Space Sciences Dictionary

4

Space Sciences

ENGLISH
FRENCH
GERMAN
SPANISH
PORTUGUESE
RUSSIAN

Dictionary 4

Earth Sciences / Solar System / Deep Space

JOSIP KLECZEK, HELENA KLECZKOVÁ

Astronomical Institute of the Czechoslovak
Academy of Sciences, Ondřejov

ELSEVIER

Amsterdam – London – New York – Tokyo 1992

Scientific Editor

Prof. RNDr. Vladimír Vanýsek, DrSc.

Reviewers

RNDr. Pavel Andrle, CSc.

Published in co-edition with ACADEMIA, Publishing House
of the Czechoslovak Academy of Sciences, Prague

Distribution of this book is being handled by the following publishers
for the U.S.A. and Canada
Elsevier Science Publishing Company, Inc.
655 Avenue of the Americas
New York, N.Y. 10017
for the East European Countries, China, Northern Corea, Cuba, Vietnam, Mongolia
Academia, Publishing House of the Czechoslovak Academy of Sciences, Prague, Czechoslovakia
for all remaining areas
Elsevier Sciences Publishers
25 Sara Burgerhartstraat
P. O. Box 211, 1000 AE Amsterdam, The Netherlands

Library of Congress Cataloging-in-Publication Data

Kleczek, Josip.
Space sciences dictionary.
In English, French, German, Spanish, Portuguese, and Russian.
Published in co-edition with Academia Pub. House of the Czechoslovak Academy of Sciences, Prague.
Includes indexes.
Contents: v. 1. Radiation/matter -- --
v. 4. Earth sciences/solar system/deep space.
1. Space sciences--Dictionaries--Polyglot.
2. Dictionaries, Polyglot, I. Kleczkova, Helena.
II. Title.
QB497.K58 1990 500.5′03
ISBN 0-444-98872-6 (v. 1)
ISBN 0-444-98816-5 (v. 4)

89-1184
CIP

Printed in Czechoslovakia

FOREWORD

The number of ways to set up a multilingual dictionary is at least as large as the number of possible positions of a Rubik cube. Kleczek succeeded in finding an excellent solution back in 1961 when his Astronomical Dictionary in Six Languages was published. It was a success at that time and was considered–at least by myself–as being the epitome of achievement in that area. This opinion was not shared by Josip Kleczek who almost immediately sat down to work on a dictionary with a much wider scope covering all space sciences, space technology and adjacent areas of basic sciences.

Within a short period–considering the task–of thirty years Kleczek has attained his aim in spite of the very fast development of space sciences and technology during that timespan. The number of hours and days devoted by the author to this work is beyond counting. And so is the number of letters exchanged with hundreds of astronomers, space scientists and experts in space applications from many countries of the world on the usage, frequency and suitability of each of the terms in each of the languages of the dictionary.

The four volumes covering wide subject areas are independent of each other, and can thus serve the specialist. Together they form a whole which is capable of serving someone with extremely wide interests.

The selection of languages was a serious problem. I know from frequent discussions with the author that oriental languages were also considered and that it was with the utmost regret that Kleczek abandoned the idea of including Chinese and Japanese. But, the necessity of producing volumes capable of being handled and available for a price accessible to users were restrictions too strong for the free spirit and enthusiasm of the author.

The dictionary is not meant to provide a norm for the terminology of the individual languages. This task is the province of national academies and scientific institutions. But Kleczek's dictionary will play an important role in normalizing and unifying terminologies. In the first place, the dictionary provides justification and guidance for work on terminology because it indicates both the necessity for standard terms and the wide scope of the project. In the second place, the dictionary reflects current usage. The author was wise enough to resist the temptation, which might have been rather strong, to coin new terms. Thus, in each language of the dictionary where the need for standard terminology is perceived, the ground has been prepared.

Users of the dictionary will find the arrangement according to subjects in the main part and to alphabetical order in the indices very convenient, whether the equivalent of a term in another language is being sought or the user is trying to express his thoughts correctly in his own language.

LUBOŠ PEREK

former Chief,
Outer Space Affairs Division
United Nations Secretariat

INTRODUCTION

The dictionary contains the terms used in space sciences, space technology, space research and allied disciplines. It consists of two main parts. In the first part the terms are listed in all six languages under eighty–six different subject sections (see page 8–9). The second part consists of complete alphabetical indices providing a quick reference to the position of any term in the first part. The whole work is thus equivalent to thirty bilingual dictionaries.

The terms have been derived from journals, monographs, textbooks and lectures; some new terms have been proposed by competent specialists. It is hoped that the dictionary will help scientists, engineers, teachers, students and other users in communicating through the language frontiers. A person who has at least a reading knowledge of any one of these languages will have no difficulty in finding his way, with the help of the dictionary, when he attempts to read scientific or technical writing in one of the other five languages.

The author is deeply obliged to all who helped him in preparing this dictionary. He acknowledges gratefully the important contributions that many scientists, engineers and professors have made. However, the final responsibility for the whole work rests with him.

Symbols used in the dictionary have a generally accepted meaning. In addition to them, ( ) and [ ] are introduced for economy. An expression in parentheses may be left out to use a shorter synonym. If a word in square brackets replaces the immediately preceding word then we get another synonym.

JOSIP KLECZEK

Astronomical Institute
Czechoslovak Academy of Sciences

251 65 Ondřejov
Czechoslovakia

CONTENTS

VOLUME IV

70 EARTH SCIENCES
71 space geodesy 36
72 geophysics 46
73 seismicity, volcanism 62
74 geology 80
75 geomagnetism 92
76 atmospheric sciences 108
77 weather 120
78 ionosphere 138
79 atmospheric optics 146
80 SOLAR SYSTEM
81 selenography 164
82 selenology 198
83 planets 232
83a planetary cosmogony 260
84 interplanetary matter 268
85 comets 280
86 meteors 294
87 meteoroids 324
88 the Sun 338
89 solar activity 364
90 DEEP SPACE
91 stars 398
92 constellations 420
93 stellar structure 438
94 stellar evolution 458
95 variable stars 470
96 stellar systems 486
97 interstellar matter 504
98 galaxies 520
99 universe 558

VOLUME I

0 RADIATION
1 electromagnetic radiation
2 photometry
3 imaging
4 telescope
5 photography
6 spectroscopy
7 spectrum
8 radio radiation
9 corpuscular radiation

10 MATTER
11 elementary particles
12 interactions
13 atomic nucleus
14 atom
15 chemical elements
16 molecules
17 macroscopic properties
18 plasma
19 magnetized plasma

VOLUME II

20 MOTION
21 coordinates
22 position
23 position determination
24 time
25 heat
26 flow
27 mechanics
28 orbits

30 SPACE FLIGHT
31 launch
32 guidance
33 flight
34 power supply
35 tracking
36 attitude
37 descent

40 DATA
41 observation
42 measurement
43 remote sensing
43a telemetry
44 data processing
45 mathematics
46 geometry
47 statistics
48 publication

VOLUME III

50 SPACE TECHNOLOGY
51 materials and structures
52 automation
53 balloons
54 rockets
55 propulsion
56 space flight centre
57 space vehicles
58 satellites
59 space applications

60 SPACE RESEARCH
61 mission
62 space cabin
63 space research
64 space astronomy
65 space biology
66 space medicine
67 exobiology
68 space law

INTRODUCTION

Ce dictionnaire comporte de nombreux termes utilisés dans le domaine des sciences et techniques spatiales, ainsi que dans les sciences associées. Il comprend deux parties: dans la première les termes sont repartis en quatre–vingt–six sections dans un ordre logique (selon le plan donné à la page 12–13). La seconde partie est un index alphabétique, qui indique pour chacune des six langues la position du terme dans la première partie. De cette façon ce dictionnaire est équivalent à trente dictionnaires bilingues. La connaissance passive d'une de ces langues suffit pour lire et comprendre la littérature concernant l'espace actuellement publiée dans les cinq autres langues.

Les termes ont été tirés de journaux, de monographies, de livres d'enseignement et de conférences. Des spécialistes compétents ont été appelés à proposer les termes manquants dans leur langue maternelle. Je voudrais exprimer ma profonde gratitude aux académiciens, professeurs, docteurs, médecins, ingénieurs, techniciens et aussi à beaucoup d'Instituts et d'Universités, à leurs bibliothèques, à l'Union Astronomique Internationale, à ses Présidents et Secrétaires Généraux, aux Sociétés Scientifiques – brièvement à tous ceux qui m'ont aidé pendant les trente années passées. Leur contribution a été fondamentale; néanmoins l'auteur tient à conserver l'entière responsabilité du résultat final.

Le sens des symboles utilisés ici est généralement conu. Pour économiser l'espace, nous avons utilisé parenthèses ou crochets. Un mot entre les parenthèses peut être omis sans changer la signification du terme. Par contre un mot entre les crochets peut remplacer le mot précédent et il en résulte un autre synonyme.

Institut astronomique
Academie tchécoslovaque des sciences
251 65 Ondřejov
Tchécoslovaquie

JOSIP KLECZEK

TABLE DES MATIÈRES

VOLUME IV

70 SCIENCES DE LA TERRE
71 géodésie spatiale . . . 36
72 géophysique . . . 46
73 sismologie, volcanisme . . . 62
74 géologie . . . 80
75 géomagnétisme . . . 92
76 sciences atmosphériques . . . 108
77 temps . . . 120
78 ionosphère . . . 138
79 optique atmosphérique . . . 146
80 SYSTÈME SOLAIRE
81 sélénographie . . . 164
82 sélénologie . . . 198
83 planètes . . . 232
83a cosmogonie planétaire . . . 260
84 matière interplanétaire . . . 268
85 comètes . . . 280
86 météores . . . 294
87 météoroïdes . . . 324
88 Soleil . . . 338
89 activité solaire . . . 364
90 ESPACE LOINTAIN
91 étoiles . . . 398
92 constellations . . . 420
93 structure stellaire . . . 438
94 évolution stellaire . . . 458
95 étoiles variables . . . 470
96 systèmes stellaires . . . 486
97 matière interstellaire . . . 504
98 galaxies . . . 520
99 univers . . . 558

VOLUME I

0 RAYONNEMENT

1 rayonnement électro-magnétique
2 photométrie
3 imagerie
4 télescope
5 photographie
6 spectroscopie
7 spectre
8 rayonnement radioélectrique
9 rayonnement corpusculaire

10 MATIÈRE

11 particules élémentaires
12 interactions
13 noyau atomique
14 atome
15 éléments chimiques
16 molécule
17 propriétés macroscopiques
18 plasma
19 plasma magnétoactif

VOLUME II

20 MOUVEMENT

21 coordonnées
22 position
23 détermination de la position
24 temps
25 chaleur
26 écoulement
27 mécanique
28 orbite

30 VOL SPATIAL

31 lancement
32 guidage
33 vol
34 système d'alimentation
35 poursuite
36 attitude
37 descente

40 DONNÉES

41 observation
42 mesure
43 télédétection
43a télémesure
44 traitement des données
45 mathématique
46 géométrie
47 statistique
48 publication

VOLUME III

50 TECHNOLOGIE SPATIALE

51 matériaux et structures
52 automatisation
53 ballons
54 fusées
55 propulsion
56 centre de vols spatiaux
57 véhicule spatial
58 satellites
59 applications spatiales

60 RECHERCHE SPATIALE

61 mission
62 cabine spatiale
63 recherche spatiale
64 astronomie spatiale
65 biologie spatiale
66 médicine spatiale
67 exobiologie
68 droit spatial

EINLEITUNG

Das vorliegende mehrsprachige Wörterbuch enthält Fachausdrücke aus Weltraumforschung, Raumfahrttechnologie, Kosmoswissenschaften und aus benachbarten Fachgebieten.

Jeder der vier Bände besteht aus zwei Teilen: im ersten Teil sind die Fachbegriffe in sechsundachtzig Abteilungen eingeordnet. Das Schema der Einordnung ist auf der Seite 16–17 dargestellt. Der zweite Teil enthält das alphabetisch geordnete Verzeichnis für jede der sechs Sprachen. Die beigefügten Zahlen weisen auf Abteilung und Nummer des Fachausdruckes im ersten Teil hin. Durch diese einfache Anordnung ersetzt das vorliegende Wörterbuch dreissig zweisprachige Wörterbücher.

Die Fachausdrücke wurden aus Zeitschriften, Monographien, Abhandlungen, Konferenzen, Symposien, Lehrbüchern und Vorträgen entnommen. In vielen Fällen wurden neue Fachausdrücke von kompetenten Fachleuten geprägt. Sogar mit passiver Kenntnis einer der sechs Sprachen kann man – mit Hilfe des Wörterbuches – das in den anderen fünf Sprachen geschriebenes Schrifttum aus Weltraumforschung, Raumfahrttechnologie und Kosmoswissenschaften verstehen.

Vielen Fachleuten aus verschiedenen Gebieten und aus vielen Ländern ist der Verfasser für ihre Hilfe bei der Zusammenstellung des Wörterbuches sehr verpflichtet. Er ist nichtsdestoweniger für die Arbeit voll verantwortlich.

Alle im Wörterbuch benutzten Symbole sind allgemein bekannt. Nur die Klammern haben folgende Bedeutung: (...) – das eingeklammerte Wort, oder den eingeklammerten Buchstaben kann man ausslassen; [...] – das eingeklammerte Wort kann des vorangehende substituiren, um ein anderes Synonymum zu bekommen.

Astronomisches Institut
Tschechoslowakische Akademie der Wissenschaften
261 65 Ondřejov
Tschechoslowakei

JOSIP KLECZEK

INHALT

BAND IV

70 GEOWISSENSCHAFTEN
71 Weltraumgeodäsie 36
72 Geophysik 46
73 Erdbebenkunde und Vulkane 62
74 Geologie 80
75 Erdmagnetismus 92
76 atmosphärische Wissenschaften 108
77 Wetter 120
78 Ionosphäre 138
79 atmosphärische Optik 146

80 SONNENSYSTEM
81 Selenographie 164
82 Selenologie 198
83 Planeten 232
83a Planetenkosmogonie 260
84 interplanetare Materie 268
85 Kometen 280
86 Meteore 294
87 Meteoroiden 324
88 Sonne 338
89 Sonnenaktivität 364

90 ERDFERNER WELTRAUM
91 Sterne 398
92 Sternbilder 420
93 Sternaufbau 438
94 Sternentwicklung 458
95 veränderliche Sterne 470
96 Sternsysteme 486
97 interstellare Materie 504
98 Galaxien 520
99 Weltall 558

BAND I

0 STRAHLUNG
1 elektromagnetische Strahlung
2 Photometrie
3 Bildererzeugung
4 Fernrohr
5 Photographie
6 Spektroskopie
7 Spektrum
8 Radiofrequenzstrahlung
9 Korpuskularstrahlung

10 MATERIE
11 Elementarteilchen
12 Wechselwirkung
13 Atomkern
14 Atom
15 chemisches Element
16 Molekül
17 makroskopische Eigenschaften
18 Plasma
19 magnetisiertes Plasma

BAND II

20 BEWEGUNG
21 Koordinaten
22 Position
23 Positionsbestimmung
24 Zeit
25 Wärme
26 Strömung
27 Mechanik
28 Bahn

30 WELTRAUMFLUG
31 Start
32 Bahnsteuerung
33 Flug
34 Energieversorgung
35 Bahnverfolgung
36 Fluglage
37 Abstieg

40 DATEN
41 Beobachtung
42 Messung
43 Fernerkundung
43a Telemetrie
44 Datenverarbeitung
45 Mathematik
46 Geometrie
47 Statistik
48 Veröffentlichung

BAND III

50 RAUMFAHRTTECHNOLOGIE
51 Weltraum-Werkstoffe
52 Automatisierung
53 Ballon
54 Raketen
55 Antrieb
56 Raumflugzentrum
57 Raumflugkörper
58 Satelliten
59 Raumfahrtanwendungen

60 WELTRAUMFORSCHUNG
61 Mission
62 Raumfahrzeugkabine
63 Weltraumforschung
64 extraterrestrische Astronomie
65 Weltraumbiologie
66 Raumfahrtmedizin
67 Exobiologie
68 Weltraumrecht

INTRODUCCIÓN

Este diccionario comprende términos utilizados en ciencias y tecnologías espaciales y asimismo en disciplinas relacionadas. Consta de dos partes: en la primera parte los términos estan dispuestos logicamente en ochenta y seis secciones (ver página 20–21); la segunda parte son índices alfabéticos para cada una de las seis lenguas. De este modo el diccionario es equivalente a treinta diccionarios bilingües. El conocimiento, aunque sea pasivo, de uno de los seis idiomas, es suficiente para leer la literatura de dichas disciplinas que se publica actualmente en los otros cinco idiomas.

Los términos fueron tomados de revistas, de monografías, de libros de enseñanza y de conferencias; especialistas competentes propusieron términos nuevos no existentes antes en su lengua materna. El autor agradece profundamente a los académicos, profesores, doctores, ingenieros, médicos, juristas, especialistas de varias disciplinas, a la Unión Astronómica Internacional y sus Presidentes y Secretarios Generales, diversos institutos, universidades, sociedades, comisiones, agrupaciones, observatorios, y todos otros los que han prestado una ayuda substancial durante los treinta años pasados. Sin embargo, la responsabilidad final pertenece al autor.

El sentido de los símbolos utilizados es bien conocido. Para ahorrar espacio, utilizamos paréntesis ( ) y corchetes [ ]. Una palabra entre paréntesis se puede omitir sin cambio del significado del término. Al contrario, una palabra entre corchetes puede reemplazar la palabra anterior y así obtenemos otro sinónimo.

Instituto astronómico
Academia checoslovaca de Ciencias
251 65 Ondřejov
Checoslovaquia

JOSIP KLECZEK

ÍNDICE DE MATERIAS

TOMO IV

70 GEOCIENCIAS
71 geodesia espacial 37
72 geofísica 47
73 sismología, volcanismo 63
74 geología 81
75 geomagnetismo 93
76 ciencias atmosféricas 109
77 tiempo 121
78 ionosfera 139
79 óptica atmosférica 147
80 SISTEMA SOLAR
81 selenografía 165
82 selenología 199
83 planetas 233
83a cosmogonía planetaria 261
84 materia interplanetaria 269
85 cometas 281
86 meteoros 295
87 meteoroides 325
88 Sol 339
89 actvidad solar 365
90 ESPACIO REMOTO
91 estrellas 399
92 constelaciones 421
93 estructura estelar 439
94 evolución estelar 459
95 estrellas variables 471
96 sistemas estelares 487
97 materia interestelar 505
98 galaxias 521
99 universo 559

TOMO I

0 RADIACIÓN

1 radiación electromagnética
2 fotometría
3 formación de imágenes
4 telescopio
5 fotografía

6 espectroscopía
7 espectro
8 radiación radio
9 radiación corpuscular

10 MATERIA

11 partículas elementales
12 interacción
13 núcleo atómico
14 átomo
15 elementos químicos

16 molécula
17 propiedades macroscópicas
18 plasma
19 plasma magnetizado

TOMO II

20 MOVIMIENTO

21 coordenadas
22 posición
23 determinación de la posición
24 tiempo
25 calor

26 flujo
27 mecánica
28 órbita

30 VUELO ESPACIAL

31 lanzamiento
32 guiado
33 vuelo

34 alimentación de energía
35 seguimiento

36 actitud
37 descenso

40 DATOS

41 observación
42 medida
43 sensoriamiento remoto
43a telemetría
44 procesamiento de datos
45 matemáticas

46 geometría
47 estadística
48 publicación

TOMO III

50 TECNOLOGÍA ESPACIAL

51 materiales de uso espacial
52 automatización
53 balón
54 cohetería
55 propulsión

56 centro de vuelos espaciales
57 vehículo espacial
58 satélites
59 aplicaciones espaciales

60 EXPLORACIÓN ESPACIAL

61 misión
62 cabina espacial
63 investigación espacial
64 astronomía espacial
65 biología espacial

66 medicina espacial
67 exobiología
68 derecho espacial

INTRODUÇÃO

Este dicionário é constituído por termos e frases utilizados em tecnologia espacial, pesquisas especiais, ciências espaciais e em disciplinas afins. Cada volume do dicionário compreende duas partes: a primeira é a lista dos termos em seis línguas, dispostos em oitenta e seis seções (página 24–25); a segunda, contém um índice aflabético para cada língua, o que facilita a procura de qualquer termo na primeira parte. Desta maneira, o dicionário é equivalente a trinta dicionários bilíngües.

Os termos e as frases provêm de diversas fontes: jornais, monografias, livros de estudo, livros de consulta, preleções e conferências; especialistas competentes propuseram termos novos que faltavam na sua língua materna.

O autor reconhece com gratidão as contribuições que outros fizeram; não obstante, ele tem toda a responsabilidade pela obra.

O sentido dos símbolos usados é bem conhecido. Para economizar espaço, temos introduzido parênteses ( ) e colchetes [ ]. A palabra entre parênteses pode omitir-se sem alteração do sentido. Pelo contrário, uma palabra entre colchetes pode substituir a palabra antecedente e assim obtemos outro sinônimo.

JOSIP KLECZEK

Instituto astronômico
Academia tchecoslovaca de ciências
251 65 Ondřejov
Tcheco-Eslováquia

ÍNDICE DE MATÉRIA

TOMO IV

70 CIÊNCIAS DA TERRA

71 geodésia espacial 37
72 geofísica 47
73 sismologia e vulcanismo 63
74 geologia 81
75 geomagnetismo 93
76 ciências da atmosfera 109
77 tempo 121
78 ionosfera 139
79 óptica atmosférica 147

80 SISTEMA SOLAR

81 selenografia 165
82 selenologia 199
83 planetas 233
83a cosmogonia planetária 261
84 matéria interplanetária 269
85 cometas 281
86 meteoros 295
87 meteoróides 325
88 Sol 339
89 atividade solar 365

90 ESPAÇO REMOTO

91 estrelas 399
92 constelações 421
93 estrutura estelar 439
94 evolução estelar 459
95 estrelas variáveis 471
96 sistemas estelares 487
97 matéria interestelar 505
98 galáxias 521
99 universo 559

TOMO I

0	RADIAÇÃO	10	MATÉRIA
1	radiação ele(c)tromagnética	11	partículas elementares
2	fotometria	12	interações
3	imageamento	13	núcleo atômico
4	telescópio	14	átomo
5	fotografia	15	elementos químicos
6	espe(c)trografia	16	moléculas
7	espe(c)tro	17	propriedades macroscópicas
8	radiação rádio	18	plasma
9	radiação corpuscular	19	plasma magnetizado

TOMO II

20	MOVIMENTO	30	VÔO ESPACIAL	40	DADOS
21	coordenadas	31	lançamento	41	observação
22	posição	32	guiamento	42	medições
23	determinação de posição	33	vôo	43	sensoriamento remoto
				43a	telemetria
24	tempo	34	alimentação de energia	44	processamento de dados
25	calor	35	rastreamento	45	matemática
26	fluxo	36	atitude	46	geometria
27	mecânica	37	descida	47	estatística
28	órbitas			48	publicação

TOMO III

50	TECNOLOGIA ESPACIAL	60	PESQUISA ESPACIAL
51	materiais e estruturas	61	missão
52	automatização	62	cabine espacial
53	balões	63	ciências espaciais
54	foguetes	64	astronomia espacial
55	propulsão	65	biologia espacial
56	centro de vôos espaciais	66	medicina espacial
57	veículos espaciais	67	exobiologia
58	satélites	68	direito espacial
59	aplicações da pesquisa espacial		

ВВЕДЕНИЕ

Настоящий Словарь содержит термины, используемые в космических науках, космической технологии, космических исследованиях и смежных дисциплинах. Он состоит из двух основных частей. В первой части во всех шести языках термины упорядочены по восьмидесяти шести разным тематическим разделам (см. стр. 28–29). Вторая часть состоит из полных алфавитных указателей, обеспечивающих быструю ссылку на место любого термина в первой части. Вся работа в этом случае эквивалентна тридцати двуязычным словарям.

Термины взяты из периодических изданий, монографий, учебников, материалов конференций, симпозиумов и школ; некоторые новые термины были предложены компетентными специалистами. Автор надеется, что Словарь поможет ученым, инженерам, преподавателям, студентам и другим пользователям при общении через языковые барьеры. Человек, который по крайней мере, может читать на каком-либо из этих языков, не встретится с трудностями, когда с помощью этого Словаря он попытается читать научные или технические тексты на любом из остающихся пяти языков.

Я выражаю большую благодарность всем, кто сделал важный вклад в эту работу, однако полную ответственность за всю работы несу я.

Сымволы, которые использованы здесь, имеют общепринятое значение. Чтобы сэкономить место, мы ввели скобки: круглие скобки обозначают, что букву или слово закрытое в них можно опустить. Слово в квадратных скобках может заместить предыдущее слово, чтобы получить новый синоним.

Астрономический Институт
ЧСАН,
251 65 Ондржейов
Чехословакия

ИОСИП КЛЕЧЕК

СОДЕРЖАНИЕ

ТОМ 4

70 НАУКИ О ЗЕМЛЕ

71 космическая геодезия 37
72 геофизика 47
73 сейсмичность и вулканизм 63
74 геология 81
75 геомагнетизм 93
76 атмосферные науки 109
77 погода 121
78 ионосфера 139
79 атмосферная оптика 147

80 СОЛНЕЧНАЯ СИСТЕМА

81 селенография 165
82 селенология 199
83 планеты 233
83а планетная космогония 261
84 межпланетная материя 269
85 кометы 281
86 метеоры 295
87 метеороиды 325
88 Солнце 339
89 солнечная активность 365

90 ДАЛЬНЫЙ КОСМОС

91 звезды 399
92 созвездия 421
93 строение звёзд 439
94 эволюция звёзд 459
95 переменные звезды 471
96 звёздные системы 487
97 межзвёздное пространство 505
98 галактики 521
99 вселенная 559

ТОМ 1

0 ИЗЛУЧЕНИЕ

1 электромагнитное излучение
2 фотометрия
3 получение изображений
4 телескоп
5 фотография
6 спектроскопия
7 спектр
8 радиоизлучение
9 корпускулярное излучение

10 МАТЕРИЯ

11 элементарные частицы
12 взаимодействия
13 атомное ядро
14 атом
15 химические элементы
16 молекула
17 макроскопические свойства
18 плазма
19 замагниченная плазма

ТОМ 2

20 ДВИЖЕНИЕ

21 координаты
22 положение
23 определение положения
24 время
25 тепло
26 течение
27 механика
28 орбита

30 КОСМИЧЕСКИЙ ПОЛЕТ

31 запуск
32 наведение
33 полёт
34 энергопитание
35 сопровождение
36 угловое пространственное положение
37 спуск

40 ДАННЫЕ

41 наблюдение
42 измерение
43 дистанционное зондирование
43а телеметрия
44 обработка данных
45 математика
46 геометрия
47 статистика
48 публикация

ТОМ 3

50 КОСМИЧЕСКАЯ ТЕХНОЛОГИЯ

51 материалы космической техники
52 автоматизация
53 баллон
54 ракетная техника
55 двигатель
56 центр космических исследований
57 космический летательный аппарат
58 спутники
59 использование космического пространства

60 КОСМИЧЕСКИЕ ИССЛЕДОВАНИЯ

61 миссия
62 кабина космического корабля
63 космические исследования
64 внеземная астрономия
65 космическая биология
66 космическая медицина
67 экзобиология
68 космическое право

ACKNOWLEDGEMENTS:

The four Volumes of the Space Sciences Dictionary would not see the light without the kind assistance of many Scientific Societies,various Departments of many Universities, National Institutions, Academic Institutes, National Committees of Astronomy, Laboratories, Observatories, and Space Agencies in each language. They are too many to be all listed here. We are deeply obliged to them for their generous help with terminology.
Our special thanks are due to many Professors, Directors, Presidents, General Secretaries, Doctors, Lecturers, Academicians, Scientists, Engineers, Technicians, Librarians and Students from many countries where the six languages are spoken (we apologize to those whose names slipped from our failing memory):

Abalakin, Viktor; Abans, Mariángela; Abdala, José; Adams, John; Adner, Heinz; Afanasjeva Praskovya, M.; Agostinho, Rui L.; Aguilar, Maria-Luisa; Alcantará, Alberto; Alcantará Gomes, Roberto; Alexander, A. F. O'D; Aleksandrov, B. M.; Aleksejev, V. M.; Alfaro, Emilio; Altavista, Carlos; Amorim Ferreira, H.; Andrillat, Henri; Andrillat, Yvette; Andrle, Pavel; Araujo Lopes Zamith, A. ; Aristidi, Eric; Arkhipova, V. P.; Armentia, Javier E.; Astapovich, I. S.; Astor, J. L.; Astronomicheskij Sovjet USSR; Astronomisches Rechen-Institut, Germany; Avignon, Yvette;

Babadzhanov, P. B.; Bagin, Petr; Baldamus, Dieter; Baldet, F.; Ballester, José L.; Baltá, José; Banin, V.G.; Barabashov, N.P.; Barreto, Jorge Muniz; Barreto, Luiz Muniz; Barroso dos Reis, Helaine; Barta-Camus, Sylvie; Bartl, Erich; Bastin, John A.; Bauriegel, Lutz; Behr, A.; Bellmann, Günter; Belorizky, D.; Benevides, P.; Bertaud, Ch.; Bezrukova, A.Ya.; Bittner, Alfred; Blagonravov, A. A.; Blaha, Milan; Blamont, Jaques; Böhme, Anna; Boishot,; Bok, Bart; Bol'shev, L. N.; Bondi, H.; Bonity, M; Boscov, Jayme; Brado, W; Braga, C. de Azevedo Coutinho; Bretagnon, P.; Bronshten, B.A.; Brown, John; Bruzek, Anton; Buchar, Emil; Budach, L.; Budějický, Miroslav; Budgen, P.; Bukovský, J.; Bulanov, S.V.; Burtsev, A.N.; Bureau des Longitudes, France; Burša, Milan; ; Calvo, Luiz Lozano; Camichel, Henri; Campos, L. M. B. C.; Capitaine, N.; Carleal, Aydano B.; Cardoso, J. Christovão; Cardús, J. O.; Carrasco, E.; Carter, L.J.; Casas, Justiniano; Catalá, Joaquín; Catalá, M. Asunción; Catalá de Alemany, José; Cavedon, M.; Čechová, Ilana; Centro Sensores Remotos, Argentine; Centro Técnico Aeroespacial, Brasil; Cepa, Jordi; Cesco, Carlos; Chapront, J.; Chavira, E.; Chugajnov, P. F.; Clauzet, L. B. F.; CNES France; CNIE Argentine; Codina Vidal, José M.; Codina, Sayd; Cohen, R. J.; Concha, Tomás Elices; Conseil International de la Langue Francaise, France; Conselho Nacional de Estatística, Brasil; Correia, Emilia; Couder, A.; Couderc, P.; Courtès, Georges; Coutrez, Raymond; Cowling, T. G.; Cox, R. R. S.; Crovisier, Jaques; da Costa Campos, L.M.B.;

da Rocha Vieira, Edmondo; da Silva, A.V.C. Simoes; da Silva, Eugenio; da Silva, C.Padilha Alves; da Silva, Licio; da Silva Machado, Luiz Eduardo; Dabbs, Joseph R.; Daene, H.; Davies, J. G.; de Abreu Coutinho, Jorge; de Albuquerque Braz, José Manuel; de Araújo, Francisco Xavier; de Assumpcao, Marcelo Sousa; de Azevedo Coutinho Braga, Carlos; de Barros, Manuel; de Barros, M. C. Pereira; de Carvalho Dias, Lindolpho; de Castro, Arcadio García; de Freitas Mourão, R. R.; de Freitas, Sergio Mange; de Jager, Cornelius; de Jesus Parada, Nelson; de la Cotardière, Philippe; de la Reza, Ramiro; de Loore, Camiel; de Mendonça, Fernando; de Rafael, Ramón; de Silva, G. A.; de Sousa, Silva A.; de Vaucouleurs, Gérard; Débarbat, Souzanne; Dědoch, Antonín; del Sola, G. Vizoso; Delahais, M.; Delcourt, J. G.; Delgado, Antonio; Delhaye, J.; Delouf, A.; Delsemme, A. H.; Delval, G.; Demaret, J.; Denis, Nelba; Denisse, J. F.; Department of Geology, Imperial College of Science and Technology, England; Departamento Nacional de Meteorologia, Brasil; Deubner, Franz; Deutsche Forschungs-und Versuchsanstalt für Luft- und Raumfahrt; Deutsche Gesellschaft für Luft- und Raumfahrtmedizin; Deutsche Gesellschaft für Weltraumforschung; Dick, J.; Dickvoss, W.; Dietzel, Ernst; Direction de la Météorologie Nationale, France, Dobrovol'skij, Oleg V.; Dokuchaeva, O.D.; Dollfus, Audouin; Dottori, Horacio; Du Puy, David L.; Dubrovskij, V.A.; Dubský, Josef; Ducati, Jorge; Dufay, J.; Dumont, Simone; Dungey, J.W.; Dunlop, Storm; Duque Simoẽs, Ilídio; Dvořák, Aleš; Dvořák, Josef;

Eckhartt, D.; Eliáš, Mojmír; Ellison, M. A.; Elsässer, H.; Enslin, H.; ESA; Escobal, P.; ESO; Estalella, Roberto; Esteve, D.; Estokas Filho, Bruno; European Space Operation Center;

Fechtig, H.; Fedynskij, V. V.; Fehrenbach, Charles; Feinstein, Alejandro; Feissel, Martin; Fernandez Fernandez, José; Fernandez Figueroa, Maria José; Ferraz Mello, Sylvio; Ferreira, G; Ferreira, H. Amorim; Fielder, George; Filloy, E. M.; Firneis, Maria; Fischer, Hans Joachim; Fischer, H. S.; Fizicheskij Institut Akademii Nauk USSR; Fossat, Eric; Fourcade, C. R.; Fraeijs de Veubeke, B.; Fricke, Walter; Froeschle, C.; Fürstenberg, F.;

Gama, Lelio I.; Gambis, Daniel; Garcia, Adriana; Garrido, Rafael; Garzón, Francisco; Gayet, Jaques; GDTA, France; Gel'frejkh, Georgij B.; Geophysics Department, Imperial College London; Gerbaldi, Michèle; Gerlach, H. A.; Gesellschaft für Weltraumforschung; Gesellschaft für Weltraumforschung und Raumfahrt der DDR; Ghielmetti, H.; Giese, Richard H.; Giovanelli, Ronald G.; Gnevyshev, Mstislav N.; Gnevysheva, Raisa S.; Goberna, J. Rafael; Goffin, Edwin; Golay, Marcel; Goldenberg, José; Golovko, Aleksej A.; Gomes, Roberto Alcantará; Gonze, R.; Gordon, Mark A.; Gorret, H.; GOSNITSIPR, Moscow; Gosudarstvennyj Astronomicheskij Institut im. Shternberka; Gosset, H.; Gougenheim, Loucienne; Goy, Gerald; Gräfendorf, Paul; Gramaccini, Fernando; Grensemann, M.G.; Grigorjev, Yurij G.; Gross, Ricardo; Grossi-Gallegos, Hugo; Groth, H.G.; Groupement pour la détection spatiale, France; Guigay,G.; Guimarães Ferreira, Pedro M.; Guinot, B.; Gullón, Enrique; Günther, O; Gusmán, E.; Guth, Vladimír; Gutiérrez-Moreno, Adelina; Guyot, E.;

Hachenberg, O.; Hälbig, Eberhard; Hardorp, H.; Harmanec, Petr; Haro, Guillermo; Harth, W.; Haug, Ulrich; Haupt, Hermann F.; Häuser, M. G.; Haynes, J. Michael; Heilmaier, Erich P.; Helin, Glo; Helin, Ronald; Hempsell, C. M.; Hernandez, Manolo; Herrera Cabello, Félix; Hey, M. H.; Heyvaerts, Jean; Hoppe, J.; Horský, Zdena; Hottes, K.; Houziaux, Léo; Hoyer, Walter; Hubený, Ivan; Hull, M. D.; Hunča, Svatopluk; Hunter, A.;

Iannini, G.M; IAU, Presidents of Commissions; IAU Secretariat since 1960; IAU – UNESCO International Schools for Young Astronomers (1967 – 1990). The greater part (more than four hundreds) of lecturers and participants; Ibañez, Miguel H.; Idlis, G. M.; Igenbergs, E.B.; IIAE, Córdoba, Argentine; IFLYSIB – U.N.L.P. Argentine; IKI, Moscow; Incertis, Fernando; INPE, Brasil; Institut d'Astronomie Spatiale de Belgique; Institut Fiziki Zemlji AN USSR, Moscow; Institut für Flugmedizin und Raumfahrt, Köln; Institut für Kosmosforschung, Berlin; Institut für Luft- und Raumfahrt, München; Institut Kosmicheskikh Issledovanij, Moscow; Institut Mediko-Biologicheskikh Problem, Moscow; Instituto Central de Física, Universidad de Concepción; Instituto de Atividades Espaciais, Brasil; Instituto de Geociencias da Universidade, Brasilia; Instituto de Investigaciones Aeronauticas y Espaciales, Argentina; Instituto de Pesquisas Espaciais, Brasil; Instituto Español de Oceanografía; Instituto Geofísico do Infante D. Luís, Lisboa; Instituto Meteorológico de Academia, Cuba; Instituto Nacional de Geofísica, Spain; Instituto Nacional de Meteorologia, Brasil; Instituto Nacional de Técnica Aeroespacial, Spain; INTA, Spain; INTA – Programas Espaciales, Madrid; Ioshpa, B. A.; IZMIRAN, Akademgorodok;

Jablonski, Francisco; Jacchia, Luigi; Jäger, Friedrich W.; Jarolím, Miroslav; Jaschek, Carlos; Jaschek, Mercedes; Jelasco, Jean Jaques; Jennings, R. E.; Jentzsch, Klaus; Johnson, Alan; Johnstone, A.; Jeřábek, Ondřej; Joly, Hubert; Jukes, Thomas H.; Juřina, Petr;

Kahn, Franz; Kaiser, Tom R.; Kaplan, S.A.; Karyagina, Zoya B.; Katasjev, L. A.; Kaufmann, Pierre; Kautzleben, H.; Kellner, H.A.; Khabibulin, Sh.T.; Khokhlova, Vyera L.; Khrgian, A. Kh.; Khromov, G. S.; Kiepenheuer, K. O.; Kiko, J.; Kippenhahn, Rudolf; Kitchin, Chris; Klein, K. E.; Kluge, Jochen; Knittl, Zdeněk; Koch Miramond, Lydie; Kockarts, G.; Koelle,; Kohler, Pierre; Kohoutek, Luboš; Kolář, Jan; Kononovich, Edward V.; Kopal, Zdeněk; Kopatz, Oswald; Kopecký, Lubomír; Kopylov, I. M.; Kostyakova, Elena V.; Kourganoff, V.; Kovalevsky, Jean; Krat, B.A.; Kratinová, Magdaléna; Krause, Fritz; Krejčí, Gustav; Kresák, Lubor; Krinov, E. L.; Křivský, Ladislav; Kříž, Antonín; Kroshkin, M. G; Kroulík, Jiří; Krüger, Albrecht; Kryukov, Aleksej Borisovich; Kuchařová Ludmila; Kukal, Zdeněk; Kukarkin, Boris V.; Kuklin, G. V.; Kunkel, W.E.; Kupryakov, Yurij Alekseyevich; Kurt, Vladimir G.; Kurth, Ljudmila; Kurzweil, J.;

Laboratoire d'Automatique et d'Analyse des Systèmes, Toulouse; Laboratoire de Physique des Matériaux, Meudon; Lála, Petr; Lambrecht, H.; Lamla, Erich; Lamzin, S. A.; Landi Dessy, Jorge; Lang, Heinrich; Lantos, Pierre; Laring, Johannes; Lauter, E.; Lebeau, André; Lebedev Physical Institute, Moscow; Ledoux, P.; Lehmann, U.; Lehnert, B.; Lemos, V.H.; Leroy, Jean Louis; Letfus, Vojtěch; Levin, Boris Yu.; Lidmilová, Pavla; Liebl, Vlastimil; Lindner, Klaus; Link, František; Lipskij, Yu. N.; London, Julius; Lopéz, R.; Lopez García, Francisco; Lorón, Mariano Martín; Lozano Calvo, Luis;

Macháček, Jakub; Macháček, Lukáš; Macháčková, Kateřina; Machado, Marcos; Maciel, Walter J.; Madeira, José A.; Magalhaes, Antonio; Magalhaes Gomes, F. A.; Malaise, Daniel J.; Maloň, St.; Malov, I.F.; Mánek, Jan; Markaryan, B. E.; Marques, E. Solano; Marques, Manuel Nunes; Martres, Marie Josephe; Martinez, Patrick; Martynov, D. Ya; Mason, R. G.; Maštalíř, Ludvík; Matematichesko-Fizicheskij Fakultet,Tekh. Universiteta, Moscow; Mattig, W.; Mayne, David Q.; McCrea, W.H.; McIntosh, Bruce A.; McNally, Derek; Meaburn, John; Medina, José; Meister, C. V.; Mejía, Gastón R.; Mekhaniko-Matemat. Fakultet Moskovskovo Gos.Universiteta; Meleshko, Gana Osipovna; Menendez Hevia, Enrique; Menendez, Luis L.; Merkulenko, . Vladimir E.; Meteorological Office,

Brecknell; Météorologie Nationale, France; Meteorologischer Dienst der DDR; Metter, F; Meyer, E.; Meyer, Friedrich; Meyer, J.; Mielke, Heinz; Mikhalkovskij, V.; Millman, Peter; Minnaert, M. G. J.; Milone, Luiy A.; Mirzoyan, L. N; Möhlmann, Dietrich; Mohr, J. M.; Molchanov, A. P.; Monsel, Philippe; Monteiro, José Fernando; Moragues, Jaime A.; Moreno, Hugo; Moreno, M. Q.; Moroz, V. I.; Morrison, Leslie V.; Moscati, Giorgio; Moskovskoye Matematicheskoe Obschestvo; Mourek, Jindřich; Muller, R.; Mãrao, Ronaldo; Museu de Ciência, São Paulo; Mustel, Evald Rudolfovich;

Nábělek, Bohumil; Nauchno-Issledovatelskij Radiofizicheskij Inst. Gorkij; Naumov, Vitalij A.; Netter, F.; Nicolet, Marcel; Nicolson, Iain; Nikolskij, Genadij M.; NIRFI, Gorkij; Nogueira Barbosa, Marcio; Notni, Peter;

Oficina Meteorológica de Chile; Ogorodnikov, Kyril P.; Oleak, H.; Onderlička, Bedřich; Oranovskij, V.E.; Orsini, Aristóteles; Orte, Alberto; Osborne, Daphne J.; Osório, José;

Pagel, Bernard; Palacios, Julio; Palouš, Jan; Parijskij, Yurij Nikolaevich; Parkinson, John H.; Partel, G.; Paul, Serge; Pavlov, N.N.; Pecina, Petr; Pecker, Jean Claude; Pekárek, Luděk; Pellas, Paul; Pelegrini, Paulo; Penaloza, Marcos A.; Pensado Iglesias, José; Percy, John R.; Perek, Luboš; Peretti Hollemaert, Alvaro; Pernegr, Jaroslav; Pešek, Rudolf; Pettauer, Thomas; Pfleiderer, Jorg; Pflug, Klaus; Philibert, J.; Pícha, Jaroslav; Pick, Monique; Pieschel, Friedemann; Pikel'ner, Salomon B.; Pinkau, K.; Pisarenko, N. P.; Pismis de Recillas, Paris; Pivoňka, Petr; Pleskač, Radek; Podolský, Jiří; Pohl, Dieter; Pollak, L. W.; Polnitzky, Gerhard; Pöppel, Wolfgang G.; Porte, J.; Poveda, Arcadio; Praderie, F.; Priesemuth, Bernhard; Priest, Eric; Prigancová, Ala; Procházka, M.; Psovskij, Yu. P.; Ptáček, Vl.; Ptitsyn, Dimitrij A.; Pudovkin, M.I.; Purgathofer, Alois;

Quast, Germano;

Rachen, Karl Heinz; Radcliffe, John A.; Rahe, Jürgen; Ramsayer, Karl; Rssbach, Erika; Raupp, Marco Antonio; Rego, Manuel; Reijnen, Elisabet; Reinhart, M.; Rému-Battiau Liliane, G. A.; Richards, H. W. P.; Richter, Nicolas; Riedel Ratisbona, Leandro; Ringuelet, Adela; Roberts, Walter Orr; Rodríguez, António; Roederer, Juan G.; Rohlfs, K.; Romana, P. Antonio; Rösch, Jean; Rose, Jochen; Roy, Archie E.; Rozhkovskij, D. A.; Rückl, Antonín; Ruprecht, Jaroslav; Ruprechtová, L.;

Sadil, Josef; Sadler, D. H.; Sadovnichij, V. A.; Sadovskij, M. A.; Safronov, V.S.; Sagner, Harald; Sahade, Jorge; Sainte Marie, M.; Sala, Oswaldo; Sanahuja, Blai; Sanchez, Francisco; Sanchez Magro, Carlos; Sanke, Heidi; Sazhin, M.V.; Scalise, Eugenio Jr.; Sharonov, V. V.; Schatzman, Evry; Scheffler, H.; Scheuer, Walter; Schevchenko, V. V.; Schmadel, L. D.; Schmidt, Hans; Schmidt, Hermann U.; Schmidt, K.H.; Schneider, Manfred; Schneider, Otto; Schneller, H.; Schönberg, E.; Schröder, Rolf; Schröter, Egon H.; Schrutka Rechtenstamm, Guntram; Schücking, E.; Schulz-Görner, J.; Scott, F.P.; Segers, Carlos; Sehnal, Ladislav; Seibold, Jorge R.; Sengbusch, Kurt; Sérsic, José L.; Servajean, Roger; Servico Meteorológico Nacional, Portugal; Sestokas Filho, Bruno; Sharov, A. S.; Sheffer, E. K.; Shcherbina-Samojlova, I. S.; Schevchenko,; Šíma, Jiří; Simon, Paul A.; Simonenko, Ala; Sisteró, Roberto F.; Sitnik, G. F.; Skuridin, G. A.; Smith, H.; Smyth, Michael J.; Sobotka, Michal; Sociedade Brasileira de Física; Société Francaise de Physique; Sommerville, William B.; Soulié, Edgar; Spänkuch, Dietrich; Stammreich, Hans; Steinlin, U. W.; Stepanov, V. E.; Stepanyan, Arnold A.; Stepanyan, Natasha N.; Šternberg, Bohumil; Stoyko, N.; Stram, Edgar; Strel'nickij, Vladimir S.; Strohmeier, W.; Strzondala, Roman; Suagher, Françoise; Sushkov, German; Švestka, Zdeněk; Swings, Paul; Sykes, John B.; Syrovatskij, S. I.;

Tabanera, Teofilo; Tandberg-Hanssen, Einar; Tasso, J. J.; Tavares, José Lopes; Tavastsherna, K. H.; Tejfel', Viktor G.; Temesváry, S.; Teryan, A.; Texereau, Jean M.; Thomas, H. C.; Tissi, Plínio; Tobyáš, Vladimír; Tobyášek, Vladislav; Torchinskaya, O.; Torres, Alberto P. C. O. Carlos; Torroja, José M.; Trellis, M.; Troche Boggino, Alexis; Troitskij, V. S.; Tschamler, Ricardo H.; Tscharnuter, Werner M.; Tunill, Reginald; Turnay, José M.; Tutukov, A. V.;

Udiás, Augustín; Urban, Jan; Urmitzer, H.;

Valníček, Boris; Vanýsek, Vladimír; Varsavsky, Carlos M.; Vaz, S.; Vazquez, Manuel; Vedrenne, G.; Velarde, M. G.; Velasco Durantez, Mariano; Vercheval, J.; Vicente, R. de Oliveira; Vidal, Codina; Vidal, Germano Seidl; Vítek, Antonín; Vitinskij, Yurij I.; Vitkevich, B.B. ; Vives, Theodoro José; Voigt, Hans H.; Vreux, Jean Marie; Vsechsvzatskij, S. K.; Vykutilová, Marie; Vystavěl, Radek;

Wallis, G.; Weberová, L.; Weidemann, Volker; Weigert, Alfred; Wellmann, Peter; Wetterdienst der DDR; Whitaker, E. A.; Wielen, Roland; Wiemann, Gero; Wilkins, George A.; Wittbrodt, H.; Wolf, Marek; Wolstencroft, Ramon D.; Wolynec, Elisa;

Yun, João Lin;

Zachoval, L.; Zadunaisky, Pedro E.; Zamorano, Jaime; Zasov, A. V.; Zavřel, Josef; Zel'manov, A.L.; Zentralinstitut für Astrophysik, Potsdam; Zentralinstitut für Kernforschung, Dresden; Zentralinstitut für Physik der Erde, Potsdam; Zentralinstitut für Solar-Terrestrische Physik, Berlin; Zheleznykh, I. M.; Zhidkov, N. P.; Zhitnik, Igor A.; Zicha, Josef; Ziegler, Reinhard;

Ondřejov, Czechoslovakia
April 1st, 1991;

Josip Kleczek

70

EARTH SCIENCES

SCIENCES DE LA TERRE

GEOWISSENSCHAFTEN

GEOCIENCIAS

CIÊNCIAS DA TERRA

НАУКИ О ЗЕМЛЕ

71	SPACE GEODESY	GÉODÉSIE SPATIALE	WELTRAUMGEODÄSIE
1	geosciences Earth sciences	sciences de la Terre *fpl*	Geowissenschaften *fpl*
2	Earth	Terre *f*	Erde *f*
3	terrestrial	terrestre	Erd- terrestrisch
4	International Union of Geodesy and Geophysics	Union Géodésique et Géophysique Internationale *f*	Internationale Union für Geodäsie und Geophysik *f*
5	Earth's surface	surface terrestre *f* surface de la Terre *f*	Erdoberfläche *f* Oberfläche der Erde *f*
6	curvature of the Earth	courbure de la Terre *f*	Erdkrümmung *f* Krümmung der Erde *f*
7	figure of the Earth shape of the Earth	figure de la Terre *f*	Erdfigur *f* Figur der Erde *f*
8	Earth's flattening	aplatissement de la Terre *m*	Erdabplattung *f*
9	equatorial bulge	bourrelet équatorial *m*	Äquatorwulst *m*
10	globe terrestrial sphere	globe terrestre *m* sphère terrestre *f*	Erdkugel *f*
11	earth ellipsoid	ellipsoïde terrestre *m*	Erdellipsoid *n*
12	polyhedron	polyèdre *m*	Polyeder *n*
13	earth spheroid	sphéroïde terrestre *m*	Erdsphäroid *n*
14	international ellipsoid	ellipsoïde international *m*	internationales Ellipsoid *n*
15	local ellipsoid	ellipsoïde local *m*	lokales Ellipsoid *n*
16	geoid	géoïde *m*	Geoid *n*
17	cogeoid compensated geoid regularized geoid	cogéoïde *m* géoïde compensé *m* géoïde régularisé *m*	Kogeoid *n* kompensiertes Geoid *n* regularisiertes Geoid *n*
18	reference ellipsoid	ellipsoïde de référence *m*	Bezugsellipsoid *n* Referenzellipsoid *n*
19	geodetic reference system	système géodésique de référence *m*	geodätisches Bezugssystem [Referenzsystem] *n*
20	geodetic datum datum	donnée géodésique *f*	geodätische Ausgangsdaten *npl* geodätischer Bezug *m*
21	arc of meridian	arc de méridien *m*	Meridianbogen *m*

geociencias *fpl* ciencias de la Tierra *fpl*	geociências *fpl* ciências da Terra *fpl*	землеведение *n* науки о Земле *fpl*	1
Tierra *f*	Terra *f*	Земля *f*	2
terrestre	terrestre	земной наземный	3
Unión Internacional de Geodesia y Geofísica *f*	União Internacional de Geodesia [Geodésia] e Geofísica *f*	Международный геодезический и геофизический союз *m*	4
superficie terrestre *f* superficie de la Tierra *f*	superfície terrestre *f* superfície da Terra *f*	земная поверхность *f* поверхность Земли *f*	5
curvatura de la Tierra *f* curvatura terrestre *f*	curvatura da Terra *f* curvatura terrestre *f*	кривизна земной поверхности *f* земная кривизна *f*	6
forma de la Tierra *f* figura de la Tierra *f*	figura da Terra *f* forma da Terra *f*	фигура Земли *f*	7
aplanamiento de la Tierra *m* achatamiento terrestre *m*	achatamento da Terra *m*	сжатие Земли *n* сплюснутость Земли *f*	8
abultamiento ecuatorial *m*	bojo [convexidade] equatorial *m*[*f*]	экваториальная выпуклость *f* экваториальное вздутие *n* экваториальный выступ *m*	9
globo terrestre *m* esfera terrestre *f*	globo terrestre *m* esfera terrestre *f*	земной шар *m*	10
elipsoide terrestre *m*	elipsóide terrestre *m*	земной эллипсоид *m*	11
poliedro *m*	poliedro *m*	полиэдр *m*	12
esferoide terrestre *m*	esferóide terrestre *m*	земной сфероид *m*	13
elipsoide internacional *m*	elipsóide internacional *m*	международный эллипсоид *m*	14
elipsoide local *m*	elipsóide local *m*	местный эллипсоид *m*	15
geoide *m*	geóide *m*	геоид *m*	16
geoide compensado *m*	geóide compensado *m* geóide regularizado *m*	когеоид *m* компенсированный геоид *m* регуляризированный геоид *m*	17
elipsoide de referencia *m*	elipsóide de referência *m*	эллипсоид относимости *m* отсчётный эллипсоид *m* референц-эллипсоид *m* эллипсоид сравнения *m*	18
sistema geodésico de referencia *m*	sistema geodésico de referência *m*	геодезическая система относимости *f*	19
datum geodésico *m* referencia geodésica *f*	datum geodésico *m* referência geodésica *f*	элементы ориентирования *mpl* исходные геодезические данные [даты]	20
arco de meridiano *m*	arco de meridiano *m*	дуга меридиана *f*	21

22	arc measurement measurement of a degree	mesure d'arc *f* mesure du degré *f*	Gradmessung *f*
23	triangulation	triangulation *f*	Triangulation *f*
24	primary triangulation first-order triangulation	triangulation fondamentale *f* triangulation du premier ordre *f*	Haupttriangulation *f* Triangulation erster Ordnung *f*
25	spatial triangulation	triangulation spatiale *f*	räumliche Triangulation *f*
26	spatial aerotriangulation [phototriangulation]	aérotriangulation [phototriangulation] spatiale *f*	räumliche Aerotriangulation [Phototriangulation] *f*
27	satellite triangulation	triangulation par satellite *f*	Satellitentriangulation *f*
28	cosmic triangulation	triangulation cosmique *f*	kosmische Triangulation *f*
29	trilateration	trilatération *f*	Trilateration *f*
30	subsatellite point	point subsatellite *m*	Subsatelliten-Punkt *m*
31	space geodesy	géodésie spatiale *f*	kosmische Geodäsie *f*
32	satellite geodesy	géodésie par satellite *f*	Satellitengeodäsie *f*
33	satellite station	station d'observations de satellites *f*	Satellitenstation *f*
34	geodetic satellite network	réseau géodésique de satellites *m*	geodätisches Satellitennetz *n*
35	simultaneous observations	observations simultanées *fpl*	Simultanbeobachtungen *fpl* gleichzeitige Beobachtungen *fpl*
36	quasi-simultaneous observations	observations quasi simultanées *fpl*	fast gleichzeitige Beobachtungen *fpl* quasisimultane Beobachtungen *fpl*
37	geodetic connection of continents	jonction géodésique des continents *f*	geodätische Verbindung von Kontinenten *f*
38	world geodetic network	réseau géodésique mondial *m*	geodätisches Weltnetz *n*
39	world geodetic system	système géodésique mondial *m*	geodätisches Weltsystem *n*
40	map projection cartographic projection	projection cartographique *f*	Kartenprojektion *f* kartographische Abbildung *f*
41	equiangular projection conformal projection	projection équiangle *f* projection conforme *f*	winkeltreue Projektion *f* konforme Projektion *f*
42	conic projection	projection conique *f*	Kegelprojektion *f*
43	cylindrical projection	projection cylindrique *f*	zylindrische Projektion *f* Zylinderprojektion *f*
44	azimuthal projection	projection azimutale *f*	azimutale Projektion *f* Azimutalprojektion *f*

medida del grado *f*	medição do grau *f*	градусное измерение *n*	22
triangulación *f*	triangulação *f*	триангуляция *f*	23
triangulación primaria *f* triangulación de primer orden *f*	triangulação de 1ª ordem *f* triangulação fundamental *f*	основная триангуляция *f* триангуляция первого класса *f*	24
triangulación espacial *f*	triangulação espacial *f*	пространственная триангуляция *f*	25
aerotriangulación [fototriangulación] espacial *f*	aerotriangulação [fototriangulação] espacial *f*	пространственная аэротриангуляция *f*	26
triangulación por satélites *f*	triangulação por satélites *f*	спутниковая триангуляция *f*	27
triangulación cósmica *f*	triangulação cósmica *f*	космическая триангуляция *f*	28
trilateración *f*	trilateração *f*	трилатерация *f*	29
punto subsatélite *m*	ponto subsatélite *m*	подспутниковая точка *f*	30
geodesia espacial *f*	geodesia [geodésia] espacial *f*	космическая геодезия *f*	31
geodesia por satélite *f*	geodesia [geodésia] por satélite *f*	спутниковая геодезия *f*	32
estación (de observación) de satélites *f*	estação (de observação) de satélites *f*	спутниковая станция *f*	33
red geodésica de satélites *f*	rede geodésica de satélites *f*	геодезическая спутниковая сеть *f*	34
observaciones simultáneas *fpl*	observações simultâneas *fpl*	синхронные наблюдения *npl*	35
observaciones c(u)asisimultáneas *fpl*	observações quasesimultâneas *fpl*	квазисинхронные наблюдения *npl* почти одновременные наблюдения *npl*	36
enlace geodésico intercontinental *f*	enlace geodésico intercontinental *f*	геодезическое соединение материков *n*	37
red geodésica mundial *f*	rede geodésica mundial *f*	мировая геодезичсская сеть *f*	38
sistema geodésico mundial *m*	sistema geodésico mundial *m*	всемирная геодезическая система *f*	39
proyección cartográfica *f*	projeção cartográfica *f*	картографическая проекция *f*	40
proyección conforme *f*	projeção conforme *f*	конформная проекция *f*	41
proyección cónica *f*	projeção cônica *f*	коническая проекция *f*	42
proyección cilíndrica *f*	projeção cilíndrica *f*	цилиндрическая проекция *f*	43
proyección azimutal *f* proyección acimutal *f*	projeção azimutal *f*	азимутальная проекция *f*	44

45	zenithal projection	projection zénithale *f*	Zenitprojektion *f*
46	gnomonic projection	projection gnomonique *f*	gnomonische Projektion [Abbildung] *f*
47	stereographic projection	projection stéréographique *f*	stereographische Projektion *f*
48	central projection	projection centrale *f*	Zentralprojektion *f*
49	map scale	échelle de la carte *f*	Kartenmasstab *m*
50	satellite altimetry	altimétrie par satellite *f*	Satellitenhöhenmessung *f* Satellitenaltimetrie *f*
51	altitude height above sea level elevation	altitude [élévation] au-dessus du niveau de la mer *f*	Meereshöhe *f* Höhe über dem Meeresniveau *f*
52	ellipsoidal height [altitude] geodetic height [altitude]	altitude ellipsoïdale *f*	ellipsoidische Höhe *f*
53	contour line isohypse	courbe de niveau *f*	Höhenlinie *f* Höhenschichtlinie *f*
54	bathymetric chart bathymetric map	carte bathymétrique *f* carte de profondeurs *f*	Tiefenkarte *f* bathymetrische Karte *f*
55	isobath bathymetric contour depth contour	courbe bathymétrique *f* isobathe *f* ligne de même profondeur *f*	Isobathe *f* Tiefenkurve *f* Tiefenlinie *f* Linie gleicher Wassertiefe *f*
56	gravity	pesanteur *f* gravité *f*	Schwere *f*
57	gravity field	champ de pesanteur *m* champ de gravité *m*	Schwerefeld *n*
58	gravity potential	potentiel de pesanteur *m* potentiel de gravité *m*	Schwerepotential *n*
59	harmonic coefficients dynamic coefficients	coefficients harmoniques *mpl*	harmonische Koeffizienten *mpl*
60	geopotential gravity potential of Earth	géopotentiel *m* potentiel terrestre *m*	Geopotential *n* Schwerepotential der Erde *n*
61	geopotential surface equipotential surface (of gravity field)	surface géopotentielle *f* surface équipotentielle (du champ de pesanteur) *f*	Geopotentialfläche *f* Äquipotentialfläche des Schwerefeldes *f* Niveaufläche (des Schwerefeldes) *f*
62	intensity of gravity	intensité de la pesanteur *f*	Schwereintensität *f* Schwerkraft *f*
63	gravimetry	gravimétrie *f*	Gravimetrie *f*
64	acceleration of gravity gravity acceleration	accélération de la pesanteur *f*	Schwerebeschleunigung *f* Schwerkraftbeschleunigung *f*

proyección zenital *f* proyección cenital *f*	projeção zenital *f*	зенитная проекция *f*	45
proyección gnomónica *f*	projeção gnomônica *f*	гномоническая проекция *f*	46
proyección estereográfica *f*	projeção estereográfica *f*	стереографическая проекция *f*	47
proyección central *f*	projeção central *f*	центральная проекция *f*	48
escala del mapa *f* escala de la carta *f*	escala do mapa *f* escala da carta *f*	масштаб карты *m*	49
altimetría por satélite(s) *f*	altimetria por satélite(s) *f*	спутниковое измерение высот *n* спутниковая альтиметрия *f*	50
altitud sobre el nivel del mar *f*	altitude sobre o nível do mar *f*	высота над уровнем моря *f*	51
altitud sobre el elipsoide (de referencia) *f*	altitude sobre o elipsóide (de referência) *f*	эллипсоидическая высота *f* высота над эллипсоидом *f*	52
curva de nivel *f*	curva de nível *f*	горизонталь *f* линия равных высот *f* изогипса *f*	53
carta batimétrica *f* mapa batimétrico *m*	carta batimétrica *f* mapa batimétrico *m*	батиметрическая карта *f* карта глубин *f*	54
isobata *f* curva batimétrica *f*	isóbata *f* curva batimétrica *f* curva de igual profundidade *f*	изобата *f* линия равных глубин *f* батиметрическая линия *f*	55
gravedad *f*	gravidade *f*	тяжесть *f* сила тяжести *f*	56
campo de gravedad *m*	campo da gravidade *m*	поле силы тяажести *n*	57
potencial de gravedad *m*	potencial da gravidade *m*	потенциал силы тяжести *m*	58
coeficientes armónicos *mpl*	coeficientes harmônicos *mpl*	Стоксовы постоянные *fpl* постоянные Стокса *fpl*	59
geopotencial *m* potencial terrestre *m*	geopotencial *m* potencial terrestre *m*	геопотенциал *m* потенциал Земли *m*	60
superficie equipotencial (de gravedad) *f* superficie geopotencial *f*	superfície equipotencial (de gravidade) *f* superfície geopotencial *f*	геопотенциальная поверхность *f* эквипотенциальная поверхность поля тяжести *f*	61
intensidad de la gravedad *f* fuerza de la gravedad *f*	intensidade da gravidade *f* força da gravidade *f*	напряжение силы тяжести *n* интенсивность силы тяжести *f*	62
gravimetría *f*	gravimetria *f*	гравиметрия *f*	63
aceleración de la gravedad *f*	aceleração da gravidade *f*	ускорение силы тяжести *n*	64

65	gal (1 cm sec^{-2})	gal *m*	Gal *n*
66	milligal (10^{-3} cm sec^{-2})	milligal *m*	Milligal *n* mGal
67	determination of gravity measurement of gravity	détermination de la pesanteur *f* mesure de la pesanteur *f*	Schweremessung *f*
68	absolute measurement of gravity	mesure absolue de la pesanteur *f*	absolute Schweremessung *f*
69	relative measurement of gravity relative gravity measurement	mesure relative de la pesanteur *f*	relative Schweremessung *f*
70	gravimeter	gravimètre *m*	Gravimeter *n*
71	vertical plumb line true normal	verticale *f* ligne de fil à plomb *f*	Lotlinie *f* Vertikale *f*
72	plumb-line deflection deflection of the vertical	déviation de la verticale *f*	Lotabweichung *f* Lotablenkung *f*
73	horizontal pendulum	pendule horizontal *m*	Horizontalpendel *n*
74	geoid undulation	ondulation du géoïde *f*	Geoidundulation *f*
75	gravity anomaly gravimetric anomaly	anomalie de la pesanteur *f* anomalie gravimétrique *f*	Schwereanomalie *f* gravimetrische Anomalie *f*
76	international gravity formula international gravimetric formula	formule internationale de la pesanteur *f* formule internationale gravimétrique *f*	internationale Schwereformel *f* internationale gravimetrische Formel *f*
77	Clairaut's theorem	théorème de Clairaut *f*	Clairautsches Theorem *n*
78	free-air reduction Faye reduction	réduction à l'air libre *f* réduction de Faye *f*	Freiluftreduktion *f* Faye-Reduktion *f*
79	gravimetric map gravity (contour) map gravimetric chart	carte de la pesanteur *f* carte gravimétrique *f*	Schwerekarte *f* gravimetrische Karte *f*
80	gravimetric network gravity network	réseau gravimétrique *m* réseau de pesanteur *m*	Schwerenetz *n* gravimetrisches Netz *n*
81	World gravity network	Réseau gravimétrique mondial *m*	Welt-Schwerenetz *n*

gal *m*	gal *m*	гал *m* гл	65
miligal *m*	miligal *m*	миллигал *m* мгал	66
determinación de la gravedad *f* medida [medición] de la gravedad *f*	determinação (da intensidade) da gravidade *f* medição [medida] da gravidade *f*	определение силы тяжести *n* измерение силы тяжести *n*	67
medida [medición] absoluta de la gravedad *f*	medição [medida] absoluta da gravidade *f*	измерение абсолютной величины силы тяжести *n* абсолютное измерение (ускорения) силы тяжести *n*	68
medida [medición] relativa de la gravedad *f*	medição [medida] relativa da gravidade *f*	относительное измерение (ускорения) силы тяжести *n* измерение относительной величины силы тяжести *n*	69
gravímetro *m*	gravímetro *m*	гравиметр *m*	70
vertical *f* plomada *f* vertical astronómica *f*	linha de prumo *f* linha vertical *f* vertical *f*	линия отвеса *f* отвесная линия *f* вертикальная линия *f*	71
desviación de la vertical *f*	desvio da vertical *m*	отклонение отвеса *n* уклонение отвеса *n* вертикальное отклонение *n*	72
péndulo horizontal *m*	pêndulo horizontal *m*	горизонтальный маятник *m*	73
ondulación del geoide *f*	ondulação do geóide *f*	уклонения геоида *npl* волнистость [ундуляция] геоида *f*	74
anomalía de la gravedad *f* anomalía gravimétrica *f*	anomalia da gravidade *f* anomalia gravimétrica *f*	аномалия силы тяжести *f* гравиметрическая аномалия *f*	75
fórmula internacional de la gravedad *f* fórmula internacional gravimétrica *f*	fórmula internacional da gravidade *f* fórmula internacional gravimétrica *f*	международная гравиметрическая формула *f* международная формула нормального значения силы тяжести *f*	76
teorema de Clairaut *m*	teorema de Clairaut *m*	теорема Клеро *f*	77
reducción al aire libre *f* reducción Faye *f*	redução ao ar livre *f* redução de Faye *f*	редукция за свободный воздух *f* редукция Фая *f*	78
mapa de gravedad *m* mapa gravimétrico *m* carta gravimétrica *f*	mapa gravimétrico *m* carta gravimétrica *f* mapa da gravidade *m*	гравиметрическая карта *f* карта силы тяжести *f*	79
red gravimétrica *f* red de gravedad *f*	rede gravimétrica *f* rede de gravidade *f*	гравиметрическая сеть сеть силы тяжести *f f*	80
Red gravimétrica mundial *f*	Rede gravimétrica mundial *f*	Мировая гравиметрическая сеть *f*	81

82	fundamental gravity [gravimetric] network	réseau gravimétrique fondamental *m*	Schweregrundnetz *n*
83	internal constitution	structure interne *f*	innerer Aufbau *m*

red gravimétrica fundamental *f*	rede gravimétrica básica *f* rede gravimétrica fundamental *f*	опорная гравиметрическая сеть *f*	82
estructura interna *f*	estrutura interna *f*	внутреннее строение *n*	83

72 GEOPHYSICS — GÉOPHYSIQUE — GEOPHYSIK

	GEOPHYSICS	GÉOPHYSIQUE	GEOPHYSIK
1	geophysics	géophysique *f*	Geophysik *f*
2	geology	géologie *f*	Geologie *f*
3	Earth's interior interior of the Earth	intérieur de la Terre *m* intérieur terrestre *m*	Erdinneres *n* Inneres der Erde *n*
4	Earth's crust	croûte terrestre *f* écorce de la Terre *f*	Erdkruste *f*
5	continental crust	croûte continentale *f*	Kontinentalkruste *f*
6	lithosphere	lithosphère *f*	Lithosphäre *f*
7	geosphere	géosphère *f*	Geosphäre *f*
8	biosphere	biosphère *f*	Biosphäre *f*
9	continent	continent *m*	Kontinent *m*
10	continental block	bloc continental *m*	Kontinentalblock *m*
11	shield	bouclier *m*	Schild *m*
12	basement basement complex	socle *m* complexe fondamental *m* complexe de base *m*	Sockel *m* Fundament *n* Grundgebirge *n*
13	transition layer transition zone	couche intermédiaire *f* zone de transition *f*	Zwischenschicht *f* Übergangszone *f*
14	Mohorovičić discontinuity Moho discontinuity crust-mantle interface crust-mantle transition	discontinuité de Mohorovičić *f* discontinuité de Moho *f* transition croûte-manteau *f*	Mohorovičić Diskontinuität *f* Moho-Diskontinuität *f* Kruste-Mantel-Grenzfläche *f*
15	mantle Earth's mantle	manteau *m* manteau terrestre *m*	Mantel *m* Erdmantel *m*
16	upper mantle	manteau supérieur *m*	oberer Erdmantel *m*
17	(Earth's) core siderosphere barysphere centrosphere	noyau terrestre *m* barysphère *f*	Erdkern *m* Kern der Erde *m* Barysphäre *f*
18	nife	nifé *m* noyau de nifé *m*	Nife *n* Nifekern *m*
19	outer core	noyau externe *m*	äusserer Kern *m*

GEOFÍSICA	GEOFÍSICA	ГЕОФИЗИКА	
geofísica *f*	geofísica *f*	геофизика *f* физика Земли *f*	1
geología *f*	geologia *f*	геология *f*	2
interior de la Tierra *m*	interior da Terra *m*	земные недра *pl* недра Земли *pl*	3
corteza [costra] terrestre *f* corteza [costra] de la Tierra *f*	crosta terrestre *f* crosta da Terra *f*	земная кора *f* кора Земли *f*	4
corteza continental *f*	crosta continental *f*	континентальная кора *f* материковая кора *f*	5
litosfera *f*	litosfera *f*	литосфера *f*	6
geosfera *f*	geosfera *f*	геосфера *f*	7
biosfera *f*	biosfera *f*	биосфера *f*	8
continente *m*	continente *m*	материк *m* континент *m*	9
bloque continental *m*	bloco continental *m*	континентальный массив *m*	10
escudo *m*	escudo *m*	щит *m* континентальная глыба *f* геоблок *m*	11
basamento *m* complejo basal *m*	basamento *m* complexo basal *m*	кристаллический фундамент *m* цоколь *m*	12
zona de transición *f* capa de transición *f*	camada de transição *f* camada intermediária *f*	зона перехода *f* переходный слой *m* переходная область [зона] *f* промежуточная область *f*	13
discontinuidad de Mohorovičić *f* discontinuidad Moho *f* transición corteza-manto *f*	discontinuidade de Mohorovičić *f* discontinuidade Moho *f* transição crosta-manto *f*	поверхность [граница] Мохоро-вичича *f* граница Мохо [M] *f* граница кора-мантия *f*	14
manto de la Tierra *m* manto terrestre *m*	manto terrestre *m* manto da Terra *m*	мантия *f* мантия Земли *f*	15
manto superior *m*	manto superior *m*	верхняя мантия *f* область B *f*	16
núcleo de la Tierra *m* núcleo terrestre *m* barisfera *f*	núcleo terrestre *m* barisfera *f* núcleo da Terra *m*	земное ядро *n* ядро Земли *n*	17
nife *m* núcleo central *m*	nife *m* núcleo central *m*	нифе *n* ядро нифе *n*	18
núcleo externo *m*	núcleo externo *m*	внешнее ядро (Земли) *n* внешняя часть ядра *f*	19

20	inner core	noyau interne *m* grain *m*	Innenkern *m* innerer Erdkern *m*
21	hydrosphere	hydrosphère *f*	Hydrosphäre *f*
22	ocean	océan *m*	Ozean *m*
23	oceanographic map oceanographic chart	carte océanographique *f*	ozeanographische Karte *f* Meereskarte *f*
24	oceanography	océanographie *f*	Ozeanographie *f* Meeresforschung *f*
25	hydrologic map hydrologic chart	carte hydrologique *f*	Gewässerkarte *f* hydrologische Karte *f*
26	mediterranean sea inland sea continental sea	mer méditerranée *f* mer intérieure *f* mer intracontinentale *f*	Mittelmeer *n* Binnenmeer *n* kontinentales Meer *n*
27	marginal sea	mer marginale *f* mer limitrophe *f*	Randmeer *n*
28	bathymetry	bathymétrie *f*	Meerestiefenmessung *f* Tiefenmessung *f*
29	echo depth sounding echo sounding	écho-sondages *fpl* sondage acoustique *f*	Tiefseelotung *f* Echolotung *f*
30	shore coast	littoral *m* côte *f*	Meeresküste *f* Küste *f*
31	shoreline coastline	ligne littorale *f* trait de côte *m*	Küstenlinie *f*
32	submerged contour	courbe de niveau (dans un lac) *f*	Höhenlinie unter Wasser *f*
33	coastal chart (UK) sea chart (USA)	carte (nautique) d'atterrissage *f*	Segelkarte *f* Seekarte *f*
34	continental shelf	plate-forme continentale *f* plateau continetal *m*	Kontinentalschelf *m* (kontinentaler) Schelf *m*
35	continental slope	pente continentale *f*	Kontinentalhang *m* Kontinentalböschung *f*
36	continental apron	talus continental *m*	Böschung am Fuss des Kontinentalhanges *f* Kontinentalfuss *m*

núcleo interno *m*	núcleo interno *m*	внутреннее ядро (Земли) *n* субъядро *n*	20
hidrosfera *f*	hidrosfera *f*	гидросфера *f* водная оболочка Земли *f*	21
océano *m*	oceano *m*	океан *m*	22
mapa oceanográfico *m*	carta oceanográfica *f*	океанографическая карта *f*	23
oceanografía *f*	oceanografia *f*	океанография *f*	24
mapa hidrográfico *m*	carta hidrográfica *f*	гидрологическая карта *f*	25
mar mediterráneo *m* mar continental *m* mar interior *m*	mar mediterrâneo *m* mar continental *m* mar interior *m*	средиземное море *n* внутриматериковое море *n* внутреннее море *n* внутриконтинентальное море *n*	26
mar marginal *m*	mar marginal *m*	окраинное море *n* эпиконтинентальное море *n* краевое море *n*	27
batimetría *f*	batimetria *f*	батиметрия *f* измерение глубин *n* глубоководный промер *m* промер глубины *m*	28
sondeos acústicos *mpl* ecosondeos *mpl* sondeos por eco *mpl* eco-sondagens *fpl*	sondagens acústicas *fpl* eco-sondagens *fpl*	измерение глубин эхолотом *n* эхолотирование *n* эхолотный промер *m*	29
litoral *m* costa *f*	litoral *m* costa *f*	берег *m*	30
línea de la costa *f*	linha litorânea *f* contorno da costa *m*	береговая линия *f*	31
curva de nivel submergida *f*	curva de nível submergida [submersa] *f*	линия равных глубин *f* подводная горизонталь *f*	32
carta de arrumbamiento *f* carta de recalar *f*	carta de marear *f* carta náutica *f*	путевая (береговая) карта *f*	33
plataforma continental *f* zócalo continental *m*	plataforma continental *f* banqueta submarina *f*	континентальная отмель *f* континентальная платформа *f* шельф *m*	34
talud continental *m*	talude continental *m*	континентальный склон *m* материковый склон *m*	35
falda continental *f* batiente continental *m*	talude continental de inclinação suave *m*	материковое подножие *n* континентальное подножие *n*	36

	GEOPHYSICS	GÉOPHYSIQUE	GEOPHYSIK
37	ocean bed ocean floor	fond des océans *m* fond océanique *m*	Ozeanboden *m* Ozeangrund *m*
38	sea floor sea bed	fond de la mer *m* fond marin *m*	Seeboden *m* Meeresboden *m* Meeresgrund *m*
39	ocean-floor spreading	expansion des fonds océaniques *f*	Verbreiterung des Ozeanbodens *f*
40	(ocean) trench (marine) trench	fosse océanique *f* fosse marine *f*	ozeanischer Graben *m* Tiefseegraben *m* Tiefseerinne *f*
41	abyssal (zone) deep ocean (2500–6000 m)	(zone) abyssale *f* abysse *m* profondeurs océaniques *fpl*	Tiefsee *f* Abyssal *n* Abyssalzone *f*
42	abyssal plain	plaine abyssale *f*	Tiefsee-Ebene *f*
43	seamount submarine mountain	mont sous-marin *m* montagne sous-marine *f*	Tiefseeberg *m* Tiefseekuppe *f* submariner Berg *m* Tiefseehügel *m*
44	guyot (submarine) tablemount flat-topped seamount	guyot *m* mont tabulaire sous-marin *m*	Guyot *m* abgestumpfter Tiefseeberg *m*
45	mid-ocean ridge	dorsale médio-océanique *f*	mittelozeanischer Rücken *m* mittelozeanische Schwelle *f*
46	submarine canyon	canyon sous-marin *m*	submarines Flusstal *n* submariner Cañon *m*
47	submarine sedimentation	sédimentation sous-marine *f*	(sub)marine Sedimentation *f* Meeresablagerung *f*
48	terrigenous deposits terrigenous sediments	dépôts terrigènes *mpl* sédiments terrigènes *mpl*	terrigene Ablagerungen [Sedimente] *f[n]pl* vom Festland stammende Ablagerungen [Sedimente] *f[n]pl*
49	turbidity current suspension current	courant de turbidité *m* courant de suspension *m*	Trübestrom *m* Suspensionsstrom *m*
50	deep-sea sediments [deposits] *pl* abyssal deposits *pl* deep-ocean deposits [sediments]	sédiments de mer profonde *mpl* dépôts abyssaux *mpl* sédiments [dépôts] de profondeurs océanigues *mpl*	Tiefseeablagerungen *fpl* Tiefseesedimente *npl* Abyssalablagerungen *fpl* Abyssalsedimente *npl*

suelo oceánico *m* fondo oceánico *m* lecho oceánico *m* fondo [suelo] del océano *m*	assoalho oceânico *m* fundo oceânico *m* leito oceânico [do oceano] *m* solo oceânico [do oceano] *m*	дно океана *n* ложе океана *n* океаническое дно *n* океанское дно *n*	37
suelo marítimo *m* lecho marítimo *m* fondo del mar *m*	solo marítimo [do mar] *m* leito marítimo *m* fundo marítimo *m*	дно моря *n* морской грунт *m* морское дно *n*	38
extensión del fondo oceánico *f*	expansão do fundo [assoalho] oceânico *f*	раздвигание дна океана *n* расширение морского дна *n*	39
fosa oceánica *f* fosa marina *f*	fossa marítima *f* fossa oceânica *f*	океанский глубоководный жёлоб [жолоб] *m* глубоководная впадина *f*	40
región abis(m)al *f* abismo *m*	abissal *m* zona abissal *f*	глубокий океан *m* абиссаль *f* абиссальная зона [равнина] *f*	41
planicie abis(m)al *f*	planície abissal *f*	абиссальная равнина *f*	42
monte submarino *m* montaña submarina *f*	monte [pico] submarino *m* montanha submarina *f*	подводная гора *f* подводная вершина *f*	43
monte tabular *m* guyot *m* mesa submarina *f*	planalto (submarino) *m* guyot *m* mesa submarina *f*	плосковершинная подводная гора *f* гайот *m* гийот *m*	44
cresta medio-oceánica *f* dorsal centro-oceánica *f*	dorsal (médio-)oceânica *f* crista médio-oceânica *f*	срединно-океанический хребет *m*	45
cañon submarino *m* desfiladero submarino *m*	canion submarino *m* canhão submarino *m*	подводный каньон *m*	46
sedimentación submarina *f*	sedimentação submarina *f*	морская седиментация *f* морское осадконакопление *n* морское осадкообразование *n* подводная аккумуляция *f*	47
depósitos terrígenos *mpl* sedimentos terrígenos *mpl*	depósitos terrígenos *mpl* sedimentos terrígenos *mpl*	терригенные осадки [отложения] *m*[*n*] *pl* осадки образованные с суши *pl*	48
corriente de turbidez [suspensión] *f* corriente de turbiedad *f*	corrente de turbidez *f* corrente de suspensão *f*	мутьевой поток *m* суспензионный поток *m*	49
sedimentos marinos profundos *mpl* sedimentos abis(m)ales *mpl* depósitos abisales *mpl* sedimentos en [de] región abis(m)al *mpl*	sedimentos marinhos profundos *mpl* sedimentos [depósitos] abissais *mpl* sedimentos [depositos] na zona abissal *mpl*	глубоководные осадки [отложения] *mpl* [*npl*] абиссальные осадки [отложения] *m*[*n*]*pl* глубинные [глубоководные] морские отложения *npl*	50

	GEOPHYSICS	GÉOPHYSIQUE	GEOPHYSIK
51	coring	carottage *m*	Kernen *n* Kerngewinnung *f*
52	core	carotte *f*	Bohrkern *m*
53	dredging	dragage *m*	Nassbaggern *n* Nassbaggerung *f* Unterwasserbaggerung *f*
54	red clay	argile rouge *f*	roter Ton *m*
54a	pelagic clay	argile pélagique *f*	pelagischer Ton *m*
55	nickel content	teneur en nickel *f*	Nickelgehalt *m*
56	extraterrestrial material	matériau extraterrestre *m*	extraterrestrisches Material *n* ausserirdisches Material *n*
57	pelagic sediments pelagic deposits	sédiments pélagiques *mpl* dépôts pélagiques *mpl*	pelagische Sedimente *npl* pelagische Ablagerungen *fpl*
58	ooze	vase *f* boue *f*	Schlick *m* Schlamm *m*
59	organogenous sediments biogenic deposits	sédiments biogéniques *mpl* dépôts organogéniques *mpl*	organogene Sedimente *npl* biogene Ablagerungen *fpl*
60	ocean current sea current	courant océanique *m* courant marin *m*	ozeanischer Strom *m* Meeresströmung *f* Meeresstrom *m*
61	subsurface (ocean) current	courant de subsurface *m*	(ozeanische) Tiefenströmung *f*
62	surface current	courant de surface *m*	Oberflächenströmung *f*
63	warm current	courant (marin) chaud *m*	Wärmestrom *m* warmer Meeresstrom *m*
64	cold current	courant (marin) froid *m*	kalter Meeresstrom *m*
65	ebb current	courant de jusant *m*	Ebbestrom *m* Ebb(e)strömung *f*
66	salinity salt content	salinité *f* teneur en sel *f*	Salzhaltigkeit *f* Salzgehalt *m* Salinität *f*
67	salinity interface	surface limite de salinité *f*	Salinitätsgrenzfläche *f*
68	chlorinity chlorine content	teneur en chlore *f* chlorinité *f*	Chlorgehalt *m* Chlormenge *f*

sondeo con taladro *m* extracción de testigos *f*	(per)furação *f* testemunho *m*	отбор керна *m* получение колонок донного грунта *n* отбор грунтов(ых проб) *m*	51
muestra obtenida con taladro *f* testigo *m*	núcleo de furação *m*	керн *m* колонка *f*	52
dragado *m* dragaje *m*	dragagem *f*	драгирование *n* драгировка *f*	53
arcilla roja *f*	argila vermelha *f*	красные (глубоководные) глины *fpl*	54
arcilla pelágica *f*	argila pelágica *f*	пелагические глины *fpl*	54a
contenido de níquel *m*	conteúdo em níquel *m*	содержание никеля *n*	55
material extraterrestre *m*	material extraterrestre *m*	внеземное вещество *n* внеземной материал *m*	56
sedimentos pelágicos *mpl* depósitos pelágicos *mpl*	sedimentos pelágicos *mpl* depósitos pelágicos *mpl*	пелагические осадки *mpl* пелагические отложения *npl* морские [океанические] осадки *mpl*	57
limo *m* lodo *m* fango *m*	lodo *m* lama *f*	ил *m*	58
sedimentos organógenos *mpl* depósitos biogénicos *mpl*	sedimentos organogênicos *mpl* depósitos biogênicos *mpl*	биогенные отложения [осадки] *n*[*м*]*pl* органические отложения *npl*	59
corriente oceánica *f* corriente marina *f*	corrente oceânica *f* corrente marinha *f*	морское течение *n* океаническое течение *n*	60
corriente sub-superficial *f*	corrente subsuperficial *f*	течение под поверхностью *n*	61
corriente superficial *f*	corrente superficial *f*	поверхностное течение *n*	62
corriente (marina) cálida *f*	corrente (marinha) quente *f*	теплое (морское) течение *n*	63
corriente (marina) fría *f*	corrente (marinha) fria *f*	холодное (морское) течение *n*	64
corriente de reflujo *f*	corrente de refluxo *f*	отливное течение *n* течение отлива *n*	65
salinidad *f* cantidad del sal *f*	salinidade *f* conteúdo do sal *m*	засолённость *f* солёность *f* минерализация *f*	66
interfase de salinidad *f*	interface de salinidade *f*	граничный слой солёности *m* граница раздела солёности *f*	67
clorinidad [cloricidad] *f* cantidad del cloro *f*	clorinidade *f* conteúdo do cloro *m*	содержание хлора *n* количество хлора *n*	68

69	space oceanography	océanographie spatiale *f*	Raumozeanographie *f*
70	satellite oceanography	océanographie par satellite *f*	Satellitenozeanographie *f*
71	oil spill	déversement de pétrole *m*	Ölauslaufverlust *m*
72	pollution event	cas de pollution *m* événement de pollution *m*	Verschmutzungsereignis *n* Verunreinigungsfall *m*
73	water pollution	pollution de l'eau *f*	Wasserverunreinigung *f*
74	pollutant	polluant *m*	Verunreinigungsstoff *m*
75	flood inundation	inondation *f*	Überflutung *f* Überschwemmung *f*
76	storm surge sea surge	onde de tempête *f*	Sturmflut *f*
77	tidal-wave damage	dégât par ondes de marées *m*	Flutwellenschaden *m* Gezeitenwellenschaden *m*
78	coastline [shoreline] altering coastline modification	changement de la ligne littorale *m* modification du trait de côte *f*	Küstenveränderung *f* Küstenlinienveränderung *f*
79	off-shore silting	sédimentation fine au large *f*	Verschlickung [Verschlammung] in äusserem Schelf *f*
80	sea discoloration	changement de coloration de la mer *m*	Farbänderung der See *f*
81	new island	nouvelle île *f*	neue Insel *f*
82	coast mapping	levé de côte *m* cartographie de littoral *f*	Küstenkartierung *f* Küstenaufnahme *f*
83	thermal mapping	tracé de température *m* cartografie thermique *f*	thermische Kartierung *f* thermische Bildaufnahne *f*
84	fish location	repérage des poissons *m*	Fischortung *f*
85	ice surveillance ice survey	surveillance des icebergs *f*	Eis(berg)überwachung *f* Eisüberwachungsdienst *m*
86	sea state motion of the sea	état de la mer *m* agitation de la mer *f*	Meereszustand *m* Seegang *m*
87	sea-level	niveau de la mer *m*	Meeresniveau *n* Meeresspiegel *m* Normalnull *f*

GEOFÍSICA	GEOFÍSICA	ГЕОФИЗИКА	
oceanografía espacial *f*	oceanografia espacial *f*	космическая океанография *f*	69
oceanografía por satélite *f*	oceanografia por satélite *f*	спутниковая океанография *f*	70
derrame de petróleo *m* vertido de petróleo *m*	derrame de óleo *m* vazamento de óleo *m*	разлитие нефти *n* утечка нефти *f*	71
suceso de contaminación *m* caso de contaminación *m*	evento de poluição *m* evento poluente *m*	случай загрязнения *m* происшествие загрязнения *n*	72
contaminación del agua *f*	poluição da água *f*	загрязнение воды *n*	73
contaminante *m*	poluente *m*	загрязнитель *m* загрязняющее вещество *n*	74
inundación *f*	inundação *f*	затопление *n* наводнение *n*	75
ola ciclónica *f* marejada *f*	vaga de tempestade *f* onda de tempestade *f*	штормовой прибой *m* штормовой прилив *m*	76
daño por ondas de marea *m* destrucción por marea *f*	dano por ondas de maré *m* destruição por maré *f* dano por maré *m*	повреждение приливной волной *n* повреждение приливо-отливной волной *n*	77
modificación del litoral *f* cambio de la línea de la costa *m*	modificação da linha litorânea *f* modificação do contorno da costa *f*	изменение береговой линии *n*	78
aterramiento [sedimentación] en mar abierto [a alta mar] *m*[*f*]	sedimentação oceânica [em mar alto] *f*	заиление в открытом море *n* заиливание во внешней зоне шельфа *n*	79
descoloración del mar *f*	descoloração do mar *f*	изменение цвета моря *n*	80
isla nueva *f*	ilha nova *f*	новый остров *m*	81
mapificación costera *f* cartografía del litoral *f*	mapeamento da costa *m* cartografia do litoral *f*	съёмка морских берегов *f* береговое картирование *n*	82
mapificación de temperatura *f* cartografía térmica *f*	traçado da temperatura *m* mapeamento térmico *m*	термическое картирование *n* картографирование температуры *n*	83
localización de la zona piscícola *f*	locação de peixes [de zona piscosa] *f*	определение хода рыбы *n* определение рыболовного места *n*	84
vigilancia de hielos *f* reconocimiento de hielos *m*	levantamento de geleiras *m* vigilancia de icebergs *f*	обследование льдов *n* съёмка льдов *f* служба наблюдения за льдами *f*	85
estado de la mar *m* agitación de la mar *f*	estado do mar *m* agitação do mar *f*	волнение *n* состояние моря *n*	86
nivel del mar *m*	nível do mar *m*	уровень моря *m*	87

	GEOPHYSICS	GÉOPHYSIQUE	GEOPHYSIK
88	upwelling	upwelling *m* poussée de l'eau *f*	Wasserauftrieb *m*
89	tides flood and ebb	marées *fpl* flot et jusant *m, m*	Gezeiten *fpl* Ebbe und Flut *f, f* Tiden *fpl*
90	high tide high water	marée haute *f*	Gezeitenhochwasser *n* Tidehochwasser *n*
91	low tide low water	marée basse *f* basse-mer *f*	Gezeitenniedrigwasser *n* Niedrigwasser *n*
92	tidal forces	forces de marée *fpl*	Gezeitenkräfte *fpl* fluterzeugende Kräfte *fpl*
93	infrasonic wave	onde infrasonore *f*	Infraschallwelle *f*
94	planetary wave	onde planétaire *f*	planetarische Welle *f*
95	tsunami tsunami wave seaquake wave	tsunami *m* raz de marée *m* onde sismique marine *f*	Tsunami *n* Seebebenwelle *f* Erdbebenflutwelle *f* seismische Meereswelle *f*
96	shoal bank	haut fond *m* barre *f* banc *m*	Untiefe *f* Schwelle *f* Sandbank *m*
97	navigation hazard	danger de navigation *m*	Navigationsgefahr *f* Schiffahrtsgefahr *f*
98	chemical pollution	pollution chimique *f*	chemische Verunreinigung *f* chemische Verschmutzung *f*
99	research buoy	bouée de recherche *f*	Forschungsboje *f*
100	automatic buoy	bouée automatique *f*	automatische Boje *f*
101	water sampler	échantillonneur d'eau *m*	Wasserprobenehmer *m* Wasserprobenentnahmegerät *n*
102	hummocky ice ice hummock pack ice	hummock de glace *m* banquise *f*	Eishügel *m* Packeis *n*
103	glaciology geology of glaciers	glaciologie *f* géologie de glaciers *f*	Glaziologie *f* Glazialgeologie *f* Gletscherkunde *f*
104	glaciation	glaciation *f*	Vereisung *f* Vergletscherung *f*
105	glacial period	période glaciaire *f*	Glazialperiode *f* Eiszeit *f* Vereisungsperiode *f*

afloramiento *m*	soerguimento *m* afloramento *m*	нагон воды *m*	88
mareas *fpl* flujo y reflujo *m, m*	marés *fpl* fluxo e refluxo *m, m*	прилив и отлив *m, m* приливы *mpl* наивысший уровень прилива *m*	89
pleamar *f* marea alta *f* alta [plena] mar *f* aguas llenas *fpl*	preamar *f* maré cheia *f* maré alta *f*	полный прилив *m* полная вода *f*	90
bajamar *f* marea baja *f*	baixa-mar *f* maré vazia *f* maré baixa *f*	меженная вода *f* низкая вода *f* малая вода *f*	91
fuerzas de marea *fpl*	forças de maré *fpl*	приливообразующие силы *fpl* приливные силы *fpl* приливо-отливные силы *fpl*	92
onda infrasónica *f*	onda infra-sônica *f*	инфразвуковая волна *f*	93
onda planetaria *f*	onda planetária *f*	планетарная волна *f*	94
tsunami *m* onda sísmica de maremoto *f* gran ola *f*	tsunami *m* onda sísmica do maremoto *f*	цунами *pl* волна цунами [тсунами] *f* сейсмическая волна моретрясения *f*	95
bajío *m* barra *f* banco *m*	baixio *m* banco *m* água rasa *f*	отмель *f* мель *f* банка *f* мелководье *n*	96
riesgos de la navegación *mpl*	riscos de navegação *mpl*	опасность для судоходства *f* навигационная опасность *f*	97
contaminación química *f*	poluição química *f*	химическое загрязнение *n*	98
boya de investigación *f*	bóia de pesquisa *f*	исследовательский буй *m*	99
boya automática *f*	bóia automática *f*	автоматический буй *m*	100
recogedor de muestras de agua *m* tomamuestras de agua *m*	recolhedor de amostras de água *m* amostrador de água *m*	пробоотбиратель воды *m* устройство для забора проб воды *n*	101
(campo de) hielo flotante *m*	elevação de gelo *f* gelo flutuante *m*	плотный лёд *m* паковый лёд *m* торос *m*	102
glaciología *f* geología de hielo *f*	glaciologia *f* geologia de gelo *f*	гляциология *f* наука о легниках *f*	103
glaciación *f*	glaciação *f*	континентальное обледенение *n* оледенение *n*	104
período glacial *m* período glaciario *m*	período glacial *m* período glaciário *m*	ледниковый период *m* гляциальный период *m*	105

	GEOPHYSICS	GÉOPHYSIQUE	GEOPHYSIK
106	glacial	glaciaire	eiszeitlich glazial
107	interglacial period	période interglaciaire *f*	Warmzeit *f* Interglazial(zeit) *n(f)* Zwischeneiszeit *f*
108	interglacial	interglaciaire	interglazial zwischeneiszeitlich
109	Holocene post-glacial epoch	époque postglaciaire *f* holocène *m*	Postglazial *n* Nacheiszeit *f*
110	post-glacial	postglaciaire	postglazial nacheiszeitlich
111	snowline snow limit	limite des neiges *f*	Schneegrenze *f*
112	glacier	glacier *m*	Gletscher *m*
113	mountain glacier	glacier de montagne *m*	Gebirgsgletscher *m*
114	alpine glacier valley glacier	glacier alpin *m* glacier de vallée *m*	alpiner Gletscher *m* Hochgebirgsgletscher *m* Talgletscher *m*
115	ice-cap	calotte glaciaire *f* couverture de glace *f*	Eiskappe *f* Eisschild *m* Eisdecke *f*
116	continental glacier	glacier continental *m*	Kontinentalgletscher *m* Inlandgletscher *m*
117	iceberg	iceberg *m*	Eisberg *m*
118	firn coarse-grained snow granular ice	neige à gros grains *f* névé *m*	Firn *m* Firnschnee *m* vereister Schnee *m*
119	firn field field of compacted snow permanent snowfield	champ de névé *m* neige éternelle *f*	Firnschneefeld *n* Firnfeld *n*
120	glacier flow	écoulement de la glace (dans un glacier) *m*	Gletscherabfluss *m* Eisbewegung im Gletscher *f* Gletscherfliessen *n*
121	plastic flow	écoulement plastique *m*	plastisches Fliessen *n*
122	zone of growth (of a glacier) zone of snow-supply	zone d'accumulation (d'un glacier) *f* région d'alimentation (d'un glacier) *f*	Gletschernährungsgebiet *n* Akkumulationsgebiet *n* Nährgebiet (eines Gletschers) *n*
123	zone of ablation	région [zone] de fonte des neiges *f* zone [région] d'ablation *f*	Ablationsgebiet *n* Abschmelzgebiet *n*

glacial glaciario	glaciário glacial	ледниковый гляциальный	106
interglaciación *f* período interglacial *m* período interglaciario *m*	interglaciário *m* período interglacial *m* período integlaciário *m*	межледниковье *n* интергляциаль(ный период) *m* межледниковый период *m*	107
interglacial integlaciario	interglacial interglaciário	межледниковый интергляциальный	108
época postglacial *f* holoceno *m*	época pós-glacial [holocena] *f* holoceno *m*	послеледниковая эпоха *f* послеледниковье *n*	109
postglacial	pós-glacial	послеледниковый постгляциальный	110
línea de nieves *f* límite de las nieves *m*	linha de neve *f* limite de neve *m*	снеговая линия *f* снеговая граница *f*	111
glaciar *m* helero *m*	geleira *f* glaciar *m*	ледник *m* глетчер *m*	112
glaciar de montaña *m*	geleira [glaciar] de montanha *f*	горный ледник *m*	113
glaciar alpino *m* glaciar de valle *m*	geleira alpina *f* glaciar de vale *m*	альпийский ледник *m* долинный ледник *m* ледник альпийского типа *m*	114
casquete glaciar *m* calota de hielo *f*	calota glacial *f* capa de gelo *f*	ледниковый купол *m* ледниковый покров *m* ледниковая шапка *f*	115
glaciar continental *m*	geleira continental *f*	континентальный ледник *m* материковый ледник *m* покровный ледник *m*	116
iceberg *m*	iceberg(ue) *m* monte de gelo flutuante *m*	айсберг *m* ледяная гора *f*	117
neviza *f* ventisquero *m* nieve granular endurecida *f*	neve granulada *f* nevado *m* neve granular (endurecida) *f*	фирн *m* зернистый лёд *m* фирновый снег *m*	118
nieves perpetuas *fpl* campo de nevero *m* (campo de) ventisquero *m*	campo de nevado *m* neve eterna *f*	фирновое поле *n*	119
movimiento del glaciar *m* flujo de glaciar *m* flujo glacial *m*	movimento glaciário *m* fluxo de geleira *m* fluxo glaciário *m*	ледниковый сток *m* стекание ледника *n* течение ледника *n* движение ледника *n*	120
flujo plástico *m*	fluxo plástico *n*	пластическое течение *n*	121
zona de alimentación *f* zona de crecimiento *f* región de acumulación de un glaciar *f*	zona de crescimento *f* zona de alimentação glaciária *f* região de acumulação de uma geleira *f*	область накопления ледника *f* зона питания *f* область питания *f* область аккумуляции *f*	122
zona de ablación *f* región de ablación *f*	zona de ablação *f* região de ablação *f*	область абляции ледника *f* зона абляции *f*	123

124	zone of dissipation	zone de fusion glaciaire *f*	Schwundgebiet *n* Zehrgebiet (eines Gletschers) *n*
125	ice tongue glacier tongue	glacier d'écoulement *m* langue glaciaire *f*	Eiszunge *f* Gletscherzunge *f*
126	paleotemperature	paléotempérature *f*	Paläotemperatur *f*
127	glacial deposits *pl*	dépôts glaciaires *mpl*	glaziale Ablagerungen *fpl* Gletscherablagerungen *fpl*
128	moraine till	moraine *f*	Moräne *f* Gletscherwall *m*
129	ground moraine	moraine de fond *f*	Grundmoräne *f*
130	terminal moraine	moraine frontale *f*	Stirnmoräne *f* Endmoräne *f*
131	lateral moraine	moraine latérale *f*	Randmoräne *f* Seitenmoräne *f*
132	indicator boulder erratic block (glacial) erratic	galet indicateur *m* erratique (glaciaire) *m* bloc erratique *m*	Leitgeschiebe *n* erratischer Block *m* Findling *m*

región de fusión de un glaciar *f* zona de disipación *f*	zona de fusão glaciária *f* zona de dissipação *f*	зона таяния *f* область стаивания ледника *f*	124
lengua de hielo *f* lengua de un glaciar *f* lengua glacial *f*	língua de gelo *f* língua glaciária *f* língua da geleira *f*	ледниковый язык *m* выводной ледник *m* стоковый ледник *m* разгрузочный ледник *m*	125
paleotemperatura *f*	paleotemperatura *f*	палеотемпература *f*	126
depósitos glaciales *mpl*	depósitos glaciais *mpl*	ледниковые отложения *npl* гляциальные отложения *npl*	127
mor(r)ena *f*	morena *f*	морена *f*	128
mor(r)ena de fondo *f*	morena de fundo *f*	основная морена *f* поддонная морена *f* донная морена *f*	129
mor(r)ena terminal *f* mor(r)ena frontal *f*	morena terminal *f* morena frontal *f*	конечная морена *f* краевая морена *f*	130
mor(r)ena lateral *f*	morena lateral *f*	береговая морена *f* боковая морена *f*	131
bloque errático *m* mole errática *f*	bloco errático *m*	ледниковый валун *m* эрратический валун *m* валун-индикатор *m* ледниковая глыба *f* ледниковый отторженец *m*	132

73	SEISMICITY VOLCANISM	SISMOLOGIE VOLCANISME	ERDBEBENKUNDE VULKANE
1	seismology	sismologie *f* séismologie *f*	Seismologie *f* Erdbebenkunde *f*
2	seismologist	sismologue *m* séismologue *m*	Seismologe *m* Erdbebenforscher *m*
2a	earthquake earth tremors	tremblement de terre *m* secousses sismiques [séismiques] *fpl* séisme *m* sisme [séïsme] *m*	Erdbeben *n*
2b	inland earthquake continental earthquake	séisme terrestre *m* tremblement de terre continental *m* sisme terrestre *m*	Landbeben *n* Binnenland-Erdbeben *n* kontinentales Beben [Erdbeben] *n*
2c	submarine earthquake seaquake	séisme sous-marin *m* tremblement de terre sous-marin *m* sisme sous-marin *m*	Seebeben *n* submarines Erdbeben *n*
2d	dislocation earthquake tectonic earthquake	séisme de dislocation *m* tremblement de terre tectonique *m* séisme tectonique *m* sisme tectonique *m*	Dislokationsbeben *n* tektonisches Erdbeben *n*
2e	volcanic earthquake	séisme volcanique *m* tremblement de terre volcanique *m*	vulkanisches Erdbeben *n*
3	seismicity	sismicité *f* séismicité *f*	Seismizität *f*
4	seismic activity	activité sismique *f* activité séismique *f*	Erdbebentätigkeit *f* seismische Aktivität *f*
5	seismologic(al) observatory	observatoire sismologique *f*	seismologisches Observatorium *n*
6	seismograph(ic) station seismic station seismologic station	station sismique *f* station séismique *f* station sismologique *f*	seismologische Station *f* Erdbebenstation *f*
7	seismograph	séismographe *m* sismographe *m*	Seismograph *m* Erdbebenschreiber *m*
8	seismogram seism(ograph)ic record	sismogramme *m* séismogramme *m*	Seismogramm *n*
9	earthquake wave seismic wave	onde séismique *f* onde sismique *f*	Erdbebenwelle *f* seismische Welle *f*
10	foreshock	secousses prémonitoires *fpl*	Vorbeben *n*
11	fore-runner (of earthquake) precursor	précurseur (de tremblement de terre) *m*	Vorläufer (des Erdbebens) *m*

sismología *f* seismología *f*	sismologia *f*	сейсмология *f* наука о землетрясениях *f*	1
sismólogo *m*	sismólogo *m*	сейсмолог *m*	2
terremoto *m* temblor de tierra *m* sismo *m* seísmo *m*	terremoto *m* tremor de terra *m* sismo *m*	землетрясение *n* удар землетрясения *m* толчок землетрясения *m*	2a
terremoto terrestre *m* terremoto continental *m*	terremoto continental *m* sismo continental *m*	материковое землетрясение *n* землетрясение с материковым очагом *n* землетрясение на суше *n*	2b
terremoto submarino *m* maremoto *m*	sismo submarino *m* terremoto submarino *m* maremoto *m*	подводное землетрясение *n* моретрясение *n*	2c
terremoto de dislocación *m* terremoto tectónico *m*	terremoto de deslocamento *m* sismo tectônico *m* terremoto tectônico *m*	дислокационное землетрясение *n* тектоническое землетрясение *n* землетрясение тектонического происхождения *n* обвальное землетрясение *n* провальное землетрясение *n*	2d
terremoto volcánico *m*	sismo vulcânico *m* terremoto vulcânico *m*	вулканическое землетрясение *n* землетрясение вулканического происхождения *n*	2e
sismicidad *f*	sismicidade *f*	сейсмичность *f*	3
actividad sísmica *f*	atividade sísmica *f*	сейсмическая активность *f*	4
observatorio sismológico *m*	observatório sismológico *m*	сейсмологическая обсерватория *f*	5
estación sismográfica *f* estación sísmica *f* estación sismológica *f*	estação sismográfica *f* estação sísmica *f* estação sismológica *f*	сейсмографическая станция *f* сейсмическая станция *f*	6
sismógrafo *m* seismógrafo *m*	sismógrafo *m* sismômetro *m*	сейсмограф *m* сейсмоприёмник *m*	7
sismograma *m* seismograma *m*	sismograma *m*	сейсмограмма *f* сейсмозапись *f*	8
onda sísmica *f* onda de terremoto *f*	onda sísmica *f* onda de terremoto *f*	сейсмическая волна *f* волна землетрясения *f*	9
sacudidas premonitoras *fpl* premonitores *mpl*	choques premonitores *mpl*	предварительные толчки *mpl* форшоки *mpl*	10
precursor (de un terremoto) *m*	prognóstico *m* precursor (do terremoto) *m*	предвестник (землетрясения) *m*	11

12	main phase	phase principale *f*	Hauptphase *f*
13	aftershock	réplique *f* secousses consécutives *fpl*	Nachbeben *n*
14	seismic ray	rayon sismique *m*	seismischer Strahl *m*
15	epicentrum epicentre	épicentre *m*	Epizentrum *n*
16	epicentral region	région de l'épicentre *f* région épicentrale *f*	Epizentralgebiet *n* Schüttergebiet *n*
17	earthquake focus hypocenter (US) seismic focus hypocentre (UK)	foyer du tremblement de terre *m* hypocentre *m*	Herd des Erdbebens *m* Bebenherd *m* Hypozentrum *n*
18	distance of the focus focal distance hypocentral distance	distance de l'hypocentre *f* distance du foyer *f*	Herdentfernung *f* Herddistanz *f* Hypozentraldistanz *f*
19	epicentral distance	distance de l'épicentre *f*	Epizentralentfernung *f* Epizentraldistanz *f* Epizentralabstand *m*
20	focal depth	profondeur du foyer *f*	Herdtiefe *f*
21	shallow earthquake	séisme [sisme] superficiel *m* tremblement de terre superficiel *m*	Oberflächenbeben *n* flaches Beben *n* Flachbeben *n*
22	normal earthquake	séisme [sisme] normal *m* tremblement de terre normal *m*	normales Erdbeben *n*
23	deep-focus earthquake deep earthquake	séisme [sisme] profond *m* tremblement de terre profond *m*	tiefes Erdbeben *n* Tiefbeben *n*
24	intermediate earthquake	séisme [sisme] intermédiaire *m* tremblement de terre intermédiaire *m*	intermediäres Erdbeben *n* mitteltiefes Erdbeben *n*
25	local earthquake	séisme [sisme] local *m* tremblement de terre local *m*	Lokalbeben *n* Ortsbeben *n*
26	near(by) earthquake	séisme [sisme] proche *m*	Nahbeben *n*
27	distant earthquake teleseism	séisme [sisme] lointain *m* téléséisme *m*	Fernbeben *n*
28	earthquake region seismic region	région sismique *f* région séismique *f*	Erdbebengebiet *n* seismisches Gebiet *n* seismische Region *f*

fase principal *f*	fase principal *f*	главная фаза *f* основная фаза *f*	12
réplicas *fpl* temblores secundarios *mpl*	réplicas *fpl* tremores secundários *mpl*	афтершок *m* последующий удар [толчок] землетрясения *m* затухающий удар [толчок] землетрясения *m*	13
rayo sísmico *m*	raio sísmico *m*	сейсмический луч *m*	14
epicentro *m*	epicentro *m*	эпицентр *m*	15
región epicentral *f* región del epicentro *f*	região epicentral *f* região epicêntrica *f*	область эпицентра *f*	16
foco del terremoto *m* foco sísmico *m* hipocentro *m*	foco do terremoto *m* foco do sismo *m* hipocentro *m*	фокус землетрясения *m* очаг землетрясения *m* сейсмический очаг *m* гипоцентр *m*	17
distancia del hipocentro *f* distancia hipocentral *f* distancia focal *f*	distância hipocentral *f* distância do foco *f* distância focal *f*	расстояние до очага землетрясения *n* гипоцентральное расстояние *n*	18
distancia del epicentro *f* distancia epicentral *f*	distância do epicentro *f* distância epicentral *f*	расстояние до эпицентра *n* эпицентральное расстояние *n*	19
profundidad del foco *f* profundidad focal *f*	profundidade do foco *f* profundidade focal *f*	глубина фокуса *f* глубина (залегания) очага *f*	20
terremoto superficial *m* sismo superficial *m* seísmo superficial *m*	sismo superficial *m* terremoto superficial *m*	мелкое землетрясение *n* неглубокое землетрясение *n*	21
terremoto normal *m* seísmo [sismo] normal *m*	sismo normal *m* terremoto normal *m*	нормальное землетрясение *n*	22
terremoto de foco profundo *m* terremoto profundo *m*	sismo profundo *m* terremoto profundo *m*	глубокофокусное землетрясение *n* глубокое землетрясение *n*	23
terremoto de profundidad focal intermedia *m* terremoto intermedio *m*	sismo intermediário *m* terremoto de profundidade intermediária *m*	промежуточное землетрясение *n* землетрясение промежуточной глубины *n* среднее землетрясение *n*	24
terremoto local *m* sismo [seísmo] local *m*	sismo local *m* terremoto local *m*	местное землетрясение *n* локальное землетрясение *n*	25
terremoto [sismo] cercano *m*	sismo [terremoto] próximo *m*	близкое землетрясение *n*	26
terremoto [sismo] lejano *m* seísmo lejano *m*	sismo distante *m* terremoto distante *m*	далекое землетрясение *n* удалённое землетрясение *n*	27
región sísmica *f* región de terremoto *f*	região sísmica *f* região de terremoto *f*	сейсмическая область *f* сейсмический район *m*	28

29	aseismic region	région aséismique *f* région asismique *f*	aseismisches Gebiet *n* erdbebenfreies Gebiet *n* aseismische [erdbebenfreie] Region *f*
30	earthquake belt	ceinture sismique *f* ceinture séismique *f*	Erdbebengürtel *m* seismischer Gürtel *m*
31	earthquake zone seismic zone	zone sismique *f* zone séismique *f*	Erdbebenzone *f* seismische Zone *f*
32	active belt	ceinture active *f*	aktiver Gürtel *m*
33	earthquake intensity	intensité sismique *f* intensité séismique *f*	Erdbebenintensität *f* Intensität eines Erdbebens *f*
34	earthquake magnitude magnitude of an earthquake	magnitude d'un tremblement de terre *f* magnitude s(é)ismique *f*	Magnitude eines Erdbebens *f* Stärke des Erdbebens *f* Erdbebenmagnitude *f*
35	slight earthquake	tremblement de terre faible *m*	leichtes Erdbeben *n*
36	moderate earthquake	tremblement de terre moyen *m*	mittelstarkes Erdbeben *n*
37	violent earthquake	violent tremblement de terre *m*	starkes Erdbeben *n*
38	damaging earthquake earthquake of magnitude IX	tremblement de terre désastreux *m*	verwüstendes Erdbeben *n*
39	destructive earthquake earthquake of magnitude X	tremblement de terre destructif *m*	zerstörendes Erdbeben *n*
40	catastrophic earthquake earthquake of magnitude XI	tremblement de terre catastrophique *m* catastrophe s(é)ismique *f*	katastrophales Erdbeben *n* Erdbebenkatastrofe *f* seismische Katastrofe *f*
41	transverse wave	onde transversale *f*	Transversalwelle *f*
42	body wave	onde de volume *f*	Raumwelle *f*
43	S wave secondary wave "undae secundae" distortional wave shear wave equivoluminar wave	onde S *f* onde secondaire *f* onde de distorsion *f* onde de cisaillement *f* onde de rotation *f*	S-Welle *f* Scherungswelle *f* Scherwelle *f* Schubwelle *f* sekundäre Welle *f* Querwelle *f* „undae secundae"
44	elastic wave	onde élastique *f*	elastische Welle *f*
45	longitudinal wave	onde longitudinale *f*	Longitudinalwelle *f*
46	P wave primary wave "undae primae" compressional wave dilational [dilatational] wave push wave pressure wave irrotational wave	onde P *f* onde primaire *f* onde de compression *f* onde de dilatation *f* onde de distension *f*	P-Welle *f* primäre Welle *f* Kompressionswelle *f* Primärwelle *f* Längswelle *f* Verdichtungswelle *f* „undae primae"

región asísmica *m* región sin terremotos *f*	região assísmica *f* região sem terremotos *f*	асейсмичная [несейсмичная] область *f* асейсмичный [несейсмичный] район *m*	29
cinturón sísmico *m* franja sísmica *f*	cinturão sísmico *m* faixa sísmica *f*	сейсмический пояс *m* пояс землетрясений *m*	30
zona sísmica *f*	zona sísmica *f*	зона землетрясения *f* сейсмическая зона *f*	31
cinturón activo *m* franja activa *f*	cinturão ativo *m* faixa ativa *f*	активный пояс *m*	32
intensidad del terremoto *f* intensidad sísmica [del seísmo] *f*	intensidade sísmica *f* intensidade de terremoto *f*	балльность землетрясения *f* интенсивность землетрясения *f*	33
magnitud de un terremoto *f* magnitud sísmica *f*	magnitude de um terremoto *f* magnitude sísmica *f*	магнитуда землетрясения *f* величина землетрясения *f* балл землетрясения *m*	34
terremoto débil *m*	terremoto fraco *m*	слабое землетрясение *n*	35
terremoto medio *m*	terremoto médio *m*	умеренное землетрясение *n*	36
terremoto intenso *m*	terremoto intenso *m*	сильное землетрясение *n*	37
terremoto desastroso *m* terremoto muy intenso *m*	terremoto desastroso *m* terremoto muito intenso *m*	разрушительное землетрясение *n* опустошительное землетрясение *n*	38
terremoto destructor *m*	terremoto destruidor *m*	уничтожающее замлетрясение *n*	39
terremoto catastrófico *m* catástrofe sísmica *f*	terremoto catastrófico *m*	катастрофическое землетрясение *n* сейсмическая катастрофа *f*	40
onda transversal *f*	onda transversal *f*	поперечная волна *f*	41
onda de volumen *f*	onda de volume *f*	объёмная волна *f*	42
onda S *f* onda de distorsión *f* secundaria *f* onda secundaria *f* onda de cizalla *f* onda equivolumétrica *f*	onda S *f* onda de distorção *f* onda secundária *f* onda de cisalhamento *f* onda lateral *f* onda eqüivolumétrica *f* onda distorcida *f* onda rotacional *f*	S волна *f* вторичная волна *f* поперечная волна землетрясения *f* волна сдвига *f*	43
onda elástica *f*	onda elástica *f*	упругая волна *f*	44
onda longitudinal *f*	onda longitudinal *f*	продольная волна *f*	45
onda P *f* primaria *f* onda de compresión *f* onda primaria *f* onda de presión *f*	onda P *f* onda primária *f* onda de compressão *f* onda de pressão *f*	P волна *f* первичная волна *f* сейсмическая продольная волна *f* продольная волна землетрясения *f* волна сжатия *f*	46

47	surface wave	onde de surface *f*	Oberflächenwelle *f*
48	Love wave Q wave "undae longae"	onde de Love *f* onde Q *f* onde de longue période *f*	Love-Welle *f* Q-Welle *f* „undae longae"
49	free oscillations of the Earth	oscillations libres de la Terre *fpl* vibrations propres de la Terre *fpl*	freie Erdschwingungen *fpl* freie Oszillationen der Erde *fpl*
50	arrival time	temps d'arrivée *m* instant d'arrivée *m*	Ankunftzeit *f* Ansatzzeit *f*
51	travel time travelling time transit time	temps de trajet *m* durée de parcours *f* durée de propagation *f*	Laufzeit *f*
52	velocity–depth curve velocity–depth section	relation [rapport] vitesse-profondeur *f*[*m*]	Geschwindigkeits-Tiefe-Kurve *f*
53	microseisms	microséismes *mpl* microsismes *mpl*	Mikroseismen *pl*
54	microseismic noise seismic noise	agitation microséismique [microsismique] *f* bruit séismique [sismique] *f*	mikroseismische Bodenunruhe *f* seismisches Rauschen *n*
55	seismic background	(bruit de) fond séismique [sismique] *m*	seismischer Hintergrund *m*
56	rigidity	rigidité *f*	Starrheit *f* Righeit *f*
57	compressibility	compressibilité *f*	Kompressibilität *f*
58	elasticity	élasticité *f*	Elastizität *f*
59	Young's modulus modulus of elasticity	module de Young *m* module d'élasticité *m*	Youngscher Modul *m* Elastizitätsmodul *m*
60	Poisson('s) ratio Poisson('s) constant Poisson('s) number	coefficient de Poisson *m* rapport de Poisson *m* constante de Poisson *f*	Poissonsches Verhältnis *n* Poissonsche Zahl *f* Poisson-Konstante *f*
61	internal force	force interne *f*	innere Kraft *f*
62	stress	contrainte *f*	Stress *m* Beanspruchung *f*
63	shear stress tangential stress	contrainte de cisaillement *f* contrainte tangentielle *f*	Scherstress *m* Scherbeanspruchung *f*
64	principal stress	contrainte principale *f*	Hauptstress *m* Hauptbeanspruchung *f*

onda superficial *f*	onda superficial *f*	поверхностная волна *f*	47
onda (de) Love *f* onda Q *f* onda larga *f*	onda de Love *f* onda Q *f*	волна Лява *f* волна Q *f*	48
vibraciones libres de la Tierra *fpl* oscilaciones libres de la Tierra *fpl*	oscilações livres da Terra *fpl* vibrações livres da Terra *fpl*	свободные колебания Земли *npl*	49
tiempo de llegada *m*	tempo de chegada *m* instante de chegada *m*	время прихода *n* время вступления волны *n*	50
tiempo de propagación *m* tiempo de recorrido *m* duración del recorrido *f*	tempo de percurso *m* duração do percurso *f*	время пробега (сейсмической волны) *n*	51
dromocrónica *f* curva de velocidad-profundidad *f* distribución de la velocidad con la profundidad *f*	curva da velocidade-profundidade *f* relação velocidade-profundidade *f* distribuição da velocidade com profundidade *f*	скоростной разрез *m* график изменения скорости с глубиной *m* соотношение скорость–глубина *f*	52
microseismos *mpl* microsismos *mpl*	microssismos *mpl* pequenos abalos sísmicos *mpl*	микроколебания *npl* микросейсмы *fpl*	53
agitación microsísmica *f* ruido sísmico *m*	agitação microssísmica *f* ruído sísmico *m*	сейсмический шум *m*	54
ruido sísmico de fondo *m*	fundo sísmico *m* ruído de fundo sísmico *m*	сейсмический фон *m*	55
rigidez *f*	rigidez *f*	жёсткость *f*	56
compresibilidad *f*	compressibilidade *f*	сжимаемость *f*	57
elasticidad *f*	elasticidade *f*	упругость *f*	58
módulo de Young *m* módulo de elasticidad *m*	módulo de Young *m* módulo de elasticidade *m*	модуль Юнга *m* модуль (продольной) упругости *m*	59
coeficiente de Poisson *m* constante de Poisson *f* relación de Poisson *f*	razão de Poisson *f* constante de Poisson *f*	коэффициент Пуассона *m* постоянная Пуассона *f* отношение Пуассона *n*	60
fuerza interna *f*	força interna *f*	внутренняя сила *f*	61
esfuerzo (interno) *m*	esforço *m*	напряжение *n* стресс *m*	62
esfuerzo cortante *m* esfuerzo de cizalla *m*	esforço de cisalhamento *m* esforço de corte *m*	срезывающее напряжение *n* касательное напряжение *n* сдвигающее напряжение *n* напряжение сдвига *n* тангенциальное напряжение *n*	63
esfuerzo principal *m*	esforço principal *m*	главное напряжение *n*	64

65	thrust	poussée *f* pression latérale *f*	Schub *m* Vortrieb *m*
66	tension	tension *f*	Dehnung *f* Zerrung *f*
67	strain	déformation *f*	Deformation *f* Strain *m* Zerrung *f*
68	distortion change in shape	distorsion *f* changement de forme *m*	Verzerrung *f* Formveränderung *f* Verformung *f*
69	stress tensor tenseur des tension	tenseur des contraintes *m*	Spannungstensor *m* Stresstensor *m*
70	strain tensor deformation tensor distortion tensor	tenseur des déformations *m*	Deformationstensor *m* Verformungstensor *m* Verzerrungstensor *m*
71	viscosity	viscosité *f*	Viskosität *f* Zähigkeit *f*
72	diastrophism crustal movements crustal motions	diastrophisme *m* mouvement de la croûte terrestre *m*	Diastrophismus *m* diastrophische Bewegungen *fpl* Krustenbewegungen *fpl* Erdkrustenbewegungen *fpl*
73	epeirogenesis epeirogenic movement epeirogeny uplift or depression	ép(é)irogenèse *f* mouvement épirogénique *m* mouvement épéirogénique *m*	Epirogenese *f* epirogenetische Bewegungen *fpl* epirogene Bewegungen *fpl*
74	orogeny orogenesis mountain building	orogenèse *f* formation des montagnes *f*	Gebirgsbildung *f* Orogenese *f*
75	orogen(et)ic belt	ceinture orogénique *f*	Orogengürtel *m* orogenetischer Gürtel *m*
75a	orogen(et)ic zone	zone orogénique *f*	orogene Zone *f* Faltungszone *f*
76	convection current	courant de convection *m*	Konvektionsstrom *m*
77	contraction theory	théorie de la contraction *f*	Kontraktionstheorie *f* Schrumpfungstheorie *f*
78	fracture rupture	fracture *f* cassure *f* rupture *f* fissure *f*	Bruch *m* Kluft *f* Riss *m* Spalte *f*
79	fracture system	système de fractures [cassures] *m* système de ruptures [fissures] *m*	Bruchsystem *n* Spaltensystem *n*

empuje *m* presión lateral *f*	empuxo *m* pressão lateral *f*	боковое давление *n* надвиг *m*	65
tensión *f*	tensão *f*	натяжение *n*	66
deformación *f*	deformação *f*	относительная деформация *f* растяжение *n*	67
distorsión *f* cambio de forma *m*	distorção *f* mudança de forma *f*	деформирование *n* изменение формы *n*	68
tensor de esfuerzos *m*	tensor dos esforços *m*	тензор напряжения *m*	69
tensor de deformaciones *m*	tensor das deformações *m*	тернзор деформации *m*	70
viscosidad *f*	viscosidade *f*	вязкость *f* вязкотекучесть *f*	71
diastrofismo *m* movimientos de la corteza terrestre *mpl*	diastrofismo *m* movimentos crustais *mpl* movimentos da crosta terrestre *mpl*	диастрофизм *m* тектоническое движение *n* движения (земной) коры *npl*	72
epirogénesis *f* movimiento epirogénico *m* epirogenia *m*	epirogênese *f* movimento epirogenético *m* epirogenia *f*	эпейрогеническое движение *n* вертикальное движение континентов *n* эпейрогенез *m*	73
orogénesis *f* formación de las montañas *f* orogenia *f*	orogênese *f* formação de montanhas *f* orogenia *f*	орогенез *m* горообразование *n*	74
franja orogénica *f*	cinturão orogênico *m* cinturão orogenético *m*	орогенический пояс *m* подвижный пояс *m*	75
zona orogénica *f*	zona orogênica *f* zona orogenética *f*	орогеническая зона *f* подвижная зона *f*	75a
corriente de convección *f*	corrente de convecção *f*	конвективное течение *n* конвекционный поток *m*	76
teoría de la contracción *f*	teoria da contração *f*	теория [гипотеза] контракции *f* контракционная гипотеза [теория] *f* теория [гипотеза] сжатия *f*	77
fractura *f* rotura *f* disyunción *f* fisura *f*	fratura *f* ruptura *f* fissura *f* disjunção *f*	трещина *f* разлом *m* разрыв *m* излом *m* расселина *f*	78
sistema de fracturas *m* sistema de roturas *m*	sistema de fraturas [fissuras] *m* sistema de rupturas *m*	система разломов [изломов] *f* система трещин [разрывов] *f*	79

80	mid-ocean fracture system	système de fractures médio-océaniques *m*	mittelozeanisches Bruchsystem *n*
81	continental fracture system	système de fractures continentales *m*	kontinentales Bruchsystem *n*
82	primary arc	arc primaire *m*	Hauptbogen *m* primärer Bogen *m*
83	secondary arc	arc secondaire *m*	Nebenbogen *m* sekundärer Bogen *m*
84	mountain arc	arc montagneux *m*	Gebirgsbogen *m*
85	mountain range	chaîne de montagnes *f*	Gebirgskette *f* Bergrücken *m* Bergkette *f* Gebirgszug *m*
86	folded mountains fold mountains	montagnes plissées *fpl*	Faltengebirge *n* Faltungsgebirge *n*
87	fold	pli *m*	Falte *f*
88	folding	plissement *m*	Faltung *f*
89	syncline synclinal fold trough	synclinal *m* pli synclinal *m*	Mulde *f* Synklinale *f* Synkline *f* Synklinalfalte *f*
90	anticline anticlinal fold	anticlinal *m* pli anticlinal *m*	Antiklinalfalte *f* Antiklinale *f* Antikline *f* Sattel *m*
91	fault paraclase dislocation	faille *f* paraclase *f* cassure (avec déplacement relatif) *f*	Verwerfung *f* Sprung *m* Verwurf *m*
92	faulting faultage	formation de failles *f* faillage *m*	Verwerfen *n* Sprungbildung *f* Verwerfungsbildung *f* Bruchbildung *f*
93	normal fault	faille normale [de distension] *f* faille directe [distensive] *f* faille régulière *f*	normale Verwerfung *f* gewöhnliche Verwerfung *f* gewöhnlicher Sprung *m*
94	reverse fault reversed fault	faille inverse [compressive] *f* faille renversée *f* faille de compression *f*	antithetische Verwerfung *f* inverse Verwerfung *f* gegensinnige Verwerfung *f* widersinnige Verwerfung *f*

sistema de fracturas oceánicas *m*	sistema de fraturas oceânicas *m*	(срединно)океан(иче)ская система разломов *f*	80
sistema de fracturas continentales *m*	sistema de fraturas continentais *m*	континентальная система разломов *f*	81
arco primario *m*	arco primário *m*	первичная дуга *f*	82
arco secundario *m*	arco secundário *m*	вторичная дуга *f*	83
arco montañoso *m* arco de montañas *m*	arco montanhoso *m*	дугообразный горный хребет *m*	84
cordillera *f* sierra *f* cadena de montañas *f*	cordilheira *f* serra *f* cadeia de montanhas *f*	горный хребет *m* горная цепъ *f* (горный) кряж *m*	85
montañas plegadas *f* montañas de plegamiento *f*	montanhas dobradas *fpl* montanhas de dobramento *fpl*	складчатые горы *fpl* складкообразные горы *fpl*	86
pliegue *m*	dobra *f*	складка *f* изгиб *m*	87
plegamiento *m*	dobramento *m*	складкообразование *n* складчатость *f* пликативная дислокация *f*	88
sinclinal *m* pliegue sinclinal *m*	sinclinal *f* dobra sinclinal *f*	синклиналь *f* мульда *f* синклинальная складка *f*	89
anticlinal *m* pliegue anticlinal *m*	anticlinal *f* dobra anticlinal *f*	антиклиналь *f* седло *n* антиклинальная складка *f*	90
falla *f* dislocación *f* paraclasa *f*	falha *f* paráclase *f* deslocamento *m*	разрыв *m* сброс *m* сдвиг *m* параклас *m*	91
formación de fallas *f* fallamiento *m* fracturación *f*	formação das falhas *f*	образование сбросов *n* сбрасывание *n* разрывное нарушение *n* сбросовая деятельность *f*	92
falla normal *f* falla directa *f* falla de distensión *f*	falha normal *f* falha direta *f* falha de distensão *f*	нормальный сброс *m* простой сброс *m* обыкновенный сброс *m* (нисходящий) сброс *m*	93
falla invertida *f* falla inversa *f* falla de compresión *f*	falha invertida *f* falha inversa *f* falha de compressão *f*	обратный сброс *m* обращённый сброс *m* взброс *m* надвиг *m*	94

95	geotectonics tectonics	géotectonique *f* tectonique *f*	Geotektonik *f* Tektonik *f*
96	tectonic map	carte tectonique *f*	tektonische Karte *f*
97	structural map	carte structurale *f*	Strukturkarte *f*
98	isostasy	isostasie *f*	Isostasie *f*
99	isostatic equilibrium	équilibre isostatique *m*	isostatisches Gleichgewicht *n*
100	isostatic adjustment isostatic compensation	ajustement isostatique *m* compensation isostatique *f*	isostatischer Ausgleich *m* isostatische Kompensation *f*
101	Archimedean principle Archimedes principle	principe d'Archimède *m* principe archimédien *m*	Archimedisches Prinzip *n* Prinzip von Archimedes *n*
102	depth of (isostatic) compensation	profondeur de compensation isostatique *f*	Tiefe der (isostatischen) Ausgleichsfläche *f* Kompensationstiefe *f* isostatische Tiefe *f*
103	continental plate continental platform	plaque continentale *f* plate-forme continentale *f* marge continentale *f*	kontinentale Platte *f* Festlandtafel *f* kontinentale Plattform *f*
104	continental drift drift of continents	dérive des continents *f*	Wanderung der Kontinente *f* Kontinentaldrift *f*
105	geosyncline geosynclinal	géosynclinal *m*	Geosynklinale *f*
106	tectonic forces *pl*	forces tectoniques *fpl*	tektonische Kräfte *fpl*
107	erosion	érosion *f*	Erosion *f* Auswaschung *f*
108	denudation	dénudation *f*	Denudation *f* Abtragung *f*
109	deposition sedimentation	sédimentation *f* déposition *f*	Ablagerung *f* Sedimentation *f*
110	volcanic activity igneous activity	activité volcanique *f*	vulkanische Tätigkeit *f* magmatische Tätigkeit *f*
111	volcano	volcan *m*	Vulkan *m*

geotectónica *f* tectónica *f*	geote(c)tônica *f* te(c)tônica *f*	геотектоника *f* тектоника *f*	95
carta tectónica *f* mapa tectónico *m*	mapa te(c)tônico *m* carta te(c)tônica *f*	тектоническая карта *f*	96
mapa estructural *m*	mapa estrutural *m*	структурная карта *f*	97
isostasia *f*	isostasia *f*	изостазия *f* уравновешивание *n*	98
equilibrio isostático *m*	equilíbrio isostático *m*	изостатическое равновесие *n*	99
ajuste isostático *m* compensación isostática *f*	ajustamento isostático *m* compensação isostática *f*	изостатическое выравнивание *n* изостатическая компенсация *f*	100
principio de Arquímedes *m* principio Arquimediano *m*	princípio de Arquimedes *m* princípio Arquimediano *m*	архимедов принцип *m* закон Архимеда *m*	101
profundidad de compensación (isostática) *f*	profundidade da compensação (isostática) *f*	глубина (уровня) (изостатической) компенсации *f*	102
plataforma continental *f* meseta continental *f*	placa continental *f* plataforma continental *f*	континентальная плита *f* континентальная платформа *f* материковая платформа *f*	103
deriva continental *f* deriva de los continentes *f*	deriva dos continentes *f* deriva continental *f*	дрейф материков [континентов] *m* (горизонтальное)движение континентов [материков] *n* [горизонтальное] перемещение континентов [материков] *n* эпейрофорез *m*	104
geosinclinal *m*	geossinclinal *m*	геосинклиналь *f*	105
fuerzas tectónicas *fpl*	forças tectônicas *fpl*	тектонические силы *fpl*	106
erosión *f*	erosão *f*	эрозия *f* смывание *n* размыв [смыв] *m*	107
denudación *f*	denudação *f* desnudamento *m*	денудация *f* снос *m*	108
sedimentación *f* depósito de sedimentos *m*	sedimentação *f* deposição *f*	нанос *m* отложение (осадочных материалов) *n* осаждение *n* намывание *n*	109
actividad volcánica *f*	atividade vulcânica *f*	вулканическая деятельность [активность] *f* магматическая деятельность *f*	110
volcán *m*	vulcão *m*	вулкан *m*	111

	SEISMICITY VOLCANISM	SISMOLOGIE VOLCANISME	ERDBEBENKUNDE VULKANE
112	volcanic eruption volcanic explosion	éruption volcanique *f* explosion volcanique *f*	Vulkaneruption *f* Vulkanausbruch *m*
113	fissure eruption linear eruption	éruption fissurale *f*	Spaltenausbruch *m* Spalteneruption *f* Lineareruption *f*
114	active volcano burning volcano fuming volcano	volcan actif *m* volcan en activité *m*	tätiger Vulkan *m* dampfender Vulkan *m*
115	extinct volcano dead volcano	volcan éteint *m* volcan mort *m*	erloschener Vulkan *m* toter Vulkan *m*
116	dormant volcano quiescent volcano inactive volcano	volcan en sommeil *m* volcan en repos *m* volcan inactif *m*	schlafender Vulkan *m* ruhender Vulkan *m* untätiger Vulkan *m*
117	submarine volcano subaquatic volcano	volcan sous-marin *m* volcan subaquatique *m*	submariner Vulkan *m* subaquatischer Vulkan *m* unterseeischer Vulkan *m*
118	volcanic cone cone of eruption	cône volcanique *m* cône d'éruption *m*	Vulkankegel *m* Eruptionskegel *m* vulkanischer Kegel *m*
119	shield volcano	volcan-bouclier *m* volcan effusif *m*	Schildvulkan *m*
120	pyroclastic cone debris cone cinder cone	cône pyroclastique *m* cône de débris *m* cône de cendres *m*	pyroklastischer Kegel *m* Schuttkegel *m* Aschenkegel *m* Schlackenkegel *m*
121	strato-cone composite cone strato-volcano	strato-cône *m* cône mixte *m* strato-volcan *m* cône composé *m*	Stratovulkan *m* Schichtvulkan *m* geschichteter Vulkan *m*
122	volcanic vent diatreme	cheminée volcanique *f* évent *m* diatrème *f*	Vulkanschlot *m* Schusskanal *m* Diatreme *f*
123	chimney	cheminée *f*	Schlot *m*
124	conduit	conduit *m*	Förderschlot *m*
125	volcanic neck	neck volcanique *m*	Schlotgang *m* Halskuffe *f*
126	crater	cratère *m*	Krater *m*
127	main crater	cratère principal *m*	Hauptkrater *m*

erupción volcánica *f* explosión volcánica *f*	erupção vulcânica *f* explosão vulcânica *f*	вулканическое извержение *n* извержение вулкана *n*	112
erupción de fisura *f* erupción fisural *f* erupción longitudinal *f*	erupção fissural *f* erupção longitudinal *f*	трещинное извержение *n* извержение из (региональной) трещины *n* линейное извержение *n*	113
volcán activo *m* volcán en actividad *m* volcán vivo *m*	vulcão ativo *m*	действующий вулкан *m*	114
volcán apagado *m* volcán extinto *m* volcán extinguido *m*	vulcão extinto *m*	потухший вулкан *m* мёртвый вулкан *m*	115
volcán inactivo *m* volcán durmiente *m*	vulcão inativo *m* vulcão dormente *m*	бездействующий вулкан *m* уснувший вулкан *m* спящий [недействующий] вулкан *m*	116
volcán submarino *m* volcán suboceánico *m*	vulcão submarino *m* vulcão suboceânico *m*	подводный вулкан *m* субмаринный вулкан *m*	117
cono volcánico *m* cono de erupción *m*	cone vulcânico *m* cone de erupção *m*	вулканический конус *m* конус вулкана *m*	118
volcán (de escudo) de lava *m* volcán Hawaiano *m*	vulcão escudo *m* vulcão tipo Hawai *m*	щитовой вулкан *m* щитовидный вулкан *m*	119
cono piroclástico *m* cono de cenizas *m* cono de escorias *m* cono de eyección *m*	cone de cinza *m* cone piroclástico *m* cone de explosão *m* cone de ejeção *m*	эксплозивный конус *m* пирокластический конус *m* пепловый конус *m* насыпной конус *m* туфовый конус *m*	120
cono de capas *m* cono mixto *m* volcán de capas *m* cono compuesto *m*	cone de estratos *m* estratovulcão *m* cone estratificado *m*	смешанный вулкан *m* стратовулкан *m* сложный вулканический конус *m* наслоенный вулкан *m*	121
boca volcánica *f* abertura volcánica *f* chimenea volcánica *f*	chaminé vulcânica *f* diatrema *f*	диатрема *f* жерло вулкана *n* горловина вулкана *f*	122
chimenea *f*	chaminé *f*	трубка взрыва *f*	123
conducto *m*	conduto *m* duto *m*	канал вулкана *m*	124
cuello volcánico *m*	garganta vulcânica *f*	жерловина (вулкана) *f* (вулканический) нэк [некк] *m*	125
cráter *m*	cratera *f*	(вулканический) кратер *m*	126
cráter principal *m*	cratera principal *f*	главный кратер *m*	127

128	lateral crater parasitic crater adventive crater	cratère latéral *m* cratère parasitaire *m* cratère adventif *m*	Seitenkrater *m* Schmarotzerkrater *m* Parasitärkrater *m* Nebenkrater *m* Adventivkrater *m*
129	summit crater	cratère sommital *m*	Gipfelkrater *m*
130	effusion of lava emission of lava lava outflow	effusion de lave *f* émission de lave *f* écoulement de lave *m*	Ausströmung von Lava *f* Effusion *f* Lavaausfluss *m* Lavaerguss *m*
131	magma	magma *m*	Magma *n*
132	extrusion	extrusion *f* expulsion *f*	Erguss *m* Extrusion *f* Auswurf *m*
133	intrusion	intrusion *f*	Intrusion *f*
134	concordant intrusion	intrusion concordante *f*	konkordante Intrusion *f*
135	sill	filon-couche *m* sill *m* lame concordante *f* lame intrusive *f*	Intrusivlager *m* Lagergang *m*
136	laccolith [laccolite]	laccolite *m*	Lakkolith *m*
137	discordant intrusion	intrusion discordante *f*	diskordante Intrusion *f* transgressive Intrusion *f*
138	dyke dike	dyke *m* filon lithique *m* filon rocheux *m*	Eruptivgang *m* Intrusivgang *m* Gesteinsgang *m*
139	batholith [batholite]	batholite *m*	Batholith *m*
140	lava flow	coulée de lave *f*	Lavastrom *m*
141	fissure flow	émission fissurale *f*	Spaltenerguss *m*
142	volcanic ash	cendre volcanique *f*	vulkanische Asche *f*
143	volcanic dust	poussière volcanique *f*	vulkanischer Staub *m*
144	volcanic bomb	bombe volcanique *f*	vulkanische Bombe *f*
145	endogenous	endogène	endogen
146	exogenous	exogène	exogen

cráter lateral *m* cráter secundario *m* cráter parasitario *m* cráter adventicio *m*	crátera lateral *f* crátera secundária *f* cratera parasita *f* cratera adventícia *f*	латеральный кратер *m* боковой кратер *m* побочный кратер *m* паразитический кратер *m*	128
cráter de cumbre *m*	cratera de vértice *f*	вершинный кратер *m* терминальный кратер *m*	129
derrame de lava *m* efusión de lava *f* emisión de lava *f* colada de lava *f*	emissão de lava *f* efusão de lava *f*	эффузия *f* истечение лавы *n* излияние лавы *n*	130
magma *m*	magma *m*	магма *f*	131
extrusión *f* expulsión *f*	extrusão *f* expulsão *f*	экструзия *f* экструзивное извержение *n*	132
intrusión *f*	intrusão *f*	интрузия *f* внедрение *n*	133
intrusión concordante *f*	intrusão concordante *f*	согласная интрузия *f* конкордантная интрузия *f*	134
capa intrusiva *f* sill *m* capa concordante *f*	camada intrusiva *f* camada concordante *f* sill *m*	пластовая интрузия *f* интрузивная залежь *f* силл *m*	135
lacolito *m*	lacólito [lacolito] *m*	лакколит *m*	136
intrusión discordante *f*	intrusão discordante *f*	несогласная интрузия *f* дискордантная интрузия *f*	137
dique *m* filón de inyección *m* filón de roca *m*	dique *m* veio mineral discordante *m*	дайка *f*	138
batolito *m*	batólito *m*	батолит *m*	139
torrente de lava *m* flujo dc lava *m*	fluxo de lava *m*	лавовый поток *m*	140
flujo fisural *m* flujo de fisura *f*	fluxo fissural *m*	истечение [излияние] из трещины *n* трещинное излияние [истечение] *n*	141
ceniza volcánica *f*	cinza vulcânica *f*	вулканический пепел *m*	142
polvo volcánico *m*	poeira vulcânica *f* pó vulcânico *m*	вулканическая пыль *f*	143
bomba volcánica *f*	bomba vulcânica *f*	вулканическая бомба *f*	144
endógeno	endógeno	эндогенный	145
exógeno	exógeno	экзогенный	146

	GEOLOGY	GÉOLOGIE	GEOLOGIE
1	rock	roche *f* roc *m*	Gestein *n* Fels(en) *m*
2	igneous rock magmatic rock	roche ignée *f* roche magmatique *f*	Erstarrungsgestein *n* Magmatit *m* magmatisches Gestein *n* Magmagestein *n* Eruptivgestein *n*
3	volcanic rock extrusive rock effusive rock	roche volcanique *f* roche extrusive *f* roche effusive *f*	vulkanisches Gestein *n* Extrusivgestein *n* Effusivgestein *n* Ergussgestein *n*
4	acid rock	roche acide *f*	saueres Gestein *n*
5	basic rock basite	roche basique *f*	basisches Gestein *n* Basit *m*
6	ultrabasic rock	roche ultra-basique *f*	ultrabasisches Gestein *n* Ultrabasit *m*
7	intrusive rock plutonic rock	roche intrusive *f* roche plutonienne *f*	Intrusivgestein *n* Plutonit *m*
8	granite	granite *m*	Granit *m*
9	basalt	basalte *m*	Basalt *m*
10	sedimentary rock	roche sédimentaire *f*	Sedimentgestein *n* Absatzgestein *n* Ablagerungsgestein *n*
11	metamorphic rock	roche métamorphique *f*	metamorphes Gestein *n* Metamorphit *m*
12	mineral	minéral *m*	Mineral *n*
13	silica silicon dioxide	silice *f* oxyde de silicium *m*	Siliziumdioxid *n* Silizium (IV)-oxid *n*
14	silicate	silicate *m*	Silikat *n*
15	feldspar (US) felspar (UK)	feldspath *m*	Feldspat *m*
16	orthoclase potash feldspar	ortho(cla)se *f* feldspath potassique *m*	Orthoklas *m* Kalifeldspat *m*
17	albite	albite *f*	Albit *m*
18	anorthite	anorthite *f*	Anorthit *m*
19	plagioclase	plagioclase *m*	Plagioklas *m*
20	amphibole	amphibole *f*	Amphibol *m*

GEOLOGÍA	GEOLOGIA	ГЕОЛОГИЯ	
roca *f* peña *f*	rocha *f*	горная порода *f* порода *f* скала *f*	1
roca ígnea *f* roca magmática *f*	rocha ígnea *f* rocha magmática *f*	магматическая порода *f*	2
roca volcánica *f* roca extrusiva *f* roca efusiva *f*	rocha vulcânica *f* rocha extrusiva *f* rocha efusiva *f*	вулканическая порода *f* излившаяся порода *f* эффузивная порода *f* вулканит *m*	3
roca ácida *f*	rocha ácida *f*	кислая порода *f*	4
roca básica *f*	rocha básica *f*	основная порода *f* базит *m*	5
roca ultrabásica *f*	rocha ultrabásica *f*	ультраосновная порода *f* ультрабазит *m* гипербазит *m*	6
roca intrusiva *f* roca plutónica *f*	rocha intrusiva *f* rocha plutônica *f*	интрузивная порода *f* глубинная порода *f* плутоническая порода *f*	7
granito *m*	granito *m*	гранит *m*	8
basalto *m*	basalto *m*	базальт *m*	9
roca sedimentaria *f*	rocha sedimentar *f*	осадочная порода *f*	10
roca metamórfica *f*	rocha metamórfica *f*	метаморфическая порода *f*	11
mineral *m*	mineral *m*	минерал *m*	12
sílice *f* dióxido de silicio *m*	sílica *f* dióxido de silício *m*	двуокись кремния *f* кремнезём *m*	13
silicato *m*	silicato *m*	силикат *m*	14
feldespato *m*	feldspato *m*	полевой шпат *m*	15
ortoclasa *f* feldespato potásico *m*	ortoclásio *m* feldspato potássico *m*	ортоклаз *m* калийный полевой шпат *m*	16
albita *f*	albita *f*	альбит *m*	17
anortita *f*	anortita *f*	анортит *m* известковый полевой шпат *m*	18
plagioclasa *f*	plagioclásio *m*	плагиоклаз *m*	19
anfíbol *m*	anfibólio *m*	амфибол *m*	20

21	pyroxene	pyroxène *m*	Pyroxen *m*
22	olivine	olivine *f*	Olivin *m*
23	mica	mica *m*	Glimmer *m*
24	muscovite white mica	muscovite *f* mica blanc *m*	Muskovit *m* Hellglimmer *m* weisser Glimmer *m*
25	biotite dark mica	biotite *f* mica noir *m*	Biotit *m* Dunkelglimmer *m* dunkler Glimmer *m*
26	oxide	oxyde *m*	Oxid *n* Oxyd *n*
27	carbonate	carbonate *m*	Karbonat *n*
28	sulphide	sulfure *m*	Sulfid *n*
28a	sulphite	sulfite *m*	Sulfit *n*
29	sulphate	sulfate *m*	Sulfat *n*
30	geochemistry	géochimie *f*	Geochemie *f*
31	cosmochemistry	cosmochimie *f*	Kosmochemie *f*
32	geochemical cycle	cycle géochimique *m*	geochemischer Zyklus *m*
33	geochemical evolution	évolution géochimique *f*	geochemische Entwicklung *f*
34	metamorphism metamorphic process	métamorphisme *m* processus métamorphique *m*	Metamorphose *f* metamorpher Prozess *m*
35	contact metamorphism	métamorphisme de contact *m*	Kontaktmetamorphose *f*
36	dynamic metamorphism dynamometamorphism	dynamo-métamorphisme *m*	Dynamometamorphose *f* dynamische Metamorphose *f*
37	thermal metamorphism	métamorphisme thermique *m*	thermische Metamorphose *f* Thermometamorfose *f*
38	diagenesis	diagenèse *f*	Diagenese *f*
39	migmatite	migmatite *f*	Migmatit *m* Mischgestein *n*
40	fractional crystallization	cristallisation fractionnée *f*	fraktionierte Kristallisation *f*
41	crystallization differentiation	différenciation par cristallisation *f*	Kristallisationsdifferentiation *f*

piroxeno *m* piroxena *f*	piroxênio *m*	пироксен *m*	21
olivina *f*	olivina *f*	оливин *m*	22
mica *f*	mica *f*	слюда *f*	23
moscovita *f* mica blanca *f*	moscovita *f* mica branca *f*	мусковит *m* обыкновенная слюда *f* белая слюда *f*	24
biotita *f* mica negra *f*	biotita *f* mica preta *f*	биотит *m* чёрная слюда *f*	25
óxido *m*	óxido *m*	окись *f* окисел *m* оксид *m* закись *f*	26
carbonato *m*	carbonato *m*	карбонат *m* углекислая соль *f*	27
sulfuro *m*	sulfeto *m* sulfureto *m*	сульфид *m* сернистый металл *m* сернистая соль *f*	28
sulfito *m*	sulfito *m*	сульфит *m* сернистокислая соль *f*	28a
sulfato *m*	sulfato *m*	сульфат *m* сернокислая соль *f*	29
geoquímica *f*	geoquímica *f*	геохимия *f*	30
cosmoquímica *f*	cosmoquímica *f*	космохимия *f*	31
ciclo geoquímico *m*	ciclo geoquímico *m*	геохимический цикл *m*	32
evolución geoquímica *f*	evolução geoquímica *f*	геохимическая эволюция *f*	33
metamorfismo *m* proceso metamórfico *m*	metamorfismo *m* processo metamórfico *m*	метаморфизм *m* метаморфический процесс *m*	34
metamorfismo de contacto *m*	metamorfismo de conta(c)to *m*	контактный метаморфизм *m*	35
metamorfismo dinámico *m* dinamometamorfismo *m*	metamorfismo dinâmico *m*	динамический метаморфизм *m* динамометаморфизм *m*	36
termometamorfismo *m* metamorfismo térmico *m*	metamorfismo térmico *m*	термальный метаморфизм *m* термометаморфизм *m*	37
diagénesis *f*	diagênese *f*	диагенез *m*	38
migmatita *f*	migmatito *m*	мигматит *m*	39
cristalización fraccionada *f*	cristalização fracionada *f*	фракционная кристаллизация *f*	40
direrenciación por cristalización *f*	diferenciação por cristalização *f*	кристаллизационная дифферен- циация *f*	41

	GEOLOGY	GÉOLOGIE	GEOLOGIE
42	recrystallization	recristallisation *f*	Rekristallisation *f* Umkristallisierung *f* Neukristallisation *f*
43	chemically active fluid	fluide chimiquement actif *m*	chemisch aktive Flüssigkeit *f* chemisch aktive Lösung *f*
44	metasomatism metasomatosis	métasomatisme *m* métasomatose *f*	Metasomatose *f* Verdrängung *f*
45	sedimentation	sédimentation *f*	Ablagerung *f* Sedimentation *f* Sedimentbildung *f*
46	sedimentary process	processus de sédimentation *m*	Sedimentationsprozess *m* Sedimentationsvorgang *m*
47	sediment	sédiment *m* dépôt *m*	Absatz *m* Sediment *n* Ablagerung *f*
48	stratified rock	roche stratifiée *f*	Schichtgestein *n*
49	lithification petrification	lithification *f* pétrification *f*	Versteinerung *f* Lithifikation *f*
50	compaction	compaction *f*	Kompaktion *f* Sedimentverfestigung *f*
51	desiccation drying	dessiccation *f* dessèchement *m*	Austrocknung *f* Austrocknen *n*
52	cementation cementing	cimentation *f* cémentation *f*	Zementation *f* Zementierung *f* Verkittung (von Gesteinen) *f*
53	foliation	foliation *f* feuilletage *m* schistosité *f*	Bänderung *f* Schieferung *f* Blätterung *f*
54	foliated rock schistous rock	roche feuilletée *f* roche schisteuse *f*	geschiefertes Gestein *n* Blattgestein *n*
55	geochemical differentiation	différenciation géochimique *f*	geochemische Differentiation *f*
56	gravitational differentiation	différenciation par gravité *f*	Schweredifferentiation *f*
57	weathering	altération (météorique) *f* désagrégation *f*	Verwitterung *f*
58	geothermics geothermy	géothermie *f*	Geothermie *f* Geothermik *f*

GEOLOGÍA	GEOLOGIA	ГЕОЛОГИЯ	
recristalización *f*	recristalização *f*	перекристаллизация *f* рекристаллизация *f*	42
fluido químicamente activo *m*	fluido quimicamente ativo *m*	химически активная жидкость *f*	43
metasomatismo *m* metasomatosis *f*	metassomatismo *m* metassomatose *f*	метасоматоз *m* метасоматизм *m*	44
sedimentación *f*	sedimentação *f*	седиментация *f* осадкообразование *n* осадконакопление *n*	45
proceso sedimentario *m*	processo sedimentar *m*	осадочный процесс *m* процесс осадкообразования *m*	46
sedimento *m* depósito *m*	sedimento *m*	осадок *m* отложение *n* отстой *m*	47
roca estratificada *f*	rocha estratificada *f*	слоистая порода *f*	48
litificación *f* petrificación *f*	litificação *f* petrificação *f*	литификация *f* окаменение *n*	49
compactación *f* consolidación *f*	compactação *f*	уплотнение осадков *n*	50
desecación *f*	dessecação *f*	обезвоживание *n* высушивание *n* усыхание *n*	51
cementación *f* aglutinación *f*	cementação *f* aglutinação *f*	цементация (осадков) *f*	52
foliación *f* esquistosidad *f*	esfoliação *f* xistosidade *f*	разлистование *n* сланцеватость *f* расслоение *n* полосчатость *f* листоватость *f*	53
roca foliada *f* roca lamelar *f* roca esquistosa *f*	rocha folheada *f* rocha xistácea *f* rocha xistóide *f*	рассланцованная порода *f* расслоенная порода *f*	54
diferenciación geoquímica *f*	diferenciação geoquímica *f*	геохимическая дифференциа-ция *f*	55
diferenciación gravitacional *f*	diferenciação gravitacional *f*	гравитационная дифференциа-ция *f*	56
meteorización *f* intemperización *f* desagregación *f*	meteorização *f* alteraçao pelo tempo *f* intemperismo *m*	выветривание *n*	57
geotermia *f*	geotermia *f*	геотермия *f* геотермика *f*	58

	GEOLOGY	GÉOLOGIE	GEOLOGIE
59	geothermal gradient geothermic step geothermic degree (of depth)	gradient géothermique *m* échelon géothermique *m* degré géothermique *m*	geothermischer Gradient *m* geothermische Tiefenstufe *f* geothermische Stufe *f*
60	heat flow flow of heat	flux de chaleur *m*	Wärmestrom *m* Wärmefluss *m*
61	radioactivity	radioactivité *f*	Radioaktivität *f*
62	source of heat	source de chaleur *f*	Wärmequelle *f*
63	original heat fossil heat	chaleur originelle *f* chaleur primitive *f*	ursprüngliche Wärme *f*
64	radioactive decay	désagrégation radioactive *f*	radioaktiver Zerfall *m*
65	disintegrate decay	(se) désintégrer se décomposer	zerfallen zersetzen sich
66	thermal history	histoire thermique *f*	thermische Geschichte *f*
67	geochronology	géochronologie *f* chronologie géologique *f*	Geochronologie *f* geologische Zeitrechnung *f*
68	geochronological scale	échelle géochronologique *f*	geochronologische Skala *f*
69	era	ère (géologique) *f*	Ära *f* Zeitalter *n*
70	Precambrian Pre-Cambrian (era) Cryptozoic (era)	Précambrien *m* ère précambrienne *f* Antécambrien *m*	Präkambrium *n* Kryptozoikum *n* Erdurzeit *f* Urzeit der Erde *f*
71	Paleozoic Palaeozoic Paleozoic era	Paléozoïque *m* ère primaire *f* Primaire *m* ère paléozoïque *f*	Paläozoikum *n* Primärzeit *f* Erdaltertum *n* paläozoische Ära *f*
72	Mesozoic Mesozoic era	Mésozoïque *m* ère secondaire *f* Secondaire *m* ère mésozoïque *f*	mesozoikum *n* Sekundärzeit *f* Erdmittelalter *n* mesozoische Ära *f*
73	Cenozoic Cainozoic Cainozoic era	Cénozoïque *m* ère cénozoïque *f* ère néozoïque *f* Néozoïque *m*	Känozoikum *n* Erdneuzeit *f* känozoische Ära *f* Neozoikum *n*
74	geological period	période géologique *f*	geologische Periode *f*
75	Azoic Early Precambrian	Azoïque *m* Précambrien inférieur *m*	Azoikum *n* Abiotikum *n*
76	Archaeozoic	Archéozoïque *m*	Archaikum *n* Archäozoikum *n*

GEOLOGÍA	GEOLOGIA	ГЕОЛОГИЯ	
grado geotérmico *m* gradiente geotérmico *m* intervalo geotérmico *m*	gradiente geotérmico *m* grau geotérmico *m* passo geotérmico *m*	геотермическая ступень *f* геотермический градиент *m*	59
flujo de calor *m*	fluxo de calor *m*	тепловой поток *m* поток тепла *m*	60
radi(o)actividad *f*	radi(o)atividade *f*	радиоактивность *f*	61
fuente de calor *f*	fonte de calor *f*	источник тепла *m* источник тепловой энергии *m*	62
calor originario *m* calor fósil *m*	calor original *m* calor fóssil *m*	первичное тепло *n*	63
desintegración radi(o)activa *f*	desintegração radi(o)ativa *f*	радиоактивный распад *m*	64
desintegrar(se) descomponerse	desintegrar(se) decair	распадаться	65
historia térmica *f*	história térmica *f*	термическая история *f* тепловая история *f*	66
geocronología *f* cronología geológica *f*	geocronologia *f* cronologia geológica *f*	геохронология *f* геологическое летоисчисление *n*	67
escala geocronológica *f*	escala geocronológica *f*	геохронологическая шкала *f* геоисторическая шкала *f*	68
era (geológica) *f*	era *f*	эра *f*	69
Precámbrico *m* era precámbrica *f* Criptozoico *m*	Pré-cambriano *m* era pré-cambriana *f* Criptozóico *m*	докембрий *m* криптозой *m* докембрийская эра [группа] *f*	70
Paleozoico *m* era paleozoica *f* Primario *m*	Paleozóico *m* era paleozóica *f* Primário *m*	палеозой *m* палеозойская эра *f* палеозойская группа *f*	71
Mesozoico *m* era mesozoica *f* Secundario *m*	Mesozóico *m* era mesozóica *f* Secundário *m*	мезозой *m* мезозойская эра *f* мезозойская группа *f*	72
Cenozoico *m* era kenozoica *f* Kenozoico *m* Neozoico *m*	Cenozóico *m* era cenozóica *f* era neozóica *f* Neozóico *m*	кайнозой *m* кайнозойская эра *f* кайнозойская группа *f* неозой *m*	73
período geológico *m*	período geológico *m*	геологический период *m*	74
Azoico *m* Agnostozoico *m* Precámbrico inferior *m*	Azóico *m* período arqueano *m* Pré-cambriano inferior *m*	азой *m* азойская эра *f* азойская группа *f*	75
Arqueozoico *m*	Arqueozóico *m*	археозой *m* археозойская эра [группа] *f*	76

	GEOLOGY	GÉOLOGIE	GEOLOGIE
77	Proterozoic Eozoic Algonkian	Protérozoïque *m* Éozoïque *m* Algonquien *m*	Proterozoikum *n* Eozoikum *n* Algonkium *n*
78	Cambrian	Cambrien *m*	Kambrium *n*
79	Ordovician	Ordovicien *m*	Ordovizium *n*
80	Silurian Gothlandian	Silurien *m* Gothlandien *m*	Silur *n* Gotlandium *n*
81	Devonian	Dévonien *m*	Devon *n*
82	Carboniferous	Carbonifère *m*	Karbon *n*
83	Permian	Permien *m*	Perm *n*
84	Triassic	Trias *m*	Trias *f*
85	Jurassic Jura	Jurassique *m*	Jura *m*
86	Cretaceous	Crétacé *m*	Kreide *f*
87	Tertiary	Tertiaire *m*	Tertiär *n*
88	Quaternary	Quaternaire *m*	Quartär *n*
89	(geologic) epoch	époque (géologique) *f*	(geologische) Epoche *f*
90	geologic age	âge géologique *m*	geologisches Alter *n*
91	age determination determination of age	détermination de l'âge *f* datation *f*	Altersbestimmung *f*
92	radiometric method radioactive method	méthode radiométrique *f* méthode radioactive *f*	radiometrische Methode *f* radioaktive Methode *f*
93	isotopic composition	composition isotopique *f*	Isotopenzusammensetzung *f*
94	isotope ratio	rapport isotopique *m*	Isotopenverhältnis *n*

GEOLOGÍA	GEOLOGIA	ГЕОЛОГИЯ	
Proterozoico *m* Eozoico *m* Algónquico *m*	Proterozóico *m* Eozóico *m* Algonquiano *m*	протерозой *m* протерозойская группа [над-эра] *f* протерозойский эон *m*	77
Cámbrico *m* Cambriano *m*	Cambriano *m*	кембрий *m* кембрийский период *m* кембрийская система *f*	78
Ordoviciano *m*	Ordoviciano *m*	ордовик *m* ордовикский период *m* ордовикская система *f*	79
Silúrico *m* Gotlandiano *m*	Siluriano *m* Gotlândico *m*	силур *m* силурийский период *m* силурийская система *f*	80
Devónico *m*	Devoniano *m*	девон *m* девонский период *m* девонская система *f*	81
Carbonífero *m*	Carbonífero *m*	карбон *m* каменноугольный период *m* каменноугольная система *f*	82
Pérmico *m*	Permiano *m*	пермь *f* пермский период *m* пермская система *f*	83
Triásico *m*	Triás(s)ico *m* Triádico *m*	триас *m* триасовый период *m* триасовая система *f*	84
Jurásico *m*	Jurássico *m*	юра *m* юрский период *m* юрская система *f*	85
Cretáceo *m*	Cretáceo *m*	мел *m* меловой период *m* меловая система *f*	86
Terciario *m*	Terciário *m*	третичный период *m* третичная система *f*	87
Cuaternario *m*	Quaternário *m*	четвертичный период *m* четвертичная система *f*	88
época (geológica) *f*	época (geológica) *f*	(геологическая) эпоха *f*	89
edad geológica *f*	idade geológica *f*	геологический возраст *m*	90
determinación de la edad *f*	determinação da idade *f*	определение возраста *n*	91
método radiométrico *m* método radiactivo *m*	método radi(o)métrico *m* método radi(o)ativo *m*	радиологический метод (определения возраста) *m*	92
composición isotópica *f*	composição isotópica *f*	изотопный состав *m*	93
relación isotópica *f* proporción isotópica *f*	razão isotópica *f* razão de isótopos *f*	изотопное отношение *n* (со)отношение изотопов *n*	94

	GEOLOGY	GÉOLOGIE	GEOLOGIE
95	radiocarbon radioactive carbon	radiocarbone *m* carbone radioactif *m*	radioaktiver Kohlenstoff *m* Radiokohlenstoff *m* Radiokarbon *n*
96	uranium method	méthode à l'uranium *f*	Uran-Methode *f*
97	radiogenic lead	plomb radiogénique *m*	radiogenes Blei *n*
98	prim(a)eval lead primary lead	plomb primitif *m*	ursprüngliches Blei *n*
99	common lead	plomb commun *m*	gemeines Blei *n* gewöhnliches Blei *n*
100	lead-lead age lead-ratio age	âge déterminé par la méthode au plomb isotopique *m* datation plomb-plomb *f* datation au plomb *f*	Bleiverhältnis-Alter *n* Blei-Blei-Alter *n*
101	uranium-lead age	âge déterminé par la méthode du rapport uranium-plomb *m* datation uranium-plomb *f*	Uran-Blei-Alter *n*

GEOLOGÍA	GEOLOGIA	ГЕОЛОГИЯ	
carbono radioactivo *m* radiocarbono *m*	carbono radi(o)ativo *m*	радиоактивный углерод *m* радиоуглерод *m*	95
método del uranio *m*	método do urânio *m*	урановый метод *m*	96
plomo radiogénico *m*	chumbo radiogênico *m*	радиогенный свинец *m* свинец радиоактивного происхождения *m*	97
plomo primitivo *m* plomo original *m*	chumbo primitivo *m* chumbo original *m*	первоначальный свинец *m*	98
plomo común *m* plomo ordinario *m*	chumbo comum *m*	обыкновенный свинец *m*	99
edad por relación plomo(-plomo) *f* determinación de la edad por el método del plomo (isotópico) *f*	idade chumbo-chumbo *f* idade de razão isotópica de chumbo *f*	возраст по соотношению изотопов свинца *m* свинцово-свинцовый возраст *m*	100
edad por relación uranio-plomo *f* determinación de la edad por la relación uranio-plomo *f*	idade urânio-chumbo *f*	ураново-свинцовый возраст *m*	101

	GEOMAGNETISM	GÉOMAGNÉTISME	ERDMAGNETISMUS
1	terrestrial electricity geoelectricity	électricité tellurique *f* géoélectricité *f*	Erdelektrizität *f* Geoelektrizität *f*
2	atmospheric electricity	électricité atmosphérique *f*	Luftelektrizität *f* atmosphärische Elektrizität *f*
3	Earth's charge	charge de la Terre *f*	Erdladung *f*
4	Earth current field electrotelluric field	champ tellurique *m* champ des courants telluriques *m*	elektrotellurisches Feld *n* Feld der Erdströme *n*
5	Earth currents telluric currents	courants telluriques *mpl*	Erdströme *mpl* tellurische Ströme *mpl*
6	stray Earth currents	courants telluriques vagabonds *mpl*	vagabundierende Erdströme *mpl*
7	magnetosphere	magnétosphère *f*	Magnetosphäre *f*
8	geomagnetic cavity	cavité géomagnétique *f*	geomagnetischer Hohlraum *m*
9	magnetopause magnetospheric boundary	magnétopause *f* frontière de la magnétosphère *f*	Magnetopause *f* Magnetosphärengrenze *f*
10	magnetospehric tail magnetotail	queue de la magnétosphère *f* queue magnétosphérique *f*	magnetosphärischer Schweif *m* Schweif der Magnetosphäre *m*
11	plasma sheet	couche de plasma *f*	Plasmaschicht *f*
12	plasmasphere protonosphere	plasmasphère *f* protonosphère *f*	Plasmasphäre *f* Protonosphäre *f* Protonensphäre *f*
13	magnetospheric cleft polar cusp	cornet magnétosphérique *m* cornet polaire *m*	cusp der Magnetosphäre *n* Horn der Magnetosphäre *n* cusp des Poles *n*
14	corotating field line	ligne de force en co-rotation *f*	mitrotierende Feldlinie *f*
15	open field line	ligne de champ non-fermée *f*	offene Feldlinie *f*
16	neutral sheet neutral surface	couche neutre *f* surface neutre *f*	Neutralschicht *f* Neutralfläche *f*
17	noon meridian	méridien de midi *m*	Mittagsmeridian *m*
18	dawn meridian	méridien de l'aube *m*	Morgenmeridian *m*
19	dusk meridian	méridien du coucher *m*	Abendmeridian *m*

electricidad de la tierra *f* electricidad terrestre *f*	ele(c)tricidade terrestre *f*	геоэлектричество *n*	1
electricidad atmosférica *f*	ele(c)tricidade atmosférica *f*	атмосферное электричество *n*	2
carga de la Tierra *f*	carga da Terra *f*	(электрический) заряд Земли *m*	3
campo de corrientes telúricas *m* campo electrotelúrico *m*	campo ele(c)trotelúrico *m* campo das correntes telúricas *m*	электротеллурическое поле *n* поле земных электрических токов *n*	4
corrientes telúricas *fpl*	correntes telúricas *fpl*	земные (электрические) токи *mpl* теллурические токи *mpl*	5
corrientes telúricas parásitas *fpl* corrientes telúricas vagabundas *fpl*	correntes telúricas vagabundas *fpl*	блуждающие токи *mpl*	6
magnetosfera *f*	magnetosfera *f*	магнитосфера *f*	7
cavidad geomagnética *f*	cavidade geomagnética *f*	геомагнитная полость (в потоке солнечной плазмы) *f*	8
magnetopausa *f* frontera magnetosférica *f*	magnetopausa *f* fronteira da magnetosfera *f*	магнитопауза *f* граница магнитосферы *f*	9
cola magnetosférica *f* cola magnética (de la Tierra) *f*	cauda da magnetosfera *f* cauda magnética (da Terra) *f*	хвост магнитосферы *m* геомагнитный шлейф *m*	10
lámina de plasma *f* hoja de plasma *f*	camada de plasma *f*	плазменный слой *m* плазменная оболочка *f*	11
plasma(e)sfera *f* protonó(e)sfera *f*	plasmosfera *f* protonosfera *f*	плазмосфера *f* протоносфера *f*	12
cuello magnetosférico *m* cuerno magnetosférico *m*	cúspide magnetosférica *f* ponta magnetosférica *f*	магнитосферный клефт *m* рог магнитосферы *m* полярный касп *m*	13
línea de fuerza corrotacional *f*	linha de força corrotativa *f*	силовая линия участвующая во вращении *f*	14
línea de campo abierta *f*	linha de campo aberta *f*	разомкнутая силовая линия *f*	15
lámina neutra *f* hoja neutra *f* superficie neutra *f*	camada neutra *f* lâmina neutra *f* superfície neutra *f*	нейтральный слой (в хвосте магнитосферы) *m*	16
meridiano de mediodía *m*	meridiano do meio-dia *m*	полуденный меридиан *m*	17
meridiano del amanecer *m* meridiano matutino *m*	meridiano do nascente *m*	утренний меридиан *m*	18
meridiano del anochecer *m* meridiano vespertino *m*	meridiano do poente *m*	вечерний меридиан *m*	19

	GEOMAGNETISM	GÉOMAGNÉTISME	ERDMAGNETISMUS
20	sunlit side day side sunlit hemisphere daylight hemisphere illuminated hemisphere	côté éclairé *m* côté jour *m* hémisphère éclairé *m*	sonnenbeschienene Seite *f* Tagseite *f* sonnenbeschienene Hemisphäre *f* beleuchtete Hemisphäre *f*
21	night side night hemisphere dark side unilluminated hemisphere	côté nuit *m* côté à l'ombre *m* hémisphère obscur *m* hémisphère nocturne *m*	Nachtseite *f* dunkle Seite *f* Nachthemisphäre *f* unbeleuchtete Hemisphäre *f*
22	antisolar direction	direction antisolaire *f*	antisolare Richtung *f* Richtung entgegen der Sonne *f*
23	(magneto)conjugate points magnetically conjugate points	points conjugués magnétiques *mpl*	magnetisch konjugierte Punkte *mpl*
24	geomagnetism terrestrial magnetism Earth's magnetism	géomagnétisme *m* magnétisme terrestre *m*	Geomagnetismus *m* Erdmagnetismus *m*
25	paleomagnetism palaeomagnetism	paléomagnétisme *m*	Paläomagnetismus *m*
26	ferromagnetism	ferromagnétisme *m*	Ferromagnetismus *m*
27	magnetic susceptibility	susceptibilité magnétique *f*	magnetische Suszeptibilität *f*
28	remanent magnetization residual magnetization remanent magnetism residual magnetism	aimantation rémanente *f* aimantation résiduelle *f* magnétisme résiduel *m*	bleibende Magnetisierung *f* remanente Magnetisierung *f* Restmagnetismus *m* remanenter Magnetismus *m* Restmagnetisierung *f*
29	induced magnetization induced magnetism	aimantation induite *f* magnétisme induit *m*	induzierte Magnetisierung *f* induzierter Magnetismus *m*
30	thermoremanent magnetization thermoresidual magnetization TRM	aimantation thermorémanente *f* aimantation thermorésiduelle *f*	thermoremanente Magnetisierung *f*
31	chemical magnetization	aimantation chimique *f*	chemische Magnetisierung *f*
32	viscous magnetization	aimantation visqueuse *f*	viskose Magnetisierung *f*
33	detrital (remanent) magnetization depositional (remanent) magnetization DRM	aimantation détritique *f* aimantation de dépôt *f*	Sedimentationsmagnetisierung *f* Ablagerungsmagnetisierung *f*
34	magnetic viscosity (of rocks)	viscosité magnétique (de roches) *f*	magnetische Viskosität *f* magnetische Zähigkeit *f*

lado diurno *m* lado iluminado *m* lado del Sol *m* lado de día *m* hemisferio alumbrado por el Sol *m* hemisferio iluminado *m*	lado do dia *m* lado iluminado pelo Sol *m* lado do Sol *m* hemisfério iluminado pelo Sol *m*	дневное полушарие (Земли) *n* дневная сторона (Земли) *f* освещённая полусфера (Земли) *f* сторона (Земли) обращённая к Солнцу *f* освещённое Солнцем полушарие *n*	20
lado de noche *m* lado o(b)scuro *m* hemisferio nocturno *m* hemisferio o(b)scuro *m*	lado da noite *m* lado escuro *m* hemisfério não-iluminado *m* hemisfério escuro *m*	ночное полушарие *n* неосвещённое полушарие *n* неосвещённая сторона *f* ночная сторона *f* теневая сторона (Земли) *f*	21
dirección antisolar *f* dirección opuesta al Sol *f*	direção anti-solar *f*	направление от Солнца *n* антисолнечное направление *n*	22
puntos magneticamente conjugados *mpl* puntos magnetoconjugados *mpl*	pontos magneticamente conjugados *mpl* pontos magnetoconjugados *mpl*	магнитосопряжённые точки *fpl*	23
geomagnetismo *m* magnetismo terrestre *m*	geomagnetismo *m* magnetismo terrestre *m*	геомагнетизм *m* земной магнетизм *m*	24
paleomagnetismo *m*	paleomagnetismo *m*	палеомагнетизм *m*	25
ferromagnetismo *m*	ferromagnetismo *m*	ферромагнетизм *m*	26
susceptibilidad magnética *f*	susce(p)tibilidade magnética *f*	магнитная восприимчивость *f*	27
magnetismo remanente *m* imanación [magnetización] remanente *f* magnetismo [magnetización] residual *m*[*f*] imanación residual *f*	magnetização residual *f* imantação residual *f* magnetismo residual *m* magnetização remanente [remanescente] *f*	остаточный магнетизм *m* остаточное намагничение *n*	28
imanación inducida *f* magnetización inducida *f* magnetismo inducido *m*	magnetização induzida *f* imantação induzida *f* magnetismo induzido *m*	индуцированное намагничение *n* индуцированная намагниченность *f*	29
magnetización termorresidual *f* imanación termorremanente *f*	magnetização termorremanente *f* magnetização termorremanescente *f*	термоостаточная намагниченность *f* термоостаточное намагничение *n*	30
imanación química *f* magnetización química *f*	magnetização química *f*	химическая намагниченность *f* химическое намагничение *n*	31
imanación viscosa *f* magnetización viscosa *f*	magnetização viscosa *f*	вязкая намагниченность *f* вязкое намагничение *n*	32
imanación sedimentaria *f* magnetización detrítica *f* magnetización sedimentaria *f*	magnetização detrítica *f* magnetização deposicional *f* magnetização sedimentar *f*	ориентационная намагниченность *f* ориентационное намагничение *n*	33
viscosidad magnética *f*	viscosidade magnética *f*	магнитная вязкость (горных пород) *f*	34

35	primary magnetization	aimantation primaire *f*	primäre Magnetisierung *f* Primärmagnetisierung *f*
36	secondary magnetization	aimantation secondaire *f*	sekundäre Magnetisierung *f* Sekundärmagnetisierung *f*
37	reversed magnetization	aimantation inverse *f* aimantation inversée *f*	umgekehrte Magnetisierung *f* inverse Magnetisierung *f*
38	self-reversal of magnetization magnetic self-reversal	auto-inversion de l'aimantation *f*	Selbstumkehr der Magnetisierung *f*
39	coercive field coercive force	champ coercitif *m* force coercitive *f*	Koerzitivfeld *n* Koerzitivkraft *f*
40	thermal demagnetization	désaimantation thermique *f*	thermische Entmagnetisierung *f*
41	Curie point Curie temperature	point de Curie *m* température de Curie *f*	Curie-Punkt *m* Curie-Temperatur *f*
42	magnetizing field	champ magnétisant *m*	Magnetisierungsfeld *n* magnetisierendes Feld *n*
43	rock magnetism	magnétisme des roches *m*	Gesteinsmagnetismus *m*
44	Earth's magnetic field terrestrial magnetic field geomagnetic field	champ magnétique terrestre *m* champ magnétique de Terre *m* champ géomagnétique *m*	erdmagnetisches Feld *n* Erdmagnetfeld *n* Magnetfeld der Erde *n* geomagnetisches Feld *n*
45	magnetic dip pole Earth's magnetic pole	pôle magnétique terrestre *m* pôle d'inclinaison *m*	magnetischer Pol (der Erde) *m* magnetischer Erdpol *m* magnetischer Neigungspol *m* Pol magnetischer Inklination *m*
46	magnetic axis	axe magnétique *m*	magnetische Achse *f*
47	terrestrial magnetic axis geomagnetic axis	axe géomagnétique *m* axe magnétique terrestre *m*	geomagnetische Achse *f* erdmagnetische Achse *f*
48	geomagnetic axis pole geomagnetic pole	pôle de l'axe géomagnétique *m* pôle de l'aimantation uniforme *m* pôle géomagnétique *m*	Pol der geomagnetischen Achse *m* geomagnetischer Achsenpol *m*
49	magnetic meridian	méridien magnétique *m*	magnetischer Meridian *m*
50	magnetic equator	équateur magnétique *m*	magnetischer Äquator *m*
51	geomagnetic coordinates	coordonées géomagnétiques *fpl*	geomagnetische [erdmagnetische] Koordinaten *fpl*
52	geomagnetic latitude	latitude géomagnétique *f*	geomagnetische [erdmagnetische] Breite *f*
53	geomagnetic longitude	longitude géomagnétique *f*	geomagnetische [erdmagnetische] Länge *f*

GEOMAGNETISMO	GEOMAGNETISMO	ГЕОМАГНЕТИЗМ	
imanación primaria *f* magnetización primaria *f*	magnetização primária *f*	первичное намагничение *n* первичная намагниченность *f*	35
magnetización secundaria *f* imanación secundaria *f*	magnetização secundária *f*	вторичное намагничение *n* вторичная намагниченность *f*	36
imanación inversa *f* magnetización inversa *f*	magnetização reversa *f* magnetização inversa *f*	обратное намагничение *n* обратная намагниченность *f*	37
autoinversión de la magnetización *f* autoinversión de la imanación *f*	auto-reversão da magnetização *f* auto-inversão da magnetização *f*	самообращение намагниченно- сти *n*	38
campo coercitivo *m* fuerza coercitiva *f*	campo coerci(ti)vo *m* força coerci(ti)va *f*	коэрцитивная сила *f* задерживающая сила *f* коэрцитивное поле *n*	39
desmagnetización térmica *f* desimantación térmica *f*	desmagnetização térmica *f*	тепловое размагничивание *n* термическое размагничение *n*	40
punto de Curie *m* temperatura de Curie *f*	ponto de Curie *m* temperatura (de) Curie *f*	точка Кюри *f* температура размагничения *f*	41
campo imanador *m* campo magnetizante *m*	campo magnetizante *m* campo iman(iz)ante *m*	поле намагничивания *n*	42
magnetismo de las rocas *m*	magnetismo das rochas *m*	магнетизм пород *m*	43
campo geomagnético *m* campo magnético terrestre *m* campo magnético de la Tierra *m*	campo geomagnético *m* campo magnético da Terra *m* campo magnético terrestre *m*	магнитное поле Земли *n* земное магнитное поле *n* геомагнитное поле *n*	44
polo de la inclinación magnética *m* polo magnético terrestre *m*	pólo magnético da Terra *m* pólo magnético de inclinação *m*	геомагнитный полюс *m* полюс наклонения *m* магнитный полюс (Земли) *m*	45
eje magnético *m*	eixo magnético *m*	магнитная ось *f*	46
eje magnético terrestre *m* eje magnético de la Tierra *m* eje geomagnético *m*	eixo magnético da Terra *m* eixo geomagnético *m*	земная магнитная ось *f* магнитная ось Земли *f* ось геодиполя *f*	47
polo del eje geomagnético *m* polo geomagnético *m*	pólo do eixo geomagnético *m* pólo geomagnético *m*	полюс однородного намагниче- ния *m*	48
meridiano magnético *m*	meridiano magnético *m*	магнитный меридиан *m*	49
ecuador magnético *m*	equador magnético *m*	магнитный экватор *m*	50
coordenadas geomagnéticas *fpl*	coordenadas geomagnéticas *mpl*	геомагнитные координаты *fpl*	51
latitud geomagnética *f*	latitude geomagnética *f*	геомагнитная широта *f*	52
longitud geomagnética *f*	longitude geomagnética *f*	геомагнитная долгота *f*	53

	GEOMAGNETISM	GÉOMAGNÉTISME	ERDMAGNETISMUS
54	(geo)magnetic elements	éléments (géo)magnétiques *mpl*	(geo)magnetische [erdmagnetische] Elemente *npl*
55	geomagnetic field intensity	intensité du champ géomagnétique *f*	Intensität des erdmagnetischen Feldes *f*
56	magnetic declination	déclinaison magnétique *f*	Missweisung *f* magnetische Deklination *f*
57	isogonal line isogonic line isogone	isogone *f* ligne isogone *f*	Isogone *f* Isolinie magnetischer Deklination *f*
58	magnetic dip magnetic inclination	inclinaison magnétique *f*	magnetische Inklination *f* erdmagnetische Inklination *f*
59	isoclinic line isocline isodip line	ligne isocline *f* isocline *f*	Isokline *f* Isolinie magnetischer Inklination *f*
60	horizontal intensity (of magnetic field) horizontal component horizontal field intensity	intensité horizontale (du champ magnétique) *f* composante horizontale magnétique *f*	Horizontalintensität (des Magnetfeldes) *f* Horizontalkomponente *f*
61	isodynamic line isodynamical line	isodyname *f* ligne isodynamique *f*	Isodyname *f*
62	vertical (field) intensity vertical component	intensité verticale *f* composante verticale *f*	Vertikalintensität *f* vertikale Komponente *f*
63	magnetic chart	carte magnétique *f*	magnetische Karte *f*
64	magnetic survey	levé magnétique *m*	magnetische Vermessung *f* magnetische Aufnahme *f* magnetische Landesvermessung *f*
65	World Magnetic Survey WMS	Levé magnétique mondial *m*	magnetische Vermessung der Erde *f*
66	multipole	multipôle *m*	Multipol *m*
67	magnetic anomaly	anomalie magnétique *f*	magnetische Anomalie *f*
68	South Atlantic Anomaly	anomalie sud-atlantique *f*	südatlantische Anomalie *f*
69	(geo)magnetic variations variations of geomagnetic field	variations (géo)magnétiques *fpl* variations du champ géomagnétique *fpl*	erdmagnetische [geomagnetische] Variationen *fpl* erdmagnetische [geomagnetische] Schwankungen *fpl* Variationen des Erdmagnetfeldes *fpl*

elementos (geo)magnéticos *mpl*	elementos (geo)magnéticos *mpl*	(гео)магнитные элементы *mpl*	54
intensidad del campo geomagnético *f*	intensidade do campo geomagnético *f*	напряжённость геомагнитного поля *f* напряжённость магнитного поля Земли *f*	55
declimación magnética *f*	declinação magnética *f*	магнитное склонение *n* отклонение магнитной стрелки от географического меридиана *n*	56
isogona *f* línea isogona *f*	linha isogônica *f* isógona *f*	изогона *f* изогональная *f* линия равного магнитного склонения *f*	57
inclinación magnética *f*	inclinação magnética *f*	магнитное наклонение *n*	58
(línea) isoclina *f* (línea) isoclinal *f*	isóclina *f* curva isoclinal *f*	изоклина *f* линия равных значений наклонения *f*	59
intensidad horizontal (del campo magnético) *f* componente horizontal *m*	intensidade horizontal (do campo magnético) *f* componente horizontal (do campo magnético) *m*	горизонтальный компонент напряжённости *m* горизонтальная составляющая магнитного поля *f*	60
isodinámica *f* línea isodinámica *f*	isodinâmica *f*	изодинама *f* линия равной напряжённости (гео)магнитного поля *f*	61
intensidad vertical *f* componente vertical *m*	intensidade vertical *f* componente vertical *m*	вертикальный компонент напряжённости *m* вертикальная составляюща геомагнитного поля *f*	62
mapa magnético *m* carta magnética *f*	carta magnética *f*	(изо)магнитная карта *f* карта элементов земного магнетизма *f*	63
levantamiento magnético *m* reconocimiento magnético *m*	levantamento magnético *m*	(гео)магнитная съёмка *f*	64
Reconocimiento magnético mundial *m* Levantamiento magnético mundial *m*	Levantamento magnético mundial *m*	Мировая магнитная съёмка *f*	65
multipolo *m*	multipolo *m*	мультиполь *m*	66
anomalía magnética *m*	anomalia magnética *f*	магнитная аномалия *m*	67
anomalía sudatlántica *f*	anomalia do Atlântico Sul *f*	южноатлантическая аномалия *f* югоатлантическая аномалия *f*	68
variaciones (geo)magnéticas *fpl* variaciones del campo geomagnético *fpl*	variações do campo geomagnético *fpl* variações geomagnéticas *fpl*	геомагнитные вариации *fpl* вариации элементов земного магнетизма *fpl*	69

	GEOMAGNETISM	GÉOMAGNÉTISME	ERDMAGNETISMUS
70	(geo)magnetic disturbances	perturbations (géo)magnétiques *fpl*	(erd)magnetische Störungen *fpl*
71	(geo)magnetic storm	orage (géo)magnétique *m*	(erd)magnetischer Sturm *m* (geo)magnetischer Sturm *m*
72	standard-type storm	orage magnétique typique *m*	Sturm des Standard-Typs *m*
73	storm with sudden commencement	orage magnétique à début brusque *m*	Sturm mit plötzlichen Beginn *m*
74	storm sudden commencement ssc sudden commencement	début brusque d'orage *m* début brusque *m*	plötzlicher Sturmanfang *m*
75	initial phase (of a magnetic storm)	phase initiale (d'un orage magnétique) *f*	Anfangsphase (eines magnetischen Sturmes) *f*
76	main phase	phase principale *f*	Hauptphase *f*
77	recovery phase	phase de retour au calme *f*	Erholungsphase *f*
78	gradually commencing storm sg	orage à début progressif *m*	allmählich beginnender Sturm *m*
79	magnetic crochet crochet	crochet magnétique *m*	hakenartige Ausbuchtung *f* hakenartige Aufwölbung *f* Haken im Magnetogramm *m*
80	magnetic bay bay disturbance	baie magnétique *f* perturbation en baie *f*	erdmagnetische Bai *f* Bai-Störung *f*
81	(geo)magnetic pulsations	pulsations (géo)magnétiques *fpl*	(erd)magnetische [geomagnetische] Pulsationen *fpl*
82	(geo)magnetic micropulsations	micropulsations (géo)magnétiques *fpl*	(erd)magnetische [geomagnetische] Mikropulsationen *fpl* Mikropulsationen des Erdfeldes *fpl*
83	magnetograph	magnétographe *m*	Magnetograph *m*
84	magnetogram magnetic record	magnétogramme *m*	Magnetogramm *n*
85	secular (geo)magnetic variations secular variation of the Earth's magnetic field	variations séculaires (géo)magnétiques *fpl* variation séculaire du champ magnétique terrestre *f*	Säkularvariationen des (erd)magnetischen Feldes *fpl*
86	geomagnetic activity	activité géomagnétique *f*	geomagnetische Aktivität *f* erdmagnetische Aktivität *f*
87	atmospheric heating	échauffement atmosphérique *m*	atmosphärische Aufheizung *f* atmosphärische Erwärmung *f*
88	infrasonic pressure waves	ondes infrasoniques (de pression) *fpl*	Infraschallwellen *fpl*

perturbaciones (geo-)magnéticas *fpl*	perturbações (geo-)magnéticas *fpl*	(гео)магнитные возмущения *npl*	70
tormenta (geo)magnética *f* tempestad (geo)magnética *f*	tempestade (geo)magnética *f*	(гео)магнитная буря *f*	71
tormenta (magnética) típica *f*	tempestade típica *f*	буря обычного типа *f*	72
tormenta (magnética) de comienzo brusco *f*	tempestade de início súbito *f*	буря с внезапным началом *f*	73
comienzo brusco *m* comienzo rápido *m*	começo brusco *m* início súbito da tempestade *m* início brusco *m*	внезапное начало (магнитной бури) *n*	74
fase inicial (de una tormenta magnética) *f*	fase inicial (de uma tempestade magnética) *f*	начальная фаза (магнитной бури) *f*	75
fase principal *f*	fase principal *f*	главная фаза *f*	76
fase de restablecimiento *f*	fase de recuperação *f*	фаза восстановления *f*	77
tormenta de comienzo gradual *f*	tempestade de começo [inicio] gradual *f*	буря с постепенным началом *f*	78
gancho magnético *m* corchete *m*	crochê magnético *m* colchete *m*	кроше *n* микробухта *f* магнитный крючок *m* магнитный крюк *m*	79
bahía magnética *f*	baía magnética *f*	магнитная бухта *f* бухтообразное возмущение *n*	80
pulsaciones (geo)magnéticas *fpl*	pulsações (geo)magnéticas *fpl*	(гео)магнитные пульсации *fpl* (гео)магнитные колебания *npl* пульсации (гео)магнитного поля *fpl*	81
micropulsaciones (geo)magnéticas *fpl*	micropulsações (geo)magnéticas *fpl*	(гео)магнитные микропульсации *fpl* микропульсации (гео)магнитного поля *fpl*	82
magnetógrafo *m*	magnetógrafo *m*	магнитограф *m*	83
magnetograma *m*	magnetograma *m*	магнитограма *f* запись магнитографа *f*	84
variación (geo)magnética secular *f* variación secular del campo geomagnético *f*	variacão secular (geo)magnética *f* variação secular do campo (geo)magnético *f*	вековые вариации геомагнитного поля *fpl* вековые магнитные изменения *npl*	85
actividad geomagnética *f*	atividade geomagnética *f*	геомагнитная активность *f*	86
calentamiento atmosférico *m*	aquecimento atmosférico *m*	нагрев атмосферы *m*	87
ondas infrasónicas de presión *fpl*	ondas infra-sônicas de pressão *fpl*	инфразвуковые волны *fpl*	88

	GEOMAGNETISM	GÉOMAGNÉTISME	ERDMAGNETISMUS
89	K-index of geomagnetic activity three-hour-range index	indice K (d'activité géomagnétique) *m* indice trihoraire *m*	K-Index (der geomagnetischen Aktivität) *m* dreistündliche Kennziffer *f*
90	planetary (magnetic) Kp-index planetary index	indice planétaire Kp *m*	planetarischer Kp-Index *m* planetarische Kennziffer *f*
91	magnetically disturbed period	période magnétiquement agitée *f*	magnetisch gestörter Zeitabschnitt *m* magnetisch gestörte Periode *f*
92	quiet day magnetically quiet day	jour (magnétiquement) calme *m*	(magnetisch)ruhiger Tag *m*
93	active day magnetically active day	jour magnétiquement actif *m* jour (magnétiquement) agité *m*	aktiver Tag *m* magnetisch aktiver Tag *m*
94	disturbed day	jour perturbé *m*	gestörter Tag *m*
95	(geo)magnetic substorm	sous-orage (géo)magnétique *m*	magnetischer Elementarsturm *m* magnetischer Teilsturm *m*
96	polar magnetic substorm	sous-orage magnétique polaire *m*	polarer magnetischer Elementarsturm *m* polarer magnetischer Teilsturm *m*
97	current system	système de courants *m*	Stromsystem *n*
98	equatorial ring current	anneau équatorial de courant *m*	äquatorialer Ringstrom *m*
99	geomagnetic dynamo theory	dynamo-théorie géomagnétique *f* théorie dynamo du champ magnétique terrestre *f*	geomagnetische Dynamotheorie *f*
100	ring currents	courants d'anneau *mpl* anneau de courants *m*	Ringströme *mpl*
101	radiation belt Van Allen belt radiation zone trapped-radiation region	ceinture de rayonnement *f* ceinture de Van Allen *f* zone de rayonnement *f* zone de piégeage *f*	Strahlungsgürtel *m* Van Allenscher Strahlungsgürtel *m* Van Allen-Gürtel *m*
102	isocount contours contours of constant intensity	lignes d'égal taux de comptage *fpl* iso-intensités *fpl*	Isozahl-Kontouren *fpl* Umgrenzungslinien gleicher Zählrate *fpl* Schichtlinien gleicher Zählrate *fpl*
103	unidirectional intensity	intensité directionnelle *f*	Zählrate für eine bestimmte Richtung *f*
104	omnidirectional intensity	intensité omnidirectionnelle *f*	Zählrate für den gesamten Raumwinkel *f*
105	outer radiation zone outer radiation belt outer Van Allen (radiation) belt	ceinture extérieure *f* zone extérieure *f* ceinture de rayonnement externe *f*	äusserer Strahlungsgürtel *m*

índice K de actividad magnética *m* índice tri-horario *m*	índice K da atividade magnética *m* índice triorário *m*	К-индекс (гео)магнитной активности *m* трёхчасовой индекс активности *m*	89
índice planetario Kp *m* índice magnético planetario *m*	índice planetário Kp *m*	планетарный индекс (гео)магнитной активности Kp *m* планетарный индекс *m*	90
período magnéticamente perturbado *m*	período magneticamente perturbado *m*	магнитно-возмущённый период *m*	91
día de calma (magnética) *m* día (magnéticamente) tranquilo *m*	dia calmo [quieto] (magnético) *m* dia (magneticamente) calmo [quieto] *m*	спокойный день *m*	92
día (magnéticamente) activo *m*	dia ativo (magnético) *m* dia (magneticamente) ativo *m*	активный день *m*	93
día perturbado *m*	dia perturbado *m*	возмущённый день *m*	94
tormenta (geo)magnética secundaria *f*	subtempestade magnética *f*	магнитная суббуря *f*	95
tormenta magnética polar secundaria *f*	subtempestade magnética polar *f*	полярная магнитная суббуря *f*	96
sistema de corrientes *m*	sistema de correntes *m*	система электрических токов *f*	97
corriente anular ecuatorial *m*	corrente anular equatorial *f*	экваториальный кольцевой ток *m*	98
teoría geomagnética de la dínamo *f* dinamoteoría geomagnética *f*	teoria dínamo geomagnética *f* teoria (de) dínamo do campo geomagnético *f*	теория геомагнитного динамо *f*	99
corrientes de anillo *mpl* corrientes anulares *mpl*	correntes anulares *fpl*	кольцевые токи *mpl*	100
cinturón de radiación *m* cinturón de Van Allen *m* zona de radiación (atrapada) *m*	cinturão de radiação *m* cinturão de Van Allen *m* zona de radiação (aprisionada) *f*	пояс ван Аллена-Вернова *m* радиационный пояс *m* радиационный пояс Земли *m* зона захваченной радиации *f*	101
contornos de isoconteos *mpl* contornos de igual contaje *mpl* líneas de igual intensidad *fpl*	eqüiescalares de igual contagem *mpl* contornos de igual contagem [intensidade] *mpl*	изолинии скорости счёта [счётчика Гейгера] *fpl* кривые равного счёта *fpl*	102
intensidad unidireccional *f*	intensidade unidirecional *f*	однонаправленная интенсивность *f*	103
intensidad omnidireccional *f*	intensidade o(m)nidirecional *f*	всенаправленная интенсивность *f*	104
región exterior de radiación *f* zona exterior de radiación *f* cinturón de radiación exterior *m*	cinturão externo *m* zona de radiação externa *f*	внешний радиационный пояс *m* второй радиационный пояс *m*	105

	GEOMAGNETISM	GÉOMAGNÉTISME	ERDMAGNETISMUS
106	inner radiation zone inner radiation belt inner Van Allen (radiation) belt	ceinture interne *f* ceinture intérieure *f* zone intérieure *f*	innerer Strahlungsgürtel *m*
107	natural radiation belt [zone]	ceinture [zone] de radiation naturelle *f* ceinture [zone] de rayonnement naturel *f*	natürlicher Strahlungsgürtel *m*
108	artificial radiation belt man-made radiation belt artificial radiation zone	zone de radiation artificielle *f* ceinture de rayonnement artificiel *f* zone de rayonnement artificiel *f*	künstlicher Strahlungsgürtel *m* durch Menschen hergestelte Strahlungszone *f* künstliche Strahlungszone *f*
109	trapping	piégeage *m* capture *f*	Einfangen *n* Einfang *m*
110	trapping region	région de capture *f*	Einfangregion *f*
111	(geomagnetically) trapped particles	particules capturées (par le champ magnétique terrestre) *fpl* particules piégées *fpl*	(geomagnetisch) eingefangene Partikeln *fpl*
112	(geomagnetically) trapped radiation	radiation capturée (par le champ magnétique terrestre) *f* rayonnement piégé *m*	(geomagnetisch) eingefangene Strahlung *f*
113	artificial trapped radiation	radiation artificielle capturée *f* radiation provoquée *f*	künstlich eingefangene Strahlung *f*
114	high-altitude nuclear explosion	explosion nucléaire à haute altitude *f*	nukleare Explosion in grosser Höhe *f*
115	albedo neutron	neutron d'albédo *m*	Albedo-Neutron *n*
116	cosmic-ray albedo neutron albedo	albédo du rayonnement cosmique *m* albédo neutronique *m*	Albedo der kosmischen Strahlung *f* Neutronenalbedo *f*
117	magnetic rigidity	rigidité magnétique *f*	magnetische Steifigkeit *f* magnetische Righeit *f*
118	particle trajectory	trajectoire d'une particule *f*	Partikelbahn *f* Teilchenbahn *f*
119	latitudinal bounce	va-et-vient latitudinal *m* rebond latitudinal *m*	Schwingung zwischen den Spiegelpunkten *f*
120	longitudinal drift drift in longitude	dérive en longitude *f*	Drift in (geomagnetischer) Länge *f* longitudinale Drift *f* longitudinale Verschiebung *f*
121	curvature of lines of force	courbure de lignes de force *f*	Krümmung der Feldlinen *f* Krümmung der Kraftlinien *f*

región interior de radiación *f* zona interior de radiación *f* cinturón de radiación interior *m*	cinturão interno *m* cinturão de radiação interno *m* zona de radiação interna *f*	внутренний радиационный пояс *m* первый радиационный пояс *m* внутренняя зона захваченной радиации *f*	106
cinturón [zona] de radiación natural *m*[*f*]	cinturão [zona] de radiação natural *m*[*f*]	естественный радиационный пояс *m*	107
cinturón de radiación artificial *m* cinturón de radiación producida por el hombre *m*	cinturão de radiação artificial *m* zona de radiação artificial *f*	искусственный радиационный пояс *m* искусственная зона радиации *f* пояс радиации искусственного происхождения *m*	108
aprisionamiento *m* captura *f*	aprisionamento *m* retenção *f* captura *f*	захват (частиц) *m*	109
región de captura *f* región de aprisionamiento *f*	região de aprisionamento *f* região de retenção *f*	область захвата (частиц) *f*	110
partículas (geomagnéticamente) atrapadas *fpl* partículas capturadas *fpl*	partículas retidas (geomagnéticamente) *fpl* partículas aprisionadas *fpl*	частицы, захваченные геомагнитным полем *fpl*	111
radiación (geomagnéticamente) atrapada *f* radiación capturada *f*	radiação retida (geomagnéticamente) *f* radiação aprisionada *f*	радиация, захваченная геомагнитным полем *f* захваченная радиация *f*	112
radiación artificial atrapada *f*	radiação artificial retida *f* radiação artificial aprisionada *f*	захваченная радиация искусственного происхождения *f*	113
explosión nuclear a gran altura *f*	explosão nuclear de altitude *f*	высотный взрыв *m*	114
neutrón de albedo *m*	nêutron de albedo *m*	альбедный нейтрон *m*	115
albedo de rayos cósmicos *m* albedo de neutrones *m*	albedo de radiação cósmica *m* albedo neutrônico *m*	альбедо нейтронов *n* нейтронное альбедо *n* альбедо космических лучей *n*	116
rigidez magnética *f*	rigidez magnética *f*	магнитная жёсткость *f*	117
trayectoria de una partícula *f*	trajetória de uma partícula *f*	траектория частицы *f*	118
rebote latitudinal *m*	oscilação latitudinal *f* movimento entre pontos espelhentos *m*	осцилляции между зеркальными точками *fpl* колебания (частицы) между зеркальными точками *npl*	119
deriva longitudinal *f* deriva en longitud (geomagnética) *f*	deriva longitudinal *f* deriva em longitude *f*	долготный дрейф *m* дрейф по долготе *m*	120
curvatura de líneas de fuerza [de líneas de campo] *f*	curvatura das linhas de campo [de força] *f*	кривизна силовых линий *f*	121

	GEOMAGNETISM	GÉOMAGNÉTISME	ERDMAGNETISMUS
122	trapped particle lifetime	durée de vie d'une particule capturée [piégée] *f*	Lebensdauer eingefangener Partikel *f*
123	particle storage storage of particles	stockage de particules *m*	Partikelspeicherung *f* Speicherung von Partikeln *f*
124	mirror point reflection point	point miroir *m* point de réflexion *m*	Spiegelpunkt *m* Reflexionspunkt *m*
125	mirror-point altitude	altitude du point miroir *f*	Höhe des Spiegelpunktes *f*

(tiempo de) vida de una partícula atrapada *(m)f* duración de una partícula capturada *f*	(tempo de) vida da partícula retida *(m)f* (tempo de) vida da partícula aprisionada *(m)f*	время жизни захваченной частицы *n*	122
almacenamiento de partículas *m* depósito de partículas *m*	acumulação de partículas *f* armazenamento de partículas *f*	накопление частиц *n*	123
punto de reflexión *m* punto espejo *m*	ponto espelhento *m* ponto de reflexão *m*	зеркальная точка *f* точка отражения *f*	124
altura del punto espejo *f* altitude del punto de reflexión *f*	altitude do ponto espelhento *f*	высота зеркальной точки *f*	125

76 ATMOSPHERIC SCIENCES / SCIENCES ATMOSPHÉRIQUES / ATMOSPHÄRISCHE WISSENSCHAFTEN

1	atmospheric sciences	sciences atmosphériques *fpl*	atmosphärische Wissenschaften *fpl*
2	environmental science ecology	science de l'environnement *f* écologie *f*	Ökologie *f* Umweltwissenschaft *f*
3	environment	environnement *m* milieu *m*	Umwelt *f*
4	atmospheric physics	physique de l'atmosphère *f*	Physik der Atmosphäre *f*
5	atmosphere of the Earth Earth's atmosphere	atmosphère terrestre *f* atmosphère de la Terre *f*	Erdatmosphäre *f* Atmosphäre der Erde *f*
6	air	air *m*	Luft *f*
7	ambient	ambiant	umgebend Umgebungs-
8	airless	vide d'air	luftleer
9	vacuum free space	espace libre *m* espace vide *m* vide *m*	Vakuum *n* freier Raum *m*
10	homogeneous atmosphere	atmosphère homogène *f*	homogene Atmosphäre *f*
11	standard atmosphere	atmosphère standard *f*	Standard-Atmosphäre *f* Normalatmosphäre *f*
12	adiabatic atmosphere	atmosphère adiabatique *f*	adiabatische Atmosphäre *f* adiabatisch geschichtete Atmosphäre *f*
13	free atmosphere	atmosphère libre *f*	freie Atmosphäre *f*
14	dry adiabatic lapse rate DALR	gradient (vertical de température) adiabatique sec *m*	trockenadiabatischer Temperaturgradient *m*
15	barometric height formula	formule barométrique *f* formule altimétrique *f*	barometrische Höhenformel *f*
16	pressure altitude	altitude-pression *f*	Höhe einer Druckfläche *f* Druckhöhe *f*
17	vertical distribution	distribution verticale *f*	Vertikalverteilung *f*
17a	height of the homogeneous atmosphere	hauteur de l'atmosphère homogène *f*	Höhe der homogenen Atmosphäre *f*
18	scale height	échelle d'altitude *f* échelle de hauteur *f* hauteur de référence *f*	Skalenhöhe *f* Masstabshöhe *f* Masstabhöhe *f*
19	height of unit decrease of pressure (in metres/millibar)	échelon hypsométrique *m*	barometrische Höhenstufe *f*
20	isothermal equilibrium	équilibre isotherme *m*	isothermes Gleichgewicht *n*

ciencias atmosféricas *fpl* ciencias da atmósfera *fpl*	ciências atmosféricas *fpl* ciências da atmosfera *fpl*	атмосферные науки *fpl* науки об атмосфере *fpl*	1
ciencia ambiental *f* ecología *f*	ciência ambiental *f* ecologia *f*	экология *f* наука об окружающей среде *f*	2
medio ambiente *m* ambiente *m*	meio ambiente *m* ambiente *m*	окружающая среда *f*	3
física atmosférica *f*	física atmosférica *f*	физика атмосферы *f*	4
atmósfera de la Tierra *f* atmósfera terrestre *f*	atmosfera terrestre *f* atmosfera da Terra *f*	атмосфера Земли *f* земная атмосфера *f*	5
aire *m*	ar *m*	воздух *m*	6
ambiente	ambiente	окружающий	7
vacío	vazio	безвоздушный	8
espacio libre *m* vacío *m*	espaço livre *m* vácuo *m*	безвоздушное пространство *n* свободное пространство *n*	9
atmósfera homogénea *f*	atmosfera homogênea *f*	однородная атмосфера *f*	10
atmósfera patrón *f* atmósfera tipo *f*	atmosfera padrão *f* atmosfera standard *f*	стандартная атмосфера *f*	11
atmósfera adiabática *f*	atmosfera adiabática *f*	адиабатическая атмосфера *f*	12
atmósfera libre *f*	atmosfera livre *f*	свободная атмосфера *f*	13
gradiente (de temperatura) adiabático seco *m*	gradiente (vertical de temperatura) adiabático seco *m* taxa de variação vertical de temperatura *f*	сухоадиабатический (вертикальный) градиент *m*	14
fórmula barométrica de altura *f* fórmula de nivelación barométrica *f*	fórmula barométrica [altimétrica] *f* fórmula de nivelamento barométrico *f*	барометрическая формула *f*	15
altura de presión *f*	altitude de pressão *f*	барометрическая высота *f* высота по давлению *f*	16
distribución vertical *f*	distribuição vertical *f*	вертикальное распределение *n*	17
altura de atmósfera homogénea *f*	altura da atmosfera homogênea *f*	высота однородной атмосферы *f*	17a
escala de alturas *f* altura a escala *f*	escala de altura *f* altura de referência *f*	шкала высот *f* масштабная высота *f* приведённая высота *f*	18
escalón barométrico *m*	escalão hipsométrico *m* escala barométrica *f*	барическая ступень *f*	19
equilibrio isotérmico *m*	equilíbrio isotérmico *m*	изотермическое равновеие *n*	20

21	adiabatic equilibrium	équilibre adiabatique *m*	adiabatisches Gleichgewicht *n*
22	convective equilibrium	équilibre convectif *m*	konvektives Gleichgewicht *n*
23	diffusion equilibrium diffusive equilibrium	équilibre de diffusion *m*	Diffusionsgleichgewicht *n*
24	escape of gases	libération des gaz *f*	Entweichen von Gasen *n*
25	top of the atmosphere	limite supérieure de l'atmosphère *f*	obere Atmosphärengrenze *f* Atmosphärenobergrenze *f*
26	critical level	niveau critique *m*	kritisches Niveau *n*
27	cone of escape	cône de libération *m*	Entweichungskegel *m*
28	tidal oscillations	oscillations de marée *fpl*	Gezeitenschwingungen *fpl*
29	semi-diurnal wave	onde semi-diurne *f*	halbtägige Welle *f*
30	diurnal wave	onde diurne *f*	ganztägige Welle *f*
31	atmospheric acoustics	acoustique atmosphérique *f*	atmosphärische Akustik *f*
32	propagation of sound	propagation du son *f*	Schallausbreitung *f*
33	sound wave	onde sonore *f*	Schallwelle *f*
34	sound ray	rayon sonore *m*	Schallstrahl *m*
35	normal sound	son normal *m*	normaler Schall *m*
36	anomalous sound	son anormal *m*	anomaler Schall *m*
37	anomalous sound propagation anomalous propagation of sound	propagation anormale du son *f*	anomale Schallausbreitung *f*
38	sound level sound intensity	niveau du son *m*	Schallpegel *m*
39	decibel	décibel *m*	Dezibel *n*
40	sound field	champ sonore *m*	Schallfeld *n*
41	zone of audibility	zone d'audibilité *f*	Hörbarkeitszone *f*
42	zone of silence	zone de silence *f*	Zone des Schweigens *f*
43	characteristic velocity	vitesse caractéristique *f*	charakteristische Geschwindigkeit *f*
44	sound barrier sonic barrier	mur du son *m*	Schallmauer *f*

equilibrio adiabático *m*	equilíbrio adiabático *m*	адиабатическое равновесие *n*	21
equilibrio convectivo *m*	equilíbrio convectivo *m*	конвективное равновесие *n*	22
equilibrio difusivo *m*	equilíbrio de difusão *m*	диффузионное равновесие *n*	23
escape de gases *m*	escapamento [escape] de gases *m*	убегание газов (из атмосферы) *n* ускользание газов (из атмосферы) *n*	24
límite superior de la atmósfera *m* techo de la atmósfera *m*	topo da atmosfera *m* limite superior da atmosfera *m*	верхняя граница атмосферы *f* потолок атмосферы *m* предельная высота атмосферы *f* вершина атмосферы *f*	25
nivel crítico *m*	nível crítico *m*	критический уровень *m*	26
cono de escape *m*	cone de escape [escapamento] *m*	конус потерь *m* конус ускользания *m* конус убегания *m*	27
oscilaciones de marea *fpl*	oscilações de maré *fpl*	приливные колебания *npl*	28
onda semidiurna *f*	onda semi-diurna *f*	полусуточная волна *f*	29
onda diurna *f*	onda diurna *f*	суточная волна *f*	30
acústica atmosférica *f*	acústica atmosférica *f*	атмосферная акустика *f*	31
propagación del sonido *f*	propagação do som *f*	распространение звука *n*	32
onda sonora *f*	onda sonora *f*	звуковая волна *f*	33
rayo sonoro *m*	raio sonoro *m*	звуковой луч *m*	34
sonido normal *m*	som normal *m*	нормальный звуковой луч *m*	35
sonido anómalo *m*	som anômalo *m*	аномальный звуковой луч *m*	36
propagación anómala del sonido *f*	propagação anômala do som *f*	аномальное распространение звука *n*	37
nivel del sonido *m*	nível do som *m* intensidade do som *f*	уровень звука *m* уровень силы звука *m*	38
decibel(io) *m*	decibel *m*	децибел *m*	39
campo sonoro *m*	campo sonoro [de som] *m*	звуковое поле *n*	40
zona de audibilidad *f*	zona de audibilidade *f*	зона слышимости *f*	41
zona de silencio *f*	zona de silêncio *f*	зона молчания *f*	42
velocidad característica *f*	velocidade característica *f*	характеристическая скорость *f*	43
barrera del sonido *f* barrera sónica *f*	barreira sonora *f* barreira do som *f*	звуковой барьер *m*	44

45	sound probe acoustic sounding	sonde acoustique *f*	Schallsonde *f*
46	hot-wire microphone	microphone à fil chaud *m*	Hitzdrahtmikrophon *n*
47	sounding	sondage *m*	Sondierung *f*
48	rocket sounding	sondage par fusées *m*	Raketensondierung *f*
49	atmospheric layer	couche atmosphérique *f*	atmosphärische Schicht *f*
50	lower atmosphere	atmosphère inférieure *f*	untere Atmosphäre *f*
51	middle atmosphere	atmosphère moyenne *f*	mittlere Atmosphäre *f*
52	upper atmosphere	atmosphère supérieure *f*	obere Atmosphäre *f*
53	vertical profile (of density, of temperature, of wind, of humidity)	profil vertical (de densité, de température, du vent, de l'humidité) *m*	Höhenprofil (der Dichte, der Temperatur, des Windes, der Feuchte) *n*
54	rise increase	incline *f* croissance *f*	Anstieg *m* Neigung *f*
55	peak value (of a layer)	maximum de la région *m*	Maximum der Schicht *n* Spitze (der Schicht) *f*
56	peak value of E-layer	maximum de la région E *m*	E-Schicht-Maximum *n*
57	decrease fall	décline *f* décroissance *f*	Abfall *m* Abnahme *f*
58	upper boundary –pause	limite supérieure *f* –pause *f* surface supérieure de... *f*	obere Grenze (einer Schicht) *f*
59	thermosphere	thermosphère *f*	Thermosphäre *f*
60	homosphere	homosphère *f*	Homosphäre *f*
61	heterosphere	hétérosphère *f*	Heterosphäre *f*
62	chemosphere	chimiosphère [chémosphère] *f*	Chemosphäre *f*
63	photochemical reaction	réaction photochimique *f*	photochemische Reaktion *f*
64	neutrosphere	neutrosphère *f*	Neutrosphäre *f*
65	homopause	homopause *f*	Homopause *f*
66	mesosphere	mésosphère *f*	Mesosphäre *f*
67	mesopause	mésopause *f*	Mesopause *f*

sonda acústica *f*	sonda acústica *f*	акустический зонд *m*	45
micrófono de hilo caliente *m* micrófono de alambre caliente *m*	microfone de fio quente *m*	электромагнитный микробарограф *m*	46
sondeo *m*	sondagem *f*	зондирование *n*	47
sondeo por cohete *m*	sondagem por foguete *f*	ракетное зондирование *n*	48
capa atmosférica *f*	camada atmosférica *f*	слой атмосферы *m* атмосферный слой *m*	49
baja atmósfera *f*	baixa atmosfera *f*	нижняя атмосфера *m*	50
atmósfera media *f*	atmosfera média *f*	средняя атмосфера *f*	51
alta atmósfera *f*	alta atmosfera *f* atmosfera superior *f*	верхняя атмосфера *f* верхние слои атмосферы *mpl*	52
perfil de altura (de densidad, de temperatura, del viento, de la humedad) *m* curva de (...) con la altura *f* perfil vertical *m*	perfil vertical (da densidade, da temperatura, do vento, da umidade) *m*	высотный профиль (плотности, температуры, ветра, влажности) *m* вертикальный профиль *m*	53
subida *f* ascendente *m*	subida *f* elevação *f* aumento *m*	подъём *m* слой повышения (температуры, ...) (вверх) *m*	54
máximo (de una capa) *m*	máximo (de uma camada) *m* valor máximo *m*	пик (слоя) *m* максимум (слоя) *m*	55
máximo de la capa E *m*	(valor) máximo da camada E *m*	максимум слоя E *m*	56
descendente *m* caída *f*	declínio *m* queda *f*	падение *n* слой понижения (температуры, ...) (вверх) *m*	57
límite superior de una capa *m* –pausa *f*	limite superior *m* pausa *f*	верхняя граница (слоя) *f*	58
termosfera *f*	termosfera *f*	термосфера *f*	59
homosfera *f*	homosfera *f*	гомосфера *f*	60
heterosfera *f*	heterosfera *f*	гетеросфера *f*	61
quimiosfera *f*	quimiosfera *f*	хемисфера *f*	62
reacción fotoquímica *f*	reação fotoquímica *f*	фотохимическая реакция *f*	63
neutrosfera *f*	neutrosfera *f*	нейтросфера *f*	64
homopausa *f*	homopausa *f*	гомопауза *f*	65
mesosfera *f*	mesosfera *f*	мезосфера *f*	66
mesopausa *f*	mesopausa *f*	мезопауза *f*	67

	ATMOSPHERIC SCIENCES	SCIENCES ATMOSPHÉRIQUES	ATMOSPHÄRISCHE WISSENSCHAFTEN
68	escape region exosphere fringe region	exosphère *f* atmosphère extérieure *f*	Exosphäre *f* äussere Atmosphäre *f*
69	ionosphere	ionosphère *f*	Ionosphäre *f*
70	stratosphere	stratosphère *f*	Stratosphäre *f*
71	stratopause	stratopause *f*	Stratopause *f*
72	turbosphere	turbosphère *f*	Turbosphäre *f*
73	geocorona	géocouronne *f*	Geokorona *f*
74	neutral hydrogen geocorona	géocouronne d'hydrogène neutre *f*	neutrale Wasserstoff-Geokorona *f*
75	protonosphere	protonosphère *f*	Protonosphäre *f* Protonensphäre *f*
76	plasmosphere outer ionosphere	plasmosphère *f* ionosphère extérieure *f*	Plasmasphäre *f* äussere Ionosphäre *f*
77	tropopause	tropopause *f*	Tropopause *f*
78	troposphere	troposphère *f*	Troposphäre *f*
79	atmospheric chemistry	chimie de l'atmosphère *f*	Chemie der Atmosphäre *f*
80	air pollution atmospheric pollution	pollution de l'air *f* pollution atmosphérique *f*	Luftverunreinigung *f* Luftverschmutzung *f*
81	pollutant	polluant *m*	Pollutant *m* Beimengung *f*
82	atmospheric pollutant	polluant atmosphérique *m*	Luftverschmutzer *m*
83	smog	smog *m*	Smog *m* Rauch mit Nebel *m*
84	photolysis	décomposition photochimique *f*	Photolyse *f*
85	(air) sampling	échantillonage (de l'air) *m* prise d'échantillons *f*	Entnahme einer Luftprobe *f* Probenentnahme *f*
86	air sample	échantillon de l'air *m*	Luftprobe *f*
87	atmospheric constituent	constituant atmosphérique *m*	Bestandteil der Atmosphäre *m*
88	trace gas	gaz en trace *m*	Spurengas *n*
89	aerosol	aérosol *m*	Aerosol *n*
90	hydrosol	hydrosol *m*	Hydrosol *n*

CIENCIAS ATMOSFÉRICAS	CIÊNCIAS ATMOSFÉRICAS	АТМОСФЕРНЫЕ НАУКИ	
exosfera *f* atmósfera externa *f*	exosfera *f* região de escape *f* atmosfera exterior *f*	экзосфера *f* область рассеяния *f* внешняя область атмосферы *f*	68
ionosfera *f*	ionosfera *f*	ионосфера *f*	69
estratosfera *f*	estratosfera *f*	стратосфера *f*	70
estratopausa *f*	estratopausa *f*	стратопауза *f*	71
turbosfera *f*	turbosfera *f*	турбосфера *f*	72
geocorona *f*	geocorona *f*	геокорона *f*	73
geocorona de hidrógeno neutro *f*	geocorona de hidrogênio neutro *f*	геокорона из нейтрального водорода *f* водородная оболочка Земли *f*	74
protonosfera *f*	protonosfera *f*	протоносфера *f*	75
plasmasfera *f* ionosfera exterior *f*	plasmosfera *f* ionosfera exterior *f*	плазмосфера *f* внешняя ионосфера *f*	76
tropopausa *f*	tropopausa *f*	тропопауза *f*	77
troposfera *f*	troposfera *f*	тропосфера *f*	78
química de la atmósfera *f*	química da atmosfera *f*	химия атмосферы *f*	79
contaminación atmosférica *f* polución del aire *f*	polu(i)ção do ar *f* polu(i)ção atmosférica *f*	загрязнение воздуха *n* загрязнение атмосферы *n*	80
contaminante *m*	poluente *m*	загрязнитель *m* загрязняющее вещество *n*	81
contaminante atmosférico *m*	poluente atmosférico *m*	загрязнитель атмосферы *m*	82
niebla espesa mezclada con humo *f* (e)smog *m*	smog *m* bruma com polu(i)ção *f*	смог *m*	83
fotolisis *f*	fotólise *f*	фотохимическое разложение *n* фотолиз *m*	84
muestreo (de aire) *m* sacar muestras	amostragem (do ar) *f* tirar amostra (do ar)	отбор проб (воздуха) *m*	85
muestra de aire *f*	amostra do ar *f*	проба воздуха *f*	86
componente atmosférico *m*	constituente [componente] atmosférico *m*	атмосферная составляющая *f*	87
gas residual *m*	gás traço *m* traço de gás *m*	микрогаз *m* малая составляющая (атмосферы) *f*	88
aerosol *m*	aerossol *m*	аэрозоль *m*	89
hidrosol *m*	hidrossol *m*	гидрозоли *pl*	90

	ATMOSPHERIC SCIENCES	SCIENCES ATMOSPHÉRIQUES	ATMOSPHÄRISCHE WISSENSCHAFTEN
91	aerosol particle	particule d'aérosol *f*	Aerosolteilchen *n*
92	hydrophilic	hydrophile	wasserfreundlich wasseraffin hydrophil
93	hydrophobic water resistant	hydrophobe	hydrophob wasserabstossend wasserabweisend wasserfeindlich
94	aerosol filter	filtre pour aérosol *m*	Aerosolfilter *n*
95	dust filter	filtre pour poussière *m*	Staubfilter *n*
96	colloid	colloïde *m*	Kolloid *n*
97	colloidal	colloïdal	kolloidal
98	colloidal particle	particule colloïdale *f*	Kolloidteilchen *n*
99	solid particle	particule solide *f*	feste Partikel *f* festes Teilchen *n*
100	liquid particle	particule liquide *f* gouttelette *f*	flüssiges Teilchen *n* Tröpfchen *n*
101	saturation	saturation *f*	Sättigung *f*
102	supersaturation	supersaturation *f*	Übersättigung *f*
103	condensation nucleus	noyau de condensation *m*	Kondensationskern *m*
104	ice nuclei icy nuclei	noyaux de glace *mpl*	Eiskerne *mpl*
105	supercooling subcooling	superrefroidissement *m* sous-refroidissement *m* surfusion *f*	Unterkühlung *f*
106	supercooled subcooled	superrefroidi surfondu	unterkühlt
107	crystallization	cristallisation *f*	Kristallisation *f*
108	crystallization nuclei	noyaux de cristallisation *mpl*	Kristallisationskerne *mpl*
109	freezing nuclei	noyaux de congélation *mpl*	Gefrierkerne *mpl* Eisnuklei *mpl*
110	sublimation nuclei	noyaux de sublimation *mpl*	Sublimationskerne *mpl*
111	ice crystal	cristal de glace *m*	Eiskristall *m*
112	diamond dust ice needles	aiguilles de glace *fpl*	Eisnadeln *fpl*

partícula de aerosol *f*	partícula de aerossol *f*	аэрозольная частица *f*	91
hidrofílico hidrófilo	hidrófilo	гидрофильный	92
hidrófobo hidrofóbico	hidrófobo hidrofóbico	гидрофобный нерастворимый	93
filtro de aerosol *m*	filtro de aerossol *m*	аэрозольный фильтр *m*	94
filtro de polvo *m*	filtro de poeiras *m*	фильтр для пыли *m*	95
coloide *m*	colóide *m*	коллоид *m*	96
coloidal	coloidal	коллоидальный коллоидный	97
partícula coloidal *f*	partícula coloidal *f*	коллоидная частица *f*	98
partícula sólida *f*	partícula sólida *f*	твёрдая частица *f*	99
partícula líquida *f* gotita *f*	partícula líquida *f* gotícula *f*	жидкая частица *f* капелька *f*	100
saturación *f*	saturação *f*	насыщение *n*	101
sobresaturación *f*	supersaturação *f*	перенасыщение *n* пересыщение *n*	102
núcleo de condensación *m*	núcleo de condensação *m*	ядро конденсации *n*	103
núcleos de hielo *mpl*	núcleos do gelo *mpl*	ледяные ядра *npl* льдообразующие ядра *npl*	104
sobreenfriamento *m f* subfusión *f* subenfriamento *m*	super-resfriamento *m* superfusão *f* sub-resfriamento *m*	переохлаждение *n*	105
sobreenfriado subfundido subenfriado	super-resfriado superfundido sub-resfriado	переохлаждённый	106
cristalización *f*	cristalização *f*	кристаллизация *f*	107
núcleos de cristalización *mpl*	núcleos de cristalização *mpl*	ядра кристаллизации *npl*	108
núcleos de congelación *mpl*	núcleos de congelamento *mpl*	ядра замерзания *npl*	109
núcleos de sublimación *mpl*	núcleos de sublimação *mpl*	ядра сублимации *npl*	110
cristal de hielo *m*	cristal de gelo *m*	ледяной кристалл *m*	111
agujas de hielo *fpl*	agulhas de gelo *fpl*	ледяные иглы *fpl*	112

113	drop	goutte *f*	Tropfen *m*
114	size of particles	grandeur des particules *f*	Partikelgrösse *f*
115	large ion Langevin ion ($25–55 . 10^{-7}$cm)	ion lourd *m* ion de Langevin *m* ion lent *m*	Grossion *n* Langevin-Ion *n*
116	Aitken particles Aitken nuclei ($10^{-2}\mu–10^{-1}\mu$)	particules d'Aitken *fpl*	Aitkensche Teilchen *npl* Aitkenkerne *mpl*
117	precipitable particles	particules précipitables *fpl*	ausfällende Teilchen *npl*
118	giant particle giant nucleus (1 μ–30 μ)	particule géante *f*	Riesenteilchen *n* Riesenkern *m*
119	coagulation	coagulation *f*	Koagulation *f*
120	coalescence	coalescence *f*	Koaleszenz *f* Zusammenfliessen *n*
121	number of particles	nombre des particules *m*	Partikelzahl *f*
122	aerology	aérologie *f*	Aerologie *f*
122a	aeronomy	aéronomie *f*	Aeronomie *f*
123	ozone layer ozonosphere	couche d'ozone *f* ozonosphère *f*	Ozonschicht *f* Ozonosphäre *f*
124	atmospheric ozone	ozone atmosphérique *m*	atmosphärisches Ozon *n*
125	atmospheric ozone content	quantité d'ozone atmosphérique *f*	Ozongehalt *m*
126	Umkehr effect	effet Götz *m* effet d'inversion *m*	Umkehreffekt *m*
127	ozone shadow	ombre de l'ozone *f*	Ozonschatten *m*

gota *f*	gota *f*	капля *f*	113
tamaño de las partículas *m*	tamanho das partículas *m* dimensão das partículas *f*	размер частиц *m*	114
ion grande *m* ion Langevin *m* ion lento *m*	íon grande *m* íon (de) Langevin *m*	тяжёлый ион *m* ион Ланжевена *m*	115
partículas (de) Aitken *fpl* núcleos (de) Aitken *mpl*	partículas (de) Aitken *fpl* núcleos (de) Aitken *mpl*	частицы Айткена *fpl* ядра Айткена *npl*	116
partículas precipitantes *fpl*	partículas precipitáveis *fpl*	высыпающиеся частицы *fpl*	117
partícula gigante *f* núcleo gigante *m*	partícula gigante *f* núcleo gigante *m*	гигантская частица *f* гигантское ядро *n*	118
coagulación *f*	coagulação *f*	коагуляция *f*	119
coalescencia *f*	coalescência *f*	коалесценция *f*	120
número de partículas *m*	número de partículas *m*	число частиц *n*	121
aerología *f*	aerologia *f*	аэрология *f*	122
aeronomía *f*	aeronomia *f*	аэрономия *f*	122a
capa de ozono *f* ozonosfera *f*	camada de ozônio [de ozone] *f* ozonosfera *f*	озонный слой *m* озоносфера *f*	123
ozono atmosférico *m*	ozônio [ozone] atmosférico *m*	атмосферный озон *m*	124
cantidad de ozono en la atmósfera *f*	teor de ozônio na atmosfera *m* conteúdo de ozônio atmosférico *f*	содержание озона (в атмосфе- ре) *n*	125
efecto de inversión *m* efecto Umkehr *m*	efeito de Umkehr [inversão] *m* efeito Götz *m*	умкер-эффект *m* эффект обращения *m*	126
sombra de ozono *f*	sombra de ozônio *f*	озонная тень *f*	127

	WEATHER	TEMPS	WETTER
1	weather	temps *m*	Wetter *n*
2	weather conditions meteorological conditions	conditions météorologiques *fpl*	Wetterverhältnisse *npl* Wetterbedingungen *fpl*
3	standard conditions	conditions normales *fpl*	Normalbedingungen *fpl*
4	ambient air	air ambiant *m*	Umgebungsluft *f*
5	weather service meteorological service	service de prévision du temps *m* service météorologique *m*	Wetterdienst *m* meteorologischer Dienst *m*
6	meteorology	météorologie *f*	Meteorologie *f*
7	dynamic(al) meteorology	météorologie dynamique *f*	dynamische Meteorologie *f*
8	synoptic meteorology	météorologie synoptique *f*	synoptische Meteorologie *f*
9	micrometeorology	micrométéorologie *f*	Mikrometeorologie *f*
10	weather report	bulletin météorologique *m* rapport météorologique *m*	Wetterbericht *m*
11	synoptic chart synoptic map	carte synoptique *f*	synoptische Karte *f*
12	weather map weather chart	carte météorologique *f* carte du temps *f*	Wetterkarte *f*
13	weather forecast weather prognosis	prévision du temps *f* prévision météorologique *f*	Wetterprognose *f* Wettervorhersage *f*
14	weather situation	situation météorologique *f* état du temps *m*	Wetterlage *f*
15	general weather situation large-scale weather situation	situation météorologique générale *f*	Grosswetterlage *f*
16	World Weather Watch WWW	Veille Météorologique Mondiale *f* VMM	Weltwetterwacht *f* WWW
17	World Meteorological Organiza-tion WMO	Organisation Météorologique Mondia-le *f* OMM	Meteorologische Weltorganisation *f* MWO Weltorganisation für Meteorologie *f*
18	world network	réseau mondial *m*	weltweites (Beobachtungs)netz *n*
19	long-range forecast	prévision à longue échéance *f*	langfristige Vorhersage *f* langfristige Prognose *f* Langfristprognose *f*
20	short-range forecast	prévision à courte échéance *f*	Kurzfristprognose *f* kurzfristige Vorhersage [Prognose] *f*

TIEMPO	TEMPO	ПОГОДА	
tiempo *m*	tempo *m*	погода *f*	1
condiciones del tiempo *fpl* condiciones meteorológicas *fpl*	condições do tempo *fpl* condições meteorológicas *fpl*	метеорологические условия *npl* метеоусловия *npl*	2
condiciones normales *fpl*	condições normais *fpl*	стандартные (атмосферные) условия *npl*	3
aire ambiente *m*	ar ambiente *m*	окружающий воздух *m*	4
servicio meteorológico *m*	serviço meteorológico *m*	служба погоды *f* служба прогнозов *f*	5
meteorología *f*	meteorologia *f*	метеорология *f*	6
meteorología dinámica *f*	meteorologia dinâmica *f*	динамическая метеорология *f* динамика и термодинамика атмосферы *f*	7
meteorología sinóptica *f*	meteorologia sinó(p)tica *f*	синоптическая метеорология *f*	8
micrometeorología *f*	micrometeorologia *f*	микрометеорология *f*	9
boletín meteorológico *m*	boletim do tempo *m* boletim meteorológico *m*	сообщение о погоде *n* бюллетень погоды *m*	10
mapa sinóptico *m* carta sinóptica *f*	mapa sinó(p)tico *m* carta sinó(p)tica *f*	синоптическая карта *f*	11
mapa del tiempo *m* carta del tiempo *f*	carta do tempo *f* mapa do tempo *m*	карта погоды *f*	12
predicción del tiempo *f* previsión del tiempo *f* prognóstico del tiempo *m*	previsão do tempo *f* prognóstico do tempo *m*	предсказание погоды *n* прогноз погоды *m*	13
estado del tiempo *m* situación meteorológica *f*	estado do tempo *m* situação meteorológica *f*	состояние погоды *n* синоптическое положение *n*	14
estado meteorológico general *m* situación general meteorológica *f* estado del tiempo a gran escala *f*	condições meteorológicas gerais *fpl* situação meteorológica geral *f* estado do tempo de grande escala *m*	макросиноптическое положение *n* общее синоптическое положение *n* общее состояние погоды *n*	15
Vigilancia Meteorológica Mundial *f* VMM	Vigilância Meteorológica Mundial *f* VMM	Всемирная служба погоды *f* ВСП	16
Organización Meteorológica Mundial *f* OMM	Organização Meteorológica Mundial *f* OMM	Всемирная метеорологическая организация *f* ВМО	17
red mundial *f*	rede mundial *f*	мировая сеть *f*	18
predicción a largo plazo *f* previsión a largo plazo *f* prognóstico a largo plazo *m*	previsão a longo prazo *f* prognóstico a longo prazo *m*	долгосрочный прогноз *m* долгосрочное предсказание *n*	19
predicción a corto plazo *f* prognóstico a corto plazo *m*	previsão a curto prazo *f* prognóstico a curto prazo *m*	краткосрочный прогноз *m* краткосрочное предсказание *n*	20

	WEATHER	TEMPS	WETTER
21	forecast map	carte prévisionnelle *f*	Vorhersagekarte *f*
22	change of weather weather change	changement de temps *m*	Wetteränderung *f*
23	deterioration of weather	aggravation du temps *f*	Wetterverschlechterung *f*
24	visibility	visibilité *f*	Sicht *f*
25	good visibility	bonne visibilité *f*	gute Sicht *f*
26	poor visibility	mauvaise visibilité *f*	schlechte Sicht *f*
27	haze (dust haze)	brume sèche *f*	Dunst *m* trockner Dunst *m*
28	meteorological element	élément météorologique *m*	meteorologisches Element *n*
29	temperature	température *f*	Temperatur *f*
30	normal temperature and pressure	température et pression normales	Normal-Temperatur-und-Druck
31	room temperature normal room temperature	température locale *f* température de chambre *f* température (normale) de la pièce *f*	Zimmertemperatur *f* normale Raumtemperatur *f*
32	ambient temperature	température ambiante *f*	Umgebungstemperatur *f*
33	Celsius (temperature) scale centigrade scale	échelle Celsius *f* échelle centigrade *f*	Celsius–Skala *f* Zentigradskala *f*
34	shade temperature temperature in the shade	température sous abri *f* température á l'ombre *f*	Temperatur im Schatten *f*
35	temperature conditions	conditions thermiques *fpl*	Temperaturverhältnisse *npl*
36	temperature inversion	inversion thermique *f* inversion de température *f*	Temperaturinversion *f*
37	atmospheric inversion	inversion atmosphérique *f*	atmosphärische Inversion *f*
38	surface [ground] inversion	inversion au sol *f*	Bodeninversión *f*
39	isothermal atmosphere	atmosphère isotherme *f*	isotherme Atmosphäre *f*
40	thermal stratification	stratification thermique *f*	Temperaturschichtung *f*
41	decrease in temperature fall in temperature temperature drop	abaissement de la température *m* baisse de la température *f*	Temperaturabnahme *f* Temperaturabfall *m*
42	rise in temperature temperature increase	hausse de la température *f* augmentation de la température *f*	Temperaturzunahme *f*
43	atmospheric pressure	pression atmosphérique *f*	Luftdruck *m* atmosphärischer Druck *m*

TIEMPO	TEMPO	ПОГОДА	77
carta de predicción *f*	carta da previsão *f*	прогностическая карта *f*	21
cambio del tiempo *m*	mudança do tempo *f*	изменение погоды *n*	22
empeoramiento del tiempo *m*	agravamento do tempo *m*	ухудшение погоды *n*	23
visibilidad *f*	visibilidade *f*	видимость *f*	24
buena visibilidad *f*	boa visibilidade *f*	хорошая видимость *f*	25
mala visibilidad *f*	má visibilidade *f*	плохая видимость *f*	26
bruma seca *f* calima *f*	bruma seca *f* névoa seca *f*	дымка *f* сухая мгла *f*	27
elemento meteorológico *m*	elemento meteorológico *m*	метеорологический элемент *m*	28
temperatura *f*	temperatura *f*	температура *f*	29
temperatura y presión normales	temperatura e presão normais	нормальные температура и давление	30
temperatura normal *f*	temperatura de sala *f* temperatura ambiente normal *f*	комнатная температура *f* нормальная комнатная температура *f*	31
temperatura ambiente *f*	temperatura ambiente *f*	окружающая температура *f*	32
escala de Celsio *f* escala centígrada *f* escala centesimal *f*	escala Celsius *f* escala centígrada *f* escala centesimal *f*	шкала Цельсия *f* стоградусная шкала *f*	33
temperatura a la sombra *f*	temperatura do ar (à sombra) *f* temperatura no abrigo *f*	температура в тени *f*	34
condiciones térmicas *fpl*	condições de temperatura *fpl*	температурные условия *npl*	35
inversión térmica *f* inversión de temperatura *f*	inversão de temperatura *f*	температурная инверсия *f*	36
inversión atmosférica *f*	inversão atmosférica *f*	атмосферная инверсия *f*	37
inversión junto al suelo *f*	inversão superficial [no solo] *f*	приземная инверсия *f*	38
atmósfera isotérmica *f*	atmosfera isotérmica *f*	изотермическая атмосфера *f*	39
estratificación térmica *f*	estratificação térmica *f*	температурная стратификация *f*	40
caída de temperatura *f* disminución de la temperatura *f*	declínio de temperatura *m* abaixamento de temperatura *m* queda de temperatura *f*	падение температуры *n*	41
subida de temperatura *f* aumento de la temperatura *m*	subida da temperatura *f* aumento da temperatura *m* elevação da temperatura *f*	повышение температуры *n*	42
presión atmosférica *f*	pressão atmosférica *f*	атмосферное давление *n* давление воздуха *n*	43

	WEATHER	TEMPS	WETTER
44	sea-level pressure	pression au niveau de la mer *f*	Druck in Meereshöhe *m*
45	reduction to sea-level	réduction au niveau de la mer *f*	Reduktion auf den Meeresspiegel *f*
46	at normal pressure	sous pression normale [ordinaire]	unter normalem Druck
47	millibar	millibar *m*	Millibar *n*
48	millimetre of mercury (column)	millimètre (de colonne) de mercure *m*	Millimeterquecksilbersäule *f*
49	sunlight	lumière solaire *f*	Sonnenlicht *n*
50	direct solar radiation	radiation solaire directe *f*	direkte Sonnenstrahlung *f*
51	insolation	insolation *f*	Einstrahlung *f* Insolation *f*
52	sunshine	rayonnement solaire incident *m*	Sonnenschein *m* Sonneneinstrahlung *f*
53	skylight sky light	lumière du ciel *f*	Himmelslicht *n*
54	radiation of the sky	radiation du ciel *f*	Himmelsstrahlung *f*
55	global radiation	rayonnement global *m*	Globalstrahlung *f*
56	Earth radiation	radiation de la Terre *f* rayonnement terrestre *m*	Erdstrahlung *f* Ausstrahlung der Erde *f*
57	(atmospheric) scattering of light	diffusion (atmosphérique) de la lumière *f*	(atmosphärische) Streuung des Lichtes *f*
58	scattering pattern function of scattering indicatrix of scattering	indicatrice de diffusion *f* fonction de diffusion *f*	Streuindikatrix *f* Streudiagramm *n* Streufunktion *f*
59	diffuse sky radiation	rayonnement diffus du ciel *m*	diffuse Himmelsstrahlung *f*
60	blueness of the sky	bleu du ciel *m* azur du ciel *m*	Himmelsblau *m*
61	sunshine recorder heliograph	héliographe *m* enregistreur du rayonnement solaire *m*	Sonnenscheinschreiber *m* Heliograph *m*
62	heat balance (of the atmosphere)	bilan thermique (de l'atmosphère) *m*	Wärmehaushalt (der Atmosphäre) *m* Wärmebilanz (der Atmosphäre) *f*
63	advection	advection *f*	Advektion *f*
64	atmospheric circulation	circulation atmosphérique *f*	atmosphärische Zirkulation *f*
65	general circulation	circulation générale *f*	allgemeine Zirkulation *f*
66	planetary circulation	circulation planétaire *f*	planetare Zirkulation *f*

TIEMPO	TEMPO	ПОГОДА	
presión a nivel del mar *m*	pressão ao nível do mar *m*	давление на уровне моря *n* давление, приведённое к уровню моря *n*	44
reducción al nivel de mar *f*	redução ao nível do mar *m*	приведение к уровню моря *n*	45
a presión normal	a pressão normal	при нормальном давлении	46
milibar *m*	milibar *m*	миллибар *m*	47
milímetro (de la columna) de mercurio *m*	milímetro (da coluna) de mercúrio *m*	миллиметр ртутного столба *m*	48
luz solar *f*	luz solar *f*	солнечный свет *m*	49
radiación solar directa *f*	radiação solar direta *f*	прямая солнечная радиация *f*	50
insolación *f*	insolação *f* tempo do Sol descoberto *m*	инсоляция *f*	51
radiación solar incidente *f*	radiação solar incidente *f*	солнечное сияние *n* солнечный свет *m*	52
luz del cielo *f*	luz do céu *f*	свечение неба *n* свет неба *m*	53
radiación del cielo *f*	radiação do céu *f*	излучение неба *n* радиация небесного свода *f*	54
radiación global *f*	radiação global *f*	суммарная радиация *f*	55
radiación de la Tierra *f* radiación terrestre *f*	radiação terrestre *f* radiação da Terra *f*	излучение Земли *n*	56
difusión (atmosférica) de luz *f*	espalhamento (atmosférico) da luz *m* difusão (atmosférica) da luz *f*	рассеяние света (в атмосфере) *n*	57
indicatriz de difusión *f* función de difusión *f*	padrão de espalhamento [difusão] *m* indicatriz de espalhamento [difusão] *f* função de espalhamento [difusão] *f*	индикатриса рассеяния *f* функция рассеяния *f*	58
radiación difusa del cielo *f*	radiação difusa do céu *f*	рассеянный свет неба *m*	59
azul del cielo *m*	azul do céu *m*	голубой цвет неба *m* синева неба *f*	60
heliógrafo *m* registrador de luz solar *m*	heliógrafo *m* registrador de brilho solar *m*	гелиограф *m*	61
balance calorífico (de la atmósfera) *m* equilibrio térmico (de la atmósfera) *m*	balanço térmico *m* balanço de calor (da atmosfera) *m*	тепловой баланс (атмосферы) *m*	62
advección *f*	advecção *f*	адвекция *f*	63
circulación atmosférica *f*	circulação atmosférica *f*	атмосферная циркуляция *f*	64
circulación general *f*	circulação geral *f*	общая циркуляция *f*	65
circulación planetaria *f*	circulação planetária *f*	планетарная циркуляция *f*	66

	WEATHER	TEMPS	WETTER
67	air mass	masse d'air *f*	Luftmasse *f*
68	front	front *m*	Front *f*
69	warm front	front chaud *m*	Warmfront *f*
70	cold front	front froid *m*	Kaltfront *f*
71	occluded front occlusion front	front occlus *m*	Okklusionsfront *f*
72	occlusion	occlusion *f*	Okklusion *f*
73	polar front	front polaire *m*	Polarfront *f*
74	frontogenesis	frontogénèse *f*	Frontogenese *f*
75	frontolysis	frontolyse *f*	Frontolyse *f*
76	cyclone depression	cyclone *m* dépression *f*	Tiefdruckgebiet *n* Zyklone *f* Tief *n*
77 78	tropical cyclone hurricane	cyclone tropical *m* ouragan *m*	tropische Zyklone *f* tropischer Wirbelsturm [Orkan] *m*
79	Coriolis force	force de Coriolis *f*	Corioliskraft *f*
80	deviating force	force déviante *f*	Ablenkungskraft *f*
81	Coriolis effect	effet Coriolis *m*	Coriolis-Effekt *m*
82	deepening of a depression	creusement d'une dépression *m*	Vertiefung einer Zyklone *f*
83	filling of a depression	comblement d'une dépression *m*	Auffüllung einer Zyklone *f*
84	anticyclone high high-pressure area	zone de haute pression *f* anticyclone *m* centre de haute pression *m*	Antizyklone *f* Hochdruckgebiet *n* Hoch *n*
85	anticyclonic	anticyclonique	antizyklonal
86	wedge of high pressure	dorsale de haute pression *f* dorsale barométrique *f*	Hochdruckkeil *m*
87	ridge of high pressure	crête de haute pression *f* crête barométrique *f*	Hochdruckrücken *m*
88	jet stream	courant-jet *m*	Strahlstrom *m* Strahlströmung *f*
89	wind	vent *m*	Wind *m*

masa de aire *f*	massa de ar *f*	воздушная масса *f*	67
frente *m*	frente *f*	фронт *m*	68
frente cálido *m*	frente quente *f*	теплый фронт *m*	69
frente frío *m*	frente fria *f*	холодный фронт *m*	70
frente ocluido *m*	frente oclusa *f*	фронт окклюзии *m* окклудированный фронт *m*	71
oclusión *f*	oclusão *f*	окклюзия *f*	72
frente polar *m*	frente polar *f*	полярный фронт *m*	73
frontogénesis *f*	frontogênese *f*	фронтогенез *m* образование фронта *n* фронтообразование *n*	74
frontólisis *f*	frontólise *f*	фронтолиз *m* размывание фронта *n*	75
depresión *f* ciclón *m*	depressão *f* ciclone *m*	область низкого давления *f* циклон *m* депрессия *f*	76
ciclón tropical *m* huracán (tropical) *m*	ciclone tropical *m* furação (tropical) *m*	тропический циклон *m* (тропический) ураган *m*	77 78
fuerza de Coriolis *f*	força de Coriolis *f*	сила Кориолиса [кориолисова] *f*	79
fuerza desviadora *f*	força desviante *f*	отклоняющая сила *f*	80
efecto Coriolis *m*	efeito Coriolis *m*	эффект Кориолиса *m*	81
profundización de una depresión *f*	aprofundamento de uma depressão *m* cavamento de uma depressão *m*	углубление циклона *m* углубление депрессии *n*	82
llenado de una depresión *m*	enchimento de uma depressão *m*	заполнение циклона *n* заполнение депрессии *n*	83
alta *f* anticiclón *m* área de alta presión *f*	anticiclone *m* alta *f* região [área] de alta pressão *f*	антициклон *m* область высокого давления *f* барометрическая высота *f*	84
anticiclónico	anticiclônico	антициклонический	85
cunã de alta presión *f* cuña anticiclónica *f*	cunha de alta pressão *f*	клин высокого давления *m*	86
cresta [dorsal] de alta presión *f*	crista anticiclônica *f* dorsal anticiclônica *f*	гребень высокого [повышенного] давления *m* барнческий гребень *m*	87
corriente en chorro *f*	corrente de ja(c)to *f*	струйное течение *n*	88
viento *m*	vento *m*	ветер *m*	89

	WEATHER	TEMPS	WETTER
90	air current	courant aérien *m*	Luftströmung *f*
91	updraught (GB) updraft (US) up-current	courant ascendant *m*	Aufwärtsströmung *f*
92	anabatic wind	vent anabatique *m*	aufsteigender Wind *m* Aufwind *m*
93	katabatic moving downward	catabatique descendant	katabatisch absteigend
94	gradient wind	vent du gradient *m*	Gradientwind *m*
95	cyclostrophic wind	vent cyclostrophique *m*	zyklostrophischer Wind *m*
96	geostrophic wind	vent géostrophique *m*	geostrophischer Wind *m*
97	sea-breeze	brise de mer *f*	Seebrise *f* Seewind *m*
98	land-breeze	brise de terre *f*	Landbrise *f* Landwind *m*
99	monsoon	mousson *f*	Monsun *m*
100	trade winds *pl* trades *pl*	vents alizés *mpl* alizés *mpl*	Passatwinde *mpl* Passate *mpl*
101	doldrums	calmes équatoriaux *mpl* zone de calmes équatoriaux *f*	Mallungen *fpl* Kalmengürtel *m*
102	horse latitudes	(latitudes des) calmes tropicaux (*fpl*) *mpl*	Rossbreiten *fpl*
103	anti-trades	contre-alizés *mpl*	Gegenpassate *mpl* Antipassate *mpl*
104	wind vane	girouette *f*	Windfahne *f*
105	blow	souffler	wehen
106	pressure gradient	gradient de pression *m*	Druckgefälle *n* Druckgradient *m*
107	wind force	force du vent *f*	Windstärke *f*
108	Beaufort scale (of wind force)	échelle de Beaufort (de la force du vent) *f*	Beaufortskala (der Windstärke) *f*
109	wind velocity (vector) wind speed (scalar)	vitesse du vent *f*	Windgeschwindigkeit *f*
110	calm	calme *m*	Windstille *f*
111	gusty wind	vent à rafales *m*	böiger Wind *m*

TIEMPO	TEMPO	ПОГОДА	
corriente de aire *f*	corrente atmosférica *f*	воздушный поток *m*	90
corriente ascendente *f*	corrente ascendente *f*	восходящее течение *n* восходящий ток конвекции *m* восходящий поток *m*	91
viento anabático *m* viento ascendente *m*	vento anabático *m* vento ascendente *m*	восходящий ветер *m* анабатический ветер *m*	92
catabático descendente	catabático descendente	катабатический нисходящий	93
viento del gradiente *m*	vento (do) gradiente *m*	градиентный ветер *m*	94
viento ciclostrófico *m*	vento ciclostrófico *m*	циклострофический ветер *m*	95
viento geostrófico *m*	vento geostrófico *m*	геострофический ветер *m*	96
brisa de mar *f* virazón *m*	brisa marítima *f*	морский бриз *m* дневной бриз *m*	97
brisa de tierra *f*	brisa terrestre *f*	береговой бриз *m* ночной бриз *m*	98
monzón *m*	monção *m*	муссон *m*	99
vientos alisios *mpl* alisios *mpl*	ventos alísios *mpl* alísios *mpl*	пассаты *mpl* пассатные ветры *mpl*	100
zona de calmas ecuatoriales *f* calmas ecuatoriales *fpl*	calmarias equatoriais *fpl* zona de calmarias equatoriais *f*	зона экваториальных штилей *f* (экваториальная) зона затишья *f*	101
(latitudes de las) calmas tropicales *fpl*	(região de) calmarias tropicais (*f*) *fpl*	конские широты *fpl* субтропические широты *fpl*	102
(vientos) contraalisios *mpl*	(ventos) anti-alísios *mpl* contra-alísios *mpl*	антипассаты *mpl*	103
veleta *f*	grimpa *f* cata-vento *m*	флюгер *m*	104
soplar	soprar	дуть	105
gradiente de presión *m* gradiente bárico [barométrico] *m*	gradiente barométrico [bárico] *m* gradiente de pressão *m*	градиент давления *m* барический градиент *m*	106
fuerza del viento *f*	força do vento *f*	сила ветра *f*	107
escala (de vientos) Beaufort *f*	escala (dos ventos) (de) Beaufort *f*	шкала Бофорта (силы ветра) *f*	108
velocidad del viento *f*	velocidade do vento *f*	скорость ветра *f*	109
calma *f*	calmaria *f*	безветрие [затишье] *n* штиль *f*	110
viento a rachas *m*	vento de rajadas *m*	порывистый ветер *m*	111

	WEATHER	TEMPS	WETTER
112	gust	rafale *f*	Windstoss *m*
113	anemometer	anémomètre *m*	Anemometer *n* Windmesser *m*
114	anemograph recording anemometer	anémographe *m* anémomètre enregistreur *m*	Anemograph *m* Windschreiber *m*
115	wind direction	direction du vent *f*	Windrichtung *f*
116	wind rose	rose des vents *f*	Windrose *f*
117	prevailing wind	vent dominant *m*	vorherrschender Wind *m*
118	atmospheric humidity atmospheric moisture	humidité de l'air *f* humidité atmosphérique *f*	Luftfeuchtigkeit *f* Luftfeuchte *f*
119	water vapour	vapeur d'eau *f*	Wasserdampf *m*
120	hygrometer	hygromètre *m*	Hygrometer [Feuchtigkeitsmesser] *n*
121	psychrometer	psychromètre *m*	Psychrometer *m*
122	absolute humidity	humidité absolue *f*	absolute Feuchte *f* absolute Feuchtigkeit *f*
123	specific humidity	humidité spécifique *f*	spezifische Feuchtigkeit *f*
124	amount of water vapour	quantité de vapeur d'eau *f*	Wasserdampfgehalt *m*
125	water-vapour pressure water-vapour tension	tension de la vapeur d'eau *f*	Wasserdampfdruck *m* Wasserdampfspannung *f*
126	extinction due to water vapour	extinction par la vapeur d'eau *f*	Wasserdampfextinktion *f*
127	water-vapour absorption	absorption par la vapeur d'eau *f* absorption de vapeur d'eau *f*	Wasserdampfabsorption *f* Absorption durch Wasserdampf *f*
127a	atmospheric extinction	extinction atmosphérique *f*	atmosphärische Extinktion *f*
128	relative humidity	humidité relative *f*	relative Feuchtigkeit *f*
129	condense	se condenser	kondensieren
130	fog	brouillard *m*	Nebel *m*
131	cloud	nuage *m*	Wolke *f*
132	state of the sky	état du ciel *m*	Himmelszustand *m*
133	cloud form	forme de nuage *f*	Wolkenform *f*
134	orographic cloud	nuage orographique *m*	orographische Wolke *f*
135	cloud type	espèce de nuage *f*	Wolkenart *f*

TIEMPO	TEMPO	ПОГОДА	
racha *f* ráfaga *f*	rajada *f*	порыв ветра *m*	112
anemómetro *m*	anemômetro *m*	анемометр *m*	113
anemógrafo *m* anemómetro registrador *m*	anemógrafo *m* anemômetro registrador *m*	анемограф *m* самопишущий анемометр *m*	114
dirección del viento *f*	direção do vento *f*	направление ветра *n*	115
rosa de los vientos *f*	rosa dos ventos *f*	роза ветров *f*	116
viento dominante *m* viento predominante *m*	vento predominante *m*	преобладающий ветер *m*	117
humedad del aire *f* humedad atmosférica *f*	umidade do ar *f* umidade atmosférica *f*	влажность воздуха *f* атмосферная влажность *f*	118
vapor de agua *m*	vapor de água *m*	водяной пар *m*	119
higrómetro *m*	higrômetro *m*	гигрометр *m*	120
psicrómetro *m*	psicrômetro *m*	психрометр *m*	121
humedad absoluta *f*	umidade absoluta *f*	абсолютная влажность *f*	122
humedad específica *f*	umidade específica *f*	удельная влажность *f*	123
cantidad del vapor de agua *f*	quantidade de vapor de água *f*	количество водяного пара *n*	124
presión del vapor de agua *f*	tensão do vapor de água *f* pressão do vapor de água *f*	давление водяного пара *n*	125
extinción del vapor de agua *f*	extinção por vapor de água *f*	экстинкция водяным паром *f*	126
absorción por vapor de agua *f*	absorção por vapor de água *f* absorção do vapor de água *f*	поглощение водяным паром *n* абсорбция водяным паром *f*	127
extinción atmosférica *f*	extinção atmosférica *f*	атмосферная экстинкция *f*	127a
humedad relativa *f*	umidade relativa *f*	относительная влажность *f*	128
condensar(-se)	condensar(se)	конденсировать конденсироваться	129
niebla *f*	nevoeiro *m*	туман *m*	130
nube *f*	nuvem *f*	облако *n*	131
estado del cielo *m*	estado do céu *m*	состояние неба *n*	132
forma de nube *f*	forma de nuvem *f*	облачная форма *f* форма облака *f*	133
nube orográfica *f*	nuvem orográfica *f*	орографическое облако *n*	134
especie de nube *f*	espécie de nuvem *f*	вид облака *m*	135

	WEATHER	TEMPS	WETTER
136	Cirrus Ci	cirrus *m*	Cirrus *m* Federwolke *f*
137	Cumulonimbus Cb	cumulonimbus *m*	Kumulonimbus *m* Gewitterwolke *f* Cumulonimbus *m*
138	Cumulus Cu	cumulus *m*	Kumulus *m* Cumulus *m* Haufenwolke *f*
139	Stratus St	stratus *m*	Stratus *m* Schichtwolke *f*
140	Altostratus As	altostratus *m*	Altostratus *m* hohe Schichtwolke *f*
141	Cirrostratus Cs	cirro-stratus *m*	Cirrostratus *m* Federschichtwolke *f*
142	Stratocumulus Sc	strato-cumulus *m*	Stratokumulus *m*
143	Cirrocumulus Cc	cirro-cumulus *m* ciel moutonné *m*	Cirrokumulus *m* Federhaufenwolke *f*
144	Altocumulus Ac	altocumulus *m*	Altokumulus *m* hohe Haufenwolke *f*
145	Nimbostratus Ns	nimbo-stratus *m* couche de nuages de pluie *f*	Nimbostratus *m* Regenschichtwolke *f*
146	Fractocumulus cumulus fractus	cumulus fractus *m*	Fraktokumulus *m*
147	cloud amount cloudiness	nébulosité *f*	Bewölkung *f* Bedeckungsgrad *m*
148	cloudy sky cloudy	ciel nuageux *m* nuageux	bewölkter Himmel *m* wolking bewölkt
149	cloudless	sans nuages serein	wolkenlos wolkenfrei
150	cloud layer	couche de nuages *f*	Wolkenschicht *f*
151	veil of clouds cloud veil	voile de nuages *m*	Wolkenschleier *m*
152	overcast sky overcast	ciel couvert *m* couvert	bedeckter Himmel *m* bedeckt
153	clear sky	ciel clair *m* ciel serein *m*	klarer Himmel *m* heiterer Himmel *m*
154	precipitation	précipitations *fpl*	Niederschlag *m*

TIEMPO	TEMPO	ПОГОДА	
cirro *m*	cirro *m*	перистое облако *n* циррус *m*	136
cúmulonimbo *m*	cúmulo-nimbo *m*	кучево-дождевое облако *n* кумулус *m*	137
cúmulo *m* nube cumuliforme *f*	cúmulo *m* nuvem cumuliforme *f*	кучевое облако *n* кумулонимбус *m*	138
estrato *m*	estrato *m*	слоистое облако *n* стратус *m*	139
altostrato *m* altoestrato *m*	alto-estrato *m*	высокослоистые облака *npl* высокое слоистое облако *n* алтостратус *m*	140
cirrostrato *m* cirroestrato *m*	cirro-estrato *m*	перисто-слоистые облака *npl* цирростратус *m*	141
estratocúmulo *m*	estrato-cúmulo *m*	слоисто-кучевые облака *npl* стратокумулус *m*	142
cirrocúmulo *m*	cirro-cúmulo *m*	перисто-кучевые облака *npl* циррокумулус *m*	143
altocúmulo *m*	alto-cúmulo *m*	высококучевые облака *npl* альтокумулус *m*	144
nimboestrato *m* capa de nubes lluviosas *f*	nimbo-estrato *m* estrato-nimbo *m*	слоисто-дождевые облака *npl* нимбостратус *m*	145
fractocúmulo *m*	fra(c)to-cúmulo *m*	разорванно-кучевое облако *n* фрактокумулус *m*	146
cantidad de nubes *f* nubosidad *f*	nebulosidade *f* quantidade de nuvens *f*	количество облаков *n* облачность *f* степень облачности *f*	147
cielo nuboso *m* nublado nuboso	ceú nublado *m* nublado	облачное небо *n* облачно	148
sereno despejado	limpo sem nuvens	безоблачно прояснено	149
capa de nubes *f*	camada de nuvens *f*	облачный слой *m*	150
velo de nubes *m*	véu de nuvens *m*	облачная пелена *f* пелена облаков *f*	151
cielo cubierto *m* cubierto	céu encoberto *m* encoberto	пасмурное небо *n* пасмурно	152
cielo despejado *m* cielo claro *m*	céu limpo *m* céu claro *m*	ясное небо *n* ясно	153
precipitaciones (atmosféricas) *fpl*	precipitação (atmosférica) *f*	(атмосферные) осадки *mpl*	154

	WEATHER	TEMPS	WETTER
155	heat of condensation	chaleur de condensation *f*	Kondensationswärme *f*
156	condensation level	niveau de condensation *m*	Kondensationsniveau *n*
157	latent heat	chaleur latente *f*	latente Wärme *f*
158	precipitation amount amount of precipitation	hauteur des précipitations *f* quantité des précipitations *f*	Niederschlagsmenge *f*
159	thunderstorm	orage (avec tonnerre) *m*	Gewitter *n*
160	atmospheric electricity	électricité atmosphérique *f*	Luftelektrizität *f*
161	atmospheric discharge	décharge atmosphérique *f*	atmosphärische Entladung *f*
162	lightning	éclair *m*	Blitz *m*
163	ball lightning	éclair en boule *m*	Kugelblitz *m*
164	bead lightning	éclair en chapelet *m*	Perlschnurblitz *m* Perlenschnurblitz *m*
165	thunder	tonnerre *m*	Donner *m*
166	orographic rain	pluie orographique *f*	Stauregen *m* orographischer Regen *m*
167	orographic precipitation	précipitation orographique *f*	orographischer Niederschlag *m*
168	total precipitation sum of precipitation	somme des précipitations *f*	Niederschlagssumme *f*
169	rain	pluie *f*	Regen *m*
170	rain-gauge	pluviomètre *m*	Regenmesser *m*
171	totalizator totalizer rain-gauge	(pluviomètre) totalisateur *m*	(Niederschlags-) Totalisator *m*
172	drizzle	bruine *f*	Nieseln *n*
173	dew	rosée *f*	Tau *m*
174	dew-point	point de rosée *m*	Taupunkt *m*
175	snow	neige *f*	Schnee *m*
176	snow cover	couche de neige *f*	Schneedecke *f*
177	hoarfrost	gelée blanche *f*	Reif *m*
178	hail	grêle *f*	Hagel *m*
179	climatology	climatologie *f*	Klimatologie *f*

TIEMPO	TEMPO	ПОГОДА	
calor de condensación *m*	calor de condensação *m*	теплота конденсации *f* тепло конденсации *n*	155
nivel de condensación *m*	nível de condensação *m*	уровень конденсации *f*	156
calor latente *m*	calor latente *m*	скрытое тепло *n* скрытая теплота *m*	157
cantidad de precipitaciones *f*	quantidade de precipitação *f* precipitação *f*	количество осадков *n*	158
tormenta (de trueno) *f*	tempestade com trovoada(s) *f*	гроза *f*	159
electricidad atmosférica *f*	ele(c)tricidade atmosférica *f*	атмосферное электричество *n*	160
descarga atmosférica *f*	descarga atmosférica *f*	атмосферный разряд *m*	161
rayo *m*	relâmpago *m* raio *m*	молния *f*	162
rayo en bola *m*	bola de fogo *f* relâmpago em bola *m*	шаровая молния *f*	163
rayo-rosario *m* rayo en forma de rosario *m*	relâmpago em rosário *m* relâmpago em contas *m* relâmpago em cadeia *m*	чёточная молния *f*	164
trueno *m*	trovão *m*	гром *m*	165
lluvia orográfica *f*	chuva orográfica *f*	орографический дождь *m*	166
precipitación orográfica *f*	precipitação orográfica *f*	орографические осадки *mpl*	167
precipitación total *f* suma de la precipitación *f*	precipitação total *f* soma das precipitações *f*	сумма осадков *f*	168
lluvia *f*	chuva *f*	дождь *m*	169
pluviómetro *m*	pluviômetro *m*	дождемер *m*	170
(pluviómetro) totalizador *m*	(pluviômetro) totalizador *m*	суммарный дождемер *m*	171
llovizna *f*	chuvisco *m*	морось *f*	172
rocío *m*	orvalho *m*	роса *f*	173
punto de rocío *m*	ponto de orvalho *m*	точка росы *f*	174
nieve *f*	neve *f*	снег *m*	175
capa de nieve *f*	cobertura de neve *f*	снежный покров *m*	176
escarcha *f*	geada (branca) *f*	иней *m*	177
granizo *m*	granizo *m*	град *m*	178
climatología *f*	climatologia *f*	климатология *f*	179

	WEATHER	TEMPS	WETTER
180	climate	climat *m*	Klima *n*
181	satellite climatology	climatologie par satellite *f*	Satellitenklimatologie *f*
182	climatic	climatique	klimatisch
183	climatic fluctuation	variation du climat *f*	Klimaschwankung *f*
184	climatic change	changement du climat *m*	Klimaänderung *f*
185	continental climate	climat continental *m*	Landklima *n* kontinentales Klima *n*
186	maritime climate	climat maritime *m*	Seeklima *n* maritimes Klima *n*
187	mountain climate	climat de montagne *m*	Höhenklima *n*
188	paleoclimate	paléoclimat *m*	Paläoklima *n*
189	glacial period	période glaciaire *f*	Eiszeit *f*
190	interglacial	interglaciaire	Zwischeneiszeit- interglazial
191	torrid zone	zone torride *f*	heisse Zone *f*
192	temperate zone	zone tempérée *f*	gemässigte Zone *f*
193	polar zone	zone polaire *f*	polare Zone *f*

TIEMPO	TEMPO	ПОГОДА	
clima *m*	clima *m*	климат *m*	180
climatología desde satélite *f*	climatologia por satélites *f*	спутниковая климатология *f*	181
climático	climático	климатический	182
variación del clima *f*	variação do clima *f* flutuação climática *f*	колебание климата *n*	183
cambio del clima *m*	mudança climática [do clima] *f*	изменение климата *n*	184
clima continental *m*	clima continental *m*	континентальный климат *m* материковый климат *m*	185
clima marítimo *m*	clima marítimo *m*	морской климат *m* приморский климат *m*	186
clima de montaña *m*	clima de montanha *m*	горный климат *m*	187
paleoclima *m*	paleoclima *m*	палеоклимат *m*	188
período glacial *m*	período glaciário *m*	ледниковый период *m*	189
interglacial	interglaciário	межледниковый	190
zona tórrida *f*	zona tórrida *f*	тропическая зона *f* жаркая зона *f*	191
zona templada *f*	zona temperada *f*	умеренная зона *f*	192
zona polar *f*	zona polar *f*	полярная зона *f*	193

	IONOSPHERE	IONOSPHÈRE	IONOSPHÄRE
1	ionosphere	ionosphère *f*	Ionosphäre *f*
2	ionospheric	ionosphérique	ionosphärisch
3	ionized layer	couche ionisée *f*	ionisierte Schicht *f*
4	ionospheric region	région ionosphérique *f*	Ionosphärenregion *f*
5	ionospheric layer	couche ionosphérique *f*	Ionosphärenschicht *f*
6	electron density	densité électronique *f*	Elektronendichte *f*
7	ionospheric inhomogeneity	inhomogénéité ionosphérique *f*	ionosphärische Inhomogenität *f*
8	electron-density inhomogeneity	inhomogéneité de densité d'électrons *f*	Elektronendichteinhomogenität *f*
9	magnetic-field inhomogeneity	inhomogénéité de champ magnétique *f*	Magnetfeldinhomogenität *f*
10	ionic composition	composition ionique *f*	Ionenzusammensetzung *f*
11	molecular composition	composition moléculaire *f*	Molekülzusammensetzung *f* molekulare Zusammensetzung *f*
12	direct wave	onde directe *f*	direkte Welle *f*
13	space wave sky wave	onde d'espace *f*	Raumwelle *f*
14	ground wave	onde de sol *f*	Bodenwelle *f*
15	transmitter	émetteur *m*	Sender *m*
16	transmitter aerial transmitting aerial transmitting antenna	antenne émettrice *f*	Sendeantenne *f*
17	receiver	récepteur *m*	Empfänger *m*
18	receiving aerial receiving antenna	antenne réceptrice *f*	Empfangsantenne *f*
19	ionospheric sounding	sondage ionosphérique *m*	Ionosphären-Sondierung *f*
20	topside sounding	sondage par le haut *m* sondage en contre-haut *m*	Sondierung von oben *f*
21	ionospheric observatory	observatoire ionosphérique *m*	Ionosphären-Observatorium *n*
22	ionosonde ionospheric recorder	ionosonde *f* sondeur ionosphérique *m*	Ionosonde *f* Ionosphärenabtaster *m*
23	ionogram frequency-height curve	ionogramme *m* enregistrement ionosphérique *m*	Ionogramm *n*

ionosfera *f*	ionosfera *f*	ионосфера *f*	1
ionosférico	ionosférico	ионосферный	2
capa ionizada *f*	camada ionizada *f*	иониз(ир)ованный слой *m*	3
región ionosférica *f*	região ionosférica *f*	область ионосферы *f*	4
capa ionosférica *f*	camada ionosférica *f*	ионосферный слой *m*	5
densidad electrónica *f*	densidade ele(c)trônica *f*	электронная плотность *f* электронная концентрация *f*	6
heterogeneidad ionosférica *f*	heterogeneidade ionosférica *f*	ионосферная неоднородность *f*	7
heterogeneidad de la densidad electrónica *f*	heterogeneidade da densidade ele(c)trônica *f*	неоднородность электронной плотности *f*	8
heterogeneidad del campo magnético *f*	heterogeneidade do campo magnético *f*	неоднородность магнитного поля *f*	9
composición iónica *f*	composição iônica *f*	ионный состав *m*	10
composición molecular *f*	composição molecular *f*	молекулярный состав *m*	11
onda directa *f*	onda direta *f*	прямая волна *f*	12
onda del espacio *f* onda reflejada *f*	onda espacial *f*	пространственная волна *f*	13
onda terrestre *f*	onda no solo *f* onda terrestre *f*	земная волна *f* приземная волна *f*	14
emisor *m* transmisor *m*	emissor *m* transmissor *m*	передатчик *m*	15
antena emisora *f* antena transmisora *f*	antena emissora *f* antena de emissão *f* antena transmissora *f*	передающая антенна *f*	16
receptor *m*	receptor *m*	приёмник *m*	17
antena receptora *f*	antena receptora *f* antena de recepção *f*	приёмная антенна *f*	18
sondeo [sondaje] ionosférico *m*	sondagem ionosférica *f*	зондирование ионосферы *n*	19
sondeo descendente *m* sondaje descendente *m*	sondagem desccendente *f*	зондирование сверху *n*	20
observatorio ionosférico *m*	observatório ionosférico *m*	ионосферная обсерватория *f*	21
ionosonda *f* registrador ionosférico *m*	ionossonda *f* registrador ionosférico *m*	ионосферный самописец *m*	22
ionograma *m*	ionograma *m* registro ionosférico *m*	высотно-частотная характеристика *f* ионограмма *f*	23

24	critical frequency	fréquence critique *f*	kritische Frequenz *f*
25	sporadic E-layer	couche E sporadique *f*	sporadische E-Schicht *f*
26	Heaviside layer Kennelly-Heaviside layer E-layer	couche d'Heaviside *f* couche E *f* couche de Kennelly-Heaviside *f*	Heaviside-Schicht *f* E-Schicht *f* Heaviside-Kennelly-Schicht *f*
27	layer height	hauteur d'une couche *f*	Schichthöhe *f*
28	virtual height equivalent height	hauteur virtuelle *f* hauteur équivalente *f*	scheinbare Höhe *f* äquivalente Höhe *f*
29	real height true height	hauteur réelle *f* hauteur vraie *f*	wahre Höhe *f*
30	extraordinary wave	onde extraordinaire *f*	ausserordentliche Welle *f*
31	ordinary wave	onde ordinaire *f*	ordentliche Welle *f*
32	ionospheric prediction ionospheric forecast	prévision ionosphérique *f*	Ionosphären-Vorhersage *f* Funkprognose *f*
33	lowest usable frequency LUF	fréquence minimum utilisable *f* FmU *f*	niedrigste brauchbare Übertragungsfrequenz *f* LUF *f*
34	propagation of radio waves radio-wave propagation	propagation des ondes radioélectriques *f*	Ausbreitung der Radiowellen *f* Funkwellenausbreitung *f*
35	direction of propagation	direction de propagation *f*	Ausbreitungsrichtung *f*
36	propagation vector	vecteur de propagation *m*	Ausbreitungsvektor *m*
37	maximum usable frequency MUF	fréquence maximum utilisable *f* FMU *f*	höchste [maximale] brauchbare [nutzbare] Übertragungsfrequenz *f*
38	skip distance silent area	portée minimum *f* largeur de la zone de silence *f*	Hopentfernung *f* Sprungentfernung *f* Zone des Schweigens *f*
39	ionospheric disturbance	perturbation ionosphérique *f*	Ionosphärenstörung *f* ionosphärische Störung *f*
40	sudden short-wave fade-out SSWF fade-out	évanouissement des ondes courtes *m*	Kurzwellentotalschwund *m* Mögel-Dellinger-Effekt *m* totaler Schwund *m*
41	sudden ionospheric disturbance SID	perturbation ionosphérique à début brusque *f*	plötzliche Ionosphärenstörung *f* plötzliche ionosphärische Störung *f*
42	sudden cosmic noise absorption SCNA	absorption brusque du bruit cosmique *f*	Sonneneruptionseffekt im kosmischen Rauschen *m* plötzliche Absorption im kosmischen Rauschen *f*

frecuencia crítica *f*	freqüência crítica *f*	критическая частота *f*	24
capa E esporádica *f*	camada E esporádica *f*	спорадический слой E *m*	25
capa Heaviside-Kennelly *f* capa E *f*	camada Heaviside-Kennelly *f* camada E *f*	слой E *m* слой (Кеннелли-)Хевисайда *m*	26
altura de una capa *f*	altura de uma camada *f* altitude de uma camada *f*	высота слоя *f*	27
altura virtual *f* altura equivalente *f*	altura [altitude] virtual *f* altura equivalente *f*	действующая высота *f* эквивалентная высота *f*	28
altura real *f*	altura verdadeira *f* altura [altitude] real *f*	действительная высота *f* истинная высота *f* реальная высота *f*	29
onda extraordinaria *f*	onda extraordinária *f*	необыкновенная волна *f*	30
onda ordinaria *f*	onda ordinária *f*	обыкновенная волна *f*	31
predicción ionosférica *f*	previsão ionosférica *f*	ионосферный прогноз *m*	32
mínima frecuencia utilizable *f* mFU *f*	freqüência mínima utilizável *f* FmU *f*	минимальная применимая частота *f* мПЧ *f*	33
propagación de las ondas de radio *f*	propagação das ondas radioelé(c)tricas *f*	распространение радиоволн *n*	34
dirección de propagación *f*	direção de propagação *f*	направление распространения *n*	35
vector de propagación *m*	ve(c)tor de propagação *m*	вектор распространения *m*	36
máxima frecuencia utilizable *f* MFU *f*	freqüência máxima utilizável *f* FMU *f*	максимальная применимая частота *f* МПЧ *f*	37
intervalo de salto *m* ancho de la zona de silencio *f*	alcance mínimo *m* área [largura] da zona de silêncio *f*	наименьшее расстояние связи *n* зона молчания *f*	38
perturbación ionosférica *f*	perturbação ionosférica *f* distúrbio ionosférico *m*	ионосферное возмущение *n*	39
desvanecimiento repentino (de ondas cortas) *m*	desaparecimento súbito das ondas curtas *m* esva(n)ecimento [enfraquecimento] súbito [brusco] (das ondas curtas) *m*	замирание коротких волн *n* внезапное поглощение коротких волн *n* эффект Деллинджера *m*	40
perturbación ionosférica repentina *f*	perturbação ionosférica brusca [de início repentino] *f* distúrbio ionosférico súbito [brusco] *m*	внезапное ионосферное возмущение *n*	41
absorción brusca del ruido cósmico *f*	absorção súbita [brusca] do ruído cósmico *f* ASRC	внезапное поглощение космического шума *n*	42

	IONOSPHERE	IONOSPHÈRE	IONOSPHÄRE
43	influx of high-energy particles invasion of high-energy particles	invasion des particules à haute énergie *f*	Einbruch hochenergetischer Partikeln *m*
44	polar cap absorption PCA	absorption ionosphérique polaire *f*	Polkappenabsorption *f* Funkwellenabsorption in der Polarkappe *f*
45	riometer relative ionospheric opacity meter	riomètre *m*	Riometer *n*
46	auroral absorption AA	absorption aurorale *f*	Polarlichtabsorption *f*
47	sudden commencement absorption SCA	absorption à début brusque *f*	plötzlicher Absroptionsanstieg *m* Absorption mit plötzlichem Anfang *f*
48	electron precipitation	précipitation d'électrons *f*	Elektronen-Abregnung *f* Elektronen-Einfall *m*
49	ionospheric storm	orage ionosphérique *m*	Ionosphärensturm *m*
50	spherics atmospherics	atmosphériques *mpl* parasites atmosphériques *mpl*	atmosphärischer Störpegel *m* atmosphärische Störungen *fpl* Spherics *pl*
51	sudden enhancement of atmospherics SEA	augmentation brusque (du niveau) des atmosphériques *f* renforcement brusque des atmosphériques *m*	plötzliche Erhöhung des atmosphärischen Störpegels *f* Sonneneruptionseffekt im atmosphärischen Störpegel *m*
52	sudden phase anomaly SPA	anomalie de phase à début brusque *f*	sonnenbedingte Phasenanomalie *f* plötzliche Phasenanomalie *f*
53	whistlers whistling atmospherics	siffleurs *mpl* sifflements *mpl*	Whistler(s) *m(pl)* atmosphärische Pfeifstörungen *fpl*
54	very low frequency emission VLF (3 000 to 30 000 Hz)	émission de très basse fréquence *f* TBF	Emission in niedriger Langwellenfrequenz *f* VLF Emission *f*
55	extremely low frequency emission ELF (300 to 3 000 Hz)	émission d'extrêmement basse fréquence *f*	Emission in extrem niedriger Langwellenfrequenz *f* ELF Emission *f*
56	group delay	délai de groupe *m* retard de groupe *m*	Gruppenverzögerung *f*
57	long whistler	siffleur long *m*	langer Whistler *m* lange Pfeifstörung *f*
58	short whistler	siffleur court *m*	kurzer Whistler *m* kurze Pfeifstörung *f*
59	nose whistler	nose whistler *m* siffleur en forme de nez *m*	Nose Whistler *m* „Nasen" – Pfeifstörung *f*

penetración de las (primeras) partículas energéticas *f*	invasão de partículas de alta energia *f*	вторжение частиц высоких энергий *n*	43
absorción de casquete polar *f* absorción ionosférica polar *f*	absorção da calota polar *f* absorção ionosférica polar *f*	поглощение в полярной шапке *n* ППШ	44
riómetro *m* medidor relativo de opacidad ionosférica *m*	riômetro *m* medidor [indicador] da opacidade ionosférica relativa *m*	риометр *m* измерительный прибор относительной непрозрачности ионосферы *m*	45
absorción auroral *f*	absorção auroral [devido às auróras] *f*	авроральное поглощение *n*	46
absorción de comienzo brusco *f*	absorção de começo brusco *f* absorção de início súbito *f*	поглощение с внезапным началом *n*	47
precipitación de electrones *f*	precipitação de elé(c)trons *f* precipitação ele(c)trônica *f*	вторжение электронов *n*	48
tempestad ionosférica *f*	tempestade ionosférica *f*	ионосферная буря *f*	49
atmosféricos *mpl* interferencias estáticas *fpl* esféricos *mpl*	atmosféricos *mpl* parasitas atmosféricos *mpl* interferências estáticas *fpl*	атмосферики *mpl* атмосферные помехи *fpl*	50
aumento brusco de atmosféricos *m*	aumento súbito [brusco] dos atmosféricos *m* reforço súbito [brusco] dos atmosféricos *m*	внезапное усиление атмосфериков *n* внезапное повышение атмосфериков *n*	51
anomalía brusca de fase *f*	anomalia súbita [brusca] de fase *f*	внезапная фазовая аномалия *f*	52
silbidos *mpl* silbidos atmosféricos *mpl*	silvos atmosféricos *mpl* assobiadores *mpl* assobios *mpl*	свисты *mpl* свистящие (атмосферики) *mpl*	53
emisión de muy baja frecuencia *f*	emissão de freqüência muito baixa *f*	эмиссия очень низких частот *f* эмиссия ОНЧ *f* ОНЧ *fpl*	54
emisión de extremadamente baja frecuencia *f*	emissão de freqüência extremamente baixa *f*	эмиссия сверхнизких частот *f* эмиссия исключительно низких частот *f* СНЧ *fpl*	55
retardo de grupo *m*	retardo de grupo *m*	групповая задержка *f*	56
silbido largo *m* silbido prolongado *m*	silvo longo *m* assobio longo *m*	длинный свистящий атмосферик *m*	57
silbido corto *m*	silvo curto *m* assobio curto *m*	короткий свистящий атмосферик *m*	58
silbido (en forma) de nariz *m*	silvo (em forma) de nariz *m*	свистящий атмосферик в форме носа *m*	59

78 IONOSPHERE

	IONOSPHERE	IONOSPHÈRE	IONOSPHÄRE
60	electron whistler	siffleur électronique *m*	Elektronenwhistler *m* Elektronen-Pfeifstörung *f*
61	proton whistler	siffleur protonique *m*	Protonenwhistler *m* Protonen-Pfeifstörung *f*
62	dawn chorus	choeur de l'aube *m*	Dämmerungschor *m*
63	tweek	twik *m*	tweek *m*
64	hiss	hiss *m*	hiss *m*

silbido electrónico *m*	silvo ele(c)trônico *m*	электронный свист *m*	60
silbido protónico *m*	silvo protônico *m*	протонный свист *m*	61
coros matutinos *mpl*	coro de alvorada *m* coro de aurora *m*	утренние хоры *mpl*	62
tvik *m*	musicais atmosféricos *mpl*	твик *m*	63
soplido *m*	hiss *m*	хис *m*	64

79 ATMOSPHERIC OPTICS

	ATMOSPHERIC OPTICS	OPTIQUE ATMOSPHÉRIQUE	ATMOSPHÄRISCHE OPTIK
1	atmospheric optics	optique atmosphérique *f*	atmosphärische Optik *f*
2	airglow	émission atmosphérique *f* luminescence atmosphérique *f* luminescence du ciel *f*	Luftleuchten *n* Himmelsleuchten *n*
3	dayglow day-time airglow	luminescence du ciel diurne *f* luminescence atmosphérique diurne *f*	Taghimmelsleuchten *n*
3a	day sky	ciel diurne *m*	Tag(es)himmel *m*
3b	northern sky	ciel boréal *m*	Nordhimmel *m*
3c	southern sky	ciel austral *m*	Südhimmel *m*
3d	starry night	nuit étoilée *f*	Sternnacht *f*
3e	moonless night	nuit sans lune *f*	mondlose Nacht *f*
3f	photometric(ally clear) night	nuit photométrique *f*	photometrische Nacht *f* photometrisch klare Nacht *f*
3g	clear night cloudless night	nuit sans nuages *f* nuit sereine *f*	wolkenfreie Nacht *f* klare Nacht *f*
4	twilight airglow twilight glow	luminescence atmosphérique crépusculaire *f* luminescence du ciel crépusculaire *f*	Dämmerungsleuchten *n*
5	nightglow night airglow nightglow emission night-sky luminescence permanent aurora	luminescence du ciel nocturne *f* émission du ciel nocturne *f* luminescence atmosphérique nocturne *f*	Nachthimmelsleuchten *n*
6	night sky	ciel nocturne *m*	Nachthimmel *m*
7	extraterrestrial light	lumière extraterrestre *f*	extraterrestrisches Licht *n*
8	light of the night sky night-sky light	lumière du ciel nocturne *f*	Nachthimmelslicht *n* Licht des Nachthimmels *n*
9	Lα-glow of the night sky Lα-emission of the night sky	emission Lα du ciel nocturne *f* luminescence Lα du ciel nocturne *f*	Lα Leuchten des Nachthimmels *n*
9a	heavenly body	corps céleste *m* astre *m*	Himmelskörper *m*
10	starlight	lumière stellaire *f*	Sternlicht *n*
11	direct starlight	lumière stellaire directe *f*	direktes Sternlicht *n*

óptica atmosférica *f*	ó(p)tica atmosférica *f*	атмосферная оптика *f*	1
luminescencia atmosférica *f* luminosidad del aire *f* claror atmosférico *m*	emissão atmosférica *f* luminescência do ar *f* brilho [clarão] atmosférico *m*	свечение атмосферы *n* собственное свечение атмосферы *n* атмосферное свечение *n*	2
luminescencia diurna *f* luminosidad diurna del aire *f* claror atmosférico diurno *m*	luminescência diurna atmosférica *f* brilho atmosférico diurno *m* luminescência diurna do ar *f*	дневное свечение атмосферы *n* (атмосферное) свечение дневного неба *n*	3
cielo diurno *m*	céu diurno *m*	дневное небо *n*	3a
cielo boreal *m*	céu boreal *m*	северное небо *n* северное полушарие неба *n*	3b
cielo austral *m*	céu austral *m*	южное небо *n* южное полушарие неба *n*	3c
noche estrallada *f*	noite de céu estrelado *f*	звёздная ночь *f*	3d
noche sin (claro de la) luna *f*	noite sem luar *f*	безлунная ночь *f*	3e
noche (de observación) fotométrica *f*	noite fotométrica(mente clara) *f*	фотометрическая ночь *f* фотометрически ясная ночь *f*	3f
noche clara *f* noche despejada *f*	noite clara *f* noite sem nuvens *f*	ясная ночь *f* безоблачная ночь *f*	3g
luminescencia crepuscular *f* luminosidad crepuscular *f* claror atmosférico crepuscular *m*	luminescência crepuscular do ar *f* brilho atmosférico crepuscular *m* emissão atmosférica crepuscular *f*	сумеречное свечение атмосферы *n* атмосферное свечение сумеречного неба *n*	4
luminescencia nocturna *f* luminosidad nocturna (del aire) *f* claror atmosférico nocturno *m* aurora permanente *f*	luminescência noturna do ar *f* emissão atmosférica noturna *f* aurora permanente *f* luminescência noturna do ar *f*	ночное свечение атмосферы *n* атмосферное свечение ночного неба *n* атмосферная составляющая свечения ночного неба *f*	5
cielo nocturno *m*	céu noturno *m*	ночное небо *n*	6
luz extraterrestre *f*	luz extraterrestre *f*	внеатмосферное [заатмосферное] свечение *n* внеатмосферный свет *m*	7
luminosidad del cielo nocturno *f* luz del cielo nocturno *f*	luminosidade do céu noturno *f* luz do céu noturno *f*	свет ночного неба *m* свечение ночного неба *n*	8
luminescencia Lα del cielo nocturno *f* emisión Lα del cielo nocturno *f*	luminescência Lα do céu noturno *f* emissão Lα do céu noturno *f*	Lα свечение ночного неба *n*	9
cuerpo celeste *m* astro *m*	corpo celeste *m* astro *m*	светило *n* (самосветящееся) небесное тело *n*	9a
luz estelar *f*	luz estelar *f*	свет звёзд *m* звёздный свет *m*	10
luz estelar directa *f*	luz estelar direta *f*	прямой свет звёзд *m*	11

12	scattered starlight	lumière stellaire diffus(é)e	gestreutes Sternlicht *n*
13	zodiacal light	lumière zodiacale *f*	Zodiakallicht *n*
14	galactic light	lumière galactique *f*	galaktisches Licht *n*
15	red line	raie rouge *f*	rote Linie *f*
16	green line	raie verte *f*	grüne Linie *f*
17	luminescent layer airglow layer luminous layer	couche lumineuse *f* couche émissive *f* couche luminescente *f*	Emissionsschicht *f*
18	band system	système de bandes *m*	Bandensystem *n*
19	Vegard-Kaplan bands	bandes de Vegard-Kaplan *mpl*	Vegard-Kaplan-Banden *fpl*
20	first positive band system of nitrogen	premier système positif de l'azote *m*	erste positive Stickstoffbanden *fpl*
21	negative nitrogen bands	système negatif de l'azote *m*	negative Stickstoffbanden *fpl*
22	afterglow	postluminescence *f*	Nachleuchten *n*
23	molecular oxygen	oxygène moléculaire *m*	molekularer Sauerstoff *m*
24	atomic oxygen	oxygène atomique *m*	atomarer Sauerstoff *m*
25	hydroxyl	hydroxyle *m* oxhydryle *m* oxhydride *m*	Hydroxyl *n*
26	altitude of the luminous layer airglow height	altitude de la couche émissive *f*	Höhe der Emissionsschicht *f*
27	excitation mechanism	mécanisme d'excitation *m*	Anregungsmechanismus *m*
28	ozone shadowing	occultation par la couche d'ozone *f*	Abschirmung durch die Ozonschicht *f*
29	mutual neutralization	neutralisation mutuelle *f*	gegenseitige Neutralisation *f*
30	triple collision	collision triple *f*	Dreierstoss *m*
31	chemoluminescence	chimiluminescence *f*	Chemolumineszenz *f*
32	twilight phenomena	phénomènes crépusculaires *mpl*	Dämmerungserscheinungen *fpl*
33	optical phenomenon	phénomène optique *m*	optische Erscheinung *f*
34	twilight	crépuscule *m*	Dämmerung *f*
35	morning twilight dawn	crépuscule du matin *m* aube *f*	Morgendämmerung *f*

luz estelar difusa *f*	luz estelar espalhada [difusa] *f*	рассеянный свет звёзд *m*	12
luz zodiacal *f*	luz zodiacal *f*	зодиакальный свет *m*	13
luz galáctica *f*	luz galáctica *f*	галактический свет *m* свет Галактики *m*	14
línea roja *f*	linha vermelha *f*	красная (спектральная) линия *f*	15
línea verde *f*	linha verde *f*	зелёная (спектральная) линия *f*	16
capa emisiva *f* capa luminosa *f*	camada luminescente *f* camada emissora (do ar) *f* camada luminosa *f*	слой свечения *m* излучающий слой *m*	17
sistema de bandas *m*	sistema de bandas *m*	система полос (в молекулярном спектре) *f*	18
bandas de Vegard-Kaplan *fpl*	bandas de Vegard-Kaplan *fpl*	полосы Вегарда-Каплана *fpl*	19
primeras bandas positivas del nitrógeno *fpl*	primeiras bandas positivas do nitrogênio *fpl*	полосы первой положитеьной системы азота *fpl*	20
bandas negativas del nitrógeno *fpl*	bandas negativas do nitrogênio *fpl*	полосы отрицательной системы азота *fpl*	21
postluminiscencia *f* resplandor residual *m*	pós-luminescência *f*	послесвечение *n*	22
oxígeno molecular *m*	oxigênio molecular *m*	молекулярный кислород *m*	23
oxígeno atómico *m*	oxigênio atômico *m*	атомный кислород *m*	24
hidroxilo *m* oxidrilo *m* oxhidrilo *m*	hidroxila *f* oxidrila *f*	гидроксил *m*	25
altura de la capa luminosa *f*	altura da camada emissora *f*	высота излучающего слоя *f* высота слоя свечения *f*	26
mecanismo de excitación *m*	mecanismo de excitação *m*	механизм возбуждения *m*	27
ocultación por la capa de ozono *f* blindaje por la capa de ozono *f*	ocultação pela camada de ozônio [ozone] *f*	экранирование озоном *n* затемнение озонным слоем *n*	28
neutralización mutua *f*	neutralização mútua *f*	взаимная нейтрализация *f*	29
colisión triple *f* choque triple *m*	choque triplo *m*	тройное столкновение *n* тройное соударение *n*	30
quimioluminiscencia *f*	quimi(o)luminescência *f*	хемилюминесценция *f*	31
fenómenos crepusculares *mpl*	fenômenos crepusculares *mpl*	сумеречные явления *npl*	32
fenómeno óptico *m*	fenômeno ó(p)tico *m*	оптическое явление *n*	33
crepúsculo *m*	crepúsculo *m*	сумерки *pl*	34
crepúsculo matutino *m* alba *f* amanecer *m*	crepúsculo matutino *m* amanhecer *m* aurora *f*	рассвет *m* утренние сумерки *pl* утренняя заря *f*	35

79	ATMOSPHERIC OPTICS	OPTIQUE ATMOSPHÉRIQUE	ATMOSPHÄRISCHE OPTIK
36	evening twilight dusk	crépuscule du soir *m* crépuscule vespéral *m*	Abenddämmerung *f* abendliches Zwielicht *n*
37	astronomical twilight	crépuscule astronomique *m*	astronomische Dämmerung *f*
38	civil twilight	crépuscule civil *m*	bürgerliche Dämmerung *f*
39	nautical twilight	crépuscule nautique *m*	nautische Dämmerung *f*
40	twilight...	crépusculaire	Dämmerungs-
41	twilight sounding	sondage crépusculaire *f*	Dämmerungssondierung *f*
42	twilight observation	observation crépusculaire *f*	Dämmerungsbeobachtung *f*
43	twilight sky	ciel crépusculaire *m*	Dämmerungshimmel *m*
44	Earth's shadow	ombre de la Terre *f*	Erdschatten *m*
45	solar depression	dépression solaire *f*	Sonnendepression *f*
46	shadow height	hauteur d'ombre *f*	Schattenhöhe *f*
47	geometrical shadow height	hauteur d'ombre géométrique *f*	geometrische Schattenhöhe *f*
48	screening height	hauteur de blindage *f*	Abschirmungshöhe *f*
49	screened shadow height true shadow height	hauteur vraie d'ombre *f*	abgeschirmte Schattenhöhe *f* wahre Schattenhöhe *f*
50	crepuscular arc	arc crépusculaire *m*	Dämmerungsbogen *m*
51	pre-twilight emission subpolar sheets	émission précrépusculaire *f*	Vordämmerungsemission *f*
52	post-twilight emission western sheets	émission postcrépusculaire *f*	Nachdämmerungsemission *f*
53	intertropical arcs	arcs intertropicaux *mpl*	innertropische Bögen *mpl*
54	purple light	lueur pourprée *f*	Purpurlicht *n*
55	sunset colours sunset glow	couleurs du couchant *fpl*	Abendrot *n* Abendröte *f*
56	aurora sunrise colours	aurore *f*	Morgenrot *n*
57	antitwilight	anticrépuscule *m*	Gegendämmerung *f*
58	green flash	rayon vert *m*	grüner Strahl *m*
59	bright night	nuit brillante *f*	helle Nacht *f*
60	aureole corona	auréole *f* couronne *f*	Strahlenkranz *m* Aureole *f* Korona *f*

crepúsculo vespertino *m* anochecer *m*	crepúsculo vespertino *m* anoitecer *m*	вечерние сумерки *pl* сумрак *m*	36
crepúsculo astronómico *m*	crepúsculo astronômico *m*	астрономические сумерки *pl*	37
crepúsculo civil *m*	crepúsculo civil *m*	гражданские сумерки *pl*	38
crepúsculo náutico *m*	crepúsculo náutico *m*	навигационные сумерки *pl*	39
crepuscular	crepuscular	сумеречный	40
sondeo crepuscular *m*	sondagem crepuscular [de crepúsculo] *f*	сумеречное зондирование *n*	41
observación del crepúsculo *f*	observação do crepúsculo *f*	наблюдение сумерек *n*	42
cielo crepuscular *m*	céu crepuscular *m*	сумеречное небо *n*	43
sombra de la Tierra *f*	sombra da Terra *f*	земная тень *f* тень Земли *f*	44
depresión solar *f*	depressão solar *f*	погружение Солнца под горизонт *n*	45
altura de la sombra *f*	altura da sombra *f*	высота тени *f*	46
altura geométrica de la sombra *f*	altura geométrica da sombra *f*	геометрическая высота тени *f*	47
altura de blindaje *f*	altura de blindagem [de abrigo] *f*	высота экранирования *f*	48
altura real de la sombra *f*	altura verdadeira da sombra *f*	экранированная высота тени *f* истинная высота тени *f*	49
arco crepuscular *m*	arco crepuscular *m*	сумеречная дуга *f*	50
emisión precrepuscular *f*	emissão pré-crepuscular *f* camadas subpolares *fpl*	предсумеречная эмиссия *f*	51
emisión postcrepuscular *f*	emissão pós-crepuscular *f* camadas ocidentais *fpl*	послесумеречная эмиссия *f*	52
arcos intertropicales *mpl*	arcos intertropicais *mpl*	внутритропические дуги *fpl*	53
luz purpúrea *f*	luz purpúrea *f*	пурпурный свет *m*	54
colores del atardecer *mpl* arrebol vespertino *m*	cores do (Sol) poente *fpl*	вечерняя заря *f*	55
aurora *f* colores de la aurora *mpl*	aurora *f* cores do (Sol) nascente *fpl*	утренняя заря *f*	56
anticrepúsculo *m*	anticrepúsculo *m* luminescência anticrepuscular *f*	противосумерки *pl*	57
rayo verde *m*	raio verde *m*	зелёный луч *m*	58
noche clara *f*	noite clara *f*	светлая ночь *f*	59
corona *f* aureola *f*	coroa *f* auréola *f*	ореол *m* венец *m*	60

79	ATMOSPHERIC OPTICS	OPTIQUE ATMOSPHÉRIQUE	ATMOSPHÄRISCHE OPTIK
61	lunar aureole lunar corona	auréole lunaire *f* couronne lunaire *f*	Mondkranz *m* Mondhof *m* Mondaureole *f*
62	solar aureole solar corona	auréole solaire *f* couronne solaire *f*	Sonnenkranz *m* Sonnenkorona *f* Sonnenaureole *f*
63	glory	gloire *f*	Glorie *f*
64	mirage	mirage *m*	Luftspiegelung *f*
65	rainbow	arc-en-ciel *m*	Regenbogen *m*
66	halo phenomenon	phénomène de halo *m*	Haloerscheinung *f*
67	halo (haloes *pl.*)	halo *m*	Halo *m*
68	lunar halo	halo lunaire *m*	Mondring *m*
69	halo of 22° 22°-halo	halo principal *m* petit halo *m*	kleiner Ring *m* 22°-Ring *m*
70	halo of 46° 46°-halo	grand halo *m* halo de 46° *m*	grosser Ring *m* 46°-Ring *m*
71	parhelion sun-dog mock sun	parhélie *m*	Nebensonne *f*
72	undersun lower parhelion subsun	parhélie inférieur *m*	Untersonne *f*
73	paraselene mock moon	parasélène *f*	Nebenmond *m*
74	circumzenithal arc circumzenithal circle	arc circumzénithal *m* cercle circumzénithal *m*	Zirkumzenitalbogen *m* Zirkumzenitalkreis *m*
75	tangential arc of halo	arc tangent au halo *m*	Berührungsbogen *m*
76	lower tangential arc	arc tangent inférieur *m*	unterer Berührungsbogen *m*
77	upper tangential arc	arc tangent supérieur *m*	oberer Berührungsbogen *m*
78	sun pillar	colonne lumineuse *f*	Lichtsäule *f*
79	light cross halo cross	croix lumineuse *f* croix de halo *f*	Lichtkreuz *n* Halokreuz *n*
80	lunar cross	croix lunaire *f*	Mondkreuz *n*

corona lunar *f*	coroa lunar *f* auréola lunar *f*	венец вокруг Луны *m* окололунный ореол *m* ореол Луны *m*	61
aureola solar *f* corona solar *f*	auréola solar *f* coroa solar *f*	венец вокруг Солнца *m* околосолнечный ореол *m* ореол Солнца *m*	62
gloria *f*	glória *f*	глория *f*	63
espejismo *m*	miragem *f*	мираж *m*	64
arco iris *m*	arco-íris *m*	радуга *f*	65
fenómeno de halo *m*	fenômeno de halo *m*	явление гало *n*	66
halo *m*	halo *m*	гало *n* (*pl.*галосы)	67
halo lunar *m*	halo lunar *m*	гало вокруг Луны *n* лунное гало *n*	68
halo principal *m* halo primario *m*	halo de 22° *m* pequeno halo *m*	малое гало *n* малое кольцо *n* малый круг *m*	69
halo grande *m* halo secundario *m*	halo de 46° *m* grande halo *m*	большое гало *n* большое кольцо *n* большой круг *m*	70
parhelio *m* falso sol *m*	parélio *m* sol falso *m*	паргелий *m* ложное солнце *n* побочное солнце *n*	71
parhelio inferior *m*	parélio inferior *m*	нижнее (побочное) солнце *n*	72
paraselene *f* paraselena *f*	parasselene *m* parasselênio *m*	ложная луна *f*	73
arco circuncenital *m* círculo circuncenital *m*	arco circunzenital *m* círculo circunzenital *m*	околозенитная дуга *f* околозенитный круг *m*	74
arco tangencial de halo *m*	arco tangencial do halo *m*	касательная дуга (к) гало *f*	75
arco tangencial inferior *m*	arco tangencial inferior *m*	нишняя касательная дуга (к гало) *f*	76
arco tangencial superior *m*	arco tangencial superior *m*	верхняя касательная дуга (к гало) *f*	77
columna luminosa *f*	coluna luminosa *f* coluna solar *f*	солнечный столб *m* световой столб *m*	78
cruz luminosa *f* cruz de halo *f*	cruz luminosa *f* cruz de halo *f*	световой крест *m*	79
cruz lunar *f*	cruz lunar *f*	лунный крест *m*	80

	ATMOSPHERIC OPTICS	OPTIQUE ATMOSPHÉRIQUE	ATMOSPHÄRISCHE OPTIK
81	solar cross	croix solaire *f*	Sonnenkreuz *n*
82	circumhorizontal arc parhelic circle	arc circumhorizontal *m* cercle parhélique *m*	Horizontalkreis *m* Horizontalring *m*
83	anthelion	anthélie *m*	Gegensonne *f*
84	paranthelion	paranthélie *m*	Nebengegensonne *f*
85	antiselene	antisélène *f*	Gegenmond *m*
86	parantiselene	parantisélène *f*	Nebengegenmond *m*
87	lateral lower tangential arc	arc tangent infralatéral *m*	seitlicher Berührungsbogen *m*
88	Bishop's ring	anneau de Bishop *m*	Bishopscher Ring *m*
89	noctilucent cloud NLC	nuage noctilucent *m* nuage nocturne lumineux *m*	leuchtende Nachtwolken *fpl*
90	iridescent cloud	nuage iridescent *m*	irisierende Wolke *f*
91	nacreous cloud mother-of-pearl cloud	nuage nacré *f*	Perlmutt(er)wolke *f*
92	blue Sun	soleil bleu *m*	blaue Sonne *f*
93	polar aurora aurora (polaris)	aurore polaire *f*	Polarlicht *n* Aurora *f*
94	radio aurora	aurore radio *f*	Radio-Polarlicht *n* Radio-Aurora *f*
95	radar aurora	aurore détectée par radar *f*	Radar-Polarlicht *n* Radar-Aurora *f*
96	low-latitude aurora	aurore de basse latitude *f*	Polarlicht in niederen Breiten *n*
97	dayside aurora	aurore de jour *f*	Polarlicht auf der Tagseite *n*
98	polar-cap aurora	aurore de calotte polaire *f*	Polarlicht in der Polkappe *n*
99	artificial aurora	aurore artificielle *f*	künstliches Polarlicht *n*
100	auroral activity	activité aurorale *f*	Polarlichtaktivität *f*

cruz solar *f*	cruz solar *f*	солнечный крест *m*	81
círculo horizontal *m* círculo azimutal *m*	arco circum-horizontal *m* círculo par(h)élico *m*	окологоризонтальная дуга *f* паргелический круг *m* горизонтальный круг *m*	82
antihelio *m*	antélio *m* anti-hélio *m*	противосолнце *n* антигелий *m* антелий *m*	83
paranthelio *m*	parantélio *m* paranti-hélio *m*	побочное противосолнце *n* парантгелий *m*	84
antiselene *f* antiselena *f*	anti-selene *m* antisselênio *m*	антиселена *f* противолуна *f*	85
parantiselene *f* parantiselena *f*	parantisselênio *m* paranti-selene *m*	побочная противолуна *f* парантиселена *f*	86
arco tangencial infralateral *m*	arco tangencial infralateral *m*	побочная нижняя касательная дуга *f*	87
anillo de Bishop *m*	anel de Bishop *m*	кольцо Бишопа *n*	88
nube nocturna luminosa *f* nube noctiluciente *f*	nuvem noctilucia [noctiluzente] *f* nuvem luminosa noturna *f*	ночное светящееся облако *n* серебристое облако *n*	89
nube iridiscente *f* nube tornasol *f*	nuvem iridescente *f*	иризирующее облако *n*	90
nube nacreada *f*	nuvem nacarada *f*	перламутровое облако *n*	91
sol azul *m*	sol azul *m*	голубое солнце *n*	92
aurora polar *f*	aurora polar *f*	полярное сияние *n*	93
aurora radio *f* radioaurora *f*	aurora rádio *f* radioaurora *f*	радиоаврора *f* полярное сияние излучающее радиоволны *n*	94
aurora radar *f*	aurora radar *f*	радарное полярное сияние *n* полярное сияние наблюденное с помощью радиолокатора *n*	95
aurora da baja latitud *f*	aurora de baixa latitude *f*	низкоширотное полярное сияние *n*	96
aurora diurna *f*	aurora diurna *f*	полярное сияние на дневной стороне *n*	97
aurora de casquete polar *f*	aurora de calota polar *f*	полярное сияние в полярной шапке *n*	98
aurora artificial *f*	aurora artificial *f*	искусственное полярное сияние *n*	99
actividad auroral *f*	atividade auroral *f*	активность полярных сияний *f*	100

101	auroral observatory	observatoire auroral *m*	Polarlichtobservatorium *n*
102	parallactic photography	photographie parallactique *f*	parallaktische Photographie *f*
103	sequence photography	photographie séquentielle *f*	Sequenzphotographie *f* fortlaufende Photographie *f*
104	all-sky camera	caméra grand champ *f* caméra (pour) tout le ciel *f*	All-Sky-Kamera *f*
105	aurora borealis	aurore boréale *f*	Nordlicht *n* nördliche Aurora *f*
106	aurora australis	aurore australe *f*	Südlicht *n* südliche Aurora *f*
107	auroral particle	particule aurorale *f*	Polarlichtteilchen *n*
108	proton aurora	aurore à protons *f*	Protonenpolarlicht *n*
109	auroral electron	électron auroral *m*	Polarlichtelektron *n*
110	electron aurora	aurore à électrons *f*	Elektronenpolarlicht *n*
111	trapped electrons	électrons piégés *mpl*	eingefangene Elektronen *npl*
112	auroral electron precipitation	précipitation aurorale d'électrons *f*	Elektronen-Abregnung im Polarlicht *f*
113	auroral electrojet	electrojet auroral *m*	Ionosphärenstrom in Polarlichtbreiten *m* Elektrojet in Polarlichtbreiten *m*
114	jet of electrons	jet d'électrons *m*	Elektronenstrom *m* Elektrojet *m*
115	electron cloud	nuage d'électrons *m*	Elektronenwolke *f*
116	electron beam	faisceau électronique *m*	Elektronenbündel *n* Elektronenstrahl *m*
117	X-ray burst	sursaut de rayonnement X *m*	Röntgen(strahlungs)-burst *m*
118	dumped electrons	électrons précipités *mpl*	Dumping-Elektronen *npl*
119	electron source	source d'électrons *f*	Elektronenquelle *f*
120	atmospheric interaction	interaction atmosphérique *f*	atmosphärische Wechselwirkung *f*
121	auroral echo	écho auroral *m*	Polarlichtecho *n*
122	auroral frequency	fréquence aurorale *f*	Polarlichthäufigkeit *f*

observatorio auroral *m*	observatório de aurora *m* observatório auroral *m*	обсерватория полярных сияний *f*	101
fotografía paraláctica *f*	fotografia paralá(c)tica *f*	параллактическая фотография *f*	102
fotografía en secuencia *f*	seqüência de fotografias *f* fotografia de seqüência *f*	последовательная фотография *f*	103
cámara fotográfica de cielo global *f*	câmara [máquina] fotográfica para todo o céu *f*	камера всего неба *f* фотокамера для небесного свода *f* камера с полем зрения, охватающим полусферу *f*	104
aurora boreal *f*	aurora boreal *f*	северное (полярное) сияние *n*	105
aurora austral *f*	aurora austral *f*	южное (полярное) сияние *n*	106
partícula auroral *f*	partícula auroral *f*	авроральная частица *f*	107
aurora protónica *f*	aurora de prótons *f*	протонное полярное сияние *n*	108
electrón auroral *m*	elé(c)tron auroral *m*	авроральный электрон *m*	109
aurora electrónica *f*	aurora de elé(c)trons *f*	электронное полярное сияние *n*	110
electrones atrapados *mpl*	elé(c)trons retidos [aprisionados] *mpl*	захваченные электроны *mpl*	111
precipitación de electrones aurorales *f*	precipitação auroral ele(c)trônica *f*	высыпание электронов *n* вторжение электронов (полярного сияния) *n*	112
electrochorro auroral *m*	ja(c)to auroral *m* ele(c)troja(c)to auroral *m*	электроструя полярного сияния *f* авроральная электроструя *f*	113
chorro de electrones *m*	ja(c)to ele(c)trônico *m*	струя электронов *f*	114
nube de electrones *f*	nuvem de elé(c)trons *f*	электронное облако *n*	115
haz electrónico *m* haz de electrones *m*	feixe ele(c)trônico *m*	электронный пучок *m* пучок электронов *m*	116
estallido de rayos X *m* brusca emisión de rayos X *f*	explosão de raios X *f* emissão brusca de raios X *f*	всплеск рентгеновского излучения *m*	117
electrones desechados *mpl*	elé(c)trons precipitados *mpl*	высыпанные электроны *mpl*	118
fuente de electrones *f*	fonte de elé(c)trons *f*	источник электронов *m*	119
interacción atmosférica *f*	interação atmosférica *f*	атмосферное взаимодействие *n* взаимодействие с атмосферой *n*	120
eco auroral *m*	eco auroral *m*	эхо полярного сияния *n*	121
frecuencia de auroras *f*	freqüência (da ocorrência) de auroras *f*	повторяемость полярных сияний *f* частота появления полярных сияний *f*	122

	ATMOSPHERIC OPTICS	OPTIQUE ATMOSPHÉRIQUE	ATMOSPHÄRISCHE OPTIK
123	auroral zone	zone aurorale *f*	Polarlichtzone *f*
124	auroral oval	ovale auroral *m*	Polarlichtoval *n*
125	zone of maximum auroral frequency	zone du maximum de fréquence aurorale *f*	Zone maximaler Polarlichthäufigkeit *f*
126	auroral region	région aurorale *f*	Polarlichtregion *f*
127	auroral cap	calotte aurorale *f*	Polarlichtkappe *f*
128	subauroral belt	ceinture subaurorale *f*	Gürtel südlich [nördlich] der Nord-[Süd]lichtzone *m*
129	minauroral belt	ceinture minaurorale *f*	Gürtel geringer Polarlichthäufigkeit *m*
130	sunlit aurora	aurore éclairée par le Soleil *f*	sonnenbelichtetes Polarlicht *n*
131	magnetic midnight	minuit magnétique *f*	magnetische Mitternacht *f*
132	geomagnetic time	temps géomagnétique *m*	geomagnetische Zeit *f*
133	terrella terrestrial globe model of the Earth	terrella *f* modèle de la Terre *m*	Terrella *f* Modell der Erde *n*
134	allowed space allowed region	espace permis *m* région permise *f*	erlaubte Räume *mpl*
135	forbidden space forbidden region	espace interdit *m* région interdite *f*	verbotene Räume *mpl*
136	auroral type	type d'aurore *m*	Art des Polarlichtes *f*
137	auroral forms	formes aurorales *fpl*	Polarlichtformen *fpl*
137a	auroral corona	couronne aurorale *f*	Polarlichtkorona *f*
138	active features	formations actives *fpl*	aktive Formen *fpl*
139	folds	plis *mpl*	Falten *fpl*
140	ribbon-like structure aurora	aurore avec structure à rubans *f*	Polarlicht mit bandförmiger Struktur *n*
141	auroral substorm	sous-orage auroral *m*	Polarlicht-Substorm *m*
142	forms without ray-structure	formes sans structure de rayons *fpl*	Formen ohne Strahlen-Struktur *fpl*
143	homogeneous quiet arc HA	arc homogène calme *m*	homogener ruhiger Bogen *m*
144	substorm	sous-orage *m*	Substorm *m*
145	hydrogen glow	lueur d'hydrogène *f*	Wasserstoffleuchten *n*

zona auroral *f*	zona auroral *f*	зона полярных сияний *m*	123
óvalo auroral *m*	oval auroral *m*	овал полярных сияний *m*	124
zona de máxima frecuencia auroral *f*	zona de freqüência máxima de auroras *f*	зона максимальной повторяемости (полярных сияний) *f*	125
región de auroras *f* región auroral *f*	região de auroras *f* região auroral *f*	область полярных сияний *f* авроральная область *f*	126
casquete auroral *f*	calota auroral *f*	шапка полярных сияний *f*	127
cinturón subauroral *m*	cinturão subauroral *m*	субавроральный пояс *m* субполярный пояс *m*	128
cinturón minauroral *m*	cinturão minauroral *m*	зона минимума полярных сияний *f*	129
aurora iluminada por el Sol *f*	aurora iluminada pelo Sol *f*	полярное сиание освещённое Солнцем *n*	130
medianoche magnética *f*	meia-noite magnética *f*	магнитная полночь *f*	131
tiempo geomagnético *m*	tempo geomagnético *m*	геомагнитное время *n*	132
terrella *f* modelo de la Tierra *m*	terrela *f* modelo da Terra *m*	терелла *f* модель Земли *f*	133
espacio permitido *m* región posible *f*	espaço permitido *m* região permitida *f*	разрешённое пространство *n* разрешённая область *f*	134
espacio prohibido *m* región prohibida *f*	espaço interdito [proibido] *m* região interdita [proibida] *f*	запрещённое пространство *n* запрещённая область *f*	135
tipo de aurora *m*	tipo de aurora *m*	тип полярного сияния *m*	136
formas aurorales *fpl*	formas de aurora *fpl*	формы полярных сияний *fpl*	137
corona auroral *f*	coroa auroral *f*	авроральная корона *f*	137a
características activas *fpl*	formações ativas *fpl*	активные образования *npl*	138
pliegues *mpl*	dobras *fpl*	складки *fpl*	139
aurora de estructura de cinta *f*	aurora com estrutura de fita *f*	полярное сияние с ленточной структурой *n*	140
subtormenta auroral *f*	subtempestade auroral *f*	авроральная суббуря *f*	141
formas sin (estructura de) rayos *fpl*	formas sem estrutura raiada *fpl*	формы без лучевой структуры *fpl*	142
arco homogéneo quieto *m*	arco homogêneo calmo *m*	однородная дуга *f*	143
subtormenta *f*	subtempestade *f*	суббуря *f*	144
luminosidad del hidrógeno *f*	luminosidade do hidrogênio *f*	водородное свечение *n*	145

79	**ATMOSPHERIC OPTICS**	**OPTIQUE ATMOSPHÉRIQUE**	**ATMOSPHÄRISCHE OPTIK**
146	homogeneous band HB	bande homogène *f*	homogene Bande *f*
146a	auroral arc	arc auroral *m*	Polarlichtbogen *m*
147	pulsating arc PA	arc pulsant *m*	pulsierender Bogen *m*
148	diffuse luminous surface DS	surface lumineuse diffuse *f* lueur diffuse *f*	diffus leuchtende Fläche *f* Nordlichtdunst *m* Polarlichtdunst *m*
149	pulsating surface PS	surface pulsante *f*	pulsierende Fläche *f*
150	band with ray-structure rayed band RB	bande rayée *f*	Bande mit Strahlenstruktur *f*
151	glow G	(faible) leur *f*	Schein *m* Leuchten *n*
152	forms with ray-structure	formes avec structure de rayons *fpl*	Formen mit Strahlen(struktur) *fpl*
153	arc with ray-structure rayed arc RA	arc rayé *m* arc avec structure de rayons *m*	Bogen mit Strahlen(struktur) *m*
154	curtains draperies D	draperies *fpl*	Draperien *fpl*
155	rays R	raies *fpl* rayons *mpl*	Strahlen *mpl*
156	corona C	couronne *f*	Nordlichtkrone *f* Korona *f*
157	flaming aurora F	aurore flamboyante *f*	flammendes Nordlicht *n*
158	pulsations	pulsations *fpl*	Pulsationen *fpl*

banda homogénea *f*	faixa homogênea *f*	однородная полоса *f*	146
arco auroral *m*	arco auroral *m*	авроральная дуга *f*	146a
arco pulsante *m*	arco pulsante *m*	пульсирующая дуга *f*	147
superficie luminosa difusa *f* velo auroral *m*	superficie luminosa difusa *f*	диффузная светящаяся поверхность *f*	148
superficie pulsante *f*	superficie pulsante *f*	пульсирующая поверхность *f*	149
banda con rayos *f* banda estriada *f*	faixa com estrutura raiada *f*	полоса с лучевой структурой *f*	150
luminosidad débil *f* resplandor *m*	clarão fraco *m*	свечение *n*	151
formas con (estructura de) rayos *fpl*	formas com estrutura raiada *fpl*	формы с лучевой структурой *fpl*	152
arco estriado *m* arco con (estructura de) rayos *m*	arco com estrutura raiada *m*	дуга с лучевой структурой *f*	153
franjas *fpl*	cortinas *fpl*	драпри *pl*	154
rayos *mpl* estrías *fpl*	raios *mpl*	лучи *mpl*	155
corona *f*	coroa *f*	корона *f*	156
aurora flameante *f*	aurora flamejante *f*	пламенное (полярное) сияние *n* полярное сияние в виде пламени *n*	157
pulsaciones *fpl*	pulsações *fpl*	пульсации *fpl*	158

80

SOLAR SYSTEM

SYSTÈME SOLAIRE

SONNENSYSTEM

SISTEMA SOLAR

SISTEMA SOLAR

СОЛНЕЧНАЯ СИСТЕМА

81 SELENOGRAPHY | SÉLÉNOGRAPHIE | SELENOGRAPHIE

	English	Français	Deutsch
1	selenography	sélénographie *f*	Selenographie *f*
2	Moon	Lune *f*	Mond *m*
3	lunar of the Moon	lunaire de la Lune	lunar Mond
4	cislunar	cislunaire	cislunar innerhalb der Mondbahn
5	circumlunar	circumlunaire	zirkumlunar
6	translunar	translunaire	translunar jenseits der Mondbahn
8	selenographic	sélénographique	selenographisch
9	selenographic coordinates	coordonées sélénographiques *fpl*	selenographische Koordinaten *fpl*
10	selenographic longitude	longitude sélénographique *f*	selenographische Länge *f*
11	selenographic latitude	latitude sélénographique *f*	selenographische Breite *f*
12	Earth-Moon system	système Terre-Lune *m*	Erde-Mond-System *n*
13	distance of the Moon from the Earth Earth-Moon distance	distance de la Lune à la Terre *f* distance Terre-Lune *f*	Entfernung des Mondes von der Erde *f* Erde-Mond Entfernung *f*
14	bound rotation synchronous rotation	rotation liée *f* rotation synchrone *f*	gebundene Rotation *f*
15	front side of the Moon near side of the Moon visible side of the Moon lunar frontside [nearside] Earthside hemisphere	face visible de la Lune *f* face antérieure de la Lune *f* hémisphère visible *m* hémisphère proche *m*	der Erde zugewandte Mondseite *f* sichtbare Seite des Mondes *f* Vorderseite des Mondes *f* (erd)zugewandte Hemisphäre [Mondseite] *f*
16	far side of the Moon rear side of the Moon Moon's far side reverse side of the Moon invisible side averted hemisphere	face cachée de la Lune *f* face opposée de la Lune *f* face arrière de la Lune *f* face invisible de la Lune *f*	Mondrückseite *f* Rückseite des Mondes *f* erdabgewandte Mondseite *f* unsichtbare Seite [Hemisphäre] des Mondes *f* (erd)abgewandte Hemisphäre *f*
17	polar region	région polaire *f*	Polargebiet *n*
18	lunation	lunaison *f*	Lunation *f*
19	phase of the Moon change of the Moon	phase de la Lune *f*	Mondphase *f*
20	waxing moon	Lune croissante *f* Lune jeune *f*	zunehmender Mond *m*

selenografía *f*	selenografia *f*	селенография *f*	1
Luna *f*	Lua *f*	Луна *f*	2
lunar	lunar da Lua	лунный	3
cislunar	cislunar	предлунный цизлунный передлунный	4
circunlunar	circunlunar circum-lunar	окололунный	5
translunar	translunar	за орбитой Луны транслунный залунный	6
selenográfico	selenográfico	селенографический	8
coordenadas selenográficas *fpl*	coordenadas selenográficas *fpl*	селенографические координаты *fpl*	9
longitud selenográfica *f*	longitude selenográfica *f*	селенографическая долгота *f*	10
latitud selenográfica *f*	latitude selenográfica *f*	селенографическая широта *f*	11
sistema Tierra-Luna *m*	sistema Terra-Lua *m*	система Земля-Луна *f*	12
distancia de la Luna a la Tierra *f* distancia Tierra-Luna *f*	distância de Lua à Terra *f* distância Terra-Lua *f*	расстояние от Земли до Луны *n*	13
rotación ligada *f* rotación forzada *f*	rotação forçada [ligada] *f* rotação síncrona [vinculada] *f*	вынужденное вращение *n* несвободное вращение *n*	14
cara visible de la Luna *f* faz visible de la Luna *f* lado visible de la Luna *m* hemisferio visible *m*	face frontal [visível] da Lua *f* lado frontal [visível] da Lua *m* hemisfério voltado para a Terra *m* hemisfério do lado da Terra *m*	видимая сторона Луны *f* передняя сторона Луны *f* ближайшая сторона Луны *f* обращённая к Земле сторона Луны *f* видимое полушарие Луны *n*	15
cara oculta de la Luna *f* parte posterior de la Luna *f* lado alejado de la Luna *m* faz invisible *f* cara opuesta de la Luna *f* hemisferio invisible *m*	lado oposto [distante] da Lua *m* face traseira da Lua *f* face [lado] não-visível *f*[*m*] face oculta da Lua *f* hemisfério não-visível *m* lado traseiro da Lua *m*	обратная сторона Луны *f* задняя сторона Луны *f* невидимая сторона Луны *f*	16
región polar *f*	região polar *f*	полярная область *f*	17
lunación *f*	lunação *f*	лунация *f*	18
fase de la Luna *f*	fase da Lua *f* mudança da Lua *f*	фаза Луны *f* лунная фаза *f*	19
Luna creciente *f*	Lua crescente *f*	растущая [нарастающая, прибывающая] Луна *f*	20

	SELENOGRAPHY	SÉLÉNOGRAPHIE	SELENOGRAPHIE
21	waning moon	Lune décroissante *f* décours de la Lune *m* Lune vieille *f*	abnehmender Mond *m*
22	new moon	nouvelle Lune *f*	Neumond *m*
23	crescent crescent moon crescent phase of the Moon	croissant lunaire *m*	Mondsichel *f*
24	lunar cusps	cornes lunaires *fpl* cornes de la Lune *fpl*	Hörnerspitzen des Mondes *fpl*
25	first quarter waxing half-moon	premier quartier *m* Lune en premier quartier *f*	erstes Viertel *n* erstes Mondviertel *n*
26	quarter half moon	quartier *m* demi-lune *f*	Mondviertel *n* Halbmond *m*
27	full moon	pleine Lune *f*	Vollmond *m*
28	phase defect	défaut de la phase *m*	Phasendefekt *m*
29	syzygies *pl* syzygy *sg*	syzygies *fpl* syzygie *f*	Syzygien *fpl* Syzygie *f*
30	last quarter third quarter waning half-moon	dernier quartier *m* Lune en dernier quartier *f*	letztes Viertel *n* letztes Mondviertel *n*
31	moon's age age of the moon	âge de la Lune *m*	Alter des Mondes *n* Mondalter *n*
32	lunar day	jour lunaire *m*	Tag auf dem Monde *m* Mondtag *m*
33	lunar night	nuit lunaire *f*	Mondnacht *f* Nacht auf dem Monde *f*
34	sunward side illuminated side sunlit side	côté ensoleillé *m* face éclairée *f* hemisphère illuminé *m*	beleuchtete Mondseite *f* Tagseite des Mondes *f*
35	earthshine earthshine on the Moon earthlight	lumière cendrée *f*	aschgraues Licht *n*
36	Moon's orbit lunar orbit	orbite de la Lune *f* orbite lunaire *f*	Mondbahn *f*
37	inclination to the lunar equator	inclinaison sur l'équateur lunaire *f*	Neigung gegen den Mondäquator *f*
38	perigee	périgée *m*	Erdnähe *f* Perigäum [Perigeum] *n*

Luna decreciente *f* Luna menguante *f*	Lua minguante *f*	убывающая Луна *f* Луна на ущербе *f* ущерб(лён)ная Луна *f*	21
Luna nueva *f* novilunio *m*	Lua nova *f* novilúnio *m*	новолуние *n* новая Луна *f*	22
creciente (lunar) *f* fase creciente de la Luna *f*	crescente (lunar) *m* fase crescente da Lua *f* Lua crescente *f*	серп Луны *m* лунный серп *m* серповидная фаза Луны *f*	23
cúspides de la Luna *fpl*	cúspides lunares *fpl* cornos lunares *mpl*	рога серпа Луны *mpl* рога Луны *mpl*	24
cuarto creciente *m* primer cuarto *m*	quarto crescente *m* primeiro quarto *m*	первая четверть *f* Луна в первой четверти *f*	25
cuarto *m* media luna *f*	quarto *m* meia-lua *f*	полумесяц *m* четверть *f* половина диска Луны *f*	26
luna llena *f* plenilunio *m*	lua cheia *f* plenilúnio *m*	полная фаза Луны *f* полнолуние *n* полная Луна *f*	27
defecto de fase *m*	defeito de fase *m*	дефект фазы *m*	28
sicigias *fpl* sicigia *f*	sizígias *fpl* sizígia *f*	сизигии *fpl*	29
cuarto menguante *m* último cuarto *m* tercer cuarto *m*	último quarto *m* quarto minguante *m* terceiro quarto *m*	последняя четверть *f* третья четверть *f*	30
edad de la Luna *f*	idade da Lua *f*	возраст Луны *m*	31
día lunar *m*	dia lunar *m*	день на Луне *m* лунные сутки *pl* лунный день *m*	32
noche lunar *f*	noite lunar *f*	лунная ночь *f* ночь на Луне *f*	33
faz iluminada de la Luna *f* lado iluminado *m* cara iluminada *f*	lado voltado para o Sol *m* face iluminada *f* lado iluminado *m*	обращёная к Солнцу сторона *f* сторона освещённая Солнцем *f* дневная сторона *f*	34
luz cenicienta *f*	luz cinzenta *f* brilho terreal (na Lua) *m*	пепельный свет Луны *m*	35
órbita de la Luna *f* órbita lunar *f*	órbita lunar *f* órbita da Lua *f*	орбита Луны *f* лунная орбита *f*	36
inclinación sobre el ecuador lunar *f*	inclinação sobre o equador lunar *f*	наклон к лунному экватору *m*	37
perigeo *m*	perigeu *m*	перигей *m*	38

39	apogee	apogée *m*	Erdferne *f* Apogäum [Apogeum] *n*
40	perilune pericynthion periselene periselenium	périsélène *m* périlune *f*	Mondnähe *f* Periselen *n*
41	apolune apocynthion aposelene aposelenium	aposélène *m* apolune *f*	Mondferne *f* Aposelen *n*
42	month	mois *m*	Monat *m*
43	anomalistic month anomalistic revolution	mois anomalistique *m* révolution anomalistique *f*	anomalistischer Monat *m* anomalistischer Umlauf *m*
44	tropical revolution tropical month	révolution tropique *f* mois tropique *m*	tropischer Umlauf *m* tropischer Monat *m*
45	sidereal revolution sidereal month	révolution sidérale *f* mois sidéral *m*	siderischer Umlauf *m* siderischer Monat *m*
46	synodic(al) revolution synodic(al) month lunar month	révolution synodique *f* mois synodique *m*	synodischer Umlauf *m* synodischer Monat *m*
47	draconic revolution draconic month nodical month draconic period	révolution draconitique *f* mois draconitique *m*	drakonitischer Umlauf *m* drakonitischer Monat *m*
48	libration	libration *f*	Libration *f*
49	libration in longitude	libration en longitude *f*	Libration in (selenographischer) Länge *f*
50	libration in latitude	libration en latitude *f*	Libration in (selenographischer) Breite *f*
51	geometrical libration optical libration	libration géométrique *f* libration optique *f*	optische Libration *f*
52	diurnal libration parallactic libration	libration diurne *f* libration parallactique *f*	tägliche Libration *f* parallaktische Libration *f*
53	physical libration	libration physique *f*	physische Libration *f*
54	lunar disk lunar [Moon's] disc	disque lunaire *m*	Mondscheibe *f*
55	lunar landscape	paysage lunaire *m*	Mondlandschaft *f*

SELENOGRAFÍA	SELENOGRAFIA	СЕЛЕНОГРАФИЯ	
apogeo *m*	apogeu *m*	апогей *m*	39
perilunio *m* pericintio *m*	pericíntio *m* perilúnio *m* perisselênio *m*	периселений *m* перилуний *m*	40
apolunio *m* apocintio *m*	apolúnio *m* apocíntio *m* aposselênio *m*	апоселений *m* аполуний *m*	41
mes *m*	mês *m*	месяц *m*	42
mes anomalístico *m* revolución anomalística *f*	mês anomalístico *m* revolução anomalística *f*	аномалистический месяц *m* аномалистический период обращения *m*	43
revolución trópica *f* mes trópico *m*	revolução trópica *f* mês trópico *m*	тропический период обращения *m* тропический месяц *m*	44
revolución sideral *f* período sidéreo *m* mes sidéreo *m*	revolução sideral *f* mês sideral *m*	звездный период обращения *m* сидерическое обращение *n* звёздный месяц *m* сидерический месяц *m*	45
revolución sinódica *f* período sinódico *m* mes lunar *m*	revolução sinódica *f* mês sinódico *m* mês lunar *m*	синодический период обращения *m* синодическое обращение *n* синодический месяц *m* лунный месяц *m*	46
revolución draconítica *f* mes draconítico *m*	revolução dracônica *f* mês nódico *m* mês dracônico *m*	драконическое обращение *n* драконический месяц *m* драконический период обращения *m*	47
libración *f*	libração *f*	либрация *f*	48
libración en longitud *f*	libração em longitude *f*	либрация по долготе *f*	49
libración en latitud *f*	libracão em latitude *f*	либрация по широте *f*	50
libración geométrica *f* libración óptica *f*	libração geométrica *f* libração ó(p)tica *f*	геометрическая либрация *f* оптическая либрация *f*	51
libración diurna *f* libration parallactique *f*	libração diurna *f* libração paralá(c)tica *f*	суточная либрация *f* параллактическая либрация *f*	52
libración física *f*	libração física *f*	физическая либрация *f*	53
disco lunar *m* disco de la Luna *m*	disco lunar *m* disco da Lua *m*	лунный диск *m* диск Луны *m*	54
paisaje lunar *m*	paisagem lunar *f* panorama lunar *m*	лунный пейзаж *m* лунный ландшафт *m*	55

	SELENOGRAPHY	SÉLÉNOGRAPHIE	SELENOGRAPHIE
56	eclipse of the Moon lunar eclipse	éclipse de Lune *f*	Mondfinsternis *f*
57	eclipsed Moon	Lune éclipsée *f*	verfinsterter Mond *m*
58	Earth's shadow umbra of the Earth	ombre de la Terre *f*	Erdschatten *m*
59	density of the shadow shadow density	densité de l'ombre *f*	Schattendichte *f* Stärke des Schattens *f*
60	Moon's right ascension	ascension droite de la Lune *f*	Rektaszension des Mondes *f*
61	Sun's right ascension	ascension droite du Soleil *f*	Rektaszension der Sonne *f*
62	Moon's declination	déclinaison de la Lune *f*	Deklination des Mondes *f*
63	Sun's declination	déclinaison du Soleil *f*	Deklination der Sonne *f*
64	penumbra of the Earth	pénombre de la Terre *f*	Halbschatten der Erde *m*
65	penumbral lunar eclipse	éclipse de Lune par la pénombre *f*	Halbschattenfinsternis *f*
66	total eclipse of the Moon	éclipse totale de Lune *f*	totale Mondfinsternis *f*
67	partial eclipse of the Moon	éclipse partielle de Lune *f*	partielle Mondfinsternis *f*
68	opposition in right ascension	opposition en ascension droite *f*	Opposition in Rektaszension *f*
69	Sun's apparent radius Sun's apparent semi-diameter	rayon apparent du Soleil *m* demi-diàmetre apparent du Soleil *m*	scheinbarer Halbmesser der Sonne *m*
70	Moon's apparent radius Moon's apparent semi-diameter	rayon apparent de la Lune *m* demi-diamètre apparent de la Lune *f*	scheinbarer Halbmesser des Mondes *m*
71	position angle	angle de position *m*	Positionswinkel *m*
72	hourly motion	mouvement horaire *m*	stündliche Bewegung *f* stündliche Änderung *f*
73	circumstances of the eclipse phases of the eclipse	phases de l'éclipse *fpl*	Bedingungen der Finsternis *fpl* Verlauf der Finsernis *m* Phasen der Finsternis *fpl*
74	beginning of the eclipse	commencement de l'éclipse *m*	Beginn der Finsternis *m* Anfang der Finsternis *m*
75	Moon enters penumbra	entrée de la Lune dans la pénombre *f*	Entritt des Mondes in den Halbschatten *m*
76	Moon enters umbra	entrée de la Lune dans l'ombre *f*	Eintritt des Mondes in den Kernschatten *m*

SELENOGRAFÍA	SELENOGRAFIA	СЕЛЕНОГРАФИЯ	
eclipse de Luna *m* eclipse lunar *m*	eclipse da Lua *m* eclipse lunar *m*	затмение Луны *n* лунное затмение *n*	56
Luna eclipsada *f*	Lua eclipsada *f*	затмившаяся Луна *f* Луна в земной тени *f*	57
sombra de la Tierra *f*	sombra da Terra *f*	земная тень *f* тень Земли *f*	58
densidad de la sombra *f*	densidade da sombra *f*	плотность тени *f*	59
ascensión recta de la Luna *f*	ascensão reta da Lua *f*	прямое восхождение Луны *n*	60
ascensión recta del Sol *f*	ascensão reta do Sol *f*	прямое восхождение Солнца *n*	61
declinación de la Luna *f*	declinação da Lua *f*	склонение Луны *n*	62
declinación del Sol *f*	declinação do Sol *f*	склонение Солнца *n*	63
penumbra de la Tierra *f*	penumbra da Terra *f*	земная полутень *f* полутень Земли *f*	64
eclipse penumbral de Luna *m* eclipse de Luna por la penumbra *m*	eclipse penumbral da Lua *m*	полутеневое лунное затмение *n*	65
eclipse total de Luna *m*	eclipse total da Lua *m*	полное лунное затмение *n* полное затмение Луны *n*	66
eclipse parcial de Luna *m*	eclipse parcial da Lua *m*	частное лунное затмение *n* частное затмение Луны *n*	67
oposición en ascensión recta *f*	oposição em ascensão reta *f*	противостаяние по нрямому восхождению *n*	68
radio aparente del Sol *m* semidiámetro aparente del Sol *m*	raio aparente do Sol *m* semidiâmetro aparente do Sol *m*	видимый радиус Сольца *m* видимый полудиаметр Солнца *m*	69
radio aparente de la Luna *m* semidiámetro aparente de la Luna *m*	raio aparente da Lua *m* semidiâmetro aparente da Lua *m*	видимый радиус Луны *m* видимый полудиаметр Луны *m*	70
ángulo de posición *m*	ângulo de posição *m*	позиционный угол *m* угол положения *m*	71
movimiento horario *m*	movimento horário *m*	часовое изменение *n* часовое движение *n*	72
fases del eclipse *fpl* circunstancias del eclipse *fpl*	fases do eclipse *fpl* circunstâncias do eclipse *fpl*	обстоятельства затмения *npl*	73
principio del eclipse *m* comienzo del eclipse *m*	começo do eclipse *m* princípio do eclipse *m*	начало затмения *n*	74
entrada de la Luna en la penumbra *f*	entrada da Lua na penumbra *f*	вступление Луны в полутень *n*	75
entrada de la Luna en la sombra *f*	entrada de Lua na sombra *f*	вступление Луны в тень *n*	76

77	total eclipse begins beginning of totality beginning of the total eclipse	début de la totalité *m* la totalité commence	Beginn der totalen Verfinsterung *m* Beginn der Totalität *m* Anfang der Totalität *m* Beginn der totalen Finsternis *m*
78	middle of the eclipse	milieu de la totalité *m* milieu de l'éclipse *m*	Mitte der Finsternis *f* gröste Phase *f*
78a	partial phase (of the eclipse)	phase partielle (de l'éclipse) *f*	partielle Phase (der Finsternis) *f*
79	magnitude of the eclipse	grandeur de l'éclipse *f*	Grösse der Finsternis *f*
79a	partial eclipse	éclipse partielle *f*	partielle Finsternis *f*
80	total eclipse ends end of totality	fin de la totalité *m*	Ende der totalen Verfinsterung *n* Ende der Totalität *n*
81	Moon leaves umbra	sortie de l'ombre *f*	Austritt des Mondes aus dem Kern-schatten *m*
82	Moon leaves penumbra	sortie de la pénombre *f*	Austritt des Mondes aus dem Halb-schatten *m*
83	end of the eclipse	fin de l'éclipse *m*	Ende der Finsternis *n*
83a	end of partial eclipse	fin de l'éclipse partielle *m*	Ende der partiellen Finsternis *n*
84	first contact	premier contact *m*	erste Berührung *f* erster Kontakt *m*
85	last contact	dernier contact *m*	letzte Berührung *f* letzter Kontakt *m*
86	eclipse is visible in...	éclipse visible à...	Finsternis ist sichtbar in...
87	eclipse is invisible in ...	éclipse invisible à...	Finsternis ist unsichtbar in ...
88	visibility of eclipse	visibilité de l'éclipse *f*	Sichtbarkeit der Finsternis *f*
89	zone of visibility (of eclipse)	zone de visibilité (de l'éclipse) *f*	Sichtbarkeitsgebiet (der Finsternis) *f*
90	shadow cone	cône d'ombre *m*	Schattenkegel *m*
91	Earth's terminator	terminateur de la Terre *m*	Erdterminator *m* Terminator des Erdschattens *m*
92	isophote isophotal contour	isophote *f* courbe isophote *f*	Isophote *f* Isophotenlinie *f*
93	enlargement of the Earth's shadow	élargissement de l'ombre de la Terre *m*	Vergrösserung des Erdschattens *f*
94	occultation of a star by the Moon lunar occultation occultation lunaire *f*	occultation d'une étoile par la Lune *f* occultation lunaire *f*	Sternbedeckung durch den Mond *f*
95	limb of the Moon lunar limb	bord de la Lune *m* bord lunaire *m*	Mondrand *m* Rand des Mondes *m*
96	bright limb	bord éclairé *m*	heller Mondrand *m* heller Rand *m*

SELENOGRAFÍA	SELENOGRAFIA	СЕЛЕНОГРАФИЯ	
principio de la totalidad *m* comienzo del eclipse total *m* eclipse total comienza	começo da totalidade *m* começo do eclipse total *m* o eclipse total começa	начало полного затмения *n* полное затмение начинается	77
medio del eclipse *m* fase máxima del eclipse *f*	meio do eclipse *m* fase máxima do eclipse *f*	середина затмения *f* момент наибольшей фазы *m*	78
fase parcial (del eclipse) *f*	fase parcial (do eclipse) *f*	частная фаза (затмения) *f*	78a
magnitud del eclipse *f*	grandeza do eclipse *f*	величина наибольшей фазы *f*	79
eclipse parcial *m*	eclipse parcial *m*	частное затмение *n*	79a
fin de la totalidad *m* fin del eclipse total *m*	fim do eclipse total *m* fim da totalidade *m*	конец полного затмения *m*	80
salida de la Luna de la sombra *f*	saída da Lua da sombra *f*	выход Луны из тени *m*	81
salida de la Luna de la penumbra *f*	saída da Lua da penumbra *f*	выход Луны из полутени *m* конец полутеневого затмения *m*	82
fin del eclipse *m*	fim do eclipse *m*	конец затмения *m*	83
fin del eclipse parcial *m*	fim do eclipse parcial *m*	конец частного затмения *m*	83a
primer contacto *m*	primeiro conta(c)to *m*	первый контакт *m*	84
último contacto *m*	último conta(c)to *m*	четвёртый контакт *m* последний контакт *m*	85
el eclipse es visible en...	o eclipse é visível em...	затмение видимо в ...	86
el eclipse es invisible en ...	o eclipse é invisível em ...	затмение невидимо в ...	87
visibilidad del eclipse *f*	visibilidade do eclipse *f*	видимость затмения *f*	88
zona de visibilidad (del eclipse) *f*	zona de visibilidade (do eclipse) *f*	область видимости (затмения) *f*	89
cono de sombra *m*	cone de sombra *m*	конус тени *m*	90
terminador de la Tierra *m* terminador terrestre *m*	terminadouro da Terra *m* terminadouro terrestre *m*	граница земной тени *f* земной терминатор *m*	91
isofota *f* curva isofota *f*	isofota *f* contorno isofótico [de isofotas] *m*	изофота *f* кривая равной освещённости [яркости] *f*	92
alargamiento de la sombra de la Tierra *m*	alargamento da sombra da Terra *m*	увеличение земной тени *n* расширение земной тени *n*	93
ocultación de una estrella por la Luna *f* ocultación lunar *f*	ocultação de uma estrela pela Lua *f* ocultação lunar *f*	покрытие звезды Луной *n*	94
borde de la Luna *m* borde lunar *m*	limbo [bordo] da Lua *m* bordo lunar *m*	край Луны *m* лимб Луны *m*	95
borde iluminado *m*	limbo iluminado [brilhante] *m* bordo claro *m*	светлый край *m* светлый лимб *m*	96

	SELENOGRAPHY	SÉLÉNOGRAPHIE	SELENOGRAPHIE
97	dark limb	bord obscur *m*	dunkler Mondrand *m* dunkler Rand *m*
98	immersion disappearance	immersion *f* disparition *f*	Eintritt des Sternes *m* Verschwinden *n*
99	emersion desoccultation de-occultation reappearance	émersion *f* réapparition *f*	Austritt des Sternes *m*
100	grazing occultation	occultation rasante *f*	streifende Bedeckung *f*
101	appulse	appulse *m*	Berührung *f*
102	occultation predictions	prédiction des occultations *f*	Voraussage der Sternbedeckungen *f* Bedeckungsvorausberechnungen *fpl*
103	visibility of an occultation	visibilité d'une occultation *f*	Sichtbarkeit der (Stern)Bedeckung *f*
104	limiting latitudes of the visibility of an occultation	latitudes limites de la visibilité d'une occultation *fpl*	Breitengrenzen der Sichtbarkeit einer Bedeckung *fpl* Grenzbreiten der Sichtbarkeit einer Bedeckung *fpl*
105	northern limiting latitude	latitude limite nord *f* limite boréale *f*	nördliche Grenzbreite *f* nördliche Breitengrenze *f*
106	southern limiting latitude	latitude limite sud *f* limite australe *f*	südliche Grenzbreite *f* südliche Breitengrenze *f*
107	occultation zone zone of occultation	zone d'occultation *f*	Sternbedeckungszone *f*
108	reduction of occultations	réduction des occultations *f*	Reduktion der Sternbedeckungen *f* Reduktion der Bedeckungen *f*
109	selenodesy	sélénodésie *f* géodésie lunaire *f*	Selenodäsie *f* Mondgeodäsie *f*
110	selenodetic	sélénodétique	selenodätisch
111	mapping of the Moon lunar mapping lunar charting lunar cartography	cartographie lunaire *f* cartographie de la Lune *f*	Mondkartographie *f* Kartographierung des Mondes *f* Mondmappierung *f*
112	lunar globe	globe lunaire *m*	Mondkugel *f* Mondglobus *m*
113	limb zone marginal zone	zone marginale *f*	Randzone *f*
114	central zone	zone centrale *f*	Zentralzone *f*
115	stereoscopic method	méthode stéréoscopique *f*	stereoskopische Methode *f*
116	shape of the Moon figure of the Moon	figure de la Lune *f*	Mondgestalt *f*

borde o(b)scuro *m*	limbo escuro *m* bordo escuro *m*	тёмный край *m* тёмный лимб *m*	97
desaparición *f* immersión *f*	imersão *f* desaparecimento *m* desaparição *f*	закрытие звезды *n* исчезновение звезды *n* иммерсия *f*	98
emersión *f* reaparición *f*	emersão *f* reaparecimento *m* reaparição *f*	появление звезды *n* открытие звезды *n*	99
ocultación rasante *f*	ocultação rasante *f*	касательное покрытие *n*	100
apulso *m*	apulso *m*	аппульс *m*	101
predicción de ocultaciones *f*	predição das ocultações *f* previsão das ocultações *f*	предвычисление покрытий *n* предсказание покрытий *n*	102
visibilidad de una ocultación *f*	visibilidade de uma ocultação *f*	видимость покрытия *f*	103
latitudes límites de visibilidad de una ocultación *fpl*	latitudes limites da visibilidade de uma ocultação *fpl*	предельные широты видимости покрытия *fpl*	104
latitud límite boreal *f* latitud límite norte *f*	latitude limite norte *f*	северная предельная широта *f*	105
latitud límite austral *f* latitud límite sur *f*	latitude limite sul *f*	южная предельная широта *f*	106
zona de ocultación *f*	zona de ocultação *f*	зона покрытия *f* пояс видимости покрытия *m*	107
reducción de ocultaciones *f*	redução das ocultações *f*	редукция наблюдений покрытий *f* обработка покрытий *f*	108
selenodesia *f* geodesia lunar *f*	selenodésia *f* geodésia [geodesia] lunar *f*	селенодезия *f* лунная геодезия *f*	109
selenodésico	selenodésico	селенодезический	110
cartografía lunar *f* mapeo de la Luna *m* cartografía de la Luna *f*	mapeamento lunar *m* mapeamento da Lua *m* cartografia lunar *f*	лунное картирование *n* составление лунной карты *n* картографирование Луны *n* лунная картография *f*	111
globo lunar *m*	globo lunar *m*	лунный шар *m* лунный глобус *m*	112
zona del borde [del limbo] *f* zona marginal *f*	zona do limbo *f* zona marginal *f*	краевая зона (Луны) *f*	113
zona central *f*	zona central *f*	центральная зона *f*	114
método estereoscópico *m*	método estereoscópico *m*	стереоскопический метод *m*	115
forma de la Luna *f* figura de la Luna *f*	forma da Lua *f* figura da Lua *f*	фигура Луны *f*	116

117	selenoid	sélénoïde *m*	Selenoid *n*
118	lunar bulge	bourrelet lunaire *m* renflement lunaire *m*	Mondwulst *m*
119	tidal bulge	bourrelet de marée *m* renflement dû aux marées *m*	Gezeitenwulst *m* Gezeitenberg *m*
120	tidal interaction	interaction due aux marées *f*	Gezeitenwechselwirkung *f*
121	lunar equator	équateur lunaire *m*	Mondäquator *m*
122	equator of illumination	équateur d'illumination *m*	Beleuchtungsequator *m*
123	lunar terminator	terminateur lunaire *m*	Mondterminator *m*
124	sunrise terminator dawn terminator morning terminator	terminateur du matin *m* terminateur matinal *m*	Morgenterminator *m*
125	sunset terminator dusk terminator evening terminator	terminateur du soir *m*	Abendterminator *m*
126	sunlit hemisphere	hémisphère ensoleilé *m* hémisphère éclairé *m* hémisphère illuminé *m*	beleuchtete Hemisphäre *f*
127	subsolar point	point subsolaire *m*	subsolarer Punkt *m*
128	sub-Earth point centre of the Moon's apparent disk	point subterrestre *m* centre du disque apparent de la Lune *m*	subterrestrischer Punkt *m*
129	sublunar point	point sublunaire *m*	sublunarer Punkt *m*
130	substellar point	point substellaire *m*	substellarer Punkt *m*
131	selenocentric lunicentric	sélénocentrique	selenozentrisch
132	selenocentric position	position sélénocentrique *f*	selenozentrische Lage *f*
133	lunar map map of the Moon lunar chart	carte sélénographique *f* carte lunaire *f* carte de la Lune *f*	Mondkarte *f*
134	physical map of the Moon (photometric map, colorimetric map, polarization map)	carte lunaire physique *f* (carte photométrique, carte colorimétrique, carte polarimétrique)	physikalische Mondkarte *f* (photometrische Karte, kolorimetrische Karte, polarimetrische Karte)

selenoide *m*	selenóide *m*	селеноид *m*	117
abultamiento lunar *m* protuberancia lunar *f*	protuberância lunar *f* bojo lunar *m*	лунный горб *m* лунный выступ *m*	118
abultamiento debido a las mareas *m* protuberancia de las mareas *f*	protuberância da maré *f* bojo devido à maré *m*	приливный горб *m* приливный выступ *m* приливное вздутие *n*	119
interacción de marea *f*	interação de maré *f*	приливное взаимодействие *n*	120
ecuador lunar *m*	equador lunar *m*	лунный экватор *m*	121
ecuador de iluminación *m*	equador de iluminação *m*	экватор освещения *m*	122
terminador lunar *m*	terminadouro lunar *m*	лунный терминатор *m*	123
terminador matutino *m*	terminadouro matutino *m* terminadouro de nascente *m* terminadouro da manhã *m*	утренний терминатор *m* терминатор на рассвете *m* терминатор при восходе Солнца *m*	124
terminador vespertino *m*	terminadouro vespertino *m* terminadouro do poente *m* terminadouro do anoitecer *m*	сумеречный терминатор *m* вечерний терминатор *m* терминатор при заходящем Солнце *m*	125
hemisferio iluminado *m*	hemisfério iluminado *m*	освещённая полусфера [сторона] *f* полушарие, освещённое Солнцем *f* сторона, обращённая к Солнцу *f*	126
punto subsolar *m*	ponto subsolar *m*	подсолнечная точка *f*	127
punto subterrestre *m* centro del disco lunar aparente *m*	ponto subterrestre *m* centro do disco aparente da Lua *m*	подземная точка *f* центр видимого диска Луны *m*	128
punto sublunar *m*	ponto sublunar *m*	подлунная точка *f* географическое место Луны *n*	129
punto subestelar *m*	ponto subestelar *m*	подзвёздная точка *f* (селенографическое, географическое место звезды)	130
selenocéntrico lunicéntrico	selenocêntrico lunicêntrico	селеноцентрический	131
posición selenocéntrica *f*	posição selenocêntrica *f*	селеноцентрическое положение *n*	132
mapa lunar *m* mapa de la Luna *m* carta lunar *f*	mapa lunar *m* mapa da Lua *m* carta lunar *f*	карта Луны *f* лунная карта *f*	133
mapa físico lunar *m* carta física lunar *f* (mapa fotométrico, mapa colorimétrico, mapa polarimétrico)	carta física lunar *f* (carta fotométrica, carta colorimétrica, carta polarimétrica) mapa físico lunar *m*	физическая карта Луны *f* (фотометрическая карта, колориметрическая карта, поляризационная карта)	134

135	lunar thermal map	carte lunaire thermique *f*	thermische Mondkarte *f*
135a	radio map radio chart	carte radio *f* carte radioélectrique *f*	Radiokarte *f*
136	lunar atlas atlas of the Moon Moon atlas	atlas lunaire *m* atlas de la Lune *m*	Mondatlas *m* Atlas des Mondes *m*
137	lunar photographic map photographic map of the Moon	carte lunaire photographique *f*	photographische Mondkarte *f*
138	lunar topographic map	carte lunaire topographique *f*	topographische Mondkarte *f*
139	isolevel line isolevel contour isohypse	courbe de niveau *f* courbe d'égale hauteur *f* isohypse *f*	Höhengleiche *f* Höhen(schicht)linie *f* Isohypse *f*
140	vertical difference	différence verticale *f*	Höhenunterschied *m*
141	isolevel map hypsometric map	carte altimètrique *f* carte hypsomètrique *f* carte en courbes de niveau *f*	Höhenlinienkarte *f* Höhenschichtenkarte *f* Höhenkarte *f*
142	hypsometric level hypsographic curve	niveau hypsométrique *m* courbe hypsographique *f*	Höhenniveau *n* hypsographische Kurve *f*
143	gravimetric map gravity contour map	carte gravimétrique *f*	gravimetrische Karte *f*
144	lunar gravitational anomaly	anomalie de la (force de) pesanteur lunaire *f*	lunare Schwereanomalie *f*
145	mascon lunar mass concentration local mass concentration lunar mascon	concentration de masses lunaires *f* mascon *m*	Maskon *n* lunare Massenkonzentration *f* Mascon *n*
146	lunar topography	topographie lunaire *f*	Mondtopographie *f*
147	lunar morphology lunar features morphology	morphologie lunaire *f*	Mondmorphologie *f*
148	lunar features lunar formations surface markings	configurations lunaires *fpl* formations lunaires *fpl* caractères lunaires *mpl* caractéristiques lunaires *fpl* taches lunaires *fpl*	Oberflächenformationen des Mondes *fpl* Mondformationen *fpl* Merkmale der Oberfläche *npl*

mapa térmico lunar *m* carta térmica lunar *f*	carta térmica lunar *f* mapa térmico lunar *m*	тепловая карта Луны *f* термическая карта Луны *f*	135
radiomapa *m* mapa radio *m*	radiomapa *m* mapa rádio *m*	радиокарта *f* карта изофот радиоизлучения *f*	135a
atlas lunar *m* atlas de la Luna *m*	atlas lunar *m* atlas da Lua *m*	лунный атлас *m* атлас Луны *m*	136
mapa fotográfico lunar *m* mapa fotográfico de la Luna *m*	mapa fotográfico lunar *m* mapa fotográfico da Lua *m*	фотографическая карта Луны *f*	137
mapa topográfico lunar *m* mapa topográfico de la Luna *m*	mapa topográfico lunar *m* mapa topográfico da Lua *m*	топографическая карта Луны *f*	138
línea de nivel *f* contorno de nivel *m* isohipsa *f*	curva [linha] de nível *f* contôrno de nível *m* isoípsa *f*	изогипса *f* горизонталь *f*	139
diferencia vertical *f*	diferença vertical *f*	различие [расхождение] по высоте *n*	140
mapa de nivel constante *m* mapa hipsométrico *m* carta hipsométrica *f*	mapa de curvas [linhas] de nível *f* mapa hipsométrico *m* carta hipsométrica *f*	гипсометрическая карта *f* карта рельефа [горизонталей] *f* карта высот *f*	141
nivel hipsométrico *m* curva hipsográfica *f*	nível hipsométrico *m* curva hipsográfica *f*	гипсометрический уровень *m* гипсометрическая горизонталь *f*	142
mapa gravimétrico *m*	mapa gravimétrico *m*	гравиметрическая карта *f* карта изолиний силы тяжести *f*	143
anomalía gravitacional lunar *f*	anomalia gravitacional lunar *f*	лунная гравитационная аномалия *f*	144
mascon *m* concentración de masa lunar *f*	mascon *m* concentração de massa lunar *f*	маскон *m* лунный маскон *m* концентрация масс в коре Луны *f* область повышенной концентрации масс *f*	145
topografía lunar *f*	topografia lunar *f*	лунная топография *f* топография Луны *f*	146
morfología lunar *f*	morfologia lunar *f*	лунная морфология *f* морфология лунных образований *f*	147
estructuras lunares *fpl* formaciones lunares *f* detalles lunares *mpl* características lunares *fpl* rasgos lunares *mpl*	formações lunares *fpl* aspectos lunares *mpl* marcas superficiais *fpl* detalhes lunares *mpl*	лунные образования *npl* образования лунной поверхности *npl* характерные черты местности *fpl* подробности лунного рельефа *fpl* особенность лунного рельефа *f* лунные формы *fpl*	148

149	topographic feature topographic unit	caractéristique topographique *f* caractère topographique *m* formation topographique *f*	topographische Formation *f* Terrainbeschaffenheit *f* Geländegestaltung *f*
150	primary feature	formation [caractéristique] primaire *f*	ursprüngliche Formation *f*
151	secondary feature	formation [caractéristique] secondaire *f*	Sekundärformation *f* sekundäre Formation *f*
152	structural forms	formes structurales *fpl*	Strukturformen *fpl* Strukturen *fpl*
153	pre-mare structure	structure prémare *f*	prae-mare Struktur *f*
154	post-mare structure	structure postmare *f*	post-mare Struktur *f*
155	ring form ring structure ring-like structure	formation circulaire *f* forme circulaire *f* structure circulaire *f*	Ringstruktur *f*
156	lunar elevation	surélévation (du terrain) lunaire *f*	Monderhöhung *f* Erhebung [Erhöhung] auf dem Mond *f*
157	lunar depression	dépression lunaire *f*	lunare Depression *f* Mondvertiefung *f*
158	highland depression continental depression	dépression continentale *f*	kontinentale Depression *f* kontinentale Vertiefung *f*
159	linear depression	dépression linéaire *f*	lineare Depression *f* lineare Vertiefung *f*
160	arcuate depression	dépression en arc *f*	mäanderförmige Depression *f* bogenförmige Vertiefung *f*
161	sinuous depression	dépression sinueuse *f*	sinusförmige Vertiefung *f* sinusförmige Depression *f*
162	terraced depression	dépression à terrasses *f*	Terrassendepression *f* terassenförmige Vertiefung *f*
163	valley	vallée *f*	Tal *n*
164	Alpine Valley	vallée des Alpes *f* vallée Alpine *f*	Alpental *n* Alpenquertal *f*
165	rill rille cleft groove furrow rima	rainure *f* sillon *m* crevasse *f* fissure *f*	Rille *f* Riss *m* Furche *f* Graben *m*

detalle topográfico *m* unidad topográfica *f* aspecto topográfico *m* formación topográfica *f*	aspecto topográfico *m* unidade topográfica *f* formação topográfica *f*	топографическая характерная черта *f* топографический элемент *m* характер местности *m*	149
característica [formación] primaria *f* rasgo primario *m*	formação primária *f* marca primária *f*	древняя характерная особенность (рельефа) *f*	150
característica secundaria *f* rasgo secundario *m* formación secundaria *f*	formação secundária *f* marca secundária *f*	поздняя характерная особенность (рельефа) *f* новейшая характерная особенность (рельефа) *f*	151
formas estructurales *fpl*	formas estruturais *fpl*	структурные формы *fpl*	152
estructura premarítima [anterior a los mares] *f*	estrutura pré-mares *f* estrutura pré-marítima *f*	доморская структура *f* п(е)редморская структура *f*	153
estructura postmarítima *f*	estrutura pós-mares [pós-marítima] *f*	послеморская структура *f*	154
forma anular *f* estructura anular *f*	forma anular *f* estrutura anular *f* forma anelar *f*	кольцевая структура *f* кольцевая форма *f*	155
elevación lunar *f*	elevação lunar *f*	лунное возвышение *n* возвышение лунной местности *n*	156
depresión lunar *f*	depressão lunar *f*	лунная впадина *f* лунная депрессия *f* лунное понижение *n*	157
depresión continental *f*	depressão continental *f*	внутриматериковая впадина [депрессия] *f*	158
depresión lineal *f*	depressão linear *f*	линейная депрессия *f* линейное понижение *n*	159
depresión arqueada *f* depresión en arco *f*	depressão arqueada *f*	арковидная депрессия *f* дугообразное понижение *n*	160
depresión sinuosa *f*	depressão sinuosa *f*	извилистая депрессия *f* извилистое понижение *n*	161
depresión aterrazada *f* depresión en terrazas *f*	depressão terraceada *f* depressão em [de] terraços *f*	террасированная депрессия *f* террасовое понижение *n*	162
valle *m*	vale *m*	долина *f*	163
valle Alpino *m*	vale dos Alpes *m* vale Alpino *m*	долина Альп *f*	164
hendidura *f* grieta *f* acanaladura *f* ranura *f* surco *m*	ranhura *f* sulco *m* fenda *f* fratura *f* greta *f* rima *f*	разлом *m* трещина *f* расселина *f* раскол *m* борозда *f* расщелина *f*	165

166	linear rille [rima, groove]	rainure linéaire *f*	lineare Rille *f*
167	sinuous rille [rima, groove]	rainure sinueuse *f*	sinusförmige Rille *f*
168	lunar rays bright rays lunar streaks	auréole *f* traînées *fpl* raies [rayons] lunaires *fpl* [*mpl*]	Mondstrahlen *mpl* Strahlen *mpl* Strahlen auf dem Mond *mpl*
169	tangential ray	raie tangentielle *f* rayon tangentiel *m*	tangentialer Strahl *m*
170	ray system system of rays	système rayonnant *m* système de rayons [de raies] *m*	Strahlensystem *m*
171	ray surface	surface de raies *f* surface de rayons *f*	mit Strahlen bedeckte Oberfläche *f* Strahlenoberfläche *f*
172	inter-ray surface	surface interraies *f*	Oberfläche zwischen den Strahlen *f*
173	surface fracture	fracture de surface *f*	Oberflächenbruch *m*
174	fracture system	système de fractures *m*	Bruchsystem *n*
175	basin	bassin *m* thallassoide *m*	Becken *n* Mulde *f*
176	volcanic dome	dôme volcanique *m*	Vulkankuppe *f* Vulkandom *m*
176a	lava dome	dôme [cône] de lave *m*	Lavakuppe *f* Lavadom *m*
177	lunar dome dome	dôme lunaire *m*	Beule *f* Dom *m* Kuppel *f* kuppelartige Bodenform *f*
178	cluster of domes	amas de dômes *m*	Anhäufung von Domen *f* Anhäufung von Kuppeln *f*
179	caldera basin of volcanic origin	caldère *f* caldeira *f* bassin volcanique *m*	Kaldera *f* Caldera *f* Vulkanbecken *n*
180	explosion caldera	caldeira d'explosion *f*	Explosionskaldera *f* Explosionscaldera *f*
181	collapse caldera	caldeira d'affaissement *f*	Einbruchkaldera *f* Einbruchcaldera *f*
182	subsidence caldera	caldeira de subsidence *f*	Senkungskaldera *f*
183	erosion caldera	caldeira d'erosion *f*	Erosionskaldera *f*
184	rupes scarp	escarpement *m*	Steilwand *f* Steilabfall *m* Böschung *f*

hendidura lineal *f*	ranhura linear *f*	прямолинейная борозда *f*	166
ranura sinuosa *f*	ranhura sinuosa *f*	извилистая борозда *f*	167
rayos lunares *mpl*	raias lunares *fpl* raios lunares *mpl*	лунные лучи *mpl* светлые лучи (на Луне) *mpl*	168
rayo tangencial *m*	raia [raio] tangencial *f*[*m*]	касательный луч *m* тангенциальный луч *m*	169
sistema de rayos *m*	sistema de raias [raios] *m*	лучевая система *f*	170
superficie de rayos *f* superficie rayada *f*	superfície de raias [raios] *f* superfície raiada *f*	лучевая поверхность *f* поверхность, покрытая лучами *f*	171
superficie entre rayos *f*	superfície entre raias [raios] *f*	межлучевая поверхность *f*	172
fractura superficial *f*	fratura superficial *f*	поверхностный разлом *m* поверхностная трещина *f*	173
sistema de fracturas *m*	sistema de fraturas *m*	система разломов *f* система трещин *f*	174
cuenca *f*	bacia *f*	талассоид *m* бассейн *m* котловина *f* впадина *f*	175
domo volcánico *m*	domo vulcânico *m* cúpula vulcânica *f*	вулканический купол *m*	176
domo de lava *m*	domo de lava *m* cúpula de lava *f*	лавовый купол *m* лавовый конус *m*	176a
domo lunar *m*	domo lunar *m*	лунный купол *m* купол *m*	177
conjunto de domos *m* cúmulo de domos *m*	aglomerado de domos *m*	скопление куполов *n*	178
caldera *f*	caldeira *f*	кальдера *f* вулканическая впадина *f*	179
caldera de explosión *f*	caldeira de explosão *f*	кальдера взрыва *f*	180
caldera colapsada *f* caldera de hundimiento *f*	caldeira de colapso *f*	кальдера оседания *f* кальдера обрушения *f*	181
caldera de subsidencia *f*	caldeira de subsidência *f*	кальдера опускания *f*	182
caldera de erosión *f*	caldeira de erosão *f*	эрозионная кальдера *f*	183
escarpa *f* declive *m*	escarpa *f*	крутой склон *m* откос *m* обрыв *m*	184

185	Straight Wall	Mur Droit *m*	Gerade Wand *f* Lange Wand *f*
186	ridge mountain ridge	crête *f* arête *f*	Bergrücken *m*
187	wrinkle ridge	crête de plissement *f*	Falte *f* Faltung (srücken) *f*(*m*)
188	mare ridge	crête de mer *f* crête de mare *f*	Mare-Kamm *m* Bergrücken im Meer *m*
189	highland	haut-plateau *m*	Hochland *n*
190	terra (pl. terrae)	terra *f*	Terra *f*
191	continent	continent *m*	Festland *n* Kontinent *m*
192	plateau (pl. plateaux)	plateau *m*	Hochebene *f*
193	lunar mountains	montagnes lunaires *f*	Mondgebirge *n*
194	mountain chain	chaîne de montagnes *f* cordillère *f*	Gebirgskette *f* Bergkette *f*
195	mountain crest	crête *f*	Gebirgskamm *m*
196	mountain region	région montagneuse *f*	Gebirgsregion *f*
197	lunar crater	cratère lunaire *m*	Mondkrater *m*
198	primary crater	cratère primaire *m*	Primärkrater *m* primärer Krater *m*
199	secondary crater	cratère secondaire *m*	Sekundärkrater *m* sekundärer Krater *m*
200	satellite crater satellitic crater	cratère satellite *m*	Satellitenkrater *m*
201	microcrater	microcratère *m*	Mikrokrater *m*
202	unnamed crater	cratère sans nom *m* cratère non-baptisé *m* cratère non-désigné *m*	namenloser Krater *m* unbenannter Krater *m*
203	flooded crater	cratère rempli *m* cratère comblé *m*	überfluteter Krater *m*
204	buried crater	cratère enseveli *m*	zugeschütteter Krater *m*

Muro Recto *m* Pared Recta *f*	Parede Direita *f* Parede Re(c)ta *f*	Прямая Стена *f*	185
cordillera *f* cresta montañosa *f*	crista (de montanhas) *f* dorsal *f*	горный гребень [кряж] *m* хребет *m*	186
cordillera de plegamiento *f* cresta de plegamiento *f*	cordilheira dobrada *f* crista dobrada [de dobra(mento)] *f*	складчатый хребет *m* складчатый гребень *m*	187
cordillera marina *f* cresta de mar(e) *f*	dorsal marítima [marinha] *f* crista de mar *f*	гребень моря *m* морской хребет *m*	188
tierra alta *f* altiplanicie *f* altiplano *m*	planalto *m* terras altas *fpl* altiplano *m*	нагорье *n* высокогорье *m*	189
terra *f* (pl. terrae)	terra *f* (pl. terras)	терра *f* грунт *m*	190
continente *m*	continente *m*	материк *m* континент *m*	191
meseta *f*	platô *m* plató *m*	плато *n* плоскогорье *n* плоская возвышенность *f*	192
montañas lunares *fpl*	montanhas lunares *fpl*	лунные горы *fpl*	193
cadena montañosa *f* cadena de montañas *f* cordillera *f*	cadeia de montanhas *f* cordilheira *f*	горная цепь *f*	194
cresta de montaña *f*	crista de montanhas [montanhosa] *f*	горный хребет *m*	195
región montañosa *f*	região montanhosa *f*	гористый район *m*	196
cráter lunar *m*	cratera lunar *f*	лунный кратер *m*	197
cráter primario *m*	cratera primária *f*	первичный кратер *m*	198
cráter secundario *m*	cratera secundária *f*	вторичный кратер *m* поздний кратер *m* побочный кратер *m*	199
cráter satélite *m*	cratera satélite *f*	кратер-спутник *m*	200
microcráter *m*	microcratera *f*	микрократер *m* микроскопический кратер *m*	201
cráter innominado *m* cráter sin nombre *m*	cratera sem nome *f* cratera inominada *f*	безымянный кратер *m*	202
cráter inundado *m*	cratera alagada [cheia] *f* cratera inundada *f*	затопленный кратер *m* заполненный кратер *m*	203
cráter sepultado *m* cráter soterrado *m* crater enterrado *m*	cratera enterrada *f*	погруженный кратер *m*	204

205	ancient crater	ancien cratère *m*	alter Krater *m*
206	young crater	cratère jeune *m*	junger Krater *m*
207	pre-mare crater	cratère prémare *m* cratère antérieur aux mers *m*	prae-mare Krater *m*
208	post-mare crater	cratère postmare cratère postérieur aux mers *m* cratère postérieur à la mare *m*	post-mare Krater *m*
209	ghost crater elementary ring	cratère fantôme *m*	Geisterkrater *m* Phantomkrater *m*
210	probable crater	cratère probable *m*	wahrscheinlicher Krater *m*
211	possible crater	cratère possible *m*	möglicher Krater *m*
212	polygonal crater	cratère polygonal *m*	vieleckiger Krater *m*
213	circular crater	cratère circulaire *m*	kreisförmiger Krater *m*
214	rayed crater crater with rays ray crater	cratère à raies *m* cratère rayonnant *m* cratère à rayons *m*	Krater mit Strahlen *m*
215	dark-haloed crater	cratère à halo sombre *m*	Krater mit dunklem Halo *m*
216	concentric crater	cratère concentrique *m*	konzentrischer Krater *m*
217	central crater	cratère central *m*	Zentralkrater *m*
218	outer crater	cratère extérieur *m*	Aussenkrater *m* äusserer Krater *m*
219	dish-like crater cup-shaped crater	cratère en coupe *m* cratère en bol *m*	schüsselförmiger Krater *m* tellerförmiger krater *m*
220	irregular crater	cratère irrégulier *m*	unregelmässiger Krater *m*
221	crater rim	lèvre d'évent *f* bord de cratère *m*	Kraterrand *m*
222	raised rim	rebord élevé *m*	erhöhter Rand *m*
223	raised-rim crater	cratère à rebord élevé *m* cratère à rebord relevé *m*	Krater mit erhöhtem Rand *m*
224	blocky rim	rempart à éboulis *m*	Blöckenrand *m*
225	blocky-rim crater blocky-rimmed crater	cratère à rempart d'eboulis *m*	Krater mit Rand aus Blöcken *m*
226	rimless crater	cratère sans rebord *m*	randloser Krater *m*

cráter antiguo *m* cráter viejo *m*	cratera antiga *f*	древний кратер *m*	205
cráter joven [juvenil, reciente] *m*	cratera jovem [nova] *f*	молодой кратер *m*	206
cráter premarítimo *m* cráter anterior al mar *m*	cratera precedente [anterior] à mar *f* cratera pré-marítima [pré-mares] *f*	кратер, возникший до образования морей *m*	207
cráter postmarítimo *m* cráter posterior al mar *m*	cratera pós-mares [pós-marítima] *f* cratera posterior à mar *f*	кратер, возникший после образования морей *m*	208
cráter fantasma *m* falso cráter *m*	cratera fantasma *f* falsa cratera *f* anel elementar *m*	кратер-призрак *m* кратер-фантом *m* реликтовый кратер *m*	209
cráter probable *m*	cratera provável *f*	вероятный кратер *m*	210
cráter posible *m*	cratera possível *f*	возможный кратер *m*	211
cráter poligonal *m*	cratera poligonal *f*	полигональный кратер *m*	212
cráter circular *m*	cratera circular *f*	круглый кратер *m*	213
cráter radiado *m* cráter con rayos *m*	cratera raiada *f* cratera com raias [raios] *f* cratera de raios *f f*	лучевой кратер *m* лучистый кратер *m*	214
cráter de halo oscuro *m*	cratera com halo escuro *f*	кратер с тёмным кольцом *m*	215
cráter concéntrico *m*	cratera concêntrica *f*	концентрический кратер *m*	216
cráter central *m*	cratera central *f*	центральный кратер *m* внутренний кратер *m*	217
cráter externo *m*	cratera externa *f*	внешний кратер *m*	218
cráter en forma de plato *m* cráter en forma de taza *m*	cratera em forma de prato *f* cratera em forma de taça *f*	кратер с плоским дном *m* чашечный кратер *m*	219
cráter irregular *m*	cratera irregular *f*	неправильный кратер *m*	220
borde del cráter *m*	borda da cratera *f* bordo da cratera *m*	край кратера *m*	221
borde levantado *m*	borda levantada *f* bordo levantado *m*	выступающий вал *m* поднятый край *m*	222
cráter con el borde levantado *m*	cratera com bordos levantados *f* cratera com bordas levantadas *f*	кратер с выступающим валом *m*	223
borde de bloques *m*	bordo com blocos *m* borda com blocos de pedra *f*	вал [край], состоящий из глыб *m*	224
cráter con el borde de bloques *m* cráter bordeado de bloques *m*	cratera de bordos com blocos *f* cratera com bordos formados por blocos *f*	кратер с валом из глыб *m*	225
cráter sin borde *m*	cratera sem bordos *f*	кратер без вала *m*	226

227	crater parameters	paramètres de cratère *mpl*	Kraterparameter *m* Kraterabmessungen *fpl*
228	vertical profile	profil vertical *m*	vertikales Profil *n*
229	crater terrace	terrasse de cratère *f*	Kraterterrasse *f*
230	slope	pente *f*	Abhang *m*
231	moderate slope	pente modérée *f*	flacher Abhang *m*
232	steep slope	pente abrupte *f*	steiler Abhang *m*
233	detrital slope	pente détritique *f*	Schutthang [Geröllabhang] *m*
234	talus scree	talus *m*	Schuttdecke *f*
235	rock debris	débris rocheux *mpl* débris de rochers *mpl*	Felstrümmer *fpl*
236	mass wasting mass wastage	perte de matière *f* dispersion de matière *f* matériaux détritiques *mpl*	Geröllmasse *f* Steinschutt *m*
237	rolling stones	pierres roulantes *fpl* avalanches *fpl*	rollende Steine *mpl*
238	explosion crater	cratère d'explosion *m*	Explosionskrater *m*
239	drainage crater	cratère à écoulement *m*	Dränungskrater *m*
240	impact crater	cratère d'impact *m*	Einschlagskrater *m*
241	crater chain	chaîne de cratères *f*	Kraterkette *f*
242	ghost ring	anneau fantôme *m*	Ringstruktur *f*
243	dimple dimple crater	petit creux *m* petite dépression *f*	Kratergrube *f*
244	summit pit (of a central peak)	cheminée du pic central *f*	Gipfelmulde (des Zentralkegels) *f*
245	cratered surface	surface cratérée *f*	mit Kratern bedeckte Oberfläche *f*
246	crater density crater number-density	densité de cratères *f*	Kraterdichte *f*

parámetros de un cráter *mpl*	parâmetros de uma cratera *mpl*	параметры кратера *mpl*	227
perfil vertical *m*	perfil vertical *m*	вертикальный профиль *f*	228
terraza del cráter *f*	terraço de (uma) cratera *m*	кратерная терраса *f*	229
pendiente *f*	declive *m*	склон *m* наклон *m*	230
pendiente moderada *f*	declive moderado *m*	умеренный наклон *m* пологий склон [наклон] *m*	231
pendiente fuerte *f* pendiente empinada *f*	declive abrupto *m*	значительный наклон *m* крутой склон *m*	232
pendiente detrítico [de derrubio] *f*	declive detrítico *m*	склон, покрытый обломками *m*	233
talud *m* detrito *m* derrubio *m*	tálus *m* talude *m* detrito *m*	скат *m* откос *m* осыпь *f*	234
fragmentos de roca *mpl*	fragmentos de rocha *mpl*	обломки пород *mpl*	235
desgaste de masa *m* remoción de masa *f* materiales detríticos *mpl*	desgaste de massa *m* materiais detríticos *mpl*	истощение массы *n* истощение породы *n*	236
piedras rodantes *fpl* cantos rodados *mpl*	brugalhau [burgalhau] *m* pedras roladas *fpl*	катящиеся камни *mpl*	237
cráter de explosión *m*	cratera de explosão *f*	кратер взрывного происхождения *m* взрывной кратер *m*	238
cráter de drenaje *m*	cratera de drenagem *f*	дренированный кратер *m* кратер, образованный при истечении лавы *m*	239
cráter de impacto *m*	cratera de impacto *f*	ударный кратер *m*	240
cadena de cráteres *f*	cadeia de crateras *f*	кратерная цепочка [цепь] *f* цепь кратеров *f*	241
anillo fantasma *m* falso anillo *m*	anel fantasma *m*	кольцо-призрак *n*	242
hoyuelo *m* pequeña depresión *f*	cova *f* pequena depressão *f*	ямочка *f* лунка *f* маленькое углубление *n*	243
hoya de la cumbre *f* depresión del pico central *f*	pequena cova (de pico central) *f* buraco no pico central *m*	вершинная впадина (центральной горки) *f*	244
superficie con cráteres *f* superficie craterizada *f*	superfície (coberta) com crateras *f*	поверхность, покрытая кратерами *f*	245
densidad de cráteres *f*	densidade de crateras *f*	плотность кратеров *f*	246

247	crater frequency	fréquence des cratères *f*	Kraterhäufigkeit *f*
248	diameter-depth relation (of craters)	relation diamètre-profondeur (des cratères) *f*	Verhältnis Durchmesser zu Tiefe *n* Beziehung Durchmesser-Tiefe *f*
249	size frequency	fréquence en fonction de la taille *f* répartition en taille *f*	Grössenhäufigkeit *f*
250	crater pit small crater craterlet	petit cratère *m* craterlet *m*	kleiner Krater *m* Kratergrube *f* Krater pit
251	crater wall	enceinte du cratère *f*	Kraterwand *f* Kraterwall *m*
252	inner crater wall	rempart intérieur du cratère *m*	Kraterinnenwand *f*
253	outer crater wall	rempart extérieur du cratère *m*	Krateraussenwand *f*
254	concentric wall	rempart concentrique *m*	konzentrischer Wall *m*
255	multiple wall	rempart multiple *m*	mehrfacher Wall *m*
256	internal terracing	terrasses intérieures *fpl*	innere Terrasse *f*
257	terraced wall	rempart en terrasse *m*	Terrassenwall *m* terrassenförmige Wand *f*
258	central peak	pic central *m* piton central *m*	Zentralberg *m* Zentralkegel *m*
258a	crater with central peak	cratère avec pic central *m*	Krater mit Zentralberg *m*
259	crater floor floor of the crater	fond du cratère *m*	Kraterboden *m*
260	flat bottom	fond plat *m*	flacher Kraterboden *m*
261	impact hypothesis	hypothèse des impacts *f*	Einschlagshypothese *f* Aufsturzhypothese *f*
262	meteoroidal theory meteori(ti)c theory	hypothèse météoroïdale *f* hypothèse météori(ti)que *f*	Meteor(it)enhypothese *f* Meteoroidenhypothese *f*
263	meteori(ti)c crater meteoroidal crater	cratère météorique *m* cratère météoroïdal *m*	Meteor(iten)krater *m* Meteoroidenkrater *m*
264	maar crater	(cratère) maar *m* cratère du type maar *m*	Maarkrater *m* Maar-ähnlicher Krater *m*
265	walled plain lunar ring	cirque lunaire *m*	Wallebene *f* Ringgebirge *n*
266	mare (pl. maria) sea	mer *f* mare *f*	Mare *n* (pl. Maria) Mondmeer *n*

SELENOGRAFÍA	SELENOGRAFIA	СЕЛЕНОГРАФИЯ	
frecuencia de cráteres *f*	freqüência de crateras *f*	частота кратеров *f*	247
relación diámetro-profundidad *f*	relação diâmetro-profundidade (das crateras) *f*	отношение диаметра к глубине *n*	248
frecuencia de tamaño *f*	freqüência de tamanho *f*	частота размеров *f* частота величин *f*	249
cráter pequeño *m* agujero crateroide *m* (p)seudocráter *m* cratercito *m*	pequena cratera *f* craterleta *f*	небольшой кратер *m* мелкий кратер *m* кратер-лунка *m*	250
pared del cráter *f*	parede da cratera *f*	стена кратера *f* вал кратера *m*	251
pared interna del cráter *f*	parede interna da cratera *f*	внутренняя стена *f* внутренний вал [склон вала] *m*	252
pared externa del cráter *f*	parede externa da cratera *f*	внешний вал [склон вала] *m*	253
pared concéntrica *f*	parede concêntrica *f*	концентрический вал *m*	254
pared múltiple *f*	parede múltipla *f*	кратный вал *m*	255
terrazas internas *fpl*	terraço interno *m*	внутреннее террасовое строение *n*	256
pared aterrazada *f* pared con terrazas *f*	parede em forma de terraço *f*	террасовый вал *m*	257
pico central *m*	pico central *m*	центральная вершина *f* центральная горка *f*	258
cráter con pico central *m*	cratera com pico central *f*	кратер с центральной горкой *m*	258a
fondo de un cráter *m*	fundo de (uma) cratera *m* piso de (uma) cratera *m*	дно кратера *n*	259
fondo llano *m*	fundo plano *m*	плоское дно *n*	260
hipótesis de impacto *f*	hipótese de impacto *f*	гипотеза падения *f* гипотеза столкновений *f*	261
hipótesis meteoroidal *f* hipótesis meteorítica *f*	hipótese meteoroídica *f* hipótese meteorítica *f*	метеоритная гипотеза *f* метеороидная гипотеза *f*	262
cráter meteórico *m* cráter meteorítico *m* cráter meteoroidal *m*	cratera meteorítica *f* cratera meteórica *f* cratera meteoroidal *f*	метеоритный кратер *m* метеороидный кратер *m*	263
cráter maar *m*	cratera (tipo) maar *f*	маар *m* маар-кратер *m*	264
circo lunar *m*	circo lunar *m* anel lunar *m*	лунный цирк *m*	265
mar *m*	mar *m* (pl. maria, mares)	море *n*	266

	SELENOGRAPHY	SÉLÉNOGRAPHIE	SELENOGRAPHIE
267	circular mare circular sea	mer circulaire *f*	kreisförmiges Mare *n* kreisförmiges Meer *n*
268	mare region	région des mare [des mers] *f* région [zone] maritime *f*	Mare-Gebiet *n* Mare-Region *f*
269	pre-mare region	région [zone] prémare *f* région [zone] antérieure aux mers *f*	prae-mare Gebiet *n* prae-mare Region *f*
270	post-mare region	région [zone] postérieure aux mers *f* région [zone] postmare *f*	post-mare Gebiet *n* post-mare Region *f*
271	continental region terra (pl. terrae)	région continentale *f* zone continentale *f*	Hochland-Gebiet *n* Terra-Gebiet *n*
272	lowland region lowlands	plaine-basse *f* terrain bas *m* région basse *f*	Tiefland-Gebiet *n* Tiefebene *f* Niederung *f*
273	Sea of Showers Sea of Rains Mare Imbrium	Mer des Pluies *f*	Regen-Meer *n*
274	Sea of Fertility Mare Fecunditatis	Mer de la Fécondité *f*	Meer der Fruchtbarkeit *n*
275	Sea of Tranquillity Mare Tranquillitatis Sea of Peace	Mer de la Tranquillité *f*	Meer der Ruhe *n*
276	Sea of Serenity Mare Serenitatis	Mer de la Sérénité *f*	Meer der Heiterkeit *n* Mer des heiteren Himmels *n*
277	Oceanus Procellarum Ocean of Storms	Océan des Tempêtes *m*	Ozean der Stürme *m*
278	Sea of Cold Mare Frigoris	Mer du Froid *f*	Meer der Kälte *n*
279	Sea of Nectar Mare Nectaris	Mer du Nectar *f*	Nektar-Meer *n*
280	Sea of Crises Mare Crisium Sea of Conflicts	Mer des Crises *f*	Meer der Gefahren *n*
281	Sea of Clouds Mare Nubium	Mer des Nuages *f*	Wolken-Meer *n* Meer der Wolken *n*
282	Sea of Moisture [Humours] Mare Humorum	Mer des Humeurs *f*	Meer der Feuchtigkeit *n*
283	Moscow Sea [Sea of Moscow] Mare Moscoviense	Mer de Moscou *f*	Moskau-Meer *n* Moskauer Meer *n*
284	Eastern Sea Mare Orientale	Mer Orientale *f*	Ostmeer *n*

mar circular *m*	mar circular *m*	круглое море *n* круговое море *n*	267
región de mares *f* región marítima *f*	região de mares *f* região marítima *f*	морской район *m* морская область *f*	268
región premarítima *f* región anterior a los mares *f*	região anterior aos mares *f* região pré-mares [pré-marítima] *f*	предморской район *m* область, возникшая до образования морей *f*	269
región postmarítima *f* región posterior a los mares *f*	região pós-mares [pós-marítima] *f* região posterior aos mares *f*	послеморской район *m* область, возникшая после образования морей *f*	270
región continental *f*	região continental *f* terra *f* (pl. terras)	материковая область *f* континентальная область *f* материковый район *m*	271
región baja *f* tierras bajas *fpl* planicies *fpl*	região baixa *f* planícies *fpl* (terra) chã *f* terras baixas *fpl*	низменность *f* низина *f* низменный район *m* низменная область *f*	272
Mar de las Lluvias *m*	Mar das Chuvas *m*	Море Дождей *n*	273
Mar de la Fecundidad *m*	Mar da Fecundidade *m* Mar da Fertilidade *m*	Море Изобилия *n* Море Плодородия *n*	274
Mar de la Tranquilidad *m*	Mar da Tranqüilidade *m*	Море Спокойствия *n*	275
Mar de la Serenidad *m*	Mar da Serenidade *m*	Море Ясности *n*	276
Océano de las Tempestades *m* Océano de las Tormentas *m*	Oceano das Tempestades *m*	Океан Бурь *m*	277
Mar de los Fríos *m* Mar del Frío *m*	Mar do Frio *m*	Море Холода *n*	278
Mar de los Néctares *m* Mar del Néctar *m*	Mar do Néctar *m* Mar dos Néctares *m*	Море Нектара *n*	279
Mar de las Crisis *m*	Mar das Crises *m*	Море Кризисов *n*	280
Mar de las Nubes *m*	Mar das Nuvens *m*	Море Облаков *n*	281
Mar de los Humores *m*	Mar dos Humores *m* Mar Úmido [da Umidade] *m*	Море Влажности *n*	282
Mar de Moscú *m*	Mar de Moscou *m* Mar Moscovita *m*	Море Москвы *n*	283
Mar Oriental *m*	Mar Oriental *m*	Море Восточное *n*	284

285	Border Sea Mare Marginis	Mer Marginale *f*	Randmeer *n*
286	Southern Sea Mare Australe	Mer Australe *f*	Südmeer *n*
287	Smyth's Sea Mare Smythii	Mer de Smyth *f*	Smyth-Meer *n*
288	Humboldt's Sea Mare Humboldtianum	Mer de Humboldt *f*	Humboldt-Meer *n*
289	Known Sea Knowledge Sea Mare Cognitum	Mer de la Connaissance *f* Mer Connue *f*	Bekanntes Meer *n* Meer des Wissens *n*
290	Sea of Waves Mare Undarum	Mer des Vagues *f*	Meer der Wellen *n*
291	Serpent Sea Mare Anguis	Mer du Serpent *f*	Meer der Schlangen *n*
292	Sea of Ingenuity Mare Ingenii	Mer des Inventions *f*	Meer der Sennsucht *n*
293	Promontorium promontory cape	promontoire *m*	Kap *m*
294	sinus gulf bay	golfe *m*	Bucht *f* Golf *m* Meerbusen *m*
295	Sinus Iridum Bay of Rainbows	Golfe des Iris *m*	Regenbogenbucht *f*
296	Sinus Aestuum Seething Bay Bay of Billows [of Heats]	Golfe Torride *m*	Bucht der Gluten *f*
297	Lacus lake	lac *m*	See *m*
298	Lacus Mortis Lake of Death	Lac de la Mort *m*	See des Todes *m*
299	Lacus Somniorum Lake of Dreams [of Sleepers] Lake of Dreamers	Lac des Rêves *m*	See der Träume *m*
300	Lacus Aetatis Summer Lake	Lac de l'Été *m*	See des Sommers *m*
301	Lacus Autumnae Autumn Lake	Lac de l'Autumne *m*	See des Herbstes *m*
302	Lacus Veris Spring Lake	Lac du Printemps *m*	See des Frühlings *m*
303	Palus marsh	marais *m*	Sumpf *m* Moor *n*

Español	Português	Русский	No.
Mar Marginal *m*	Mar da Margem *m* Mar Marginal *m*	Море Краевое *n*	285
Mar Austral *m*	Mar Austral *m*	Море Южное *n*	286
Mar de Smyth *m*	Mar de Smyth *m*	Море Смита *n*	287
Mar de Humboldt *m*	Mar de Humboldt *m*	Море Гумбольдта *n*	288
Mar Conocido *m* Mar del Conocimiento *m*	Mar Conhecido *m*	Море Познанное *n* Море Познаний *n*	289
Mar de las Ondas *m*	Mar das Ondas *m*	Море Волн *n*	290
Mar de las Serpientes *m*	Mar das Serpentes *m*	Море Змеи *n*	291
Mar de la Ingenuidad *m*	Mar da Engenhosidade *m* Mar do Engenho *m*	Море Мечты *n*	292
cabo *m* promontorio *m*	cabo *m* promontório *m*	мыс *m*	293
bahía *f* golfo *m*	golfo *m* baía *f*	залив *m*	294
Bahía del Arco Iris *f*	Golfo do Arco-Iris *m* Baía do Arco-Iris *f*	Залив Радуги *m*	295
Golfo Tórrido *m*	Baía dos Calores [Ardores] *f* Golfo Tórrido *m* Baía Tórrida *f*	Залив Зноя *m* Залив Волнений *m*	296
lago *m*	lago *m*	озеро *n*	297
Lago de la Muerte *m*	Lago da Morte *m*	Озеро Смерти *n*	298
Lago de los Sueños *m*	Lago dos Sonhos *m*	Озеро Сновидений *n*	299
Lago de Verano *m*	Lago do Verão *m*	Озеро Лета *n*	300
Lago de Otoño *m*	Lago do Outono *m*	Озеро Осени *n*	301
Lago de Primavera *m*	Lago da Primavera *m*	Озеро Весны *n*	302
pantano *m*	pântano *m*	болото *n*	303

304	Palus Epidemiarum Marsh of Diseases [Epidemics]	Marais des Épidemies *m*	Sumpf der Epidemien *m*
305	Palus Putredinis Marsh of Decay	Marais de la Putréfaction *m*	Sumpf des Zerfalls *m*
306	Palus Somnii Marsh of Sleep	Marais du Sommeil *m*	Sumpf des Schlafes *m*

Pantano de las Epidemias *m*	Pântano das Epidemias *m*	Болото Эпидемий *n*	304
Pantano de la Decadencia *m* Pantano de la Putrefacción *m*	Pântano da Putrefação *m* Pântano da Decadência *m*	Болото Гниения *n* Болото Гнилое *n*	305
Pantano del Sueño *m*	Pântano do Sono *m*	Болото Сна *n* Болото Сонное *n*	306

	SELENOLOGY	SÉLÉNOLOGIE	SELENOLOGIE
1	selenology	sélénologie *f*	Selenologie *f*
2	lunar sciences	sciences lunaires *fpl*	Mondkunde *f*
3	lunar research	recherche lunaire *f*	Mondforschung *f*
4	lunar-based astronomy	astronomie depuis une base lunaire *f*	Astronomie von der Mondoberfläche *f*
5	lunar module LM	module lunaire *m* section lunaire *f* véhicule d'alunissage [d'atterrissage] *m*	Mondmodul *m* Mondlandegerät *n* Mondlandeaggregat *n*
6	command module	section de commande *f* module [véhicule] de commande *m*	Befehlseinheit *f* Steuermodul *m*
7	service module	module [véhicule] de service *m*	Service Modul *n* Bedienungsmodul *n* Kommandomodul *n*
8	ascent	ascension *f*	Aufstieg *m*
9	ascending trajectory	trajectoire ascendante *f*	Aufstiegsbahn *f*
10	path correction flight-path correction	correction de trajectoire *f*	Bahnkorrektur *f* Kurskorrektur *f*
11	circumlunar trajectory	trajectoire circumlunaire *f*	Mondumlaufbahn *f*
12	interplanetary trajectory	trajectoire interplanétaire *f*	interplanetare Bahn *f*
13	lunar landing	atterrissage lunaire [sur la Lune] *f* alunissage *m*	Mondlandung *f*
14	landing site	site d'alunissage *m* site d'atterrissage lunaire [sur la Lune] *m*	Landeplatz *m* Landepunkt *m* Mondlandungsplatz *m*
15	proposed landing site	site d'atterrissage proposé *m* site d'alunissage sélectionné [choisi] *m*	vorgesehener Landeplatz *m*
16	touchdown	instant d'atterrissage *m* impact *m* instant d'alun(iss)age *m*	Landung *f* Landungsmoment *m*
17	lunar base Moon base	base lunaire *f* base sur la Lune *f*	Mondbasis *f*
18	manned lunar base	base lunaire habitée *f*	bemannte Mondbasis *f*

SELENOLOGÍA	SELENOLOGIA	СЕЛЕНОЛОГИЯ	
selenología *f*	selenologia *f*	селенология *f*	1
ciencias lunares *fpl*	ciências lunares *fpl*	науки о Луне *fpl*	2
investigación lunar *f*	pesquisa lunar *f*	исследование Луны *n* изучение Луны *n*	3
astronomía hecha desde la Luna *f* astronomía de base lunar *f*	astronomia de [com] base lunar *f*	астрономия, базирующаяся на наблюдениях с Луны *f*	4
módulo lunar *m* módulo de alunizaje *m*	módulo lunar *m* módulo de alun(iss)agem *m*	лунный отсек *m*	5
módulo de comando *m* módulo de mando *m*	módulo de comando *m*	отсек управления *m* командный отсек *m* командный модуль *m*	6
módulo de servicio *m*	módulo de serviço *m*	служебный отсек *m* отсек со вспомогательными системами *m*	7
ascenso *m*	subida *f*	подьём *m* выход на орбиту *m*	8
trayectoria ascendente *f*	trajetória ascendente *f*	восходящая траектория *f* траектория подъёма *f*	9
corrección de la trayectoria (del vuelo) *f*	correção de trajetória *f* correção de trajetória do vôo *f*	коррекция траектории (полёта) *f*	10
trayectoria circunlunar *f*	trajetória circunlunar *f*	селеноцентрическая траектория *f* траектория облёта Луны *f* траектория вокруг Луны *f*	11
trayectoria interplanetaria *f*	trajetória interplanetária *f*	межпланетный участок (траектории) *m* межпланетная траектория *f*	12
alunizaje *m*	alunissagem *f* alunagem *f*	посадка на Луну *f* посадка на поверхность Луны *f* прилунение *n*	13
lugar de alunizaje *m* sitio de alunizaje *m*	local de pouso (na Lua) *m* sítio de pouso (na Lua) *m* sítio de alun(iss)agem *m* local da alun(iss)agem *m*	место посадки *n* место прилунения *n*	14
sitio propuesto de alunizaje *m*	sítio proposto de pouso *m* local proposto da alun(iss)agem *m*	предполагаемое место посадки *n*	15
impacto *m* instante de alunizaje *m* toma de contacto *f*	impacto *m* instante de alun(iss)agem *m*	момент посадки *m* приземление *n* прилунение *n*	16
base lunar *f* base en la Luna *f*	base lunar *f* base na Lua *f*	лунная база [станция] *f* база на Луне *f*	17
base lunar tripulada *f*	base lunar habitada *f* base lunar tripulada *f*	обитаемая лунная база [станция] *f*	18

19	automatic lunar station	station lunaire automatique *f*	automatische Station auf dem Mond *f*
20	lunar observatory observatory on the Moon	observatoire lunaire *f* observatoire sur la Lune *f*	Observatorium auf dem Mond *n*
21	lunar lander	véhicule d'alunissage *m* véhicule d'atterrissage lunaire [sur la Lune] *m*	Mondlandefähre *f*
22	landing leg	pied d'atterrissage *m*	Landebein *n*
23	footpad	coupelle du pied d'atterrissage *f*	Landefuss (dämpfer) *m*
24	shock-absorber	amortisseur *m*	Dämpfer *m*
25	aluminium honeycomb	structure d'aluminium en nid d'abeille *f*	Aluminium-Wabenblech *n*
26	lunar [Moon's] photograph picture of the Moon	photo(graphie) lunaire *f* image photographique de la Lune *f*	Mondphotographie *f* Mondaufnahme *f* Mondfoto *n*
27	panoramic photo(graph)	prise de vue panoramique *f*	Rundbildaufnahme *f* Panoramaaufnahme *f*
28	vertical photo(graph)	photographie verticale *f*	Senkrechtaufnahme *f* Nadiraufnahme *f*
29	oblique photo(graph)	prise de vue oblique *f*	Schrägaufnahme *f*
30	photographic survey photographic surveying	relevé photographique *m*	photographische Vermessung [Übersicht] *f* Aufnahme des Geländes *f*
31	orbital reconnaisance	reconnaissance orbitale *f*	orbitale Erkundung *f* Erkundung von der Umlaufbahn aus *f*
32	surface reconnaisance	reconnaissance à la surface *f* observation de surface *f*	Oberflächenerkundung *f*
33	manned landing	atterrissage humain *m*	bemannte Landung *f*
34	stay time duration of stay	durée du séjour *f*	Aufenthalt *m* Aufenthaltszeit *f* Aufenthaltsdauer *f*
35	traverse traverse survey	excursion *f* cheminement *m*	Exkursion *f* Poligonaufnahme *f* Poligonieren *n*

estación lunar automática *f*	estação lunar automática *f*	автоматическая лунная станция *f*	19
observatorio lunar *m*	observatório lunar *m*	лунная обсерватория *f* обсерватория на Луне *f*	20
vehículo de alunizaje *m* alunizador *m*	dispositivo de pouso lunar *m* dispositivo de alun(iss)agem *m*	лунный посадочный аппарат *m* аппарат для посадки на Луну *m*	21
pie del vehículo de alunizaje *m* pata del alunizador *f*	perna para pouso *f*	посадочная нога *f*	22
amortiguador del pie de alunizaje *m*	amortecedor de pouso *m*	посадочная опора *f* пятка посадочной ноги *f*	23
amortiguador *m*	amortecedor *m*	амортизатор *m*	24
panal de aluminio *m*	alumínio alveolado [celular, en forma de fovos de colmeia] *m*	сотовый алюминий *m* ячеистый алюминий *m*	25
fotografía lunar *f* fotografía de la Luna *f* foto lunar *f*	fotografia lunar *f* fotografia da Luna *f* foto da Lua *f*	фотография лунной поверхности *f* фотография Луны *f*	26
fotografía panorámica *f*	fotografia panorâmica *f*	панорамный снимок *m* широкоугольная фотография *f*	27
fotografía vertical *f*	fotografia vertical *f*	горизонтальная фотография *f* фотография в отвесном направлении *f*	28
fotografía oblícua *f*	fotografia oblíqua *f*	фотография под углом наклона *f* фотография в наклонном направлении *f*	29
levantamiento fotográfico *m* reconocimiento fotográfico *m*	levantamento fotográfico *m*	фотографический обзор *m* фотографическое обозрение *n* фотографическая съёмка *f*	30
reconocimiento orbital *m* reconocimiento desde la órbita *m*	reconhecimento orbital *m*	орбитальная разведка *f* орбитальная рекогносцировка *f*	31
reconocimiento de la superficie *m* reconocimiento en la superficie *m*	reconhecimento da superfície *m* reconhecimento superficial [na superfície] *m*	рекогносцировка поверхности *f* первоначальная разведка *f* предварительное исследование (поверхности) *n*	32
desembarco tripulado *m* aterrizaje tripulado *m*	pouso controlado por ser humano *m*	пилотируемая посадка *f*	33
tiempo de permanencia *m*	tempo de permanência *m*	время пребывания (на Луне) *n*	34
exploración (de reconocimiento) *f* travesía de levantamiento *f*	excursão *f* (levantamento de) travessia *f* levantamento poligonal *m*	маршрутная съёмка *f* траверс *m* курс *m* маршрутный обзор *m*	35

	SELENOLOGY	SÉLÉNOLOGIE	SELENOLOGIE
36	scientific traverse	excursion scientifique *f*	wissenschaftliche Exkursion *f*
37	traverse route	itinéraire *m*	Exkursionsroute *f*
38	footprints	empreintes de pas *fpl*	Fusstapfen *fpl*
39	selenophysical survey	levé sélénophysique *m* relevé sélénophysique *m*	selenophysikalische Durchmusterung [Aufnahme, Vermessung] *f*
40	selenophysical monitoring network	cadastrage sélénophysique *m*	selenophysikalisches Beobachtungsnetz *n*
41	automated instrument	instrument automatisé *m*	automatisiertes Gerät *n*
42	lunar-surface experiment	appareils déposés sur la Lune *mpl* expérience sur la Lune *f*	Mondoberflächenexperiment *n*
43	retroreflector laser retroreflector	réflecteur laser *m*	Laser-Reflektor *m*
44	portable magnetometer	magnétomètre portatif *m*	tragbares Magnetometer *n*
45	selenomagnetic field	champ sélénomagnétique *m* champ magnétique lunaire *m*	selenomagnetisches Feld *n* magnetisches Feld des Mondes *n*
46	solar-wind spectrometer	spectromètre de vent solaire *m*	Sonnenwindspektrometer *n*
47	selenology lunar geology	sélénologie *f* géologie lunaire *f*	Selenologie *f* Mondkunde *f* Mondgeologie *f*
48	selenologic(al)	sélénologique	selenologisch
49	selenologist	sélénologue *m*	Selenolog *m*
50	selenologic survey geologic survey of the Moon	relevé sélénologique *m* levé sélénologique *m*	selenologische Aufnahme *f* selenologische Vermessung *f*
51	selenologic(al) mapping	cartographie sélénologique *f*	selenologische Kartographie *f*
52	petrographic analysis	analyse pétrographique *f*	petrographische Analyse *f*
53	mineralogic(al) analysis	analyse minéralogique *f*	mineralogische Analyse *f*
54	Moon's crust lunar crust	croûte lunaire *f* croûte de la Lune *f*	Mondkruste *f*

exploración científica *f* travesía científica *f*	excursão científica *f* travesia científica *f*	научный траверс [обзор] *m* научная съёмка *f* маршрутный обзор [траверс] *m*	36
ruta de exploración *f*	itinerário da travesia *m* rota cruzada de excursão *f*	трасса маршрута *f* трасса прохождения *f*	37
huellas del pie *fpl* pisadas *fpl*	pegadas *fpl*	следы (ног) *mpl*	38
reconocimiento [levantamiento] selenofísico *m*	levantamento selenofísico *m*	селенофизическая съёмка *f*	39
red monitora selenofísica *f* red de vigilancia selenofísica *f*	rede de vigilancia lunar *f* rede de referência selenofísica *f*	селенофизическая сеть регистрации *f*	40
instrumento automatizado *m*	instrumento automatizado *m*	автоматизированный инструмент *m*	41
experimento en la superficie lunar *m*	experiência na superfície lunar *f*	эксперимент на лунной поверхности *m*	42
retroreflector *m* retroreflector laser *m*	retrorrefletor *m* retrorrefletor laser *m*	лазерный ретрорефлектор *m* отражатель *m* лазерный отражатель *m*	43
magnetómetro portátil *m*	magnetômetro portátil *m*	переносный магнитометр *m* портативный магнитометр *m*	44
campo selenomagnético *m* campo magnético lunar *m*	campo selenomagnético *m* campo magnético lunar *m*	селеномагнитное поле *n* магнитное поле Луны *n*	45
espectrómetro del viento solar *m*	espe(c)trômetro de vento solar *m*	спектрометр солнечного ветра *m*	46
selenología *f* geología lunar *f*	selenologia *f* geologia lunar *f*	селенология *f* лунная геология *f*	47
selenológico	selenológico	селенологический	48
selenólogo *m*	selenólogo *m*	селенолог *m*	49
relevamiento [cartografiado] selenológico *m* reconocimiento selenológico *m*	levantamento selenológico *m* levantamento geológico da Lua *m*	селенологический обзор *m* селенологическая съёмка *f* селенологическое обозрение *n* геологическая съёмка Луны *f*	50
mapeo selenológico *m* cartográfia selenológica *f*	mapeamento selenológico *m*	селенологическое картирование *n* составление селенологической карты *n*	51
análisis petrográfico *m*	análise petrográfica *f*	петрографический анализ *m*	52
análisis mineralógico *m*	análise mineralógica *f*	минералогический анализ *m*	53
corteza lunar *f* corteza de la Luna *f*	crosta lunar *f* crosta da Lua *f*	лунная кора *f* кора Луны *f*	54

55	lunar surface Moon's surface surface of the Moon	surface lunaire *f* surface de la Lune *f*	Mondoberfläche *f*
56	highland surface	(surface de) haut-plateau *f*	Hochlandoberfläche *f*
57	mare surface	surface de mare *f* surface de plaine *f*	Mare-Oberfläche *f* Meeresoberfläche *f*
58	lunar material	matériau lunaire *m*	Mondmaterial *n* lunares Material *n*
59	lunar surface material	matériau lunaire de surface *m*	lunares Oberflächenmaterial *n*
60	core sample	carotte-échantillon *m* échantillon carotté [de carotte] *m*	Kernprobe *f*
61	bore hole	trou de forage *m*	Bohrloch *n*
62	lunar soil	sol lunaire *m*	Mondboden *m* Mondgrund *m*
63	cohesive soil	sol cohésif *m* sol cohérent *m*	bindiger Boden *m* kohäsiver Boden *m*
64	bearing strength	capacité de portance *f* force portante *f*	Tragfähigkeit *f* Tragkraft *f*
65	lunar rock moonrock	roche lunaire *f* roc lunaire *m*	Mondgestein *m*
66	crystalline rock	roche cristalline *f*	kristallines Gestein *n*
67	equigranular rock	roche équigrainue *f*	gleichmässig körniges Gestein *n*
68	rock fragment	fragment rocheux *m*	Gesteinbruchstück *n*
69	lunar sample sample of lunar material	échantillon lunaire *m*	Mondprobe *f* Probe des Mondmaterials *f*
70	returned lunar sample	échantillon lunaire rapporté *m*	zurückgebrachte Mondprobe *f*
71	sample return	retour d'échantillons *m*	Bergung von Proben *f*
72	sample collection	prise d'échantillons *f*	Probennahme *f*

SELENOLOGÍA	SELENOLOGIA	СЕЛЕНОЛОГИЯ	
superficie lunar *f* superficie de la Luna *f*	superfície lunar *f* superfície da Lua *f*	лунная поверхность *f* поверхность Луны *f*	55
tierras altas *fpl* superficie alta *f* superficie de altiplanicie *f*	superfície de planalto *f*	высокогорье *n* гористая поверхность *f* материковая поверхность *f* плоскогорье *n*	56
superficie del mar *f*	superfície do mar *f*	поверхность моря *f*	57
material lunar *m*	material lunar *m*	лунное вещество *n* лунный материал *m*	58
material de la superficie lunar *m*	material da superfície lunar *m*	материал лунной поверхности *m*	59
testigo *m* muestra del núcleo *m*	amostra do núcleo *f*	образец керна [из пробоотборника] *m* керн *m* буровой образец *m*	60
agujero taladrado *m*	furo *m* perfuração *f*	буровая скважина *f* высверленное отверстие *n*	61
suelo lunar *m*	solo lunar *m*	лунный грунт *m* лунная почва *f* лунный почвенный слой *m*	62
suelo cohesivo *m* suelo consistente *m*	solo coesivo *m* solo coerente *m*	связный грунт *m* связная почва *f*	63
capacidad portadora *f* capacidad sustentadora *f*	força portante *f* capacidade de suporte *f*	несущая способность *f*	64
roca lunar *f*	rocha lunar *f*	лунный камень *m* лунная порода *f* лунная скала *f*	65
roca cristalina *f*	rocha cristalina *f*	кристаллическая порода *f*	66
roca equigranular *f*	rocha equigranular *f*	равномерно зернистая порода *f* гранитная порода *f*	67
fragmento rocoso *m*	fragmento de rocha *m*	обломок (горной) породы *m*	68
muestra lunar *f* muestra de material lunar *f*	amostra lunar *f* amostra de material lunar *f*	лунный образец *m* образец лунного вещества *m* лунная проба *f*	69
muestra lunar devuelta *f*	amostra lunar de retorno [devolvida] *f*	привезённый лунный образец *m* лунная проба, доставленная на Землю *f* образец лунного грунта, (доставленный на Землю *m*	70
retorno de muestras *m*	retorno de amostras *m*	доставка образца *f*	71
colección de muestras *f*	coleção de amostras *f*	собирание образцов *n*	72

73	microstructure	microstructure *f*	Mikrostruktur *f*
74	selenochemistry Moon's chemistry	sélénochimie *f* chimie de la Lune *f*	Selenochemie *f* Mondchemie *f*
75	chemical analysis	analyse chimique *f*	chemische Analyse *f*
76	alpha-scattering instrument	expérience de diffusion (des particules) alpha *f*	Alpha-Streuinstrument *n*
77	non-destructive γ-ray spectrometry	spectroscopie gamma non-destructive *f*	zerstörungsfreie Gamma-Spektroskopie *f*
78	soil sampling	prélèvement d'échantillons du sol *m* prise d'échantillons du sol *f*	Proben(n)ahme von Boden *f* Bodenproben(n)ahme *f* Probe(n)entnahme vom Grund *f*
79	soil processor	analyseur du sol *m*	Bodenuntersuchungsgerät *n*
80	sample-collection tool (surface) sampler	outil de prélèvement d'échantillons *m*	Gerät zur Entnahme von Proben *n*
81	sample container	récipient des échantillons *m* conteneur [container] d'échantillons *m*	Probenbehälter *m*
82	subsurface investigation	recherche en profondeur *f* investigation souterraine *f*	Untersuchung unterhalb der Oberfläche *f*
83	(surface) penetrometer soil penetrometer	pénétromètre *m*	Penetrometer *n*
84	lunar basalt	basalte lunaire *m*	Mondbasalt *m* lunarer Basalt *m*
85	lunar breccia	brèche lunaire *f*	Mondbreccie *f* Mondbrekzie *f*
86	lunar microbreccia	microbrèche lunaire *f*	lunare Mikrobreccie *f* lunare Mikrobrekzie *f*
87	anorthosite	anorthosite *f*	Anorthosit *m*
88	plagioclase	plagioclase *m*	Plagioklas *m*
89	olivine	olivine *f*	Olivin *m* Peridot *m*
90	pigeonite	pigeonite *f*	Pigeonit *m*

microestructura *f*	microestrutura *f*	микроструктура *f*	73
selenoquímica *f* química lunar *f*	selenoquímica *f* química lunar *f*	селенохимия *f* лунная химия *f* химия Луны *f*	74
análisis químico *m*	análise química *f*	химический анализ *m*	75
instrumento para dispersión alfa *m*	instrumento de espalhamento alfa *m*	эксперимент рассеяния альфа-частиц *m*	76
espectrometría gamma no destructiva *f*	espe(c)trometria não-destrutiva de raios gama *f*	недеструктивная γ-спектроскопия *f* неразрушающая гамма-спектроскопия *f*	77
colección de muestras de suelos *f*	coleta [coleção] de amostras de solo *f* amostragem do solo *f*	забор проб лунного грунта *m* отбор образца грунта *m* забор проб грунта *m*	78
procesador de suelos *m*	processador de solo *m*	прибор, определяющий характеристики грунта *m* прибор, испытывающий грунт [почвы] *m*	79
herramienta recogedora de muestras *f* muestreador de superficie *m*	ferramenta de coleta de amostras *f*	пробоотборник *m* пробоотбиратель *m* прибор для отбирания проб *m* орудие для собирания образцов *n*	80
recipiente de muestras *m* contenedor de muestras *m*	contenedor de amostras *m* recipiente que contém amostras *m*	контейнер для образцов *m*	81
investigación subsuperficial *f* investigación del subsuelo *f*	investigacão subsuperficial *f*	подповерхностное исследование *n*	82
penetrómetro (de suelos) *m* sonda de suelos *f*	penetrômetro de solo *m* penetrômetro (de superfície) *m* sonda de solos *f*	пенетрометр *m* прибор для измерения проницаемости грунта *m*	83
basalto lunar *m*	basalto lunar *m*	лунный базальт *m*	84
brecha lunar *f*	brecha lunar *f*	лунная брекчия *f* обломочная порода *f*	85
microbrecha lunar *f*	microbrecha lunar *f*	лунная микробрекчия *f*	86
anortosita *f*	anortosita (B) *f* anortosite (P) *f*	анортозит *m* известковый полевой шпат *m*	87
plagioclasa *f*	plagioclásio *m*	плагиоклаз *m*	88
olivina *f* olivino *m*	olivina *f*	оливин *m*	89
pigeonita *f*	pigeonita (B) *f* pigeonite (P) *f*	пижонит *m* пиджонит *m*	90

91	ilmenite	ilménite *f*	Ilmenit *m*
92	magnetite	magnétite *f* fer magnétique *m*	Magnetit *m* Magneteisenerz *n*
93	feldspar	feldspath *m*	Feldspat *m*
94	troilite	troïlite *f*	Troilit *m*
95	crystobalite	cristobalite *f*	Cristobalit *m*
96	lunar glass	verre lunaire *m*	Mondglas *n*
97	glassy fragment	fragment vitreux *m*	glasiges Bruchstück *n* glasartiges Fragment *n*
98	igneous intrusion	intrusion de roches ignées *f*	Intrusion der eruptiven Gesteine *f* magmatische Intrusion *f*
99	rock-forming mineral rock constituent	constituant des roches *m*	gesteinsbildendes Mineral *n* Gesteinsbestandteil *m*
100	mineralogical constituent	constituant minéralogique *m*	mineralogische Komponente *f*
101	petrographic clasification	classification pétrographique *f*	petrographische Einteilung *f*
102	texture	texture *f* structure *f*	Struktur *f* Textur *f* Gefüge *n*
103	hummocky material	matériau de monticule *m*	höckeriges Material *n* rundhöckerähnliches Material *n* maulwurfshaufenähnliche Gebilde *npl*
104	regolith fragmental debris layer fragmented debris layer	régolite *m* surface fragmentée *f*	Regolith *m* lockere Bodenschicht *f*
105	subsoil	sous-sol *m*	Untergrund *m*
106	surface roughness	rugosité de la surface *f*	Oberflächenrauheit *f* Unebenheit der Oberfläche *f* Rauhigkeit der Oberfläche *f*
107	roughness factor	coefficient de rugosité *m*	Rauhigkeitsbeiwert *m*
108	soil characteristic	caractéristique du sol *f*	Bodeneigenschaft *f*
109	fines fines particles	particules fines *fpl* poussière *f* particules à grain fin *fpl*	feinkörnige Partikeln *fpl* Teilchen mit weniger als 1 mm Durchmesser *npl*

SELENOLOGÍA	SELENOLOGIA	СЕЛЕНОЛОГИЯ	
ilmenita *f*	ilmenita (B) *f* ilmenite (P) *f*	ильменит *m* титанистый железняк *m*	91
magnetita *f*	magnetita (B) *f* magnetite (P) *f*	магнетит *m* магнитный железняк *m*	92
feldespato *m*	feldspato *m*	полевой шпат *m*	93
troilita *f*	troilita (B) *f* troilite (P) *f*	тройлит *m*	94
cristobalita *f*	cristobalita (B) *f* cristobalite (P) *f*	кристобалит *m*	95
vidrio lunar *m*	vidro lunar *m*	лунное стекло *n*	96
fragmento vítreo *m*	fragmento vitrificado *m*	стекловатый обломок *m* стекловидный фрагмент *m*	97
intrusión ígnea *f* intrusión magmática *f*	intrusão ígnea *f* intrusão magmática *f*	магматическая интрузия *f* пирогенная интрузия *f*	98
mineral que constituye la roca *m* mineral que forma la roca *m*	constituinte mineral de uma rocha *m* mineral constituinte de rocha *m*	породообразующий минерал *m* породосоставляющая часть *f*	99
componente [constituyente] mineralógico *m*	constituinte mineralógico *m*	минералогический компонент *m* минералогическая составная часть *f*	100
clasificación petrográfica *f*	classificação petrográfica *f*	петрографическая классификация *f*	101
textura *f*	textura *f*	структура *f* текстура *f* строение *n*	102
material de montículo *m*	material de cômoros *m*	бугристый материал *m*	103
regolito *m* capa de detrito fragmentario *f*	regolito *m* (camada de) detritos fragmentários (*f*) *mpl* rególito *m*	(лунный) реголит *m* (лунный) раздробленный слой *m* покрывающая порода *f* поверхностная рыхлая порода *f*	104
subsuelo *m*	subsolo *m*	подгрунт *m*	105
rugosidad superficial *f* aspereza superficial *f*	rugosidade superficial *f* aspereza superficial *f*	шероховатость поверхности *f* неровность поверхности *f*	106
coeficiente de rugosidad *m* coeficiente de aspereza *m*	coeficiente de rugosidade *m*	коэффициент шероховатости *m*	107
característica del suelo *f*	característica do solo *f*	характеристика грунта *f*	108
partículas finas *fpl* partículas de „finos" *fpl*	partículas finas *fpl* partículas de "fines" *fpl*	мельчайшие частицы *fpl* частицы мелочи *fpl* мелкодисперсный материал *m*	109

	SELENOLOGY	SÉLÉNOLOGIE	SELENOLOGIE
110	fines grain	grain de poussière *m* grain fin *m*	Gruskorn *n*
111	porous substance	substance poreuse *f* matière poreuse *f*	poröse Substanz *f* poröser Stoff *m*
112	lunar fines	particules lunaires fines *fpl* arène lunaire *f*	Mondstaub *m* lunarer Grus *m*
113	grain size	dimension des grains *f* taille des grains *f*	Korngrösse *f*
114	fine-grained	à grain fin	feinkörnig
115	fine-grained matrix	matrice à grain fin *f* liant à grain fin *m*	feinkörnige Grundmasse *f*
116	medium-grained	à grain moyen	mittelkörnig von mittlerer Korngrösse
117	coarse-grained	à gros grain	grobkörnig von grober Korngrösse
118	matrix	matrice *f* liant *m*	Grundmasse *f* Bindemittel *n*
119	dust-free	dépourvu de poussière libre de poussière	staubfrei
120	dust-covered	couvert de poussière	staubbedeckt
121	lunar dust sample	échantillon de poussière lunaire *f*	Probe des Mondstaubes *f*
122	dust layer	couche de poussière *f*	Staubschicht *f*
123	lunar particle	particule lunaire *f*	Mondteilchen *n*
124	size distribution	distribution des dimensions *f*	Grössenverteilung *f*
125	impact dust	poussière d'impact *f*	Aufschlagsstaub *m* Einsturzstaub *m*
126	volcanic dust	poussière volcanique *f* cendre volcanique fine *f*	vulkanischer Staub *m* vulkanische Asche *f*
127	glassy spherules	sphérules vitreuses *fpl*	glasige Kügelchen *npl*

grano fino *m* grano de "finos" *m*	grão fino *m* grão de "fines" *m*	зерно [гранула] мелкодисперсного материала *n[f]* зерно мелочи *n*	110
substancia porosa *f*	su(b)stância porosa *f*	пористое вещество *n*	111
"finos" lunares *mpl*	poeira fina lunar *f* pó fino lunar *m*	мелкодисперсный лунный материал *m* лунная мелочь *f* тонкозернистые частицы лунного грунта *fpl*	112
tamaño de grano *m*	tamanho dos grãos *m*	размер зёрен *m* величина зёрен *f*	113
de grano fino	de granulação fina	мелкозернистый тонкозернистый с зёрнами величиной менее одного миллиметра	114
matriz de grano fino *f*	matriz de granulação fina *f*	мелкозернистое вяжущее вещество *n*	115
con grano de tamaño medio granulado medio	de granulação média de grãos médios	среднезернистый	116
de grano grueso	de grãos grossos de granulação grossa	крупнозернистый грубозернистый	117
matriz *f*	matriz *f*	основная масса *f* материнская порода *f* цементирующая среда *f* вяжущее [цементирующее] вещество *n*	118
libre de polvo	livre de poeira livre de pó	свободный от пыли обеспыленный	119
cubierto de polvo	coberto de poeira coberto de pó	покрытый пылью	120
muestra de polvo lunar *f*	amostra de poeira lunar *f* amostra de pó lunar *f*	образец лунной пыли *m*	121
capa de polvo *f*	camada de poeira *f*	слой пыли *m* пылевой слой *m*	122
partícula lunar *f*	partícula lunar *f*	лунная частица *f*	123
distribución de tamaños *f*	distribuição de tamanhos *f*	распределение по размерам *n*	124
polvo de impactos *m*	poeira de impacto *f* pó de impacto *m*	пыль ударного происхождения *f* пыль, образованная ударом *f*	125
polvo volcánico *m*	poeira vulcânica *f* pó vulcânico *m*	вулканическая пыль *f* вулканический пепел *m*	126
esférulas vítreas *fpl*	esférulas vitrificadas *fpl*	стеклянные шарики *mpl* стеклянные сферулы *fpl* стеклянные гранулы *fpl*	127

128	pebble	gravier *m* caillou *m* galet *m*	Geröll *n*
129	pellet	bille *f* boulette *f*	kleiner Ballen *m* Kügelchen *n*
130	boulder block of rock	pierre *f* rocher *m* bloc rocheux *m*	loser Block *m* Gesteinsblock *m* Felsblock *m*
131	angular rock angular block	pierre anguleuse *f* roche anguleuse *f*	eckiger Stein *m* kantiges Gestein *n*
132	graben	fossé *m* graben *m* fossé d'effondrement *m*	Graben *m* Tiefscholle *f*
133	stratification	stratification *f*	Schichtung *f* Lagerung *f*
134	horizontal layering	stratification horizontale *f*	horizontale Schichtung *f*
135	soil strength	résistance du sol *f*	Bodenfestigkeit *f*
136	in-situ analysis	analyse "in situ" *f* analyse sur place *f*	in-situ Untersuchung *f* Untersuchung an Ort und Stelle *f*
137	adhesive	adhésif	adhäsiv
138	cohesive	cohésif	kohäsiv
139	rock field stone field	étendue de pierres *f* champ de pierres *m* champ de roches *m*	Trümmerfeld *n* Felsenmeer *n*
140	isotopic composition	composition isotopique *f*	isotopische Zusammensetzung *f*
141	elemental abundances	abondance(s) des éléments *f(pl)*	Elementenhäufigkeit *f* Häufigkeit chemischer Elemente *f*
142	Mössbauer spectrography	spectrographie (de) Mössbauer *f*	Mössbauer-Spektrographie *f*
143	microprobe analysis	analyse par microsonde *f*	Mikroproben-Analyse *f*

guijarro *m*	calhau *m* seixo *m*	галька *f* окатный обломок *m* голыш *m* булыжник *m*	128
gránulos *mpl*	pelota *f*	дробинка *f* шарик *m*	129
bloque *m* bloque de rocas *m*	matação *m* pedregulho *m* bloco de rocha *m*	валун *m* глыба *f* блок *m*	130
bloque anguloso *m*	bloco anguloso *m* rocha angulosa *f*	скала [блок, утёс] с острыми углами *f*[*m*]	131
fosa *f* graben *m*	graben *m* fossa *f*	грабен *m* сбросовый ров *m* сбросовая мульда *f*	132
estratificación *f*	estratificação *f*	слоистость *f* стратификация *f* наслоение *n* напластование *n*	133
estratificación horizontal *f*	estratificação horizontal *f*	горизонтальная слоистость *f* прокладка *f* слоистость в горизонтальном направлении *f*	134
resistencia del suelo *f*	resistência do solo *f*	прочность грунта [почвы] *f*	135
análisis in situ *m*	análise local *f*	анализ на месте *m*	136
adhesivo	adesivo	липкий пристающий	137
cohesivo	coesivo	связующий вязкий связный	138
campo de rocas *m*	campo de rochas *m* campo de pedras *m*	каменное поле *n* каменное море *n* поле камней *n*	139
composición isotópica *f*	composição isotópica *f*	изотопный состав *m*	140
abundancia de los elementos *f*	abundâncias elementares *fpl*	распространенность элементов *f* обилие химических элементов *n*	141
espectrografía Mössbauer *f*	espe(c)trografia Mössbauer *f*	ядерная резонансная спектрография *f* мёссбауэровская спектрография *f* ЯМР – спектрография *f*	142
análisis de microprueba *m*	análise de microsonda *f*	микроанализ *m* анализ микроскопических проб *m*	143

144	major elements	éléments dominants *mpl*	häufige Elemente *npl*
145	trace element	élement-trace *f*	Spurenelement *n*
146	primordial isotope	isotope primordial *m*	Urisotop *n* primäres Isotop *n*
147	cosmogenic isotope cosmic-ray produced isotope	isotope cosmogénique *m* isotope produit par les rayons cosmiques *m*	kosmogones Isotop *n* durch kosmische Strahlung erzeugtes Isotop *n*
148	trapped solar-wind gases	gaz capturés du vent solaire *m*	eingefangene Gase des Sonnenwindes *npl*
149	lunar activity	activité lunaire *f*	Mondaktivität *f*
150	lunar process	processus lunaire *m*	Mondprozess *m*
151	selenophysics physics of the Moon	sélénophysique *f* physique de la Lune *f*	Mondphysik *f* Physik des Mondes *f*
152	soil creep	glissement du sol *m*	Bodenkriechen *n*
153	lunar transient transient lunar phenomenon transient lunar event	phénomène lunaire bref *m* phénomène lunaire transitoire *m*	vorübergehende Erscheinung auf dem Mond *f*
154	volcanic activity	activité volcanique *f*	vulkanische Tätigkeit *f*
155	triggered volcanism	activité volcanique déclenchée *f*	ausgelöste vulkanische Tätigkeit *f*
156	meteorite-induced volcanism	activité volcanique déclanchée par chute météoritique *f*	durch Meteoriteneinschlag ausgelöste vulkanische Tätigkeit *f*
157	volcanism along fractures	activité volcanique le long de fractures *f*	vulkanische Tätigkeit längs der Brüche *f*
158	internal activity	activité interne *f*	innere Tätigkeit *f*
159	volcanic feature volcanic formation volcanic marking	trait volcanique *m* caractère volcanique *m* formation volcanique *f*	vulkanisches Kennzeichen *n* vulkanische Formation *f*
160	volcanic cone	cône volcanique *m*	Vulkankegel *m*
161	maar	maar *m*	Maar *n*
162	volcanic vent	évent *m* cheminée de volcan *f*	Vulkanschlot *m* Vulkanesse *f* Kraterschlot *m*

elementos mayores *mpl*	elementos principais *mpl*	главные элементы *mpl* преобладающие элементы *mpl*	144
elemento de trazas *m*	elemento em traços *m* traços de elemento *mpl*	рассеянный элемент *m* микроэлемент *m*	145
isótopo primordial *m*	isótopo primordial *m*	изначальный [первичный] изотоп *m* примордиальный изотоп *m*	146
isótopo cosmogénico *m* isótopo producido por rayos cósmicos *m*	isótopo cosmogênico *m* isótopo produzido por raios cósmicos *m*	космогенный изотоп *m* изотоп, образованный космическими лучами *m*	147
gases atrapados del viento solar *mpl*	gases captados do vento solar *mpl*	захваченные газы солнечного ветра *mpl*	148
actividad lunar *f*	atividade lunar *f*	лунная активность *f*	149
proceso lunar *m*	processo lunar *m*	лунный процесс *m* процесс на Луне *m*	150
selenofísica *f* física de la Luna *f*	selenofísica *f* física lunar [da Lua] *f*	селенофизика *f* физика Луны *f*	151
asentamiento de tierras *m*	deslizamento [arrasto] do solo *m*	ползучесть грунта *f*	152
fenómeno lunar transitorio *m*	fenômeno lunar transiente *m* fenômeno lunar transitório *m*	преходящий лунный процесс *m* временное лунное явление *n*	153
actividad volcánica *f*	atividade vulcânica *f*	вулканическая активность *f* вулканическая деятельность *f*	154
volcanismo inducido *m*	vulcanismo induzido *m*	вулканизм, вызванный ... *m* вулканизм, обусловленный ... *m*	155
volcanismo inducido por un meteorito *m* volcanismo inducido por impacto meteorítico *m*	vulcanismo induzido por um meteorito *m* vulcanismo producido por queda de um meteorito *m*	вулканизм, вызванный падением метеорита *m*	156
volcanismo a lo largo de fracturas *m*	vulcanismo ao longo de fraturas *m*	вулканизм вдоль трещин [разлома, разрыва] *m*	157
actividad interna *f*	atividade interna *f*	внутренняя активность *f*	158
rasgo volcánico *m* carácter volcánico *m* formación volcánica *f*	aspe(c)to vulcânico *m* formação vulcânica *f* marca vulcânica *f*	вулканический признак *m* вулканическая особенность *f* вулканическое образование *n*	159
cono volcánico *m*	cone vulcânico *m*	вулканический конус *m*	160
maar *m*	maar *m*	маар *m* провал взрыва *m* кратерное озеро *n*	161
abertura volcánica *f* chimenea volcánica *f*	descarga vulcânica *f* chaminé *f* conduto vulcânico *m*	жерло вулкана *n* канал вулкана *m* диатрема *f*	162

	SELENOLOGY	SÉLÉNOLOGIE	SELENOLOGIE
163	fumarole	fumerolle *f*	Fumarole *f*
164	fumarolic activity	activité de fumerolles *f*	Fumarolentätigkeit *f*
165	ejecta ejectamenta	éjecta *mpl* matières éjectées *fpl* éjections *fpl* déjections *fpl* rejets *mpl*	Auswürflinge *mpl* ausgeworfene Massen *fpl*
166	volcanic ejecta	éjecta volcaniques *mpl* matières volcaniques éjectées *fpl*	vulkanische Auswürflinge *mpl*
167	impact ejecta	éjecta d'impact *mpl* éjections de choc *fpl* éjections d'impact *fpl*	Einschlagsauswürflinge *mpl* Stossauswürflinge *mpl* beim Einschlag ausgeworfene Massen *fpl*
168	volcanic ash	cendre volcanique *f*	vulkanische Asche *f*
169	lapilli volcanic pellets	lapilli *mpl*	Lapilli *mpl* Rapilli *mpl*
170	volcanic eruption	éruption volcanique *f*	Vulkaneruption *f* Vulkanausbruch *m*
171	extrusion of lava	extrusion de lave *f* épanchement de lave *m*	Lavaextrusion *f* Ausdrücken der Lava *n*
172	effusion of lava	effusion de lave *f* écoulement de lave *m*	Lavaerguss *m*
173	lava flow	coulée de lave *f*	Lavastrom *m* Lavaflut *f*
174	ash flow	écoulement de cendre *m* coulée de cendre *f*	Aschenfluss *m* Aschenfliessen *n* Aschenstrom *m*
175	lava sheet	nappe de lave *f* couche de lave *f*	Lavadecke *f*
176	cool lava solidified lava	lave refroidie *f* lave figée *f*	erkaltete Lava *f* erstarrte Lava *f*
177	lava cavern lava cave	grotte de lave *f*	Lavahöhle *f*
178	lava blisters	vésicules de lave *fpl*	Lavablasen *fpl*

fumarola *f*	fumarola *f*	фумарола *f* суфляр *m*	163
actividad fumarólica *f*	atividade fumarólica *f*	фумарольная активность *f*	164
eyectos *mpl* materiales expulsados *mpl* materiales proyectados *mpl* material eyectado *m*	ejetos *mpl* ejetamentos *mpl*	выбросы *mpl* продукты извержения *mpl* изверженный материал *m*	165
materiales volcánicos eyectados [proyectados] *mpl* eyectos volcánicos *mpl*	ejetos vulcânicos *mpl* materiais vulcânicos ejetados *mpl* material lanzado pelos vulcões *m*	вулканические выбросы *mpl* изверженный материал *m*	166
eyectos de impacto *mpl* material eyectado por un impacto *m*	ejetos de impacto *mpl*	материал, изверженный [выброшенный] при соударении *m* материал, изверженный соударением *m* ударом изверженный материал *m* материал, изверженный в результате удара *m*	167
ceniza volcánica *f*	cinza vulcânica *f*	вулканический пепел *m*	168
lapilli *mpl*	lapíli *mpl* pedrinhas vulcânicas *fpl* ejetólitos *mpl*	лапилли *pl* частицы лавы, выбрасываемые вулканом *fpl*	169
erupción volcánica *f*	erupção vulcânica *f*	вулканическое извержение *n*	170
extrusión de lava *f* expulsión de lava *f*	extrusão de lava *f* expulsão de lava *f*	экструзия *f* выталкивание лавы *n*	171
efusión de lava *f*	efusão de lava *f*	эффузия *f* истечение лавы *n*	172
colada de lava *f*	fluxo de lava *m*	лавовый поток *m* излияние лавы *n*	173
corriente de cenizas *f* flujo de cenizas *m*	fluxo de cinza *m*	поток пепла *m* течение пепла *n* скольжение пепла *n*	174
capa de lava *f*	camada de lava *f*	лавовый покров *m*	175
lava enfriada *f* lava solidificada *f*	lava arrefecida *f* lava solidificada *f*	застывшая лава *f* затвердевшая лава *f* отвердевшая лава *f*	176
caverna de lava *f* cueva de lava *f*	caverna de lava *f*	лавовая каверна *f* лавовая пещера *f* лавовый грот *m*	177
burbujas de lava *fpl*	bolhas de lava *fpl*	лавовые пузыри [волдыри] *mpl*	178

	SELENOLOGY	SÉLÉNOLOGIE	SELENOLOGIE
179	laccolith	laccolit(h)e *f*	Lakkolith *n*
180	vesicle	vésicule *f* vacuole *f* bulle *f*	Blase *f* Gasblase im Gestein *f*
181	vesicular blistered vuggy porous	vésiculaire vésiculeux bulleux à bulles	blasig porös Blasen-
182	vesiculation	vésiculation *f*	Blasigkeit *f* Entstehung von Gasblasen *f*
183	lava lake	lac de lave *m*	Lavasee *m*
184	lava field	champ de lave *m*	Lavafeld *n*
185	lava plain	plaine de lave *f*	Lavaebene *f*
186	lava tube	tube de lave *m*	Lavaröhre *f*
187	lava tongue	langue de lave *f*	Lavazunge *f*
188	fluidity of the lava lava fluidity	fluidité de la lave *f* coulage de la lave *m*	Fluidität der Lava *f*
189	flow stage of magma	état de coulage [fluidité] du magma *m*	Flusszustand des Magmas *m*
190	consolidation of magma	consolidation du magma *f*	Erstarrung des Magmas *f*
191	flow texture fluidal texture	texture d'écoulement [fluidale] *f* structure d'écoulement [fluidale] *f*	Fluidaltextur *f* Fliesstruktur *f*
192	flow channel	canal d'écoulement *m*	Fliessrinne *f* Flussrinne *f*
193	flow front	front de la coulée *m*	Strömungsfront *f*
194	flow lines line of flow stream line	ligne d'écoulement *f*	Strömungslinie *f* Fliesslinie *f* Stromlinie *f*
195	mass movements mass wastage downslope creep	mouvements de masse *mpl*	Massenbewegungen *fpl*
196	slide sliding	éboulement *m* glissement *m*	Abgleitung *f* Abschiebung *f*
197	slide track	trace d'éboulement *f*	Gleitbahn *f*

lacolito *m*	lacólito *m* lacolito *m*	лакколит *m*	179
vesícula *f*	vesícula *f*	пузырёк *m* полость в минерале *f*	180
vesicular vesiculoso	vesicular poroso vesiculoso	пузырчатый ячеистый пористый кавернозный	181
vesiculación *f*	vesiculação *f*	образование пузырьков (во время застывания) *n* пористость *f*	182
lago de lava *m*	lago de lava *m*	лавовое озеро *n*	183
campo de lava *m*	campo de lava *m*	лавовое поле *n*	184
planicie de lava *f*	planície de lava *f*	лавовая равнина *f*	185
tubo de lava *m*	tubo de lava *m*	лавовая трубка *f*	186
lengua de lava *f*	língua de lava *f*	лавовый язык *m*	187
fluidez de la lava *f*	fluidez da lava *f* fluência de lava *f*	текучесть лавы *f*	188
estado de fluidez del magma *m*	estado de fluidez [fluência] *m*	текучая фаза магмы *f*	189
consolidación del magma *f*	consolidação do magma *f*	отверде(ва)ние лавы *n* затвердение [консолидация] магмы *n*[*f*]	190
textura fluidal *f* textura del flujo *f*	textura de fluxo *f* textura fluidal *f*	текстура течения *m* флуидальная текстура *f* структура потока *f*	191
canal de flujo *m*	canal de fluxo *m*	канал потока *m*	192
frente de flujo *f*	frente do fluxo *f*	фронт потока *m* лоб течения *m*	193
línea da flujo *f*	linha do fluxo *f* linha de corrente *f*	линия потока *f* линия течения *f*	194
movimiento de masa *m*	movimento(s) de massa *m*(*pl*) rastejamento (encosta abaixo) *m*	движение масс *n* сползание *n* оползание *n*	195
deslizamiento *m*	escorregamento *m* corrediça *f* deslizamento *m*	скольжение *n* сдвиг *m* оползень *m*	196
pista de deslizamiento *f*	traço de escorregamento *m* pista de deslizamento *f*	путь скольжения *m* путь обвала *m*	197

	SELENOLOGY	SÉLÉNOLOGIE	SELENOLOGIE
198	slumping (of walls) downward slipping	glissement de remparts *m* éboulement de remparts *m*	Herabgleiten *n* Gleiten *n* Rutschung *f* Rutschen *n*
199	slump material (that has) slipped down from rock slopes	affaissement *m* effondrement *m* éboulis *m*	Rutsch *m*
200	tectonic origin	origine tectonique *f*	tektonischer Ursprung *m* tektonische Entstehung *f*
201	tectonic process	processus tectonique *m*	tektonischer Prozess *m*
202	tectonic map	carte tectonique *f*	tektonische Karte *f*
203	structural map	carte structurale *f*	Strukturkarte *f*
204	local faulting	formation des failles locales *f* faillage local *m*	lokale Sprungbildung *f* lokale Verwerfungsbildung *f*
205	regional faulting	formation des failles régionales *f*	regionale Sprungbildung *f*
206	tension crack	crique de tension *f*	Spannungsriss *m*
207	tension fracture	fissure de tension *f* fissure d'extension *f*	Tensionsspalte *f* Ausdehnungsspalte *f*
208	linear fracture plane fracture	fracture linéaire *f*	gerader Bruch *m* ebener Bruch *m*
209	concentric fracture	fracture concentrique *f*	konzentrischer Bruch *m*
210	crustal stress	contrainte de la croûte *f*	Krustenspannung *f*
211	crater-margin fracture	fracture au rebord d'un cratère *f* fracture au rempart d'un cratère *f*	Spalte [Bruch] am Kraterrande *f*[*m*]
212	fault crustal dislocation	faille *f* rupture *f* dislocation *f*	Verwerfung *f* Bruch *m* Sprung *m*
213	lunar seismology seismology of the Moon	séismologie lunaire *f* sismologie lunaire *f*	Mondseismologie *f*
214	moonquake	seisme [tremblement] de Lune *m*	Mondbeben *n*
215	seismic event	événement sismique *m*	Bebenereignis *n*

SELENOLOGÍA	SELENOLOGIA	СЕЛЕНОЛОГИЯ	
deslizamiento de paredes *m* corrimiento de paredes *m*	deslizamento (de paredes) *m* escorregamento (de paredes) *m* desmoronamento *m*	оползание стенок *n*	198
hundimiento *m* desplome *m*	afundamento *m*	оползень *m* обвал *m* оседание пластов *n*	199
origen tectónico *m*	origem tectônica *f*	тектоническое происхождение *n*	200
proceso tectónico *m*	processo tectônico *m*	тектонический процесс *m*	201
mapa tectónico *m*	mapa tectônico *m*	тектоническая карта *f*	202
mapa estructural *m*	mapa estrutural *m*	структурная карта *f*	203
faulting local *m* formación de una falla local *f*	falhamento local *m* formação de falha local *f*	местное образование сбросов *n* местное сбрасывание *n* местная сбросовая деятельность *f*	204
formación de una falla regional *f*	falhamento regional *m* formação de falha regional *f*	региональное образование сбросов *n* региональное сбрасывание *n* региональная сбросовая деятельность *f*	205
grieta de tensión *f*	fenda de tensão *f*	трещина растяжения *f*	206
fractura de tensión *f*	fratura de tensão *f*	разрыв растяжения *m* излом растяжения *m*	207
fractura lineal *f*	fratura linear *f* fratura plana *f*	ровный излом *m*	208
fractura concéntrica *f*	fratura concêntrica *f*	концентрический излом *m* концентрическая трещина *f*	209
tensión cortical *f*	tensão crustal [da crosta] *f*	напряжение в коре *n* стресс в породах коры *m*	210
fractura de borde de cráter *f*	fratura da margem de cratera *f*	трещина вдоль края кратера *f*	211
falla *f* dislocación cortical *f*	falha *f* deslocamento da crosta *m*	сброс *m* разрыв *m* дислокация *f* разлом *m* сдвиг коры *m*	212
sismología lunar *f* sismología de Luna *f*	sismologia lunar *f* sismologia da Lua *f*	сейсмология Луны *f* лунная сейсмология *f*	213
lunamoto *m*	lunamoto *m*	лунотрясение *n*	214
suceso sísmico *m*	evento sísmico *m*	сейсмическое событие *n* сейсмическое явление *n*	215

216	seismic sounding	sondage sismique *m*	seismische Sondierung *f*
217	active seismology	séismologie active *f* sismométrie active *f*	aktive Seismologie *f*
218	passive seismology	séismologie passive *f*	passive Seismologie *f*
219	seismic experiment	expérience sismique *f* appareil sismique *m*	seismischer Versuch *m*
220	active seismic experiment	expérience sismique active *f* appareil de sismologie active *m*	aktiver seismischer Versuch *m*
221	passive seismometer	séismomètre passif *m* sismomètre passif *m*	passives Seismometer *n*
222	active seismometer	séismomètre actif *m* sismomètre actif *m*	aktives Seismometer *n*
223	seismic echo	écho sismique *m*	seismisches Echo *n*
224	impact seismogram	séismogramme d'un impact *m* sismogramme d'un impact *m*	Aufschlagsseismogramm *n* Aufsturzseismogramm *n*
225	impacting body	corps d'impact *m* corps impactant *m*	aufschlagender Körper *m* aufstürzender Körper *m*
226	man-made impact artificial impact	impact artificiel *m*	künstlicher Aufschlag *m* künstlicher Aufsturz *m*
227	natural fall	chute naturelle *f*	natürlicher Fall *m*
228	impact feature	formation [caractère] d'impact *f* caractéristique d'impact *f*	Einschlagformation *f* Aufschlagspur *f*
229	impact origin	formation par impact *f* origine d'impact *f*	Entstehung durch Einschlag *f* Entstehung durch Einsturz *f*
230	cometary impact	impact cométaire *m* impact d'une comète *m*	Kometenstoss *m*
231	surface activity	activité superficielle *f* activité de surface *f*	Tätigkeit der Oberfläche *f* Oberflächenaktivität *f*
232	cratering formation of craters	cratérisation *f* formation de cratères *f*	Kraterbildung *f*
233	cratering rate	taux de la formation de cratères *m* taux de la cratérisation *m*	Kraterbildungrate *f*
234	impact cratering	formation de cratères par impact *f*	Einschlagskraterbildung *f* Kraterbildung beim Einsturz *f*
235	meteoroidal bombardement meteoritic bombardement	bombardement par météorites [par météoroïdes] *m*	Meteoritenaufprall(en) *n* haufiger Einschlag von Meteoriten *m*

sondeo sísmico *m*	sondagem sísmica *f*	сейсмическое зондирование *n*	216
sismología activa *f*	sismologia ativa *f*	активная сейсмология *f*	217
sismología pasiva *f*	sismologia passiva *f*	пассивная сейсмология *f*	218
experimento sísmico *m*	experiência sísmica *f* experimento sísmico *m*	сейсмический эксперимент *m*	219
experimento sísmico activo *m*	experiência sísmica ativa *f* experimento sísmico ativo *m*	активный сейсмический эксперимент *m*	220
sismómetro pasivo *m*	sismômetro passivo *m*	пассивный сейсмометр *m*	221
sismómetro activo *m*	sismômetro ativo *m*	активный сейсмометр *m*	222
eco sísmico *m*	eco sísmico *m*	сейсмическое эхо *n*	223
sismograma de impacto *m*	sismograma de impacto *m*	сейсмограмма падения (тела на Луну) *f*	224
cuerpo impactante *m*	corpo de impacto *m*	ударяющее тело *n* упавшее тело *n*	225
impacto artificial *m*	impacto artificial *m*	искусственный удар *m*	226
caída natural *f*	queda [impacto] natural *f*[*m*]	естественное падение *n*	227
rasgo de impacto *m*	formação de impacto *f* figura [característica] de impacto *f*	ударное образование *n*	228
origen por impacto *m* formación por impacto *f*	origem de impacto *f*	происхождение ударом *n* ударное происхождение *n*	229
impacto cometario *m*	impacto cometário *m*	столкновение кометы *n* падение кометы *n* столкновение с кометой *n*	230
actividad superficial *f*	atividade superficial *f*	поверхностная активность *f*	231
formación de cráteres *f* craterización *f*	formação de crateras *f* craterização *f*	образование кратеров *n*	232
velocidad de craterización *f* velocidad de formación de cráteres *f*	taxa da formação de crateras *f* taxa da craterizacão *f*	частота образования кратеров *f* скорость образования кратеров *f*	233
craterización por impacto *f* formación de cráteres por impacto *f*	formação de crateras por impacto *f*	образование кратеров ударом *n* ударное образование кратеров *n*	234
bombardeo meteorítico *m* bombardeo meteoroidal *m*	bombardeamento meteoroidal *m* bombardeamento meteorítico *m* bombardeio meteoroidal [meteorítico] *m*	бомбардировка метеорными телами *f* бомбардировка метеороидами *f* метеоритная бомбардировка *f*	235

236	meteoroid impact meteoroidal impact meteoritic impact	impact d'un météoroïd *m* chute d'une météorite *f*	Meteoroidensturz *m* Meteoritenaufschlag *m* Aufprall eines Meteoriten *m* Meteoriteneinschlag *m*
237	shock metamorphism	métamorphisme de choc *m*	Stossmetamorphose *f*
238	impact lithification impact petrification	lithification par impact *f* pétrification par impact *f*	Stosslithifikation *f* Stossversteinerung *f*
239	pulverization	pulvérisation *f*	Zerstäubung *f* Pulverisierung *f* Pulverisieren *n*
240	particle bombard(e)ment	bombardement par les particules *m*	Beschuss mit Partikeln *m*
241	surface erosion	érosion de surface *f*	Erosion der Oberfläche *f*
242	meteoroid erosion meteoritic erosion	érosion par météoroïdes (par météorites) *f* érosion météoritique *f*	Abtragung durch Meteoroiden *f* Meteoritenerosion *f*
243	gas erosion	érosion gazeuse *f*	Gaserosion *f*
244	space weathering	désagrégation en milieu spatial *m* altération en environnement cosmique *f* altération par exposition à l'espace *f*	Verwitterung im kosmischen Raum *f*
245	solar-wind sputtering	pulvérization par le vent solaire *f* atomisation par le vent solaire *f*	Zerstäubung durch den Sonnenwind *f* Sonnenwindzerstäubung *f*
246	(lunar) photoelectron layer	couche de photoélectrons *f*	Photoelektronenschicht *f*
247	lunar flare (transient) luminous event	éruption lunaire *f*	Mondflare *f* vorübergehendes Aufleuchten *n* vorübergehende Leuchterscheinung *f*
248	lunar luminescence	luminescence lunaire *f*	Mondlumineszenz *f*
249	thermoluminescence	thermoluminescence *f*	Thermolumineszenz *f*
250	cosmic-ray irradiation	irradiation par rayons cosmiques *f*	Bestrahlung durch kosmische Strahlung *f*
251	particle track track	trace d'une particule *f*	Teilchenspur *f* Partikelspur *f*
252	fossil (nuclear) track	trace nucléaire fossile *f*	fossile Kernspur *f*
253	track density	densité de traces *f*	Dichte der Kernspuren *f*
254	cosmic-ray spallation	spallation (provoquée) par rayons cosmiques *f*	Kern(zer)splitterung durch kosmische Strahlung *f* Zersplitterung durch kosmische Strahlung *f*

SELENOLOGÍA	SELENOLOGIA	СЕЛЕНОЛОГИЯ	
impacto meteoroidal *m* impacto meteorítico *m*	impacto meteoroidal *m* impacto meteorítico *m*	удар метеорного тела *m* удар метеорита *m* удар метеороида *m*	236
metamorfismo de choque *m*	metamorfismo de choque *m*	ударный метаморфизм *m*	237
litificación por impacto *f* petrificación por impacto *f*	litificação por impacto *f* petrificação por impacto *f*	окаменение в результате удара *n* ударная литификация *f*	238
pulverización *f*	pulverização *f*	распыление *n* испыление *n* пульверизация *f*	239
bombardeo de [por] partículas *m*	bombardeamento de [por] partículas *m*	бомбардировка частицами *f*	240
erosión superficial *f*	erosão da superfície *f*	поверхностная эрозия *f*	241
erosión por meteoroides *f* erosión meteorítica *f*	erosão meteoroidal *f* erosão meteorítica *f*	метеороидная эрозия *f* метеоритная эрозия *f*	242
erosión gaseosa *f*	erosão gasosa *f*	газовая эрозия *f*	243
meteorización espacial *f*	meteorização espacial *f* desagregação espacial *f*	пространственное выветривание *n*	244
pulverización por viento solar *f* atomización por viento solar *f*	pulverização pelo vento solar *f*	распыление солнечным ветром *n* атомизация солнечным ветром *f*	245
capa fotoelectrónica *f*	camada fotoele(c)trônica *f*	фотоэлектронный слой *m*	246
destello lunar *m* fenómeno luminoso transitorio *m*	erupção lunar *f*	лунная вспышка *f*	247
luminescencia lunar *f*	luminescência lunar *f*	лунная люминесценция *f*	248
termoluminescencia *f*	termoluminescência *f*	термолюминесценция *f*	249
irradiación por radiación cósmica *f* irradiación de rayos cósmicos *f*	irradiação de [por] raios cósmicos *f*	облучение космическими лучами *n*	250
traza de una partícula *f*	traço de partícula *m*	след частицы *m* трек (частицы) *m*	251
traza nuclear fósil *f* paleotraza nuclear *f*	traço nuclear fóssil *m*	первичный ядерный след *m* первичный трек ядра *m*	252
densidad de trazas *f*	densidade de traços *f*	плотность треков [следов] *f*	253
astillado por radiación cósmica *f*	espalação por raios cósmicos *f*	растрескивание космическими лучами *n* скалывание ядра космическими лучами *n*	254

255	cosmogenic radionuclide bombard(e)ment-produced radionuclide	radioisotope cosmogénique *m*	kosmogenes Radionuklid *n*
256	primordial radionuclide	radioélément primitif *m*	radioaktives Urnuklid *n*
257	chemical evolution	évolution chimique *f*	chemische Entwicklung *f*
258	internal process endogenous process	processus interne [endogène] *m* activité interne *f*	innerer Prozess *m* innerer Vorgang *m*
259	endogenous endogenic	endogène	endogen innenbürtig
260	thermal history	histoire thermique *f* évolution thermique passée *f*	thermische Geschichte *f*
261	degassing outgassing	dégazage *m*	Entgasung *f* Gasaustreibung *f*
262	volatile substance volatile	substance volatile *f* volatil *m*	flüchtiger Stoff *m* Flüchtige *n*
263	defluidization	défluidisation *f*	Defluidisation *f*
264	Moon's age	âge de la Lune *m*	Alter des Mondes *n*
265	dating of lunar events	datation des événements lunaires *f* détermination de l'âge *f*	Datierung der Mondereignisse *f* Altersbestimmung *f*
266	structural relations	relations structurales *fpl*	Strukturbeziehungen *fpl*
267	stratigraphic age	âge stratigraphique *m*	stratigraphisches Alter *n*
268	relative age	âge relatif *m*	relatives Alter *n*
269	absolute age	âge absolu *m*	absolutes Alter *n*
270	thorium-lead dating	datation par le rapport thorium--plomb *f*	Thorium-Blei-Datierung *f*
271	strontium age rubidium-strontium age	âge de [par le] strontium *m* âge déterminé par le rapport rubidium--strontium *m*	Strontiumalter *n* Rubidium-Strontium Alter *n*
272	neap tide dead tide	marée de morte eau *f* marée de quadrature *f* petite marée *f*	Nippflut *f* taube Flut *f*

radionúclido cosmogénico *m* radionúclido producido por bombardeo *m* radionucleido cosmogénico *m*	radionuclídeo cosmogênico *m* radionuclídeo produzido por bombardeio [bombardeamento] *m*	космогенический радионуклид *m* радиоизотоп, вызванный бомбардировкой космических лучей *m*	255
radionúclido primordial *m* radionucleido primitivo *m*	radionuclídeo primordial *m*	первичный радиоактивный изотоп *m* изначальный радионуклид *m*	256
evolución química *f*	evolução química *f*	химическая эволюция *f*	257
proceso interno *m* proceso endógeno *m*	processo interno processo endógeno *m*	внутренний процесс *m* эндогенный процесс *m*	258
endógeno	endógeno	эндогенный [глубинного происхождения]	259
historia térmica *f*	história térmica *f*	термическая история *f*	260
desgasificación *f*	desprendimento [escape] de gás *m* ação de degasar *f*	обезгаживание *n* дегазация *f*	261
substancia volátil *f* (fase) volátil *f*	su(b)stância volátil *f* volátil *m*	летучее вещество *n* летучее *n*	262
desfluidización *f*	de(s)fluidização *f* de(s)fluid(ific)ação *f*	обезвоживание *n* дефлюидизация *n*	263
edad de la Luna *f*	idade da Lua *f*	возраст Луны *m*	264
datación de sucesos lunares *f* fechado de sucesos lunares *m*	datação de eventos lunares *f*	датирование лунных событий *n* установление периодов в лунной истории *n* определение возраста *n*	265
relaciones estructurales *fpl*	relações estruturais *fpl*	структурные отношения *npl*	266
edad estratigráfica *f*	idade estratigráfica *f*	стратиграфический возраст *m*	267
edad relativa *f*	idade relativa *f*	относительный возраст *m*	268
edad absoluta *f*	idade absoluta *f*	абсолютный возраст *m*	269
datación por torio-plomo *f* fechado torio-plomo *m*	datação tório-chumbo *f* datando por meio da relação chumbo-tório *m*	торий-свинцовый метод определения возраста *m*	270
edad de estroncio *f* edad rubidio-estroncio *f*	idade de estrôncio *f* idade (de) rubídio-estrôncio *f*	рубидий-стронциевый возраст *m* возраст по радиогенному стронцию *m*	271
mareas de cuadratura *fpl* mareas bajas *fpl* aguas muertas *fpl* marea muerta *f*	maré de águas mortas *f* maré de quadratura *f* maré de quarto *f* maré morta *f*	квадратурный прилив *m*	272

273	syzygial tide (high water of) spring tide big tide total tide	marée de syzygie *f* marée de vive-eau *f* grande marée *f* hautes eaux de syzygie *pl*	Springflut *f* Springtide *f* Syzygienhochwasser *n*
274	tidal interaction	interaction due aux marées *f*	Gezeitenwechselwirkung *f*
275	Earth-induced tides Earth-produced tides	marée produite par la Terre *f*	erdinduzierte Gezeiten *pl*
276	tidal effect	effet de marée *m*	Gezeiteneffekt *m* Gezeitenphänomen *n*
277	tidal deformation	déformation due à la marée *f*	Gezeitendeformation *f*
278	tidal bulge	renflement dû aux marées *m*	Gezeitenberg *m* Flutberg *m*
279	tidal current tidal stream	courant de marée *m* flux de marée *m*	Gezeitenstrom *m* Gezeitenströmung *f* Flutströmung *f* Tidenstrom *m*
280	flood afflux flood-tide current	flot *m* flux *m* marée montante *f*	Flut *f* Flutstrom *m* Anlauf *m*
281	high tide high water	marée haute *f*	Gezeitenhochwasser *n* Tidenhochwasser *n*
282	ebb reflux falling tide	jusant *m* reflux *m* marée descendante *f*	Ebbe *f* Ebbestrom *m* Abstrom *m* Rückfluss *m*
283	low tide low water	marée basse *f* basse-mer *f*	Gezeitenniedrigwasser *n* Niedrigwasser *n*
284	tide-raising body	corps producteur de marée *m*	Gezeiten verursachender Körper *m*
285	energy dissipation	dissipation de l'énergie *f*	Dissipation der Energie *f*
286	tidal dissipation	dissipation (par effet) de marée *f*	Gezeitendissipation *f*
287	tidal heat	chaleur de marée *f*	Gezeitenwärme *f*
288	tidal torque	couple freinant de marée *m* couple dû à la marée *m*	Moment des Gezeitenkräftepaares *n*
289	tidal force tide-raising force tide-generating force	force de marée *f* force marémotrice *f*	Gezeitenkraft *f* Flutkraft *f* gezeitenerzeugende Kraft *f*
290	tidal day lunar day	jour de marée *m*	Tidentag *m*

SELENOLOGÍA	SELENOLOGIA	СЕЛЕНОЛОГИЯ	
mareas vivas *fpl* mareas de sicigias *fpl* mareas altas *fpl* aguas vivas *fpl* marea equinoccial *f*	maré de sizígias *f* maré alta de sizígia *f* maré de águas vivas *f* água-viva *f*	сизигийный прилив *m* сизигийная полная вода *f*	273
interacción de marea *f*	interação de marés *f*	приливное взаимодействие *n*	274
mareas inducidas por la Tierra *fpl*	marés induzidas pela Terra *fpl*	прилив, вызванный Землей *m*	275
efecto de marea *m*	efeito de maré *m*	приливный эффект *m* приливо-отливное явление *n*	276
deformación de marea *f*	deformação de maré *f*	приливная деформация *f*	277
ola de marea *f*	saliência [protuberância] da maré alta *f*	приливный горб *m* приливная выпуклость *f*	278
flujo de marea *m*	fluxo de maré *m*	приливно-отливное течение *n* приливное течение *n*	279
flujo *m* marea entrante *f*	enchente *f* enchente da maré *f* maré montante *f*	прилив *m*	280
alta [plena] mar *f* marea alta *f* pleamar *f* aguas llenas *fpl*	preamar *f* maré alta *f* maré cheia *f*	полный прилив *m* полная вода *f* наивысший уровень прилива *m*	281
marea menguante *f* reflujo *m* marea vacilante *f*	maré descendente *f* refluxo da maré *m* vazante da maré *f*	отлив *m*	282
marea baja *f* bajamar *f*	maré vazia *f* maré baixa *f* baixa-mar *f*	малая вода *f* меженная вода *f* низкая вода *f*	283
cuerpo causante de marea *m*	corpo produtor de maré *m*	приливообразующее тело *n*	284
disipación de la energía *f*	dissipação da energia *f*	диссипация энергии *f* рассеяние энергии *n*	285
disipación de marea *f*	dissipação da maré *f*	приливная диссипация *f*	286
calor de marea *m*	calor da maré *m*	теплота, вызванная приливом *f*	287
par de marea *m* cupla de marea *f*	torque [conjugado] da maré *m* binário da maré *m*	приливный момент вращения *m* приливная пара сил *f*	288
fuerza de marea *f*	força de maré *f* força geradora da maré *f*	приливная сила *f* приливообразующая сила *f* приливо-отливная сила *f*	289
día de mareas *m* día lunar *m*	dia de maré *m* dia lunar *m*	приливные сутки *pl*	290

291	tidal wave	onde de marée *f*	Gezeitenwoge *f* Gezeitenwelle *f* Flutwelle *f*
292	tidal amplitude tidal range	amplitude des marées *f* marnage *m*	Gezeitenhub *m* Tidenhub *m* Tidenstieg *m*
293	tidal potential	potentiel de marée *m*	Gezeitenpotential *n*
294	atmospheric tides air tides	marées atmosphériques *fpl*	atmosphärische Gezeiten *pl* Atmosphärengezeiten *pl*
295	geomagnetic tides	marées géomagnétiques *fpl*	erdmagnetische Gezeiten *pl*
296	earth tides terrestrial tides	marées terrestres *fpl*	Erdgezeiten *pl* Gezeiten des Erdkörpers *pl*
297	crustal tides	marées de la croûte *fpl*	Erdkrustengezeiten *pl*
298	gravitational tides	marées gravitationnelles *fpl*	Gravitationsgezeiten *pl*
299	thermal tides	marées thermiques *fpl*	thermische Gezeiten *pl*
300	lunar time	temps lunaire *m*	Mondzeit *f*
301	apparent lunar time	temps lunaire vrai *m*	wahre Mondzeit *f*
302	mean lunar time	temps lunaire moyen *m*	mittlere Mondzeit *f*
303	(meridian) transit of the Moon lunar transit culmination of the Moon	passage de la Lune par le méridien *m* culmination de la Lune *f*	Mond(meridian)durchgang *m* Mondkulmination *f*
304	mean lunar day	jour lunaire moyen *m*	mittlerer Mondtag *m*
305	tidal theory	théorie des marées *f*	Gezeitentheorie *f*
306	forced tidal oscillations	oscillation forcée de marée *f*	erzwungene Gezeitenschwingungen *fpl*
307	lunisolar component	composante lunisolaire *f*	Mond-Sonnentidenkomponente *f*
308	lunar tides	marées lunaires *fpl*	Mondgezeiten *pl* Mondtiden *pl*
309	sea tides	marées maritimes *fpl*	Meeresgezeiten *pl*
310	ocean tides	marées océaniques *fpl*	ozeanische Gezeiten *pl*
311	tide gauge mareograph	indicateur d'amplitudes des marées *m* marégraphe *m*	Gezeitenhubmesser *m* Mareograph *m*
312	tidal friction	friction (résultant) des marées *f*	Gezeitenreibung *f*
313	retardation of Earth's rotation	ralentissement de la rotation de la Terre *m*	Verzögerung der Erdrotation *f*

onda de marea *f*	onda de maré *f*	приливная волна *f*	291
rango de las mareas *m* amplitud de las mareas *f*	amplitude da maré *f* extensão da maré *f*	амплитуда прилива *f* подъём прилива *m* величина прилива *f*	292
potencial de marea *m*	potencial da maré *m*	приливный потенциал *m*	293
mareas atmosféricas *fpl*	marés atmosféricas *fpl*	атмосферные приливы и отливы *mpl*	294
mareas geomagnéticas *fpl*	marés geomagnéticas *fpl*	геомагнитные приливы и отливы *mpl*	295
mareas terrestres *fpl*	marés terrestres *fpl* marés sólidas *fpl*	земные приливы *mpl*	296
mareas de la corteza *fpl*	marés da crosta *fpl*	приливы земной коры *mpl*	297
mareas gravitacionales *fpl*	marés gravitacionais *fpl*	гравитационные приливы *mpl*	298
mareas térmicas *fpl*	marés térmicas *fpl*	термические приливы *mpl*	299
tiempo lunar *m*	tempo lunar *m*	лунное время *n*	300
tiempo lunar aparente *m*	tempo lunar aparente *m*	истинное лунное время *n*	301
tiempo lunar medio *m*	tempo lunar médio *m*	среднее лунное время *n*	302
tránsito de la Luna *m* tránsito lunar *m* paso de la Luna por el meridiano *m* culminación de la Luna *f*	trânsito da Lua *m* trânsito lunar *m* culminação da Lua *f*	кульминация Луны *f* прохождение Луны через меридиан *n*	303
día lunar medio *m*	dia lunar médio *m*	средние лунные сутки *pl*	304
teoría de las mareas *f*	teoria das marés *f*	теория приливов *f*	305
oscilaciones forzadas de las mareas *fpl* oscilaciones de marea inducidas *fpl*	oscilações forçadas da maré *fpl*	вынужденные приливные колебания *npl*	306
componente lunisolar *m*	componente lunissolar *m*	лунно-солнечная составляющая *f* лунно-солнечный компонент *m*	307
mareas lunares *fpl*	marés lunares *fpl*	лунные приливы *mpl*	308
mareas de mar *fpl*	marés marítimas *fpl*	морские приливы *mpl*	309
mareas oceánicas *fpl*	marés oceânicas *fpl*	океанские приливы *mpl*	310
indicador de mareas *m* mareómetro *m* mareógrafo *m* marégrafo *m*	mareômetro *m* mareógrafo *m* marégrafo *m* marêmetro *m*	приливомер *m* мареограф *m*	311
fricción de la marea *f*	atrito [fricção] das marés *m*[*f*]	приливное трение *n* трение приливной волны *n*	312
retardo de la rotación terrestre *m*	atraso da rotação da Terra *m*	замедление вращения Земли *n*	313

83 PLANETS

	PLANETS	PLANÈTES	PLANETEN
1	planet	planète *f*	Planet *m* Wandelstern *m*
2	planetary system	système planétaire *m*	Planetensystem *n*
3	Bode('s) law law of planetary distances Titius-Bode formula [law]	loi de Titius-Bode [de Bode] *f* loi de distances planétaires [des planètes] *f*	Bodesches Gesetz *n* Abstandsgesetz der Planeten *n* Titius-Bodesche Regel *f*
4	planetary physics planetophysics physics of planets	planétophysique *f* physique planétaire *f* physique des planètes *f*	Planetenphysik *f* Physik der Planeten *f*
5	planetology planetary geology	planétologie *f* géologie des planètes *f*	Planetologie *f* planetarische Geologie *f* Planetengeologie *f*
6	planetary	planétaire	Planeten- planetar(isch)
7	planetographic	planétographique	planetographisch
8	planetographic coordinates	coordonées planétographiques *fpl*	planetographische Koordinaten *fpl*
9	planetocentric (orbit, coordinates)	planétocentrique (orbite, coordonnées)	planetozentrisch (Umlaufbahn, Koordinaten)
10	planetocentric system (of reference, of coordinates)	système planétocentrique (de coordonnées, de reférence) *m*	planetozentrisches System (von Koordinaten) *n* Bezugssystem *n*
11	solar planet planet of the Sun planet of the solar system	planète solaire *f* planète du Soleil *f* planète du système solaire *f*	Sonnenplanet *m* Planet der Sonne *m* Planet des Sonnensystems *m*
12	solar system	système solaire *f*	Sonnensystem *n*
13	major planets	planètes principales *fpl* grosses planètes *fpl* planètes majeures *fpl*	grosse Planeten *mpl* Hauptplaneten *mpl*
14	inferior planet inner planet	planète inférieure *f* planète intérieure *f*	innerer Planet *m* unterer Planet *m* sonnennaher Planet *m*
15	superior planet outer planet	planète supérieure *f* planète extérieure *f*	äusserer Planet *m* oberer Planet *m* sonnenferner Planet *m*
16	drawing of a planet	dessin d'une planète *m*	Zeichnung eines Planeten *f* Planetenzeichnung *f*
17	satellite	satellite *m*	Satellit *m* Trabant *m*
18	natural satellite	satellite naturel *m*	natürlicher Satellit *m*
19	satellite system	système de satellites *m*	Satellitensystem *n*

planeta *m*	planeta *m*	планета *f*	1
sistema planetario *m*	sistema planetário *m*	планетная система *f*	2
ley de Bode *f* ley de distancias planetarias *f* fórmula de Titius-Bode *f*	lei de Bode *f* lei das distâncias planetárias *f* fórmula de Titius-Bode *f*	закон Боде *m* закон планетных расстояний *m* правило Тициуса-Боде *n*	3
física planetaria *f* física de los planetas *f*	planetofísica *f* física planetária *f* física dos planetas *f*	физика планет *f* планетофизика *f*	4
planetología *f* geología planetaria *f*	planetologia *f* geologia planetária *f* geologia dos planetas *f*	планетная геология *f* геология планет *f* планетология *f*	5
planetario	planetário	планетный	6
planetográfico	planetográfico	планетографический	7
coordenadas planetográficas *fpl*	coordenadas planetográficas *fpl*	планетографические координаты *fpl*	8
planetocéntrico (órbita, coordenadas)	planetocêntrico (órbita, coordenadas)	планетоцентрический (орбита, координаты)	9
sistema planetocéntrico (de coordenadas, de referencia) *m*	sistema planetocêntrico (de coordenadas, de referência) *m*	планетоцентрическая система (координат) *f*	10
planeta solar *m* planeta del Sol *m* planeta del sistema solar *m*	planeta solar *m* planeta do Sol *m* planeta do sistema solar *m*	планета солнечной системы *f* планета Солнца *f* солнечная планета *f*	11
sistema solar *m*	sistema solar *m*	солнечная система *f*	12
grandes planetas *mpl* planetas mayores *mpl* planetas principales *mpl*	grandes planetas *mpl* planetas maiores *mpl* planetas principais *mpl*	большие планеты *fpl* главные планеты *fpl*	13
planeta inferior *m* planeta interior *m*	planeta inferior *m* planeta interior *m*	внутренняя планета *f* нижняя планета *f*	14
planeta superior *m* planeta exterior *m*	planeta superior *m* planeta exterior *m* planeta externo *m*	внешняя планета *f* верхняя планета *f*	15
dibujo de un planeta *m*	desenho de um planeta *m*	рисунок планеты *m*	16
satélite *m*	satélite *m*	спутник *m*	17
satélite natural *m*	satélite natural *m*	естественный спутник *m*	18
sistema de satélites *m*	sistema de satélites *m*	система спутников *f* спутниковая система *f*	19

20	innermost satellite	satellite le plus proche (de sa planète) *m*	der innerste Satellit *m*
21	irregular satellite retrograde satellite	satellite irrégulier *m* satellite rétrograde *m*	unregelmässiger Satellit *m* rückläufiger Satellit *m* Satellit auf retrograder Bahn *m*
22	rocky satellite	satellite rocheux *m* satellite lithique *m*	Felssatellit *m* vorwiegend aus Felsen bestehender Satellit *m*
23	icy satellite	satellite de glace *m*	vorwiegend aus Eis bestehender Satellit *m* Eissatellit *m*
24	solar perturbations (on satellites)	perturbations solaires (sur les satellites) *fpl*	Störungen (der Satellitenbewegung) durch die Sonne *fpl*
25	mutual perturbations (among satellites)	perturbations mutuelles (entre les satellites) *fpl*	gegenseitige Störungen (zwischen den Satelliten) *fpl*
26	satellite survey satellite monitoring	surveillance des satellites *f* levé [observation] de satellites *m*[*f*]	Satellitenüberwachung *f* Überwachen von Satelliten *n*
27	satellite search	recherche des satellites *f*	Aufsuchen von Satelliten *n*
28	planetary motion motion of planets	mouvement planétaire *m* mouvement des planètes *m*	Planetenbewegung *f*
28a	aphelion	aphélie *m*	Aphel *n*
29	planetary orbit orbit of a planet	orbite planétaire *f* orbite d'une planète *f*	Planetenbahn *f* Bahn eines Planeten *f*
29a	perihelion	périhélie *m*	Perihel *n*
30	planetary rotation rotation of planets	rotation planétaire *f* rotation de planètes *f*	Planetenrotation *f* Drehung der Planeten *f*
31	rotation period period of rotation spin period axial period	période de rotation *f* durée de rotation *f* période diurne *f*	Rotationsperiode *f* Rotationszeit *f*
32	velocity of rotation rotational velocity rotation rate [speed] spin rate	vitesse de rotation *f*	Rotationsgeschwindigkeit *f* Umdrehungsgeschwindigkeit *f* Drehungsgeschwindigkeit *f*
33	rotation axis axis of rotation	axe de rotation *m*	Rotationsachse *f* Umdrehungsachse *f*
34	axis direction orientation of the axis	direction de l'axe *f* orientation de l'axe *f*	Achsenrichtung *f* Richtung der Achse *f*
35	axial inclination axial tilt	inclinaison de l'axe *f*	Achsenneigung *f* Neigung der Achse *f*

satélite más interior *m*	satélite mais interior *m*	ближайший спутник *m*	20
satélite irregular *m* satélite retrógrado *m*	satélite irregular *m* satélite retrógrado *m*	иррегулярный спутник *m* спутник с обратным движением *m* обратный спутник *m*	21
satélite rocoso *m*	satélite rochoso *m*	каменный спутник *m* спутник из скальных пород *m* спутник из каменных пород *m*	22
satélite de hielo *m*	satélite de gelo *m*	ледяной спутник *m*	23
perturbaciones solares (en los satélites) *fpl*	perturbações solares (em satélites) *fpl*	возмущения, вызванные Солнцем (в движении спутников) *npl*	24
perturbaciones mutuas (entre satélites) *fpl*	perturbações mútuas (entre satélites) *fpl*	взаимные возмущения (в движении спутников) *npl*	25
patrullaje [levantamiento] de satélites *m* vigilancia [monitoraje] de satélites *f*	pesquisa [levantamento, patrulhamento] de satélites *f*[*m*] patrulha [monitoração] de satélites *f*	наблюдения спутников *npl* слежение за спутниками *n*	26
búsqueda de satélites *f*	procura [busca] de satélites *f*	поиск спутников *m*	27
movimiento planetario *m* movimiento de planetas *m*	movimento planetário *m* movimento de planetas *m*	планетное движение *n* движение планет *n*	28
afelio *m*	afélio *m*	афелий [апогелий] *m*	28a
órbita planetaria *f* órbita de un planeta *f*	órbita planetária *f* órbita de um planeta *f*	планетная орбита *f* орбита планеты *f*	29
perihelio *m*	periélio *m*	перигелий *m*	29a
rotación planetaria *f* rotación de planetas *f*	rotação planetária *f* rotação dos planetas *f*	вращение планет *n* планетное вращение *n*	30
período de rotación *m* duración de la rotación *f*	periódo de rotação *m* duração da rotação *f*	период (осевого) вращения *m* вращательный период *m*	31
velocidad de rotación *f* velocidad de giro *f*	velocidade de rotação *f*	скорость вращения *f* вращательная скорость *f*	32
eje de rotación *m*	eixo de rotação *m*	ось вращения *f*	33
dirección del eje *f*	direção do eixo *f*	направление оси *n*	34
inclinación del eje *f*	inclinação axial *f* inclinação do eixo *f*	наклон оси *m* угол наклона оси *m*	35

36	axis pole axial pole pole of the rotation axis	pôle de rotation *m* pôle d'axe *m*	Achsenpol *m* Rotationspol *m*
37	synchronous rotation synchronous spin	rotation synchrone *f*	gebundene Rotation *f*
38	non-synchronous rotation [spin] asynchronous rotation	rotation non-synchrone *f* rotation asynchrone *f*	nichtgebundene Rotation *f* ungebundene Rotation *f*
39	direct rotation	rotation directe *f*	rechtläufige [direkte] Rotation *f*
40	retrograde rotation	rotation rétrograde *f*	retrograde Rotation *f* rückläufige Rotation *f*
41	apparent motion	mouvement apparent *m*	scheinbare Bewegung *f*
42	direct	direct	direkt rechtläufig
43	direct motion	mouvement direct *m*	rechtläufige Bewegung *f*
44	station of a planet	station d'une planète *f*	Stillstand eines Planeten *m*
45	stationary point stationary position	point stationnaire *m*	stationärer Punkt *m*
46	retrograde	rétrograde	retrograd rückläufig
47	retrograde motion	mouvement rétrograde *m*	rückläufige Bewegung *f*
48	deferent	déférent *m*	Deferent *m* Deferenzkreis *m*
49	epicycle	épicycle *m*	Epizykel *m*
50	planetary configuration aspect relative position of planets	configuration des planètes *f* configuration planétaire *f* aspect *m*	Konfiguration der Planeten *f* (Planeten-)Konstellation *f*
51	transit of a planet across the solar disk	passage d'une planète devant le (disque du) Soleil *m*	Vorübergang eines Planeten vor der Sonnenscheibe *m* Vorübergang eines Planeten vor der Sonne *m*
52	black drop	goutte noire *f*	schwarzer Tropfen *m*
53	shadow transit	passage de l'ombre *m*	Vorübergang des Schattens (des Trabanten, des Satelliten) *m*
54	satellite transit (across the planet's disk)	passage du satellite *m*	Vorübergang des Trabanten (vor der Planetenscheibe) *m* Vorübergang des Satelliten *m*
55	conjunction	conjonction *f*	Konjuktion *f*
56	superior conjunction	conjonction supérieure *f*	obere Konjunktion *f*

polo del eje *m* polo de rotación *m*	pólo de rotação *m* pólo de eixo *m*	полюс оси *m* полюс (оси) вращения *m*	36
rotación sincrónica *f*	rotação síncrona *f* rotação sincrônica *f*	синхронное вращение *n*	37
rotación no sincrónica *f* rotación asincrónica *f*	rotação assíncrona *f* rotação não-sincrônica *f*	асинхронное вращение *n*	38
rotación directa *f*	rotação direta *f*	прямое вращение *n*	39
rotación retrógrada *f*	rotação retrógrada *f*	обратное вращение *n*	40
movimiento aparente *m*	movimento aparente *m*	видимое движение *n*	41
directo	direto	прямой	42
movimiento directo *m*	movimento direto *m*	прямое движение *n*	43
estación de un planeta *f*	estação de um planeta *f*	стояние планеты *n*	44
punto estacionario *m* posición estacionaria *f*	posição estacionária *f* ponto estacionário *m*	точка стояния *f*	45
retrógado	retrógrado	обратный попятный	46
movimiento retrógrado *m*	movimento retrógrado *m*	обратное движение *n* попятное движение *n*	47
deferente *m* círculo deferente *m*	deferente *m*	деферент *m*	48
epiciclo *m*	epiciclo *m*	эпицикл *m*	49
configuración planetaria *f* constelación *f* aspecto *m*	configuração planetária *f* aspe(c)to dos planetas *m* posição relativa dos planetas *f*	конфигурация планет *f* планетная конфигурация *f* аспекты планет *mpl*	50
paso de un planeta por el disco solar *m* tránsito de un planeta sobre el disco solar *m*	trânsito de um planeta diante do Sol *m* passagem de um planeta diante do disco solar *f*	прохождение планеты по диску Солнца *n*	51
gota negra *f*	gota negra *f*	чёрная капля *f*	52
tránsito de la sombra *m*	passagem da sombra *f*	прохождение тени (спутника по диску планеты) *n*	53
paso de un satélite (a través del disco de un planeta) *m* tránsito del satélite *m*	passagem do satélite *f* trânsito de um satélite (diante do disco do planeta) *m*	прохождение спутника (по диску планеты) *n*	54
conjunción *f*	conjunção *f*	соединение *n*	55
conjunción superior *f*	conjunção superior *f*	верхнее соединение *n*	56

57	inferior conjunction	conjonction inférieure *f*	untere Konjunktion *f*
58	triple conjunction	conjonction triple *f*	dreifache Konjunktion *f*
59	opposition	opposition *f*	Opposition *f*
60	elongation	élongation *f*	Elongation *f*
61	eastern elongation	élongation orientale *f*	östliche Elongation *f*
62	western elongation	élongation occidentale *f*	westliche Elongation *f*
63	greatest elongation	élongation maximale *f*	grösste Elongation *f*
64	quadrature	quadrature *f*	Quadratur *f*
65	eastern quadrature	quadrature orientale *f*	östliche Quadratur *f*
66	western quadrature	quadrature occidentale *f*	westliche Quadratur *f*
67	illuminated hemisphere sunlit hemisphere day hemisphere sunlit side illuminated side daylight hemisphere	hémisphère éclairé *m* hémisphère diurne *m* hémisphère ensoleillé *m* côté illuminé *m* côté jour *m*	beleuchtete Hemisphäre *f* Tagseite *f* Tagesseite *f* sonnenbeschienene Hemisphäre *f* helle Seite *f*
68	non-illuminated hemisphere dark hemisphere antisolar hemisphere dark side night hemisphere nocturnal hemisphere unilluminated hemisphere	hémisphère obscur *m* hémisphère nocturne *m* côté nocturne *m* côté nuit *m* côté non-éclairé *m* côté à l'ombre *m*	dunkle Hemisphäre *f* unbeleuchtete Hemisphäre *f* Nachtseite *f* Nachthemisphäre *f* dunkle Seite *f*
69	terminator shadow line	terminateur *m*	Terminator *m* Schattengrenze *f*
70	sunrise terminator morning terminator dawn terminator	terminateur du matin *m* terminateur matinal *m*	Morgenterminator *m*
71	dusk terminator sunset terminator evening terminator	terminateur du soir *m* terminateur vesperal *m*	Abendterminator *m*
72	phase angle	angle de phase *m*	Phasenwinkel *m*
73	zero phase	phase nulle *f*	Nullphase *f*
74	phase curve	courbe de phase *f*	Phasenkurve *f*

conjunción inferior *f*	conjunção inferior *f*	нижнее соединение *n*	57
conjunción triple *f*	conjunção tripla *f*	тройное соединение *n*	58
oposición *f*	oposição *f*	противостояние *n* оппозиция *f*	59
elongación *f*	elongação *f*	элонгация *f*	60
elongación este *f* elongación oriental *f*	elongação oriental *f*	восточная элонгация *f*	61
elongación occidental *f* elongación oeste *f*	elongação ocidental *f*	западная элонгация *f*	62
máxima elongación *f*	máxima elongação *f*	наибольшая элонгация *f*	63
cuadratura *f*	quadratura *f*	квадратура *f*	64
cuadratura oriental *f*	quadratura oriental *f*	восточная квадратура *f*	65
cuadratura occidental *f*	quadratura ocidental *f*	западная квадратура *f*	66
hemisferio iluminado *m* hemisferio diurno *m* lado diurno *m* lado iluminado *m* lado del Sol *m* hemisferio de dia *m*	hemisfério iluminado *m* hemisfério diurno *m* lado iluminado *m* hemisfério voltado para o Sol *m*	полушарие освещённое Солнцем *n* дневное полушарие *n* освещённое полушарие *n* освещённая полусфера *f* солнечная сторона *f* дневная сторона [полусфера] *f*	67
hemisferio no iluminado *m* hemisferio oscuro [obscuro] *m* hemisferio nocturno *m* lado de noche *m* lado obscuro *m* hemisferio de noche *m*	hemisfério não-iluminado *m* hemisfério noturno *m* hemisfério escuro *m* hemisfério obscuro *m* lado da noite *m* lado escuro *m*	ночное полушарие *n* неосвещёное полушарие *n* теневая сторона *f* неосвещённая полусфера *f* ночная полусфера *f*	68
terminador *m* línea de sombra *f*	terminadouro *m* círculo de iluminação *m* terminadoiro *m*	терминатор *m* граница тени *f*	69
terminador matutino *m*	terminadouro matinal [da manhã] *m* terminadoiro matutino [de nascente] *m*	утренний терминатор *m* терминатор на рассвете *m* терминатор при восходе Солнца *m*	70
terminador vespertino *m*	terminadouro da tarde [do poente] *m* terminadouro vespertino [do anoitecer] *m* terminadoiro vespertino *m*	вечерний терминатор *m* сумеречный терминатор *m* терминатор при заходящем Солнце *m*	71
ángulo de fase *m*	ângulo de fase *m*	угол фазы *m* фазовый угол *m*	72
fase cero *f*	fase zero *f* fase nula *f*	нулевая фаза *f*	73
curva de fase *f*	curva de fase *f*	фазовая кривая *f*	74

	PLANETS	PLANÈTES	PLANETEN
75	opposition effect	effet d'opposition *m*	Oppositionseffekt *m*
76	Mie scattering	diffusion de Mie *f*	Mie-Streuung *f*
77	phase coefficient	coefficient de phase *m*	Phasenkoeffizient *m*
78	phase function	fonction de phase *f*	Phasenfunktion *f*
79	phase integral	intégrale de phase *f*	Phasenintegral *n*
80	polarimetric measurements	mesures polarimétriques *fpl*	polarimetrische Messungen *fpl* Polarisationsmessungen *fpl*
81	polarization curve	courbe de polarisation *f*	Polarisationskurve *f*
82	visual polarimeter	polarimètre visuel *m*	visuelles Polarimeter *n*
83	photoelectric polarimeter	polarimètre photoélectrique *m*	lichtelektrisches Polarimeter *n*
84	photographic polarimeter	polarimètre photographique *m*	photographisches Polarimeter *n*
85	mean magnitude mean opposition magnitude	magnitude moyenne *f* magnitude moyenne à l'opposition *f*	mittlere Helligkeit *f* mittlere Sterngrösse *f* Sterngrösse in der mittleren Opposition *f* mittlere Grössenklasse *f*
86	absolute magnitude of a planet	magnitude absolue d'une planète *f*	absolute Helligkeit eines Planeten *f* absolute Grössenklasse eines Planeten *f* absolute Sterngrösse eines Planeten *f*
87	special colo(u)r-index	indice de couleur spécial *m*	spezieller Farbenindex *m*
88	yellow colo(u)r-index	indice de couleur pour le jaune *m*	Gelb(farben)index *m*
89	Lambert's albedo	albédo de Lambert *m*	Lambert(sche) Albedo *f*
90	brightness factor	facteur de brillance *m*	Helligkeitsfaktor *m*
90a	mean albedo	albédo moyen *m*	mittlere Albedo *f*
90b	total albedo	albédo total [global] *m*	Totalalbedo *f*
91	Bond's albedo spherical albedo	albédo de Bond *m* albédo sphérique *m*	Bond(sche) Albedo *f* sphärische Albedo *f*
92	geometric(al) albedo	albédo géométrique *m*	geometrische Albedo *f*
93	monochromatic albedo	albédo monochromatique *m*	monochromatische Albedo *f*

efecto de oposición *m*	efeito de oposição *m*	эффект противостояния *m*	75
difusión de Mie *f*	difusão [espalhamento] de Mie *f*	рассеяние Ми *n*	76
coeficiente de fase *m*	coeficiente de fase *m*	коэффициент фазы *m* фазовый коэффициент *m*	77
función de fase *f*	função de fase *f*	фазовая функция *f*	78
integral de fase *f*	integral de fase *m*	фазовый интеграл *m*	79
medidas polarimétricas *fpl*	medidas polarimétricas *fpl*	поляриметрические измерения *npl*	80
curva de polarización *f*	curva de polarização *f*	кривая хода поляризации *f*	81
polarímetro visual *m*	polarímetro visual *m*	визуальный поляриметр *m*	82
polarímetro fotoeléctrico *m*	polarímetro fotoelé(c)trico *m*	фотоэлектрический поляриметр *m*	83
polarímetro fotográfico *m*	polarímetro fotográfico *m*	фотографический поляриметр *m*	84
magnitud media *f* magnitud en oposición media *m*	magnitude média *f* magnitude média na oposição *f*	средняя звёздная величина *f* звёздная величина в средней оппозиции *f*	85
magnitud absoluta de un planeta *f*	magnitude absoluta de um planeta *f*	абсолютная (звёздная) величина планеты *f*	86
índice de color especial *m*	índice de cor especial *m*	специальный показатель цвета *m*	87
índice de color amarillo *m*	índice de cor (para o) amarelo *m*	показатель желтизны *m*	88
albedo de Lambert *m*	albedo de Lambert *m*	ламбертово альбедо *n* плоское альбедо *n* истинное альбедо *n* альбедо по Ламберту *n* альбедо Ламберта *n*	89
factor de brillo *m*	fator de brilho *m* fator de brilhância *m*	яркостный фактор *m* видимое альбедо *n*	90
albedo medio *m*	albedo médio *m*	среднее альбедо *n*	90a
albedo total *m*	albedo total *m*	полное альбедо *n*	90b
albedo de Bond *m* albedo esférico *m*	albedo de Bond *m* albedo esférico *m*	альбедо по Бонду *n* альбедо Бонда *n* сферическое альбедо *n*	91
albedo geométrico *m*	albedo geométrico *m*	геометрическое альбедо *n*	92
albedo monocromático *m*	albedo monocromático *m*	монохроматическое альбедо *n*	93

	PLANETS	PLANÈTES	PLANETEN
94	radio-frequency albedo	albédo radioélectrique *m* albédo du rayonnement radio *m* radio-albédo *m*	Radiostrahlungsalbedo *f* Radio(frequenz)albedo *f*
95	cloud albedo albedo of the clouds	albédo des nuages *m*	Wolkenalbedo *f*
96	brightness coefficient	coefficient de brillance *m*	Helligkeitskoefficient *m*
97	surface of a planet planetary surface	surface d'une planète *f* surface planétaire *f*	Planetenoberfläche *f* Oberfläche eines Planeten *f*
98	solid surface	surface solide *f*	feste Oberfläche *f*
99	surface roughness	rugosité de surface *f*	Oberflächenrauheit *f* Oberflächenrauhigkeit *f*
100	porosity	porosité *f*	Porosität *f*
101	layered surface	surface stratifieé *f*	geschichtete Oberfläche *f*
102	rough surface	surface rugueuse *f*	rauhe Oberfläche *f*
103	integrated photometry	photométrie globale *f*	Integralphotometrie *f*
104	detailed photometry	photométrie détaillée *f*	Detailphotometrie *f* Ausschnittsphotometrie *f*
105	surface feature surface marking surface formation	tache de la surface *f* détail de la surface *m* formation de la surface *f* configuration de la surface *f*	Oberflächengebilde *n* Oberflächenformation *f* Merkmal der Oberfläche *n*
106	permanent feature permanent marking	détail permanent *m* formation permanente *f* tache permanente *f*	beständiges Gebilde *n* beständige Formation *f*
107	atmospheric feature	formation atmosphérique *f*	atmosphärisches Gebilde *n*
108	nomenclature of Martian features	nomenclature des configurations martiennes *f*	Nomenklatur von Marsformationen *f* Benennung von Marsformationen *f*
109	figure of a planet shape of a planet	figure d'une planète *f* aspect [forme] d'une planète *m*	Gestalt eines Planeten *f* Figur eines Planeten *f*
110	solid body	corps solide *m*	fester Körper *m*
111	rigid body	corps rigide *m*	starrer Körper *m*
112	equilibrium figure equilibrium shape hydrostatic equilibrium figure	figure d'équilibre (hydrostatique) *f* forme d'équilibre *f*	Gleichgewichtsform *f* Gleichgewichtsgestalt *f* Gleichgewichtsfigur *f*

albedo de radiofrecuencia *m* radioalbedo *m*	albedo de radiofreqüências *m* radioalbedo *m*	альбедо в области радиочастот *n* радиоальбедо *n*	94
albedo de las nubes *m*	albedo das nuvens *m*	альбедо облаков *n*	95
coeficiente de brillo *m*	coeficiente de brilhância [brilho] *m*	коэффициент яркости *m*	96
superficie de un planeta *f* superficie planetaria *f*	superfície de um planeta *f* superfície planetária *f*	поверхность планеты *f* планетная поверхность *f*	97
superficie sólida *f*	superfície sólida *f*	твёрдая поверхность *f*	98
aspereza de superficie *f* rugosidad superficial *f*	aspereza [rugosidade] superficial *f* aspereza de superfície *f*	шероховатость поверхности *f*	99
porosidad *f*	porosidade *f*	пористость *f* скважность *f*	100
superficie en capas *f* superficie estratificada *f*	superfície em camadas *f* superfície estratificada *f*	слоистая поверхность *f*	101
superficie rugosa *f*	superfície áspera *f*	шероховатая поверхность *f*	102
fotometría integrada *f*	fotometria integrada *f*	интегральная фотометрия *f*	103
fotometría detallada *f*	fotometria detalhada [pormenorizada] *f*	детальная фотометрия *f*	104
detalle de la superficie *m* formación de la superficie *f* característica de la superficie *f*	formação [aspecto, pormenor, detalhe, característica] da superfície *f*[*m*] marca [forma] superficial *f*	подробность рельефа *f* образование на поверхности *n* деталь на поверхности *f* особенность поверхности *f* деталь поверхности *f*	105
característica permanente *f* detalle permanente *m* formación permanente *f*	formação [forma, característica] permanente *f* marca [detalhe, pormenor] permanente *f*(*m*)	постоянное образование *n* постоянная деталь *f* постоянная подробность *f*	106
formación atmosférica *f*	formação atmosférica *f*	атмосферное образование *n*	107
nomenclatura de los detalles marcianos *f* nomenclatura de las formaciones marcianas *f*	nomenclatura das formações marsianas *f* nomenclatura dos detalhes de Marte *f*	номенклатура марсианских образований *f* номенклатура марсианских деталей *f*	108
forma de un planeta *f* figura de un planeta *f*	figura de um planeta *f* forma de um planeta *f*	фигура планеты *f* форма планеты *f*	109
cuerpo sólido *m*	corpo sólido *m*	твёрдое тело *n*	110
cuerpo rígido *m*	corpo rígido *m*	жёсткое тело *n* абсолютно твёрдое тело *n*	111
figura de equilibrio *f* forma de equilibrio (hidrostático) *f*	figura de equilibrio (hidrostático) *f*	фигура равновесия *f* гидростатическая фигура *f* фигура гидростатического равновесия *f*	112

	PLANETS	PLANÈTES	PLANETEN
113	flattening oblateness	aplatissement *m*	Abplattung *f*
114	polar flattening polar compression	aplatissement polaire *m*	polare Abplattung *f*
115	equatorial radius equatorial semidiameter	rayon équatorial *m*	Äquatorhalbmesser *m* Äquatorradius *m*
116	polar radius	rayon polaire *m*	Polradius *m* Polhalbmesser *m*
116a	mean radius	rayon moyen *m*	mittlerer Halbmesser *m*
117	apparent diameter angular diameter	diamètre apparent *m* diamètre angulaire *m*	scheinbarer Durchmesser *m* Winkeldurchmesser *m*
118	linear diameter true diameter	diamètre réel *m*	wahrer Durchmesser *m* linearer Durchmesser *m*
119	terrestrial planets Earth-like planets minor planets	planètes terrestres *fpl* planètes telluriques *fpl* planètes de type terrestre *fpl*	erdähnliche Planeten *mpl* terrestrische Planeten *mpl* erdartige Planeten *mpl*
120	Mercury	Mercure *m*	Merkur *m*
121	day-time observation	observation de jour *f* observation diurne *f*	Tagesbeobachtung *f*
122	hypothetical planet	planète hypothétique *f*	hypothetischer Planet *m*
123	intramercurial planet intramercurian planet Vulcan	planète intramercurielle *f* Vulcain *m*	intramerkurieller Planet *m* Vulkan *m*
124	transplutonian planet planet X *m*	planète transplutonienne *f* planète X *f*	transplutonischer Planet *m* Transpluto *m*
125	Venus	Vénus *f*	Venus *f*
126	Cytherean of Venus	cythérien vénusien de Vénus	Venus-
127	morning star	étoile du matin *f*	Morgenstern *m*
128	evening star	étoile du soir *f* étoile du berger *f*	Abendstern *m*
129	crescent (of) Venus	croissant de Vénus *m*	Venussichel *m*
130	cusps horns	cornes *fpl* extrémités du croissant *fpl*	Hörner *npl* Hörnerspitzen *fpl*

Español	Português	Русский	№
achatamiento *m*	achatamento *m*	сжатие *n* сплющенность *f*	113
achatamiento polar *m*	achatamento polar *m*	полярное сжатие *n*	114
radio ecuatorial *m* semidiámetro ecuatorial *m*	raio equatorial *m* semidiâmetro equatorial *m*	экваториальный радиус *m*	115
radio polar *m*	raio polar *m*	полярный радиус *m*	116
radio medio *m*	raio médio *m*	средний радиус *m*	116a
diámetro aparente *m* diámetro angular *m*	diâmetro aparente *m* diâmetro angular *m*	видимый диаметр *m* угловой диаметр *m*	117
diámetro verdadero *m* diámetro lineal *m*	diâmetro linear *m* diâmetro verdadeiro *m* diâmetro real *m*	истинный диаметр *m* линейный диаметр *m*	118
planetas terrestres *mpl* planetas telúricos *mpl*	planetas terrestres *mpl* planetas telúricos *mpl*	земные планеты *fpl* планеты земного типа *fpl* планеты земной группы *fpl* планеты типа Земли *fpl*	119
Mercurio *m*	Mercúrio *m*	Меркурий *m*	120
observación de día *f*	observação diurna *f*	дневное наблюдение *n*	121
planeta hipotético *m*	planeta hipotético *m*	гипотетическая планета *f*	122
planeta intramercurial *m* Vulcano *m*	planeta intramercurial *m* Vulcano *m* Vulcão *m*	интрамеркуриальная планета *f* Вулкан *m*	123
planeta transplutoniano *m* planeta X *m*	planeta transplutoniano *m* planeta X *m*	транcплутоновая планета *f* заплутоновая планета *f*	124
Venus *f*	Vênus *f*	Венера *f*	125
de Venus venusino	de Vênus Venusiano	венерианский Венеры	126
estrella [lucero] de la mañana *f* estrella del alba *f* lucero matutino *m* lucero miguero *m*	estrela da manhã *f* estrela d'alva *f* estrela matutina *f* papa-ceia *f*	утренняя звезда *f*	127
estrella vespertina *f* estrella de la tarde *f* lucero vespertino *m* lucero de la tarde *m*	estrela Vespertina *f* estrela da tarde *f* estrela do pastor *f* Vésper(o) *m*	вечерняя звезда *f*	128
creciente de Venus *f*	crescente de Vênus *m*	серп Венеры *m*	129
cuernos *mpl* cúspides *fpl*	cúspides *fpl* pontas *fpl*	рога *mpl* рога серпа *mpl*	130

	PLANETS	PLANÈTES	PLANETEN
131	extension of the horns of Venus extension of the crescent of Venus	allongement des cornes de Vénus *m*	Übergreifen der Venus-Hörnerspitzen *n*
132	dichotomy of Venus first or last quarter	dichotomie de Vénus *f* premier ou dernier quartier *m*	Venus-Dichotomie *f*
133	Mars	Mars *m*	Mars *m*
134	areophysics	aréophysique *f*	Marsphysik *f* Areophysik *f*
135	Martian areo-	martien aréo-	Mars- Areo-
136	areographic chart	carte aréographique *f*	Marskarte *f*
137	areographic	aréographique	areographisch areiographisch
138	areocentric (orbit, coordinates)	aréocentrique (orbite, coordonnées)	areozentrisch (Bahn, Koordinaten) areiozentrisch Marsmittelpunkt-
139	areographic longitude	longitude aréographique *f*	areographische Länge *f*
140	areographic latitude	latitude aréographique *f*	areographische Breite *f*
141	apareon apereon apomartium apomortian	apomars *m* apoastre martien *m*	Marsferne *f* Apare(i)on *n*
142	periareon perimartium perimartian	périmars *m* périastre martien *m*	Periareon *n* Marsnähe *f*
143	polar cap polar hood	calotte polaire *f*	Polkappe *f* Polkalotte *f*
144	solid carbon dioxide carbon-dioxide ice dry ice	neige carbonique *f* anhydride carbonique solide *m*	festes Kohlendioxid *n* Kohlensäureschnee *m* Trockeneis *n*
145	waxing of the polar cap	croissance [développement] de la calotte polaire *f*	Zunehmen der Polkappe *n*
146	waning of the polar cap	décroissance [régression] de la calotte polaire *f*	Abnehmen der Polkappe *n*
147	melting rate	vitesse de fusion *f*	Abschmelzgeschwindigkeit *f*
148	laminar structure	structure laminaire *f*	laminare Struktur *f*
149	lamina (pl. laminae) thin layer	feuille *f* couche (extrêment) mince *f* feuillet *m*	Folie *f* dünne Schicht *f*

prolongación de los cuernos de Venus *f*	prolongamento do crescente de Vênus *m*	удлинение рогов (серпа) Венеры *n*	131
dicotomía de Venus *f* mitad del disco de Venus *f*	dicotomia de Vênus *f* metade do disco de Vênus *f*	дихотомия Венеры *f* первая или последняя четверть *f*	132
Marte *m*	Marte *m*	Марс *m*	133
areofísica *f*	areofísica *f*	физика Марса *f*	134
marciano areo-	marciano areo-	арео- марсианский марсов	135
mapa del Marte *m*	mapa de Marte *m*	карта Марса *f*	136
areográfico	areográfico	ареографический	137
areocéntrico (órbita, coordenadas)	areocêntrico (órbita, coordenadas)	ареоцентрический (орбита, координаты)	138
longitud areográfica *f*	longitude areográfica *f*	ареографическая долгота *f*	139
latitud areográfica *f*	latitude areográfica *f*	ареографическая широта *f*	140
apareo *m*	apáreon *m*	апоарий *m* апомарсий *m*	141
periáreo *m*	periáreon *m*	периарий *m* перимарсий *m*	142
casquete polar *m*	calota polar *f*	полярная шапка *f*	143
anhídrido carbónico sólido *m* hielo seco *m*	dióxido de carbono sólido *m* anidrido carbônico sólido *m* gelo seco *m*	твёрдая двуокись углерода *f* твёрдая углекислота *f* сухой лед *m*	144
crecimiento del casquete polar *m*	crescimento da calota polar *m*	нарастание полярной шапки *n*	145
decrecimiento del casquete polar *m*	decrescimento da calota polar *m*	убывание полярной шапки *n*	146
velocidad de fusión *f*	velocidade [taxa] de fusão *f*	скорость таяния *f*	147
estructura laminar *f*	estrutura laminar *f*	ламинарная структура *f* слоистая структура *f*	148
lámina *f* plancha delgada *f*	lâmina *f* camada [chapa] delgada *f*	тонкий слой *m* тонкая пластинка *f* лист *m*	149

150	hoarfrost	gelée blanche *f*	Reif *m*
151	polar fringe	frange polaire *f*	Saum der Polkappe *m*
152	permafrost permanently frozen (sub)soil pergelisol	permafrost *m* sol perpétuellement gelé *m* tjäle *m* pergélisol *m* congélation perpétuelle *f*	Permafrost *m* ewiger Bodenfrost *m* Dauerfrostboden *m*
153	river bed	lit fluvial *m*	Flussbett *n*
154	canyon	cañon *m* canyon *m*	Schlucht *f* Klamm *f*
155	dark area	région sombre *f* tache [aire] sombre *f*	dunkles Gebiet *n* dunkle Fläche *f*
156	sea mare (pl. maria)	mer *f*	Meer *n*
157	canals of Mars Martian canals	canaux de Mars *mpl* canaux martiens *mpl*	Marskanäle *mpl*
158	lacus lake	lac *m*	See *m*
159	continent	continent *m*	Kontinent *m* Festland *n*
160	light region bright region	région claire *f* région brillante *f*	helle Gegend *f* helle Region *f*
161	desert	désert *m*	Wüste *f*
162	dune	dune *f*	Düne *f*
163	wind erosion	érosion éolienne *f*	Winderosion *f*
164	chaotic terrain	terrain chaotique *m* terrain désordonné *m*	chaotisches Terrain *n* ungeordnetes Gebiet *n*
165	violet layer	couche violette *f*	violette Schicht *f*
166	seasonal changes	changements saisonniers *mpl*	jahreszeitliche Änderungen *fpl*
167	diurnal changes	changements diurnes *mpl*	tägliche Änderungen [Schwankungen] *fpl*
168	Phobos	Phobos *m*	Phobos *m*
169	Deimos	Deimos *m*	Deimos *m*
170	Jupiter	Jupiter *m*	Jupiter *m*
171	Jovian	Jovien de Jupiter	Jupiter-

escarcha *f*	geada (branca) *f*	иней *m*	150
borde de los casquetes polares *m* franja polar *f*	orla da calota polar *f* franja polar *f*	кайма полярной шапки *f* ободок полярной шапки *m*	151
tierra permanentemente congelada *f*	solo permanentemente gelado *m*	вечная мерзлота *f*	152
álveo *m* lecho fluvial *m*	álveo *m* leito de um rio *m*	русло *n*	153
cañon *m*	canhão *m* barranco *m*	каньон *m* узкое ущелье *n*	154
región o(b)scura *f*	região escura *f*	тёмная область *f*	155
mar *m*, *f*	mar *m*	море *n*	156
canales de Marte *mpl*	canais de Marte *mpl*	каналы Марса *mpl* марсианские каналы *mpl*	157
lago *m*	lago *m*	озеро *n*	158
continente *m*	continente *m*	материк *m* континент *m*	159
región clara *f* región brillante *f*	região clara *f* região brilhante *f*	светлая область *f*	160
desierto *m*	deserto *m*	пустыня *f*	161
duna *f*	duna *f*	дюна *f*	162
erosión eólica *f*	erosão eóli(c)a *f*	ветровая эрозия *f*	163
terreno caótico *m* terreno desordenado *m*	terreno caótico *m* terreno desordenado *m*	хаотическая местность *f* беспорядочная поверхность *f*	164
capa violeta *f* estrato violáceo *m*	camada violeta *f*	фиолетовый слой *m* голубая мгла *f*	165
cambios estacionales *mpl*	variações estacionais [sazonais] *fpl*	сезонные изменения *npl*	166
cambios diurnos *mpl*	variações diurnas *fpl*	суточные изменения *npl*	167
Fobos *m*	Fobos *m*	Фобос *m*	168
Deimos *m*	Deimos *m*	Деймос *m*	169
Júpiter *m*	Júpiter *m*	Юпитер *m*	170
joviano de Júpiter	de Júpiter joviano	юпитеров йови- [юпитерианский]	171

	PLANETS	PLANÈTES	PLANETEN
172	perijove Jovian periapse [pericentre, periapsis]	périjove *m* périjovium *m* périastre jovien [de Jupiter] *m*	Perijovium *n* Jupiternähe *f*
173	apojove Jovian apoapse [apoapsis, apocentre]	apojove *m* apojovium *m* apoastre jovien [de Jupiter] *m*	Apojovium *f* Jupiterferne *f*
174	giant planet Jovian planet	planète géante *f* planète Jovienne *f*	Riesenplanet *m* jupiterähnlicher Planet *m*
175	jovigraphic coordinates	coordonées jovigraphiques [zenographiques] *fpl*	jovigraphische Koordinaten *fpl*
176	jovicentric coordinates	coordonnées jovicentriques [zenocentriques] *fpl*	jovizentrische Koordinaten *fpl*
177	longitude system	système de longitudes *m*	Längensystem *n* Rotationssystem *n*
178	equatorial longitude system longitude system I	système de longitudes équatorial *m* système I (de longitudes) *m*	äquatoriales Längensystem *n* Längensystem I *n* Rotationssystem I *n*
179	longitude system II system for intermediate latitudes	système II (de longitudes) *m* système des latitudes moyennes *m* système des latitudes temperées *m*	Längensystem II *n* Längensystem für mittlere Breiten *n* Rotationssystem II *n*
180	active region (on Jupiter)	région active (sur Jupiter) *f*	aktives Gebiet (des Jupiters) *n*
181	Jupiter's sporadic radio emissions	emissions radio sporadiques de Jupiter *fpl*	sporadische Radioemission des Jupiters *f*
182	Jovian radio burst	sursaut radioélectrique de Jupiter *m*	Radioausbruch des Jupiters *m*
183	decametric burst	sursaut décamétrique *m*	Dekameterwellenausbruch *m*
184	low-frequency bursts	sursauts de basse fréquence *mpl*	Ausbrüche bei niedriger Frequenz *mpl*
185	Jovian noise storm	tempête de bruit de Jupiter *f*	Rauschsturm des Jupiters *m*
186	decametric radio source	source décamétrique *f* radiosource décamétrique *f*	dekametrische Quelle *f* dekametrische Radioquelle *f*
187	zones currents	zones *fpl* courants *mpl*	Zonen *fpl* Strömungen *fpl*
188	equatorial zone	zone équatoriale *f*	Äquatorzone *f*
189	temperate zone	zone tempérée *f*	gemässigte Zone *f*
190	dusky belt belt	bande *f* bande sombre *f*	Band *n* Streifen *m*

perijove *m* perijoviano *m* pericentro joviano *m*	perijove *m* periastro joviano *m*	перийовий *m*	172
apojove *m* apojoviano *m* apocentro joviano *m*	apojove *m*	апойовий *m*	173
planeta gigante *m* planeta joviano *m*	planeta gigante *m* planeta joviano *m*	планета-гигант *f* планета типа Юпитера *f*	174
coordenadas jovigráficas *fpl*	coordenadas jovigráficas *fpl*	йовиграфические координаты *fpl* юпитерографические координаты *fpl*	175
coordenadas jovicéntricas *fpl*	coordenadas jovicêntricas *fpl*	йовицентрические координаты *fpl* юпитероцентрические координаты *fpl*	176
sistema de longitudes *m*	sistema de longitudes *m*	система долгот *f*	177
sistema ecuatorial de longitudes *m* sistema I de longitudes *m*	sistema equatorial de longitudes *m* sistema de longitudes I *m*	экваториальная система долгот *f* система долгот I *f*	178
sistema II de longitudes *m* sistema para latitudes intermedias *m*	sistema de longitudes II *m* sistema para latitudes intermédias *m*	система долгот для умеренных широт *f* система долгот II *f*	179
región activa (de Júpiter) *f*	região ativa (em Júpiter) *f*	активная область (на Юпитере) *f*	180
radioemisiones esporádicas de Júpiter *fpl*	emissões de rádio esporádicas de Júpiter *fpl*	спорадическое радиоизлучение Юпитера *n*	181
sobresalto [emisión repentina] en radioondas de Júpiter *m*[*f*]	erupção em radiofreqüências de Júpiter *f* evento rádio de Júpiter *m*	радиовсплеск Юпитера *m* радиовплеск на Юпитере *m*	182
sobresalto en ondas decamétricas *m*	erupção em ondas decamétricas *f*	декаметровый (радио)всплеск *m*	183
sobresaltos de baja frecuencia *mpl*	erupções de baixa freqüência *fpl*	низкочастотные (радио)всплески *mpl*	184
tormenta de ruido de Júpiter *f*	tempestade de ruídos em Júpter *f*	шумовая буря на Юпитере *f*	185
fuente decamétrica *f* radiofuente decamétrica *f*	fonte decamétrica *f* radiofonte decamétrica *f*	источник [радиоисточник] декаметрового излучения *m*	186
zonas *fpl* corrientes *fpl*	zonas *fpl*	зоны *fpl*	187
zona ecuatorial *f*	zona equatorial *f*	экваториальная зона *f*	188
zona templada *f*	zona temperada *f*	умеренная зона *f*	189
banda (obscura) *f* franja (oscura) *f*	banda *f* faixa escura *f*	полоса *f* тёмная полоса *f* пояс *m*	190

191	equatorial belt	bande équatoriale *f*	Äquatorband *n* Äquatorstreifen *m*
192	polar region	région polaire *f*	polares Gebiet *n* Polargebiet *n*
193	Red Spot Great Red Spot	Tache Rouge *f* Grande Tache Rouge *f*	Roter Fleck *m* Grosser Roter Fleck *m*
194	South Tropical Disturbance	(Grande) Perturbation Australe *f*	Südliche tropische Störung *f* Grosse südliche Störung *f* Südliches Störungsgebiet *n*
195	White Oval	Ovale Blanc *m*	Weisses Oval *n*
196	model planet model of a planet planetary model	modèle planétaire *m* modèle d'une planète *m*	Modellplanet *m* Planetenmodell *n*
197	hydrogen planet hydrogen model (of a planet)	planète d'hydrogène *f* modèle d'hydrogène (d'une planète) *m*	Wasserstoffmodell (eines Planeten) *m* Wasserstoffplanet *m*
198	a planet's own radiation	radiation propre d'une planète *f*	Eigenstrahlung eines Planeten *f*
199	heat source source of heat	source de chaleur *f*	Wärmequelle *f*
200	primordial heat	chaleur primordiale *f*	Urwärme *f*
201	cooling time	temps de refroidissement *m*	Kühlungszeit *f*
202	metallic hydrogen	hydrogène metallique *m*	metallischer Wasserstoff *m*
203	gravitational shrinkage gravitational contraction	contraction gravitationnelle *f* rétrécissement gravitationnel *m*	Schwerekontraktion *f* Schwerezusammenziehung *f* Gravitationszusammenschrumpfung *f*
204	Galilean satellite	satellite Galiléen *m*	Galileischer Statellit *m*
205	Io	Io *f*	Jo *f*
206	Europa	Europe *f*	Europa *f*
207	Ganymede	Ganymède *m*	Ganymed *m*
208	Callisto	Callisto *f*	Callisto *f*
209	Saturn	Saturne *m*	Saturn *m*
210	perisaturnian perisaturnium Saturnian periapse [periapsis, pericentre]	périsaturnium *m* périastre saturnien *m*	Perisaturnium *n* Saturnnähe *f*
211	aposaturnian aposaturnium Saturnian apoapse [apoapsis, apocentre]	aposaturnium *m* apoastre saturnien *m*	Aposaturnium *n* Saturnferne *f*

banda ecuatorial *f* franja ecuatorial *f*	banda equatorial *f* faixa equatorial *f*	экваториальная полоса *f* экваториальный пояс *m*	191
región polar *f*	região polar *f*	полярная область *f*	192
Mancha Roja *f* Gran Mancha Roja *f*	Mancha Vermelha *f* Grande Mancha Vermelha *f*	Красное пятно *n* Большое красное пятно *n*	193
(Gran) Perturbación Austral *f* Perturbación Tropical Sur *f*	(Grande) Perturbação Austral *f*	Южное тропическое возмущение *n* Большое южное возмущение *n*	194
Óvalo Blanco *m*	Oval Branca *f*	белый овал *m*	195
modelo de un planeta *m*	planeta modelo *m* modelo de um planeta *m*	модель планеты *f*	196
planeta de hidrógeno *m* modelo de hidrógeno (de un planeta) *m*	planeta de hidrogênio *m* modelo de hidrogênio (de um planeta) *m*	водородная планета *f* водородная модель (планеты) *f*	197
radiación propia de un planeta *f*	radiação própria de um planeta *f*	собственное излучение планеты *n*	198
fuente de calor *f*	fonte de calor *f*	источник тепла *m*	199
calor primordial *m*	calor primordial *m*	первичное тепло *n*	200
tiempo de enfriamiento *m*	tempo de arrefecimento [resfriamento] *m*	время охлаждения *n*	201
hidrógeno metálico *m*	hidrogênio metálico *m*	металлический водород *m*	202
contracción gravitacional *f*	contração gravitacional *f*	гравитационное сжатие *n* гравитационная усадка *f*	203
satélite Galileano *m*	satélite Galileano *m*	Галилеев(ский) спутник *m*	204
Io *f*	Io *f*	Ио *f*	205
Europa *f*	Europa *f*	Европа *f*	206
Ganímedes *m*	Ganimedes *m*	Ганимед *m*	207
Calisto *f*	Calisto *f*	Каллисто *f*	208
Saturno *m*	Saturno *m*	Сатурн *m*	209
perisaturno *m* pericentro saturnal *m*	perissatúrnio *m* perissaturno *m*	перисатурний *m*	210
aposaturno *m* apocentro saturnal *m*	apossatúrnio *m* apossaturno *m*	апосатурний *m*	211

	PLANETS	PLANÈTES	PLANETEN
212	saturnographic	saturnographique	saturnographisch
213	saturnocentric	saturnocentrique	saturnozentrisch
214	White Spot	Tache Blanche *f* Tache Claire *f*	Weisser Fleck *m*
215	Saturn's ring Saturn's rings	anneau de Saturne *m* anneaux de Saturne *mpl*	Saturnring *m* Saturnringe *mpl*
216	ring plane	plan de l'anneau *m*	Ringebene *f*
217	ring disappearance	disparition de l'anneau *f*	Verschwinden des Ringes *n*
218	edgewise ring ring seen edge-on	anneau vu par la tranche *m* anneau vu de profil *m*	Ring, von der Kante gesehen *m*
219	ring reappearance	réapparition de l'anneau *f*	Wiedererscheinen des Ringes *n*
220	Saturn's ring system	système des anneaux de Saturne *m*	Ringsystem des Saturns *n*
221	crape ring crêpe ring gauze ring dusky ring ring C	anneau de crêpe *m* anneau intérieur *m* anneau sombre *m* anneau transparent *m* anneau C *m*	Kreppring *m* Florring *m* dunkler Ring *m* C-Ring *m* Innenring *m*
222	middle ring main ring ring B	anneau médian *m* anneau B *m*	B-Ring *m* mittlerer Ring *m*
223	division	division *f*	Teilung *f*
224	Cassini's division Cassini division	division de Cassini *f*	Cassinische Teilung *f*
225	outer ring ring A	anneau extérieur *m* anneau A *m*	Aussenring *m* äusserer Ring *m* A-Ring *m*
226	inner radius interior radius	rayon intérieur *m* demi-diamètre intérieur *m*	Innenhalbmesser *m* innerer Halbmesser *m*
227	outer edge	bord extérieur *m*	Aussenrand *m* äusserer Rand *m*
228	outer radius exterior radius	rayon extérieur *m* demi-diamètre extérieur *m*	Aussenhalbmesser *m* äusserer Halbmesser *m*
229	thickness of rings	épaisseur des anneaux *f*	Dicke der Ringe *f*
230	hydrogen torus	tore d'hydrogène *m*	Wasserstofftorus *m*
231	equatorial plane plane of the equator	plan équatorial *m* plan de l'équateur *m*	Äquatorebene *f*

saturnográfico	saturnigráfico	сатурнографический	212
saturnocéntrico	saturnicêntrico	сатурноцентрический	213
Mancha Blanca *f*	Mancha Branca *f* Mancha Clara *f*	Белое пятно *n*	214
anillo de Saturno *m* anillos de Saturno *mpl*	anel de Saturno *m* anéis de Saturno *mpl*	кольцо Сатурна *n* кольца Сатурна *npl*	215
plano del anillo *m*	plano do anel *m*	плоскость кольца *f*	216
desaparición del anillo *f*	desaparecimento [desaparição] do anel *m[f]*	исчезновение колец *n*	217
anillo de canto *m* anillo visto de canto *m*	anel visto de perfil *m*	кольцо, видимое с ребра *n*	218
reaparición del anillo *f*	reaparecimento [reaparição] do anel *m[f]*	появление кольца *n*	219
sistema de los anillos de Saturno *m*	sistema dos anéis de Saturno *m*	система сатурновых колец *f* система колец Сатурна *f*	220
anillo nebuloso *m* anillo interior *m* anillo o(b)scuro *m* anillo C *m*	anel de crepe *m* anel interior *m* anel transparente *m* anel C *m*	креповое кольцо *n* кольцо C *n*	221
anillo central *m* anillo medio *m* anillo B *m*	anel médio *m* anel central *m* anel B *m*	среднее кольцо *n* кольцо B *n*	222
división *f*	divisão *f*	деление *n* щель (между кольцами) *f*	223
división de Cassini *f*	divisão de Cassini *f*	деление Кассини *n* щель Кассини *f*	224
anillo exterior *m* anillo externo *m* anillo A *m*	anel exterior *m* anel A *m*	внешнее кольцо *n* кольцо A *n*	225
radio interior [interno] *m* semidiámento interior [interno] *m*	raio interno [interior] *m* semidiâmetro interior *m*	внутренний радиус *m*	226
borde exterior *m*	bordo exterior *m*	внешний край *m*	227
radio exterior [externo] *m* semidiámetro exterior [externo] *m*	raio externo [exterior] *m* semidiâmetro exterior *m*	внешний радиус *m*	228
espesor de los anillos *m*	espessura dos anéis *f*	толщина кольца *f*	229
toro de hidrógeno *m*	toro de hidrogênio *m*	водородный тор *m*	230
plano ecuatorial *m* plano del ecuador *m*	plano equatorial *m* plano do equador *m*	экваториальная плоскость *f* плоскость экватора *f*	231

232	invariable plane Laplacian plane	plan invariable *m* plan de Laplace *m*	invariable Ebene *f* Laplacesche Ebene *f*
233	Titan	Titan *m*	Titan *m*
234	Uranus	Uranus *m*	Uranus *m*
235	Ariel	Ariel *m*	Ariel *m*
236	Neptune	Neptune *m*	Neptun *m*
237	transneptunian planet	planète transneptunienne *f*	transneptunischer Planet *m*
238	planet search	recherche des planètes *f*	Planetensuche *f*
239	Pluto	Pluton *m*	Pluto *m*
240	former satellite	ancien satellite *m*	ehemaliger Satellit *m*
241	atmosphere	atmosphère *f*	Atmosphäre *f*
242	atmospheric	atmosphérique	atmosphärisch
243	atmosphere of a planet planetary atmosphere	atmosphère planétaire *f* atmosphère d'une planète *f*	Planetenatmosphäre *f* Atmosphäre eines Planeten *f*
244	primitive [primordial] atmosphere	atmosphère primitive *f*	ursprüngliche Atmosphäre *f*
245	planetary corona	couronne planétaire *f*	Planetenkorona *f*
246	hydrogen corona	couronne d'hydrogène *f*	Wasserstoffkorona *f*
247	model of an atmosphere model atmosphere	modèle d'atmosphère *m* atmosphère modèle *f*	Atmosphärenmodell *n* Modellatmosphäre *f*
248	cloud layer	couche nuageuse *f*	Wolkenschicht *f*
249	haze layer	couche de brume *f*	Dunstschicht *f*
250	haze veil	brume *f* voile *m*	Dunst *m* Trübung *f*
251	turbid medium	milieu trouble *m*	trübes Medium *n*
252	turbidity turbidness	trouble *m* turbidité *f*	Trübung *f* Trübheit *f*
253	cloud-covered	couvert de nuages nuageux	mit Wolken bedeckt bewölkt
254	clouds	nuages *mpl*	Wolken *fpl*
255	dust storm	tempête de poussière *f*	Staubsturm *m*
256	dust cloud	nuage de poussière *m*	Staubwolke *f*
258	scattering layer	couche diffusante *f*	streuende Schicht *f*

plano invariable *m* plano Laplaciano *m*	plano invariável *m* plano de Laplace *m*	неизменяемая плоскость *f* плоскость Лапласа *f*	232
Titán *m*	Titã *m*	Титан *m*	233
Urano *m*	Urano *m*	Уран *m*	234
Ariel *m*	Ariel *m*	Ариэль *m*	235
Neptuno *m*	Netuno *m*	Нептун *m*	236
planeta transneptuniano *m*	planeta transnetuniano *m*	транснептуновая планета *f*	237
búsqueda de planetas *f*	procura de planetas *f*	поиск планет *m*	238
Plutón *m*	Plutão *m*	Плутон *m*	239
antiguo satélite *m*	antigo [inicial] satélite *m*	бывщий спутник *m*	240
atmósfera *f*	atmosfera *f*	атмосфера *f*	241
atmosférico	atmosférico	атмосферный	242
atmósfera planetaria *f* atmósfera de un planeta *f*	atmosfera de um planeta *f* atmosfera planetária *f*	атмосфера планеты *f* планетная атмосфера *f*	243
atmósfera primitiva *f*	atmosfera primitiva *f*	первичная атмосфера *f*	244
corona planetaria *f*	coroa planetária *f*	планетная корона *f*	245
corona de hidrógeno *f*	coroa de hidrogênio *f*	водородная корона *f*	246
modelo de atmosfera *m*	modelo de atmosfera *m*	модель атмосферы *f* атмосферная модель *f*	247
capa de nubes *f*	camada de nuvens *f*	облачный слой *m*	248
capa de bruma *f*	camada de bruma *f*	слой дымки *m* слой тумана *m*	249
bruma *f* velo *m*	bruma *f* véu *m* névoa *f*	дымка *f* туманные образования *npl* вуаль *f*	250
medio turbio *m*	meio turvo *m*	мутная среда *f*	251
turbiedad *f* enturbiamiento *m*	turvação *f* turvamento *m*	мутность *f* помутнение *n*	252
cubierto de nubes nublado	coberto de nuvens nublado	покрытый облаками облачный	253
nubes *fpl*	nuvens *fpl*	облака *npl*	254
tormenta de polvo *f*	tempestade de poeira [pó] *f*	пылевая буря *f*	255
nube de polvo *f*	nuvem de poeira [pó] *f*	пылевое облако *n*	256
capa de difusión *f* capa difusora *f*	camada difusora *f* camada de espelhamento *f*	рассеивающий слой *m*	258

259	temperature profile (of the atmosphere)	profil de température (de l'atmosphère) *m*	Temperaturprofil (der Atmosphäre) *n*
260	aeolosphere model aelospheric model	modèle éolosphérique *m* modèle éolien *m*	Äolosphärenmodell *n*
261	ionospheric model	modèle ionosphérique *m*	Ionosphärenmodell *n*
262	greenhouse model	modèle à l'effet de serre *m*	Glashausmodell *n* Treibhausmodell *n*
263	greenhouse effect	effet de serre *m*	Glashauseffekt *m*
264	transparency window transmission window	fenêtre de transparence *f*	Durchlässigkeitsfenster *n* Transparenzfenster *n*
265	atmospheric constituent	constituant atmosphérique *m*	atmosphärische Komponente *f* atmosphärischer Bestandteil *m*
266	trace concentration	concentration de traces *f* microconcentration	Spurenkonzentration *f*

perfil de temperatura (de la atmósfera) *m*	perfil de temperatura (da atmosfera) *m*	температурный профиль (атмосферы) *m*	259
modelo eolosférico *m*	modelo eolosférico *m*	эолосферная модель *f*	260
modelo ionosférico *m*	modelo ionosférico *m*	ионосферная модель *f*	261
modelo de invernadero *m*	modelo de estufa *m*	парниковая модель *f* тепличная модель *f*	262
efecto de invernadero *m*	efeito (de) estufa *m*	тепличный эффект *m* парниковый эффект *m* оранжерейный эффект *m*	263
ventana de transparencia *f* ventana de transmisión *f*	janela de transmissão *f*	окно прозрачности *n*	264
componente atmosférico *m* constituyente atmosférico *m*	componente atmosférico *m* constituinte atmosférico *m*	атмосферная составляющая *f*	265
concentración de trazas [vestigios] *f*	concentração de traços [vestígios] *f*	следы *mpl* незначительная концентрация *f* микроконцентрация *f*	266

83a	PLANETARY COSMOGONY	COSMOGONIE PLANÉTAIRE	PLANETENKOSMOGONIE
1	planetary cosmogony	cosmogonie planétaire *f*	Planetenkosmogonie *f* Kosmogonie der Planeten *f* planetarische Kosmogonie *f*
2	law of planetary distances	loi des distances des planètes *f*	Gesetz der Planetenentfernungen *n* Titius-Bodesche Regel *f*
3	coplanarity of orbits	coplanarité des orbites *f*	Ko(m)planarität der Bahnen *f*
4	moment of momentum angular momentum	moment angulaire *m* moment cinétique *m* moment de la quantité de mouvement *m*	Impulsoment *n* Drehimpuls *m* Drall *m* Schwung *m*
5	rotation(al) moment torque	moment de rotation (axiale) *m* moment du couple *m*	Drehmoment *n* Rotationsmoment *n* Moment des Kräftepaares *n*
6	orbital angular momentum	moment orbital *m* quantité de mouvement orbital *f*	Bahndrehimpuls *m*
7	nebular hypothesis	hypothèse nébulaire *f*	Nebularhypothese *f*
8	rotational instability	instabilité rotationnelle *f* instabilité de rotation *f*	Rotationsinstabilität *f*
9	separation (from the Sun)	séparation (du Soleil) *f*	Abtrennung (von der Sonne) *f*
10	nebular ring	anneau nébulaire *m*	Nebelring *m*
11	vortex hypothesis Weizsäcker('s) turbulence theory	théorie des tourbillons *f* théorie de Weizsäcker *f*	Wirbelhypothese *f* Weizsäckersche Turbulenztheorie *f*
12	cosmic-scale turbulence large-scale turbulence	turbulence (à l'échelle) cosmique *f* turbulence à grande échelle *f*	Turbulenz mit kosmischen Ausmass *f* gross-ausgedehnte Turbulenz *f*
13	system of vortices vortex system	système de tourbillons *m*	Wirbelsystem *n*
14	encounter hypothesis collision hypothesis catastrophic hypothesis	hypothèse de marée *f* hypothèse catastrophique *f*	Begegnungshypothese *f* Stosshypothese *f* Gezeitenhypothese *f*
15	tidal action	action de marée *f*	Gezeitenwirkung *f*
16	stellar collision collision between stars	collision stellaire *f* collision entre étoiles *f*	Sternzusammenstoss *m* Zusammenstoss der Sterne *m*
17	head-on collision central collision	collision centrale *f* collision directe *f*	direkter Zusammenstoss *m* zentraler Zusammenstoss *m*
18	capture hypothesis	hypothèse de capture *f*	Einfangshypothese *f*
19	capture	capture *f*	Einfang *m*
20	accretion theory	théorie de l'accrétion *f*	Akkretionstheorie *f* Aufsammlungstheorie *f*

cosmogonía planetaria *f*	cosmogonia planetária *f*	планетная космогония *f* космогония планетной системы *f*	1
ley de distancias planetarias *f* ley de Titius (y) Bode *f*	lei das distancias planetárias *f* lei de Titius (e) Bode *f*	закон планетных расстояний *m* зуакон Тициуса-Боде *m*	2
coplanareidad de órbitas *f*	coplanaridade das órbitas *f*	компланарность орбит *f*	3
ímpetu angular *m* momento angular *m* momento cinético *m* momento de la cantidad de movimiento *m*	momento angular *m* momento cinético *m* quantidade de movimento angular *f*	момент количества движения *m* угловой момент *m*	4
momento rotacional *m* momento de par (de fuerzas) *m*	momento de rotação *m* quantidade de movimento de rotação *f* torque *m*	момент вращения *m*	5
momento angular orbital *m*	momento orbital *m* quantitade de movimento orbital *f*	орбитальный момент количества движения *m*	6
hipótesis nebular *f*	hipótese nebular *f*	небулярная гипотеза *f*	7
inestabilidad rotacional *f* inestabilidad de rotación *f*	instabilidade rotacional *f* instabilidade de rotação *f*	ротационная неустойчивость *f*	8
separación (desde el Sol) *f*	separação (do Sol) *f*	отделение (от Солнца) *n*	9
anillo nebular *m*	anel nebular *m*	туманное кольцо *n*	10
hipótesis de los vórtices *f* hipótesis de Weizsäcker *f* hipótesis de los torbellinos *f*	hipótese dos turbilhões *f* hipótese de Weizsäcker *f* hipótese dos vórtices *f*	вихревая теория *f* турбулентная теория *f* теория Вейцзеккера *f*	11
turbulencia (a escala) cósmica *f* turbulencia a gran escala *f*	turbulência (em escala) cósmica *f* turbulência em grande escala *f*	турбулентность космического масштаба *f* турбулентность большого масштаба *f*	12
sistema de torbellinos *m* sistema de vórtices *m*	sistema de turbilhoẽs *m* sistema de vórtices *m*	система вихрей *f*	13
hipótesis del choque *f* hipótesis del encuentro *f* hipótesis catastrófica *f*	hipótese de colisão *f* hipótese catastrófica *f* hipótese de maré *f*	гипотеза столкновения *f* приливная гипотеза *f* катастрофическая гипотеза *f*	14
(acción de) marea *f*	(efeito de) maré *f*	приливное воздействие *n*	15
colisión estelar *f* colisión de estrelas *f*	colisão estelar *f* colisão de estrelas *f*	звёздное столкновение *n*	16
colisión frontal *f* colisión central *f*	colisão central *f* colisão frontal *f*	столкновение в лоб *n* центральное столкновение *n*	17
hipótesis de la captura *f*	hipótese de captura *f*	гипотеза захвата *f*	18
captura *f*	captura *f*	захват *m*	19
teoría de la acreción *f*	teoria de acresção [acréscimo] *f*	теория наращивания *f* теория акреции *f*	20

21	primitive nebula pre-solar nebula	nébuleuse primitive *f* nébuleuse protosolaire *f* nébuleuse présolaire *f*	Urnebel *m* Sonnenurnebel *m*
22	flattening	aplatissement *m*	Abplattung *f*
23	solar nebula	nébuleuse solaire *f*	Sonnennebel *m*
24	protoplanetary cloud preplanetary cloud	nuage protoplanétaire *m*	protoplanetare Wolke *f*
25	gas-dust matter gas-dust mixture	mélange gaz-poussière *m* matière gaz-poussière *f*	gas-und staubförmige Materie *f* Gas-Staub-Mischung *f*
26	condensation	condensation *f*	Kondensation *f*
27	solid particle	particule solide *f*	festes Teilchen *n* feste Partikel *f*
28	dust cloud	nuage de poussière *m*	Staubwolke *f*
29	atmophilic elements	éléments atmophiles *mpl*	atmophile Elemente *npl*
30	lithophilic elements	éléments lithophiles *mpl*	lithophile Elemente *npl*
31	chalcophilic elements	éléments chalcophiles *mpl*	chalkophile Elemente *npl*
32	agglomeration	agglomération *f*	Agglomeration *f*
33	accretion	accrétion *f*	Materieakkretion *f* Materieaufsammlung *f*
34	accumulation	accumulation *f*	Anhäufung *f* Akkumulation *f*
35	planetesimals	planétesimales *mpl*	Planetesimale *npl*
36	embryo of a planet	embryon d'une planète *m*	Embryo eines Planeten *n*
37	growth of a planet planetary growth	croissance d'une planète *f*	Wachsen eines Planeten *n*
38	mass increase	accroissement de masse *m* gain de masse *m*	Vergrösserung der Masse *f*
39	Protosun	Soleil primitif *m* Protosoleil *m*	Ursonne *f*
40	protoplanet	protoplanète *f*	Urplanet *m* Protoplanet *m*
41	sweeping out of gases	expulsion des gaz *f* balayage des gas *m*	Ausfegen von Gasen *n*

COSMOGONÍA PLANETARIA	COSMOGONIA PLANETÁRIA	ПЛАНЕТНАЯ КОСМОГОНИЯ	
nebulosa primitiva *f* nebulosa presolar *f*	nebulosa primitiva *f* nebulosa pré-solar *f* nebulosa proto-solar *f*	досолнечная туманность *f* протосолнечная туманность *f*	21
achatamiento *m*	achatamento *m*	уплощение *n* сплющивание *n*	22
nebulosa solar *f*	nebulosa solar *f*	солнечная туманность *f*	23
nube protoplanetaria *f* nube preplanetaria *f*	nuvem protoplanetária *f*	протопланетное облако *n*	24
materia de gas y polvo *m* mezcla de gas y polvo *f*	mistura de gás e poeira *f*	смесь газа и пыли *f* газово-пылевое вещество *n*	25
condensación *f*	condensação *f*	конденсация *f* сгущение *n* уплотнение *n*	26
partícula sólida *f*	partícula sólida *f*	твёрдая частица *f*	27
nube de polvo (cósmico) *f*	nuvem de poeira *f*	пылевое облако *n*	28
elementos atmófilos *mpl*	elementos atmófilos *mpl*	атмофилы *mpl* атмофильные элементы *mpl*	29
elementos litófilos *mpl*	elementos litófilos *mpl*	литофилы *mpl* литофильные элементы *mpl*	30
elementos calcófilos *mpl*	elementos calcófilos *mpl*	халькофилы *mpl* халькофильные элементы *mpl*	31
aglomeración *f*	aglomeração *f*	скопление *n* агломерация *f*	32
acreción *f* acrecimiento *m*	acréscimo *m* acrescimento [acresção] *m* [*f*]	акреция *f* наращивание *n*	33
acumulación *f*	acumulação *f*	аккумуляция *f* накопление *n*	34
planetesimales *fpl*	planetesimais *fpl*	планетезимали *mpl*	35
embrión de un planeta *m*	embrião de um planeta *m*	зародыш планеты *m*	36
crecimiento planetario *m* crecimiento de un planeta *m*	crescimento planetário *m* crescimento de um planeta *m*	рост планеты *m*	37
incremento de masa *m* aumento de masa *m*	incremento de massa *m* aumento de massa *m*	прирост массы *m* увеличение массы *n*	38
Protosol *m*	Sol primitivo *m* Protosol *m*	Протосолнце *n*	39
protoplaneta *m*	protoplaneta *m*	протопланета *f*	40
expulsión de gases *f*	expulsão de gases *f*	выметание газов *n*	41

42	decrease of mass	diminution de la masse *f* perte de masse *f*	Massenverminderung *f*
43	incandescent gases	gaz incandescents *mpl*	glühende Gase *npl*
44	initial atmosphere protoatmosphere	atmosphère initiale *f* proto-atmosphère *f*	Uratmosphäre *f*
45	early atmosphere	atmosphère primitive *f*	Atmosphäre im Anfangszustand *f*
46	reducing atmosphere	atmosphère réductrice *f*	reduzierende Atmosphäre *f*
47	oxidizing atmosphere	atmosphère oxydante *f*	oxydierende Atmosphäre *f*
48	separation of chemical elements	séparation des éléments chimiques *f*	Trennung der chemischen Elemente *f*
49	chemical evolution	évolution chimique *f*	chemische Entwicklung *f*
50	chemical differentiation	différenciation chimique *f*	chemische Differentiation *f* chemische Aufteilung *f*
51	fractionation of chemical elements	fractionnement des éléments chimiques *m*	Fraktionierung der chemischen Elemente *f*
52	volatile substance	substance volatile *f*	flüchtiger Stoff *m*
53	non-volatile substances	substance non-volatile *f*	nichtflüchtiger Stoff *m*
54	gravitational differentiation	différenciation gravitationnelle *f*	Differenzierung im Schwerefeld *f* Sedimentation im Schwerefeld *f*
55	refractory material material difficult to melt [to fuse] not-easily-fusible material not-easily-fusible substance substance difficult to fuse	substance difficilemet fusible *f* substance réfractaire *f* matière réfractaire *f*	schwer schmelzbarer Stoff *m* schwer schmelzender Stoff *m* feuerfester Stoff *m*
56	easily fusible substance	substance facilement fusible *f*	leicht schmelzbarer Stoff *m*
57	cooling	refroidissement *m*	Abkühlung *f*
58	phase transition	transition de phase *f*	Phasenübergang *m*
59	solidification	solidification *f*	Erstarrung *f*
60	thermal history	histoire thermique *f* évolution thermique passée *f*	thermische Geschichte *f*
61	thermal evolution	évolution thermique *f*	thermische Entwicklung *f*
62	radiogenic heat	chaleur radiogénique *f* chaleur (de désintégration) radioactive *f*	die durch Radioaktivität erzeugte Wärme *f* radiogene Wärme *f*

disminución de masa *f*	decréscimo de massa *f*	уменьшение массы *n* убыль массы *f*	42
gases incandescentes *mpl*	gases incandescentes *mpl*	раскалённые газы *mpl*	43
atmósfera inicial *f* protoatmósfera *f*	atmosfera inicial *f* proto-atmosfera *f*	первоначальная атмосфера *f* протоатмосфера *f*	44
atmósfera primitiva *f*	atmosfera primitiva *f*	ранняя атмосфера *f*	45
atmósfera reductora *f*	atmosfera redutora *f*	восстановительная атмосфера *f*	46
atmósfera oxidante *f*	atmosfera oxidante *f*	окислительная атмосфера *f*	47
separación de los elementos químicos *f*	separação dos elementos químicos *f*	разделение химических элементов *n*	48
evolución química *f*	evolução química *f*	химическая эволюция *f*	49
diferenciación química *f*	diferenciação química *f*	химическая дифференциация *f*	50
fraccionamiento de los elementos químicos *m*	fracionamento dos elementos químicos *m*	фракционирование химических элементов *n*	51
substancia volátil *f*	su(b)stância volátil *f*	летучее вещество *n*	52
substancia no volátil *f*	su(b)stância não-volátil *f*	нелетучее вещество *n*	53
diferenciación gravitacional *f*	diferenciação gravitacional *f*	гравитационная дифференциация *f*	54
material difícil de fundir *m* substancia difícil de fundir *f* substancia refractaria *f* material refractario *m*	su(b)stancia dificilmente fusível *f* su(b)stância refratária *f* material refratário *m*	тугоплавкое вещество *n*	55
substancia fácilmente fundible *f*	su(b)stância facilmente fusível *f*	легкоплавкое вещество *n*	56
enfriamiento *m*	arrefecimento *m* resfriamento *m*	охлаждение *n* остывание *n*	57
transición de fase *f*	transição de fase *f* mudança de fase *f*	фазовое превращение *n* фазовый переход *m* превращение фаз *n*	58
solidificación *f*	solidificação *f*	затвердевание *n* отверде(ва)ние *n* затвердение *n*	59
historia térmica *f*	história térmica *f*	тепловая история *f* термическая история *f*	60
evolución térmica *f*	evolução térmica *f*	тепловая эволюция *f* термическая эволюция *f*	61
calor radiogénico *m* calor radi(o)activo *m*	calor devido à radioatividade *m* calor radiogênico *m*	радиоактивное тепло *n* радиогенное тепло *n*	62

63	Olbers' hypothetical planet	planète hypothétique (de Olbers) *f*	hypothetischer (Olbersscher) Planet *m*
64	molten globe	globe fondu *m* globe en fusion *m*	flüssiger Erdball *m*
65	tidal evolution	évolution sous l'effet des marées *f*	Gezeitenentwicklung *f*
66	tidal deformation	deformation due aux marées *f*	Gezeitendeformation *f*
67	tidal bulge	renflement dû aux marées *m* bourrelet de marées *m*	Gezeitenberg *m* Gezeitenwulst *m*
68	age of the Earth	âge de la Terre *m*	Erdalter *n*
69	geological age	âge géologique *m*	geologisches Alter *n*
70	absolute age	âge absolu *m*	absolutes Alter *n*
71	age determination dating	détermination de l'age *f* datation *f*	Altersbestimmung *f* Datierung *f*
72	cosmochronology	cosmochronologie *f*	Kosmochronologie *f*
73	nucleochronology isotopic dating radioactive age determination	datation par les isotopes *f* datation isotopique [radioactive] *f*	Altersbestimmung mittels Isotopen *f* radioaktive Altersbestimmung *f* absolute Altersbestimmung *f*
74	short-lived isotope short-life isotope	isotope de courte période *m*	kurzlebiges Isotop *n*
75	long-lived isotope long-life isotope	isotope de longue période *m*	langlebiges Isotop *n*
76	lead isotope	isotope du plomb *m*	Bleiisotop *n*
77	geological period	période géologique *f*	geologische Periode *f*
78	ice age glacial period	période glaciaire *f* période glaciale *f*	Eiszeit *f*
79	paleomagnetism	paléomagnétisme *m*	Paläomagnetismus *m*
80	rock magnetism	magnétisme des roches *m*	Gesteins-Magnetismus *m*
81	fragmentation	fragmentation *f*	Zersplitterung *f*
82	fragmentation of asteroids	fragmentation des astéroides *f*	Zersplitterung der Asteroiden *f*
83	fragments	fragments *mpl*	Bruchstücke *npl*

planeta hipotético (de Olbers) *m*	planeta hipotético de Olbers *m*	гипотетическая планета (Ольберса) *f*	63
globo fundido *m* globo derretido *m*	globo fundido *m* globo derretido *m*	расплавленный земной шар *m*	64
evolución por (efecto de) las mareas *f* evolución debida a las mareas *f*	evolução por efeito de marés *f* evolução devida à maré *f*	приливная эволюция *f*	65
deformación por mareas *f* deformación debida a las mareas *f*	deformação devida à maré *f* deformação por maré *f*	приливная деформация *f*	66
abultamiento debido a las mareas *m*	bojo devido à maré *m* protuberância da maré *f*	приливный горб *m* приливный выступ *m*	67
edad de la Tierra *f*	idade da Terra *f*	возраст Земли *m*	68
edad geológica *f*	idade geológica *f*	геологический возраст *m*	69
edad absoluta *f*	idade absoluta *f*	абсолютный возраст *m*	70
determinación de la edad *f* datación *f*	datação *f* determinação de idade *f*	определение возраста *n*	71
cosmocronología *f*	cosmocronologia *f*	космохронология *f* определение возраста космических тел *n*	72
datación por isótopos *f* datación radiactiva *f* datación isotópica *f*	datação por isótopos *f* datação radi(o)ativa *f* datação absoluta *f*	определение возраста с помощью изотопов *n*	73
isótopo de vida corta *m*	isótopo de vida curta *m*	короткоживущий изотоп *m*	74
isótopo de largo período *m*	isótopo de vida longa *m*	долгоживущий изотоп *m*	75
isótopo del plomo *m*	isótopo do chumbo *m*	изотоп свинца *m*	76
período geológico *m*	período geológico *m*	геологический период *m*	77
período glacial *m*	época glacial *f* período glacial *m*	ледниковый период *m*	78
paleomagnetismo *m*	paleomagnetismo *m*	палеомагнетизм *m*	79
magnetismo de rocas *m*	magnetismo das rochas *m*	магнетизм горных пород *m*	80
fragmentación *f*	fragmentação *f*	дробление *n*	81
fragmentación de asteroides *f*	fragmentação dos asteróides *f*	дробление астероидов *n*	82
fragmentos *mpl*	fragmentos *mpl*	обломки *mpl* осколки *mpl* фрагменты *mpl*	83

84	INTERPLANETARY MATTER	MATIÈRE INTERPLANÉTAIRE	INTERPLANETARE MATERIE
1	interplanetary matter	matière interplanétaire *f*	interplanetare Materie *f*
2	interplanetary space	espace interplanétaire *m*	interplanetarer Raum *m*
3	geocentric distance	distance géocentrique *f*	geozentrische Distanz *f* geozentrische Entfernung *f*
4	heliocentric distance	distance héliocentrique *f*	heliozentrische Distanz *f* heliozentrische Entfernung *f*
5	Earth's orbit	orbite de la Terre *f* orbite terrestre *f*	Erdbahn *f*
6	beyond the Earth's orbit	au delà de l'orbite terrestre	jenseits der Erdbahn ausserhalb der Erdbahn
7	in the vicinity of the Earth's orbit	au voisinage de l'orbite terrestre	in der Nähe der Erdbahn
8	interplanetary	interplanétaire	interplanetar
9	interplanetary magnetic field	champ magnétique interplanétaire *m*	interplanetares magnetisches Feld *n*
10	(magnetic) field direction	direction du champ (magnétique) *f*	Richtung des (Magnet)feldes *f*
11	magnetic field intensity (magnetic) field strength	intensité du champ (magnétique) *f*	(Magnet-)Feldstärke *f* magnetische Feldstärke *f*
12	magnetic field irregularity	irrégularité du champ magnétique *f*	Ungeordnetheit [Irregularität] des Magnetfeldes *f* Unregelmässigkeit des Magnetfeldes *f*
13	small-scale irregularity	irrégularité à petite échelle *f*	kleinräumige Ungeordnetheit *f* kleinräumige Irregularität *f*
14	(magnetic) field variability	variabilité du champ (magnétique) *f*	Variabilität des (Magnet-)Feldes *f*
15	sector structure	structure en secteur (du champ interplanétaire) *f*	Sektorstruktur *f*
16	interplanetary (magnetic) sector	secteur (magnétique) interplanétaire *m*	interplanetarer (Magnet-)Sektor *m*
17	corotating structure (of magnetic field)	structure en corotation (du champ magnétique) *f*	mitrotierende Struktur (des Magnetfeldes) *f*
18	sector boundary	limite de secteur *f*	Sektorgrenze *f*

MATERIA INTERPLANETARIA	MATÉRIA INTERPLANETÁRIA	МЕЖПЛАНЕТНАЯ МАТЕРИЯ	
materia interplanetaria *f*	matéria interplanetária *f*	межпланетная материя *f* межпланетное вещество *n*	1
espacio interplanetario *m*	espaço interplanetário *m*	межпланетное пространство *n*	2
distancia geocéntrica *f*	distância geocêntrica *f*	геоцентрическое расстояние *n*	3
distancia heliocéntrica *f*	distância heliocêntrica *f*	гелиоцентрическое расстояние *n*	4
órbita de la Tierra *f* órbita terrestre *f*	órbita da Terra *f* órbita terrestre *f*	орбита Земли *f* земная орбита *f*	5
mas allá de la órbita terrestre	além da órbita da Terra além da órbita terrestre	вне орбиты Земли вне земной орбиты	6
en las cercanías [en las proximidades, en la vecindad] de la órbita terrestre	próximo à órbita terrestre na proximidade da órbita da Terra	вблизи орбиты Земли вблизи земной орбиты	7
interplanetario	interplanetário	межпланетный	8
campo magnético interplanetario *m*	campo magnético interplanetário *m*	межпланетное магнитное поле *n*	9
dirección del campo (magnético) *f*	direção do campo (magnético) *f*	направление (магнитного) поля *n*	10
intensidad del campo (magnético) *f*	intensidade do campo (magnético) *f*	напряжённость магнитного поля *f* сила [магнитного] поля *f*	11
irregularidad del campo magnético *f*	irregularidade do campo magnético *f*	неоднородность магнитного поля *f*	12
irregularidad a pequeña escala *f*	irregularidade de pequena escala *f*	маломасштабная неоднородность *f* мелкомасштабная неоднородность *f*	13
variabilidad del campo (magnético) *f*	variabilidade do campo (magnético) *f*	переменность (магнитного) поля *f*	14
estructura en sector *f*	estrutura setorial [de setor] *f*	секторная структура *f*	15
sector (magnético) interplanetario *m*	setor (magnético) interplanetário *m*	межпланетный (магнитный) сектор *m* сектор межпланетного поля *m*	16
estructura en corrotación *f* estructura corrotacional *f*	estrutura em corrotação *f* estrutura corrotacional *f*	совместно вращающаяся структура *f* коротирующая структура (магнитного поля) *f* совращающаяся структура *f*	17
límite del sector *m m*	limite do setor *m*	граница сектора *f* секторная граница *f*	18

19	magnetic loop	boucle magnétique *f*	magnetische Schleife *f*
20	solar interplanetary loop solar interplanetary tongue	boucle solaire interplanétaire *f* langue solaire interplanétaire *f*	solare interplanetare Schleife *f*
21	interplanetary loop interplanetary tongue	boucle interplanétaire *f* langue interplanétaire *f*	interplanetare Schleife *f*
22	hydromagnetic cavity magnetic cavity	cavité hydromagnétique *f* cavité magnétique *f*	hydromagnetischer Hohlraum *m* magnetischer Hohlraum *m*
23	geomagnetic cavity	cavité géomagnétique *f*	geomagnetischer Hohlraum *m*
24	planetary cavity	cavité planétaire *f*	planetarer Hohlraum *m*
25	interplanetary cavity	cavité interplanétaire *f*	interplanetarer Hohlraum *m*
26	(cosmic-ray) modulation region	région de la modulation (des rayons cosmiques) *f*	Modulationsgebiet (der kosmischen Strahlung) *n*
27	cosmic-ray diffusion	diffusion du rayonnement cosmique *f*	Diffusion der kosmischen Strahlung *f*
28	solar-wind cavity	cavité du vent solaire *f*	Sonnenwind-Hohlraum *m*
29	solar plasma	plasma solaire *m*	Sonnenplasma *n*
30	plasma cloud	nuage de plasma *m*	Plasmawolke *f*
31	particle cloud	nuage de particules *m*	Partikelwolke *f*
32	magnetized solar plasma	plasma solaire magnétisé [aimanté] *m*	magnetisiertes Sonenplasma *n*
33	stream of solar plasma	courant de plasma solaire *m*	solarer Plasmastrom *m* Strom des Sonnenplasmas *m*
34	Archimedes spiral Archimedean spiral	spirale d'Archimède *f* spirale archimédienne *f*	Archimedische Spirale *f*
35	solar corpuscular stream	courant corpusculaire solaire *m*	Sonnenkorpuskularstrom *m*
35a	solar-wind stream	courant du vent solaire *m*	Sonnenwindstrom *m*
36	radio scintillation	radioscintillation *f* scintillation des radiosources *f*	Radioszintillation *f*
37	interplanetary radio scintillation	scintillation (des radiosources) interplanétaire *f*	interplanetare Radioszintillation *f*
38	burst of solar plasma	sursaut [éjection] de plasma solaire *m*[*f*]	Ausbruch des solaren Plasmas *m*
39	interplanetary plasma	plasma interplanétaire *f*	interplanetares Plasma *n*
40	permanent solar wind	vent solaire permanent *m*	permanenter Sonnenwind *m*

MATERIA INTERPLANETARIA	MATÉRIA INTERPLANETÁRIA	МЕЖПЛАНЕТНАЯ МАТЕРИЯ	
bucle magnético *m*	laço [arco] magnético *m*	магнитная петля *f*	19
bucle solar-interplanetario *m* lengua solar-interplanetaria *f*	laço [arco] solar interplanetário *m* língua solar interplanetária *f*	солнечная межпланетная петля *f*	20
bucle interplanetario *m* lengua interplanetaria *f*	laço [arco] interplanetário *m* língua interplanetária *f*	межпланетная петля *f* межпланетный выступ *m*	21
cavidad hidromagnética *f* cavidad magnética *f*	cavidade hidromagnética *f* cavidade magnética *f*	гидромагнитная полость *f* магнитная полость *f*	22
cavidad geomagnética *f*	cavidade geomagnética *f*	геомагнитная полость *f*	23
cavidad planetaria *f*	cavidade planetária *f*	планетная полость *f*	24
cavidad interplanetaria *f*	cavidade interplanetária *f*	межпланетная полость *f*	25
región de modulación (de la radiación cósmica) *f*	região de modulação (da radiação cósmica) *f*	область модуляции (космических лучей) *f*	26
difusión de los rayos cósmicos *f*	difusão dos raios cósmicos *f*	диффузия космических лучей *f*	27
cavidad del viento solar *f*	cavidade do vento solar *f*	полость солнечного ветра *f*	28
plasma solar *m*	plasma solar *m*	солнечная плазма *f*	29
nube de plasma *f*	nuvem de plasma *f*	облако плазмы *n* плазменное облако *n*	30
nube de partículas *f*	nuvem de partículas *f*	облако частиц *n*	31
plasma solar magnetizado *m*	plasma solar magnetizado *m*	замагниченная [намагниченная] солнечная плазма *f*	32
corriente de plasma solar *f*	corrente de plasma solar *f*	поток солнечной плазмы *m*	33
espiral de Arquímedes *f* espiral Arquimediana *f*	espiral de Arquimedes *f* espiral Arquimediana *f*	архимедова спираль *f* спираль Архимеда *f*	34
corriente corpuscular solar *f*	corrente corpuscular solar *f*	солнечный корпускулярный поток *m* поток солнечных корпускул *m*	35
corriente de viento solar *f*	corrente de vento solar *f*	поток солнечного ветра *m*	35a
radiocentelleo *m* centelleo [titileo] radio *m*	radiocintilação *f*	мерцание радиоисточников *n* радиомерцуание *n*	36
radiocentelleo interplanetario *m* centelleo interplanetario en radio *m* titileo interplanetario en radio *m*	cintilação interplanetária das radiofontes *f* radiocintilação interplanetária *f*	межпланетное мерцание радиоисточников *n* межпланетно обусловленное мерцание радиоисточников *n*	37
explosión [eyección] de plasma solar *f*	erupção [ejeção] de plasma solar *f*	выброс солнечной плазмы *m*	38
plasma interplanetario *m*	plasma interplanetário *m*	межпланетная плазма *f*	39
viento solar permanente *m*	vento solar permanente *m*	постоянный солнечный ветер *m*	40

84	INTERPLANETARY MATTER	MATIÈRE INTERPLANÉTAIRE	INTERPLANETARE MATERIE
41	quiescent solar wind quiet solar wind	vent solaire continu *m* vent solair calme *m*	ruhiger Sonnenwind *m*
42	burst of solar wind solar-wind burst	sursaut du vent solaire *m*	Ausbruch des Sonnenwindes *m*
43	(lunar) libration cloud	nuage de libration (de la Lune) *m*	(lunare) Librationswolke *f*
44	interplanetary storm	tempête interplanétaire *f*	interplanetarer Sturm *m*
45	interplanetary disturbance	perturbation interplanétaire *f*	interplanetare Störung *f*
46	flare-associated disturbance	perturbation associée à une éruption *f* perturbation due à une éruption *f*	einer Eruption zugeordnete Störung *f*
47	transit time time of transit Sun-Earth	temps de parcours Soleil-Terre *m*	Laufzeit (Sonne-Erde) *f*
48	transport of magnetic field magnetic-field transport	transport de champ magnétique *m*	Transport des Magnetfeldes *m*
49	collisionless shock-wave	onde de choc sans collisions *f*	stossfreie Schockwelle *f*
50	collisionless plasma	plasma sans collisions *m*	stossfreies Plasma *n*
51	interplanetary dust	poussière interplanétaire *f*	interplanetarer Staub *m*
52	mass distribution mass-frequency function	distribution des masses *f* fonction de fréquences de masses *f*	Massenverteilung *f* Massen-Frequenz-Funktion *f*
53	Earth-orbiting dust circumterrestrial dust	poussière circulant autour de la Terre *f* poussière circumterrestre *f*	die Erde umkreisender Staub *m* zirkumterrestrischer Staub *m*
54	zodiacal cloud	nuage zodiacal *m*	Zodiakalwolke *f*
55	solar-terrestrial relations Sun-Earth relations terrestrial effects of solar activity solar-terrestrial relationships	relations entre les phénomènes solaires et terrestres *fpl* effets terrestres de l'activité solaire *mpl* relations Soleil-Terre *fpl* phénomènes héliogéophysiques *mpl*	solar-terrestrische Beziehungen *fpl* solar-terrestrische Erscheinungen *fpl* terrestrische Effekte der Sonnenaktivität *npl* heliogeophysikalische Effekte *npl*
56	solar-terrestrial physics physics of solar-terrestrial relations	héliogéophysique *f* physique des relations Soleil-Terre *f*	Heliogeophysik *f* Physik der solar-terrestrischen Beziehungen *f*
57	cosmic geophysics	géophysique cosmique *f*	kosmische Geophysik *f*
58	passage of interplanetary discontinuity	passage d'une discontinuité interplanétaire *m*	Durchgang [Durchlaufen] einer interplanetaren Diskontinuität *m*[*n*]
59	proton event	événement à protons *m*	Protonereignis *n*

Español	Português	Русский	№
viento solar quiescente *m* viento solar en calma *m*	vento solar quiescente *m* vento solar calmo *m*	спокойный солнечный ветер *m*	41
explosión de viento solar *m*	erupção de vento solar *f*	вспышка солнечного ветра *f* всплеск солнечного ветра *m*	42
nube de libración (lunar) *f*	nuvem de libração (lunar) *f*	либрационное облако (Луны) *n*	43
tormenta interplanetaria *f*	tempestade interplanetária *f*	межпланетная буря *f*	44
disturbio interplanetario *m* perturbación interplanetaria *f*	perturbação interplanetária *f*	межпланетное возмущение *n*	45
disturbio asociado a una fulguración *m* perturbación asociada a una fulguración *f*	perturbação associada a uma fulguração *f*	возмущение, обусловленное вспышкой *n*	46
tiempo del tránsito (Sol-Tierra) *m*	tempo de trânsito (Sol-Terra) *m*	время пробега *n* время пробега (корпускул) от Солнца к Земле *n*	47
transporte del campo magnético *m*	transporte do campo magnético *m*	перенос магнитного поля *m*	48
onda de choque sin colisiones *f*	onda de choque sem colisões *f* onda de choque acolisional *f*	бесстолкновительная ударная волна *f*	49
plasma sin colisiones *m*	plasma sem colisões *m* plasma acolisional *m*	бесстолкновительная плазма *f*	50
polvo interplanetario *m*	poeira interplanetária *f* pó interplanetário *m*	межпланетная пыль *f*	51
distribución de las masas *f* función de frequencia de las masas *f*	distribuição das massas *f* função da freqüência de massa *f*	распределение по массе *n*	52
polvo orbitante (alrededor de la Tierra) *m* polvo circunterrestre *m*	poeira orbitante (ao redor da Terra) *f* poeira circunterrestre *f* pó circum-terrestre *m*	пыль, движущаяся вокруг Земли *f*	53
nube zodiacal *f*	nuvem zodiacal *f*	зодиакальное облако *n*	54
relaciónes heliogeofísicas *fpl* relación entre los fenómenos solares y terrestres *f* efectos terrestres de la actividad solar *mpl* relaciones Sol-Tierra *fpl*	relações solares-terrestres *fpl* efeitos terrestres da atividade solar *mpl* relações Sol-Terra *fpl* relações helio-geofísicas *fpl*	земные проявления солнечной активности *npl* солнечно-земные связи *fpl* связь между солнечными и земными явлениями *f* связь Солнце-Земля *f*	55
heliogeofísica *f* física de las relaciones Sol-Tierra *f*	helio-geofísica *f* física das relações Sol-Terra *f*	гелиогеофизика *f* физика солнечно-земных связей *f*	56
geofísica cósmica *f*	geofísica cósmica *f*	космическая геофизика *f*	57
paso de una discontinuidad interplanetaria *m*	passagem de discontinuidade interplanetária *f*	прохождение разрыва в межпланетной среде *n*	58
suceso protónico *m*	evento protônico *m* evento de protons *m*	выброс протонов *m* протонная вспышка *f*	59

	INTERPLANETARY MATTER	MATIÈRE INTERPLANÉTAIRE	INTERPLANETARE MATERIE
60	helium-enriched plasma	plasma enrichi d'hélium *m*	heliumangereichertes Plasma *n*
61	zodiacal light	lumière zodiacale *f*	Tierkreislicht *n* Zodiakallicht *n*
62	false zodiacal light	fausse lumière zodiacale *f*	falsches Zodiakallicht *n*
63	zodiacal band	bande (de la lumière) zodiacale *f*	Zodiakalband *n* Lichtbrücke *f*
64	counterglow gegenschein	lueur antisolaire *f* gegenschein *m*	Gegenschein *m*
65	tail of the Earth	queue de la Terre *f*	Erdschweif *m*
66	zodiacal particles	particules zodiacales *fpl*	Zodiakallichtteilchen *npl*
67	Poynting-Robertson effect	effet Poynting-Robertson *m*	Poynting-Robertson Effekt *m*
68	asteroid minor planet planetoid	astéroïde *m* petite planète *f* planétoïde *m*	Asteroid *m* kleiner Planet *m* Planetoid *m*
69	closest approach	passage au plus près *m* passage à distance minimale *m* approche minimale *f*	grösste Annäherung *f*
69a	artificial asteroid man-made asteroid	astéroïde artificiel *m*	künstlicher Asteroid *m*
70	ring of asteroids asteroid ring minor planet belt asteroid belt belt of planetoids	anneau des astéroïdes *m* ceinture des petites planètes *f* ceinture d'astéroïdes *f*	Asteroidenring *m* Asteroidengürtel *m* Ring kleiner Planeten *m*
71	ring asteroid asteroid [planetoid, minor planet] of the belt	asteroïde de l'anneau *m* planétoïde de l'anneau *m* petite planète de l'anneau *f*	Ringasteroid *m* Ringplanetoid *m*
72	gravitational crushing gravitational compression	écrasement gravitationnel *m*	Zerdrücken durch Gravitation *n* Schwerezerquetschung (eines Asteroiden) *f*
73	asteroid family	famille des astéroïdes *f*	Asteroidenfamilie *f*
74	Mars (-family) asteroid	astéroïde de (la famille de) Mars *m* planétoïde (de la famille) de Mars *m*	Marsasteroid *m* Asteroid der Marsfamilie *m*
75	Mars-crossing asteroid Marscrosser Mars-orbit crosser	astéroïde traversant l'orbite de Mars *m* planétoïde traversant l'orbite de Mars *m*	Marsbahn kreuzender Asteroid *m*
76	Earth-crossing asteroid Earth crosser Earth-orbit crosser	astéroïde traversant l'orbite de la Terre *m*	Erdbahn kreuzender Asteroid *m*

plasma enriquecido en hélio *m*	plasma enriquecido de hélio *m*	плазма, обогащённая гелием *f*	60
luz zodiacal *f*	luz zodiacal *f*	зодиакальный свет *m*	61
luz zodiacal falsa *f*	falsa luz zodiacal *f*	ложный зодиакальный свет *m*	62
banda (de la luz) zodiacal *f*	banda (da luz) zodiacal *f* faixa zodiacal *f*	зодиакальная полоса *f*	63
contrabrillo zodiacal *m* luz antisolar *f*	contrabrilho zodiacal *m* brilho anti-solar *m*	противосияние *n*	64
cola de la Tierra *f*	cauda da Terra *f*	хвост Земли *m*	65
partículas zodiacales *fpl*	partículas zodiacais *fpl*	зодиакальные частицы *fpl* частицы зодиакального облака *fpl*	66
efecto Poynting-Robertson *m*	efeito (de)Poynting-Robertson *m*	эффект Пойнтинга-Робертсона *m*	67
asteroide *m* planeta menor *m* pequeño planeta *m* planetoide *m* planetilla *f*	asteróide *m* pequeno planeta *m* planetóide *m*	астероид *m* малая планета *f* планетоид *m*	68
aproximación cercana *f*	aproximação máxima *f*	теснейшее сближение *n*	69
asteroide artificial *m* planetoide hecho por el hombre *m*	asteróide artificial *m* planetóide feito pelo homem *m*	искусственный астероид *m* искусственная малая планета *f*	69a
anillo de asteroides *m* anillo asteroidal *m* anillo de planetas menores *m* cinturón de asteroides *m*	anel de asteróides *m* cinturão dos pequenos planetas *m* anel asteroidal *m* cinturão de asteróides *m*	кольцо астероидов *n* кольцо малых планет *n* пояс малых планет *m* пояс астероидов *m*	70
asteroide del anillo *m* asteroide del cinturón *m* planetoide del anillo *m*	asteróide do anel [cinturão] *m* planetóide do cinturão [anel] *m*	астероид кольца *m* малая планета кольца *f* планетоид пояса *m*	71
aplastamiento gravitacional *m*	esmagamento gravitacional *m*	гравитационное разрушение *n*	72
familia de los asteroides *f*	família de asteróides *f*	семейство астероидов *n* семейство малых планет *n*	73
asteroide marciano *m* planetoide de (la familia de) Marte *m*	asteróide (da família) de Marte *m* asteróide marciano *m*	астероид (семейства) Марса *n*	74
asteroide que cruza la órbita de Marte *m*	asteróide cruzante à órbita de Marte *m*	астероид [планетоид], пересекающий орбиту Марса *m*	75
asteroide que cruza la órbita terrestre [de la Tierra] *m*	asteróide cruzante à órbita de Terra *m*	астероид [малая планета], пересекающий орбиту Земли *m*	76

77	cigar-shaped asteorid	astéroïde en forme de cigare *m*	zigarrenförmiger Asteroid *m*
78	close-approach asteroid Earth-grazing asteroid Earth grazer	astéroïde de courte approche *m* astéroïde d'approche serrée *m*	Asteroid mit dichter Annäherung *m*
79	binary asteroid double planetoid [asteroid]	astéroïde double *m* planétoïde binaire *m*	Doppelasteroid *m* Doppelplanetoid *m*
80	Amor(-type) asteroid	astéroïde de type Amor *m*	Amor-ähnlicher Asteroid [Planetoid] *m*
80a	Apollo(-type) asteroid	astéroïde de type Apollo *m*	Apollo-Asteroid *m*
81	Apollo	Apollon *m*	Apollo *m*
82	Ceres	Cérès *f*	Ceres *f*
83	Hermes	Hermès *m*	Hermes *m*
84	Icarus	Icare *m*	Ikarus *m*
85	Juno	Junon *f*	Juno *f*
86	Pallas	Pallas *f*	Pallas *f*
87	Vesta	Vesta *f*	Vesta *f*
88	commensurability gap gap of commensurability	manque de commensurabilité *m* lacune *f* lacune de commensurabilité *f*	Lücke *f* Kommensurabilitätslücke *f*
89	Kirkwood gaps Hornstein-Kirkwood gaps	lacunes de Kirkwood *fpl* manques de Kirkwood *mpl*	Kirkwood-Lücken *fpl*
90	group of minor planets asteroid group asteroidal groups	groupe astéroïdal *m* groupe de petites planètes *m* groupe d'astéroïdes *m*	Asteroidengruppe *f* Gruppe kleiner Planeten *f* Planetoidengruppe *f*
91	Trojan group Trojans Trojan group of asteroids	groupe troyen *m* petites planètes troyennes *fpl* groupe troyen d'astéroïdes *m*	Trojaner *pl* Trojaner-Asteroidengruppe *f*
91a	libration asteroid	astéroïde de libration *m*	Librationsasteroid *m*
92	ephemeris of minor planet positions predictions of minor planet positions	éphémérides des petites planètes *fpl*	Ephemeriden von kleinen Planeten *fpl*
93	photometric diameter	diamètre photométrique *m*	photometrischer Durchmesser *m*
94	asteroid survey	surveillance des astéroïdes *f*	Überwachung von Asteroiden *f*
95	catalog(u)ed asteroid numbered planetoid	astéroïde catalogué *m* planétoïde numéroté *m*	katalogisierter Asteroid *m* numerierter Asteroid *m*

asteroide en forma de cigarro *m*	asteróide em forma de charuto *m*	астероид в форме сигары *m* сигарообразный астероид *m*	77
asteroide de aproximación cercana (a la Tierra) *m*	asteróide rasante (à Terra) *m* asteróide de grande aproximação (à Terra) *m*	астероид тесного сближения *m* планетоид [малая планета], тесно сближающийся [...щаяся] *m[f]*	78
asteroride binario *m* planetoide doble *m*	asteróide binário *m* planetóide duplo *m*	двойной астероид *m* двойной планетоид *m*	79
objeto [asteroide] tipo Amor *m*	asteróide [planetóide] tipo Amor *m*	объект [астероид] типа Амур *m*	80
asteroide [planetoide] tipo Apolo *m*	objeto [asteróide] tipo Apolo *m*	астероид типа Аполлон *m*	80a
Apolo *m*	Apolo *m*	Аполлон *m*	81
Ceres *f*	Ceres *f*	Церера *f*	82
Hermes *m*	Hermes *m*	Гермес *m*	83
Ícaro *m*	Ícaro *m*	Икар *m*	84
Juno *f*	Juno *f*	Юнона *f*	85
Palas *f*	Palas *f*	Паллада *f*	86
Vesta *f*	Vesta *f*	Веста *f*	87
laguna *f* laguna de conmensurabilidad *f*	falha *f* lacuna *f* lacuna de comensurabilidade *f*	пробел (соизчеримости) *m* люк *m* промежуток (соизнеримости) *m*	88
lagunas de Kirkwood *fpl*	lacunas de Kirkwood *fpl*	промежутки Кирквуда *mpl* люки Кирквуда *mpl*	89
grupo asteroidal *m* grupo de los asteroides *m* grupo de los planetas menores *m*	grupo de asteróides [planetóides] *m* grupo de pequenos planetas *m* grupo asteroidal *m*	группа малых планет *f* группа астероидов *f* группа планетноидов *f*	90
asteroides troyanos *mpl* Troyanos *mpl* grupo troyano *m*	grupo troiano *m* asteróides troianos *mpl* Troianos *mpl*	Троянцы *mpl* группа Троянцев *f*	91
asteroide de libración *m*	asteróide em [de] libração *m*	либрирующий астероид *m*	91a
predicciones de las posiciones de pequeños planetas *fpl* efemérides de los pequeños planetas *fpl*	efeméride(s) dos pequenos planetas *f(pl)*	предвычисление положений малых планет *n* эфемериды малых планет *fpl*	92
diámetro fotométrico *m*	diâmetro fotométrico *m*	фотометрический диаметр *m*	93
patrullaje de asteroides *m*	levantamento [patrulhamento] dos asteróides *m* patrulha dos asteróides *f*	обзор наблюдений малых планет *m* обозрение астероидов *n*	94
asteroide catalogado *m* asteroide numerado *m*	asteróide [planetóide] catalogado *m* asteróide [planetóide] numerado *m*	каталогизированная малая планета *f* нумерованный [каталогизированный] астероид *m*	95

96	uncatalogued object uncataloged object (US)	objet non-catalogué *m* objet non-numéroté *m*	nichtkatalogisiertes Objekt *n* nichtnumeriertes Objekt *n*
97	lost asteroid lost planetoid	astéroïde perdu *m* planétoide perdu *m*	verlorener Asteroid *n* verlorener Planetoid *m*
98	sample-return mission to an asteroid	mission (spatiale) vers un astéroïde pour ramener des échantillons *f*	Raumsonde zu einem Asteroiden die Asteroidenmaterie zur Erde zurückbringt *f*
99	asteroid mission mission to an asteroid	mission (spatiale) vers un astéroïde *f*	Asteroidenmission *f* Mission [Raumsonde] zu einem Asteroiden *f*
100	asteroid stream asteroidal (jet) stream	courant d'astéroïdes *m*	Asteroidenstrom *m*
101	commensurability (asteroid) group	groupe astéroïdal de commensurabilité *m*	Kommensurabilitäts-(Asteroiden)gruppe *f*
102	orientation of rotation axis	direction de l'axe de rotation *f*	Orientierung der Drehachse *f*
103	geometrical cross-section	section (efficace) géométrique *f*	geometrischer Querschnitt *m*
104	gravitational cross-section	section (efficace) gravitationnelle (de collision) *f*	Gravitationsquerschnitt *m*
105	erosive collision	collision érosive *f*	erosiver Zusammenstoss *m*
106	catastrophic collision	collision catastrophique *f*	katastrophischer Zusammenstoss *m*
107	collisional break-up	rupture par collision *f*	Zerfall durch Stoss *m*
108	asteroid fragmentation fragmentation of asteroids	fragmentation des astéroïdes *f* désintégration des astéroïdes *f*	Zerplatzen der Asteroiden *n*
109	outgassing of a cometary nucleus	dégazage d'un noyau cométaire *m*	Ausgasen eines Kometenkernes *n*

objeto no catalogado *m* objeto no numerado *m*	objeto não-catalogado *m* objeto não-numerado *m*	некаталогизированный [ненумерованный] объект *m*	96
asteroide perdido *m*	asteróide perdido *m*	потерянный астероид *m* потерянная малая планета *f*	97
misión a un asteroide con retorno de muestras *f*	missão a um asteróide com retorno de amostras *f*	полёт к астероиду с возвращением образцов на Землю *m*	98
misión a un asteroide *f*	missão a um asteróide *f*	полёт к астероиду *m* полёт к малой планете *m*	99
corriente de asteroides *f*	corrente de asteróides *f*	поток астероидов *m* поток малых планет *m*	100
grupo (asteroidal) de conmensurabilidad *m*	grupo (asteroidal) de comensurabilidade *m*	группа соизмеримых астероидов *f*	101
orientación del eje de rotación *f*	orientação do eixo de rotação *f*	направление оси вращения *n*	102
sección (eficaz) geométrica *f*	se(c)ção de choque geométrica *f* se(c)ção (eficaz) geométrica *f*	геометрическое сечение столкновения *n*	103
sección (eficaz) gravitacional *f*	se(c)ção (eficaz) gravitacional *f* se(c)ção de choque gravitacional *f*	гравитационное сечение столкновения *n*	104
colisión erosiva *f*	colisão erosiva *f*	эрозионное столкновение *n*	105
colisión catastrófica *f*	colisão catastrófica *f*	катастрофическое столкновение *n*	106
ruptura por colisión *f*	ruptura colisional *f*	разрыв [разрушение] ударением [ударом] *f*[*n*] распад при столкновении *m*	107
fragmentación de los asteroides *f*	fragmentação dos asteróides *f* desintegração dos asteróides *f*	раздробление малых планет *n* фрагментация астероидов *f*	108
desgasificación del un núcleo cometario *f*	degasação de um núcleo cometário *f*	выделение газов из кометного ядра *n* дегазация кометного ядра *f*	109

85 COMETS — COMÈTES — KOMETEN

	COMETS	COMÈTES	KOMETEN
1	comet	comète *f*	Komet *m* Haarstern *m* Schweifstern *m*
2	cometary	cométaire	kometarisch Kometen-
3	periodic comet	comète périodique *f*	periodischer Komet *m*
4	elliptical-orbit comet elliptic comet	comète elliptique *f*	elliptischer Komet *m* Komet mit elliptischer Bahn *m*
5	short-period comet	comète à courte période *f*	kurzperiodischer Komet *m*
6	long-period comet	comète à longue période *f*	langperiodischer Komet *m*
7	parabolic-orbit comet parabolic comet	comète parabolique *f*	parabolischer Komet *m* Komet mit parabolischer Bahn *m*
8	hyperbolic-orbit comet hyperbolic comet	comète hyperbolique *f*	hyperbolischer Komet *m* Komet mit hyperbolischer Bahn *m*
9	return of a comet cometary return	retour d'une comète *m*	Wiederkehr eines Kometen *m*
10	non-periodic comet	comète non-périodique *f*	nichtperiodischer Komet *m*
11	predicted comet	comète prédite *f*	vorausberechneter Komet *m*
12	eclipse comet comet discovered during an eclipse	comète de l'éclipse *f* comète découverte pendant l'éclipse *f*	Finsterniskomet *m*
13	Sun-grazing comet sungrazer	comète qui rase le Soleil *f*	die Sonne(atmosphäre) berührender Komet *m*
14	artificial comet	comète artificielle *f*	künstlicher Komet *m*
14a	telescopic comet	comète télescopique *f*	teleskopischer Komet *m*
15	search for comets	recherche des comètes *f*	Kometensuche *f* Aufsuchung der Kometen *f*
16	discovery of a comet comet discovery	découverte d'une comète *f*	Kometenentdeckung *f* Entdeckung eines Kometen *f*
17	rediscovery	rédécouverte *f*	Wiederentdeckung *f*
18	apparition of a comet	apparition d'une comète *f*	Erscheinen eines Kometen *n*
19	first apparition (of a comet)	première apparition (d'une comète) *f*	erstes Erscheinen (eines Kometen) *n*
20	astronomical telegram	télégramme astronomique *m*	astronomisches Telegramm *n*
21	discoverer	auteur de la découverte *m*	Entdecker *m*

cometa *m*	cometa *m*	комета *f*	1
cometario	cometário	кометный	2
cometa periódico *m*	cometa periódico *m*	периодическая комета *f*	3
cometa de órbita elíptica *m* cometa elíptico *m*	cometa elíptico *m* cometa elítico *m* cometa com órbita elí(p)tica *m*	эллиптическая комета *f* комета с эллиптической орбитой *f*	4
cometa de corto período *m*	cometa de curto período *m*	короткопериодическая комета *f*	5
cometa de largo período *m*	cometa de longo período *m*	долгопериодическая комета *f*	6
cometa de órbita parabólica *m* cometa parabólico *m*	cometa parabólico *m* cometa com órbita parabólica *m*	параболическая комета *f* комета с параболической орбитой *f*	7
cometa de órbita hiperbólica *m* cometa hiperbólico *m*	cometa hiperbólico *m* cometa com órbita hiperbólica *m*	гиперболическая комета *f* комета с гиперболической орбитой *f*	8
retorno de un cometa *m*	retorno de um cometa *m*	возвращение кометы *n*	9
cometa no periódico *m*	cometa não-periódico *m*	непериодическая комета *f*	10
cometa predicho *m*	cometa previsto [predito] *m*	предвычисленная комета *f* предсказанная комета *f*	11
cometa de eclipse *m*	cometa de eclipse *m*	затменная комета *f* комета, открытая при затмении *f*	12
cometa rasante al Sol *m*	cometa rasante ao Sol *m*	задевающая Солнце комета *f* комета, тесно сближающаяся с Солнцем *f*	13
cometa artificial *m*	cometa artificial *m*	искусственная комета *f*	14
cometa telescópico *m*	cometa telescópico *m*	телескопическая комета *f*	14a
búsqueda de cometas *f*	busca de cometas *f* procura de cometas *f*	поиски комет *mpl*	15
descubrimiento de un cometa *m*	descoberta de um cometa *f*	открытие кометы *n*	16
redescubrimiento *m*	redescoberta *f*	переоткрытие *n*	17
aparición de un cometa *f*	aparição [aparecimento] de um cometa *f*	появление кометы *n*	18
primera aparición (de un cometa) *f*	primeira aparição (de um cometa) *f*	первое появление (кометы) *n*	19
telegrama astronómico *m*	telegrama astronômico *m*	астрономическая телеграмма *f*	20
descubridor *m*	autor da descoberta *m* descobridor *m*	первооткрыватель (кометы) *m* открыватель (кометы) *m*	21

	COMETS	COMÈTES	KOMETEN
22	computer	calculateur *m*	Berechner *m*
23	communicator	expéditeur *m*	Berichter *m*
24	nature of the object	nature de l'objet *f*	Art des Objektes *f*
25	diffuse object	objet diffus *m*	diffuses Objekt *n*
26	approximate position	position approchée *f*	genäherter Ort *m* genäherte Position *f*
27	predicted position calculated position	position prédite *f* position calculée *f*	vorausberechnete Position *f* vorhergesagter Ort *m* vorhergesagte Position *f* vorherberechnete Position *f*
28	accurate position	position précise *f*	genauer Ort *m* genaue Position *f*
29	daily motion	mouvement diurne *m*	tägliche Bewegung *f*
30	check number	nombre de contrôle *m*	Kontrollzahl *f*
31	designation of a comet	désignation d'une comète *f*	Bezeichnung eines Kometen *f*
32	preliminary designation	désignation provisoire *f*	vorläufige Bezeichnung *f*
33	definitive designation	désignation définitive *f*	endgültige Bezeichnung *f*
34	structure of comets cometary structure	structure des comètes *f* structure cométaire *f*	Aufbau der Kometen *m* Kometenaufbau *m*
35	head of a comet cometary head	tête d'une comète *f* tête cométaire *f*	Kopf eines Kometen *m* Kometenkopf *m*
36	cometary nucleus nucleus of a comet	noyau cométaire *m* noyau d'une comète *m*	Kometenkern *m* Kern eines Kometen *m*
37	central condensation	condensation centrale *f*	zentrale Verdichtung *f*
38	starlike nucleus	noyau (quasi) stellaire *m* noyau en forme d'étoile *m*	sternförmiger Kern *m*
39	solid nucleus	noyau solide *m*	fester Kern *m*
40	ice nucleus icy nucleus	noyau de glace *m*	Eiskern *m*
41	conglomerate	conglomérat *m*	Konglomerat *n*
42	ice	glace *f*	Eis *n*
43	icy conglomerate	conglomérat de glace *m*	Eiskonglomerat *n*
44	amorphous ice	glace amorphe *f*	amorphes Eis *n*

calculista *m*, *f*	calculador *m* calculista *m*	вычислитель *m*	22
comunicador *m* comunicante *m*	comunicador *m*	отправитель *m*	23
naturaleza del objeto *f*	natureza do objeto *f*	природа объекта *f*	24
objeto difuso *m*	objeto difuso *m*	диффузный объект *m*	25
posición aproximada *f*	posição aproximada *f*	приблизительное положение *n*	26
posición predicha *f* posición calculada *f*	posição prevista *f* posição calculada *f* posição predita *f*	предвычисленное положение *n* предсказанное положение *n*	27
posición precisa *f*	posição precisa *f*	точное положение *n*	28
movimiento diario *m*	movimento diário *m*	суточное движение *n*	29
número de control *m*	número de verificação [de controle] *m*	контрольное число *n*	30
designación de un cometa *f* nombre de un cometa *m*	designação de um cometa *f* nome de um cometa *m*	обозначение кометы *n*	31
designación preliminar *f* designación previa *f*	designação provisória *f* designação preliminar *f*	предварительное обозначение *n*	32
designación definitiva *f*	designação definitiva *f*	окончательное обозначение *n*	33
estructura de los cometas *f* estructura cometaria *f*	estrutura dos cometas *f* estrutura cometária *f*	строение комет *f* структура комет *f*	34
cabeza de un cometa *f* cabeza cometaria *f*	cabeça de um cometa *f* cabeça cometária *f*	голова кометы *f* кометная голова *f*	35
núcleo cometario *m* núcleo de un cometa *m*	núcleo de um cometa *m* núcleo cometário *m*	ядро кометы *n* кометное ядро *n*	36
condensación central *f*	condensação central *f*	центральное уплотнение *n* центральное сгущение *n*	37
núcleo cuasiestelar *m* núcleo esteliforme *m*	núcleo (quase-)estelar *m* núcleo esteliforme *m*	звёздообразное ядро *n* звёздоподобное ядро *n*	38
núcleo sólido *m*	núcleo sólido *m*	твёрдое ядро *n*	39
núcleo de hielo *m*	núcleo de gelo *m*	ледяное ядро *n*	40
conglomerado *m*	conglomerado *m*	конгломерат *m*	41
hielo *m*	gelo *m*	лёд *m*	42
conglomerado de hielo *m*	conglomerado de gelo *m*	ледяной конгломерат *m*	43
hielo amorfo *m*	gelo amorfo *m*	аморфный лёд *m*	44

45	frozen gases	gaz congelés *mpl*	gefrorene Gase *npl*
46	cubic ice	glace cubique *f*	kubisches Eis *n*
47	water ice H_2O ice	glace d'eau *f*	Wassereis *n*
48	solid particles	particules solides *fpl* corpuscules solides *fpl*	feste Partikeln *fpl* feste Teilchen *npl*
49	refractory solid	solide réfractaire *m*	feuerfester Feststoff *m*
50	porous meteoroid	météoroïde poreux *m*	poröses Meteoroid *n*
51	clathrate	clathrate *m*	Käfigeinschlussverbindung *f*
52	icy grain	grain de glace *m*	Eiskorn *n*
53	silicate dust	poussière de silicate *f*	Silikatstaub *m*
54	rocky grain	grain pierreux *m*	Gesteinskorn *n*
55	jet ray filament	rayon *m* jet *m* filament *m*	Ausströmung *f* Strahl *m*
56	non-gravitational motion	mouvement non-gravitationnel *m*	nicht durch Gravitation verursachte Bewegung *f*
57	rectilinear jet	jet rectiligne *m*	geradlinige [gerichtete] Ausströmung *f* rückwärts gerichteter Strahl *m*
58	curved jet	jet incurvé *m*	gekrümmter Strahl *m*
59	spiral jet	jet en spirale *m*	spiralförmiger Strahl *m*
60	sector fan	secteur *m* éventail *m*	Sektor *m* Fächer *m*
61	straight fan	éventail droit *m*	gerader Fächer *m*
62	curved fan	éventail incurvé *m*	gekrümmter Fächer *m*
63	inverted fan	éventail inversé *m*	umgekehrter Fächer *m*
64	coma (pl. comae)	chevelure *f*	Koma *f* Hülle des Kometenkerns *f*

Spanish	Portuguese	Russian	No.
gases congelados *mpl*	gases congelados *mpl*	замерзшие [замороженные] газы *mpl*	45
hielo cúbico *m*	gelo cúbico *m*	кубический лёд *m*	46
hielo de agua *m*	gelo de água *m*	водяной лёд *m*	47
partículas sólidas *fpl* corpúsculos sólidos *mpl*	partículas sólidas *fpl* corpúsculos sólidos *mpl*	твёрдые частицы *fpl*	48
material refractario *m*	su(b)stância sólida refratária *f*	тугоплавкое твёрдое вещество *n*	49
meteoroide poroso *m*	meteoróide poroso *m*	пористый метеороид *m*	50
clatrato *m*	clatrato *m*	клатрат *m*	51
grano de hielo *m*	grão de gelo *m*	ледяная частица *f*	52
polvo de silicato *m*	poeira de silicato *f* pó de silicato *m*	силикатная пыль *f*	53
grano de roca *m*	grão rochoso *m*	каменистая частица *f*	54
rayo *m* chorro *m* filamento *m*	ja(c)to *m* filamento *m*	выброс *m* излияние *n*	55
movimiento no gravitacional *m* movimiento no gravitatorio *m*	movimento não-gravitacional *m*	негравитационное движение *n*	56
chorro rectilíneo *m* rayo rectilíneo *m*	ja(c)to retilíneo *m*	прямолинейное излияние *n* прямолинейный выброс *m*	57
chorro curvo *m* rayo curvado *m*	ja(c)to curvado *m*	искривлённый выброс *m* искривлённое излияние *n* кривящийся выброс *m*	58
chorro en espiral *m*	ja(c)to em espiral *m*	спиральное излияние *n* спиральный выброс *m*	59
sector *m* abanico *m*	leque *m* se(c)tor *m*	веерообразное истечение *n* веерообразный выброс *m*	60
abanico recto *m*	leque [se(c)tor] reto *m*	прямое веерообразное истечение *n* прямой веерообразный выброс *m*	61
abanico curvo *m* abanico curvado *m*	leque curvado *m* se(c)tor curvado *m*	искривлённое веерообразное истечение *n* (искривлённый) веерообразный выброс *m*	62
abanico invertido *m*	leque [se(c)tor] invertido *m*	перевернутое [обращённое] веерообразное истечение *n* обращённый веерообразный выброс *m*	63
coma *f* cabellera *f*	coma *f* cabeleira *f*	кома *f* оболочка головы кометы *f*	64

	COMETS	COMÈTES	KOMETEN
65	molecular coma inner coma photochemical coma	chevelure moléculaire *f* chevelure intérieure *f*	molekulare Koma *f* innere Koma *f*
66	radical coma visible coma	chevelure de radicaux *f* chevelure visible *f*	Radikalkoma *f* sichtbare Koma *f*
67	atomic coma UV coma	chevelure atomique *f* chevelure ultraviolette *f*	atomare Koma *f* Ultraviolettkoma *f*
68	haloes of a comet	halos d'une comète *mpl*	Kometenhalos *mpl*
69	envelopes of a comet shells of a comet	enveloppes d'une comète *fpl*	Kometenenveloppen *fpl*
70	tail of a comet cometary tail	queue d'une comète *f* queue cométaire *f*	Kometenschweif *m*
71	cometary tail type	type de la queue cométaire *m*	Kometenschweiftyp(us) *m*
72	straight tail	queue rectiligne *f*	gerader Schweif *m*
73	curved tail	queue incurvée *f*	gekrümmter Schweif *m*
74	fan-like fan-shaped	en évantail	fächerförmig fächerartig
75	anomalous tail	queue anomale *f*	anomaler Schweif *m*
76	multiple tail	queue multiple *f*	mehrfachar Schweif *m*
77	main tail	queue principale *f*	Hauptschweif *m*
78	tail directed toward the Sun sunward tail antitail	queue dirigée vers le Soleil *f* antiqueue *f*	zur Sonne gerichteter Schweif *m* Gegenschweif *m*
79	dust tail	queue de poussières *f*	Staubschweif *m*
80	dust content	teneur en poussières *f*	Staubgehalt *m*
81	(tail) directed away from the Sun	(queue) dirigée à l'opposé du Soleil *f*	von der Sonne abgewandter Schweif *m*
82	ion tail	queue d'ions *f*	Ionenschweif *m*
83	cometary form shape of a comet	forme cométaire *f*	Kometenform *f*
84	mechanical theory of cometary forms	théorie mécanique des formes cométaires *f*	mechanische Theorie der Kometenformen *f*
85	fountain model	modèle (en forme de jet) de fontaine *m*	Fontänenmodell *n*
86	tail formation	formation de la queue *f*	Schweifbildung *f*
87	repulsive force	force répulsive *f* force de répulsion *f*	Repulsivkraft *f* Repulsionskraft *f* repulsive Kraft *f*

coma molecular *f* coma interior *f*	coma molecular *f* coma interior *f*	внутренняя кома *f* молекулярная кома *f* фотохимическая кома *f*	65
coma de radicales *f* coma visible *f*	coma de radicais *f* coma visível *f*	видимая кома *f* свободнорадикальная кома *f* кома свободных радикалов *f*	66
coma atómica *f* coma ultravioleta *f*	coma atômica *f* coma ultravioleta *f*	атомарная кома *f* ультрафиолетовая кома *f*	67
halos de un cometa *mpl*	halos de um cometa *mpl*	галосы кометы *mpl*	68
capas envolventes de un cometa *fpl* envolturas de un cometa *fpl*	invólucros de um cometa *mpl*	оболочки кометы *fpl*	69
cola de un cometa *f* cola cometaria *f*	cauda de um cometa *f* cauda cometária *f*	хвост кометы *m* кометный хвост *m*	70
tipo de la cola cometaria *m*	tipo de cauda cometária *m*	тип кометного хвоста *m*	71
cola recta *f*	cauda ret(ilíne)a *f*	прямой хвост *m*	72
cola curvada *f*	cauda curvada *f*	изогнутый хвост *m*	73
en forma de abanico	em forma de leque	веерообразный	74
cola anómala *f*	cauda anômala *f*	аномальный хвост *m*	75
cola múltiple *f*	cauda múltipla *f*	многолучевой хвост *m* кратный хвост *m*	76
cola principal *f*	cauda principal *f*	главный хвост *m*	77
cola dirigida hacia el Sol *f* anticola *f*	cauda dirigida para o Sol *f* anticauda *f*	хвост, направленный к Солнцу *m* аномальный хвост *m*	78
cola de polvo *f*	cauda de pó *f*	пылевой хвост *m*	79
contenido de polvo *m*	conteúdo de pó *m*	пылесодержание *n* содержание пыли *n*	80
cola en dirección opuesta al Sol *f*	cauda em direção oposta ao Sol *f*	(хвост) направленный от Солнца *m*	81
cola de iones *f* cola iónica *f*	cauda iônica *f*	ионный хвост *m*	82
forma de un cometa *f*	forma de um cometa *f*	форма кометы *f* кометарная [кометная] форма *f*	83
teoría mecánica de la forma de cometas *f*	teoria mecânica da forma dos cometas *f*	механическая теория кометных форм *f*	84
modelo de surtidor *m*	modelo em chafariz *m*	фонтанная модель *f*	85
formación de la cola *f*	formação de cauda *f*	образование хвоста *n*	86
fuerza repulsiva *f* fuerza de repulsión *f*	força repulsiva *f* força de repulsão *f*	отталкивающая [отталкивательная] сила *f* сила отталкивания *f*	87

	COMETS	COMÈTES	KOMETEN
88	effective acceleration	accéleration effective *f*	effektive Beschleunigung *f*
89	ejection of particles	éjection des particules *f*	Ausschleuderung von Partikeln *f* Auswurf von Partikeln *m*
90	velocity of ejection	vitesse d'éjection *f*	Auswurfgeschwindigkeit *f*
91	syndyname syndyne	syndyname *f*	Syndyname *f*
92	synchrone	synchrone *f*	Synchrone *f*
93	terminal synchrone	synchrone terminale *f*	letzte Synchrone *f*
94	complete synchrone	synchrone complète *f*	vollständige Synchrone *f*
95	knot tail condensation	condensation dans la queue *f*	Knoten *m* Schweifkondensation *f*
96	origin of comets	origine des comètes *f*	Ursprung der Kometen *f* Herkunft der Kometen *f*
97	Oort's cloud (of comets) Öpik-Oort cloud	nuage (de comètes) d'Oort *m*	Oortsche Kometenwolke *f*
98	ejection theory	théorie de l'éjection *f*	Auswurfstheorie *f*
99	capture theory	théorie de la capture *f*	Einfangtheorie *f*
100	comet cloud	nuage de comètes *m* essaim de comètes *m*	Kometenwolke *f*
101	capture of comets	capture des comètes *f*	Kometeneinfang *m* Einfang der Kometen *m*
102	family of comets	famille de comètes *f*	Kometenfamilie *f*
103	Jupiter's family Jovian family	famille de Jupiter *f*	Kometenfamilie des Jupiters *f*
104	physics of comets	physique des comètes *f*	Physik der Kometen *f*
105	comet group	groupe de comètes *m*	Kometengruppe *f*
106	photometric parameters	paramètres photométriques *mpl*	photometrische Parameter *mpl*
107	intrinsic brightness total brightness	éclat global *m* éclat total *m* éclat intrinsèque *m*	Gesamthelligkeit *f* Eigenhelligkeit *f*
108	reduced brightness	éclat réduit *m*	reduzierte Helligkeit *f*
109	absolute brightness (of a comet)	éclat absolu (d'une comète) *m*	absolute Helligkeit (eines Kometen) *f*

Español	Português	Русский	№
aceleración efectiva *f*	aceleração efetiva *f*	эффективное ускорение *n*	88
eyección de partículas *f*	ejeção de partículas *f*	выброс частиц *m* извержение частиц *n*	89
velocidad de eyección *f*	velocidade de ejeção *f*	скорость выброса *f* скорость извержения *f*	90
sindínama *f*	sindínama *f*	синдинама *f*	91
síncrona *f*	síncrona *f*	синхрона *f*	92
síncrona terminal *f*	síncrona terminal *f*	концевая синхрона *f*	93
síncrona completa *f*	síncrona completa *f*	полная синхрона *f*	94
condensación en la cola *f*	condensação na cauda *f*	узел в хвосте *m* облачное образование в хвосте *n*	95
origen de los cometas *m*	origem dos cometas *f*	происхождение комет *n*	96
nube de cometas de Oort *f* nube de Oort *f*	nuvem dos cometas de Oort *f* nuvem de Oort *f*	облако Оорта [Эпика-Оорта] *n* облако комет *n*	97
teoría de eyección *f*	teoria de ejeção *f*	гипотеза извержения *f*	98
teoría de captura *f*	teoria de captura *f*	гипотеза захвата *f* гипотеза увлечения *f*	99
nube de cometas *f*	nuvem de cometas *f*	кометное облако *n*	100
captura de cometas *f*	captura de cometas *f*	захват комет *m*	101
familia de cometas *f*	família de cometas *f*	семейство комет *n*	102
familia de Júpiter *f*	família de Júpiter *f*	семейство (комет) Юпитера *n*	103
física de los cometas *f*	física dos cometas *f*	физика комет *f*	104
grupo de cometas *m*	grupo de cometas *m*	группа комет *f*	105
parámetros fotométricos *mpl*	parâmetros fotométricos *mpl*	фотометрические параметры *mpl*	106
brillo intrínseco *m* brillo total *m*	brilhância global [total] *f* brilho global [total] *m* brilhância intrínseca *f*	интегральный блеск *m* собственный блеск *m*	107
brillo reducido *m*	brilho reduzido *m* brilhância reduzida *f*	приведённый блеск *m*	108
brillo absoluto (de un cometa) *m*	brilhância absoluta *f* brilho absoluto (de um cometa) *m*	абсолютный блеск (кометы) *m*	109

110	cometary activity	activité cométaire *f*	Kometentätigkeit *f* Kometenaktivität *f*
111	splitting of a comet	fission d'une comète *f*	Kometenaufspaltung *f*
112	tidal splitting	fission due aux marées *f*	Gezeitenaufspaltung *f*
113	water model	modèle d'eau (d'une comète) *m*	Wassermodell (eines Kometen) *n*
114	disruption of a comet cometary disruption	rupture d'une comète *f* fragmentation d'une comète *f*	Zerstörung eines Kometen *f* Zerplatzen eines Kometen *n*
115	disintegration of a comet	désagrégation d'une comète *f* désintégration d'une comète *f*	Auflösung eines Kometen *f* Zerfall eines Kometen *m*
116	nuclear disintegration disintegration of a cometary nucleus	désagrégation du noyau cométaire *f* fragmentation du noyau cométaire *f* désintégration du noyau cométaire *f*	Zerstückelung des Kometenkernes *f* Zerkleinerung des Kometenkernes *f* Auflösung des Kometenkernes *f*
117	cohesive strength	force cohésive *f*	Kohäsionsfestigkeit *f*
118	tidal disintegration tidal disruption	désagrégation due aux marées *f* fragmentation [rupture] due aux marées *f*	Gezeitenzerstückelung *f* Gezeitenzerplatzen *n*
119	rotational break-up	désagrégation rotationnelle *f*	Auseinanderbrechen infolge von Rotation *n* Rotationszerfall *m*
120	adsorptive power adsorption power	pouvoir adsorbant *m* adsorptivité *f*	Adsorptionskraft *f* Adsorptionsvermögen *n*
121	adsorption equilibrium	équilibre adsorptif *m*	Adsorptionsgleichgewicht *n*
122	adsorption potential	potentiel d'adsorption *m*	Adsorptionspotential *n*
123	desorption	désorption *f*	Desorption *f*
124	heat of desorption	chaleur de désorption *f*	Desorptionswärme *f*
125	comet [cometary] outburst outburst of a comet comet burst	augmentation brusque de l'éclat *f* explosion d'une comète *f*	Ausbruch eines Kometen *m*
126	brightness outburst	sursaut de l'éclat *m*	Helligkeitsausbruch *m*
127	secular decrease of brightness	décroissance séculaire de l'éclat *f* diminution séculaire de l'éclat *f*	säkulare Helligkeitsabnahme *f*
128	adsorption	adsorption *f*	Adsorption *f*
129	adsorb	adsorber	adsorbieren
130	adsorbent	adsorbant *m* corps adsorbant *m*	Adsorbens *n* absorbierender Stolff *m*

actividad cometaria *f*	atividade cometária *f*	кометная активность *f*	110
escisión de un cometa *f* ruptura de un cometa *f*	cisão de um cometa *f*	расщепление кометы *n*	111
escisión por marea *f*	cisão de [causada por] maré *f*	приливное расщепление *n*	112
modelo de agua *m*	modelo de água (de cometas) *m*	водяная модель (кометы) *f*	113
fraccionamiento de un cometa *m* disrupción de un cometa *f*	dirupção [ruptura] de um cometa *f* desagregação de um cometa *f*	распад кометы *m*	114
desintegración de un cometa *f*	desintegração de um cometa *f*	разрушение кометы *n* дезинтеграция кометы *f*	115
desintegración del núcleo cometario *f*	desintegração do núcleo cometário *f*	разрушение кометного ядра *n* дезинтеграция кометного ядра *f*	116
fuerza cohesiva *f* potencia cohesiva *f*	força de coesão *f* poder de coesão *m*	когезионная прочность *f*	117
disrupción por marea *f* desintegración por efecto de marea *f*	dirupção [ruptura] por (efeito de) maré *f*	приливное разрушение *n*	118
desintegración rotacional *f*	desagregação rotacional *f*	вращательная дезинтеграция *f*	119
poder adsorbente *m* poder de adsorción *m*	poder adsorvente *m* poder de adsorção *m*	адсорбционная способность *f* адсорбируемость *f*	120
equilibrio de adsorción *m*	equilíbrio de adsorção *m*	адсорбционное равновесие *n*	121
potencial de adsorción *m*	potencial de adsorção *m*	потенциал адсорбции *m* адсорбционный потенциал *m*	122
desorción *f*	dessorção *f*	десорбция *f*	123
calor de desorción *m*	calor de dessorção *m*	теплота десорбции *f*	124
explosión de un cometa *f*	explosão de um cometa *f*	внезапное усиление блеска (кометы) *n* вспышка кометы *f*	125
explosion de brillo *f*	explosão de brilho *f*	вспышка блеска *f*	126
disminución secular del brillo *f*	diminuição secular do brilho *f*	вековое падение блеска *n* вековое понижение яркости *n*	127
adsorción *f*	adsorção *f*	адсорбция *f* поверхностное поглощение *n*	128
adsorber	adsorver	адсорбировать	129
adsorbente *m* materia adsorbente *f*	adsorvente *m* corpo adsorvente *m*	адсорбент *m* адсорбирующее вещество *n*	130

131	adsorbate adsorbed matter	adsorbat *m* substance adsorbée *f*	Adsorbat *n* adsorbierter Stoff *m*
132	fluorescent bands	bandes de fluorescence *fpl*	Fluoreszenzbanden *fpl*
133	Swan bands	bandes de Swan *fpl*	Swanbanden *fpl*
134	unidentified bands	bandes non-identifiées *fpl*	unidentifizierte Banden *fpl*
135	parent molecule	molécule-mère *f* molécule parente *f*	Muttermolekül *n*
135a	daughter molecule	molécule-fille *f*	Tochtermolekül *n*
136	radical	radical *m*	Radikal *n*
137	free radical	radical libre *m*	freies Radikal *n*
138	sublimation	sublimation *f*	Sublimation *f*
139	heat of sublimation	chaleur de sublimation *f*	Sublimationswärme *f*
140	sputtering	pulvérisation *f*	Zerstäubung *f*
141	volatile substance	substance volatile *f*	flüchtiger Stoff *m*
142	non-volatile substance	substance non-volatile *f*	nichtflüchtiger Stoff *m*
143	production rate formation rate	taux de production *m* vitesse de production *f*	Erzeugungsrate *f* Entstehungsrate *f*
144	gas-production rate	taux de production de gaz *m*	Gaserzeugungsrate *f*
145	mass emission	émission de masse *f*	Massenemission *f*
146	mass-emission rate	taux d'émission de masse *m*	Massenemissionsrate *f*
147	inactive comet	comète inactive *f*	nichtaktiver [inaktiver] Komet *m*
148	deactivation (of a comet)	désactivation (d'une comète) *f*	Inaktivierung (eines Kometen) *f* Deaktivierung (eines Kometen) *f*
149	active comet live comet	comète active *f*	aktiver Komet *m*
150	extinct comet	comète éteinte *f*	erloschener Komet *m*

adsorbato *m* materia adsorbida *f*	adsorvato *m* matéria adsorvida *f*	адсорбат *m* адсорбированное вещество *n*	131
bandas de fluorescencia *fpl*	bandas de fluorescência *fpl*	флуоресцентные (молекуляр-ные) полосы *fpl*	132
bandas de Swan *fpl*	bandas de Swan *fpl*	полосы Свана *fpl*	133
bandas no identificadas *fpl*	bandas não-identificadas *fpl*	неотож(д)ествлённые полосы *fpl*	134
molécula madre *f*	molécula-mãe *f*	родительская молекула *f* материнская молекула *f*	135
molécula hija *f*	molécula-filha *f*	дочерняя молекула *f*	135a
radical *m*	radical *m*	радикал *m*	136
radical libre *m*	radical livre *m*	свободный радикал *m*	137
sublimación *f*	sublimação *f*	сублимация *f* возгонка *f*	138
calor de sublimación *m*	calor de sublimação *m*	теплота сублимации *f*	139
pulverización *f*	pulverização *f*	распыление *n*	140
sustancia volátil *f*	substância volátil *f*	летучее вещество *n*	141
sustancia no volátil *f*	substância não-volátil *f*	нелетучее вещество *n*	142
tasa de producción *f* tasa de formación *f*	taxa de produção *f* taxa de formação *f*	скорость образования *f* скорость выделения *f*	143
tasa de producción de gases *f*	taxa de produção de gases *f*	скорость выделения газов *f* скорость газовыделения *f*	144
emisión de masa *f*	emissão de massa *f*	испускание [эмиссия] массы *n*[*f*]	145
tasa de emisión de masa *f*	taxa de emissão de massa *f*	скорость эмиссии [испускания] массы *f*	146
cometa inactivo *m*	cometa inativo *m*	неактивная комета *f*	147
desactivación (de un cometa) *f*	desativação de um cometa *f*	дезактивизация кометы *f*	148
cometa activo *m* cometa vivo *m*	cometa ativo *m* cometa vivo *m*	действующая комета *f* активная комета *f*	149
cometa extinto *m*	cometa extinto *m*	истощившаяся комета *f*	150

86 METEORS | MÉTÉORES | METEORE

1	meteor	météore *m*	Meteor *n*, *m*
2	shooting star	étoile filante *f*	Sternschnuppe *f*
3	meteor astronomy	astronomie météorique *f*	Meteorastronomie *f*
4	meteor- meteoric	météorique	meteorisch Meteor-
5	passage of a meteor appearance of a meteor	passage d'un météore *m* chute d'un météore *f* apparition d'un météore *f*	Meteorerscheinung *f*
6	meteor observation	observation de météores *f*	Meteorbeobachtung *f*
7	meteor observer	observateur de météores *m*	Meteorbeobachter *m*
8	brightness of a meteor meteor brightness	éclat d'un météore *m*	Helligkeit eines Meteors *f* Meteorhelligkeit *f*
9	superbolide giant bolide [fireball]	très gros bolide *m* bolide géant *m*	Überbolid *m* Riesenmeteor *n*
10	great fireball [bolide] bright fireball [bolide]	gros bolide *m* bolide brillant *m*	grosser Bolid [Feuerball] *m* heller Bolid [Feuerball] *m*
11	bolide fireball	grand météore *m* bolide *m*	Feuerball *m* Feuerkugel *f* Bolid *m*
12	bright meteor	météore brillant *m*	helles Meteor *n*
13	photographic meteor	météore photographique *m* météore photographié *m*	photographisches Meteor *n* photographiertes Meteor *n*
14	ordinary meteor	météore ordinaire *m*	gewöhnliches Meteor *n*
15	faint meteor	météore faible *m*	schwaches Meteor *n*
16	visual meteor	météore visuel *m*	freisichtbares [freisichtiges] Meteor *n* visuelles Meteor *n*
17	meteor with train	météore avec traînée *m* étoile filante traînée *f*	Meteor mit Schweif *n*
18	telescopic meteor	météore télescopique *m*	teleskopisches Meteor *n* Telemeteor *n*
19	faint telescopic meteor	météore télescopique faible *m*	schwaches Telemeteor *n* schwaches teleskopisches Meteor *n*
20	ultratelescopic meteor micrometeor very faint telescopic meteor	météore ultratélescopique *m* météore télescopique très faible *m* micrométéore *m*	ultrateleskopisches Meteor *n* Mikrometeor *n* sehr schwaches Telemeteor *n*
21	artificial meteor	météore artificiel *m*	künstliches Meteor *n*
22	day-time meteor	météore diurne *m*	Tagesmeteor *n*

METEOROS	METEOROS	МЕТЕОРЫ	86
meteoro *m*	meteoro *m*	метеор *m*	1
estrella fugaz *f*	estrela cadente [filante, fugaz] *f*	падающая звезда *f*	2
astronomía meteórica *f*	astronomia meteórica *f*	метеорная астрономия *f*	3
meteórico	meteórico	метеорный	4
paso de un meteoro *m* caída de un meteoro *f*	aparecimento de um meteoro *m* passagem de um meteoro *f* queda [aparição] de um meteoro *f*	полёт метеора *m* пролёт метеора *m* падение метеора *n*	5
observación de meteoros *f*	observação de meteoros *f*	наблюдение метеоров *n*	6
observador de meteoros *m*	observador de meteoros *m*	наблюдатель метеоров *m* метеорщик *m*	7
brillo de un meteoro *m*	brilho de um meteoro *m*	блеск метеора *m* яркость метеора *f*	8
bólido muy grande *m* bólido gigante *m*	bólido gigante *m* bólido muito grande *m*	сверхболид *m* очень большой болид *m*	9
bólido grande *m* bólido brillante *m*	bólido grande *m* bólido brilhante *m*	яркий болид *m* большой болид *m*	10
bólido *m*	bólido *m* bola [globo] de fogo *f*[*m*]	болид *m*	11
meteoro brillante *m*	meteoro brilhante *m*	яркий метеор *m*	12
meteoro fotográfico *m* meteoro fotografiado *m*	meteoro fotográfico *m* meteoro fotografado *m* *m*	фотографический [сфотографированный] метеор *m*	13
meteoro ordinario *m*	meteoro ordinário *m*	обычный метеор *m*	14
meteoro débil *m*	meteoro fraco *m*	слабый метеор *m*	15
meteoro visual *m* meteoro visible *m*	meteoro visual *m* meteoro visível (a olho nu) *m*	визуальный метеор *m*	16
meteoro con rastro *m*	meteoro com rast(r)o [esteira] *m*	метеор со следом *m*	17
meteoro telescópico *m*	meteoro telescópico *m*	телескопический метеор *m* телеметеор *m*	18
meteoro telescópico débil *m*	meteoro telescópico fraco *m*	слабый телескопический метеор *m* слабый телеметеор *m*	19
meteoro ultratelescópico *m* micrometeoro *m* meteoro telescópiso muy débil *m*	meteoro ultratelescópico *m* micrometeoro *m* meteoro telescópico muito fraco *m*	ультрателеметеор *m* микрометеор *m* очень слабый телеметеор *m*	20
meteoro artificial *m*	meteoro artificial *m*	искусственный метеор *m*	21
meteoro diurno *m*	meteoro diurno *m*	дневной метеор *m*	22

23	sporadic meteor	météore sporadique *m*	sporadisches Meteor *n*
24	shower meteor	météore d'essaim *m* météore appartenant à un essaim *m*	Strommeteor *n*
25	non-shower meteor	météore du champ *m* météore n'appartenant pas à un essaim *m*	Nichtstrommeteor *n* Feldmeteor *n*
26	noctilucent cloud	nuage nocturne lumineux *m*	leuchtende Nachtwolke *f*
27	light strip luminous band	bande de lumière *f* bande lumineuse *f*	Leuchtstreifen *m*
28	bright night sky	ciel nocturne lumineux *m*	heller Nachthimmel *m*
29	meteor trail (on plate)	trace photographique du météore *f* trace météorique sur photographie [plaque photographique] *f*	Meteorspur (auf der Platte) *f*
30	trajectory of a meteor meteor trajectory meteor path	trajectoire d'un météore (dans l'atmosphère) *f*	Meteorbahn *f* Meteortrajektorie (in der Atmosphäre) *f*
31	true path	trajectoire réelle *f*	räumliche Bahn *f*
32	apparent path (on the celestial sphere)	trajectoire apparente (sur le ciel) *f*	scheinbare Bahn (auf der Himmelssphäre) *f*
33	distorted path wavy path	trajectoire avec distorsion *f* trajectoire ondulante *f*	schlangenförmige [verwundene, wellenförmige] Meteorbahn *f*
34	stationary meteor	météore stationnaire *m*	stationäres Meteor *n*
35	appearance of a meteor	apparition d'un météore *f*	Aufleuchten eines Meteors *n*
36	appearance point beginning point start point	point d'apparition *m* point initial *m*	Aufleuchtpunkt *m* Anfangspunkt *m*
37	disappearance of a meteor extinction of a meteor	disparition d'un météore *f* extinction d'un météore *f*	Erlöschen eines Meteors *n*
38	disappearance point end point	point d'extinction *m* point de disparition *m*	Punkt des Erlöschens *m* Hemmungspunkt *m* Endpunkt *m*
39	height of appearance point appearance height beginning height	altitude du point d'apparition *f* altitude d'apparition *f*	Anfangshöhe *f* Aufleuchthöhe *f* Höhe des Aufleuchtens [des Anfangspunktes] *f*
40	height of disappearance height of end point end height	altitude de disparition *f* altitude finale *f* altitude (du point) d'extinction *f*	Endhöhe *f* Höhe des Endpunktes *f* Höhe des Erlöschens *f*
41	geocentric orbit	orbite géocentrique *f*	geozentrische Bahn *f*

meteoro esporádico *m*	meteoro esporádico *m*	спорадический метеор *m*	23
meteoro de enjambre *m*	meteoro pertencente a um enxame *m* meteoro de enxame *m* meteoro de [pertenecente a] chuva *m*	поточный метеор *m* метеор, принадлежащий (к) потоку *m* метеор потока *m*	24
meteoro solitario *m* meteoro no perteneciente a un enjambre *m*	meteoro não-pertencente a um enxame *m*	метеор, не принадлежащий (к) потоку *m*	25
nube noctiluciente *f*	nuvem noctilúcia [noctiluzente] *f*	серебристое облако *n*	26
banda luminosa *f* faja luminosa *f*	banda luminosa *f* faixa luminosa *f*	светящаяся полоса *f*	27
cielo nocturno luminoso *m*	céu noturno luminoso *m*	яркое ночное небо *n*	28
traza meteórica (sobre la placa fotográfica) *f* traza fotográfica de un meteoro *f*	rast(r)o de um meteoro (na chapa fotográfica) *m*	снимок метеора (на пластинке) *m* след метеора на пластинке *m*	29
rastro de un meteoro (en la atmósfera) *m* trayectoria de un meteoro *f*	trajetória de um meteoro (na atmosfera) *f*	траектория метеора (в атмосфере) *f* путь метеора *m*	30
trayectoria verdadera *f*	trajetória real *f*	истинный путь *m* истинная траектория *f*	31
trayectoria aparente (en la esfera celeste) *f*	trajetória aparente (no céu) *f*	видимый путь на (небесной) сфере *m*	32
trayectoria distorsionada *f* trayectoria ondulada *f*	trajetória distorcida *f* trajetória ondulada *f*	волнистая траектория *f* змеевидная траектория *f*	33
meteoro estacionario *m*	meteoro estacionário *m*	стационарный метеор *m*	34
aparición de un meteoro *f*	aparição de um meteoro *f*	появление метеора *n*	35
punto de aparición *m* punto inicial *m*	ponto de aparição *m* início de trajetória aparente *m*	точка появленияа *f*	36
desaparición de un meteoro *f* extinción de un meteoro *f*	extinção de um meteoro *f* desaparição de um meteoro *f*	погасание метеора *n* исчезновение метеора *n*	37
punto de extinción *m* punto de desaparición *m*	ponto de extinção *m* fim da trajetória aparente *m* ponto de desaparição *m*	точка исчезновения *f* точка погасания *f* конец видимого пути *m*	38
altura [altitud] del punto de aparición *f* altura de aparición *f*	altura do ponto de aparição *f* altura de aparição *f*	высота появления метеора *f* высота точки появления *f*	39
altura (del punto) de extinción *f* altitud (del punto) de desaparición *f*	altura (do ponto) de extinção *f* altura (do ponto) de desaparição *f*	высота погасания *f* высота точки исчезновения *f* высота исчезновения *f*	40
órbita geocéntrica *f*	órbita geocêntrica *f*	геоцентрическая орбита *f*	41

	METEORS	MÉTÉORES	METEORE
42	heliocentric orbit	orbite héliocentrique *f*	heliozentrische Bahn *f*
43	interstellar meteor hyperbolic meteor	météore interstellaire *m* météore hyperbolique *m*	interstellares Meteor *n* hyperbolisches Meteor *n*
44	interplanetary meteor solar-system meteor	météore interplanétaire *m* météore du système solaire *m*	interplanetares Meteor *n* Sonnensystem-Meteor *n*
45	cometary meteor	météore d'origine cométaire *m* météore cométaire *m*	kometarisches Meteor *n* Kometenmeteor *n*
46	ecliptical meteor	météore écliptique *m*	ekliptikales Meteor *n*
47	path length length of a meteor path	longueur de la trajectoire d'un météore *f*	Länge der Meteorbahn *f* Bahnlänge des Meteors *f*
48	angular path length	longueur angulaire de la trajectoire *f*	Winkellänge der Meteorbahn *f*
49	moment [time] of appearance	instant d'apparition *m*	Zeitpunkt des Erscheinens *m*
50	duration of flight	durée du parcours d'un météore *f*	Flugdauer *f*
51	curvature of the path	courbure de la trajectoire *f*	Bahnkrümmung *f*
52	zenith attraction	attraction zénithale *f*	Zenitattraktion *f*
53	radiant meteor radiant	radiant *m* radiant météorique *m*	(Meteor)radiant *m* Strahlungspunkt *m*
54	group radiant	radiant de groupe *m*	Gruppenradiant *m*
55	individual radiant	radiant individuel *m*	individueller Radiant *m*
56	apparent radiant	radiant apparent *m*	scheinbarer Radiant *m*
57	diurnal aberration	aberration diurne *f*	tägliche Aberration *f*
58	corrected radiant geocentric radiant	radiant corrigé *m* radiant géocentrique *m*	korrigierter Radiant *m* geozentrischer Radiant *m*
59	true radiant	radiant vrai *m*	wahrer Radiant *m*
60	radiant position	position du radiant *f*	Radiantenort *m* Position des Radianten *f*
61	curvature of the Earth's surface	courbure de la surface terrestre *f*	Krümmung der Erdoberfläche *f*
62	correction for the curvature (of the Earth's surface)	correction de courbure (de la surface terrestre) *f*	Korrektur für die Krümmung (der Erde) *f*
63	radiant density	densité des radiants *f*	Radiantendichte *f*
64	apex	apex *m*	Apex *m*
65	antapex	ant(i)apex *m*	Antapex *m*

órbita heliocéntrica *f*	órbita heliocêntrica *f*	гелиоцентрическая орбита *f*	42
meteoro interestelar *m* meteoro hiperbólico *m*	meteoro interestelar *m* meteoro hiperbólico *m*	межзвёздный метеор *m* гиперболический метеор *m* галактический метеор *m*	43
meteoro interplanetario *m* meteoro del sistema solar *m*	meteoro interplanetário *m* meteoro do sistema solar *m*	межпланетный метеор *m* метеор солнечной системы *m*	44
meteoro de origen cometario *m* meteoro cometario *m*	meteoro de origem cometária *m* meteoro cometário *m*	метеор кометного происхождения *m* кометный метеор *m*	45
meteoro eclíptico *m*	meteoro eclíptico *m*	эклиптический метеор *m*	46
longitud de la trayectoria *f*	comprimento da trajetória de um meteoro *m*	длина пути метеора *f*	47
longitud angular de la trayectoria *f*	comprimento angular da trajetória *m*	угловая длина пути *f*	48
tiempo [instante] de aparición *m*	hora [instante] de aparição *f*[*m*]	момент появления *m* время появления *n*	49
duración del recorrido meteórico *f*	duração do percurso de um meteoro *f*	продолжительность полёта *f* длительность полёта *f*	50
curvatura de la trayectoria *f*	curvatura da trajetória *f*	кривизна пути *f*	51
atracción zenital *f*	atração zenital *f*	зенитное притяжение *n*	52
radiante *m* radiante meteórico *m*	radiante *m* radiante meteórico *m*	радиант *m* метеорный радиант *m*	53
radiante de grupo *m*	radiante de grupo *m*	групповой радиант *m*	54
radiante individual *m*	radiante individual *m*	индивидуальный радиант *m*	55
radiante aparente *m*	radiante aparente *m*	видимый радиант *m*	56
aberración diurna *f*	aberração diurna *f*	суточная аберрация *f*	57
radiante corregido *m* radiante geocéntrico *m*	radiante corrigido *m* radiante geocêntrico *m*	геоцентрический радиант *m* исправленный радиант *m*	58
radiante verdadero *m*	radiante verdadeiro *m*	истинный радиант *m*	59
posición del radiante *f*	posição do radiante *f*	положение радианта *n*	60
curvatura de la superficie terrestre *f*	curvatura da superficie da Terra *f*	кривизна земной поверхности *f*	61
corrección por la curvatura (de la superficie terrestre) *f*	correção de [para a] curvatura (da superficie da Terra) *f*	поправка [исправление] за кривизну земной поверхности *f*[*n*]	62
densidad de radiantes *f*	densidade dos radiantes *f*	плотность радиантов *f*	63
ápex *m*	ápex *m*	апекс *m*	64
antiápex *m*	antiápex *m*	антиапекс *m*	65

	METEORS	MÉTÉORES	METEORE
66	elevation of the apex apex height [elevation]	hauteur de l'apex *f*	Höhe des Apex *f*
67	elongation of the apparent radiant	élongation du radiant apparent *f*	Elongation des scheinbaren Radianten *f*
68	elongation of the true radiant	élongation du radiant vrai *f*	Elongation des wahren Radianten *f*
69	daily motion of the radiant	mouvement diurne du radiant *m*	tägliche Verschiebung des Radianten *f* tägliche Radiantenverschiebung *f*
70	meteor light curve	courbe de lumière d'un météore *f*	Lichtkurve eines Meteors *f* Helligkeitskurve eines Meteors *f*
71	apparent magnitude of a meteor	magnitude apparente d'un météore *f*	scheinbare Grössenklasse eines Meteors *f*
72	absolute magnitude of a meteor	magnitude absolue d'un météore *f*	absolute Grössenklasse eines Meteors *f*
73	apparent meteor brightness	éclat apparent d'un météore *m*	scheinbare Meteorhelligkeit *f*
74	absolute meteor brightness	éclat absolu d'un météore *m*	absolute Meteorhelligkeit *f*
75	progressive fading progressive extinction	disparition progressive *f* extinction progressive *f*	langsames Erlöschen *n*
76	sudden disappearance sudden extinction	extinction soudaine *f*	plötzliches Erlöschen *n*
77	spindle-shaped meteor	météore en forme de fuseau *m* météore fusiforme *m*	spindelförmiges Meteor *n*
78	maximum brightness maximum light	maximum de luminosité *m* maximum d'éclat *m* luminosité maximale *f*	maximale Helligkeit *f* Maximum der Helligkeit *n* Lichtmaximum *n*
79	light increase	augmentation d'éclat *f* accroissement de luminosité *m*	Helligkeitsanstieg *m*
80	meteor flare	éruption lumineuse d'un météore *f*	Lichtausbruch eines Meteors *m*
81	explosion of a meteor	explosion d'un météore *f*	Meteorexplosion *f*
82	disruption of a meteor	rupture d'un météore *f* éclatement (d'un météore) *m* désintégration d'un météore *f*	Zerspringen eines Meteors *n* Zerplatzen eines Meteors *n*
83	fragmentation of a meteor	fragmentation d'un météore *f*	Zerbröckeln [Aufbrechen, Zerbröcklung] eines Meteors *n(n, f)*
84	sparks meteor sparks	étincelles (d'un météore) *fpl*	Funkensprühen *n* Funken *mpl*
85	meteor with irregular light-curve flickering meteor	météore à courbe de lumière irrégulière *m* météore présentant des variations de luminosité *m* météore pulsant *m*	flackerndes Meteor *n* flimmerndes Meteor *n*

altura del ápex *f*	altura do ápex *f*	высота апекса *f*	66
elongación del radiante aparente *f*	elongação do radiante aparente *f*	элонгация видимого радианта *f*	67
elongación del radiante verdadero *f*	elongação do radiante verdadeiro *f*	элонгация истинного радианта *f*	68
movimiento diurno del radiante *m*	movimento diário do radiante *m*	суточное смещение радианта *n*	69
curva de luz de un meteoro *f*	curva de luz de um meteoro *f*	кривая блеска метеора *f*	70
magnitud aparente de un meteoro *f*	magnitude aparente de um meteoro *f*	видимая (звёздная) величина метеора *f*	71
magnitud absoluta de un meteoro *f*	magnitude absoluta de um meteoro *f*	абсолютная (звёздная) величина метеора *f*	72
brillo aparente de un meteoro *m*	brilho aparente de um meteoro *m*	видимый блеск метеора *m*	73
brillo absoluto de un meteoro *m*	brilho absoluto de um meteoro *m*	абсолютный блеск метеора *m*	74
extinción gradual *f* desaparición gradual *f*	extinção progressiva *f* desaparição gradual *f*	плавное [тупое] погасание *n* постепенное погасание *n*	75
extinción instantánea *f* desaparición instantánea *f*	desaparecimento súbito *m* extinção [desaparição] repentina *f*	резкое [мгновенное] погасание *n* мгновенное исчезновение *n*	76
meteoro fusiforme *m*	meteoro (com rast(r)o) fusiforme *m*	веретенообразный метеор *m*	77
máximo brillo *m*	brilho máximo *m*	максимальный блеск *m* максимальная сила света *f*	78
aumento de brillo *m* incremento de luminosidad *m*	aumento de brilho *m* aumento de luminosidade *m*	усиление блеска *n*	79
erupción luminosa de un meteoro *f*	erupção luminosa de um meteoro *f*	вспышка метеора *f*	80
explosión de un meteoro *f*	explosão de um meteoro *f*	взрыв метеора *m*	81
disrupción de un meteoro *f*	cisão [ruptura] de um meteoro *f*	разрушение метеора *n* распад метеора *m*	82
fragmentación de un meteoro *f*	fragmentação de um meteoro *f*	дробление метеора *n*	83
centellas (de un meteoro) *fpl* chispas (de un meteoro) *fpl*	faíscas (de um meteoro) *fpl* chispas (de um meteoro) *fpl*	искры (метеора) *fpl*	84
meteoro pulsante *m* meteoro con variaciones irregulares de brillo *m*	meteoro de luz pulsante *m* meteoro com variações luminosas *m*	мерцающий метеор *m*	85

86	meteor velocity	vitesse d'un météore *f*	Meteorgeschwindigkeit *f*
87	fast meteor	météore rapide *m*	schnelles Meteor *n*
88	slow meteor	météore lent *m*	langsames Meteor *n*
89	angular velocity	vitesse angulaire *f*	Winkelgeschwindigkeit *f*
90	instantaneous angular velocity	vitesse angulaire instantanée *f*	momentane [augenblickliche] Winkelgeschwindigkeit *f*
91	mean angular velocity	vitesse angulaire moyenne *f*	mittlere Winkelgeschwindigkeit *f*
92	initial velocity velocity of entry	vitesse d'entrée *f* vitesse à l'entrée dans l'atmosphère *f*	Eintrittsgeschwindigkeit *f* Eindringgeschwindigkeit *f* Anfangsgeschwindigkeit *f*
93	velocity outside the atmosphere	vitesse en dehors de l'atmosphère *f*	ausseratmosphärische Geschwindigkeit *f*
94	atmospheric velocity velocity in the atmosphere	vitesse dans l'atmosphère *f*	atmosphärische Geschwindigkeit *f* Meteorgeschwindigkeit in der Atmosphäre *f*
95	geocentric meteor velocity geocentric velocity of meteor	vitesse géocentrique d'un météore *f*	geozentrische Meteorgeschwindigkeit *f*
96	heliocentric velocity	vitesse héliocentrique *f*	heliozentrische Geschwindigkeit *f*
97	velocity at the mid-point of the trail	vitesse à mi-trajectoire *f*	Geschwindigkeit im Mittelpunkt der Bahn *f*
98	average velocity mean velocity	vitesse moyenne *f*	mittlere Geschwindigkeit *f*
99	instantaneous velocity	vitesse instantanée *f*	momentane Geschwindigkeit *f* augenblickliche Geschwindigkeit *f*
100	apparent deceleration	ralentissement apparent *m* freinage apparent *m* décélération apparent *m*	scheinbare Geschwindikeitsabnahme *f* scheinbare Abbremsung *f* scheinbare Verzögerung *f*
101	real deceleration	décélération réelle *f* freinage réel *m* relentissement réel *m*	wahre Verzögerung *f* wirkliche Abbremsung *f* wahre Geschwindigkeitsabnahme *f*
102	meteor nucleus	noyau d'un météore *m*	Meteorkern *m*
103	meteor train train of a meteor	traînée météorique *f*	Schweif eines Meteors *m* Meteorschweif *m*
104	persistent train	traînée météorique persistante *f*	beständiger Meteorschweif *m*
105	noctilucent train	traînée visible de nuit *f*	Nachtschweif *m* nachts sichtbarer Schweif *m*
106	day-time train	traînée visible de jour *f*	Tagschweif *m* am Tage sichtbarer Schweif *m*
107	spectrum of a meteor train	spectre d'une traînée météorique *m*	Schweifspektrum *n*

velocidad de un meteoro *f*	velocidade de um meteoro *f*	скорость метеора *f*	86
meteoro rápido *m*	meteoro rápido *m*	быстрый метеор *m*	87
meteoro lento *m*	meteoro lento *m*	медленный метеор *m*	88
velocidad angular *f*	velocidade angular *f*	угловая скорость *f*	89
velocidad angular instantánea *f*	velocidade angular instantânea *f*	мгновенная угловая скорость *f*	90
velocidad angular media *f*	velocidade angular média *f*	средняя угловая скорость *f*	91
velocidad inicial *f* velocidad de entrada en la atmósfera *f*	velocidade da entrada na atmosfera *f*	скорость входа *f* начальная скорость *f* скорость влета в атмосферу *f*	92
velocidad fuera de la atmósfera *f*	velocidade fora da atmosfera *f*	внеатмосферная скорость *f*	93
velocidad en la atmósfera *f* velocidad atmosférica *f*	velocidade na atmosfera *f* velocidade atmosférica *f*	атмосферная скорость *f* скорость в атмосфере *f*	94
velocidad geocéntrica de un meteoro *f*	velocidade geocêntrica de um meteoro *f*	геоцентрическая скорость метеора *f*	95
velocidad heliocéntrica *f*	velocidade heliocêntrica *f*	гелиоцентрическая скорость *f*	96
velocidad en el punto medio de la trayectoria *f*	velocidade no ponto médio da trajetória *f*	скорость в середине пути *f*	97
velocidad media *f*	velocidade média *f*	средняя скорость *f*	98
velocidad instantánea *f*	velocidade instantânea *f*	мгновенная скорость *f*	99
deceleración aparente *f* aceleración negativa aparente *f*	desaceleração aparente *f* redução aparente de velocidade *f*	видимое торможение *n* видимое замедление *n*	100
deceleración real *f* aceleración negativa real *f*	desaceleração real *f* redução real da velocidade *f*	истинное замедление *n* истинное торможение *n*	101
núcleo de un meteoro *m*	núcleo de um meteoro *m*	ядро метеора *n*	102
estela de un meteoro *f* estela meteórica *f*	rasto [rastro] de um meteoro *m* rast(r)o meteórico *m*	след метеора *m* метеорный след *m*	103
estela meteórica persistente *f*	rast(r)o meteórico persistente *m*	стойкий след метеора *m*	104
estela visible de noche *f*	rast(r)o noctilúcio [noctilucente] *m*	ночной след *m*	105
estela visible de día *f*	rast(r)o visível de dia *m*	дневной след *m*	106
espectro de una estela meteórica *m*	espe(c)tro do rast(r)o de um meteoro *m*	спектр метеорного следа *m*	107

108	self-luminous train	traînée lumineuse (par elle-même) *f* traîmée à lumière propre *f*	selbstleuchtender Schweif *m* Leuchtschweif *m*
109	unstable train	traînée instable *f*	instabiler Meteorschweif *m*
110	wake of a meteor	sillage d'un météore *m* queue d'un météore *f*	Schweifansatz *m* Schweifspur *f*
111	duration of train visibility persistence of a train	durée de la visibilité d'une traînée *f* persistence d'une traînée *f*	Sichtbarkeitsdauer eines Meteor-schweifs *f*
112	short-duration train short-enduring meteor train	traînée de courte durée *f*	kurzlebiger [kurzdauernder] Meteor-schweif *m*
113	long-enduring meteor train long-duration train	traînée météorique de longue durée *f*	langdauernder [langlebiger] Meteor-schweif *m*
114	drift of the train	déplacement de la traînée *m*	Meteorschweifverschiebung *f*
115	dust train	traînée de poussières *f*	Staubschweif *m*
116	gaseous train	traînée gazeuse *f*	Gasschweif *m*
117	passage of a fireball appearance of a fireball	passage d'un bolide *m* chute d'un bolide *f* apparition d'un bolide *f*	Feuerkugelerscheinung *f*
118	daylight fireball	bolide visible de jour *m*	Tagesbolid *m*
119	detonating fireball sound-emitting fireball	bolide détonant *m*	detonierendes Meteor *n* Feuerkugel mit Donner *f*
120	area of audibility	aire [región] d'audibilité *f* zone d'audibilité *f*	Schallfeld *n*
121	point of disruption	point de rupture *m*	Punkt des (Meteor-)Zerplatzens *m*
122	point of explosion	point d'explosion *m*	Punkt der Meteorexplosion *m*
123	stopping point	point d'arrêt *m*	Hemmungspunkt *m*
124	impact point point of impact	point de chute *m*	Aufschlagspunkt *m*
125	simultaneous observation	observation parallèle *f* observation simultanée *f*	Parallelbeobachtung *f* korrespondierende Beobachtung *f* gleichzeitige Beobachtung *f*
126	area of visibility	région de visibilité *f*	Sichtbarkeitsgebiet *n*
127	point of clearest vision	point de la meilleure vision *m* point de vision fovéale *m*	Augenruhepunkt *m*

METEOROS	METEOROS	МЕТЕОРЫ	
estela luminosa *f* estela con luz propia *f*	rast(r)o luminoso [com luz própria] *m*	светящийся след *m* самосветящийся след *m*	108
estela inestable *f*	rast(r)o instável *m*	неустойчивый след *m*	109
estela muy corta de un meteoro *f* cola de meteoro *f*	esteira de um meteoro *f* cauda de um meteoro *m*	хвост метеора *m* кильватер *m* весьма кратковременный след *m*	110
duración de la visibilidad de una estela *f* persistencia de una estela *f*	duração da visibilidade do rast(r)o *f*	продолжительность видимости следа *f*	111
estela de corta duración *f*	rasto [rastro] de curta duração *m*	кратковременный след *m*	112
estela de larga duración *f*	rast(r)o de longa duração *m*	длительный след *m* долговременный след *m*	113
desplazamiento de la estela *m*	deslocamento do rast(r)o *m*	дрейф метеорного следа *m*	114
estela de polvo *f*	rast(r)o de poeira *m*	пылевой след *m*	115
estela gaseosa *f*	rast(r)o gasoso *m*	газовый след *m*	116
paso de un bólido *m* caída de un bólido *f*	aparecimento de um bólido *m* queda de um bólido *f* passagem de um bólido *f*	пролёт болида *m* полёт болида *m* падение болида *n*	117
bólido visíble de día *m*	bólido visível de dia *m*	дневной болид *m*	118
bólido detonante *m*	bólido detonante [sonoro] *m* bólido produzindo efeitos sonoros *m*	звуковой болид *m* детонирующий болид *m*	119
área de audibilidad *f* zona de audibilidad *f*	área de audibilidade *f*	область слышимости звуков *f*	120
punto de disrupción *m*	ponto de ruptura *m* ponto de cisão *m*	точка распада *f*	121
punto de explosión *m*	ponto de explosão *m*	точка взрыва *f*	122
punto de parada *m*	ponto de parada *m*	точка задержки *f* область задержки *f*	123
punto de impacto *m* punto de caída *m*	ponto de impacto *m* ponto de queda *m*	точка удара *f* место удара *n*	124
observación simultánea *f*	observação simultânea *f*	параллельное наблюдение *n* корреспондирующее наблюдение *n*	125
área [región] de visibilidad *f*	área de visibilidade *f*	область видимости *f*	126
punto de visión foveal *m*	ponto de visão foveal *m* ponto de máxima visão *m*	точка фиксации *f* центр фиксации *m*	127

	METEORS	MÉTÉORES	METEORE
128	peripheral vision parafoveal vision side vision extrafoveal vision averted vision	vision périphérique *f* vision scotopique *f* vision de coté *f* vision extrafovéale *f*	indirektes Sehen *n* extrafoveales Sehen *n* seitliches Sehen *n*
129	reticle	réticule *f*	Netz *n*
130	rotating mirror	miroir tournant *m*	rotierender Spiegel *m*
130a	rocking mirror	miroir oscillant *m*	vibrierender Spiegel *m* beweglicher Spiegel *m*
131	meteor pair double meteor	météore double *m*	Doppelmeteor *n*
132	double counting method	méthode du double comptage *f*	Methode der Doppelzählung *f*
133	multiple meteor	météore multiple *m*	mehrfaches Meteor *n*
134	meteor activity	activité météorique *f*	meteorische Aktivität *f*
135	frequency of meteors hourly rate of meteors rate of shooting stars number of meteors per hour	fréquence de météores *f* nombre horaire (des météores) *m* fréquence horaire des météores *f*	Frequenz der Meteore *f* stündliche Häufigkeit *f* stündliche Anzahl von Meteoren *f* Häufigkeit von Meteoren *f*
136	background meteor activity	activité des météores sporadiques *f*	Aktivität der Feldmeteore *f*
137	diurnal variation daily variation	variation diurne *f*	tägliche Variation *f*
138	yearly variation	variation annuelle *f*	jährliche Variation *f*
139	seasonal variation	variation saisonnière *f*	jahreszeitliche Variation *f*
140	meteor photography photographic observation of meteors	photographie des météores *f* observation photographique des météores *f*	Meteorphotographie *f* photographische Beobachtung von Meteoren *f*
141	meteor camera	chambre pour la photographie des météores *f*	Meteorkamera *f*
142	guided camera driven camera	chambre entraînée *f* chambre guidée *f*	nachgeführte Kamera *f*
143	undriven camera	chambre stationnaire *f* chambre non-entraînée *f* chambre fixe *f*	feste Kamera *f* unbewegliche Kamera *f* nichtnachgeführte Kamera *f*
144	meteor patrol	surveillance météorique *f*	Meteorüberwachung *f*
145	patrol camera	chambre (à grand champ) pour la détection des météores *f*	Überwachungskamera *f*

visión periférica *f* visión parafoveal *f* visión extrafoveal *f*	visão periférica *f* visão indireta *f* visão extrafoveal *f*	периферическое зрение *n*	128
retículo *m*	retículo *m*	сетка *f*	129
espejo rotatorio espejo giratorio *m*	espelho giratório *m*	вращающееся зеркало *n*	130
espejo oscilante *m*	espelho oscilante *m*	вибрирующее зеркало *n*	130a
meteoro doble *m*	meteoro duplo *m*	двойной метеор *m*	131
método del doble conteo *m*	método de contagem dupla *m*	метод двойного счёта *m*	132
meteoro múltiple *m*	meteoro múltiplo *m*	кратный метеор *m*	133
actividad meteórica *f*	atividade meteórica *f*	метеорная активность *f*	134
frecuencia horaria de los meteoros *f* número de meteoros por hora *m*	freqüência horária de meteoros *f* número de meteoros por hora *m*	часовое число (метеоров) *n* частота метеоров *f* количество метеоров в один час *n*	135
actividad de los meteoros esporádicos *f*	atividade dos meteoros esporádicos *f*	активность метеорного фона *f* фоновая активность *f*	136
variación diurna *f*	variação diária *f*	суточная вариация *f*	137
variación anual *f*	variação anual [ânua] *f*	годичная вариация *f*	138
variación estacional *f*	variação estacional [sazonal] *f*	сезонная вариация *f*	139
fotografía de meteoros *f* observación fotográfica de meteoros *f*	fotografia de meteoros *f* observação fotográfica dos meteoros *f*	фотографирование метеоров *n* метеорная фотография *f*	140
cámara meteórica *f* cámara para la fotografía de los meteoros *f*	câmara meteórica *f* câmara [máquina] para a fotografia de meteoros *f*	метеорная камера *f*	141
cámara guiada *f* cámara móvil *f*	câmara fotográfica guiada *f* câmara fotográfica móvel *f*	гидируемая камера *f* подвижная камера *f*	142
cámara fija *f* cámara no guiada *f*	câmara fotográfica fixa *f* câmara não-guiada *f* câmara não-móvel *f*	неподвижная камера *f* фиксированная камера *f* негидируемая камера *f*	143
patrulla(je) [vigilancia] meteórica *f* batería de cámaras meteóricas para detectar meteoros *f*	vigilância (para a detecção) de meteoros *f* patrulhamento de meteoros *m*	метеорный патруль *m*	144
cámara de patrulla(je) [vigilancia] *f* cámara para la detección de meteoros *f*	câmara de patrulhamento *f* câmara de vigilância *f*	патрульная камера *f*	145

146	meteor spectrograph	spectrographe pour les météores *m*	Meteorspektrograph *m*
147	rotating sector rotating shutter	secteur tournant *m*	rotierender Sektor *m*
148	moving-blade shutter	obturateur à lames mobiles *m*	Sektor mit beweglichem Unterbrecher *m* unterbrechender Sektor *m*
149	two-blade shutter	obturateur à deux lames *m*	Doppelunterbrecher *m*
150	shutter break	interruption par l'obturateur *f*	Sektorunterbrechung *f*
151	meteor plate	cliché météorique *m*	Meteorplatte *f*
151a	meteor photograph	photographie d'un météore *f*	Meteorphotographie *f*
152	star trail	trace stellaire *f*	(photographische) Sternspur *f*
153	meteor trail	trace météorique *f*	(photographische) Meteorspur *f*
154	trailing velocity	vitesse de trace *f*	Spurbildungsgeschwindigkeit *f* Spurgeschwindigkeit *f*
155	doubly photographed meteor meteor photographed from two stations	météore photographié de deux stations *m*	Doppelaufnahme *f* photographische Doppelaufnahme *f*
156	plate centre [center]	centre de la plaque *m*	Plattenmitte *f* Plattenzentrum *n*
157	field correction	correction de champ *f*	Feldkorrektion *f* Feldkorrektur *f*
158	meteor station	station d'étude des météores *f* station météorique *f*	Meteorstation *f*
159	physical theory of meteors	théorie physique des météores *f*	physikalische Theorie der Meteore *f*
160	meteoroid	météoroïde *m*	Meteoroid *n*
160a	meteor body meteoric body meteoric particle	corps météorique *m* particule météorique *f*	Meteorkörper *m* meteorisches Teilchen *n* meteorischer Körper *m* meteorische Partikel *f*
161	entry into the atmosphere atmospheric entry	entrée dans l'atmosphère *f*	Eindringen in die Atmosphäre *n* Eintritt in die Atmosphäre *m*
162	atmospheric path trajectory in the atmosphere	trajectoire dans l'atmosphère *f* trajectoire atmosphérique *f*	atmosphärische Bahn *f*
163	mass of the meteor(ic) body meteor mass	masse d'une particule météorique *f* masse d'un météore *f*	Masse des Meteorkörpers *f* Masse des Meteors *f* Masse des meteorischen Teilchens *f*

espectrógrafo para (la observación de) meteoros *m*	espe(c)trógrafo para meteoros *m*	метеорный спектрограф *m*	146
sector giratorio *m* obturador giratorio *m*	obturador rotativo *m*	обтюратор *m* вращающийся сектор *m*	147
obturador de lámina móvil *m*	obturador de lâmina móvel *m*	обтюратор с подвижным крылом *m*	148
obturador de dos láminas *m*	obturador de duas lâminas *m*	двухлопастный обтюратор *m*	149
interrupción por el obturador *f*	interrupção [corte] pelo obturador *f*[*m*]	перерыв (от обтюратора) *m*	150
placa meteórica *f*	chapa [placa] meteórica *f*	метеорная пластинка *f*	151
fotografía de un meteoro *f*	fotografia de um meteoro *f*	метеорный снимок *m*	151a
traza estelar *f*	rast(r)o de uma estrela (na chapa fotográfica) *m*	след звезды (на пластинке) *m* звёздный след *m*	152
traza meteórica *f*	rast(r)o de um meteoro (na chapa fotográfica) *m*	снимок метеора (на пластинке) *m*	153
velocidad en la traza *f*	velocidade sobre o rast(r)o (na chapa fotográfica) *f*	скорость перемещения изображения (на пластинке) *f*	154
meteoro fotografiado desde dos estaciones *m*	meteoro fotografado de dois locais *m*	базисный метеор *m* метеор, снятый из двух пунктов *m*	155
centro de la placa *m*	centro da chapa fotográfica *m*	центр пластинки *m*	156
corrección de campo *f*	correção de campo *f*	поправка поля *f*	157
estación meteórica *f* estación para la observación de meteoros *f*	estação meteórica *f* estação para a observação de meteoros *f*	метеорная станция *f*	158
teoria física de los meteoros *f*	teoria física dos meteoros *f*	физическая теория метеоров *f*	159
meteoroide *m*	meteoróide *m*	метеороид *m*	160
partícula meteórica *f* cuerpo meteórico *m*	corpo meteórico *m* partícula meteórica *f*	метеорное тело *n* метеорная частица *f*	160a
entrada en la atmósfera *f*	entrada na atmosfera *f*	вход в атмосферу *n* вход в плотные слои атмосферы *m*	161
trayectoria atmosférica *f* trayectoria en la atmósfera *f*	trajetória na atmosfera *f*	атмосферная траектория *f* траектория в атмосфере Земли *f* путь в атмосфере *m*	162
masa del cuerpo meteórico *f* masa meteórica *f* masa de la partícula meteórica *f*	massa da partícula meteórica *f* massa do meteoro *f* massa meteórica *f*	масса метеорного тела *f* масса метеорной частицы *f* метеорная масса *f*	163

164	original mass initial mass	masse originelle *f* masse initiale *f*	Anfangsmasse *f*
165	mass distribution	distribution des masses *f*	Massenverteilung *f*
166	brightness distribution	distribution selon éclat *f*	Helligkeitsverteilung *f*
167	variable mass	masse variable *f*	veränderliche Masse *f*
168	mass loss	perte de masse *f*	Massenabnahme *f* Massenverlust *m* Massenschwund *m*
169	mass equation mass-loss equation	équation de masse *f* équation pour la perte de masse *f*	Massengleichung *f* Massenabnahmegleichung *f*
170	compression cap air cap gas cap	zone de compression *f* coiffe d'air *f* coiffe gazeuse *f*	Verdichtungskappe *f* Kompressionskappe *f* Verdichtungszone *f*
171	shielding	blindage *m* effet d'écran *m*	Abschirmung *f*
172	air particle air molecule	particule d'air *f* molécule d'air *f*	Luftpartikel *n* Luftmolekül *n*
173	friction	frottement *m*	Reibung *f*
174	heating of the meteor body	échauffement du corps météorique *m*	Erwärmung des meteorischen Körpers *f* Erhitzung des Meteorkörpers *f*
175	shape of the meteor body form of the meteoric particle	forme du corps météorique *f* forme de la particule météorique *f*	Gestalt des Meteorkörpers *f* Gestalt des meteorischen Körpers *f*
176	head-on cross-section	section frontale *f*	stirnseitiger Querschnitt *m*
177	shape factor form factor geometry factor	coefficient de forme *m* facteur de forme *m* facteur géométrique *m*	Gestaltfaktor *m* Formfaktor *m* Formparameter *m* Geometriefaktor *m*
178	drag air resistance aerodynamic resistance	traînée *f* résistance de l'air *f* résistance aérodynamique *f*	Rücktrieb *m* Luftwiderstand *m*
179	drag equation resistance equation	équation de traînée *f* équation de résistance *f*	Widerstandsgesetz *n*
180	drag coefficient	coefficient de traînée *m*	Widerstandsbeiwert *m*
181	fusion melting	fusion *f*	Aufschmelzung *f* Schmelzen *n*
182	ablation	ablation *f*	Ablation *f* Massenabnahme *f* Abschmelzung *f* Abtragung *f*

masa inicial *f* masa original *f*	massa inicial *f* massa original *f*	начальная масса *f*	164
distribución de masas *f* distribución según la masa *f*	distribuição das massas *f* distribuição segundo a massa *f*	распределение по массе *n*	165
distribución según el brillo *f*	distribuição de brilho [brilhância] *f*	распределение по яркости *n*	166
masa variable *f*	massa variável *f*	переменная масса *f*	167
pérdida de masa *f*	perda de massa *f*	убыль массы *f* потеря массы *f* уменьшение массы *n*	168
ecuación de (la) pérdida de masa *f* ecuación de masa *f*	equação da perda de massa *f* equação de massa *f*	уравнение потери массы *n* уравнение убыли массы *n*	169
zona de compresión *f* casquete de compresión *f*	coifa de compressão *f* zona de compressão *f* calota de compressão *f*	шапка сжатого воздуха *f* газовая подушка *f*	170
blindaje *m* efecto de pantalla *m*	blindagem *f*	экранирование *n* загораживание *n* экранировка *f*	171
partícula de aire *f* molécula de aire *f*	partícula de ar *f* molécula de ar *f*	частица воздуха *f* молекула воздуха *f*	172
fricción *f*	atrito *m*	трение *n*	173
calentamiento de la partícula meteórica *m* calentamiento del cuerpo meteórico *m*	aquecimento do corpo meteórico *m* aquecimento da partícula metéorica *m*	разогрев метеорного тела *m* разогрев метеорной частицы *m*	174
forma del cuerpo meteórico *f* forma de la partícula metéorica *f*	forma do corpo meteórico *f* forma da partícula meteórica *f*	форма метеорного тела *f* форма метеорной частицы *f*	175
sección frontal *f*	se(c)ção frontal *f*	лобовое [миделево] сечение *n* мидель *m*	176
coeficiente de forma *m* factor de forma *m* factor geométrico *m*	coeficiente de forma *m* fator de forma *m* fator geométrico *m*	коэффициент формы *m* геометрический фактор *m*	177
arrastre *m* resistencia del aire *f* resistencia aerodinámica *f*	resistência do ar *f* resistência aerodinâmica *f* arrast(r)o (atmosférico) *m*	лобовое сопротивление *n* увлечение *n* аэродинамическое [воздушное] сопротивление *n*	178
ecuación de la resistencia *f* ley de la resistencia *f*	equação de resistência [arrast(r)o] *f* lei de resistência *f*	уравнение торможения *n* закон сопротивления *m*	179
coeficiente de resistencia *m* coeficiente de arrastre *m*	coeficiente de resistência [arrast(r)o] *m*	коэффициент сопротивления *m* коэффициент увлечения *m*	180
fusión *f*	fusão *f*	плавление *n* расплавление *n*	181
ablación *f*	ablacão *f*	сдувание расплавленного слоя *n* абляция [процесс абляции] *f*[*m*] унос массы *m* унос вещества с поверхности *m*	182

183	ablation process	mécanisme d'ablation *m* mécanisme de la perte de masse *m*	Ablationsprozess *m* Prozess der Massenabnahme *m*
184	heat of ablation	chaleur d'ablation *f*	Ablationswärme *f* Abschmelzungswärme *f*
185	intensive vaporization	vaporisation intense *f*	intensive [lebhafte] Verdampfung *f*
186	disintegration of a meteor [meteoric] body	désintégration d'un corps météorique *f*	Zerbrechen eines Meteorkörpers *n*
187	meteor disintegration	désintégration d'un météore *f*	Meteorzerfall *m*
188	fragmentation of a meteor [meteoric] body meteor fragmentation	fragmentation d'un corps météorique *f* fragmentation météorique *f*	Zerbrechen [Zerkleinerung, Zerstückelung, Auflösung, Aufbrechen] eines Meteorkörpers *n* [*f, f, f, n*]
189	pulverization of a meteor [meteoric] body	pulvérisation d'un corps météorique *f*	Zerstäubung eines Meteorkörpers *f*
190	dustball	boule de poussière *f* météore de poussière *f*	Staubball *m* Staubmeteor *n*
191	fragile	fragile	bröckelig brüchig
192	fragility	fragilité *f*	Brüchigkeit *f* Sprödigkeit *f*
193	meteor height	altitude des météores *f*	Meteorhöhe *f* Höhe der Meteore *f*
194	height equation	équation d'altitude *f*	Höhengleichung *f*
195	meteor luminosity	luminosité d'un météore *f*	Leuchtkraft eines Meteors *f*
196	luminosity equation	équation de luminosité *f*	Leuchtkraftgleichung *f*
197	luminosity factor	facteur de luminosité *m*	Lichtausbeute *f*
198	evaporation of a meteor body vapo(u)rization of a meteor body	vaporisation d'un corps météorique *f* évaporisation d'un corps météorique *f*	Verdampfen eines Meteorkörpers *n* Verdampfung eines Meteorkörpers *f*
199	heat of evaporation energy of vapo(u)rization	chaleur de vaporisation *f* énergie de vaporisation *f*	Verdampfungsenergie *f* Verdampfungswärme *f*
200	vapo(u)rization equation evaporation equation	équation de vaporisation *f*	Verdampfungsgleichung *f*
201	heat-transfer coefficient	coefficient de transmission calorifique *m*	Wärmeübertragungskoeffizient *m*
202	heat transfer transmission of heat	transfert [transport] de chaleur *m* transmission de chaleur *f*	Wärmeübertragung *f* Wärmeübergang *m* Wärmetransport *m*

proceso de ablación *m*	processo de ablação *m*	процесс уноса массы *m* процесс абляции *m*	183
calor de ablación *m*	calor de ablação *m*	теплота абляции *f* тепло абляции *n*	184
vaporización intensa *f*	vaporização intensa *f*	интенсивное испарение *n*	185
desintegración de un cuerpo meteórico *f*	desintegração de um corpo meteórico *f*	распад метеорного тела *m* распад метеорной частицы *m*	186
desintegración de un meteoro *f*	desintegração de um meteoro *f*	распад метеора *m*	187
fragmentación de un cuerpo meteórico *f*	fragmentação de um corpo meteórico *f*	дробление метеорного тела *n* дробление метеора *n*	188
pulverización de un cuerpo meteórico *f*	pulverização de um corpo meteórico *f*	распыление метеорного тела *n* распыление метеорной частицы *n*	189
bola de polvo *f* meteoro de polvo *m*	bola de poeira *f* meteoro de pó *m*	пылевой метеор *m* пылевой комок *m*	190
frágil quebradizo	frágil quebradiço	ломкий хрупкий	191
fragilidad *f*	fragilidade *f*	ломкость *f* хрупкость *f*	192
altura de los meteoros *f* altitud de los meteoros *f*	altura dos meteoros *f* altitude dos meteoros *f*	высота метеоров *f*	193
ecuación de altura *f*	equação de altura *f*	уравнение высоты *n*	194
luminosidad de un meteoro *f*	luminosidade de um meteoro *f*	светимость метеора *f*	195
ecuación de la luminosidad *f*	equação de luminosidade *f*	уравнение светимости *n*	196
factor de luminosidad *m*	fator de luminosidade *m*	коэффициент эффективности излучения *m* коэффициент светимости *m*	197
vaporización de un cuerpo meteórico *f* evaporación de una partícula meteórica *f*	vaporização de um corpo meteórico *f* evaporacão de uma partícula meteórica *f*	испарение метеорного тела *n* испарение четеорной частицы *n*	198
calor de vaporización *m* energía de vaporización *f*	calor de vaporização *m* energia de vaporização *f*	(удельная) теплота испарения *f* энергия испарения *f*	199
ecuación de vaporización *f*	equação de vaporização *f*	уравнение испарения *n*	200
coeficiente de transmisión del calor *m*	coeficiente de transmissão de calor *m*	коэффициент теплопередачи *m*	201
transmisión de calor *f* transferencia de calor *f*	transmissão de calor *f* transferência de calor *f*	теплопередача *f* перенос тепла *m*	202

203	preheating	préchauffage *m*	Vorwärmung *f*
204	melting temperature melting point fusing temperature	température de fusion *f* point de fusion *m*	Schmelztemperatur *f* Schmelzpunkt *m* Fliesspunkt *m*
205	thermal stress	contrainte thermique *f*	Wärmespannung *f*
206	thermal shock	choc thermique *m*	Wärmeschock *m* Wärmestoss *m*
207	meteor radiation	rayonnement d'un météore *m*	Meteorstrahlung *f*
208	transformation of energy	transformation de l'énergie *f*	Energieumsatz *m* Umwandlung der Energie *f*
209	ionization efficiency	rendement d'ionisation *m*	Ionisierungsausbeute *f*
210	luminous efficiency light-output ratio	rendement lumineux *m* taux de débit lumineux *m* efficacité lumineuse *f*	Lichtausbeute *f* Lichtleistung *f*
211	radiation process radiative process	processus de rayonement *m* processus radiatif *m*	Strahlungsvorgang *m* Strahlungsprozess *m*
212	recombination radiation free-bound radiation	rayonnement de recombinaison *f* rayonnement libre-lié *m*	Rekombinationsstrahlung *f* Frei-Gebunden-Strahlung *f*
213	deionization recombination	désionisation recombinaison *f*	Deionisierung *f* Entionisierung *f*
214	dissociative recombination	recombinaison dissociative *f*	dissoziative Rekombination *f*
215	meteor shower meteoric shower	pluie météorique *f* pluie d'étoiles filantes *f*	Meteorschauer *f* Sternschnuppenfall *m* Sternschnuppenschwarm *m*
216	meteor stream	courant météorique *m* essaim météorique *m*	Meteorstrom *m*
217	knot within a meteor stream condensation within a meteor stream	noeud d'un courant météorique *m* condensation dans un courant météorique *f*	Verdichtung innerhalb eines Meteorstromes *f*
218	night-time stream night-time shower	courant nocturne *m* essaim nocturne *m* pluie de météores nocturnes *f*	Nachtstrom *m*
219	day-time stream day-time shower daylight stream	courant diurne *m* essaim diurne *m* courant (visible) de jour *m*	Tagesstrom *m* Tagstrom *m* Tages-Meteorstrom *m*
220	planetary stream	courant planétaire *m*	planetarischer Strom *m*
221	ecliptical stream	courant écliptique *m*	Ekliptikalstrom *m*
222	cometary stream	courant cométaire *m*	kometarischer Strom *m*
223	active stream	courant actif *m*	tätiger Strom *m*

precalentamiento *m*	preaquecimento *m*	предварительный нагрев *m*	203
temperatura de fusión *f* punto de fusión *m*	temperatura de fusão *f* ponto de fusão *m*	температура таяния *f* температура плавления *f*	204
esfuerzo térmico *m*	esforço térmico *m*	термическое напряжение *n*	205
choque térmico *m*	choque térmico *m*	тепловой удар *m*	206
radiación de un meteoro *f*	radiação de um meteoro *f*	излучение метеора *n*	207
transformación de la energía *f*	transformação de energia *f*	превращение энергии *n*	208
eficiencia de ionización *f*	rendimento de ionização *m*	эффективность ионизации *f*	209
eficiencia luminosa *f* eficacia luminosa *f*	rendimento luminoso *m* eficiência luminosa *f*	световая отдача *f* световая эффективность *f* эффективность излучения *f*	210
proceso de radiación *m* proceso radiativo *m*	processo de radiação *m* processo radiativo *m*	процесс излучения *m* радиационный процесс *m*	211
radiación de recombinación *f* radiación libre-ligado *f*	radiação de recombinação *f* radiação livre-ligado *f*	рекомбинационное излучение *n* свободно-связанное излучение *n*	212
desionización *f* recombinación *f*	de(s)ionização *f* recombinação *f*	деионизация *f* рекомбинация *f*	213
recombinación disociativa *f*	recombinação dissociativa *f*	диссоциативная рекомбинация *f*	214
lluvia meteórica *f* lluvia de estrellas (fugaces) *f*	chuva de meteoros *f* chuva de estrelas cadentes *f*	метеорный дождь *m* звёздный дождь *m*	215
corriente meteórica *f* enjambre meteórico *m*	corrente meteórica *f* enxame meteórico *m*	метеорный поток *m*	216
condensación en una corriente meteórica *f*	condensação de [em] uma corrente meteórica *f*	метеорный рой *m* рой метеорных тел *m*	217
corriente nocturna *f* enjambre nocturno *m*	corrente noturna *f* enxame noturno *m*	ночной поток *m* ночной рой *m*	218
corriente diurna *f* enjambre diurno *m*	corrente diurna *f* enxame diurno *m*	дневной поток *m* дневной рой *m*	219
corriente planetaria *f*	corrente planetária *f*	поток планетного происхождения *m*	220
coriente eclíptica *f*	corrente eclíptica *f f*	эклиптический поток *m*	221
corriente cometaria *f*	corrente cometária *f*	кометный поток *m*	222
corriente activa *f*	corrente ativa *f*	активный метеорный поток *m*	223

224	interstellar stream	courant interstellaire *m*	interstellarer Strom *m*
225	temporary stream	courant temporaire *m*	temporärer Strom *m*
226	sporadic stream	courant sporadique *m*	sporadischer Strom *m*
227	permanent stream	courant permanent *m*	permanenter Strom *m*
228	periodic stream	courant périodique *m*	periodischer Strom *m*
229	stream activity	activité d'un courant *f*	Stromaktivität *f*
230	index of stream activity	indice de l'activité d'un courant *m*	Index der Stromaktivität *m*
231	relative stream activity	activité relative d'un courant *f*	relative Stromaktivität *f*
232	meteor background	fond météorique *m*	meteorischer Hintergrund *m*
233	sporadic rate	taux sporadique *m*	Rate sporadischer Meteore *f*
234	meteor component of night sky radiation	composante météorique de la radiation du ciel nocturne *f*	meteorische Komponente des Nachthimmelleuchtens *f* meteorisch bedingter Hintergrund *m*
235	sharp maximum	maximum aigu *m*	spitzes Maximum *n* scharfes Maximum *n*
236	flat maximum	maximum plat *m*	flaches Maximum *n* breites Maximum *n*
237	radiant of meteor stream shower radiant	radiant d'un courant météorique *m* radiant d'un essaim météorique *m*	Stromradiant *m* Meteorschauerradiant *m* Radiant eines Meteorstromes *m*
238	area of radiation	aire de radiation *f*	Radiationsfläche *f* Radiationsfeld *n*
239	subcentre secondary radiant	radiant secondaire *m*	Nebenradiant *m* Teilradiant *m*
240	stream branch	branche du courant *f*	Zweig eines Stromes *m* Stromzweig *m*
241	stationary radiant	radiant stationnaire *m*	stationärer Radiant *m*
242	displacement of the radiant radiant displacement	déplacement du radiant *m*	Verschiebung des Radianten *f* Radiantenverschiebung *f*
243	meteor stream structure	structure du courant météorique *f*	Struktur des Meteorstromes *f*
244	central condensation	condensation centrale *f*	zentrale Verdichtung *f*
245	age of a stream	âge d'un courant *m*	Alter eines Stromes *n*
246	age of a meteor shower	âge d'une pluie météorique *m* âge d'un essaim météorique *m*	Alter eines Meteorschauers *n*
247	Quadrantids	Quadrantides *fpl*	Quadrantiden *fpl*

corriente interestelar *f*	corrente inter(e)stelar *f*	межзвёздный поток *m*	224
corriente temporal *f*	corrente temporária *f*	временный поток *m*	225
corriente esporádica *f*	corrente esporádica *f*	спорадический поток *m*	226
corriente permanente *f*	corrente permanente *f*	постоянный поток *m*	227
corriente periódica *f*	corrente periódica *f*	периодический поток *m*	228
actividad de una corriente *f*	atividade de uma corrente *f*	активность потока *f*	229
índice de actividad de una corriente *m*	índice da atividade da corrente *m*	индекс активности потока *m*	230
actividad relativa de la corriente *f*	atividade relativa da corrente *f*	относительная активность потока *f*	231
fondo meteórico *m*	fundo meteórico *m*	метеорный фон *m*	232
frecuencia de los meteoros esporádicos *f*	freqüência dos meteoros esporádicos *f* taxa esporádica *f*	частота спорадических метеоров *f*	233
componente meteórica de la radiación del cielo nocturno *f*	componente meteórico da radiação do céu noturno *m*	метеорная составляющая свечения ночного неба *f*	234
máximo agudo *m*	máximo agudo *m*	острый максимум *m*	235
máximo llano *m*	máximo chatado *m*	пологий максимум *m* плоский максимум *m*	236
radiante de una corriente meteórica *m* radiante de una lluvia meteórica *m*	radiante de um enxame meteórico *m* radiante de uma corrente meteórica *m*	радиант метеорного потока *m*	237
área de radiación *f*	área de radiação *f*	площадь радиации *f*	238
radiante secundario *m*	radiante secundário *m*	побочный радиант *m*	239
rama de una corriente *f*	ramo de uma corrente *m*	ветвь потока *f*	240
radiante estacionario *m*	radiante estacionário *m*	стационарный радиант *m*	241
desplazamiento del radiante *m*	deslocamento do radiante *m*	смещение радианта *n*	242
estructura de la corriente meteórica *f*	estrutura da corrente meteórica *f*	структура метеорного потока *f*	243
condensación central *f*	condensação central *f*	центральное сгущение *n*	244
edad de una corriente *f*	idade de uma corrente *f*	возраст потока *m*	245
edad de una lluvia meteórica *f* edad de un enjambre *f*	idade da uma chuva meteórica *f* idade de um enxame de meteoros *f*	возраст роя *m*	246
Cuadrántidas *fpl*	Quadrântidas *fpl* Quadrantídeos *mpl*	Квадрантиды *fpl*	247

248	Lyrids	Lyrides *fpl*	Lyriden *fpl*
249	Aquarids	Aquarides *fpl*	Aquariden *fpl*
250	Pisces Australids	essaim d'α Piscis Austrini *m* Piscides *fpl*	Pisciden *fpl*
251	Perseids	Perséides *fpl*	Perseiden *fpl*
252	Giacobinids Draconids	Giacobinides *fpl* Draconides *fpl*	Giacobiniden *fpl* Draconiden *fpl*
253	Orionids	Orionides *fpl*	Orioniden *fpl*
254	Taurids	Taurides *fpl*	Tauriden *fpl*
255	Leonids November Swarm	Léonides *fpl*	Leoniden *fpl*
256	Andromedids Bielids	Andromédides *fpl* Bielides *fpl*	Andromediden *fpl* Bieliden *fpl*
257	Geminids	Géminides *fpl*	Geminiden *fpl*
258	December Ursids Ursids	Ursides de décembre *fpl*	Dezember-Ursiden *fpl*
259	comet associated with a shower	comète associée à un essaim *f*	erzeugender Komet *m* mit einem Schauer in Zusammenhang stehender Komet *m*
260	parent comet	comète génératrice *f* comète mère *f*	Ursprungskomet *m*
261	decay of a cometary nucleus	désagrégation d'un noyau cométaire *f*	Zerfall eines Kometenkerns *m* Auflösung eines Kometenkerns *f*
262	ejection from a comet	éjection depuis une comète *f*	Materiausstoss aus dem Kometen *m* Auswurf aus einem Kometen *m*
263	space density	densité spatiale *f*	Raumdichte *f*
264	density of a meteor stream	densité d'un courant météorique *f*	Meteorstromdichte *f*
265	radar observations radar echo observations	observations par radar *fpl* observations par échos de radar *fpl*	Radarbeobachtungen *fpl* Radioecho-Methode *f* Radioecho-Beobachtung *f*
266	range indicator	indicateur de distance *m*	Entfernungsanzeiger *m*

Líridas *fpl*	Líridas *fpl* Lirídeos *mpl*	Лириды *fpl*	248
Acuáridas *fpl*	Acuáridas *fpl* Aquarídeos *mpl*	Аквариды *fpl*	249
Austrálidas *fpl*	Austrálidas *fpl* Pisces Australídeos *mpl*	Писциды *fpl*	250
Perseidas *fpl*	Persêidas *fpl* Perseídeos *mpl*	Персеиды *fpl*	251
Giacobínidas *fpl* Dracónidas *fpl*	Giacobínidas *fpl* Giacobinídeos *mpl*	Джакобиниды *fpl* Дракониды *fpl*	252
Oriónidas *fpl*	Oriônidas *fpl* Orionídeos *mpl*	Ориониды *fpl*	253
Táuridas *fpl*	Táuridas *fpl* Taurídeos *mpl*	Тауриды *fpl*	254
Leónidas *fpl*	Leônidas *fpl* Leonídeos *mpl*	Леониды *fpl*	255
Andromédidas *fpl* Biélidas *fpl*	Andromédidas *fpl* Bielídios *mpl* Andromedídeos *mpl*	Андромедиды *fpl* Биелиды *fpl*	256
Gemínidas *fpl*	Gemínidas *fpl* Geminídeos *mpl*	Геминиды *fpl*	257
Úrsidas de diciembre *fpl*	Úrsidas de dezembro *pl* Ursídeos *mpl*	Декабрьские Урсиды *fpl*	258
cometa asociado a un enjambre *m*	cometa associado a um enxame *m*	родительская комета *f* комета, связанная с роём *f*	259
cometa padre *m* cometa generador *m*	cometa pai do enxame [da chuva meteórica] *m* cometa gerador *m*	материнская комета *f* комета-родоначальница *f* родительская комета *f*	260
desintegración del núcleo de un cometa *f*	desagregação do núcleo dum cometa *f*	распад кометного ядра *m*	261
eyección por un cometa *f*	ejeção de um cometa *f*	выброс из кометы *m*	262
densidad espacial *f*	densidade espacial *f*	пространственная плотность *f*	263
densidad de una corriente meteórica *f*	densidade de uma corrente meteórica *f*	(пространственная) плотность метеорного потока *f*	264
observaciones con radar *fpl*	observações por radar *fpl*	радиолокационные наблюдения *npl* наблюдения радиоэха *npl*	265
indicador de distancia *m*	indicador de alcance *m*	индикатор дальности *m* указатель расстояния *m*	266

267	radio meteor	radio-météore *m*	Radiometeor *n*
268	radio magnitude	magnitude radio-électrique *f*	Radio-Sterngrösse *f*
269	transmitter	émetteur *m*	Sender *m*
270	transmitter power	puissance de l'émetteur *f*	Sendeleistung *f*
271	peak power	puissance de crête *f*	Spitzenleistung *f*
272	mean power	puissance moyenne *f*	mittlere (Sende)leistung *f*
273	transmitter frequency	fréquence de l'émetteur *f*	Sendefrequenz *f*
274	carrier frequency	fréquence porteuse *f*	Trägerfrequenz *f*
275	modulation frequency	fréquence de modulation *f*	Modulationsfrequenz *f*
276	carrier carrier wave	porteuse *f* onde porteuse *f*	Träger *m* Trägerwelle *f*
277	continuous wave (c.w.)	onde continue *f* signal continu *m*	kontinuierliche Welle *f* Dauerstrich *m*
278	pulsed signals	signaux discontinus *mpl* impulsions *fpl*	Impulssignale *npl*
279	pulse repetition frequency recurrence frequency recurrence rate pulse repetition rate	fréquence d'impulsions *f* taux de récurrence *m* fréquence de récurrence *f* taux de répétition (des impulsions) *m*	Impulsfrequenz *f* Modulationsfrequenz *f* Tastfrequenz *f* Wiederholungsfrequenz *f*
280	pulse width pulse duration pulse length	durée d'impulsion *f* largeur d'impulsion *f* longueur d'impulsion *f*	Impulsbreite *f* Impulsdauer *f* Impulslänge *f* Pulsdauer *f*
281	meteor ionization	ionisation par les météores *f*	meteorische Ionisation *f*
282	ionized column column of ionization	colonne ionisée *f* colonne d'ionisation *f*	ionisierte Luftsäule *f* Ionisationsschlauch *m*
283	line density electron line density	densité électronique linéaire *f* densité linéaire des électrons *f*	lineare Elektronendichte *f*
284	underdense of [with] line density smaller than 10^{12} cm^{-1}	sous-dense (de) densité linéaire inférieure à 10^{12} cm^{-1}	unterdicht mitlinearer Elektronendichte kleiner als 10^{12} cm^{-1}

radiometeoro *m*	radiometeoro *m*	радиометеор *m* метеор, наблюдённый [наблюдаемый] радиолокатором *m*	267
radiomagnitud *f*	radiomagnitude *f*	радиозвёздная величина *f* звёздная величина по радионаблюдениям *f*	268
transmisor *m*	transmissor *m*	передатчик *m*	269
potencia del transmisor *f*	potência do transmissor *f*	мощность передатчика *f*	270
potencia de pico *f*	potência máxima *f*	пиковая мощность *f*	271
potencia media *f*	potência média *f*	средняя мощность *f*	272
frecuencia del transmisor *f*	freqüência do transmissor *f*	частота передатчика *f*	273
frecuencia portadora *f*	freqüência (da onda) portadora *f*	несущая частота *f*	274
frecuencia moduladora *f*	freqüência moduladora *f*	частота модуляции *f*	275
portadora *f* onda portadora *f*	portadora *f* onda portadora *f*	несущая волна *f*	276
onda continua *f*	onda contínua *f*	незатухающая волна *f* непрерывное излучение *n*	277
señales pulsadas *fpl* señales intermitentes *fpl*	sinais pulsantes *mpl* sinais pulsados *mpl*	импульсное излучение *n* импульсы *mpl*	278
frecuencia de los pulsos *f* frecuencia de recurrencia *f* tasa de recurrencia *f* tasa de repetición de pulsos *f*	freqüência dos pulsos *f* freqüência de recorrência *f* taxa de recorrência *f* taxa de repetição dos pulsos *f*	частота импульсов *f* частота повторения (импульсов) *f* скорость повторения (импульсов) *f*	279
duración de pulsos *f* ancho de pulsos *m* anchura del pulso *f*	duração dos pulsos *f* largura do pulso *f*	длительность импульса *f* ширина импульса *f* продолжительность импульса *f* длина импульса *f*	280
ionización por los meteoros *f*	ionização pelos meteoros *f*	метеорная ионизация *f*	281
columna ionizada *f*	coluna ionizada *f*	иониз(ир)ованный цилиндр *m* столб иониз(ир)ованного газа *m* иониз(ир)ованная колонна *f*	282
densidad electrónica lineal *f*	densidade ele(c)trônica linear *f*	линейная [погонная] плотность электронов *f* линейная [погонная] электронная плотность *f*	283
subdenso (de) densidad lineal inferior a 10^{12} cm^{-1}	subdenso (com) densidade linear inferior a 10^{12} cm^{-1}	ненасыщенный недоуплотнённый с линейной электронной плотностью меньше 10^{12} $см^{-1}$	284

285	over-dense of [with] line density larger than 10^{12} cm^{-1}	superdense (de) densité linéaire supérieure à 10^{12} cm^{-1}	überdicht mit linearer Elektronendichte grösser als 10^{12} cm^{-1}
286	ionic diffusion diffusion of ions	diffusion des ions *f* diffusion ionique *f*	Diffusion der Ionen *f* Ionendiffusion *f*
287	atmospheric diffusion coefficient	coefficient de diffusion atmosphérique *m*	Diffusionskoeffizient (in der Atmosphäre) *m*
288	range radar range	distance *f* portée (du radar) *f*	Entfernung *f* Abstand *m* Reichweite *f* Radarreichweite *f*
289	maximum range	distance maximale *f* distance limite *f*	Grenzentfernung *f* maximaler Abstand *m*
290	echo pulse echo signal	impulsion d'écho *f* signal d'écho *m*	Echoimpuls *m* Rücklaufimpuls *m* Rückstrahlimpuls *m*
291	echo time	temps d'écho *m*	Laufzeit *f*
292	oscilloscope oscillograph	oscilloscope *m* oscillographe *m*	Oszilloskop *n* Oszillograph *m*
293	cathode ray tube	tube cathodique *m*	Kathodenstrahlröhre *f*
294	Fresnel zones	zones de Fresnel *fpl*	Fresnelsche Zonen *fpl*
295	radio [radar] echo radar response	écho radio-électrique *m* écho radar [radio] *m*	Radioecho *n* Radarecho *n*
296	meteor echo	écho météorique *m*	Meteorecho *n*
297	meteoric head echo	écho météorique de tête *m*	Frontalecho *n*
298	trail echo	écho de traînée *m*	Schweifecho *n*
299	echo duration	durée de l'écho météorique *f*	Echodauer *f*
300	short-duration meteor echo	écho météorique de courte durée *m*	kurzzeitiges Meteorecho *n*
301	long-duration meteor echo	écho météorique de longue durée *m*	langdauerndes Meteorecho *n*

sobredenso (de) densidad lineal superior a 10^{12} cm^{-1}	sobredenso (com) densidade linear superior a 10^{12} cm^{-1} superdenso	пересыщенный переуплотнённый с электронной плотностью больше 10^{12} эл. $см^{-1}$	285
difusión ionica *f* difusión de los iones *f*	difusão iônica *f* difusão dos íons *f*	ионная диффузия *f* диффузия ионов *f*	286
coeficiente de difusión atmosférica *m*	coeficiente de difusão atmosférica *m*	коэффициент атмосферной диффузии *m*	287
distancia *f* alcance (del radar) *m*	distância *f* alcance (do radar) *m*	дальность *m* дальность [расстояние] действия (радиолокатора) *f*[*n*]	288
distancia máxima *f*	distância máxima *f*	максимальная дальность *f* предельная дальность *f*	289
pulso del eco *m* pulso de retorno *m*	pulso do eco *m*	отражённый импульс *m* отражённый сигнал *m* эхо-сигнал *m*	290
tiempo de eco *m*	tempo (de recepção) do eco *m*	запаздывание эха *n*	291
osciloscopio *m* oscilógrafo *m*	osciloscópio *m* oscilógrafo *m*	осциллоскоп *m* осциллограф *m*	292
tubo de rayos catódicos *m*	tubo [lâmpada] de raios catódicos *m*[*f*]	катодная трубка *f* электронно-лучевая трубка *f* катодно-лучевая трубка *f*	293
zonas de Fresnel *fpl*	zonas de Fresnel *fpl*	зоны Френеля *fpl*	294
radioeco *m* eco de radar [radio] *m*	eco radioelé(c)trico [rádio] *m* radioeco *m*	радиоэхо *n* эхо-сигнал *m*	295
eco meteórico *m*	eco meteórico *m*	метеорное эхо *n* эхо от метеорного следа *n*	296
eco frontal meteórico *m*	eco meteórico frontal *m*	головное эхо (от) метеора *n*	297
eco de estela *m*	eco do rast(r)o *m*	эхо от следа *n*	298
duración del radioeco *f* duración del eco *f*	duração do eco *f* duracão do radioeco *f*	продолжительность метеорного эха *f*	299
eco meteórico de corta duración *m*	eco meteórico de curta duração *m*	кратковременное метеорное эхо *n*	300
eco meteórico de larga duración *m*	eco meteórico de longa duração *m*	длительное метеорное эхо *n*	301

87 METEOROIDS | MÉTÉOROÏDES | METEOROIDEN

	METEOROIDS	MÉTÉOROÏDES	METEOROIDEN
1	meteorite	météorite *f*	Meteorit *m*
2	meteoritic meteoritical	météoritique	meteoritisch Meteoriten-
3	micrometeorites	micrométéorites *fpl*	Mikrometeorite(n) *mpl*
4	macrometeorite	macrométéorite *f*	Makrometeorit *m*
5	of extraterrestrial origin of cosmic origin	d'origine extraterrestre d'origine cosmique	extraterrestrischen Ursprungs kosmischen Ursprungs
6	meteoritics meteoritic astronomy	astronomie météoritique *f*	Meteoritenkunde *f* Meteoritenastronomie *f*
7	meteorite investigator meteoriticist	spécialiste des météorites *m*	Meteoritenforscher *m*
8	meteorite collection collection of meteorites	collection de météorites *f*	Meteoritensammlung *f*
9	meteoritic matter	matière météoritique *f*	meteoritische Materie *f*
10	meteoritic dust	poussière météoritique *f*	Meteoritenstaub *m* meteoritischer Staub *m*
11	meteoric dust	poussière météorique *f*	meteorischer Staub *m* Meteorstaub *m*
12	collection of meteoric dust dust collection	collection de poussières météoriques *f*	Aufsammeln meteorischen Staubes *n* Sammeln meteoritischen Staubes *n*
13	meteoric-dust collector	collecteur de poussière météorique *m*	Staubkollektor *m* Kollektor des meteorischen Staubes *m*
14	sticky slide	plaque gluante *f*	Haftplatte *f*
15	deep-sea deposits	dépôts des mers profondes *mpl*	Tiefseesedimente *npl*
16	oceanographic soundings	sondages océanographiques *mpl* prélèvements océanographiques *mpl*	Tiefseelotungen *fpl*
17	annual deposit	dépôt annuel *m*	jährlicher Niederschlag *m* jährliche Niederschlagsmenge *f*
18	increase in mass of the Earth	accroissement de la masse de la Terre *m*	Massenzuwachs der Erde *m*
19	fall of a meteorite meteoritic fall	chute d'une météorite *f* chute météoritique *f*	Meteoritenfall *m*
20	destruction by a meteorite	destruction par une météorite *f* catastrophe produite par une météorite *f*	Zerstörung durch einen Meteoriten *f*
21	smoke train	traînée de fumée *f*	Rauchschweif *m*
22	dust train	traînée de poussière *f*	Staubschweif *m*

METEOROIDES	METERÓIDES	МЕТЕОРОИДЫ	
meteorito *m*	meteorito *m*	метеорит *m*	1
meteorítico	meteorítico	метеоритный	2
micrometeoritos *mpl*	micrometeoritos *mpl*	микрометеориты *mpl*	3
macrometeorito *m*	macrometeorito *m*	макрометеорит *m*	4
de origen extra-terrestre de origen cósmico	de origem extraterrestre de origem cósmica	внеземного происхождения космического происхождения	5
astronomía meteorítica *f*	ciência dos meteoritos *f* astronomia meteorítica *f*	метеоритика *f*	6
especialista en meteoritos *m*	especialista em meteoritos *m*	исследователь метеоритов *m* метеоритолог *m*	7
colección de meteoritos *f*	coleção de meteoritos *f*	коллекция метеоритов *f* собрание метеоритов *n*	8
materia meteorítica *f*	matéria meteorítica *f*	метеоритное вещество *n*	9
polvo meteorítico *m*	poeira meteorítica *f* pó meteorítico *m*	метеоритная пыль *f*	10
polvo meteórico *m*	poeira meteórica *f* pó meteórico *m*	метеорная пыль *f*	11
recolección de polvo meteórico *f*	recolha [recolhimento, coleta] de poeira meteórica *f* [*m*, *f*]	сбор метеорной пыли *m*	12
recolector de polvo meteórico *m*	recolhedor de poeira meteórica *m* coletor de poeira meteórica *m*	оборудование для сбора метеорной пыли *n*	13
placa adhesiva *f*	chapa [placa] adesiva *f*	клейкая пластинка *f*	14
sedimentos marinos *mpl* depósitos en el fondo del mar *mpl*	sedimentos marinhos *mpl* depósitos marinhos *mpl*	глубоководный осадок [ил] *m* глубоководные отложения *npl*	15
sondeos marinos *mpl*	sondagens oceanográficas *fpl*	взятие донной пробы *n*	16
depósito anual *m*	depósito anual *m*	годовой осадок *m*	17
aumento de masa de la Tierra *m*	aumento da massa da Terra *m*	прирост массы Земли *m* увеличение массы Земли *n*	18
caída de un meteorito *f* caída meteorítica *f*	queda de um meteorito *f* queda meteorítica *f*	падение метеорита *n*	19
destrucción por un meteorito *f* catástrofe causada por un meteorito *f*	destruição por um meteorito *f* catástrofe pela queda de um meteorito *f*	разрушение от метеорита *n* попадание метеорита *n*	20
cola de humo *f*	cauda de fumo [fumaça] *f* rast(r)o de fumaça *m*	дымовой след *m*	21
cola de polvo *f*	cauda [rast(r)o] de poeira *f* [*m*]	пылевой след *m*	22

	METEOROIDS	MÉTÉOROÏDES	METEOROIDEN
23	dust cloud	nuage de poussière *m*	Staubwolke *f*
24	destruction	destruction *f*	Zerstörung *f*
25	fragmentation	fragmentation *f*	Zertrümmerung *f*
26	atmospheric passage passage through atmosphere	traversée de l'atmosphère *f*	Atmosphärendurchgang *m*
27	thermal stress	contrainte thermique *f*	Wärmespannung *f* thermische Spannung *f*
28	pulverization	pulvérisation *f*	Zerstäubung *f* Pulverisierung *f*
29	spherules spherical particles	sphérules *fpl* particules sphériques *fpl*	Kügelchen *npl* sphärische Partikeln *fpl*
30	explosion	explosion *f*	Explosion *f*
31	region of visibility area of visibility	région de visibilité *f* aire de visibilité *f*	Sichtbarkeitsgebiet *n* Sichtweite *f*
32	area of audibility	région d'audibilité *f* aire d'audibilité *f*	Hörbarkeitsgebiet *n* Hörweite *f*
33	meteorite impact meteoritic impact	impact d'une météorite *m* impact météoritique *m*	Meteoritenimpakt *m* Aufschlag eines Meteoriten *m*
34	meteoritic seism meteoritic seismic wave	séïsme (d'origine) météoritique *m* séïsme par la chute d'une météorite *m*	Beben meteoritischen Ursprungs *n* meteoritisches Beben *n*
35	destruction of surface	destruction de la surface *f*	Oberflächenzerstörung *f*
36	fall of a single meteorite single fall	chute unique *f*	Einzelfall *m*
37	meteorite shower	averse de météorites *f* averse météoritique *f*	Meteoritenschauer *m* meteoritischer Schauer *m*
38	area of impact area of fall area over which meteorites are scattered	aire [région] de dispersion de la pluie météoritique *f* zone [aire] de chute *f* aire de répartition *f* surface de distribution *f*	Fallgebiet eines Meteoritenschauers *n* Streufeld eines Meteoritenschauers *n*
39	farthest point of area of impact latest point of area of impact	partie de la région de dispersion située en avant *f* dernier point d'impact *m*	Kopfteil des Streufeldes *m*
40	nearest point of area of impact earliest point of area of impact	partie de la région de dispersion située en arrière *f* premier point d'impact *m*	rückwärtiges Streufeld *n* rückwärtiger Teil des Streufeldes *m*
41	largest fragment [particle] principal mass main mass	le plus grand fragment *m* masse principale *f*	Hauptmeteorit *m* Hauptmasse *f*

nube de polvo *m*	núvem de poeira *f*	пылевое облако *n*	23
destrucción *f*	destruição *f*	разрушение *n*	24
fragmentación *f*	fragmentação *f*	дробление *n* фрагментация *f*	25
paso por la atmósfera *m*	passagem pela atmosfera *f*	пролёт через атмосферу *m*	26
tensión térmica *f* esfuerzo térmico *m*	esforço térmico *m* tensão térmica *f*	термическое напряжение *n* тепловое напряжение *n*	27
pulverización *f*	pulverização *f*	распыление *n*	28
esférulas *fpl* partículas esféricas *fpl*	esférulas *fpl* partículas esféricas *fpl*	шарики *mpl* сферические частицы *fpl* сферулы *fpl*	29
explosión *f*	explosão *f*	взрыв *m*	30
región de visibilidad *f* área de visibilidad *f*	região de visibilidade *f* área de visibilidade *f*	область видимости *f* оптическое поле *n*	31
región de audibilidad *f* área de audibilidad *f*	região de audibilidade *f* área de audibilidade *f*	область слышимости *f* акустическое поле *n*	32
impacto de un meteorito *m* impacto meteorítico *m*	impacto de um meteorito *m* impacto meteorítico *m*	удар метеорита (о поверхность Земли) *m* клевок *m*	33
seísmo meteorítico *m*	sismo meteorítico *m*	метеоритная сейсма *f*	34
destrucción de la superficie *f*	destruição da superfície *f*	разрушение поверхности *n*	35
caída aislada *f* caída de un meteorito aislado *f*	queda única *f* queda de um único meteorito *f*	одиночное падение (метеорита) *n*	36
lluvia meteorítica *f* lluvia de meteoritos *f*	chuva meteorítica *f* chuva de meteoritos *f*	метеоритный дождь *m* дождь метеоритов *m*	37
región de dispersión (de los meteoritos) *f* área de caída *f* área de impacto (de la lluvia meteorítica) *f*	região do impacto (dos meteoritos) *f* área de queda *f* área do impacto (da chuva meteorítica) *f*	эллипс рассеяния *m* область падения метеоритного дождя *f* область рассеяния метеоритного дождя *f*	38
punto extremo de la región de dispersión *m*	ponto mais afastado da área do impacto *m* último ponto da área do impacto *m*	головная часть области [эллипса] рассеяния *f*	39
punto más próximo de la región de dispersión *m*	ponto mais próximo da área do impacto *m* primeiro ponto da área do impacto *m*	тыловая часть области (эллипса) рассеяния *f*	40
el mayor meteorito *m* masa principal *f*	maior meteorito *m* massa principal *f* fragmento principal *m*	главная масса *f* наибольшая масса *f*	41

	METEOROIDS	MÉTÉOROÏDES	METEOROIDEN
42	individual specimen	spécimen *m* échantillon *m*	individuelles Exemplar *n* Einzelexemplar *n*
43	meteorite fragment fragment of a meteorite	fragment d'une météorite *m* fragment météoritique *m*	Bruchstück eines Meteoriten *m* Meteoritenbruchstück *m*
44	finding of a meteorite meteorite find	découverte d'une météorite *f*	Meteoritenfund *m*
45	search for meteorites	recherche [ramassage] de météorites *f*[*m*]	Suche nach Meteoriten *f*
47	circumstances of the fall	circonstances de la chute *fpl*	Umstände des Falls *mpl*
48	eye-witness of the fall	témoin oculaire de la chute *m*	Augenzeuge des Falls *m*
49	suspected fall of a meteorite	chute suspectée de météorite *f*	vermuteter Meteoritenfall *m*
50	place of a meteoritic fall	lieu d'une chute (météoritique) *m*	Fallort des Meteoriten *n*
51	time [moment] of a meteoritic fall	instant [moment] d'une chute (météoritique) *m*	Zeitpunkt des Falls *m*
52	finding place place of discovery	lieu de découverte *m*	Fundstelle *f* Fundort *m*
53	meteoritic crater meteorite crater meteor crater	cratère météoritique *m* cratère météorique *m*	Meteoritenkrater *m* Meteorkrater *m*
54	terrestrial meteoritic crater	cratère météoritique terrestre *f*	irdischer Meteoritenkrater *m*
55	fossil meteoritic crater	cratère météoritique fossile *f*	fossiler Meteoritenkrater *m*
56	crater field	champ de cratères *m*	Kraterfeld *n*
57	main crater	cratère principal *m*	Hauptkrater *m*
58	meteorite pit meteorite hole hole of impact	trou météoritique *m* trou creusé par la météorite *m* cavité creusée par la météorite *f*	Meteoritengrube *f* Einschlaggrube *f* Kratergrube *f*
59	classification of meteorites	classification des météorites *f*	Klassifizierung der Meteoriten *f*
60	stony meteorite meteoritic stone aerolite aerolith	météorite pierreuse *f* aérolithe *m*	Steinmeteorit *m* Aerolith *m* Meteorstein *m*
61	chondrite	chondrite *f*	Chondrit *m*
62	carbonaceous meteorite	météorite carbonée *f*	Kohlenstoffmeteorit *m*
63	carbonaceous chondrite	chondrite carbonée *f*	kohliger Chondrit *m*

ejemplar individual *m* muestra individual *m*	espécime(n) individual *m* exemplar individual *m*	индивидуальный экземпляр *m*	42
fragmento de un meteorito *m*	fragmento de um meteorito *m*	осколок метеорита *m* обломок метеорита *m* метеоритный обломок *m*	43
hallazgo de un meteorito *m* descubrimiento de un meteorito *m*	descoberta de um meteorito *f* achado de um meteorito *m*	находка метеорита *f*	44
búsqueda de meteoritos *f*	busca de meteoritos *f*	поиски метеоритов *mpl* сбор метеоритов *m*	45
circunstancias de la caída *fpl*	circunstâncias da queda *fpl*	обстоятельства падения *npl*	47
testimonio visual de la caída *m*	testemunha ocular de queda *f*	очевидец падения *m*	48
caída supuesta de un meteorito *m*	queda pressuposta de um meteorito *m*	предполагаемое падение метеорита *n* подозреваемое падение метеорита *n*	49
lugar de caída (de un meteorito) *m*	local da queda de um meteorito *m*	место падения метеорита *n*	50
instante de caída *m* momento de caída de un meteorito *m*	instante da queda de um meteorito *m*	время падения метеорита *n*	51
sitio del hallazgo (de un meteorito) *m* lugar del hallazgo *m*	local de encontro *m* local da descoberta *m*	место находки (метеорита) *n* место обнаружения (метеорита) *n*	52
cráter meteorítico *m* cráter meteórico *m*	cratera meteorítica *f* cratera meteórica *f*	метеоритный кратер *m* метеорный кратер *m*	53
cráter meteorítico terrestre *m*	cratera meteorítica terrestre *f*	земной метеоритный кратер *m*	54
cráter meteorítico fósil *m*	cratera meteorítica fóssil *f*	древний метеоритный кратер *m*	55
campo de cráteres *m*	campo de crateras *m*	кратерное поле *n*	56
cráter principal *m*	cratera principal *f*	главный кратер *m*	57
agujero [hoyo] producido por el meteorito *m* hoyo meteorítico *m* agujero meteorítico *m*	cova meteorítica *f* buraco do impacto *m*	метеоритная лунка *f* метеоритная яма *f* воронка *f*	58
clasificación de los meteoritos *f*	classificação dos meteorítos *f*	классификация метеоритов *f*	59
meteorito pétreo *m* aerolito *m* piedra meteórica *f* astrolito *m*	meteorito lítico *m* pedra meteórica *f* aerólito *m* pedra meteorítica *f*	каменный метеорит *m* аэролит *m*	60
condrito *m*	condrito *m*	хондрит *m*	61
meteorito carbonáceo *m*	meteorito carbonáceo *m*	углистый метеорит *m*	62
condrito carbonáceo *m*	condrito carbonáceo *m*	углистый хондрит *m*	63

	METEOROIDS	MÉTÉOROÏDES	METEOROIDEN
64	carbonaceous matter	matière carbonée *f*	Kohlenstoffmaterie *f*
65	meteoritic diamonds	diamants météoritiques *mpl*	meteoritische Diamanten *mpl*
66	organic inclusion	inclusion organique *f*	organische Einschlüsse *mpl*
67	organic remains in meteorites	traces organiques dans les météorites *fpl*	organische Spuren in Meteoriten *fpl*
68	meteoritic organic matter	matière météoritique organique *f*	meteoritische organische Materie *f*
69	organic acid	acide organique *m*	organische Säure *f*
70	amino acid	aminoacide *m* acide aminé *m*	Aminosäure *f*
71	amino group amino [amin-, -amine]	groupe amine *m* amine *f*	Aminogruppe *f*
72	α-aminopropionic acid alanine	acide α-amino-propionique *m* alanine *f*	α-Aminopropionsäure *f* Alanin *n*
73	extraterrestrial origin	origine extraterrestre *f*	ausserirdischer Ursprung *m*
73a	extraterrestrial space	espace extraterrestre *m*	ausserirdischer Raum *m*
74	terrestrial origin	origine terrestre *f*	irdischer Ursprung *m*
75	non-carbonaceous chondrite	chondrite non-carbonée *f*	nichtkohliger Chondrit *m*
76	chondrule	chondre *m*	Chondre *f*
77	achondrite achondritic meteorite	achondrite *f* météorite achondritiqe *f*	Achondrit *m* achondritischer Meteorit *m*
78	siderolite	sidérolite *f*	Siderolith *m*
79	mesosiderite	mésosidérite *f*	Mesosiderit *m*
80	stony-iron meteorite lithosiderite	lithosidérite *f*	Lithosiderit *m*
81	pallasite	pallasite *f*	Pallasit *m*
82	iron meteorite siderite	météorite ferreuse [métallique] *f* sidérite *f* holosidères *m*	Eisenmeteorit *m* Siderit *m*
83	meteori(ti)c iron	fer météori(ti)que *m*	Meteor(iten)eisen *n*
84	hexahedrite	hexaédrite *f*	Hexaedrit *m*
85	troilite	troïlite *f*	Troilit *m*
86	octahedrite	octaédrite *f*	Oktaedrit *m*
87	ataxite	ataxite *f*	Ataxit *m*

materia carbonácea *f* materia carbónica *f*	material carbonáceo *m* material carbônico *m*	углеродистое вещество *n* углистое вещество *n*	64
diamantes meteoríticos *mpl*	diamantes meteoríticos *mpl*	метеоритные алмазы *mpl*	65
inclusión orgánica *f*	inclusão orgânica *f*	органическое включение *n*	66
restos orgánicos en meteoritos *mpl*	traços orgânicos em meteoritos *mpl*	органические остатки в метеоритах *mpl*	67
materia orgánica meteorítica *f*	matéria orgânica meteorítica *f*	метеоритное органическое вещество *n*	68
ácido orgánico *m*	ácido orgânico *m*	органическая кислота *f*	69
aminoácido *m* ácido amínico *m*	aminoácido *m* ácido aminado *m*	аминокислота *f*	70
grupo amino *m* amino *m*	grupamento amino *m* amino *m*	аминогруппа *f*	71
ácido α-aminopropiónico *m* alanina *f*	ácido α-aminopropiônico *m* alanina *f*	аланин *m* α-аминопропионовая кислота *f*	72
origen extraterrestre *m*	origem extraterrestre *f*	внеземное происхождение *n*	73
espacio extraterrestre *m*	espaço extraterrestre *m*	внеземное пространство *n*	73a
origen terrestre *m*	origem terrestre *f*	земное происхождение *n*	74
condrito no carbonáceo *m*	condrito não-carbonáceo *m*	неуглистый хондрит *m*	75
cóndrulo *m*	côndrulo *m*	хондра *f*	76
meteorito acondrítico *m* acondrito *m*	acondrito *m* meteorito acondrítico *m*	ахондрит *m*	77
siderolito *m*	siderólito *m*	сидеролит *m*	78
mesosiderito *m*	mesossiderito *m*	мезосидерит *m*	79
litosiderito *m*	meteorito ferrolítico *m* litossiderito *m*	каменножелезный метеорит *m* литосидерит *m*	80
pallasito *m*	pallasito *m*	палласит *m*	81
meteorito férrico *m* siderito *m*	meteorito férrico *m* siderito *m*	железный метеорит *m* сидерит *m*	82
hierro meteórico [meteorítico] *m*	ferro meteórico [meteorítico] *m*	метеоритное железо *n*	83
hexaedrito *m*	hexaédrita *f*	гексаэдрит *m*	84
troilita *f*	troilita *f*	троилит *m*	85
octaedrito *m*	octaedrita *f* octaedrito *m*	октаэдрит *m*	86
ataxito *m*	ataxita *f* ataxito *m*	атаксит *m*	87

88	nickel-poor ataxite	ataxite pauvre en nickel *f*	nickelarmer Ataxit *m*
89	nickel-rich ataxite	ataxite riche en nickel *f*	nickelreicher Ataxit *m*
90	nickel iron	ferro-nickel *m*	Nickeleisen *n*
91	trias	trias *m*	Triade *f*
92	kamacite	kamacite *f*	Kamazit *m*
93	taenite	taenite *f*	Taenit *m*
94	plessite	plessite *f*	Plessit *m* Fülleisen *n*
95	tektite	tectite *f*	Tektit *m*
96	splash-form tektite drop-form tektite	tectite en forme de goutte *f*	Tropfentektit *m*
97	microtektite	microtectite *f*	Mikrotektit *m*
98	australite	australite *f*	Australit *m*
99	moldavite	moldavite *f*	Moldavit *m*
100	structure of meteorites	structure des météorites *f*	Struktur der Meteoriten *f*
101	polished surface	surface polie *f*	polierte Fläche *f*
102	etching of a polished surface	morsure d'une surface polie *f* attaque chimique *f*	Ätzen einer polierten Fläche *n*
103	Neumann lines	lignes de Neumann *fpl*	Neumannsche Linien *fpl*
104	lamellae	lamelles *fpl*	Balken *mpl*
105	Widmanstätten figures [pattern]	figures de Widmanstätten *fpl*	Widmanstättensche Figuren *fpl*
106	meteorite surface	surface d'une météorite *f*	Oberfläche eines Meteoriten *f*
107	primary surface first type surface	surface primaire *f* surface du premier type *f*	ursprüngliche Oberfläche *f* primäre Oberfläche *f*
108	secondary surface second type surface	surface secondaire *f* surface du second type *f*	sekundäre Oberfläche *f*
109	crusted surface	surface recouverte d' une croûte *f*	mit Schmelzkruste bedeckte Oberfläche *f*
110	fusion crust	croûte de fusion *f*	Schmelzkruste *f*

ataxito pobre en níquel *m*	ataxita [ataxito] pobre em [de] níquel *f*[*m*]	бедный никелем атаксит *m*	88
ataxito rico en níquel *m*	ataxita rica em [de] níquel *f*	богатый никелем атаксит *m*	89
ferroníquel *m*	ferro-níquel *m*	никелистое железо *n*	90
tríada *f*	tríade *m*	триада *f*	91
kamacita *f* camacita *f*	camacita *f* kamacita *f*	камасит *m*	92
taenita *f*	t(a)enita *f*	тэнит *m*	93
plessita *f*	plessita *f*	плессит *m*	94
tectita *f*	tectito *m* tectita *f*	тектит *m*	95
tectita en forma de gota *f*	tectita [tectito] em forma de gota *f* [*m*]	каплеобразный тектит *m*	96
microtectita *f*	microtectito *m* microtectita *f*	микротектит *m*	97
australita *f*	australito *m* australita *f*	австралит *m*	98
moldavita *f*	moldavito *m* moldavita *f*	молдавит *m*	99
estructura de los meteoritos *f*	estrutura dos meteoritos *f*	структура метеоритов *f*	100
superficie pulida *f*	superfície polida *f*	полированная поверхность *f*	101
tratamiento con ácido (de una superficie pulida) *m* ataque químico *m*	cauterização *f*	травление полированной поверхности *n*	102
rayas de Neumann *fpl*	linhas de Neumann *fpl*	неймановы линии *fpl* линии Неймана *fpl*	103
laminillas *fpl*	lamelas *fpl*	балки *fpl*	104
figuras de Widmanstaetten *fpl*	figuras de Widmanstaetten *fpl*	видманщтеттеновы фигуры *fpl*	105
superficie de un meteorito *f*	superfície de um meteorito *f*	поверхность метеорита *f*	106
superficie primaria *f* superficie original *f*	superfície primária *f* superfície original *f*	первичная поверхность *f* поверхность первого рода *f*	107
superficie secundaria *f*	superfície secundária *f*	вторичная поверхность *f* поверхность второго рода *f*	108
superficie con costra *f*	superfície encrostada *f*	поверхность с корой плавления *f*	109
costra producida por la fusión *f*	crosta de fusão *f*	кора плавления *f*	110

	METEOROIDS	MÉTÉOROÏDES	METEOROIDEN
111	veins	veines *fpl*	Adern *fpl*
112	oriented meteorite	météorite orientée *f*	orientierter Meteorit *m*
113	oriented form	forme orientée *f*	orientierte Form *f*
114	front surface	surface [partie] antérieure *f* surface [partie] avant *f*	Vorderseite *f* Frontteil *m*
115	rear surface back surface	surface [partie] arrière *f* surface [partie] postérieure *f*	Rückseite *f*
116	atmospheric treatment atmospheric action	action de l'atmosphère *f* action atmosphérique *f*	atmosphärische Einwirkung *f*
117	cavity	cavité *f*	Aushöhlung *f*
118	surface cavity	cavité superficielle *f*	Oberflächenaushöhlung *f*
119	inner cavity	cavité interne *f*	innere Aushöhlung *f*
120	radiation track track of a particle particle track	trace nucléaire *f* trace d'une particule *f*	Partikeleinschlag-Spur *f* Teilchenspur *f*
121	fossil track	trace fossile *f*	fossile Spur *f*
122	regmaglypts piezoglypts	regmaglyptes *mpl* piézoglyptes *mpl*	Rhegmaglypten *fpl* Piezoglypten *fpl*
123	metorite parent body parent body	corps parental *m* météorite parente *f*	meteoritischer Mutterkörper *m* Muttermeteorit *m*
124	break-up of parent body disruption of parent body	fragmentation du corps parental *f* éclatement [cassure] de la météorite [du corps] parent *m*[*f*]	Zersplitterung des Mutterkörpers *f*
125	collisional breakup collisional disruption	fragmentation par collision *f* cassure par collision *f*	Kollisionszersplitterung *f*
126	age determination dating age measurement	détermination de l'âge *f* datation *f* mesure de l'âge *f*	Altersbestimmung *f* Datierung *f*
127	age of meteorites	âge des météorites *f*	Alter der Meteorite(n) *n*
128	helium method (of age determination)	méthode à l'hélium (pour les mesures d'âges) *f*	Heliummethode (der Altersbestimmung) *f*
129	isotopic composition	composition isotopique *f*	isotopische Zusammensetzung *f* Isotopenzusammensetzung *f*
130	isotope dating isotopic dating	datation isotopique *f* datation par les isotopes *f*	Altersbestimmung mit Hilfe von Isotopen *f* Altersbestimmung mittels Isotopen *f*

vetas *fpl*	veios *mpl*	жилки *fpl*	111
meteorito orientado *m*	meteorito orientado *m*	ориентированный метеорит *m*	112
forma orientada *f*	forma orientada *f*	ориентированная форма *f*	113
superficie anterior *f* superficie frontal *f*	superfície anterior *f* superfície frontal *f*	передняя поверхность *f* лобовая поверхность *f*	114
superficie posterior *f*	superfície posterior *f*	тыловая поверхность *f* тыловая часть *f*	115
acción de la atmósfera *f*	ação da atmosfera *f*	воздействие атмосферы *n* атмосферная обработка *f*	116
cavidad *f*	cavidade *f*	каверна *f* полость *f*	117
cavidad superficial *f*	cavidade superficial *f*	поверхностная полость *f*	118
cavidad interna *f*	cavidade interior *f*	внутренняя полость *f*	119
traza nuclear *f* traza de una partícula *f*	traço de radiação *m* traço nuclear *m*	трек [след] радиационного разрушения *m* след частицы *m*	120
traza fósil *f*	traço fóssil *m*	древний трек *m* древний след *m*	121
regmaglyptos *mpl* piezoglyptos *mpl*	regmaglitas *mpl* piezoglitas *mpl*	пьезоглипты *mpl* регмаглипты *mpl*	122
meteorito original *m* cuerpo madre *m*	meteorito original *m* corpo parente [origem] do meteorito *m*	исходное тело *n* материнское тело *n* метеоритное родительское тело *n*	123
disrupción del cuerpo original *f*	ruptura do corpo parente [original] *f*	распад родительского тела *m* разрушение материнского тела *n*	124
disrupción por colisión *f*	ruptura por colisão *f*	столкновительный распад *m* столкновительное разрушение *n*	125
determinación de la edad *f*	determinação da idade *f* datação *f*	определение возраста *n* датирование *n* измерение возраста *n*	126
edad de los meteoritos *f*	idade dos meteoritos *f*	возраст метеоритов *m*	127
método del helio (de la determinación de la edad) *m*	método do hélio (de determinação da idade) *m*	гелиевый метод (определения возраста) *m*	128
composición isotópica *f*	composição isotópica *f*	изотопный состав *m* изотопический состав *m*	129
determinación isotópica de la edad *f*	determinação da idade isotópica [por isótopos] *f* datação isotópica *f*	изотопическое определение возраста *n* определение возраста изотопным методом *n*	130

131	dating by radioactivity radioactive dating	datation radioactive [par radioactivité] *f* détermination de l'âge par radioactivité *f*	radioaktive Altersbestimmung *f*
132	radiogenic noble gas	gaz rare d'origine radioactive *m* gaz rare radiogénique *m*	radiogenes Edelgas *n* durch radioaktiven Zerfall entstandenes Edelgas *n*
133	primordial noble gas	gaz rare primordial *m*	Uredelgas *n*
134	gas retention	rétention gazeuse *f* rétention de gaz *f*	Gasretention *f* Gaszurück(be)haltung *f*
135	gas-retention age retention age	âge de la rétention (gazeuse) *m*	Gasretentionsalter *n* Gasgehaltalter *n*
136	outgassing degassing	dégazage *m*	Entgasung *f* Gasaustreibung *f*
137	solidification age	âge de solidification *m*	Erstarrungsalter *n*
138	fractionation	fractionnement *m*	Fraktionierung *f*
139	cosmic-ray exposure age exposure age radiation age irradiation age	âge d'exposition (au rayonnement cosmique) *m*	Bestrahlungsalter *n*
140	microphotograph photomicrograph	microphotographie *f*	Mikroaufnahme *f*
141	track density	densité de traces *f*	Spurendichte *f* Dichte der Kernspuren *f*
142	space erosion	érosion dans l'éspace (cosmique) *f* érosion cosmique *f*	Erosion im Weltraum *f* kosmische Erosion *f*

determinación de la edad por radiactividad *f*	determinação da idade por radioatividade *f*	радиоактивное датирование *n* определение возраста по радиоактивности *n* датирование по радиоизотопам *n*	131
gas noble radiogénico *m*	gás nobre radiogênico *m*	радиогенный благородный газ *m* благородный газ радиоактивного происхождения *m*	132
gas noble primordial *m*	gás nobre primordial *m*	первичный благородный газ *m*	133
retención de gas *f*	retencão de gás *f*	удержание газа *n* удерживание газа *n*	134
edad de rentención de gas *f*	idade de retenção de gás *f*	возраст, определённый по количеству остаточных газов *m*	135
desgasificación *f*	degasificação *f* degasação *f*	дегазация *f* обезгаживание *n* выделение газов *n*	136
edad de solidificación *f*	idade de solidificação *f*	возраст затвердевания *m*	137
fraccionamiento *m*	fracionamento *m*	фракционирование *n*	138
edad de (ir)radiación *f* edad de exposición (a los rayos cósmicos) *f*	idade de exposição (aos raios cósmicos) *f* idade de radiação *f*	экспозиционный возраст *m* возраст, определённый по экспонированию космическими лучами *m*	139
microfotografia *f*	microfotografia *f*	микрофотоснимок *m* микроснимок *m* микрофотография *f*	140
densidad de trazas *f*	densidade dos traços *f*	плотность следов *f* плотность треков *f*	141
erosión en el espacio cósmico *f* erosión cósmica *f*	erosão no espaço cósmico *f* erosão cósmica *f*	эрозия в космическом пространстве *f*	142

88 THE SUN — SOLEIL — SONNE

	THE SUN	SOLEIL	SONNE
1	Sun	Soleil *m*	Sonne *f*
2	optical Sun visible Sun	Soleil optique *m* Soleil visible *m*	optische Sonne *f* sichtbare Sonne *f*
3	radio Sun	Soleil radio *m*	Radiosonne *f*
4	X-ray Sun	Soleil (en rayons) X *m*	Röntgensonne *f*
5	solar helio-	solaire	Sonnen- solar Helio-
6	solar physics heliophysics	physique solaire *f* héliophysique *f*	Sonnenphysik *f* Physik der Sonne *f*
7	solar observatory	observatoire solaire *f*	Sonnenobservatorium *n*
8	sunrise	lever du Soleil *m*	Sonnenaufgang *m*
9	sunshine	ensoleillement	Sonnenschein *m*
10	duration of sunshine	durée d'ensoleillement *f*	Sonnenscheindauer *f*
11	sunset	coucher du Soleil *m*	Sonnenuntergang *m*
12	heliometer	héliomètre *m*	Heliometer *n*
13	screen	écran (de projection) *m*	Projektionsschirm *m* Sonnenprojektionsschirm *m*
14	eclipse	éclipse *f*	Finsternis *f*
15	solar eclipse eclipse of the Sun	éclipse du Soleil *f* éclipse solaire *f*	Sonnenfinsternis *f*
15a	annular eclipse of the Sun	éclipse annulaire de Soleil *f*	ringförmige Sonnenfinsternis *f*
16	partial solar eclipse	éclipse partielle de Soleil *f*	partielle Sonnenfinsternis *f*
16a	central eclipse	éclipse centrale *f*	zentrale Finsternis *f*
17	total solar eclipse total eclipse of the Sun	éclipse totale de Soleil *f*	totale Sonnenfinsternis *f*
17a	artificial solar eclipse	éclipse artificielle de Soleil *f*	künstliche Sonnenfinsternis *f*
18	annular eclipse	éclipse annulaire *f*	ringförmige Finsternis *f*
19	southern limit of eclipse	limite australe de l'éclipse *f*	südliche Grenze der Finsternis *f*

Sol *m*	Sol *m*	Солнце *n*	1
Sol óptico *m* Sol visible *m*	Sol ó(p)tico *m* Sol visível *m*	видимый [оптический] диск *m* оптическое Солнце *n* видимый диск *m*	2
Sol radio *m* radiosol *m*	rádio-Sol [Sol-rádio] *m* radiossol *m*	радиосолнце *n* Солнце в радиоволнах *n*	3
Sol (en rayos) X *m*	Sol (em raios) X *m*	рентгеновское Солнце *n*	4
solar helio-	solar heli(o)	солнечный гелио-	5
física solar *f* heliofísica *f*	física solar *f* heliofísica *f*	физика Солнца *f* солнечная физика *f* гелиофизика *f*	6
observatorio solar *m*	observatório solar *m*	солнечная обсерватория *f*	7
salida del Sol *f*	nascer do Sol *m*	восход Солнца *m*	8
insolación *f*	luz solar *f* insolação *f*	солнечное сияние *n* инсоляция *f* облучение Солнцем *n* солнечный свет *m*	9
duración de la insolación *f*	duração da insolação *f*	продолжительность солнечного сияния *f*	10
puesta del Sol *f*	pôr [ocaso] do Sol *m*	заход Солнца *m*	11
heliómetro *m*	heliômetro *m*	гелиометр *m*	12
pantalla *f*	tela *f* quadro de projeção *m*	экран для наблюдения Солнца *m*	13
eclipse *m*	eclipse *m*	затмение *n*	14
eclipse solar *m* eclipse del Sol *m*	eclipse solar *m* eclipse do Sol *m*	солнечное затмение *n* затмение Солнца *n*	15
eclipse anular del Sol *m*	eclipse anular do Sol *m*	кольцеобразное [кольцевое] затмение Солнца *n*	15a
eclipse parcial del Sol *m*	eclipse parcial do Sol *m*	частное солнечное затмение *n*	16
eclipse central *m*	eclipse central *m*	центральное затмение *n*	16a
eclipse total del Sol *m*	eclipse total do Sol *m*	полное солнечное затмение *n*	17
eclipse artificial del Sol *m*	eclipse artificial do Sol *m*	искусственное затмение Солнца *n*	17a
eclipse anular *m*	eclipse anular *m*	кольцеобразное [кольцевое] затмение *n*	18
límite austral [sur] del eclipse *m*	limite sul do eclipse *m*	южная граница затмения *f*	19

20	northern limit of eclipse	limite boréale de l'éclipse *f*	nördliche Grenze der Finsternis *f*
21	eclipse ends at sunrise	fin au lever du Soleil *f*	Ende bei Sonnenaufgang *n*
22	eclipse begins at sunset	commencement au coucher du Soleil *m*	Anfang bei Sonnenuntergang *m*
23	ionospheric eclipse	éclipse ionosphérique *f*	ionosphärische Finsternis *f*
24	corpuscular eclipse	éclipse corpusculaire *f*	korpuskulare Finsternis *f*
25	flash spectrum	spectre-éclair *m*	Flash-Spektrum *n*
26	duration of total eclipse length [duration] of totality	durée de la totalité *f*	Dauer der totalen Sonnenfinsternis *f* Dauer der Totalität *f* Finsternisdauer *f*
27	moving plate	plaque mobile *f*	bewegliche (photographische) Platte *f*
28	eclipse chart	carte de (visibilité de) l'éclipse *f*	Finsterniskarte *f*
29	path of the total phase	zone de totalité *f*	Zone der totalen Finsternis *f* Totalitätszone *f*
30	solar-eclipse expedition	expédition d'éclipse *f*	Sonnenfinsternisexpedition *f*
31	shadow bands	ombres volantes *fpl*	fliegende Schatten *mpl*
32	Baily's beads	grains de Baily *mpl*	Perl(en)schnur-Phänomen *n*
33	smoked glass	verre fumé *m*	Rauchglas *n* Russglas *n* geschwärztes Glas *n*
34	sunspot drawing	dessin des taches *m*	Fleckenzeichnung *f*
35	solar instruments	instruments solaires *mpl*	Sonneninstrumente *npl*
36	heliograph	héliographe *m*	Heliograph *m*
37	white-light photograph	photographie en lumière blanche *f*	Integralaufnahme *f* photographische Aufnahme im weissen Licht *f*
38	monochromatic telescope monochromatic heliograph	héliographe monochromatique *m*	monochromatischer Heliograph *m*
39	pyrheliometer	pyrhéliomètre *m*	Pyrheliometer *n*
40	water-flow pyrheliometer	pyrhéliomètre à courant d'eau *m*	Wasserstrom-Pyrheliometer *n*

SOL	SOL	СОЛНЦЕ	
límite boreal [norte] del eclipse *m*	limite norte do eclipse *m*	северная граница затмения *f*	20
fin del eclipse a la salida del Sol *m*	eclipse termina ao nascer do Sol	конец (затмения) при восходе *m*	21
comienzo del eclipse a la puesta del Sol *m*	eclipse começa ao pôr do Sol	начало затмения при заходе Солнца *n* кривая начала при заходе *f*	22
eclipse ionosférico *m*	eclipse ionosférico *m*	ионосферное затмение *n*	23
eclipse corpuscular *m*	eclipse corpuscular *m*	корпускулярное затмение *n*	24
espectro relámpago *m*	espectro relâmpago *m*	спектр вспышки *m*	25
duración de la totalidad *f*	duração do eclipse total *f* duração da totalidade *f*	продолжительность полного затмения *f*	26
placa móvil *f*	chapa (fotográfica) móvel *f*	движущаяся фотопластинка *f*	27
carta del eclipse *f*	carta do eclipse *f*	карта солнечного затмения *f*	28
zona de totalidad *f*	faixa de totalidade *f*	полоса полного затмения *f* полоса полной фазы *f*	29
expedición para observar un eclipse Solar [del Sol] *f*	expedição de eclipse solar *f*	экспедиция для наблюдения солнечного затмения *f* солнечная экспедиция *f*	30
sombras volantes *fpl* bandas de sombra *fpl*	sombras volantes *fpl* faixas sombreadas *fpl*	бегущие тени *fpl*	31
perlas de Baily *fpl*	grãos de Baily *mpl*	чётки Бейли *fpl*	32
cristal ahumado *m* vidrio ahumado *m*	vidro enfumaçado *m* vidro fosco *m*	дымчатое стекло *n*	33
dibujo de las manchas solares *m* mapa de las manchas solares *m*	desenho das manchas *m*	зарисовка Солнца *f* чертёж Солнца *m*	34
instrumentos solares *mpl*	instrumentos solares *mpl*	солнечные инструменты *mpl*	35
heliógrafo *m*	heliógrafo *m*	гелиограф *m* фотогелиограф *m*	36
fotografía en luz blanca *f*	fotografia em luz branca *f*	фотография в белом свете *f* фотография в интегральном свете *f*	37
heliógrafo monocromático *m*	heliógrafo monocromático *m*	монохроматический гелиограф *m*	38
pirheliómetro *m*	pireliômetro *m*	пиргелиометр *m*	39
pirheliómetro de corriente de agua *m*	pireliômetro a corrente de água *m*	пиргелиометр с текущей водой *m* водоструйный пиргелиометр *m*	40

	THE SUN	SOLEIL	SONNE
41	water-stir pyrheliometer	pyrhéliomètre à circulation d'eau *m*	Wasserrühr-Pyrheliometer *n*
42	silver-disk pyrheliometer	pyrhéliomètre à disque d'argent *m*	Silberscheibenpyrheliometer *n*
43	pyranometer	pyranomètre *m*	Pyranometer *n*
44	siderostat	sidérostat *m*	Siderostat *m*
45	heliostat	héliostat *m*	Heliostat *m*
46	coelostat	coelostat *m*	Coelostat *m* Zölostat *m*
47	horizontal coelostat	coelostat horizontal [renvoyant le rayonnement horizont]alement *m*	Horizontalzölostat *m* Horizontalcoelostat *m*
48	vertical coelostat	coelostat vertical [renvoyant le rayonnement verticalement] *m*	Vertikalzölostat *m* Vertikalcoelostat *m*
49	solar telescope	télescope solaire *m*	Sonnenteleskop *n*
50	automatic solar telescope	télescope solaire automatique *m*	automatisches Sonnenteleskop *n*
51	Sun seeker guider	chercheur [guide] solaire *m* lunette guide *f*	Sonnensucher *m*
52	balloon-borne solar telescope	télescope solaire embarqué sur ballon *m*	Ballon-Sonnenteleskop *n*
53	domeless solar refractor	réfracteur solaire sans coupole *m*	kuppelloser Sonnenrefraktor *m*
54	tower telescope solar tower	tour solaire *f* télescope solaire vertical *m*	Turmteleskop *n*
55	horizontal solar telescope	télescope solaire (à axe optique) horizontal *m*	horizontales Sonnenteleskop *n*
56	high-altitude observatory	observatoire de haute montagne *f* observatoire d'altitude *f*	Höhenstation *f* Höhenobservatorium *n*
57	spectrograph recording velocities spectro-enregistreur des vitesses	spectro-enregistreur des vitesses *m*	registrierender Geschwindigkeitsmesser *m* spectro-enregistreur des vitesses *m*
58	prominence spectroscope	spectroscope pour les protubérances *m* spectroscope protubéranciel *m*	Protuberanzenspektroskop *n*
59	coronagraph	coronographe *m*	Koronograph *m*

SOL	SOL	СОЛНЦЕ	88
pirheliómetro de circulación de agua *m*	pireliômetro a circulação de agua *m*	водяной пиргелиометр с перемешиванием *m*	41
pirheliómetro de disco de plata *m*	pireliômetro a disco de prata *m*	пиргелиометр с серебряным диском *m*	42
piranómetro *m*	piranômetro *m*	пиранометр *m*	43
sideróstato *m*	sideróstato *m*	сидеростат *m*	44
helióstato *m*	helióstato *m*	гелиостат *m*	45
celóstato *m*	celóstato *m*	целостат *m*	46
celóstato horizontal *m*	celóstato horizontal *m*	горизонтальный целостат *m*	47
celóstato vertical *m*	celóstato vertical *m*	вертикальный целостат *m*	48
telescopio solar *m*	telescópio solar *m*	солнечный телескоп *m*	49
telescopio solar automático *m*	telescópio solar automático *m*	автоматический [автоматизированный] солнечный телескоп *m*	50
buscador del Sol *m*	localizador do Sol *m* buscador solar *m*	солнечный искатель [фотогид] *m* искатель Солнца *m*	51
telescopio solar llevado por [embarcado en] un globo *m*	telescópio solar colocado em um balão *m*	стратосферный солнечный телескоп *m* солнечный телескоп, (помещённый) на баллоне *m*	52
refractor solar sin cúpula *m*	refrator solar sem cúpula *m*	рефрактор без башни *m* солнечный телескоп без купола *m*	53
torre solar *f*	torre solar *f*	башенный телескоп *m* солнечная башня *f*	54
telescopio solar horizontal *m*	telescópio solar horizontal *m*	горизонтальный солнечный телескоп *m*	55
observatorio de gran altitud *m* observatorio de alta montaña *m*	observatório de grande altitude *m*	высокогорная обсерватория *f* высокогорная станция *f*	56
registrador espectroscópico de velocidades *m*	espe(c)troregistrador de velocidades *m*	спекторегистратор скоростей *m*	57
espectroscopio de protuberancias *m*	espe(c)troscópio de protuberâncias *m*	протуберанц-спектроскоп *m*	58
coronógrafo *m*	coronógrafo *m*	коронограф *m* внезатменный коронограф *m*	59

	THE SUN	SOLEIL	SONNE
60	externally occulted coronagraph externally occulting coronagraph	coronographe à occultation externe *m* coronographe à disque occulteur externe *m*	Koronograph mit äusserer Abdeckung [Abblendung] *m* Koronograph mit äusserer Blende [Abdeckscheibe] *m*
61	occulting disk occulting disc	disque d'occultation *m* disque occulteur *m*	Blende *f* Abdeckscheibe *f*
62	reflecting cone	cône réfléchissant *m*	reflektierender Kegel *m*
63	truncated cylinder	cylindre tronqué *m*	abgestumpfter Zylinder *m*
64	light trap	piège à lumière *m*	Lichtfalle *f*
65	curved slit	fente courbe *f*	gekrümmter Spalt *m*
66	K-coronameter	coronomètre K *m* polarimètre coronal *m*	K-Koronometer *n*
67	spectrohelioscope	spectrohélioscope *m*	Spektrohelioskop *n*
68	rotating prism Anderson prism	prisme rotatif *m* prisme d'Anderson *m*	rotierendes Prisma *n* Anderson-Prisma *n*
69	line-shifter	dispositif modifiant la longuer d'onde *m* dispositif de sélection de λ *m*	Linienschieber *m*
70	oscillating slit	fente oscillante *f*	oszillierender Spalt *m*
71	spectroheliograph	spectrohéliographe *m*	Spektroheliograph *m*
72	spectroheliogram	spectrohéliogramme *m*	Spektroheliogramm *n*
73	chromospheric telescope	télescope chromosphérique *m*	Chromosphären-Teleskop *n*
74	X-ray heliogram	héliogramme en rayons X *m*	Röntgenheliogramm *n* Röntgenstrahlen-Heliogramm *n* X-Strahlen-Heliogramm *n*
74a	solar spectrum	spectre solaire *m*	Sonnenspektrum *n*
75	spectrobolometer	spectrobolomètre *m*	Spektrobolometer *n*
76	spectrobologram	spectrobologramme *m*	Spektrobologramm *n* Bolographenkurve *f*
77	solar spectrograph	spectrographe solaire *m*	Sonnenspektrograph *m*
78	horizontal spectrograph	spectrographe horizontal *m*	Horizontalspektrograph *m*
79	vertical spectrograph	spectrographe vertical *m*	Vertikalspektrograph *m*
80	step weakener step wedge [attenuator]	atténuateur à échelons *m*	Stufenkeil *m*
81	spectroheliocinematograph	spectro-héliocinématographe *m*	Spektroheliokinematograph *m*

SOL	SOL	СОЛНЦЕ	
coronógrafo con disco externo *m* coronógrafo ocultado externamente *m*	coronógrafo de ocultação exterior *m* coronógrafo ocultado exteriormente *m*	коронограф с внешним затмевающим диском *m* коронограф с внешним затмением *m*	60
disco de ocultamiento *m* disco ocultante [ocultador] *m*	disco ocultante *m* disco de ocultação *m*	искусственная луна *f* затмевающий диск *m*	61
cono reflectante *m*	cone refletor *m*	отражающий конус *m*	62
cilindro truncado *m*	cilindro truncado [troncado] *m*	усечённый [косо срезанный] цилиндр *m*	63
trampa para la luz *f*	armadilha de luz [para a luz] *f*	световая ловушка *f*	64
rendija curvada *f* ranura curvada *f*	fenda curvada *f* fenda curva *f*	кривая щель *f*	65
coronómetro K *m* polarímetro coronal *m*	K-coronômetro *m* polarímetro coronal *m*	К-корономeтр *m*	66
espectrohelioscopio *m*	espe(c)tro(h)elioscópio *m*	спектрогелиоскоп *m*	67
prisma de Anderson *m* prisma rotante *m*	prisma rotativo *m* prisma de Anderson *m*	вращающаяся призма *f* призма Андерсона *f*	68
dispositivo de desplazamiento de la raya *m*	desviador de raia *m* deslocador de raia *m*	лайншифтер *m* сместитель линий *m* устройство для смещения линий *n*	69
ranura oscilante *f* rendija oscilante *f*	fenda oscilante *f*	колеблющаяся щель *f*	70
espectroheliógrafo *m*	espe(c)tro(h)eliógrafo *m*	спектрогелиограф *m*	71
espectroheliograma *m*	espe(c)tro(h)eliograma *m*	спектрогелиограмма *f*	72
telescopio cromosférico *m*	telescópio cromosférico *m*	хромосферный телескоп *m*	73
heliograma en rayos X *m*	heliograma em raios X *m*	рентгеновская гелиограмма *f* гелиограмма в рентгеновских лучах *f*	74
espectro solar *m*	espe(c)tro solar *m*	солнечный спектр *m*	74a
espectrobolómetro *m*	espe(c)trobolômetro *m*	спектроболометр *m*	75
espectrobologrаma *m*	espe(c)trobolograma *m*	спектроболограмма *f*	76
espectrógrafo solar *m*	espe(c)trógrafo solar *m*	солнечный спектрограф *m*	77
espectrógrafo horizontal *m*	espe(c)trógrafo horizontal *m*	горизонтальный спектрограф *m*	78
espectrógrafo vertical *m*	espe(c)trógrafo vertical *m*	вертикальный спектрограф *m*	79
cuña escalonada *f* atenuador escalonado *m*	redutor em escada *m* atenuador escalonado *m*	ступенчатый ослабитель *m*	80
espectroheliocinematógrafo *m*	espe(c)tro(h)eliocinematógrafo *m*	спектрогелиокинематограф *m*	81

82	solar magnetograph	magnétographe solaire *m*	Sonnenmagnetograph *m*
83	longitudinal magnetograph magnétographe vectoriel *m*	magnétographe longitudinal *m* magnétographe du champ longitudinal *m*	Longitudinalmagnetograph *m* Magnetograph für das Longitudinalfeld *m*
84	vector magnetograph	magnétographe (pour la détermination) du vecteur champ *m* magnétographe vectoriel *m*	Vektormagnetograph *m*
85	videomagnetograph	vidéomagnétographe *m*	Videomagnetograph *m*
86	electrooptic modulation	modulation électrooptique *f*	elektrooptische Modulation *f*
87	mica strip	lame de mica *f* bandelette de mica *f*	Glimmerstreifen *m*
88	beam-splitter beam splitting prism	diviseur [séparateur] de faisceau *m* prisme diviseur de faisceau *m* dédoubleur de faisceau *m*	Strahlenteilungsprisma *n m* Bildteiler(prisma) *m*(*n*) Prisma zur Bildteilung *n* Strahlteiler *m*
89	σ-component	composante σ *f*	σ-Komponente *f*
90	compensation for Doppler shifts	compensation de l'effet Doppler-Fizeau *f*	Kompensation für Doppler-Verschiebungen *f* Dopplerkompensation *f*
91	Doppler compensator	compensateur d'effet Doppler-Fizeau *m*	Doppler-Kompensator *m*
92	compensating plate	lame de compensation *f*	Kompensationsplatte *f*
93	magnetoheliograph Leighton's magnetoheliograph	magnétohéliographe *m*	Magnetoheliograph *m* Leighton-Magnetoheliograph *m*
94	photographic subtraction	soustraction photographique *f*	photographische Subtraktion *f*
95	radioheliograph	radiohéliographe *m*	Radioheliograph *m*
96	radioheliogram	radiohéliogramme *m*	Radioheliogramm *n*
97	radio interferometer	interféromètre radio(astronomique) *m*	Radiointerferometer *n*
98	two-element interferometer	interféromètre à deux antennes *m*	Zweielement-Interferometer *n*
99	aerial [antenna] spacing spacing of antennae [aerials]	espacement des antennes *m* distance des antennes *f*	Antennenabstand *m* Antennendistanz *f* Zwischenraum *m*
100	interferometer base	base de l'interféromètre *f*	Interferometerbasis *f*
101	zero spacing	espacement nul *m*	ohne Zwischenraum
102	fixed base-line fixed spacing	base fixe *f* espacement fixe *m*	feste Basislänge *f* fester Zwischenraum *m*

magnetógrafo solar *m*	magnetógrafo solar *m*	солнечный магнитограф *m*	82
magnetógrafo longitudinal *m* magnétografo del campo longitudinal *m*	magnetógrafo longidutinal *m* magnetógrafo do campo longidutinal *m*	магнитограф для продольной составляющей *m* магнитограф продольного поля *m*	83
magnetógrafo vectorial *m*	magnetógrafo ve(c)torial *m*	векторный магнитограф *m*	84
videomagnetógrafo *m*	videomagnetógrafo *m*	видеомагнитограф *m*	85
modulación electro-óptica *f*	modulação ele(c)troó(p)tica *f*	электрооптическая модуляция *f*	86
lámina de mica *f* tira de mica *f*	lamínula [lâmina] de mica *f* folha de mica *f*	полоска из слюды *f* полоска поляроидной мозаики *f*	87
divisor de haz *m* separador de haces *m* prisma divisor de haz *m* separador de haz *m*	separador do feixe luminoso *m* prisma separador do feixe luminoso *m*	светоделительная призма *f* призма разделяющая пучок *f* делитель пучка *m* расщепляющая пучок призма *f*	88
componente σ *m*	componente σ *m*	σ-компонент *m*	89
compensación para los desplazamientos Doppler *f*	compensação do efeito Doppler *f*	компенсация доп(п)леровских сдвигов *f*	90
compensador Doppler *m*	compensador Doppler *m*	компенсатор доп(п)леровских сдвигов *m* доп(п)леровский компенсатор *m*	91
placa compensadora *f*	lamínula [lâmina] de compensação *f*	компенсирующая пластинка *f*	92
magnetoheliógrafo *m* magnetoheliógrafo de Leighton *m*	magneto(h)eliógrafo *m* magneto(h)eliógrafo de Leighton *m*	магнитогелиограф *m* лейтоновский магнитогелиограф *m*	93
subtracción fotográfica *f*	subtração fotográfica *f*	фотографическое вычитание *n*	94
radioheliógrafo *m*	radioeliógrafo *m*	радиогелиограф *m*	95
radioheliograma *m*	radioeliograma *m*	радиогелиограмма *f*	96
radiointerferómetro *m*	radiointerferômetro *m*	радиоинтерферометр *m*	97
interferómetro de dos elementos *m*	interferômetro de dois elementos *m*	двухантенный [двухэлементный] интерферометр *m*	98
espaciamiento de antenas *m* separación entre antenas *f*	espaçamento entre antenas *m* distância de antenas *f*	расстояние между антеннами *n* межантенное расстояние *n* расстояние антенн *n*	99
base del interferómetro *f*	base do interferometro *f*	база [базис] интерферометра *f*	100
espaciamiento cero *m*	espaçamento nulo *m*	нулевое расстояние (антенн) *n*	101
línea-base fija *f* espaciamiento fijo *m*	(linha de) base fixa *f* espaçamento fixo *m*	неизменяющаяся база *f* постоянное расстояние антенн *n*	102

103	variable base-line variable spacing	base variable *f* espacement variable *m*	veränderliche Basislänge *f* veränderlicher Zwischenraum *m*
104	long-base-line interferometer long-base interferometer	interféromètre à grande base *m*	Langbasis-Interferometer *n*
105	sweep in position	balayage en position *m*	Schwenkung der Stellung *f*
106	sea interferometer cliff(-top) interferometer	interféromètre marin *m* interféromètre à une antenne *m*	Kliff-Interferometer *n*
107	variable base-line interferometer variable-spacing interferometer	interféromètre à base variable *m* interféromètre à espacement variable *m*	Interferometer mit veränderlicher Basis [Basislänge] *n*
108	movable element	antenne mobile *f*	bewegliches Element *n*
109	swept-lobe interferometer phase-swept interferometer	interféromètre à balayage de lobe [de phase] *m* interféromètre à lobe balayé *m*	Interferometer mit schwenkbarer Richtkeule *n* Swept-Lobe-Interferometer *n* Phasendrehinterferometer *n*
110	sweep-frequency interferometer	interféromètre à balayage de fréquence *m*	Interferometer mit schneller Frequenzänderung *n*
111	sweep in wavelength	balayage en longueur d'onde *n*	(schnelle) Wellenlängenänderung *f*
112	phase-switching	commutation de phase *f*	Phasenschaltung *f*
113	phase-switching interferometer	interféromètre à commutation de phase *m*	Interferometer mit Phasenschaltung *n*
114	phase-changing switch	changeur de phase *m*	Phasenumschalter *m*
115	phase shifter phase-changing unit	déphaseur *m*	Phasenschieber *m*
116	radio-link interferometer	interféromètre à liaison radioélectrique *m*	Interferometer mit Radioverbindung *n*
117	path difference difference in optical path	différence de marche (de rayons) *f* différence de chemin optique *f*	Wegunterschied *m* Gangunterschied *m* Gangdifferenz *f*
118	aperture synthesis	synthèse d'ouverture *f*	Apertursynthese *f*
119	synthesis system	système à synthèse d'ouverture *m*	Synthesesystem *n*

línea-base variable *f* espaciamiento variable *m*	(linha de) base variável *f* espaçamento variável *m*	переменная база *f* переменное расстояние (антенн) *n*	103
interferómetro de línea-base larga *m* interferómetro de larga base *m*	interferômetro de longa (linha de) base *m*	длиннобазисный интерферометр *m* интерферометр с большим базисом *m* интерферометр с большой базой *m*	104
barrido en posición *m*	varredura [varrimento] em posição *f*	развертка в положении *f* сканирование по небу *m*	105
interferómetro de acantilado *m* interferómetro marítimo *m*	interferômetro marítimo *m*	морской интерферометр *m* наскальный интерферометр *m*	106
interferómetro de base variable *m*	interferômetro de base variável *m*	интерферометр с переменным базисом радиоинтерферометр с переменной базой *m*	107
elemento movible *m*	elemento móvel *m*	подвижный элемент *m*	108
interferómetro de lóbulo móvil *m* interferómetro con barrido de lóbulo [de fase] *m*	interferômetro por [a] varredura de lóbulo *m* interferômetro por [a] varredura [varrimento] de [em] fase *m*	интерферометр с качающейся диаграммой *m* интерферометр с фазовой перестройкой *m*	109
interferómetro de barrido en frecuencia *m*	interferômetro a varredura [varrimento] em freqüência *m*	интерферометр с частотной развёрткой *m* интерферометр со сканированием по частоте *m*	110
barrido en longitud de onda *m*	varredura [varrimento] em comprimento de onda *f*	сканирование по длине волны *n*	111
conmutación de fase *f*	comutação de fase *m*	переключение фазы *m*	112
interferómetro de conmutación de fase *m*	interferômetro à comutação de fase *m*	интерферометр с переключением фазы *m* интерферометр Райла *m*	113
conmutador de fase *m*	conversor [comutador] de fase *m*	переключатель фазы *m*	114
desfasador *m* cambiador de fase *m*	unidade de mudança de fase *f* de(s)fasador *m*	фазовращатель *m* фазосдвигающее устройство *n*	115
interferómetro de enlace radioeléctrico *m*	interferômetro por ligação radioelé(c)trica *m*	интерферометр с радиосвязью *m*	116
diferencia de recorrido óptico *f* diferencia de camino óptico *f*	diferença de trajeto *f* diferença de caminho ó(p)tico *f*	разность хода лучей *f* разность путей *f*	117
síntesis de abertura *f*	síntese de abertura *f*	апертурный синтез *m* синтез апертуры *m*	118
sistema de síntesis *m*	sistema sintetizador [de sintese] *m* sistema de sintetização *m*	система апертурного синтеза *f*	119

	THE SUN	SOLEIL	SONNE
120	interference pattern interference image	système de franges d'interférence *m* figure d'interférence	Interferenzmuster *m* Interferenzfigur *f*
121	Mills'cross (aerial) pencil-beam interferometer	croix de Mills *f* interféromètre à lobe étroit *m* interféromètre en pinceau (fin) *m* interféromètre à pinceau *m*	Mills-Kreuz-Antenne *f* Millssches Kreuz *n* Strahlungempfänger mit extrem schmaler Richtkeule *m* scharfbündelndes [starkbündelndes] Interferometer *n*
122	fan-beam intereferometer	interféromètre à lobe en éventail *m* interféromètre à faisceau en éventail *m*	Fächerkeulen-Interferometer *n* Fächerbündel-Interferometer *n*
123	multiantenna interferometer multiwave interferometer multielement interferometer Christiansen interferometer grating interferometer	interféromètre à antennes multiples *m* interféromètre à réseau *m* interféromètre de Christiansen *m*	Gitterinterferometer *n* Gitter-Interferometer *n* Vielelement-Interferometer *n* Vielfachinterferometer *n*
124	dynamic spectrograph radio spectrograph radio spectrometer swept-frequency spectrograph solar radio spectroscope	spectrographe dynamique *m* radiospectrographe *m* spectromètre *m* spectrographe à balayage de fréquence *m*	dynamischer Spektrograph *m* Radiofrequenz-Spektrometer *n* Frequenzdrift-Spektrograph *m*
125	dynamic spectrum dynamic spectrogram	spectre dynamique *m* spectrogramme dynamique *m*	dynamisches Spektrum *n* dynamisches Spektrogramm *n*
126	quarter-wave plate Venetian blind $\lambda/4$ plate	grille quart d'onde *f* lame quart d'onde *f* vénitienne *f* lame $\lambda/4$ *f*	Viertelwellenlängenplatte *f* Viertelwellenplatte *f* $\lambda/4$- Platte *f*
127	rotation of the Sun solar rotation	rotation du Soleil *f* rotation solaire *f*	Rotation der Sonne *f* Sonnenrotation *f*
128	differential rotation	rotation différentielle *f*	differentielle Rotation *f*
129	period of solar rotation	période de la rotation solaire *f*	Rotationsperiode der Sonne *f*
130	isorotation	isorotation *f*	Isorotation *f*
131	equatorial acceleration	accélération équatoriale *f*	äquatoriale Beschleunigung *f*
132	recurrence	récurrence *f*	Wiederkehr *f*
133	return of a spot group	retour d'un groupe *m*	Wiederkehr der Gruppe *f*

sistema de franjas de interferencia *m* imagen [figura] de interferencia *f*	diagrama [figura] de interferência *m*[*f*] sistema de franjas de interferência *m*	интерференционная картина *f* интерференционное изображение *n*	120
(antena en) cruz de Mills *f* radiómetro de lóbulo estrecho *m* interferómetro de haz en pincel *m* interferómetro superdireccional *m*	antena cruzada Mills *f* interferômetro a feixe estreito *m* interferômetro de lóbulo estreito *m* interferômetro a feixe de lapiz *m*	интерферометр с карандашной диаграммой направленности *m* интерферометр с карандашным пучком *m* крестообразный интерферометр *m* крест Миллса *m*	121
interferómetro de lóbulo en abanico *m* interferómetro de haz en abanico *m*	interferômetro de lóbulo em leque *m* interferômetro produzindo feixe em leque *m*	интерферометр с веерообразной диаграммой направленности *m*	122
interferómetro de antenas múltiples *m* interferómetro de Christiansen *m* interferómetro de red *m*	interferômetro com antenas múltiplas *m* interferômetro de Christiansen *m* interferômetro a multi-elementos *m* interferômetro multionda *m* interferômetro em [à] rede *m*	многоэлементный (радио)интерферометр *m* многоантенный [многозеркальный] интерферометр *m* многолучевой интерферометр *m* (радио)интерферометр Христиансена *m*	123
radioespectrógrafo de barrido (de frecuencia) *m* radioespectrómetro *m* radioespectrógrafo (dinámico) *m* espectrómetro *m*	radioespe(c)trógrafo *m* espe(c)trógrafo dinâmico *m* radiospe(c)trômetro *m* espe(c)trômetro *m* radioespe(c)troscópio solar *m*	динамический спектрограф *m* динамический спектрометр *m* радиоспектрометр *m* солнечный радиоспектроскоп *m*	124
espectro dinámico *m* espectrograma dinámico *m*	espe(c)tro dinâmico *m* espe(c)trograma dinâmico *m*	динамический спектр *m* динамическая спектрограмма *f*	125
lámina de un cuarto de onda *f* lámina $\lambda/4$ *f* persiana veneciana *f* placa de un cuarto de onda *f*	lâmina (de um) quarto de onda *f* placa quarto de onda *f* janela veneziana *f*	пластинка в четверть волны *f* четвертьволновая пластика *f*	126
rotación del Sol *f* rotación solar *f*	rotação do Sol *f* rotação solar *f*	вращение Солнца *n* солнечное вращение *n*	127
rotación diferencial *f*	rotação diferencial *f*	дифференциальное вращение *n*	128
período de la rotación del Sol *m*	período da rotação solar *m*	период вращения Солнца *m*	129
isorrotación *f*	isorrotação *f*	изоротация *f*	130
aceleración ecuatorial *f*	aceleração equatorial *f*	экваториальное ускорение *n*	131
recurrencia *f*	recorrência *f*	рекуррентность *f* возвращение *n* возврат *m*	132
retorno de un grupo de manchas *m*	retorno de um grupo de manchas *m*	рекуррентность [возвращение] группы пятен *f*[*n*] возврат группы пятен *m*	133

134	solar radiation	rayonnement solaire *m*	Sonnenstrahlung *f*
135	direct solar radiation	rayonnement solaire direct *m*	direkte Sonnenstrahlung *f*
136	diffuse sky radiation	rayonnement diffus du ciel *m* rayonnement diffusé par le ciel *m*	Streustrahlung des Himmels *f*
137	Fraunhofer spectrum	spectre de Fraunhofer *m*	Fraunhoferspektrum *n*
138	photometric atlas	atlas photométrique *m*	photometrischer Atlas *m*
139	Fraunhofer line	raie de Fraunhofer *f*	Fraunhoferlinie *f*
140	D-line	raie D *f*	D-Linie *f*
140a	atmospheric line telluric line	raie atmosphérique *f* raie tellurique *f*	atmosphärische Spektrallinie *f* tellurische Spektrallinie *f*
141	brightness distribution	distribution [répartition] de la brillance *f*	Helligkeitsverteilung *f*
142	limb darkening	assombrissement au bord *m* assombrissement vers le bord *m*	Randverdunkelung *f*
143	law of darkening	loi d'assombrissement *f*	Randverdunkelungsgesetz *n*
144	coefficient of darkening	coefficient d'assombrissement *m*	Randverdunkelungskoeffizient *m*
145	centre-limb variation centre-to-limb variation	variation centre-bord *f*	Mitte-Rand-Variation *f*
146	total radiation	rayonnement total *m*	Gesamtstrahlung *f*
147	integrated radiation	rayonnement intégré *m*	integrierte Strahlung *f*
148	solar constant	constante solaire *f*	Solarkonstante *f*
149	outside the Earth's atmosphere	hors de l'atmosphère en dehors de l'atmosphère terrestre	ausserhalb der Erdatmosphäre
150	corpuscular radiation particle radiation	rayonnement corpusculaire *m*	Teilchenstrahlung *f* Partikelstrahlung *f* Korpuskularstrahlung *f*
151	particle emission	émission de particules *f*	Teilchenemission *f*
152	corpuscular emission	émission corpusculaire *f*	Korpuskularemission *f*
153	corpuscle	corpuscule *m*	Korpuskel *n*
154	particle	particule *f*	Teilchen *n* Partikel *f*

radiación solar *f*	radiação solar *f*	солнечное излучение *n* солнечная радиация *f* излучение Солнца *n*	134
radiación solar directa *f*	radiação solar direta *f*	прямая солнечная радиация *f* прямое солнечное излучение *n*	135
radiación difusa del cielo *f*	radiação difusa do céu *f*	диффузное излучение неба *n* рассеяное излучение неба *n*	136
espectro de Fraunhofer *m*	espe(c)tro de Fraunhofer *m*	фраунгоферов спектр *m*	137
atlas fotométrico *m*	atlas fotométrico *m*	фотометрический атлас *m*	138
línea de Franhofer *f*	linha [raia] de Fraunhofer *f*	фраунгоферова линия *f*	139
línea D *f*	linha [raia] D *f*	D-линия *f*	140
línea atmosférica *f* línea telúrica *f*	linha [raia] atmosférica *f* linha [raia] telúrica *f*	атмосферная (спектральная) линия *f* теллурическая (спектральная) линия *f*	140a
distribución de brillo *f*	distribuição de brilhância *f*	распределение яркости *n*	141
o(b)scurecimiento del borde *m* oscurecimiento del limbo *m*	obscurecimento de limbo [bordo] *m* escurecimento de limbo *m*	потемнение к краю (диска) *n* потемнение диска к краю *n*	142
ley de o(b)scurecimiento *f*	lei de obscurecimento *f*	закон потемнения *m*	143
coeficiente de o(b)scurecimiento *m*	coeficiente de obscurecimento *m*	коэффициент потемнения *m*	144
variación centro-limbo *f* variación centro-borde *f*	variação centro-limbo *f* variação centro-ao-limbo *f*	изменение от центра к краю *n*	145
radiación total *f*	radiação total *f*	полное излучение *m*	146
radiación integral *f*	radiação integrada *f*	интегральное излучение *n*	147
constante solar *f*	constante solar *f*	солнечная постоянная *f*	148
fuera de la atmósfera terrestre	fora da atmosfera terrestre exterior à atmosfera terrestre	за пределами земной атмосферы вне земной атмосферы	149
radiación corpuscular *f*	radiação coruscular *f* radiação de partículas *f*	корпускулярное излучение *n* корпускулярная радиация *f*	150
emisión de partículas *f*	emissão de partículas *f*	испускание [эмиссия] частиц *f*	151
emisión corpuscular *f*	emissão corpuscular *f*	корпускулярная эмиссия *f* корпускулярное излучение *n*	152
corpúsculo *m*	corpúsculo *m*	корпускула *f*	153
partícula *f*	partícula *f*	частица *f*	154

	THE SUN	SOLEIL	SONNE
155	corpuscular	corpusculaire	korpuskular
156	solar wind	vent solaire *m*	Sonnenwind *m* solarer Wind *m*
157	solar-wind density	densité du vent solaire *f*	Dichte des Sonnenwindes *f*
158	solar-wind velocity	vitesse du vent solaire *f*	Geschwindigkeit des Sonnenwindes *f*
159	hydromagnetic expansion	expansion hydromagnétique *f*	hydromagnetische Expansion *f*
160	corpuscular stream	courant corpusculaire *m*	Korpuskularstrom *m*
161	ejection of particles from (the Sun)	éjection solaire de particules *f* éjection de particules (du Soleil) *f*	Partikelauswurf (aus der Sonne) *m* Partikelejektion *f*
162	burst of particles	émission brusque de particules *f*	Teilchenausbruch *m* Partikelausbruch *m*
163	solar atmosphere atmosphere of the Sun	atmosphère solaire *f* atmosphère du Soleil *f*	Sonnenatmosphäre *f* Atmosphäre der Sonne *f*
164	solar surface	surface solaire *f*	Sonnenoberfläche *f*
165	outer solar atmosphere higher layers of the solar atmosphere	couches élevées de l'atmosphère solaire *fpl* couches externes [extérieures] de l'atmosphère solaire *fpl*	höhere Schichten der Sonnenatmosphäre *fpl* äussere Sonnenatmosphäre *f*
166	lower solar atmosphere lower layers of the solar atmosphere	couches basses de l'atmosphère solaire *fpl* couches profondes de l'atmosphère solaire *fpl*	tiefere Schichten der Sonnenatmosphäre *fpl* tiefere Sonnenatmosphäre *f*
167	photosphere	photosphère *f*	Photosphäre *f*
168	subphotospheric layer	couche subphotosphérique *f*	subphotosphärische Schicht *f*
169	undisturbed photosphere quiet photosphere	photosphère non-perturbée *f* photosphère calme *f*	ungestörte Photosphäre *f* ruhige Photosphäre *f*
170	photospheric	photosphérique	photosphärisch
171	visible disc [disk]	disque (en lumière) visible *m*	sichtbare Scheibe *f*
172	center [centre] of the visible disc [disk]	centre du disque visible *m*	Zentrum der sichtbaren Scheibe *n*
173	on the disc [disk]	sur le disque	auf der Scheibe
174	reversing layer	couche renversante *f*	umkehrende Schicht *f*
175	absorbing layer	couche absorbante *f*	absorbierende Schicht *f*
176	photospheric granulation	granulation photosphérique *f*	photosphärische Granulation *f*
177	intergranular area intergranular region	région intergranulaire *f*	intergranulares Gebiet *n*

corpuscular	corpuscular	корпускулярный	155
viento solar *m*	vento solar *m*	солнечный ветер *m*	156
densidad del viento solar *f*	densidade do vento solar *f*	плотность солнечного ветра *f*	157
velocidad del viento solar *f*	velocidade do vento solar *f*	скорость солнечного ветра *f*	158
expansión hidromagnética *f*	expansão hidromagnética *f*	гидромагнитное расширение *n*	159
corriente corpuscular *f*	torrente [corrente] corpuscular *f*	корпускулярный поток *m* поток корпускул *m*	160
eyección de partículas (por el Sol) *f*	ejeção de partículas (do Sol) *f*	испускание частиц (из Солнца) *n* эжекция частиц (из Солнца) *f*	161
emisión (brusca) de partículas *f* tormenta de partículas *f*	explosão de partículas *f* emissão brusca de partículas *f*	(внезапный) выброс частиц *m*	162
atmósfera solar *f* atmósfera del Sol *f*	atmosfera solar *f* atmosfera do Sol *f*	солнечная атмосфера *f* атмосфера Солнца *f*	163
superficie solar *f*	superfície solar *f*	поверхность Солнца *f*	164
capas superiores de la atmósfera solar *fpl*	atmosfera solar exterior *f* camadas (mais) elevadas da atmosfera solar *fpl*	верхние слои солнечной атмосферы *mpl*	165
capas bajas de la atmósfera solar *fpl*	atmosfera solar inferior *f* camadas mais baixas da atmosfera solar *fpl*	нижние слои солнечной атмосферы *mpl*	166
fotosfera *f*	fotosfera *f*	фотосфера *f*	167
capa subfotosférica *f*	camada subfotosférica *f*	подфотосферный слой *m*	168
fotosfera no perturbada *f* fotosfera tranquila [inactiva] *f*	fotosfera não-perturbada *f* fotosfera calma *f*	невозмущённая фотосфера *f* спокойная фотосфера *f*	169
fotosférico	fotosférico	фотосферный	170
disco visible *m*	disco visível *m*	видимый диск *m*	171
centro del disco visible *m*	centro do disco visível *m*	центр видимого диска *m*	172
en [sobre] el disco	no disco	на диске	173
capa inversora [de inversión] *f*	camada de inversão *f*	обращающий слой *m*	174
capa absorbente *f*	camada absorvente *f*	поглощающий слой *m*	175
granulación fotosférica *f*	granulação fotosférica *f*	фотосферная грануляция *f*	176
región intergranular *f* área intergranular *f*	região intergranular *f* área intergranular *f*	межгранульная область *f* межгранульный промежуток *m*	177

	THE SUN	SOLEIL	SONNE
178	granule granular element	granule *m*	Granulum *n* Granule *f*
179	facular granule	granule faculaire *m*	Fackelgranule *f* Fackelgranulum *n*
180	umbral granule	granule dans l'ombre *m* granule d'ombre *m*	Umbragranule *f* Umbragranulum *n*
181	photospheric granule	granule photosphérique *m*	photosphärisches Granulum *n* photosphärische Granule *f*
182	supergranule supergranular element	supergranule *m*	Supergranulum *n* Supergranule *f* supergranulares Element *n*
183	supergranulation	supergranulation *f*	Supergranulation *f*
184	horizontal motion	mouvement horizontal *m*	Horizontalbewegung *f*
185	vertical motion	mouvement vertical *m*	Vertikalbewegung *f*
186	five-minute oscillations	oscillations (à période) de 5 minutes *fpl*	Fünf-Minuten-Oszillation *f*
187	velocity field	champ de vitesses *m*	Geschwindigkeitsfeld *n*
188	small-scale motion	mouvement à petite échelle *m*	Mikrobewegung *f*
189	large-scale motion	mouvement à grande échelle *m*	Makrobewegung *f*
190	inclined inhomogeneity	inhomogénéité oblique *f*	geneigte Inhomogenität *f*
191	chromosphere	chromosphère *f*	Chromosphäre *f*
192	upper chromosphere	chromosphère supérieure *f*	höhere Chromosphäre *f*
193	lower chromosphere	basse chromosphère *f*	untere Chromosphäre *f*
194	chromospheric	chromosphérique	chromosphärisch
195	spicule	spicule *f*	Spikulum *n* Spikule *f*
196	two-component model	modèle à deux composantes *m*	Zwei-Komponenten-Modell *n*
197	two-temperature model (of the chromosphere)	modèle à deux températures *m*	Zwei-Temperaturen-Modell (der Chromosphäre) *n*
198	interspicular region	région interspiculaire *f*	interspikulares Gebiet *n*
199	cold region	région froide *f*	kaltes Gebiet *n*
200	hot region	région chaude *f*	heisses Gebiet *n*
201	gravity wave	onde de gravité *f*	Schwerewelle *f*

SOL	SOL	СОЛНЦЕ	
gránulo *m* elemento [célula] granular *m*[*f*]	grânulo *m* elemento granular *m*	гранула *f*	178
gránulo facular *m*	grânulo facular *m*	факельная гранула *f*	179
gránulo de la umbra *m* gránulo umbral *m*	grânulo da umbra *m*	гранула тени *f*	180
gránulo fotosférico *m*	grânulo fotosférico *m*	фотосферная гранула *f*	181
supergranulo *m* elemento supergranular *m* célula supergranular *f*	supergrânulo *m* elemento supergranular *m*	сверхгранула *f* супергранула *f* супергранулярный [супергрануляционный] элемент *m*	182
supergranulación *f*	supergranulação *f*	супергрануляция *f*	183
movimiento horizontal *m*	movimento horizontal *m*	горизонтальное движение *n*	184
movimiento vertical *m*	movimento vertical *m*	вертикальное движение *n*	185
oscilaciones de 5 minutos *fpl*	oscilacões de cinco minutos *fpl*	пятиминутные колебания *npl* пятиминутные осцилляции *fpl*	186
campo de velocidades *m*	campo de velocidades *m*	поле скоростей *n*	187
movimiento en [a] pequeña escala *m*	movimento de pequena escala *m*	мелкомасштабное движение *n*	188
movimiento en [a] gran escala *m*	movimento de grande escala *m*	крупномасштабное движение *n*	189
inhomogeneidad inclinada *f*	heterogeneidade inclinada *f*	наклонная неоднородность *f*	190
cromosfera *f*	cromosfera *f*	хромосфера *f*	191
alta cromosfera *f*	alta cromosfera *f*	верхняя хромосфера *f*	192
baja cromosfera *f*	baixa cromosfera *f*	нижняя хромосфера *f*	193
cromosférico	cromosférico	хромосферный	194
espícula *f*	espícula *f*	спикула *f*	195
modelo de dos componentes *m*	modelo de dois componentes *m*	двухкомпонентная модель *f*	196
modelo de dos temperaturas (de la cromósfera) *m*	modelo de duas temperaturas (da cromosfera) *m*	двухтемпературная модель (хромосферы) *f*	197
región interespicular *f*	região interspicular *f*	межспикульная область *f* межспикульный промежуток *m*	198
región fría *f*	região fria *f*	холодная область *f*	199
región caliente *f*	região quente *f*	горячая область *f*	200
onda de gravedad *f*	onda de gravidade *f*	волна тяготения *f*	201

	THE SUN	SOLEIL	SONNE
202	height of the chromosphere	hauteur de la chromosphère *f*	Höhe der Chromosphäre *f*
203	chromospheric fine structure	structure fine de la chromosphère *f*	chromosphärische Feinstruktur *f*
204	spatial resolution	résolution spatiale *f*	räumliche Auflösung *f*
205	time resolution	résolution temporale *f*	zeitliche Auflösung *f* Zeitauflösung *f*
206	macroscopic inhomogeneities	inhomogénéités macroscopiques *fpl*	makroskopische Inhomogenitäten *fpl*
207	large-scale structure	structure à grande échelle *f*	Makrostruktur *f* makroskopische Struktur *f*
208	small-scale structure	structure à petite échelle *f*	mikroskopische Struktur *f* Feinstruktur *f*
209	fibrils chromospheric striation	fibrilles *fpl* fibres de la chromosphère *fpl* petites fibres chromosphériques *fpl*	fadenähnliche Strukturen *fpl* Fäden *fpl* Fadenstruktur *f* chromosphärische Fasern *fpl* Fibrillen *fpl*
210	mottles	tachetures *fpl* structures élémentaires chromosphériques *fpl*	Mottles *pl*
211	bright mottles	tachetures brillantes *fpl*	helle Mottles *pl*
212	dark mottles	tachetures sombres *fpl*	dunkle Mottles *pl*
213	heating of the chromosphere	chauffage de la chromosphère *m*	Aufheizung der Chromosphäre *f*
213a	heating of the corona	chauffage de la couronne *m*	Aufheizung der Korona *f*
214	heating mechanism	mécanisme de chauffage *m*	Aufheizungsmechanismus *m*
215	generation of sound	production d'ondes sonores *f*	Schallerzeugung *f*
216	energy generation	production d'énergie *f*	Energieerzeugung *f*
217	energy propagation	propagation de l'énergie *f*	Energieausbreitung *f*
218	non-radiative energy	énergie non-radiative *f*	andere Energieform als Strahlung *f* Nichtstrahlungsenergie *f*
219	acoustic energy sound energy	énergie acoustique *f* énergie sonore *f*	akustische Energie *f* Schallenergie *f*
220	acoustic [sound] noise	bruit acoustique *m*	akustisches Rauschen *n*
221	flux of acoustic energy sound energy flux	flux d'énergie acoustique *m*	Schallenergiefluss *m* Fluss akustischer Energie *m*
222	dissipation of waves	dissipation d'ondes *f*	Wellendissipation *f*
223	support of the chromosphere	support de la chromosphère *m* soutien de la chromosphère *m*	Tragen der Chromosphäre *n* Aufrechterhaltung der Chromosphäre *f*

SOL	SOL	СОЛНЦЕ	
altura de la cromosfera *f*	altura da cromosfera *f*	высота хромосферы *f*	202
estructura fina de la cromosfera *f*	estrutura fina da cromosfera *f*	тонкая структура хромосферы *f*	203
resolución espacial *f*	resolucão espacial *f*	пространственное разрешение *n*	204
resolución temporal *f* resolución en el tiempo *m*	resolução temporal *f*	временное разрешение *n*	205
inhomogeneidades macroscópicas *fpl*	inomogeneidades macroscópicas *fpl*	макроскопические неоднородности *fpl*	206
estructura en [a] gran escala *f*	estrutura de grande escala *f*	крупномасштабная структура *f*	207
estructura en pequeña escala *f*	estrutura de pequena escala *f*	мелкомасштабная структура *f*	208
fibrillas (cromosféricas) *fpl* estrías cromosféricas *fpl* estriado cromosférico *m*	fibrilas *fpl* estriação cromosférica *f* estrias cromosféricas *fpl* estrutura estriada [fibrilar] *f*	хромосферные волокна *npl* фибриллы *fpl* нитеподобные структуры *fpl*	209
motas *fpl* moteado *m*	pintas *fpl* jaspes *mpl*	узелки *mpl*	210
motas brillantes *fpl*	pintas [jaspes] brilhantes *f[m]pl*	яркие узелки *mpl*	211
motas o(b)scuras *fpl*	jaspes escuros [obscuros] *mpl* pintas escuras [obscuras] *fpl*	тёмные узелки *mpl*	212
calentamiento de la cromosfera *m*	aquecimento da cromosfera *m*	нагрев хромосферы *m*	213
calentamiento de la corona *m*	aquecimento da coroa *m*	нагрев короны *m*	213b
mecanismo de calentamiento *m*	mecanismo de aquecimento *m*	механизм нагрева *m*	214
generación de sonido *f*	produção de som *f*	генерация звука *f*	215
generación de energía *f*	produção de energia *f*	генерация энергии *f*	216
propagación de la energía *f*	propagação de energia *f*	распространение энергии *n*	217
energía no radiativa *f*	energia não-radiativa *f*	нелучевая энергия *f* неизлучаемая энергия *f*	218
energía acústica *f*	energia acústica *f*	звуковая энергия *f* акустическая энергия *f*	219
ruido acústico *m*	ruído acústico *m*	акустический шум *m*	220
flujo de energía acústica *m*	fluxo de energia acústica *m* fluxo de energia sonora *m*	поток звуковой энергии *m* поток акустической энергии *m*	221
disipación de ondas *f*	dissipação de ondas *f*	диссипация волн *f*	222
soporte de la cromosfera *m*	suporte da cromosfera *m*	поддержание хромосферы *n*	223

	THE SUN	SOLEIL	SONNE
223a	solar oscillations	oscillations solaires *fpl*	Sonnenschwingungen *fpl* Oszillationen der Sonne *fpl*
223b	solar seismology helioseismology	sismologie solaire *f* helios(é)ismologie *f*	Sonnenseismologie *f* Helioseismologie *f*
224	transition region transition layer chromosphere-corona transition region	région de transition *f* zone [région] de transition chromosphère-couronne *f*	Übergangsschicht *f*
225	solar corona	couronne solaire *f*	Sonnenkorona *f*
226	eclipse observation	observation pendant [d'] une éclipse *f*	Finsternisbeobachtung *f*
227	out-of-eclipse observation	observation en dehors d'une éclipse *f*	Beobachtung ausserhalb von Finsternissen *f*
227a	artificial eclipse	éclipse artificielle *f*	künstliche Finsternis *f*
228	coronal	coronal	Korona- koronal
229	inner corona	couronne interne *f*	innere Korona *f*
230	middle corona	couronne moyenne *f*	mittlere Korona *f*
231	outer corona	couronne extérieure *f*	äussere Korona *f*
232	expanding corona	couronne en expansion *f*	expandierende Korona *f*
233	electron density	densité électronique *f*	Elektronendichte *f*
234	emission measure	mesure d'émission *f*	Emissionsmass *n*
235	Abel equation	équation d'Abel *f*	Abelsche Gleichung *f*
236	scale height	échelle de hauteur *f*	Skalenhöhe *f* Äquivalenthöhe *f*
237	white-light corona white corona	couronne blanche *f*	weisse Korona *f*
238	monochromatic corona	couronne monochromatique *f*	monochromatische Korona *f*
239	radio corona	couronne radio *f*	Radiokorona *f*
240	X-ray corona Röntgen corona	couronne de rayons X *f* courone (dans le domaine) X *f*	Röntgenkorona *f* X-Strahlen-Korona *f*
241	Fraunhofer corona F corona	couronne de Fraunhofer *f* couronne F *f*	Fraunhoferkorona *f* F-Korona *f*
241a	emission corona E corona	couronne d'émission *f* couronne E *f*	Emissionskorona *f* E-Korona *f*

SOL	SOL	СОЛНЦЕ	
oscilaciones solares *fpl*	oscilações solares *fpl*	солнечные колебания *npl* солнечные осцилляции *fpl*	223a
sismología solar *f* heliosismología *f*	sismologia solar *f* heliosismologia *f*	солнечная сейсмология *f* гелиосейсмология *f*	223b
región de transición *f* región de transición entre cromosfera y corona *f*	região de transição *f* região de transição cromosfera-coroa *f* camada de transição *f*	переходная область *f* переходный слой (между хромосферой и короной) *m*	224
corona solar *f*	coroa solar *f*	солнечная корона *f*	225
observación durante el eclipse *f*	observação de eclipse *f*	затменные наблюдения *npl*	226
observación fuera del eclipse *f*	observação fora do eclipse *f*	внезатменное наблюдение *n*	227
eclipse artificial *m*	eclipse artificial *m*	искусственное затмение *n*	227a
coronal	coronal	корональный	228
corona interior [interna] *f*	coroa interna *f*	внутренняя корона *f*	229
corona media *f*	coroa média *f*	средняя корона *f*	230
corona exterior *f*	coroa externa *f*	внешняя корона *f*	231
corona en expansión *f*	coroa em expansão *f*	расширяющаяся корона *f*	232
densidad electrónica *f*	densidade ele(c)trônica *f*	электронная плотность *f*	233
medida de la emisión *f*	medida de emissão *f*	мера эмиссии *f*	234
ecuación de Abel *f*	equação de Abel *f*	уравнение Абеля *n* абелево уравнение *n*	235
altura de escala *f*	altura de escala *f*	шкала высот *f* высота однородной атмосферы *f*	236
corona blanca *f* corona de luz blanca *f*	coroa branca *f* coroa em luz branca *f*	белая корона *f* корона в белом свете *f*	237
corona monocromática *f*	coroa monocromática *f*	монохроматическая корона *f*	238
corona radio(eléctrica) *f*	radiocoroa *f*	радиокорона *f*	239
corona en rayos X *f* corona Roentgen *f*	coroa de raios X *f* coroa Roentgen *f*	рентгеновская корона *f* корона в рентгеновских лучах *f*	240
corona de Fraunhofer *f* corona F *f*	coroa de Fraunhofer *f* coroa F *f*	фраунгоферова корона *f* F-корона *f* F-компонента короны *f*	241
corona de emisión *f* corona E *f*	coroa de emissão *f* coroa E *f*	эмиссионная (линейчатая) корона *f* излучающая корона *f* E-корона *f*	241a

242	K corona electron corona	couronne K *f* couronne électronique *f*	K-Korona *f* Elektronenkorona *f*
243	smearing of Fraunhofer lines washing out of Franhofer lines	effacement de raies de Fraunhofer *m* dilution de raies de Fraunhofer *f* raies devenant floues *fpl*	Verwischung der Fraunhoferlinien *f* Verschmierung der Fraunhoferlinien *f*
244	maximum corona	couronne de maximum *f*	Maximumskorona *f*
245	intermediate corona	couronne intermédiaire *f*	Übergangstyp der Korona *m*
245a	coronal hole	trou coronal *m*	koronales Loch *n*
246	minimum corona	couronne de minimum *f*	Minimumskorona *f*
247	flattening of the corona	aplatissement de la couronne *f*	Abplattung der Korona *f*
248	structural feature structural formation	caractère structural *m* structure caractéristique *f*	Strukturmerkmal *n* Strukturelement *m*
249	arch structure	structure en arche *f*	Bogenstruktur *f*
250	mass balance	bilan de masse *m*	Massenbilanz *f*
251	energy balance	bilan d'énergie *m* bilan énergétique *m*	Energiebilanz *f*
252	radiation balance	bilan de radiation *m*	Strahlungsgleichgewicht *n*
253	energy gain	gain d'énergie *m*	Energiegewinn *n*
254	energy loss	perte d'énergie *f*	Energieverlust *m* Energieabgabe *f*
255	heliosphere	héliosphère *f*	Heliosphäre *f*
256	heliospheric	héliosphérique	heliosphärisch
257	heliospheric cavity	cavité hélisphérique *f*	heliosphärischer Hohlraum *m*

corona K *f* corona de electrones *f*	coroa K *f* coroa de elé(c)trons *f*	К-корона *f* электронная корона *f*	242
desdibujamiento de las líneas de Fraunhofer *m*	extenuação das linhas de Fraunhofer *f*	размазывание фраунгоферовых линий *n*	243
corona de máximo *f* corona máxima *f*	coroa de máximo *f* coroa máxima *f*	корона в период максимума *f* максимальная корона *f*	244
corona intermedia *f* corona de tipo intermedio *f*	coroa intermediária *f*	промежуточная корона *f*	245
agujero coronal *m*	buraco coronal *m*	корональная дыра *f*	245a
corona de mínimo *f* corona mínima *f*	coroa mínima *f* coroa de mínimo *f*	корона в период минимума *f* минимальная корона *f*	246
achatamiento de la corona *m*	achatamento da coroa *m*	сжатие короны *n* сплюснутость короны *f*	247
detalle estructural *m* característica estrutural *f*	formação [detalhe, pormenor] estrutural *f*[*m*] estrutura característica *f*	структурное образование *n*	248
estructura en arco *f*	estrutura em arco *f*	арочная структура *f*	249
equilibrio [balance] de masa *m*	equilíbrio de massa *m*	баланс массы *m*	250
equilibrio energético *m* balance de energía *m*	equilíbrio de energia *m* equilíbrio energético *m*	энергетический баланс *m* баланс энергии *m*	251
balance de radiación *m* equilibrio de radiación *m*	equilíbrio de radiação *m* equilíbrio radiativo *m*	радиационный баланс *m*	252
ganancia de energía *f*	ganho de energia *m*	выигрыш энергии *m*	253
pérdida de energía *f*	perda de energia *f*	потеря энергии *f*	254
heliosfera *f*	heliosfera *f*	гелиосфера *f*	255
heliosférico	heliosférico	гелиосферный	256
cavidad heliosférica *f*	cavidade heliosférica *f*	гелиосферная полость *f*	257

89	SOLAR ACTIVITY	ACTIVITÉ SOLAIRE	SONNENAKTIVITÄT
1	solar activity	activité solaire *f*	Sonnenaktivität *f* Sonnentätigkeit *f*
2	solar maximum maximum of solar activity	maximum solaire *m* maximum de l'activité solaire *m*	Sonnenmaximum *n* Maximum der Sonnentätigkeit (Sonnenaktivität) *n*
3	optical activity	activité optique *f* manifestation optique de l'activité *f*	optische Aktivität *f* Aktivität im optischen Spektralbereich *f*
4	solar minimum minimum of solar activity	minimum solaire *m* minimum de l'activité solaire *f*	Sonnenminimum *n* Minimum der Sonnenaktivität (-tätigkeit) *n*
5	magnetic activity	activité (du champ) magnétique *f*	magnetische Aktivität *f* Aktivität des Magnetfeldes *f*
6	radio activity activity in radio waves	activité radioélectrique *f* manifestation radio de l'activité *f*	Aktivität im Radio(wellen)gebiet *f*
7	corpuscular activity	activité corpusculaire *f* manifestation corpusculaire de l'activité *f*	korpuskulare Aktivität *f*
8	index of solar activity solar index	indice d'activité solaire *m* indice solaire *m*	Index der Sonnenaktivität *m* Sonnenindex *m*
9	solar data	données solaires *fpl*	Sonnendaten *npl*
10	Quiet Sun Year Year of the Quiet Sun	Année du Soleil Calme *f*	Jahr der ruhigen Sonne *n*
11	quiet Sun undisturbed Sun	Soleil calme *m* Soleil non-perturbé *m*	ruhige Sonne *f* ungestörte Sonne *f*
12	active Sun disturbed Sun	Soleil actif *m* Soleil perturbé *m*	aktive Sonne *f* gestörte Sonne *f*
13	solar phenomenon solar feature	phénomène solaire *m* structure solaire *f* manifestation de l'activité solaire *f*	Sonnenphänomen *n* Sonnenerscheinung *f*
14	active feature	phénomène actif *m*	aktive Erscheinung *f* aktives Phänomen *n*
15	short-lived phenomenon	phénomène de courte durée *m*	kurzlebige Erscheinung *f*
16	solar survey	surveillance [observation systématique] du Soleil *f* patrouille [service] du Soleil *f*[*m*]	Sonnenüberwachung *f*
17	heliographic map heliographic chart	carte héliographique *f*	heliographische Karte *f*
18	daily map (of solar activity) daily solar map daily map of the Sun	carte journalière du Soleil *f*	Tageskarte der Sonne *f*

actividad solar *f*	atividade solar *f*	солнечная активность *f* активность Солнца *f*	1
máximo solar *m* máximo de la actividad solar *m*	máximo solar *m* máximo da atividade solar *m*	солнечный максимум *m* максимум солнечной активности *m*	2
actividad óptica *f*	atividade ó(p)tica *f*	активность в оптическом диапазоне *f*	3
mínimo solar *m* mínimo de la actividad solar *m*	mínimo solar *m* mínimo da atividade solar *m*	солнечный минимум *m* минимум солнечной активности *m*	4
actividad magnética *f*	atividade magnética *f*	магнитная активность *f* активность магнитных полей *f*	5
actividad radioeléctrica *f* actividad en radiación radio *f*	atividade rádio *f* atividade em ondas radioelé(c)tricas *f*	активность в радиодиапазоне [в радиоволнах] *f*	6
actividad corpuscular *f*	atividade corpuscular *f*	корпускулярная активность *f*	7
índice de actividad solar *m* índice solar *m*	índice de atividade solar *m* índice solar *m*	индекс солнечной активности *m* солнечный индекс *m*	8
datos solares *mpl*	dados solares *mpl*	солнечные данные *pl*	9
Año del Sol calmo [quieto, tranquilo, en calma, inactivo] *m*	Ano do Sol calmo *m* Ano do Sol quieto *m*	Год Спокойного Солнца *m*	10
Sol calmo [quieto, tranquilo, no perturbado, inactivo] *m* Sol en calma *m*	Sol calmo *m* Sol quieto [não-perturbado] *m*	спокойное Солнце *n* невозмущённое Солнце *n*	11
Sol. activo *m* Sol perturbado *m*	Sol ativo *m* Sol perturbado *m*	активное Солнце *n* возмущённое Солнце *n*	12
fenómeno solar *m*	fenômeno solar *m* característica solar *f*	солнечное явление [образование] *n* проявление солнечной активности *n*	13
fenómeno activo *m*	fenômeno ativo *m* característica ativa *f*	активное образование *n* активное явление *n*	14
fenómeno de vida corta *m* fenómeno de corta duración *m*	fenômeno de curta duração *m*	короткоживущее явление [образование] *n*	15
patrullaje solar *m* observación regular del Sol *f*	patrulha do Sol *f* observação regular do Sol *f*	служба Солнца *f*	16
mapa heliográfico *m*	carta heliográfica *f*	гелиографическая карта *f*	17
mapa diario del Sol *m*	carta diária do Sol *f*	ежедневная карта Солнца *f*	18

19	polar map	carte polaire *f*	Polarkarte *f*
20	synoptic map of the photosphere	carte synoptique de la photosphère *f*	synoptische Karte der Photosphäre *f* synoptische Photosphärenkarte *f*
21	synoptic map of the chromosphere	carte synoptique de la chromosphère *f*	synoptische Karte der Chromosphäre *f* synoptische Chromosphärenkarte *f*
22	synoptic map of the corona	carte synoptique de la couronne *f*	synoptische Karte der Korona *f*
23	velocity map	carte de vitesses *f*	Geschwindigkeitskarte *f*
24	solar disc solar disk	disque solaire *m*	Sonnenscheibe *f*
25	on the disk [disc]	sur le disque	auf der Scheibe
26	optical disk [disc] visible disk [disc]	disque visible *m*	sichtbare (Sonnen)scheibe *f*
27	visible hemisphere	hemisphère visible *m*	sichtbare Halbkugel *f* sichtbare Hemisphäre *f*
28	solar limb limb of the Sun	bord (du disque) solaire *m* limbe solaire *m* bord apparent solaire *m*	Sonnenrand *m* Rand der Sonne *m*
29	on the limb	au bord	am Rand
30	above the limb	au-dessus du bord	über dem Rand
31	limb passage	passage au bord *m*	Randübertritt *m*
32	extreme limb	extrême bord *m*	aüsserster Rand *m*
33	west limb	bord ouest *m*	Westrand *m*
34	east limb	bord est *m*	Ostrand *m*
35	centre [center] of the (solar) disk [disc]	centre du disque (solaire) *m*	Mitte der Sonnenscheibe *f* Scheibenmitte *f*
36	distance from the centre [center] of the disk [disc]	distance au centre du disque *f*	Abstand von der Scheibenmitte [Sonnenmitte] *m*
37	heliocentric angle	angle héliocentrique *m*	heliozentrischer Winkel *m*
38	heliographic position	position héliographique *f*	heliographische Lage *f*
39	heliographic coordinates	coordonnées héliographiques *fpl*	heliographische Koordinaten *fpl*
40	heliographic longitude	longitude héliographique *f*	heliographische Länge *f*
41	active longitude	longitude active *f*	aktive Länge *f*

mapa polar *m*	carta polar *f*	полярная карта *f*	19
mapa sinóptico de la fotosfera *m*	carta sinótica da fotosfera *f*	синоптическая карта фотосферы *f*	20
mapa sinóptico de la cromosfera *m*	carta sinótica da cromosfera *f*	синоптическая карта хромосферы *f*	21
mapa sinóptico de la corona *m*	carta sinótica da coroa *f*	синоптическая карта короны *f*	22
mapa de velocidades *f*	carta das velocidades *f*	карта скоростей *f*	23
disco solar *m* disco del Sol *m*	disco solar *m* disco do Sol *m*	солнечный диск *m* диск Солнца *m*	24
sobre el disco	no disco	на диске	25
disco visible *m*	disco visível *m*	видимый диск *m* оптический диск *m*	26
hemisferio visible *m*	hemisfério visível *m*	видимое полушарие *n* видимая полусфера *n*	27
limbo solar *m* limbo del Sol *m*	limbo [bordo] solar *m* limbo [bordo] do Sol *m*	край (диска) Солнца *m* солнечный край [лимб] *m* край солнечного диска *m*	28
sobre el limbo	no limbo [bordo]	на краю на лимбе	29
encima del limbo	acima do limbo [bordo]	над краем над лимбом	30
paso por [sobre] el limbo *m*	passagem no limbo [bordo] *m*	прохождение через край диска *n*	31
limbo extremo *m*	limbo [bordo] extremo *m*	внешний край *m* экстремальный край [лимб] *m*	32
limbo oeste *m*	limbo [bordo] oeste [do oeste] *m*	западный край [лимб] *m*	33
limbo este *m*	limbo [bordo] leste [do leste] *m*	восточный край [лимб] *m*	34
centro del disco (solar) *m*	centro do disco (solar) *m*	центр (солнечного) диска *m*	35
distancia al centro del disco *f*	distância ao centro do disco *f*	расстояние от центра диска *n* гелиоцентрическое расстояние *n*	36
ángulo heliocéntrico *m*	angulo heliocêntrico *m*	гелиоцентрический угол *m*	37
posición heliográfica *f*	posição heliográfica *f*	гелиографическое положение *n*	38
coordenadas heliográficas *fpl*	coordenadas heliográficas *fpl*	гелиографические координаты *fpl*	39
longitud heliográfica *f*	longitude heliográfica *f*	гелиографическая долгота *f* гелиодолгота *f*	40
longitud activa *f*	longitude ativa *f*	активная долгота *f*	41

	SOLAR ACTIVITY	ACTIVITÉ SOLAIRE	SONNENAKTIVITÄT
42	heliographic latitude	latitude héliographique *f*	heliographische Breite *f*
43	central meridian	méridien central *m*	Zentralmeridian *m*
44	cross the central meridian	passer au méridien central	durch den Zentralmeridian gehen
45	C. M. P. central-meridian passage	passage au méridien central *m*	Durchgang durch den Zentralmeridian *m* Zentralmeridiandurchgang *m*
46	distance from the central meridian	distance au méridien central *f*	Abstand vom Zentralmeridian *m*
47	central zone	zone centrale *f*	Zentralzone *f*
48	(heliographic) longitude distribution distribution in (heliographic) longitude	distribution en longitude (héliographique) *f* distribution longitudinale *f*	Verteilung nach (heliographischer) Länge *f* Längenverteilung *f*
49	east-west asymmetry	asymétrie [dissymétrie] est-ouest *f*	Ost-West-Asymmetrie *f*
50	north-south asymmetry	asymétrie [dissymétrie] nord-sud *f*	Nord-Süd-Asymmetrie *f*
51	(heliographic) latitude distribution distribution in (heliographic) latitude	distribution en latitude (héliographique) *f* distribution latitudinale *f*	Verteilung nach (heliographischer) Breite *f* Breitenverteilung *f*
52	low-latitude	de basse latitude	in niedrigen Breiten
53	high-latitude	de haute latitude	in hohen Breiten
54	spot zone royal zone activity zone	zone d'apparition [de formation] des taches *f* zone active [tachée] *f* zone royale *f*	Fleckenzone *f* Aktivitätszone *f* Zone der Aktivität *f*
55	visibility function	fonction de visibilité *f*	Sichtbarkeitsfunktion *f*
56	undisturbed photospheric region quiet photospheric region	région photosphérique calme *f* région photosphérique non-perturbée *f*	ungestörtes Gebiet der Photosphäre *n* ruhiges Gebiet der Photosphäre *n*
57	active solar region AR region of activity	région active *f* région d'activité *f*	Aktivitätsgebiet *n* aktives Gebiet *n*
58	polar region	région polaire *f*	Polregion *f* polares Gebiet *n*
59	equatorial region	région équatoriale *f*	Äquatorregion *f* äquatoriales Gebiet *n*
60	centre [center] of activity active centre [center]	centre d'activité *m* centre actif *m*	Aktivitätszentrum *n* Zentrum der Sonnentätigkeit *n*

latitud heliográfica *f*	latitude heliográfica *f*	гелиографическая широта *f* гелиоширота *f*	42
meridiano central *m*	meridiano central *m*	центральный меридиан *m*	43
cruzar el meridiano central	passar pelo meridiano central	проходить через центральный меридиан	44
paso por el meridiano central *m*	passagem pelo meridiano central *f*	прохождение через центральный меридиан *n*	45
distancia al meridiano central *f*	distância ao meridiano central *f*	расстояние от центрального меридиана *n*	46
zona central *f*	zona central *f*	центральная зона *f*	47
distribución en longitud (heliográfica) *f* distribución longitudinal *f*	distribuição em longitude (heliográfica) *f* distribuição longitudinal *f*	распределение по (гелиографической) долготе *n* долготное распределение *n*	48
asimetría este-oeste *f*	assimetria este-oeste *f*	восточно-западная асимметрия *f*	49
asimetría norte-sur *f* asimetría norte-sud *f* (Am)	assimetria norte-sul *f*	северо-южная асимметрия *f*	50
distribución en latitud (heliográfica) *f* distribución latitudinal *f*	distribuição em latitude (heliográfica) *f* distribuição latitudinal *f*	распределение по широте *n* широтное распределение *n*	51
de latitud baja	de baixa latitude	низкоширотный	52
de latitud alta	de alta latitude	высокоширотный	53
zona de manchas *f* zona de actividad *f* zona real *f*	zona de manchas *f* zona ativa *f* zona régia *f*	зона пятнообразования *f* королевская зона *f* зона пятен *f*	54
función de visibilidad *f*	função de visibilidade *f*	функция видимости *f*	55
región fotosférica no perturbada *f*	região calma da fotosfera *f*	невозмущённая область фотосферы *f*	56
región activa solar *f* RA región de actividad *f*	região (solar) ativa *f* região de atividade *f*	солнечная активная область *f* активная область *f* область активности *f*	57
región polar *f*	região polar *f*	полярная область *f*	58
región ecuatorial *f*	região equatorial *f*	экваториальная область *f*	59
centro de actividad *m* centro activo *m*	centro de atividade *m* centro ativo *m*	центр активности *m* активный центр *m*	60

61	M-region	région M *f*	M-Gebiet *n*
62	sunspot activity spot activity	activité des taches solaires *f*	(Sonnen-) Fleckenaktivität *f* Fleckentätigkeit *f*
63	eruption hypothesis	hypothèse d'éruption *f*	Eruptionshypothese *f*
64	forecast of solar activity prediction of solar activity	prévision de l'activité solaire *f* prédiction de l'activité solaire *f*	Voraussage der Sonentätigkeit *f* Voraussage der Sonnenaktivität *f*
65	sunspot number Wolf (sunspot) number (Zürich) relative sunspot number	nombre relatif (des taches solaires) *m* nombre de Wolf *m*	Relativzahl *f* Fleckenrelativzahl *f* Sonnenfleckenrelativzahl *f*
66	sunspot frequency	fréquence d'apparition [de formation] des taches *f*	Fleckenhäufigkeit *f*
67	area of sunspots spot area	aire des taches *f*	Fleckenfläche *f*
68	millionths of the visible hemisphere	millionièmes de l'hémisphère visible *fpl*	Millionstel der sichtbaren Halbkugel *npl*
69	millionth of the solar disk [disc]	millionième du disque solaire *f*	Millionstel der Sonnenscheibe *n*
70	11-year period eleven-year period	période undécennale *f* période de onze ans *f*	elfjährige Periode *f*
71	solar cycle cycle of solar activity	cycle d'activité solaire *m* cycle solaire *m*	Sonnenzyklus *m* Zyklus der Sonnentätigkeit *m*
72	sunspot cycle	cycle des taches solaires *m*	(Sonnen-)Fleckenzyklus *m*
73	11-year cycle eleven-year cycle	cycle undécennal *m* cycle de onze ans *m*	elfjähriger Zyklus *m* Elfjahreszyklus *m*
74	migration of spot zone	déplacement des zones royales *m* déplacement des zones de formation des taches *m*	Zonenwanderung *f* Fleckenzonenwanderung *f* Wanderung der Aktivitätszonen *f*
75	Spörer's law law of latitudes	loi de Spoerer *f* loi de déplacement des zones royales *f*	Spörersches Gesetz *n* Gesetz der Fleckenzonenwanderung *n* Spoerersches Gesetz *n*
76	butterfly diagram Maunder diagram	diagramme (en) papillon *m* diagramme de Maunder *m*	Schmetterlingsdiagramm *n*
77	even spot cycle	cycle pair *m*	geradzahliger Zyklus *m* gerader Zyklus *m*
78	odd spot cycle	cycle impair *m*	ungerad(zahlig)er Zyklus *m*

región M *f*	região M *f*	М область *f*	61
actividad de las manchas (solares) *f*	atividade das manchas (solares) *f*	активность солнечных пятен *f* пятнообразовательная деятельность *f*	62
hipótesis de la erupción *f*	hipótese de erupção *f*	взрывная гипотеза *f*	63
predicción de la actividad solar *f* previsión de la actividad solar *f*	predicão da atividade solar *f* previsão da atividade solar *f*	прогноз солнечной активности [деятельности] *m* предсказание солнечной активности [деятельности] *n*	64
número relativo de manchas *m* número de Wolf *m*	número relativo (de manchas) *m* número de Wolf *m*	число Вольфа *n* относительное число (солнечных пятен) *n* цюрихское число пятен *n*	65
frecuencia de las manchas *f*	freqüência das manchas *f*	частота солнечных пятен *f*	66
área de las manchas *f*	área das manchas *f*	площадь пятен *f* занятнённая площадь *f*	67
millonésimas de hemisferio visible *fpl*	milionésimos do hemisfério visível *mpl*	миллионные доли видимой полусферы *fpl*	68
millonésima de disco solar *f*	milionésimo do disco solar *m*	миллионная доля солнечного диска *f*	69
período undecenal *m* período de once años *m*	período undecenal *m* período de onze anos *m*	11-летний период *m* одиннадцатилетний период *m*	70
ciclo de actividad solar *m* ciclo solar *m*	ciclo solar *m* ciclo da atividade solar *m*	солнечный цикл *m* цикл солнечной активности *m*	71
ciclo de las manchas (solares) *m*	ciclo das manchas (solares) *m*	цикл солнечных пятен *m* цикл пятнообразовательной деятельности *m*	72
ciclo undecenal *m* ciclo de once ãnos *m*	ciclo undecenal *m* ciclo de onze anos *m*	одиннадцатилетний цикл *m* 11-летний цикл *m*	73
desplazamiento [migración] de la zona de las manchas *m* corrimiento de la zona de actividad *m*	deslocamento da zona de manchas *m* migração da zona ativa *f*	смещение зоны пятен *n* смещение [дрейф] зоны пятнообразования *n*	74
ley de Spoerer *f* ley de las latitudes *f*	lei de Spoerer *f* lei das latitudes *f*	закон Шпёрера *m* закон зон *m* широтный дрейф *m*	75
diagrama (de) mariposa *m* diagrama de Maunder *m*	diagrama de [em] borboleta *m* diagrama de Maunder *m*	бабочки Маундера *fpl* диаграмма бабочек [широта-время] *f*	76
ciclo solar par *m*	ciclo solar par *m*	чётный цикл *m*	77
ciclo solar impar *m*	ciclo solar ímpar *m*	нечётный цикл *m*	78

	SOLAR ACTIVITY	ACTIVITÉ SOLAIRE	SONNENAKTIVITÄT
79	phase of the cycle	phase du cycle *f*	Phase des Zyklus *f*
80	epoch of minimum	époque de minimum *f*	Epoche des Minimums *f*
81	epoch of maximum	époque de maximum *f*	Epoche des Maximums *f*
82	beginning of the cycle	début du cycle *m*	Beginn des Zyklus *m*
83	ascending phase of the solar cycle rising phase	phase ascendante du cycle solaire *f*	aufsteigende Phase des Sonnenzyklus *f*
84	descending phase of the solar cycle declining phase	phase de déclin du cycle solaire *f*	absteigende Phase des Sonnenzyklus *f*
85	end of the cycle	fin du cycle *f*	Ende des Zyklus *n*
86	(period of) overlap of two cycles	(période de) recouvrement de deux cycles *f*	Überlappung von zwei Zyklen *f* Überdeckung von zwei Zyklen *f*
87	duration of the cycle	durée du cycle *f*	Dauer des Zyklus *f*
88	sunspot	tache solaire *f*	Sonnenfleck *m*
89	Wilson effect	effet Wilson *m*	Wilson-Effekt *m*
90	spotted area area of sunspots	aire tachée *f* aire des taches *f*	die von Flecken bedeckte Fläche *f* Fleckenfläche *f*
91	apparent sunspot area	aire projetée des taches *f*	scheinbare Fleckenfläche *f*
92	foreshortening (of sunspots)	dimension réduite par (effet de) perspective *f* raccourcissement perspectif (des taches) *m*	perspektivische Verkürzung (der Flecken) *f*
93	corrected area area corrected for foreshortening reduced area	aire calculée [réelle] *f* aire corrigée *f* aire corrigée de la perspective *f*	korrigierte (Flecken)fläche *f* Fläche korrigiert wegen perspektivischer Verkürzung *f*
94	total area	aire totale (des taches) *f*	Gesamtfläche *f*
95	spot umbra sunsport umbra	ombre (d'une tache) *f*	Umbra (eines Flecks) *f*
96	(bright) umbral dot	point (brillant) de l'ombre *m* point ombral *m*	(heller) Umbrapunkt *m*
97	(sunspot) penumbra spot penumbra	pénombre (d'une tache) *f*	Penumbra (eines Flecks) *f*
98	penumbral filament penumbral fibril	filament pénombral *m* filament de pénombre *m*	Penumbra-Filament *n* Penumbra-Faser *f*

fase del ciclo *f*	fase do ciclo *f*	фаза цикла *f*	79
época de mínimo *f*	época de mínimo *f*	эпоха минимума *f*	80
época de máximo *f*	época de máximo *f*	эпоха максимума *f*	81
comienzo del ciclo *m*	início do ciclo *m*	начало цикла *n*	82
fase ascendente del ciclo solar *f* fase creciente *f*	fase ascendente do ciclo solar *f* fase crescente *f*	фаза возрастания [роста] солнечного цикла *f* восходящая фаза [ветвь] солнечного цикла *f* ветвь роста активности *f*	83
fase descendente del ciclo solar *f* fase declinante *f*	fase descendente do ciclo solar *f*	фаза спада (активности) солнечного цикла *f* нисходящая фаза [ветвь] солнечного цикла *f* ветвь спада активности *f*	84
fin del ciclo *m*	fim do ciclo *m*	конец цикла *m*	85
(período de) superposición de dos cilos *f*	(período de) sobreposição [superposição] de dois ciclos *f*	интервал [период] наложения двух (соседних) циклов *m* период [эпоха, интервал] перекрытия двух циклов *m*[*f*, *m*]	86
duración del ciclo *f*	duração do ciclo *f*	продолжительность цикла *f*	87
mancha solar *f*	mancha solar *f*	солнечное пятно *n*	88
efecto Wilson *m*	efeito Wilson *m*	эффект Вильсона *m*	89
área con las manchas *f* área manchada *f*	área manchada *f* área com manchas *f*	запятнённая площадь *f* площадь с пятнами *f*	90
área aparente de las manchas *f*	área aparente das manchas *f*	видимая площадь пятен *f*	91
escorzo (de las manchas) *m*	encurtamento (das manchas) *m* achatamento (das manchas) por perspectiva *m*	(перспективное) сокращение пятен *n*	92
área reducida *f* área corregida *f*	área reduzida *f* área corrigida de perspectiva *f*	площадь пятен, исправленная за перспективное сокращение *f*	93
área total (de las manchas) *f*	área total (das manchas) *f*	суммарная [общая] площадь пятен *f*	94
umbra (de la mancha) *f*	sombra da mancha *f* umbra (da mancha) *f*	тень пятна *f* ядро пятна *n*	95
punto (brillante) umbral *m* punto de umbra *m*	ponto de sombra [de umbra] *m* ponto umbral *m*	(яркая) точка (в) тени (пятна) *f*	96
penumbra (de una mancha) *f*	penumbra (da mancha) *f*	полутень (пятна) *f*	97
filamento de penumbra *m* fibrilla de penumbra *f*	filamento da penumbra *m* fibrila penumbral *f*	волокно полутени *n* полутеневое волокно *n*	98

	SOLAR ACTIVITY	ACTIVITÉ SOLAIRE	SONNENAKTIVITÄT
99	interfilamentary space	espace interfilamentaire *m*	Interfilamentraum *m* interfilamentarer Raum *m*
100	penumbral waves	ondes pénombrales *fpl*	Penumbra-Wellen *fpl*
101	bright ring	anneau brillant *m*	heller Ring *m*
102	sunspot model	modèle de tache *m*	Fleckenmodell *n*
103	Evershed effect	effet Evershed *m*	Evershed-Effekt *m*
104	bright bridge	pont lumineux *m*	Lichtbrücke *f*
105	spot group group of sunspots	groupe de tache (solaires) *m*	Fleckengruppe *f* Sonnenfleckengruppe *f*
106	new-cycle group	groupe du nouveau cycle *f*	Gruppe des neuen Zyklus *f*
107	old-cycle group	groupe du cycle précédent *m*	Gruppe des alten [vorigen] Zyklus *f*
108	recurrent group	groupe récurrent *m*	wiederkehrende Gruppe *f* rekurrente Gruppe *f*
109	magnetic classification	classification magnétique *f*	magnetische Klassifikation *f*
110	unipolar group	groupe unipolaire *m*	unipolare Gruppe *f*
111	bipolar group	groupe bipolaire *m*	bipolare Gruppe *f*
112	complex group	groupe complexe *m*	komplexe Gruppe *f*
113	leading spot leader preceding spot *p* spot western spot	tache de tête *f* tache ouest *f* tache-leader *f* *p*-tache solaire *f*	vorangehender Fleck *m* *P*-Fleck *m* vorderer Fleck *m* westlicher Fleck *m* führender Fleck *m*
114	follower *f* spot eastern spot trailing spot following spot trailer (spot)	tache est *f* tache de queue *f* tache arrière *f* tache postérieure *f* *f*-tache solaire *f*	nachfolgender Fleck *m* *F*-Fleck *m* östlicher Fleck *m* folgender Fleck *m*
115	major spot	tache principale *f*	Hauptfleck *m*
116	single spot	tache isolée *f* tache unique *f*	Einzelfleck *m*
117	proper motion of sunspots	mouvement propre des taches *m*	Eigenbewegung der Flecken *f*
118	development of a spot group	évolution d'un groupe (des taches) *f*	Entwicklung der Gruppe *f* Entwicklung einer Fleckengruppe *f*
119	development curve (of a spot group)	courbe d'évolution (d'un groupe) *f*	Entwicklungskurve (einer Fleckengruppe) *f*

espacio interfilamentario *m*	espaço interfilamentar *m*	пространство между волокнами *n*	99
ondas penumbrales *fpl*	ondas penumbrais [na penumbra] *fpl*	волны в полутени *fpl*	100
anillo brillante *m*	anel brilhante *m*	яркое [светлое] кольцо *n*	101
modelo de manchas *m*	modelo de mancha *m*	модель пятна *f*	102
efecto Evershed *m*	efeito Evershed *m*	эффект Эвершеда *m*	103
puente brillante *m*	ponte luminosa *f* ponte brilhante *f*	светлый мост *m* яркая перемычка (в пятне) *f*	104
grupo de manchas (solares) *m*	grupo de manchas (solares) *m*	группа солнечных пятен *f* группа пятен *f*	105
grupo del nuevo ciclo *m*	grupo do novo ciclo *m*	группа нового цикла *f*	106
grupo del viejo ciclo [del ciclo precedente, anterior] *m*	grupo do ciclo anterior [precedente] *m*	группа старого цикла *f* группа предыдущего цикла *f*	107
grupo recurrente *m*	grupo recorrente *m*	рекуррентная група *f*	108
clasificación magnética *f*	classificação magnética *f*	магнитная классификация *f*	109
grupo unipolar *m*	grupo unipolar *m*	униполярная группа *f*	110
grupo bipolar *m*	grupo bipolar *m*	биполярная группа *f*	111
grupo complejo *m*	grupo complexo *m*	сложная группа *f* комплексная группа *f*	112
mancha delantera *f* mancha precedente *f* mancha del oeste *f* mancha *p* *f*	mancha da frente *f* mancha de oeste *f* mancha precedente *f* mancha *p* *f*	головное пятно *n* ведущее пятно *n* лидер (группы) *m* *p*-пятно *n*	113
mancha trasera *f* mancha zaguera *f* mancha siguiente *f* mancha del este *f* mancha *f* *f*	mancha de cauda *f* mancha de este *f* mancha seguinte *f* mancha *f* *f*	хвостовое пятно *n* последующее пятно *n* следующее пятно *n* заднее пятно *n* *f*-пятно *n*	114
mancha principal *f*	mancha principal *f*	главное пятно *n*	115
mancha aislada *f*	mancha isolada *f*	одиночное пятно *n*	116
movimiento propio de las manchas *m*	movimento próprio das manchas *m*	собственное движение пятен *n*	117
desarrollo [evolución] de un grupo de manchas *m[f]*	desenvolvimento [evolução] de um grupo de manchas *m[f]*	развитие группы пятен *n*	118
curva de desarrollo [evolución] (de un grupo) *f*	curva de evolução (de um grupo) *f*	кривая развития (группы пятен) *f*	119

	SOLAR ACTIVITY	ACTIVITÉ SOLAIRE	SONNENAKTIVITÄT
120	sunspot-evolution table	tableau d'évolution des taches *m*	Fleckenentwicklungstabelle *f*
121	birth of a spot formation of a spot birth of a sunspot	formation d'une tache *f* naissance d'une tache *f*	Bildung eines Flecks *f* Entstehung eines Flecks *f* Fleckenbildung *f*
122	pore	pore *m*	Pore *f*
123	lifetime of a spot group	durée de vie (d'un groupe) *f*	Lebensdauer (einer Fleckengruppe) *f*
124	long-lived	à longue durée de vie	langlebig
125	short-lived	à courte durée de vie	kurzlebig
126	decay of a sunspot group	désagregation [décroissance] du groupe de taches *f* désintégration d'un groupe *f*	Auflösung einer Fleckengruppe *f* Zerfall einer Fleckengruppe *m*
127	Zürich classification	classification de Zürich *f*	Züricher Klassifikation *f*
128	general magnetic field (of the Sun)	champ magnétique général (du Soleil) *m*	allgemeines Magnetfeld (der Sonne) *n*
129	solar magnetic fields	champs magnétiques du Soleil *mpl*	Magnetfelder auf der Sonne *npl* solare Magnetfelder *npl*
130	photospheric magnetic field	champ magnétique photosphérique *m*	photosphärisches Magnetfeld *n*
131	magnetic (fields) map	carte (du champ) magnétique *f*	Magnetfeldkarte *f* magnetishe Karte *f*
132	map the magnetic field	faire la carte des champs magnétiques	das Magnetfeld kartographieren
133	high-resolution map	carte à haute résolution *f*	Karte hoher Auflösung *f*
134	isogauss line isogauss contour	ligne isogauss *f* courbe de niveau magnétique *f*	Isogauss-Linie *f* Isogauss-Kontur *f*
135	longitudinal component	composante longitudinale *f*	Longitudinalkomponente *f*
136	transversal component	composante transversale *f*	Transversalkomponente *f*
137	transversal field	champ transversal [transverse] *m*	Transversalfeld *n*
138	longitudinal field	champ longitudinal *m*	Longitudinalfeld *n*
139	magnetic vector	vecteur champ magnétique *m*	magnetischer Vektor *m*
140	field configuration structure of the field	configuration du champ *f*	Feldstruktur *f*
141	background (magnetic) field	champ (magnétique) de fond *m* valeurs „plancher" du champ magnétique *fpl*	Hintergrundfeld *n* Hintergrund-Magnetfeld *n*

cuadro [tabla] de evolución de los grupos *m(f)*	quadro de evolução das manchas *m*	таблица развития солнечных пятен *f*	120
formación de una mancha *f* nacimiento de una mancha *m*	formação de uma mancha *f* nascimento de uma mancha *m*	рождение пятна *n* возникновение пятна *n* образование пятна *n*	121
poro *m*	poro *m*	пора *f*	122
duración de un grupo *f*	duração de [vida de] um grupo de manchas *f*	продолжительность жизни группы пятен *f*	123
de vida larga de larga duración	de longa duração a longo prazo	долгоживущий	124
de vida corta de corta duración	de pequena duração a curto prazo	короткоживущий	125
disolución de un grupo de manchas *f*	decaimento de um grupo de manchas *m*	распад группы пятен *m* гибель группы пятен *f*	126
clasificación de Zürich *f*	classificação de Zürich *f*	цюрихская классификация *f*	127
campo magnético general (del Sol) *m*	campo magnético geral (do Sol) *m*	общее магнитное поле (Солнца) *n*	128
campos magnéticos solares *mpl*	campos magnéticos solares *mpl*	солнечные магнитные поля *npl*	129
campo magnético fotosférico *m*	campo magnético fotosférico *m*	фотосферное магнитное поле *n*	130
mapa magnético *m* mapa de campos magnéticos *m*	carta magnética *f* carta de campos magnéticos *f*	карта магнитного поля *f*	131
hacer el mapa (del campo) magnético	regist(r)ar o campo magnético fazer a carta magnética	построить карту магнитного поля	132
mapa de alta resolución *m*	carta de grande [alta] resolução *f*	карта с высоким разрешением *f*	133
línea de isogauss *f* contorno de isogauss *m*	(risca) isogáussica *f* contorno isogáussico *m*	изолиния магнитного поля *f* изогаусса *f*	134
componente longitudinal *m*	componente longitudinal *m*	продольная составляющая *f* продольный компонент *m*	135
componente transversal *m*	componente transversal *m*	поперечная составляющая *f* поперечный компонент *m*	136
campo transversal *m*	campo transversal *m*	поперечное поле *n*	137
campo longitudinal *m*	campo longitudinal *m*	продольное поле *n*	138
vector magnético *m*	ve(c)tor magnético *m*	(полный) вектор магнитного поля *m*	139
configuración del campo *f* estructura del campo *f*	configuração do campo *f* estrutura do campo *f*	структура поля *f* конфигурация поля *f*	140
campo (magnético) de fondo *m*	campo (magnético) de fundo *m*	фоновое (магнитное) поле *n*	141

142	large-scale (magnetic) field	champ (magnétique) à grande échelle *m*	grossmasstäbliches (magnetisches) Feld *n* grossskaliges Feld [Magnetfeld] *n*
143	small-scale (magnetic) field	champ (magnétique) à petite échelle *m*	kleinmasstäbliches Feld *n* kleinskaliges (magnetisches) Feld *n* Feinstruktur-Magnetfeld *n*
144	magnetic region	région magnétique *f*	magnetisches Gebiet *n* magnetische Region *f*
145	magnetic tube of force	tube de force magnétique *m*	magnetische Kraftröhre *f*
146	magnetic rope	corde magnétique *f* tube de force en forme de corde *m*	magnetischer Strang *m*
147	magnetic buoyancy	poussée des forces magnétiques *f* flottabilité magnétique *f*	magnetischer Auftrieb *m*
148	magnetic polarity	polarité magnétique *f*	magnetische Polarität *f*
149	northern polarity positive magnetic polarity	polarité nord *f* polarité N *f* polarité positive *f*	nördliche Polarität *f* N-Polarität *f* positive Polarität *f*
150	southern polarity negative magnetic polarity	polarité sud *f* polarité S *f* polarité négative *f*	südliche Polarität *f* S-Polarität *f* negative Polarität *f*
151	neutral line line separating opposite polarities zero line (of magnetic fields)	ligne neutre *f* ligne de séparation des polarités (opposées) *f* ligne d'inversion des polarités *f*	neutrale Linie *f* magnetische Null-Linie *f* Trennungslinie der Polaritäten *f*
152	neutral point	point neutre *m*	neutraler Punkt *m*
153	reversal of magnetic polarity	inversion (alternance) de la polarité *f* changement de polarité *m*	Wechsel der magnetischen Polarität *m* Umkehrung der magnetischen Polarität *f*
154	law of polarity Hale's law	loi (de l'alternance) de polarité *f* loi de Hale *f*	Polaritätsgesetz *n* Halesches Gesetz *n*
155	opposite polarity	polarité opposée *f*	entgegengesetzte Polarität *f*
155a	unipolar magnetic region	région magnétique unipolaire *f*	unipolares magnetisches Gebiet *n*
156	(magnetic-)field amplification	amplification du champ (magnétique) *f*	Feldverstärkung *f*
157	random walk random motion	marche [trajet] aléatoire *f*[*m*] marche [mouvement] au hasard *f*[*m*]	(zufällige) Irrfahrt *f* Zufallsbewegung *f*
158	annihilation of magnetic fields	annihilation des champs magnétiques *f*	Feldvernichtung *f* Feldannihilation *f*
159	invisible sunspot	tache invisible *f*	unsichtbarer Fleck *m*
160	magnetic foreshortening	réduction [diminution apparente] centre-bord du champ magnétique *f*	Mitte-Rand Variation der magnetischen Feldstärke *f*

campo (magnético) de macroestructura *m* campo de gran escala *m*	campo de grande escala *m* campo de conjunto *m*	крупномасштабное [магнитное] поле *n*	142
campo (magnético) de microestructura *m* campo de pequeña escala *m*	campo (magnético) local *m* campo de pequena escala *m*	мелкомасштабное (магнитное) поле *n*	143
región magnética *f*	região magnética *f*	область магнитного поля *f* магнитная область *f*	144
tubo de fuerza magnética *m*	tubo de força magnética *m*	магнитная силовая трубка *f*	145
cuerda magnética *f*	corda magnética *f*	жгут магнитного поля *m*	146
flotación magnética *f*	flotação magnética *f*	магнитная подъёмная сила *f* магнитная плавучесть *f*	147
polaridad magnética *f*	polaridade magnética *f*	магнитная полярность *f*	148
polaridad norte *f* polaridad magnética positiva *f*	polaridade norte *f* polaridade magnética positiva *f*	северная полярность *f* положительная магнитная полярность *f*	149
polaridad sur *f* polaridad magnética negativa *f*	polaridade sul *f* polaridade magnética negativa *f*	южная полярность *f* отрицательная магнитная полярность *f*	150
línea neutra *f* línea de separación de polaridades opuestas *f*	linha neutra *f* linha de separação de polaridades opostas *f*	линия раздела полярностей *f* нейтральная линия *f* нулевая линия (магнитного поля) *f*	151
punto neutro *m*	ponto neutro *m*	нейтральная точка *f*	152
alternación [inversión] de polaridad magnética *f*	inversão [alternação] da polaridade magnética *f*	обращение [изменение] магнитной полярности *n*	153
ley (de la alternación) de polaridad *f* ley de Hale *f*	lei (de alternação) de polaridade *f* lei de Hale *f*	закон чередования полярности *m* закон Хэйла *m*	154
polaridad opuesta *f*	polaridade oposta *f*	противоположная полярность *f*	155
región magnética unipolar *f*	região magnética unipolar *f*	униполярная магнитная область *f*	155a
amplificación del campo (magnético) *f*	amplificação do campo (magnético) *f*	усиление (магнитного) поля *n*	156
movimiento aleatorio [al azar] *m* paseo al azar *m*	percurso aleatório *m* movimento aleatório [ao acaso] *m*	беспорядочное движение *n* случайное блуждание *n*	157
aniquilación de campos magnéticos *f*	aniquilação de campos magnéticos *f*	аннигиляция магнитных полей *f*	158
mancha invisible *f*	mancha invisível *f*	невидимое пятно *n*	159
variación del campo magnético del centro al limbo *f*	encurtamento magnético (por perspectiva) *m*	магнитное сокращение *n*	160

160a	bipolar magnetic region	région magnétique bipolaire *f*	bipolares magnetisches Gebiet *n*
161	facula photospheric faculae *pl*	facule *f* facules photosphériques *fpl*	(Sonnen-)Fackel *f* photosphärisches Fackelgebiet *n*
162	polar faculae	facules polaires *fpl*	polare Fackeln *fpl*
163	facular granules facular granulation	granules faculaires *mpl* granulation faculaire *f*	Fackelgranulen *fpl* Feckelgranulation *f*
164	plages *pl* plage areas *pl* chromospheric faculae *pl*	plage faculaire *f* facules chromosphériques *fpl*	chromospärische Fackeln *fpl* Plage *f*
165	optical plage	région faculaire observée en rayonnement visible *f*	optische Fackel *f* optische Plage *f*
166	X-ray plage	région faculaire observée en rayonnement X *f*	Röntgen(strahlen)fackel *f* Röntgenplage *f*
167	flocculus flocculi *pl*	flocule *f* flocculus *m* flocculi *pl*	Flocculus *n* Flocculi *npl* Flocke *f* Flocken *fpl*
168	hydrogen flocculi	flocules d'hydrogène *fpl* flocculi d'hydrogène *mpl*	Wasserstoff-Flocken *fpl* Wasserstoff-Flocculi *fpl*
169	bright hydrogen flocculi	flocules brillantes d'hydrogène *fpl* flocculi brillants d'hydrogène *mpl*	helle Wasserstoff-Flocculi *fpl*
170	calcium flocculi	flocules de calcium ionisé *fpl* flocculi de calcium ionisé *mpl*	Kalziumflocculi *npl* Kalziumflocken *fpl*
171	chromospheric network	réseau chromosphérique *m*	chromosphärisches Netzwerk *n*
172	chromospheric mottles fine mottles	tachetures chromosphériques *fpl* structures fines (chromosphériques) *fpl*	chromosphärische Mottles *npl*
173	striation	striation *f* stries *fpl*	Faserung *f* Riefung *f*
174	striation pattern	structure striée *f*	Fasermuster *n*
175	flare chromospheric flare solar flare	éruption *f* éruption chromosphérique *f* éruption solaire *f*	Flare *m* chromosphärische Eruption *f* (Sonnen-)Eruption *f*
176	flare activity	activité éruptive *f* activité d'éruptions chromosphériques *f*	Flare-Aktivität *f* Eruptionstätigkeit *f* Flare-Tätigkeit *f*
177	proton-flare project	campagne d'observation des éruptions à protons *f*	Proton-Flare-Projekt *n*

región magnética bipolar *f*	região magnética bipolar *f*	биполярная магнитная область *f*	160a
fácula *f* fáculas fotosféricas *fpl*	fácula *f* fáculas fotosféricas *fpl*	факел *m* фотосферные факелы *mpl*	161
fáculas polares *fpl*	fáculas polares *fpl*	полярные факелы *mpl*	162
gránulos faculares *mpl* granulación facular *f*	grânulos faculares *mpl* granulação facular *f*	факельные гранулы *fpl* факельная грануляция *f*	163
fáculas cromosféricas *fpl* playas *fpl*	fáculas cromosféricas *fpl* plagas [praias] *fpl* áreas de plagas [praias] *fpl*	флоккульные поля *npl* флоккульные площадки *fpl* хромосферные факелы *mpl*	164
fácula óptica *f* playa óptica *f*	plaga ó(p)tica *f* fácula ó(p)tica *f* praia ó(p)tica *f*	флоккульное поле в оптическом диапазоне *n* флоккулы в оптическом диапазоне *mpl*	165
fácula en rayos X *f* playa en rayos X *f*	plaga em raios X *f* fácula em raios X *f* praia em raios X *f*	флоккульное поле в рентгеновских лучах *n* рентгеновские флоккулы *mpl*	166
flóculo *m* flóculos *mpl*	flóculo *m* flóculos *mpl*	флоккул *m* флоккулы *mpl*	167
flóculos de hidrógeno *mpl*	flóculos de hidrogênio *mpl*	водородные флоккулы *mpl*	168
flóculos brillantes de hidrógeno *mpl*	flóculos brilhantes de hidrogênio *mpl*	яркие водородные флоккулы *mpl*	169
flóculos de calcio *mpl*	flóculos de cálcio *mpl*	кальциевые флоккулы *mpl*	170
reticulado cromosférico *m*	rede cromosférica *f*	хромосферная сетка *f*	171
motas cromosféricas *fpl* moteado cromosférico *m*	pintas cromosféricas *fpl*	узлы хромосферной сетки *mpl* хромосферные узелки *mpl*	172
estriación *f* estrías *fpl*	estriamento *m* estriação *f*	хромосферные волоконца *npl* бороздчатость хромосферы *f*	173
estructura de estriación *f* estructura de estrías *f*	configuração estriada [de estrias] *f* configuração [estrutura] de estriamento *f*	волоконцевая структура *f* волоконная структура *f*	174
fulguración *f* fulguración cromosférica *f* fulguración [erupción] solar *f*	fulguração *f* erupção cromosférica *f* erupção solar *f*	вспышка *f* хромосферная вспышка *f* солнечная вспышка *f*	175
actividad eruptiva *f* actividad de fulguraciones *f* actividad de erupciones *f*	atividade em erupções [fulgurações] *f* atividade eruptiva [fulgurante] *f* atividade de fulgurações *f*	вспышечная активность *f*	176
proyecto de fulguraciones protónicas *m*	projeto de erupções [fulgurações] protônicas [de protons] *m*	проект „протонная вспышка" *m* программа наблюдений протонных вспышек *f*	177

	SOLAR ACTIVITY	ACTIVITÉ SOLAIRE	SONNENAKTIVITÄT
178	homologous flare	éruption homologue *f*	homologer Flare *m*
179	sympathetic flare	éruption par contagion *f*	sympathetischer Flare *m* gleichzeitiger Flare *m*
180	radio flare	éruption raio(électrique) *f* éruption ayant une composante radio *f*	Radioflare *m*
181	optical flare visible flare	éruption dans le domaine visible *f* manifestation visible d'une éruption *f*	sichtbarer Flare *m* optischer Flare *m*
182	white-light flare white flare	éruption (visible) en lumière blanche *f*	Flare im weissen Licht *m* Integrallicht-Eruption *f*
183	X-ray flare	éruption en rayons X *f* manifestation X d'une éruption *f*	X-Strahlen-Flare *m* Röntgen(strahlen)flare *m*
183a	X-ray bright point	point brillant en rayos rayons X *m*	heller X-Strahlen-Punkt *m*
184	X-ray burst solar X-ray event X-ray event	sursaut de l'émission X *m*	X-Strahlen-Burst *m* Röntgenstrahlungsausbruch *m*
185	proton flare	éruption à protons *f*	Protonen-Flare *m* Protonen-Eruption *f*
185a	coronal transient	transitoire coronal *m*	koronaler Transient *m* vorübergehendes koronales Event *n*
186	non-proton flare	éruption sans émission de protons *f*	Nicht-Protonen-Eruption [Flare] *f*[*m*]
187	flare ribbon	filet brillant d'une éruption *m* fil d'une éruption *m*	Eruptionsband *n* Flare-Band *n*
187a	two-ribbon flare	éruption à deux fils *f* éruption en double filet *f*	Zweibänderflare *m* Flare mit zwei Faden *m*
188	flare-associated	associé à une éruption	Flare-verbunden
189	flare nimbus dark nimbus	nuage absorbant observé lors d'une éruption *m*	Flare-Wolke *f* (Flare-verbundene) dunkle Wolke *f*
190	flare surge	jet de matiére accompagnant une éruption *m* surrection d'une éruption *f*	Eruptionsauswurf *m* Flare-Surge *f*
191	energetic particle event high-energy particle event	événement à particules de haute énergie *m*	Ereignis (hoch)energetischer Partikeln *n* Event hochenergetischer Teilchen *n*
192	solar proton event proton event	événement solaire à protons *m*	Protonen-Event *n* Protonen-Ereignis *n*

fulguración [erupción] homóloga *f*	erupção [fulguração] homóloga *f*	гомологическая вспышка *f*	178
fulguración [erupción] simpatética *f*	erupção [fulguração] simpatética *f*	симпатическая вспышка *f*	179
fulguración [erupción] radio(-eléctrica) *f*	erupção [fulguração] rádio *f*	радиовспышка *f* вспышка в радиоизлучении *f*	180
fulguración [erupción] visible *f* fulguración [erupción] óptica *f*	erupcão [fulguração] visível *f* erupção [fulguração] ó(p)tica *f*	оптическая вспышка *f* вспышка в видимом диапазоне *f*	181
fulguración [erupción] en luz blanca *f*	erupcão [fulguração] em luz branca *f* erupção [fulguração] branca *f*	вспышка в белом свете *f*	182
fulguración [erupción] en rayos X *f*	erupcão [fulguração] de raios X *f* erupção em raios X *f* fulguração em raios X *f*	рентгеновская вспышка *f* вспышка в рентгеновских лучах *f*	183
punto brillante en rayos X *m*	ponto brilhante em raios X *m*	яркая рентгеновская точка *f*	183a
sobresalto en rayos X *m* suceso en rayos X *m*	explosão de raios X *f* evento solar em raios X *m* evento em raios X *m*	всплеск рентгеновского излучения *m* событие в (солнечном) рентгене *n* событие в рентгеновском излучении *n*	184
fulguración [erupción] protónica *f* fulguración [erupción] a protones *f*	erupção [fulguração] de prótons *f* erupção [fulguração] protônica *f*	протонная вспышка *f*	185
(fénomeno) transitorio coronal *m*	transiente coronal *m* (fenômeno) transitório coronal *m* evento transitório coronal *m*	корональное транзиентное явление *n* корональный транзиент *m*	185a
fulguración [erupción] aprotónica [sin protones] *f*	erupção [fulguração] não-protônica [sem prótons] *f*	непротонная вспышка *f*	186
cinta [banda] de fulguración [erupción] *f*	filamento de uma erupção *m* fita eruptiva *f*	(яркая) лента вспышки *f* вспышечная лента *f*	187
fulguración [erupción] con dos cintas *f*	erupção com duas fitas *f* fulguração com duas fitas *f*	двухленточная вспышка *f* двухниточная вспышка *f*	187a
asociado a una fulguración [erupción]	associado à erupção [fulguração]	связанный со вспышкой	188
nimbo de la fulguración [erupción] *m* nimbo o(b)scuro *m*	nimbo de erupção [fulguracão] *m* nimbo escuro *m*	вспышечный нимб *m* тёмный лимб вспышки *m*	189
surtidor de fulguración [erupción] *m* oleada de fulguración [erupción] *f*	surgimento de uma erupção *m* surgimento de uma fulguração *m*	послевспышечный выброс *m* вспышечный возвратный выброс *m*	190
suceso de partículas energéticas *m* suceso de partículas de alta energía *m*	evento a partículas de alta energia *m* fenômeno de partículas energéticas *f*	появление частиц высокой энергии *n* появление энергичных частиц *n*	191
suceso (solar) con eyección de protones *m*	evento [fenômeno] solar de prótons [protões] *m*	протонное событие *n* протонное явление *n*	192

	SOLAR ACTIVITY	ACTIVITÉ SOLAIRE	SONNENAKTIVITÄT
193	arrival at the Earth	arrivée sur la Terre *f*	Aufreffen auf die Erde *n* Ankunft auf der Erde *f*
194	solar cosmic-ray event	événement solaire à rayons cosmiques *m*	Ereignis solarer kosmischer Strahlen *n*
195	subflare	sous-éruption *f*	Subflare *m*
196	stored energy	énergie emmagasinée *f* énergie accumulée *f*	gespeicherte Energie *f*
197	available energy	énergie disponible *f*	verfügbare Energie *f*
198	trigger	déclencher	auslösen
199	triggering trigger action	déclenchement *m*	Auslösung *f*
200	moustaches *pl* Ellerman bomb	moustaches *mpl* bombe d'Ellerman *f*	Moustaches *pl* Ellerman-Bombe *f*
201	bright point	point brillant *m*	heller Punkt *m*
202	limb flare	éruption chromosphérique au bord *f*	Randflare *m* Randeruption *f* Eruption am Sonnenrand *f*
203	flare behind the limb	éruptión au delà du limbe *f* éruption derrière le bord *f*	Flare [Eruption] jenseits des Randes *m[f]*
204	disk [disc] flare	éruption sur le disque *f*	Flare [Eruption] auf der Scheibe *m[f]*
205	importance of a flare class of a flare	importance d'une éruption *f* classe d'éruption *f*	Importanz einer Eruption [eines Flares] *f* Intensitätsklasse *f*
206	flare of importance 2 class 2 flare	éruption d'importance 2 *f* éruption de classe 2 *f*	Flare der Importanz 2 *m* Eruption der Klasse 2 *f*
207	area of a flare	aire de l'éruption *f*	Flare-Fläche *f* Eruptionsfläche *f* Fläche einer Eruption [eines Flares] *f*
208	effective line width	largeur effective de la raie *f*	effektive Linienbreite *f*
209	lifetime of a flare duration of a flare	durée d'une éruption *f*	Eruptionsdauer *f* Lebensdauer eines Flares *f*
210	flash-phase flare flash explosive phase eruptive phase	phase explosive *f*	Flash-Phase *f* eruptive Phase *f* Explosionsphase *f*
211	central Hα intensity	intensité au centre de la raie Hα *f*	Zentralintensität der Hα-Linie *f*

llegada a la Tierra *f*	chegada à Terra *f*	приход на Землю *m*	193
suceso solar de rayos cósmicos *m*	evento de raios cósmicos solares *m*	вспышка солнечных космических лучей *f* событие в солнечных космических лучах *n*	194
subfulguración [subcrupción] *f*	sub-erupcão *f*	субвспышка *f*	195
energía acumulada *f* energía almacenada *f*	energia acumulada *f* energia armazenada *f*	запасённая энергия *f* накопленная энергия *f*	196
energía disponible *f*	energia disponível *f*	имеющаяся энергия *f*	197
gatillar disparar	disparar dar início	приводить в действие запустить	198
acción de disparo *f*	engatilhamento *m* ação de disparar [disparo] *f*	спусковое [триггерное] действие *n* спуск [триггер] *m*	199
bigotes *mpl* bomba de Ellerman *f*	bigodes *mpl* bomba de Ellerman *f*	усы *pl* бомба Эллермана *f*	200
punto brillante *m*	ponto brilhante *m*	яркая точка *f*	201
fulguración [erupción] en el limbo *f*	erupção [fulguração] no bordo *f* erupção [fulguração] no limbo *f*	лимбовая вспышка *f* вспышка на краю диска *f*	202
fulguración [erupción] detrás del limbo *f*	erupção [fulguração] por detrás do limbo [bordo] *f*	залимбовая вспышка *f*	203
fulguración [erupción] en el disco *f*	erupção sobre o disco [no disco] *f*	вспышка на диске *f*	204
importancia de una fulguración [erupción] *f* clase de una fulguración [erupción] *f*	importância de uma erupção [fulguração] *f* classe de uma erupção [fulguração] *f*	балл вспышки *m* мощность вспышки *f*	205
fulguración [erupción] de importancia 2 *f* fulguración [erupción] de clase 2 *f*	erupção de importância 2 *f* erupção [fulguração] de classe 2 *f*	вспышка балла 2 *f*	206
área de una fulguración [erupción] *f*	área de uma erupção *f*	площадь вспышки *f*	207
ancho efectivo de la línea *m*	largura efetiva de uma raia [linha] *f*	эффективная ширина линии *f*	208
duración de una fulguración [erupción] *f*	duração de uma erupção [fulguração] *f*	продолжительность (жизни) вспышки *f*	209
fase de flash *f* fase eruptiva *f* fase explosiva *f*	fase eruptiva *f* fase explosiva *f* fase „flash“ *f*	взрывная фаза *f* фаза всплеска *f* резкое увеличение яркости *n* возгорание вспышки *n*	210
intensidad en el centro de (la línea) Hα *f*	intensidade central da (raia) Hα *f* intensidade no centro da (raia) Hα *f*	центральная интенсивность линии Hα *f*	211

	SOLAR ACTIVITY	ACTIVITÉ SOLAIRE	SONNENAKTIVITÄT
212	brightness of a flare	éclat de l'éruption *m*	Flare-Helligkeit *f* Helligkeit der Eruption *f*
213	brightening of a plage	renforcement d'éclat [de brillance] d'une plage faculaire *m*	Aufhellung eines Fackelgebietes *f* Aufleuchten einer Plage *n*
214	ejection of matter flare ejection	éjection de matière *f* éjection associée à une éruption *f*	Materieauswurf *m* Flare-Auswurf *m*
215	activation of a filament	activation d'un filament *f*	Filamentaktivierung *f*
216	winking filament	filament oscillant *m*	wedelndes Filament *n*
217	frequency of flare occurence	fréquence des éruptions *f*	Eruptionshäufigkeit *f* Häufigkeit der Flares *f*
219	flare patrol	surveillance des éruptions *f* observation permanente [systématique, réguliére] des éruptions *f*	Flare-Überwachung *f* Überwachung der Eruptionen *f* Überwachen der Flares *n*
220	cinematographic (flare) patrol	surveillance [observation] cinématographique des éruptions *f*	kinematographische Flare-Überwachung *f*
221	automatic flare patrol	surveillance [observation] automatique des éruptions *f*	automatische Flare-Überwachung *f*
222	flare-like prominence	protubérance à apparence d'éruption *f*	Flare-ähnliche Protuberanz *f*
223	prominence	protubérance *f*	Protuberanz *f*
224	filament dark flocculus dark hydrogen flocculus dark marking dark absorption marking	filament *m* plage faculaire obscure *f* plage faculaire sombre *f* structure absorbante (en $H\alpha$, K_3) *f*	Filament *n* dunkler Flocculus *m* dunkle Flocke *f*
225	magnetic shielding	blindage magnétique *m* formation d'un écran magnétique *f* protection magnétique *f*	magnetische Abschirmung *f*
226	bright rim of a filament bright edge of a filament	bord brillant d'un filament *m* marge brillante d'un filament *f*	heller Saum eines Filaments *m*
227	low-latitude prominence	protubérance de basse latitude *f*	Protuberanz in niedriger Breite *f*
228	equatorial prominence	protubérance équatoriale *f*	äquatoriale Protuberanz *f* Protuberanz der Äquatorzone *f*
229	high-latitude prominence	protubérance de haute latitude *f*	Protuberanz in hoher Breite *f*

brillo de una fulguración [erupción] *m*	brilho de uma erupção *m* brilhância de uma fulguração *f*	яркость вспышки *f*	212
intensificación del brillo de una playa [fácula] *f*	aumento de brilho du uma praia [plaga] *m*	поярчание флоккула *n*	213
eyección de materia *f*	ejeção de matéria *f*	выброс вещества *m* вспышечный выброс *m*	214
activación de un filamento *f*	ativação de um filamento *f*	активизация волокна *f*	215
filamento tremolante *m* filamento oscilante *m*	filamento oscilante [tremulante] *m*	мигающее волокно *n* осциллирующее волокно *n*	216
frecuencia de las fulguraciones *f* frecuencia de las erupciones *f*	freqüência de erupções *f* freqüência de fulgurações *f*	частота появления вспышек *f*	217
vigilancia de fulguraciones *f* seguimiento de fulguraciones *m* monitoreo [monitoraje] de fulguraciones *m*	observação regular das erupções *f* patrulha(mento) de erupções [fulgurações] monitoramento de erupções *m*	патруль вспышек *m*	219
seguimiento [vigilancia, patrullaje] cinematográfico de fulguraciones *m*[*fm*]	patrulha cinematográfica das erupções [fulgurações] *f*	фотографический патруль вспышек *f*	220
seguimiento [vigilancia, patrullaje] automático de fulguraciones *m* [*fm*]	patrulha automática das erupções [fulgurações]	автоматический патруль вспышек *m*	221
protuberancia semejante a una fulguración [erupción] *f* protuberancia de tipo fulguración	protuberância semelhante a erupção [fulguração] *f*	вспышечноподобный протуберанец *m* вспышкоподобный протуберанец *m*	222
protuberancia *f*	protuberância *f*	протуберанец *m*	223
filamento *m* flóculo o(b)scuro *m* flóculo o(b)scuro de hidrógeno *m*	filamento *m* flóculo escuro *m* flóculo escuro de hidrogênio *m*	волокно *n* тёмный флоккул *m* тёмный водородный флоккул *m*	224
blindaje magnético *m* protección magnética *f*	blindagem magnética *f*	магнитное экранирование *n*	225
borde brillante de un filamento *m*	bordo brilhante de um filamento *m*	яркое окаймление волокна *n* эмиссионная кайма волокна *f* яркая кайма волокна *f*	226
protuberancia de latitud baja *f*	protuberância de baixa latitude *f*	низкоширотный протуберанец *m*	227
protuberancia ecuatorial *f*	protuberância equatorial *f*	экваториальный протуберанец *m* протуберанец экваториальной зоны *f*	228
protuberancia de la latitud alta *f*	protuberância de alta latitude *f*	высокоширотный протуберанец *m*	229

No.	SOLAR ACTIVITY	ACTIVITÉ SOLAIRE	SONNENAKTIVITÄT
230	polar prominence	protubérance polaire *f*	polare Protuberanz *f* Protuberanz der Polarzone *f*
231	poleward motion of prominences poleward migration (of prominences)	mouvement des protubérances vers les pôles *m* remontée des filaments vers les pôles *f*	polwärtige Wanderung der Protuberanzen *f* Poldrift der Protuberanzen *f*
232	polar crown (of filaments)	couronne polaire (des filaments) *f*	polare Filamentkrone *f*
233	prominence area	aire des protubérances *f*	Protuberanzenfläche *f*
234	base of a prominence	base d'une protubérance *f*	Protuberanzenbasis *f*
235	prominence unit	unité de protubérance *f*	Protuberanzeneinheit *f*
236	daily areas of prominences	aires journalières des protubérances *fpl*	tägliche Protuberanzenflächen *fpl*
237	prominence frequency	fréquence des protubérances *f*	Protuberanzenhäufigkeit *f*
238	active prominence	protubérance active *f*	aktive Protuberanz *f*
239	prominence activation activation of a (quiescent) prominence	activation d'une protubérance (quiescente) *f*	Aktivierung einer (ruhenden) Protuberanz *f*
240	active prominence region	région à [de] protubérances actives *f*	aktives Protuberanzen-Gebiet *n*
241	centre [center] of attraction	centre d'attraction *m*	Anziehungszentrum *n*
242	motion pictures of prominences prominence movie [film]	film de protubérances *m*	Protuberanzenfilm *m*
243	interacting prominences interactive prominences	protubérances interactives *fpl*	interaktive Protuberanzen *fpl* Protuberanzen die aufeinander wirken *fpl*
244	prominence knot knot of a prominence	nœud d'une protubérance *m* nodosité dans une protubérance *f*	Protuberanzenknoten *m*
245	prominence thread prominence streamer	filet protubérantiel *m* cordon protubérantiel *m*	Protuberanzenfaden *m*
246	common active prominence	protubérance active commune *f*	gewöhnliche aktive Protuberanz *f*
247	coronal active prominence	protubérance active coronale *f*	koronale aktive Protuberanz *f*
248	eruptive prominence	protubérance éruptive *f*	eruptive Protuberanz *f*
249	quasi-eruptive prominence	protubérance quasi-éruptive *f*	quasi-eruptive Protuberanz *f*
250	common eruptive prominence	protubérance éruptive commune *f*	gewöhnliche eruptive Protuberanz *f*

protuberancia polar *f*	protuberância polar *f*	полярный протуберанец *m* протуберанец полярной зоны *m*	230
desplazamiento de las protuberancias hacia el polo *m* migración de protuberancias hacia el polo *f*	migração de protubérâncias para os pólos *f*	смещение протуберанцев к полюсам *n* полярный дрейф протуберанцев *m*	231
corona polar (de los filamentos) *f*	coroa polar (de filamentos) *f*	полярный венец волокон *m* полярная крона волокон *f*	232
área de las protuberancias *f*	área das protuberâncias *f*	площадь протуберанцев *f*	233
base de una protuberancia *f*	base de uma protuberância *f*	основание протуберанца *f* базис протуберанца *m*	234
unidad de protuberancia *f*	unidade de protuberância *f*	протуберанц-единица *f*	235
áreas diarias de protuberancias *fpl*	áreas diárias das protuberâncias *fpl*	ежедневные площади протуберанцев *fpl* среднесуточные площади протуберанцев *fpl*	236
frecuencia de las protuberancias *f*	freqüência de protuberâncias *f*	частота протуберанцев *f*	237
protuberancia activa *f*	protuberância ativa *f*	активный протуберанец *m*	238
activación de una protuberancia (quiescente) *f*	ativação de uma protuberância (quiescente) *f*	активизация спокойного протуберанца *f*	239
región de protuberancias activas *f*	região ativa de protuberâncias *f*	область активных протуберанцев *f*	240
centro de atracción *m*	centro de atração *m*	центр притяжения *m*	241
película (cinematográfica) de protuberâncias *m*	filme (cinematográfico) de protuberâncias *m*	кинофильм протуберанцев *m*	242
protuberancias interactivas *fpl*	protuberâncias interativas *fpl*	взаимодействующие протуберанцы *mpl*	243
nudo de una protuberancia *m* condensación de una protuberancia *f*	nó de uma protuberância *m*	узел протуберанца *m* сгусток протуберанца *m*	244
fibra [filete] de protuberancia *f* columna de una protuberancia *f*	fibra de uma protuberância *f* filamento de protuberância *m*	струя [лента] протуберанца *f* волокно протуберанца *f*	245
protuberancia activa ordinaria *f*	protuberância ativa comum *f*	обычный активный протуберанец *m*	246
protuberancia activa coronal *f*	protuberância ativa coronal *f*	корональный активный протуберанец *m*	247
protuberancia eruptiva *f*	protuberância eruptiva *f*	эруптивный протуберанец *m*	248
protuberancia casi-eruptiva *f*	protuberância quase-eruptiva *f*	квазиэруптивный протуберанец *m*	249
protuberancia eruptiva ordinaria *f*	protuberância eruptiva comum *f*	обычный эруптивный протуберанец *m*	250

	SOLAR ACTIVITY	ACTIVITÉ SOLAIRE	SONNENAKTIVITÄT
251	eruptive arch	arche éruptive *f*	eruptiver Bogen *m*
252	sunspot prominence spot prominence	protubérance de tache *f*	Fleckenprotuberanz *f*
253	cap Prominence	protubérance en coiffe *f*	cap Protuberanz *f* Protuberanz in Form einer kleinen Ausbuchtung *f* aktive Fleckenprotuberanz *f*
254	active sunspot prominence	protubérance active de tache *f*	aktive Fleckenprotuberanz *f*
255	surge	surge *f* jet de matière *m* surrection *f*	Surge *f* Spritz(fleck)protuberanz *f*
256	dark surge	jet de matière absorbante *m* surge sombre *f*	dunkle Surge *f* dunkle Spritz(fleck)protuberanz *f*
257	bright surge	jet de matière visible par son émission *m* surge brillante *f*	helle Surge *f* helle Spritz(fleck)protuberanz *f*
258	spray spray prominence	protubérance en éventail *f*	Spray *f* Fächerprotuberanz *f*
259	ejection	éjection *f*	Auswurf *m*
260	secondary prominence	protubérance secondaire *f*	sekundäre Protuberanz *f*
261	coronal cloud	nuage coronal *m*	koronale Wolke *f*
262	tornado prominence	protubérance tourbillonnaire *f*	Tornado-Protuberanz *f*
263	quiescent prominence	protubérance quiescente *f*	ruhende Protuberanz *f*
264	coronal prominence	protubérance coronale *f*	koronale Protuberanz *f*
265	metallic prominence	protubérance à raies métalliques *f*	Protuberanz mit Metallinien *f*
266	hydrogen prominence	protubérance d'hydrogène *f*	Wasserstoffprotuberanz *f*
267	hydrogen filament	filament d'hydrogène *m*	Wasserstoff-Filament *n*
268	metallic filament	filament métallique *m*	Metall-Filament *n*
269	funnel funnel prominence	protubérance en entonnoir *f*	Trichter *m* Trichter-Protuberanz *f*
270	loop prominence loops	protubérance en boucle *f* protubérance en arche *f*	Bogenprotuberanz *f* Loopprotuberanz *f*

arco eruptivo *m*	arco eruptivo *m*	эруптивная арка *f*	251
protuberancia de mancha *f*	protuberância de mancha *f*	протуберанец типа солнечных пятен *m*	252
protuberancia (en forma) de casquete *f*	protuberância de calota *f*	колпачковый протуберанец *m* протуберанец типа колпак *m*	253
protuberancia activa de mancha *f*	protuberância ativa de mancha *f*	активный протуберанец типа солнечных пятен *m*	254
surtidor *m* protuberancia en chorro *f* oleada *f*	surgimento *m* surge *m*	возвратный выброс *m* возвратный протуберанец *m*	255
surtidor o(b)scuro *m* chorro o(b)scuro *m* eyección de materia o(b)scura [absorbente] *f*	surgimento escuro *m* ejeção da matéria absorvente *f*	тёмный возвратный выброс [протуберанец] *m*	256
surtidor brillante *m* chorro brillante *m* eyección de materia brillante *f*	surgimento brilhante *m* ejeção da matéria brilhante *f*	яркий возвратный выброс [протуберанец] *m*	257
protuberancia en abanico *f*	protuberância em leque *f* esguicho *m*	веерообразный выброс *m* быстрый вспышечный выброс *m* струйчатый выброс *m*	258
eyección *f*	ejeção *f*	выброс *m* извержение *n*	259
protuberancia secundaria *f*	protuberância secundária *f*	вторичный протуберанец *m*	260
nube coronal *f*	nuvem coronal *f*	корональное облако *m*	261
protuberancia (en forma) de tornado *f*	protuberância turbilionária *f*	протуберанец торнадо *m*	262
protuberancia quiescente *f*	protuberância quiescente *f*	спокойный протуберанец *m*	263
protuberancia coronal *f*	protuberância coronal *f*	корональный протуберанец *m*	264
protuberancia con líneas (de emisión) metálicas *f*	protuberância de linhas metálicas *f*	металлический протуберанец *m*	265
protuberancia de hidrógeno *f*	protuberância de hidrogênio *f*	водородный протуберанец *m*	266
filamento de hidrógeno *m*	filamento de hidrogênio *m*	водородное волокно *n*	267
filamento metálico *m*	filamento metálico *m*	металлическое волокно *n*	268
embudo *m* protuberancia (en forma) de embudo *f*	funil *m* protuberância (em forma) de funil *f*	воронка *f* воронкообразный протуберанец *m*	269
protuberancia (en forma) de arco *f* bucles *mpl* lazos *mpl*	protuberância (em forma) de anel *f* protuberância (em forma) de arco *f*	петельный [петлеобразный] протуберанец *m* петли *fpl*	270

271	coronal rain	pluie coronale *f*	koronaler Regen *m* Koronaregen *m*
272	tree trunks	(protubérance en forme de) troncs d'arbres (*f*) *mpl*	Baumstämme *mpl* Baumstämme-Protuberanz *f*
273	trees	(en forme d') arbres *mpl*	Bäume *mpl*
274	hedgerow	(en forme de) haie *f*	Zaunreihe *f*
275	suspended cloud	(en forme de) nuage suspendu *m*	schwebende Wolke *f*
276	mound	(en forme de) tas *m*	Damm *n*
277	puff	(en forme de) bouffée *f*	Hauch *m* Bausch *m*
278	rising prominence	protubérance ascendante *f*	aufsteigende Protuberanz *f*
279	sudden disappearance (of a filament) S.D.F.	disparition brusque *f* D.B.	plötzliche Auflösung *f* D.B. plötzliches Verschwinden *n*
280	dissolution of a prominence	dissolution d'une protubérance *f*	Auflösung einer Protuberanz *f*
281	coronal condensation	condensation coronale *f*	koronale Kondensation *f* Koronakondensation *f*
282	permanent coronal condensation	condensation coronale permanente *f*	permanente koronale Kondensation *f* permanente Koronakondensation *f*
283	sporadic coronal condensation	condensation coronale sporadique *f*	sporadische koronale Kondensation *f* sporadische Koronakondensation *f*
284	green coronal region	région coronale (émettant la raie) verte *f*	grüne Koronaregion *f* grünes Koronagebiet *n*
285	red coronal region	région coronale (émettant la raie) rouge *f*	rote Koronaregion *f* rotes Koronagebiet *n*
286	coronal active region	région active de la couronne *f*	aktive Koronaregion *f* aktives Koronagebiet *n*
287	coronal ray coronal streamer	jet coronal *m*	Koronastrahl *m*
288	monochromatic coronal ray	jet coronal monochromatique *m*	monochromatischer Koronastrahl *m*

lluvia coronal *f*	chuva coronal *f*	корональный дождь *m*	271
troncos *mpl* protuberancia (en forma) de troncos de arboles *f*	protuberância (em forma) de tronco de árvore *f*	стволы деревьев *mpl*	272
(en forma de) árboles *mpl*	(com forma de) árvores *mpl*	деревья *npl*	273
(en forma de) seto *m* cerco *m* (Am)	(com forma de) alameda *f* cerca viva *f*	(живая) изгородь *f*	274
(en forma de) nube suspendida *f*	nuvem suspensa *f*	висящее облако *n* взвешенное облако *n*	275
(en forma de) montículo *m* terraplén *m*	(com forma de) escarpa *f*	холм *m*	276
(en forma de) soplo *m* bocanada *f*	(com forma de) sopro *m*	облако дыма *n* клуб дыма *m*	277
protuberancia ascendente *f*	protuberância ascendente *f*	поднимающийся протуберанец *m*	278
desaparición brusca *f*	desaparecimento súbito (de um filamento) *m*	внезапное изчезновение (волокна) *n*	279
disolución de una protuberancia *f*	dissolução de uma protuberância *f*	рассасывание протуберанца *n*	280
condensación coronal *f*	condensação coronal *f*	корональная конденсация *f*	281
condensación coronal permanente *f*	condensação coronal permanente *f*	неизменная корональная конденсация *f* перманентная корональная конденсация *f*	282
condensación coronal esporádica *f*	condensação coronal esporádica *f*	спорадическая корональная конденсация *f*	283
región verde de la corona *f*	região verde da coroa *f*	зелёная корональная область *f* область свечения зелёной корональной линии *f*	284
región roja de la corona *f*	região vermelha da coroa *f*	красная корональная область *f* область свечения красной корональной линии *f*	285
región activa coronal *f*	região coronal ativa *f*	корональная активная область *f*	286
rayo coronal *m* corriente coronal *f*	raio coronal *m* corrente coronal *m*	корональный луч *m*	287
rayo coronal monocromático *m*	raio coronal monocromático *m*	монохроматический корональный луч *m* корональный луч в монохроматическом свете *m*	288

289	highly ionized	hautement ionisé	hochionisiert
290	coronal line	raie coronale *f*	Koronalinie *f*
291	green line	raie verte *f*	grüne Koronalinie *f*
292	yellow line	raie jaune *f*	gelbe Koronalinie *f*
293	red line	raie rouge *f*	rote Koronalinie *f*
294	plumes polar rays	plumes polaires *fpl*	Polarstrahlen *mpl* polare Strahlen *mpl* polare Büschel *mpl*
295	fan helmet	éventail coronal *m* jet en (forme de) bulbe *m*	koronaler Fächer *m*
296	fan ray	jet en (forme d') éventail *m* jet d'éventail *m*	Fächerstrahl *m*
297	dark domes (around prominences)	cavité sombre (autour des protubérances) *f*	dunkler Raum (um Protuberanzen) *m*
298	coronal arch	arche coronale *f*	koronaler Bogen *m* Koronabogen *m*
299	arch system	système d'arches *m*	Bogensystem *n*

altamente [fuertemente] ionizado	altamente ionizado	высокоионизованный	289
línea coronal *f*	raia [linha] coronal *f*	корональная линия *f*	290
línea verde *f*	raia [linha] verde *f*	зелёная линия *f*	291
línea amarilla *f*	raia [linha] amarela *f*	жёлтая линия *f*	292
línea roja *f*	raia [linha] vermelha *f*	красная линия *f*	293
rayos polares *mpl* plumas polares *fpl*	raios polares *mpl* plumas *fpl*	полярные щеточки *fpl* полярные лучи *mpl*	294
bulbo coronal *m* abanico coronal *m* casco coronal *m*	leque coronal *m* elmo coronal *m*	опахало *n* шлем *m*	295
rayo de un bulbo [abanico] coronal *m*	raio de leque coronal *m*	опахальный луч *m* луч опахала *m*	296
bóveda obscura (alrededor de las protuberancias) *f*	domos escuros (em torno das protuberâncias) *mpl*	тёмные купола (вокруг протуберанцев) *mpl*	297
arco coronal *m* bucle coronal *m*	arco coronal *m*	корональная дуга *f*	298
sistema de arcos (coronales) *m* sistema de bucles *m*	sistema de arcos *m*	система дуг *f* арочная система *f*	299

90

DEEP SPACE

ESPACE LOINTAIN

ERDFERNER WELTRAUM

ESPACIO REMOTO

ESPAÇO REMOTO

ДАЛЬНЫЙ КОСМОС

91	STARS	ÉTOILES	STERNE
1	star	étoile *f*	Stern *m*
1a	starry sky	ciel étoilé *m*	Sternhimmel *m*
2	stellar astronomy	astronomie stellaire *f*	Stellarastronomie *f*
3	distance	distance *f*	Entfernung *f* Abstand *m* Distanz *f*
4	stellar parallax	parallaxe stellaire *f*	Sternparallaxe *f*
5	trigonometric parallax	parallaxe trigonométrique *f*	trigonometrische Parallaxe *f*
6	annual parallax	parallaxe annuelle *f*	jährliche Parallaxe *f*
7	cluster parallax moving-cluster parallax	parallaxe de groupe *f* parallaxe de courant stellaire *f*	Sternstromparallaxe *f* Gruppenparallaxe *f*
8	statistical parallax	parallaxe statistique *f*	statistische Parallaxe *f*
8a	Cepheid parallax	parallaxe des [par les] céphéides *f*	Cepheiden-Parallaxe *f*
9	secular parallax	parallaxe séculaire *f*	säkulare Parallaxe *f* Säkularparallaxe *f*
10	relative (trigonometric) parallax	parallaxe (trigonométrique) relative *f*	relative (trigonometrische) Parallaxe *f*
11	absolute parallax	parallaxe absolue *f*	absolute Parallaxe *f*
12	dynamic(al) parallax	parallaxe dynamique *f* parallaxe hypothétique *f*	dynamische Parallaxe *f* hypothetische Parallaxe *f*
13	Strömgren parallax parallax of Strömgren spheres	parallaxe de Strömgren *f* parallaxe des sphères de Strömgren *f*	Parallaxe aus dem Strömgrenradius *f* Parallaxe der Strömgrensphären *f*
14	photometric parallax	parallaxe photométrique *f*	photometrische Parallaxe *f*
15	spectroscopic parallax	parallaxe spectroscopique *f*	spektroskopische Parallaxe *f*
16	distance modulus	module de distance *m*	Entfernungsmodul *m*
17	parsec pc	parsec *m*	Parsec *n* Parsek *n*
18	light year light years l.y. ly lt-yr	année-lumière *f* années-lumière *fpl* année de lumière *f* anné-de-lumière *f* al a.l.	Lichtjahr *n* Lj

ESTRELLAS	ESTRELAS	ЗВЕЗДЫ	
estrella *f*	estrela *f*	звезда *f*	1
cielo estrellado *m*	céu estrelado *m*	звёздное небо *n*	1a
astronomía estelar *f*	astronomia estelar *f*	звёздная астрономия *f*	2
distancia *f*	distância *f*	расстояние *n*	3
paralaje estelar *f*	paralaxe estelar *f*	звёздный параллакс *m* параллакс звезды *m*	4
paralaje trigonométrica *f*	paralaxe trigonométrica *f*	тригонометрический параллакс *m*	5
paralaje anual *f*	paralaxe anual [ânua] *f*	годичный параллакс *m*	6
paralaje de cúmulo (en movimiento) *f* paralaje de grupo *f*	paralaxe de corrente estelar *f* paralaxe de cúmulo móvel *f* paralaxe de grupo *f*	групповой параллакс *m* параллакс движущихся скоплений *m*	7
paralaje estadística *f*	paralaxe estatística *f*	статистический параллакс *m*	8
paralaje cefeídica *f*	paralaxe cefêidica *f* paralaxe de cefeidas *f* paralaxe cefeídica *f*	цефеидный параллакс *m* параллакс, вычисленный по цефеидам *m* параллакс цефеид *m*	8a
paralaje secular *f*	paralaxe secular *f*	вековой параллакс *m*	9
paralaje (trigonométrica) relativa *f*	paralaxe (trigonométrica) relativa *f*	относительный (тригонометрический) параллакс *m*	10
paralaje absoluta *f*	paralaxe absoluta *f*	абсолютный параллакс *m*	11
paralaje dinámica *f*	paralaxe dinâmica *f*	динамический параллакс *m*	12
paralaje de Strömgren *f* paralaje de esferas de Strömgren *f*	paralaxe de Strömgren *f* paralaxe das esferas de Strömgren *f*	параллакс Стрёмгрена *m* параллакс сферы Стрёмгрена *m*	13
paralaje fotométrica *f*	paralaxe fotométrica *f*	фотометрический параллакс *m*	14
paralaje espectroscópica *f*	paralaxe espe(c)troscópica *f*	спектральный параллакс *m* спектроскопический параллакс *m*	15
módulo de distancia *m*	módulo de distância *m*	модуль расстояния *m*	16
parsec *m*	parsec *m*	парсек *m* пс, пк	17
año luz *m* año-luz *m* año de luz *m* AL	ano-luz *m* anos-luz *mpl*	световой год *m*	18

	STARS	ÉTOILES	STERNE
19	proper motion	mouvement propre *m*	Eigenbewegung *f*
20	annual proper motion	mouvement propre annuel *m*	jährliche Eigenbewegung *f*
21	proper motion in right ascension	mouvement propre en ascension droite *m*	Eigenbewegung in Rektaszension *f*
22	proper motion in declination	mouvement propre en déclinaison *m*	Eigenbewegung in Deklination *f*
23	apex of the Sun solar apex	apex du Soleil *m* apex solaire *m*	Apex der Sonne(nbewegung) *m* Zielpunkt der Sonnenbewegung *m*
24	radial velocity line-of-sight velocity	vitesse radiale *f* vitesse le long du rayon visuel *f*	Radialgeschwindigkeit *f* Geschwindigkeit in Beobachtungsrichtung [im Sehstrahl] *f*
25	Doppler shift	déplacement Doppler *m* déplacement Doppler-Fizeau *m*	Dopplerverschiebung *f*
26	positive velocity recession velocity	vitesse positive *f* vitesse de fuite *f*	positive Geschwindigkeit *f* Fluchtgeschwindigkeit *f*
27	negative velocity approach velocity	vitesse négative *f* vitesse d'approche *f*	negative Geschwindigkeit *f* Annäherungsgeschwindigkeit *f*
28	line of sight line of vision line-of-sight	rayon visuel *m* ligne de vision *f* axe de visée *m*	Sehstrahl *m* Visionsradius *m* Visionslinie *f*
29	in the line-of-sight	le long du rayon visuel en visibilité directe	auf dem Sehstrahl in Beobachtungsrichtung
30	spatial velocity space velocity	vitesse spatiale *f*	Raumgeschwindigkeit *f*
31	peculiar motion peculiar velocity	mouvement particulier *m* vitesse particulière *f*	Pekuliarbewegung *f*
32	stereocomparator	stéréo-comparateur *m*	Stereokomparator *m*
33	parallactic motion	mouvement parallactique *m*	parallaktische Bewegung *f*
34	blink microscope blink comparator	microscope à papillotement *m* blink microscope *m* blink comparateur *m*	Blinkkomparator *m* Blinkmikroskop *n*
35	stellar rotation rotation of stars	rotation stellaire *f* rotation des étoiles *f*	Rotation der Sterne *f* Sternrotation *f*
36	non-uniform rotation	rotation non-uniforme *f*	ungleichförmige Rotation *f*
37	rotating star	étoile en rotation *f* étoile tournante *f*	rotierender Stern *m*
38	non-rotating star	étoile sans rotation *f* étoile non-tournante *f*	nichtrotierender Stern *m*

ESTRELLAS	ESTRELAS	ЗВЕЗДЫ	
movimiento propio *m*	movimento próprio *m*	собственное движение *n*	19
movimiento propio anual *m*	movimento próprio anual *m*	годичное собственное движение *n*	20
movimiento propio en ascensión recta *m*	movimento próprio em ascensão reta *m*	собственное движение по прямому восхождению *n*	21
movimiento propio en declinación *m*	movimento próprio em declinação *m*	собственное движение по склонению *n*	22
ápex del Sol *m* ápex solar *m*	ápex solar *m* ápex do Sol *m*	солнечный апекс *m* апекс (движения) Солнца *m*	23
velocidad radial *f* velocidad en la visual *f*	velocidade radial *f* velocidade na linha de visada [visão] *f*	лучевая скорость *f* скорость вдоль луча зрения *f*	24
desplazamiento Doppler *m* corrimiento Doppler *m*	deslocamento Doppler *m* desvio Doppler *m*	доп(п)леровское смещение *n*	25
velocidad positiva *f* velocidad de recesión *f*	velocidade positiva *f* velocidade de recessão *f* velocidade de afastamento *f*	положительная скорость *f* скорость удаления *f*	26
velocidad negativa *f* velocidad de aproximación *f*	velocidade negativa *f* velocidade de aproximação *f*	отрицательная скорость *f* скорость приближения *f*	27
línea de la visual *f* visual *f* dirección de la visual *f*	linha de visada *f* raio de visada *m* linha de visão *f*	луч зрения *m*	28
en la visual en la dirección de la visual	na linha de visada na linha de visão	вдоль луча зрения	29
velocidad espacial *f*	velocidade espacial *f*	пространственная скорость *f*	30
movimiento peculiar *m*	movimento peculiar *m*	пекулярное движение *n*	31
estereocomparador *m*	estereocomparador *m*	стереокомпаратор *m*	32
movimiento paraláctico *m*	movimento paralá(c)tico *m*	параллактическое движение *n*	33
microscopio "blink" *m* microscopio de centelleo *m* microscopio de salto *m*	microscópio "blink" *m* microscópio de cintilação *m*	блинк-компаратор *m* блинк-микроскоп *m*	34
rotación estelar *f* rotación de las estrellas *f*	rotação estelar *f* rotação das estrelas *f*	звёздное вращение *n* вращение звёзд *n*	35
rotación no uniforme *f*	rotação não-uniforme *f*	неоднородное вращение *n*	36
estrella en rotación *f* estrella rotante *f*	estrela em rotação *f* estrela rotante *f*	вращающаяся звезда *f*	37
estrella sin rotación estrella no rotante *f*	estrela sem rotação *f* estrela não-rotante *f*	невращающаяся звезда *f*	38

	STARS	ÉTOILES	STERNE
39	rapidly rotating star	étoile en rotation rapide *f*	schnell rotierender Stern *m*
40	pole-on star	étoile observée dans la direction de son pôle *f*	Stern in Pol-Aufsicht *m* Stern mit Rotationsachse in Richtung des Sehstrahls *m*
41	gravity darkening	assombrissement gravitationnel *m* assombrissement de [par] gravité *m*	Gravitationsverdunk(e)lung *f*
42	rotor rotating body rotator	rotateur *m* corps tournant *m* rotor *m*	Rotator *m* rotierender Körper *m*
43	rapid rotor rapid rotator	rotateur rapide *m* rotor rapide *m*	schnell rotierender Körper *m*
44	rotational distortion (of velocity curve) Rossiter effect rotational effect (on velocity curve)	distorsion rotationnelle (d'une courbe de vitesse) *f* effet Rossiter *m* effet rotationnel (sur la courbe de vitesse) *m*	Rotationsverzerrung *f* Rossiter-Effekt *m*
45	braking of the rotation	freinage de la rotation *m*	Abbremsung der Rotation *f*
46	braking mechanism	mécanisme de freinage *m*	Bremsmechanismus *m*
47	magnetic braking	freinage magnétique *m*	magnetische Abbremsung *f*
48	rotation(al) velocity [speed] velocity of rotation	vitesse de rotation *f* vitesse rotatoire *f*	Rotationsgeschwindigkeit *f* Dreh(ungs)geschwindigkeit *f*
49	equatorial velocity	vitesse équatoriale *f*	äquatoriale Geschwindigkeit *f*
50	inclination of the rotation axis to . . .	inclinasion de l'axe de rotation sur . . . *f*	Neigung der Rotationsachse gegen . . . *f*
51	magnetic star	étoile magnétique *f*	magnetischer Stern *m*
52	oblique rotator	rotateur oblique *m*	schiefer Rotator *m*
53	fossil magnetic field relic magnetic field	champ magnétique fossile *m* champ magnétique résiduel *m*	fossiles Magnetfeld *n* übrigbleibendes Magnetfeld *n*
54	brightness	éclat *m*	Helligkeit *f*
55	stellar brightness	éclat stellaire *m*	Sternhelligkeit *f*
56	stellar magnitude magnitude	magnitude *f* magnitude stellaire *f*	Grössenklasse *f* Sterngrösse *f* Magnitudo *f*
57	magnitude system	système de magnitudes *m*	Grössenklassensystem *n*

ESTRELLAS	ESTRELAS	ЗВЕЗДЫ	
estrella en rotación rápida *f*	estrela em [de] rotação rápida *f*	быстровращающаяся звезда *f*	39
estrella vista por el polo *f* estrella con el eje de rotación en la dirección del observador *f*	estrela com o pólo à vista *f* estrela vista do pólo *f*	звезда, видимая с полюса *f*	40
o(b)scurecimiento gravitacional *m*	obscurecimento gravitacional *m* escurecimento por gravidade *m*	гравитационное потемнение *n*	41
rotor *m* cuerpo giratorio *m* cuerpo rotante *m*	rotor *m* corpo em rotação *m* corpo rotante *m*	ротатор *m* вращающееся тело *n*	42
rotor rápido *m*	rotor rápido *m*	быстрый ротатор *m*	43
distorsión rotacional (de la curva de velocidad) *f* efecto Rossiter *m*	distorção rotacional (da curva de velocidade) *f* efeito Rossiter *m*	вращательная дисторсия *f* влияние вращения на кривую лучевой скорости *m* вращательный эффект *m*	44
frenado de la rotación *m*	frenagem [fre(n)amento] da rotação *f*[*m*]	торможение вращения *n*	45
mecanismo de frenado *m*	mecanismo de frenagem [fre(n)amento, travagem] *m*	механизм торможения *m*	46
frenado magnético *m*	frenagem magnética *f* fre(n)amento magnético *m*	магнитное торможение *n*	47
velocidad de rotación *f* velocidad rotacional *f*	velocidade de rotação *f* velocidade rotacional *f*	скорость вращения *f* вращательная скорость *f*	48
velocidad ecuatorial *f*	velocidade equatorial *f*	экваториальная скорость *f*	49
inclinación del eje de rotación con respecto a . . . *f*	inclinação do eixo de rotação em relação a . . . *f*	наклонение оси вращения к . . . *n* наклон оси вращения к . . . *m*	50
estrella magnética *f*	estrela magnética *f*	магнитная звезда *f*	51
rotor oblicuo *m*	rotor oblíquo *m*	наклонный ротатор *m*	52
campo magnético fósil *m* campo magnético residual *m*	campo magnético fóssil *m*	реликтовое магнитное поле *n* изначальное магнитное поле *n*	53
brillo *m*	brilho *m*	блеск *m* яркость *f*	54
brillo estelar *m*	brilho estelar *m*	звёздная яркость *f* блеск звезды *m*	55
magnitud estelar *f*	magnitude estelar *f*	звёздная величина *f* яркость звезды *f*	56
sistema de magnitudes *m*	sistema de magnitudes *m*	система звёздных величин *f*	57

	STARS	ÉTOILES	STERNE
58	Pogson's equation magnitude-intensity relation	équation de Pogson *f* relation magnitude-intensité *f*	Pogsonsche Gleichung *f* Beziehung zwischen Grössenklassen und Intensitäten *f*
59	magnitude difference	différence de magnitudes *f*	Grössenklassendifferenz *f* Grössenklassenunterschied *m*
60	intensity ratio illuminance ratio	rapport d'intensité *m* rapport des éclairements (lumineux) *m*	Intensitätsverhältnis *n* Verhältnis der Beleuchtungsstärken *n*
61	apparent magnitude apparent brightness	magnitude apparente *f* éclat apparent *m*	scheinbare Sternhelligkeit *f* scheinbare Helligkeit *f* scheinbare Sterngrösse *f*
62	limiting magnitude limiting brightness	magnitude limite *f* éclat limite *m*	Grenzhelligkeit *f* Grenzgrösse *f*
63	zenithal magnitude zenith magnitude	magnitude réduite au zénith *f* magnitude zénithale *f*	Zenithelligkeit *f* Sterngrösse im Zenit *f*
64	absolute magnitude absolute brightness	magnitude absolue *f* éclat absolu *m*	absolute Sterngrösse *f* absolute (Stern)helligkeit *f*
65	visual magnitude visual brightness	magnitude visuelle *f* éclat visuel *m*	visuelle Helligkeit *f* visuelle Sterngrösse *f*
66	photographic magnitude photographic brightness	magnitude photographique *f* éclat photographique *m*	photographische Helligkeit *f* photographische Sterngrösse *f*
67	electrographic magnitude electrographic brightness	magnitude électro(no)graphique *f* éclat électrographique *m* éclat électronographique *m*	elektrographische Helligkeit *f* elektronographische Sterngrösse *f*
68	bolometric magnitude bolometric brightness	magnitude bolométrique *f* éclat bolométrique *m*	bolometrische Helligkeit *f* bolometrische Sterngrösse *f*
69	photovisual magnitude photovisual brightness	magnitude photovisuelle *f* éclat photovisuel *m*	photovisuelle Helligkeit *f* photovisuelle Sterngrösse *f*
70	radiometric magnitude radiometric brightness	magnitude radiométrique *f* éclat radiométrique *m*	radiometrische Helligkeit *f* radiometrische Sterngrösse *f*
71	infrared magnitude infrared brightness	magnitude infrarouge *f* éclat infrarouge *m*	infrarote Helligkeit *m* infrarote Sterngrösse *f*
72	colour system (GB) color system (US)	système de couleurs *m* système à bande passante (étroite ou large) *m*	Farbsystem *n*
73	colour equation color equation	équation de couleur *f*	Farbgleichung *f*

ecuación de Pogson *f* relación magnitud-intensidad *f*	equação de Pogson *f* relação magnitude-intensidade *f*	уравнение [формула] Погсона *n(f)* зависимость звёздной величины от освещённости *f*	58
diferencia de magnitudes *f*	diferença de magnitudes *f*	разность звёздных величин *f*	59
relación [razón] de intensidades *f* razón de iluminancias *f*	relação [razão] de intensidades *f* razão de iluminâncias *f*	отношение интенсивностей *n* отношение освещённостей *n*	60
magnitud aparente *f* brillo aparente *m*	magnitude aparente *f* brilho aparente *m*	видимая (звёздная) величина *f* видимий блеск звезды *m*	61
magnitud límite *f* brillo límite *m*	magnitude limite *f* brilho limite *m*	предельная (звёздная) величина *f* предельный блеск *m*	62
magnitud reducida al zenit [cenit] *f* magnitud cenital [zenital] *f*	magnitude zenital *f* magnitude reduzida ao zênite *f*	зенитная (звёздная) величина *f* звёздная величина в зените *f*	63
magnitud absoluta *f* brillo absoluto *m*	magnitude absoluta *f* brilho absoluto *m*	абсолютная (звёздная) величина *f* абсолютный блеск *m*	64
magnitud visual *f* brillo visual *m*	magnitude visual *f* brilho visual *m*	визуальная (звёздная) величина *f* визуальный блеск *m*	65
magnitud fotográfica *f* brillo fotográfico *m*	magnitude fotográfica *f* brilho fotográfico *m*	фотографическая (звёздная) величина *f* фотографический блеск *m*	66
magnitud electrográfica *f* brillo electrográfico *m*	magnitude ele(c)trográfica *f* brilho ele(c)trográfico *m*	электро(но)графическая (звёздная) величина *f* электро(но)графический блеск *m*	67
magnitud bolométrica *f* brillo bolométrico *m*	magnitude bolométrica *f* brilho bolométrico *m*	болометрическая (звёздная) величина *f* болометрический блеск *m*	68
magnitud fotovisual *f* brillo fotovisual *m*	magnitude fotovisual *f* brilho fotovisual *m*	фотовизуальная величина *f* фотовизуальный блеск *m*	69
magnitud radiométrica *f* brillo radiométrico *m*	magnitude radiométrica *f* brilho radiométrico *m*	радиометрическая (звёздная) величина *f* радиометрический блеск *m*	70
magnitud infrarroja *f* brillo infrarrojo *m*	magnitude infravermelha *f* brilho infravermelho *m*	инфракрасная (звёздная) величина *f* инфракрасный блеск *m*	71
sistema de colores *m*	sistema de cores *m*	цветовая система *f*	72
ecuación de color *f*	equação de cor *f*	цветовое уравнение *n* уравнение цвета *n*	73

	STARS	ÉTOILES	STERNE
74	intrinsic colo(u)r	couleur intrinsèque *f*	Eigenfarbe *f*
75	intrinsic colo(u)r index	indice de couleur intrinsèque *m*	Eigenfarb(en)index *m*
76	colo(u)r index	indice de couleur *m*	Farbenindex *m* Farbindex *m*
77	colo(u)r excess	excés de couleur *m*	Farb(en)exzess *m*
78	ultraviolet excess UV excess	excès ultraviolet *m* excès UV *m*	Ultraviolett-Überschuss *m* UV-Exzess *m* ultravioletter Überschuss *m*
79	infrared excess IR excess	excès infrarouge *m* excès IR *m*	infraroter Überschuss *m* IR-Exzess *m* Infrarot-Überschuss *m*
80	bolometric correction	correction bolométrique *f*	bolometrische Korrektion *f*
81	heat index	indice de chaleur *m*	Wärmeindex *m*
82	faint star	étoile faible *f*	schwacher Stern *m*
83	bright star	étoile brillante *f*	heller Stern *m*
84	halo star	étoile du halo *f*	Halostern *m* Stern der Halopopulation *m*
85	disk [disc] star	étoile du disque (galactique) *f*	Stern der Scheibenpopulation *m* Scheibenstern *m*
86	non-spherical star	étoile non-sphérique *f*	nichtsphärischer Stern *m*
87	North Polar Sequence Polar Sequence	séquence polaire (internationale) *f* séquence polaire nord *f*	(internationale) Polsequenz *f* Nordpolarsequenz *f*
88	stellar spectrum (pl. stellar spectra)	spectre stellaire *m*	Sternspektrum *n* (pl. Sternspektren)
89	normal stellar spectrum	spectre stellaire normal *m*	normales Sternspektrum *n*
90	spectral classification	classification spectrale *f*	Spektralklassifikation *f* Klassifikation der Sternspektren *f*
91	Harvard system of spectral classification Harvard classification	système de classification spectrale de Harvard *m* clasification de Harvard *f* système Harvard des spectres *m*	Harvard-Klassifikation *f* Harvard-System der Spektralklassen *n*
92	peculiar star star with a peculiar spectrum	étoile particulière *f* étoile à particularités spectrales *f* étoile à spectre particulier	Stern mit pekuliarem Spektrum *m* Stern mit Besonderheiten (im Spektrum) *m*

color intríseco *m*	cor intrínseca *f*	истинный цвет *m*	74
índice intrínseco de color *m*	índice de cor intrínseco *m*	истинный [собственный] показатель цвета *m*	75
índice de color *m*	índice de cor *m*	показатель цвета *m* колор-индекс *m*	76
exceso de color *m*	excesso de cor *m*	избыток цвета *m*	77
exceso ultravioleta *m* exceso UV *m*	excesso (de) ultravioleta *m* excesso UV *m*	ультрафиолетовый избыток *m* УФ-избыток *m* избыток УФ-излучения *m*	78
excesso infrarrojo exceso (de) IR *m*	excesso (de) infravermelho *m* excesso (de) IV *m*	инфракрасный избыток *m* ИК-избыток *m* избыток инфракрасного излучения *m*	79
corrección bolométrica *f*	correção bolométrica *f*	болометрическая поправка *f*	80
índice de calor *m*	índice de calor *m*	тепловой показатель *m* показатель тепла *m*	81
estrella débil *f*	estrela fraca *f*	слабая звезда *f*	82
estrella brillante *f*	estrela brilhante *f*	яркая звезда *f*	83
estrella del halo *f*	estrela do halo *f*	звезда (населения) гало *f*	84
estrella del disco *f*	estrela do disco *f*	звезда (населения) диска *f*	85
estrella no esférica *f*	estrela anesférica estrela não-esférica *f*	несферическая звезда *f*	86
secuencia polar (internacional) *f* secuencia boreal *f*	seqüência boreal *f* seqüência polar internacional [norte] *f*	СПН *m* Северный Полярный Ряд *m* Международный Полярный Ряд *m*	87
espectro estelar *m*	espe(c)tro estelar *m*	спектр звезды *m* звёздный спектр *m*	88
espectro estelar normal *m*	espe(c)tro estelar normal *m*	нормальный звёздный спектр *m* неискажённый звёздный спектр *m*	89
clasificación espectral *f*	classificação espe(c)tral *f*	спектральная классификация *f*	90
sistema de clasificación espectral de Harvard *m* clasificación de Harvard *f*	sistema de classificação espe(c)tral de Harvard *m* classificação (espe(c)ral) de Harvard *f*	гарвардская (спектральная) классификация *f* гарвардская система спектральных классов *f*	91
estrella peculiar *f* estrella con peculiaridades espectrales *f* estrella con espectro peculiar *f*	estrela peculiar *f* estrela com peculiaridades espe(c)trais *f* estrela com espe(c)tro peculiar *f*	пекулярная звезда *f* звезда с пекулярным спектром *f*	92

93	spectral peculiarity	particularité spectrale *f*	spektrale Besonderheit *f*
94	spectral sequence	séquence spectrale *f*	Spektralsequenz *f*
95	normal star common star	étoile normale *f* étoile courante [banale] *f*	normaler Stern *m* gewöhnlicher Stern *m*
96	anomalous star	étoile ano(r)male *f*	anomaler Stern *m*
97	red star	étoile rouge *f*	roter Stern *m*
98	yellow	jaune	gelb
99	blue	bleu	blau
100	intrinsic property	propriété intrinsèque *f*	stellare Eigenschaft *f*
101	star type stellar type	type d'étoile *m* type stellaire *m*	Sterntyp *m*
102	early-type stars early stars	étoiles des premiers types *fpl*	Sterne frühen Typs *mpl* frühe Sterntypen *mpl*
103	late-type stars late stars *fpl*	étoiles des types avancés *fpl* étoiles des derniers types spectraux *fpl* étoiles tardives *fpl*	Sterne späten Typs *mpl* späte Sterntypen *mpl*
104	solar-type star	étoile de type solaire *f*	Stern vom (Spektral-) Typ der Sonne *m*
105	two-dimensional spectral classification	classification spectrale bidimensionnelle *f*	zweidimensionale Spektralklassifikation *f*
106	three-dimensional spectral classification	classification spectrale tridimensionelle *f*	dreidimensionale Spektralklassifikation *f*
107	spectral class	classe spectrale *f*	Spektralklasse *f*
108	spectral type	type spectral *m*	Spektraltyp *m* Spektraltypus *m*
109	colo(u)r temperature	température de couleur *f*	Farbtemperatur *f*
110	classification criteria (sg. criterion)	critères de classification *mpl*	Klassifikationskriterien *npl* Klassifizierungskriterien *npl* (sg. ... kriterium)
111	absolute gradient	gradient absolu *m*	absoluter Gradient *m*
112	spectrophotometric gradient	gradient spectrophotométrique *m*	spektrophotometrischer Gradient *m*
113	effective temperature	température effective *f*	effektive Temperatur *f*

ESTRELLAS	ESTRELAS	ЗВЕЗДЫ	
peculiaridad espectral *f*	peculiaridade espe(c)tral *f*	спектральная особенность *f*	93
secuencia espectral *f*	seqüência espe(c)tral *f*	спектральная последовательность *f*	94
estrella normal *f* estrella común *f*	estrela normal *f* estrela comum *f*	нормальная звезда *f* обычная звезда *f*	95
estrella anómala *f*	estrela anômala *f*	аномальная звезда *f*	96
estrella roja *f*	estrela vermelha *f*	красная звезда *f*	97
amarillo	amarelo	жёлтый	98
azul	azul	голубой	99
propiedad intrínseca *f*	propriedade intrínseca *f*	собственное свойство (звезды) *n* подлинная особенность *f*	100
tipo estelar *m* tipo de estrellas *m*	tipo estelar *m* tipo de estrelas *m*	тип звёзд *m* звёздный тип *m*	101
estrellas de tipos tempranos *fpl* estrellas de los primeros tipos *fpl*	estrelas dos primeiros tipos *fpl* estrelas dos tipos iniciais *fpl* estrelas anteriores *fpl*	звёзды ранних спектральных типов (классов) *fpl* ранние звёзды *fpl*	102
estrellas de tipos tardíos *fpl* estrellas de los últimos tipos *fpl*	estrellas dos tipos finais *fpl* estrelas dos últimos tipos *fpl* estrelas de tipo avançado [tardio] *fpl* estrelas posteriores *fpl*	звёзды поздних спектральных типов [классов] *fpl* поздние звёзды *fpl*	103
estrella de tipo solar *f*	estrela de tipo solar *f*	звезда солнечного типа *f* звезда типа Солнца *f*	104
clasificación espectral bidimensional *f*	classificação espe(c)tral bidimensional *f*	двумерная спектральная классификация *f*	105
clasificación espectral tridimensional *f*	classificação espe(c)tral tridimensional *f*	трёхмерная спектральная классификация *f*	106
clase espectral *f*	classe espe(c)tral *f*	спектральный класс *m*	107
tipo espectral *m*	tipo espe(c)tral *f*	спектральный тип *m*	108
temperatura de color *f*	temperatura de cor *f*	цветовая температура *f*	109
criterios de clasificación *mpl*	critérios de classificação *mpl*	критерии классификации *mpl* классификационные критерии *mpl*	110
gradiente absoluto *m*	gradiente absoluto *m*	абсолютный градиент *m*	111
gradiente espectrofotométrico *m*	gradiente espe(c)trofotométrico *m*	спектрофотометрический градиент *m*	112
temperatura efectiva *f*	temperatura efetiva *f*	эффективная температура *f*	113

	STARS	ÉTOILES	STERNE
114	brightness temperature (at a given wavelength λ)	température de brillance monochromatique (pour la longueur d'onde λ) *f*	Strahlungstemperatur (für eine gegebene Wellenlänge λ) *f*
115	kinetic temperature	température cinétique *f*	kinetische Temperatur *f*
116	electron temperature	température électronique *f*	Elektronentemperatur *f*
117	ionization temperature	température d'ionisation *f*	Ionisationstemperatur *f*
118	excitation temperature	température d'excitation *f*	Anregungstemperatur *f*
119	rotation temperature	température de rotation *f*	Rotationstemperatur *f*
120	vibration temperature	température de vibration *f*	Schwingungstemperatur *f*
120a	dissociation temperature	température de dissociation *f*	Dissoziationstemperatur *f*
121	stellar mass	masse stellaire *f*	Sternmasse *f*
122	solar mass M⊙	masse solaire *f* masse du Soleil *f*	Sonnenmasse *f* Masse der Sonne *f*
123	solar units (M⊙, L⊙)	unités solaires *fpl*	Sonneneinheiten *fpl*
124	low-mass star light star	étoile de masse faible *f* étoile légère *f*	Stern (von) kleiner Masse *m* massearmer Stern *m*
125	intermediate-mass star middle-weight star	étoile de masse intermédiaire *f*	Stern mittlerer Masse *m*
126	massive star heavy-weight star	étoile massive *f*	massereicher Stern *m* Stern (von) grosser Masse *m*
127	supermassive star	étoile supermassive *f* étoile de très grande masse *f*	supermassiver Stern *m*
128	luminosity	luminosité *f*	Leuchtkraft *f*
129	mass-luminosity relation mass-luminosity diagram	relation masse-luminosité *f* diagramme masse-luminosité *m*	Masse-Leuchtkraft-Beziehung *f* Masse-Leuchtkraft-Diagramm *n*
130	Hertzsprung-Russell-diagram spectrum-luminosity relation HR diagram	diagramme de Hertzsprung-Russell *m* relation luminosité-type spectral *f* diagramme luminosité-type spectral *m*	Hertzsprung-Russell-Diagramm *n* Spektrum-Leuchtkraft-Beziehung *f* HR-Diagramm *n*
131	main sequence principal sequence	séquence principale *f* série principale *f*	Hauptreihe *f* Hauptsequenz *f*
132	horizontal branch	branche horizontale *f*	Horizontalast *m*

ESTRELLAS	ESTRELAS	ЗВЕЗДЫ	
temperatura de brillo (a la longitud de onda λ) *f*	temperatura de brilho monocromático (para o comprimento λ) *f*	яркостная температура (для данной длины волны λ) *f*	114
temperatura cinética *f*	temperatura cinética *f*	кинетическая температура *f*	115
temperatura electrónica *f*	temperatura ele(c)trônica *f*	электронная температура *f*	116
temperatura de ionización *f*	temperatura de ionização *f*	ионизационная температура *f*	117
temperatura de excitación *f*	temperatura de excitação *f*	температура возбуждения *f*	118
temperatura de rotación *f*	temperatura de rotação *f*	вращательная температура *f*	119
temperatura de vibración *f*	temperatura de vibração *f*	колебательная температура *f*	120
temperatura de dissociación *f*	temperatura de dissociação *f*	диссоциационная температура *f*	120a
masa estelar *f*	massa estelar *f*	масса звезды *f* звёздная масса *f*	121
masa solar *f* masa del Sol *f*	massa solar *f* massa do Sol *f*	масса Солнца *f* солнечная масса *f*	122
unidades solares *fpl*	unidades solares *fpl*	единицы Солнца *fpl* солнечные единицы *fpl* (в единицах Солнца)	123
estrella de masa pequeña *f* estrella liviana [poco masiva] *f*	estrela de pequena massa *f* estrela leve [poco massiva] *f*	звезда малой массы *f*	124
estrella de masa intermedia *f*	estrela de massa intermédia [intermediária, moderada] *f*	звезда средней массы *f* звезда промежуточной массы *f*	125
estrella masiva *f*	estrela massiva *f*	массивная звезда *f* звезда большой масы *f*	126
estrella supermasiva *f*	estrela supermassiva *f*	сверхмассивная звезда *f*	127
luminosidad *f*	luminosidade *f*	светимость *f*	128
relación masa-luminosidad *f* diagrama masa-luminosidad *m*	relação massa-luminosidade *f* diagrama massa-luminosidade *m*	соотношение масса-светимость *n* зависимость масса-светимость *f* диаграмма масса-светимость *f*	129
diagrama de Hertzsprung-Russell *m* relación espectro-luminosidad *f* diagrama H-R *m*	diagrama de Hertzsprung-Russell *m* relação luminosidade-tipo espe(c)tral *f* diagrama H-R *m*	диаграмма Герцшпрунга-Рессела *f* зависимость спектр-светимость *f* диаграмма спектр-абсолютная (звёздная) величина *f* диаграмма спектр-светимость *f*	130
secuencia principal *f*	seqüência principal *f* série principal *f*	главная последовательность *f*	131
rama horizontal *f*	ramo horizontal *m*	горизонтальная ветвь *f*	132

133	subluminous star underluminous star	étoile sous-lumineuse *f* étoile de faible luminosité *f*	unterluminöser Stern *m* unterleuchtkräftiger Stern *m* Stern zu geringer Leuchtkraft *m*
134	overluminous star superluminous star	étoile surlumineuse *f* étoile superlumineuse *f*	überluminöser Stern *m* überleuchtkräftiger Stern *m* Stern zu hoher Leuchtkraft *m*
135	dwarf dwarf star	naine *f* étoile naine *f*	Zwerg *m* Zwergstern *m*
136	red dwarf	naine rouge *f*	roter Zwerg *m*
137	white dwarf	naine blanche *f*	weisser Zwerg *m*
138	Sirius companion	compagnon de Sirius *m*	Sirius-Begleiter *m*
138a	brown dwarf	naine brune *f*	brauner Zwerg *m*
139	black dwarf	naine noire *f*	schwarzer Zwerg *m*
140	subdwarf	sous-naine *f*	Unterzwerg *m*
141	horizontal-branch star	étoile de la branche horizontale *f*	Horizontalast-Stern *m*
142	main-sequence star	étoile de la série principale *f*	Hauptreihenstern *m* Stern der Hauptreihe *m* Stern der Hauptsequenz *m*
142a	blue giant	géante bleue *f*	blauer Riese *m*
143	giant giant star	géante *f* étoile géante *f*	Riese *m* Riesenstern *m*
143a	early-type giant	géante des premiers types *f*	Riese frühen Typs *m* früher Riesenstern *m*
144	Hertzsprung gap	lacune d'Hertzsprung *f*	Hertzsprung-Lücke *f*
144a	late-type giant	géante des types avancés *f*	Riese(nstern) späten Typs *m*
145	supergiant (star)	supergéante *f*	Überriese *m*
146	luminosity class	classe de luminosité *f*	Leuchtkraftklasse *f*
147	most luminous supergiants hypergiants Ia	supergéantes les plus brillantes *fpl* hypergéantes *fpl*	hellste Überriesen *mpl*

ESTRELLAS	ESTRELAS	ЗВЕЗДЫ	
estrela subluminosa *f*	estrela subluminosa *f*	недосветимая звезда *f* звезда пониженной [ослабленной] светимости *f* звезда с ослабленной светимостью *f*	133
estrella superluminosa *f*	estrela superluminosa *f*	пересветимая звезда *f* звезда повышенной [избыточной] светимости *f* звезда с избыточной светимостью *f*	134
enana *f* estrella enana *f*	anã *f* estrela anã *f*	звезда карлик *f* карлик(овая звезда) *m(f)*	135
enana roja *f*	anã vermelha *f*	красный карлик *m*	136
enana blanca *f*	anã branca *f*	белый карлик *m*	137
compañero de Sirio *m*	companheira de Sírio [Sirius] *f*	спутник Сириуса *m*	138
enana marrón [parda] *f*	anã parda [marrom] *f*	коричневый карлик *m*	138a
enana negra *f*	anã preta *f* anã negra *f*	чёрный карлик *m*	139
subenana *f*	subanã *f*	субкарлик *m*	140
estrella de la rama horizontal *f*	estrela do ramo horizontal *f*	звезда горизонтальной ветви *f*	141
estrela de la secuencia principal *f*	estrela da seqüência principal *f*	звезда главной последовательности *f* звезда на главной последовательности *f*	142
gigante azul *f*	gigante azul *f*	голубой гигант *m*	142a
gigante *f* estrella gigante *f*	gigante *f* estrela gigante *f*	гигант *m* звезда гигант *f*	143
gigante de tipos tempranos *f*	gigante dos tipos iniciais *f*	гигант ранних спектральных классов *m*	143a
laguna de Hertzsprung *f*	lacuna [falha] de Hertzsprung *f*	пробел Герцшпрунга *m* разрыв Герцшпрунга *m*	144
gigante de tipos tardíos *f*	gigante dos tipos finais *f*	гигант поздних спектральных классов *m*	144a
supergigante *f*	supergigante *f*	сверхгигант *m*	145
clase de luminosidad *f*	classe de luminosidade *f*	класс светимости *m*	146
supergigantes más luminosas *fpl* supergigantes más brillantes *fpl*	supergigantes (mais) brilhantes *fpl* supergigantes (mais) luminosas *fpl* hipergigantes *fpl*	наиболее яркие сверхгиганты *mpl* ярчайшие сверхгиганты *mpl*	147

148	less luminous supergiants Ib	supergéantes faibles *fpl*	schwächere Überriesen *mpl*
149	bright giants II	géantes brillantes *fpl*	helle Riesen *mpl*
150	normal giants III	géantes normales *fpl*	normale Riesen *mpl*
151	subgiants IV	sous-géantes *fpl*	Unterriesen *mpl*
152	neutron star	étoile neutronique *f* étoile à neutrons *f*	Neutronenstern *m*
152a	red gigant	géante rouge *f*	roter Riese(nstern) *m*
152b	giant branch	branche des géantes *f*	Riesenast *m*
153	superdense star	étoile superdense *f*	überdichter Stern *m*
154	hyperon star hyperonic star	étoile à hypérons *f* étoile hypéronique *f*	Hyperonenstern *m*
155	hot star	étoile chaude *f*	heisser Stern *m*
156	cool star	étoile froide *f*	kühler Stern *m*
157	infrared star	étoile infrarouge *f*	Infrarotstern *m*
158	radio star	étoile radio *f*	Radiostern *m*
159	emission-line star	étoile à raies d'émission *f*	Emissionsstern *m* Stern mit Emissionslinien *m*
160	Wolf-Rayet star WR star W star	étoile de Wolf-Rayet *f* étoile WR *f* étoile W *f*	Wolf-Rayet-Stern *m* WR-Stern *m* W-Stern *m*
161	contracting star	étoile en contraction *f*	kontrahierender Stern *m*
162	expanding star	étoile en expansion *f*	expandierender Stern *m*
163	intergalactic star	étoile intergalactique *f*	intergalaktischer Stern *m*
164	imploding star collapsing star collapsar	étoile en implosion [en effondrement] *f* étoile implosive *f* étoile (en cours) d'écroulement *f*	kollabierender Stern *m* zusammenfallender Stern *m*
165	exploding star	étoile en explosion *f* étoile explosive *f*	explodierender Stern *m*
166	zero-temperature star	étoile au zéro absolu *f* étoile à température nulle *f*	Stern mit Temperatur Null *m*
167	dead star stellar corpse	étoile morte *f*	abgestorbener Stern *m*

ESTRELLAS	ESTRELAS	ЗВЕЗДЫ	
supergigantes débiles *fpl* supergigantes subluminosas *fpl*	supergigantes subluminosas [fracas] *fpl* supergigantes menos luminosas [brilhantes] *fpl*	слабые сверхгиганты *mpl* менее яркие сверхгиганты *mpl*	148
gigantes luminosas *fpl* gigantes brillantes *fpl*	gigantes brilhantes *fpl* gigantes luminosas *fpl*	яркие гиганты *mpl*	149
gigantes normales *fpl*	gigantes normais *fpl*	нормальные гиганты *mpl*	150
subgigantes *fpl*	subgigantes *fpl*	субгиганты *mpl*	151
estrella de neutrones *f* estrella neutrónica *f*	estrela de nêutrons *f* estrela neutrônica *f*	нейтронная звезда *f*	152
gigante roja *f*	gigante vermelha *f*	красный гигант *m*	152a
rama de las gigantes *f*	ramo das gigantes *m*	ветвь гигантов *f*	152b
estrella superdensa *f*	estrela superdensa *f*	сверхплотная звезда *f*	153
estrella de hiperones *f* estrella hiperónica *f*	estrela de híperons *f* estrela hiperônica *f*	гиперонная звезда *f*	154
estrella caliente *f*	estrela quente *f*	горячая звезда *f*	155
estrella fría *f*	estrela fria *f*	холодная звезда *f*	156
estrella infrarroja *f*	estrela infravermelha *f*	инфракрасная звезда *f*	157
radioestrella *f*	radioestrela [radiestrela] *f*	радиозвезда *f*	158
estrella con líneas de emisión *f*	estrela com linhas de emissão *f* estrela com raias de emissão *f*	звезда с эмиссионными линиями *f* эмиссионная звезда *f*	159
estrella Wolf-Rayet *f* estrella WR *f*	estrela Wolf-Rayet *f* estrela WR *f*	звезда (типа) Вольфа-Райе *f*	160
estrella en contracción *f*	estrela em contração *f*	сжимающаяся звезда *f*	161
estrella en expansión *f*	estrela em expansão *f*	расширяющаяся звезда *f*	162
estrella intergaláctica *f*	estrela intergalática *f*	межгалактическая звезда *f*	163
estrella colapsante *f* estrella en implosión *f* estrella implosiva *f* estrella en colapso *f*	estrela em implosão *f* estrela em colapso *f* estrela implosiva *f* estrela colapsante *f*	коллапсирующая звезда *f* спадающая звезда *f* спавшая [сжатая] звезда *f* сколлапсировавшая звезда *f*	164
estrella en explosión *f* estrella explosiva *f*	estrela em explosão *f* estrela explosiva *f*	взрывающаяся звезда *f*	165
estrella de temperatura cero *f*	estrela a temperatura zero *f* estrela a zero absoluto *f*	звезда нулевой температуры *f*	166
estrella muerta *f*	estrela morta *f*	угасшая [умершая] звезда *f* мёртвая звезда *f*	167

	STARS	ÉTOILES	STERNE
168	chemical composition (of a star)	composition chimique (d'une étoile) *f*	chemische Zusammensetzung (eines Sternes) *f*
169	chemical abundance abundance of an element	abondance d'un élement *f* abondance chimique *f*	Häufigkeit eines Elementes *f* chemische Häufigkeit *f*
170	metal abundance metallicity	abondance métallique *f* métallicité *f*	Metallhäufigkeit *f*
171	natural abundance	abondance naturelle *f*	natürliche Häufigkeit *f*
172	normal abundance	abondance normale *f*	normale Häufigkeit *f*
173	anomalous abundance	abondance anormale *f*	anomale Häufigkeit *f*
174	relative abundances of elements	abondances relatives des éléments *fpl*	relative Häufigkeiten von Elementen *fpl*
175	abundance ratio	rapport d'abondance *m*	Häufigkeitsverhältnis *n*
176	hydrogen-metal ratio metal-to-hydrogen ratio	rapport hydrogène-métaux *m*	Häufigkeitsverhältnis von Wasserstoff zu den Metallen *n* Wasserstoff-Metall-Verhältnis *n*
177	total abundance	abondance totale *f*	totale Häufigkeit *f*
178	isotope effect isotopic effect	effet isotopique *m*	Isotopie-Effekt *m*
179	isotopic abundance	abondance isotopique *f*	Isotopenhäufigkeit *f*
180	betatron effect	effet bêtatron *m*	Betatron-Effekt *m*
181	surface nuclear reaction	réaction nucléaire en surface *f*	Kernreaktion an der Sternoberfläche *f*
182	accelerated particle	particule accélérée *f*	beschleunigtes Teilchen *n*
183	abundance anomalies	anomalies de l'abondance *fpl*	Häufigkeitsanomalien *fpl*
184	over-abundant element	élément surabondant *m*	überhäufiges Element *n* Element von zu grosser Häufigkeit *n*
185	over-abundance	surabondance *f*	Überhäufigkeit *f*
186	under-abundant element	élément sous-abondant *m*	Element von zu geringer Häufigkeit *n* unterhäufiges Element *n*

ESTRELLAS	ESTRELAS	ЗВЕЗДЫ	
composición química (de una estrella) *f*	composição química (de uma estrela) *f*	химическая состав (звезды) *m*	168
abundancia de un elemento *f* abundancia química *f*	abundância de um elemento *f* abundância química *f*	распространённость элемента *f* обилие [содержание] элемента *n*	169
abundancia de metales *f* abundancia metálica *f*	metalicidade *f* abundância metálica *f*	металличность *f* содержание металлов *n*	170
abundancia natural *f*	abundância natural *f*	распространённость в природе *f* естественная распространённость *f*	171
abundancia normal *f*	abundância normal *f*	нормальное содержание [обилие] *n*	172
abundancia anómala *f*	abundância anômala *f*	аномальное содержание [обилие] *n*	173
abundancias relativas de los elementos *fpl*	abundâncias relativas dos elementos *fpl*	относительное содержание элементов *n* относительная рапространённость элементов *f*	174
relación [razón] de abundancia *f*	razão de abundância *f*	отношение содержаний *n*	175
relación [razón] hidrógeno-metales *f*	razão hidrogênio-metais *f*	отношение (содержаний) водорода к металлам *n*	176
abundancia total *f*	abundância total *f*	общее содержание *n* полное обилие *n*	177
efecto isotópico *m*	efeito isotópico *m*	изотопический эффект *m*	178
abundancia isotópica *f*	abundância isotópica *f*	распространённость изотопов *f*	179
efecto betatrón *m*	efeito bétatron *m*	бетатронный эффект *m*	180
reacción nuclear superficial *f*	reação nuclear superficial *f*	ядерная реакция на поверхности (звёзд) *f*	181
partícula acelerada *f*	partícula acelerada *f*	ускоренная частица *f*	182
anomalías de abundancia *fpl*	anomalias de abundância *fpl*	аномалии обилия (в содержании) *fpl*	183
elemento superabundante *m*	elemento superabundante *m*	элемент с повышенным обилием *m* элементы с избыточным содержанием *m*	184
superabundancia *f*	superabundância *f*	избыток содержания [обилия] *m*	185
elemento subabundante *n* elemento deficiente *m*	elemento subabundante *m* elemento em deficiência *m* elemento deficiente *m*	элемент с недостатком обилия [содержания] *m* элемент с пониженным обилием [содержанием] *m*	186

187	under-abundance	sous-abondance *f*	Unterhäufigkeit *f*
188	carbon star	étoile à carbone *f* étoile carbonée *f*	Kohlenstoffstern *m*
189	nitrogen star	étoile azotée *f* étoile à azote *f*	Stickstoffstern *m*
190	silicon star	étoile à silicium *f*	Silizium-Stern *m*
191	hydrogen-poor star hydrogen-deficient star	étoile pauvre en hydrogène *f*	wasserstoffarmer Stern *m*
192	helium-poor star helim-deficient star	étoile pauvre en hélium *f*	heliumarmer Stern *m*
193	helium-rich star	étoile riche en hélium *f*	heliumreicher Stern *m*
194	carbon-poor star carbon-deficient star	étoile pauvre en carbone *f*	kohlenstoffarmer Stern *m*
195	metallic-line star metal-line star	étoile à raies métalliques *f*	Metallinienstern *m*
196	metal-rich star	étoile riche en métaux *f*	metallreicher Stern *m*
197	metal-poor star metal-deficient star	étoile pauvre en métaux *f*	metallarmer Stern *m*

subabundancia *f* deficiencia (en abundancia) *f*	subabundância *f* deficiência em abundância *f*	недостаток [недостаточность] содержания [обилия] *m[f]*	187
estrella de carbono *f*	estrela carbonada *f* estrela de carbono *f*	углеродная звезда *f*	188
estrella de nitrógeno *f*	estrela nitrogenada [de nitrogênio] *f*	азотная звезда *f*	189
estrella de silicio *f*	estrela de silício *f*	кремниевая звезда *f*	190
estrella pobre en hidrógeno *f* estrella deficiente en hidrógeno *f*	estrela pobre em hidrogênio *f* estrela deficiente em hidrogênio *f*	звезда с дефицитом водорода *f* звезда с низким содержанием водорода *f* звезда бедная водородом *f*	191
estrella pobre en helio *f* estrella deficiente en helio *f*	estrela pobre em hélio *f* estrela deficiente em hélio *f*	звезда с дефицитом гелия *f* звезда с низким содержанием гелия *f* звезда бедная гелием *f*	192
estrella rica en helio *f*	estrela rica em hélio *f*	звезда с большим содержанием гелия *f*	193
estrella deficiente en carbono *f* estrella pobre en carbono *f*	estrela pobre em carbono *f* estrela deficiente em carbono *f*	звезда с дефицитом углерода *f* звезда с низким содержанием углерода *f* звезда бедная углеродом *f*	194
estrella con líneas metálicas *f*	estrela com linhas [raias] metálicas *f*	металлическая звезда *f* звезда с усиленными линиями металлов *f*	195
estrella rica en metales *f*	estrela rica em metais *f*	звезда с большим содержанием металлов *f*	196
estrella pobre en metales *f* estrella deficiente en metales *f*	estrela pobre em metais *f* estrela deficiente em metais *f*	звезда с дефицитом металлов *f* звезда с недостаточным содержанием металлов *f* звезда бедная металлами *f*	197

1	constellation	constellation *f*	Sternbild *n*
1a	summer constellation	constellation d'été *f*	Sommersternbild *n*
1b	winter constellation	constellation d'hiver *f*	Wintersternbild *n*
1c	spring constellation	constellation de printemps *f*	Frühlingssternbild *n*
1d	autumn [fall] constellation	constellation d'automne *f*	Herbststernbild *n*
2	star	étoile *f*	Stern *m*
2a	circumpolar star	étoile circumpolaire *f*	Zirkumpolarstern *m*
3	stellar	stellaire	Stern-
3a	asterism group of stars	astérisme *m* groupe d'étoiles *f*	Asterismus *m*
4	fixed star	étoile fixe *f*	Fixstern *m*
4a	circumpolar constellation	constellation circumpolaire *f*	Zirkumpolarsternbild *n*
5	zodiacal constellation	constellation zodiacale *f*	Tierkreis–Sternbild *n*
5a	zodiac	zodiaque *m*	Tierkreis *m*
6	zodiacal signs	signes du zodiaque *mpl*	Tierkreiszeichen *npl*
7	celestial globe	globe céleste *m*	Himmelsglobus *m*
8	celestial sphere	sphère céleste *f*	Himmelskugel *f* Himmelssphäre *f*
9	star map [chart] map [chart] of sky	carte céleste *f* carte du ciel *f*	Sternkarte *f* Himmelskarte *f*
10	revolving planisphere revolving star map revolving star chart	carte planiciel [des étoiles tournante] *f* carte céleste mobile [tournante] *f* planisphère mobile *m* carte stellaire tournante *f*	drehbare Sternkarte *f*
11	sky survey	surveillance du ciel *f* cartographie du ciel *f*	Durchmusterung des Himmels *f*
12	star atlas	atlas du ciel *m* atlas céleste *m*	Sternatlas *m* Himmelsatlas *m*
12a	Messier catalogue	catalogue de Messier *m*	Messier-Katalog *m*
13	star catalogue catalogue of stars	catalogue stellaire *m* catalogue d'étoiles *m*	Sternkatalog *m*
14	Andromeda (gen.-dae) And Chained Woman	Andromeda *f* Andromède *f* Femme Enchaînée *f*	Andromeda *f* Andromeda *f*
15	Antlia (gen.-liae) Ant Air Pump	Antlia *f* Machine Pneumatique *f*	Antlia *f* Luftpumpe *f*

constelación *f*	constelação *f*	созвездие *n*	1
constelación de verano *f*	constelação de verão *f*	созвездие летнего неба *n*	1a
constelación de invierno *f*	constelação de inverno *f*	созвездие зимнего неба *n*	1b
constelación primaveral *f*	constelação primaveral *f*	весеннее созвездие *n*	1c
constelación autumnal *f*	constelação outonal *f*	осеннее созвездие *n*	1d
estrella *f*	estrela *f*	звезда *f*	2
estrella circumpolar *f*	estrela circumpolar *f*	околополярная [около-полюсная, близполюсная] звезда *f*	2a
estelar	estelar	звёздный	3
asterismo *m*	asterismo *m*	астеризм *m* группа звёзд *f*	3a
estrella fija *f*	estrela fixa *f*	неподвижная звезда *f*	4
constelación circumpolar *f*	constelação circumpolar *f*	околополярное [околополюсное, близполюсное] созвездие *n*	4a
constelación zodiacal *f*	constelação zodiacal *f*	зодиакальное созвездие *n*	5
zodíaco *m*	zodíaco *m*	зодиак *m*	5a
signos del zodíaco *mpl*	signos do zodíaco *mpl*	зодиакальные знаки *mpl* зодиакальные обозначения *npl*	6
globo celeste *m*	globo celeste *m*	небесный глобус *m*	7
esfera celeste *f*	esfera celeste *f*	небесная сфера *f*	8
carta celeste *f* carta del cielo *f*	carta celeste *f* carta do céu *f*	звёздная карта *f* карта неба *f*	9
carta celeste giratoria *f* planisferio *m*	carta celeste giratória *f* planisfério celeste *m*	подвижная звёздная карта *f* подвижная карта неба *f*	10
relevamiento celeste *m* relevamiento del cielo *m*	levantamento celeste *m* levantamento do céu *m*	обзор небесной сферы *m* обозрение неба *n*	11
atlas celeste *m* atlas del cielo *m*	atlas celeste *m* atlas do céu *m*	звёздный атлас *m* атлас неба *m*	12
catálogo de Messier *m*	catálogo de Messier *m*	каталог Мессье *m*	12a
catálogo estelar *m* catálogo de estrellas *m*	catálogo estelar *m* catálogo de estrelas *m*	каталог звёзд *m* звёздный каталог *m*	13
Andromeda *f* Andrómeda *f*	Andromeda *f* Andrômeda *f*	Андромеда *f*	14
Antlia *f* Máquina Neumática *f*	Antlia *f* Máquina [Bomba] Pneumática *f* Máquina [Bomba] de Ar *f*	Насос *m*	15

16	Apus (gen.-podis) Aps Bird of Paradise	Apus *m* Oiseau de Paradis *m*	Apus *m* Paradiesvogel *m*
17	Aquarius (gen.-rii) Aqr Water Bearer Water Carrier Waterman	Aquarius *m* Verseau *m* Porteur d'eau *m*	Aquarius *m* Wassermann *m*
18	Aquila (gen.-lae) Aql Eagle	Aquila *f* Aigle *m*	Aquila *f* Adler *m*
19	Ara (gen.-rae) Ara Altar	Ara *f* Autel *m*	Ara *f* Altar *m*
20	Aries (gen.-ietis) Ari Ram	Aries *m* Bélier *m*	Aries *m* Widder *m*
21	Auriga (gen.-gae) Aur Charioteer	Auriga *m* Cocher *m*	Auriga *m* Fuhrmann *m*
22	Boötes (gen.-tis) Boo Herdsman Kite Ploughman Bear Driver Bull Driver	Boötes *m* Bouvier *m* Conducteur de troupeau *m* Conducteur de boeuf *m* Conducteur de taureau *m*	Boötes *m* Bärenhüter *m*
23	Caelum (gen.-aeli) Cae Chisel Sculptor's Tool	Caelum *n* Burin du Graveur *m* Ciseau du Graveur *m* Outil du Sculpteur *m*	Caelum *n* Grabstichel *m*
24	Camelopardalis *f* Cam Camelopardus (gen.-di) Giraffe *m*	Camelopardalis *f* Girafe *f*	Camelopardalis *f* Giraffe *f*
25	Cancer (gen.-cri) Cnc Crab	Cancer *m* Cancer *m*	Cancer *m* Krebs *m*
26	Canes Venatici (gen. -num -corum) CVn Hunting Dogs	Canes Venatici *mpl* Chiens de Chasse *mpl*	Canes Venatici *mpl* Jagdhunde *mpl*
27	Canis Major (gen. -is -ris) CMa Big Dog Greater Dog	Canis Major *m* Grand Chien *m*	Canis Major *m* Grosser Hund *m*
28	Canis Minor (gen. -is -ris) CMi (Small) Dog	Canis Minor *m* Petit Chien *m*	Canis Minor *m* Kleiner Hund *m*

CONSTELACIONES	CONSTELAÇÕES	СОЗВЕЗДИЯ	
Apus *m* Ave del Paraíso *f*	Apus *m* Ave do Paraíso *f*	Райская Птица *f*	16
Aquarius *m* Acuario *m*	Aquarius *m* Aquário *m*	Водолей *m*	17
Aquila *f* Águila *f*	Aquila *f* Águia *f*	Орёл *m*	18
Ara *f* Altar *m*	Ara *f* Altar *m*	Жертвенник *m*	19
Aries *m* Carnero *m*	Aries *m* Carneiro *m*	Овен *m*	20
Auriga *m* Cochero *m*	Auriga *m* Cocheiro *m*	Возничий *m*	21
Boötes *m* Boyero *m*	Boötes *m* Boieiro *m* Boiadeiro *m*	Волопас *m*	22
Caelum *n* Buril *m* Cincel *m*	Caelum *n* Buril *m* Cinzel *m*	Резец *m*	23
Camelopardalis *f* Jirafa *f*	Camelopardalis *f* Girafa *f*	Жираф *m*	24
Cancer *m* Cangrejo *m*	Cancer *m* Caranguejo *m*	Рак *m*	25
Canes Venatici *mpl* Lebreles *mpl* Perros de Caza *mpl*	Canes Venatici *mpl* Cães de Caça *mpl* Cães Caçadores *mpl*	Гончие Псы *mpl* Гончие Собаки *fpl*	26
Canis Major *m* Can Mayor *m*	Canis Major *m* Cão Maior *m*	Большой Пёс *m*	27
Canis Minor *m* Can Menor *m*	Canis Minor *m* Cão Menor *m*	Малый Пёс *m*	28

	CONSTELLATIONS	CONSTELLATIONS	STERNBILDER
29	Capricornus (gen. -ni) Cap (Sea-) Goat He-Goat	Capricornus *m* Capricorne *m*	Capricornus *m* Steinbock *m*
30	Carina (gen. -nae) Car Keel	Carina *f* Carène *f*	Carina *f* Kiel des Schiffes *m*
31	Cassiopeia (gen. -peiae) Cas Lady in Chair	Cassiopeia *f* Cassiopée *f*	Cassiopeia *f* Kassiopeia *f*
32	Centaurus (gen. -ri) Cen Centaur	Centaurus *m* Centaure *m*	Centaurus *m* Kentaur *m* Zentaur *m*
33	Cepheus (gen. -phei) Cep King	Cepheus *m* Céphée *m* Le Roi *m*	Cepheus *m* Kepheus *m*
34	Cetus (gen. -ti) Cet Sea-Monster Whale	Cetus *m* Baleine *f* Monstre Marin *m*	Cetus *m* Walfisch *m*
35	Chamaeleon (gen.-ntis) Cha Chamaeleon	Chamaeleon *m* Caméléon *m*	Chamaeleon *m* Chamäleon *n*
36	Circinus (gen.-ni) Cir Compasses	Circinus *m* Compas *m*	Circinus *m* Zirkel *m*
37	Columba (gen.-bae) Col Dove	Columba *f* Colombe *f*	Columba *f* Taube *f*
38	Coma Berenices (gen.-mae -cis) Com Berenice's Hair	Coma Berenices *f* Chévelure de Bérénice *f*	Coma Berenices *f* Haar der Berenike *n*
39	Corona Australis (gen. -nae -lis) CrA Corona Austrina (gen. -nae -nae) (Southern) Crown S Crown	Corona Australis *f* Couronne Australe *f* Couronne du Sud *f*	Corona Australis *f* Südliche Krone *f*
40	Corona Borealis (gen.-nae -lis) CrB Northern Crown N Crown	Corona Borealis *f* Couronne Boréale *f* Couronne du Nord *f*	Corona Borealis *f* Nördliche Krone *f*
41	Corvus (gen.-vi) Crv Crow	Corvus *m* Corbeau *m*	Corvus *m* Rabe *m*
42	Crater (gen.-eris) Crt Cup	Crater *m* Coupe *f*	Crater *m* Pokal *m* Becher *m*

CONSTELACIONES	CONSTELAÇÕES	СОЗВЕЗДИЯ	
Capricornus *m* Capricornio *m*	Capricornus *m* Capricôrnio *m* Cabra *f*	Козерог *m*	29
Carina *f* Quilla *f*	Carina *f* Quilha *f* Carena *f* Querena *f*	Киль *m*	30
Cassiopeia *f* Casiopea *f*	Cassiopeia *f* Cassiopéia *f*	Кассиопея *f*	31
Centaurus *m* Centauro *m*	Centaurus *m* Centauro *m*	Центавр *m*	32
Cepheus *m* Cefeo *m*	Cepheus *m* Cefeu *m*	Цефей *m*	33
Cetus *m* Ballena *f*	Cetus *m* Baleia *f*	Кит *m*	34
Chamaeleon *m* Camaleón *m*	Chamaeleon *m* Camaleão *m*	Хамелеон *m*	35
Circinus *m* Compás *m*	Circinus *m* Compasso *m*	Циркуль *m*	36
Columba *f* Paloma *f*	Columba *f* Pomba *f*	Голубь *m*	37
Coma Berenices *f* Cabellera de Berenice *f*	Coma Berenices *f* Cabeleira de Berenice *f*	Волосы Вероники *mpl*	38
Corona Australis *f* Corona Austral *f*	Corona Australis *f* Coroa Austral *f*	Южная Корона *f*	39
Corona Borealis *f* Corona Boreal *f*	Corona Borealis *f* Coroa Boreal *f*	Северная Корона *f*	40
Corvus *m* Cuervo *m*	Corvus *m* Corvo *m*	Ворон *m*	41
Cráter *m* Copa *f*	Crater *m* Taça *f*	Чаша *f*	42

43	Crux (gen.-ucis) Cru Southern Cross S Cross	Crux *f* Croix du Sud *f*	Crux *f* Kreuz des Südens *n*
44	Cygnus (gen.-gni) Cyg Swan Northern Cross	Cygnus *m* Cygne *m* Croix du Nord *f*	Cygnus *m* Schwan *m* Kreuz des Nordens *n*
45	Delphinus (gen.-ni) Del Dolphin Job's Coffin	Delphinus *m* Dauphin *m*	Delphinus *m* Delphin *m*
46	Dorado (gen.-dus) *m* Dor Swordfish Gold-fish Dorado Fish	Dorado *m* Dorade *f* Daurade *f* Poisson Doré *m*	Dorado *m* Schwertfisch *m*
47	Draco (gen.-onis) Dra Dragon	Draco *m* Dragon *m*	Draco *m* Drache *m*
48	Equuleus (gen.-lei) Equ Colt Small Horse Little Horse	Equuleus *m* Petit Cheval *m* Poulain *m* Le Petit du Cheval *m*	Equuleus *m* Kleines Pferd *n* Pferdchen *n*
49	Eridanus (gen.-ni) Eri River Eridanus	Eridanus *m* Eridan *m* Le Fleuve Eridan *m*	Eridanus *m* Eridanus-Fluss *m* Fluss Eridanus *m*
50	Fornax (gen.-acis) For Furnace	Fornax *f* Fourneau *m* Fourneau Chimique *m*	Fornax *f* Chemischer Ofen *m*
51	Gemini (gen.-norum) Gem (Heavenly) Twins	Gemini *mpl* Gémeaux *mpl*	Gemini *mpl* Zwillinge *mpl*
52	Grus (gen.-ruis) Gru Crane	Grus *f* Grue *f*	Grus *f* Kranich *m*
53	Hercules (gen.-lis) Her Kneeling Man Kneeling Giant Kneeler	Hercules *m* Hercule	Hercules *m* Herkules *m*
54	Horologium (gen.-gii) Hor Clock	Horologium *n* Pendule Astronomique *f* Horloge *f*	Horologium *n* Pendeluhr *f*
55	Hydra (gen.-drae) Hya Water Snake Water Monster Monster	Hydra *f* Hydre Femelle *f* Serpent d'Eau *m* Monstre Aquatique *m*	Hydra *f* Nördliche Wasserschlange *f* Weibliche Wasserschlange *f*

Crux *f* Cruz *f* Cruz del Sur *f*	Crux *f* Cruz do Sul *f* Cruzeiro do Sul *m* Cruzeiro *m*	Крест *m* Южный Крест *m*	43
Cygnus *m* Cisne *m* Cruz del Norte *f*	Cygnus *m* Cisne *m* Cruz do Norte *f*	Лебедь *m*	44
Delphinus *m* Delfín *m*	Delphinus *m* Golfinho *m* Delfim *m*	Дельфин *m*	45
Dorado *m* Dorada *f*	Dorado *m* Dourado *m* Saipé *m* Dourada *f*	Золотая Рыба *f*	46
Draco *m* Dragón *m*	Draco *m* Dragão *m*	Дракон *m*	47
Equuleus *m* Caballo Menor *m* Caballito *m*	Equuleus *m* Potro *m* Cavalinho *m*	Малый Конь *m* Жеребёнок *m*	48
Eridanus *m* (Río) Erídano *m*	Eridanus *m* Erídano *m*	Эридан *m*	49
Fornax *f* Horno *m*	Fornax *f* Forno *m* Fornalha *f*	Печь *f*	50
Gemini *mpl* Gemelos *mpl*	Gemini *mpl* Gêmeos *mpl*	Близнецы *mpl*	51
Grus *f* Grulla *f*	Grus *f* Grou *m* Grua *f*	Журавль *m*	52
Hercules *m* Hércules *m*	Hercules *m* Hércules *m*	Геркулес *m*	53
Horologium *n* Reloj (de Péndulo) *m*	Horologium *n* Relógio *m*	Часы *pl*	54
Hydra *f* Hidra (Hembra) *f* Serpiente de Agua *f*	Hydra *f* Hidra (Fêmea) *f*	Гидра *f*	55

56	Hydrus (gen.-dri) Hyi Sea Serpent	Hydrus *m* Hydre Mâle *f* Hydre Australe *f*	Hydrus *m* Südliche Wasserschlange *f* Männliche Wasserschlange *f*
57	Indus (gen.-di) Ind Indian	Indus *m* Indien *m*	Indus *m* Inder *m*
58	Lacerta (gen.-tae) Lac Lizard	Lacerta *f* Lézard *m*	Lacerta *f* Eidechse *f*
59	Leo (gen.-onis) Leo Lion	Leo *m* Lion *m*	Leo *m* Löwe *m*
60	Leo Minor (gen.-onis -ris) LMi Small Lion Little Lion Lesser Lion	Leo Minor *m* Petit Lion *m* Le Petit du Lion *m*	Leo Minor *m* Kleiner Löwe *m*
61	Lepus (gen.-poris) Lep Hare	Lepus *m* Lièvre *m*	Lepus *m* Hase *m*
62	Libra (gen.-rae) Lib Scales Balance	Libra *f* Balance *f*	Libra *f* Waage *f*
63	Lupus (gen.-pi) Lup Wolf	Lupus *m* Loup *m*	Lupus *m* Wolf *m*
64	Lynx (gen.-ncis) Lyn Lynx	Lynx *m, f* Lynx *m*	Lynx *m, f* Luchs *m*
65	Lyra (gen.-rae) Lyr Lyre Harp	Lyra *f* Lyre *f*	Lyra *f* Leier *f* Leyer *f*
66	Mensa (gen.-sae) Men Table (Mountain)	Mensa *f* Table (Montagne) *f*	Mensa *f* Tafelberg *m* Tisch *m*
67	Microscopium (gen.-pii) Mic Microscope	Microscopium *n* Microscope *m*	Microscopium *n* Mikroskop *n*
68	Monoceros (gen.-rotis) *m* Mon Unicorn	Monoceros *m* Licorne *f*	Monoceros *m* Einhorn *n*
69	Musca (gen.-cae) Mus Fly	Musca *f* Mouche *f*	Musca *f* Fliege *f*

CONSTELACIONES	CONSTELAÇÕES	СОЗВЕЗДИЯ	
Hydrus *m* Hidra Macho *f* Hidra Austral *f*	Hydrus *m* Hidra Macho *f* Hidra Austral *f*	Южная Гидра *f* Южный Змей *m*	56
Indus *m* Indio *m*	Indus *m* Índio *m*	Индеец *m*	57
Lacerta *f* Lagarto *m*	Lacerta *f* Lagarto *m* Lagartixa *f*	Ящерица *f*	58
Leo *m* León *m*	Leo *m* Leão *m*	Лев *m*	59
Leo Minor *m* León Menor *m*	Leo Minor *m* Leão Menor *m*	Малый Лев *m*	60
Lepus *m* Liebre *f*	Lepus *m* Lebre *f*	Заяц *m*	61
Libra *f* Balanza *f*	Libra *f* Balança *f*	Весы *pl*	62
Lupus *m* Lobo *m*	Lupus *m* Lobo *m*	Волк *m*	63
Lynx *m, f* Lince *m*	Lynx *m, f* Lince *m*	Рысь *m*	64
Lyra *f* Lira *f*	Lyra *f* Lira *f*	Лира *f*	65
Mensa *f* (Montaña de la) Mesa *f*	Mensa *f* Mesa *f*	Столовая Гора *f*	66
Microscopium *n* Microscopio *m*	Microscopium *n* Microscópio *m*	Микроскоп *m*	67
Monoceros *m* Unicornio *m*	Monoceros *m* Unicórnio *m*	Единорог *m*	68
Musca *f* Mosca *f*	Musca *f* Mosca *f*	Муха *f*	69

70	Norma (gen.-mae) Nor Square (Level)	Norma *f* Equerre *f* Règle *f*	Norma *f* Winkelmass *n*
71	Octans (gen.-ntis) *m* Oct Octant	Octans *m* Octant *m*	Octans *m* Oktant *m*
72	Ophiuchus (gen.-chi) Oph Serpent Bearer Serpentarius	Ophiuchus *m* Ophiuchus *m* Serpentaire *m*	Ophiuchus *m* Schlangenträger *m*
73	Orion (gen.-nis) Ori Hunter	Orion *m* Orion *m* Le Chasseur *m*	Orion *m* Himmelsjäger *m*
74	Pavo (gen.-vonis) Pav Peacock	Pavo *m* Paon *m*	Pavo *m* Pfau *m*
75	Pegasus (gen.-si) Peg Winged Horse	Pegasus *m* Pégase *m* Cheval Ailé *m*	Pegasus *m* Pegasos *m* Geflügeltes Ross *n*
76	Perseus (gen.-sei) Per Champion	Perseus *m* Persée *m* Le Vainqueur *m*	Perseus *m* Perseus *m*
77	Phoenix (gen.-nicis) Phe Phoenix	Phoenix *m* Phénix *m*	Phoenix *m* Phoenix *m*
78	Pictor (gen.-ris) Pic Painter's Easel	Pictor *m* Chevalet du Peintre *m*	Pictor *m* Maler *m* Malerstaffelei *f*
79	Pisces (gen.-cium) Psc Fishes	Pisces *mpl* Poissons *mpl*	Pisces *mpl* Fische *mpl*
80	Piscis Austrinus (gen. -is -ni) PsA Southern Fish S Fish	Piscis Austrinus *m* Poisson Austral *m*	Piscis Austrinus *m* Südlicher Fisch *m*
81	Puppis (gen.-ppis) Pup Poop Ship's Stern	Puppis *f* Poupe *f* L'Arrière du Bateau *m*	Puppis *f* Hinterdeck *n* Hinterteil des Schiffes *m* Achterschiff *n*
82	Pyxis (gen.-xidis) Pyx Compass Ship's Compass Mariner's Compass	Pyxis *f* Boussole *f* Compas du Bateau *m* Compas du Marin *m*	Pyxis *f* Schiffskompass *m*
83	Reticulum (gen.-li) Ret Net	Reticulum *n* Réticule *m* Réticule Astronomique *m*	Reticulum *n* Netz *n*

CONSTELACIONES	CONSTELAÇÕES	СОЗВЕЗДИЯ	
Norma *f* Escuadra *f*	Norma *f* Régua (paraláctica) *f* Esquadro *m*	Наугольник *m*	70
Octans *m* Octante *m*	Octans *m* Octante *m* Oitante *m*	Октант *m*	71
Ophiuchus *m* Ofiuco *m* Serpentario *m*	Ophiuchus *m* Ofiúco *m* Serpentário *m*	Змееносец *m*	72
Orion *m* Orión *m*	Orion *m* Órion *m* (Grande) Caçador *m*	Орион *m*	73
Pavo *m* Pavo (Real) *m*	Pavo *m* Pavão *m*	Павлин *m*	74
Pegasus *m* Pegaso *m*	Pegasus *m* Pégaso *m*	Пегас *m*	75
Perseus *m* Perseo *m*	Perseus *m* Perseu *m*	Персей *m*	76
Phoenix *m* (Ave) Fénix *f*	Phoenix *m* Fênix *f*	Феникс *m*	77
Pictor *m* Pintor *m* Caballete del Pintor *m*	Pictor *m* Pintor *m* Cavalete do Pintor *m*	Живописец *m*	78
Pisces *mpl* Peces *mpl*	Pisces *mpl* Peixes *mpl*	Рыбы *fpl*	79
Piscis Austrinus *m* Pez Austral *m*	Piscis Austrinus *m* Peixe Austral *m*	Южная Рыба *f*	80
Puppis *f* Popa *f*	Puppis *f* Popa *f*	Корма *f*	81
Pyxis *f* Brújula *f*	Pyxis *f* Bússola *f*	Компас *m*	82
Reticulum *n* Réticulo *m*	Reticulum *n* Retículo *m*	Сетка *f*	83

CONSTELLATIONS		CONSTELLATIONS	STERNBILDER
84	Sagitta (gen.-tae) Sge Arrow	Sagitta *f* Flèche *f*	Sagitta *f* Pfeil *m*
85	Sagittarius (gen.-rii) Sgr Archer	Sagittarius *m* Sagittaire *m*	Sagittarius *m* Schütze *m*
86	Scorpio (gen.-onis) Scorpius (gen.-pii) Sco Scorpion	Scorpio *m* Scorpius *m* Scorpion *m*	Scorpio *m* Scorpius *m* Skorpion *m*
87	Sculptor (gen.-ris) Scl Sculptor	Sculptor *m* Atelier du Sculpteur *m*	Sculptor *m* Bildhauer *m* Bildhauerwerkstatt *m*
88	Scutum (gen.-ti) Sct Shield	Scutum *n* Écu de Sobiesky *m*	Scutum *n* Schild *m* Sobieskischer Schild *m*
89	Serpens (Caput + Cauda) (gen.-ntis) Ser Serpent (Head and Tail)	Serpens *m, f* Serpent *m* (Tête et Queue)	Serpens *m, f* Schlange *f* (Kopf und Schwanz)
90	Sextans (gen.-ntis) Sex Sextant	Sextans *m* Sextant *m*	Sextans *m* Sextant *m*
91	Taurus (gen.-ri) Tau Bull	Taurus *m* Taureau *m*	Taurus *m* Stier *m*
92	Telescopium (gen.-pii) Tel Telescope	Telescopium *n* Télescope *m*	Telescopium *n* Fernrohr *n*
93	Triangulum (gen.-li) Tri Triangle	Triangulum *n* Triangle *m*	Triangulum *n* Dreieck *n*
94	Triangulum Australe (gen.-li -lis) TrA Southern Triangle S Triangle	Triangulum Australe *n* Triangle Austral *m*	Triangulum Australe *n* Südliches Dreieck *n*
95	Tucana (gen.-nae) Tuc Toucan	Tucana *f* Toucan *m*	Tucana *f* Tukan *m*
96	Ursa Major (gen.-sae -ris) Uma Big Bear Great Bear Plough Charles's Wain Big Dipper	Ursa Major *f* Grand Chariot *m* Grande Ourse *f* Grande Louche *f*	Ursa Major *f* Grosser Bär *m*

CONSTELACIONES	CONSTELAÇÕES	СОЗВЕЗДИЯ	
Sagitta *f* Flecha *f*	Sagitta *f* Flecha *f* Seta *f*	Стрела *f*	84
Sagittarius *m* Sagitario *m*	Sagittarius *m* Sagitário *m* Arqueiro *m*	Стрелец *m*	85
Scorpio *m* Scorpius *m* Escorpión *m*	Scorpio *m* Scorpius *m* Escorpião *m*	Скорпион *m*	86
Sculptor *m* Escultor *m* Taller del Escultor *m*	Sculptor *m* Escultor *m* Atelier do Escultor *m*	Скульптор *m*	87
Scutum *n* Escudo *m* Escudo de Sobieski *m*	Scutum *n* Escudo *m* Escudo de Sobieski *m*	Щит *m*	88
Serpens *m, f* Serpiente *f* (Cabeza y Cola)	Serpens *m, f* Serpente *f* (Cabeça e Cauda)	Змея *f* (Голова и Хвост)	89
Sextans *m* Sextante *m*	Sextans *m* Sextante *m*	Секстант *m*	90
Taurus *m* Toro *m*	Taurus *m* Touro *m*	Телец *m*	91
Telescopium *n* Telescopio *m*	Telescopium *n* Telescópio *m*	Телескоп *m*	92
Triangulum *n* Triángulo *m*	Triangulum *n* Triângulo *m*	Треугольник *m*	93
Triangulum Australe *n* Triángulo Austral *m*	Triangulum Australe *n* Triângulo Austral *m*	Южный Треугольник *m*	94
Tucana *f* Tucán *m*	Tucana *f* Tucano *m*	Тукан *m*	95
Ursa Major *f* Osa Mayor *f* Carro *m*	Ursa Major *f* Ursa Maior *f*	Большая Медведица *f* Большой Ковш *m*	96

97	Ursa Minor (gen.-sae -ris) UMi Little Bear Small Bear Little Dipper	Ursa Minor *f* Petite Ourse *f* Petit Chariot *m* Petite Louche *f*	Ursa Minor *f* Kleiner Bär *m*
98	Vela (gen.-lorum) Vel Sails Ship's Sails	Vela *npl* Voile *f* Voiles du Navire *fpl*	Vela *npl* Segel (des Schiffes) *n*
99	Virgo (gen.-ginis) Vir Virgin	Virgo *f* Vierge *f*	Virgo *f* Jungfrau *f*
100	Volans (gen.-ntis) Vol Piscis Volans Flying Fish	Volans *m* Poisson Volant *m*	Volans *m* Fliegender Fisch *m*
101	Vulpecula (gen.-lae) Vul Little Fox Small Fox	Vulpecula *f* Petit Renard *m*	Vulpecula *f* Fuchs *m* Füchslein *n* Füchschen *n*
102	Achernar [α Eri]	Achernar *m*	Achernar *m*
103	Albireo [α Cyg]	Albireo *m*	Albireo *m*
104	Alcor [80 UMa]	Alcor *m*	Alkor *m* Reiterlein *n*
105	Algol [β Per] Demon Star	Algol *m* Tête de Méduse *f*	Algol *m*
106	Altair [α Agl]	Altaïr *m*	A(l)tair *m*
107	Antares [α Sco]	Antarès *m*	Antares *m*
108	Betelgeuse [α Ori]	Bételgeuse *f*	Betelgeuze [Beteigeuze] *m*
109	Canopus [α Car]	Canopus *m*	Kanopus *m*
110	Capella [α Aur]	Capella *f* La Chèvre *f*	Kapella *f*
111	Castor [α Gem]	Castor *m*	Kastor *m*
112	Deneb [α Cyg]	Deneb *m*	Deneb *m*
113	Denebola [β Leo]	Denebola *f*	Denebola *f*
114	Dubhe [α UMa]	Dubhe *m*	Dubhe *m*
115	Etamin [γ Dra]	Etamin *m*	Etamin *m*
116	Fomalhaut [α PsA]	Fomalhaut *f*	Fomalhaut
117	Gemma [α CrB] Alphecca [Alphekka]	Perle *f* Margarita *f*	Gemma *f*

CONSTELACIONES	CONSTELAÇÕES	СОЗВЕЗДИЯ	
Ursa Minor *f* Osa Menor *f*	Ursa Minor *f* Ursa Menor *f*	Малая Медведица *f* Малый Ковш *m*	97
Vela *npl* Vela *f*	Vela *npl* Velas *fpl* Vela *f*	Паруса *mpl*	98
Virgo *f* Virgen *f*	Virgo *f* Virgem *f*	Дева *f*	99
Volans *m* Pez Volador *m*	Volans *m* Peixe-voador *m*	Летучая Рыба *f*	100
Vulpécula *f* Zorra *f* Vulpeja *f*	Vulpecula *f* Raposa *f* Raposinha *f*	Лисичка *f*	101
Achernar *m*	Aquernar [Achernar] *m*	Ахернар *m*	102
Albireo *m*	Albíreo *m*	Альбирео *m*	103
Alcor *m*	Alcor *m*	Алькор *m*	104
Algol *m*	Algol *m*	Алголь *m*	105
Altair *m*	Altair *m*	Альтаир *m*	106
Antares *m*	Antares *m*	Антарес *m*	107
Betelgeuse *m*	Betelgeuse *m*	Бетельгейзе *m*	108
Canopus *m*	Canopo [Canopéia] *m* Canopus *m*	Канопус *m*	109
Capella *f* La Cabra *f*	Capella *f* Cabra *f*	Капелла *f*	110
Cástor *m*	Castor *m*	Кастор *m*	111
Deneb *m*	Deneb *m*	Денеб *m*	112
Denébola *f*	Denébola *f*	Денебола *f*	113
Dubhe *m*	Dubhe *m*	Дубхе *m*	114
Etamín *m*	Etamin *m*	Этамин *m*	115
Fomalhaut	Fomalhaut	Фомальгаут *m*	116
Gemma *f*	Gemma *f* Alfeca [Pérola] *f*	Гемма *f*	117

118	Hamal [α Ari]	Hamal *m*	Hamal *m*
119	Markab [α Peg]	Markab *m*	Markab *m*
120	Mizar [ζ UMa]	Mizar *m*	Mizar *m*
121	Polaris [α UMi] Polestar North Star	Polaire *f* Polaris *f* Étoile polaire *f*	Polarstern *m*
122	Pollux [β Gem]	Pollux *m*	Pollux *m*
123	Procyon [α CMi]	Procyon *m*	Prokyon *m*
124	Proxima Centauri	Proxima du Centaure *f*	Proxima Centauri *f*
125	Ras Alhague [α Oph]	Ras Alhague *m*	Ras Alhague *m*
126	Regulus [α Leo]	Régulus *m*	Regulus *m*
127	Rigel [β Ori]	Rigel *m*	Rigel *m*
128	Sirius [α CMa] Dog Star	Sirius *m* Étoile du Chien *f*	Sirius *m* Hundsstern *m*
129	Spica [α Vir]	Épi *f*	Spika *f*
129a	Orion's Belt Three Wise Men	Baudrier d'Orion *m* Trois Rois *mpl*	Oriongürtel *m* Jakobsstab *m*
130	Sword of Orion	Epée d'Orion *f*	Schwertgehänge des Orions *n*
131	Thuban [α Dra]	Thuban *m*	Thuban *m*
132	Vega [α Lyr]	Véga *f*	Wega *f*

CONSTELACIONES	CONSTELAÇÕES	СОЗВЕЗДИЯ	
Hamal *m*	Hamal *m*	Хамаль *m*	118
Markab *m*	Markab *m*	Маркаб *m*	119
Mizar *m*	Mizar *m*	Мицар *m*	120
Polar *f* Estrella polar *f*	Polar *f* Estrela polar *f*	Полярная *f*	121
Pólux *m*	Pólux *m*	Поллукс *m*	122
Proción *m*	Prócion *m* Procião *m*	Процион *m*	123
Próxima Centauri *f*	Próxima Centauri *f*	Проксима Центавра *f*	124
Ras Alhague *m*	Ras Alhague *m*	Рас-Альхаг *m*	125
Régulo *m*	Régulo *m*	Регул *m*	126
Rigel *m*	Rigel *m*	Ригель *m*	127
Sirius *m* Sirio *m*	Sirius *m* Sírio *m*	Сириус *m*	128
La Espiga *f* Spica *f*	Espiga *f* Spica *f*	Спика *f*	129
Cinturón de Orión *m* Tahalí de Orión *m* Tres Marías *fpl*	Cinto [Cinturão] de Órion *m* Três Marias *fpl* Três Irmãs *fpl*	Пояс Ориона *m*	129a
Enjambre de Orión *m* Puñal de Orión *m*	Espada do Órion *f* Três Reis Magos *mpl*	Меч Ориона *m*	130
Thuban *m*	Tubã *m*	Тубан *m*	131
Vega *f*	Vega *f*	Вега *f*	132

93	STELLAR STRUCTURE	STRUCTURE STELLAIRE	STERNAUFBAU
1	stellar structure	structure stellaire *f*	Sternaufbau *m* Sternstruktur *f*
2	stellar atmosphere	atmosphère stellaire *f*	Sternatmosphäre *f*
3	outer layers of a star	couches extérieures d'une étoile *fpl*	Aussenschichten eines Sternes *fpl*
4	stellar corona	couronne stellaire *f*	Sternkorona *f*
5	shell star	étoile à enveloppe *f*	Stern mit Hülle *m* Hüllenstern *m*
6	star with an extended atmosphere	étoile à atmosphère étendue *f*	Stern mit einer ausgedehnten Atmosphäre *m*
7	star with an extended envelope	étoile à enveloppe étendue *f*	Stern mit einer ausgedehnten Hülle *m*
8	extended envelope	enveloppe étendue *f*	ausgedehnte Hülle *f*
9	extended atmosphere	atmosphère étendue *f*	ausgedehnte Atmosphäre *f*
10	circumstellar shell	enveloppe stellaire [circumstellaire] *f*	zirkumstellare Hülle *f*
11	stationary shell	enveloppe stationnaire *f*	stationäre Hülle *f*
12	expanding shell	enveloppe en expansion *f*	ausdehnende Hülle *f* expandierende Hülle *f*
13	mass outflow	flux de masse (vers l'extérieur) *m*	Massenausfluss *m*
14	stellar wind	vent stellaire *m*	Sternwind *m* stellarer Wind *m*
15	mass ejection	éjection de masse *f*	Massenausstoss *m*
16	(purely) scattering atmosphere	atmosphère purement diffusante *f*	(rein) streuende Atmosphäre *f*
17	model atmosphere	modèle d'atmosphère *m*	Atmosphärenmodell *n*
18	blanketed model	modèle avec [tenant compte de] l'effet de serre *m*	Modell mit (Berücksichtigung der) Linienabsorption *n*
19	unblanketed model	modèle sans [ne tenant pas compte de] l'effet de serre *m*	Modell ohne (Berücksichtigung der) Linienabsorption *n*
20	blanketing	effet de serre *m* couverture *f*	Blanketing *n* Glashauswirkung *f*
21	plane-parallel atmosphere	atmosphère stratifiée en couches planes parallèles *f*	planparallele Atmosphäre *f*
22	semi-infinite atmosphere	atmosphère semi-infinie *f*	halbunendliche Atmosphäre *f*
23	finite atmosphere	atmosphère finie *f*	endliche Atmosphäre *f*

estructura estelar *f*	estrutura estelar *f*	строение звёзд *n*	1
atmósfera estelar *f*	atmosfera estelar *f*	звёздная атмосфера *f*	2
capas exteriores de una estrella *fpl*	camadas exteriores de uma estrela *fpl*	внешние слои звезды *mpl* наружные слои звезды *mpl*	3
corona estelar *f*	coroa estelar *f*	звёздная корона *f*	4
estrella de envoltura *f*	estrela com invólucro [envelope, envoltória] *f*	звезда с оболочкой *f*	5
estrella con atmósfera extendida *f*	estrela com atmosfera extensa *f*	звезда с протяжённой атмосферой *f*	6
estrella con envoltura extendida *f*	estrela com invólucro [envoltória, envelope] extenso *f*	звезда с протяжённой оболочкой *f*	7
envoltura extendida *f*	invólucro [envelope] extenso *m*	протяжённая оболочка *f*	8
atmósfera extendida *f*	atmosfera extensa *f*	протяжённая атмосфера *f*	9
envoltura circumestelar *f*	invólucro [envoltória] circunstelar [circum-estelar] *m* [*f*]	оболочка звезды *f*	10
envoltura estacionaria *f*	invólucro estacionário *m* envoltória estacionária *f*	стационарная оболочка *f*	11
envoltura en expansión *f*	invólucro [envoltória] em expansão *m* [*f*]	расширяющаяся оболочка *f*	12
flujo de masa (hacia el exterior) *m*	efluxo de massa *m* fluxo de massa (para o exterior) *m*	истечение вещества *n* истечение массы *n*	13
viento estelar *m*	vento estelar *m*	звёздный ветер *m*	14
eyección de masa *f*	ejeção de massa *f*	выброс массы [вещества] *m*	15
atmósfera (puramente) difusora *f*	atmosfera de difusão (pura) *f* atmosfera difusora *f*	(чисто) рассеивающая атмосфера *f*	16
modelo de atmósfera *m*	modelo de atmosfera *m*	модель атмосферы *f*	17
modelo con blanketing *m*	modelo com [de] cobertura *m*	модель с экранированием *f*	18
modelo sin blanketing *m*	modelo descoberto [sem cobertura] *m*	модель без экранирования *f*	19
blanketing *m*	cobertura *f*	экранирование (линиями) *n*	20
atmósfera plano-paralela [estratificada en capas plano-paralelas] *f*	atmosfera plano-paralela [de camadas plano-paralelas] *f*	плоско-параллельная атмосфера *f*	21
atmósfera semiinfinita *f*	atmosfera semi-infinita *f*	полубесконечная атмосфера *f*	22
atmófera finita *f*	atmosfera finita *f*	атмосфера конечной толщины *f* конечная атмосфера *f*	23

	STELLAR STRUCTURE	STRUCTURE STELLAIRE	STERNAUFBAU
24	spherical atmosphere	atmosphère sphérique *f*	kugelsymmetrische Atmosphäre *f*
25	isothermal atmosphere	atmosphère isotherme [isothermique] *f*	isotherme Atmosphäre *f*
26	grey atmosphere (UK) gray atmosphere (US)	atmosphère grise *f*	graue Atmosphäre *f*
27	non-grey atmosphere (UK) non-gray atmosphere (US)	atmospheère non-grise *f*	nicht-graue Atmosphäre *f*
28	electron atmosphere	atmosphère d'électrons *f* atmosphère électronique *f*	Elektronenatmosphäre *f*
29	radiative transfer transfer of radiation	transfert du rayonnement *m*	Strahlungstransport *m*
30	radiation field	champ de rayonnement *m*	Strahlungsfeld *n*
31	diffuse radiation field	champ de rayonnement diffus *m*	diffuses Strahlungsfeld *n*
32	transfer equation	équation de transfert *f*	Transportgleichung *f*
33	equation of radiative transfer [transport]	équation de transfert radiatif *f*	Strahlungstransportgleichung *f*
34	source function	fonction source *f*	Ergiebigkeit *f*
35	thermodynamic(al) equilibrium TE	équilibre thermodynamique *m* ET	thermodynamisches Gleichgewicht *n* TE
36	deviations from TE deviations from thermodynamic equilibrium departure from thermodynamic equilibrium	écarts (par rapport) à l'équilibre thermodynamique *mpl* écarts à l'ET *mpl*	Abweichungen vom thermodynamischen Gleichgewicht *fpl* Abweichungen vom TE *fpl*
37	dilution factor	facteur de dilution *m*	Verdünnungsfaktor *m*
38	diluted radiation dilute radiation	rayonnement dilué *m*	verdünnte Strahlung *f*
39	local thermodynamic equilibrium LTE	équilibre thermodynamique local *m* ETL	lokales thermodynamisches Gleichgewicht *n* LTE
40	deviations from LTE deviations from local thermodynamic equilibrium	écarts à l'ETL *mpl* écarts à l'équilibre thermodynamique local *mpl*	Abweichungen vom lokalen thermodynamischen Gleichgewicht *fpl* Abweichungen vom LTE *fpl*
41	detailed balancing microreversibility	bilan detaillé *m* microréversibilité *f*	detailiertes Gleichgewicht *n* Gleichgewicht der Elementarprozesse *n* Mikroreversibilität *f*
42	statistical equilibrium	équilibre statistique *m*	statistisches Gleichgewicht *n*

atmósfera esférica *f*	atmosfera esférica *f*	сферическая атмосфера *f*	24
atmósfera isotérmica *f*	atmosfera isotérmica *f*	изотермическая атмосфера *f*	25
atmósfera gris *f*	atmosfera cinza *f*	серая атмосфера *f*	26
atmósfera no gris *f*	atmosfera não-cinza *f*	несерая атмосфера *f*	27
atmósfera de electrones *f* atmósfera electrónica *f*	atmosfera ele(c)trônica *f* atmosfera de elé(c)trons *f*	атмосфера, состоящая [составленная] из электронов *f* электронная атмосфера *f*	28
transporte de radiación *f* transporte radiativo *m*	transferência radiativa [da radiação] *f* transporte radiativo [da radiação] *m*	перенос излучения *m*	29
campo de radiación *m*	campo de radiação *m*	поле излучения *m*	30
campo de radiación difusa *m*	campo de radiação difusa *m*	поле диффузного излучения *n*	31
ecuación del transporte *f*	equação de transferência [transporte] *f*	уравнение переноса *n*	32
ecuación del transporte radiativo *f*	equação de transferência radiativa [de transporte radiativo] *f*	уравнение лучистого переноса *n* уравнение переноса излучения *n*	33
función fuente *f*	função fonte [origem] *f*	функция источника [отдачи] *f*	34
equilibrio termodinámico *m* ET	equilíbrio termodinâmico *m* ET	термодинамическое равновесие *n*	35
desviaciones del equilibrio termodinámico *fpl* desviaciones del ET *fpl*	desvios do equilíbrio termodinâmico *mpl* desvios do ET *mpl* afastamentos do ET *mpl*	отклонения от термодинамического равновесия *npl* отклонения от ТР *npl*	36
factor de dilución *m*	fator [coeficiente] de diluição *m*	коэффициент дилюции *m* фактор разрежения [дилюции] *m*	37
radiación diluída *f*	radiação diluída *f*	дилютированное [расслабленное] излучение *n*	38
equilibrio termodinámico local *m* ETL	equilíbrio termodinâmico local *m* ETL	локальное термодинамическое равновесие *n* ЛТР	39
desviaciones del ETL *fpl* desviaciones del equilibrio termodinámico local *fpl*	desvios do ETL *mpl* desvios do equilíbrio termodinâmico local *mpl* afastamentos do ETL *mpl*	отклонения от (состояния) локального термодинамического равновесия *npl* отклонения от ЛТР *npl*	40
balance detallado *m* equilibrio detallado *m* microrreversibilidad *f*	balanço [balanceamento] detalhado [pormenorizado] *m* microrreversibilidade *f* equilíbrio pormenorizado *m*	детальный баланс *m* детальное балансирование *n* детальное равновесие *n* детальное уравновешивание *n*	41
equilibrio estadístico *m*	equilíbrio estatístico *m*	статистическое равновесие *n*	42

43	generalized equilibrium non-LTE (equilibrium) non-thermodynamic(al) equilibrium	équilibre général *m* équilibre hors ETL *m*	verallgemeinertes Gleichgewicht *n* nicht-LTE-Gleichgewicht *n*
44	statistical equilibrium equations	équations de l'équilibre statistique *fpl*	Gleichungen des statistischen Gleichgewichtes *fpl*
45	two-level atom	atome à deux niveaux *m*	(Modell-)Atom mit zwei Energieniveaus *n*
46	multilevel atom	atome à plusieurs niveaux *m*	(Modell-)Atom mit vielen Energieniveaus *n*
47	integrated flux	flux intégré *m*	Gesamtstrahlungsstrom *m*
48	radiative equilibrium	équilibre radiatif *m*	Strahlungsgleichgewicht *n*
49	continuity equation	équation de continuité *f*	Kontinuitätsgleichung *f*
50	Milne-Eddington approximation	approximation de Milne-Eddington *f*	Milne-Eddington(sche) Näherung *f*
51	Chandrasekhar approximation	approximation de Chandrasekhar *f*	Chandrasekhar(sche) Näherung *f*
52	integro-differential equation	équation intégro-différentielle *f*	Integro-Differentialgleichung *f*
53	integral equation of Schwarzschild and Milne	équation intégrale de Schwarzschild et Milne *f*	Schwarzschild-Milne(sche) Integralgleichung *f*
54	Laplace transform Laplace transformation	transformation laplacienne [de Laplace] *f*	Laplace-Transformation *f* Laplacesche Transformation *f*
55	Legendre polynomials Legendre coefficients zonal harmonic (functions)	polynômes de Legendre *mpl* fonctions (harmoniques) zonales *fpl*	Legendresche Polynome *npl* zonale Kugelfunktionen *fpl* Legendre-Polynome *npl*
56	iterative method	méthode itérative *f*	Iterationsmethode *f*
57	variational method	méthode variationnelle *f*	Variationsmethode *f*
58	weight function	fonction poids *f*	Gewichtsfunktion *f*
59	method of weight functions	méthode des fonctions poids *f*	Methode der Gewichtsfunktionen *f*
60	saturation function	fonction de saturation *f*	Sättigungsfunktion *f*
61	principle of invariance	principe d'invariance *m*	Invariànzprinzip *n*
62	exponential integral function	fonction intégro-exponentielle *f*	Integralexponentialfunktion *f*
63	line formation	formation de raies *f*	Linienentstehung *f*
64	effective depth of line formation	profondeur effective de formation des raies *f*	effektive Tiefe der Linienentstehung *f*

equilibrio generalizado *m* equilibrio diferente del ETL *m*	equilíbrio generalizado *m* equlíbrio diferente do ETL *m* equilíbrio fora do ETL *m*	обобщённое равновесие *n* не ЛТР *n* отсутствие ЛТР *n*	43
ecuaciones del equilibrio estadístico *fpl*	equações do equilíbrio estatístico *fpl*	уравнения стационарности *npl*	44
átomo de dos niveles *m*	átomo de dois níveis *m*	двухуровенный атом *m*	45
átomo de niveles múltiples *m*	átomo de níveis múltiplos *m*	атом с несколькими уровнями *m*	46
flujo integrado *m*	fluxo integrado *m*	полный [интегральный] поток *m*	47
equilibrio radiativo *m*	equilíbrio radiativo *m*	лучистое равновесие *n*	48
ecuación de continuidad *f*	equação de continuidade *f*	уравнение непрерывности *n*	49
aproximación de Milne-Eddington *f*	aproximação de Milne-Eddington *f*	приближение Милна-Эддингтона *n*	50
aproximación de Chandrasekhar *f*	aproximação de Chandrasekhar *f*	приближение Чандрасекара *n*	51
ecuación integro-diferencial *f*	equação íntegro-diferencial *f*	интегро-дифференциальное уравнение *n*	52
ecuación integral de Schwarzschild y Milne *f*	equação integral de Schwarzschild e Milne *f*	интегральное уравнение Шварцшильда-Милна *n*	53
transformación laplaciana [de Laplace] *f*	transfomação de Laplace *f*	преобразование Лапласа *n* лапласово преобразование *n*	54
polinomios [coeficientes] de Legendre *mpl* armónicos zonales *mpl*	polinômios [coeficientes] de Legendre *mpl* harmônicas zonais *fpl*	полиномы Лежандра *mpl* зональные гармонические (функции) *fpl*	55
método de iteración *m*	método iterativo *m*	метод итераций *m*	56
método variacional *m*	método variacional *m*	вариационный метод *m*	57
función de peso *f*	função (de) peso [ponderação] *f*	весовая функция *f*	58
método de las funciones de peso *m*	método das funções (de) peso [ponderação] *f*	метод весовых функций *m*	59
función de saturacion *f*	função (de) saturação *f*	функция насыщения *f*	60
principio de invariancia *m*	princípio de invariância *m*	принцип инвариантности *m*	61
función integro-exponencial *f*	função integro-exponencial *f*	интегрально-показательная функция *f*	62
formación de las líneas *f*	formação de linhas [raias] *f*	образование линий *n*	63
profundidad efectiva de formación de las líneas *f*	profundidade efetiva da formação das raias [linhas] *f*	эффективная глубина испускания излучения *f* эффективная глубина образования линий *f*	64

65	hydrogen ion	ion (d')hydrogène *m*	Wasserstoffion *n*
66	negative hydrogen ion	ion négatif (d')hydrogène *m*	negatives Wasserstoffion *n*
67	blanketing effect	effet de couverture [de serre] *m*	Blanketingeffekt *m* Blanketing-Effekt *m*
68	curve of growth	courbe de croissance *f*	Wachstumskurve *f*
69	empirical curve of growth	courbe de croissance empirique *f*	emipirische Wachstumskurve *f*
70	theoretical curve of growth	courbe de croissance théorique *f*	theoretische Wachstumskurve *f*
71	internal structure of stars stellar structure internal constitution	structure interne des étoiles *f* structure stellaire *f* constitution interne des étoiles *f*	innerer Aufbau der Sterne *m* Sternaufbau *m*
72	stellar interior	intérieur stellaire *m* intérieur des étoiles *m*	das Sterninnere *n* das Innere der Sterne *n*
73	gas sphere	sphère gazeuse *f*	Gaskugel *f*
74	equilibrium conditions	conditions d'équilibre *fpl*	Gleichgewichtsbedingungen *fpl*
75	hydrostatic equilibrium	équilibre hydrostatique [mécanique] *m*	hydrostatisches Gleichgewicht *n*
76	hydrostatic equation	équation de l'équilibre mécanique *f* équation de l'équilibre hydrostatique *f*	Gleichung des hydrostatischen Gleichgewichts *f* hydrostatische Gleichung *f*
77	hydrostatic pressure	pression hydrostatique *f*	hydrostatischer Druck *m*
78	thermal equilibrium	équilibre thermique *m*	thermisches Gleichgewicht *n*
79	convective equilibrium	équilibre convectif *m*	Konvektionsgleichgewicht *n* konvektives Gleichgewicht *n*
80	surface (boundary) condition	condition à la surface *f*	Oberflächenbedingung *f*
81	central (boundary) condition	condition au centre *f*	Zentrumsbedingung *f*
82	uniqueness of the solution	unicité de la solution *f*	Eindeutigkeit der Lösung *f*
83	Vogt-Russell theorem Russell-Vogt theorem uniqueness theorem	théorème de Vogt-Russell *m* théorème d'unicité *m*	Eindeutigkeitssatz (des Sternaufbaus) *m* Vogt-Russell-Theorem *n*
84	centre of the star	centre de l'étoile *m*	Sternzentrum *n* Sternmittelpunkt *m*
85	distance from the centre of the star	distance au centre de l'étoile *f*	Entfernung [Abstand] vom Sternmittelpunkt *f*
86	mean density	densité moyenne *f*	mittlere Dichte *f*
87	central density	densité centrale *f* densité au centre *f*	Zentraldichte [Mittelpunktdichte] *f* zentrale Dichte *f*

ion de hidrógeno *m*	íon de hidrogênio *m*	водородный ион *m*	65
ion negativo del hidrógeno *m*	íon negativo de hidrogênio *m*	отрицательный ион водорода *m*	66
efecto blanketing *m*	efeito de cobertura *m*	покровный эффект *m* эффект экранирования *m*	67
curva teórica de crecimiento *f*	curva de crescimento *f*	кривая роста *f*	68
curva de crecimiento empírica *f*	curva empírica de crescimento *f*	эмпирическая кривая роста *f*	69
curva de crecimiento teórica *f*	curva teórica de crescimento *f*	теоретическая кривая роста *f*	70
constitución interna de las estrellas *f* estructura estelar *f*	estrutura interna das estrelas *f* estrutura estelar *f* constituição interna das estrelas *f*	внутреннее строение звёзд *n*	71
interior estelar *m* interior de estrellas *m*	interior estelar *m* interior das estrelas *m*	звёздные недра *pl* недра звёзд *pl*	72
esfera gaseosa *f*	esfera gasosa *f*	газовой шар *m*	73
condiciones de equilibrio *fpl*	condições de equilíbrio *fpl*	условия равновесия *npl*	74
equilibrio hidrostático *m*	equilíbrio hidrostático *m*	гидростатическое равновесие *n*	75
ecuación del equilibrio hidrostático *f* ecuación hidrostática *f*	equação do equilíbrio hidrostático *f* equação hidrostática *f*	гидростатическое уравнение *n* уравнение гидростатики [гидростатического равновесия] *n*	76
presión hidrostática *f*	pressão hidrostática *f*	гидростатическое давление *n*	77
equilibrio térmico *m*	equilíbrio térmico *m*	тепловое равновесие *n*	78
equilibrio convectivo *m*	equilíbrio convectivo *m*	конвективное равновесие *n*	79
condición en la superficie *f*	condição na [à] superfície *f*	граничное условие на поверхности *n*	80
condición (de contorno) en el centro *f*	condição no centro *f*	граничное условие в центре *n*	81
unicidad de la solución *f*	unicidade da solução *f*	однозначность решения *f* единственность решения *f*	82
teorema de Vogt-Russel *m* teorema de Russell-Vogt *m* teorema de unicidad *m*	teorema de Vogt-Russell *m* teorema de Russell-Vogt *m* teorema da unicidade *m*	теорема Рессела-Фогта *f* теорема однозначности *f*	83
centro de la estrella *m*	centro da estrela *m*	центр звезды *m*	84
distancia desde el centro de la estrella *f*	distância do centro da estrela *f*	расстояние от центра звезды *n*	85
densidad media *f*	densidade média *f*	средняя плотность *f*	86
densidad central *f* densidad en el centro *f*	densidade central *f* densidade no centro *f*	центральная плотность *f* плотность в центре *f*	87

	STELLAR STRUCTURE	STRUCTURE STELLAIRE	STERNAUFBAU
88	surface temperature	température superficielle *f*	Oberflächentemperatur *f*
89	central temperature	température centrale *f* température au centre *f*	Zentraltemperatur *f* Mittelpunkttemperatur *f* zentrale Temperatur *f*
90	gas pressure particle pressure	pression gazeuse *f* pression de particules *f*	Gasdruck *m* Teilchendruck *m*
91	radiation pressure photon pressure	pression de radiation *f* pression de photons *f*	Strahlungsdruck *m* Photonendruck *m*
92	Poynting vector	vecteur de Poynting *m*	Poyntingvektor *m* Poynting-Vektor *m*
93	total pressure	pression totale *f*	Gesamtdruck *m*
94	gradient	gradient *m*	Gradient *m*
95	pressure gradient	gradient de pression *m*	Druckgradient *m*
96	temperature gradient	gradient de température *m* gradient thermique *m*	Temperaturgradient *m* Temperaturgefälle *n*
97	radiative temperature gradient	gradient de température radiatif *m*	radiativer Temperaturgradient *m*
98	adiabatic temperature gradient	gradient adiabatique de température *m*	adiabatischer Temperaturgradient *m*
99	superadiabatic gradient	gradient superadiabatique *m*	überadiabatischer Gradient *m*
100	mean molecular weight	poids moléculaire moyen *m*	mittleres Molekulargewicht *n*
101	Russell mixture	mélange de Russell *m*	Russellmischung *f*
102	Kramers' law	loi de Kramers *f*	Kramerssches Gesetz *n*
103	guillotine factor	facteur (de) guillotine *m*	Guillotinefaktor *m*
104	Gaunt factor	facteur de Gaunt *m*	Gauntfaktor *m*
105	energy generation energy production	génération d'énergie *f* production d'énergie *f*	Energieproduktion *f* Energieerzeugung *f* Energiefreisetzung *f*
106	free-free absorption	absorption libre-libre *f*	frei-freie Absorption *f*
107	bound-free absorption	absorption lié-libre *f*	gebunden-freie Absorption *f*

temperatura superficial *f*	temperatura superficial *f*	поверхностная температура *f*	88
temperatura central *f* temperatura en el centro *f*	temperatura central *f* temperatura no centro *f*	центральная температура *f* температура в центре *f*	89
presión del gas *f* presión de partículas *f*	pressão gasosa [do gás] *f* pressão de partículas *f*	газовое давление *n* давление газа *n* давление частиц *n*	90
presión de la radiación *f* presión de fotones *f*	pressão da radiação *f* pressão dos fótons *f*	лучистое [радиационное] давление *n* давление излучения [фотонов] *n*	91
vector de Poynting *m*	ve(c)tor de Poynting *m*	вектор Пойнтинга [Умова-Пойнтинга] *m*	92
presión total *f*	pressão total *f*	полное давление *n*	93
gradiente *m*	gradiente *m*	градиент *m*	94
gradiente de presión *m*	gradiente de pressão *m*	градиент давления *m*	95
gradiente térmico *m* gradiente de temperatura *m*	gradiente de temperatura *m* gradiente térmico *m*	температурный градиент *m* градиент температуры *m*	96
gradiente radiativo de temperatura *m*	gradiente radiativo de temperatura *m*	градиент температуры при лучистом равновесии *m*	97
gradiente adiabático de temperatura *m*	gradiente adiabático de temperatura *m*	адиабатический температурный градиент *m*	98
gradiente superadiabático *m*	gradiente superadiabático *m*	сверхадиабатический градиент *m*	99
peso molecular medio *m*	peso molecular médio *m*	средний молекулярный вес *m*	100
mezcla de Russell *f*	mistura de Russell *f*	ресселова смесь *f*	101
ley de Kramers *f*	lei de Kramers *f*	закон Крамерса *m*	102
factor de guillotina *m*	fator de guilhotina *m*	гильотинный множитель *m* поправка к формуле Крамерса *f*	103
factor de Gaunt *m*	fator de Gaunt *m*	гаунтовский поправочный множитель *m*	104
generación de energía *f* producción de energía *f*	geração de energia *f* produção de energia *f*	выделение энергии *n* генерация энергии *f* энерговыделение *n*	105
absorción libre-libre *f*	absorção livre-livre *f*	свободно-свободная абсорбция *f* свободно-свободное поглощение *n*	106
absorción ligada-libre *f*	absorção ligada-livre *f*	связанно-свободная абсорбция *f* связанно-свободное поглощение *n*	107

108	bound-bound absorption line absorption	absorption lié-lié *f* absorption dans la raie *f*	gebunden-gebundene Absorption *f* Linienabsorption *f*
109	energy output energy release	débit d'énergie *m* rendement énergétique *m*	Energieabgabe *f* Energieausstoss *m*
110	source of energy energy source	source d'énergie *f*	Energiequelle *f*
111	energy transport energy transfer	transport d'énergie *m* transfert d'énergie *m*	Energietransport *m*
112	energy flux	flux d'énergie *m*	Energiefluss *m* Energiestrom *m*
113	radiation flux	flux de rayonnement *m*	Strahlungsfluss *m*
114	emergent radiative flux	flux de rayonnement émergent *m*	austretender Strahlungsfluss *m*
115	convective energy flux convective flux	flux d'énergie convectif *m* flux convectif *m*	konvektiver Energiefluss *m*
116	total energy flux	flux d'énergie total *m*	Gesamtenergiefluss *m*
117	contraction	contraction *f*	Kontraktion *f*
118	gravitational contraction	contraction gravitationnelle *f*	Gravitationskontraktion *f*
119	gravitational energy	énergie gravitationnelle *f*	Gravitationsenergie *f*
120	rotational energy	énergie de rotation *f*	Rotationsenergie *f*
121	atomic energy	énergie atomique *f*	Atomenergie *f*
122	nuclear energy	énergie nucléaire *f*	Kernenergie *f*
123	stellar model	modèle stellaire *m* modèle d'étoile *m*	Sternmodell *n*
124	stellar matter stellar material	matière stellaire *f*	stellare Materie *f* Sternmaterie *f*
125	mixture of gas and radiation	mélange de gaz et de rayonnement *f*	Mischung von Gas und Strahlung *f*
126	polytrope	polytrope *m*	Polytrope *f*
127	polytropic gaseous sphere polytropic gas sphere	sphère de gaz polytropique *f*	polytrope Gaskugel *f*
128	polytropic index	indice polytropique *m* index polytropique *m*	Polytropenindex *m* Index der Polytrope *m*
129	polytropic model	modèle polytropique *m*	polytropes Modell *n*

absorción ligada-ligada *f* absorción de línea *f*	absorção ligada-ligada *f* absorção de linha [de raia] *f*	поглощение в линии *n* поглощение в результате связанно-связанных переходов *n* абсорбция в линии *f*	108
rendimiento energético *m* liberación de energía *f*	rendimento energético *m* liber(t)ação de energia *f*	выход энергии *m* энергетический выход *m*	109
fuente de energía *f*	fonte de energia *f*	источник энергии *m*	110
transporte de energía *m*	transporte de energia *m* transferência de energia *f*	перенос энергии *m*	111
flujo de energía *m*	fluxo de energia *m*	поток энергии *m*	112
flujo de radiación *m*	fluxo de radiação *m*	поток излучения *m*	113
flujo radiativo emergente *m*	fluxo radiativo emergente *m*	выходящий поток излучения *m*	114
flujo de energía convectivo *m* flujo convectivo *m*	fluxo de energia convectivo *m* fluxo convectivo *m*	конвективный поток энергии *m*	115
flujo total de energía *m*	fluxo total de energia *m*	полный поток энергии *m*	116
contracción *f*	contração *f*	сжатие *n*	117
contracción gravitacional *f*	contração gravitacional *f*	гравитационное сжатие *n*	118
energía gravitacional *f*	energia gravitacional *f*	гравитационная энергия *f*	119
energía rotacional *f*	energia rotacional [de rotação] *f*	энергия вращения *f* вращательная энергия *f*	120
energía atómica *f*	energia atômica *f*	атомная энергия *f*	121
energía nuclear *f*	energia nuclear *f*	ядерная энергия *f*	122
modelo estelar *m* modelo de estrella *m*	modelo estelar *m* modelo da estrela *m*	модель звезды *f* звёздная модель *f*	123
materia estelar *f*	matéria estelar *f* material estelar *m*	звёздное вещество *n* вещество звезды *n*	124
mezcla de gas y radiación *f*	mistura de gás e radiação *f*	смесь излучения и газа *f* излучение и газ *n*, *m*	125
polítropo *m*	politrópica *f*	политропа *f*	126
esfera de gas politrópica *f*	esfera gasosa politrópica *f*	политропный газовый шар *m* политропная газообразная сфера *f*	127
índice politrópico *m*	índice politrópico *m*	индекс политропы *m* показатель политропы *m*	128
modelo politrópico *m*	modelo politrópico *m*	политропная модель *f*	129

130	Emden's equation	équation d'Emden *f*	Emdensche (Differential)gleichung *f*
131	homologous	homologue	homolog
132	homology transformation homologous transformation	transformation homologue *f*	homologe Transformation *f*
133	non-dimensional variables	variables non-dimensionnelles [sans dimensions] *fpl*	dimensionslose Veränderliche *fpl*
134	radiative layer radiative zone	zone radiative *f* couche radiative *f*	radiative Schicht *f* Strahlungszone *f*
135	radiative envelope	enveloppe radiative *f*	Hülle im Strahlungsgleichgewicht *f*
136	convective envelope	enveloppe convective *f*	konvektive Hülle *f*
137	Eddington model standard model	modèle d'Eddington *m* modèle standard *m*	Eddingtonsches Modell Standardmodell *n*
138	point-source model	modèle à source ponctuelle *m*	Punktquellenmodell *n*
139	inhomogeneous stellar model	modèle stellaire non-homogène *m*	inhomogenes Sternmodell *n*
140	composite model	modèle mixte [composite] *m*	zusammengesetztes Modell *n*
141	interface	interface *f*	Übergangszone *f*
142	fitting condition	condition de raccordement *f*	Anpassungsbedingung *f* Fit-Bedingung *f*
143	shell-source model	modèle à sources en couches *m*	Schalenquellenmodell *n*
144	hydrogen burning zone hydrogen shell	zone de combustion de l'hydrogène *f* couche d'hydrogène *f*	Zone des Wasserstoffsbrennens *f* Wasserstoffschale *f*
145	helium burning zone helium shell	zone de combustion de l'hélium *f* couche d'hélium *f*	Zone des Heliumbrennens *f* Heliumschale *f*
146	double energy source	source d'énergie double *f*	doppelte Energiequelle *f*
147	convection	convection *f*	Konvektion *f*
148	convective	convectif	konvektiv
149	convective zone convective layer	zone convective *f* couche convective *f*	Konvektionszone *f* konvektive Schicht *f*
150	inward stream descending stream falling current	courant descendant *m* courant tombant *m* courant vers l'intérieur *m*	absinkender Strom *m* fallender Strom *m*

ecuación de Emden *f*	equação de Emden *f*	уравнение Эмдена *n*	130
homólogo	homólogo	гомологический	131
transformación homológica *f*	transformação homóloga *f*	гомологическое преобразование *n*	132
variables adimensionales *fpl*	variáveis adimensionais *fpl*	безразмерные переменные *fpl*	133
capa radiativa *f* zona radiativa *f*	camada radiativa *f* zona radiativa *f*	зона лучистого равновесия *f*	134
envoltura radiativa *f*	envelope [invólucro] radiativo *m* envoltória radiativa *f*	оболочка в лучистом равновесии *f* лучистая оболочка *f*	135
envoltura convectiva *f*	envelope [invólucro] convectivo *m* envoltória convectiva *f*	оболочка в конвективном равновесии *f* конвективная оболочка *f*	136
modelo de Eddington *m* modelo estándar *m*	modelo de Eddington *m* modelo padrão *m*	модель Эддингтона *f* стандартная модель *f*	137
modelo de fuente puntual *m*	modelo de fonte pontual *m*	модель с точечным источником *f*	138
modelo estelar no homogéneo [inhomogéneo] *m*	modelo estelar inomogêneo [não-homogêneo) *m*	неоднородная модель звезды *f*	139
modelo compuesto *m*	modelo composto *m*	составная модель *f*	140
interface *f* zona común *f*	interface *f* interzona *f*	промежуточная поверхность *f*	141
condición de ajuste *f* condición de empalme *f*	condição de ajuste *f*	условие согласования *n* условие сшивки *n*	142
modelo de fuente en capa *m*	modelo de fonte em camada *m*	модель с источниками в оболочке *f*	143
capa (de combustión) de hidrógeno *f* zona (de combustión) de hidrógeno *f*	zona de (queima [de combustão] do) hidrogênio *f* camada de (queima do) hidrogênio *f*	зона горения водорода *f* водородная оболочка *f*	144
capa (de combustión) de helio *f* zona (de combustión) de helio *f*	zona de queima do hélio *f* camada de combustão do hélio *f*	зона горения гелия *f* гелиевая оболочка *f*	145
fuente doble de energía *f*	fonte dupla de energia *f*	модель двойного источника *f* модель с двумя источниками *f*	146
convección *f*	convecção *f*	конвекция *f*	147
convectivo	convectivo	конвективный	148
zona convectiva *f* capa convectiva *f*	zona convectiva [de convecção] *f* camada convectiva *f*	конвективная зона *f*	149
corriente descendente *f*	corrente descendente *f*	нисходящий конвективный поток *m* падающий поток *m*	150

151	outward stream rising current ascending current	courant ascendant *m* courant vers l'extérieur *m* courant surgissant *m*	aufsteigender Strom *m*
152	turbulent convection	convection turbulente *f*	turbulente Konvektion *f*
153	laminar convection	convection laminaire *f*	laminare Konvektion *f*
154	local instability	instabilité locale *f*	lokale Instabilität *f*
155	convective instability	instabilité convective *f*	konvektive Instabilität *f*
156	Schwarzschild inequality	inégalité de Schwarzschild *f*	Schwarzschild(sche) Ungleichung *f*
157	convective temperature gradient	gradient convectif de température *m*	konvektiver Temperaturgradient *m*
159	hydrogen convection [convective] zone	zone convective d'hydrogène *f*	Wassertoffkonvektionszone *f*
160	semi-convective zone	zone semi-convective *f*	halbkonvektive Zone *f* semikonvektive Zone *f*
161	acoustic energy flux	flux d'énergie acoustique *m*	Schallenergiefluss *m* akustischer Energiefluss *m*
162	mixing of stellar matter	brassage de la matière stellaire *m*	Durchmischung der Sternmaterie *f*
163	meridional [meridian] circulation	circulation méridionale [méridienne] *f*	meridionale Zirkulation *f*
164	core of a star stellar core	noyau stellaire *m* noyau d'une étoile *m*	Kern des Sternes *m* Sternkern *m*
165	convective [convection] core	noyau convectif *m*	konvektiver Kern *m*
166	radiative core	noyau radiatif *m*	radiativer Kern *m*
167	helium core	noyau d'hélium *m*	Heliumkern *m*
168	isothermal core	noyau isotherme *m*	isothermer Kern *m*
169	inhomogeneous core	noyau non-homogène *m*	inhomogener Kern *m*
170	degenerate core	noyau dégénéré *m*	entarteter Kern *m*
171	non-degenerate core	noyau non-dégénéré *m*	nichtentarteter Kern *m*
172	degeneracy	dégénérescence *f*	Entartung *f*
173	partial degeneracy	dégénérescence partielle *f*	teilweise Entartung *f*
174	degree of degeneracy	degré de dégénérescence *m*	Entartungsgrad *m*
175	complete degeneracy	dégénérescence complète *f*	vollständige Entartung *f*

corriente ascendente *f*	corrente ascendente *f*	восходящий конвективный поток *m*	151
convección turbulenta *f*	convecção turbulenta *f*	турбулентная конвекция *f*	152
convección laminar *f*	convecção laminar *f*	ламинарная конвекция *f*	153
inestabilidad local *f*	instabilidade local *f*	локальная неустойчивость [нестабильность] *f*	154
inestabilidad convectiva *f*	instabilidade convectiva *f*	конвективная неустойчивость [нестабильность] *f*	155
desigualdad de Schwarzschild *f*	desigualdade de Schwarzschild *f*	неравенство Шварцшильда *n*	156
gradiente de temperatura convectivo *m*	gradiente de temperatura convectivo *m*	конвективный температурный градиент *m*	157
zona convectiva de hidrógeno *f*	zona convectiva [de convecção] do hidrogênio *f*	конвективная водородная зона *f*	159
zona semiconvectiva *f* capa semiconvectiva *f*	zona semiconvectiva *f* camada semiconvectiva *f*	полуконвективная зона *f*	160
flujo de energía acústica *m*	fluxo de energia acústica *m* fluxo de energia sonora *m*	поток звуковой энергии *m*	161
mezcla de la materia estelar *f*	mistura da matéria estelar *f*	перемешивание звёздного вещества *n*	162
circulación meridional *f*	circulação meridional *f*	меридиональная циркуляция *f*	163
núcleo estelar *m* núcleo de una estrella *m*	núcleo estelar *m* núcleo de uma estrela *m*	ядро звезды *n* звёздное ядро *n*	164
núcleo convectivo *m*	núcleo convectivo *m*	конвективное ядро *n*	165
núcleo radiativo *m*	núcleo radiativo *m*	ядро в лучистом равновесии *n*	166
núcleo de helio *m*	núcleo de hélio *m*	гелиевое ядро *n*	167
núcleo isotérmico *m*	núcleo isotérmico *m*	изотермическое ядро *n*	168
núcleo no homogéneo *m*	núcleo inomogêneo [não-homogêneo] *m*	неоднородное ядро *n*	169
núcleo degenerado *m*	núcleo degenerado *m*	вырожденное ядро *n*	170
núcleo no degenerado *m*	núcleo não-degenerado *m*	невырожденное ядро *n*	171
degeneración *f*	degenerescência *f* degeneração *f*	вырождение *n*	172
degeneración parcial *f*	degenerescência parcial *f*	частичное вырождение *n*	173
grado de degeneración *m*	grau de degenerescência *m*	степень вырождения *f*	174
degeneración total *f*	degenerescência total [completa] *f*	полное вырождение *n*	175

176	relativistic degeneracy	dégénérescence relativiste *f*	relativistische Entartung *f*
177	non-relativistic degeneracy	dégénérescence non-relativiste *f*	nichtrelativistische Entartung *f*
178	degenerate	dégénéré	entartet
179	degenerate gas	gaz dégénéré *m*	entartetes Gas *n*
180	degenerate star	étoile dégénérée *f*	entarteter Stern *m*
181	criterion of degeneracy degeneracy criterion	critère [condition] de dégénérescence *m[f]*	Entartungskriterium *n*
182	non-degenerate gas	gaz non-dégénéré *m*	nichtentartetes Gas *n*
183	partially degenerate	partiellement dégénéré	teilweise entartet
184	highly degenerate	fortement dégénéré	hochentartet
185	completely degenerate	complètement dégénéré	vollständig entartet
186	zero-temperature model (of a star)	modèle (d'étoile) au zéro absolu *m*	Modell (eines Sternes) mit Temperatur Null *n*
187	zero-point (kinetic) energy	énergie (cinétique) au point zéro *f* énergie au zéro absolu *f*	(kinetische) Nullpunktsenergie *f*
187a	degeneracy pressure	pression de dégénérescence *f*	Entartungsdruck *m*
188	degeneracy temperature	température (limite) de dégénérescence *f*	Entartungstemperatur *f*
189	critical mass (of white dwarfs) mass limit Chandrasekhar limit	masse critique (des naines blanches) *f* masse limite *f* limite de Chandrasekhar *f*	kritische Masse (weisser Zwerge) *f* Chandrasekharsche Grenzmasse *f*
190	subnuclear density	densité subnucléaire *f*	subnukleare Dichte *f*
191	neutron degeneracy	dégénérescence de neutrons *f* dégénérescence neutronique *f*	Neutronengasentartung *f*
192	neutrons gas	gaz de neutrons *m* gaz neutronique *m*	Neutronengas *n*
193	hyperon gas hyperonic gas	gaz d'hypérons *m* gaz hypéronique *m*	Hyperonengas *n*
194	baryon gas baryonic gas	gaz de baryons *m* gaz baryonique *m*	Baryonengas *n*
195	neutron star	étoile à neutrons *f* étoile neutronique *f*	Neutronenstern *m*
196	hyperon star hyperonic star	étoile à hypérons *f* étoile hypéronique *f*	Hyperonenstern *m*

degeneración relativista *f*	degeneração [degenerescência] relativista *f*	релятивистское [релятивистическое] вырождение *n*	176
degeneración no relativista *f*	degeneração [degenerescência] não-relativista *f*	нерелятивистское вырождение *n*	177
degenerado	degenerado	вырожденный	178
gas degenerado *m*	gás degenerado *m*	вырожденный газ *m*	179
estrella degenerada *f*	estrela degenerada *f*	вырожденная звезда *f*	180
criterio de degeneración *m*	critério de degenerescência [degeneração] *m*	критерий вырождения *m*	181
gas no degenerado *m*	gás não-degenerado *m*	невырожденный газ *m*	182
parcialmente degenerado	parcialmente degenerado	частично вырожденный	183
altamente degenerado	altamente [fortemente] degenerado	вырожденный в высокой степени	184
totalmente [completamente] degenerado	completamente [totalmente] degenerado	полностью вырожденный	185
modelo (de estrella) de temperatura nula *m*	modelo (de uma estrela) de temperatura zero *m*	модель (звезды) с нулевой температурой *f*	186
energía (cinética) del punto cero *f*	energia (cinética) do ponto zero *f*	энергия при нулевой температуре *f* нулевая энергия *f*	187
pressión de degeneración *f*	pressão de degenerescência *f*	давление вырождения *n*	187a
temperatura de degeneración *f*	temperatura de degenerescência *f*	температура вырождения *f*	188
masa crítica (de las enanas blancas) *f* límite de masa *m* límite de Chandrasekhar *m*	massa crítica (das anãs brancas) *f* massa limite *f* (massa) limite de Chandrasekhar (*f*) *m*	критическая масса (белых карликов) *f* предел Чандрасекара *m* граница Чандрасекара *f*	189
densidad subnuclear *f*	densidade subnuclear *f*	субядерная плотность *f*	190
degeneración (del gas) de neutrones *f*	degenerescência (do gás) de nêutrons [neutrões] *f*	нейтронное вырождение *n* вырождение нейтронного газа *n*	191
gas de neutrones *m* gas neutrónico *m*	gás de nêutrons [neutrões] *m* gás neutrônico *m*	нейтронный газ *m*	192
gas hiperónico *m* gas de hiperones *m*	gás de híperons [hiperões] *m* gás hiperônico *m*	гиперонный газ *m*	193
gas bariónico *m* gas de bariones *m*	gás de bárions [bariões] *m* gás bariônico *m*	барионный газ *m*	194
estrella de neutrones *f* estrella neutrónica *f*	estrela de nêutrons [neutrões, neutrônica] *f*	нейтронная звезда *f*	195
estrella de hiperones *f* estrella hiperónica *f*	estrela de híperons [hiperões, hiperônica] *f*	гиперонная звезда *f*	196

197	superdense star	étoile hyperdense *f*	überdichter Stern *m*
198	relativistic stellar model	modèle relativiste d'étoile *m*	relativistisches Sternmodell *n*
199	non-degenerate outer layers	couches extérieures non-dégénérées *fpl*	nichtentartete Aussenschichten *fpl*
200	mass-radius relation	relation masse-rayon *f*	Masse-Radius-Beziehung *f*
201	cooling time	temps de refroidissement *m*	Abkühlungszeit *f*
202	cooling track cooling line	chemin [courbe] de refroidissement *m[f]* trajet de refroidissement *m*	Abkühlungsweg *m* Abkühlungskurve *f*
203	cooling rate	vitesse [taux] de refroidissement *f[m]*	Abkühlungsgeschwindigkeit *f*
204	stability	stabilité *f*	Stabilität *f*
205	stability condition	condition de stabilité *f*	Stabilitätsbedingung *f*
206	stability of stars	stabilité des étoiles *f*	Stabilität der Sterne *f*
207	dynamical stability	stabilité dynamique *f*	dynamische Stabilität *f*
208	stable	stable	stabil
209	thermally stable	thermiquement stable	thermisch stabil
210	unstable	instable	instabil
211	instability	instabilité *f*	Instabilität *f*
212	secular instability	instabilité séculaire *f*	säkulare Instabilität *f*
213	pulsational instability	instabilité pulsationnelle *f*	Pulsationsinstabilität *f*
214	vibrational instability	instabilité vibrationnelle *f*	vibrationelle Instabilität *f*
215	radial perturbations	perturbations radiales *fpl*	radiale Störungen *fpl*
216	rotational instability	instabilité rotationnelle *f*	Rotationsinstabilität *f*
217	tidal instability	instabilité due aux [à l'égard des] marées *f*	Gezeiteninstabilität *f*

estrella superdensa *f*	estrela superdensa *f*	сверхплотная звезда *f*	197
modelo estelar relativista *m*	modelo estelar relativístico [relativista] *m*	релятивистская звёздная модель *f*	198
capas exteriores no degeneradas *fpl*	camadas exteriores não-degeneradas *fpl*	невырожденные внешние слои *mpl*	199
relación masa-radio *f*	relação massa-raio *f*	зависимость масса-радиус *f*	200
tiempo de enfriamento *m*	tempo de resfriamento [arrefecimento] *m*	время охлаждения *n*	201
trayectoria de enfriamiento *f* traza de enfriamiento *f*	traço [trajetória, linha] de resfriamento [arrefecimento] *m*[*f*]	трек [кривая] охлаждения *m* [*f*] след охлаждения *m*	202
velocidad de enfriamiento *f*	taxa de resfriamento [arrefecimento] *f*	скорость охлаждения *f*	203
estabilidad *f*	estabilidade *f*	устойчивость [стабильность] *f*	204
condición de estabilidad *f*	condição de estabilidade *f*	условие устойчивости *n*	205
estabilidad de las estrellas *f*	estabilidade das estrelas *f*	устойчивость звёзд *f*	206
estabilidad dinámica *f*	estabilidade dinâmica *f*	динамическая устойчивость *f*	207
estable	estável	устойчивый	208
térmicamente estable	termicamente estável	термически устойчивый	209
inestable	instável	неустойчивый	210
inestabilidad *f*	instabilidade *f*	неустойчивость *f*	211
inestabilidad secular *f*	instabilidade secular *f*	вековая неустойчивость *f*	212
inestabilidad pulsante [de pulsación] *f*	instabilidade pulsacional [de pulsação] *f*	пульсационная неустойчивость *f*	213
inestabilidad vibracional [de vibración] *f*	instabilidade vibracional [de vibração] *f*	колебательная [вибрационная] неустойчивость *f*	214
perturbaciones radiales *fpl*	perturbações radiais *fpl*	радиальные возмущения *npl*	215
inestabilidad rotacional *f*	instabilidade rotacional [da rotação] *f*	вращательная неустойчивость *f*	216
inestabilidad de [por] marea *f*	instabilidade de marés *f*	приливная неустойчивость *f*	217

	STELLAR EVOLUTION	ÉVOLUTION STELLAIRE	STERNENTWICKLUNG
1	evolution of stars stellar evolution	évolution des étoiles *f* évolution stellaire *f*	Sternentwicklung *f* Entwicklung der Sterne *f*
2	cosmogony	cosmogonie *f*	Kosmogonie *f*
3	cosmogonic cosmogonical	cosmogonique	kosmogonisch
4	cosmogonist	spécialiste de cosmogonie *m* cosmogoniste *m*	Kosmogonist *m* Fachmann für Kosmogonie *m*
5	cosmogonical problem	problème cosmogonique *m*	kosmogonisches Problem *n*
6	hypothesis	hypothèse *f*	Hypothese *f*
7	stellar cosmogony	cosmogonie stellaire *f*	Sternkosmogonie *f*
8	cosmogonic hypothesis	hypothèse cosmogonique *f*	kosmogonische Hypothese *f*
9	origin	origine *f*	Herkunft *f* Ursprung *m*
10	formation	formation *f*	Entstehung *f*
11	star formation formation of stars	formation des étoiles *f*	Sternentstehung *f*
11a	starburst violent star formation	formation violente des étoiles *f* générescence d'étoiles *f*	heftige Sternentstehung *f*
12	rate of star formation birth rate of stars	taux de formation des étoiles *m* vitesse de formation des étoiles *f*	Sternentstehungsrate *f*
13	pre-stellar	préstellaire	prästellar
14	pre-stellar evolution (of galaxies)	évolution préstellaire (des galaxies) *f*	prästellare Entwicklung (der Galaxien) *f*
15	pre-stellar body	corps préstellaire *m*	prästellarer Körper *m*
16	pre-stellar matter	matière préstellaire *f*	prästellare Materie *f*
17	primordial galactic cloud protogalactic cloud	nuage galactique primitif [primordial] *m*	galaktische Urwolke *f* protogalaktische Wolke *f*
18	protogalaxy	protogalaxie *f*	Protogalaxie *f*
19	halo phase halo stage	phase de halo *f* stade (de) halo *m*	Halophase *f* Halostadium *n*
20	contraction phase contracting phase contracting stage	phase de contraction *f* stade de contraction *m*	Kontraktionsphase *f* Kontraktionsstadium *n*
21	intermediate phase intermediate stage	phase intermédiaire *f* stade intermédiaire *m*	Zwischenphase *f* Zwischenstadium *n*

EVOLUCIÓN ESTELAR	EVOLUÇÃO ESTELAR	ЭВОЛЮЦИЯ ЗВЁЗД	
evolución de las estrellas *f* evolución estelar *f*	evolução das estrelas *f* evolução estelar *f*	эволюция звёзд *f* звёздная эволюция *f*	1
cosmogonía *f*	cosmogonia *f*	космогония *f*	2
cosmogónico	cosmogônico	космогонический	3
especialista en cosmogonía *m* cosmogonista *m*	especialista em cosmogonia *m* cosmogonista *m*	космогонист *m*	4
problema cosmogónico *m*	problema cosmogônico *m*	космогоническая проблема *f*	5
hipótesis *f*	hipótese *f*	гипотеза *f*	6
cosmogonía estelar *f*	cosmogonia estelar *f*	звёздная космогония *f*	7
hipótesis cosmogónica *f*	hipótese cosmogônica *f*	космогоническая гипотеза *f*	8
origen *m*	origem *f*	возникновение *n* происхождение *n*	9
formación *f* creación *f*	formação *f* criação *f*	образование *n* формирование *n*	10
formación de estrellas *f* creación de estrellas *f*	formação estelar [das estrelas] *f* criação das estrelas *f*	звёздообразование *n* образование звёзд *n*	11
formación violenta de estrellas *f*	formação violenta de estrelas *f* pulso de formação estelar *m*	бурное звёздообразование *n* бурное образование звёзд *n*	11a
tasa [velocidad] de formación de estrellas *f*	taxa de formação estelar [de estrelas] *f* taxa de nascimento de estrelas *f*	темп звёздообразования *m* скорость звёздообразования *f*	12
preestelar	pré-estelar	дозвёздный	13
evolución preestelar (de las galaxias) *f*	evolução pré-estelar (das galáxias) *f*	дозвёздная эволюция (галактик) *f*	14
cuerpo preestelar *m*	corpo pré-estelar *m*	дозвёздное тело *n*	15
materia preestelar *f*	matéria pré-estelar *f* material pré-estelar *m*	дозвёздное вещество *n* дозвёздная материя *f*	16
nube galáctica primitiva *f* nube protogaláctica *f*	nuvem galáctica primoridal *f* nuvem protogaláctica *f*	догалактическое облако *n* протогалактическое облако *n*	17
protogalaxia *f*	protogaláxia *f*	протогалактика *f*	18
fase(de) halo *f* estadio de halo *m*	fase de halo *f* estádio [estágio] de halo *m*	стадия [фаза] гало *f* этап гало *m*	19
fase de contracción *f* estadio contractivo *m*	fase de contração *f* estádio [estágio] de contração *m*	стадия [фаза] сжатия *f* этап сжатия *m*	20
fase intermedia *f* estadio intermedio *m*	fase intermediária *f* estágio [estádio] intermediário *m*	промежуточная стадия [фаза] *f* промежуточный этап *m*	21

	STELLAR EVOLUTION	ÉVOLUTION STELLAIRE	STERNENTWICKLUNG
22	disk [disc] phase disk [disc] stage	stade (de) disque *m* phase (de) disque *f*	Scheibenphase *f* Scheibenstadium *n*
23	generation of stars stellar generation	génération d'étoiles *f* génération stellaire *f*	Sterngeneration *f* Generation der Sterne *f*
24	first-generation star	étoile de la première génération *f*	Stern der ersten Generation *m*
25	common origin	origine commune *f*	gemeinsamer Ursprung *m*
26	pre-main-sequence evolution pre-main-seguence phase of evolution	évolution avant la série [séquence] principale *f*	Vorhauptreihenentwicklung *f*
27	evolutionary change	changement évolutif *m*	Entwicklungsänderung *f*
28	evolutionary phase evolutionary stage	phase évolutive *f* stade d'évolution *m*	Entwicklungsphase *f* Entwicklungsstadium *n*
29	evolutionary track evolutionary line	trajet évolutif *m* chemin évolutif *m*	Entwicklungsweg *m* Entwicklungskurve *f*
30	evolutionary sequence	séquence d'évolution *f*	Entwicklungsreihe *f*
31	cosmic cloud nebular cloud	nuage cosmique *m* nuage nébulaire *m*	kosmische Wolke *f* Nebelwolke *f*
32	gaseous mass	masse gazeuse *f*	Gasmasse *f*
33	self-gravitating	soumis à sa propre gravitation autogravitant	selbstgravitierend
34	self-gravitation	gravitation propre *f* autogravitation *f*	Eigengravitation *f*
35	gravitational instability	instabilité gravitationnelle *f*	Gravitationsinstabilität *f*
36	contraction	contraction *f*	Kontraktion *f*
37	gravitational contraction Helmholtz contraction	contraction gravitationnelle *f* contraction gravifique [d'Helmholtz] *f*	Gravitationskontraktion *f* Helmholtzsche Kontraktion *f*
38	fragmentation of a cosmic cloud	fragmentation d'un nuage cosmique *f*	Fragmentation einer kosmischen Wolke *f*
39	free fall	chute libre *f*	freier Fall *m*
40	quasi-static contraction	contraction quasi statique *f* contraction quasi stationnaire *f*	quasistatische Kontraktion *f* quasistationäre Kontraktion *f*
41	thermal pressure	pression thermique *f*	thermischer Druck *m*
42	non-thermal pressure	pression non-thermique *f*	nichtthermischer Druck *m*
43	hydrodynamic phase hydrodynamic stage	phase hydrodynamique *f* stade hydrodynamique *m*	hydrodynamische Phase *f* hydrodynamisches Stadium *n*

fase (de) disco *f* estadio (de) disco *m*	fase de disco *f* estágio [estádio] de disco *m*	дисковая стадия [фаза] *f* дисковый этап *m*	22
generación de estrellas *f* generación estelar *f*	geração de estrelas *f* geração estelar *f*	поколение звёзд *n* звёздное поколение *n*	23
estrella de primera generación *f*	estrela da primeira geração *f*	звезда первого поколения *f*	24
origen común *m*	origem comum *f*	совместное происхождение *n*	25
evolución preenana *f* evolución anterior a la secuencia principal *f* evolución presecuencia principal *f*	evolução pré-seqüência [pré-série] principal *f* evolução anterior à seqüência [série] principal *f*	эволюция [развитие] до главной последовательности *f*[*n*]	26
cambio evolutivo *m*	mudança evolutiva *f*	эволюционное изменение *n*	27
fase de evolución *f* estadio evolutivo *m*	fase evolutiva *f* estádio [estágio] evolutivo *m*	эволюционная фаза *f* фаза развития *f* стадия эволюции *f*	28
trayectoria evolutiva *f* traza evolutiva *f*	trajetória [linha] evolutiva *f* traço [caminho] evolutivo *m*	путь [трек] развития *m* эволюционный трек [путь] *m*	29
secuencia evolutiva *f*	seqüência evolutiva *f*	эволюционная последовательность *f*	30
nube cósmica *f*	nuvem cósmica *f*	космическое облако *n*	31
masa gaseosa *f*	massa gasosa *f*	газовая масса *f* масса газа *f*	32
autogravitante autogravitatorio	autogravitante de [por] gravitação própria	самогравитирующий гравитирующий	33
autogravitación *f* gravitación propia *f*	autogravitação *f* gravitação própria *f*	собственное притяжение *n* самогравитация *f*	34
inestabilidad gravitacional *f*	instabilidade gravitacional *f*	гравитационная неустойчивость *f*	35
contracción *f*	contração *f*	сжатие *n*	36
contracción gravitacional *f* contracción (de) Helmholtz *f*	contração gravitacional *f* contração devida à gravitação *f*	гравитационное сжатие *n* гравитационное самосжатие *n*	37
fragmentación de una nube cósmica *f*	fragmentação de uma nuvem cósmica *f*	дробление космического облака *n*	38
caída libre *f*	queda livre *f*	свободное падение *n*	39
contracción cuasiestática *f* contracción cuasiestacionaria *f*	contração quase-estática *f* contração quase-estacionária *f*	квазистатическое сжатие *n* квазистационарное сжатие *n*	40
presión térmica *f*	pressão térmica *f*	тепловое давление *n*	41
presión no térmica *f*	pressão não-térmica [atérmica] *f*	нетепловое давление *n*	42
fase hidrodinámica *f* estadio hidrodinámico *m*	fase hidrodinâmica *f* estágio [estádio] hidrodinâmico *m*	гидродинамическая стадия *f* гидродинамическая фаза *f*	43

44	Hayashi phase Hayashi stage	phase (de) Hayashi *f* stade (de) Hyashi *m*	Hayashi-Phase *f* Hayashi-Stadium *n*
45	protostar	proto-étoile *f*	Urstern *m* Protostern *m*
46	protostellar	protostellaire	protostellar
47	T-Tauri stage	stade T-Tauri *m*	T-Tauri Stadium *n*
48	circumstellar cloud	nuage circumstellaire *m*	zirkumstellare Wolke *f*
49	pre-main-sequence stage	stade avant la série [séquence] principale *m*	Vorhauptreihenstadium *n*
50	nuclear evolution thermonuclear evolution	évolution nucléaire *f* évolution thermonucléaire *f*	nukleare Entwicklung *f* thermonukleare Entwicklung *f*
51	Urca process	processus Urca *m*	Urca-Prozess *m*
52	nucleosynthesis element formation synthesis of the elements nuclear synthesis nucleogenesis	nucléosynthèse *f* formation des éléments *f* nucléogenèse *f* synthèse des éléments *f*	Nukleosynthese *f* Kernaufbau *m* Elementaufbau *m* Elementsynthese *f* Entstehung der Elemente *f* Element(en)entstehung *f* Elementbildung *f*
53	thermonuclear reaction rate	taux des réactions thermonucléaires *m* vitesse de réaction *f*	thermonukleare Reaktionsrate *f* Reaktionsgeschwindigkeit *f*
54	differential cross-section partial cross-section	section efficace différentielle *f* section efficace partielle *f*	differentieller Wirkungsquerschnitt *m* Partialquerschnitt *m* partieller Wirkungsquerschnitt *m*
55	hydrogen ignition	démarrage de la combustion de l'hydrogène *m*	Wasserstoff-Zündung *f*
56	hydrogen burning	combustion de l'hydrogène *f*	Wasserstoffbrennen *n*
57	explosive nucleosynthesis explosive nucleogenesis	nucléosynthèse explosive *f* nucléogenèse explosive *f*	explosive Nukleosynthese *f* explosiver Kernaufbau *m*
58	non-explosive nucleosyntheis non-explosive nucleogenesis	nucléosynthèse non-explosive *f* nucléogenèse non-explosive *f*	nichtexplosive Nukleosynthese *f* nichtexplosiver Kernaufbau *m*
59	zero-age star	étoile d'âge zéro *f*	Stern vom Alter Null *m*
60	zero-age main sequence zero-age line	séquence principale initiale *f* ligne d'âge zéro *f*	Nullalter-Hauptreihe *f* Alter-Null-Hauptreihe *f*
61	age of stars	âge des étoiles *m*	Alter der Sterne *n*
62	age determination determination of age	détermination de l'âge *f* datation *f*	Altersbestimmung *f*
63	young star	étoile jeune *f*	junger Stern *m*

fase (de) Hayashi *f* estadio (de) Hayashi *m*	fase (de) Hayashi *f* estágio [estádio] (de) Hayashi *m*	эволюционная стадия Хаяши *f* эволюционный этап Хаяши *m*	44
protoestrella *f*	proto-estrela *f*	протозвезда *f* зарождающаяся звезда *f*	45
protoestelar	proto-estelar *f*	протозвёздный	46
estadio [fase] T-Tauri *m*[*f*]	estágio [fase] T-Tauri *m*	стадия Т-Тельца *f*	47
nube circumestelar *f*	nuvem circunstelar [circum-estelar] *f*	околозвёздное облако *n*	48
estadio presecuencia principal *m* estadio de preenana *m*	estágio anterior à seqüência [série] principal *m* estádio pré-seqüência [pré-série] principal *m*	стадия [этап] до главной последовательности *f*[*m*]	49
evolución nuclear *f* evolucién termonuclear *f*	evolução nuclear *f* evolução termonuclear *f*	ядерная эволюция *f* термоядерная эволюция *f*	50
proceso Urca *m*	processo Urca *m*	Урка-процесс *m*	51
nucleosíntesis *f* formación de elementos *f* síntesis de elementos *f* síntesis nuclear *f* nucleogénesis *f*	nucleossíntese *f* formação dos elementos *f* síntese dos elementos *f* gênese dos elementos *f* nucleogênese *f*	нуклеосинтез *m* происхождение атомных ядер *n* образование элементов *n* синтез элементов *m* ядерный синтез *m* нуклеогенез(ис) *m*	52
velocidad de reacción termonuclear *f* tasa de reacción termonuclear *f*	taxa de reação termonuclear *f*	скорость термоядерной реакции *f*	53
sección eficaz diferencial *f*	se(c)ção eficaz diferencial *f* se(c)ção de choque diferencial [parcial] *f*	дифференциальное сечение *n* парциальное сечение *n*	54
ignición del hidrógeno *f*	ignição do hidrogênio *f*	загорание водорода *n*	55
combustión [quemazón] del hidrógeno *f*	queima [combustão] do hidrogênio *f*	горение водорода *n*	56
nucleosíntesis explosiva *f* nucleogénesis explosiva *f*	nucleossíntese explosiva *f* nucleogênese explosiva *f*	взрывной нуклеосинтез [ядерный синтез, нуклеогенез] *m*	57
nucleosíntesis no explosiva *f* nucleogénesis no explosiva *f*	nucleossíntese não-explosiva *f* nucleogênese não-explosiva *f*	невзрывной нуклеосинтез [ядерный синтез, нуклеогенез] *m*	58
estrella de edad cero *f*	estrela de idade zero [nula] *f*	звезда нулевого возраста *f*	59
secuencia principal de edad cero *f*	seqüência [linha, traço] principal de idade zero [nula] *f*	главная последовательность [линия] нулевого возраста *f*	60
edad de las estrellas *f*	idade das estrelas *f*	возраст звёзд *m*	61
determinación de la edad *f*	determinação da idade *f* datação *f*	определение возраста *n*	62
estrella joven *f*	estrela jovem *f*	молодая звезда *f*	63

64	old star	étoile vieille *f*	alter Stern *m*
65	time-scale	échelle de temps *f*	Zeitskala *f* Zeitmasstab *m*
66	short time-scale	échelle courte d'évolution *f*	kurze Zeitskala *f*
67	long time-scale	échelle longue d'évolution *f*	lange Zeitskala *f*
68	post-main-sequence stage post-main-sequence (phase of) evolution	stade après la série [séquence] principale *m* phase d'évolution après la série [séquence] principale *f*	Nachhauptreihenstadium *n* (Entwicklungs)phase nach der Hauptreihe *f*
69	exhaustion of hydrogen	épuisement de l'hydrogène *m*	Abbrennen des Wasserstoffs *n*
70	core contraction	contraction du noyau *f*	Kernkontraktion *f*
71	helium ignition	démarrage de la combustion de l'hélium *m*	Heliumzündung *f*
72	mass loss	perte de masse *f*	Massenverlust *m*
73	dissipation	dissipation *f*	Dissipation *f*
74	evolved star	étoile évoluée *f*	entwickelter Stern *m*
75	evolve	évoluer	(sich) entwickeln
76	thermal dissipation	dissipation termique *f*	thermische Dissipation
77	mass ejection ejection of matter	éjection de matière [de masse] *f*	Materieausstoss *m* Auswurf von Masse [Materie] *m*
78	helium burning	combustion de l'hélium *f*	Heliumbrennen *n*
79	helium flash explosive helium burning	combustion explosive de l'hélium *f* flash de l'hélium *m*	Heliumflash *m* explosives Heliumbrennen *n*
80	carbon ignition	démarrage de la combustion du carbone *m*	Kohlenstoffzündung *f*
81	carbon burning	combustion du carbone *f*	Kohlenstoffbrennen *n*
82	oxygen burning	combustion de l'oxygène *f*	Sauerstoffbrennen *n*
83	silicon burning	combustion du silicium *f*	Siliziumbrennen *n*
84	neutron capture	capture de neutron(s) *f* capture neutronique *f*	Neutroneneinfang *m*
85	neutron source	source de neutrons *f* source neutronique *f*	Neutronenquelle *f*
86	neutron flux neutron flux density	flux de neutrons *m* densité du flux neutronique *f*	Neutronenflussdichte *f* Neutronenfluss *m*

estrella vieja *f*	estrela velha *f*	старая звезда *f*	64
escala de tiempo *f*	escala de tempo *f*	шкала времени *f* масштаб времени *m*	65
escala de tiempo corta *f*	escala curta de tempo *f*	короткая шкала времени *f*	66
escala de tiempo larga *f*	escala longa de tempo *f*	длинная шкала времени *f*	67
estadio postenana [postsecuencia principal] *m* evolución posterior a secuencia principal *f*	estágio posterior à seqüência [série] principal *m* estágio [estádio, fase] pós-seqüência principal *m*[*f*]	стадия [этап] после главной последовательности *f*[*m*]	68
agotamiento del hidrógeno *m* exhaustión del hidrógeno *f*	esgotamento do hidrogênio *m* exaustão do hidrogênio *f*	исчерпание водорода *n*	69
contracción del núcleo *f*	contração do núcleo *f*	сжатие ядра *n*	70
ignición de helio *f*	ignição do hélio *f*	загорание гелия *n*	71
pérdida de masa *f*	perda de massa *f*	потеря массы *f*	72
disipación *f*	dissipação *f*	диссипация *f*	73
estrella evolucionada *f*	estrela evoluída *f*	(про)эволюционировавшая звезда *f*	74
evolucionar	evolver evoluir	эволюционировать	75
disipación térmica *f*	dissipação térmica *f*	термическая диссипация *f*	76
eyección de materia [masa] *f*	ejeção de massa [matéria] *f*	выброс [сброс] вещества *m* выброс массы [материи] *m*	77
combustión [quemazón] del helio *f*	queima [combustaõ] do hélio *f*	горение гелия *n*	78
combustión [quemazón] explosiva de helio *f*	queima explosiva do hélio *f* lampejo [estelho] de hélio *m*	гелиевая вспышка *f* взрывное горение гелия *n*	79
ignición del carbono *f*	ignição do carbono [carbônio] *f*	загорание углерода *n*	80
combustión [quemazón]del carbono *f*	queima do carbono [carbônio] *f*	горение углерода *n*	81
combustión [quemazón] del oxígeno *f*	queima [combustão] do oxigênio *f*	горение кислорода *n*	82
combustión [quemazón] del silicio *f*	queima [combustão] do silício *f*	горение кремния *n*	83
captura de neutrones *f* captura neutrónica *f*	captura de nêutrons [neutrões] *f* captura neutrônica *f*	захват нейтронов *m* нейтронный захват *m*	84
fuente de neutrones *f* fuente neutrónica *f*	fonte de nêutrons [neutrões] *f* fonte neutrônica *f*	источник нейтронов *m* нейтронный источник *m*	85
flujo de neutrones *m* densidad del flujo neutrónico *f*	fluxo de nêutrons *m* densidade do fluxo neutrônico *f*	нейтронный поток *m* плотность нейтронного потока *f*	86

87	neutron-capture reaction	réaction de capture neutronique *f*	Neutroneneinfangreaktion *f*
88	slow neutron absorption *s*-process slow process slow neutron-capture process	absorption lente des neutrons *f* processus *s m* processus lent *m* processus de capture neutronique lente *m* absorption neutronique lente *f*	langsame Neutronenabsorption *f* *s*-Prozess *m* langsamer Einfang von Neutronen *m*
89	rapid neutron absorption *r*-process rapid process rapid neutron-capture process	absorption neutronique rapide *f* processus *r m* processus rapide *m* processus de capture neutronique rapide *m*	schnelle Neutronenabsorption *f* *r*-Prozess *m* schneller Einfang von Neutronen *m*
90	proton capture	capture de proton(s) *f* capture protonique *f*	Protoneneinfang *m*
91	*p*-process	processus *p m*	*p*-Prozess *m*
92	neutrino reaction	réaction à neutrinos *f* réaction neutrinique *f*	Neutrinoreaktion *f*
93	photo-neutrino	photo-neutrino *m*	Photoneutrino *n*
94	neutrino radiation neutrino emission	rayonnement de neutrinos *m* rayonnement neutrinique *m*	Neutrinostrahlung *f*
95	neutrino opacity	opacité neutrinique *f*	Neutrinoopazität *f*
96	neutrino luminosity	luminosité neutrinique *f*	Neutrinoleuchtkraft *f*
97	neutrino cooling cooling by neutrino emission	refroidissement neutrinique [par neutrinos] *m* refroidissement par emission de neutrinos *m*	Neutrinokühlung *f* Abkühlung mittels Neutrinos *f*
98	radiative cooling radiation cooling	refroidissement radiatif *m* refroidissement par rayonnement *m*	Strahlungskühlung *f* Strahlungsabkühlung *f*
99	equilibrium process *e*-process	processus d'équilibre *m* processus *e m*	Gleichgewichtsprozess *m* *e*-Prozess *m*
100	spallation (reaction)	(réaction de) spallation *f*	(Kern-)Zersplitterung *f* Spaltungsvorgang *m* Spallation(sreaktion) *f*
101	photodisintegration (of a nucleus)	photo-désintegration (d'un noyau) *f*	Photospaltung *f* Photozerlegung (des Kernes) *f*
102	nuclear fragment	fragment nucléaire *m*	Kernfragment *n*
103	nuclear (statistical) equilibrium	équilibre (statistique) nucléaire *m*	nukleares (statistisches) Gleichgewicht *n*
104	equilibrium ratio	rapport d'équilibre *m*	Gleichgewichtsverhältnis *n*

reacción de captura de neutrones *f*	reação de captura de nêutrons [neutrões] *f*	реакция захвата нейтрона *f* реакция нейтронного захвата *f*	87
absorción lenta de neutrones *f* proceso *s m* proceso lento de captura de neutrones *m*	absorção lenta de nêutrons [neutrões] *f* processo *s m* processo lento de captura de nêutrons [neutrões] *m*	медленный захват нейтронов *m* *s*-процесс *m*	88
absorción rápida de neutrones *f* proceso *r m* proceso rápido de captura de neutrones *m*	absorção rápida de nêutrons [neutrões] *f* processo *r m* processo rápido de captura de nêutrons [neutrões] *m*	*r*-процесс *m* быстрый захват нейтронов *m* быстрый нейтронный захват *m*	89
captura de protones *f* captura protónica *f*	captura de protons [protões] *f*	захват протонов *m* протонный захват *m*	90
proceso *p m*	processo *p m*	*p*-процесс *m*	91
reacción neutrínica [de neutrinos] *f*	reação neutrínica [de neutrinos] *f*	нейтринная реакция *f*	92
fotoneutrino *m*	fotoneutrino *m*	фотонейтрино *n*	93
radiación neutrínica [de neutrinos] *f*	radiação neutrínica [de neutrinos] *f*	нейтринное излучение *n*	94
opacidad neutrínica *f*	opacidade neutrínica *f*	нейтринная непрозрачность *f*	95
luminosidad neutrínica *f*	luminosidade de neutrino *f* luminosidade neutrínica *f*	нейтринная светимость *f*	96
enfriamiento neutrínico [por neutrinos] *m*	resfriamento [arrefecimento] neutrínico [por neutrinos] *m*	нейтринное охлаждение *n* охлаждение за счёт нейтринного излучения *n*	97
enfriamiento radiativo *m*	resfriamento [arrefecimento] radiativo *m* resfriamento [arrefecimento] por radiação *m*	радиативное охлаждение *n* охлаждение излучением *n*	98
proceso de equilibrio *m* proceso *e m*	processo de equilíbrio *m* processo *e m*	равновесный процесс *m* *e*-процесс *m*	99
(reacción de) astillamiento (*f*) *m* reacción de espalación [astillado] *f*	(reação de) espalação *f*	реакция скалывания *f* (ра)скалывание *n*	100
fotodesintegración (del núcleo) *f*	fotodesintegração (do núcleo) *f*	фоторасщепление (ядра) *n* фотодезинтеграция *f*	101
fragmento nuclear *m*	fragmento nuclear *m*	осколок ядра *m*	102
equilibrio (estadístico) nuclear *m*	equilíbrio (estatístico) nuclear *m*	(статистическое) ядерное равновесие *n*	103
relación de equilibrio *f*	razão [relação] de equilíbrio *f*	равновесное отношение *n*	104

105	iron-group elements [nuclei] iron-peak elements [nuclei]	éléments du groupe du fer *mpl* noyaux du pic du fer *mpl*	Elemente der Eisengruppe *npl*
106	isotopic abundance	abondance isotopique *f* abondance des isotopes *f*	Isotopenhäufigkeit *f* Häufigkeit der Isotope *f*
107	cosmic(al) abundance of the elements	abondance cosmique des éléments *f*	kosmische Häufigkeit der Elemente *f*
108	abundance curve	courbe d'abondances *f*	Häufigkeitskurve *f*
109	collapse gravitational collapse catastrophic contraction	écroulement [effondrement] (catastrophique) *m* contraction catastrophique *f* écroulement [collapse] gravitationnel *m*	Kollaps *m* Gravitationskollaps *m* Zusammenbruch *m*
110	stellar core implosion	implosion du noyau stellaire *f*	Kernimplosion *f*
111	collapsed star	étoile écroulée [effondrée, collapsée] *f*	kollabierter Stern *m*
112	collapsing star	étoile en cours d'écroulement [d'effondrement] *f* étoile collapsante *f*	kollabierender Stern *m* (in) sich zusammenstürzender Stern *m*
113	gravitational radius Schwarzschild radius	rayon gravitationnel *m* rayon de Schwarzschild *m*	Gravitationsradius *m* Schwarzschildscher Radius *m* Schwarzschild-Radius *m*
114	black hole	trou noir *m*	schwarzes Loch *n*
115	rate of stellar extinction rate of star deaths	taux de mortalité des étoiles *m* taux d'extinction stellaire *m*	Stern-Sterberate *f* Sternsterberate *f*
116	singularity singular point	singularité *f* point singulier *m*	Singularität *f* singulärer Punkt *m*
117	naked singularity	singularité nue *f*	nackte Singularität *f*
117a	non-spherical [aspherical] collapse	écroulement [collapse] non-sphérique *m*	nichtsphärischer Kollaps *m*
118	evaporation of black holes	évaporation de trous noirs *f*	Verdampfung schwarzer Löcher *f*

elementos del grupo del hierro *mpl* elementos del pico del hierro *mpl*	elementos do grupo do ferro *mpl*	элементы группы железа *mpl* элементы железного пика *mpl*	105
abundancia isotópica *f*	abundância isotópica *f* abundância de isótopos *f*	распространённость изотопов *f* изотопическая распространённость *f*	106
abundancia cósmica de los elementos *f*	abundância cósmica dos elementos *f*	космическая распространённость элементов *f* космическое обилие элементов *n*	107
curva de abundancia *f*	curva de abundância *f*	кривая распространённости (элементов) *f*	108
colapso *m* colapso gravitacional *m* contracción catastrófica *f*	colapso *m* colapso gravitacional *m* contração catastrófica *f*	коллапс *m* гравитационный коллапс *m* катастрофическое сжатие *n* катастрофическое самосжатие *n*	109
implosión del núcleo estelar *f*	implosão do núcleo estelar *f*	имплозия [коллапс] звёздного ядра *f* [*m*]	110
estrella colapsada *f*	estrela colapsada *f*	сколлапсировавшая звезда *f*	111
estrella en colapso *f*	estrela em colapso *f*	коллапсирующая звезда *f*	112
radio gravitacional *m* radio de Schwarzschild *m*	raio gravitacional *m* raio de Schwarzschild *m*	гравитационный радиус *m* шварцшильдовский радиус *m* радиус Шварцшильда *m*	113
agujero negro *m*	buraco negro *m*	чёрная дыра *f*	114
tasa de desaparición de estrellas *f* velocidad de desaparición de estrellas *f*	taxa de morte das estrelas *f* taxa de extinção [desaparecimento] estelar *f*	скорость умирания звёзд *f* темп умирания звёзд *m*	115
singularidad *f* punto singular *m*	singularidade *f* ponto singular *m*	особенность [сингулярность] *f* особая точка *f*	116
singularidad desnuda *f*	singularidade desnuda *f*	голая особенность *f* голая сингулярность *f*	117
colapso non esférico *m*	colapso não-esférico *m*	несферический коллапс *m*	117a
evaporación de agujeros negros *f*	evaporação de buracos negros *f*	испарение чёрных дер *f*	118

95	VARIABLE STARS	ÉTOILES VARIABLES	VERÄNDERLICHE STERNE
1	variable star variable	étoile variable *f* variable *f*	veränderlicher Stern *m* Veränderlicher *m*
2	non-stationary star	étoile non-stationnaire *f*	nichtstationärer Stern *m*
3	stationary star	étoile stationnaire *f*	stationärer Stern *m*
4	stellar variability	variabilité stellaire *f*	Sternveränderlichkeit *f*
5	suspected variable star suspected of variability	étoile soupçonnée variable *f* étoile suspectée de variabilité *f*	vermuteter Veränderlicher *m* ein der Veränderlichkeit verdächtig(t)er Stern *m* verdächtig(t)er Veränderlicher *m*
6	light variation	variation de lumière *f* variation lumineuse *f*	Lichtwechsel *m*
7	brightness variation	variation d'éclat *f*	Helligkeitsänderung *f*
8	stable star	étoile stable *f*	stabiler Stern *m*
9	unstable star non-stable star	étoile instable *f*	instabiler Stern *m*
10	step method	méthode d'Argelander *f* méthode des degrés *f*	Stufenschätz(ungs)methode *f* Argelandermethode *f*
11	estimate of brightness	estimation d'éclat *f*	Helligkeitsschätzung *f*
12	estimate	estimer	schätzen
13	over-estimate	surestimer	überschätzen
14	underestimate	sous-estimer	unterschätzen
15	chart of the variable	carte (des environs) de l'étoile variable *f* carte de la variable *f*	Umgebungskarte des Veränderlichen *f*
16	comparison star	étoile de comparaison *f*	Vergleichsstern *m*
17	light-curve	courbe de lumière *f* courbe d'éclat *f*	Lichtkurve *f*
18	rectification (of light-curves)	rectification (des courbes de lumière) *f*	Rektifikation [Berichtigung] (der Lichtkurven) *f*
19	mean light-curve	courbe de lumière moyenne *f* courbe moyenne d'éclat *f*	mittlere Lichtkurve *f*
20	radial-velocity curve	courbe de vitesses radiales *f*	Radialgeschwindigkeitskurve *f*
21	colo(u)r curve	courbe de couleur *f*	Farb(enindex)kurve *f*
22	photoelectric light-curve	courbe de lumière photoélectrique *f*	photoelektrische Lichtkurve *f*
23	photographic light-curve	courbe de lumière photographique *f*	photographische Lichtkurve *f*

ESTRELLAS VARIABLES	ESTRELAS VARIÁVEIS	ПЕРЕМЕННЫЕ ЗВЕЗДЫ	95
estrella variable *f* variable *f*	estrela variável *f* variável *f*	переменная звезда *f* переменная *f*	1
estrella no estacionaria *f*	estrela não-estacionária *f*	нестационарная звезда *f*	2
estrella estacionaria *f*	estrela estacionária *f*	стационарная звезда *f*	3
variabilidad estelar *f*	variabilidade estelar *f*	звёздная переменность *f*	4
supuesta variable *f* variable sospechosa *f* estrella sospechosa de variabilidad *f*	variável suspeita *f* estrela suspeita de variabilidade *f*	звезда, заподозренная [подозреваемая] в переменности *f* кандидат в переменные *m*	5
variación de luz *f*	variação de luz *f* variação luminosa *f*	световые изменения *npl* изменения яркости *npl*	6
variación de brillo *f*	variação de brilho *f*	изменение блеска *n*	7
estrella estable *f*	estrela estável *f*	устойчивая звезда *f*	8
estrella no estable *f* estrella inestable *f*	estrela instável *f*	неустойчивая звезда *f*	9
método de pasos *m* método de Argelander *m*	método dos passos *m* método de Argelander *m*	степенной способ *m* способ степеней *m*	10
estimación de brillo *f*	estimativa de brilho *f*	оценка блеска *f*	11
estimar	estimar	оцени(ва)ть	12
sobreestimar	sobreestimar superestimar	переоценить	13
subestimar	subestimar	недооценить	14
carta de la variable *f*	carta da variável *f* mapa da variável *f*	карта (окрестностей) переменной звезды *f*	15
estrella de comparación *f*	estrela de comparação *f*	звезда сравнения *f*	16
curva de luz *f* curva de brillo *f*	curva de luz *f* curva de brilho *f*	кривая блеска *f* кривая изменения блеска *f*	17
rectificación (de la curva de luz) *f*	retificação (da curva de luz) *f*	поверка (кривой блеска) *f* ректификация *f*	18
curva media de luz *f*	curva média de luz *f* curva média de brilho *f*	средняя кривая блеска *f*	19
curva de velocidades radiales *f*	curva de velocidades radiais *f*	кривая лучевых скоростей *f*	20
curva de color *f*	curva de cor *f*	кривая изменения цвета *f*	21
curva fotoeléctrica de luz *f*	curva fotoelé(c)trica de luz *f* curva de luz fotoelé(c)trica *f*	фотоэлектрическая кривая блеска *f*	22
curva fotográfica de luz *f*	curva fotográfica de luz *f*	фотографическая кривая блеска *f*	23

24	visual light-curve	courbe de lumière visuelle *f*	visuelle Lichtkurve *f*
25	photometric elements	éléments photométriques *mpl*	photometrische Elemente *npl* Systemkonstanten *fpl*
26	primary minimum principal minimum	minimum primaire *m* minimum principal *m*	Hauptminimum *n* primäres Minimum *n*
27	secondary minimum	minimum secondaire *m*	Nebenminimum *n* sekundäres Minimum *n*
28	rise to maximum increase to maximum	montée vers le maximum *f*	Anstieg zum Maximum *m* Aufstieg zum Maximum *m*
29	rate of rise to maximum	taux de montée vers le maximum *m*	Steilheit des Anstiegs *f* relative Anstiegsdauer *f* Anstiegsgeschwindigkeit *f*
30	pre-maximum	prémaximum *m*	Prämaximum [Praemaximum] *n*
31	primary maximum principial maximum	maximum primaire *m* maximum principal *m*	Hauptmaximum *n* primäres Maximum *n*
32	secondary maximum	maximum secondaire *m*	Nebenmaximum *n* sekundäres Maximum *n*
33	post-maximum	postmaximum *m*	Postmaximum *n*
34	ascending branch rising branch	branche ascendante *f* montée de la courbe de lumière *f*	ansteigender Ast *m* Anstieg der Lichtkurve *m*
35	descending branch decreasing branch decline of light-curve	branche descendante *f* branche décroissante *f* déclin de la courbe de lumière *m*	absteigender Ast *m* Abfall der Lichtkurve *m*
36	rate of decline (from maximum)	taux de déclin *m* vitesse de déclin (à partir du maximum) *f*	Steilheit de Abstieges *f* Abfallgeschwindigkeit *f* relative Abfallsdauer *f*
37	hump (on [of] the curve)	bosse [extremum] (de la courbe) *f*	Buckel (der Kurve) *m*
38	amplitude (of brightness variation)	amplitude (de variation de l'éclat) *f*	Amplitude *f* Helligkeitsamplitude *f*
39	period duration of the period period time	période *f* durée de la période *f*	Periode *f* Periodenlänge *f* Periodendauer *f*
40	primary period	période principale *f*	Hauptperiode *f*
41	secondary period	période secondaire *f*	sekundäre Periode *f*
42	beat period	période de battement *f*	Schwebungsperiode *f*
43	change of period	variation de la période *f*	Periodenänderung *f*
44	secondary fluctuations superposed fluctuations	fluctuations secondaires *fpl* fluctuations superposées *fpl*	sekundäre Schwankungen *fpl* überlagerte Schwankungen *fpl*
45	light equation	équation de lumière *f*	Lichtgleichung *f*

curva visual de luz *f*	curva visual de luz *f*	визуальная кривая блеска *f*	24
elementos fotométricos *mpl*	elementos fotométricos *mpl*	фотометрические элементы *mpl*	25
mínimo primario *m* mínimo principal *m*	mínimo primário *m* mínimo principal *m*	главный минимум *m*	26
mínimo secundario *m*	mínimo secundário *m*	вторичный минимум *m*	27
aumento [subida] al máximo *m*[*f*] aumento de brillo *m*	crescimento para o máximo *m* subida [ascensão] para o máximo *f*	подъём к максимуму *m*	28
ritmo de aumento hacia el máximo *m* tasa [rapidez] de crecimiento hacia el máximo *f*	taxa de subida para o máximo *f* inclinação do ramo ascendente (da curva de luz) *f*	скорость подъёма к максимуму *f*	29
premáximo *m*	pré-máximo *m*	(время) до максимума *n*	30
máximo primario *m* máximo principal *m*	máximo primário *m* máximo principal *m*	главный максимум *m*	31
máximo secundario *m*	máximo secundário *m*	вторичный максимум *m*	32
postmáximo *m*	pós-máximo *m*	(время) после максимума *n*	33
rama ascendente *f* rama creciente *f*	ramo ascendente *m* trecho ascendente *m*	восходящая ветвь *f*	34
rama descendente *f* rama decreciente *f*	ramo descendente *m* trecho descendente *m*	нисходящая ветвь *f*	35
velocidad de decrecimiento *f* tasa de decrecimiento *f*	taxa de decréscimo *f* inclinação do ramo descendente *f*	скорость падения (блеска) от максимума *f*	36
joroba (de la curva) *f*	protuberância (da curva) *f*	горб кривой (блеска) *m*	37
amplitud (de variación de brillo) *f*	amplitude (da variação de brilho) *f*	амплитуда (изменения блеска) *f*	38
período *m* tiempo del período *m*	período *m* duração do período *f* tempo de período *m*	период *m* длина периода *f* длительность периода *f*	39
período primario *m*	período principal *m*	главный [основной] период *m*	40
período secundario *m*	período secundário *m*	вторичный период *m*	41
período de pulsación *m*	período de batimento *m*	период биений *m*	42
cambio de período *m*	variação do período *f*	изменение периода *n*	43
fluctuaciones secundarias *fpl* fluctuaciones sobrepuestas *fpl*	flutuações secundárias *fpl* flutuações sobrepostas [superpostas] *fpl*	вторичные колебания *mpl* вторичные флуктуации *fpl*	44
ecuación de luz *f*	equação da luz *f*	световое уравнение *n*	45

	VARIABLE STARS	ÉTOILES VARIABLES	VERÄNDERLICHE STERNE
46	classification of variable stars	classification des (étoiles) variables *f*	Klassifikation der Veränderlichen *f*
47	... -type variable ... variables *pl* (Algol variables, Beta Lyrae variables etc)	variable du type . . . *f*	Veränderlicher der ... Klasse *m* ... Veränderlicher *m* ... Stern *m* Veränderlicher vom ... Typus *m*
48	spectrum variable spectral variable	étoile à spectre variable *f* variable spectrale *f*	Spektrumveränderlicher *m* Stern mit veränderlichem Spektrum *m*
49	spectral variations	variations spectrales *fpl* variations du spectre *fpl*	spektrale Änderungen *fpl* spektrale Veränderungen *fpl*
50	symbiotic star	étoile symbiotique *f*	symbiotischer Stern *m*
51	pseudovariable false [improper] variable	pseudovariable *f*	Pseudo-Veränderlicher *m*
52	magnetic variable	variable magnétique *f*	Magnetfeld-Veränderlicher *m*
53	magnetic variations	variations magnétiques *fpl*	magnetische Variationen *fpl*
54	stellar magnetism	magnétisme stellaire *m*	Sternmagnetismus *m*
55	periodic magnetic variable (α)	variable magnétique regulière *f*	periodischer [regelmässiger] magnetischer Veränderlicher *m*
56	irregular (magnetic) variable with reversing polarity (β)	variable (magnétique) irrégulière à polarité alternative [réversible] *f*	unregelmässiger (magnetischer) Veränderlicher mit wechselnder Polarität *m*
57	irregular (magnetic) variable with constant polarity (γ)	variable (magnétique) à polarité constante *f*	unregelmässiger (magnetischer) Veränderlicher mit konstanter Polarität *m*
58	catalaog(ue) of stars suspected of variability	catalogue des étoiles suspectées [soupçonnées] de variabilité *m*	Katalog vermuteter veränderlicher Sterne *m*
59	effective field strength effective field (intensity)	champ effectif *m* intensité du champ effective *f* force de champ effective *f*	effektives Feld *n* effektive Feldintensität [Feldstärke] *f*
60	oblique rotator	rotateur oblique *m*	schiefer Rotator *m*
61	hydromagnetic oscillator magnetohydrodynamic oscillator	oscillateur hydromagnétique *m* oscillateur magnétohydrodynamique *m*	hydromagnetischer Oszillator *m* magnetohydrodynamischer Oszillator *m*
62	fossil magnetic field prim(a)eval magnetic field primitive magnetic field relic magnetic field	champ magnétique primitif *m* champ magnétique résiduel *m* champ magnétique fossile *m*	ursprüngliches Magnetfeld *n* primordiales Magnetfeld *n* übrigbleibendes Magnetfeld *n*
63	internal field	champ interne *m*	inneres Feld *n*
64	surface field	champ superficiel *m*	Oberflächenfeld *n*

ESTRELLAS VARIABLES	ESTRELAS VARIÁVEIS	ПЕРЕМЕННЫЕ ЗВЕЗДЫ	
clasificación de (estrellas) variables *f*	classificação das (estrelas) variáveis *f*	классификация переменных (звёзд) *f*	46
tipo de variable ... *m* variable tipo ... *f*	variável tipo ... *f* variável ... *f*	переменная типа ... *f* переменная звезда типа ... *f* переменная звезда класса ... *f*	47
variable espectroscópica *f* variable de espectro *f* variable espectral *f*	variável espe(c)troscópica *f* variável espe(c)tral *f*	спектрально-переменная (звезда) *f* спектральная переменная *f*	48
variaciones espectrales *fpl* variaciones del espectro *fpl*	variações espe(c)trais *fpl* variações do espe(c)tro *fpl*	спектральные изменения *npl* вариации спектра *fpl*	49
estrella simbiótica *f*	estrela simbiótica *f*	симбиотическая звезда *f*	50
seudovariable [pseudovariable] *f* falsa variable *f*	pseudovariável *f* variável imprópria *f*	псевдопеременная звезда *f*	51
variable magnética *f*	variável magnética *f*	магнитно-переменная (звезда) *f* магнитная переменная (звезда) *f*	52
variaciones magnéticas *fpl*	variações magnéticas *fpl*	магнитные изменения *npl*	53
magnetismo estelar *m*	magnetismo estelar *m*	звёздный магнетизм *m*	54
variable magnética regular [periódica] *f*	variável magnética regular [periódica] *f*	периодическая магнитно-переменная (звезда) *f*	55
variable (magnética) irregular con polaridad alterna(nte) *f*	variável (magnética) irregular de polaridade alternada *f*	неправильная (магнитная) переменная с переменной полярностью *f*	56
variable (magnética) irregular con polaridad constante *f*	variável (magnética) irregular de polaridade constante *f*	неправильная (магнитная) переменная с постоянной полярностью *f*	57
catálogo de estrelas sospechosas de variabilidad *m* catálogo de supuestas variables *m*	catálogo das estrelas suspeitas de variabilidade *m* catálogo das variáveis suspeitas *m*	каталог звёзд, заподозренных [подозреваемых] в переменности *m*	58
campo efectivo *m* intensidad efectiva del campo (magnético) *f*	campo efetivo *m* intensidade efetiva do campo (magnético) *f*	эффективная напряжённость магнитного поля *f*	59
rotor oblícuo *m*	rotor inclinado [oblíquo] *m*	наклонный ротатор *m*	60
oscilador hidromagnético *m* oscilador magneto-hidrodinámico *m*	oscilador hidromagnético *m* oscilador magneto-hidrodinâmico *m*	гидромагнитный осциллятор *m* магнитогидродинамический осциллятор *m*	61
campo magnético fósil *m* campo magnético primitivo *m* campo magnético remanente *m*	campo magnético fóssil *m* campo magnético primitivo *m* campo magnético remane(sce)nte *m*	изначальное магнитное поле *n* первоначальное магнитное поле *n* реликтовое магнитное поле *n*	62
campo interno *m*	campo interno *m*	внутреннее поле *n*	63
campo superficial *m*	campo superficial *m*	поверхностное поле *n*	64

65	rotational variable	variable rotationnelle *f*	Rotationsveränderlicher *m*
66	gaseous ring	anneau gazeux *m*	Gasring *m*
67	gaseous stream	courant gazeux *m*	Gasstrom *m*
68	flare star flare variable	variable éruptive *f* étoile [variable] à éruption *f* variable à flare *f*	Eruptionsveränderlicher *m* Flare-Stern *m* Flackerstern *m*
69	flare	éruption *f* augmentation brusque d'éclat *f* sursaut lumineux [d'éclat] *m*	kurzzeitiges Aufleuchten *n* Eruption *f* Flare *m*
70	star flare (US) stellar flare	éruption stellaire *f*	Sterneruption *f*
71	flare activity	activité éruptive *f* activité d'éruptions *f*	Eruptionstätigkeit *f* Flare-Aktivität *f*
72	geometrical variable	variable géométrique *f*	geometrischer Veränderlicher *m*
73	photometric binary (star)	binaire photométrique *f*	photometrischer Doppelstern *m*
74	eclipsing binary eclipsing variable	binaire à éclipse *f* variable à éclipses *f*	Bedeckungsveränderlicher *m*
75	depth of eclipse	profondeur (du minimum) de l'éclipse *f*	Tiefe (des Minimums) der Finsternis *f*
76	atmospheric eclipse	éclipse atmosphérique *f*	Verfinsterung der Atmosphäre *f*
77	Algol-type eclipsing binary Algol variable Beta Persei star	algolide *f* variable du type Algol *f*	Algolstern *m* Algolveränderlicher *m* Bedeckungsveränderlicher vom Algol--Typ(us) *m*
78	duration of the eclipse	durée de l'éclipse *f*	Dauer der Bedeckung *f*
79	ellipsoidal variable	variable ellipsoïdale *f*	ellipsoidischer Veränderlicher *m*
80	occultation	occultation *f*	Bedeckung *f*
81	transit	passage *m* transit *m*	Vorübergang *m*
82	reflection effect (in binaries) reflexion effect	effet de réflexion (dans les binaires) *m*	Reflexionseffekt (bei Doppelsternen) *m*
83	veiled variable veil variable occultation variable R Coronae Borealis variable	variable à occultation *f* variable du type R Coronae Borealis *f* variable à obscurcissement *f*	Verhüllungsveränderlicher *m* R Coronae Borealis Stern *m*
84	veiling	obscurcissement *m*	Verhüllung *f*
85	nebular variable *f*	variable nébulaire *f*	Nebelveränderlicher *m*

variable rotacional *f*	variável rotacional *f*	вращающаяся переменная *f*	65
anillo gaseoso *m*	anel gasoso *m*	газовое кольцо *n*	66
corriente gaseosa *f*	corrente gasosa *f*	поток газа *m*	67
estrella fulgurante *f* estrella ráfaga *f* variable de fulguraciones *f*	variável eruptiva [fulgurante] *f* estrela eruptiva [fulgurante] *f* variável de erupções *f*	вспыхивающая звезда *f* эруптивная звезда *f* эруптивная переменная *f*	68
fulguración *f* erupción *f* ráfaga *f*	erupção *f* aumento brusco de brilho *m* fulguração *f*	вспышка *f* эрупция *f*	69
erupción estelar *f* fulguración estelar *f*	erupção estelar *f* fulguração estelar *f*	вспышка звезды *f* звёздная вспышка *f*	70
actividad eruptiva *f* actividad de fulguraciones *f*	atividade em erupções [fulgurações] *f* atividade eruptiva [fulgurante] *f*	вспышечная активность *f*	71
variable geométrica *f*	variável geométrica *f*	геометрическая переменная *f*	72
(estrella) binaria fotométrica *f*	binária fotométrica *f*	фотометрическая двойная *f*	73
binaria de eclipse *f* variable de eclipse [eclipsante] *f*	binária eclipsante *f* variável eclipsante [de eclipse] *f*	затменно-переменная (звезда) *f* затменная переменная (звезда) *f*	74
profundidad del eclipse *f*	profundidade do eclipse *f*	глубина затмения *f*	75
eclipse atmosférico *m*	eclipse atmosférico *m*	атмосферное затмение *n*	76
binaria de eclipse tipo Algol *f* variable Algol *f* binaria eclipsante tipo Algol *f*	variável tipo Algol *f* binária eclipsante tipo Algol *f*	затменная переменная типа Алголя *f* алголь *m*	77
duración del eclipse *f*	duração do eclipse *m*	продолжительность затмения *f*	78
variable elipsoidal *f*	variável elipsoidal *f*	эллипсоидальная переменная *f*	79
ocultación *f*	ocultação *f*	покрытие *n*	80
tránsito *m* pasaje *m*	trânsito *m* passagem *f*	прохождение *n*	81
efecto de reflexión (en binarias) *m*	efeito de reflexão (nas binárias) *m*	эффект отражения (в двойных) *m*	82
variable por ocultación *f* variable tipo R Coronae Borealis *f* variable velada *f* variable de velo *f*	variável de velamento *f* variável tipo R Coronae Borealis *f* variável por ocultação *f* variável por obscurecimento *f*	загораживаемая переменная (звезда) *f* экранируемая [вуалируемая] переменная (звезда) *f* переменная типа R Coronae Borealis *f* затмеваемая переменная (звезда) *f*	83
efecto de velo *m*	velamento [obscurecimento] *m*	вуалирование *n*	84
variable nebular *f*	variável nebular *f*	небулярная переменная (звезда) *f*	85

	VARIABLE STARS	ÉTOILES VARIABLES	VERÄNDERLICHE STERNE
86	extrinsic variable	variable extrinsèque *f*	scheinbar Veränderlicher *m* uneigentlicher Veränderlicher *m*
87	intrinsically variable star intrinsic variable	variable intrinsèque *f*	eigentlicher Veränderlicher *m* physischer Veränderlicher *m*
88	periodic variable	variable périodique *f*	periodischer Veränderlicher *m*
89	short-period variable	variable à courte période *f*	Kurzperiodischer *m* kurzperiodischer Veränderlicher *m*
90	ultra-short-period variable	variable à période ultracourte *f*	Veränderlicher mit ultrakurzer Periode *m*
91	long-period variable	variable à longue période *f*	Langperiodischer *m* langperiodischer Veränderlicher *m*
92	Cepheid variable Cepheid δ-Cephei type variable	céphéïde *f* variable du type δ Céphée *f*	Cepheid *m* Delta-Cephei-Stern *m* Delta-Cephei Veränderlicher *m* δ-Cephei-Stern *m*
93	RR Lyrae star cluster-type variable cluster variable	variable du type RR Lyrae *f* variable d'amas *f* céphéïde à courte période *f*	RR Lyrae-Stern *m* RR Lyrae Veränderlicher *m* Haufenveränderlicher *m*
94	classical cepheid type I cepheid	céphéïde classique *f* céphéïde à longue période *f*	klassischer Cepheid *m* klassischer Delta Cephei-Stern *m*
95	pulsating star	étoile pulsante *f*	pulsierender Stern *m*
96	radial pulsations	pulsations radiales *fpl*	radiale Pulsationen *fpl*
97	non-radial pulsations	pulsations non-radiales *fpl*	nichtradiale Pulsationen *fpl*
98	adiabatic pulsations	pulsations adiabatiques *fpl*	adiabatische Pulsationen *fpl*
99	non-adiabatic pulsations	pulsations non-adiabatiques *fpl*	nichtadiabatische Pulsationen *fpl*
100	period of pulsations	période de pulsation *f*	Pulsationsperiode *f*
101	zonal harmonics Legendre coefficients Legendre polynomials	polynômes de Legendre *mpl* harmoniques zonaux *mpl* fonctions sphériques primitives *fpl*	Legendresche Polynome *npl* zonale Kugelfunktionen *fpl*
102	tesseral harmonics non-zonal harmonics	fonctions harmoniques tessérales *fpl* harmoniques tesseraux *mpl*	tesserale Kugelfunktionen *fpl* nichtzonale Kugelfunktionen *fpl*
103	period-luminosity relation period-luminosity curve	relation période-luminosité *f* courbe période-luminosité *f*	Perioden-Leuchtkraft-Beziehung *f* Perioden-Helligkeits-Beziehung *f*

variable extrínseca *f*	variável extrínseca *f*	нефизическая [несобственная] переменная (звезда) *f*	86
estrella intrínsecamente variable *f* variable intrínseca *f*	variável intrínseca *f* estela intrinsecamente variável *f*	физическая переменная (звезда) *f* собственная переменная (звезда) *f*	87
variable periódica *f*	variável periódica *f*	периодическая переменная (звезда) *f*	88
variable de corto período *f*	variável de curto período *f*	короткопериодическая переменная (звезда) *f*	89
variable de período ultracorto *f*	variável de período ultracurto *f*	сверхкороткопериодическая переменная (звезда) *f*	90
variable de largo período *f*	variável de longo período *f*	долгопериодическая переменная (звезда) *f*	91
variable cefeida *f* cefeida *f* variable tipo δ Cephei *f*	cefeida *f* variável tipo cefeida *f* variável tipo δ Cephei *f*	цефеида *f* звезда типа дельта Цефея *f* переменная типа дельта Цефея *f*	92
estrella RR Lyrae *f* variable de tipo cúmulo *f* variable de cúmulo *f* cefeida de corto período *f*	variável tipo RR Lyrae *f* variável de cúmulo *f* cefeida de curto período *f*	звезда типа RR Лиры *f* короткопериодическая цефеида *f* переменная (звезда) типа скоплений *f* анталголь *m*	93
cefeida clásica *f* cefeida de largo período *f*	cefeida clássica *f* cefeida de longo período *f*	классическая цефеида *f* долгопериодическая цефеида *f*	94
estrella pulsante *f*	estrela pulsante *f*	пульсирующая звезда *f*	95
pulsaciones radiales *fpl*	pulsações radiais *fpl*	радиальные пульсации [колебания] *fpl*[*npl*]	96
pulsaciones no radiales *fpl*	pulsações não-radiais *fpl*	нерадиальные колебания [пульсации] *npl*[*fpl*]	97
pulsaciones adiabáticas *fpl*	pulsações adiabáticas *fpl*	адиабатические пульсации [колебания] *fpl*[*npl*]	98
pulsaciones no adiabáticas *fpl*	pulsações não-adiabáticas *fpl*	неадиабатические пульсации [колебания] *fpl*[*npl*]	99
período de las pulsaciones *m*	período de pulsação *m*	период пульсаций *m*	100
polinomios de Legendre *mpl* coeficientes de Legendre *mpl*	polinômios de Legendre *mpl* funções esféricas zonais *fpl* harmônicos zonais *mpl*	полиномы Лежандра *mpl* многочлены Лежандра *mpl* зональные гармоники *fpl*	101
funciones esféricas teserales *fpl*	(funções) harmônicas tesserais *fpl* funções harmônicas não-zonais *fpl*	тессеральные гармоники *fpl* незональные гармоники *fpl*	102
curva período-luminosidad *f* relación período-luminosidad *f*	relação período-luminosidade *f* curva período-luminosidade *f*	зависимость [кривая] период-светимость *f*	103

104	zero point of the period-luminosity relation	zéro de la relation période-luminosité *m*	Nullpunkt der Perioden-Leuchtkraft-Beziehung *m*
105	semi-regular variable	variable semi-régulière *f*	halbregelmässig(er) Veränderlicher *m*
106	semi-regular red variable	variable semi-régulière rouge *f*	roter halbregelmässig(er) Veränderlicher *m*
107	irregular variable	variable irrégulière *f*	unregelmässig(er) Veränderlicher *m*
108	cyclic(al) variable	variable cyclique *f*	zyklisch(er) Veränderlicher *m*
109	cataclysmic variable	variable cataclysmique *f*	Ausbruchsveränderlicher *m* kataklysmisch(er) Veränderlicher *m*
110	nova (pl. novae)	nova *f* (pl. novae)	Nova (pl. Novae) *f* Neuer Stern *m*
111	typical nova	nova typique *f*	typische Nova *f*
112	dwarf nova U Geminorum star	nova naine *f*	Zwergnova *f*
112a	hot spot	point chaud *m*	heisse Stelle *f* heisser Fleck *m*
113	galactic nova	nova galactique *f*	galaktische Nova *f*
114	extragalactic nova	nova extragalactique *f*	extragalaktische Nova *f*
115	principal spectrum	spectre principal *m*	Hauptspektrum *n*
116	recurrent nova	nova récurrente *f* nova à répétition *f*	wiederkehrende Nova *f* rekurrierende [wiederholte] Nova *f*
117	symbiotic spectrum	spectre symbiotique *m*	symbiotisches Spektrum *n*
118	nova-like (variable) nova-like star	novoïde *f* étoile de type nova *f*	novaähnlicher Veränderlicher *m* Novaähnlicher (Stern) *m*
119	fast nova	nova rapide *f*	schnelle Nova *f*
120	slow nova	nova lente *f*	langsame Nova *f*
121	nova outburst nova explosion	explosion d'une nova *f*	Aufleuchten [Helligkeitsausbruch] einer Nova *n*[*m*] Novaausbruch *m*
122	pre-nova	prénova *f*	Praenova *f* Pränova *f*
123	post-nova	postnova *f*	Postnova *f*

punto cero de la relación período-luminosidad *m*	(ponto) zero da relação período-luminosidade *m*	нуль-пункт зависимости период--светимость *m*	104
variable semirregular *f*	variável semi-regular *f*	полуправильная переменная (звезда) *f*	105
variable roja semirregular *f*	variável semi-regular vermelha *f*	красная полуправильная переменная (звезда) *f*	106
variable irrregular *f*	variável irregular *f*	неправильная переменная (звезда) *f*	107
variable cíclica *f*	variável cíclica *f*	циклическая переменная (звезда) *f*	108
variable cataclísmica *f*	variável cataclísmica *f*	катаклизмическая пременная (звезда) *f* взрывная переменная (звезда) *f*	109
nova (pl. novas) *f*	nova (pl. novas, novae) *f*	новая (звезда) *f* (pl. новые звёзды)	110
nova típica *f*	nova típica *f*	типичная [обыкновенная] новая (звезда) *f*	111
nova enana *f*	nova anã *f*	карликовая новая (звезда) *f* звезда типа U Близнецов *f*	112
mancha caliente *f*	mancha quente *f*	горячее пятю [место] *n*	112a
nova galáctica *f*	nova galáctica *f*	галактическая новая (звезда) *f*	113
nova extragaláctica *f*	nova extragaláctica *f*	внегалактическая новая (звезда) *f*	114
espectro principal *m*	espe(c)tro principal *m*	главный спектр *m*	115
nova recurrente *f*	nova recorrente *f*	повторная новая (звезда) *f*	116
espectro simbiótico *m*	espe(c)tro simbiótico *m*	симбиотический [комбинированный] спектр *m*	117
novoide *f* variable parecida à nova *f*	novóide *f* estrela parecida à nova *f*	новоподобная переменная (звезда) *f*	118
nova rápida *f*	nova rápida *f*	быстрая новая (звезда) *f*	119
nova lenta *f*	nova lenta *f*	медленная новая (звезда) *f*	120
explosión de una nova *f*	explosão de uma nova *f*	вспышка новой (звезды) *f* взрыв новой (звезды) *f*	121
prenova *f*	pré-nova *f*	новая (звезда) до вспышки *f* пре(д)новая *f*	122
postnova *f*	pós-nova *f*	бывшая новая (звезда) *f*	123

124	ejection of matter mass ejection	éjection de matière *f* éjection de masse *f*	Auswurf von Materie [von Masse] *m* Materieausstoss *m*
125	nebular stage	phase nébulaire *f*	Nebelstadium *n*
126	supernova	supernova *f*	Supernova *f*
127	galactic supernova	supernova galactique *f*	galaktische Supernova *f*
128	extragalactic supernova	supernova extragalactique *f*	extragalaktische Supernova *f*
129	pre-supernova	présupernova *f*	Präsupernova *f* Praesupernova *f*
130	supernova of type I	supernova de type I *f*	Supernova vom Typ(us) I *f*
131	supernova outburst supernova explosion SN event	explosion d'une supernova *f*	Supernovaausbruch *m* Aufleuchten einer Supernova *n*
132	chain of supernovae	supernovae en chaîne *fpl*	Kette von Supernovae *f*
133	remnant of a supernova explosion supernova remnant relics [remains] of a supernova SNR	restes d'explosion de supernova *mpl* restes d'une supernova *mpl*	Überreste der Supernova *mpl* Supernovarest *m*
134	post-supernova past supernova	ancienne supernova *f*	ehemalige Supernova *f* Postsupernova *f*
135	Tycho's star Tychonic supernova	étoile de Tycho *f* supernova de Tycho *f*	Tychonischer Stern *m* Tycho's Supernova *f*
136	supernova search	recherche de supernovae *f*	Supernova-Suche *f*
137	pulsar	pulsar *m*	Pulsar *m*
138	X-ray pulsar	pulsar à rayons X *m*	Pulsar mit Röntgenstrahlung *m* Röntgenpulsar *m*
139	gamma-ray pulsar	pulsar à rayons gamma *m*	Pulsar mit Gammastrahlung *m* Gamma-Pulsar *m*
140	optical pulsar	pulsar optique *m*	optischer Pulsar *m*
140a	binary pulsar	pulsar binaire *m*	Doppelpulsar *m* Pulsarpaar *n*
141	optical variability	variabilité optique *f*	optische Veränderlichkeit *f*
142	Crab pulsar	pulsar (de la Nébuleuse) du Crabe *m*	Krebsnebelpulsar *m*
143	precursor	précurseur *m*	Vorläufer *m*

expulsión [eyección] de materia *f* expulsión [eyección] de masa *f*	ejeção de massa [de matéria] *f*	выброс материи [вещества] *m* сброс материи *m*	124
etapa [fase] nebular *f*	fase nebular *f*	небулярная стадия *f*	125
supernova *f*	supernova *f*	сверхновая (звезда) *f*	126
supernova galáctica *f*	supernova galáctica *f*	галактическая сверхновая (звезда) *f*	127
supernova extragaláctica *f*	supernova extragaláctica *f*	внегалактическая сверхновая (звезда) *f*	128
presupernova *f*	pré-supernova *f*	сверхновая (звезда) до вспышки *f* предсверхновая *f*	129
supernova de tipo I *f*	supernova (de) tipo I *f*	сверхновая (звезда) типа I *f*	130
explosión de una supernova *f*	explosão de uma supernova *f*	вспышка сверхновой (звезды) *f* взрыв сверхновой (звезды) *m*	131
cadena de supernovas *f*	cadeia de supernovas *f*	цепь сверхновых *f*	132
restos de una supernova *mpl*	remanescentes de uma supernova *mpl* restos (da explosão) de uma supernova *mpl*	остаток сверхновой (звезды) *m* остаток вспышки сверхновой (звезды) *m*	133
ex-supernova *f* antigua supernova *f*	ex-supernova *f* supernova passada *f* pós-supernova *f*	сверхновая (звезда) после вспышки *f* бывшая сверхновая (звезда) *f*	134
estrella de Tycho *f* supernova de Tycho *f*	estrela de Tycho *f* supernova de Tycho *f*	звезда Тихо Браге *f* сверхновая Тихо Браге *f*	135
búsqueda de supernovas *f*	busca de supernovas *f*	поиски сверхновых звёзд *mpl*	136
pulsar *m*	pulsar *m*	пульсар *m*	137
pulsar de rayos X *m*	pulsar de raios X *m*	рентгеновский пульсар *m*	138
pulsar de rayos gamma *m*	pulsar de raios gama *m*	гамма-пульсар *m*	139
pulsar óptico *m*	pulsar ó(p)tico *m*	оптический пульсар *m*	140
pulsar binario *m*	pulsar binário *m*	двойной пульсар *m*	140a
variabilidad óptica *f*	variabilidade ó(p)tica *f*	оптическая переменность *f*	141
pulsar del Cangrejo *m*	pulsar do Caranguejo *m*	пульсар (в) Крабовидной туманности *m*	142
precursor *m*	precursor *m*	предшественник *m* предвестник *m*	143

144	principal pulse main pulse	pulsation principale *f*	Hauptpuls *m* Hauptpulsation *f*
145	subpulse	sous-pulsation *f*	Nebenpuls *m* Nebenpulsation *f*
146	isolated pulse	pulsation isolée *f*	Einzelpuls *m* vereinzelte Pulsation *f*
147	pulsar magnetosphere	magnétosphère de pulsar *f*	Pulsarmagnetosphäre *f* Magnetosphäre des Pulsars *f*
148	superstrong magnetic field	champ magnétique très intense *m*	überstarkes Magnetfeld *n* superstarkes Magnetfeld *n*
149	superfluid	superfluide *m*	Supraflüssigkeit *f*
150	superconducting nucleus	noyau supraconducteur *m* noyau superconducteur *m*	supraleitender Kern *m* Supraleitungskern *m*
151	solid crust	croûte solide *f*	feste Kruste *f* feste Schale *f*
152	crustquakes	tremblements de la croûte *mpl* secousse de la croûte *f*	Krustenbeben *n* Erschütterung der Schale *f*
153	starquake	secousse stellaire *f* tremblement stellaire *m*	Sternbeben *n* Erschütterung des Sterns *f*
154	glitch sudden jump (in pulsar frequency)	sursaut soudain *m*	Periodensprung *m* abrupte Änderung der Pulsperiode *f*
155	light cylinder	cylindre de lumière *m*	Lichtzylinder *m*
156	light cone ray cone	cône lumineux *m* cône de lumière *m*	Lichtkegel *m* Strahlenkegel *m*

pulsación principal pulso principal *m*	pulso principal *m*	главный импульс *m*	144
subpulsación *f* subpulso *m*	subpulso *m*	подимпульс *m* субимпульс *m*	145
pulsación aislada *f* pulso aislado *m*	pulso isolado *m*	изолированный импульс *m*	146
magnetosfera de un pulsar *f*	magnetosfera de um pulsar *f*	магнитосфера пульсара *f*	147
campo magnético ultraintenso *m* campo magnético muy intenso *m*	campo magnético superintenso *m* campo magnético muito intenso *m*	сверхсильное магнитное поле *n* очень сильное магнитное поле *n*	148
superfluido *m*	superfluido *m*	сверхтекучая жидкость *f*	149
núcleo superconductor *m*	núcleo superconductor *m*	сверхпроводящее ядро *n*	150
corteza sólida *f*	crosta sólida *f*	твёрдая кора *f*	151
temblor de la corteza *m*	abalos [tremor] da crosta *mpl* sismos da crosta *mpl*	трясение [сотрясение] коры *n* коротрясение *n*	152
temblor de la estrella *m*	abalos [tremor] da estrela *mpl* sismos estelares *mpl*	звездотрясение *n*	153
deslizamiento en frecuencia *m* salto brusco *m*	salto repentino *m* salto brusco *m*	скачкообразное изменение периода *n*	154
cilindro de luz *m*	cilindro de luz *m*	световой цилиндр *m*	155
cono de luz *m*	cone de luz *m* cone luminoso *m*	световой конус *m* конус световых лучей *m*	156

	STELLAR SYSTEMS	SYSTÈMES STELLAIRES	STERNSYSTEME
1	stellar system system of stars	système stellaire *m* système d'étoiles *m*	Sternsystem *n* System von Sternen *n*
2	single star	étoile simple *f*	Einzelstern *m*
3	double star binary star binary	(étoile) double *f* binaire *f* étoile binaire *f*	Doppelstern *m* Sternpaar *n*
4	triple star	étoile triple *f* système triple *m*	Dreifachstern *m* Dreifachsternsystem *n*
5	multiple star	étoile multiple *f*	Mehrfachstern *f*
6	optical double star	(étoile) double optique *f*	optischer Doppelstern *m*
7	physical double star binary star binary	(étoile) double physique *m* binaire *f* (étoile) binaire *f*	Doppelsternsystem *n* physischer Doppelstern *m* physisches Sternpaar *n*
8	visual double star	(étoile) double visuelle *f*	visueller Doppelstern *m*
9	wide double star	binaire à grande séparation *f*	weites Sternpaar *n* weiter Doppelstern *m*
10	unresolved binary	(étoile) double non-résolue *f* binaire non-résolue *f*	nicht aufgelöster Doppelstern *m*
11	photometric binary	binaire photométrique *f*	photometrischer Doppelstern *m*
12	astrometric binary	binaire astrométrique *f*	astrometrischer Doppelstern *m*
13	star with a dark companion	étoile avec un compagnon obscur *f*	Stern mit einem unsichtbaren Begleiter *m*
14	spectroscopic binary spectrum binary	binaire spectroscopique *f* étoile double spectroscopique *f*	spektroskopischer Doppelstern *m* Spektrum-Doppelstern *m*
15	single-line (spectroscopic) binary	binaire (spectroscopique) à raies simples *f* étoile double (spectroscopique) à un seul spectre *f*	(spektroskopischer) Doppelstern mit Linien einer Komponente *m*
16	single spectrum	spectre unique *m*	Einzelspektrum *n*
17	double-line (spectroscopic) binary	binaire (spectroscopique) à raies doubles *f* étoile double (spectroscopique) à deux spectres *f*	(spektroskopischer) Doppelstern mit Linien beider Komponenten *m*
18	double spectrum	spectre dédoublé *m*	Doppelspektrum *n*

sistema estelar *m* sistema de estrellas *m*	sistema estelar *m* sistema de estrelas *m*	звёздная система *f* система звёзд *f*	1
estrella aislada *f* estrella simple *f*	estrela única *f* estrela simples *f*	одиночная звезда *f*	2
estrella doble *f* binaria *f* doble *f*	estrela dupla *f* binária *f* dupla *f*	двойная звезда *f* двойная *f*	3
estrella triple *f* sistema triple *m*	estrela tripla *f* estrela tríplice *f* sistema triplo *m*	тройная звезда *f* тройная система *f*	4
estrela múltiple *f*	estrela múltipla *f*	кратная звезда *f*	5
(estrella) doble óptica *f*	(estrela) dupla ó(p)tica *f*	оптическая двойная (звезда) *f*	6
(estrella) doble física *f* estrella binaira *f* binaria *f*	(estrela) dupla física *f* estrela binária *f* binária *f*	физическая двойная (звезда) *f* двойная (звезда) *f* физическая пара *f*	7
(estrella) doble visual *f*	(estrela) dupla visual *f* binária visual *f*	визуально двойная (звезда) *f* визуальная пара *f*	8
(estrella) doble de gran separación *f*	(estrela) dupla de grande separação *f* binária afastada *f*	широкая двойная (звезда) *f* широкая пара *f*	9
binaria irresoluble *f* binaria no resuelta *f*	(estrela) dupla não-resolvida *f* binária [dupla] não-resolúvel *f* binária [dupla] não-resolvida *f*	неразрешимая [неразрешённая] двойная (звезда) *f*	10
binaria fotométrica *f*	binária fotométrica *f* dupla fotométrica *f*	фотометрическая двойная (звезда) *f*	11
binaria astrométrica *f*	binária astrométrica *f* dupla astrométrica *f*	астрометрическая двойная (звезда) *f*	12
estrella con un compañero invisible *f*	estrela com uma companheira invisível *f*	звезда с тёмным спутником *f*	13
binaria espectroscópica *f* binaria espectral *f*	(estrela) dupla espe(c)troscópica *f* binária espe(c)troscópica *f*	спектрально-двойная (звезда] *f* спектральная [спектроскопическая] двойная (звезда) *f*	14
binaria (espectroscópica) con espectro simple *f* binaria espectroscópica con líneas simples *f*	dupla espe(c)troscópica com espe(c)tro simples *f* binária espe(c)troscópica com raias [linhas] simples *f*	(спектрально-)двойная с одиночными линиями в спектре *f*	15
espectro simple *m*	espe(c)tro simples *m*	одиночный спектр *m*	16
binaria espectroscópica con líneas dobles *f*	binária espe(c)troscópica de linhas [raias] duplas *f* binária espe(c)troscópica com linhas [raias] duplas *f*	(спектрально-)двойная с двойными линиями в спектре *f*	17
espectro doble *m*	espe(c)tro duplo *m*	двойной спектр *m*	18

19	composite spectrum	spectre composite *m*	gemischtes [zusammengesetztes, überlagertes] Spektrum *n*
20	symbiotic star	étoile symbiotique *f*	symbiotischer Stern *m*
21	accretion disk accretion disc	disque d'accrétion *m*	Akkretionsscheibe *f* Aufsammlungsscheibe *f*
22	periodic line shift	déplacement périodique des raies *m* décalage périodique des raies *m*	periodische Linienverschiebung *f*
23	close binary	(étoile)double serrée *f* binaire serrée *f*	enger Doppelstern *m* enges Paar *n*
24	superposition of spectra	superposition des spectres *f* recouvrement des spectres *m*	Überlagerung von Spektren *f* Überdeckung von Spektren *f*
25	gravity darkening	assombrissement de gravité *m* assombrissement par gravité *m*	Gravitationsverdunk(e)lung *f* Schwereverdunk(e)lung *f*
26	reflection effect	effet de réflexion *m*	Reflexionseffekt *m*
27	critical surface contact surface	surface critique (de Roche) *f* surface de contact *f*	Kontaktfläche *f* kritische Äquipotentialfläche *f*
28	Roche lobe	lobe de Roche *m*	Rochescher Lappen *m* Rochesche Grenze *f*
29	contact binary	binaire en contact *f*	Kontaktdoppelstern *m*
30	tidal distortion	distorsion par effet de marée *f*	Gezeitendeformation *f*
31	conical point inner Lagrangian point first Lagrangian point	point conique *m* point interne de Lagrange *m* premier point de Lagrange *m*	innerer Lagrange-Punkt *m* erster Lagrange-Punkt *m*
32	common envelope (of a binary)	enveloppe commune (d'une double) *f*	gemeinsame Hülle (eines Doppelsterns) *f*
33	mass transfer (from one component to the other)	transfert de masse (d'une composante à l'autre) *m*	Massentransport (von einer Komponente zu der anderen) *m* Massenübertragung (von einem Stern zum anderen) *f* Massenaustausch *m*
34	semi-detached binary	double semi-détachée *f* binaire presque séparée *f* double semi-séparée *f*	halbgetrennter Doppelstern *m* halbgetrenntes Sternpaar *n*
35	detached binary	double séparée *f* binaire séparée *f*	getrennter Doppelstern *m* getrenntes Sternpaar *n*
36	component	composante *f*	Komponente *f*

espectro compuesto *m*	espe(c)tro composto *m*	сложный спектр *m*	19
estrella simbiótica *f*	estrela simbiótica *f*	симбиотическая звезда *f*	20
disco de acrecimiento *m* disco de acreción *m*	disco de acrescimento *m* disco de acrescão *m*	аккреционный диск *m*	21
desplazamiento periódico de líneas *m*	deslocamento periódico das linhas [raias] *m*	периодическое смещение линий *n*	22
estrella binaria cerrada *f* binaria próxima [cercana] *f*	dupla estreita [cerrada] *f* binária cerrada [estreita] *f*	тесная двойная звезда *f* тесная пара *f*	23
sobreposición de los espectros *f* superposición de los espectros *f*	sobreposição dos espe(c)tros *f* superposição dos espe(c)tros *f*	наложение спектров *n* перекрытие спектров *n*	24
o(b)scurecimiento gravitacional *m* o(b)scuración por gravedad *f*	obscurecimento por gravidade *m* escurecimento por gravidade *m*	гравитационное потемнение *n*	25
efecto de reflexión *m*	efeito de reflexão *m*	эффект отражения *m*	26
superficie crítica *f* superficie de contacto *f*	superfície crítica *f* superfície de conta(c)to *f*	критическая [контактная] поверхность (Роша) *f* критическая эквипотенциальная поверхность *f*	27
lóbulo de Roche *m*	lóbulo de Roche *m*	лепесток Роша *m*	28
binaria de contacto *f* doble de contacto *f*	dupla de [em] conta(c)to *f* binária em [de] conta(c)to *f*	контактная двойная *f*	29
distorsión de marea *f*	distorção de maré *f*	приливное искажение *n* приливная деформация [дисторсия] *f*	30
punto cónico *m* punto interior de Lagrange *m* primer punto Lagrangiano *m*	ponto cônico *m* ponto interior de Lagrange *m* primeiro ponto de Lagrange *m*	коническая точка *f* первая критическая точка *f* внутренняя [первая] точка Лагранжа *f*	31
envolvente común (de una binaria) *f*	envelope [invólucro] comum (de uma binária) *m*	общая оболочка (двойной звезды) *f*	32
transporte de masa (de una componente a otra) *m*	transferência de massa (de um componente para outro) *f* transporte de massa *m*	перенос массы (с одной компоненты на другую) *m*	33
binaria semiseparada *f*	dupla semidestacada *f* binária semi-separada *f* binária semidesligada *f*	полуразделённая двойная *f*	34
binaria separada *f*	dupla separada [destacada] *f* binária destacada [desligada] *f*	разделённая двойная *f*	35
componente *m, f*	componente *m, f*	составляющая *f* компонент *m*	36

37	primary primary component	étoile principale *f* composante principale *f*	Hauptstern *m* Hauptkomponente *f*
38	secondary secondary component companion	secondaire *f* composante secondaire *f* compagnon *m*	Nebenkomponente *f* Begleiter *m* Sekundärkomponente *f*
39	companion star	étoile compagnon *f*	Sternbegleiter *m* begleitender Stern *m*
40	threshold curve of the secondary component threshold of the secondary	courbe de seuil de la secondaire *f* seuil du compagnon *m*	Schwellenkurve der Sekundärkomponente [des Begleiters] *f* Grenzkurve der Sekundärkomponente [des Begleiters] *f*
41	invisible companion	compagnon invisible *m*	unsichtbarer Begleiter *m*
42	dark companion dark secondary	compagnon obscur *m* secondaire obscure *f*	dunkler Begleiter *m* dunkle Sekundärkomponente *f*
43	planetary companion	compagnon planétaire *m*	planetarischer Begleiter *m*
44	non-luminous	non-lumineux	nichtleuchtend
45	self-luminous	lumineux par lui-même à lumière propre	selbstleuchtend leuchtend mit Eigenlicht
46	photocentre (UK) photocenter (US)	centre photométrique *m*	Lichtzentrum *n*
47	photocentric orbit	orbite photocentrique *f*	photozentrische Bahn *f*
48	apparent orbit	orbite apparente *f*	scheinbare Bahn *f*
49	micrometric measurement	mesure micrométrique *f*	mikrometrische Messung *f*
50	tangent plane plane of the sky	plan tangent à la sphère céleste *m* plan du ciel *m*	Tangentialebene an der Himmelssphäre *f*
51	inclination of orbital plane inclination of orbit	inclinaison du plan orbital *f* inclinaison d'une orbite *f*	Neigung der Bahnebene *f* Neigung der Bahn *f*
52	angular separation angular distance	distance angulaire *f* séparation angulaire *f*	Winkelabstand *m* Winkelentfernung *f*
53	true orbit	orbite réelle *f*	wahre Bahn *f*
54	relative orbit	orbite relative *f*	relative Bahn *f*
55	absolute orbit	orbite absolue *f*	absolute Bahn *f*
56	spectroscopic orbit spectrographic orbit	orbite spectroscopique *f* orbite spectrographique *f*	spektroskopische Bahn *f* spektrographische Bahn *f*

estrella primaria *f* componente primaria *f* componente [estrella] principal *m*[*f*]	primário *m* componente principal [primário] *m* estrela principal [primária] *f*	центральная звезда *f* главный компонент *m* главная звезда *f*	37
componente secundaria *f* compañero *m* compañera *f*	componente secundário *m* companheira *f* companheiro *m*	спутник *m*	38
estrella compañera *f*	estrela companheira *f*	звезда-спутник *f*	39
curva umbral de la secundaria *f* umbral de la secundária [compañera] *m*	curva limiar do componente secundário *f* limiar do secundário *m*	предельная кривая спутника (двойной) *f* порог разрешения спутника *m*	40
componente [compañera] invisible *m*[*f*]	companheira invisível *f*	невидимый спутник *m*	41
componente o(b)scuro [obscura] *m*[*f*] compañera o(b)scura *f*	companheira obscura [escura] *f* componente obscuro [escuro] *m*	тёмный спутник *m*	42
compañera planetaria *f*	companheira planetária *f*	планета-спутник *f* планетарный [планетный] спутник *m*	43
no luminoso sin luz propia	não-luminoso sem luz própria	несветящийся несамосветящийся	44
autoluminoso con luz propia	autoluminoso luminoso por si próprio com luz própria	самосветящийся светящийся	45
fotocentro *m* centro fotométrico *m*	fotocentro *m* centro fotométrico *m*	фотоцентр *m*	46
órbita fotocéntrica *f*	órbita fotocêntrica *f*	фотоцентрическая орбита *f*	47
órbita aparente *f*	órbita aparente *f*	видимая орбита *f*	48
medición micrométrica *f*	medida micrométrica *f*	микрометрическое измерение *n*	49
plano tangencial [tangente] a la esfera celeste *m* plano del cielo *m*	plano do céu *m* plano tangente à esfera celeste *m*	плоскость перпендикулярная к лучу зрения *f* картинная плоскость *f*	50
inclinación del plano orbital *f* inclinación de la órbita *f*	inclinação do plano orbital *f* inclinação da órbita *f*	наклонение плоскости орбиты *n*	51
distancia angular *f*	distância angular *f* separação angular *f*	угловое расстояние (между компонентами) *n*	52
órbita verdadera *f*	órbita verdadeira *f*	истинная орбита *f*	53
órbita relativa *f*	órbita relativa *f*	относительная орбита *f*	54
órbita absoluta *f*	órbita absoluta *f*	абсолютная орбита *f*	55
órbita espectroscópica *f* órbita espectrográfica *f*	órbita espe(c)troscópica *f* órbita espe(c)trográfica *f*	спектроскопическая орбита *f* спектрографическая орбита *f*	56

	STELLAR SYSTEMS	SYSTÈMES STELLAIRES	STERNSYSTEME
57	astrometric orbit	orbite astrométrique *f*	astrometrische Bahn *f*
58	periastron	périastre *m*	Periastron *n*
59	longitude of periastron	longitude du périastre *f*	Periastronlänge *f* Länge des Periastrons *f*
60	epoch of periastron time of periastron passage	instant du passage au périastre *m* époque du passage au périastre *f*	Periastronzeit *f* Zeit des Durchganges durch das Periastron *f* Durchgang(szeit) durch das Periastron *m(f)*
61	apastron	apoastre *m*	Ap(o)astron *n*
62	orbital motion	mouvement orbital *m*	Bahnbewegung *f*
63	orbital velocity	vitesse orbitale *f*	Bahngeschwindigkeit *f*
64	dimensions of the system	dimensions du système *fpl*	Dimensionen des Systems *fpl*
65	magnitude difference brightness difference	différence de magnitude *f* différence d'éclat *f*	Helligkeitsdifferenz *f* Helligkeitsunterschied *m*
66	separation of components component separation distance between components	séparation des composantes *f* distances entre les composantes *f*	Abstand der Komponenten *m*
67	mass ratio	rapport des masses *m*	Massenverhältnis *n*
68	mass function	fonction des masses *f*	Massenfunktion *f*
69	mass determination	détermination de la masse *f*	Massenbestimmung *f*
70	variable mass	masse variable *f*	veränderliche Masse *f*
71	apsidal motion rotation of the line of apsides advance of periastron	mouvement de la ligne des apsides *m* avance du périastre *m* rotation de la ligne des apsides *f*	Drehung der Apsidenlinie *f* Periastrondrehung *f* Apsidenbewegung *f*
72	capture theory	théorie de la capture *f*	Einfangstheorie *f*
73	star capture	capture d'étoile *f*	Sterneinfang *m*
74	three-body encounter triple collision three-body collision	rencontre de trois corps *f* triple collision *f* collision à trois corps *f*	Dreikörperzusammentreffen *n* Dreikörperbegegnung *f* Dreikörperstoss *m* Dreierstoss *m*
75	fission theory	théorie de la fission *f*	Spaltungstheorie *f*
76	separate-nuclei theory theory of multiple formation	théorie de la condensation en plusieurs centres *f* théorie des noyaux séparés *f* théorie de la formation multiple *f*	Theorie der Bildung getrennter Kerne *f* Theorie getrennter Kerne *f*
77	star chain	chaîne stellaire *f*	Sternkette *f*

órbita astrométrica *f*	órbita astrométrica *f*	астрометрическая орбита *f*	57
periastro *m*	periastro *m*	периастр *m* периастрий *m*	58
longitud del periastro *f*	longitude do periastro *f*	долгота периастра *f*	59
instante [momento] del paso por el periastro *m* tiempo [época] del (paso por el) periastro *m[f]*	época do periastro *f* instante [tempo] da passagem pelo periastro *m*	момент прохождения через периастр *m*	60
ap(o)astro *m*	ap(o)astro *m*	апоастр *m* апоастрий *m*	61
movimiento orbital *m*	movimento orbital *m*	орбитальное движение *n*	62
velocidad orbital *f*	velocidade orbital *f*	орбитальная скорость *f*	63
dimensiones del sistema *fpl*	dimensões do sistema *fpl*	размеры системы *mpl*	64
diferencia de magnitud *f* diferencia de brillo *f*	diferença de magnitude *f* diferença de brilho *f*	разность блеска [яркости] *f* разность звёздных величин *f*	65
distancia entre las componentes *f*	distância entre os componentes *f* separação dos componentes *f*	расстояние между компонентами *n*	66
relación de (las) masas *f*	razão [relação] das massas *f*	отношение масс *n*	67
función de masa [de las masas] *f*	função de massa [das massas] *f*	функция масс *f*	68
determinación de la masa *f*	determinação da massa *f*	определение массы *n*	69
masa variable *f*	massa variável *f*	переменная масса *f*	70
rotación de la línea de los ápsides *f* movimiento apsidal *m* avance del periastro *m*	rotação da linha dos apsides *f* movimento apsidal *m* avanço do periastro *m*	вращение линии апсид *n* движение линии апсид *n* апсидальное движение *n*	71
teoría de captura *f*	teoria da captura *f*	теория захвата *f*	72
captura de estrellas *f*	captura de estrelas *f*	захват звезды *m*	73
colisión triple *f* encuentro triple *m* encuentro de tres cuerpos *m* colisión de tres cuerpos *f*	encontro de três corpos *m* colisão tripla *f* colisão de três corpos *f*	тройное столкновение *n* тройное сближение *n* тройное соударение *n*	74
teoría de la fisión *f*	teoria da fissão [divisão] *f*	теория деления *f*	75
teoría de núcleos separados *f* teoría de formación múltiple *f*	teoria dos núcleos separados *f* teoria da formação múltipla *f*	теория образования отдельных ядер *f*	76
cadena de estrellas *f* cadena estelar *f*	cadeia estelar *f* cadeia de estrelas *f*	звёздная цепочка *f* цепочка звёзд *f*	77

	STELLAR SYSTEMS	SYSTÈMES STELLAIRES	STERNSYSTEME
78	ring of stars star ring	anneau d'étoiles *m* anneau stellaire *m*	Sternring *m*
79	star cluster	amas d'étoiles *m* amas stellaire *m*	Sternhaufen *m*
80	structural index of a cluster	indice de structure d'un amas *m*	Strukturindex eines Sternhaufens *m*
81	double cluster	amas double *m*	Doppelhaufen *m*
82	cluster member member of the cluster cluster star	membre de l'amas *m* étoile d'un amas *f* étoile membre de l'amas *f*	Sternhaufenmitglied *n* Haufenmitglied *n* Haufenstern *m*
83	physical membership	appartenance physique (à l'amas) *f*	physische Mitgliedschaft *f*
84	field star background star non-member star non-cluster star non-member of the cluster	étoile du fond stellaire *f* étoile de champ *f* étoile non-membre de l'amas *f* étoile n'appartenant pas à l'amas *f* non-membre de l'amas *m*	Feldstern *m* Hintergrundstern *m* nicht zum Haufen gehörender Stern *m*
85	foreground star	étoile située devant [en avant] *f*	Vordergrundstern *m*
86	probable member of the cluster	membre probable de l'amas *m*	wahrscheinliches Mitglied des Haufens *n*
87	uncertain member of the cluster	membre incertain de l'amas *m*	unsicheres Mitglied des Sternhaufens *n*
88	stellar association	association stellaire *f*	Sternassoziation *f*
89	O-association	association O *f*	O-Assoziation *f*
90	T-association	association T *f*	T-Assoziation *f*
91	nucleus of the association	noyau de l'association *m*	Kern der Assoziation *m*
92	Trapezium in Orion	Trapèze d'Orion *m*	Trapez im Orion *n*
93	young star	étoile jeune *f*	junger Stern *m*
94	expansion velocity	vitesse d'expansion *f*	Expansionsgeschwindigkeit *f*
95	star formation formation of stars	formation des étoiles *f*	Sternentstehung *f*
96	protostar	proto-étoile *f*	Protostern *m*
97	moving cluster moving star cluster	amas en mouvement *m* courant stellaire *m*	Bewegungssternhaufen *m* Bewegungshaufen *m*

anillo de estrellas *m* anillo estelar *m*	anel de estrelas *m* anel estelar *m*	звёздное кольцо *n* кольцо звёзд *n*	78
cúmulo estelar *m*	aglomerado estelar *m* cúmulo [enxame] estelar *m*	звёздное скопление *n* скопление звёзд *n*	79
índice estructural de un cúmulo *m*	índice estrutural de um cúmulo [aglomerado] *m*	структурный индекс скопления *m*	80
cúmulo doble *m*	cúmulo [aglomerado] duplo *m* enxame duplo *m*	двойное скопление *n*	81
miembro del cúmulo *m* estrella del cúmulo *f*	membro do aglomerado [cúmulo] *m* estrela pertencente ao cúmulo *f* estrela do cúmulo [aglomerado] *f* estrela membro do cúmulo *f*	член скопления *m* звезда скопления *f* звезда, принадлежащая скоплению *f*	82
pertenecencia física al cúmulo *f* membrecía del cúmulo *f*	pertinência física ao cúmulo [ao aglomerado] *f*	физическая принадлежность *f* физическое членство *n*	83
estrella de campo *f* estrella de fondo *f* estrella no perteneciente al cúmulo *f*	estrela do campo *f* estrela do fundo *f* estrela não-membro do cúmulo *f* estrela não-pertencente ao cúmulo *f* estrela não-pertencente ao sistema *f*	звезда фона [поля] *f* нечлен скопления *m* звезда, не принадлежащая скоплению *f* звезда не является членом скопления *f*	84
estrella de primer plano *f*	estrela preposta *f* estrela de primeiro plano *f*	звезда переднего фона *f* передняя звезда *f*	85
miembro probable del cúmulo *m*	membro provável do aglomerado [cúmulo] *m*	возможный член скопления *m*	86
miembro incierto del cúmulo *m*	membro incerto do aglomerado [cúmulo] *m*	недостоверный [неуверенный, неопределённый] член скопления *m*	87
asociación estelar *f*	associação estelar *f*	звёздная ассоциация *f*	88
asociación O *f*	associação O *f*	О-ассоциация *f*	89
asociación T *f*	associação T *f*	Т-ассоциация *f*	90
núcleo de la asociación *m*	núcleo da associação *m*	ядро ассоциации *n*	91
El Trapecio de Orión *m*	Trapézio do Órion *m*	Трапеция Ориона *f*	92
estrella joven *f*	estrela jovem *f*	молодая звезда *f*	93
velocidad de expansión *f*	velocidade de expansão *f*	скорость расширения *f*	94
formación de estrellas creación de estrellas *f*	formação estelar [de estrelas] *f* criação de estrelas *f*	образование звёзд *n* звёздообразование *n*	95
proto-estrella *f*	proto-estrela *f*	протозвезда *f*	96
cúmulo en movimiento *m*	cúmulo móvel (de estrelas) *m* aglomerado [enxame] móvel *m* cúmulo em movimento *m*	движущееся (звёздное) скопление *n* местный звёздный поток *m*	97

	STELLAR SYSTEMS	SYSTÈMES STELLAIRES	STERNSYSTEME
98	convergent point vertex	point de convergence *m* vertex *m*	Konvergenzpunkt *m* Vertex (pl. Vertizes) *m*
99	open star cluster galactic star cluster	amas ouvert *m* amas galactique *m*	offener Sternhaufen *m* galaktischer Sternhaufen *m*
100	star-rich cluster rich cluster of stars	amas riche en étoiles *m*	sternreicher Haufen *m*
101	concentration index	indice de concentration *m*	Konzentrationsindex *m*
102	star-poor cluster poor cluster of stars	amas pauvre en étoiles *m*	sternarmer Haufen *m*
103	dissociation of a galactic cluster cluster disintegration cluster dissolution	dissociation d'un amas galactique *f* désintégration d'un amas *f*	Auflösung eines offenen Sternhaufens *f* Sternhaufenauflösung *f*
104	dissociation time dissolution time dispersion time	durée de dissociation *f* temps de dissipation *m* temps de désintégration *m*	Auflösungszeit *f*
105	disruption of a cluster	désintégration [désagrégation] d'un amas *f*	Zerstörung eines Sternhaufens *f*
106	galactic tidal field	champ de marées galactique *m*	galaktisches Gezeitenfeld *n*
107	tidal instability	instabilité due aux marées *f*	Gezeiteninstabilität *f*
108	evaporation of a cluster	évaporation d'un amas *f*	Verdampfung eines Sternhaufens *f*
109	evaporation time	durée d'évaporation *f*	Verdampfungszeit *f* Verdampfungsdauer *f*
110	Pleiades Seven Sisters	Pléiades *fpl* Poussinière *f*	Plejaden *pl* Siebengestirn *n* Kuckuckgestirn *n* Gluckhenne *f*
111	Praesepe Beehive	Praesepe *pl*	Praesepe *pl* Krippe *f*
111a	Hyades *pl*	Hyades *fpl*	Hyaden *fpl*
112	metal-rich cluster	amas stellaire riche en métaux *m*	metallreicher Sternhaufen *m*
113	metal-poor cluster	amas stellaire pauvre en métaux *m*	metallarmer Sternhaufen *m*
114	globular cluster globular star cluster	amas globulaire *m*	Kugelsternhaufen *m* Kugelhaufen *m* kugelförmiger Sternhaufen *m*

punto de convergencia *m* vértice *m*	ponto de convergência *m* vértice *m*	радиант *m* вертекс *m*	98
cúmulo (estelar) abierto *m* cúmulo galáctico *m*	cúmulo [aglomerado] aberto *m* cúmulo [aglomerado] galáctico *m*	рассеянное звёздное скопление *n* открытое звёздное скопление *n* галактическое звёздное скопление *n*	99
cúmulo estelar rico *m*	cúmulo [aglomerado] rico de estrelas *m* cúmulo (estelar) rico *m*	скопление богатое звёздами *n* богатое звёздное скопление *n*	100
índice de concentración *m*	índice de concentração *m*	индекс концентрации *m*	101
cúmulo estelar pobre *m*	cúmulo [aglomerado] pobre (de [em] estrelas) *m*	скопление бедное звёздами *n* бедное звёздное скопление *n*	102
desintegración [disolución] de un cúmulo galáctico *f* disgregación de un cúmulo abierto *f*	dissipação [dissociação] de um cúmulo galáctico *f* desmembramento do enxame [cúmulo] galáctico *m*	распад открытого скопления *m* разрушение галактического скопления *n*	103
tiempo de disgregación *m* tiempo de disolución *m* tiempo de desintegración *m*	tempo de dissociação [desintegração] *m* duração do desmembramento *f* tempo de dissipação *m*	время распада *n*	104
desagregación de un cúmulo *f* disgregación de un cúmulo *f*	desagregação de um cúmulo [aglomerado] *f*	разрушение скопления *n*	105
campo de marea galáctico *m*	campo de maré galáctico *m*	галактическое приливное поле *n*	106
inestabilidad de marea *f*	instabilidade de maré *f*	приливная неустойчивость *f*	107
evaporación de un cúmulo *f*	evaporação de um cúmulo [aglomerado] *f*	испарение скопления *n*	108
tiempo de evaporación *m*	tempo de evaporação *m* duração de evaporação *f*	срок испарения *m* продолжительность испарения *f*	109
Pléyadas *fpl* Pléyades *fpl* Las Siete Cabrillas *fpl*	Plêiades *fpl* Sete cabrinhas *fpl* Sete-estrelo *m*	Плеяды *pl* Стожары *pl*	110
Pesebre *m* Presepio *m*	Presépio *m*	Ясли *pl*	111
Híades *fpl* Híadas *fpl*	Híades *fpl*	Гиады *pl*	111a
cúmulo rico en metales *m*	cúmulo [aglomerado, enxame] rico em metais *m*	скопление с высоким содержанием металлов *n*	112
cúmulo pobre en metales *m*	cúmulo [aglomerado, enxame] pobre em metais *m*	скопление с низким содержанием металлов *n*	113
cúmulo globular *m*	cúmulo globular *m* aglomerado globular *m* enxame globular *m*	шаровое звёздное скопление *n* шарообразное звёздное скопление *n*	114

115	angular diameter apparent diameter	diamètre angulaire *m* diamètre apparent *m*	Winkeldurchmesser *m* scheinbarer Durchmesser *m*
116	photometric distance	distance photométrique *f*	photometrische Entfernung *f*
117	apparent modulus	module apparent *m*	scheinbarer Entfernungsmodul *m*
118	true modulus	module vrai *m*	wahrer Entfernungsmodul *m*
119	distribution in space spatial distribution	distribution dans l'espace *f* distribution spatiale *f*	Verteilung im Raum *f* Raumverteilung *f* räumliche Verteilung *f*
120	structural parameter	paramètre de structure *m*	Strukturparameter *m*
122	ellipticity of a globular cluster	ellipticité d'un amas globulaire *f*	Elliptizität eines Kugelsternhaufens [Kugelhaufens] *f*
123	flattening of a globular cluster	aplatissement d'un amas globulaire *m*	Abplattung eines Kugelsternhaufens [Kugelhaufens] *f*
124	density distribution	distribution de la densité *f*	Dichteverteilung *f*
125	isopleth	isoplèthe *f*	Isoplethe *f*
126	central condensation	condensation centrale *f*	zentrale Verdichtung *f*
127	concentration	concentration *f*	Konzentration *f*
128	degree of concentration	degré de concentration *m*	Konzentrationsgrad *m*
129	high-concentration cluster highly concentrated cluster	amas à forte concentration *m*	Sternhaufen mit hoher Konzentrati-on *m*
130	core of a cluster	noyau d'un amas *m*	Sternhaufenkern *m*
131	background density	densité du fond (stellaire) *f*	Hintergrunddichte *f*
132	apparent surface density	densité de surface apparente *f*	scheinbare Flächendichte *f*
133	space density real density volume density	densité spatiale *f* densité réelle *f* densité volumique *f*	räumliche Dichte *f* wahre Dichte *f*
134	integrated properties	propriétés intégrées *fpl* propriétés globales *fpl*	integrale Eigenschaften *fpl* globale Eigenschaften *fpl*
135	integrated brightness total brightness	éclat global *m* éclat intégré *m* éclat total *m*	Gesamthelligkeit *f* integrale Helligkeit *f* Integralhelligkeit *f*

diámetro angular *m* diámetro aparente *m*	diâmetro angular *m* diâmetro aparente *m*	угловой диаметр *m* видимый диаметр *m*	115
distancia fotométrica *f*	distância fotométrica *f*	фотометрическое расстояние *n*	116
módulo aparente (de distancia) *m*	módulo aparente (de distância) *m*	видимый модуль (расстояния) *m*	117
módulo verdadero *m*	módulo verdadeiro *m*	истинный модуль (расстояния) *m*	118
distribución espacial *f* distribución en el espacio *f*	distribuição espacial *f* distribuição no espaço *f*	пространственное распределение *n* распределение в пространстве *n*	119
parámetro estructural *m*	parâmetro estrutural *m*	структурный параметр *m*	120
elipticidad de un cúmulo globular *f*	elipticidade de um cúmulo [aglomerado] globular *f*	эллиптичность шарового скопления *f*	122
achatamiento de un cúmulo globular *m*	achatamento de um cúmulo [aglomerado] globular *m*	сжатие шарового скопления *n* сплюснутость шарового скопления *f*	123
distribución de densidad *f*	distribuição de densidade *f*	распределение плотности *n*	124
isopleta *f*	isopleta *f*	изолиния плотности *f* изоплета *f*	125
condensación central *f*	condensação central *f*	центральное сгущение *n* центральное уплотнение *n*	126
concentración *f*	concentração *f*	концентрация *f* сгущение *n*	127
grado de concentración *m*	grau de concentração *m*	степень концентрации *f* степень сгущения *f*	128
cúmulo de alta concentración *m*	cúmulo [aglomerado] de grande concentração *m*	скопление с сильным сгущением к центру *n* сильно сконцентрированное скопление *n*	129
núcleo de un cúmulo *m*	núcleo de um cúmulo [aglomerado] *m*	ядро скопления *n*	130
densidad del fondo *f*	densidade de fundo (estelar) *f*	плотность (звёздного) фона *f*	131
densidad superficial aparente *f*	densidade superficial aparente *f*	видимая поверхностная плотность *f*	132
densidad espacial *f* densidad real *f*	densidade espacial *f* densidade real *f*	пространственная плотность *f* истинная плотность *f* объёмная плотность *f*	133
propiedades integrales *fpl* propiedades globales *fpl*	propriedades integrais *fpl* propriedades globais *fpl*	интегральные свойства *npl* суммарные свойства *npl*	134
brillo integrado *m* brillo integral *m* brillo total *m*	brilho global *m* brilho integral *m* brilho total *m*	интегральная яркость *f* суммарный блеск *m* общий блеск *m*	135

	STELLAR SYSTEMS	SYSTÈMES STELLAIRES	STERNSYSTEME
136	integrated magnitude total magnitude	magnitude totale *f* magnitude globale *f* magnitude intégrée *f*	Gesamtgrösse *f* Integralgrösse *f* integrale Magnitudo *f*
137	integrated spectrum composite spectrum	spectre global *m* spectre composite *m*	Gesamtspektrum *n* integrales Spektrum *n*
138	integrated colo(u)r index composite colo(u)r index	indice de couleur global *m* indice de couleur composite *m*	integraler Farb(en)index *m* Gesamtfarb(en)index *m*
139	total mass of the cluster	masse globale de l'amas *f* masse totale de l'amas *f*	Gesamtmasse des Haufens *f* Integralmasse des Haufens *f*
140	colo(u)r system	système de couleurs *m*	Farbsystem *n*
141	three-colo(u)r photometry	photométrie en trois couleurs *f*	Dreifarbenphotometrie *f*
142	effective wavelength	longueur d'onde effective *f*	effektive Wellenlänge *f*
143	multicolo(u)r photometry	photométrie à plusieurs couleurs *f*	Mehrfarbenphotometrie *f*
144	UBV system ultraviolet-blue-visual colo(u)r system ultraviolet-blue-yellow colo(u)r system	système U, B, V *m*	U, B, V-System *n*
145	blue-minus-yellow colo(u)r index B–V colo(u)r index	indice de couleur B–V *m* indice B–V *m*	Farb(en)index B–V *m*
146	ultraviolet-minus-blue colo(u)r index U–B colo(u)r index	indice de couleur U–B *m* indice U–B *m*	Farb(en)index U–B *m*
147	two-colo(u)r diagram	diagramme à deux couleurs *m*	Zweifarben-Diagramm *n*
148	six-colo(u)r photometry	photométrie en six couleurs *f*	Sechsfarben-Photometrie *f*
149	reddening line	raie de rougissement *f*	Verfärbungslinie *f* Rötungslinie *f*
150	colo(u)r-magnitude diagram	diagramme couleur-magnitude *m*	Farbenhelligkeitsdiagramm *n*
151	evolution diagram evolutionary diagram	diagramme d'évolution *m* diagramme évolutif *m*	Entwicklungsdiagramm *n*
152	zero-age sequence zero-age main sequence	séquence d'âge zéro *f* série principale d'âge zéro *f*	Alter-Null-Hauptreihe *f* Hauptreihe für das Alter Null *f*
153	cluster dynamics	dynamique des amas stellaires *f*	Dynamik von Sternhaufen *f*
154	relativistic star cluster	amas stellaire relativiste *m*	relativistischer Sternhaufen *m*

magnitud total *f* magnitud integral *f* magnitud integrada *f*	magnitude total *f* magnitude global *f* magnitude integral *f*	интегральная [общая] звёздная величина *f* суммарная звёздная величина *f*	136
espectro integral [compuesto] *m* espectro integrado *m*	espe(c)tro integral *m* espe(c)tro global [composto]	интегральный спектр *m* суммарный спектр *m*	137
índice de color integral *m* índice de color compuesto *m*	índice de cor global *m* índice de cor integral *m* índice de cor composto *m*	интегральный [общий] показатель цвета *m* суммарный показатель цвета *m*	138
masa total del cúmulo *f* masa integral del cúmulo *f*	massa total do cúmulo [aglomerado] *f* massa global [integral] do cúmulo [aglomerado] *f*	полная масса скопления *f* общая масса скопления *f* суммарная масса скопления *f*	139
sistema de colores *m*	sistema de cores *m*	система цветов *f*	140
fotometría en tres colores *f*	fotometria em três cores *f*	трёхцветная фотометрия *f*	141
longitud de onda efectiva *f*	comprimento de onda efetivo *m*	эффективная длина волны *f*	142
fotometría multicolor *f*	fotometria multicor [multicolor] *f*	многоцветная фотометрия *f*	143
sistema U, B, V *m*	sistema U, B, V *m*	система U, B, V *f*	144
índice de color B–V *m* índice de azul menos amarillo *m*	índice (de cor) B–V *m* índice de cor azul menos amarelo [visual] *m*	показатель цвета B–V *m* показатель цвета в системе B–V *m*	145
índice de color U–B *m* índice de ultravioleta menos azul *m*	índice (de cor) U–B *m* índice (de cor) ultravioleta menos azul *m*	показатель цвета U–B *m* ультрафиолетовый показатель цвета *m*	146
diagrama color-color *m* diagrama en dos colores *m*	diagrama bicolor *m* diagrama a duas cores *m*	двухцветная диаграмма *f*	147
fotometría en seis colores *f*	fotometria em seis cores *f*	шестицветная фотометрия *f*	148
línea de enrojecimiento *f*	linha de avermelhamento *f*	линия покраснения *f*	149
diagrama color-magnitud *m*	diagrama cor-magnitude *m*	диаграмма цвет-величина *f* диаграмма показатель цвета-– звёздная вечина *f*	150
diagrama de evolución *m*	diagrama de evolução *m* diagrama evolutivo *m*	эволюционная диаграмма *f* диаграмма развития *f*	151
secuencia de edad cero *f* secuencia principal de edad cero *f*	seqüência de idade zero *f*	последовательность нулевого возраста *f*	152
dinámica de cúmulos estelares *f*	dinâmica dos cúmulos [aglomerados] estelares *f*	динамика звёздных скоплений *f*	153
cúmulo de estrellas relativista *m*	cúmulo [aglomerado] de estrelas relativista [relativístico] *m*	релятивистское звёздное скопление *n*	154

155	escape velocity from a system	vitesse de libération d'un système *f*	Entweichgeschwindigkeit *f*
156	gravitationally bound [stable] system system with negative total energy Ep + Ek < 0	système lié par gravitation *m* système à énergie totale négative *m* système gravitationnellement lié *m*	System mit negativer Gesamtenergie *n* gebundenes System *n*
157	compact system	système compact *m*	kompaktes System *n*
158	gravitationally unbound [unstable] system system with positive total energy $E_p + E_k > 0$	système à énergie totale positive *m* système gravitationnellement libre [non-lié] *m*	System mit positiver Gesamtenergie *n* nichtgebundenes System *n*
159	escape of a star from the system	évasion d'une étoile du système *f*	Entweichen eines Sternes aus dem System *n*
160	escaper (US) ecaping star escapee runaway star	étoile d'évasion *f* étoile à vitesse de libération *f* étoile en voie d'échappement *f*	entweichender Stern *m* Fluchtstern *m* Ausreisser *m*
160a	gravitational potential	potentiel gravitationnel *m* potentiel de gravitation *m*	Gravitationspotential *n*
161	dynamical evolution	évolution dynamique *f*	dynamische Entwicklung *f*

velocidad de escape desde un sistema *f*	velocidade de escape de um sistema *f*	скорость убегания [ускользания] из системы *f*	155
sistema ligado por gravitación *m* sistema con energía total negativa *m* sistema gravitatoriamente ligado *m*	sistema ligado por gravitação *m* sistema com energia total negativa *m* sistema estável (gravitacionalmente) *m*	гравитационно связанная система *f* система с отрицательной полной энергией *f*	156
sistema compacto *m*	sistema compacto *m*	компактная система *f*	157
sistema con energía total positiva *m* sistema gravitatoriamente libre *m*	sistema com energia total positiva *m* sistema gravitacionalmente livre *m* sistema instável (gravitacionalmente) *m*	гравитационно несвязанная система *f* система с положительной полной энергией *f*	158
escape [fuga] de una estrella del sistema *m(f)*	escape de uma estrela do sistema *m* fuga de uma estrela do sistema *f*	освобождение звезды из системы *n* уход звезды из системы *m*	159
estrella fugada *f* estrella escapante *f* estrella en fuga *f*	estrela escapante *f* estrela de [em] escape *f* estrela em afastamento *f* estrela fugitiva *f* fugitiva *f*	вылетевшая (из системы) звезда *f* освобождающаяся звезда *f* ускользающая звезда *f* уходящая звезда *f*	160
potential gravitacional *m* potencial gravitatorio *m*	potencial gravitacional *m* potencial de gravitação *m*	гравитационный потенциал *m*	160a
evolución dinámica *f*	evolução dinâmica *f*	динамическая эволюция *f*	161

97 INTERSTELLAR MATTER | MATIÈRE INTERSTELLAIRE | INTERSTELLARE MATERIE

	INTERSTELLAR MATTER	MATIÈRE INTERSTELLAIRE	INTERSTELLARE MATERIE
1	interstellar matter	matière interstellaire *f*	interstellare Materie *f*
2	interstellar	interstellaire	interstellar
3	circumstellar	circumstellaire	zirkumstellar
4	interstellar space	espace interstellaire *m*	interstellarer Raum *m*
5	extragalactic space	espace extragalactique *m*	extragalaktischer Raum *m* aussergalaktischer Raum *m*
6	intergalactic	intergalactique	intergalaktisch
7	intergalactic space	espace intergalactique *m*	intergalaktischer Raum *m*
8	interstellar medium	milieu interstellaire *m*	interstellares Medium *n*
9	interstellar gas	gaz interstellaire *m*	interstellares Gas *n*
10	interstellar molecule	molécule interstellaire *f*	interstellares Molekül *n*
11	molecular cloud	nuage moléculaire *m*	Molekularwolke *f*
12	maser cloud masering cloud masing cloud	nuage à maser *m*	Maserwolke *f* Masermolekularwolke *f*
13	natural maser cosmic maser astrophysical maser celestial maser	maser natural *m* maser cosmique *m*	natürlicher Maser *m* natürlicher Maserverstärker *m* kosmicher Maser *m*
13a	giant molecular cloud GMC	nuage moléculaire géant *m*	Riesenmolekularwolke *f*
14	interstellar maser	maser interstellaire *m*	interstellarer Maser *m*
15	population inversion	inversion de population *f*	Besetzungsinversion *f*
16	pumping mechanism	mécanisme de pompage *m*	Pumpmechanismus *m*
17	radiative pumping optical pumping photon pumping	pompage optique *m* pompage par rayonnement *m* pompage photonique *m*	optisches Pumpen *n* Pumpen durch Strahlung *n*
18	infrared pumping	pompage (par rayonnement) infrarouge *m*	Pumpen durch infrarote Strahlung *n*
18a	ultraviolet pumping	pompage (par rayonnement) ultraviolet *m*	Pumpen durch ultraviolette Strahlung *n*
19	chemical pumping	pompage chimique *m*	chemisches Pumpen *n*
20	collisional pumping	pompage par collision *m* pompage par choc *m*	Stosspumpen *n*

materia interestelar *f*	matéria inter(e)stelar *f*	межзвёздная материя *f* межзвёздное вещество *n*	1
interestelar	inter(e)stelar	межзвёздный	2
circumestelar	circum-estelar circunstelar	околозвёздный	3
espacio interestelar *m*	espaço inter(e)stelar *m*	межзвёздное пространство *n*	4
espacio extragaláctico *m*	espaço extragaláctico *m*	внегалактическое простран-ство *n*	5
intergaláctico	intergaláctico	межгалактический	6
espacio intergaláctico *m*	espaço intergaláctico *m*	межгалактическое простран-ство *n*	7
medio interestelar *m*	meio inter(e)stelar *m*	межзвёздная среда *f*	8
gas interestelar *m*	gás inter(e)stelar *m*	межзвёздный газ *m*	9
molécula interestelar *f*	molécula inter(e)stelar *f*	межзвёздная молекула *f*	10
nube molecular *f*	nuvem molecular *f*	молекулярное облако *n*	11
nube de máser *f*	nuvem maser *f*	мазерное облако *n*	12
máser natural *m* máser cósmico *m*	maser natural *m* maser cósmico *m*	естественный мазер *m* космический мазер *m* астрофизический мазер *m*	13
nube molecular gigante *f*	nuvem molecular gigante *f*	гигантское молекулярное обла-ко *n*	13a
máser interestelar *m*	maser inter(e)stelar *m*	межзвёздный мазер *m*	14
inversión de poblaciones *f*	inversão de população *f*	инверсия населённостей (уро-вней *f*	15
mecanismo de bombeo *m*	mecanismo de bombeamento *m*	механизм накачки *m*	16
bombeo óptico *m* bombeo radiativo *m*	bombeamento radiativo *m* bombeamento ó(p)tico *m* bombeamento por radiação *m*	радиативная накачка *f* оптическая накачка *f* фотонная накачка *f*	17
bombeo infrarrojo *m*	bombeamento infravermelho *m*	накачка инфракрасным излуче-нием *f* инфракрасная накачка *f*	18
bombeo ultravioleta *m*	bombeamento ultravioleta *m*	накачка ультрафиолетовым из-лучением *f* ультрафиолетовая накачка *f*	18a
bombeo químico *m*	bombeamento químico *m*	химическая накачка *f*	19
bombeo colisional *m* bombeo por choque *m*	bombeamento colisional *m*	накачка столкновениями *f* столкновительная накачка *f*	20

	INTERSTELLAR MATTER	MATIÈRE INTERSTELLAIRE	INTERSTELLARE MATERIE
21	polyatomic molecule	molécule polyatomique *f*	vielatomiges Molekül *n*
22	hydroxyl OH	hydroxyle *m*	Hydroxyl *n* Hydroxylgruppe *f*
23	methylidyne CH	méthylidine *n*	Methylidin *n*
24	cyanogen C_2N_2	cyanogène *m*	Zyan *n* Dizyan *n*
25	ammonia NH_3	ammoniac *m*	Ammoniak *n*
26	hydrogen oxide water H_2O	oxyde d'hydrogène *m* eau *f*	Wasserstoffoxid *n* Wasser *n*
27	formaldehyde H_2CO	formaldédyde *m* aldéhyde formique *m* méthanal *m*	Formaldehyd *m* Ameisensäurealdehyd *m* Methanal *n*
28	carbon monoxide CO	oxyde de carbone *m*	Kohlenmonoxid *n*
29	hydrogen cyanide HCN	cyanure d'hydrogène *m*	Wasserstoffzyanid *n*
30	cyanoacetylene HC_3N	cyanoéthylène *m*	Zyanazetylen *n*
31	molecular hydrogen H_2	hydrogène moléculaire *m*	molekularer Wasserstoff *m*
32	methanol methyl alcohol wood spirit CH_3OH	alcool méthylique *m* méthanol *m*	Methylakohol *m* Methanol *n*
33	formic acid HCOOH	acide formique *m*	Ameisensäure *f*
34	carbon monosulfide CS	monosulfure de carbone *m*	Kohlenstoffmonosulfid *n*
35	formamide NH_2CHO	formamide *m* amide formique *m*	Formamid *n* Ameisensäureamid *n*
36	carbonyl sulfide OCS	sulfure de carbonyle *m*	Karbonyl-Sulfid *n*
37	silicon monoxide SiO	monoxyde de silicium *m*	Siliziummonoxid *n*
38	methyl cyanide CH_3CN	cyanure de méthyle *m* acétonitrile *m*	Methylzyanid *n* Azetonitril *n*
39	isocyanic acid HNCO	acide isocyanique *m*	Isozyansäure *f*

molécula poliatómica *f*	molécula poliatômica *f*	многоатомная молекула *f*	21
oxidrilo *m* hidroxilo *m*	oxidrila *f* hidroxila *f*	гидроксил *m* гидроксильная группа *f*	22
metilideno *m*	metilidino *m*	метин *m* метенил *m*	23
cianógeno *m*	cianogênio *m*	дициан *m*	24
amoníaco *m*	amoníaco *m*	аммиак *m*	25
óxido de hydrógeno *m* agua *f*	óxido de hidrogênio *m* água *f*	окись водорода *m* вода *m*	26
formaldehído *m* metanal *m* aldehído fórmico *m*	formaldeído *m* aldeído fórmico *m* metanal *m*	формальдегид *m* муравьиный альдегид *m*	27
monóxido de carbono *m*	monóxido de carbono *m*	окись углерода *m* угарный газ *m*	28
cianuro de hidrógeno *m* ácido cianhídrico *m*	ácido cianídrico *m*	цианид водорода *m* водородный цианид *m*	29
cianoacetileno *m*	cianoacetileno *m*	цианацетилен *m*	30
hidrógeno molecular *m*	hidrogênio molecular *m*	молекулярный водород *m*	31
alcohol metílico *m* metanol *m*	álcool metílico *m* metanol *m*	метанол *m* метиловый спирт *m*	32
ácido fórmico *m*	ácido fórmico *m*	муравьиная кислота *f*	33
monosulfuro de carbono *m*	monossulfeto de carbono *m*	односернистый углерод *m* моносульфид углерода *m*	34
formamida *f* metanamida *f*	formamida *f* metanamida *f*	формамид *m* амид муравьиной кислоты *m*	35
sulfuro de carbonilo *m*	sulfeto de carbonila *m*	карбонил-сульфид *m* карбонильный сульфид *m*	36
monóxido de silicio *m*	monóxido de silício *m*	одноокись кремния *f* моноокись кремния *f*	37
cianuro de metilo *m* acetonitrilo *m*	cianeto de metila *m* acetonitrilo *m*	метилцианид *m* цианистый метил *m* ацетонитрил *m*	38
acido isociánico *m*	ácido isociánico *m*	изоциановая кислота *f*	39

40	hydrogen isocyanide HNC	carbylamine d'hydrogène *f* isonitrile d'hydrogène *m*	Wasserstoffisozyanid *n* Wasserstoffkarbylamin *n*
41	methylacetylene CH_3CCH	acétylène de méthyle *m* éthyne méthylique *m* méthylacétylène *m*	Methylazetylen *n* Methyläthin *n*
42	acetaldehyde CH_3CHO	aldéhyde acétique *m* acétaldéhyde *m*	Azetaldehyd *m*
43	thioformaldehyde H_2CS	thio-formaldéhyde *m*	Thioformaldehyd *m*
44	hydrogen sulfide H_2S	sulfure d'hydrogène *m*	Schwefelwasserstoff *m*
45	methanimine CH_2NH	méthane-imine *f*	Methanimin *n*
46	interstellar hydrogen	hydrogène interstelaire *m*	interstellarer Wasserstoff *m*
47	neutral hydrogen	hydrogène neutre *m*	neutraler Wasserstoff *m*
48	ionization front	front d'ionisation *m*	Ionisationsfront *f*
49	H I region neutral hydrogen region	région d'hydrogène neutre *f* région H I *f*	Gebiet neutralen Wasserstroffs *n* H I-Gebiet *n* H I-Region *f*
50	H II region ionized hydrogen region	région H II *f* région d'hydrogène ionisé *f*	H II-Gebiet *n* Gebiet ionisierten Wasserstoffs *n* H II-Region *f*
51	compact H II region	région H II compacte *f*	kompaktes H II-Gebiet *n*
52	cosmic dust	poussière cosmique *f*	kosmischer Staub *m*
53	galactic dust	poussière galactique *f*	galaktischer Staub *m*
54	interstellar dust	poussière interstellaire *f*	interstellarer Staub *m*
55	circumstellar dust	poussière circumstellaire *f*	zirkumstellarer Staub *m*
56	diffuse galactic light	lumière galactique diffus(é)e *f*	diffuses galaktisches Licht *n*
57	smoke	fumée *f*	Rauch *m*
58	rarefaction wave	onde de raréfaction *f*	Verdünnungswelle *f*
59	interstellar dust grains interstellar grains	grains de poussière interstellaire *mpl* grains interstellaires *mpl*	interstellare Staubkörner *npl*

isocianuro de hidrógeno *m* isonitrilo de hidrógeno *m*	isocianeto de hidrogênio *m* isonitrilo de hidrogênio *m*	изонитрил водорода *m* карбиламин водорода *m*	40
metilacetileno *m* propino *m*	acetileno de metilo *m* metilacetileno *m*	метилацетилен *m* метилэтин *m*	41
acetaldehído *m* aldehído acético *m*	aldeído acético *m* acetaldeído *m*	уксусный альдегид *m* ацетальдегид *m*	42
tioformaldehído *m*	tioformaldeído *m*	тиоформальдегид *m*	43
ácido sulfhídrico *m* sulfuro de hidrógeno *m*	ácido sulfídrico *m* sulfeto de hidrogênio *m*	сернистый водород *m* сероводород *m*	44
metanimina *f*	metanimina *f*	метанимин *m* метиламин *m*	45
hidrógeno interestelar *m*	hidrogênio inter(e)stelar *m*	межзвёздный водород *m*	46
hidrógeno neutro *m*	hidrogênio neutro *m*	нейтральный [неиониз(ир)ованный] водород *m*	47
frente de ionización *m*	frente de ionização *f*	ионизационный фронт *m* фронт ионизации *m*	48
región de hidrógeno neutro *f* región H I *f*	região de hidrogênio neutro *f* região H I *f*	область H I *f* область нейтрального водорода *f*	49
región H II *f* región de hidrógeno ionizado *f*	região H II *f* região de hidrogênio ionizado *f*	область иониз(ир)ованного водорода *f* область H II *f*	50
región H II compacta *f*	região H II compacta *f*	компактная область H II *f*	51
polvo cósmico *m*	poeira cósmica *f* pó cósmico *m*	космическая пыль *f*	52
polvo galáctico *m*	poeira galáctica *f* pó galáctico *m*	галактическая пыль *f*	53
polvo interestelar *m*	pó [poeira] inter(e)stelar *m*[*f*]	межзвёздная пыль *f*	54
polvo circumestelar *m*	pó [poeira] circunstelar [circum-estelar] *m*[*f*]	околозвёздная пыль *f*	55
luz galáctica difusa *f*	luz galáctica difusa *f*	диффузный галактический свет *m*	56
humo *m*	fumaça *f* fumo *m*	дым *m*	57
onda de rarefacción *f*	onda de rarefação *f*	волна разрежения *f*	58
granos interestelares *mpl* granos de materia interestelar *mpl*	grãos inter(e)stelares *mpl* grãos de matéria inter(e)stelar *mpl* grãos de poeira inter(e)stelar *mpl*	межзвёздные пылинки *fpl* межзвёздные пылевые частицы *fpl*	59

	INTERSTELLAR MATTER	MATIÈRE INTERSTELLAIRE	INTERSTELLARE MATERIE
60	grain formation formation of grains	formation des grains *f*	Entstehung von Staubkörnern *f*
61	grain growth	croissance des grains *f*	Wachstum der Staubkörner *n*
62	star spots	taches stellaires *fpl*	Sternflecken *mpl*
63	primitive stellar nebula parent stellar nebula stellar nebula	nébuleuse stellaire primitive *f* nébuleuse génératrice *f* nébuleuse mère *f*	Sternnurnebel *m* Sternnebel *m* stellarer Urnebel *m*
64	refractory elements	éléments réfractaires *mpl*	feuerfeste Elemente *npl*
65	molecular aggregate	agrégat moléculaire *m*	Molekül-Aggregat *n*
66	condensation temperature	température de condensation *f*	Kondensationstemperatur *f*
67	nucleus of condensation	noyau de condensation *m*	Kondensationskern *m*
68	interstellar particles	particules interstellaires *fpl*	interstellare (Staub)partikeln *fpl*
69	solid particle	particule solide *f*	festes Teilchen *n* feste Partikel *f*
70	core-mantle particle	particule cœur-manteau *f* particule avec envenloppe *f*	Teilchen mit Mantel *n* Partikel mit Mantel *f*
71	graphite core	cœur de graphite *m*	Graphitkern *m*
72	ice mantle icy mantle	enveloppe de glace *f* enveloppe glacée *f*	Eismantel *m*
73	dirty ice	glace sale *f*	schmutziges Eis *n*
74	iron grain	grain de fer *m*	Eisenpartikel *f*
75	graphite grain	grain de graphite *m*	Graphitpartikel *f*
76	silicate grain	grain de silicate *m*	Silikatpartikel *f*
77	ice-coated grain	grain enveloppé de glace *m*	eisüberzogenes Teilchen *n*
77a	polymer grain	grain polymérique *m*	Polymerenteilchen *n*
78	alignment of grains	alignement de grains *m*	Ausrichtung von Teilchen *f*
79	destruction mechanism	mécanisme de destruction *m*	Zerstörungsmechanismus *m*
80	interstellar extinction	extinction interstellaire *f*	interstellare Extinktion *f* interstellare Lichtschwächung *f*
81	extinction curve	courbe d'extinction *f*	Extinktionskurve *f*
82	total extinction	extinction totale *f*	Gesamtextinktion *f*

formación de granos *f*	formação de grãos *f*	образование пылинок *n*	60
crecimiento de granos *m*	crescimento de grãos *m*	рост пылевых частиц *m*	61
manchas estelares *fpl*	manchas estelares *fpl*	звёздные пятна *npl*	62
nebulosa primitiva [generadora] *f* nebulosa preestelar *f* nebulosa primigenia *f*	nebulosa primitiva *f* nebulosa pré-estelar *f* nebulosa-mãe *f*	дозвёздная туманность *f* родительская туманность *f* материнская туманность *f*	63
elementos refractarios *mpl*	elementos refratários *mpl*	огнеупорные элементы *mpl*	64
agregado molecular *m*	agregado molecular *m*	молекулярный агрегат *m*	65
temperatura de condensación *f*	temperatura de condensação *f*	температура конденсации *f*	66
núcleo de condensación *m*	núcleo de condensação *m*	ядро конденсации *n*	67
partículas interestelares *fpl*	partículas inter(e)stelares *fpl*	межзвёздные частицы *fpl*	68
partícula sólida *f*	partícula sólida *f*	твёрдая частица *f*	69
partícula (del) núcleo-manto *f*	partícula (do) núcleo-manto *f*	частица, содержащая ядро [сердцевину] и оболочку *f* частица ядро-оболочка *f*	70
núcleo de grafito *m*	núcleo de grafite *m*	графитовое ядро *n* углеродная сердцевина *f*	71
manto de hielo *m*	manto de gelo *m*	ледяная оболочка *f*	72
hielo sucio *m*	gelo sujo *m*	грязный [загрязнённый] лёд *m*	73
grano de hierro *m*	grão de ferro *m*	железная пылинка *f*	74
grano de grafito *m*	grão de grafite *m*	графитовая пылинка *f*	75
grano de silicato *m*	grão de silicato *m*	силикатная пылинка *f*	76
grano recubierto de hielo *m*	grão recoberto de gelo *m*	частица с ледяной оболочкой *f*	77
grano polimérico *m*	grão de polímero *m*	полимерная частица *f*	77a
alineamiento de granos *m*	alinhamento de grãos *m*	(преимущественная) ориентация частиц [пылинок] *f* упорядоченность частиц *f*	78
mecanismo de destrucción *m*	mecanismo de destruição *m*	механизм разрушения [деструкции] *m*	79
extinción inter(e)stelar *f*	extinção inter(e)stelar *f*	межзвёздная экстинкция *f* межзвёздное ослабление света *n*	80
curva de extinción *f*	curva de extinção *f*	кривая экстинкции *f* кривая ослабления света *f*	81
extinción total *f*	extinção total *f*	полная экстинкция *f*	82

83	selective extinction	extinction sélective *f*	selektive Extinktion *f*
84	interstellar absorption	absorption interstellaire *f*	interstellare Absorption *f*
85	absorbing matter	matière absorbante *f*	absorbierende Materie *f*
86	interstellar obscuration	obscurcissement interstellaire *m*	interstellare Verdunklung *f*
87	general interstellar absorption	absorption interstellaire générale *f*	allgemeine interstellare Absorption *f*
87a	total [integral] interstellar absorption	absorption interstellaire totale *f*	totale interstellare Absorption *f*
88	zone of avoidance	zone d'absence *f*	nebelfreie Zone *f*
89	galactic absorbing layer	couche absorbante galactique *f*	galaktische Absorptionsschicht *f*
90	selective interstellar absorption	absorption interstellaire sélective *f* absorption différentielle *f*	selektive interstellare Absorption *f*
91	interstellar reddening space reddening	rougissement interstellaire *m*	interstellare Verfärbung *f* interstellare Rötung *f*
92	reddening law law of interstellar reddening	loi du rougissement *f* loi du rougissement interstellaire *f*	Verfärbungsgesetz *n*
93	reddening curve	courbe de rougissement *f*	Verfärbungskurve *f* Kurve interstellarer Verfärbung *f*
94	reddened star	étoile rougie *f*	rot verfärbter Stern *m*
95	unreddened star	étoile non-rougie *f*	nichtverfärbter Stern *m*
96	interstellar line	raie interstellaire *f*	interstellare Linie *f*
97	circumstellar line	raie (d'origine) circumstellaire *f*	zirkumstellare Linie *f*
98	interstellar radio line microwave interstellar line	raie radio(électrique) interstellaire *f* raie d'hyperfréquence interstellaire *f* raie micro-ondes interstellaire *f*	interstellare Mikrowellenlinie *f*
99	optical interstellar line	raie optique interstellaire *f*	optische interstellare Linie *f*
100	diffuse interstellar band diffuse interstellar feature	bande interstellaire diffuse *f*	diffuse interstellare Bande *f*
101	interstellar cloud	nuage interstellaire *m*	interstellare Wolke *f*
102	circumstellar cloud	nuage circumstellaire *m*	zirkumstellare Wolke *f*
103	H I cloud neutral hydrogen cloud	nuage d'hydrogène neutre *m* nuage H I *m*	H I Wolke *f* Wolke neutralen Wasserstoffes *f*

extinción selectiva *f*	extinção seletiva *f*	селективное ослабление *n* избирательная экстинкция *f*	83
absorción interestelar *f*	absorção inter(e)stelar *f*	межзвёздное поглощение *n*	84
materia absorbente *f*	matéria absorvente *f*	поглощающее вещество *n*	85
o(b)scurecimiento interestelar *m*	obscurecimento inter(e)stelar *m* escurecimento inter(e)stelar *m*	экранирование межзвёздным веществом *n*	86
absorción interestelar general *f*	absorção inter(e)stelar geral *f*	общее межзвёздное поглощение *n*	87
absorción interestelar total *f*	absorção inter(e)stelar total *f*	полное межзвёздное поглощение *n*	87a
zona de ausencia *f*	zona de ausência [exclusão] *f*	зона избегания *f*	88
estrato galáctico absorbente *m* capa galáctica absorbente *f*	camada galáctica absorvente *f*	галактический поглощающий слой *m*	89
absorción interestelar selectiva *f* absorción diferencial *f*	absorção inter(e)stelar seletiva *f* absorção diferencial *f*	избирательное межзвёздное поглощение *n* дифференциальное поглощение *n*	90
enrojecimiento interestelar *m*	avermelhamento inter(e)stelar *m*	межзвёздное покраснение *n*	91
ley de enrojecimiento *f*	lei de avermelhamento *f*	закон (межзвёздного) покраснения *m*	92
curva de enrojecimiento *f*	curva de avermelhamento *f*	кривая покраснения *f*	93
estrella enrojecida *f*	estrela avermelhada *f*	покрасневшая звезда *f*	94
estrella no enrojecida *f*	estrela não-avermelhada *f*	непокрасневшая звезда *f*	95
línea interestelar *f*	linha [raia] inter(e)stelar *f*	межзвёздная линия *f*	96
línea circumestelar *f*	linha [raia] circum-estelar [circunstelar] *f*	околозвёздная линия *f*	97
línea interestelar en radio *f* línea de microondas interstelar *f*	linha [raia] inter(e)stelar rádio [em microondas] *f*	микроволновая межзвёздная линия *f* межзвёздная радиолиния *f*	98
línea interestelar óptica *f*	linha [raia] inter(e)stelar ó(p)tica *f*	оптическая межзвёздная линия *f*	99
banda difusa interestelar *f*	banda difusa inter(e)stelar *f*	диффузная межзвёздная полоса *f*	100
nube interestelar *f*	nuvem inter(e)stelar *f*	межзвёздное облако *n*	101
nube circumestelar *f*	nuvem circum-estelar [circunstelar] *f*	околозвёздное облако *n*	102
nube H I *f* nube de hidrógeno neutro *f*	nuvem H I *f* nuvem de hidrogênio neutro *f*	облако нейтрального водорода *n* облако H I *n*	103

97	INTERSTELLAR MATTER	MATIÈRE INTERSTELLAIRE	INTERSTELLARE MATERIE
104	absorbing cloud	nuage absorbant *m*	Absorptionswolke *f*
105	interstellar polarization	polarisation interstellaire *f*	interstellare Polarisation *f*
106	interstellar magnetic field	champ magnétique interstellaire *m*	interstellares Magnetfeld *n*
107	dark cloud	nuage obscur *m*	Dunkelwolke *f*
108	dark matter	matière obscure *f*	dunkle Materie *f*
109	luminous matter	matière lumineuse *f*	leuchtende Materie *f*
109a	interstellar scintillation	scintillation interstellaire *f*	interstellare Szintillation *f*
109b	interstellar radio scintillation	scintillation des radiosources interstellaire *f*	interstellare Radioszintillation *f*
110	galactic nebula	nébuleuse galactique *f*	galaktischer Nebel *m*
111	Messier number	numéro de Messier *m*	Nummer im Messier-Katalog *f* Messier-Nummer *f*
112	gaseous nebula	nébuleuse gazeuse *f*	Gasnebel *m*
113	gas cloud	nuage gazeux *m*	Gaswolke *f*
114	dust cloud	nuage de poussière *m*	Staubwolke *f*
115	dust nebula	nébuleuse de poussière *f*	Staubnebel *m*
116	dust shell	enveloppe de poussière *f*	Staubhülle *f*
117	circumstellar dust cloud circumstellar dust shell	nuage de poussière circumstellaire *m* enveloppe de poussière circumstellaire *f*	zirkumstellare Staubhülle *f* zirkumstellare Staubwolke *f*
118	depolarization	dépolarisation *f*	Depolarisation *f*
119	gas-and-dust nebula	nébuleuse de gaz et de poussière *f*	Gas- und Staubnebel *m* gas-staubförmiger Nebel *m*
120	gas-to-dust ratio	rapport du gaz à la poussière *m*	Gas-Staub-Verhältnis *n*
121	bright nebula	nébuleuse lumineuse *f* nébuleuse brillante *f*	leuchtender Nebel *m* heller Nebel *m*
122	diffuse nebula	nébuleuse diffuse *f*	diffuser Nebel *m*
123	emission nebula	nébuleuse à émission *f*	Emissionsnebel *m*
124	isophotal contours (of a nebula)	(contours) isophotes (d'une nébuleuse) (*mpl*) *fpl*	Konturen gleicher Beleuchtungsstärke (eines Nebels) *fpl*

nube absorbente *f*	nuvem absorvente *f*	поглощающее облако *n*	104
polarización interestelar *f*	polarização inter(e)stelar *f*	межзвёздная поляризация *f*	105
campo magnético interestelar *m*	campo magnético inter(e)stelar *m*	межзвёздное магнитное поле *n*	106
nube oscura [obscura] *f*	nuvem escura [obscura] *f*	тёмное облако *n*	107
materia oscura [obscura] *f*	matéria escura [obscura] *f*	тёмное вещество *n*	108
materia luminosa *f*	matéria luminosa *f*	светящееся вещество *n*	109
centelleo interestelar *m*	cintilação inter(e)stelar *f*	межзвёздное мерцание *n*	109a
radiocentelleo interestelar *m* centelleo radio interestelar *m*	cintilação inter(e)stelar das radiofontes *f*	межзвёздное мерцание радиоисточников *n*	109b
nebulosa galáctica *f*	nebulosa galáctica *f*	галактическая туманность *f*	110
número (en el catálogo) de Messier *m*	número de Messier *m*	номер по каталогу Мессье *m*	111
nebulosa gaseosa *f*	nebulosa gasosa *f*	газовая туманность *f*	112
nube gaseosa *f*	nuvem de gás *f*	газовое облако *n*	113
nube de polvo *f*	nuvem de poeira *f* nuvem de pó *f*	пылевое облако *n*	114
nebulosa pulverulenta *f* nebulosa de polvo *f*	nebulosa de poeira *f* nebulosa de pó *f*	пылевая туманность *f*	115
envoltura de polvo *f* envoltura pulverulenta *f*	invólucro de poeira *m* invólucro de pó *m*	пылевая оболочка *f*	116
envoltura circunestelar pulverulenta *f* envoltura circunestelar de polvo *f*	invólucro circum-estelar de pó *m* nuvem de poeira circunstelar *f* invólucro circunstelar de pó *m*	околозвёздная пылевая оболочка *f* околозвёздное пылевое облако *n*	117
depolarización *f*	de(s)polarização *f*	деполяризация *f*	118
nebulosa de gas y polvo *f* nebulosa gaseoso-polvorienta *f*	nebulosa de gas e poeira *f* nebulosa de gás e pó *f*	газово-пылевая туманность *f*	119
relación [razón] del gas al polvo *f*	razão gás-poeira [entre gás e pó] *f*	отношение масс газа и пыли *n* отношение газа к пыли *n*	120
nebulosa luminosa *f* nebulosa brillante *f*	nebulosa luminosa *f* nebulosa brilhante *f*	яркая туманность *f* светлая туманность *f*	121
nebulosa difusa *f*	nebulosa difusa *f*	диффузная туманность *f*	122
nebulosa de emisión *f*	nebulosa de emissão *f*	эмиссионная туманность *f* туманность с эмиссионным спектром *f*	123
contornos isofotales (de una nebulosa) *mpl*	isótofas [contornos isófotos] (de uma nebulosa) *fpl* [*mpl*]	изофоты туманности *fpl*	124

125	exciting star	étoile excitatrice *f*	anregender Stern *m* leuchtanregender Stern *m*
126	accidental connection accidental association	connexion accidentelle *f* association accidentelle *f*	zufälliger Zusammenhang *m* zufällige Beziehung *f*
127	genetic connection genetic association	connexion génétique *f* association génétique *f*	genetischer Zusammenhang *m* genetische Beziehung *f*
128	reflection nebula reflecting nebula	nébuleuse à réflexion *f* nébuleuse réfléchissante *f*	Reflexionsnebel *m*
129	filamentary nebula	nébuleuse à filaments *f* nébuleuse filamenteuse *f*	Fasernebel *m* filamentartiger Nebel *m*
130	filamentary structure	structure filamenteuse *f*	Filamentstruktur *f* faserige Struktur *f* Faserstruktur *f*
131	filament of a nebula	filament d'une nébuleuse *m*	Faser eines (Gas)nebels *f*
132	dark nebula	nébuleuse obscure *f*	Dunkelnebel *m*
133	globule	globule *m*	Globule *f*
134	Crab Nebula	Nébuleuse du Crabe *f*	Krebsnebel *m* Krabbennebel *m* Crabnebel *m*
135	Horsehead Nebula Dark Bay	Tête de Cheval *f*	Pferdekopfnebel *m*
136	Orion Nebula	(Grande) Nébuleuse d'Orion *f*	(Grosser) Orionnebel *m*
137	Loop Nebula Network Nebula Veil Nebula Looped Nebula Cygnus Loop	Nébuleuse de la Dentelle *f* La Dentelle du Cygne *f*	Netzwerknebel *m* Schleiernebel *m*
138	Coal Sack	Sac à Charbon *m*	Kohlensack *m*
139	Owl Nebula	Nébuleuse du Hibou *f*	Eulennebel *m*
140	North America Nebula	Nébuleuse America *f* Nébuleuse Amérique du Nord *f*	Nordamerikanebel *m*
141	Trifid Nebula	Nébuleuse Trifide *f*	Trifidnebel *m*
142	cometary nebula comet-like nebula	nébuleuse cométaire *f* nébuleuse cométoïde *f*	kometenähnlicher Nebel *m*
143	Horseshoe Nebula Swan Nebula Omega Nebula	Nébuleuse du Fer à Cheval *f* Nébuleuse Oméga *f*	Hufeisennebel *m* Omeganebel *m*
144	planetary nebula	nébuleuse planétaire *f*	planetarischer Nebel *m*
145	angular diameter	diamètre angulaire *m*	Winkeldurchmesser *m*

estrella excitante *f* estrella excitadora *f*	estrela excitadora *f* estrela excitante *f*	возбуждающая звезда *f*	125
relación accidental *f* asociación accidental *f*	conexão acidental *f* associação [relação] acidental *f*	случайная связь *f* случайная ассоциация *f*	126
relación genética *f* asociación genética *f*	conexão genética *f* relação [associação] genética *f*	генетическая связь *f*	127
nebulosa de reflexión *f* nebulosa reflectante *f*	nebulosa refletora *f* nebulosa de reflexão *f*	отражательная туманность *f* отражающая туманность *f*	128
nebulosa filamentosa *f* nebulosa con filamentos *f*	nebulosa filamentosa [filamentar] *f* nebulosa de filamentos *f*	волокнистая туманность *f*	129
estructura filamentosa *f*	estrutura filamentosa [filamentar] *f*	волокнистая структура *f*	130
filamento de una nebulosa *m*	filamento de uma nebulosa *m*	волокно туманности *n*	131
nebulosa obscura [oscura] *f*	nebulosa escura [obscura] *f*	тёмная туманность *f*	132
glóbulo *m*	glóbulo *m*	глобула *f*	133
Nebulosa del Cangrejo *f*	Nebulosa do Caranguejo *f*	Крабовидная туманность *f* Ракообразная туманность *f* Краб *m*	134
Cabeza de Caballo *f* Bahía O(b)scura *f*	Cabeça de Cavalo *f*	Конская голова *f*	135
Nebulosa de Orión *f*	Nebulosa de Orion *f*	Большая туманность Ориона *f*	136
Nebulosa del Encaje *f* Nebulosa del Velo *f*	Nebulosa de Renda *f* Nebulosa do Laço *f* Nebulosa do Véu *f* Véu da Noiva *m*	Волокнистая туманность *f* Туманность Рыбачья Сеть *f* Петля в Лебеде *f*	137
Saco [Bolsa] de Carbón *m*	Saco de Carvão *m*	Угольный мешок *m*	138
Nebulosa del Búho *f*	Nebulosa da Coruja [do Mocho] *f*	Туманность Сова *f*	139
Nebulosa Norteamérica *f*	Nebulosa América do Norte *f*	Туманность Северная Америка *f*	140
Nebulosa Trífida *f*	Nebulosa Trífida *f*	Тройная туманность *f* Трёхраздельная туманность *f*	141
nebulosa cometaria *f* nebulosa cometoide *f*	nebulosa cometária *f* nebulosa cometóide *f*	кометообразная туманность *f* кометарная туманность *f*	142
Nebulosa Omega *f* Nebulosa Herradura *f*	Nebulosa Ômega *f* Nebulosa Ferradura *f*	Туманность Омега *f* Туманность Подкова *f*	143
nebulosa planetaria *f*	nebulosa planetária *f*	планетарная туманность *f*	144
diámetro angular *m*	diâmetro angular *m*	угловой диаметр *m*	145

	INTERSTELLAR MATTER	MATIÈRE INTERSTELLAIRE	INTERSTELLARE MATERIE
146	ring ...	annulaire	ringförmig Ring ...
147	ring nebula	nébuleuse annulaire *f*	Ringnebel *m*
148	central star nucleus of the planetary nebula	étoile centrale *f* noyau de la nébuleuse planétaire *m*	Zentralstern *m* Kern des planetarischen Nebels *m*
149	internal motions	mouvements internes *mpl*	innere Bewegungen *fpl*
150	interstellar gas dynamics	dynamique des gaz interstellaires *f*	interstellare Gasdynamik *f*
151	interstellar wind	vent interstellaire *m*	interstellarer Wind *m*
152	relativistic gas	gaz relativiste *m*	relativistisches Gas *n*
153	relativistic electrons	électrons relativistes *mpl*	relativistische Elektronen *npl*
154	stratification	stratification *f*	Schichtung *f*
155	two-photon emission	émission (simultanée) de deux photons *f*	Zwei-Photonen-Ausstrahlung *f* Zwei-Photonen-Emission *f*
156	theory of cycles	théorie des cycles *f*	Zyklentheorie *f*
157	splitting of lines by expansion doubling of lines	dédoublement des raies par expansion *m*	Aufspaltung der Linien durch Expansion *f*
158	monochromatic images	images monochromatiques *fpl*	monochromatische Bilder *npl*

anular de anillo	anular anelar	кольцеобразный кольцевой	146
nebulosa anular *f*	nebulosa anular *f*	кольцевая [кольцеобразная] туманность *f*	147
estrella central *f* núcleo de la nebulosa planetaria *m*	estrela central *f* núcleo da nebulosa planetária *m*	центральная звезда *f* ядро планетарной туманности *n*	148
movimientos internos *mpl*	movimentos internos *mpl*	внутрение движения *npl*	149
dinámica del gas interestelar *f*	dinâmica do gás inter(e)stelar *f*	межзвёздная газодинамика *f*	150
viento interestelar *m*	vento inter(e)stelar *m*	межзвёздный ветер *m*	151
gas relativista *m*	gás relativístico [relativista] *m*	релятивистский газ *m*	152
electrones relativistas *mpl*	elé(c)trons relativísticos [relativistas] *mpl*	релятивистские электроны *mpl*	153
estratificación *f*	estratificação *f*	стратификация *f*	154
emisión de dos fotones *f*	emissão (simultânea) de dois fótons *f*	двухфотонное излучение *n*	155
teoría de los ciclos *f*	teoria dos ciclos *f*	теория циклов *f*	156
desdoblamiento de las líneas por expansión *m*	desdobramento das linhas [raias] por expansão *m*	расщепление линий вследствие расширения *n*	157
imágenes monocromáticas *fpl*	imagens monocromáticas *fpl*	монохроматические изображения *npl*	158

	GALAXIES	GALAXIES	GALAXIEN
1	Galaxy Milky Way Galaxy Milky Way	Galaxie *f* Voie Lactée *f* notre Galaxie *f*	Galaxis *f* Milchstrasse *f* Milchstrassensystem *n*
2	galactic	galactique	galaktisch
3	galactic system	système galactique *m*	galaktisches System *n*
4	structure of the Galaxy structure of the Milky Way galactic structure	structure de la Galaxie *f* structure de la Voie Lactée *f*	Aufbau des Milchstrassensystems *m* Struktur der Galaxis *f*
5	bifurcation	bifurcation *f*	Gabelung *f* Abzweigung *f*
5a	ramification	branchement *m*	Verzweigung *f*
6	star cloud	nuage d'étoiles *m*	Sternwolke *f*
7	spiral	spir(al)e *f*	Spirale *f*
8	spiral structure	structure spirale [spiralée, en spirale] *f*	Spiralstruktur *f*
9	centre [center] of the Galaxy galactic centre [center]	centre galactique *m* centre du système galactique *m*	Zentrum des Milchstrassensystems *n* galaktisches Zentrum *n*
10	galactocentric	galactocentrique	galaktozentrisch
11	galactocentric distance distance from the galactic center [centre]	distance au centre de la Galaxie *f* distance galactocentrique *f*	galaktozentrische Entfernung *f* Entfernung zum Zentrum der Milchstrasse *f*
12	galactocentric coordinates	coordonnées galactocentriques *fpl*	galaktozentrische Koordinaten *fpl*
13	region of the galactic centre [center] central region of the Galaxy	région du centre galactique *f* région centrale de la Galaxie *f*	Gebiet des galaktischen Zentrums *n* Zentralgebiet der Galaxis *n*
14	galactic anticentre (UK) galactic anticenter (US) towards the anticenter	anticentre galactique *m* en direction de l'anticentre galactique	galaktisches Antizentrum *n* in der Richtung des galaktischen Antizentrums
15	(galactic) anticentre [anticenter] region	région de l'anticentre galactique *f*	Gebiet des galaktischen Antizentrums *n*
16	central bulge nuclear bulge	bulbe central *m* bulbe nucléaire *m* renflement galactique *m*	zentraler Wulst *m* Kernverdickung *f* Zentralkörper *m*
17	nuclear region	région nucléaire *f* région du noyau (galactique) *f*	Kernbereich *m* Kerngebiet *n*
18	nucleus of the Galaxy galactic nucleus	noyau de la Galaxie *m* noyau (central) galactique *m*	Kern der Galaxis *m* galaktischer Kern *m*
19	starlike nucleus	noyau quasi stellaire *m*	sternförmiger Kern *m*

GALAXIAS	GALÁXIAS	ГАЛАКТИКИ	
Galaxia *f* Vía Láctea *f* Camino de Santiago *m*	Galáxia *f* Via Láctea [Via-lá(c)tea] *f* Caminho [Estrada, Carreira] de São Tiago *m*[*f*]	Галактика *f* наша галактика *f* Млечный Путь *m*	1
galáctico	galáctico	галактический	2
sistema galáctico *m*	sistema galáctico *m*	галактическая система *f*	3
estructura de la Vía Láctea *f* estructura galáctica *f* estructura de la Galáxia *f*	estrutura da Via Láctea *f* estrutura galáctica *f* estructura da Galáxia *f*	строение Галактики *n* структура Млечного Пути *f*	4
bifurcación *f*	bifurcação *f*	раздвоение *n* бифуркация *f*	5
ramificación *f*	ramificação *f*	разветвление *n*	5a
nube estelar *f*	nuvem estelar [de estrelas] *f*	звёздное облако *n*	6
espiral *f*	espiral *f*	спираль *f*	7
estructura espiral *f*	estrutura espiral *f*	спиральная структура *f*	8
centro galáctico *m* centro de la Galaxia *m*	centro galáctico [da Via Láctea] *m* centro do sistema galáctico *m*	центр Галактики *m* галактический центр *m*	9
galactocéntrico	galactocêntrico	галактоцентрический	10
distancia galactocéntrica *f* distancia al centro galáctico *f*	distância galactocêntrica *f* distância ao centro da Galáxia *f*	галактоцентрическое расстояние *n* расстояние до центра Галактики *n*	11
coordenadas galactocéntricas *fpl*	coordenadas galactocêntricas *fpl*	галактоцентрические координаты *fpl*	12
región del centro galáctico *f* región central de la Galaxia *f*	região do centro galáctico *f* região central da Galáxia *f*	центральная область Галактики *f* область галактического центра *f*	13
anticentro galáctico *m* en dirección al anticentro	anticentro galáctico *m* em direção ao anticentro	галактический антицентр *m* в направлении на антицентр *m*	14
región del anticencentro (galáctico) *f*	região do anticentro (galáctico) *f*	область (галактического) антицентра *f*	15
núcleo central *m* abultamiento central *m* bulbo central *m*	bojo galáctico *m* protuberância central *f*	центральное утолщение *n* балдж *m*	16
región nuclear *f* región del núcleo *f*	região do núcleo *f* região nuclear *f*	ядерная область *f* область ядра *f*	17
núcleo de la Galaxia *m* núcleo galáctico *m*	núcleo da Galáxia *m* núcleo galáctico *m*	ядро Галактики *n* галактическое ядро *n*	18
núcleo cuasiestelar *m* núcleo esteliforme *m*	núcleo quase-estelar [esteliforme] *m*	звёздоподобное ядро *n*	19

20	peculiar nucleus	noyau particulier *m*	pekuliarer Kern *m*
21	compact object	objet compact *m*	kompaktes Objekt *n*
22	supermassive body supermassive object	corps supermassif *m* objet supermassif *m*	supermassiver Körper *m* supermassives Objekt *n*
22a	supermassive black hole	trou noir supermassif *m*	supermassives schwarzes Loch *n*
23	collapsed object	objet effondré *m*	kollabiertes Objekt *n*
23a	galatic black hole	trou noir galactique *m*	galaktisches schwarzes Loch *n*
24	white hole	trou blanc *m*	weisses Loch *n*
25	magnetoid rotating magnetoplasm(at)ic body spinar	magnétoïde *m* objet de plasma magnétisé en rotation *m*	Magnetoid *n* rotierender magnetoplasmatischer Körper *m* Spinar *m*
26	non-elastic star collision	collision inélastique des étoiles *f*	unelastischer Stoss zwischen Sternen *m*
26a	galactic arm	bras galactique *m*	galaktischer Arm *m*
27	spiral arm	bras spiral *m*	Spiralarm *m*
28	trailing	enroulement à la traîne *m*	nachziehende [nachschleppende, feuerradähnliche] Rotation *f* Rotation mit konvexer Seite voraus *f*
29	trailing arm	bras à la traîne *m* bras enroulé *m*	nachgezogener [nachgeschleppter, nachlaufender] Arm *m*
30	outward winding of spiral arms	déroulement des bras spiraux (dans le sens de la rotation) *m*	Öffnung der Spiralwindungen nach aussen *f* Öffnung der Spiralwindungen in Rotationsrichtung *f*
31	unwinding arm leading arm	bras déroulé *m*	vorlaufender Arm *m* sich öffnender Arm *m*
32	angular length	longueur angulaire *f*	Windungslänge *f*
33	characteristic angle	angle caractéristique *m*	charakteristischer Winkel *m*
34	Perseus arm	bras de Persée *m*	Perseus-Arm *m*
35	Sagittarius arm	bras du Sagittaire *m*	Sagittarius-Arm *m*

núcleo peculiar *m*	núcleo peculiar *m*	пекулярное ядро *n*	20
objeto compacto *m*	objeto compacto *m*	компактный объект *m*	21
cuerpo supermasivo *m* objeto supermasivo *m*	corpo supermassivo *m* objeto supermassivo *m*	сверхмассивное тело *n* сверхмассивный объект *m*	22
agujero negro supermasivo *m*	buraco negro supermassivo *m*	сверхмассивная чёрная дыра *f*	22a
objeto colapsado *m*	objeto colapsado *m*	сколлапсированный объект *m* сколлапсировавший объект *m*	23
agujero negro galáctico *m*	buraco negro galáctico *m*	галактическая чёрная дыра *f*	23a
agujero blanco *m*	buraco branco *m*	белая дыра *f*	24
magnetoide *m* cuerpo magnetoplasmático rotatorio *m*	magnetóide *m* corpo rotatório [girante] de magnetoplasma *m*	магнитоид *m* вращающееся магнитоплазменное тело *n*	25
colisión inelástica de estrellas *f*	colisão inelástica de estrelas *f*	неупругое столкновение звёзд *f*	26
brazo galáctico *m*	braço galáctico *m*	галактический рукав *m* галактическая ветвь *f*	26a
brazo espiral *m*	braço espiral *m*	спиральный рукав *m* спиральная ветвь *f*	27
arrolamiento [enrollamiento] (de los brazos espirales) *m*	enrolamento [enroscamento] (dos braços espirais) *m*	закручивание ветвей *n* скручивание галактических рукавов *n*	28
brazo enrollado *m*	braço enrolado *m* braço enroscado *m*	закручивающаяся [раскрученная] галактическая ветвь *f* закручивающийся [закручиванный, закручиваемый, закрученный] галактический рукав *m*	29
desenrollamiento (de los brazos espirales) *m*	desenrolamento [desenroscamento] dos braços espirais) *m*	развёртывание ветвей *n* раскручивание [раскручение] рукавов *n*	30
brazo desenrollado *m*	braço desenrolado *m* braço desenroscado *m*	раскручивающаяся [закручиваемая, лидирующая] ветвь *f* раскручиваемый [раскручивающийся, раскручиванный, раскрученный] рукав *m*	31
longitud angular *f*	comprimento angular *m*	длина в угловой мере *f*	32
ángulo característico *m*	ângulo característico *m*	характеристический угол *m*	33
brazo de Perseo *m*	braço de Perseu *m*	ветвь Персея *f* Персеев рукав *m*	34
brazo de Sagitario *m*	braço de Sagitário *m*	рукав [ветвь] Стрельца *m*[*f*]	35

36	Orion arm	bras d'Orion *m*	Orion-Arm *m*
37	3-kpc arm three-kiloparsec arm	bras de 3 kpc *m* bras de trois kiloparsec *m*	Drei-Kiloparsec-Arm *m*
38	expanding arm	bras expansif [en expansion] *m*	expandierender Arm *m*
39	spiral tracer spiral-arm tracer	traceur [indicateur] de la structure spirale *m*	Spiralarm-Indikator *m*
40	interarm (space, region, gas ...) between arms	(espace, région, gaz ...) entre deux bras spiraux	Zwischen-Arm- ... (Raum, Gebiet, Gas ...) zwischen Armen
41	ring pattern ring structure	structure annulaire *f* forme en anneau *f*	Ringform *f* Ringstruktur *f*
42	one-arm(ed) galaxy unilateral galaxy	galaxie à un (seul) bras *f*	einarmige Galaxie *f* Einarmgalaxie *f*
43	two-armed galaxy two-arm system	galaxie à deux bras *f*	zweiarmige Galaxie *f* Zweiarmgalaxie *f*
44	multiple-arm(ed) galaxy multiple-arm system multiply-armed galaxy	galaxie à plusieurs bras *f* galaxie à bras multiples *f*	mehrarmige Galaxie *f* Mehrarmgalaxie *f*
45	galactic halo	halo galactique *m*	galaktischer Halo *m*
46	hot halo	halo chaud *m*	heisser Halo *m*
47	cool halo	halo froid *m*	kühler Halo *m*
48	magnetic halo	halo magnétique *m*	magnetischer Halo *m*
49	galactic corona	couronne galactique *f*	galaktische Korona *f*
50	galactic magnetic field	champ magnétique galactique *m*	galaktisches Magnetfeld *n*
51	magnetic confinement of cosmic rays containment of cosmic radiation	confinement magnétique des rayons cosmiques *m*	magnetische Einschliessung der kosmischen Strahlung *f*
52	galactic disc [disk]	disque galactique *m*	galaktische Scheibe *f*
53	galactic plane	plan galactique *m* plan principal de Galaxie *m*	galaktische Ebene [Hauptebene] *f* Milchstrassenebene *f*
54	galactic distribution	distribution galactique *f*	galaktische Verteilung *f*
55	concentration toward the galactic plane	concentration vers le plan galactique *f*	Konzentration zur galaktischen Ebene *f*
56	zone of avoidance	zone d'absence *f*	nebelfreie Zone *f*

Español	Português	Русский	№
brazo de Orión *m*	braço de Orion *m*	Орионов рукав *m* рукав [ветвь] ассоциации Ориона *m*[*f*]	36
brazo de 3 kpc *m* brazo de tres kiloparsec *m*	braço de 3 kpc *m* braço de três kiloparsec *m*	3 кпс рукав *m* рукав на растоянии 3 кпс от центра Галактики *m*	37
brazo en expansión *m*	braço em expansão *m*	расширяющийся рукав *m*	38
trazador de estructura espiral *m* indicador de estructura espiral *m*	traçador da estrutura espiral *m* indicador da estrutura espiral *m*	индикатор спиральной структуры *m*	39
(espacio, región, gas ...) entre los brazos	(espaço, região, gás ...) entre os braços	(пространство, область, газ ...) между спиральными рукавами межрукавный	40
forma anular *f* estructura anular *f*	estrutura anelar [anular] *f* forma em [de] anel *f*	кольцевая форма *f* кольцевая структура *f*	41
galaxia con brazo único *f* galaxia de un solo brazo *f*	galáxia de um (só) braço *f* galáxia com um braço *f*	однорукавная галактика *f*	42
galaxia de dos brazos *f*	galáxia de dois braços *f*	двухрукавная галактика *f*	43
galaxia de varios brazos *f* galaxia de brazos múltiples *f*	galáxia de vários braços *f* galáxia de múltiplos braços *f*	многорукавная галактика *f*	44
halo galáctico *m*	halo galáctico *m*	галактическое гало *n*	45
halo caliente *m*	halo quente *m*	горячее гало *n*	46
halo frío *m*	halo frio *m*	холодное гало *n*	47
halo magnético *m*	halo magnético *m*	магнитное гало *n*	48
corona galáctica *f*	coroa galáctica *f*	галактическая корона *f*	49
campo magnético galáctico *m*	campo magnético galáctico *m*	галактическое магнитное поле *n*	50
confinamiento magnético de los rayos cósmicos *m*	confinamento magnético dos raios cósmicos *m*	магнитное удержание космического излучения *n* удержание космических лучей магнитным полем *n*	51
disco galáctico *m*	disco galáctico *m*	галактический диск *m*	52
plano galáctico *m* plano principal de la Galaxia *m*	plano galáctico *m* plano principal da Galáxia *m*	основная плоскость Галактики *f* галактическая плоскость *f*	53
distribución galáctica *f*	distribuição galáctica *f*	галактическое распределение *n*	54
concentración hacia el plano galáctico *f*	concentração em direção ao [para o] plano galáctico *f*	концентрация к плоскости галактики *f*	55
zona de ausencia *f*	zona de ausência [exclusão] *f*	зона избегания *f*	56

57	galactic windows windows in the zone of avoidance	fenêtres galactiques *fpl* fenêtres de la zone d' absence *fpl*	galaktische Fenster *npl* Fenster in der nebelfreien Zone *npl*
58	solar neighbo(u)rhood	voisinage du Soleil *m*	Sonnenumgebung *f* Umgebung der Sonne *f*
59	in the neighbo(u)rhood of the Sun	au voisinage du Soleil	in der Sonnenumgebung
60	nearby star	étoile proche *f*	naher Stern *m*
61	local system	système local *m*	lokales (Stern)system *n*
62	solar motion	mouvement solaire *m*	Sonnenbewegung *f*
63	standard solar motion	mouvement solaire standard *m*	Standard-Sonnenbewegung *f*
64	solar apex apex of the solar motion	apex solaire *m* apex du Soleil *m*	Sonnenapex *m* Apex der Sonne *m*
65	solar antapex	antiapex solaire *m*	Antapex der Sonne *m*
66	basic solar motion	mouvement solaire fondamental *m*	fundamentale Sonnenbewegung *f* Sonnenbewegung um galaktisches Zentrum *f*
67	stellar [star] population	population stellaire *f*	Sternpopulation *f*
68	population II type II population	population II *f* population de type II *f*	Population II *f* Typ-II-Sternpopulation *f*
69	halo population halo-type population	population du type halo *f* population du halo *f*	Halopopulation *f*
70	intermediate population	population intermédiaire *f*	intermediäre Population *f* Zwischenpopulation *f*
71	population I type I population	population I *f* population de type I *f*	Population I *f* Typ-I-Sternpopulation *f*
72	disk population disk-type population	population du disque *f* population (de) disque *f*	Scheibenpopulation *f* Population der galaktischen Scheibe *f*
73	extreme population I spiral-arm population	population I extrême *f* population des bras spiraux *f*	extreme Population I *f* Spriralarm-Population *f*
74	subsystem	sous-système *m*	Untersystem *n*
75	system of the globular clusters	système des amas globulaires *m*	System der Kugelhaufen *n*
76	galactic concentration (toward the center, toward the principal plane)	concentration galactique *f* (vers le centre, vers le plan fondamental)	galaktische Konzentration *f* (zum Zentrum, zur Hauptebene)

ventanas galácticas *fpl* ventanas en la zona de ausencia *fpl*	janelas galácticas *fpl* janelas na zona de ausência *fpl*	окна видимости *npl* галактические коридоры видимости *mpl*	57
vecindad solar *m* proximidades del Sol *fpl*	vizinhança solar [do Sol] *f* proximidades do Sol *fpl*	окрестности Солнца *fpl*	58
en la vecindad solar	na vizinhança do Sol	в окрестностях Солнца	59
estrella próxima [vecina] *f*	estrela próxima [vizinha] *f*	близкая звезда *f*	60
sistema local *m*	sitema local *m*	местная система *f*	61
movimiento solar *m*	movimento solar *m*	движение Солнца *n* солнечное движение *n*	62
movimiento solar estándar *m*	movimento solar padrão *m*	стандартное солнечное движение *n*	63
ápex solar *m* ápex del Sol *m*	ápex solar *m* ápex do Sol *m*	солнечный апекс *m* апекс движения Солнца *m*	64
antiápex solar *m*	antiápex solar *m*	солнечный антиапекс *m*	65
movimiento solar básico *m*	movimento solar básico *m*	основное солнечное движение *n* движение Солнца вокруг центра Галактики *n*	66
población estelar *f*	população estelar *f*	звёздное население *n*	67
población II *f* población del tipo II *f*	populacão II *f* população de tipo II *f*	население типа II *n* население второго типа *n*	68
población del halo *f*	população do halo *f*	сферическая составляющая *f* население гало *n*	69
población intermedia *f*	população intermediária *f*	промежуточная составляющая *f* промежуточное население *n*	70
población I *f* población del tipo I *f*	população I *f* população de tipo I *f*	плоская составляющая *f* население первого типа [типа I] *n*	71
población del disco *f*	população do disco *f*	население галактического диска *n*	72
población I extrema *f* población de los brazos espirales *f*	população I extrema *f* população dos braços espirais *f*	экстремальное население первого типа *n* население спиральных рукавов *n*	73
subsistema *m*	subsistema *m*	подсистема *f* субсистема *f*	74
sistema dc cúmulos globulares *m*	sistema dos cúmulos globulares *m*	система шаровых скоплений *f*	75
concentración galáctica *f* (hacia el centro, hacia el plano principal)	concentração galáctica *f* (em direção ao centro, em direção ao plano fundamental)	галактическая концентрация *f* (к центру, к основной плоскости) *f*	76

	English	French	German
77	stellar statistics	statistique stellaire *f*	Stellarstatistik *f*
78	selected areas areas of Kapteyn Kapteyn areas	régions sélectionnées *fpl* aires choisies *fpl* aires de Kapteyn *fpl*	selected Areas *npl* Kapteynsche Eichfelder *npl* Kapteynsche Felder *npl*
79	star counts	dénombrements d'étoiles *mpl*	Sternzählung *f*
80	star number	nombre des étoiles *m*	Sternzahl *f*
81	selection effect	effet de sélection *m*	Auswahleffekt *m*
82	star density	densité stellaire *f*	Sterndichte *f*
83	stellar space density	densité spatiale des étoiles *f*	räumliche Sterndichte *f*
84	cubic parsec (per cubic parsec)	parsec-cube *m* (par parsec-cube)	Kubikparsec *n* (pro Kubikparsec)
85	projected density surface density	densité projetée *f* densité de la distribution projetée (sur plan galactique) *f*	projizierte Dichte *f* flächenhafte Dichte *f*
86	space distribution spatial distribution	distribution dans l'espace *f* distribution spatiale *f*	räumliche Verteilung *f* Raumverteilung *f* Verteilung im Raum *f*
87	density gradient	gradient de densité *m*	Dichtegradient *m*
88	density function	fonction de densité *f*	Dichtefunktion *f*
89	luminosity function	fonction de luminosité *f*	Leuchtkraftfunktion *f*
90	kinematic properties	propriétés cinématiques *fpl*	kinematische Eigenschaften *fpl*
91	residual velocity	vitesse résiduelle *f*	Restgeschwindigkeit *f*
92	distribution of the residual velocities	distribution des vitesses résiduelles *f*	Verteilung der Geschwindigkeitskomponenten *f* Verteilung von Restgeschwindigkeiten *f*
93	randomly distributed	distribué au hasard distribué d'une manière aléatoire	zufällig verteilt
94	random motion	mouvement aléatoire *m* mouvement au hasard *m*	Zufallsbewegung *f* (stochastische) Irrfahrt *f* stochastische Bewegung *f*
95	ordered motion	mouvement ordonné *m*	geordnete Bewegung *f*
96	Maxwellian distribution of velocities	distribution maxwellienne des vitesses *f*	Maxwellsche Geschwindigkeitsverteilung *f*
97	velocity space	espace des vitesses *m*	Geschwindigkeitsraum *m*
98	star stream	courant d'étoiles *m* courant stellaire *m*	Sternstrom *m*

estadística estelar *f*	estatística estelar *f*	звёздная статистика *f*	77
áreas seleccionadas *fpl* áreas de Kapteyn *fpl*	áreas selecionadas *fpl* áreas escolhidas *fpl* áreas de Kapteyn *fpl*	избранные площадки *fpl* избранные площади *fpl* площадки Каптейна *fpl*	78
conteos [recuentos] estelares *mpl*	contagem de estrelas *f*	подсчёт звёзд *n*	79
número de estrellas *m*	número de estrelas *m*	число звёзд *n*	80
efecto de selección *m*	efeito de seleção *f*	избирательный эффект *m*	81
densidad estelar *f*	densidade estelar *f*	плотность звёзд *f*	82
densidad estelar espacial *f*	densidade espacial das estrelas *f*	пространственная плотность звёзд *f*	83
parsec cúbico *m* (por parsec cúbico)	parsec cúbico *m* (por parsec cúbico)	кубический парсек *m* (на один кубический парсек)	84
densidad proyectada *f* densidad superficial *f*	densidade projetada *f* densidade superficial *f*	поверхностная плотность *f* проекционная плотность *f*	85
distribución espacial *f* distribución en el espacio *f*	distribuição no espaço *f* distribuição espacial *f*	пространственное распределение *n* распределение в пространстве *n*	86
gradiente de densidad *m*	gradiente de densidade *m*	градиент плотности *m*	87
función de densidad *f*	função de densidade *f*	функция плотности *f*	88
función de luminosidad *f*	função de luminosidade *f*	функция светимости *f*	89
propiedades cinemáticas *fpl*	propriedades cinemáticas *fpl*	кинематические свойства *npl*	90
velocidad residual *f*	velocidade residual *f*	остаточная скорость *f*	91
distribución de velocidades residuales *f*	distribuição das velocidades residuais *f*	распределение остаточных скоростей *n*	92
distribuido al azar distribuido aleatoriamente	distribuido ao acaso distribuido aleatoriamente [casualmente]	случайно распределённый	93
movimiento aleatorio *m* movimiento al azar [casual] *m*	movimento aleatório *m* movimento casual *m* movimento ao acaso *m*	беспорядочное [хаотическое] движение *n* случайное блуждание *n*	94
movimiento ordenado *m*	movimento ordenado *m*	упорядоченное движение *n*	95
distribución maxwelliana de velocidades *f*	distribuição maxwelliana das velocidades *f*	максвелловское распределение скоростей *n*	96
espacio de velocidades *m*	espaço das velocidades *m*	пространство скоростей *n*	97
corriente estelar *f*	corrente de estrelas *f* corrente estelar *f*	поток звёзд *m* звёздный поток *m*	98

99	two-stream hypothesis	hypothèse des deux courants stellaires *f*	Zweistromtheorie *f*
100	ellipsoidal distribution of velocities	distribution ellipsoïdale des vitesses *f*	ellipsoidische Geschwindigkeitsverteilung *f* ellipsoidische Verteilung der Sterngeschwindigkeiten *f*
101	velocity ellipsoid	ellipsoïde des vitesses *m*	Geschwindigkeitsellipsoid *n*
102	coefficients of the velocity ellipsoid	coefficients [constantes] de l'ellipsoïde des vitesses *mpl [fpl]*	Koeffizienten des Geschwindigkeitsellipsoids *mpl*
103	velocity dispersion dispersion of velocities	dispersion des vitesses *f*	Geschwindigkeitsstreuung *f* Geschwindigkeitsdispersion *f*
104	asymmetric drift	courant asymétrique *m*	asymmetrische Geschwindigkeitsverteilung *f*
105	vertex (pl. vertexes, vertices)	vertex *m*	Vertex (pl. Vertizes) *m*
106	antivertex	antivertex *m*	Antivertex *m*
107	galactic rotation	rotation galactique *f*	galaktische Rotation *f*
108	sense of rotation	sens de la rotation *m*	Rotationssinn *m*
109	rotation curve	courbe de rotation *f*	Rotationskurve *f*
110	rotation law	loi de rotation *f*	Rotationsgesetz *n*
111	solid rotation solid-body rotation	rotation rigide [du solide] *f* rotation du corps solide [rigide] *f*	starre Rotation *f* Festkörperrotation *f* Rotation eines starren Körpers *f*
112	Keplerian rotation Kepler's rotation	rotation keplérienne *f* rotation de Kepler *f*	Keplersche Rotation *f* Rotation gemäss den Keplerschen Gesetzen *f*
113	differential rotation of the Galaxy differential galactic rotation	rotation différentielle de la Galaxie *f* rotation galactique différentielle *f*	differentielle galaktische Rotation *f* differentielle Rotation der Galaxis *f*
114	field of differential motions	champ des mouvements différentiels *m*	differentielles Bewegungsfeld *n* Feld der differentiellen Bewegung *n*
115	differential-motion term	terme de mouvement différentiel *m*	Glied der differentiellen Bewegung *n*
116	galactic rotation term	terme de rotation galactique *m*	Glied der galaktischen Rotation *n*
117	double wave	onde double *f*	Doppelwelle *f*
118	Oort constants	constantes de Oort *fpl*	Oortsche Konstanten *fpl*
119	K term	terme K *m*	K-Term *m*

hipótesis de las dos corrientes *f*	hipótese das duas correntes *f*	гипотеза двух потоков *f*	99
distribución elipsoidal de velocidades *f*	distribuição elipsoidal das velocidades *f*	эллипсоидальное распределение скоростей *n*	100
elipsoide de velocidad(es) *m*	elipsóide das velocidades *m*	эллипсоид скоростей *m*	101
coeficientes [constantes] del elipsoide de velocidades *mpl*[*fpl*]	coeficientes [constantes] do elipsóide de velocidades *mpl*[*fpl*]	коэффициенты [постоянные] эллипсоида скоростей *mpl* [*fpl*]	102
dispersión de las velocidades *f*	dispersão de velocidades *f*	дисперсия скоростей *f*	103
desplazamiento asimétrico *m*	desvio assimétrico *m* deslocamento assimétrico *m*	асимметрический [асимметричный] дрейф *m* асимметрическое распределение скоростей *n*	104
vértice *m*	vértice *m*	вертекс *m*	105
antivértice *m*	antivértice *m*	антивертекс *m*	106
rotación galáctica *f*	rotação galáctica *f*	галактическое вращение *n*	107
sentido de la rotación *m*	sentido da rotação *m*	направление вращения *n*	108
curva de rotación *f*	curva de rotação *f*	кривая вращения *f*	109
ley de rotación *f*	lei de rotação *f*	закон вращения *m*	110
rotación rígida *f* rotación de cuerpo rígido *f*	rotação rígida *f* rotação de corpo rígido *f*	твердотельное вращение *n*	111
rotación kepleriana *f* rotación de Kepler *f*	rotação kepleriana *f* rotação de Kepler *f*	кеплеровское вращение *n*	112
rotación diferencial de la Galaxia *f* rotación galáctica diferencial *f*	rotação diferencial da Galáxia *f* rotação galáctica diferencial *f*	дифференциальное вращение Галактики *n* дифференциальное галактическое вращение *n*	113
campo de movimientos diferenciales *m*	campo dos movimentos diferenciais *m*	поле дифференциальных скоростей [движений] *n*	114
término del movimiento diferencial *m*	termo de movimento diferencial *m*	дифференциальный член *m*	115
término de la rotación galáctica *m*	termo da rotação galáctica *m*	член галактического вращения *m*	116
onda doble *f*	onda dupla *f*	двойная волна *f*	117
constantes de Oort *fpl*	constantes de Oort *fpl*	постоянные Оорта *fpl*	118
término K *m*	termo K *m*	K-член *m*	119

120	stellar dynamics	dynamique stellaire *f*	Stellardynamik *f*
121	stellar encounter	rencontre d'étoiles *m*	Sternbegegnung *f*
122	number of encounters	nombre de rencontres *m*	Anzahl der Sternbegegnungen *f*
123	deflection of a star	déviation (du mouvement) d'une étoile *f*	Ablenkung eines Sternes *f*
124	cumulative effect	effet cumulatif *m*	kumulativer Effekt *m*
125	energy exchange	échange d'énergie *m*	Energieaustausch *m*
126	exchanged energy	énergie échangée *f*	ausgetauschte Energie *f*
127	relaxation	relaxation *f*	Relaxation *f*
128	time of relaxation relaxation time	temps de relaxation *m* période de relaxation *f*	Relaxationszeit *f* Erholungszeit *f*
129	rate of relaxation relaxation velocity	vitese de relaxation *f* taux de relaxation *m*	Relaxationsgeschwindigkeit *f*
130	mean time of relaxation	temps moyen de relaxation *m*	mittlere Relaxationszeit *f*
131	collisional relaxation	relaxation par collisions *f*	Stossrelaxation *f*
132	collisionless relaxation	relaxation sans collisions *f*	stossfreie Relaxation *m*
133	star of the system member of the system	étoile du système *f* membre du système *m* étoile membre du système *f*	Systemstern *m* Stern des Systems *m* Systemmitglied *n*
134	interloper	étoile étrangère au système *f* (étoile) non-membre du système *f* étoile interlope *f*	systemfremder Stern *m* Fremdstern *m*
135	runaway star escaping star escaper escapee	étoile qui s'est enfuie *f* étoile en voie d'échappement *f* étoile en libération [évasion] *f*	entweichender Stern *m* Ausreisser *m* Fluchtstern *m*
136	reference frame frame of reference system of reference	système de repère *m* système de référence *m* référentiel *m*	Bezugssystem *n* Referenzsystem *n*
137	fundamental frame of reference fundamental standard of rest	système de référence fondamental *m* système de référence général *m*	fundamentales Bezugssystem *n*
138	rotating frame of reference	système de référence en rotation *m*	rotierendes Bezugssystem *n*
139	local frame of reference	système de référence local *m*	lokales Bezugssystem *n*
140	steady state	état stationnaire *m*	stationärer Zustand *m*

dinámica estelar *f*	dinâmica estelar *f*	звёздная динамика *f*	120
encuentro estelar *m*	encontro estelar [de estrelas] *m*	звёздное сближение *n*	121
número de encuentros *m*	número de encontros *m*	число сближений *n*	122
desviación de una estrella *f*	desvio [deflexão] (do movimento) de uma estrela *m*[*f*]	отклонение звезды *n*	123
efecto acumulativo *m*	efeito cumulativo *m*	кумулятивный эффект *m*	124
intercambio de energía *m*	troca de energia *f*	обмен энергии *m*	125
energía intercambiada *f*	energia trocada *f*	обмениваемая энергия *f*	126
relajación *f*	relaxação *f*	релаксация *f*	127
tiempo de relajación *m*	tempo de relaxação *m*	время релаксации *n* период релаксации *m*	128
velocidad de relajación *f* tasa de relajación *f*	velocidade de relaxação *f* taxa de relaxação *f*	скорость релаксации *f*	129
tiempo medio de relajación *m*	tempo médio de relaxação *m*	среднее время релаксации *n*	130
relajación por colisiones *f*	relaxação colisional [por colisões] *f*	релаксация за счёт столкновений *f* столкновительная релаксация *f*	131
relajación sin colisiones *f*	relaxação acolisional [sem colisões] *f*	бесстолкновительная релаксация *f*	132
estrella del sistema *f* miembro del sistema *f* estrella perteneciente al sistema *f*	estrela do sistema *f* estrela pertencente ao sistema *f* membro do sistema *m*	звезда системы *f* член системы *m* звезда, принадлежащая системе *f*	133
estrella intrusa *f* estrella no perteneciente al sistema *f*	estrela estranha [intrusa] *f* estrela não-pertencente ao sistema *f*	звезда, не принадлежащая системе *f*	134
estrella fugada [en fuga] *f* estrella escapante [que escapa] *f*	estrela escapante [que escapa] *f* estrela em fuga *f* (estrela) fugitiva *f*	убегающая [уходящая] звезда *f* вылетевшая (из системы) звезда *f*	135
sistema de referencia *m*	sistema de referência *m* referencial *m*	система отсчёта *f* система координат *f*	136
sistema fundamental de referencia *m*	sistema fundamental de referência *m* referencial padrão *m*	фундаментальная система отсчёта *f*	137
sistema de referencia en rotación [rotante] *m*	referencial [sistema de referência] em rotação [rotante] *m*	вращающаяся система отсчёта *f*	138
sistema local de referencia *m*	sistema local de referência *m*	местная система отсчёта *f*	139
estado estacionario *m*	estado estacionário *m*	стационарное состояние *n*	140

141	system in a steady state	système en état stationnaire *m*	System im stationären Zustand *n*
142	stellar hydrodynamics	hydrodynamique stellaire *f*	stellare Hydrodynamik *f*
143	distribution function	fonction de distribution *f*	Verteilungsfunktion *f*
144	six-dimensional phase space	espace des phases à six dimensions *m*	sechsdimensionaler Phasenraum *m*
145	gravitational field	champ gravitationnel *m* champ de gravitation *m*	Gravitationsfeld *n*
146	galactic field galactic gravitational field	champ de force galactique *m* champ gravitationnel galactique *m*	galaktisches Feld *n* galaktisches Gravitationsfeld *n*
147	stationary potential	potential stationnaire *m*	stationäres Potential *n*
148	statistically steady state	état statistiquement stationnaire *m*	statistisch stationärer Zustand *m*
149	statistical equilibrium	équilibre statistique *m*	statistisches Gleichgewicht *n*
150	invariable in time constant with the time	invariable dans le temps constant avec temps	zeitlich unveränderlich unveränderlich in der Zeit zeitlich konstant
151	in a non-steady state	dans un état non-stationnaire	im nichtstationären Zustand
152	system in a non-steady state	système dans un état non-stationnaire *m*	System in einem nichtstationären Zustand *n*
153	axially symmetric(al) potential	potentiel à symétrie axiale *m*	axialsymmetrisches Potential *n*
154	potential with cylindrical symmetry	potentiel à symétrie cylindrique *m*	Potential mit Zylindersymmetrie *n* zylindersymmetrisches Potential *n*
155	cylindrical symmetry	symétrie cylindrique *f* cylindrosymétrie *f*	Zylindersymmetrie *f* zylindrische Symmetrie *f*
156	cylindrical coordinates *pl*	coordonnées cylindriques *fpl*	Zylinderkoordinaten *fpl* zylindrische Koordinaten *fpl*
157	axially symmetric(al) system	système à symétrie axiale *m*	axialsymmetrisches System *n*
158	oblate spheroid	sphéroïde aplati *m*	abgeplattetes Sphäroid *n*
159	spheroidal system	système sphéroïdal *m*	sphäroidisches System *n*
160	spherical symmetry	symétrie sphérique *f*	Kugelsymmetrie *f* sphärische Symmetrie *f*
161	helical symmetry	symétrie hélicoïdale *f*	Schraubensymmetrie *f*
162	symmetry condition	condition de symétrie *f*	Symmetriebedingung *f*
163	galactic model	modèle de Galaxie *m*	galaktisches Modell *n*
164	highly oblate system	système très aplati *m*	stark abgeplattetes System *n*

Español	Português	Русский	Nº
sistema (en estado) estacionario *m*	sistema (em estado) estacionário *m*	система в стационарном состоянии *f*	141
hidrodinámica estelar *f*	hidrodinâmica estelar *f*	звёздная гидродинамика *f*	142
función de distribución *f*	função de distribuição *f*	функция распределения *f*	143
espacio de las fases 6-dimensional [hexadimensional] *m*	espaço de fases a seis dimensões *m*	шестимерное фазовое пространство *n*	144
campo gravitational *m*	campo gravitacional *m* campo gravífico *m*	гравитационное поле *n* поле тяготения *n*	145
campo galáctico *m* campo gravitacional galáctico *m*	campo galáctico *m* campo gravitacional galáctico *m*	галактическое поле *n* галактическое гравитационное поле *n*	146
potencial estacionario *m*	potencial estacionário *m*	стационарный потенциал *m*	147
estado estadísticamente estacionario *m*	estado estatísticamente estacionário *m*	статистически стационарное состояние *n*	148
equilibrio estadístico *m*	equilíbrio estatístico *m*	статистическое равновесие *n*	149
invariable en el tiempo invariable con el tiempo	invariável no [com o] tempo constante no tempo invariante no tempo	неизменный во времени неизменяющийся со временем неизменяемый во времени	150
en estado no estacionario	em estado não-estacionário	в нестационарном состоянии	151
sistema en estado no estacionario *m*	sistema em estado não-estacionário *m*	система в нестационарном состоянии *f*	152
potencial axisimétrico *m*	potential axialmente simétrico *m*	потенциал с осевой симметрией *m*	153
potencial de simetría cilíndrica *m*	potencial de simetria cilíndrica *m*	потенциал с цилиндрической симметрией *m*	154
simetría cilíndrica *f*	simetria cilíndrica *f*	цилиндрическая симметрия *f*	155
coordenadas cilíndricas *fpl*	coordenadas cilíndricas *fpl*	цилиндрические координаты *fpl*	156
sistema axisimétrico *m*	sistema de simetria axial *m*	система с осевой симметрией *f*	157
esferoide achatado *m*	esferóide achatado *m*	сжатый сфероид *m*	158
sistema esferoidal *m*	sistema esferoidal *m*	сфероидальная система *f*	159
simetría esférica *f*	simetria esférica *f*	сферическая симметрия *f*	160
simetría helicoidal *f*	simetria helicoidal *f*	винтовая симметрия *f*	161
condición de simetría *f*	condição de simetria *f*	условие симметрии *n*	162
modelo galáctico *m*	modelo galáctico *m*	модель Галактики *f*	163
sistema muy achatado *m*	sistema muito achatado *m*	сильно сжатая система *f*	164

165	stellar system formed by superposition	système stellaire résultant de la superposition (de sous-systèmes) *m* système stellaire formé par superposition *m*	durch Superposition (von Untersystemen) gebildetes Sternsystem *n*
166	collisional system	système à collisions *m* système collisionnel *m*	durch Stösse beherrschtes System *n* Stossystem *n*
167	collisionless system	système sans collisions *m*	stossfreies System *n*
168	collective interaction	interaction collective *f*	kollektive Wechselwirkung *f*
169	self-gravitating system	système autogravitant *m* système soumis à sa propre gravitation [à la autogravitation] *m*	selbstgravitierendes System *n* System mit Eigengravitation *n*
170	stable solution	solution stable *f*	stabile Lösung *f*
171	unstable system	système instable *m*	instabiles System *n*
172	stellar content	contenu stellaire *m*	Sterninhalt *m*
173	theory of spiral structure theory of spiral-arm formation	théorie de la structure spirale *f* théorie de la formation des bras spiraux *f*	Theorie der Spiralstruktur *f* Theorie der Spiralarmentstehung *f*
174	(density) wave theory (of spiral structure)	théorie d'ondes de densité (de la structure spiralée) *f*	Dichtewellentheorie (der Spiralstruktur) *f*
175	density wave galactic density wave	onde de densité *f* onde de densité galactique *f*	(galaktische) Dichtewelle *f*
176	potential trough	auge [creux] de potentiel *f*	Potentialmulde *f*
177	characteristic envelope	enveloppe caractéristique *f*	charakteristische Enveloppe *f*
178	turning point	point de rebroussement *m*	Wendepunkt *m*
179	ejection process of ejection	éjection *f* processus d'éjection *m*	Ejektion *f* Ejektionsvorgang *m*
180	asymptotic to the circular orbits	asymptotique aux orbites circulaires	asymptotisch zu Kreisbahnen
181	virial theorem law of virial	théorème du viriel *m* loi virielle *f* équation du viriel *f*	Virialsatz *m* Virialgleichung *f*
182	Lagrangian identities	identités de Lagrange *fpl*	Lagrangesche Identitäten *fpl*
183	generalized Lagrangian identities	identités de Lagrange généralisées *fpl*	verallgemeinerte Lagransche Identitäten *fpl*
184	galactic orbit of a star	orbite galactique d'une étoile *f*	galaktische Bahn eines Sternes *f*
185	three-dimensional orbit	orbite à trois dimensions *f*	dreidimensionale Bahn *f*

sistema estelar formado por superposición [sobreposición] *m*	sistema estelar formado por superposição [sobreposição] *m*	звёздная система, образованная наложением *f*	165
sistema colisional *m* sistema con colisiones *m*	sistema colisional *m* sistema com colisões *m*	столкновительная система *f*	166
sistema sin colisiones *m*	sistema acolisional [sem colisões] *m*	бесстолкновительная система *f*	167
interacción colectiva *f*	interação coletiva *f*	коллективное взаимодействие *n*	168
sistema autogravitante *m* sistema autogravitatorio *m*	sistema autogravitante *m* sistema de gravitação própria *m*	самогравитирующая система *f* гравитирующая система *f*	169
solución estable *f*	solução estável *f*	устойчивое решение *n*	170
sistema inestable *m*	sistema instável *m*	неустойчивая система *f*	171
contenido estelar *m*	conteúdo estelar *m*	звёздное содержание *n*	172
teoría de la estructura espiral *f* teoría de los brazos espirales *f*	teoria da estrutura espiral *f* teoria (de formação) dos braços espirais *f*	теория спиральной структуры *f* теория образования спиральных рукавов [ветвей] *f*	173
teoría de ondas de densidad (de la estructura espiral) *f*	teoria das ondas de densidade (da estrutura espiral) *f*	волновая теория спиральной структуры *f*	174
onda de densidad (galáctica) *f*	onda de densidade (galáctica) *f*	волна плотности [сжания] *f* галактическая волна плотности [сжатия] *f*	175
zanja de potencial *f*	vala [cocho] de potencial *f*[*m*]	потенциальное корыто *n*	176
envolvente característica *f*	invólucro característico *m*	характеристическая огибающая *f*	177
punto de regresión *m* punto de retorno *m*	ponto de regressão [retorno] *m* ponto de retrocesso *m*	точка возврата *f*	178
eyección *f* proceso de eyección *m*	ejeção *f* processo de ejeção *m*	выбрасывание *n* процесс выбрасывания *m*	179
asintótico a las órbitas circulares	assintótico [assimptótico] às órbitas circulares	асимптотический к круговым орбитам	180
teorema del virial *m* ecuación del virial *f*	teorema do virial *m* equação do virial *f*	теорема вириала *f* вириальная теорема *f* теорема о вириале *f*	181
identidades lagrangianas *fpl*	identidades lagrangianas [de Lagrange] *fpl*	тождества Лагранжа *npl*	182
identidades lagrangianas generalizadas *fpl*	identidades lagrangianas [de Lagrange] generalizadas *fpl*	обобщённые тождества Лагранжа *npl*	183
órbita galáctica de una estrella *f*	órbita galáctica de uma estrela *f*	галактическая орбита звезды *f*	184
órbita tridimensional *f*	órbita tridimensional *f*	трёхмерная орбита *f* пространственная орбита *f*	185

186	third integral of motion	troisième intégrale du mouvement *m*	drittes Bewegungsintegral *n*
187	*z*-oscillation oscillation perpendicular to the galactic plane	oscillation en *z* *f* oscillation perpendiculaire au plan galactique *f*	Oszillation in *z*-Richtung *m* Oszillation senkrecht zur galaktischen Ebene *f*
188	*z*-motion motion perpendicular to the galactic plane	mouvement suivant la composante *z* *m* mouvement perpendiculaire au plan galactique *m*	*z*-Bewegung *f* Bewegung senkrecht zur galaktischen Ebene *f*
189	galactic pericenter [pericentre, perifocus, periapsis, periapse] pericentron perigalacticon perigalacticum	péricentre galactique *m* périastre galactique *m* périapside galactique *f* apside inférieur galactique *f*	Perigalaktikum *n* galaktische Periapsis [Periapside] *f* galaktische Brennpunktsnähe *f* galaktischer Perifokus *m*
190	galactic apocenter [apocentre, apofocus, apoapse, apoapsis] apocentron apogalacticum apogalacticon	apocentre galactique *m* apoastre galactique *m* apoapside galactique *f* apside supérieure galactique *f*	Apogalaktikum *n* galaktische Apoapsis [Apoapside] *f* galaktische Brennpunktsferne *f* galaktischer Apofokus *m*
191	high-velocity star	étoile à grande vitesse *f*	Schnelläufer *m*
192	low-velocity star	étoile de faible vitesse *f*	Stern mit geringer Geschwindigkeit *m* Langsamläufer *m*
193	stability of an orbit	stabilité d'une orbite *f*	Stabilität einer Bahn *f*
194	escape of stars from a cluster	évasion d'étoiles d'un amas *f*	Entweichen von Sternen aus einem Haufen *n*
195	disintegration of clusters	désintégration des amas *f*	Auflösung von Haufen *f*
196	tidal field	champ des forces de marée *m*	Gezeitenfeld *n*
197	critical density	densité critique *f*	kritische Dichte *f*
198	dynamical stability	stabilité dynamique *f*	dynamische Stabilität *f*
199	coefficient of stability	coefficient de stabilité *m*	Stabilitätskoeffizient *m*
200	primitive orbit	orbite primitive *f*	ursprüngliche Bahn *f*
201	galaxy	galaxie *f*	Galaxie *f*
202	extragalactic nebula	nébuleuse extragalactique *f*	extragalaktischer [aussergalaktischer] Nebel *m*
203	X-ray galaxy	galaxie à rayons X *f* galaxie roentgen *f*	Röntgengalaxie *f*
204	optical galaxy	galaxie optique *f* galaxie observable en lumière (visible) *f*	optische Galaxie *f* Galaxie im sichtbaren Licht *f*
205	infrared galaxy	galaxie infrarouge *f* galaxie observable en infrarouge *f*	Infrarotgalaxie *f* infrarote Galaxie *f*

tercera integral del movimiento *f*	terceira integral do movimento *f*	третий интеграл движения *m*	186
oscilación en *z f* oscilación perpendicular al plano galáctico *f*	oscilação (em) *z f* oscilação perpendicular ao plano galáctico *f*	*z*-колебания *npl* колебания перпендикулярные плоскости Галактики *npl*	187
movimiento en *z m* movimiento perpendicular al plano galáctico *m*	movimento (em) *z m* movimento perpendicular ao plano galáctico *m*	движение перпендикулярное плоскости Галактики *n* движение в направлении *z n*	188
pericentro galáctico *m* perigaláctico *m* periápside galáctico *m* perifoco galáctico *m*	pericentro galáctico *m* perigaláctico *m* periapside galáctica *f* perifoco galáctico *m*	перигалактий *m* галактический перицентр *m* галактическая периапсида *f*	189
apocentro galáctico *m* apogaláctico *m* ap(o)ápside galáctico *m* apofoco galáctico *m*	apocentro galáctico *m* apogaláctico *m* apoapside galáctica *f* apofoco galáctico *m*	апогалактий *m* галактический апоцентр *m* галактическая апоапсида *f*	190
estrella de alta velocidad *f*	estrela de alta velocidade *f*	быстродвижущаяся звезда *f*	191
estrella de baja velocidad *f*	estrela de baixa velocidade *f*	звезда с малой скоростью *f* медленная звезда *f*	192
estabilidad de una órbita *f*	estabilidade de uma órbita *f*	устойчивость орбиты *f*	193
escape de estrellas de un cúmulo *m*	fuga [escape] de estrelas de um cúmulo [aglomerado] *f*[*m*]	освобождение [ускользание, вылет, убегание] звёзд из скопления *n*	194
desintegración de cúmulos *f*	desintegração de cúmulos *f*	разрушение [распад] скоплений *n*[*m*]	195
campo de marea *m*	campo de forças de maré *m*	поле приливных сил *n*	196
densidad crítica *f*	densidade crítica *f*	критическая плотность *f*	197
estabilidad dinámica *f*	estabilidade dinâmica *f*	динамическая устойчивость *f*	198
coeficiente de estabilidad *m*	coeficiente de estabilidade *m*	коэффициент устойчивости *m*	199
órbita primitiva [original] *f*	órbita primitiva *f*	первичная орбита *f*	200
galaxia *f*	galáxia *f*	галактика *f*	201
nebulosa extragaláctica *f*	nebulosa extragaláctica *f*	внегалактическая туманность *f*	202
galaxia de rayos X *f* galaxia en radiación X *f*	galáxia de raios X *f* galaxia em raios X *f*	рентгеновская галактика *f*	203
galaxia óptica *f* galaxia observable en luz *f*	galáxia ó(p)tica *f* galáxia visível *f* galáxia observável em luz *f*	оптическая галактика *f* галактика наблюдаемая в свете *f*	204
galaxia infrarroja *f* galaxia observable en enfrarrojo *f*	galáxia infravermelha *f* galáxia observável no infravermelho *f*	инфракрасная галактика *f*	205

206	radio galaxy	galaxie radio *f* radiogalaxie *f*	Radiogalaxie *f*
207	extragalactic anagalactic	extragalactique anagalactique	extragalaktisch aussergalaktisch
208	extragalactic astronomy	astronomie extragalactique *f*	extragalaktische Astronomie *f*
209	extragalactic object	objet extragalactique *m*	extragalaktisches Objekt *n*
210	extragalactic system	système extragalactique *m*	extragalaktisches System *n*
211	neigbo(u)ring galaxy	galaxie voisine *f*	Nachbargalaxie *f*
212	Magellanic Clouds	Nuages de Magellan *mpl*	Magellansche Wolken *fpl*
213	Large Magellanic Cloud LMC	Grand Nuage de Magellan *m*	Grosse Magellansche Wolke *f*
214	Small Magellanic Cloud SMC	Petit Nuage de Magellan *m*	Kleine Magellansche Wolke *f*
215	Andromeda Nebula	Nébuleuse d'Andromède *f*	Andromedanebel *m*
216	companion of a galaxy	compagnon d'une galaxie *m* galaxie compagnon [satellite] *f*	Begleiter einer Galaxie *m*
217	elliptical companion (of a galaxy)	compagnon elliptique (d'une galaxie) *m*	elliptischer Begleiter (einer Galaxie) *m*
218	optical companion (of a galaxy)	compagnon optique (d'une galaxie) *m*	optischer Begleiter (einer Galaxie) *m*
219	physical companion (of a galaxy)	compagnon physique (d'une galaxie) *m*	physischer Begleiter (einer Galaxie) *m*
220	dwarf galaxy	galaxie naine *f*	Zwerggalaxie *f*
221	normal galaxy	galaxie normale *f*	normale Galaxie *f*
222	giant galaxy	galaxie géante *f*	Riesengalaxie *f*
223	supergiant galaxy	galaxie supergéante *f*	Überriesengalaxie *f*
224	structural features	propriétés structurales *fpl* caractères structuraux *mpl*	Struktureinzelheiten *fpl* Struktureigenschaften *fpl*
225	structural type	type structural [de structure] *m*	Strukturtyp *m*
226	classification of galaxies	classification des galaxies *f*	Galaxienklassifikation *f*
227	Hubble sequence of galaxies	séquence des galaxies de Hubble *f*	Hubblesche Galaxienfolge *f* Hubble-Sequenz (der Galaxien) *f*

radiogalaxia *f*	radiogaláxia *f*	радиогалактика *f*	206
extragaláctico anagaláctico	extragaláctico anagaláctico	внегалактический анагалактический	207
astronomía extragaláctica *f*	astronomia extragaláctica *f*	внегалактическая астрономия *f*	208
objeto extragaláctico *m*	objeto extragaláctico *m*	внегалактический объект *m*	209
sistema extragaláctico *m*	sistema extragaláctico *m*	внегалактическая система *f*	210
galaxia vecina *f*	galáxia vizinha *f*	соседняя галактика *f*	211
Nubes Magallánicas [de Magallanes] *fpl*	Nuvens de Magalhães *fpl*	Магеллановы Облака *npl*	212
Gran Nube Magallánica *f* Nube Mayor de Magallanes *f*	Grande Nuvem de Magalhães *f*	Большое Магелланово Облако *n*	213
Pequeña Nube Magallánica *f* Nube Menor de Magallanes *f*	Pequena Nuvem de Magalhães *f*	Малое Магелланово Облако *n*	214
Nebulosa de Andrómeda *f*	Nebulosa de Andrômeda *f*	Туманность Андромеды *f*	215
compañera de una galaxia *f*	companheira de uma galáxia *f* companheiro de uma galáxia *m*	спутник галактики *m* галактика-спутник *f*	216
compañera elíptica (de una galaxia) *f*	companheiro elíptico (de uma galáxia) *m* companheira elíptica *f*	эллиптическая галактика-спутник *f* эллиптический спутник галактики *m*	217
compañera óptica (de una galaxia) *f*	companheira ó(p)tica *f* companheiro ó(p)tico (de uma galáxia) *m*	оптический спутник (галактики) *m*	218
compañera física (de una galaxia) *f*	companheira física *f* companheiro físico (de uma galáxia) *m*	физический спутник (галактики) *m*	219
galaxia enana *f*	galáxia anã *f*	карликовая галактика *f* галактика-карлик *f*	220
galaxia normal *f*	galáxia normal *f*	нормальная [обычная] галактика *f*	221
galaxia gigante *f*	galáxia gigante *f*	гигантская галактика *f* галактика-гигант *f*	222
galaxia supergigante *f*	galáxia supergigante *f*	сверхгигантская галактика *f* галактика-сверхгигант *f*	223
características estructurales *fpl* propiedades estructurales *fpl*	características estruturais *fpl* propriedades estruturais *fpl*	структурные особенности *fpl* структурные детали *fpl*	224
tipo estructural *m*	tipo estrutural *m*	структурный тип *m*	225
clasificación de galaxias *f*	classificação de galáxias *f*	классификация галактик *f*	226
sucesión [secuencia] de galaxias de Hubble *f*	seqüência de galáxias de Hubble *f*	хаббловская последовательность (типов) галактик *f*	227

	GALAXIES	GALAXIES	GALAXIEN
228	morphological type	type morphologique *m*	morphologischer Typ *m*
229	Hubble type (of a galaxy)	type (galactique) de Hubble *m*	Hubblescher Typ (einer Galaxie) *m*
230	galactic population population of a galaxy	peuplement d'une galaxie *m* population d'une galaxie *f*	Population einer Galaxie *f* galaktische Sternpopulation *f*
231	stellar content (of a galaxy)	contenu stellaire (d'une galaxie) *m*	stellarer Bestandteil (einer Galaxie) *m* Sterngehalt (einer Galaxie) *m*
232	dust component	composante de poussière *f*	Staubkomponente *f*
233	mass-to-light ratio mass-luminosity ratio	rapport masse-luminosité *m*	Masse-Leuchtkraft-Verhältniss *n*
234	optical feature	propriété optique *f*	optische Eigenschaft *f* optische Einzelheit *f*
235	radio feature	propriété en ondes radioélectriques *f*	Struktureinzelheit im Radiowellenbe-reich *f*
236	regular galaxy (regular nebula)	galaxie régulière *f* (nébuleuse régulière *f*)	regelmässige Galaxie *f* (regelmässiger Nebel) *m*
237	irregular galaxy (Irr) (irregular nebula)	galaxie irrégulière *f* (nébuleuse irrégulière *f*)	irreguläre Galaxie *f* unregelmässige Galaxie *m* (unregelmässiger Nebel) *m*
238	gas-rich galaxy	galaxie riche en gaz *f*	gasreiche Galaxie *f*
239	spindle-shaped galaxy acicular galaxy needle-shaped galaxy cigar-shaped galaxy	galaxie en fuseau *f* galaxie en aiguille *f* galaxie en cigare *f* (nébuleuse en fuseau *f*)	spindelförmige Galaxie *f* nadelförmige Galaxie *f* cigarrenförmige Galaxie *f*
240	sequence of regular galaxies	séquence des galaxies régulières *f*	Folge der regelmässigen Galaxien *f*
241	elliptical galaxy ellipsoidal galaxy	galaxie elliptique *f* galaxie ellipsoïdale *f*	elliptische Galaxie *f* ellipsoidale Galaxie *f*
242	dwarf elliptical galaxy dwarf elliptical	galaxie elliptique naine *f*	elliptische Zwerggalaxie *f*
243	giant elliptical galaxy giant elliptical	galaxie elliptique géante *f*	elliptische Riesengalaxie *f*
244	spherical galaxy E0 elliptical galaxy spheroidal galaxy	galaxie sphérique *f* galaxie sphéroïdale *f* galaxie elliptique E0 *f*	kugelförmige Galaxie *f* sphäroidale Galaxie *f*
245	lenticular galaxy	galaxie lenticulaire *f*	linsenförmige Galaxie *f*
246	edge-on galaxy galaxy seen edge-on	galaxie par la tranche *f* galaxie vue de profil *f*	Galaxie von der Seite gesehen *f* von der Kante gesehene Galaxie *f*

tipo morfológico *m*	tipo morfológico *m*	морфологический тип *m*	228
tipo (galáctico) de Hubble *m*	tipo (galáctico) de Hubble *m*	хаббловский тип (галактики) *m*	229
población galáctica *f* población de una galaxia *f*	população galáctica *f* população de uma galáxia *f*	население галактики *n* галактическое население *n*	230
contenido estelar (de una galaxia) *m*	conteúdo estelar (de uma galáxia) *m*	звёздный состав (галактики) *m*	231
componente de polvo *f*	componente de poeira *m*	пылевой компонент *m* пылевая составляющая *f*	232
razón masa-luminosidad *f*	razão massa-luminosidade *f*	отношение масса-светимость *n*	233
característica óptica *f* propiedad óptica *f*	característica ó(p)tica *f* propriedade ó(p)tica *f*	оптическая особенность *f* оптическая деталь *f*	234
característica (en ondas de) radio *f* propiedad en ondas de radio *f*	característica rádio *f* propriedade rádio *f*	деталь в радиоизлучении *f* радиоособенность *f*	235
galaxia regular *f* (nebulosa regular *f*)	galáxia regular *f* (nebulosa regular *f*)	правильная галактика *f* (правильная туманность *f*)	236
galaxia irregular *f* (nebulosa irregular *f*)	galáxia irregular *f* (nebulosa irregular *f*)	неправильная галактика *f* (неправильная туманность *f*)	237
galaxia rica en gas *f*	galáxia rica em gás *f*	галактика с большим содержанием газа *f*	238
galaxia en forma de huso *f* galaxia fusiforme *f*	galáxia fusiforme *f* galáxia (em forma) de fuso *f* nebulosa em fuso *f*	веретенообразная галактика *f* иглообразная галактика *f* сигарообразная галактика *f*	239
secuencia de galaxias regulares *f*	seqüência das galáxias regulares *f*	последовательность правильных галактик *f*	240
galaxia elíptica *f* galaxia elipsoidal *f*	galáxia elíptica *f* galáxia elipsoidal *f*	эллиптическая галактика *f* эллипсоидальная галактика *f*	241
galaxia elíptica enana *f*	galáxia elíptica anã *f*	карликовая эллиптическая галактика *f*	242
galaxia elíptica gigante *f*	galáxia elíptica gigante *f*	гигантская эллиптическая галактика *f*	243
galaxia esférica *f* galaxia tipo E0 *f* galaxia esferoidal *f*	galáxia esférica *f* galáxia esferoidal *f* galáxia de tipo E0 *f*	сферическая галактика *f* сфероидальная галактика *f*	244
galaxia lenticular *f*	galáxia lenticular *f*	линзообразная [линзовая, линзовидная] галактика *f* чечевицеобразная галактика *f*	245
galaxia vista de canto *f* galaxia vista de perfil *f*	galáxia vista de perfil *f*	галактика, видимая с ребра *f* галактика, наблюдаемая с ребра *f*	246

247	face-on galaxy galaxy seen face-on	galaxie vue de face *f*	"Pol-on" Ansicht der Galaxie *f*
248	transition-type galaxy	galaxie de transition *f*	Galaxie vom Übergangstyp *f*
249	spiral galaxy spiral (spiral nebula)	galaxie spiralée *f* (galaxie) spirale *f* (nébuleuse spirale) *f*	Spiralgalaxie *f* Spirale *f* (Spiralnebel) *m*
250	normal spiral non-barred spiral	spirale normale *f* spirale non-barrée *f*	normale Spirale *f* balkenlose Spirale *f*
251	flat system	système plat *m* système aplati *m*	flaches System *n* abgeplattetes System *n*
252	highly flattened galaxy	galaxie fortement aplatie *f*	stark abgeplattete Galaxie *f*
253	barred galaxy barred spiral	galaxie barrée *f* spirale barrée *f*	Balkenspirale *f* Spiralgalaxie mit Balken *f*
254	galactic bar	barre galactique *f*	galaktischer Balken *m*
255	sequence of spirals	séquence des spirales *f*	Sequenz [Folge] der Spiralgalaxien *f*
256	peculiar galaxy	galaxie particulière [anomale] *f*	pekuliare Galaxie *f*
256a	gravitational lens	lentille gravitationnelle *f*	Gravitationslinse *f*
257	peculiarity	particularité *f*	Besonderheit *f*
258	singular galaxy	galaxie singulière *f*	singuläre Galaxie *f*
259	ring galaxy	galaxie annulaire *f*	Ringgalaxie *f*
260	active galaxy	galaxie active *f*	aktive Galaxie *f*
261	galactic activity	activité galactique *f*	galaktische Aktivität *f*
262	Seyfert galaxy	galaxie de Seyfert *f*	Seyfert-Galaxie *f*
263	N–type galaxy	galaxie du type N *f*	N-Galaxie *f*
264	compact galaxy	galaxie compacte *f* galaxie condensée *f*	Kompaktgalaxie *f* kompakte Galaxie *f*
265	blue compact galaxy BCG	galaxie bleue compacte *f*	blaue kompakte Galaxie *f* blaue Kompaktgalaxie *f*
265a	Markarian galaxy	galaxie de Markarian *f*	Markarian-Galaxie *f*
266	Haro (blue) galaxy	galaxie de Haro *f*	Haro-Galaxie *f*

galaxia vista de frente *f*	galáxia vista de frente *f*	галактика, видимая плашмя *f* галактика, наблюдаемая в плане *f*	247
galaxia de transición *f*	galáxia de transição *f*	галактика переходного типа *f*	248
galaxia espiral *f* espiral *f* (nebulosa espiral) *f*	galáxia espiral *f* espiral *f* (nebulosa espiral *f*)	спиральная галактика *f* спираль *f* (спиральная туманность *f*)	249
espiral normal *f* espiral no barreada *f*	espiral normal *f* espiral não-barrada *f*	нормальная [обычная] спираль *f* спираль без перемычки *f*	250
sistema aplastado *m* sistema [achatado] aplanado *m*	sistema chato *m* sistema achatado *m*	плоская система *f* сплюснутая система *f*	251
galaxia altamente achatada *f*	galáxia muito achatada *f*	сильно сплюснутая галактика *f*	252
galaxia de [con] barra *f* galaxia barreada *f* espiral barreada *f*	galáxia barrada *f* espiral barrada *f*	спиральная галактика с перемычкой *f* пересечённая спиральная галактика *f*	253
barra galáctica *f*	barra galáctica *f*	(галактическая) перемычка *f*	254
secuencia de espirales *f*	seqüência das espirais *f*	последовательность спиральных галактик *f*	255
galaxia peculiar *f*	galáxia peculiar *f*	пекулярная галактика *f*	256
lente gravitacional *f*	lente gravitacional *f*	гравитационная линза *f*	256a
peculiaridad *f*	peculiaridade *f*	пекулярность *f*	257
galaxia singular *f*	galáxia singular *f*	сингулярная [особенная] галактика *f*	258
galaxia anular *f*	galáxia anular *f*	кольцеобразная [кольцевая] галактика *f*	259
galaxia activa *f*	galáxia ativa *f*	активная галактика *f*	260
actividad galáctica *f*	atividade galáctica *f*	галактическая активность *f*	261
galaxia Seyfert *f*	galáxia de Seyfert *f*	сейфертовская галактика *f* галактика Сейферта *f*	262
galaxia de tipo N *f*	galáxia do tipo N *f*	N-галактика *f* галактика типа N *f*	263
galaxia compacta *f*	galáxia compacta *f*	компактная галактика *f*	264
galaxia azul compacta *f*	galáxia compacta azul *f*	голубая компактная галактика *f*	265
galaxia Markarian *f*	galáxia de Markarian *f*	галактика Маркаряна *f*	265a
galaxia Haro *f*	galáxia (azul) de Haro *f*	галактика Аро *f*	266

267	recurrent activity	activité récurrente *f*	wiederholte Aktivität *f* rekurrierende Aktivität *f*
268	exploding galaxy	galaxie en explosion *f*	explodierende Galaxie *f*
269	eruptive galaxy	galaxie éruptive *f*	eruptive Galaxie *f*
270	post-eruptive galaxy	galaxie post-éruptive *f*	Galaxie nach der Eruption *f*
271	activity of the galactic nucleus nuclear activity of a galaxy	activité du noyau galactique *f*	Aktivität des galaktischen Kernes *f* Kernaktivität einer Galaxie *f*
272	active galactic nucleus	noyau galactique actif *m*	aktiver galaktischer Kern *m*
273	compact source	source compacte *f*	kompakte Strahlungsquelle *f*
274	ejection of matter expulsion of matter (from the galactic nucleus)	éjection de matière (du noyau galactique) *f*	Auswurf von Materie *m* Materieauswurf (aus dem galaktischen Kern) *m*
275	outflow of matter (from the galactic nucleus)	écoulement de matière (du noyau galactique) *m*	Ausfluss von Materie (aus dem galaktischen Kern) *m*
276	continuous outflow	écoulement continu *m*	kontinuierlicher Ausfluss *m* ununterbrochener Ausfluss *m*
277	galactic wind	vent galactique *m*	galaktischer Wind *m*
278	jet (from a galactic nucleus) jet ejection	jet (du noyau galactique) *m* éjection d'un jet *f*	Jet *m* Materiestrahl (aus dem galaktischen Kern) *m*
279	optical jet	jet optique *m*	optischer Jet *m*
280	tail of a galaxy	queue galactique *f*	Schweif einer Galaxie *m*
281	tadpole galaxy	galaxie du type têtard *f*	Kaulquappen-Galaxie *f*
282	recurrent ejection	éjection récurrente *f*	wiederholter Ausbruch *m*
283	galaxy-nucleus explosion explosion of galactic nucleus	explosion d'un noyau galactique *f*	Explosion eines galaktischen Kernes *f*
284	parent galaxy	galaxie mère *f* galaxie parente *f*	Muttergalaxie *f* erzeugende Galaxie *f*
285	ejection of radio clouds	éjection des nuages (à émission) radio *f*	Auswurf von Radiowolken *m*
286	radio cloud radio lobe	nuage (à) radio *m* lobe à émission radio *m*	Radiowolke *f* Radioflügel *m*
287	structure element structural element	élément structural *m* élément de structure *m*	Strukturelement *n* Strukturmerkmal *n*

actividad recurrente *f*	atividade recorrente *f*	рекуррентная активность *f* повторяющаяся активность *f*	267
galaxia en explosión *f*	galáxia explosiva [em explosão] *f*	взрывающаяся галактика *f*	268
galaxia eruptiva *f*	galáxia eruptiva *f*	эруптивная галактика *f*	269
galaxia posteruptiva *f* galaxia después de explosión *f*	galáxia pós-eruptiva [depois de uma erupção) *f*	послеэруптивная галактика *f* постэруптивная галактика *f*	270
actividad del núcleo galáctico *f*	atividade do núcleo galáctico *f*	активность галактического ядра *f*	271
núcleo galáctico activo *m*	núcleo galáctico ativo *m*	активное галактическое ядро *f*	272
fuente compacta (de radiación) *f*	fonte compacta *f*	компактный источник *m*	273
eyección [expulsión] de la materia (desde el núcleo galáctico) *f*	ejeção da matéria *f* expulsão da matéria (do núcleo galáctico) *f*	выбрасывание вещества *n* выброс материи (из галактического ядра) *m*	274
derrame [flujo] de materia hacia afuera (desde el núcleo galáctico) *m*	escoamento [efluxo] da matéria (do núcleo galáctico) *m*	истечение вещества (из галактического ядра) *n*	275
derrame continuo *m* flujo contínuo hacia afuera *m*	escoamento [efluxo] contínuo *m*	непрерывное истечение *n*	276
viento galáctico *m*	vento galáctico *m*	галактический ветер *m*	277
chorro desde un núcleo galáctico *m* chorro galáctico *m*	ja(c)to (do núcleo galáctico) *m* ja(c)to galáctico *m*	выброс струи (из галактического ядра) *m* галактический выброс *m*	278
chorro óptico *m*	ja(c)to ó(p)tico *m*	оптическая струя *f* оптический выброс *m*	279
cola galáctica *f*	cauda galáctica *f*	хвост галактики *m*	280
galaxia tipo renacuajo *f*	galáxia tipo girino *f*	головастикообразная галактика *f* головастиковая галактика *f* галактика типа голова-хвост *f*	281
eyección recurrente *f*	ejeção recorrente *f*	рекуррентный выброс *m*	282
explosión de un núcleo galáctico *f*	explosão de um núcleo galáctico *f*	взрыв галактического ядра *m*	283
galaxia madre *f*	galáxia mãe *f*	родительская галактика *f*	284
eyección de las radionubes *f*	ejeção das radionuvens *f*	выброс радиоизлучающих облаков *m*	285
radionube *f* radiolóbulo *m*	radionuvem *f* radiolóbulo *m*	радиооблако *n* радиокомпонент *m*	286
elemento estructural *m*	elemento estrutural *m*	структурный элемент *m* деталь структуры *f*	287

288	dark lanes	bandes sombres *fpl* traînées sombres *fpl*	dunkle Streifen *mpl*
289	dust lane	traînée de poussière *f*	Staubstreifen *m*
290	dust blob	globule de poussière *m*	Staubglobule *f* Staubverdichtung *f*
291	system of rings	système d'anneaux *m*	System von Ringen *n*
292	circumnuclear region	région circumnucléaire *f*	zirkumnuklearer Bereich *m*
293	central region of a galaxy	région centrale d'une galaxie *f*	Zentralgebiet einer Galaxie *n* galaktisches Zentralgebiet *n*
294	rim of a galaxy	bord d'une galaxie *m*	Rand einer Galaxie *m*
295	apparent flattening	aplatissement apparent *m*	scheinbare Abplattung *f*
296	real [true] flattening	aplatissement réel *m*	wahre Abplattung *f*
297	red shift redshift	décalage vers le rouge *m* déplacement vers le rouge *m*	Rotverschiebung *f*
298	recession of galaxies recession of the nebulae	fuite de galaxies *f* récession de galaxies *f*	Fluchtbewegung der Galaxien *f* Nebelflucht *f*
299	Hubble('s) law	loi de Hubble *f*	Hubblesches Gesetz *n* Hubble-Gesetz *n*
300	velocity-distance relation	relation vitesse-distance *f*	Geschwindigkeit-Entfernungs-Beziehung *f*
301	relation of redshift and apparent magnitude redshift-apparent magnitude relation redshift-magnitude diagram Hubble diagram	diagramme déplacement vers le rouge-magnitude apparente *m* diagramme de Hubble *m*	Rotverschiebung-Helligkeits-Beziehung *f* Rotverschiebung-Helligkeits-Diagramm *n* Hubble-Diagramm *n*
302	Hubble constant constant of recession	constante de récession *f* constante de Hubble *f*	Hubblesche Konstante *f* Hubble-Konstante *f*
303	distance scale	échelle de distance *f*	Entfernungsskala *f*
304	the brightest star (of a galaxy)	étoile la plus brillante (d'une galaxie) *f*	der hellste Stern (einer Galaxie) *m*
305	double galaxy twin galaxy	galaxie double *f* galaxie jumelle *f*	Doppelgalaxie *f* Galaxienpaar *n*
306	interconnected galaxies	galaxies interconnectées *fpl*	verbundene Galaxien *fpl*
307	luminous [bright, light] bridge	pont lumineux *m*	Lichtbrücke *f*
308	multiple galaxy	galaxie multiple *f*	mehrfache Galaxie *f* Mehrfachgalaxie *f*

bandas obscuras *fpl* bandas oscuras *fpl*	bandas escuras *fpl* faixas escuras *fpl*	тёмные полосы *fpl*	288
franja de polvo *f* faja [banda] de polvo *f*	banda de poeira *f* faixa de poeira *f*	пылевая полоса *f* пылевая прослойка *f*	289
glóbulo de polvo *m*	glóbulo de poeira *m*	пылевое сгущение *n* пылевая глобула *f*	290
sistema de anillos *m*	sistema de anéis *m*	система колец *f*	291
región circunnuclear *m*	região circunuclear [circum-nuclear] *f*	околоядерная область *f*	292
región central de una galaxia *f*	região central de uma galáxia *f*	центральная область галактики *f*	293
borde de una galaxia *m*	borda [bordo] de uma galáxia *f*[*m*]	край галактики *m*	294
aplanamiento [achatamiento] aparente *m*	achatamento aparente *m*	видимое сжатие *n*	295
aplanamiento [achatamiento] real *m*	achatamento real *m*	истинное сжатие *n*	296
corrimiento hacia el rojo *m* desplazamiento hacia el rojo *m*	deslocamento para o vermelho *m* desvio para o vermelho *m*	красное смещение *n*	297
recesión de las galaxias *f* fuga de las galaxias *f*	recessão das galáxias *f* fuga das galáxias *f*	(взаимное) удаление галактик *n* разбегание [разлёт] галактик *n*[*m*]	298
ley de Hubble *f*	lei de Hubble *f*	закон Хаббла *m* закон красного смещения *m*	299
relación velocidad-distancia *f*	relação velocidade-distância *f*	зависимость скорость-расстояние *f*	300
diagrama magnitud aparente-desplazamiento hacia el rojo *m* diagrama de Hubble *m*	diagrama deslocamento [desvio] para o vermelho-magnitude *m* diagrama de Hubble *m* relação [diagrama] magnitude - desvio para o vermelho *f*[*m*]	диаграмма красное смещение – – звёздная величина *f* диаграмма Хаббла *f*	301
constante de Hubble *f* constante de recesión *f*	constante de Hubble *f* constante da recessão *f*	постоянная Хаббла *f* параметр расширения Хаббла *m*	302
escala de distancias *f*	escala de distâncias *f*	шкала расстояний *f*	303
la estrella más brillante (de una galaxia) *f*	estrela mais brilhante (de uma galáxia) *f*	ярчайщая звезда (галактики) *f*	304
galaxia doble *f*	galáxia dupla *f*	двойная галактика *f*	305
galaxias interconectadas *fpl*	galáxias interconectadas *fpl*	взаимосвязанные галактики *fpl*	306
puente brillante *m*	ponte brilhante [luminosa] *f*	яркий мост *m*	307
galaxia múltiple *f*	galáxia múltipla *f*	кратная галактика *f*	308

309	Stephan's Quintet	Quintette de Stephan *m*	Stephans Quintett *n*
310	interacting galaxies	galaxies en interaction *fpl* galaxies en action réciproque *fpl*	wechselwirkende Galaxien *fpl*
311	tidal interaction	interaction de marées *f*	Gezeitenwechselwirkung *f*
312	criterion of tidal instability	critère d'instabilité due aux marées *m*	Kriterium der Gezeiteninstabilität *n*
313	tidal instability	instabilité due aux [à l'égard des] marées *f*	Gezeiteninstabilität *f*
314	tidal disturbance tidal perturbation	perturbation de marées *f* perturbation par effet de marée *f*	Gezeitenstörung *f*
315	disturbed galaxy	galaxie perturbée *f*	gestörte Galaxie *f*
316	distorted galaxy	galaxie distordue *f*	verzerrte Galaxie *f*
317	disrupted galaxy	galaxie désintégrée *f* galaxie fragmentée *f*	aufgelöste Galaxie *f* zerfallene Galaxie *f*
318	rupture of a (galactic) nucleus division of a (galactic) nucleus	rupture du noyau (galactique) *f*	Zerreissen des (galaktischen) Kernes *n*
319	multiple-centre [-center] galaxy multiple-nucleus galaxy	galaxie à noyau multiple *f*	Galaxie mit mehrfachem Kern *f*
320	(close) passage of two galaxies	passage (proche) de deux galaxies *m*	(naher) Vorbeigang [Vorübergang] zweier Galaxien *m*
321	galaxies in collision colliding galaxies	galaxies en collision *fpl*	kollidierende Galaxien *fpl* zusammenstossende Galaxien *fpl*
322	collision of galaxies	collision de [entre] galaxies *f*	Zusammenstoss von Galaxien *m*
323	mathematical simulation [modelling]	simulation mathématique *f*	mathematische Nachahmung [Simulation] *f*
324	mathematical model	modèle mathématique *m*	mathematisches Modell *n*
325	group of galaxies	groupe de galaxies *m*	Galaxiengruppe *f*
326	Local Group of galaxies	Groupe Local de galaxies *m*	Lokale Galaxiengruppe *f* Lokale Gruppe von Galaxien *f*
327	local galaxy galaxy of the Local Group	galaxie locale *f* galaxie du Groupe Local *f*	lokale Galaxie *f* Galaxie der Lokalen Gruppe *f*
328	chain of galaxies	chaîne de galaxies *f*	Galaxienkette *f*
329	cluster of galaxies (cluster of nebulae)	amas de galaxies *m* (nid de nébuleuses *m*)	Galaxienhaufen *m* (Nebelhaufen *m*)
330	globular cluster of galaxies	amas globulaire de galaxies *m*	kugelförmiger Galaxienhaufen *m*

Quinteto de Stephan *m*	Quinteto de Stephan *m*	квинтет Стефана *m*	309
galaxias interactuantes *fpl*	galáxias interativas *fpl*	взаимодействующие галактики *fpl*	310
interacción de marea *f*	interação de maré *f*	приливное взаимодействие *n*	311
criterio de inestabilidad por marea *m*	critério de instabilidade de maré *m*	критерий приливной неустойчивости *m*	312
inestabilidad por marea *f*	instabilidade de maré *f*	приливная неустойчивость *f*	313
perturbación de marea *f*	perturbação de maré *f*	приливное возмущение *n*	314
galaxia perturbada *f*	galáxia perturbada *f*	возмущённая галактика *f*	315
galaxia deformada *f* galaxie distorsionada *f*	galáxia deformada *f* galáxia distorcida *f*	деформированная галактика *f* искажённая галактика *f*	316
galaxia disgregada *f* galaxia desintegrada *f*	galáxia desintegrada *f*	разрушенная галактика *f* распавшаяся галактика *f*	317
ruptura de un núcleo (galáctico) *f*	ruptura de um núcleo (galáctico) *f*	деление ядра *n* разрыв [взрыв] галактического ядра *m*	318
galaxia con núcleo multiple *f*	galáxia com núcleo múltiplo *f*	галактика с кратным ядром *f*	319
cruzamiento [pasaje próximo] de dos galaxias *m*	passagem (próxima) de duas galáxias *f*	(близкое) прохождение двух галактик *n*	320
galaxias en colisión *fpl* galaxias colisionantes *fpl*	galáxias em colisão *fpl*	сталкивающиеся галактики *fpl*	321
colisión de galaxias *f*	colisão de galáxias *f*	столкновение галактик *n*	322
simulación matemática *f*	simulação matemática *f*	математическое моделирование *n*	323
modelo matemático *m*	modelo matemático *m*	математическая модель *f*	324
grupo de galaxias *m*	grupo de galáxias *m*	группа галактик *f*	325
Grupo Local de galaxias *m*	Grupo Local de galáxias *m*	Местная Группа галактик *f*	326
galaxia local *f* galaxia del Grupo Local *f*	galáxia local *f* galáxia do Grupo Local *f*	галактика Местной Группы *f* местная галактика *f*	327
cadena de galaxias *f*	cadeia de galáxias *f*	цепочка [цепь] галактик *f*	328
cúmulo de galaxias *m* (cúmulo de nebulosas *m*)	aglomerado [cúmulo] de galáxias *m* (cúmulo de nebulosas *m*)	скопление галактик *n* (скопление внегалактических туманностей *n*)	329
cúmulo globular de galaxias *m*	aglomerado [cúmulo] globular de galáxias *m*	шаровое скопление галактик *n*	330

331	poor cluster of galaxies	amas pauvre en galaxies *m*	galaxienarmer Haufen *m*
332	rich cluster of galaxies	amas riche en galaxies *m*	galaxienreicher Haufen *m* reicher Galaxienhaufen *m*
333	open cluster of galaxies	amas ouvert de galaxies *m*	offener Galaxienhaufen *m*
334	compact cluster of galaxies	amas compact de galaxies *m*	kompakter Galaxienhaufen *m*
335	very compact cluster of galaxies	amas très compact de galaxies *m*	sehr kompakter Galaxienhaufen *m*
336	regular cluster of galaxies	amas régulier de galaxies *m*	regelmässiger Galaxienhaufen *m*
337	irregular cluster of galaxies	amas irrégulier de galaxies *m*	unregelmässiger Galaxienhaufen *m*
338	cluster member [galaxy] member galaxy	galaxie de l'amas *f* galaxie membre de l'amas *m*	Galaxienhaufenmitglied *n* Haufengalaxie *f*
339	brightest member brightest cluster-galaxy first-ranked cluster-galaxy	galaxie la plus brillante de l'amas *f*	hellste (Haufen-)Galaxie *f* hellstes Galaxienhaufenmitglied *n*
340	field galaxy background galaxy non-cluster galaxy	galaxie de champ *f* (galaxie) non-membre de l'amas (*f*)*m* galaxie du fond (du ciel) *f*	Feldgalaxie *f* Hintergrundgalaxie *f* nicht zum Haufen gehörende Galaxie *f*
341	cluster population	population de l'amas *f* population galactique *f*	Galaxienhaufenpopulation *f*
342	structural index of a cluster	indice structural de l'amas *m*	Strukturindex eines Galaxienhaufens *m*
343	galaxy count	dénombrement de galaxies *m*	Galaxienzählung *f*
344	density of galaxies	densité galactique *f* densité de galaxies *f*	Galaxiendichte *f*
345	Virgo cluster (of galaxies)	amas (de galaxies) de la Vierge *m*	Virgo-Haufen *m* Galaxienhaufen im Virgo *m*
346	supercluster of galaxies cluster of clusters supergalaxy	superamas de galaxies *m* amas d'amas de galaxies *m* supergalaxie *f*	Superhaufen *m* Supergalaxie *f* Supergalaxienhaufen *m*
347	Local Supercluster (of galaxies) LSC Virgo Supercluster	Supergalaxie Locale *f* Superamas Local *m* Superamas de la Vierge *m*	Lokaler Superhaufen *m* Lokale Supergalaxie *f* Virgo-Supergalaxie *f*
348	intergalactic bridge	pont (lumineux) intergalactique *m*	intergalaktische Materiebrücke [Lichtbrücke] *f*

GALAXIAS	GALÁXIAS	ГАЛАКТИКИ	
cúmulo pobre de galaxias *m*	aglomerado [cúmulo] pobre de [em] galáxias *m*	бедное скопление галактик *n*	331
cúmulo rico de galaxias *m*	aglomerado [cúmulo] rico de [em] galáxias *m*	богатое скопление галактик *n*	332
cúmulo abierto de galaxias *m*	aglomerado [cúmulo] aberto de galáxias *m*	рассеянное скопление галактик *n*	333
cúmulo compacto de galaxias *m*	aglomerado [cúmulo] compacto de galáxias *m*	компактное скопление галактик *n*	334
cúmulo muy compacto de galaxias *m*	aglomerado [cúmulo] muito compacto de galáxias *m*	очень компактное скопление галактик *n*	335
cúmulo regular de galaxias *m*	aglomerado [cúmulo] regular de galáxias *m*	правильное скопление галактик *n*	336
cúmulo irregular de galaxias *m*	aglomerado [cúmulo] irregular de galáxias *m*	неправильное скопление галактик *n*	337
galaxia del cúmulo *f* miembro del cúmulo *m*	galáxia do cúmulo *f*	член скопления галактик *m* галактика скопления *f*	338
galaxia más brillante del cúmulo *f* miembro más brillante del cúmulo *m*	galáxia mais brilhante do cúmulo [do aglomerado] *f*	ярчайший член скопления *m* ярчайшая галактика скопления *f*	339
galaxia de campo *f* galaxia de fondo *f* galaxia no perteneciente al cúmulo *f*	galáxia do campo *f* galáxia não-membro do cúmulo [aglomerado] *f* galáxia não-pertencente ao cúmulo [aglomerado] *f*	галактика поля [фона] *f* галактика, не принадлежащая скоплению *f* нечлен скопления галактик *m*	340
población del cúmulo *f*	população do cúmulo [aglomerado] *f*	населённость скопления *f*	341
índice estructural del cúmulo *m*	índice estrutural do cúmulo [do aglomerado] *m*	структурный показатель [индекс] скопления *m*	342
cuenta [recuento] de galaxias *f* [*m*]	contagem de galáxias *f*	подсчёт галактик *m*	343
densidad galáctica *f* densidad de galaxias *f*	densidade de galáxias *f* densidade galáctica *f*	плотность галактик *f*	344
cúmulo (de galaxias) de la Virgen [de Virgo] *m*	cúmulo [aglomeroado] (de galáxias) da Virgem *m*	скопление галактик в Деве *n*	345
supercúmulo de galaxias *m* cúmulo de cúmulos (de galaxias) *m* supergalaxia *f*	superaglomerado de galáxias *m* supercúmulo de galáxias *m* cúmulo dos cúmulos de galáxias *m* supergaláxia *f*	сверхскопление галактик *n* скопление скоплений *n* сверхгалактика *f*	346
Supercúmulo (de galaxias) Local *m* Supergalaxia Local *f* Supergalaxia de la Virgen *f*	Superaglomerado Local *m* Supercúmulo Local de galáxias *m* Supergaláxia Local *f* Supergaláxia da Virgem *f*	Местная сверхгалактика *f* Сверхгалактика *f* Сверхгалактика Девы *f* Местное сверхскопление *n*	347
puente intergaláctico *m*	ponte intergaláctica *f*	межгалактический (яркий) мост *m*	348

349	intergalactic matter	matière intergalactique *f*	intergalaktische Materie *f*
350	intergalactic star	étoile intergalactique *f*	intergalaktischer Stern *m*
351	circumgalactic gas	gaz circumgalactique *m*	zirkumgalaktisches Gas *n*
351a	intergalactic gas	gaz intergalactique *m*	intergalaktisches Gas *n*
352	intracluster gas	gaz dans un amas de galaxies *m* gaz intergalactique d'un amas *m*	intergalaktisches Gas im Haufen *n*
353	intercluster gas gas between clusters	gaz intergalactique entre amas *m*	Zwischenhaufen-Gas *n* intergalaktisches Gas zwischen Haufen *n*
354	intergalactic gas cloud	nuage de gaz intergalactique *m*	intergalaktische Gaswolke *f*
355	capture of intergalactic matter	capture de matière intergalactique *f*	Einfang von intergalaktischer Materie *m*
356	infall of gas (from intergalactic space)	afflux de gaz (de l'espace intergalactique) *m*	Einfall von Gas (aus dem intergalaktischen Raum) *m*
357	metagalaxy	métagalaxie *f*	Metagalaxis *f*
358	quasi-stellar source QSS	source quasi stellaire *f*	quasistellare Quelle *f*
359	quasi-stellar object QSO	objet quasi stellaire *m* objet (d'apparence) stellaire *m*	quasistellares Objekt *n*
360	quasi-stellar radio source radio QSO quasar	radiosource quasi stellaire *f* quasar (radio-électrique) *m* quasi-étoile *f*	quasistellare Radioquelle *f* Quasar *m*
361	quasi-stellar galaxy quasag QSG	galaxie quasi stellaire *f* quasag *m*	quasistellare Galaxie *f* Quasag *m*
362	quiet QSO radio quiet QSO radio quiet object radio quiet quasar	QSO calme *m* QSO non-radioémetteur *m* objet non-radioémetteur *m* quasar sans rayonnement hertzien *m* objet [quasar, QSO] radio-calme *m*	radioruhiges quasistellares Objekt *n* radioruhiger Quasar *m*
363	non-velocity redshift non-Doppler redshift	déplacement vers le rouge non-Doppler(ien) *m* déplacement non-Doppler-Fizeau *m*	Rotverschiebung die nicht auf dem Dopplereffekt beruht *f* Rotverschiebung die nicht auf dem Hubble-Effekt beruht *f*
363a	cosmological redshift	déplacement [décalage] cosmologique *m*	kosmologische Rotverschiebung *f*
364	multiple redshift	décalage multiple vers le rouge *m*	vielfache Rotverschiebung *f*

materia intergaláctica *f*	matéria intergaláctica *f*	межгалактическая материя *f*	349
estrella intergaláctica *f*	estrela intergaláctica *f*	межгалактическая звезда *f*	350
gas circungaláctico *m*	gás circungaláctico *m*	окологалактический газ *m*	351
gas intergaláctico *m*	gás intergaláctico *m*	межгалактический газ *m*	351a
gas intergaláctico en cúmulo *m* gas intracumular *m*	gás intergaláctico em um cúmulo *m* gás intracúmulo *m*	межгалактический газ внутри скопления *m*	352
gas intergaláctico entre cúmulos *m* gas entrecúmulos *m*	gás intergaláctico entre cúmulos *m* gás intercúmulo *m*	газ между скоплениями галактик *m*	353
nube de gas intergaláctico *f*	nuvem de gás intergaláctico *f*	облако межгалактического газа *n*	354
captura de (la) materia intergaláctica *f*	captura da matéria intergaláctica *f*	захват межгалактической материи *m*	355
caída de gas (del espacio intergaláctico) *f*	afluxo [queda] de gás (do espaço intergaláctico) *m*[*f*]	натекание [захват, аккреция, падение] газа (из межгалактического пространства) *n*[*m,f,n*]	356
metagalaxia *f*	metagaláxia *f*	метагалактика *f*	357
fuente cuasi-estelar *f*	fonte quase-estelar *f*	квазизвёздный источник (излучения) *m*	358
objeto cuasi-estelar *m*	objeto quase-estelar *m*	квазизвёздный объект *m*	359
fuente (radioeléctrica) cuasi-estelar *f* cuásar *m* radiofuente cuasi-estelar *f*	fonte de rádio quase-estelar *f* quasar *m* radiofonte quase-estelar *f*	квазизвёздный радиоисточник *m* квазар *m* радиоквазар *m*	360
galaxia cuasi-estelar *f* cuásag *m*	galáxia quase-estelar *f* quasag *m*	квазизвёздная галактика *f* квазаг *f*	361
QSO inactivo en radio *m* objeto silencioso en radio *m* QSO quiescente en radio *m* cuásar silencioso *m*	QSO calmo *m* quasar calmo *m* QSO radiocalmo *m* quasar radiocalmo *m*	спокойный [тихий] квазизвёздный объект *m* квазизвёздный объект без радиоизлучения *m* радиоспокойный квазар *m*	362
desplazamiento hacia el rojo no debido a velocidad *m*	deslocamento para o vermelho não-proveniente de velocidade *m*	недопплеровское красное смещение *n*	363
desplazamiento cosmológico *m*	deslocamento [desvio] cosmológico *m*	космологическое красное смещение *n*	363a
desplazamiento múltiple hacia el rojo *m*	deslocamento [desvio] múltiplo para o vermelho *m*	многократное красное смещение *n*	364

365	zero-velocity wavelength	longueur d'onde au repos *f*	Ruh(e)wellenlänge *f* Nullgeschwindigkeitswellenlänge *f*
366	redshifted line	raie déplacée vers le rouge *f*	rotverschobene Linie *f*
367	anomalous redshift discrepant redshift	décalage anormal vers le rouge *m* déplacement aberrant vers le rouge *m*	anomale Rotverschiebung *f*
368	local object (<10 Mpc)	objet local *m*	lokales Objekt *n*
369	cosmological object >10 Mpc)	objet cosmologique *m*	kosmologisches Objekt *n*
370	cosmological density	densité cosmologique *f*	kosmologische Dichte *f*

longitud de onda en reposo *f* longitud de onda de velocidad cero *f*	comprimento de onda em repouso *m* comprimento de onda de velocidade zero [nula] *m*	длина волны для нулевой скорости *f*	365
línea desplazada hacia el rojo *f*	linha deslocada para o vermelho *f*	линия с красным смещением *f* линия, смещенная к красному концу (спектра) *f*	366
desplazamiento anómalo hacia el rojo *m*	deslocamento [desvio] anômalo [discrepante] para o vermelho *m*	аномальное красное смещение *n*	367
objeto local *m*	objeto local *m*	близкий объект *m*	368
objeto cosmológico *m*	objeto cosmológico *m*	космологический объект *m*	369
densidad cosmológica *f*	densidade cosmológica *f*	космологическая плотность *f*	370

99 UNIVERSE | UNIVERS | WELTALL

	UNIVERSE	UNIVERS	WELTALL
1	universe	univers *m*	All [Weltall, Universum] *n*
2	cosmos	cosmos *m*	Kosmos *m*
3	world	monde *m*	Welt *f*
4	finiteness of the universe	caractère fini de l'univers *m* finitude de l'univers *f*	Endlichkeit des Weltalls *f*
5	radius of the universe	rayon de l'univers *m*	Weltradius *m*
6	infinity of the universe	caractère infini de l'univers *m* infinitude de l'univers *f*	Unendlichkeit des Weltalls *f*
7	cosmic matter cosmic material	matière cosmique *f* matériau cosmique *m*	kosmische Materie *f* kosmisches Material *n*
8	distribution of matter	distribution de matière *f*	Materieverteilung *f*
9	macrostructure	macrostructure *f*	Makrostruktur *f*
10	horizon of the universe	horizon de l'univers *m* horizon cosmique *m*	Horizont des Weltalls *m*
11	origin of the universe origin of the world	origine de l'univers *f* formation du monde *f*	Entstehung des Weltalls *f* Ursprung der Welt *m* Weltentstehung *f*
12	age of the universe	âge de l'univers *m*	Weltalter *n* Alter des Weltalls *n*
13	eternity of the universe	éternité de l'univers *f*	Ewigkeit des Weltalls [der Welt] *f*
14	heat death of the universe	mort thermique de l'univers *f*	Wärmetod des Weltalls [der Welt] *m*
15	theory of the heat death of the universe	théorie de la mort thermique de l'univers *f*	Theorie des Wärmetodes des Weltalls [der Welt] *f*
16	cosmology	cosmologie *f*	Kosmologie *f*
17	observational cosmology	cosmologie observationnelle *f*	beobachtende Kosmologie *f* Beobachtungskosmologie *f*
18	cosmologic(al)	cosmologique	kosmologisch
19	cosmologist expert in cosmology	cosmologue *m* cosmologiste *m*	Kosmologe *m*
20	world system cosmological system	système du monde *m* système cosmologique *m*	Weltsystem *n* Weltbild *n*
21	geocentric system	système géocentrique *m*	geozentrisches System *n*
22	Ptolemaic system	système ptolémaïque *m* système de Ptolémée *m*	Ptolemäisches Weltsystem *n* Ptolemäisches System *n*
23	heliocentric system	système héliocentrique *m*	heliozentrisches Weltsystem *n*

universo *m*	universo *m*	вселенная *f*	1
cosmos *m*	cosmo [cosmos] *m*	космос *m*	2
mundo *m*	mundo *m*	мир *m*	3
finitud del universo *f*	finitude do universo *f*	конечность вселенной *f*	4
radio del universo *m*	raio do universo *m*	радиус вселенной *m*	5
infinitud [infinidad] del universo *f*	infinidade do universo *f*	бесконечность вселенной *f*	6
materia cósmica *f* material cósmico *m*	matéria cósmica *f* material cósmico *m*	космическое вещество *n* космическая материя *f*	7
distribución de la materia *f*	distribuição da matéria *f*	распределение материи *n*	8
macroestructura *f*	macroestrutura *f*	макроструктура *f*	9
horizonte del universo *m* horizonte cósmico *m*	horizonte do universo *m* horizonte cósmico *m*	горизонт вселенной *m*	10
origen del universo *m* principio del mundo *m*	origem do universo *f* origem do mundo *f* formação do mundo *f*	происхождение вселенной *n* происхождение мира *n* начало вселенной *n*	11
edad del universo *f*	idade do universo *f*	возраст вселенной *m*	12
eternidad del universo *f*	eternidade do universo *f*	вечность вселенной *f*	13
muerte térmica del universo *f*	morte térmica do universo *f*	тепловая смерть вселенной *f*	14
teoría de la muerte térmica del universo *f*	teoria da morte térmica do universo *f*	теория тепловой смерти вселенной *f*	15
cosmología *f*	cosmologia *f*	космология *f*	16
cosmología observacional *f*	cosmologia observacional *f*	наблюдательная космология *f*	17
cosmológico	cosmológico	космологический	18
cosmólogo *m*	cosmólogo *m* cosmologista *m*	космолог *m*	19
sistema del mundo *m* sistema cosmológico *m*	sistema do mundo *m* sistema cosmológico *m*	система мира *f* космологическая [мировая] система *f*	20
sistema geocéntrico *m*	sistema geocêntrico *m*	геоцентрическая система *f*	21
sistema ptolemaico *m* sistema ptolomeico *m*	sistema ptolemaico *m* sistema ptolomaico *m*	птолемеева система *f* система Птолемея *f*	22
sistema heliocéntrico *m*	sistema heliocêntrico *m*	гелиоцентрическая система *f*	23

	UNIVERSE	UNIVERS	WELTALL
24	Copernican system system of Copernicus	système de Copernic *m* système copernicien *m*	Kopernikanisches System [Weltsystem] *n* Copernicanisches System *n*
25	Newtonian cosmology	cosmologie newtonienne *f*	Newtonsche Kosmologie *f*
26	kinematical cosmology	cosmologie cinématique *f*	kinematische Kosmologie *f*
27	non-relativistic cosmology	cosmologie non-relativiste *f*	nichtrelativistische Kosmologie *f*
28	relativistic cosmology	cosmologie relativiste *f*	relativistische Kosmologie *f*
29	cosmic(al) constant cosmologic(al) constant	constante cosmique *f* constante cosmologique *f*	kosmische Konstante *f* kosmologische Konstante *f*
30	cosmological principle	principe cosmologique *m*	kosmologisches Prinzip *n* Weltpostulat *n*
31	restricted cosmological principle	principe cosmologique restreint *m*	eingeschränktes kosmologisches Prinzip *n*
32	perfect cosmological principle	principe cosmologique parfait *m*	vollkommenes [perfektes] kosmologisches Prinzip *n* erweitertes Weltpostulat *n*
33	coherency	cohérence *f*	Kohärenz *f*
34	coherency postulate Weyl's postulate	postulat de cohérence *m* postulat de Weyl *m*	Kohärenz-Postulat *n* Weylsches Postulat *n*
35	relativity of inertia	relativité de l'inertie *f*	Relativität der Trägheit *f*
36	mathematical postulate of relativity of inertia	postulat mathématique de la relativité de l'inertie *m*	mathematisches Postulat der Relativität der Trägheit *n*
37	Mach's principle	principe de Mach *m*	Machsches Prinzip *n*
38	cosmological paradoxes	paradoxes cosmologiques *mpl*	kosmologische Paradoxa *npl*
39	gravitational paradox Seeliger's paradox	paradoxe de gravitaton *m* paradoxe de Seeliger *m*	Gravitationsparadoxon *n* Seeligersches Paradoxon *n*
40	Olbers' paradox photometrical paradox Cheseaux' and Olbers' paradox	paradoxe d'Olbers *m* paradoxe photométrique *m* paradoxe de Cheseaux et d'Olbers *m*	Olberssches Paradoxon *n* photometrisches Paradoxon *n*
41	mass horizon	horizon des masses *m*	Massenhorizont *m*
42	singular state	singularité *f*	Singularität *f*
43	event horizon	horizon des événements *m*	Ereignishorizont *m*
44	comoving coordinate systems	systèmes de coordonnées liés à la matière *mpl* coordonnées comobiles *fpl*	mitschwimmende Koordinatensysteme *npl* mitbewegte Koordinatensysteme *npl*
45	Lambert's hypothesis	hypothèse de Lambert *f*	Lambertsche Hypothese *f*

sistema de Copérnico *m* sistema copernicano *m*	sistema copernicano *m* sistema de Copernico *m*	система Коперника *f* коперниковская [коперникова] система *f*	24
cosmología newtoniana *f*	cosmologia newtoniana *f*	ньютоновская космология *f*	25
cosmología cinemática *f*	cosmologia cinemática *f*	кинематическая космология *f*	26
cosmología no relativista *f*	cosmologia não-relativista *f*	нерелятивистская космология *f*	27
cosmología relativista *f*	cosmologia relativista *f*	релятивистская космология *f*	28
constante cósmica *f* constante cosmológica *f*	constante cósmica *f* constante cosmológica *f*	космическая постоянная *f* космологическая постоянная *f*	29
principio cosmológico *m*	princípio cosmológico *m*	космологический принцип *m*	30
principio cosmológico restringido *m*	princípio cosmológico restrito *m*	ограниченный космологический принцип *m*	31
principio cosmológico perfecto *m*	princípio cosmológico perfeito *m*	совершенный космологический принцип *m*	32
coherencia *f*	coerência *f*	когерентность *f*	33
postulado de coherencia *m* postulado de Weyl *m*	postulado de coerência *m* postulado de Weyl *m*	постулат когерентности *m* постулат Вейля *m*	34
relatividad de la inercia *f*	relatividade da inércia *f*	относительность инерции *f*	35
postulado matemático de la relatividad de la inercia *m*	postulado matemático da relatividade da inércia *m*	математический постулат относительности инерции *m*	36
principio de Mach *m*	princípio de Mach *m*	принцип Маха *m*	37
paradojas cosmológicas *fpl*	paradoxos cosmológicos *mpl*	космологические парадоксы *mpl*	38
paradoja gravitatoria *f* paradoja de Seeliger *f*	paradoxo gravitacional *m* paradoxo de Seeliger *m* paradoxo de gravitação *m*	гравитационный парадокс *m* парадокс Зеелигера *m* парадокс Неймана-Зеелигера *m*	39
paradoja de Olbers *f* paradoja fotométrica *f* paradoja de Cheseaux y Olbers *f*	paradoxo de Olbers *m* paradoxo fotométrico *m* paradoxo de Cheseaux e de Olbers *m*	парадокс Ольберса *m* фотометрический парадокс *m* парадокс Шезо-Ольберса *m*	40
horizonte de masas *m*	horizonte de massas *m*	горизонт масс *m*	41
singularidad *f*	singularidade *f*	сингулярное состояние *n* особое состояние *n*	42
horizonte de los sucesos *m*	horizonte de eventos [acontecimentos] *m*	горизонт событий *m*	43
sistemas de referencia invariablemente unidos a... *mpl* sistemas de coordenadas comóviles *mpl*	sistemas de coordenadas associados à matéria *mpl*	сопутствующие координатные системы *fpl*	44
hipotésis de Lambert *f*	hipótese de Lambert *f*	гипотеза Ламберта *f*	45

	UNIVERSE	UNIVERS	WELTALL
46	cosmological model model of the universe world model	modèle cosmologique *m* modèle d'univers *m*	Weltmodell *n* Modell der Welt [des Weltalls] *n* kosmologisches Modell *n*
47	hierarchic structure (of the universe)	structure hiérarchique (de l'univers) *f* hiérarchie de structure (de l'univers) *f*	hierarchische Struktur *f* hierarchischer Aufbau (der Welt) *m*
48	hierarchic model hierarchic universe modèle hiérarchique *m* univers hiérarchique *m*	modèle hiérarchisé *m* univers hiérarchisé *m* modèle hiérarchique *m* univers hiérarchique *m*	hierarchisches Modell des Universums *n* Modell einer hierarchischen Welt *n* hierarchische Welt *f* hierarchisches Weltall *n*
49	empty model empty universe	modèle vide *m* univers vide *m*	Modell einer leeren Welt *n* leerer Kosmos *m* leeres Universum *n* leeres Weltall *n*
50	static model static universe	modèle statique *m* univers statique *m*	statisches Weltmodell *n* Modell einer statischen Welt *n* statischer Kosmos *m* statische Welt *f* statisches Weltall [Universum] *n*
51	non-static model non-static universe	modèle non-statique *m* univers non-statique *m*	nichtstatisches Weltmodell *n* Modell einer nichtstatischen Welt *n* nichtstatische Welt *f* nichtstatisches Weltall [Universum] *n*
52	stationary model stationary universe steady-state universe steady-state model	modèle stationnaire *m* univers stationnaire *m*	stationäres Weltmodell *n* Modell einer stationären Welt *n* stationäre Welt *f* stationäres Weltall [Universum] *n* stationärer Kosmos *m*
53	non-stationary model non-stationary universe	modèle non-stationnaire *m* univers non-stationnaire *m*	nichtstationäres Universum [Weltall, Weltmodell] *n* Modell einer nichtstationären Welt *n*
54	evolving universe	univers en évolution *m*	Entwicklungsuniversum *n* sich entwickelndes Universum [Weltall] *n* Entwicklungskosmos *m*
55	expanding model expanding universe	modèle en expansion *m* univers en expansion *m*	Modell einer expandierenden Welt *n* expandierendes Weltall [Weltmodell] *n* expandierender Kosmos *m*
56	contracting model contracting universe	modèle en contraction *m* univers en contraction *m*	Modell einer kontrahierenden Welt *n* kontrahierendes Weltall [Weltmodell] *n* kontrahierender Kosmos *m*
57	oscillating model oscillating universe	modèle oscillant *m* univers oscillant *m*	oszillierendes Weltall [Universum, Weltmodell] *n* Modell einer oszillierenden Welt *n* oszillierender Kosmos *m*
58	oscillating model of the first kind oscillating universe of the first kind	modèle oscillant de première espèce *m* univers oscillant de première espèce *m*	Modell einer oszillierenden Welt erster Art *n* oszillierendes Weltall erster Art *n* oszillierender Kosmos erster Art *m*

modelo cosmológico *m* modelo del universo *m*	modelo cosmológico *m* modelo de universo *m*	модель вселенной *f* космологическая модель *f*	46
estructura jerárquica (del universo) *f* hierarquia do universo *f*	estrutura hierárquica (do universo) *f* hierarquia do universo *f*	иерархическая структура (вселенной) *f*	47
modelo jerárquico *m* universo jerárquico *m*	modelo hierárquico *m* universo hierárquico *m*	иерархическая модель *f* иерархическая вселенная *f*	48
modelo vacío *m* universo vacío *m*	modelo vazio *m* universo vazio *m*	пустая модель *f* пустая вселенная *f*	49
modelo estático *m* universo estático *m*	modelo estático *m* universo estático *m*	статическая модель *f* статическая вселенная *f*	50
modelo no estático *m* universo no estático *m*	modelo não-estático *m* universo não-estático *m*	нестатическая модель *f* нестатическая вселенная *f*	51
modelo estacionario *m* universo estacionario *m*	modelo estacionário *m* universo estacionário *m*	стационарная модель *f* стационарная вселенная *f*	52
modelo no estacionario *m* universo no estacionario *m*	modelo não-estacionário *m* universo não-estacionário *m*	нестационарная модель *f* нестационарная вселенная *f*	53
universo en evolución *m*	universo em evolução *m*	эволюционирующая вселенная *f*	54
modelo (del universo) en expansión *m* universo en expansión *m*	modelo em expansão *m* universo em expansão *m*	расширяющаяся модель *f* расширяющаяся вселенная *f*	55
modelo (del universo) en contracción *m* universo en contracción *m*	modelo em contração *m* universo em contração *m*	сжимающаяся модель *f* сжимающаяся вселенная *f*	56
modelo oscilatorio *m* universo oscilatorio *m*	modelo pulsante *m* universo pulsante *m* universo oscilante *m*	осциллирующая модель *f* осциллирующая вселенная *f* пульсирующая вселенная *f* пульсируящая модель *f*	57
modelo oscilatorio de primera especie *m* universo oscilatorio de primera especie *m*	modelo pulsante de primeira espécie *m* universo pulsante de primeira espécie *m*	осциллирующая модель первого рода *f* осциллирующая вселенная первого рода *f*	58

	UNIVERSE	UNIVERS	WELTALL
59	asymptotic model asymptotic universe	modèle asymptotique *m* univers asymptotique *m*	asymptotisches Weltmodell [Weltall] *n* Modell einer asymptotischen Welt *n* asymptotischer Kosmos *m*
60	asymptotic model of the first kind asymptotic universe of the first kind	modèle asymptotique de première espèce *m* univers asymptotique de première espèce *m*	Modell einer asymptotischen Welt erster Art *n* asymptotisches Weltmodel [Weltall] erster Art *n*
61	monotonic model monotonic universe	modèle monotone *m* univers monotone *m*	Modell einer monotonen Welt *n* monotones Weltall [Weltmodell] *n* monotoner Kosmos *m*
62	monotonic model of the first kind monotonic universe of the first kind	modèle monotone de première espèce *m* univers monotone de première espèce *m*	Modell einer monotonen Welt erster Art *n* monotones Weltall [Weltmodell] erster Art *n*
63	isotropic model isotropic universe	modèle isotrope *m* univers isotrope *m*	isotropes Weltmodell [Weltall] *n* Modell einer isotropen Welt *n* isotroper Kosmos *m*
64	homogeneous model homogeneous universe	modèle homogène *m* univers homogène *m*	Modell einer homogenen Welt *n* homogenes Weltmodell [Weltall] *n* homogener Kosmos *m*
65	anisotropic model anisotropic universe	modèle anisotrope *m* univers anisotrope *m*	anisotropes Weltmodell [Weltall] *n* Modell einer anisotropen Welt *n* anisotroper Kosmos *m*
66	non-homogeneous model non-homogeneous universe	modèle non-homogène *m* univers non-homogène *m*	inhomogenes Weltmodell [Weltall] *n* Modell einer inhomogenen Welt *n* inhomogener Kosmos *m*
67	Charlier('s) model Charlier('s) universe	modèle de Charlier *m* univers de Charlier *m*	Charliersches Weltmodell *n* Charliersches Weltall [Universum] *n*
68	Einstein('s) model Einstein('s) universe	modèle d'Einstein *m* univers d'Einstein *m*	Einsteins Weltmodell [Weltall] *n* Einstein-Universum *n*
69	De Sitter('s) model De Sitter('s) universe	modèle de De Sitter *m* univers de De Sitter *m*	De Sitter-Weltmodell [Weltall] *n* De Sitter-Universum *n*
70	model [universe] of Einstein and De Sitter Einstein-De Sitter universe [model]	modèle d' Einstein-De Sitter *m* univers d'Einstein-De Sitter *m*	Weltmodell von Einstein und De Sitter *n* Einstein-De Sitter Universum [Weltall] *n*
71	Friedmann models Friedmann universes	modèles de Friedmann *mpl* univers de Friedmann *mpl*	Friedmannsche Weltmodelle *npl* Friedmannsche Welten *fpl*
72	open model open universe	modèle ouvert *m* univers ouvert *m*	offener Kosmos *m* offenes kosmologisches Modell *n* Modell einer offenen Welt *n* offenes Weltall [Universum] *n*

modelo asintótico *m* universo asintótico *m*	modelo assintótico [assimptótico] *m* universo assintótico [assimptótico] *m*	асимптотическая модель *f* асимптотическая вселенная *f*	59
modelo asintótico de primera especie *m* universo asintótico de primera especie *m*	modelo assintótico de primeira espécie *m* universo assintótico de primeira espécie *m*	асимптотическая модель первого рода *f* асимптотическая вселенная первого рода *f*	60
modelo monótono *m* universo monótono *m*	modelo monotônico *m* universo monotônico *m*	монотонная модель *f* монотонная вселенная *f*	61
modelo monótono de primera especie *m* universo monotóno de primera especie *m*	modelo monotônico de primeira espécie *m* universo monotônico de primeira espécie *m*	монотонная модель первого рода *f* монотонная вселенная первого рода *f*	62
modelo isótropo *m* universo isótropo *m*	modelo isótropo *m* universo isótropo *m* modelo [universo] isotrópico *m*	изотропная модель *f* изотропная вселенная *f*	63
modelo homogéneo *m* universo homogéneo *m*	modelo homogêneo *m* universo homogêneo *m*	однородная модель *f* однородная вселенная *f*	64
modelo anisótropo *m* universo anisótropo *m*	modelo anisotrópico [anisótropo, não-isotrópico] *m* universo anisotrópico [anisótropo, não-isotrópico] *m*	анизотропная модель *f* анизотропная вселенная *f*	65
modelo no homogéneo *m* universo no homogéneo *m* modelo inhomogéneo *m*	modelo inomogêneo *m* modelo não-homogêneo *m* universo inomogêneo [não-homogêneo] *m*	неоднородная модель *f* неоднородная вселенная *f*	66
modelo de Charlier *m* universo de Charlier *m*	modelo de Charlier *m* universo de Charlier *m*	модель Шарлье *f* вселенная Шарлье *f*	67
modelo de Einstein *m* universo de Einstein *m*	modelo de Einstein *m* universo de Einstein *m*	модель Эйнштейна *f* вселенная Эйнштейна *f*	68
modelo de De Sitter *m* universo de De Sitter *m*	modelo de De Sitter *m* universo de De Sitter *m*	модель Де Ситтера *f* вселенная Де Ситтера *f*	69
modelo de Einstein y De Sitter *m* universo de Einstein y De Sitter *m*	modelo de Einstein e De Sitter *m* universo de Einstein e De Sitter *m*	модель Эйнштейна – Де Ситтера *f* вселенная Эйнштейна – Де Ситтера *f*	70
modelos de Friedmann *mpl* universos de Friedmann *mpl*	modelos de Friedmann *mpl* universos de Friedmann *mpl*	модели Фридмана *fpl* вселенные Фридмана *fpl*	71
modelo abierto *m* universo abierto *m*	modelo aberto *m* universo aberto *m*	открытая модель *f* открытая вселенная *f*	72

73	closed model closed universe	modèle clos *m* univers fermé *m* modèle fermé *m* univers clos *m*	geschlosssenes kosmologisches Modell *n* Modell einer geschlosenen Welt *n* geschlossenes Weltall [Universum] *n* geschlossener Kosmos *m*
74	expansion of the universe	expansion de l'univers *f*	Expansion des Weltalls *f*
75	expansion rate velocity of expansion recession velocity	vitesse d'expansion *f* vitesse de récession *f* taux d'expansion *m*	Ausdehnungsrate *f* Ausdehnungsgeschwindigkeit *f* Expansionsrate *f* Fluchtgeschwindigkeit *f* Expansionsgeschwindigkeit *f*
76	redshift-distance relation	relation décalage vers le rouge-distance *f* relation distance - décalage spectral *f*	Rotverschiebung-Entfernungs-Beziehung *f*
77	Hubble('s) constant expansion constant H	constante de récession *f* constante de Hubble *f*	Hubble-Konstante *f* Hubblekonstante *f* Hubblesche Konstante *f*
78	expansion time	temps d'expansion *m*	Expansionszeit (des Weltalls) *f* Ausdehnungszeit *f*
79	characteristic age characteristic time Hubble (expansion) time H_0^{-1}	âge caractéristique *m* temps de Hubble *m* temps caractéristique *m* temp d'expansion de Hubble *m*	Hubble-Zeit *f* Hubblesche Zeit *f* charakteristische Zeit *f* Hubblesche Expansionszeit *f*
80	Friedmann (expansion) time	temps (d'expansion) de Friedmann *m*	Friedmannsche Expansionszeit *f*
81	accelerated expansion	expansion accélérée *f*	beschleunigte Expansion *f* beschleunigte Ausdehnung *f*
81a	scale factor cosmic scale factor $R(t)$	fonction $R(t)$ *f* rayon de l'univers *m* facteur d'expansion *m* facteur d'échelle (cosmique) *m*	Skalenfaktor *m* Expansionsfaktor *m*
82	deceleration of the expansion retardation of the expansion	ralentissement de l'expansion *m* décélération de l'expansion *f*	Verzögerung der Expansion *f* Ausdehnungsverzögerung *f*
83	deceleration parameter	paramètre de décélération *m*	Dezelerationsparameter (q) *m* Verzögerungsparameter *m*
84	mean [average] density of matter mean (mass) density	densité moyenne de matière *f* densité moyenne (de masse) *f*	mittlere Materiedichte *f* mittlere (Massen)dichte *f*
85	radiation density	densité de radiation *f*	Strahlungsdichte *f*
85a	critical density	densité critique *f*	kritische Dichte *f*
86	luminous matter	matière lumineuse *f*	leuchtende Materie *f*
87	missing mass	masse manquante *f*	fehlende Masse *f*
88	hidden mass	masse cachée *f*	verborgene Masse *f*

modelo cerrado *m* universo cerrado *m*	modelo fechado *m* universo fechado *m*	закрытая модель *f* закрытая вселенная *f* замкнутая модель *f* замкнутая вселенная *f*	73
expansión del universo *f*	expansão do universo *f*	расширение вселенной *n*	74
tasa de expansión *f* velocidad de expansión *f* tasa de recesión *f* velocidad de recesión *f*	taxa [razão] de expansão *f* velocidade de expansão *f* velocidade de recessão *f* taxa de recessão *f*	скорость расширения *f*	75
relación entre corrimiento hacia el rojo y la distancia *f*	relação desvio para o vermelho – distância *f* relação de Hubble *f*	зависимость красное смещение– – расстояние *f*	76
constante de Hubble *f* constante de expansión *f*	constante de Hubble *f* constante de expansão *f*	постоянная Хаббла *f* хаббловская постоянная *f*	77
tiempo de expansión *m*	tempo de expansão *m*	время расширения *n*	78
edad característica *f* tiempo de Hubble *m* tiempo característico *m*	tempo de Hubble *m* tempo característico *m* idade característica *f*	хаббловское время *n* характеристическое время *n* хаббловский возраст *m* характерный возраст *m*	79
tiempo de Friedmann *m*	tempo de Friedmann *m*	фридмановское время (расширения) *n*	80
expansión acelerada *f*	expansão acelerada *f*	ускоренное расширение *n*	81
factor de escala *m* función $R(t)$ *f* factor de la expansión cósmica *m*	função [fator] $R(t)$ *f*[*m*] fator de escala *m* fator da expansão cósmica *m*	масштабный фактор *m* фактор $R(t)$ *m* функция $R(t)$ *f*	81a
desaceleración de la expansión *f* retardación de la expansión *f*	desaceleração da expansão *f* retardo [retardação] da expansão *m*[*f*]	замедление расширения *n*	82
parámetro de desaceleración *m*	parâmetro de desaceleração *m*	параметр замедления *m*	83
densidad media de la materia *f* densidad media (de masa) *f*	densidade média da matéria *f* densidade média (de massa) *f*	средняя плотность вещества *f* средняя плотность массы *f*	84
densidad de radiación *f*	densidade da radiação *f*	плотность излучения *f*	85
densidad crítica *f*	densidade crítica *f*	критическая плотность *f*	85a
materia luminosa *f*	matéria luminosa *f*	светящееся вещество *n* светящаяся материя *f*	86
masa ausente [faltante] *f*	massa ausente [faltante] *f*	недостающаяся масса *f*	87
masa escondida *f*	massa escondida *f*	скрытая масса *f*	88

89	creation of matter	création de matière *f*	Erzeugung von Materie *f* Materieerzeugung *f*
90	unique creation	création unique *f*	einmalige Schöpfung *f*
91	continuous creation continual creation	création continue *f*	kontinuierliche Erzeugung *f* kontinuierliche Schöpfung *f*
92	relativistic universe	univers relativiste *m*	relativistisches Weltall *n*
93	classical universe Newtonian universe	univers classique *m* univers newtonien *m*	klassisches Weltall *n* Newtonsches Universum *n*
94	special theory of relativity restricted theory of relativity	relativité restreinte *f* théorie de la relativité restreinte *f*	spezielle Relativitätstheorie *f*
95	relativistic	relativiste	relativistisch
96	non-relativistic	non-relativiste	nichtrelativistisch
97	non-relativistic approximation	approximation non-relativiste *f*	nichtrelativistische Näherung *f*
98	inertial system	système inertiel [d'inertie] *m*	Inertialsystem *n*
99	relative motion	mouvement relatif *m*	Relativbewegung *f*
100	space	espace *m*	Raum *m*
100a	metric space	espace métrique [distancié] *m*	metrischer Raum *m*
101	Euclidean	euclidien	euklidisch
102	non-Euclidean	non-euclidien	nichteuklidisch
103	finite space	espace fini *m*	endlicher Raum *m*
104	infinite space	espace infini *m*	unendlicher Raum *m*
105	Lobachevski('s) space	espace de Lobatchevski *m*	Lobatschewskischer Raum *m*
106	hyperbolic space	espace hyperbolique *m*	hyperbolischer Raum *m*
107	Euclidean space Euclid's space parabolic space flat space	espace euclidien *m* espace parabolique *m* espace plat *m*	flacher Raum *m* Euklidischer Raum *m* parabolischer Raum *m*
108	Euclidean geometry	géométrie euclidienne *f* géométrie d'Euclide *f*	Euklidische [euklidische] Geometrie *f*
109	non-Euclidean geometry	géométrie non-euclidienne *f*	nichteuklidische Geometrie *f*

creación de (la) materia *f*	criação da matéria *f*	творение материи *n* образование вещества [материи] *n*	89
creación única *f*	criação única *f*	единый акт творения *m*	90
creación continua *f*	criação contínua *f*	непрерывное творение *n*	91
universo relativístico *m* universo relativista *m*	universo relativista *m*	релятивистская вселенная *f*	92
universo clásico *m* universo newtoniano *m*	universo clássico *m* universo newtoniano *m*	классическая вселенная *f* ньютоновская вселенная *f*	93
(teoría de la) relatividad especial *f* (teoría de la) relatividad restringida *f*	teoria da relatividade restrita *f* relatividade especial *f*	специальная [частная] теория относительности *f* СТО	94
relativista relativístico	relativista relativístico	релятивистский	95
no relativista	não-relativístico não-relativista	нерелятивистский	96
aproximación no relativista *f*	aproximação não-relativística *f* aproximaçao não-relativista *f*	нерелятивистское приближение *n*	97
sistema inercial *m*	sistema inercial *m*	инерциальная система отсчёта *f*	98
movimiento relativo *m*	movimento relativo *m*	относительное движение *n*	99
espacio *m*	espaço *m*	пространство *n*	100
espacio métrico *m*	espaço métrico *m*	метрическое пространство *n*	100a
euclídeo euclidiano	euclidiano	Евклида [Эвклида] евклидов [эвклидов]	101
no euclídeo	não-euclidiano	неевклидов неэвклидов	102
espacio finito *m*	espaço finito *m*	конечное пространство *n*	103
espacio infinito *m*	espaço infinito *m*	бесконечное пространство *n*	104
espacio de Lobachevski *m*	espaço de Lobatchevski *m*	пространство Лобачевского *n*	105
espacio hiperbólico *m*	espaço hiperbólico *m*	гиперболическое пространство *n*	106
espacio euclídeo *m* espacio parabólico *m* espacio plano *m*	espaço euclidiano *m* espaço parabólico *m* espaço plano *m*	пространство Эвклида [Евклида] *n* плоское пространство *n* параболическое пространство *n*	107
geometría euclídea *f*	geometria euclidiana *f*	геометрия Эвклида [Евклида] *f* евклидова [эвклидова] геометрия *f*	108
geometría no euclídea *f*	geometria não-euclidiana *f*	неевклидова геометрия *f* неэвклидова геометрия *f*	109

110	elliptic space Newcombian space	espace elliptique *m* espace de Newcomb *m*	elliptischer Raum *m* Newcombscher Raum *m*
111	spherical space	espace sphérique *m*	sphärischer Raum *m*
112	Riemannian space Riemann('s) space	espace riemannien *m* espace de Riemann *m*	Riemannscher Raum *m*
113	Riemannian geometry	géométrie riemannienne *f*	Riemannsche Geometrie *f*
114	curved space non-Euclidean space	espace courbe *m* espace non-euclidien *m*	gekrümmter Raum *m* nichteuklidischer Raum *m*
115	curvature of space	courbure de l'espace *f*	Krümmung des Raumes *f* Raumkrümmung *f*
116	absolute space	espace absolu *m*	absoluter Raum *m*
117	absolute time	temps absolu *m*	absolute Zeit *f*
118	Newtonian time	temps de Newton *m*	Newtonsche Zeit *f*
119	space-time spacetime	espace-temps *m*	Raum-Zeit *f*
120	space-time...	spatio-temporel	raumzeitlich
121	four-dimensional	à quatre dimensions quadridimensionnel	vierdimensional
122	Minkowski(an) space-time Minkowski(an) space Minkowski(an) four-space Minkowski universe	espace-temps de Minkowski *m* espace d'événements *m* espace minkowskien *m* 4-espace de Minkowski *m* univers de Minkowski *m*	Minkowskische Raum-Zeit *f* Minkowskische Welt *f* Minkowskischer Raum *m* Raum-Zeit-Welt *f* Minkowskis vierdimensionale Welt *f*
123	time coordinate (timelike coordinate)	coordonnée de temps *f* coordonné de genre temps *f*	Zeitkoordinate *f* (zeitartige Koordinate) *f*
124	space coordinate spatial coordinate	coordonnée d'espace [spatiale] *f* coordonnée de position *f*	Raumkoordinate *f* Ortskoordinate *f*
125	event world point space-time point	événement *m* point de l'espace-temps *m* point d'univers *m*	Ereignis *n* Weltpunkt *m* Raumzeitpunkt *m* raum-zeitliches Ereignis *n*
126	adjacent event	événement adjacent *m* événement infiniment voisin *m*	benachbartes Ereignis *n* unendlich benachbarter Weltpunkt *m*
127	simultaneity	simultanéité *f*	Gleichzeitigkeit *f*
128	simultaneous events	événements simultanés *mpl*	gleichzeitige Ereignisse *npl*
129	propagation of light	propagation de la lumière *f*	Ausbreitung des Lichtes *f* Fortpflanzung des Lichtes *f* Lichtfortpflanzung *f*

espacio elíptico *m* espacio de Newcomb *m*	espaço elíptico *m* espaço de Newcomb *m*	эллиптическое пространство *n* пространство Ньюкома *n*	110
espacio esférico *m*	espaço esférico *m*	сферическое пространство *n*	111
espacio de Riemann *m* espacio riemanniano *m*	espaço de Riemann *m* espaço riemanniano *m*	пространство Римана *n* риманово пространство *n*	112
geometría de Riemann *f*	geometria riemanniana *f*	риманова геометрия *f*	113
espacio curvo *m* espacio no euclídeo *m*	espaço curvo *m* espaço não-euclidiano *m*	неевклидово [неэвклидово] пространство *n* искривлённое пространство *n*	114
curvatura del espacio *f*	curvatura do espaço *f*	кривизна пространства *f*	115
espacio absoluto *m*	espaço absoluto *m*	абсолютное пространство *n*	116
tiempo absoluto *m*	tempo absoluto *m*	абсолютное время *n*	117
tiempo newtoniano *m*	tempo newtoniano *m*	ньютоновское время *n*	118
espacio-tiempo *m*	espaço-tempo *m*	пространство-время *n*	119
espaciotemporal	espaço-temporal	пространственно-временной	120
de cuatro dimensiones cuadridimensional tetradimensional	a quatro dimensões quadri(di)mensional tetradimensional	четырёхмерный четырёхразмерный	121
espacio de Minkowski *m* espacio-tiempo de Minkowski *m* espacio de sucesos *m*	espaço de Minkowski *m* espaço-tempo de Minkowski *m* espaço dos eventos *m* espaço dos acontecimentos *m*	пространство Минковского *n* четырёхмерный мир *m* пространство – время Минковского *n*	122
coordenada de tiempo *f* coordenada temporal *f*	coordenada (do) tempo *f* coordenada temporal *f*	временная координата *f* координата времени *f*	123
coordenada de espacio *f* coordenada espacial *f*	coordenada do espaço *f* coordenada espacial *f*	пространственная координата *f* координата пространства *f*	124
suceso *m* punto del espacio-tiempo *m* punto del universo *m*	evento *m* acontecimento *m* ponto do espaço-tempo *m* ponto de universo *m*	событие *n* точка пространства-времени *f* мировая точка *f*	125
suceso adyacente *m* suceso infinitamente próximo *m*	evento adjacente *m* acontecimento adjacente [infinitamente próximo] *m*	смежное событие *n*	126
simultaneidad *f*	simultaneidade *f*	одновременность *f*	127
sucesos simultáneos *mpl*	eventos [acontecimentos] simultâneos *mpl*	одновременные события *npl*	128
propagación de la luz *f*	propagação da luz *f*	распространение света *n*	129

	UNIVERSE	UNIVERS	WELTALL
130	Michelson experiment Michelson-Morley experiment	expérience de Michelson (et Morley) *f* expérience de Michelson-Morley *f*	Michelson-Versuch *m* Michelsonscher Versuch *m* Michelson-Morley-Versuch *m*
131	Galileo('s) transformation Galilean transformation	transformation de Galilée *f* transformation galiléenne *f*	Galilei-Transformation *f* Galileische Transformation *f*
132	Lorentz transformation	transformation de Lorentz *f*	Lorentz-Transformation *f*
133	invariance under a transformation	invariance dans une transformation *f*	Invarianz gegenüber einer Transformation *f*
134	Lorentz contraction Lorentz-FitzGerald contraction FitzGerald-Lorentz contraction relativistic contraction	contraction de Lorentz *f* contraction des longueurs *f* contraction relativiste *f*	Lorentzsche Kontraktion *f* Lorentz-Kontraktion *f* Lorentz-Verkürzung *f* relativistische Verkürzung *f*
135	Lorentz contraction factor $\sqrt{1 - (v^2/c^2)}$	facteur de contraction de Lorentz *m*	Lorentzscher Faktor *m* Kontraktionsfaktor *m*
136	relativistic velocity	vitesse relativiste *f*	relativistische Geschwindigkeit *f*
137	time dilatation dila(ta)tion of time	dilatation du temps *f*	Zeitdilatation *f* Zeitdehnung *f*
138	four-vector world vector four-dimensional vector	vecteur à quatre composantes *m* vecteur quadridimensionnel *m* quadrivecteur *m*	Vierervektor *m* vierdimensionaler Vektor *m*
139	proper time	temps propre *m*	Eigenzeit *f*
140	relativistic correction	correction relativiste *f*	relativistische Korrektur *f*
141	relativistic mass	masse relativiste *f*	relativistische Masse *f*
142	relativistic mass increase	croissance relativiste de la masse *f* augmentation relativiste de masse *f*	relativistische Massenzunahme *f* relativistischer Massenzuwachs *m*
143	rest mass	masse au repos *f*	Ruh(e)masse *f*
144	rest energy	énergie au repos *f*	Ruh(e)energie *f*
145	principle of equivalence of mass and energy mass-energy equivalence principle $E = m \cdot c^2$	principe d'équivalence de masse et énergie *m* équivalence de la masse et de l'énergie *f*	Äquivalenzprinzip von Masse und Energie *n* Masse-Energie-Äquivalenzprinzip *n*
146	general theory of relativity general relativity	relativité générale [généralisée] *f* théorie de la relativité générale [généralisée] *f*	allgemeine Relativitätstheorie *f*
147	non-inertial system	système [réferentiel] non-inertiel *m*	Nichtinertialsystem *n*
148	inertial mass	masse inertielle *f* masse d'inertie *f*	träge Masse *f*

experimento de Michelson (y Morley) *m*	experiência de Michelson *f* experiência de Michelson-Morley *f*	опыт Майкельсона *m* опыт Майкельсона – Морли *m*	130
transformación de Galileo *f* transformación galileana *f*	transformação de Galileu *f* transformação galileana *f*	преобразование Галилея *n* галилеевское преобразование *n*	131
transformación de Lorentz *f*	transformação de Lorentz *f*	преобразование Лоренца *n*	132
invariancia respecto de una transformación *f*	invariância em uma transformação *f*	инвариантность относительно преобразования *f*	133
contracción de Lorentz *f* contracción longitudinal *f* contracción relativista *f*	contração de Lorentz *f* contração de comprimentos *f* contração relativística *f* contração relativista *f*	лоренцево сокращение *n* сокращение длин *n* релятивистское сокращение *n*	134
factor de contracción de Lorentz *m*	fator de contração de Lorentz *m*	коэффициент сокращения (Лоренца) *m*	135
velocidad relativística *f* velocidad relativista *f*	velocidade relativística *f* velocidade relativista *f*	релятивистская скорость *f*	136
dilatación del tiempo *f*	dilatação do tempo *f*	замедление (течения) времени *n*	137
tetravector *m* cuadrivector *m* cuatrovector *m*	quadrive(c)tor *m* tetrave(c)tor *m*	четырёхмерный вектор *m* мировой вектор *m*	138
tiempo propio *m*	tempo próprio *m*	собственное время *n*	139
corrección relativística *f* corrección relativista *f*	correção relativística *f* correção relativista *f*	релятивистская поправка *f*	140
masa relativística *f* masa relativista *f*	massa relativística *f* massa relativista *f*	релятивистская масса *f*	141
aumento relativista de la masa *m*	aumento relativista da massa *m* aumento relativístico da massa *m*	релятивистское возрастание массы *n* релятивистское увеличение [приращение] массы *n*	142
masa en reposo *f*	massa de [em] repouso *f*	масса покоя *f*	143
energía en reposo *f*	energia de [em] repouso *f*	энергия покоя *f*	144
equivalencia de masa y energía *f* principio de la equivalencia de masa y energía *m*	equivalência [eqüivalência] de massa e energia *f* princípio da equivalência de massa e energia *m*	принцип эквивалентности массы и энергии *m*	145
teoría general de la relatividad *f* relatividad general *f*	teoria da relatividade generalizada *f* (teoria da) relatividade geral *f*	общая теория относительности *f* ОТО	146
sistema no inercial *m*	sistema não-inercial *m*	неинерциальная система *f*	147
masa inercial *f*	massa inercial *f*	инертная масса *f* инерционная масса *f*	148

	UNIVERSE	UNIVERS	WELTALL
149	gravitational mass	masse pesante *f* masse gravitationnelle *f*	schwere Masse *f*
150	(Einstein's) equivalence principle	principe d'équivalence (d'Einstein) *m*	(Einsteinsches) Äquivalenzprinzip *n*
151	metric	métrique *f*	Metrik *f*
152	Schwarzschild metric	métrique de Schwarzschild *f*	Schwarzschildsche Metrik *f*
152a	Robertson-Walker metric	métrique de Robertson-Walker *f*	Robertson-Walker-Metrik *f*
153	geodesic geodesic line	géodésique *f* ligne géodesique *f*	geodätische Linie *f* Geodät(isch)e *f*
154	world line	ligne d'univers *f*	Weltlinie *f*
155	field equations	équations du champ *fpl*	Feldgleichungen *fpl*
156	Einstein('s) field equations	équations de champ d'Einstein *fpl*	Einsteinsche Feldgleichungen *fpl*
156a	cosmological term	terme cosmologique *m*	kosmologisches Glied *n* kosmologischer Term *m*
157	tensor equation	équation tensorielle *f*	Tensorgleichung *f*
157a	fundamental tensor metric tensor	tenseur fondamental *m* tenseur métrique *m*	Masstensor *m* Grundtensor *m* metrischer Tensor *m* Fundamentaltensor *m*
158	energy-momentum tensor energy tensor matter tensor	tenseur d'énergie *m* tenseur d'énergie-impulsion *m* tenseur de matière *m*	Energietensor *m* Energie-Impuls-Tensor *m* Materietensor *m*
159	curvature tensor Riemannian curvature tensor Riemann-Christoffel tensor	tenseur de courbure *m* tenseur de Riemann-Christoffel *m* tenseur de courbure de Riemann *m*	Krümmungstensor *m* Riemann-Christoffelscher Tensor *m*
160	relativistic advance of the perihelion relativistic motion of the perihelion	mouvement relativiste du périhélie *m* avance (relativiste) du périhélie *f*	relativistische Perihelbewegung *f* relativistische Periheldrehung *f*
161	relativistic deflection of light bending of light in a gravitational field	déviation des rayons lumineux par le champ de gravitation *f* déviation (relativiste) de la lumière *f*	Lichtablenkung im Schwerefeld *f*
162	gravitational redshift gravitational displacement of spectral lines relativistic displacement of spectral lines	décalage gravifique [gravitationnel] (vers le rouge) *m* déplacement relativiste des raies spectrales vers le rouge *m* déplacement vers le rouge dans le champ de gravitation *m*	Rotverschiebung der Linien im Schwerefeld *f* Gravitationsverschiebung nach Rot *f* Gravitationsrotverschiebung *f*
163	gravitational radius Schwarzschild('s) radius	rayon gravitationnel *m* rayon de Schwarzschild *m*	Gravitationsradius *m* Schwarzschildscher Radius *m* Schwarzschild-Radius *m*

masa gravitatoria *f*	massa gravitacional *f*	гравитационная масса *f* тяжёлая масса *f*	149
principio de equivalencia (de Einstein) *m*	princípio de eqüivalência (de Einstein) *m*	принцип эквивалентности (Эйнштеина) *m*	150
métrica *f*	métrica *f*	метрика *f* мероопределение *n*	151
métrica de Schwarzschild *f*	métrica de Schwarzschild *f*	шварцшильдовская метрика *f*	152
métrica de Robertson-Walker *f*	métrica de Robertson-Walker *f*	метрика Робертсона-Уокера *f*	152a
línea geodésica *f* geodésica *f*	geodésica *f* curva [linha] geodésica *f*	геодезическая линия *f* кратчайшая (линия) *f*	153
línea de universo *f*	linha de universo *f*	мировая линия *f*	154
ecuaciones del campo *fpl*	equações de campo *fpl*	уравнения поля *npl*	155
ecuaciones del campo de Einstein *fpl*	equações de campo de Einstein *fpl*	эйнштейновские уравнения поля *npl*	156
término cosmológico *m*	termo cosmológico *m*	космологический член (λ) *m*	156a
ecuación tensorial *f*	equação tensorial *f*	тензорное уравнение *n*	157
tensor fundamental *m* tensor métrico *m*	tensor fundamental *m* tensor métrico *m*	фундаментальный тензор *m* метрический тензор *m*	157a
tensor de energía *m* tensor de materia *m* tensor energía-ímpetu *m*	tensor de energia *m* tensor de matéria *m* tensor energia-momento *m*	тензор энергии-импульса *n* тензор энергии *n* тензор материи *m*	158
tensor de curvatura *m* tensor de Riemann *m*	tensor de curvatura *m* tensor de Riemann *m*	тензор кривизны *m* тензор Римана *m* тензор Римана-Кристоффеля *m*	159
avance relativista del perihelio *m* corrimiento relativista del perihelio *m*	movimento relativista do periélio *m* avanço relativista do periélo *m*	релятивистское движение перигелия *n*	160
desviación relativista de los rayos luminosos *f* desviación relativista de la luz *f*	desvio [deflexão] da luz pelo campo de gravidade *m*[*f*] desvio [deflexão] relativista da luz *m*[*f*]	отклонение луча света в гравитационном поле *n* гравитационное отклонение света *n*	161
corrimiento gravitatorio del espectro hacia el rojo *m* desplazamiento gravitacional hacia el rojo *m* corrimiento relativista de las rayas espectrales *m*	deslocamento (relativista) das raias [linhas] do espe(c)tro para o vermelho *m* desvio [deslocamento] gravitacional para o vermelho *m*	гравитационное красное смещение *n*	162
radio gravitacional *m* radio de Schwarzschild *m*	raio gravitacional *m* raio de Schwarzschild *m*	шварцшильдовский радиус *m* гравитационный радиус *m*	163

	UNIVERSE	UNIVERS	WELTALL
164	gravitational wave	onde gravitationnelle *f* onde de gravitation *f*	Gravitationswelle *f*
165	gravitational radiation	rayonnement gravitationnel *m*	Gravitationsstrahlung *f*
166	hot universe hot cosmological model	univers chaud *m* modèle d'univers chaud *m* modèle cosmologique chaud *m*	heisses Universum *n* heisses Weltmodell *n* heisses Weltall *n* heisses kosmologisches Modell *n*
166a	cosmological era	ère cosmologique *f*	kosmologische Ära *f*
167	Big Bang primordial explosion initial explosion	explosion primordiale *f* explosion primitive de l'univers *f* explosion initiale *f* (modèle cosmologique du) big bang *m*	Urknall *m* Urexplosion *f* Big Bang *m*
168	fireball prim(a)eval fireball	boule de feu primordiale *f* univers primordial *m*	Feuerball *m* frühes Universum *n* primordialer Feuerball *m*
169	prim(a)eval atom primitive atom	atome primitif *m* atome unique *m*	Uratom *n*
169a	quark soup	soupe de quarks *f*	Quark-Suppe *f*
170	ylem primordial matter	ylem *m* plasma primordial *m* matière primordiale *f*	Urmaterie *f* Urstoff *m* Urplasma *n*
170a	early universe initial universe	univers primordial *m* univers initial *m*	Urkosmos *m* Urweltall *n*
170b	inflationary universe inflatory universe	univers inflatoire *m* univers d'inflation *m* univers inflationnaire *m*	inflationäres Weltall [Universum] *n*
170c	(cosmological) inflation extremely rapid expansion (of the universe)	inflation (cosmologique) *f* expansion ultrarapide (de l'univers) *f*	(kosmologische) Inflation *f* ultraschnelle Expansion (des Weltalls) *f*
170d	inflationary era inflatory era	ère d'inflation *f* ère inflationnaire *f* ère inflatoire *f*	Inflationsära *f* inflationäre Ära *f*
170e	era of superunified force Planck era	ère de la force superunifiée *f* ère de Planck *f*	Ära der Supergravitation *f* Plancksche Ära *f*
170f	Planck (fundamental) mass (2×10^{-5} g)	masse (fondamentale) de Planck *f*	Plancksche (Elementar)masse *f*
170g	Compton wavelength	longueur d'onde Compton *f*	Compton-Wellenlänge *f*
170h	Planck (fundamental) length (10^{-33} cm)	longueur (fondamentale) de Planck *f*	Plancksche (Elementar)länge *f*
170i	Planck time (10^{-43} second)	temps de Planck *m*	Plancksche Zeit *f*

onda gravitacional *f* onda de gravitación *f*	onda gravitacional *f* onda de gravitação *f*	гравитационная волна *f*	164
radiación gravitacional *f*	radiação gravitacional *f*	гравитационное излучение *n*	165
universo caliente *m* modelo cosmológico caliente *m*	universo quente *m* modelo cosmológico quente *m*	горячая вселенная *f* горячая модель вселенной *f* горячая космологическая модель *f*	166
era cosmológica *f*	era cosmológica *f*	космологическая эра *f*	166a
Gran Explosión *f* explosión primigenia *f* esplosión primordial *f* explosión inicial *f* big bang *m*	explosão primordial *f* explosão inicial *f* big-bang *m* bigue-bangue *m*	Большой взрыв *m* биг-бэнг *m* первоначальный взрыв *m* начало расширения *n*	167
bola de fuego *f* bola de fuego primordial *f* destello primigenio *m*	bola de fogo *f* bola de fogo primordial *f*	файербол *m* огненный шар *m*	168
átomo primitivo *m*	átomo primitivo *m*	первоначальный атом *m* первоатом [праатом] *m*	169
sopa de quarks *f*	sopa de quarks *f*	кварковый суп *m*	169a
materia primigenia *f* plasma primordial *m*	matéria primordial *f* plasma primordial *m*	первичное вещество *n* илем *m*	170
universo primordial *m* universo inicial *m*	universo primordial *m* universo inicial *m*	ранняя вселенная *f* первоначальная вселенная *f*	170a
universo inflacionario *m*	universo inflacionário *m*	раздувающаяся вселенная *f* раздувающаяся космологическая модель *f*	170b
inflación (cosmológica) *f* expansión ultrarrápida (del universo) *f*	inflação (cosmológica) *f* expansão ultra-rápida (do universo) *f*	(сверхбыстрое) космологическое раздувание *n* инфляция (вселенной) *f*	170c
era de inflación (cosmológica) *f*	era da inflação (cosmológica) *f*	эра [стадиа] (космологического) раздувания *f* инфляционная стадия *f*	170d
era de la fuerza superunificada *f* era de Planck *f*	era da supergravitação *f* era de Planck *f*	планковская эра *f*	170e
masa (elemental) de Planck *f*	massa (elementar) de Planck *f*	(элементарная) масса Планка *f* планковская масса *f*	170f
longitud de onda (de) Compton *f*	comprimento da onda (de) Compton *m*	комптоновская длина волны *f*	170g
longitud (elemental) de Planck *f*	comprimento (elementar) de Planck *m*	(элементарная) длина Планка *f*	170h
tiempo de Planck *m*	tempo de Planck *m*	планковское время *n* время Планка *n*	170i

	UNIVERSE	UNIVERS	WELTALL
170j	gauge theory	théorie de jauge *f*	Eichtheorie *f* „gauge" Theorie *f*
170k	vacuum energy	énergie du vide *f*	Vakuumenergie *f*
170l	Higgs field	champ de Higgs *m*	Higgssches Feld *n*
170m	quantum field	champ quantique *m*	Quantenfeld *n*
170n	superheavy particle	particule superlourde *f*	überschwere Partikel *f* überschweres Teilchen *n*
170o	magnetic monopole	monopôle magnétique *m*	magnetischer Monopol *m*
170p	phase transition phase change	transition de phase *f* changement de phase *m*	Phasenübergang *m* Phasenumwandlung *f*
170q	superunification	superunification *f*	Supervereinheitlichung *f*
170r	superunified force [interaction] supergravitation	interaction [force] superunifiée *f* supergravitation *f*	supervereinheitliche Wechselwirkung [Kraft] *f* Supergravitation *f*
170s	symmetry breaking breaking of symmetry	brisure de la symétrie *f* violation de la symétrie *f*	Symmetriebrechung *f*
170t	CP violation violation of charge parity	violation de la parité de charge *f* violation [non-conservation] CP *f*	Verletzung der Ladungsparität *f* Verletzung der CP-Invarianz *f*
170u	grand unification	grande unification *f*	grosse Vereinheitlichung *f*
170v	grand unified force [interaction]	grande force [interaction] unifiée *f*	grosse vereinheitlichte Wechselwirkung *f*
170w	unification of interactions	unification des interactions *f*	Vereinheitlichung der Wechselwirkungen *f*
170x	electroweak interaction [force] unified force	interaction [force] électrofaible *f* force unifiée *f*	elektroschwache Wechselwirkung *f* vereinheitlichte Wechselwirkung *f*
170y	primordial fluctuations	fluctuations primordiales *fpl*	primitive [ursprüngliche] Fluktuationen *fpl*
170z	clustering of matter	agrégation [conglomération] de la matière *f*	Klumpung der Materie *f*
171	large-scale structure	structure à grande échelle *f*	grossräumige Struktur *f*
171a	cellular structure cell structure	structure cellulaire *f*	Blasenstruktur *f* Zellenstruktur *f*
171b	primordial pancakes	grandes structures aplaties *fpl* crêpes primordiales *fpl*	(ursprüngliche) Plinsen *fpl*
171c	cosmic string	structure filamentaire [allongée] *f* corde cosmique *f*	kosmisches Filament *n*

teoría calibracional *f* teoría de medida *f*	teoria de calibre [calibração] *f* teoria de gauge *f*	калибровочная теория *f*	170j
energía del vacío *f*	energia do vácuo *f*	вакуумная энергия *f* энергия вакуума *f*	170k
campo de Higgs *m*	campo de Higgs *m*	поле Хиггса *n*	170l
campo cuántico *m*	campo quântico *m*	квантовое поле *n*	170m
partícula superpesada *f*	partícula superpesada *f*	сверхтяжёлая частица *f*	170n
monopolo magnético *m*	monopolo magnético *m*	магнитный монополь *m*	170o
transición de fase *f*	transição de fase *f*	фазовый переход *m*	170p
superunificación *f*	superunificação *f*	сверхвеликое объединение *n* сверхобъединение *n*	170q
interacción [fuerza] superunificada *f* supergravitación *f*	interação [força] superunificada *f* supergravitação *f*	сверхобъединённое взаимодей- ствие *n*	170r
violación de simetría *f* quebramiento de simetría *m*	violação de simetria *f* quebra de simetria *f*	нарушение симметрии *n*	170s
violación de la paridad de carga *f*	violação da paridade de carga *f*	нарушение зарядовой чётно- сти *n*	170t
gran unificación *f*	grande unificação *f*	великое объединение *n*	170u
gran interacción [fuerza] unificada *f*	grande interação [força] unificada *f*	великое объединённое взаимо- действие *n*	170v
unificación de las interacciones *f*	unificação das interações *f*	объединение взаимодействий *n*	170w
interacción [fuerza] electrodébil *f* fuerza unificada *f*	interação [força] ele(c)trofraca *f* força unificada *f*	электрослабое взаимодей- ствие *n* объединённое взаимодействие *n*	170x
fluctuaciones primordiales *fpl*	flutuações primordiais *fpl*	первичные флуктуации *fpl*	170y
aglomeración de la materia *f*	aglomeração da matéria *f*	скучивание вещества *n*	170z
estructura en [a] gran escala *f*	estrutura em grande escala *f*	крупномасштабная структура *f*	171
estructura celular *f*	estrutura celular *f*	ячеистая структура *f*	171a
hojuelas primordiales *fpl*	panquecas primordiais *fpl*	первичные блины *mpl*	171b
cuerda cósmica *f*	corda cósmica *f*	космическая струна *f*	171c

171d	void	vide *m*	Leere *f*
171e	evolution of the universe	évolution de l'univers *f*	Entwicklung der Welt *f* Entwicklung des Weltalls *f*
172	radiation era	ère de rayonnement *f* ère du plasma photonique *f*	Strahlungsära *f*
173	radiation coupling	couplage de rayonnement *m*	Strahlungskopplung *f*
173a	annihilation era	ère d'annihilation *f*	Annihilationsära *f* Zerstrahlungsära *f*
174	hadron era	ère hadronique *f*	Hadronenära *f*
175	cosmological synthesis of elements cosmological nucleosynthesis Big-Bang nucleosynthesis primordial nucleosynthesis	nucléosynthèse cosmologique *f* nucléosynthèse primordiale *f* nucléosynthèse cosmique *f* nucléogenèse cosmique [cosmologique] *f*	kosmologische Nukleosynthese *f* Urknall-Nukleosynthese *f*
176	lepton era	ère leptonique *f*	Leptonenära *f*
177	decoupling of radiation matter-radiation decoupling	découplage du rayonnement *m* découplage entre matière et rayonnement *m*	Entkopplung der Strahlung (von Materie) *f*
177a	decoupling era	ère [époque] de découplage *f*	Entkopplungsära *f*
178	stellar era	ère stellaire *f* ère de matière *f* période [époque] matérielle *f*	Materieära *f* Sternära *f*
179	cosmological horizon	horizon cosmologique *m*	kosmologischer Horizont *m*
180	contraction of the universe	contraction de l'univers *f*	Kontraktion des Weltalls *f*
181	collapse of the universe	effondrement de l'univers *m*	Zusammensturz des Weltalls *m*
182	Big Collapse final collapse Big Crunch	effondrement final *m* Grand effondrement *m* collapse final *m*	Endzusammensturz *m* Grosser Zusammensturz *m*
183	end of the world	fin du monde *f*	Weltuntergang *m* Ende der Welt *n*

vacío *m*	vazio *m*	пустота *f*	171d
evolución del universo *f*	evolução do universo *f*	развитие вселенной *n* эволюция вселенной *f*	171e
era radiativa *f* era del plasma radiativo *f*	era da radiação *f* era do plasma fotônico *f*	эра фотонной плазмы *f* эра излучения [радиации] *f* эра радиационно-доминированной плазмы *f* радиационная эра *f*	172
acoplamiento de radiación *m* acoplamiento por radiación *m*	acoplamento de radiação *m* acoplamento por radiação *m*	радиоционная связь *f* связ посредством радиации *f* связ радиации [излучения] с веществом *f*	173
era de aniquilación *f*	era de aniquilação *f* era de desmaterialização *f*	анигиляционная эра *f* эра аннигиляции *f*	173a
era hadrónica *f*	era hadrônica *f*	адронная эра *f*	174
nucleosíntesis cosmológica *f* nucleosíntesis primordial *f* nucleogénesis primordial *f* nucleogénesis cosmológica *f*	nucleossíntese cosmológica *f* nucleossíntese primordial nucleogênese cosmológica *f* nucleogênese primordial *f*	космологический синтез ядер *m* космологический нуклеосинтез *m* первичный нуклеосинтез *m*	175
era leptónica *f*	era leptônica *f*	лептонная эра *f*	176
desacoplamiento de la radiación *m* desacoplamiento entre materia y radiación *m*	desacoplamento de radiação *m* desacoplamento entre matéria e radiação *m*	отрыв излучения от вещества *m* развязывание излучения *n* разделение излучения и вещества *n*	177
era del desacoplamiento *f*	era do desacoplamento *f*	эра [эпоха] разделения [отрыва, развязывания] излучения *f*	177a
era estelar *f*	era estelar *f*	звёздная эра *f* эра вещества *f*	178
horizonte cosmológico *m*	horizonte cosmológico *m*	космологический горизонт *m*	179
contracción del universo *f*	contração do universo *f*	сжатие вселенной *n*	180
colapso del universo *m*	colapso do universo *m*	коллапс вселенной *m*	181
colapso final *m* Gran Colapso *m*	colapso final *m* Grande colapso *m*	Большой коллапс *m*	182
fin del mundo *m*	fim do mundo *m*	конец мира *m*	183

ALPHABETIC INDICES

INDEXES ALPHABETIQUES

ALPHABETISCHE INDEXE

ÍNDICES ALFABÉTICOS

ÍNDICES ALFABÉTICOS

АЛЬФАВИТНЫЕ УКАЗАТЕЛИ

ENGLISH INDEX

Index anglais

Englischer Index

Índice inglés

Índice inglês

Английский указатель

ALPHABET:

a	h	o	v
b	i	p	w
c	j	q	x
d	k	r	y
e	l	s	z
f	m	t	
g	n	u	

EN

A

AA 78–46
Abel equation 88–235
aberration, diurnal 86–57
ablation 86–182
heat of~ 86–184
~ process 86–183
zone of ~ 72–123
above the limb 89–30
absolute age 82–269
~ brightness (of a comet) 85–109
~ gradient 91–111
~ humidity 77–122
~ magnitude 91–64
~ magnitude of a meteor 86–72
~ magnitude of a planet 83–86
~ measurement of gravity 71–68
~ meteor brightness 86–74
~ orbit 96–55
~ parallax 91–11
~ space 99–116
~ time 99–117
absorbing cloud 97–104
~ layer 88–175
galactic ~ layer 97–89
~ matter 97–85
absorption, auroral 78–46
bound–bound ~ 93–108
bound–free ~ 93–107
free–free ~ 93–106
general interstellar ~ 97–87
integral interstelar~ 97–87a
interstellar~ 97–84
line ~ 93–108
polar cap ~ 78–44
rapid neutron ~ 94–89
selective interstellar ~ 97–90
slow neutron ~ 94–88
sudden commencement ~ 78–47
sudden cosmic noise ~ 78–42
total interstellar~ 97–87a
water–vapour ~ 77–127
abundance anomalies 91–183
anomalous ~ 91–173
chemic(al) ~ 91–169
cosmic(al) ~ of the elements 94–107
~ curve 94–108
elemental ~ 82–141
isotopic ~ 91–179
metal ~ 91–170
natural ~ 91–171
normal ~ 91–172
~ of an element 91–169
relative ~ of elements 91–174
~ ratio 91–175
total ~ 91–177
abyssal 72–41
~ deposits 72–50
~ plain 72–42
~ zone 72–41
Ac 77–144
accelerated expansion 99–81
~ particle 91–182
acceleration, effective 85–88
equatorial ~ 88–131
gravity ~ 71–64
~ of gravity 71–64
accidental association 97–126
~ connection 97–126
accretion 83a–33
~ disc 96–21
~ disk 96–21
~ theory 83a–20
accumulation 83a–34
accurate position 85–28
acetaldehyde 97–42
Achernar 92–102
achondrite 87–77
acicular galaxy 98–239
acid, formic 97–33
isocyanic 97–39
organic 87–69
~ rock 74–4
acoustic energy 88–219
~ energy flux 93–161
flux of ~ energy 88–221
~ noise 88–220
~ sound 88–220
~ sounding 76–45
acoustics, atmospheric 76–31
action, tidal 83a–15
trigger ~ 89–199
activation of a filament 89–215
~ of a quiescent prominence 89–239
prominence ~ 89–239
active belt 73–32
~ comet 85–149
coronal ~ region 89–286
~ day 75–93
~ feature 89–14
~ features 79–138
~ galactic nucleus 98–272
~ galaxy 98–260
~ longitude 89–41
magnetically ~ day 75–93
~ prominence 89–238
~ prominence region 89–240
~ region 89–57
~ region on Jupiter 83–180
~ seismic experiment 82–220
~ seismology 82–217
~ seismometer 82–222
~ solar region 89–57
~ stream 86–223
~ Sun 89–12
~ sunspot prominence 89–254
~ volcano 73–114
activity, auroral 79–100
background meteor ~ 86–136
center of ~ 89–60
centre of ~ 89–60
cometary ~ 85–110
corpuscular ~ 89–7
flare ~ 89–176, 95–71
fumarolic ~ 82–164
galactic ~ 98–261
geomagnetic ~ 75–86
igneous ~ 73–110
~ in radio waves 89–6
index of the stream ~ 86–230
internal ~ 82–158
lunar ~ 82–149
magnetic ~ 89–5
meteor ~ 86–134
minimum of solar ~ 89–4
nuclear ~ of a galaxy 98–271
~ of the galactic nucleus 98–271
optical ~ 89–3
radio ~ 89–6
recurrent ~ 98–267
region of ~ 89–57
relative stream ~ 86–321
seismic ~ 73–4
solar ~ 89–1
spot ~ 89–62
stream ~ 86–229
sunspot ~ 89–62
surface ~ 82–231
volcanic ~ 73–110
~ zone 89–54
adhesive 82–137
adiabatic atmosphere 76–12
~ equilibrium 76–21
~ pulsations 95–98
~ temperature gradient 93–98
adjacent event 99–126
adjustment, isostatic 73–100
adsorb 85–129
adsorbate 85–131
adsorbed matter 85–131
adsorbent 85–130
adsorption 85–128
~ equilibrium 85–121
~ potential 85–122
~ power 85–120
advance of periastron 96–71
relativistic ~ of the perihelion 99–160
advection 77–63
adventive crater 73–128
aeolosphere model 83–260
aerial, Mills' cross 88–121
receiving ~ 78–18
~ spacing 88–99
transmitter ~ 78–16
transmitting ~ 78–16
aerodynamic resistance 86–178
aerolite 87–60
aerolith 87–60
aerology 76–122
aeronomy 76–122a
aerosol 76–89
~ filter 76–94
~ particle 76–91
aerotriangulation, spatial 71–26
afflux 82–280

EN

afterglow 79–22
aftershock 73–13
age, absolute 82–269
cosmic–ray exposure ~ 87–139
~ determination 74–91
determination of ~ 74–91
exposure ~ 87–139
gas–retention ~ 87–135
geologic ~ 74–90
ice ~ 83a–78
irradiation ~ 87–139
lead–lead ~ 74–100
lead–ratio ~ 74–100
~ measurement 87–126
Moon's ~ 81–31
~ of a meteor shower 86–246
~ of a stream 86–245
~ of meteorites 87–127
~ of stars 94–61
~ of the Earth 83a–68
~ of the Moon 81–31
~ of the universe 99–12
radiation ~ 87–139
relative ~ 82–268
retention ~ 87–135
rubidium–strontium ~ 82–271
solidification ~ 87–137
stratigraphic ~ 82–267
uranium–lead ~ 74–101
agglomeration 83a–32
aggregate, molecular 97–65
air 76–6
ambient ~ 77–4
~ cap 86–170
~ current 77–90
~ mass 77–67
~ molecule 86–172
~ particle 86–172
~ pollution 76–80
~ Pump 92–15
~ resistance 86–178
~ sample 76–86
~ sampling 76–85
~ tides 82–294
airglow 79–2
day–time ~ 79–3
~ height 79–26
night ~ 79–5
twilight ~ 79–4
airless 76–8
Aitken nuclei 76–116
~ particles 76–116
alanine 87–72
albedo, Bond's 83–91
cloud ~ 83–95
cosmic–ray ~ 75–116
geometric ~ 83–92
Lambert's ~ 83–89
mean ~ 83–90a
monochromatic ~ 83–93
~ neutron 75–115
neutron ~ 75–116
~ of the clouds 83–95
radio–frequency ~ 83–94
spherical ~ 83–91
total ~ 83–90b
Albireo 92–103
albite 74–17
alcohol, methyl 97–32
Alcor 92–104
Algol 92–105
~ variable 95–77
Algol–type eclipsing binary 95–77
Algonkian 74–77
alignment of grains 97–78
allowed region 79–134
~ space 79–134
all–sky camera 79–104
α–aminopropionic acid 87–72
alpha–scattering instrument 82–76
Alphecca 92–117
alpine glacier 72–114
Alpine Valley 81–164
Altair 92–106
Altar 92–19
altering 72–78
coastline ~ 72–78
altimetry, satellite 71–50
altitude 71–51
ellipsoidal ~ 71–52
geodetic ~ 71–52
mirror point ~ 75–125
~ of the luminous layer 79–26
pressure ~ 76–16
Altocumulus 77–144
Altostratus 77–140
aluminium honeycomb 82–25
ambient 76–7
~ air 77–4
~ temperature 77–32
amino acid 87–70
~ group 87–71
ammonia 97–25
amor-type asteroid 84–80
amorphous ice 85–44
amount, cloud 77–147
~ of precipitation 77–158
~ of water vapour 77–124
precipitation ~ 77–158
amphibole 74–20
amplitude of brightness variation 95–38
tidal ~ 82–292
anabatic wind 77–92
anagalactic 98–207
analysis, chemical 82–75
in–situ ~ 82–136
microprobe ~ 82–143
mineralogic ~ 82–53
petrographic ~ 82–52
and 92–14
Anderson prism 88–68
Andromeda 92–14
~ Nebula 98–215
Andromedids 86–256
anemograph 77–114
anemometer 77–113
recording ~ 77–114
angle, characteristic 98–33
heliocentric ~ 89–37
phase ~ 83–72
position ~ 81–71
angular block 82–131
~ diameter 83–117, 96–115
~ distance 96–52
instantaneous ~ velocity 86–90
~ length 98–32
mean ~ velocity 86–91
~ momentum 83a–4
orbital ~ momentum 83a–6
~ path length 86–48
~ rock 82–131
~ separation 96–52
~ velocity 86–89
anisotropic model 99–65
~ universe 99–65
annihilation era 99–173a
~ of magnetic fields 89–158
annual deposit 87–17
~ parallax 91–6
~ proper motion 91–20
annular eclipse 88–18
anomalous abundance 91–173
~ propagation of sound 76–37
~ redshift 98–367
~ sound 76–36
~ sound propagation 76–37
~ star 91–96
~ tail 85–75
anomaly, abundance 91–183
gravimetric ~ 71–75
gravity ~ 71–75
lunar gravitational ~ 81–144
magnetic ~ 75–67
South Atlantic ~ 75–68
sudden phase ~ 78–52
anorthite 74–18
anorthosite 82–87
Ant 92–15
antapex 86–65
solar ~ 98–65
Antares 92–107
antenna, receiving 78–18
transmitting ~ 78–16
anthelion 79–83
anticenter = anticentre
galactic ~ 98–14
~ region 98–15
towards the ~ 98–14
anticlinal fold 73–90
anticline 73–90
anticyclone 77–84
anticyclonic 77–85
antiselene 79–85
antisolar direction 75–22
~ hemisphere 83–68
antitail 85–78

anti-trades 77–103
antitwilight 79–57
antivertex 98–106
Antlia 92–15
anular eclipse of the Sun 88–15a
apareon 83–141
apastron 96–61
apereon 83–141
aperture synthesis 88–118
apex 86–64
~ elevation 86–66
~ height 86–66
~ of the solar motion 98–64
~ of the Sun 91–23
solar ~ 91–23
aphelion 83–28a
apoapse 83–173
apoapsis 83–173
apocentre = apocenter 98–190
galactic ~ 98–190
Jovian ~ 83–173
Saturnian ~ 83–211
apocentron 98–190
apocynthion 81–41
apofocus 98–190
apogalacticon 98–190
apogalacticum 98–190
apogee 81–39
apojove 83–173
Apollo 84–81
~ asteroid 84–80
apolune 81–41
apomartian 83–141
apomartium 83–141
aposaturnian 83–211
aposaturnium 83–211
aposelene 81–41
aposelenium 81–41
apparent brightness 91–61
~ deceleration 86–100
~ diameter 96–115
~ flattening 98–295
~ lunar time 82–301
~ magnitude 91–61
~ magnitude of a meteor 86–71
~ meteor brightness 86–73
~ modulus 96–117
~ motion 83–41
~ orbit 96–48
~ path 86–32
~ path on the celestial sphere 86–32
~ radiant 86–56
~ sunspot area 89–91
~ surface density 96–132
apparition, first 85–19
~ of a comet 85–18
appearance height 86–39
height of ~ point 86–39
moment of ~ 86–49
~ of a fireball 86–117
~ of a meteor 86–35
~ point 86–36
approach, closest 84–69
~ velocity 91–27
approximate position 85–26
approximation, Chandrasekhar 93–51
Milne–Eddington ~ 93–50
non–relativistic ~ 99–97
appulse 81–101
apron, continental 72–36
Aps 92–16
apsidal motion 96–71
Apus 92–16
Aql 92–18
Aqr 92–17
Aquarids 86–249
Aquarius 92–17
Aquila 92–18
AR 89–57
Ara 92–19
arc, circumhorizontal 79–82
circumzenithal ~ 79–74
crepuscular ~ 79–50
homogeneous quiet ~ 79–143
intertropical ~ 79–53
lateral lower tangential ~ 79–87
lower tangential ~ 79–76
~ measurement 71–22
~ of meridian 71–21
primary ~ 73–82
rayed ~ 79–153
secondary ~ 73–83
tangential ~ of halo 79–75
upper tangential ~ 79–77
~ with ray–structure 79–153
arch, coronal 89–298
eruptive ~ 89–251
~ structure 88–249
~ system 89–299
Archaeozoic 74–76
Archer 92–85
Archimedean principle 73–101
~ spiral 84–34
Archimedes principle 73–101
~ spiral 84–34
area, apparent sunspot 89–91
corrected ~ 89–93
~ corrected for foreshortening 89–93
dark ~ 83–155
high–pressure ~ 77–84
intergranular ~ 88–177
~ of audibility 86–120, 87–32
~ of fall 87–38
~ of flare 89–207
~ of impact 87–38
~ of radiation 86–238
~ of sunspots 89–67
~ of visibility 87–31
~ over which meteorites are scattered 87–38
prominence ~ 89–233
reduced ~ 89–93
silent ~ 78–38
spot ~ 89–67
spotted ~ 89–90
total ~ 89–94
areas of Kapteyn 98–78
plage ~ 89–164
selected ~ 98–78
areo– 83–135
areocentric 83–138
~ coordinates 83–138
~ orbit 83–138
areographic 83–137
~ chart 83–136
~ latitude 83–140
~ longitude 83–139
areophysics 83–134
Ari 92–20
Ariel 83–235
Aries 92–20
arm, expanding 98–38
galactic ~ 98–26a
leading ~ 98–31
Orion ~ 98–36
outward winding of spiral ~ 98–30
Perseus ~ 98–34
Sagittarius ~ 98–35
spiral ~ 98–27
3 kpc ~ 98–37
three–kiloparsec ~ 98–37
trailing ~ 98–29
unwinding ~ 98–31
arrival at the Earth 89–193
~ time 73–50
Arrow 92–84
artificial aurora 79–99
~ impact 82–226
~ meteor 86–21
~ radiation belt 75–108
~ radiation zone 75–108
~ trapped radiation 75–113
As 77–140
ascending branch 95–34
~ current 93–151
~ phase of the solar cycle 89–83
~ trajectory 82–9
ascent 82–8
aseismic region 73–29
ash flow 82–174
volcanic ~ 73–142, 82–168
aspect 83–50
association, accidental 97–126
genetic ~ 97–127
nucleus of the ~ 96–91
O– ~ 96–89
stellar ~ 96–88
T– ~ 96–90
asterism 92–3a
asteroid = planetoid = minor planet 84–68
Amor–type ~ 84–80
Apollo ~ 84–80a
Apollo-type ~ 84–80a
artificial ~ 84–69a
~ belt 84–70

EN

binary ~ 84–79
catalogued ~ 84–95
cigar-shaped ~ 84–77
close-approach ~ 84–78
commensurability ~ group 84–101
double ~ 84–79
Earth-crossing ~ 84–76
Earth-grazing ~ 84–78
~ family 84–73
~ fragmentation 84–108
~ group 84–90
libration ~ 84–91a
lost ~ 84–97
man-made ~ 84–69a
Mars ~ 84–74
Mars-crossing ~ 84–75
Mars-family ~ 84–74
~ mission 84–99
mission to an ~ 84–99
numbered ~ 84–99
~ of the belt 84–71
~ ring 84–70
ring ~ 84–71
~ stream 84–100
~ survey 84–94
asteroidal jet stream 84–100
~ stream 84–100
asteroids, fragmentation of 83a–82
ring of ~ 84–70
Trojan group of ~ 84–91
astrometric binary 96–12
~ orbit 96–57
astronomical telegram 85–20
~ twilight 79–37
astronomy, extragalactic 98–208
lunar-based ~ 82–4
meteor ~ 86–3
meteoritic ~ 87–6
stellar ~ 91–2
astrophysical maser 97–13
asymmetric drift 98–104
asymmetry, east-west 89–49
north-south ~ 89–50
asymptotic model 99–59
~ model of the first kind 99–60
~ to the circular orbits 98–180
~ universe 99–59
asynchronous rotation 83–38
ataxite 87–87
nickel-poor ~ 87–88
nickel-rich ~ 87–89
atlas, lunar 81–136
Moon ~ 81–136
~ of the Moon 81–136
photometric ~ 88–138
star ~ 92–12
atmophilic elements 83a–29
atmosphere 83–241
adiabatic ~ 76–12
early ~ 83a–45
Earth's ~ 76–5
electron ~ 93–28
extended ~ 93–9
finite ~ 93–23
free ~ 76–13
gray ~ 93–26
grey ~ 93–26
height of the homogenous ~ 76–17a
homogeneous ~ 76–10
initial ~ 83a–44
isothermal ~ 77–39, 93–25
lower ~ 76–50
middle ~ 76–51
model ~ 93–17
model of ~ 83–247
non-gray ~ 93–27
~ of a planet 83–243
~ of the Earth 76–5
~ of the Sun 88–163
oxidizing ~ 83a–47
plane-parallel ~ 93–21
planetary ~ 83–243
primitive ~ 83–244
primordial ~ 83–244
purely scattering ~ 93–16
reducing ~ 83a–46
scattering ~ 93–16
semi-infinite ~ 93–22
solar ~ 88–163
spherical ~ 93–24
standard ~ 76–11
stellar ~ 93–2
top of the ~ 76–25
upper ~ 76–52
atmospheric 83–242
~ action 87–116
~ acoustics 76–31
~ chemistry 76–79
~ circulation 77–64
~ constituent 83–265
~ diffusion coefficient 86–287
~ discharge 77–161
~ eclipse 95–76
~ electricity 75–2
~ entry 86–161
~ feature 83–107
~ heating 75–87
~ humidity 77–118
~ interaction 79–120
~ inversion 77–37
~ layer 76–49
~ line 88–140a
~ moisture 77–118
~ optics 79–1
~ ozone 76–124
~ ozone content 76–125
~ passage 87–26
~ path 86–162
~ physics 76–4
~ pollutant 76–82
~ pollution 76–80
~ pressure 77–43
~ scattering of light 77–57
~ sciences 76–1
~ tides 82–294
~ treatment 87–116
~ velocity 86–94
atmospherics 78–50
sudden enhancement of ~ 78–51
whistling ~ 78–53
atom, multilevel 93–46
prim(a)eval ~ 99–169
primitive ~ 99–169
two-level ~ 93–45
atomic coma 85–67
~ energy 93–121
~ oxygen 79–24
attenuator, step 88–80
attraction, centre of 89–241
zenith ~ 86–52
audibility, area of 87–32
zone of ~ 76–41
Aur 92–21
aureole 79–60
lunar ~ 79–61
solar ~ 79–62
Auriga 92–21
aurora 79–56, 79–93
artificial ~ 79–99
~ australis 79–106
~ borealis 79–105
dayside ~ 79–97
electron ~ 79–110
flaming ~ 79–157
low-latitude ~ 79–96
permanent ~ 79–5
polar ~ 79–93
polar-cap ~ 79–98
~ polaris 79–93
proton ~ 79–108
radar ~ 79–95
radio ~ 79–94
ribbon-like structure ~ 79–140
sunlit ~ 79–130
auroral absorption 78–46
~ activity 79–100
~ arc 79–146a
~ cap 79–127
~ corona 79–137a
~ echo 79–121
~ electrojet 79–113
~ electron 79–109
~ electron precipitation 79–112
~ forms 79–137
~ frequency 79–122
~ observatory 79–101
~ oval 79–124
~ particle 79–107
~ region 79–126
~ substorm 79–141
~ type 79–136
~ zone 79–123
australite 87–98
automated instrument 82–41
automatic flare patrol 89–221
~ lunar station 82–19

~ solar telescope 88–50
available energy 89–197
average density of matter 99–84
~ velocity 86–98
averted hemisphere 81–16
~ vision 86–128
avoidance, zone of 98–56
axial inclination 83–35
~ pole 83–36
~ tilt 83–35
axially symmetric(al) potential 98–153
~ symmetric(al) system 98–157
axis direction 83–34
geomagnetic ~ 75–47
geomagnetic ~ pole 75–48
magnetic ~ 75–46
~ of rotation 83–33
~ pole 83–36
rotation ~ 83–33
terrestrial magnetic ~ 75–47
azimuthal projection 71–44
Azoic 74–75

B

back surface 87–115
background density 96–131
~ galaxy 98–340
meteor ~ 86–232
~ meteor activity 86–136
seismic ~ 73–55
~ star 96–84
Baily's beads 88–32
Balance 92–62
balance, energy 88–251
mass ~ 88–250
radiation ~ 88–252
balancing, detailed 93–41
ball lightning 77–163
balloon–borne solar telescope 88–52
band, diffuse interstellar 97–100
homogeneous ~ 79–146
luminous ~ 86–27
rayed ~ 79–150
~ system 79–18
~ with ray structure 79–150
zodiacal ~ 84–63
bands, fluorescent 85–132
negative nitrogen ~ 79–21
shadow ~ 88–31
Swan ~ 85–133
unidentified ~ 85–134
Vegard–Kaplan ~ 79–19
bank 72–96
bar, galactic 98–254
barometric height formula 76–15
barred galaxy 98–253
~ spiral 98–253
barrier, sonic 76–44
sound ~ 76–44
baryon gas 93–194
baryonic gas 93–194
barysphere 72–17
basalt 74–9
lunar ~ 82–84
base, interferometer 88–100
lunar ~ 82–17
manned lunar ~ 82–18
Moon ~ 82–17
~ of a prominence 89–234
base–line, fixed 88–102
variable ~ 88–103
basement 72–12
~ complex 72–12
basic rock 74–5
~ solar motion 98–66
basin 81–175
~ of volcanic origin 81–179
basite 74–5
batholite 73–139
batholith 73–139
bathymetric chart 71–54
~ contour 71–55
~ map 71–54
bathymetry 72–28
bay 81–294
~ disturbance 75–80
magnetic ~ 75–80
~ of Billows 81–296
~ of Heats 81–296
~ of Rainbows 81–295
BCG 98–265
bead lightning 77–164
beads, Baily's 88–32
beam, electron 79–116
~ splitting prism 88–88
beam–splitter 88–88
Bear Driver 92–22
bearing strength 82–64
beat period 95–42
Beaufort scale 77–108
~ scale of wind force 77–108
bed, ocean 72–37
river ~ 83–153
sea ~ 72–38
Beehive 96–111
beginning height 86–39
~ of the cycle 89–82
~ of the eclipse 81–74
~ of the total eclipse 81–77
~ of totality 81–77
~ point 86–36
belt 83–190
active ~ 73–32
asteroid ~ 84–70
dusky ~ 83–190
equatorial ~ 83–191
man–made radiation ~ 75–108
minauroral ~ 79–129
minor planet ~ 84–70
~ of planetoids 84–70
orogen(et)ic ~ 73–75
radiation ~ 75–101
subauroral ~ 79–128
van Allen ~ 75–101
bending of light in a gravitational field 99–161
Berenice's Hair 92–38
Beta Persei star 95–77
Beta Lyrae variables 95–47
betatron effect 91–180
Betelgeuse 92–108
beyond the Earth's orbit 84–6
Bielids 86–256
bifurcation 98–5
Big Bang 99–167
~ Bang nucleosynthesis 99–175
Big Bear 92–96
Big Collapse 99–182
Big Crunch 99–182
Big Dipper 92–96
Big Dog 92–27
big tide 82–273
binary 96–3, 96–7
Algol–type eclipsing ~ 95–77
~ asteroid 84–79
astrometric ~ 96–12
close ~ 96–23
contact ~ 96–29
detached ~ 96–35
double–line ~ 96–17
double–line spectroscopic ~ 96–17
eclipsing ~ 95–74
photometric ~ 96–11
photometric ~ star 95–73
~ pulsar 95–140a
semi–detached ~ 96–34
single–line ~ 96–15
single–line spectroscopic ~ 96–15
spectroscopic ~ 96–14
spectrum ~ 96–14
~ star 96–7, 96–3
unresolved ~ 96–10
biogenic deposits 72–59
biosphere 72–8
biotite 74–25
bipolar group 89–111
~ magnetic region 89–160a
Bird of Paradise 92–16
birth of a spot 89–121
~ of a sunspot 89–121
~ rate of stars 94–12
Bishop's ring 79–88
black drop 83–52
~ dwarf 91–139
galactic ~ hole 98–23a
~ hole 94–114
supermassive ~ hole 98–22a
blanketed model 93–18
blanketing 93–20
~ effect 93–67
blind, Venetian 88–126
blink comparator 91–34
~ microscope 91–34
blistered 82–181
blisters, lava 82–178

EN

blob, dust 98–290
block, angular 82–131
continental ~ 72–10
erratic ~ 72–132
~ of rock 82–130
blocky rim 81–224
blow 77–105
blue 91–99
~ compact galaxy 98–265
~ giant 91–142a
~ minus – yellow colo(u)r index 96–145
~ Sun 79–92
blueness of the sky 77–60
Bode's law 83–3
body, heavenly 79–9a
impacting ~ 82–225
meteor ~ 86–160a
meteoric ~ 86–160a
meteorite parent ~ 87–123
parent ~ 87–123
pre–stellar ~ 94–15
rigid ~ 83–111
rotating ~ 91–42
rotating magnetoplasm(at)ic ~ 98–25
solid ~ 83–110
supermassive ~ 98–22
tide–raising ~ 82–284
~ wave 73–42
bolide 86–11
giant ~ 86–9
bolometric brightness 91–68
~ correction 91–80
~ magnitude 91–68
bomb, Ellerman 89–200
volcanic ~ 73–144
bombardment, meteoritic 82–235
~ meteoridal 82–235
particle ~ 82–240
~ produced radionuclide 82–255
Bond's albedo 83–91
Boo 92–22
Boötes 92–22
Border Sea 81–285
bore hole 82–61
boulder 82–130
indicator ~ 72–132
bound rotation 81–14
boundary, magnetospheric 75–9
~ sector 84–18
upper ~ 76–58
bound–bound absorption 93–108
bound–free absorption 93–107
braking, magnetic 91–47
~ mechanism 91–46
~ of the rotation 91–45
branch, ascending 95–34
decreasing ~ 95–35
descending ~ 95–35
giant 91–152b
horizontal ~ 91–132
rising ~ 95–34
breaking of symmetry 99–170s
symmetry ~ 99–170s
break–up, collisional 84–107
~ of parent body 87–124
rotational ~ 85–119
breccia, lunar 82–85
bridge, bright 89–104, 98–307
intergalactic ~ 98–348
light ~ 98–307
luminous ~ 98–307
bright bridge 89–104, 98–307
~ edge of a filament 89–226
~ fireball 86–10
~ giants 91–149
~ hydrogen flocculi 89–169
~ limb 81–96
~ meteor 86–12
~ mottles 88–211
~ nebula 97–121
~ night 79–59
~ night sky 86–28
~ point 89–201
~ rays 81–168
~ region 83–160
~ rim of a filament 89–226
~ ring 89–101
~ star 91–83
~ surge 89–257
brightening of a plage 89–213
brightest galaxy 98–339
~ member 98–339
~ star of a galaxy 98–304
brightness 91–54
absolute ~ 91–64
absolute ~ of a comet 85–109
absolute meteor ~ 86–74
apparent ~ 91–61
apparent meteor ~ 86–73
bolometric ~ 91–68
~ coefficient 83–96
~ difference 96–65
~ distribution 86–166, 88–141
electrographic ~ 91–67
estimate of ~ 95–11
~ factor 83–90
infrared 91–71
integrated ~ 96–135
intrinsic ~ 85–107
limiting ~ 91–62
maximum ~ 86–78
meteor ~ 86–8
~ of a flare 89–212
~ of a meteor 86–8
~ outburst 85–126
photographic ~ 91–66
photovisual ~ 91–69
radiometric ~ 91–70
reduced ~ 85–108
secular decrease of ~ 85–127
stellar ~ 91–55
~ temperature 91–114
total ~ 85–107
~ variation 95–7
visual ~ 91–65
brown dwarf 91–138
bulge, central 98–16
equatorial ~ 71–9
lunar ~ 81–118
nuclear ~ 98–16
tidal ~ 81–119, 83a–67
Bull 92–91
Bull–Driver 92–22
buoy, automatic 72–100
research ~ 72–99
buried crater 81–204
burning, carbon 94–81
explosive helium ~ 94–79
helium ~ 94–78
helium ~ zone 93–145
hydrogen ~ 94–56
hydrogen ~ zone 93–144
oxygen ~ 94–82
silicon ~ 94–83
~ volcano 73–114
burst, comet 85–125
decametric ~ 83–183
Jovian radio ~ 83–182
low–frequency ~ 83–184
~ of particles 88–162
~ of solar plasma 84–38
~ of solar wind 84–42
solar-wind ~ 84–42
star ~ 94–11a
X–ray ~ 79–117, 89–184
butterfly diagram 89–76
B–V colo(u)r index 96–145

C

C 79–156
Cae 92–23
Caelum 92–23
Cainozoic 74–73
~ era 74–73
calcium flocculi 89–170
calculated position 85–27
caldera 81–179
collapse ~ 81–181
erosion ~ 81–183
explosion ~ 81–180
subsidence ~ 81–182
Callisto 83–208
calm 77–110
Cam 92–24
Cambrian 74–78
Camelopardalis 92–24
Camelopardus 92–24
camera, all–sky 79–104
driven ~ 86–142
guided ~ 86–142
meteor ~ 86–141
patrol ~ 86–145
undriven ~ 86–143
canals of Mars 83–157

Martian ~ 83–157
Cancer 92–25
Canes Venatici 92–26
Canis Major 92–27
~ Minor 92–28
Canopus 92–109
canyon 83–154
submarine ~ 72–46
Cap 92–29
cap, air 86–170
auroral ~ 79–127
compression ~ 86–170
gas ~ 86–170
polar ~ 83–143
~ prominence 89–253
cape 81–293
Capella 92–110
Capricornus 92–29
capture 83a–19
~ hypothesis 83a–18
neutron ~ 94–84
~ of comets 85–101
~ of intergalactic matter 98–355
star ~ 96–73
~ theory 85–99, 96–72
Car 92–30
carbon burning 94–81
~ ignition 94–80
~ monosulfide 97–34
~ monoxide 97–28
radioactive ~ 74–95
solid ~ dioxide 83–144
~ star 91–188
carbonaceous chondrite 87–63
~ matter 87–64
~ meteorite 87–62
carbonate 74–27
carbon–deficient star 91–194
carbon–dioxide ice 83–144
Carboniferous 74–82
carbon–poor star 91 194
carbonyl sulfide 97–36
Carina 92–30
carrier 86–276
~ frequency 86–274
~ wave 86–276
cartographic projection 71–40
Cas 92–31
Cassini division 83–224
Cassiopeia 92–31
Castor 92–111
cataclysmic variable 95–109
catalog (USA) = catalogue (UK)
Messier ~ 92–12a
~ of stars 92–13
~ of stars suspected of variability 95–58
star ~ 92–13
catalog(u)ed asteroid 84–95
catastrophic collision 84–106
~ contraction 94–109
~ earthquake 73–40
~ hypothesis 83a–14
cathode ray tube 86–293
cave, lava 82–177
cavern, lava 82–177
cavity 87–117
geomagnetic ~ 75–8, 84–23
heliospheric ~ 88–257
hydromagnetic ~ 84–22
inner ~ 87–119
interplanetary ~ 84–25
magnetic ~ 84–22
planetary ~ 84–24
solar–wind ~ 84–28
surface ~ 87–118
Cb 77–137
Cc 77–143
celestial globe 92–7
~ maser 97–13
~ sphere 92–8
cell structure 99–171a
cellular structure 99–171a
Celsius scale 77–33
~ temperature scale 77–33
cementation 74–52
cementing 74–52
Cen 92–32
Cenozoic 74–73
Centaur 92–32
Centaurus 92–32
center (USA) = centre (UK)
centigrade scale 77–33
central boundary condition 93–81
~ bulge 98–16
~ collision 83a–17
~ condensation 85–37
~ condition 93–81
~ density 93–87
~ eclipse 88–16a
~ peak 81–258
~ projection 71–48
~ region of a galaxy 98–293
~ region of the Galaxy 98–13
~ star 97–148
~ temperature 93–89
~ zone 81–114, 89–47
central–meridian passage 89–45
centre boundary condition 93–81
galactic ~ 98–9
~ of activity 89–60
~ of attraction 89–241
~ of the disk 89–35
~ of the Galaxy 98–9
~ of the Moon's apparent disk 81–128
~ of the solar disk 89–35
~ of the star 93–84
~ of the visible disk 88–172
centre–limb variation 88–145
centre–to–limb variation 88–145
centrosphere 72–17
Cep 92–33
Cepheid 95–92
classical ~ 95–94
~ parallax 91–8a
type I ~ 95–94
~ variable 95–92
Cepheus 92–33
Ceres 84–82
Cet 92–34
Cetus 92–34
Cha 92–35
chain, mountain 81–194
~ of galaxies 98–328
~ of supernovae 95–132
star ~ 96–77
Chained Woman 92–14
chalcophilic elements 83a–31
Chamaeleon 92–35
Champion 92–76
Chandrasekhar approximation 93–51
~ limit 93–189
change, climatic 77–184
evolutionary ~ 94–27
~ in shape 73–68
~ of period 95–43
~ of the Moon 81–19
~ of weather 77–22
seasonal ~ 83–166
weather ~ 77–22
channel, flow 82–192
chaotic terrain 83–164
characteristic age 99–79
~ angle 98–33
~ envelope 98–177
soil ~ 82–108
~ time 99–79
~ velocity 76–43
charge, Earth's 75–3
violation of ~ parity 99–170t
Charioteer 92–21
Charles's Wain 92–96
Charlier's model 99–67
~ universe 99–67
chart, areographic 83–136
bathymetric ~ 71–54
coastal 72–33
gravimetric ~ 71–79
hydrologic ~ 72–25
lunar ~ 81–133
magnetic ~ 75–63
oceanographic ~ 72–23
~ of sky 92–9
~ of the variable 95–15
revolving star ~ 92–10
sea ~ 72–33
star ~ 92–9
synoptic ~ 77–11
weather ~ 77–12
check number 85–30
chemical abundance 91–169
~ analysis 82–75
~ composition of a star 91–168
~ differentiation 83a–50
~ evolution 83a–49

fractionation of ~ elements 83a–51
~ pollution 72–98
~ pumping 97–19
separation of ~ elements 83a–48
chemically active fluid 74–43
chemistry, atmospheric 76–79
chemoluminescence 79–31
chemosphere 76–62
Cheseaux' and Olbers' paradox 99–40
chimney 73–123
Chisel 92–23
chlorine content 72–68
chlorinity 72–68
chondrite 97–61
carbonaceous ~ 87–63
non–carbonaceous ~ 87–75
chondrule 87–76
chorus, dawn 78–62
Christiansen interferometer 88–123
chromosphere 88–191
height of the ~ 88–202
lower ~ 88–193
support of the ~ 88–223
upper ~ 88–192
chromosphere–corona transition region 88–224
chromospheric 88–194
~ faculae 89–164
~ fine structure 88–203
~ flare 89–175
~ mottles 89–172
~ network 89–171
~ telescope 88–73
Ci 77–136
cigar–shaped asteroid 84–77
~ galaxy 98–239
cinder cone 73–120
cinematographic flare patrol 89–220
~ patrol 89–220
Cir 92–36
Circinus 92–36
circle, circumzenithal 79–74
parhelic ~ 79–82
circular mare 81–267
circulation, atmospheric 77–64
general ~ 77–65
meridian ~ 93–163
meridional ~ 93–163
planetary ~ 77–66
circumgalactic gas 98–351
circumhorizontal arc 79–82
circumlunar 81–5
~ trajectory 82–11
circumnuclear region 98–292
circumpolar constellation 92–4a
~ star 92–2a
circumstances of the eclipse 81–73
~ of the fall 87–47
circumstellar 97–3
~ cloud 94–48, 97–102
~ dust 97–55
~ dust cloud 97–117
~ dust shell 97–117
~ line 97–97
~ shell 93–10
circumzenithal arc 79–74
~ circle 79–74
Cirrocumulus 77–143
Cirrostratus 77–141
Cirrus 77–136
cislunar 81–4
civil twilight 79–38
Clairaut's theorem 71–77
class 89–205
luminosity ~ 91–146
~ of a flare 89–205
spectral ~ 91–107
~ 2 flare 89–206
classical cepheid 95–94
~ universe 99–93
classification criteria 91–110
Harvard ~ 91–91
Harvard system of spectral ~ 91–91
~ of galaxies 98–226
~ of meteorites 87–59
~ of variable stars 95–46
petrographic ~ 82–101
spectral ~ 91–90
three–dimensional spectral ~ 91–106
two–dimensional spectral ~ 91–105
Zurich ~ 89–127
clathrate 85–51
clay, pelagic 72–54a
red ~ 72–54
clear night 79–3g
~ sky 77–153
clearest vision, point of 86–127
cleft 81–165
magnetospheric ~ 75–13
cliff interferometer 88–106
cliff–top interferometer 88–106
climate 77–180
continental ~ 77–185
maritime ~ 77–186
mountain ~ 77–187
climatic 77–182
~ change 77–184
~ fluctuation 77–183
climatology 77–179
satellite ~ 77–181
Clock 92–54
close binary 96–23
~ passage of two galaxies 98–320
close–approach asteroid 84–78
closed model 99–73
~ universe 99–73
closest approach 84–69
cloud 77–131
absorbing ~ 97–104
~ albedo 83–95
albedo of the ~ 83–95
~ amount 77–147
circumstellar ~ 94–48, 97–102
circumstellar dust ~ 97–117
comet ~ 85–100
coronal ~ 89–261
cosmic ~ 94–31
dark ~ 97–107
dust ~ 83–256, 87–23, 97–114
~ form 77–133
gas ~ 97–113
giant molecular ~ 97–13a
H I ~ 97–103
intergalactic gas ~ 98–354
interstellar ~ 97–101
iridescent ~ 79–90
~ layer 77–150, 83–248
libration ~ 84–43
lunar libration ~ 84–43
maser ~ 97–12
masering ~ 97–12
masing ~ 97–12
molecular ~ 97–11
mother–of–pearl ~ 79–91
nacreous ~ 79–91
nebular ~ 94–31
neutral hydrogen ~ 97–103
noctilucent ~ 79–89, 86–26
Oort's ~ of comets 85–97
orographic ~ 77–134
particle ~ 84–31
plasma ~ 84–30
preplanetary ~ 83a–24
primordial galactic ~ 94–17
protogalactic ~ 94–17
protoplanetary ~ 83a–24
radio ~ 98–286
star ~ 98–6
suspended ~ 89–275
~ type 77–135
~ veil 77–151
zodiacal ~ 84–54
cloud–covered 83–253
cloudiness 77–147
cloudless 77–149
~ night 97–3g
clouds 83–254
veil of ~ 77–151
cloudy 77–148
~ sky 77–148
cluster disintegration 96–103
~ dissolution 96–103
double ~ 96–81
~ dynamics 96–153
first–ranked ~ galaxy 98–339
galactic star ~ 96–99
~ galaxy 98–338
globular ~ 96–114
globular ~ of galaxies 98–330
globular star ~ 96–114
high concentration ~ 96–129
highly concentrated ~ 96–129
irregular ~ of galaxies 98–337
~ member 96–82
member of the ~ 96–82
metal–poor ~ 96–113

metal–rich ~ 96–112
moving ~ 96–97
moving star ~ 96–97
~ of clusters 98–346
~ of domes 81–178
~ of galaxies 98–329
~ of nebulae 98–329
open ~ of galaxies 98–333
open star ~ 96–99
~ parallax 91–7
poor ~ of galaxies 98–331
poor ~ of stars 96–102
~ population 98–341
regular ~ of galaxies 98–336
rich ~ of galaxies 98–332
rich ~ of stars 96–100
~ star 96–82
star ~ 96–79
star–poor ~ 96–102
star–rich ~ 96–100
~ variable 95–93
very compact ~ of galaxies 98–335
Virgo ~ 98–345
Virgo ~ of galaxies 98–345
clustering of matter 99–170z
cluster–type variable 95–93
CMa 92–27
CMi 92–28
C.M.P. 89–45
Cnc 92–25
CO 97–28
coagulation 76–119
Coal Sack 97–138
coalescence 76–120
coarse–grained 82–117
~ snow 72–118
coast 72–30
~ mapping 72–82
coastal chart 72–33
coastline 72–31
~ altering 72–78
~ modification 72–78
coefficient, brightness 83–96
drag ~ 86–180
heat–transfer ~ 86–201
~ of darkening 88–144
~ of stability 98–199
phase ~ 83–77
coefficients, dynamic 71–59
harmonic ~ 71–59
Legendre ~ 95–101
~ of the velocity ellipsoid 98–102
coelostat 88–46
horizontal ~ 88–47
vertical ~ 88–48
coercive field 75–39
~ force 75–39
cogeoid 71–17
coherency 99–33
~ postulate 99–34
cohesive 82–138
~ soil 82–63
~ strength 85–117
Col 92–37
cold current 72–64
~ region 88–199
collapsar 91–164
collapse 94–109
Big ~ 99–182
~ caldera 81–181
final ~ 99–182
gravitational ~ 94–109
~ of the universe 99–181
collapsed object 98–23
~ star 94–111
collapsing star 91–164
collection, dust 87–12
meteoric–dust ~ 87–12
meteorite ~ 87–8
~ of meteorites 87–8
collective interaction 98–168
collector, meteoric–dust 87–13
colliding galaxies 98–321
collision, between stars 83a–16
catostrophic ~ 84–106
central ~ 83a–17
erosive ~ 84–105
galaxies in ~ 98–321
head–on ~ 83a–17
~ hypothesis 83a–14
non–elastic star ~ 98–26
~ of galaxies 98–322
stellar ~ 83a–16
three–body ~ 96–74
triple ~ 96–74
collisional break–up 84–107
~ pumping 97–20
~ relaxation 98–131
~ system 98–166
collisionless plasma 84–50
~ relaxation 98–132
~ shock–wave 84–49
~ system 98–167
colloid 76–96
colloidal 76–97
~ particle 76–98
colo(u)r curve 95–21
~ equation 91–73
~ excess 91–77
~ index 91–76
integrated ~ index 96–138
intrinsic ~ 91–74
intrinsic ~ index 91–75
~ –magnitude diagram 96–150
six – ~ plotometry 96–148
special ~ index 83–87
~ system 91–72, 96–140
~ temperature 91–109
two – ~ diagram 96–147
ultraviolet–blue–visual ~ system 96–144
ultraviolet–blue–yellow ~ system 96–144
yellow ~ index 83–88
colorimetric map 81–134
Colt 92–48
Columba 92–37
column, ionized 86–282
~ of ionization 86–282
Com 92–38
coma 85–64
atomic ~ 85–67
~ Berenices 92–38
inner ~ 85–65
molecular ~ 85–65
photochemical ~ 85–65
radical ~ 85–66
UV ~ 85–67
visible ~ 85–66
comet 85–1
absolute brightness of a ~ 85–109
active ~ 85–149
apparition of a ~ 85–18
artificial ~ 85–14
~ associated with a shower 86–259
~ burst 85–125
designation of a ~ 85–31
~ discovered during eclipse 85–12
~ discovery 85–16
discovery of a ~ 85–16
disruption of a ~ 85–114
eclipse ~ 85–12
elliptic ~ 85–4
elliptical–orbit ~ 85–4
envelopes of a ~ 85–69
extinct ~ 85–150
first apparition of a ~ 85–19
~ group 85–105
hyperbolic ~ 85–8
hyperbolic–orbit ~ 85–8
inactive ~ 85–147
live ~ 85–149
long–period ~ 85–6
non–periodic ~ 85–10
~ outburst 85–125
outburst of a ~ 85–125
parabolic ~ 85–7
parabolic–orbit ~ 85–7
parent ~ 86–260
periodic ~ 85–3
predicted ~ 85–11
return of a ~ 85–9
shape of a ~ 85–83
shells of a ~ 85–69
short–period ~ 85–5
Sun–grazing ~ 85–13
tail of a ~ 85–70
telescopic ~ 85–14a
cometary 85–2
~ activity 85–110
~ disruption 85–114
~ form 85–83
~ head 85–35
~ impact 82–230
~ meteor 86–45
~ nebula 97–142

~ nucleus 85–36
~ return 85–9
~ stream 86–222
~ structure 85–34
~ tail 85–70
~ tail type 85–71
comet–like nebula 97–142
comets, family of 85–102
origin of ~ 85–96
physics of ~ 85–104
search for ~ 85–15
structure of ~ 85–34
command module 82–6
commencement, sudden 75–74
sudden ~ absorption 78–47
sudden ~ storm 75–74
commensurability asteroid group 84–101
~ gap 84–88
~ group 84–101
common envelope 96–32
~ envelope of a binary 96–32
~ eruptive prominence 89–250
~ lead 74–99
~ origin 94–25
~ star 91–95
communicator 85–23
comoving coordinate systems 99–44
compact cluster of galaxies 98–334
~ galaxy 98–264
~ H II region 97–51
~ object 98–21
~ source 98–273
~ system 96–157
compaction 74–50
companion 96–38
dark ~ 96–42
elliptical ~ 98–217
elliptical ~ of a galaxy 98–217
invisible ~ 96–41
~ of a galaxy 98–216
optical ~ 98–218
optical ~ of a galaxy 98–218
physical ~ 98–219
physical ~ of a galaxy 98–219
planetary ~ 96–43
Sirius ~ 91–138
~ star 96–39
comparator, blink 91–34
comparison star 95–16
Compass 92–82
Mariner's ~ 92–82
Ship's ~ 92–82
Compasses 92–36
compensated geoid 71–17
compensating plate 88–92
compensation, depth of 73–102
depth of isostatic ~ 73–102
~ for Doppler shifts 88–90
isostatic ~ 73–100
compensator, Doppler 88–91
complete degeneracy 93–175
~ synchrone 85–94
completely degenerate 93–185
complex, basement 72–12
~ group 89–112
component 96–36
dust ~ 98–232
longitudinal ~ 89–135
lunisolar ~ 82–307
primary ~ 96–37
secondary ~ 96–38
~ separation 96–66
σ–~ 88–89
transversal ~ 89–136
vertical ~ 75–62
composite colour index 96–138
~ cone 73–121
~ model 93–140
~ spectrum 96–19
composition, chemical 91–168
chemical ~ of a star 91–168
ionic ~ 78–10
isotopic ~ 82–140, 87–129
molecular ~ 78–11
compressibility 73–57
compression cap 86–170
gravitational ~ 84–72
polar ~ 83–114
compressional wave 73–46
Compton wavelength 99–170g
computer 85–22
concentration 96–127
degree of ~ 96–128
galactic ~ 98–76
galactic ~ toward the centre 98–76
galactic ~ toward the principal plane 98–76
high– ~ cluster 96–129
~ index 96–101
local mass ~ 81–145
lunar mass ~ 81–145
~ toward the galactic plane 98–55
trace ~ 83–266
concentric fracture 82–209
~ wall 81–254
concordant intrusion 73–134
condensation 83a–26
central ~ 85–37, 86–244, 96–126
coronal ~ 89–281
heat of ~ 77–155
~ level 77–156
~ nucleus 76–103
nucleus of ~ 97–67
permanent coronal ~ 89–282
tail ~ 85–95
~ temperature 97–66
~ within a meteor stream 86–217
condense 77–129
condition, central 93–81
fitting 93–142
stability ~ 93–205
standard ~ 77–3
surface ~ 93–80
conditions, equilibrium 93–74
meteorological ~ 77–2
temperature ~ 77–35
weather ~ 77–2
conduit 73–124
cone, cinder 73–120
composite ~ 73–121
debris ~ 73–120
light ~ 95–156
~ of eruption 73–118
~ of escape 76–27
pyroclastic ~ 73–120
ray ~ 95–156
reflecting ~ 88–62
strato ~ 73–121
volcanic ~ 73–118, 82–160
confinement of cosmic rays 98–51
magnetic ~ of cosmic rays 98–51
conformal projection 71–41
conglomerate 85–41
conic projection 71–42
conical point 96–31
conjunction 83–55
inferior ~ 83–57
superior ~ 83–56
triple ~ 83–58
connection, accidental 97–126
genetic ~ 97–127
consolidation of magma 82–190
constant, cosmic(al) 99–29
cosmologic(al) ~ 99–29
Hubble ~ 98–302
~ of recession 98–302
Oort ~ 98–118
Poisson ~ 73–60
solar ~ 88–148
~ with the time 98–150
constellation 92–1
autumn ~ 92–1d
circumpolar ~ 92–4a
fall ~ 92–1d
spring ~ 92–1c
summer ~ 92–1a
winter ~ 92–1b
zodiacal ~ 92–5
constituent, atmospheric 83–265
mineralogical ~ 82–100
rock ~ 82–99
constitution, internal 71–83, 93–71
contact binary 96–29
first ~ 81–84
last ~ 81–85
~ metamorphism 74–35
~ surface 96–27
container, sample 82–81
containment of cosmic radiation 98–51
content, chlorine 72–68
nickel ~ 72–55
salt ~ 72–66
continent 72–9, 81–191, 83–159
drift of ~ 73–104
continental apron 72–36
~ block 72–10
~ climate 77–185
~ crust 72–5

~ drift 73–104
~ earthquake 73–2b
~ fracture system 73–81
~ glacier 72–116
~ plate 73–103
~ platform 73–103
~ region 81–271
~ sea 72–26
~ shelf 72–34
~ slope 72–35
continual creation 99–91
continuity equation 93–49
continuous creation 99–91
~ outflow 98–276
~ wave 86–277
contour, bathymetric 71–55
depth ~ 71–55
isocount ~ 75–102
isophotal ~ 97–124
~ line 71–53
~ of constant intensity 75–102
submerged ~ 72–32
contracting model 99–56
~ phase 94–20
~ stage 94–20
~ star 91–161
~ universe 99–56
contraction 93–117, 94–36
~ catastrophic ~ 94–109
core ~ 94–70
gravitational ~ 93–118, 94–37
Helmholtz ~ 94–37
Lorentz ~ 99–134
Lorentz ~ factor 99–135
Lorentz–FitzGerald ~ 99–134
~ of the universe 99–180
~ phase 94–20
quasi–static ~ 94–40
relativistic ~ 99–134
~ theory 73–77
convection 93–147
~ core 93–165
~ current 73–76
hydrogen ~ zone 93–159
laminar ~ 93–153
turbulent ~ 93–152
convective 93–148
~ core 93–165
~ energy flux 93–115
~ envelope 93–136
~ equilibrium 76–22, 93–79
~ flux 93–115
hydrogen ~ zone 93–159
~ instability 93–155
~ layer 93–149
semi– ~ zone 93–160
~ temperature gradient 93–157
~ zone 93–149
convergent point 96–98
cool halo 98–47
~ lava 82–176
~ star 91–156
cooling 83a–57
~ by neutrino emission 94–97
~ line 93–202
neutrino ~ 94–97
radiation ~ 94–98
radiative ~ 94–98
~ rate 93–203
~ time 83–201
~ track 93–202
coordinate, areocentric 83–138
comoving ~ systems 99–44
cylindrical ~ 98–156
galactocentric ~ 98–12
geomagnetic ~ 75–51
heliographic ~ 89–39
iovicentric ~ 83–176
jovigraphic ~ 83–175
planetographic ~ 83–8
selenographic ~ 81–9
space ~ 99–124
time ~ 99–123
timelike ~ 99–123
Copernican system 99–24
coplanarity of orbits 83a–3
core 72–52
~ contraction 94–70
convection ~ 93–165
convective ~ 93–165
degenerate ~ 93–170
Earth's ~ 72–17
graphite ~ 97–71
helium ~ 93–167
inhomogeneous ~ 93–169
inner ~ 72–20
isothermal ~ 93–168
~ –mantle particle 97–70
non–degenerate ~ 93–171
~ of a cluster 96–130
~ of a star 93–164
outer ~ 72–19
radiative ~ 93–166
~ sample 82–60
stellar ~ 93–164
coring 72–51
Coriolis effect 77–81
~ force 77–79
corona 79–60, 79–156
~ Australis 92–39
~ Austrina 92–39
~ Borealis 92–40
E ~ 88–241a
electron ~ 88–242
emission ~ 88–241a
expanding ~ 88–232
F ~ 88–241
flattening of the ~ 88–247
Fraunhofer ~ 88–241
galactic ~ 98–49
heating of the ~ 88–213a
hydrogen ~ 83–246
inner ~ 88–229
intermediate ~ 88–245
K ~ 88–242
lunar ~ 79–61
maximum ~ 88–244
middle ~ 88–230
minimum ~ 88–246
monochromatic ~ 88–238
outer ~ 88–231
planetary ~ 83–245
radio ~ 88–239
solar ~ 79–62, 88–225
stellar ~ 93–4
synoptic map of the ~ 89–22
white ~ 88–237
white light ~ 88–237
X–ray ~ 88–240
coronagraph 88–59
externally occulted ~ 88–60
externally occulting ~ 88–60
coronal 88–228
~ active region 89–286
~ arch 89–298
green ~ region 89–284
~ hole 88–245a
~ line 89–290
monochromatic ~ ray 89–288
~ prominence 89–264
~ rain 89–271
~ ray 89–287
red ~ region 89–285
sporadic ~ condensation 89–283
~ streamer 89–287
~ transient 89–185a
coronameter, K 88–66
corotating field line 75–14
~ structure 84–17
~ structure of magnetic field 84–17
corpse, stellar 91–167
corpuscle 88–153
corpuscular 88–155
~ activity 89–7
~ emission 88–152
~ radiation 88–150
~ stream 88–160
corrected area 89–93
~ radiant 86–58
correction, bolometric 91–80
~ for the curvature 86–62
~ for the curvature of the Earth's surface ~ 86–62
relativistic ~ 99–140
Corvus 92–41
cosmic(al) abundance of the elements 94–107
~ cloud 94–31
~ constant 99–29
~ dust 97–52
~ geophysics 84–57
~ maser 97–13
~ material 99–7
~ matter 99–7
~ origin 87–5
~ scale factor 99–81a

EN

~ scale turbulence 83a–12
~ string 99–171c
~ triangulation 71–28
cosmic–ray albedo 75–116
~ diffusion 84–27
~ event 89–194
~ exposure age 87–139
~ irradiation 82–250
~ modulation region 84–26
~ produced isotope 82–147
solar ~ event 89–194
~ spallation 82–254
cosmochemistry 74–31
cosmochronology 83a–72
cosmogenic isotope 82–147
~ radionuclide 82–255
cosmogonic = cosmogonical 94–3
~ hypothesis 94–8
~ problem 94–5
cosmogonist 94–4
cosmogony 94–2
planetary ~ 83a–1
stellar ~ 94–7
cosmological 99–18
~ constant 99–29
~ density 98–370
~ era 99–166a
~ horizon 99–179
hot ~ model 99–166
~ inflation 99–170c
~ model 99–46
~ nucleosynthesis 99–175
~ object 98–369
~ paradoxes 99–38
perfect ~ principle 99–32
~ principle 99–30
~ redshift 98–363a
restricted ~ principle 99–31
~ synthesis of elements 99–175
~ system 99–20
~ term 99–156a
cosmologist 99–19
cosmology 99–16
kinematical ~ 99–26
Newtonian ~ 99–25
non–relativistic ~ 99–27
observational ~ 99–17
relativistic ~ 99–28
cosmos 99–2
count, galaxy 98–343
star ~ 98–79
counterglow 84–64
coupling, radiation 99–173
cover, snow 77–176
CP violation 99–170t
CrA 92–39
Crab 92–25
~ Nebula 97–134
~ pulsar 95–142
crack, tension 82–206
Crane 92–52
crape ring 83–221
Crater 92–42
crater 73–126
adventive ~ 73–128
ancient ~ 81–205
blocky–rim ~ 81–225
blocky–rimmed ~ 81–225
buried ~ 81–204
central ~ 81–217
~ chain 81–241
circular ~ 81–213
concentric ~ 81–216
cup–shaped ~ 81–219
dark–haloed ~ 81–215
~ density 81–246
dimple ~ 81–243
dish–like ~ 81–219
drainage ~ 81–239
explosion ~ 81–238
~ field 87–56
flooded ~ 81–203
~ floor 81–259
fossil meteoritic ~ 87–55
~ frequency 81–247
ghost ~ 81–209
impact ~ 81–240
inner ~ wall 81–252
irregular ~ 81–220
lateral ~ 73–128
lunar ~ 81–197
maar ~ 81–264
main ~ 73–127
~ –margin fracture 82–211
meteor ~ 87–53
meteoric ~ 81–263
meteorite ~ 87–53
meteoritic ~ 87–53
meteoroidal ~ 81–263
~ number density 81–246
outer ~ 81–218
outer ~ wall 81–253
~ parameters 81–227
parasitic ~ 73–128
~ pit 81–250
polygonal ~ 81–212
possible ~ 81–211
post–mare ~ 81–208
pre–mare ~ 81–207
primary ~ 81–198
probable ~ 81–210
raised–rim ~ 81–223
ray ~ 81–214
rayed ~ 81–214
~ rim 81–221
rimless ~ 81–226
satellite ~ 81–200
satellitic ~ 81–200
secondary ~ 81–199
small ~ 81–250
summit ~ 73–129
~ terrace 81–229
terrestrial meteoritic ~ 87–54
unnamed ~ 81–202
~ wall 81–251
~ with central peak 81–258a
~ with rays 81–214
young ~ 81–206
cratered surface 81–245
cratering 82–232
impact ~ 82–234
~ rate 82–233
craterlet 81–250
CrB 92–40
creation, continual 99–91
continuous ~ 99–91
~ of matter 99–89
unique ~ 99–90
creep, downslope 82–195
soil ~ 82–152
crêpe ring 83–221
crepuscular arc 79–50
crescent 81–23
extension of the ~ of Venus 83–131
~ Moon 81–23
~ of Venus 83–129
~ phase of the Moon 81–23
~ Venus 83–129
crest 81–195
mountain ~ 81–195
Cretaceous 74–86
criteria, classification 91–110
criterion, degeneracy 93–181
~ of degeneracy 93–181
~ of tidal instability 98–312
critical density 99–85a
~ frequency 78–24
~ level 76–26
~ mass 93–189
~ surface 96–27
crochet 75–79
magnetic ~ 75–79
cross, halo 79–79
light ~ 79–79
lunar ~ 79–80
Northern ~ 92–44
S ~ 92–43
solar ~ 79–81
Southern ~ 92–43
~ the central meridian 89–44
crosser, Earth– 84–76
Earth–orbit ~ 84–76
Mars– ~ 84–75
Mars–orbit ~ 84–75
cross–section, differential 94–54
geometrical ~ 84–103
gravitational ~ 84–104
head–on ~ 86–176
partial ~ 94–54
Crow 92–41
Crown 92–39
N ~ 92–40
Northern ~ 92–40
S ~ 92–39
Southern ~ 92–39
crown, polar 89–232

Crt 92–42
Cru 92–43
Crunch, Big 99–182
crushing, gravitational 84–72
crust, continental 72–5
Earth's ~ 72–4
fusion ~ 87–110
lunar ~ 82–54
Moon's ~ 82–54
solid ~ 95–151
crustal dislocation 82–212
~ motions 73–72
~ movements 73–72
~ stress 82–210
~ tides 82–297
crusted surface 87–109
crust-mantle interface 72–14
~ transition 72–14
crustquakes 95–152
Crux 92–43
CrV 92–41
Cryptozoic 74–70
~ era 74–70
crystal, ice 76–111
crystalline rock 82–66
crystallization 76–107
~ differentiation 74–41
fractional ~ 74–40
~ nuclei 76–108
crystobalite 82–95
CS 97–34
Cs 77–141
Cu 77–138
cubic ice 85–46
~ parsec 98–84
culmination of the Moon 82–303
cumulative effect 98–124
Cumulonimbus 77–137
Cumulus 77–138
~ fractus 77–146
Cup 92–42
cup-shaped crater 81–219
Curie point 75–41
~ temperature 75–41
current, air 77–90
ascending ~ 93–151
cold ~ 72–64
convection ~ 73–76
Earth ~ field 75–4
ebb ~ 72–65
equatorial ring ~ 75–98
falling ~ 93–150
flood-tide ~ 82–280
ocean ~ 72–60
rising ~ 93–151
sea ~ 72–60
subsurface ~ 72–61
subsurface ocean ~ 72–61
surface ~ 72–62
suspension ~ 72–49
~ system 75–97
tidal ~ 82–279
turbidity ~ 72–49
up- ~ 77–91
warm ~ 72–63
currents 83–187
Earth ~ 75–5
stray Earth ~ 75–6
telluric ~ 75–5
curtains 79–154
curvature of lines of force 75–121
~ of space 99–115
~ of the Earth 71–6
~ of Earth's surface 86–61
~ of the path 86–51
~ tensor 99–159
curve, abundance 94–108
colo(u)r ~ 95–21
empirical ~ of growth 93–69
extinction ~ 97–81
frequency-height ~ 78–23
light ~ 95–17
meteor light ~ 86–70
~ of growth 93–68
period-luminosity ~ 95–103
phase ~ 83–74
polarization ~ 83–81
reddening ~ 97–93
rotation ~ 98–109
theoretical ~ of growth 93–70
threshold ~ 96–40
threshold ~ of the secondary component 96–40
velocity-depth ~ 73–52
curved fan 85–62
~ jet 85–58
~ slit 88–65
~ space 99–114
~ tail 85–73
cusp, polar 75–13
cusps 83–130
lunar ~ 81–24
CVn 92–26
cyanide, hydrogen 97–29
~ methyl 97–38
cyanoacetylene 97–30
cyanogen 97–24
cycle, beginning of 89–82
duration of ~ 89–87
11-year ~ 89–73
eleven-year ~ 89–73
end of the ~ 89–85
even spot ~ 89–77
geochemical ~ 74–32
odd spot ~ 89–78
~ of solar activity 89–71
phase of the ~ 89–79
solar ~ 89–71
sunspot ~ 89–72
cyclic(al) variable 95–108
cyclone 77–76
tropical ~ 77–77
cyclostrophic wind 77–95
Cyg 92–44
Cygnus 92–44
~ Loop 97–137
cylinder, light 95–155
truncated ~ 88–63
cylindrical coordinates 98–156
~ projection 71–43
~ symmetry 98–155
Cytherean 83–126

D

D 79–154
daily areas of prominences 89–236
~ map of solar activity 89–18
~ map of the Sun 89–18
~ motion 85–29
~ motion of the radiant 86–69
~ solar map 89–18
~ variation 86–137
DALR 76–14
damage, tidal-wave 71–77
damaging earthquake 73–38
dark absorption marking 89–224
~ area 83–155
~ Bay 97–135
~ cloud 97–107
~ companion 96–42
~ domes 89–297
~ flocculus 89–224
~ hemisphere 83–68
~ hydrogen flocculus 89–224
~ lanes 98–288
~ limb 81–97
~ marking 89–224
~ matter 97–108
~ mica 74–25
~ mottles 88–212
~ nebula 97–132
~ side 75–21
~ surge 89–256
darkening, coefficient of 88–144
gravity ~ 91–41, 96–25
law of ~ 88–143
limb ~ 88–142
dark-haloed crater 81–215
dating 83a–71, 87–126
~ by radioactivity 87–131
isotope ~ 87–130
isotopic ~ 87–130
~ of lunar events 82–265
radioactive ~ 87–131
thorium-lead ~ 82–270
datum 71–20
geodetic ~ 71–20
dawn 79–35
~ chorus 78–62
~ meridian 75–18
~ terminator 81–124
day, active 75–93
disturbed ~ 75–94
~ hemisphere 83–67
lunar ~ 81–32

magnetically active ~ 75–93
magnetically quiet ~ 75–92
mean lunar ~ 82–304
quiet ~ 75–93
~ side 75–20
~ sky 79–3a
tidal ~ 82–290
dayglow 79–3
daylight fireball 86–118
~ hemisphere 75–20
~ stream 86–219
dayside aurora 79–97
day–time airglow 79–3
~ meteor 86–22
~ observation 83–121
~ shower 86–219
~ stream 86–219
~ train 86–106
De Sitter's model 99–69
~ universe 99–69
deactivation of a comet 85–148
dead star 91–167
~ tide 82–272
~ volcano 73–115
debris cone 73–120
decametric burst 83–183
~ radio source 83–186
decay 74–65
~ of a cometary nucleus 86–261
~ of a sunspot group 89–126
radioactive ~ 74–64
deceleration, apparent 86–100
~ of the expansion 99–82
~ parameter 99–83
real ~ 86–101
December Ursids 86–258
decibel 76–39
declination, magnetic 75–56
Moon's ~ 81–62
Sun's ~ 81–63
decline of light–curve 95–35
declining phase 89–84
decoupling era 99–177a
matter-radiation ~ 99–177
~ of radiation 99–177
decrease 76–57
~ in temperature 77–41
~ of mass 83a–42
decreasing branch 95–35
deep earthquake 73–23
~ ocean 72–41
deepening of a depression 77–82
deep–focus earthquake 73–23
deep-ocean deposits 72–50
~ sediments 72–50
deep–sea deposits 87–50
~ sediments 72–50
deferent 83–48
definite designation 85–33
deflection of a star 98–123
~ of the vertical 71–72
plumb–line ~ 71–72
defluidization 82–263
deformation tensor 73–70
tidal ~ 82–277, 83a–66
degassing 82–261, 87–136
degeneracy 93–172
complete ~ 93–175
~ criterion 93–181
criterion of ~ 93–181
degree of ~ 93–174
neutron ~ 93–191
non–relativistic ~ 93–177
partial ~ 93–173
~ pressure 93–187a
relativistic ~ 93–176
~ temperature 93–188
degenerate 93–178
completely ~ 93–185
~ core 93–170
~ gas 93–179
highly ~ 93–184
non- ~ gas 93–182
partially ~ 93–183
~ star 93–180
degree, geothermic 74–59
geothermic ~ of depth 74–59
measurement of a ~ 71–22
~ of concentration 96–128
~ of degeneracy 93–174
Deimos 83–169
deionization 86–213
Del 92–45
delay, group 78–56
Delphinus 92–45
demagnetization, thermal 75–40
Demon Star 92–105
Deneb 92–112
Denebola 92–113
density, apparent surface 96–132
average ~ of matter 99–84
background ~ 96–131
central ~ 93–87
cosmological ~ 98–370
crater ~ 81–246
critical ~ 98–197
~ distribution 96–124
electron ~ 78–6, 88–233
electron line ~ 86–283
~ function 98–88
~ gradient 98–87
line ~ 86–283
mean ~ 93–86
mean ~ of matter 99–84
~ of a meteor stream 86–264
~ of galaxies 98–344
~ of the shadow 81–59
projected ~ 98–85
radiant ~ 86–63
radiation ~ 99–85
real ~ 96–133
shadow ~ 81–59
space ~ 86–263, 96–133
star ~ 98–82
stellar space ~ 98–83
subnuclear ~ 93–190
surface ~ 98–85
track ~ 82–253, 87–141
volume ~ 96–133
~ wave 98–175
~ wave theory 98–174
denudation 73–108
de–occultation 81–99
departure from thermodynamic equilibrium 93–36
depolarization 97–118
deposit, annual 87–17
deposition 73–109
depositional magnetization 75–33
~ remanent magnetization 75–33
deposits, abyssal 72–50
biogenic ~ 72–59
deep-ocean ~ 72–50
deep–sea ~ 87–15
glacial ~ 72–127
pelagic ~ 72–57
terrigenous ~ 72–48
depression 77–76
arcuate ~ 81–160
continental ~ 81–158
deepening of a ~ 77–82
filling of a ~ 77–83
highland ~ 81–158
linear ~ 81–159
lunar ~ 81–157
sinuous ~ 81–161
solar ~ 79–45
terraced ~ 81–162
depth contour 71–55
effective ~ of line formation 93–64
focal ~ 73–20
~ of compensation 73–102
~ of eclipse 95–75
~ of isostatic compensation 73–102
~ of line formation 93–64
descending branch 95–35
~ phace of the solar cycle 89–84
~ stream 93–150
desert 83–161
desiccation 74–51
designation, definitive 85–33
~ of a comet 85–31
preliminary ~ 85–32
desoccultation 81–99
desorption 85–123
heat of ~ 85–124
destruction 87–24
~ by a meteorite 87–20
~ mechanism 97–79
~ of surface 87–35
destructive earthquake 73–39
detached binary 96–35
detailed balancing 93–41
~ photometry 83–104
deterioration of weather 77–23
determination, age 74–91

mass ~ 96–69
~ of age 74–91
~ of gravity 71–67
detonating fireball 86–119
detrital magnetization 75–33
~ remanent magnetization 75–33
~ slope 81–233
development curve 89–119
~ curve of a spot group 89–119
~ of a spot group 89–118
deviating force 77–80
deviations from local thermodynamic equilibrium 93–40
~ from LTE 93–40
~ from TE 93–36
~ from thermodynamic equilibrium 93–36
Devonian 74–81
dew 77–173
~ point 77–174
diagenesis 74–38
diagram, butterfly 89–76
colo(u)r–magnitude ~ 96–150
evolution ~ 96–151
evolutionary ~ 96–151
Hertzsprung–Russell ~ 91–130
H–R ~ 91–130
mass–luminosity ~ 91–129
redshift–magnitude ~ 98–301
two–colo(u)r ~ 96–147
diameter, angular 83–117, 96–115
apparent ~ 83–117, 96–115
linear ~ 83–118
photometric ~ 84–93
true ~ 83–118
diameter–depth relation 81–248
diamond dust 76–112
diastrophism 73–72
diatreme 73–122
dichotomy of Venus 83–132
difference, brightness 96–65
~ in optical path 88–117
magnitude ~ 91–59, 96–65
vertical ~ 81–140
differential cross-section 94–54
~ galactic rotation 98–113
~ -motion term 98–115
~ rotation 88–128
~ rotation of the Galaxy 98–113
differentiation, chemical 83a–50
crystallization ~ 74–41
geochemical ~ 74–55
gravitational ~ 74–56, 83a–54
diffuse galactic light 97–56
~ interstellar band 97–100
~ interstellar feature 97–100
~ luminous surface 79–148
~ nebula 97–122
~ object 85–25
~ radiation field 93–31
~ sky radiation 77–59, 88–136
diffusion coefficient 86–287
atmospheric ~ coefficient 86–287
cosmic–ray ~ 84–27
~ equilibrium 76–23
ionic ~ 86–286
~ of ions 86–286
diffusive equilibrium 76–23
dike 73–138
dila(ta)tion, time 99–137
dila(ta)tional wave 73–46
dilute radiation 93–38
diluted radiation 93–38
dilution factor 93–37
dimensions of the system 96–64
dimple 81–243
~ crater 81–243
dioxide, solid carbon 83–144
dip, magnetic 75–58
magnetic ~ pole 75–45
direct 83–42
~ motion 83–43
~ rotation 83–39
~ solar radiation 77–50
~ starlight 79–11
~ wave 78–12
direction, antisolar 75–22
axis ~ 83–34
~ of propagation 78–35
wind ~ 77–115
dirty ice 97–73
disappearance 81–98
height of ~ 86–40
~ of a meteor 86–37
~ point 86–38
ring ~ 83–217
sudden ~ 86–76, 89–279
disc = disk 96–21
~ accretion 96–21
~ flare 89–204
centre of the ~ 89–35
centre of the solar ~ 89–35
~ flare 89–204
galactic ~ 98–52
lunar ~ 81–54
occulting ~ 88–61
on the ~ 88–173
optical ~ 89–26
~ phase 94–22
~ population 98–72
solar ~ 89–24
~ stage 94–22
~ star 91–85
~ -type population 98–72
visible ~ 88–171, 89–26
discharge, atmospheric 77–161
discoloration, sea 72–80
discontinuity, Moho 72–14
Mohorovičić ~ 72–14
passage of interplanetary ~ 84–58
discordant intrusion 73–137
discoverer 85–21
discovery, comet 85–16
~ of a comet 85–16
place of ~ 87–52
discrepant redshift 98–367
dish-like crater 81–219
disintegrate 74–65
disintegration, cluster 96–103
meteor ~ 86–187
nuclear ~ 85–116
~ of a comet 85–115
~ of a cometary nucleus 85–116
~ of clusters 98–195
~ of meteor body 86–186
tidal ~ 85–118
disk = disc 88–173
dislocation 73–91
~ earthquake 73–2a
dispersion of velocities 98–103
~ time 96–104
velocity ~ 98–103
displacement of spectral lines 99–162
gravitational ~ of spectral lines 99–162
~ of the radiant 86–242
radiant ~ 86–242
relativity ~ 99–162
disrupted galaxy 98–317
disruption, collisional 87–125
cometary ~ 85–114
~ of a cluster 96–105
~ of a comet 85–114
~ of a meteor 86–82
~ of parent body 87–124
point of ~ 86–121
tidal ~ 85–118
dissipation 94–73
energy ~ 82–285
~ of waves 88–222
thermal ~ 94–76
tidal ~ 82–286
zone of ~ 72–124
dissociation of a galactic cluster 96–103
~ temperature 91–120a
~ time 96–104
dissociative recombination 86–214
dissolution, cluster 96–103
~ of a prominence 89–280
~ time 96–104
distance 91–3
angular ~ 96–52
~ between components 96–66
Earth–Moon ~ 81–13
epicentral ~ 73–19
focal ~ 73–18
~ from the central meridian 89–46
~ from the centre of the disc 89–36
~ from the centre of the star 93–85
~ from the galactic centre 98–11
galactocentric ~ 98–11
geocentric ~ 84–3
heliocentric ~ 84–4
hypocentral ~ 73–18

~ modulus 91–16
~ of the focus 73–18
~ of the Moon from the Earth 81–13
photometric ~ 96–116
~ scale 98–303
skip ~ 78–38
distant earthquake 73–27
distorted galaxy 98–316
~ path 86–33
distortion 73–68
rotational ~ 91–44
rotational ~ of velocity curve 91–44
~ tensor 73–70
tidal ~ 96–30
distortional wave 73–43
distributed, randomly 98–93
distribution, brightness 86–166
density ~ 96–124
~ function 98–143
galactic ~ 98–54
heliographic latitude ~ 89–51
heliographic longitude ~ 89–48
~ in heliographic latitude 89–51
~ in heliographic longitude 89–48
~ in latitude 89–51
~ in longitude 89–48
~ in space 96–119
latitude ~ 89–51
longitude ~ 89–48
mass ~ 86–165
mass–frequency ~ 84–52
Maxwellian ~ of velocities 98–96
~ of matter 99–8
~ of residual velocities 98–92
size ~ 82–124
space ~ 98–86
spatial ~ 96–119
disturbance, bay 75–80
flare–associated ~ 84–46
geomagnetic ~ 75–70
interplanetary ~ 84–45
ionospheric ~ 78–39
magnetic ~ 75–70
South Tropical ~ 83–194
sudden ionospheric ~ 78–41
tidal ~ 98–314
disturbed day 75–94
~ galaxy 98–315
~ Sun 89–12
diurnal aberration 86–57
~ changes 83–167
~ libration 81–52
~ variation 86–137
~ wave 76–30
division 83–223
Cassini ~ 83–224
~ of a galactic nucleus 98–318
~ of a nucleus 98–318
D–line 88–140
Dog Star 92–128
doldrums 77–101
Dolphin 92–45
dome 81–177
lava ~ 81–176a
lunar ~ 81–177
volcanic ~ 81–176
domeless solar refractor 88–53
domes, cluster of 81–178
dark ~ 89–297
dark ~ around prominences 89–297
Doppler compensator 88–91
~ shift 91–25
Dor 92–46
Dorado 92–46
~ Fish 92–46
dormant volcano 73–116
dot, umbral 89–96
double cluster 96–81
~ counting method 86–132
~ energy source 93–146
~ galaxy 98–305
~ meteor 86–131
optical ~ star 96–6
physical ~ star 96–7
~ planetoid 84–79
~ spectrum 96–18
~ star 96–3
visual ~ star 96–8
~ wave 98–117
wide ~ star 96–9
double–line binary 96–17
~ spectroscopic binary 96–17
doubling of lines 97–157
doubly photographed meteor 86–155
Dove 92–37
downslope creep 82–195
downward slipping 82–198
Dra 92–47
Draco 92–47
draconic month 81–47
~ period 81–47
~ revolution 81–47
Draconids 86–252
drag 86–178
~ coefficient 86–180
~ equation 86–179
Dragon 92–47
drainage crater 81–239
draperies 79–154
drawing of a planet 83–16
sunspot ~ 88–34
dredging 72–53
drift, asymmetric 98–104
continental ~ 73–104
~ in longitude 75–120
longitudinal ~ 75–120
~ of continents 73–104
~ of the train 86–114
driven camera 86–142
drizzle 77–172
DRM 75–33
drop 76–113
black ~ 83–52
temperature ~ 77–41
drop–form tektite 87–96
dry adiabatic lapse rate 76–14
~ ice 83–144
drying 74–51
DS 79–148
Dubhe 92–114
dune 83–162
duration of cycle 89–87
~ of eclipse 95–78
~ of flare 89–209
~ of flight 86–50
~ of stay 81–34
~ of the period 95–39
~ of sunshine 88–10
~ of total eclipse 88–26
~ of train visibility 86–111
dusk 79–36
~ meridian 75–19
~ terminator 81–125
dusky belt 83–190
~ ring 83–221
dust blob 98–290
circumstellar ~ 97–55
circumterrestrial ~ 84–53
~ cloud 83–256, 97–114
~ collection 87–12
~ component 98–232
~ content 85–80
cosmic ~ 97–52
diamond ~ 76–112
Earth–orbiting ~ 84–53
~ filter 76–95
galactic ~ 97–53
~ haze 77–27
impact ~ 82–125
interplanetary ~ 84–51
interstellar ~ 97–54
interstellar ~ grains 97–59
~ lane 98–289
~ layer 82–122
lunar ~ sample 82–121
meteoric ~ 87–11
meteoritic ~ 87–10
~ nebula 97–115
~ shell 97–116
~ storm 83–255
~ tail 85–79
~ train 86–115, 87–22
volcanic ~ 73–143, 82–126
dustball 86–190
dust–covered 82–120
dust–free 82–119
dwarf 91–135
black ~ 91–139
brown ~ 91–138a
~ elliptical 98–242
~ elliptical galaxy 98–242
~ galaxy 98–220
~ nova 95–112
red ~ 91–136
~ star 91–135
white ~ 91–137

dyke 73–138
dynamic(al) coefficient 71–59
~ evolution 96–161
~ metamorphism 74–36
~ meteorology 77–7
~ parallax 91–12
~ spectrogram 88–125
~ spectrograph 88–124
~ spectrum 88–125
~ stability 93–207, 98–198
dynamics, cluster 96–153
interstellar gas ~ 97–150
stellar ~ 98–120
dynamometamorphism 74–36

E

E layer 78–26
Eagle 92–18
early atmosphere 83a–45
~ Precambrian 74–75
~ stars 91–102
~ universe 99–170a
early–type giant 91–143a
~ stars 91–102
Earth 71–2
age of ~ 83a–68
atmosphere of ~ 76–5
~ current field 75–4
~ currents 75–5
curvature of ~ 71–6
~ ellipsoid 71–11
figure of ~ 71–7
~ grazer 84–78
interior of ~ 72–3
model of ~ 79–133
penumbra of ~ 81–64
~ radiation 77–56
~ sciences 71–1
shape of ~ 71–7
~ spheroid 71–13
tail of ~ 84–65
~ tides 82–296
~ tremors 73–2a
umbra of ~ 81–58
Earth–crossing asteroid 84–76
Earth–grazer 84–78
Earth–grazing asteroid 84–78
Earth–induced tides 82–275
earthlight 81–35
Earth–like planet 83–119
Earth–Moon distance 81–13
Earth–Moon system 81–12
Earth–orbiting dust 84–53
Earth–produced tides 82–275
earthquake 73–2a
~ belt 73–30
catastrophic ~ 73–40
continental ~ 73–2b
damaging ~ 73–38
deep ~ 73–23
deep–focus ~ 73–23
destructive ~ 73–39
dislocation ~ 73–2d
distant ~ 73–27
~ focus 73–17
inland ~ 73–2b
~ intensity 73–33
intermediate ~ 73–24
local ~ 73–25
~ magnitude 73–34
magnitude of an ~ 73–34
moderate ~ 73–36
near ~ 73–26
nearby ~ 73–26
normal ~ 73–22
~ of magnitude... 73–38
~ region 73–28
shallow ~ 73–21
slight ~ 73–35
submarine ~ 73–2c
tectonic ~ 73–2d
violent ~ 73–37
volcanic ~ 73–2e
~ wave 73–9
~ zone 73–31
Earth's atmosphere 76–5
beyond the ~ orbit 84–6
~ charge 75–3
~ core 72–17
~ crust 72–4
curvature of the ~ surface 86–61
~ flattening 71–8
in the vicinity of the ~ orbit 84–7
~ interior 72–3
~ magnetic field 75–44
~ magnetic pole 75–45
~ magnetism 75–24
~ mantle 72–15
~ orbit 84–5
~ shadow 79–44
~ surface 71–5
~ terminator 81–91
earthshine 81–35
~ on the Moon 81–35
Earthside hemisphere 81–15
east limb 89–34
eastern elongation 83–61
~ quadrature 83–65
~ Sea 81–284
~ spot 89–114
east–west asymmetry 89–49
ebb 82–282
~ current 72–65
flood and ~ 72–89
echo, auroral 79–121
~ depth sounding 72–29
~ duration 86–299
meteor ~ 86–296
meteoric head ~ 86–297
~ pulse 86–290
radar ~ observations 86–265
radio ~ 86–295
seismic ~ 82–223
~ signal 86–290
~ sounding 72–29
~ time 86–291
trail ~ 86–298
eclipse 88–14
annular ~ 88–18
artificial solar ~ 88–17a
atmospheric ~ 95–76
beginning of the ~ 81–74
beginning of the total ~ 81–77
~ begins at sunset 88–22
central ~ 88–16a
~ chart 88–28
circumstances of the ~ 81–73
~ comet 85–12
corpuscular ~ 88–24
depth of ~ 95–75
duration of ~ 95–78
end of ~ 81–83
end of partial ~ 81–83a
~ ends at sunrise 88–21
ionospheric ~ 88–23
~ is invisible in... 81–87
~ is visible in... 81–86
lunar ~ 81–56
magnitude of ~ 81–79
middle of ~ 81–78
northern limit of ~ 88–20
~ observation 88–226
~ of the Moon 81–56
~ of the Sun 88–15
partial ~ of the Moon 81–67
partial solar ~ 88–16
penumbral lunar ~ 81–65
phases of the ~ 81–73
solar ~ 88–15
southern limit of ~ 88–19
total ~ begins 81–77
total ~ ends 81–80
total ~ of the Moon 81–66
total ~ of the Sun 88–17
total solar ~ 88–17
visibility of ~ 81–88
eclipsed Moon 81–57
eclipsing binary 95–74
~ variable 95–74
ecliptical meteor 86–46
~ stream 86–221
ecology 76–2
Eddington model 93–137
edge, outer 83–227
edge–on galaxy 98–246
galaxy seen ~ 98–246
ring seen ~ 83–218
edgewise ring 83–218
effect, betatron 91–180
blanketing ~ 93–67
cumulative ~ 98–124
Coriolis ~ 77–81
Evershed ~ 89–103
greenhouse ~ 83–263
isotope ~ 91–178

EN

isotopic ~ 91–178
opposition ~ 83–75
Poynting–Robertson ~ 84–67
reflection ~ 96–26
reflection ~ in binaries 95–82
Rossiter ~ 91–44
rotational ~ 91–44
rotational ~ on velocity curve 91–44
selection ~ 98–81
tidal ~ 82–276
Umkehr ~ 76–125
Wilson ~ 89–89
effective acceleration 85–88
~ depth 93–64
~ depth of line formation 93–64
~ field strength 95–59
~ line width 89–208
~ temperature 91–113
~ wavelength 96–142
effusion of lava 73–130, 82–172
effusive rock 74–3
Einstein's equivalence principle 99–150
~ field equations 99–156
~ model 99–68
~ universe 99–68
ejecta 82–165
impact ~ 82–167
volcanic ~ 82–166
ejectamenta 82–165
ejection 89–259, 98–179
flare ~ 89–214
~ from a comet 86–262
mass ~ 93–15, 94–77
~ of matter 89–214, 95–124
~ of particles 85–89
~ of radio clouds 98–285
process of ~ 98–179
recurrent ~ 98–282
~ theory 85–98
velocity of ~ 85–90
elastic wave 73–44
elasticity 73–58
modulus of ~ 73–59
E–layer 78–26
peak value of ~ 76–56
electricity, atmospheric 75–2
terrestrial ~ 75–1
electrographic brightness 91–67
~ magnitude 91–67
electrojet, auroral 79–113
electron atmosphere 93–28
~ aurora 79–110
auroral ~ 79–109
~ beam 79–116
~ cloud 79–115
~ corona 88–242
~ density 78–6, 88–233
~ line density 86–283
~ precipitation 78–48
~ source 79–119
~ temperature 91–116
~ whistler 78–60
electrons, dumped 79–118
jet of ~ 79–114
relativistic ~ 97–153
trapped ~ 79–111
electroweak force 99–170x
~ interaction 99–170x
element formation 94–52
meteorological ~ 77–28
over-abundant ~ 91–184
structural ~ 98–287
trace ~ 82–145
under-abundant ~ 91–186
elements, atmophilic 83a–29
chalcophilic ~ 83a–31
geomagnetic ~ 75–54
iron-group ~ 94–105
iron-peak ~ 94–105
lithophilic ~ 83a–30
magnetic ~ 75–54
major ~ 82–144
photometric ~ 95–25
refractory ~ 97–64
elliptical companion 98–217
ellipsoid, earth 71–11
international ~ 71–14
local ~ 71–15
reference ~ 71–18
velocity ~ 98–101
ellipsoidal altitude 71–52
~ galaxy 98–241
~ height 71–52
~ variable 95–79
ellipticity of a globular cluster 96–122
elongation 83–60
eastern ~ 83–60
greatest ~ 83–63
~ of the apparent radiant 86–67
~ of the true radiant 86–68
western ~ 83–62
embryo of a planet 83a–36
Emden's equation 93–130
emergent radiative flux 93–114
emersion 81–99
emission corona 88–241a
corpuscular ~ 88–152
extremely low frequency ~ 78–55
Jupiter's sporadic radio ~ 83–181
Lα–~ of the night sky 79–9
mass ~ 85–145
~ measure 88–234
~ nebula 97–123
neutrino ~ 94–94
nightglow ~ 79–5
~ of lava 73–130
particle ~ 88–151
post–twilight ~ 79–52
pre–twilight ~ 79–51
two–photon ~ 97–155
very low frequency ~ 78–54
emission–line star 91–159
empirical curve of growth 93–69
empty model 99–49
~ universe 99–49
encounter hypothesis 83a–14
stellar ~ 98–121
encounters, number of 98–122
end height 86–40
~ of the cycle 89–85
~ of the eclipse 81–83
~ of the world 99–183
~ of totality 81–80
~ point 86–38
endogenous 73–145, 82–259
~ process 82–258
energetic particle event 89–191
energy, acoustic 88–219
atomic ~ 93–121
available ~ 89–197
~ balance 88–251
~ dissipation 82–285
double ~ source 93–146
~ exchange 98–125
exchanged ~ 98–126
~ flux 93–112
~ gain 88–253
~ generation 88–216, 93–105
gravitational ~ 93–119
~ loss 88–254
non–radiative ~ 88–218
nuclear ~ 93–122
~ of vaporization 86–199
~ output 93–109
~ production 93–105
~ propagation 88–217
~ release 93–109
rest ~ 99–144
rotational ~ 93–120
sound ~ 88–219
sound ~ flux 88–221
~ source 93–110
source of ~ 93–110
stored ~ 89–196
~ tensor 99–158
total ~ flux 93–116
~ transfer 93–111
transformation of ~ 86–208
~ transport 93–111
vacuum ~ 99–170k
zero–point ~ 93–187
zero–point kinetic ~ 93–187
energy–momentum tensor 99–158
enlargement of the Earth's shadow 81–93
entry, atmospheric 86–161
~ into the atmosphere 86–161
velocity of ~ 86–92
envelope, characteristic 98–177
common ~ 96–32
common ~ of a binary 96–32
convective ~ 93–136
extended ~ 93–8
~ of a comet 85–69
radiative ~ 93–135
star with an extended ~ 93–7

environment 76–3
environmental science 76–2
EO elliptical galaxy 98–244
Eozoic 74–77
epeirogenesis 73–73
epeirogenic movement 73–73
epeirogeny 73–73
ephemeris of minor planet positions 84–92
epicentral distance 73–19
~ region 73–16
epicentre 73–15
epicentrum 73–15
epicycle 83–49
epoch 74–89
~ of maximum 89–81
~ of minimum 89–80
~ of periastron 96–60
post-glacial ~ 72–109
e-process 94–99
Equ 92–48
equation, Abel 88–235
colo(u)r ~ 91–73
continuity ~ 93–49
drag ~ 86–179
Einstein field ~ 99–156
Emden's ~ 93–130
evaporation ~ 86–200
field ~ 99–155
height ~ 86–194
hydrostatic ~ 93–76
integral ~ of Schwarzschild and Milne 93–53
integro–differential ~ 93–52
light ~ 95–45
mass ~ 86–169
mass-loss ~ 86–169
~ of radiative transfer 93–33
~ of radiative transport 93–33
Pogson's ~ 91–58
resistance ~ 86–179
statistical equilibrium ~ 93–44
tensor ~ 99–157
transfer ~ 93–32
vapo(u)rization ~ 86–200
equator, lunar 81–121
magnetic ~ 75–50
~ of illumination 81–122
plane of the ~ 83–231
equatorial acceleration 88–131
~ belt 83–191
~ bulge 71–9
~ longitude system 83–178
~ plane 83–231
~ prominence 89–228
~ radius 83–115
~ region 89–59
~ ring current 75–98
~ semidiameter 83–115
~ velocity 91–49
~ zone 83–188
equiangular projection 71–41
equigranular rock 82–67
equilibrium, adiabatic 76–21
adsorption ~ 85–121
~ conditions 93–74
convective ~ 93–79
diffusion ~ 76–23
diffusive ~ 76–23
~ figure 83–112
generalized ~ 93–43
hydrostatic ~ 93–75
isostatic ~ 73–99
isothermal ~ 76–20
non–LTE ~ 93–43
non–thermodynamic(al) ~ 93–43
nuclear ~ 94–103
nuclear statistical ~ 94–103
~ process 94–99
radiative ~ 93–48
~ ratio 94–104
statistical ~ 93–42
thermal ~ 93–78
thermodynamic(al) ~ 93–35
equipotential surface 71–61
~ surface of gravity field 71–61
equivalence principle 99–150
Einstein's ~ principle 99–150
mass–energy ~ principle 99–145
equivalent height 78–28
equivoluminar wave 73–43
Equuleus 92–48
era 74–69
annihilation ~ 99–173a
cosmological ~ 99–166a
decoupling ~ 99–177a
hadron ~ 99–174
inflationary ~ 99–170d
inflatory ~ 99–170d
lepton ~ 99–176
~ of superunified force 99–170e
Planck ~ 99–170e
radiation ~ 99–172
stellar ~ 99–178
Eri 92–49
Eridanus 92–49
River ~ 92–49
erosion 73–107
~ caldera 81–183
gas ~ 82–243
meteoritic ~ 82–242
meteoroid ~ 82–242
space ~ 87–142
surface ~ 82–241
wind ~ 83–163
erosive collision 84–105
erratic 72–132
~ block 72–132
glacial ~ 72–132
eruption, cone of 73–118
fissure ~ 73–113
~ hypothesis 89–63
linear ~ 73–113
volcanic ~ 73–112, 82–170
eruptive arch 89–251
~ galaxy 98–269
~ phase 89–210
~ prominence 89–248
escape, cone of 76–27
~ of a star 96–159
~ of a star from a system 96–159
~ of gases 76–24
~ of stars from a cluster 98–194
~ region 76–68
~ velocity from a system 96–155
escapee 96–160
escaper 96–160, 98–135
escaping star 96–160, 98–135
estimate 95–12
~ of brightness 95–11
Etamin 92–115
etching of a polished surface 87–102
eternity of the universe 99–13
Euclidean 99–101
~ geometry 99–108
~ space 99–107
Europa 83–206
evaporation 86–198
~ equation 86–200
~ of a cluster 96–108
~ of a meteor body 86–198
~ of black holes 94–118
~ time 96–109
even spot cycle 89–77
evening star 83–128
~ terminator 81–125
~ twilight 79–36
event 99–125
adjacent ~ 99–126
energetic particle ~ 89–191
high–energy particle ~ 89–191
~ horizon 99–43
luminous ~ 82–247
proton ~ 89–192
seismic ~ 82–215
simultaneous ~ 99–128
SN ~ 95–131
solar proton ~ 89–192
solar X–ray ~ 89–184
transient luminous ~ 82–247
X–ray ~ 89–184
Evershed effect 89–103
evolution, chemical 82–257
~ diagram 96–151
dynamical ~ 96–161
nuclear ~ 94–50
~ of stars 94–1
~ of the universe 99–171e
post–main–sequence ~ 94–68
pre–main–sequence ~ 94–26
pre–stellar ~ 94–14
pre–stellar ~ of galaxies 94–14
stellar ~ 94–1
thermal ~ 83a–61
thermonuclear ~ 94–50
tidal ~ 83a–65

EN

evolutionary change 94–27
~ diagram 96–151
~ line 94–29
~ sequence 94–30
~ stage 94–28
~ track 94–29
evolve 94–75
evolved star 94–74
evolving universe 99–54
excess, colo(u)r 91–77
infrared ~ 91–79
IR ~ 91–79
ultraviolet ~ 91–78
UV ~ 91–78
exchange, energy 98–125
exchanged energy 98–126
excitation mechanism 79–27
~ temperature 91–118
exciting star 97–125
exhaustion of hydrogen 94–69
exogenous 73–146
exosphere 76–68
expanding arm 98–38
~ corona 88–232
~ model 99–55
~ shell 93–12
~ star 91–162
~ universe 99–55
expansion, accelerated 99–81
~ constant 99–77
extremely rapid ~ 99–170c
hydromagnetic ~ 88–159
~ of the universe 99–74
~ rate 99–75
~ time 99–78
~ velocity 96–94
velocity of ~ 99–75
expedition, solar–eclipse 88–30
experiment, active seismic 82–220
lunar–surface ~ 82–42
Michelson ~ 99–130
Michelson–Morley ~ 99–130
seismic ~ 82–219
expert in cosmology 99–19
exploding galaxy 98–268
~ star 91–165
explosion 87–30
~ caldera 81–180
~ crater 81–238
high–altitude nuclear ~ 75–114
initial ~ 99–167
~ of galactic nucleus 98–283
~ of a meteor 86–81
point of ~ 86–122
primordial ~ 99–167
supernova ~ 95–131
volcanic ~ 73–112
explosive helium burning 94–79
~ nucleogenesis 94–57
~ nucleosynthesis 94–57
~ phase 89–210
exponential integral function 93–62
exposure age 87–139
cosmic ray ~ ~ 87–139
expulsion of matter 98–274
~ of matter from galactic nucleus 98–274
extended atmosphere 93–9
~ envelope 93–8
star with an ~ atmosphere 93–6
extension of the crescent of Venus 83–131
~ of the horns of Venus 83–131
exterior radius 83–228
externally occulted coronagraph 88–60
~ occulting coronagraph 88–60
extinct comet 85–150
~ volcano 73–115
extinction, atmospheric 77–127a
~ curve 97–81
~ due to water vapour 77–126
interstellar ~ 97–80
~ of a meteor 86–37
progressive ~ 86–75
rate of stellar ~ 94–115
selective ~ 97–83
sudden ~ 86–76
total ~ 97–82
extrafoveal vision 86–128
extragalactic 98–207
~ astronomy 98–208
~ nebula 98–202
~ nova 95–114
~ object 98–209
~ space 97–5
~ supernova 95–128
~ system 98–210
extraordinary wave 78–30
extraterrestrial light 79–7
~ material 72–56
extreme limb 89–32
~ population I 98–73
extremely low frequency emission 78–55
~ rapid expansion 99–170c
extrinsic variable 95–86
extrusion 73–132
~ of lava 82–171
extrusive rock 74–3
eye–witness of the fall 87–48

F

F 79–157
F corona 88–241
f spot 89–114
face-on galaxy 98–247
factor, brightness 83–90
cosmic-scale ~ 99–81a
dilution ~ 93–37
form ~ 86–177
Gaunt ~ 93–104
geometric ~ 86–177
geometry ~ 86–177
guillotine ~ 93–103
Lorentz contraction ~ 99–135
luminosity ~ 86–197
roughness ~ 82–107
scale ~ 99–81a
shape ~ 86–177
facula 89–161
chromospheric ~ 89–164
photospheric ~ 89–161
polar ~ 89–162
facular granule 88–179
~ granulation 89–163
fade–out 78–40
sudden short-wave ~ 78–40
fading, progressive 86–75
faint meteor 86–15
~ star 91–82
~ telescopic meteor 86–19
fall 76–57
area of ~ 87–38
circumstances of a ~ 87–47
eye–witness of a ~ 87–48
free ~ 94–39
~ in temperature 77–41
meteoritic ~ 87–19
natural ~ 82–227
~ of a meteorite 87–19
single ~ 87–36
suspected ~ of a meteorite 87–49
falling current 93–150
~ tide 82–282
false variable 95–51
~ zodiacal light 84–62
family, asteroid 84–73
Jovian ~ 85–103
Jupiter's ~ 85–103
~ of comets 85–102
fan 85–60, 89–295
curved ~ 85–62
inverted ~ 85–63
~ ray 89–296
straight ~ 85–61
fan–beam interferometer 88–122
fan–like 85–74
fan–shaped 85–74
far side of the Moon 81–16
farthest point of area of impact 87–39
fast meteor 86–87
~ nova 95–119
fault 73–91, 82–212
normal ~ 73–93
reverse ~ 73–94
reversed ~ 73–94
faultage 73–92
faulting 73–92
local ~ 82–204
regional ~ 82–205
Faye reduction 71–78
feature, active 89–14
atmospheric ~ 83–107
impact ~ 82–228
lunar ~ 81–148
optical ~ 98–234

permanent ~ 83–106
primary ~ 81–150
radio ~ 98–235
secondary ~ 81–151
solar ~ 89–13
structural ~ 98–224
surface ~ 83–105
topographic ~ 81–149
volcanic ~ 82–159
feldspar (US) = felspar (UK) 74–15
potash ~ 74–16
ferromagnetism 75–26
fibrils 88–209
penumbral ~ 89–98
field, background 89–141
background magnetic ~ 89–141
coercive ~ 75–39
~ configuration 89–140
co-rotating ~ lines 75–14
~ correction 86–157
crater ~ 87–56
effective ~ strength 95–59
~ equations 99–155
firn ~ 72–119
fossil magnetic ~ 91–53
galactic ~ 98–146
galactic gravitational ~ 98–146
galactic magnetic ~ 98–50
galactic tidal ~ 96–106
~ galaxy 98–340
general magnetic ~ 89–128
gravitational ~ 98–145
gravity ~ 71–57
Higgs ~ 99–170e
internal ~ 95–63
interplanetary magnetic ~ 84–9
interstellar magnetic ~ 97–106
large-scale ~ 89–142
lava ~ 82–184
longitudinal ~ 89–138
magnetic ~ amplification 89–156
magnetic ~ direction 84–10
magnetic ~ intensity 84–11
magnetic ~ irregularity 84–12
magnetic ~ strength 84–11
magnetic ~ variability 84–14
magnetizing ~ 75–42
~ of compacted snow 72–119
~ of differential motions 98–114
open ~ line 75–15
photospheric magnetic ~ 89–130
quantum ~ 99–170m
radiation ~ 93–30
relic magnetic ~ 95–62
rock ~ 82–139
selenomagnetic ~ 82–45
solar magnetic ~ 89–129
~ star 96–84
stone ~ 82–139
structure of the ~ 89–140
superstrong magnetic ~ 95–148
surface ~ 95–64
tidal ~ 98–196
transport of magnetic ~ 84–48
transversal ~ 89–137
velocity ~ 88–187
figure, equilibrium 83–112
~ of the Earth 71–7
~ of the Moon 81–116
~ of the planet 83–109
figures, Widmanstätten 87–105
filament 85–55, 89–224
activation of a ~ 89–215
bright edge of a ~ 89–226
bright rim of a ~ 89–226
hydrogen ~ 89–267
metallic ~ 89–268
~ of a nebula 97–131
penumbral ~ 89–98
winking ~ 89–216
filamentary nebula 97–129
~ structure 97–130
filling of a depression 77–83
film, prominence 89–242
filter, aerosol 76–94
dust ~ 76–95
final collapse 99–182
find, meteorite 87–44
finding of meteorite 87–44
~ place 87–52
fine mottles 89–172
chromospheric ~ structure 88–203
fine-grained 82–114
~ matrix 82–115
fines 82–109
~ grain 82–110
~ particles 82–109
lunar ~ 82–112
finite atmosphere 93–23
~ space 99–103
finiteness of the universe 99–4
fireball 86–11, 99–168
appearance of a ~ 86–117
bright ~ 86–10
daylight ~ 86–118
detonating ~ 86–119
great ~ 86–10
passage of a ~ 86–117
prim(a)eval ~ 99–168
sound-emitting ~ 86–119
firn 72–118
~ field 72–119
first apparition of a comet 85–19
~ contact 81–84
~ positive band system 79–20
~ quarter 81–25
first-generation star 94–24
first-order triangulation 71–24
first-ranked cluster galaxy 98–339
Fish, Dorado 92–46
Flying ~ 92–100
S ~ 72–80
Southern ~ 72–80
fish location 72–84
Fishes 92–79
fission theory 96–75
fissure eruption 73–113
~ flow 73–141
fitting condition 93–142
FitzGerald-Lorentz contraction 99–134
fixed base-line 88–102
~ star 92–4
flaming aurora 79–157
flare 89–175, 95–69
~ activity 89–176, 95–71
area of a ~ 89–207
automatic ~ patrol 89–221
behind the limb ~ 89–203
brightness of ~ 89–212
cinematographic ~ patrol 89–220
chromospheric ~ 89–175
class 2 ~ 89–206
disc ~ 89–204
duration of a ~ 89–209
~ ejection 89–214
~ flash 89–210
frequency of ~ occurence 89–217
homologous ~ 89–178
lifetime of a ~ 89–209
limb ~ 89–202
meteor ~ 86–80
~ nimbus 89–189
non-proton ~ 89–186
~ of importance 2 89–206
optical ~ 89–181
~ patrol 89–219
proton ~ 89–185
proton ~ project 89–177
radio ~ 89–180
~ ribbon 89–187
solar ~ 89–175
~ star 95–68
star ~ 95–70
stellar ~ 95–70
~ surge 89–190
sympathetic ~ 89–179
two-ribbon ~ 89–187
~ variable 95–68
visible ~ 89–181
white ~ 89–182
white-light ~ 89–182
X-ray ~ 89–183
flare-associated 89–188
~ disturbance 84–46
flare-like prominence 89–222
flash, flare 89–210
green ~ 79–58
helium ~ 94–79
~ spectrum 88–25
flash-phase 89–210
flat bottom 81–260
~ maximum 86–236
~ space 99–107
~ system 98–251
flattening 83–113, 83a–22
apparent ~ 98–295

Earth's ~ 71–8
~ of the corona 88–247
~ of the globular cluster 96–123
polar ~ 83–114
real ~ 98–296
true ~ 98–296
flat-topped seamount 72–44
flickering meteor 86–85
flight–path correction 82–10
flocculus 89–167
bright hydrogen ~ 89–169
calcium ~ 89–170
dark ~ 89–224
dark hydrogen ~ 89–224
hydrogen ~ 89–168
flood 82–280
~ and ebb 72–89
flooded crater 81–203
flood–tide current 82–280
floor, crater 81–259
ocean ~ 72–37
~ of a crater 81–259
sea ~ 72–38
flow, ash 82–174
~ channel 82–192
fissure ~ 73–141
~ front 82–193
glacier ~ 72–120
heat ~ 74–60
lava ~ 73–140, 82–173
~ lines 82–194
~ of heat 74–60
plastic ~ 72–121
~ stage of magma 82–189
~ texture 82–191
fluctuation, climatic 77–183
primordial ~ 99–170y
secondary ~ 95–44
superposed ~ 95–44
fluid, chemically active 74–43
fluidal texture 82–191
fluidity, lava 82–188
~ of the lava 82–188
fluorescent bands 85–132
flux, acoustic energy 93–161
convective ~ 93–115
convective energy ~ 93–115
emergent radiative ~ 93–114
energy ~ 93–112
integrated ~ 93–47
neutron ~ 94–86
neutron ~ density 94–86
~ of acoustic energy 88–221
radiation ~ 93–113
total energy ~ 93–116
Fly 92–69
Flying Fish 92–100
focal depth 73–20
~ distance 73–18
focus, distance of the 73–18
earthquake ~ 73–17
seismic ~ 73–17
fog 77–130
fold 73–87
anticlinal ~ 73–90
~ mountains 73–86
synclinal ~ 73–89
folded mountains 73–86
folding 73–88
folds 79–139
foliated rock 74–54
foliation 74–53
follower 89–114
following spot 89–114
Fomalhaut 92–116
footpad 82–23
footprints 82–38
For 92–50
forbidden region 79–135
~ space 79–135
force, coercive 75–39
Coriolis ~ 77–79
deviating ~ 77–80
electroweak ~ 99–170x
grand unified ~ 99–170v
internal ~ 73–61
repulsive ~ 85–87
superunified ~ 99–170r
tectonic ~ 73–106
tidal ~ 82–289
tide–generating ~ 82–289
unified ~ 99–170x
wind ~ 77–107
forced tidal oscillations 82–306
forecast, ionospheric 78–31
long–range ~ 77–19
~ map 77–21
~ of solar activity 89–64
short–range ~ 77–20
weather ~ 77–13
foreground star 96–85
fore–runner 73–11
~ of earthquake 73–11
foreshock 73–10
foreshortening 89–92
area corrected for ~ 89–93
magnetic ~ 89–160
~ of sunspots 89–92
form, auroral 79–137
cometary ~ 85–83
~ of meteoric particle 86–175
~ with ray–structure 79–152
~ without ray–structure 79–142
formaldehyde 97–27
formamide 97–35
formation 94–10
grain ~ 97–60
line ~ 93–63
lunar ~ 81–148
~ of craters 82–232
~ of grains 97–60
~ of spots 89–121
~ of stars 96–95
star ~ 94–11
surface ~ 83–105
tail ~ 85–86
theory of multiple ~ 96–76
former satellite 83–240
formic acid 97–33
forms, structural 81–152
formula, barometric height 76–15
international gravimetric ~ 71–76
international gravity ~ 71–76
Titius–Bode ~ 83–3
Fornax 92–50
fossil heat 74–63
~ magnetic field 91–53
~ meteoritic crater 87–55
~ nuclear track 82–252
~ track 87–121
fountain model 85–85
four–dimensional 99–121
~ vector 99–138
four–vector 99–138
Fox, Little 92–101
Small ~ 92–101
fractional crystallization 74–40
fractionation 87–138
fractocumulus 77–146
fracture 73–78
concentric ~ 82–209
continental ~ system 73–81
crater–margin ~ 82–211
linear ~ 82–208
mid-ocean ~ system 73–80
plane ~ 82–208
surface ~ 81–173
~ system 73–79
tension ~ 82–207
fragile 86–191
fragility 86–192
fragment 83a–83
largest ~ 87–41
meteorite ~ 87–43
nuclear ~ 94–102
~ of a meteorite 87–43
rock ~ 82–68
fragmentation 83a–81
asteroid ~ 84–108
meteor ~ 86–188
~ of asteroids 83a–82
~ of cosmic cloud 94–38
~ of meteor 86–83
~ of meteor body 86–188
frame of reference 98–136
fundamental ~ of refence 98–137
local ~ of reference 98–139
reference ~ 98–136
rotating ~ of reference 98–138
Fraunhofer corona 88–241
~ line 88–139
~ spectrum 88–137
free atmosphere 76–13
~ fall 94–39
~ radical 85–137
~ oscillations 73–49

~ oscillations of the Earth 73–49
free–air reduction 71–78
free–bound radiation 86–212
free–free absorption 93–106
frequency, auroral 79–122
carrier ~ 86–274
crater ~ 81–247
critical ~ 78–24
lowest usable ~ 78–33
maximum usable ~ 78–37
modulation ~ 86–275
~ of flare occurence 89–217
~ of meteors 86–135
prominence ~ 89–237
pulse repetition ~ 86–279
recurrence ~ 86–279
size ~ 81–249
sunspot ~ 89–66
transmitter ~ 86–273
Fresnel zones 86–294
friction 86–173
tidal ~ 81–312
Friedmann expansion time 99–80
~ model 99–71
~ time 99–80
~ universe 99–71
fringe, polar 83–151
~ region 76–68
front 77–68
cold ~ 77–70
ionization ~ 97–48
occluded ~ 77–71
occlusion ~ 77–71
polar ~ 77–73
~ side of the Moon 81–15
~ surface 87–114
warm ~ 77–69
frontogenesis 77–74
frontolysis 77–75
frozen gases 85–45
full moon 81–27
fumarole 82–163
fumarolic activity 82–164
fuming volcano 73–114
function, density 98–88
distribution ~ 98–143
exponential integral ~ 93–62
luminosity ~ 98–89
mass ~ 96–68
~ of scattering 77–58
phase ~ 83–78
saturation ~ 93–60
source ~ 93–34
visibility ~ 89–55
weight ~ 93–58
fundamental frame of reference 98–137
~ gravimetric network 71–82
~ gravity network 71–82
~ standard of rest 98–137
~ tensor 99–157a
funnel 89–269
~ prominence 89–269
Furnace 92–50
furrow 81–165
fusing temperature 86–204
fusion 86–181
~ crust 87–110

G

G 79–151
gain, energy 88–253
gal 71–65
galactic 98–2
~ absorbing layer 97–89
active ~ nucleus 98–272
~ activity 98–261
~ anticenter (USA) 98–14
~ anticentre 98–14
~ apocenter (USA) 98–190
~ apocentre 98–190
~ arm 98–26a
~ bar 98–254
~ black hole 98–23a
~ center (USA) 98–9
~ centre 98–9
~ concentration 98–76
~ concentration toward the centre 98–76
~ concentration toward the principal plane 98–76
~ corona 98–49
~ density–wave 98–175
differential ~ rotation 98–113
diffuse ~ light 97–56
~ disc 98–52
~ disk 98–52
dissociation of a ~ cluster 96–103
~ distribution 98–54
~ dust 97–53
~ field 98–146
~ gravitational field 98–146
~ halo 98–45
~ light 79–14
~ magnetic field 98–50
~ model 98–163
~ nebula 97–110
~ nova 95–113
~ nucleus 98–18
~ orbit 98–184
~ orbit of a star 98–184
~ pericenter 98–189
~ pericentre 98–189
~ plane 98–53
~ population 98–230
~ rotation 98–107
~ rotation term 98–116
~ star cluster 96–99
~ structure 98–4
~ supernova 95–127
~ system 98–3
~ tidal field 96–106
~ wind 98–277
~ windows 98–57
galactocentric 98–10
~ coordinates 98–12
~ distance 98–11
galaxies, chain of 98–328
classification of ~ 98–226
cluster of ~ 98–329
colliding ~ 98–321
collision of ~ 98–322
compact cluster of ~ 98–334
density of ~ 98–344
globular cluster of ~ 98–330
group of ~ 98–325
Hubble sequence of ~ 98–227
~ in collision 98–321
interacting ~ 98–310
interconnected ~ 98–306
Local Group of ~ 98–326
open cluster of ~ 98–333
passage of two ~ 98–320
poor cluster of ~ 98–331
recession of ~ 98–298
regular cluster of 98–336
rich cluster of ~ 98–332
sequence of regular ~ 98–240
supercluster of ~ 98–346
very compact cluster of ~ 98–335
Virgo cluster of ~ 98–345
galaxy 98–201
acicular ~ 98–239
active ~ 98–260
background ~ 98–340
barred ~ 98–253
blue compact ~ 98–265
brightest cluster ~ 98–339
central region of a ~ 98–293
cigar–shaped ~ 98–239
cluster ~ 98–338
compact ~ 98–264
companion of a ~ 98–216
~ count 98–343
differential rotation of ~ 98–113
disrupted ~ 98–317
distorted ~ 98–316
disturbed ~ 98–315
double ~ 98–305
dwarf ~ 98–220
dwarf elliptical ~ 98–242
edge–on ~ 98–246
ellipsoidal ~ 98–241
elliptical ~ 98–241
elliptical companion of a ~ 98–217
E0 elliptical ~ 98–244
eruptive ~ 98–269
exploding ~ 98–268
explosion of a ~ nucleus 98–283
face–on ~ 98–247
field ~ 98–340
first–ranked cluster ~ 98–339
gas–rich ~ 98–238
giant ~ 98–222
giant elliptical ~ 98–243
Haro ~ 98–266

Haro blue ~ 98–266
highly flattened ~ 98–252
infrared ~ 98–205
irregular ~ 98–237
lenticular ~ 98–245
local ~ 98–327
Markarian ~ 98–265a
member ~ 98–338
multiple ~ 98–308
multiple-arm ~ 98–44
multiple-armed ~ 98–44
multiple-centre ~ 98–319
multiple-nucleus ~ 98–319
multiply-armed ~ 98–44
needle-shaped ~ 98–239
neighbouring ~ 98–211
non-cluster ~ 98–340
normal ~ 98–221
N-type ~ 98–263
nuclear activity of a ~ 98–271
~ nucleus explosion 98–283
~ of the Local Group 98–327
one-arm ~ 98–42
one-armed ~ 98–42
optical companion of a ~ 98–218
optical ~ 98–204
parent ~ 98–284
peculiar ~ 98–256
physical companion of a ~ 98–219
population of a ~ 98–230
post-eruptive ~ 98–270
quasi-stellar ~ 98–361
radio ~ 98–206
regular ~ 98–236
rim of a ~ 98–294
ring ~ 98–259
~ seen edge-on 98–246
~ seen face-on 98–247
Seyfert ~ 98–262
singular ~ 98–258
spherical ~ 98–244
spheroidal ~ 98–244
spindle-shaped ~ 98–239
spiral ~ 98–249
supergiant ~ 98–223
tadpole ~ 98–281
tail of ~ 98–280
transition-type ~ 98–248
twin ~ 98–305
two-arm ~ 98–43
two-armed ~ 98–43
unilateral ~ 98–42
X-ray ~ 98–203
Galaxy 98–1
center of the ~ 98–9
centre of the ~ 98–9
Milky Way ~ 98–1
nucleus of the ~ 98–18
structure of the ~ 98–4
Galilean satellite 83–204
~ transformation 99–131
Galileo('s) transformation 99–131
gamma-ray pulsar 95–139
Ganymede 83–207
gap, commensurability 84–88
Hertzsprung ~ 91–144
Hornstein-Kirkwood ~ 84–89
Kirkwood ~ 84–89
~ of commensurability 84–88
gas, baryon 93–194
baryonic ~ 93–194
~ between clusters 98–353
~ cap 86–170
circumgalactic ~ 98–351
~ cloud 98–113
degenerate ~ 93–179
~ erosion 82–243
hyperon ~ 93–193
hyperonic ~ 93–193
incandescent ~ 83a–43
intercluster ~ 98–353
intergalactic ~ cloud 98–354
interstellar ~ 97–9
intracluster ~ 98–352
neutron ~ 93–192
non-degenerate ~ 93–182
polytropic ~ sphere 93–127
~ pressure 93–90
primordial noble ~ 87–133
~ production rate 85–144
radiogenic noble ~ 87–132
relativistic ~ 97–152
~ retention 87–134
~ sphere 93–73
sweeping out of ~ 83a–41
trace ~ 76–88
trapped solar-wind ~ 82–148
gas-and-dust nebula 97–119
gas-dust matter 83a–25
~ mixture 83a–25
gaseous mass 94–32
~ nebula 97–112
~ ring 95–66
~ sphere 93–127
~ stream 95–67
~ train 86–116
gas-production rate 85–144
gas-retention age 87–135
gas-rich galaxy 98–238
gas-to-dust ratio 97–120
gauge theory 99–170j
tide ~ 82–311
Gaunt factor 93–104
gauze ring 83–221
gegenschein 84–64
Gem 92–51
Gemini 92–51
Geminids 86–257
Gemma 92–117
general circulation 77–65
~ interstellar absorption 97–87
~ magnetic field 89–128
~ magnetic field of the Sun 89–128
~ relativity 99–146
~ theory of relativity 99–146
~ weather situation 77–15
generalized Lagrangian identities 98–183
generation of energy 88–216
~ of sound 88–215
~ of stars 94–23
genetic association 97–127
~ connection 97–127
geocentric distance 84–3
~ meteor velocity 86–95
~ orbit 86–41
~ radiant 86–58
~ system 99–21
~ velocity of meteor 86–95
geochemical differentiation 74–55
~ evolution 74–33
geochemistry 74–30
geochronological scale 74–68
geochronology 74–67
geocorona 76–73
neutral hydrogen ~ 76–74
geodesic 99–153
~ line 99–153
geodesy, satellite 71–32
space ~ 71–31
geodetic altitude 71–52
~ connection of continents 71–37
~ datum 71–20
~ height 71–52
~ reference system 71–19
~ satellite network 71–34
world ~ network 71–38
world ~ system 71–39
geoelectricity 75–1
geoid 71–16
compensated ~ 71–17
regularized ~ 71–17
~ undulation 71–74
geologic(al) age 74–90
~ epoch 74–89
~ period 74–74
~ survey of the Moon 82–50
geology 72–2
lunar ~ 82–47
~ of glaciers 72–103
geomagnetic activity 75–86
~ axis pole 75–48
~ cavity 75–8
~ coordinates 75–51
~ disturbances 75–70
~ dynamo theory 75–99
~ elements 75–54
~ field 75–44
~ field intensity 75–55
K-index of ~ activity 75–89
~ latitude 75–52
~ longitude 75–53
~ micropulsations 75–82
~ pulsations 75–81
secular ~ variations 75–85
~ storm 75–71

~ substorm 75–95
~ tides 82–295
~ time 79–132
~ variations 75–69
variations of ~ field 75–69
geomagnetically trapped particles 75–111
~ trapped radiation 75–112
geomagnetism 75–24
geometric(al) albedo 83–92
~ cross-section 84–103
~ factor 86–177
~ libration 81–51
~ shadow height 79–47
~ variable 95–72
geometry, Euclidean 99–108
~ factor 86–177
non-Euclidean ~ 99–109
Riemannian ~ 99–113
geophysics 72–1
cosmical ~ 84–57
geopotential 71–60
~ surface 71–61
geosciences 71–1
geosphere 72–7
geostrophic wind 77–96
geosynclinal 73–105
geosyncline 73–105
geotectonics 73–95
geothermal gradient 74–59
geothermic degree of depth 74–59
~ step 74–59
geothermics 74–58
geothermy 74–58
ghost ring 81–242
Giacobinids 86–252
giant 91–143
blue ~ 91–142a
~ bolide 86–7
~ branch 91–152b
bright ~ 91–149
early-type ~ 91–143a
~ elliptical 98–243
~ elliptical galaxy 98–243
~ galaxy 98–222
late-type ~ 91–144a
~ molecular cloud 97–13a
normal ~ 91–150
~ nucleus 76–118
~ particle 76–118
~ planet 83–174
red ~ 91–152a
~ star 91–143
Giraffe 92–24
glacial 72–106
~ deposits 72–127
~ erratic 72–132
~ period 72–105
glaciation 72–104
glacier 72–112
alpine ~ 72–114
continental ~ 72–116
~ flow 72–120
mountain ~ 72–113
~ tongue 72–125
valley ~ 72–114
glaciology 72–103
glass, lunar 82–96
smoked ~ 88–33
glassy fragment 82–97
~ spherules 82–127
glitch 95–154
global radiation 77–55
globe 71–10
celestial ~ 92–7
lunar ~ 81–112
molten ~ 83a–64
terrestrial ~ 79–133
globular cluster 96–114
flattening of a ~ cluster 96–123
~ cluster of galaxies 98–330
~ star cluster 96–114
system of ~ clusters 98–75
globule 97–133
glory 79–63
glow 79–151
sunset ~ 79–55
twilight ~ 79–4
GMC 97–13a
gnomonic projection 71–46
Goat 92–29
He- ~ 92–29
Sea- ~ 92–29
Gold-fish 92–46
good visibility 77–25
Gothlandian 74–80
graben 82–132
gradient 93–94
absolute ~ 91–111
adiabatic temperature ~ 93–98
convective temperature ~ 93–157
density ~ 98–87
pressure ~ 77–106, 93–95
radiative temperature ~ 93–97
spectrophotometric ~ 91–112
superadiabatic ~ 93–99
temperature ~ 93–96
~ wind 77–94
grain, fines 82–110
~ formation 97–60
graphite ~ 97–75
~ growth 97–61
ice-coated ~ 97–77
icy ~ 85–52
polymer ~ 97–77a
rocky ~ 85–54
silicate ~ 97–76
~ size 82–113
grains, alignment of 97–78
interstellar dust ~ 97–59
grand unification 99–170u
~ unified force 99–170v
~ unified interaction 99–170v
granite 74–8
granular ice 72–118
granulation, facular 89–163
granule 88–178
facular ~ 88–179
photospheric ~ 88–181
umbral ~ 88–180
graphite core 97–71
~ grain 97–75
grating interferometer 88–123
gravimeter 71–70
gravimetric anomaly 71–75
~ chart 71–79
~ formula 71–76
~ map 71–79
~ network 71–80
gravimetry 71–63
gravitational collapse 94–109
~ contraction 93–118
~ cross-section 84–104
~ crushing 84–72
~ differentiation 74–56
~ displacement 99–162
~ displacement of spectral lines 99–162
~ energy 93–119
~ field 98–145
~ instability 94–35
~ lens 98–256a
lunar ~ anomaly 81–144
~ mass 99–149
~ paradox 99–39
~ potential 96–160a
~ radiation 99–165
~ radius 94–113
~ redshift 99–162
~ shrinkage 83–203
~ tides 82–298
~ wave 99–164
gravitationally bound system 96–156
~ stable system 96–156
~ unbound system 96–158
~ unstable system 96–158
gravity 71–56
absolute measurement of ~ 71–68
~ acceleration 71–64
acceleration of ~ 71–64
~ anomaly 71–75
~ contour map 71–79
~ darkening 91–41, 96–25
determination of ~ 71–67
~ field 71–57
intensity of ~ 71–62
~ map 71–79
~ network 71–80
~ of the Earth 71–60
~ potential 71–58
relative ~ measurement 71–69
relative measurement of ~ 71–69
~ wave 88–201
World ~ network 71–81
gray atmosphere 93–26
grazer, Earth 84–78
grazing occultation 81–100

Great Bear 92–96
great fireball 86–10
Great Red Spot 83–193
Greater Dog 92–27
greatest elongation 83–63
green coronal region 89–284
~ flash 79–58
~ line 79–16, 89–291
greenhouse effect 83–263
~ model 83–262
grey atmosphere 93–26
groove 81–165
linear ~ 81–166
sinuous ~ 81–167
ground inversion 77–38
~ moraine 72–129
~ wave 78–14
group, amino 87–71
asteroidal ~ 84–90
bipolar ~ 89–111
comet ~ 85–105
commensurability asteroid ~ 84–101
commensurability ~ 84–101
complex ~ 89–112
~ delay 78–56
new–cycle ~ 89–106
~ of galaxies 98–325
~ of minor planets 84–90
~ of spots 89–105
~ of stars 92–3a
old–cycle ~ 89–107
~ radiant 86–54
recurrent ~ 89–108
unipolar ~ 89–110
growth, curve of 93–68
empirical curve of ~ 93–69
grain ~ 97–61
~ of a planet 83a–37
planetary ~ 83a–37
theoretical curve of ~ 93–70
Gru 92–52
Grus 92–52
guided camera 86–142
guider 88–51
guillotine factor 93–103
gulf 81–294
gust 77–112
gusty wind 77–111
guyot 72–44

H

H I region 97–49
H II region 97–50
HA 79–143
hadron era 99–174
hail 77–178
halo 79–67
cool ~ 98–47
~ cross 79–79
galactic ~ 98–45
hot ~ 98–46
lunar ~ 79–68
magnetic ~ 98–48
~ of 46° 79–70
~ of 22° 79–69
~ phase 94–19
~ phenomenon 79–66
~ population 98–69
~ stage 94–19
~ star 91–84
tangential arc of ~ 79–75
haloes 79–67
~ of a comet 85–68
halo–type population 98–69
Hamal 92–118
Hare 92–61
harmonic coefficients 71–59
harmonics, non–zonal 95–102
tesseral ~ 95–102
zonal ~ 95–101
Haro blue galaxy 98–266
~ galaxy 98–266
Harp 92–65
Harvard classification 91–91
~ system of spectral classification 91–91
Hayashi phase 94–44
~ stage 94–44
hazard, navigation 72–97
haze 83–250
dust ~ 77–27
~ layer 83–249
head of a comet 85–35
Head and Tail of Serpent 92–89
~ of Serpent 92–89
head–on collision 83a–17
~ cross–section 86–176
heat balance 77–62
~ balance of atmosphere 77–62
~ death of the universe 99–14
~ flow 74–60
flow of ~ 74–60
fossil ~ 74–63
~ index 91–81
latent ~ 77–157
~ of ablation 86–184
~ of condensation 77–155
~ of desorption 85–124
~ of evaporation 86–199
~ of sublimation 85–139
original ~ 74–63
primordial ~ 83–200
radiogenic ~ 83a–62
~ source 83–199
source of ~ 83–199
theory of ~ death 99–15
tidal ~ 82–287
~ transfer 86–202
transmission of ~ 86–202
heating, atmospheric 75–87
~ mechanism 88–214
~ of the chromosphere 88–213
~ of the corona 88–213a
~ of the meteor body 86–174
heat–transfer coefficient 86–201
Heavenly Twins 92–51
Heaviside layer 78–26
Heaviside–Kennelly layer 78–26
heavy–weight star 91–126
hedgerow 89–274
He–Goat 92–29
height above sea level 71–51
airglow ~ 79–26
apex ~ 86–66
appearance ~ 86–39
ellipsoidal ~ 71–52
end ~ 86–40
~ equation 86–194
equivalent ~ 78–28
geodetic ~ 71–52
geometrical shadow ~ 79–47
layer ~ 78–27
meteor ~ 86–193
~ of appearance point 86–39
~ of chromosphere 88–202
~ of disappearance 86–40
~ of end point 86–40
~ of the homogeneous atmosphere 76–17a
~ of unit decrease of pressure 76–19
real ~ 78–29
scale ~ 76–18
screened shadow ~ 79–49
screening ~ 79–48
shadow ~ 79–46
true ~ 78–29
true shadow ~ 79–49
virtual ~ 78–28
helical symmetry 98–161
helio– 88–5
heliocentric angle 89–37
~ distance 84–4
~ orbit 86–42
~ system 99–23
~ velocity 86–96
heliogram, X–ray 88–74
heliograph 77–61, 88–36
monochromatic ~ 88–38,
heliographic chart 89–17
~ coordinates 89–39
~ latitude 89–42
~ latitude distribution 89–51
~ longitude 89–40
~ longitude distribution 89–48
~ map 89–17
~ position 89–38
heliometer 88–12
heliophysics 88–6
heilioseismology 88–223b
heliosphere 88–255
heliospheric 88–256
~ cavity 88–257
heliostat 88–45
helium burning 94–78
~ burning zone 93–145

~ core 93–167
explosive ~ burning 94–79
~ flash 94–79
~ ignition 94–71
~ method 87–128
~ method of age determination 87–128
~ shell 93–145
helium–deficient star 91–192
helium–enriched plasma 84–60
helium–poor star 91–192
helium–rich star 91–193
helmet 89–295
Helmholtz contraction 94–37
hemisphere, antisolar 83–68
averted ~ 81–16
dark ~ 83–68
day ~ 83–67
daylight ~ 75–20, 83–67
Earthside ~ 81–15
illuminated ~ 75–20, 83–67
night ~ 75–21
nocturnal ~ 83–68
non–illuminated ~ 83–68
sunlit ~ 75–20, 83–67
unilluminated ~ 75–21, 83–68
visible ~ 89–27
Her 92–53
Hercules 92–53
Herdsman 92–22
Hermes 84–83
Hertzsprung gap 91–144
Hertzsprung–Russell diagram 91–130
heterosphere 76–61
hexahedrite 87–84
hidden mass 99–88
hierarchic model 99–48
~ structure 99–47
~ structure of the universe 99–47
~ universe 99–48
Higgs field 99–170e
high 77–84
~ concentration cluster 96–129
~ tide 72–90, 82–281
~ water 72–90
~ water of spring tide 82–273
high–altitude nuclear explosion 75 114
~ observatory 88–56
high–energy particle event 89–191
highland 81–189
~ depression 81–168
~ surface 82–56
high–latitude 89–53
~ prominence 89–229
highly concentrated cluster 96–129
~ degenerate 93–184
~ flattened galaxy 98–252
~ ionized 89–289
~ oblate system 98–164
high–pressure area 77–84
high–resolution map 89–133
high–velocity star 98–191
hiss 78–64
history, thermal 74–66
hoarfrost 77–177, 83–150
hole, black 94–114
bore ~ 82–61
coronal ~ 88–245a
galactic black ~ 98–23a
meteorite ~ 87–58
~ of impact 87–58
supermassive black ~ 98–22a
white ~ 98–24
Holocene 72–109
homogeneous atmosphere 76–10
~ band 79–146
~ model 99–64
~ quiet arc 79–143
~ universe 99–64
homologous 93–131
~ flare 89–178
~ transformation 93–132
homology transformation 93–132
homopause 76–65
homosphere 76–60
honeycomb, aluminium 82–25
Hor 92–54
horizon, cosmological 99–179
event ~ 99–43
mass ~ 99–41
~ of universe 99–10
horizontal branch 91–132
~ coelostat 88–47
~ layering 82–134
~ motion 88–184
~ pendulum 71–73
~ solar telescope 88–55
~ spectrograph 88–78
horizontal–branch star 91–141
horns 83–130
extension of the ~ of Venus 83–131
Hornstein–Kirkwood gaps 84–89
Horologium 92–54
horse latitudes 77–102
Horsehead Nebula 97–135
Horseshoe Nebula 97–143
hot cosmological model 99–166
~ halo 98–46
~ region 88–200
~ spot 95–112a
~ star 91–155
~ universe 99–166
hot–wire microphone 76–46
hourly motion 81–72
~ rate 86–135
~ rate of meteors 86–135
H–R diagram 91–130
Hubble('s) constant 99–77
~ expansion time 99–79
~ law 98–299
~ sequence of galaxies 98–227
~ time 99–79
~ type 98–229
~ type of a galaxy 98–229
Humboldtianum, Mare 81–288
Humboldt's Sea 81–288
humidity, absolute 77–122
atmospheric ~ 77–118
relative ~ 77–128
specific ~ 77–123
hummock, ice 72–102
hummocky ice 72–102
~ material 82–103
hump on a curve 95–37
Hunter 92–73
Hunting Dogs 92–26
hurricane 77–78
Hya 92–55
Hyades 96–111a
Hydra 92–55
hydrodynamic phase 94–43
~ stage 94–43
hydrodynamics, stellar 98–142
hydrogen burning 94–56
~ burning zone 93–144
~ convection zone 93–159
~ convective zone 93–159
~ corona 83–246
~ cyanide 97–29
exhaustion of ~ 94–69
~ filament 89–267
~ glow 79–145
~ ignition 94–55
interstellar ~ 97–46
~ ion 93–65
~ isocyanide 97–40
metallic ~ 83–202
~ model 83–197
~ model of a planet 83–197
molecular ~ 97–31
negative ~ ion 93–66
neutral ~ 97–47
neutral ~ region 97–49
~ oxide 97–26
~ planet 83–197
~ prominence 89–266
~ shell 93–144
~ sulfide 97–44
~ torus 83–230
hydrogen–deficient star 91–191
hydrogen–metal ratio 91–176
hydrogen–poor star 91–191
hydrologic chart 72–25
~ map 72–25
hydromagnetic cavity 84–22
~ expansion 88–159
~ oscillator 95–61
hydrophilic 76–92
hydrophobic 76–93
hydrosol 76–90
hydrosphere 72–21
hydrostatic equation 93–76
~ equilibrium 93–75
~ equilibrium figure 83–112
~ pressure 93–77
hydroxyl 79–25, 97–22

Hydrus 92–56
hygrometer 77–120
Hyi 92–56
hyperbolic comet 85–8
~ meteor 86–43
~ space 99–106
hypergiants 91–147
hyperon gas 93–193
~ star 93–196
hyperonic gas 93–193
~ star 93–196
hypocenter 73–17
hypocentral distance 73–18
hypocentre 73–17
hypothesis 94–6
capture ~ 83a–18
catastrophic ~ 83a–14
collision ~ 83a–14
cosmogonic ~ 94–8
encounter ~ 83a–14
eruption ~ 89–63
Lambert's ~ 99–45
nebular ~ 83a–7
two–stream ~ 98–99
vortex ~ 83a–11
hypothetical planet 83–122
hypsographic curve 81–142
hypsometric level 81–142
~ map 81–141

I

Icarus 84–84
ice 85–42
~ age 83a–78
amorphous ~ 85–44
carbon–dioxide ~ 83–144
~ crystal 76–111
cubic ~ 85–46
dirty ~ 97–73
dry ~ 83–144
granular ~ 72–118
~ hummock 72–102
hummocky ~ 72–102
~ mantle 97–72
~ needles 76–112
pack of ~ 72–102
~ nucleus 85–40
~ surveillance 72–85
~ survey 72–85
water ~ 85–47
iceberg 72–117
ice–cap 72–115
ice–coated grain 97–77
icy conglomerate 85–43
~ grain 85–52
~ mantle 97–72
~ nucleus 85–40
~ satellite 83–23
identities, Lagrangian 98–182
igneous activity 73–110
~ intrusion 82–98
~ rock 74–2
ignition, carbon 94–80
helium ~ 94–71
hydrogen ~ 94–55
illuminated hemisphere 75–20
~ side 81–34
illuminance ratio 91–60
illumination, equator of 81–122
ilmenite 82–91
image, interference 88–120
images, monochromatic 97–158
immersion 81–98
impact, area of 87–38
artificial ~ 82–226
cometary ~ 82–230
~ crater 81–240
~ cratering 82–234
~ dust 82–125
earliest point of area of ~ 87–40
~ ejecta 82–167
~ hypothesis 81–261
~ lithification 82–238
man–made ~ 82–226
meteorite ~ 87–33
meteoritic ~ 87–33
nearest point of area of ~ 87–40
~ origin 82–229
~ petrification 82–238
~ point 86–124
point of ~ 86–124
~ seismogram 82–224
impacting body 82–225
imploding star 91–164
implosion of stellar core 94–110
importance of a flare 89–205
improper variable 95–51
inactive comet 85–147
~ volcano 73–116
incandescent gases 83a–43
inclination, axial 83–35
magnetic ~ 75–58
~ of orbite 96–51
~ of orbital plane 96–51
~ of rotation axis to … 91–50
~ to lunar equator 81–37
inclined inhomogeneity 88–190
inclusion, organic 87–66
increase 76–54
~ in mass of the Earth 87–18
light ~ 86–79
mass ~ 83a–38
~ to maximum 95–28
Ind 92–57
index, blue minus yellow colo(u)r 96–145
B–V colo(u)r ~ 96–145
color ~ 91–76
colour ~ 91–76
composite colo(u)r ~ 96–138
concentration ~ 96–101
heat ~ 91–81
integrated colo(u)r ~ 96–138
~ of solar activity 89–8
~ of stream activity 86–230
planetary ~ 75–90
planetary K_p ~ 75–90
planetary magnetic K_p ~ 75–90
polytropic ~ 93–128
solar ~ 89–8
structural ~ of cluster 96–80
three–hour–range ~ 75–89
U–B colo(u)r ~ 96–146
ultraviolet minus blue colo(u)r ~ 96–146
yellow colo(u)r ~ 83–88
Indian 92–57
indicator boulder 72–132
range ~ 86–266
indicatrix of scattering 77–58
individual radiant 86–55
~ specimen 87–42
induced magnetism 75–29
Indus 92–57
inequality, Schwarzschild 93–156
inertial mass 99–148
~ system 99–98
infall of gas 98–356
~ of gas from intergalactic space 98–356
inferior conjunction 83–57
~ planet 83–14
infinite space 99–104
infinity of the universe 99–6
inflation 99–170c
cosmological ~ 99–170c
inflationary era 99–170d
~ universe 99–170b
inflatory era 99–170d
~ universe 99–170b
influx of high-energy particles 78–43
infrared brightness 91–71
~ excess 91–79
~ galaxy 98–205
~ magnitude 91–71
~ pumping 97–18
~ star 91–157
infrasonic pressure wave 75–88
~ wave 72–93
inhomogeneity, electron–density 78–8
inclined ~ 88–190
macroscopic ~ 88–206
magnetic–field ~ 78–9
inhomogeneous core 93–169
~ stellar model 93–139
initial atmosphere 83a–44
~ explosion 99–167
~ mass 86–164
~ phase 75–75
~ universe 99–170a
~ velocity 86–92
inland earthquake 73–2b
~ sea 72–26
inner cavity 87–119
~ coma 85–65
~ core 72–20

~ corona 88–229
~ crater wall 81–252
~ Lagrangian point 96–31
~ planet 83–14
~ radiation belt 75–106
~ radiation zone 75–106
~ radius 83–226
~ Van Allen belt 75–106
~ Van Allen radiation belt 75–106
innermost satellite 83–20
in–situ analysis 82–136
insolation 77–51
instability 93–211
convective ~ 93–155
criterion of tidal ~ 98–312
gravitational ~ 94–35
local ~ 93–154
pulsational ~ 93–213
rotational ~ 83a–8, 93–216
secular ~ 93–212
tidal ~ 98–313
vibrational ~ 93–214
instantaneous angular velocity 86–90
~ velocity 86–99
instrument, alpha–scattering 82–76
automated ~ 82–41
integral equation of Schwarzschild and Milne 93–53
phase ~ 83–79
third ~ of motion 98–186
integrated brightness 96–135
~ colo(u)r index 96–138
~ flux 93–47
~ magnitude 96–136
~ properties 96–134
~ radiation 88–147
~ spectrum 96–137
integro–differential equation 93–52
intensity, central H_α 89–211
contours of constant ~ 75–102
earthquake ~ 73–33
geomagnetic field ~ 75–55
horizontal ~ 75–60
horizontal field ~ 75–60
horizontal ~ of magnetic field 75–60
~ of gravity 71–62
omnidirectional ~ 75–104
~ ratio 91–60
sound ~ 76–38
unidirectional ~ 75–103
vertical ~ 75–62
vertical field ~ 75–62
interacting galaxies 98–310
~ prominences 89–243
interaction, atmospheric 79–120
collective ~ 98–168
electroweak ~ 99–170x
grand unified ~ 99–170v
superunified ~ 99–170r
tidal ~ 98–311
interactive prominences 89–243
interarm 98–40
~ gas 98–40
~ region 98–40
~ space 98–40
intercluster gas 98–353
interconnected galaxies 98–306
interface 93–141
crust–mantle ~ 72–14
salinity ~ 72–67
interference image 88–120
~ pattern 88–120
interferometer base 88–100
Christiansen ~ 88–123
cliff ~ 88–106
cliff–top ~ 88–106
fan–beam ~ 88–122
grating ~ 88–123
long–base ~ 88–104
long–base–line ~ 88–104
multiantenna ~ 88–123
multielement ~ 88–123
multiwave ~ 88–123
pencil–beam ~ 88–121
phase–swept ~ 88–109
phase–switching ~ 88–113
radio ~ 88–97
radio–link ~ 88–116
sea ~ 88–106
sweep–frequency ~ 88–110
swept–lobe ~ 88–109
two–element ~ 88–98
variable base–line ~ 88–107
variable-spacing ~ 88–107
interfilamentary space 89–99
intergalactic 97–6
~ brige 98–348
capture of ~ matter 98–355
~ gas 98–351a
~ gas cloud 98–354
~ matter 98–349
~ space 97–7
~ star 91–163, 98–350
interglacial 72–108, 77–190
~ period 72–107
intergranular area 88–177
~ region 88–177
interior of the Earth 72–3
~ radius 83–226
stellar ~ 93–72
interloper 98–134
intermediate corona 88–245
~ earthquake 73–24
~ phase 94–21
~ population 98–70
~ stage 94–21
intermediate–mass star 91–125
internal activity 82–158
~ constitution 71–83, 93–71
~ field 95–63
~ force 73–61
~ motions 97–149
~ process 82–258
~ structure 93–71
~ structure of stars 93–71
~ terracing 81–256
international ellipsoid 71–14
~ gravity formula 71–76
International Union of Geodesy and Geophysis 71–4
interplanetary 84–8
~ cavity 84–25
~ discontinuity 84–58
~ disturbance 84–45
~ dust 84–51
~ loop 84–21
~ magnetic field 84–9
~ magnetic sector 84–16
~ matter 84–1
~ meteor 86–44
passage of ~ discontinuity 84–58
~ plasma 84–39
~ radio scintillation 84–37
~ sector 84–16
~ space 84–2
~ storm 84–44
~ tongue 84–21
~ trajectory 82–12
inter–ray surface 81–172
interspicular region 88–198
interstellar 97–2
~ absorption 97–84
~ band 97–100
~ cloud 97–101
diffuse ~ band 97–100
diffuse ~ feature 97–100
~ dust 97–54
~ dust grain 97–59
~ extinction 97–80
~ feature 97–100
~ gas 97–9
~ gas dynamics 97–150
general ~ absorption 97–87
~ grains 97–59
~ hydrogen 97–46
law of ~ reddening 97–92
~ line 97–96
~ magnetic field 97–106
~ maser 97–14
~ matter 97–1
~ medium 97–8
~ meteor 86–43
microwave ~ line 97–98
~ molecule 97–10
~ obscuration 97–86
optical ~ line 97–99
~ particles 97–68
~ polarization 97–105
~ radio line 97–98
~ radio scintillation 97–109b
~ reddening 97–91
~ scintillation 97–109a
selective ~ absorption 97–90
~ space 97–4
~ stream 86–224
~ wind 97–151

intertropical arcs 79–53
intracluster gas 98–352
intramercurial planet 83–123
intramercurian planet 83–123
intrinsic brightness 85–107
~ colo(u)r 91–74
~ colo(u)r index 91–75
~ property 91–100
~ variable 95–87
intrinsically variable star 95–87
intrusion 73–133
concordant ~ 73–134
discordant ~ 73–137
igneous ~ 82–98
intrusive rock 74–7
inundation 72–75
invariable in time 98–150
~ plane 83–232
invariance, principle of 93–61
~ under a transformation 99–133
invasion of high-energy particles 78–43
inversion, atmospheric 77–37
ground ~ 77–38
population ~ 97–15
surface ~ 77–38
temperature ~ 77–36
inverted fan 85–63
investigation, subsurface 82–82
invisible companion 96–41
~ side 81–16
~ side of the Moon 81–16
~ sunspot 89–159
inward stream 93–150
Io 83–205
ion, hydrogen 93–65
Langevin ~ 76–115
large ~ 76–115
negative hydrogen ~ 93–66
~ tail 85–82
ionic composition 78–10
~ diffusion 86–286
ionization, column of 86–282
~ efficiency 86–209
~ front 97–48
meteor ~ 86–281
~ temperature 91–117
ionized column 86–282
highly ~ 89–289
~ hydrogen region 97–50
~ layer 78–3
ionogram 78–23
ionosphere 76–69, 78–1
outer ~ 76–76
ionospheric 78–2
~ disturbance 78–39
~ eclipse 88–23
~ forecast 78–32
~ inhomogeneity 78–7
~ layer 78–5
~ model 83–261
~ observatory 78–21
~ prediction 78–32
~ recorder 78–22
~ region 78–4
~ sounding 78–19
~ storm 78–49
IR excess 91–79
iridescent cloud 79–90
iron grain 97–74
meteoric ~ 87–83
~ meteorite 87–82
meteoritic ~ 87–83
nickel ~ 87–90
iron-groupe elements 94–105
~ nuclei 94–105
iron-peak elements 94–105
~ nuclei 94–105
Irr 98–237
irradiation age 87–139
cosmic–ray ~ 82–250
irregular cluster of galaxies 98–337
~ crater 81–220
~ galaxy 98–237
~ magnetic variable with constant polarity 95–57
~ magnetic variable with reversing polarity 95–56
~ nebula 98–237
~ satellite 83–21
~ variable 95–107
~ variable with constant polarity 95–57
~ variable with reversing polarity 95–56
irregularity, small–scale 84–13
irrotational wave 73–46
island, new 72–81
isobath 71–55
isocline 75–59
isoclinic line 75–59
isocount contours 75–102
isocyanic acid 97–39
isocyanide, hydrogen 97–40
isodip line 75–59
isodynamic(al) line 75–61
isogauss contour 89–134
~ line 89–134
isogonal line 75–57
isogone 75–57
isogonic line 75–57
isohypse 71–53, 81–139
isolated pulse 95–146
isolevel contour 81–139
~ line 81–139
~ map 81–141
isophotal contour 81–92
isophote 81–92
isopleth 96–125
isorotation 88–130
isostasy 73–98
isostatic adjustment 73–100
~ compensation 73–100
~ equilibrium 73–99
isothermal atmosphere 77–39
~ core 93–168
~ equilibrium 76–20
isotope, cosmic–ray produced 82–147
cosmogenic ~ 82–147
~ dating 87–130
~ effect 91–178
lead ~ 83a–76
long–life ~ 83a–75
long–lived ~ 83a–75
primordial ~ 82–146
~ ratio 74–94
short–life ~ 83a–74
short–lived ~ 83a–74
isotopic abundance 94–106
~ composition 74–93, 82–140
~ dating 87–130, 83a–73
~ effect 91–178
isotropic model 99–63
~ universe 99–63
iterative method 93–56

J

jet 85–55, 98–278
asteroidal ~ stream 84–100
curved ~ 85–58
~ ejection 98–278
~ from a galactic nucleus 98–278
~ of electrons 79–114
optical ~ 98–279
rectilinear ~ 85–57
spiral ~ 85–59
~ stream 77–88
Job's Coffin 92–45
Jovian 83–171
~ apoapse 83–173
~ apoapsis 83–173
~ apocentre 83–173
~ family 85–103
~ noise storm 83–185
~ periapse 83–172
~ periapsis 83–172
~ pericentre 83–172
~ radio burst 83–182
jovicentric coordinates 83–176
jovigraphic coordinates 83–175
jump, sudden 95–154
Juno 84–85
Jupiter 83–170
Jupiter's sporadic radio emission 83–181
Jura 74–85
Jurassic 74–85

K

K corona 88–242
katabatic 77–93
kamacite 87–92
Kapteyn areas 98–78
K–coronameter 88–66
Keel 92–30
Keplerian rotation 98–112

Kepler's rotation 98–112
K–index of geomagnetic activity 75–89
kinetic temperature 91–115
kinematic properties 98–90
kinematical cosmology 99–26
King 92–33
Kirkwood gaps 84–89
Kite 92–22
Kneeler 92–53
Kneeling Giant 92–53
~ Man 92–53
knot 85–95
~ of a prominence 89–244
prominence ~ 89–244
~ within a meteor stream 86–217
Known Sea 81–289
Kramers' law 93–102
K–term 98–119

L

Lac 92–58
laccolite 73–136
laccolith 73–136, 82–179
Lacerta 92–58
lacus 83–158
Lacus 81–297
~ Aetatis 81–300
~ Autumnae 81–301
~ Mortis 81–298
~ Somniorum 81–299
~ Veris 81–302
Lady in Chair 92–31
Lagrangian identities 98–182
first ~ point 96–31
inner ~ point 96–31
lake 83–158
lava ~ 82–183
Lake 81–297
Autumn ~ 81–301
~ of Death 81–298
~ of Dreamers 81–299
~ of Dreams 81–299
~ of Sleepers 81–299
Spring ~ 81–302
Summer ~ 81–300
Lambert's albedo 83–89
~ hypothesis 99–45
lamellae 87–104
lamina 83–149
laminar convection 93–153
~ structure 83–148
land–breeze 77–98
lander, lunar 82–21
landing leg 82–22
lunar ~ 82–13
manned ~ 82–33
proposed ~ site 82–15
~ site 82–14
lane, dark 98–288
dust ~ 98–289
Langevin ion 76–115
lapilli 82–169
Laplace transform 93–54
~ transformation 93–54
Laplacian plane 83–232
large ion 76–115
Large Magellanic Cloud 98–213
large–scale field 89–142
~ motion 88–189
~ structure 88–207, 99–171
~ weather situation 77–15
largest fragment 87–41
~ particle 87–41
laser retroreflector 82–43
last contact 81–85
late stars 91–103
latent heat 77–157
lateral crater 73–128
~ lower tangential arc 79–87
~ moraine 72–131
latest point of area of impact 87–39
late–type giant 91–144a
~ stars 91–103
latitude, areographic 83–140
geomagnetic ~ 75–52
heliographic ~ 89–42
heliographic ~ distribution 89–51
horse ~ 77–102
northern limiting ~ 81–105
selenographic ~ 81–11
southern limiting ~ 81–106
latitudinal bounce 75–119
lava blisters 82–178
~ cave 82–177
~ cavern 82–177
cool ~ 82–176
~ dome 81–176a
effusion of ~ 73–130, 82–172
emission of ~ 73–130
extrusion of ~ 82–171
~ field 82 184
~ flow 73–140
~ fluidity 82–188
fluidity of ~ 82–188
~ lake 82–183
~ outflow 73–130
~ plain 82–185
~ sheet 82–175
solidified ~ 82–176
~ tongue 82–187
~ tube 82–186
law, Bode's 83–3
Hale's ~ 89–154
Hubble's ~ 98–299
Kramers' ~ 93–102
~ of darkening 88–143
~ of interstellar reddening 97–92
~ of latitudes 89–75
~ of planetary distances 83–3
~ of polarity 89–154
reddening ~ 97–92
Spörer's ~ 89–75
Titius–Bode ~ 83–3
layer, absorbing 88–175
airglow ~ 79–17
atmospheric ~ 76–49
cloud ~ 77–150, 83–248
convective ~ 93–149
dust ~ 82–122
E ~ 78–26
fragmental debris ~ 82–104
galactic absorbing ~ 97–86
haze ~ 83–249
Heaviside ~ 78–26
Heaviside–Kennelly ~ 78–26
~ height 78–27
ionized ~ 78–3
ionospheric ~ 78–5
luminescent ~ 79–17
luminous ~ 79–17
radiative ~ 93–134
reversing ~ 88–174
scattering ~ 83–258
thin ~ 83–149
transition ~ 72–13, 88–224
violet ~ 83–165
layered surface 83–101
layering, horizontal 82–134
lead, common 74–99
~ isotope 83a–76
primary ~ 74–98
prim(a)eval ~ 74–98
radiogenic ~ 74–97
leader 89–113
leading arm 98–31
~ spot 89–113
lead–lead age 74–100
lead–ratio age 74–100
leg, landing 82–22
Legendre coefficients 93–55
~ polynomials 93–55
Leighton's magnetoheliograph 88–93
length, angular 98–32
~ of meteor path 86–47
~ of totality 88–26
path ~ 86–47
Planck ~ 99–170h
Planck fundamental ~ 99–170h
lens, gravitational 98–256a
lenticular galaxy 98–245
Leo 92–59
~ Minor 92–60
Leonids 86–255
Lep 92–61
lepton era 99–176
Lepus 92–61
less luminous supergiant 91–148
Lesser Lion 92–60
level, condensation 77–156
critical ~ 76–26
hypsometric ~ 81–142
sound ~ 76–38
Lib 92–62
Libra 92–62
libration 81–48

EN

~ asteroid 84–91a
~ cloud 84–43
diurnal ~ 81–52
geometrical ~ 81–51
~ in latitude 81–50
~ in longitude 81–49
lunar ~ cloud 84–43
optical ~ 81–51
parallactic ~ 81–52
physical ~ 81–53
lifetime of flare 89–209
~ of spot group 89–123
trapped particle ~ 75–122
light bridge 98–307
~ cone 95–156
~ cross 79–79
~ cylinder 95–155
diffuse galactic ~ 97–56
~ equation 95–45
extraterrestrial ~ 79–7
false zodiacal ~ 84–62
galactic ~ 79–14
~ increase 86–79
night-sky ~ 79–8
~ of the night sky 79–8
~ output ratio 86–210
purple ~ 79–54
~ region 83–160
sky ~ 77–53
~ star 91–124
~ strip 86–27
~ trap 88–64
~ variation 95–6
~ year 91–18
zodiacal ~ 79–13, 84–61
light–curve 95–17
mean ~ 95–19
photoelectric ~ 95–22
photographic ~ 95–23
visual ~ 95–24
lightning 77–162
ball ~ 77–163
bead ~ 77–164
light-output ratio 86–210
limb, above the 89–30
bright ~ 81–96
dark ~ 81–97
~ darkening 88–142
east ~ 89–34
extreme ~ 89–32
~ flare 89–202
lunar ~ 81–95
~ of the Moon 81–95
on the ~ 89–29
~ passage 89–31
solar ~ 89–28
west ~ 89–33
~ zone 81–113
limit, Chandrasekhar 93–189
mass ~ 93–189
limiting brightness 91–62
~ latitude 81–104
~ magnitude 91–62
northern ~ latitude 81–105
~ latitude of the visibility of occultation 81–104
southern ~ latitude 81–106
line absorption 93–108
atmospheric ~ 88–140a
circumstellar ~ 97–97
contour ~ 71–53
cooling ~ 93–202
coronal ~ 89–290
~ density 86–283
effective depth of ~ formation 93–64
effective ~ width 89–208
evolutionary ~ 94–29
~ formation 93–63
flow ~ 82–194
Fraunhofer ~ 88–139
geodesic ~ 99–153
green ~ 79–16, 89–291
interstellar ~ 97–96
interstellar radio ~ 97–98
isoclinic ~ 75–59
isodip ~ 75–59
isodynamic(al) ~ 75–61
isogauss ~ 89–134
isogonal ~ 75–57
isogonic ~ 75–57
isolevel ~ 81–139
microwave interstellar ~ 97–98
neutral ~ 89–151
~ of flow 82–194
~ of sight 91–28
~ of vision 91–28
optical interstellar ~ 97–99
plumb ~ 71–71
red ~ 79–15
reddening ~ 96–149
redshifted ~ 98–366
~ separating opposite polarities 89–151
shadow ~ 83–69
telluric ~ 88–140a
world ~ 99–154
yellow ~ 89–292
zero ~ 89–151
zero ~ of magnetic fields 89–151
zero–age ~ 94–60
linear depression 81–159
~ diameter 83–118
~ eruption 73–113
~ fracture 82–208
~ groove 81–166
~ rill(e) 81–166
~ rima 81–166
line–of–sight 91–28
in the ~ 91–29
~ velocity 91–24
lines of force 75–121
curvature of ~ of force 75–121
Neumann ~ 87–103
line–shifter 88–69
Lion 92–59
Lesser ~ 92–60
Little ~ 92–60
Small ~ 92–60
liquid particle 76–100
lithification 74–49
impact ~ 82–238
lithophilic elements 83a–30
lithosiderite 87–80
lithosphere 72–6
Little Bear 92–97
~ Dipper 92–97
~ Fox 92–101
~ Horse 92–48
~ Lion 92–60
live comet 85–149
Lizard 92–58
LM 82–5
LMC 98–213
LMi 92–60
Lobachevski('s) space 99–105
lobe, Roche('s) 96–28
local earthquake 73–25
~ ellipsoid 71–15
~ faulting 82–204
~ frame of reference 98–139
~ galaxy 98–327
galaxy of the ~ group 98–327
~ group of galaxies 98–326
~ instability 93–154
~ mass concentration 81–145
~ object 98–368
~ system 98–61
~ supercluster 98–347
~ supercluster of galaxies 98–347
~ thermodynamic equilibrium 93–39
location, fish 72–84
long time–scale 94–67
~ whistler 78–57
long–base interferometer 88–104
long–base–line interferometer 88–104
long–duration train 86–113
long–enduring meteor train 86–113
longitude, active 89–41
areographic ~ 83–139
~ distribution 89–48
drift in ~ 75–120
equatorial ~ system 83–178
geomagnetic ~ 75–53
heliographic ~ 89–40
heliographic ~ distribution 89–48
~ of periastron 96–59
selenographic ~ 81–10
~ system 83–177
~ system I 83–178
~ system II 83–179
longitudinal component 89–135
~ drift 75–120
~ field 89–138
~ magnetograph 88–83
~ wave 73–45
long–life isotope 83a–75

long–lived 89–124
~ isotope 83a–75
long–period comet 85–6
~ variable 95–91
long–range forecast 77–19
loop, interplanetary 84–21
magnetic ~ 84–19
~ prominence 89–270
Loop(ed) Nebula 97–137
loops 89–270
Lorentz contraction 99–134
~ contraction factor 99–135
~ transformation 99–132
Lorentz–FitzGerald contraction 99–134
loss, energy 88–254
mass ~ 94–72
lost asteroid 84–97
~ planetoid 84–97
Love wave 73–48
low tide 72–91
~ water 72–91
lower atmosphere 76–50
~ chromosphere 88–193
~ parhelion 79–72
~ tangential arc 79–76
lowest usable frequency 78–33
low–frequency bursts 83–184
lowland region 81–272
lowlands 81–272
low–latitude 89–52
~ aurora 79–96
~ prominence 89–227
low–mass star 91–124
low–velocity star 98–192
LSC 98–347
LTE 93–39
deviations from ~ 93–40
lt-yr 91–18
LUF 78–33
luminescence, lunar 82–248
night-sky ~ 79–5
luminescent layer 79–17
luminosity 91–128
~ class 91–146
~ equation 86–196
~ factor 86–197
~ function 98–89
meteor ~ 86–195
neutrino ~ 94–96
luminous band 86–27
~ bridge 98–307
~ efficiency 86–210
~ event 82–247
~ layer 79–17
~ matter 97–109, 99–86
transient ~ event 82–247
lunar 81–3
~ activity 82–149
~ atlas 81–136
~ aureole 79–61
automatic ~ station 82–19
~ base 82–17
~ breccia 82–85
~ bulge 81–118
~ cartography 81–111
~ chart 81–133
~ charting 81–111
~ corona 79–61
~ crater 81–197
~ cross 79–80
~ crust 82–54
~ cusps 81–24
dating of ~ events 82–265
~ day 81–32, 82–290
~ depression 81–157
~ disc 81–54
~ disk 81–54
~ dome 81–177
~ dust sample 82–121
~ eclipse 81–56
~ elevation 81–156
~ equator 81–121
~ feature 81–148
~ fines 82–112
~ flare 82–247
~ formations 81–148
~ geology 82–47
~ glass 82–96
~ globe 81–112
~ halo 79–68
~ lander 82–21
~ landing 82–13
~ landscape 81–55
~ libration cloud 84–43
~ limb 81–95
~ luminescence 82–248
~ map 81–133
~ mapping 81–111
~ mascon 81–145
~ mass concentration 81–145
~ material 82–58
~ microbreccia 82–86
~ module 82–5
~ night 81–33
~ observatory 82–20
~ occultation 81–94
~ orbit 81–36
~ particle 82–123
~ phenomenon 82–153
~ photograph 82–26
~ photographic map 81–137
~ process 82–150
~ rays 81–168
~ research 82–3
returned ~ sample 82–70
~ ring 81–265
~ rock 82–65
~ sample 82–69
sample of ~ material 82–69
~ sciences 82–2
~ seismology 82–213
~ soil 82–62
~ streaks 81–168
~ surface 82–55
~ surface material 82–59
~ terminator 81–123
~ thermal map 81–135
~ tides 82–308
~ time 82–300
~ topographic map 81–138
~ topography 81–146
transient ~ event 82–153
transient ~ phenomenon 82–153
~ transit 82–303
lunar–based astronomy 82–4
lunar–surface experiment 82–42
lunation 81–18, 81–46
lunicentric 81–131
lunisolar component 82–307
Lup 92–63
Lupus 92–63
l. y. 91–18
Lyn 92–64
Lynx 92–64
Lyr 92–65
Lyra 92–65
Lyre 92–65
Lyrids 86–248

M

maar 82–161
~ crater 81–264
Mach('s) principle 99–37
macrometeorite 87–4
macroscopic inhomogeneities 88–206
macrostructure 99–9
Magellanic Cloud 98–212
Large ~ Cloud 98–213
Small ~ Cloud 98–214
magma 73–131
consolidation of ~ 82–190
flow stage of ~ 82–189
magmatic rock 74–2
magnetic active day 75–93
~ activity 89–5
~ anomaly 75–67
~ axis 75–46
~ bay 75–80
bipolar ~ region 89–160a
~ braking 91–47
~ buoyancy 89–147
~ cavity 84–22
~ chart 75–63
~ classification 89–109
~ confinement of cosmic rays 98–51
~ crochet 75–79
~ declination 75–56
~ dip 75–58
~ dip pole 75–45
~ disturbances 75–70
Earth's ~ pole 75–54
~ elements 75–54
~ equator 75–50
~ foreshortening 89–160
~ halo 98–48

~ inclination 75–58
interplanetary ~ sector 84–16
~ loop 84–19
~ meridian 75–49
~ micropulsations 75–82
~ midnight 79–131
~ monopole 99–170o
periodic ~ variable 95–55
polar ~ substorm 75–96
~ polarity 89–148
positive ~ polarity 89–149
~ quiet day 75–92
~ record 75–84
~ region 89–144
reversal of ~ polarity 89–153
~ rigidity 75–117
~ rope 89–146
~ self–reversal 75–38
~ shielding 89–225
~ star 91–51
~ substorm 75–95
superstrong ~ field 95–148
~ survey 75–64
~ susceptibility 75–27
~ tube of force 89–145
~ variable 9552
~ variations 75–69, 95–53
~ vector 89–139
~ viscosity 75–34
World ~ Survey 75–65
magnetically active day 75–93
~ conjugate points 75–23
~ disturbed period 75–91
~ quiet day 75–92
magnetic field amplification 89–156
annihilation of ~ 89–158
~ ~ direction 84–10
Earth's ~ ~ 75–44
fossil ~ ~ 91–53
galactic ~ ~ 49–50
general ~ ~ 89-128
general ~ ~ of the Sun 89–128
horizontal intensity of ~ ~ 75–60
~ ~ inhomogeneity 78–9
~ ~ intensity 84–11
interplanetary ~ ~ 84–9
interstellar ~ ~ 97–106
~ ~ irregularity 84–12
photospheric ~ ~ 89–130
prim(a)eval ~ ~ 95–62
primitive ~ ~ 95–62
relic ~ ~ 91–53
solar ~ ~ 89–129
~ ~ strength 84–11
superstrong ~ ~ 95–148
~ ~ variability 84–14
magnetism, Earth's 75–24
induced ~ 75–29
remanent ~ 75–28
residual ~ 75–28
rock ~ 75–43, 83a–80
stellar ~ 95–54
terrestrial ~ 75–24
magnetite 82–92
magnetization, chemical 75–31
depositional ~ 75–33
depositional remanent ~ 75–33
detrital ~ 75–33
detrital remanent ~ 75–33
induced ~ 75–29
primary ~ 75–35
remanent ~ 75–28
residual ~ 75–28
reversed ~ 75–37
secondary ~ 75–36
self–reversal ~ 75–38
thermoremanent ~ 75–30
thermoresidual ~ 75–30
viscous ~ 75–32
magnetized solar plasma 84–32
magnetizing field 75–42
magnetoconjugate points 75–23
magnetogram 75–84
magnetograph 75–83
longitudinal ~ 88–83
solar ~ 88–82
vector ~ 88–84
magnetoheliograph 88–93
Leighton's ~ 88–93
magnetohydrodynamic oscillator 95–61
magnetoid 98–25
magnetometer, portable 82–44
magnetopause 75–9
magnetosphere 75–7
pulsar ~ 95–147
magnetospheric boundary 75–9
~ cleft 75–13
~ tail 75–10
magnetotail 75–10
magnitude 91–67
absolute ~ 91–64
absolute ~ of a meteor 86–72
absolute ~ of a planet 83–86
apparent ~ 91–61
apparent ~ of a meteor 86–71
bolometric ~ 91–68
~ difference 91–59, 96–65
earthquake ~ 73–34
electrographic ~ 91–67
infrared ~ 91–71
integrated ~ 96–136
limiting ~ 91–62
mean ~ 83–85
~ of earthquake 73–34
~ of eclipse 81–79
photographic ~ 91–66
photovisual ~ 91–69
radio ~ 86–268
radiometric ~ 91–70
stellar ~ 91–56
~ system 91–57
total ~ 96–136
visual ~ 91–65
zenith ~ 91–63
zenithal ~ 91–63
magnitude–intensity relation 91–58
main crater 73–127, 87–57
~ mass 87–41
~ phase 73–12, 75–76
~ pulse 95–144
~ ring 83–222
~ sequence 91–131
~ tail 85–77
zero–age ~ sequence 96–152
main–sequence star 91–142
major elements 82–144
~ planets 83–13
~ spot 89–115
man–made impact 82–226
~ radiation belt 75–108
manned landing 82–33
~ lunar base 82–18
mantle 72–15
Earth's ~ 72–15
ice ~ 97–72
icy ~ 97–72
upper ~ 72–16
map, bathymetric 71–54
colorimetric ~ 81–134
daily ~ of solar activity 89–18
daily ~ of the Sun 89–18
daily solar ~ 89–18
forecast ~ 77–21
gravimetric ~ 71–79
gravity ~ 71–79
gravity contour ~ 71–79
heliographic ~ 89–17
high–resolution ~ 89–133
hydrologic ~ 72–25
hypsometric ~ 81–141
isolevel ~ 81–141
lunar ~ 81–133
lunar photographic ~ 81–137
lunar thermal ~ 81–135
lunar topographic ~ 81–138
magnetic ~ 89–131
magnetic fields ~ 89–131
oceanographic ~ 72–23
~ of the Moon 81–133
~ of the sky 92–9
photographic ~ of the Moon 81–137
photometric ~ 81–134
physical ~ 81–134
polar ~ 89–19
polarization ~ 81–134
~ projection 71–40
radio ~ 81–135a
~ scale 71–49
star ~ 92–9
structural ~ 73–97, 82–203
synoptic ~ 77–11
synoptic ~ of chromosphere 89–21
synoptic ~ of photosphere 89–20
tectonic ~ 73–96, 82–202
velocity ~ 89–23
weather ~ 77–12

mapping, coast 72–82
lunar ~ 81–111
~ of the Moon 81–111
selenologic ~ 82–51
thermal ~ 72–83
mare 81–266, 83–156
~ Anguis 81–291
~ Australe 81–286
circular ~ 81–267
~ Cognitum 81–289
~ Crisium 81–280
~ Fecunditatis 81–274
~ Frigoris 81–278
~ Humboldtianum 81–288
~ Humorum 81–282
~ Imbrium 81–273
~ Ingenii 81–292
~ Marginis 81–285
~ Moscoviense 81–283
~ Nectaris 81–279
~ Nubium 81–281
~ Orientale 81–284
~ region 81–268
~ ridge 81–188
~ Serenitatis 81–276
~ Smythii 81–287
~ surface 82–57
~ Tranquillitatis 81–275
~ Undarum 81–290
mareograph 82–311
marginal sea 72–27
~ zone 81–113
maria 81–266
Mariner's Compass 92–82
maritime climate 77–186
Markab 92–119
Markarian galaxy 98–265a
marking, absorption 89–224
dark ~ 89–224
dark absorption ~ 89–224
permanent ~ 83–106
surface ~ 81–148, 83–105
volcanic ~ 82–159
Mars 83–133
~ asteroid 84–74
Mars–crosser 84–75
Mars–crossing asteroid 84–75
Mars–family asteroid 84–74
Mars–orbit crosser 84–75
marsh 81–303
~ of Decay 81–305
~ of Diseases 81–304
~ of Epidemics 81–304
~ of Sleep 81–306
martian 83–135
~ canals 83–157
mascon 81–145
lunar ~ 81–145
maser, astrophysical 97–13
celestial ~ 97–13
~ cloud 97–12
cosmic ~ 97–13
interstellar ~ 97–14
natural ~ 97–13
masering cloud 97–12
masing cloud 97–12
mass, air 77–67
~ balance 88–250
critical ~ 93–189
critical ~ of white dwarfs 93–189
decrease of ~ 83a–42
~ determination 96–69
~ distribution 84–52
~ ejection 93–15, 94–77
~ emission 85–145
~ equation 86–169
~ function 96–68
gaseous ~ 94–32
gravitational ~ 99–149
hidden ~ 99–88
~ horizon 99–41
~ increase 83a–38
increase in ~ of the Earth 87–18
inertial ~ 99–148
initial ~ 86–164
~ limit 93–189
local ~ concentration 81–145
~ loss 86–168, 94–72
main ~ 87–41
meteor ~ 86–163
missing ~ 99–87
~ movements 82–195
~ of meteoric body 86–163
original ~ 86–164
~ outflow 93–13
Planck ~ 99–170f
Planck fundamental ~ 99–170f
principal ~ 87–41
~ ratio 96–67
relativistic ~ 99–141
relativistic ~ increase 99–142
rest ~ 99–143
solar ~ 91–122
stellar ~ 91–121
total ~ of a cluster 96–139
~ transfer 96–33
~ transfer from one component to the other 96–33
variable ~ 86–167, 96–70
~ wastage 82–195
~ wasting 81–236
mass–emission rate 85–146
mass–energy equivalence principle 99–145
mass–frequency function 84–52
massive star 91–126
mass-loss equation 86–169
mass–luminosity diagram 91–129
~ ratio 98–233
~ relation 91–129
mass–radius relation 93–200
mass–to–light ratio 98–233
material, cosmic 99–7
~ difficult to fuse 83a–55
~ difficult to melt 83a–55
extraterrestrial ~ 72–56
lunar ~ 82–58
lunar surface ~ 82–59
not–easily–fusible 83a–55
refractory ~ 83a–55
~ slipped down from slopes 82–199
stellar ~ 93–124
mathematical model 98–324
~ model(l)ing 98–323
~ postulate of relativity of inertia 99–36
~ simulation 98–323
matrix 82–118
matter, absorbing 97–85
average density of ~ 99–84
capture of intergalactic ~ 98–355
carbonaceous ~ 87–64
clustering of ~ 99–170z
cosmic ~ 99–7
creation of ~ 99–89
dark ~ 97–108
distribution of ~ 99–8
ejection of ~ 89–214, 95–124
gas–dust ~ 83a–25
intergalactic ~ 98–349
interplanetary ~ 84–1
interstellar ~ 97–1
luminous ~ 97–109, 99–86
mean density of ~ 99–84
meteoritic ~ 87–9
meteoritic organic ~ 87–68
mixing of stellar ~ 93–162
pre–stellar ~ 94–16
primordial ~ 99–170
stellar ~ 93–124
~ tensor 99–158
matter-radiation decoupling 99–177
Maunder diagram 89–76
maximum brightness 86–78
~ corona 88–244
epoch of ~ 89–81
flat ~ 86–236
increase to ~ 95–28
~ light 86–78
~ of solar activity 89–2
primary ~ 95–31
principal ~ 95–31
~ range 86–289
rise to ~ 95–28
secondary ~ 95–32
sharp ~ 86–235
solar ~ 89–2
~ usable frequency 78–37
Maxwellian distribution of velocities 98–96
mean albedo 83–90a
~ angular velocity 86–91
~ density 93–86
~ density of matter 99–84
~ light–curve 95–19
~ lunar day 82–304

~ magnitude 83–85
~ mass density 99–84
~ molecular weight 93–100
~ opposition magnitude 83–85
~ power 86–272
~ radius 83–116a
~ time of relaxation 98–130
~ velocity 86–98
measure, emission 88–234
measurement, absolute gravity 71–68
age ~ 87–126
arc ~ 71–22
micrometric ~ 96–49
~ of a degree 71–22
~ of gravity 71–67
polarimetric ~ 83–80
relative gravity ~ 71–69
relative ~ of gravity 71–69
mechanical theory of cometary forms 85–84
mechanism, braking 91–46
destruction ~ 97–79
excitation ~ 79–27
heating ~ 88–214
pumping ~ 97–16
mediterranean sea 72–26
medium, interstellar 97–8
turbid ~ 83–251
medium–grained 82–116
melting 86–181
~ point 86–204
~ rate 83–147
~ temperature 86–204
member, brightest 98–339
cluster ~ 96–82, 98–338
~ galaxy 98–338
~ of the cluster 96–82
~ of the system 98–133
probable ~ of the cluster 96–86
uncertain ~ of the cluster 96–87
membership, physical 96–83
Men 92–66
Mensa 92–66
Mercury 83–120
meridian, arc of 71–21
central ~ 89–43
~ circulation 93–163
cross the central ~ 89–44
dawn ~ 75–18
dusk ~ 75–19
magnetic ~ 75–49
noon ~ 75–17
~ transit 82–303
~ transit of the Moon 82–303
meridional circulation 93–163
mesopause 76–67
mesosiderite 87–79
mesosphere 76–66
Mesozoic 74–72
~ era 74–72
Messier catalogue 92–12a
~ number 97–111
metagalaxy 98–357
metal abundance 91–170
metal-deficient star 91–197
metallic filament 89–268
~ hydrogen 83–202
~ prominence 89–265
metallicity 91–170
metallic–line star 91–195
metal–line star 91–195
metal–poor cluster 96–113
~ star 91–197
metal–rich cluster 96–112
~ star 91–196
metal–to–hydrogen ratio 91–176
metamorphic process 74–34
~ rock 74–11
metamorphism 74–34
contact ~ 74–35
dynamic ~ 74–36
shock ~ 82–237
thermal ~ 74–37
metasomatism 74–44
metasomatosis 74–44
meteor 86–1
absolute ~ brightness 86–74
absolute magnitude of a ~ 86–72
~ activity 86–134
age of a ~ shower 86–246
apparent ~ brightness 86–73
apparent magnitude of a ~ 86–71
appearance of a ~ 86–35
artificial ~ 86–21
~ astronomy 86–3
~ background 86–232
background ~ activity 86–136
~ body 86–160a
bright ~ 86–12
~ brightness 86–8
brightness of a ~ 86–8
~ camera 86–141
cometary ~ 86–45
~ component of night sky radiation 86–234
condensation within a ~ stream 86–217
~ crater 87–53
day–time ~ 86–22
density of a ~ stream 86–264
disappearance of a ~ 86–37
~ disintegration 86–187
disintegration of a ~ body 86–186
disruption of a ~ 86–82
double ~ 86–131
doubly photographed ~ 86–155
~ echo 86–296
ecliptical ~ 86–46
evaporation of a ~ body 86–198
explosion of a ~ 86–81
extinction of a ~ 86–37
faint ~ 86–15
faint telescopic ~ 86–19
fast ~ 86–87
~ flare 86–80
flickering ~ 86–85
~ fragmentation 86–188
fragmentation of a ~ 86–83
fragmentation of a ~ body 86–188
~ height 86–193
hyperbolic ~ 86–43
interplanetary ~ 86–44
interstellar ~ 86–43
~ ionization 86–281
knot within a ~ stream 86–217
length of a ~ path 86–47
~ light curve 86–70
long duration ~ echo 86–301
long enduring ~ train 86–113
~ luminosity 86–195
~ mass 86–163
multiple ~ 86–133
non–shower ~ 86–25
~ nucleus 86–102
~ observation 86–6
~ observer 86–7
ordinary ~ 86–14
~ pair 86–131
passage of a ~ 86–5
~ path 86–30
~ patrol 86–144
~ photograph 86–151a
~ photographed from two stations 86–155
photographic ~ 86–13
~ photography 86–140
~ plate 86–151
pulverization of a ~ body 86–189
~ radiant 86–53
~ radiation 86–207
radio ~ 86–267
shape of the ~ body 86–175
short duration ~ echo 86–300
short-enduring ~ train 86–112
~ shower 86–215
shower ~ 86–24
slow ~ 86–88
solar–system ~ 86–44
~ sparks 86–84
~ spectrograph 86–146
spectrum of a ~ train 86–107
spindle–shaped ~ 86–77
sporadic ~ 86–23
~ station 86–158
stationary ~ 86–34
~ stream 86–216
~ stream structure 86–243
telescopic ~ 86–18
~ trail 86–29, 86–153
~ trail on plate 86–29
~ train 86–103
train of a ~ 86–103
~ trajectory 86–30
trajectory of a ~ 86–30
ultratelescopic ~ 86–20
vapo(u)rization of a ~ body 86–198

~ velocity 86–86
very faint telescopic ~ 86–20
visual ~ 86–16
wake of a ~ 86–110
~ with irregular light-curve 86–85
~ with train 86–17
meteoric 86–4
~ body 86–160a
collection of ~ dust 87–12
~ crater 81–263
~ dust 87–11
~ head echo 86–297
~ iron 87–83
~ particle 86–160a
~ shower 86–215
meteoric–dust collector 87–13
meteorite 87–1
achondritic ~ 87–77
carbonaceous ~ 87–62
~ collection 87–8
~ crater 87–53
destruction by a ~ 87–20
fall of a ~ 87–19
~ find 87–44
finding of a ~ 87–44
~ fragment 87–43
fragment of a ~ 87–43
~ hole 87–58
~ investigator 87–7
iron ~ 87–82
moment of a ~ fall 87–51
oriented ~ 87–112
~ parent body 87–123
~ pit 87–58
~ shower 87–37
stony ~ 87–60
stony–iron ~ 87–80
~ surface 87–106
suspected fall of a ~ 87–49
time of a ~ fall 87–51
meteorite–induced volcanism 82–156
meteorites, age of 87–127
classification of ~ 87–59
collection of ~ 87–8
search for ~ 87–45
structure of ~ 87–100
meteoritic = meteoritical 87–2
~ astronomy 87–6
~ bombardement 82–235
~ crater 87–53
~ diamonds 87–65
~ dust 87–10
~ erosion 82–242
~ fall 87–19
fossil ~ crater 87–55
~ impact 82–236
~ iron 87–83
~ matter 87–9
~ organic matter 87–68
place of a ~ fall 87–50
~ seism 87–34
~ stone 87–60
~ theory 81–262
time of a ~ fall 87–51
meteoriticist 87–7
meteoritics 87–6
meteoroid 86–160
~ erosion 82–242
~ impact 82–236
meteoroidal bombardment 82–235
~ crater 81–263
~ impact 82–236
~ theory 81–262
meteorological conditions 77–2
~ element 77–28
~ service 77–5
meteorology 77–6
dynamic(al) ~ 77–7
synoptic(al) ~ 77–8
methanimine 97–45
methanol 97–32
method, double counting 86–132
iterative ~ 93–56
radioactive ~ 74–92
radiometric ~ 74–92
step ~ 95–10
stereoscopic ~ 81–115
uranium ~ 74–96
variational ~ 93–57
methyl alcohol 97–32
~ cyanide 97–38
methylacetylene 97–41
methylidyne 97–23
metric 99–151
Robertson-Walker ~ 99–152a
Schwarzschild ~ 99–152
~ space 99–100a
~ tensor 99–157a
Mic 92–67
mica 74–23
dark ~ 74–25
~ strip 88–87
white ~ 74–24
Michelson experiment 99–130
Michelson–Morley experiment 99–130
microbreccia, lunar 82–86
microcrater 81–201
micrometeor 86–20
micrometeorite 87–3
micrometeorology 77–9
micrometric measurement 96–49
microphone, hot–wire 76–46
microphotograph 87–140
microprobe analysis 82–143
micropulsations, geomagnetic 75–82
magnetic ~ 75–82
microreversibility 93–41
Microscope 92–67
blink ~ 91–34
Microscopium 92–67
microseismic noise 73–54
microseisms 73–53
microstructure 82–73
microtektite 87–97
middle atmosphere 76–51
~ corona 88–230
~ of eclipse 81–78
~ ring 83–222
middle-weight star 91–125
midnight, magnetic 79–131
mid–ocean fracture system 73–80
~ ridge 72–45
mid–point of trail 86–97
velocity at ~ 86–97
Mie scattering 83–76
migmatite 74–39
migration of spot zone 89–74
poleward ~ 89–231
poleward ~ of prominences 89–231
Milky Way 98–1
~ Way Galaxy 98–1
structure of ~ Way 98–4
millibar 77–47
milligal 71–66
millimeter of mercury column 77–48
millionth of the solar disc 89–69
~ of the visible hemisphere 89–68
Mills' cross 88–121
~ cross aerial 88–121
Milne–Eddington approximation 93–50
minauroral belt 79–129
mineral 74–12
rock–forming ~ 82–99
mineralogic(al) analysis 82–53
~ constituent 82–100
minimum corona 88–246
epoch of ~ 89–80
~ of solar activity 89–4
primary ~ 95–26
principal ~ 95–26
secondary ~ 95–27
solar ~ 89–4
Minkowski(an) four–space 99–122
~ space 99–122
~ space–time 99–122
~ universe 99–122
Minkowskian space 99–122
minor planet = asteroid = planetoid 84–68
~ planet belt 84–70
~ planet of the belt 84–71
mirage 79–64
mirror point 75–124
~ point altitude 75–125
rocking ~ 86–130
rotating ~ 86–130
missing mass 99–87
mission, asteroid 84–99
sample–return ~ 84–98
sample–return ~ to an asteroid 84–98
~ to an asteroid 84–99
mixing of stellar matter 93–162
mixture, gas–dust 83a–25
~ of gas and radiation 93–125
Russel ~ 93–101
Mizar 92–120

EN

mock moon 79–73
~ sun 79–71
model, aeolosphere 83–260
anisotropic ~ 99–65
asymptotic ~ of the first kind 99–59
~ atmosphere 83–247
blanketed ~ 93–18
Charlier's ~ 99–67
closed ~ 99–73
composite ~ 93–140
contracting ~ 99–56
cosmological ~ 99–46
De Sitter('s) ~ 99–69
Eddington ~ 93–137
Einstein('s) ~ 99–68
Einstein–De Sitter ~ 99–70
empty ~ 99–49
expanding ~ 99–55
fountain ~ 85–85
Friedmann ~ 99–71
galactic ~ 98–163
greenhouse ~ 83–262
hierarchic ~ 99–48
homogeneous ~ 99–64
hot cosmological ~ 99–166
hydrogen ~ 83–197
hydrogen ~ of a planet 83–197
inhomogeneous stellar ~ 93–139
ionospheric ~ 83–261
isotropic ~ 99–63
mathematical ~ 98–324
monotonic ~ 99–61
monotonic ~ of the first kind 99–62
non–homogeneous ~ 99–66
non-static ~ 99–51
non-stationary ~ 99–53
~ of an atmosphere 83–247
~ of Earth 79–133
~ of Einstein and De Sitter 99–70
~ of the universe 99–46
open ~ 99–72
oscillating ~ 99–57
oscillating ~ of the first kind 99–58
~ planet 83–196
point-source ~ 93–138
polytropic ~ 93–129
relativistic stellar ~ 93–198
shell-source ~ 93–143
standard ~ 93–137
static ~ 99–50
stationary ~ 99–52
steady-state ~ 99–52
stellar ~ 93–123
sunspot ~ 89–102
two-component ~ 88–196
two-temperature ~ 88–197
unblanketed ~ 93–19
water ~ 85–113
world ~ 99–46
zero-temperature ~ 93–186
zero-temperature ~ of a star 93–186
model(l)ing, mathematical 98–323
moderate earthquake 73–36
~ slope 81–231
modification, coast-line 72–78
modulation, cosmic ray 84–26
cosmic-ray ~ region 84–26
~ frequency 86–275
module, command 82–6
lunar ~ 82–5
service ~ 82–7
modulus, apparent 96–117
distance ~ 91–16
~ of elasticity 73–59
true ~ 96–118
Young's ~ 73–59
Moho 72–14
~ discontinuity 72–14
Mohorovičić discontinuity 72–14
moisture, atmospheric 77–118
moldavite 87–99
molecular aggregate 97–65
~ cloud 97–11
~ coma 85–65
~ composition 78–11
~ hydrogen 97–31
mean ~ weight 93–100
~ oxygen 79–23
molecule, air 86–172
daughter ~ 85–135a
interstellar ~ 97–10
polyatomic ~ 97–21
molten globe 83a–64
moment of appearance 86–49
~ of momentum 83a–4
rotational ~ 83a–5
momentum, angular 83a–4
moment of ~ 83a–4
orbital angular ~ 83a–6
Mon 92–68
monitoring network 82–40
salellite ~ 83–26
selenophysical ~ network 82–40
Monoceros 92–68
monochromatic albedo 83–93
~ corona 88–238
~ coronal ray 89–288
~ heliograph 88–38
~ images 97–158
~ telescope 88–38
monopole, magnetic 99–170o
monosulfide, carbon 97–34
monotonic model 99–61
~ universe 99–61
~ universe of the first kind 99–62
monoxide, carbon 97–28
silicon ~ 97–37
monsoon 77–99
Monster 92–55
month 81–42
anomalistic ~ 81–43
draconic ~ 81–47
nodical ~ 81–47
sidereal ~ 81–45
synodic(al) ~ 81–46
tropical ~ 81–44
Moon = moon 81–2
age of ~ 81–31
~ atlas 81–136
~ base 82–17
change of the ~ 81–19
crescent ~ 81–23
culmination of the ~ 82–303
distance of the ~ from the Earth 81–13
eclipse of the ~ 81–56
eclipsed ~ 81–57
~ enters penumbra 81–75
~ enters umbra 81–76
far side of the ~ 81–16
figure of the ~ 81–116
full ~ 81–27
half ~ 81–26
invisible side of the ~ 81–16
~ leaves penumbra 81–82
~ leaves umbra 81–81
limb of the ~ 81–95
mock ~ 79–73
new ~ 81–22
partial eclipse of the ~ 81–67
phase of the ~ 81–19
photographic map of the ~ 81–137
physics of the ~ 82–151
picture of the ~ 82–26
rear side of the ~ 81–16
reverse side of the ~ 81–16
shape of the ~ 81–116
total eclipse of the ~ 81–66
waning ~ 81–21
waning half ~ 81–30
waxing ~ 81–20
waxing half ~ 81–25
moonless night 79–3e
moonquake 82–214
moonrock 82–65
Moon's age 81–31
~ apparent radius 81–70
~ apparent semi-diameter 81–70
~ chemistry 82–74
~ crust 82–54
~ declination 81–62
~ disc 81–54
~ disk 81–54
~ far side 81–16
~ orbit 81–36
~ photograph 82–26
~ right ascension 81–60
~ surface 82–55
moraine 72–128
ground ~ 72–129
lateral ~ 72–131
terminal ~ 72–130
morning star 83–127
~ terminator 83–70
~ twilight 79–35
morphological type 98–228

morphology, lunar 81–147
lunar features ~ 81–147
Moscow Sea 81–283
Mössbauer spectrography 82–142
most luminous supergiant 91–147
mother-of-pearl cloud 79–91
motion, annual proper 91–20
apex of the solar ~ 98–64
apparent ~ 83–41
apsidal ~ 96–71
basic solar ~ 98–66
crustal ~ 73–72
daily ~ 85–29
daily ~ of the radiant 86–69
direct ~ 83–43
horizontal ~ 88–184
hourly ~ 81–72
internal ~ 97–149
large-scale ~ 88–189
non-gravitational ~ 85–56
~ of planets 83–28
~ of sea 72–86
orbital ~ 96–62
ordered ~ 98–95
parallactic ~ 91–33
peculiar ~ 91–31
~ perpendicular to the galactic plane 98–188
~ pictures of prominences 89–242
planetary ~ 83–28
poleward ~ 89–231
poleward ~ of prominences 89–231
proper ~ 91–19
proper ~ in declination 91–22
proper ~ in right ascension 91–21
random ~ 98–94
relative ~ 99–99
relativistic ~ of perihelion 99–160
retrograde ~ 83–47
small-scale ~ 88–188
solar ~ 98–62
standard solar ~ 98–63
third integral of ~ 98–186
vertical ~ 88–185
z ~ 98–188
mottles 88–210
bright ~ 88–211
chromospheric ~ 89–172
dark ~ 88–212
fine ~ 89–17
movable element 88–108
mound 89–276
mountain arc 73–84
~ building 73–74
~ chain 81–194
~ climate 77–187
~ crest 81–195
fold ~ 73–86
folded ~ 73–86
~ glacier 72–113
lunar ~ 81–193
~ range 73–85
~ region 81–196
~ ridge 81–186
submarine ~ 72–43
moustaches 89–200
movements, crustal 73–72
epeirogenic ~ 73–73
mass ~ 82–195
movie, prominence 89–242
moving cluster 96–97
~ plate 88–27
~ star cluster 96–97
moving-blade shutter 86–148
moving-cluster parallax 91–7
MUF 78–37
multiantenna interferometer 88–123
multicolo(u)r photometry 96–143
multielement interferometer 88–123
multilevel atom 93–46
multiple galaxy 98–308
~ meteor 86–133
~ redshift 98–364
~ star 96–5
~ tail 85–76
~ wall 81–255
multiple-arm(ed) galaxy 98–44
~ system 98–44
multiple-centre galaxy 98–319
multiple-nucleus galaxy 98–319
multiply-armed galaxy 98–44
multipole 75–66
multiwave interferometer 88–123
Mus 92–69
Musca 92–69
muscovite 74–24
mutual neutralization 79–29
~ perturbations 83–25
~ perturbations among satellites 83–25

N

nacreous cloud 79–91
naked singularity 94–117
natural abundance 91–171
~ fall 82–227
~ maser 97–13
~ radiation belt 75–107
~ radiation zone 75–107
~ satellite 83–18
nature of the object 85–24
nautical twilight 79–39
navigation hazard 72–97
near earthquake 73–26
~ side of the Moon 81–15
nearby earthquake 73–26
~ star 98–60
nearest point of area of impact 87–40
nebula, Andromeda 98–215
bright ~ 97–121
cometary ~ 97–142
comet-like ~ 97–142
Crab ~ 97–134
dark ~ 97–132
diffuse ~ 97–122
dust ~ 97–115
emission ~ 97–123
extragalactic ~ 98–202
filament of a ~ 97–131
filamentary ~ 97–129
galactic ~ 97–110
gas-and-dust ~ 97–119
gaseous ~ 97–112
Horsehead ~ 97–135
Horseshoe ~ 97–143
irregular ~ 98–237
Loop ~ 97–137
Looped ~ 97–137
Network ~ 97–137
North America ~ 97–140
nucleus of a planetary ~ 97–148
Omega ~ 97–143
Orion ~ 97–136
Owl ~ 97–139
parent stellar ~ 97–63
planetary ~ 97–144
pre-solar ~ 83a–21
primitive ~ 83a–21
primitive stellar ~ 97–63
reflecting ~ 97–128
reflection ~ 97–128
regular ~ 98–236
ring ~ 97–147
solar ~ 83a–23
spiral ~ 98–249
Swan ~ 97–143
Trifid ~ 97–141
Veil ~ 97–137
nebulae, cluster of 98–329
nebular cloud 94–31
~ hypothesis 83a–7
~ ring 83a–10
~ stage 95–125
~ variable 95–85
neck, volcanic 73–125
needles, ice 76–112
needle-shaped galaxy 98–239
negative hydrogen ion 93–66
~ magnetic polarity 89–150
~ nitrogen bands 79–21
~ velocity 91–27
neighbo(u)rhood of the Sun 98–59
solar ~ 98–58
neighbo(u)ring galaxy 98–211
Neptune 83–236
Net 92–83
network, chromospheric 89–171
fundamental gravimetric ~ 71–82
fundamental gravity ~ 71–82
gravimetric ~ 71–80
gravity ~ 71–80
~ nebula 97–137
world ~ 77–18
world geodetic ~ 71–38
world gravity ~ 71–81

EN

neutral hydrogen 97–47
~ hydrogen cloud 97–103
~ hydrogen geocorona 76–74
~ hydrogen region 97–49
~ line 89–151
~ point 89–152
~ sheet 75–16
~ surface 75–16
neutrino cooling 94–97
~ emission 94–94
~ luminosity 94–96
~ opacity 94–95
photo– ~ 94–93
~ radiation 94–94
~ reaction 94–92
neutron albedo 75–116
albedo ~ 75–115
~ capture 94–84
~ capture reaction 94–87
~ degeneracy 93–191
~ flux 94–86
~ flux density 94–86
~ gas 93–192
~ source 94–85
~ star 91–152, 93–195
neutrosphere 76–64
new island 72–81
~ moon 81–22
Newcombian space 99–110
new–cycle group 89–106
Newtonian cosmology 99–25
~ time 99–118
~ universe 99–93
nickel content 72–55
~ iron 87–90
nickel–poor ataxite 87–88
nickel–rich ataxite 87–89
nife 72–18
night airglow 79–5
bright ~ 79–59
bright ~ sky 86–28
clear ~ 79–3g
~ hemisphere 83–68
lunar ~ 81–33
moonless ~ 79–3e
photometric ~ 79–3f
~ side 75–21
~ sky 79–6
starry ~ 79–3d
nightglow 79–5
~ emission 79–5
night–sky light 79–8
~ luminescence 79–5
night–time shower 86–218
~ stream 86–218
nimbostratus 77–145
nimbus, dark 89–189
flare ~ 89–189
nitrogen star 91–189
noctilucent cloud 79–89
~ train 86–105
nocturnal hemisphere 83–68
nodical month 81–47
noise, acoustic 88–220
microseismic ~ 73–54
seismic ~ 73–54
sound ~ 88–220
nomenclature of martian features 83–108
non–adiabatic pulsations 95–99
non–barred spiral 98–250
non–carbonaceous chondride 87–75
non–cluster galaxy 98–340
~ star 96–84
non–degenerate core 93–171
~ gas 93–182
~ outer layers 93–199
non–destructive γ–ray spectrometry 82–77
non–dimensional variable 93–133
non–Doppler redshift 98–363
non–elastic star collision 98–26
non–Euclidean 99–102
~ geometry 99–109
~ space 99–114
non–explosive nucleogenesis 94–58
~ nucleosynthesis 94–58
non–gravitational motion 85–56
non–gray atmosphere 93–27
non–grey atmosphere 93–27
non–homogeneous model 99–66
~ universe 99–66
non–illuminated hemisphere 83–68
non–inertial system 99–147
non–LTE (equilibrium) 93–43
non–luminous 96–44
non–member 96–84
~ of a cluster 96–84
~ star 96–84
non–periodic comet 85–10
non–proton flare 89–186
non–radial pulsations 95–97
non–radiative energy 88–218
non–relativistic 99–96
~ approximation 99–97
~ cosmology 99–27
~ degeneracy 93–177
non–rotating star 91–38
non–shower meteor 86–25
non–spherical star 91–86
non–stable star 95–9
non–static model 99–51
~ universe 99–51
non–stationary model 99–53
~ star 95–2
~ universe 99–53
non–steady state 98–151
non–synchronous rotation 83–38
~ spin 83–38
non–thermal pressure 94–42
non–thermodynamic equilibrium 93–43
non–thermodynamical equilibrium 93–43
non–uniform rotation 91–36
non–velocity redshift 98–363
non–volatile substance 83a–53
non–zonal harmonics 95–102
noon meridian 75–17
Nor 92–70
Norma 92–70
normal abundance 91–172
~ earthquake 73–22
~ fault 73–93
~ galaxy 98–221
~ giants 91–150
~ pressure 77–46
~ room temperature 77–31
~ sound 76–35
~ spiral 98–250
~ star 91–95
~ stellar spectrum 91–89
~ temperature and pressure 77–30
North America Nebula 97–140
~ Polar Sequence 91–87
~ Star 92–121
Northern Cross 92–44
~ Crown 92–40
~ limit of eclipse 88–20
~ limiting latitude 81–105
~ polarity 89–149
north–south asymmetry 89–50
nose whistler 78–59
not–easily fusible material 83a–55
nova 95–110
dwarf ~ 95–112
~ explosion 95–121
extragalactic ~ 95–114
fast ~ 95–119
galactic ~ 95–113
~ outburst 95–121
recurrent ~ 95–116
slow ~ 95–120
typical ~ 95–111
nova–like 95–118
~ star 95–118
~ variable 95–118
November swarm 86–255
N–type galaxy 98–263
nuclear activity of a galaxy 98–271
~ bulge 98–16
~ disintegration 85–116
~ energy 93–122
~ equilibrium 94–103
~ evolution 94–50
~ fragment 94–102
high–altitude ~ explosion 75–114
~ region 98–17
~ statistical equilibrium 94–103
~ synthesis 94–52
nuclei, Aitken 76–116
crystallization ~ 76–108
freezing ~ 76–109
iron-groupe ~ 94–105
iron-peak ~ 94–105
sublimation ~ 76–110
nucleochronology 83a–73
nucleogenesis 94–52

exlosive ~ 94–57
non-explosive ~ 94–58
nucleosynthesis 94–52
Big–Bang ~ 99–175
cosmological ~ 99–175
explosive ~ 94–57
non–explosive ~ 94–58
primordial ~ 99–175
nucleus, active galactic 98–272
activity of galactic ~ 98–271
cometary ~ 85–36
condensation ~ 76–103
decay of cometary ~ 86–261
disintegration of cometary ~ 85–116
galactic ~ 98–18
giant ~ 76–118
ice ~ 85–40
icy ~ 85–40
meteor ~ 86–102
~ of a comet 85–36
~ of an association 96–91
~ of condensation 97–67
~ of the Galaxy 98–18
~ of the planetary nebula 97–148
peculiar ~ 98–20
solid ~ 85–39
starlike ~ 85–38, 98–19
superconducting ~ 95–150
number, check 85–30
Messier ~ 97–111
~ of encounters 98–122
~ of meteors per hour 86–135
~ of particles 76–121
Poisson's ~ 73–60
star ~ 98–80
sunspot ~ 89–65
Wolf sunspot ~ 89–65
Zürich relative sunspot ~ 89–65
numbered planetoid 84–95

O

O–association 96–89
object, collapsed 98–23
compact ~ 98–21
cosmological ~ 98–369
diffuse ~ 85–25
extragalactic ~ 98–209
nature of the ~ 85–24
quasi-stellar ~ 98–359
radio quiet ~ 98–362
supermassive ~ 98–22
uncatalogued ~ 84–96
oblate spheroid 98–158
highly ~ system 98–164
oblateness 83–113
oblique photo 82–29
~ photograph 82–29
~ rotator 91–52, 95–60
obscuration, interstellar 97–86
observation, day–time 83–121
eclipse ~ 88–226
meteor ~ 86–6
out–of–eclipse ~ 88–227
radar ~ 86–265
radar echo ~ 86–265
simultaneous ~ 71–35, 86–125
twilight ~ 79–42
observational cosmology 99–17
observatory, auroral 79–101
high–altitude ~ 88–56
ionospheric ~ 78–21
lunar ~ 82–20
~ on the Moon 82–20
seismologic(al) ~ 73–5
solar ~ 88–7
observer, meteor 86–7
occlusion 77–72
occultation 95–80
grazing ~ 81–100
limiting latitudes of visibility of ~ 81–104
lunar ~ 81–94
~ of a star by the Moon 81–94
~ predictions 81–102
reduction of ~ 81–108
visibility of an ~ 81–103
~ zone 81–107
zone of ~ 81–107
occulting disc 88–61
ocean 72–22
~ bed 72–37
~ current 72–60
deep ~ 72–41
~ floor 72–37
~ of Storms 81–277
~ tides 82–310
~ trench 72–40
ocean–floor spreading 72–39
oceanographic chart 72–23
~ map 72–23
~ sounding 87–16
oceanography 72–24
satellite ~ 72–70
space ~ 72–69
Oceanus Procellarum 81–277
OCS 97–36
Oct 92–71
octahedrite 87–86
Octans 92–71
Octant 92–71
off–shore silting 72–79
oil spill 72–71
Olbers' hypothetical planet 83a–63
~ paradox 99–40
old star 94–64
old–cycle group 89–107
olivine 74–22
Omega Nebula 97–143
omnidirectional intensity 75–104
one–arm galaxy 98–42
one–armed galaxy 98–42
Oort's cloud 85–97
~ cloud of comets 85–97
~ constants 98–118
ooze 72–58
opacity, neutrino 94–95
open cluster of galaxies 98–333
~ model 99–72
~ star cluster 96–99
~ universe 99–72
Oph 92–72
Ophiuchus 92–72
opposite polarity 89–155
opposition 83–59
~ effect 83–75
~ in right ascension 81–68
mean ~ magnitude 83–85
optical activity 89–3
~ companion 98–218
~ companion of a galaxy 98–218
~ disc 89–26
~ disk 89–26
~ double star 96–6
~ feature 98–234
~ flare 89–181
~ galaxy 98–204
~ interstellar line 97–99
~ jet 98–279
~ libration 81–51
~ phenomenon 79–33
~ plage 89–165
~ pulsar 95–140
~ pumping 97–17
~ Sun 88–2
~ variability 95–141
optics, atmospheric 79–1
orbit, absolute 96–55
apparent ~ 96–48
astrometric ~ 96–57
beyond the Earth's ~ 84–6
Earth's ~ 84–5
galactic ~ of a star 98–184
geocentric ~ 86–41
heliocentric ~ 86–42
lunar ~ 81–36
Moon's ~ 81–36
~ of a planet 83–29
photocentric ~ 96–47
planetary ~ 83–29
primitive ~ 98–200
relative ~ 96–54
spectrographic ~ 96–56
spectroscopic ~ 96–56
three–dimensional ~ 98–185
true ~ 96–53
orbital motion 96–62
~ reconnaissance 82–31
~ velocity 96–63
orbits, coplanarity of 83a–3
ordered motion 98–95
ordinary wave 78–31
Ordovician 74–79
organic acid 87–69
~ inclusion 87–66
~ remains in meteorites 87–67

organogenous sediments 72–59
Ori 92–73
orientation of rotation axis 84–102
oriented form 87–113
~ meteorite 87–112
origin 94–9
common ~ 94–25
cosmic ~ 87–5
extraterrestrial ~ 87–73
impact ~ 82–229
~ of comets 85–96
~ of the universe 99–11
~ of the world 99–11
terrestrial ~ 87–74
original heat 74–63
~ mass 86–164
Orion 92–73
~ arm 98–36
~ Nebula 97–136
Orionids 86–253
Orion's Belt 92–129a
orogenesis 73–74
orogen(et)ic belt 73–75
orogenic zone 73–75a
orogeny 73–74
orographic cloud 77–134
~ precipitation 77–167
~ rain 77–166
orthoclase 74–16
oscillating model 99–57
~ slit 88–70
~ universe 99–57
~ universe of the first kind 99–58
oscillation perpendicular to galactic plane 98–187
z– ~ 98–187
oscillations, five–minute 88–186
forced tidal ~ 82–306
free ~ of the Earth 73–49
solar ~ 88–223a
tidal ~ 76–28
oscillator, hydromagnetic 95–61
magnetohydrodynamic ~ 95–61
oscillograph 86–292
oscilloscope 86–292
outburst, brightness 85–126
comet ~ 85–125
nova ~ 95–121
supernova ~ 95–131
outer core 72–19
~ corona 88–231
~ crater 81–218
~ edge 83–227
~ ionosphere 76–76
~ layers of a star 93–3
~ planet 83–15
~ radiation belt 75–105
~ radiation zone 75–105
~ radius 83–228
~ ring 83–225
~ van Allen (radiation) belt 75–105
outflow, continuous 98–276
lava ~ 73–130
mass ~ 93–13
~ of matter from galactic nucleus 98–275
outgassing 82–261, 87–136
~ of cometary nucleus 84–109
out–of–eclipse observation 88–227
output, energy 93–109
outward stream 93–151
~ winding of spiral arms 98–30
oval, auroral 79–124
white ~ 83–195
over–abundance 91–185
over–abundant element 91–184
overcast 77–152
over–dense 86–285
over–estimate 95–13
overlap of two cycles 89–86
period of ~ of two cycles 89–86
over–luminous star 91–134
Owl Nebula 97–139
oxide = oxid 74–26
hydrogen ~ 97–26
oxidizing atmosphere 83a–47
oxygen, atomic 79–24
~ burning 94–82
molecular ~ 79–23
ozone, atmospheric 76–124
atmospheric ~ content 76–125
~ layer 76–123
~ shadow 76–127
~ shadowing 79–28
ozonosphere 76–123

P

p spot 89–113
P wave 73–46
pack ice 72–102
Painter's Easel 92–78
pair, meteor 86–131
pal(a)eoclimate 77–188
pal(a)eomagnetism 75–25, 83a–79
pal(a)eotemperature 72–126
Pal(a)eozoic 74–71
~ era 74–71
Pallas 84–86
pallasite 87–81
Palus 81–303
~ Epidemiarum 81–304
~ Putredinis 81–305
~ Somnii 81–306
pancakes, primordial 99–171b
panoramic photograph 82–27
parabolic comet 85–7
~ space 99–107
paraclase 73–91
Seeliger's ~ 99–39
paradox, Cheseaux' and Olbers' 99–40
cosmological ~ 99–38
gravitational ~ 99–39
Olbers' ~ 99–40
photometrical ~ 99–40
Seeliger's ~ 99–39
parafoveal vision 86–128
parallactic libration 81–52
~ motion 91–33
~ photography 79–102
parallax, absolute 91–11
annual ~ 91–6
Cepheid ~ 91–8a
cluster ~ 91–7
dynamic(al) ~ 91–12
moving–cluster ~ 91–7
~ of Strömgren spheres 91–13
photometric ~ 91–14
relative (trigonometric) ~ 91–10
secular ~ 91–9
spectroscopic ~ 91–15
statistical ~ 91–8
stellar ~ 91–4
Strömgren ~ 91–13
trigonometric ~ 91–5
parameter, crater 81–227
deceleration ~ 99–83
photometric ~ 85–106
structural ~ 96–120
paranthelion 79–84
parantiselene 79–86
paraselene 79–73
parasitic crater 73–128
parent body 87–123
break–up of ~ body 87–124
~ comet 86–260
disruption of ~ body 87–124
~ galaxy 98–284
meteorite ~ body 87–123
~ molecule 85–135
~ stellar nebula 97–63
parhelic circle 79–82
parhelion 79–71
lower ~ 79–72
parsec 91–17
cubic ~ 98–84
per cubic ~ 98–84
partial cross–section 94–54
~ degeneracy 93–173
~ eclipse of the Moon 81–67
end of ~ eclipse 81–83a
~ phase 81–78a
~ phase of the eclipse 81–78a
~ solar eclipse 88–16
partially degenerate 93–183
particle 88–154
accelerated ~ 91–182
aerosol ~ 76–91
air ~ 86–172
auroral ~ 79–107
~ bombardment 82–240
~ cloud 84–31
colloidal ~ 76–98
core–mantle ~ 97–70
~ emission 88–151
form of a meteoric ~ 86–175

geomagnetically trapped ~ 75–111
giant ~ 76–118
high–energy ~ event 89–191
largest ~ 87–41
liquid ~ 76–100
lunar ~ 82–123
meteoric ~ 86–160a
~ pressure 93–90
~ radiation 88–150
solid ~ 76–99, 97–69
~ storage 75–123
superheavy ~ 99–170n
~ track 82–251
~ trajectory 75–118
particles, Aitken 76–116
ejection of ~ 85–89
ejection of ~ from 88–161
ejection of ~ from the Sun 88–161
fines ~ 82–109
influx of high–energy ~ 78–43
interstellar ~ 97–68
number of ~ 76–121
precipitable ~ 76–117
size of ~ 76–114
solid ~ 85–48
spherical ~ 87–29
storage of ~ 75–123
trapped ~ 75–111
zodiacal ~ 84–66
passage, atmospheric 87–26
central–meridian ~ 89–45
close ~ of two galaxies 98–320
limb ~ 89–31
~ of a fireball 86–117
~ of a meteor 86–5
~ of interplanetary discontinuity 84–58
~ of two galaxies 98–320
~ through atmosphere 87–26
passive seismology 82–218
~ seismometer 82–221
past supernova 95–134
path, apparent 86–32
angular ~ length 86–48
atmospheric ~ 86–162
~ correction 82–10
~ difference 88–117
difference in optical ~ 88–117
distorted ~ 86–33
~ length 86–47
meteor ~ 86–30
~ of the total phase 88–29
true ~ 86–31
wavy ~ 86–33
patrol, automatic flare 89–221
~ camera 86–145
cinematic flare ~ 89–220
flare ~ 89–219
pattern, interference 88–120
ring ~ 98–41
scattering ~ 77–58
striation ~ 89–174
Widmanstätten ~ 87–105
pause 76–58
Pav 92–74
Pavo 92–74
pc 91–17
PCA 78–44
Peacock 92–74
peak, central 81–258
~ power 86–271
~ value 76–55
~ value of a layer 76–55
pebble 82–128
peculiar galaxy 98–256
~ motion 91–31
~ nucleus 98–20
~ star 91–92
~ velocity 91–31
peculiarity 98–257
spectral ~ 91–93
Peg 92–75
Pegasus 92–75
pelagic clay 72–54a
~ deposits 72–57
~ sediments 72–57
pellet 82–129
volcanic ~ 82–169
pencil–beam interferometer 88–121
pendulum, horizontal 71–73
penetrometer, soil 82–83
surface ~ 82–83
penumbra of the Earth 81–64
Moon enters ~ 81–75
Moon leaves ~ 81–82
spot ~ 89–97
sunspot ~ 89–97
penumbral fibril 89–98
~ filament 89–98
~ lunar eclipse 81–65
~ waves 89–100
Per 92–76
perfect cosmological principle 99–32
pergelisol 83–152
periapse 83–172
periapsis 83–172
Jovian ~ 83–172
Saturnian ~ 83–210
periareon 83–142
periastron 96–58
advance of ~ 96–71
epoch of ~ 96–60
longitude of ~ 96–59
relativistic advance of ~ 99–160
time of ~ passage 96–60
pericenter = pericentre 98–189
galactic ~ 98–189
Jovian ~ 83–172
Saturnian ~ 83–210
pericentron 98–189
pericynthion 81–40
perifocus, galactic 98–189
perigalacticon 98–189
perigalacticum 98–189
perigee 81–38
perihelion 83–29a
perijove 83–172
perilune 81–40
perimartian 83–142
perimartium 83–142
period 95–39
axial ~ 83–31
beat ~ 95–42
change of ~ 95–43
draconic ~ 81–47
duration of the ~ 95–39
eleven-year ~ 89–70
geological ~ 74–74
glacial ~ 72–105
interglacial ~ 72–107
magnetically disturbed ~ 75–91
~ of overlap of two cycles 89–86
~ of pulsations 95–100
~ of rotation 83–31
~ of solar rotation 88–129
primary ~ 95–40
secondary ~ 95–41
spin ~ 83–31
~ time 95–39
periodic comet 85–3
~ line shift 96–22
~ magnetic variable 95–55
~ stream 86–228
~ variable 95–88
period–luminosity curve 95–103
~ relation 95–103
peripheral vision 86–128
perisaturnian 83–210
perisaturnium 83–210
periselene 81–40
periselenium 81–40
permafrost 83–152
permanent aurora 79–5
~ coronal condensation 89–282
~ feature 83–106
~ marking 83–106
~ snowfield 72–119
~ stream 86–227
permanently frozen soil 83–152
~ frozen subsoil 83–152
Permian 74–83
Perseids 86–251
Perseus 92–76
~ arm 98–34
persistence of a train 86–111
persistent train 86–104
perturbations, radial 93–215
solar ~ 83–24
solar ~ on satellites 83–24
tidal ~ 98–314
petrification 74–49
impact ~ 82–238
petrographic analysis 82–52
~ classification 82–101
phase angle 83–72
~ change 99–170p

~ changing switch 88–114
~ changing unit 88–115
~ coefficient 83–77
contracting ~ 94–20
contraction ~ 94–20
crescent ~ of the Moon 81–23
~ curve 83–74
declining ~ 89–84
~ defect 81–28
disc ~ 94–22
disk ~ 94–22
eruptive ~ 89–210
evolutionary ~ 94–28
explosive ~ 89–210
~ function 83–78
halo ~ 94–19
Hayashi ~ 94–44
initial ~ 75–75
initial ~ of a magnetic storm 75–75
~ integral 83–79
intermediate ~ 94–21
main ~ 73–12, 75–76
~ of the cycle 89–79
~ of the eclipse 81–73
~ of the Moon 81–19
recovery ~ 75–77
rising ~ 89–83
~ shifter 88–115
~ transition 83a–58,99–170p
zero ~ 83–73
phase–swept interferometer 88–109
phase–switching 88–112
~ interferometer 88–113
Phe 92–77
phenomena, twilight 79–32
phenomenon, halo 79–66
optical ~ 79–33
solar ~ 89–13
Phobos 83–168
Phoenix 92–77
photocenter 96–46
photocentre 96–46
photocentric orbit 96–47
photochemical coma 85–65
~ reaction 76–63
photodisintegration 94–101
~ of nucleus 94–101
photoelectric light–curve 95–22
~ polarimeter 83–83
photoelectron layer 82–246
lunar ~ layer 82–246
photograph, lunar 82–26
meteor ~ 86–151a
oblique ~ 82–29
panoramic ~ 82–27
vertical ~ 82–28
white–light ~ 88–37
photographic brightness 91–66
~ light-curve 95–23
lunar ~ map 81–137
~ magnitude 91–66
~ map of the Moon 81–137
~ meteor 86–13
~ polarimeter 83–84
~ subtraction 88–94
~ survey 82–30
~ surveying 82–30
photography, meteor 86–140
parallactic ~ 79–102
sequence ~ 79–103
photolysis 76–84
photometric(al) atlas 88–138
~ binary 96–11
~ binary star 95–73
~ diameter 84–93
~ distance 96–116
~ elements 95–25
~ map 81–134
~ paradox 99–40
~ parallax 91–14
photometry, detailed 83–104
integrated ~ 83–103
multicolo(u)r ~ 96–143
six–colo(u)r ~ 96–148
three–colo(u)r ~ 96–141
photomicrograph 87–140
photon pressure 93–91
~ pumping 97–17
photo–neutrino 94–93
photosphere 88–167
synoptic map of the ~ 89–20
undisturbed ~ 88–169
photospheric 88–170
~ faculae 89–161
~ granulation 88–176
~ granule 88–181
~ magnetic field 89–130
quiet ~ region 89–56
undisturbed ~ region 89–56
phototriangulation 71–26
spatial ~ 71–26
physical companion 98–219
~ compamion of a galaxy 98–219
~ double star 96–7
~ libration 81–53
~ map 81–134
~ membership 96–83
~ theory of meteors 86–159
physics, atmospheric 76–4
~ of comets 85–104
~ of planets 83–4
~ of solar–terrestrial relations 84–56
~ of the Moon 82–151
solar ~ 88–6
solar–terrestrial ~ 84–56
Pic 92
Pictor 92–78
picture of the Moon 82–26
piezoglypts 87–122
pigeonite 82–90
Pisces 92–79
Pisces Australids 86–250
Piscis Austrinus 92–80
~ Volans 92–100
pit, crater 81–250
meteorite ~ 87–58
summit ~ 81–244
summit ~ of central peak 81–244
place, finding 87–52
~ of discovery 87–52
~ of meteoritic fall 87–50
plage 89–164
~ areas 89–164
brightening of a ~ 89–213
optical ~ 89–165
X–ray ~ 89–166
plagioclase 74–19
plain, abyssal 72–42
lava ~ 82–185
walled ~ 81–265
Planck era 99–170e
~ fundamental length 99–170h
~ fundamental mass 99–170f
~ length 99–170h
~ mass 99–170f
~ time 99–170i
plane, concentration toward the galactic 98–55
equatorial ~ 83–231
~ fracture 82–208
galactic ~ 98–53
invariable ~ 83–232
Laplacian ~ 83–232
~ of the equator 83–231
~ of the sky 96–50
ring ~ 83–216
tangent ~ 96–50
plane-parallel atmosphere 93–21
planet 83–1
absolute magnitude of a ~ 83–86
drawing of a ~ 83–16
Earth–like ~ 83–119
figure of a ~ 83–109
giant ~ 83–174
growth of a ~ 83a–37
hydrogen ~ 83–197
hydrogen model of a ~ 83–197
hypothetical ~ 83–122
inferior ~ 83–14
inner ~ 83–14
intramercurial ~ 83–123
intramercurian ~ 83–123
Jovian ~ 83–174
major ~ 83–13
minor ~ 84–68
model ~ 83–196
model of a ~ 83–196
Olbers' hypothetical ~ 83a–63
orbit of a ~ 83–29
~ of the solar system 83–11
~ of the Sun 83–11
outer ~ 83–15
own radiation of a ~ 83–198
~ search 83–238
rotation of ~ 83–30
shape of ~ 83–109

solar ~ 83–11
superior ~ 83–15
terrestrial ~ 83–119
transneptunian ~ 83–237
transplutonian ~ 83–124
planetary 83–6
~ atmosphere 83–243
~ cavity 84–24
~ circulation 77–66
~ companion 96–43
~ configuration 83–50
~ corona 83–245
~ cosmogony 83a–1
~ geology 83–5
~ growth 83a–37
~ index 75–90
~ K_p index 75–90
~ magnetic index 75–90
~ model 83–196
~ nebula 97–144
nucleus of a ~ nebula 97–148
~ orbit 83–29
~ physics 83–4
~ rotation 83–30
~ stream 86–220
~ surface 83–97
~ system 83–2
~ wave 72–94
planetesimals 83a–35
planetocentric 83–9
~ system 83–10
~ system of coordinates 83–10
~ system of reference 83–10
planetographic 83–7
~ coordinates 83–8
planetoid = asteroid = minor planet 84–68
double ~ 84–73
lost ~ 84–97
numbered 84–95
~ of the belt 84–71
planetoids, belt of 84–70
planetology 83–5
planetophysics 83–4
plasma cloud 84–30
collisionless ~ 84–50
helium–enriched ~ 84–60
interplanetary ~ 84–39
magnetized solar ~ 84–32
~ sheet 75–11
solar ~ 84–29
stream of solar ~ 84–33
plasmasphere 76–76
plastic flow 72–121
plate center 86–156
~ centre 86–156
compensating ~ 88–92
continental ~ 73–103
λ/4 ~ 88–126
meteor ~ 86–151
moving ~ 88–27
plateau 81–192
platform, continental 73–103
Pleiades 96–110
plessite 87–94
Plough 92–96
Ploughman 92–22
plumb line 71–71
plumb–line deflection 71–72
plumes 89–294
Pluto 83–239
plutonic rock 74–7
Pogson's equation 91–58
point, appearence 86–36
beginning ~ 86–36
bright ~ 89–201
conical ~ 96–31
convergent ~ 96–98
Curie ~ 75–41
dew ~ 77–174
disappearance ~ 86–38
end ~ 86–38
first Lagrangian ~ 96–31
height of appearance ~ 86–39
height of end ~ 86–40
impact ~ 86–124
inner Lagrangian ~ 96–31
magnetically conjugate ~ 75–23
magnetoconjugate ~ 75–23
melting ~ 86–204
neutral ~ 89–152
~ of clearest vision 86–127
~ of disruption 86–121
~ of explosion 86–122
~ of impact 86–124
reflection ~ 75–124
singular ~ 94–116
space–time ~ 99–125
start ~ 86–36
stationary ~ 83–45
stopping ~ 86–123
subsatellite ~ 71–30
world ~ 99–125
X–ray bright ~ 89–183a
points, magnetically conjugate 75–23
point–source model 93–138
Poisson's constant 73–60
~ number 73–60
~ ratio 73–60
polar aurora 79–93
~ cap 83–143
~ cap absorption 78–44
~ cap aurora 79–98
~ compression 83–114
~ crown 89–232
~ crown of filaments 89–232
~ faculae 89–162
~ flattening 83–114
~ fringe 83–151
~ front 77–73
~ hood 83–143
~ magnetic substorm 75–96
~ map 89–19
~ prominence 89–230
~ radius 83–116
~ rays 89–294
~ region 81–17, 83–192
~ sequence 91–87
~ zone 77–193
polarimeter, photoelectric 83–83
photographic ~ 83–84
visual ~ 83–82
polarimetric measurements 83–80
Polaris 92–121
polarity, magnetic 89–148
negative magnetic ~ 89–150
northern ~ 89–149
opposite ~ 89–155
positive magnetic ~ 89–149
southern ~ 89–150
polarization curve 83–81
interstellar ~ 97–105
~ map 81–134
pole, axial 83–36
axis ~ 83–36
Earth's magnetic ~ 75–45
geomagnetic ~ 75–48
geomagnetic axis ~ 75–48
magnetic dip ~ 75–45
~ of the rotation axis 83–36
pole–on star 91–40
Polestar 92–121
poleward migration 89–231
~ migration of prominences 89–231
~ motion of prominences 89–231
polished surface 87–101
pollutant 72–74, 76–81
atmospheric ~ 76–82
pollution, air 76–80
atmospheric ~ 76–80
chemical ~ 72–98
~ event 72–72
water ~ 72–73
Pollux 92–122
polyatomic molecule 97–21
polygonal crater 81–212
polyhedron 71–12
polymer grain 97–77a
polynomials, Legendre(') 93–55
polytrope 93–126
polytropic gas sphere 93–127
~ index 93–128
~ model 93–129
Poop 92–81
poor cluster of galaxis 98–331
~ visibility 77–26
population, cluster 98–341
disc ~ 98–72
disc–type ~ 98–72
disk ~ 98–72
disk–type ~ 98–72
extreme ~ I 98–73
galactic ~ 98–230
halo ~ 98–69
halo–type ~ 98–69
intermediate ~ 98–70

EN

~ inversion 97–15
~ of a galaxy 98–230
~ I 98–71
spiral–arm ~ 98–73
star ~ 98–67
stellar ~ 98–67
~ II 98–68
type I ~ 98–71
type II ~ 98–68
pore 89–122
porosity 83–100
porous 82–181
~ meteoroid 85–50
~ substance 82–111
portable magnetometer 82–44
position, accurate 85–28
~ angle 81–71
approximate ~ 85–26
calculated ~ 86–27
heliographic ~ 89–38
predicted ~ 85–27
radiant ~ 86–60
selenocentric ~ 81–132
positive magnetic polarity 89–149
~ velocity 91–26
possible crater 81–211
post–eruptive galaxy 98–270
post–glacial 72–110
~ epoch 72–109
post–main–sequence evolution 94–68
~ phase of evolution 94–68
~ stage 94–68
post–mare crater 81–208
~ region 81–270
~ structure 81–154
post–maximum 95–33
post–nova 95–123
post-supernova 95–134
post–twilight emission 79–52
postulate, coherency 99–34
Weyl's ~ 99–34
potash feldspar 74–16
potential, adsorption 85–122
axially symmetric(al) ~ 98–153
gravitational ~ 96–160a
gravity ~ 71–58
stationary ~ 98–147
tidal ~ 82–293
~ trough 98–176
~ with cylindrical symmetry 98–154
power, adsorption 85–120
adsorptive ~ 85–120
mean ~ 86–272
transmitter ~ 86–270
Poynting vector 93–92
Poynting–Robertson effect 84–67
Praesepe 96–111
Precambrian = Pre–Cambrian 74–70
Early ~ 74–75
~ era 74–70
preceding spot 89–113
precipitable particles 76–117
precipitation 77–154
~ amount 77–158
amount of ~ 77–158
auroral electron ~ 79–112
electron ~ 78–48
orographic ~ 77–167
sum of ~ 77–168
total ~ 77–168
precursor 73–11, 95–143
predicted comet 85–11
prediction, ionospheric 78–32
occultation ~ 81–102
~ of minor planet positions 84–92
~ of solar activity 89–64
preheating 86–203
preliminary designation 85–32
pre–main–sequence evolution 94–26
~ phase of evolution 94–26
~ stage 94–49
pre–mare crater 81–207
~ region 81–269
~ structure 81–153
pre–maximum 95–30
pre–nova 95–122
pre–planetary cloud 83a–24
pre–solar nebula 83a–21
pressure altitude 76–16
at normal ~ 77–46
atmospheric ~ 77–43
degeneracy ~ 93–187a
gas ~ 93–90
~ gradient 77–106
hydrostatic ~ 93–77
non–thermal ~ 94–42
normal ~ 77–46
normal temperature and ~ 77–30
particle ~ 93–90
photon ~ 93–91
radiation ~ 93–91
ridge of high ~ 77–87
sea–level ~ 77–44
thermal ~ 94–41
total ~ 93–93
water–vapo(u)r ~ 77–125
~ wave 73–46
wedge of high ~ 77–86
pre–stellar 94–13
~ body 94–15
~ evolution 94–14
~ evolution of galaxies 94–14
~ matter 94–16
pre–supernova 95–129
pre–twilight emission 79–51
prevailing wind 77–117
prim(a)eval atom 99–169
~ fireball 99–168
~ lead 74–98
~ magnetic field 95–62
primary 96–37
~ arc 73–82
~ component 96–37
~ crater 81–198
~ feature 81–150
~ lead 74–98
~ magnetization 75–35
~ maximum 95–31
~ minimum 95–26
~ period 95–40
~ surface 87–107
~ triangulation 71–24
~ wave 73–46
primitive atmosphere 83–244
~ atom 99–169
~ magnetic field 95–62
~ nebula 83a–21
~ orbit 98–200
~ stellar nebula 97–63
primordial atmosphere 83–244
~ explosion 99–167
~ fluctuation 99–170y
~ galactic cloud 94–17
~ heat 83–200
~ isotope 82–146
~ matter 99–170
~ noble gas 87–133
~ nucleosynthesis 99–175
~ pancakes 99–171b
~ radionuclide 82–256
principal mass 87–41
~ maximum 95–31
~ minimum 95–26
~ pulse 95–144
~ sequence 91–131
~ spectrum 95–115
~ stress 73–64
principle, cosmological 99–30
Einstein's equivalence ~ 99–150
equivalence ~ 99–150
Mach's ~ 99–37
mass–energy equivalence ~ 99–145
~ of equivalence of mass and energy 99–145
~ of invariance 93–61
perfect cosmological ~ 99–32
restricted cosmological ~ 99–31
prism, Anderson 88–68
beam splitting ~ 88–88
rotating ~ 88–68
probable crater 81–210
~ member of cluster 96–86
problem, cosmogonic(al) 94–5
process, ablation 86–183
e– ~ 94–99
endogenous ~ 82–258
equilibrium ~ 94–99
internal ~ 82–258
lunar ~ 82–150
metamorphic ~ 74–34
~ of ejection 98–179
p– ~ 94–91
r– ~ 94–89
radiation ~ 86–211
radiative ~ 86–211
rapid neutron-capture ~ 94–89

s– ~ 94–88
sedimentary ~ 74–46
slow neutron-capture ~ 94–88
tectonic ~ 82–201
Urca ~ 94–51
processor, soil 82–79
Procyon 92–123
production, energy 93–105
~ rate 85–143
profile, temperature 83–259
temperature ~ of atmosphere 83–259
vertical ~ 81–228
vertical ~ of density 76–53
prognosis, weather 77–13
progressive extinction 86–75
~ fading 86–75
project, proton–flare 89–177
projected density 98–85
projection, azimuthal 71–44
cartographic ~ 71–40
central ~ 71–48
conformal ~ 71–41
conic ~ 71–42
cylindrical ~ 71–43
equiangular ~ 71–41
gnomonic ~ 71–46
map ~ 71–40
stereographic ~ 71–47
zenithal ~ 71–45
prominence 89–223
~ activation 89–239
active ~ 89–238
active sunspot ~ 89–254
~ area 89–233
base of ~ 89–234
cap ~ 89–253
common active ~ 89–246
common eruptive ~ 89–250
coronal ~ 89–264
coronal active ~ 89–247
dissolution of ~ 89–280
equatorial ~ 89–228
eruptive ~ 89–248
~ film 89–242
flare–like ~ 89–222
~ frequency 89–237
funnel ~ 89–269
high–latitude ~ 89–229
hydrogen ~ 89–266
~ knot 89–244
knot of ~ 89–244
loop ~ 89–270
low–latitude ~ 89–227
metallic ~ 89–265
~ movie 89–242
polar ~ 89–230
quiescent ~ 89–263
rising ~ 89–278
secondary ~ 89–260
~ spectroscope 88–58
spot ~ 89–252
spray ~ 89–258
~ stream 89–245
sunspot ~ 89–252
~ thread 89–245
tornado ~ 89–262
~ unit 89–235
prominences, daily areas of ~ 89–236
interacting ~ 89–243
interactive ~ 89–243
motion pictures of ~ 89–242
poleward migration of ~ 89–231
poleward motion of ~ 89–231
Promontorium 81–293
promontory 81–293
propagation, direction of 78–35
anomalous ~ of sound 76–37
anomalous sound ~ 76–37
energy ~ 88–217
~ of light 99–129
~ of radio waves 78–34
~ of sound 76–32
radio–wave ~ 78–34
~ vector 78–36
proper motion 91–19
annual ~ motion 91–20
~ motion in declination 91–22
~ motion in right ascension 91–21
~ motion of sunspots 89–117
~ time 99–139
properties, integrated 96–134
intrinsic ~ 91–100
kinematic ~ 98–90
proposed landing site 82–15
Proterozoic 74–77
protoatmosphere 83a–44
protogalactic cloud 94–17
protogalaxy 94–18
proton aurora 79–108
~ capture 94–90
~ event 89–192
~ flare 89–185
~ whistler 78–61
proton–flare project 89–177
protonosphere 75–12, 76–75
protoplanet 83a–40
protoplanetary cloud 83a–24
protostar 94–45, 96–96
protostellar 94–46
Protosun 83a–39
Proxima Centauri 92–124
PS 79–149
PsA 92–80
Psc 92–79
pseudovariable 95–51
psychrometer 77–121
Ptolematic system 99–22
puff 89–277
pulsar 95–137
binary ~ 95–140a
Crab ~ 95–142
gamma–ray ~ 95–139
~ magnetosphere 95–147
optical ~ 95–140
X–ray ~ 95–138
pulsating arc 79–147
~ star 95–95
~ surface 79–149
pulsational instability 93–213
pulsations 79–158
adiabatic ~ 95–98
geomagnetic ~ 75–81
magnetic ~ 75–81
non–adiabatic ~ 95–99
non–radial ~ 95–97
period of ~ 95–100
radial ~ 95–96
pulse duration 86–280
echo ~ 86–290
isolated ~ 95–146
~ length 86–280
main ~ 95–144
principal ~ 95–144
~ repetition frequency 86–279
~ repetition rate 86–279
~ width 86–280
pulsed signals 86–278
pulverization 82–239, 87–28
~ of a meteor body 86–189
pumping, chemical 97–19
collisional ~ 97–20
infrared ~ 97–18
~ mechanism 97–16
optical ~ 97–17
photon ~ 97–17
radiative ~ 97–17
ultraviolet ~ 97–18a
Pup 92–81
Puppis 92–81
purely scattering atmosphere 93–16
purple light 79–54
push wave 73–46
pyranometer 88–43
pyrheliometer 88–39
silver–disc ~ 88–42
water–flow ~ 88–40
water–stir ~ 88–41
pyroclastic cone 73–120
pyroxene 74–21
Pyx 92–82
Pyxis 92–82

Q

Q wave 73–48
QSG 98–361
QSO 98–359
quiet ~ 98–362
radioquiet ~ 98–362
QSS 98–358
Quadrantids 86–247
quadrature 83–64
eastern ~ 83–65
western ~ 83–66
quantum field 99–170m
quark soup 99–169a

EN

quarter 81–26
first ~ 81–25
first or last ~ 83–132
last ~ 81–30
third ~ 81–30
quarter–wave plate 88–126
quasag 98–361
quasar 98–360
radio quiet ~ 98–362
quasi–eruptive prominence 89–249
quasi–simultaneous observations 71–36
quasi–static contraction 94–40
quasi–stellar galaxy 98–361
~ object 98–359
~ radio source 98–360
~ source 98–358
Quaternary 74–88
quiescent prominence 89–263
activation of ~ prominence 89–239
~ solar wind 84–41
~ volcano 73–116
quiet day 75–92
magnetically ~ day 75–92
~ photospheric region 89–56
~ QSO 98–362
radio ~ object 98–362
radio ~ QSO 98–362
radio ~ quasar 98–362
~ solar wind 84–41
~ Sun 89–11
~ Sun Year 89–10

R

R 79–155
R Coronae Borealis variable 95–83
RA 79–153
radar aurora 79–95
~ echo 86–295
~ echo observations 86–265
~ range 86–288
~ response 86–295
radial perturbations 93–215
~ pulsations 95–96
~ velocity 91–24
radial–velocity curve 95–20
radiant 86–53
apparent ~ 86–56
corrected ~ 86–58
~ density 86–63
~ displacement 86–242
displacement of ~ 86–242
geocentric ~ 86–58
individual ~ 86–55
meteor ~ 86–53
~ of meteor stream 86–237
~ position 86–60
secondary ~ 86–239
shower ~ 86–237
stationary ~ 86–241
true ~ 86–59
radiation age 87–139
area of ~ 86–238
artificial ~ belt 75–108
artificial ~ zone 75–108
~ balance 88–252
~ belt 75–101
~ cooling 94–98
corpuscular ~ 88–150
~ coupling 99–173
decoupling of ~ 99–177
~ density 99–85
diffuse ~ field 93–31
diffuse sky ~ 77–59
dilute ~ 93–38
diluted ~ 93–38
direct solar ~ 77–50
Earth ~ 77–56
~ era 99–172
~ field 93–30
~ flux 93–113
free–bound ~ 86–212
geomagnetically trapped ~ 75–112
global ~ 77–55
gravitational ~ 99–165
inner Van Allen ~ belt 75–106
integrated ~ 88–147
man–made ~ belt 75–108
meteor ~ 86–207
neutrino ~ 94–94
~ of the sky 77–54
particle ~ 88–150
planet's own ~ 83–198
~ pressure 93–91
~ process 86–211
recombination ~ 86–212
solar ~ 88–134
total ~ 88–146
~ track 87–120
transfer of ~ 93–29
trapped ~ region 75–101
trapped ~ zone 75–101
radiative cooling 94–98
~ core 93–166
~ envelope 93–135
equation of ~ transfer 93–33
~ equilibrium 93–48
~ layer 93–134
~ pumping 97–17
~ transfer 93–29
~ zone 93–134
radical 85–136
~ coma 85–66
free ~ 85–137
radio activity 89–6
activity in ~ waves 89–6
~ aurora 79–94
~ chart 81–135a
~ cloud 98–286
~ corona 88–239
decametric ~ source 83–186
~ echo 86–295
ejection of ~ clouds 98–285
~ feature 98–235
~ flare 89–180
~ galaxy 98–206
~ interferometer 88–97
interplanetary ~ scintillation 84–37
interstellar ~ line 97–98
interstellar ~ scintillation 97–109b
Jovian ~ burst 83–182
~ lobe 98–286
~ magnitude 86–268
~ map 81–135a
~ meteor 86–267
~ QSO 98–360
quasistellar ~ source 98–360
~ quiet object 98–362
~ quiet QSO 98–362
~ quiet quasar 98–362
~ scintillation 84–36
solar ~ spectroscope 88–124
~ spectrograph 88–124
~ spectrometer 88–124
~ star 91–158
~ Sun 88–3
radioactive age determination 83a–73
~ carbon 74–95
~ dating 87–131
~ decay 74–64
~ method 74–92
radioactivity 74–61
dating by ~ 87–131
radiocarbon 74–95
radio–frequency albedo 83–94
radiogenic heat 83a–62
~ lead 74–97
~ noble gas 87–132
radioheliogram 88–96
radio–link interferometer 88–116
radiometric method 74–92
radionuclide, bombardment–produced 82–255
cosmogenic ~ 82–255
primordial ~ 82–256
radio–wave propagation 78–34
radius, apparent 81–70
equatorial ~ 83–115
exterior ~ 83–228
gravitational ~ 94–113, 99–163
inner ~ 83–226
interior ~ 83–226
mean ~ 83–116a
Moon's apparent ~ 81–70
~ of the universe 99–5
outer ~ 83–228
polar ~ 83–116
Schwarzschild('s) ~ 94–113, 99–163
Sun's apparent ~ 81–69
rain 77–169
coronal ~ 89–271
orographic ~ 77–166
rainbow 79–65
rain–gauge 77–170
totalizer ~ 77–171
raised rim 81–222

raised–rim crater 81–223
Ram 92–20
ramification 98–5a
random motion 89–157
~ walk 89–157
randomly distributed 98–93
range 86–288
maximum ~ 86–289
mountain ~ 73–85
radar ~ 86–288
tidal ~ 82–292
rapid neutron absorption 94–89
~ neutron–capture process 94–89
~ process 94–89
~ rotator 91–43
~ rotor 91–43
rapidly rotating star 91–39
rarefaction wave 97–58
Ras Alhague 92–125
rate, cooling 93–203
cratering ~ 82–233
formation ~ 85–143
gas–production ~ 85–144
hourly ~ 86–135
hourly ~ of meteors 86–135
mass–emission ~ 85–146
melting ~ 83–147
~ of decline 95–36
~ of decline from maximum 95–36
~ of relaxation 98–129
~ of rise to maximum 95–29
~ of shooting stars 86–135
~ of star deaths 94–115
~ of star formation 94–12
~ of stellar extinction 94–115
production ~ 85–143
pulse repetition ~ 86–279
recurrence ~ 86–279
rotation ~ 83–32
spin ~ 83–32
sporadic ~ 86–233
thermonuclear reaction ~ 94–53
ratio, abundance 91–175
equilibrium ~ 94–104
gas–to–dust ~ 97–120
hydrogen–metal ~ 91–176
illuminance ~ 91–60
intensity ~ 91–60
isotope ~ 74–94
mass ~ 96–67
mass–luminosity ~ 98–233
mass–to light ~ 98–233
metal–to–hydrogen ~ 91–176
Poisson('s) ~ 73–60
ray 85–55, 89–287
~ cone 95–156
coronal ~ 89–287
~ crater 81–214
fan ~ 89–296
seismic ~ 73–14
sound ~ 76–34
~ surface 81–171
~ system 81–170
tangential ~ 81–169
rayed arc 79–153
~ band 79–150
~ crater 81–214
rays 79–155
bright ~ 81–168
crater with ~ 81–214
lunar ~ 81–168
polar ~ 89–294
ray–structure, arc with 79–153
band with ~ 79–150
forms with ~ 79–152
forms without ~ 79–142
RB 79–150
reaction, neutrino 94–92
neutron–capture ~ 94–87
photochemical ~ 76–63
spallation ~ 94–100
surface nuclear ~ 91–181
real deceleration 86–101
~ density 96–133
~ flattening 98–296
~ height 78–29
reappearance 81–99
rear side of the Moon 81–16
receiver 78–17
receiving aerial 78–18
~ antenna 78–18
recession of galaxies 98–298
~ of nebulae 98–298
~ velocity 91–26, 99–75
recombination, dissociative 86–214
~ radiation 86–212
reconnaissance 82–31
orbital ~ 82–31
surface ~ 82–32
recorder, ionospheric 78–22
sunshine ~ 77–61
recording anemometer 77–114
recovery phase 75–77
recrystallization 74–42
rectification 95–18
~ of light–curves 95–18
rectilinear jet 85–57
recurrence 88–132
~ frequency 86–279
~ rate 86–279
recurrent activity 98–267
~ ejection 98–282
~ group 89–108
~ nova 95–116
red coronal region 89–285
~ dwarf 91–136
~ giant 91–152a
Great ~ Spot 83–193
~ line 79–15, 89–293
semiregular ~ variable 95–106
~ star 91–97
Red Spot 83–193
reddened star 97–94
reddening curve 97–93
interstellar ~ 97–91
~ law 97–92
law of interstellar ~ 97–92
~ line 96–149
space ~ 97–91
rediscovery 85–17
redshift = red shift 98–297
anomalous ~ 98–367
discrepant ~ 98–367
gravitational ~ 99–162
multiple ~ 98–364
non–Doppler ~ 98–363
non–velocity ~ 98–363
relation of ~ and apparent magnitude 98–301
redshift–apparent magnitude relation 98–301
redshift–distance relation 99–76
redshifted line 98–366
redshift–magnitude diagram 98–301
reduced area 89–93
~ brightness 85–108
reducing atmosphere 83a–46
reduction, Faye 71–78
free–air ~ 71–78
~ to sea–level 77–45
reference ellipsoid 71–18
~ frame 98–136
geodetic ~ system 71–19
local frame of ~ 98–139
system of ~ 98–136
reflecting cone 88–62
~ nebula 97–128
reflection effect 96–26
~ nebula 97–128
~ point 75–124
reflux 82–282
refractor, domeless solar 88–53
refractory elements 97–64
~ material 83a–55
~ solid 85–49
region, active 83–180, 89–57
active ~ on Jupiter 83–180
active prominence ~ 89–240
active solar ~ 89–57
allowed ~ 79–134
anticenter ~ 98–15
aseismic ~ 73–29
auroral ~ 79–126
bipolar magnetic ~ 89–160a
bright ~ 83–160
central ~ of a galaxy 98–293
central ~ of the Galaxy 98–13
circumnuclear ~ 98–292
cold ~ 88–199
compact H II ~ 97–51
continental ~ 81–271
coronal active ~ 89–286
cosmic–ray modulation ~ 84–26
earthquake ~ 73–28
epicentral ~ 73–16
equatorial ~ 89–59

escape ~ 76–68
forbidden ~ 79–135
fringe ~ 76–68
green coronal ~ 89–284
H–I ~ 97–49
H–II ~ 97–50
hot ~ 88–200
ionized hydrogen ~ 97–50
interarm ~ 98–40
interspicular ~ 88–198
ionospheric ~ 78–4
light ~ 83–160
lowland ~ 81–272
M ~ 89–61
magnetic ~ 89–144
mare ~ 81–268
modulation ~ 84–26
mountain ~ 81–196
neutral hydrogen ~ 97–49
nuclear ~ 98–17
~ of activity 89–57
~ of galactic center 98–13
~ of visibility 87–31
polar ~ 81–17, 83–192
post–mare ~ 81–270
pre–mare ~ 81–269
quiet photospheric ~ 89–56
red coronal ~ 89–285
seismic ~ 73–28
transition ~ 88–224
trapping ~ 75–110
undisturbed photospheric ~ 89–56
unipolar magnetic ~ 89–155a
regional faulting 82–205
regmaglypts 87–122
regolith 82–104
regular cluster of galaxies 98–336
~ galaxy 98–236
~ nebula 98–236
regularized geoid 71–17
Regulus 92–126
relation, diameter–depth 81–248
diameter–depth ~ of craters 81–248
magnitude–intensity ~ 91–58
mass–luminosity ~ 91–129
mass–radius ~ 93–200
~ of redshift and apparent magnitude 98–301
period–luminosity ~ 95–103
redshift–apparent magnitude ~ 98–301
redshift–distance ~ 99–76
solar–terrestrial ~ 84–55
spectrum–luminosity ~ 91–130
velocity–distance ~ 98–300
relations, solar–terrestrial 84–55
structural ~ 82–266
Sun–Earth ~ 84–55
relationships, solar–terrestrial 84–55
relative abundances of elements 91–174
~ age 82–268
~ gravity measurement 71–69
~ humidity 77–128
~ measurement of gravity 71–69
~ motion 99–99
~ orbit 96–54
~ position of planets 83–50
~ stream activity 86–231
~ sunspot number 89–65
~ trigonometric parallax 91–10
Zürich ~ sunspot number 89–65
relativistic 99–95
~ advance of perihelion 99–160
~ contraction 99–134
~ correction 99–140
~ cosmology 99–28
~ deflection of light 99–161
~ degeneracy 93–176
~ electrons 97–153
~ gas 97–152
~ mass 99–141
~ mass increase 99–142
~ motion of perihelion 99–160
~ star cluster 96–154
~ stellar model 93–198
~ universe 99–92
~ velocity 99–136
relativity displacement of spectral lines 99–162
general ~ 99–146
general theory of ~ 99–146
~ of inertia 99–35
relaxation 98–127
collisional ~ 98–131
collisionless ~ 98–132
mean time of ~ 98–130
rate of ~ 98–129
~ time 98–128
time of ~ 98–128
~ velocity 98–129
release, energy 93–109
relic magnetic field 95–62
relics of a supernova 95–133
remains of a supernova 95–133
organic ~ in meteorites 87–67
remanent magnetism 75–28
~ magnetization 75–28
remnant of a supernova explosion 95–133
supernova ~ 95–133
repetition rate, pulse 86–279
report, weather 77–10
repulsive force 85–87
research buoy 72–99
lunar ~ 82–3
residual magnetism 75–28
distribution of ~ velocities 98–92
~ magnetization 75–28
~ velocity 98–91
resistance, aerodynamic 86–178
air ~ 86–178
~ equation 86–179
resolution, spatial 88–204
time ~ 88–205
response, radar 86–295
rest energy 99–144
~ mass 99–143
restricted cosmological principle 99–31
~ theory of relativity 99–94
Ret 92–83
retardation of Earth's rotation 82–313
~ of expansion 99–82
retention age 87–135
gas ~ 87–134
gas ~ age 87–135
reticle 86–129
Reticulum 92–83
retrograde 83–46
~ motion 83–47
~ rotation 83–40
~ satellite 83–21
retroreflector 82–43
laser ~ 82–43
return of a comet 85–9
~ of a spot group 88–133
returned lunar sample 82–70
reversal of magnetic polarity 89–153
reverse fault 73–94
~ side 81–16
~ side of the Moon 81–16
reversed fault 73–94
~ magnetization 75–37
reversing layer 88–174
revolution, draconic 81–47
sidereal ~ 81–45
synodic(al) ~ 81–46
tropical ~ 81–44
revolving planisphere 92–10
~ star chart 92–10
~ star map 92–10
ribbon, flare 89–187
ribbon–like structure 79–140
~ structure aurora 79–140
rich cluster of galaxies 98–332
ridge 81–186
mare ~ 81–188
mid–ocean ~ 72–45
mountain ~ 81–186
~ of high pressure 77–87
wrinkle ~ 81–187
Riemann–Christoffel tensor 99–159
Riemannian curvature tensor 99–159
~ geometry 99–113
~ space 99–112
Rigel 92–127
right ascension, Moon's 81–60
Sun's ~ ascension 81–61
rigid body 83–111
rigidity 73–56
magnetic ~ 75–117
rille = rill 81–165
linear ~ 81–166
sinuous ~ 81–167
rim, blocky 81–224
bright ~ 89–226
bright ~ of a filament 89–226

crater ~ 81–221
~ of a galaxy 98–294
rima 81–165
linear ~ 81–166
sinuous ~ 81–167
rimless crater 81–226
ring... 97–146
~ A 83–225
~ asteroid 84–71
asteroid ~ 84–70
~ B 83–222
Bishop's ~ 79–88
bright ~ 89–101
~ C 83–221
crape ~ 83–221
crêpe ~ 83–221
~ currents 75–100
~ disappearance 83–217
dusky ~ 83–221
edgewise ~ 83–218
elementary ~ 81–209
equatorial ~ current 75–98
~ form 81–155
~ galaxy 98–259
gaseous ~ 95–66
gauze ~ 83–221
ghost ~ 81–242
lunar ~ 81–265
main ~ 83–222
middle ~ 83–222
~ nebula 97–147
nebular ~ 83a–10
~ of asteroids 84–70
~ of stars 96–78
outer ~ 83–225
~ pattern 98–41
~ plane 83–216
~ reappearance 83–219
Saturn's ~ 83–215
Saturn's ~ system 83–220
~ seen edge–on 83–218
star ~ 96–78
~ structure 81–155, 98–41
ring–like structure 81–155
rings, Saturn's 83–215
system of ~ 98–291
thickness of ~ 83–229
riometer 78–45
rise 76–54
~ in temperature 77–42
rate of ~ to maximum 95–29
~ to maximum 95–28
rising branch 95–34
~ current 93–151
~ phase 89–83
~ prominence 89–278
river bed 83–153
River Eridanus 92–49
Robertson–Walker metric 99–152a
Roche lobe 96–28
rock 74–1
acid ~ 74–4
angular ~ 82–131
basic ~ 74–5
block of ~ 82–130
~ constituent 82–99
crystalline ~ 82–66
~ debris 81–235
effusive ~ 72–3
equigranular ~ 82–67
extrusive ~ 74–3
~ field 82–139
foliated ~ 74–54
~ fragment 82–68
igneous ~ 74–2
intrusive ~ 74–7
lunar ~ 82–65
magmatic ~ 74–2
~ magnetism 75–43, 83a–80
metamorphic ~ 74–11
plutonic ~ 74–7
schistous ~ 74–54
sedimentary ~ 74–10
stratified ~ 74–48
ultrabasic ~ 74–6
volcanic ~ 74–3
rocket sounding 76–48
rock–forming mineral 82–99
rocking mirror 86–130
rocky satellite 83–22
rolling stones 81–237
room temperature 77–31
rope, magnetic 89–146
rose, wind 77–116
Rossiter effect 91–44
rotating body 91–42
~ frame of reference 98–138
~ magnetoplasmati(ati)c body 98–25
~ mirror 86–130
~ prism 88–68
~ sector 86–147
~ shutter 86–147
~ star 91–37
rotation, asynchronous 83–38
~ axis 83–33
axis of ~ 83–33
bound ~ 81–14
braking of the rotation 91–45
~ curve 98–109
differential ~ 88–128
differential galactic ~ 98–113
direct ~ 83–39
galactic ~ 98–107
inclination of ~ axis 91–50
Keplerian ~ 98–112
Kepler's ~ 98–112
~ law 98–110
non–synchronous ~ 83–38
non–uniform ~ 91–36
~ of line of apsides 96–71
~ of planets 83–30
~ of stars 91–35
~ of the Sun 88–127
orientation of ~ axis 84–102
period of ~ 83–31
~ period 83–31
planetary ~ 83–30
pole of ~ axis 83–36
~ rate 83–32
retardation of Earth's ~ 82–313
retrograde ~ 83–40
solar ~ 88–127
solid ~ 98–111
solid–body ~ 98–111
~ speed 83–32
stellar ~ 91–35
synchronous ~ 81–14, 83–37
~ temperature 91–119
~ velocity 91–48
velocity of ~ 83–32
rotational break–up 85–119
~ distortion 91–44
~ distortion of velocity curve 91–44
~ effect 91–44
~ effect on velocity curve 91–44
~ energy 93–120
~ instability 83a–8, 93–216
~ moment 83a–5
~ speed 91–48
~ variable 95–65
~ velocity 83–32, 91–48
rotator 91–42
oblique ~ 91–52, 95–60
rapid ~ 91–43
rotor 91–42
rapid ~ 91–43
rough surface 83–102
roughness factor 82–107
surface ~ 82–106, 83–99
route, traverse 82–37
royal zone 89–54
RR Lyrae star 95–93
R(t) 99–81a
rubidium–strontium age 82–271
runaway star 98–135
rupes 81–184
rupture 73–78
~ of galactic nucleus 98–318
Russell mixture 93–101
Russell–Vogt theorem 93–83

S

S Cross 92–43
S Crown 92–39
S Fish 92–80
S Triangle 92–94
S wave 73–43
Sagitta 92–84
Sagittarius 92–85
~ arm 98–35
Sails 92–98
salinity 72–66
~ interface 72–67
salt content 72–66
sample, air 76–86

~ collection 82–72
~ container 82–81
core ~ 82–60
lunar ~ 82–69
lunar dust ~ 82–121
~ of lunar material 82–69
~ return 82–71
sample–collection tool 82–80
sampler 82–80
surface ~ 82–80
water ~ 72–101
sample–return mission 84–98
sampling, air 76–85
soil ~ 82–78
satellite 83–17
~ altimetry 71–50
~ climatology 77–181
~ crater 81–200
former ~ 83–240
Galilean ~ 83–204
~ geodesy 71–32
geodetic ~ network 71–34
icy ~ 83–23
innermost ~ 83–20
irregular ~ 83–21
~ monitoring 83–26
natural ~ 83–18
~ oceanography 72–70
retrograde ~ 83–21
rocky ~ 83–22
~ search 83–27
~ station 71–33
~ survey 83–26
~ system 83–19
~ transit 83–54
~ triangulation 71–27
saturation 76–101
~ function 93–60
Saturn 83–209
Saturnian apoapse 83–211
~ apoapsis 83–211
~ apocentre 83–211
~ periapse 83–210
~ periapsis 83–210
~ pericentre 83–210
saturnocentric 83–213
saturnographic 83–212
Saturn's rings 83–215
scale, Beaufort 77–108
Beaufort ~ of wind force 77–108
Celsius ~ 77–33
Celsius temperature ~ 77–33
centigrade ~ 77–33
cosmic ~ factor 99–81a
distance ~ 98–303
~ factor 99–81a
geochronological ~ 74–68
~ height 76–18, 88–236
map ~ 71–49
small ~ structure 88–208
Scales 92–62
scarp 81–184
scattered starlight 79–12
scattering, atmospheric 77–57
function of ~ 77–58
indicatrix of ~ 77–58
~ layer 83–258
Mie ~ 83–76
~ of light 77–57
~ pattern 77–58
purely ~ atmosphere 93–16
schistous rock 74–54
Schwarzschild inequality 93–156
~ radius 99–163
science, environmental 76–2
sciences, atmospheric 76–1
Earth ~ 71–1
lunar ~ 82–2
scientific traverse 82–36
scintillation, interplanetary 84–37
interplanetary radio ~ 84–37
interstellar ~ 97–109a
interstellar radio ~ 97–109b
radio ~ 84–36
Scl 92–87
SCNA 78–42
Sco 92–86
Scorpio 92–86
Scorpion 92–86
Scorpius 92–86
scree 81–234
screen 88–13
screened shadow height 79–49
screening height 79–48
Sct 92–88
Sculptor 92–87
Sculptor's Tool 92–23
Scutum 92–88
SDF 89–279
sea 81–266, 83–156
Border ~ 81–285
~ chart 72–33
circular ~ 81–267
continental ~ 72–26
~ current 72–60
~ discoloration 72–80
Eastern ~ 81–284
~ floor 72–38
Humboldt's ~ 81–288
inland ~ 72–26
~ interferometer 88–106
Knowledge ~ 81–289
Known ~ 81–289
marginal ~ 72–27
mediterranean ~ 72–26
Moscow ~ 81–283
motion of the ~ 72–86
~ of Clouds 81–281
~ of Cold 81–278
~ of Conflicts 81–280
~ of Crises 81–280
~ of Fertility 81–274
~ of Humours 81–283
~ of Ingenuity 81–292
~ of Moisture 81–282
~ of Moscow 81–283
~ of Nectar 81–279
~ of Peace 81–275
~ of Rains 81–273
~ of Serenity 81–276
~ of Showers 81–273
~ of Tranquillity 81–275
~ of Waves 81–290
Serpent ~ 81–291
Smyth's ~ 81–287
Southern ~ 81–286
~ state 72–86
~ surge 72–76
~ tides 82–309
SEA 78–51
seabed 72–38
sea–breeze 77–97
Sea–Goat 92–29
sea–level 72–87
~ pressure 77–44
reduction to ~ 77–45
Sea–Monster 92–34
seamount 72–43
flat-topped ~ 72–44
seaquake 73–2c
~ wave 72–95
search for comets 85–15
~ for meteorites 87–45
planet ~ 83–238
satellite ~ 83–27
supernova ~ 95–136
Sea–Serpent 92–56
seasonal changes 83–166
~ variations 86–139
second type surface 87–108
secondary 96–38
~ arc 73–83
~ component 96–38
~ crater 81–199
~ feature 81–151
~ fluctuations 95–44
~ magnetization 75–36
~ maximum 95–32
~ minimum 95–27
~ period 95–41
~ prominence 89–260
~ radiant 86–239
~ surface 87–108
threshold of ~ 96–40
~ wave 73–43
section, velocity–depth 73–52
sector 85–60
~ boundary 84–18
interplanetary ~ 84–16
interplanetary magnetic ~ 84–16
rotating ~ 86–147
~ structure 84–15
secular decrease of brightness 85–127
~ instability 93–212
~ geomagnetic variations 75–85
~ magnetic variations 75–85

~ parallax 91–9
~ variations of the Earth's magnetic field 75–85
sediment 74–47
deep-ocean ~ 72–50
deep-sea ~ 72–50
sedimentary process 74–46
~ rock 74–10
sedimentation 73–109, 74–45
submarine ~ 72–47
sediments, deep-sea 72–50
organogenous ~ 72–59
pelagic ~ 72–57
terrigenous ~ 72–48
seeker, Sun 88–51
Seeliger's paradox 99–39
Seething Bay 81–296
seism, meteoritic 87–34
seismic activity 73–4
~ background 73–55
~ echo 82–223
~ event 82–215
~ experiment 82–219
~ focus 73–17
~ noise 73–54
~ ray 73–14
~ region 73–28
~ sounding 82–216
~ station 73–6
~ wave 73–9
seismicity 73–3
seismogram 73–8
impact ~ 82–224
seismograph 73–7
seismographic record 73–8
~ station 73–6
seismologic(al) observatory 73–5
~ station 73–6
seismologist 73–2
seismology 73–1
active ~ 82–217
lunar ~ 82–213
~ of the Moon 82–213
passive ~ 82–218
solar ~ 88–223b
seismometer, active 82–222
passive ~ 82–221
selected areas 98–78
selection effect 98–81
selective extinction 97–83
~ interstellar absorption 97–90
selenocentric 81–131
~ position 81–132
selenochemistry 82–74
selenodesy 81–109
selenodetic 81–110
selenographic 81–8
~ coordinates 81–9
~ latitude 81–11
~ longitude 81–10
selenography 81–1
selenoid 81–117
selenologic(al) 82–48
~ mapping 82–51
~ survey 82–50
selenologist 82–49
selenology 82–1
selenomagnetic field 82–45
selenophysical monitoring network 82–40
~ survey 82–39
selenophysics 82–151
self-gravitating 94–33
self-gravitation 94–34
~ system 98–169
self-luminous 96–45
~ train 86–108
self-reversal, magnetic 75–38
~ of magnetization 75–38
semi-convective zone 93–160
semi-detached binary 96–34
semidiameter, equatorial 83–115
Moon's apparent ~ 81–70
Sun's apparent ~ 81–69
semi-diurnal wave 76–29
semi-infinite atmosphere 93–22
semi-regular red variable 95–106
~ variable 95–105
sense of rotation 98–108
separate-nuclei theory 96–76
separation, angular ~ 96–52
component ~ 96–66
~ from the Sun 83a–9
~ of chemical elements 83a–48
~ of components 96–66
sequence, evolutionary 94–30
Hubble ~ of galaxies 98–227
main ~ 91–131
North Polar ~ 91–87
~ of regular galaxies 98–240
~ of spirals 98–255
~ photography 79–103
Polar ~ 91–87
principal ~ 91–131
spectral ~ 91–94
zero-age ~ 96–152
zero-age main ~ 94–60
Ser 92–89
Serpens 92–89
~ Caput 92–89
~ Cauda 92–89
Serpent 92–89
~ Bearer 92–72
Sea ~ 92–56
Serpent('s) Head 92–89
~ Tail 92–89
Serpentarius 92–72
service, meteorological 77–5
~ module 82–7
weather ~ 77–5
Seven Sisters 96–110
Sex 92–90
Sextans 92–90
Sextant 92–90
Seyfert galaxy 98–262
Sge 92–84
Sgr 92–85
shade temperature 77–34
shadow bands 88–31
~ cone 81–90
~ density 81–59
density of ~ 81–59
Earth's ~ 79–44
~ height 79–46
~ line 83–69
ozone ~ 76–127
~ transit 83–53
shallow earthquake 73–21
shape, equilibrium 83–112
~ factor 86–177
~ of a comet 85–83
~ of meteor body 86–175
~ of the Earth 71–7
~ of the Moon 81–116
~ of the planet 83–109
sharp maximum 86–235
shear stress 73–63
~ wave 73–43
sheet, lava 82–175
neutral ~ 75–16
plasma ~ 75–11
subpolar ~ 79–51
shelf, continental 72–34
shell, circumstellar 93–10
circumstellar dust ~ 97–117
dust ~ 97–116
expanding ~ 93–12
helium ~ 93–145
hydrogen ~ 93–144
~ star 93–5
stationary ~ 93–11
shells of a comet 85–69
shell-source model 93–143
shield 72–11
~ volcano 73–119
Shield 92–88
shielding 86–171
magnetic ~ 89–225
shift, Doppler 91–25
periodic line ~ 96–22
red ~ 98–297
shifter, phase 88–115
Ship's Compass 92–82
~ Sails 92–98
~ Stern 92–81
shoal 72–96
shock metamorphism 82–237
thermal ~ 86–206
shock-absorber 82–24
shock-wave, collisionless 84–49
shooting stars, rate of 86–135
shore 72–30
shoreline 72–31
~ altering 72–78
~ modification 72–78
short time-scale 94–66

~ whistler 78–58
short–duration meteor echo 86–300
~ train 86–112
short-enduring meteor train 86–112
short–lived = short–life 89–125
~ isotope 83a–74
~ phenomenon 89–15
short–period comet 85–5
~ variable 95–89
short–range forecast 77–20
shower, age of meteor 86–246
comet associated with ~ 86–259
~ meteor 86–24
meteor ~ 86–215
meteorite ~ 87–37
~ radiant 86–237
shrinkage, gravitational 83–203
shutter break 86–150
moving–blade ~ 86–148
rotating ~ 86–147
two–blade ~ 86–149
SID 78–41
side, dark 75–21, 83–68
illuminated ~ 81–34
Moon's far ~ 81–16
night ~ 75–21
rear ~ of the Moon 81–16
sunlit ~ 81–34
sunward ~ 81–34
~ vision 86–128
sidereal month 81–45
siderite 87–82
siderolite 87–78
siderolith 87–78
siderosphere 72–17
siderostat 88–44
sight, line of 91–28
σ–component 88–89
signal, echo 86–290
pulsed ~ 86–278
signs, zodiacal 92–6
silent area 78–38
silica 74–13
silicate 74–14
~ dust 85–53
~ grain 97–76
silicon burning 94–83
~ dioxide 74–13
~ monoxide 97–37
~ star 91–190
sill 73–135
silting, off–shore 72–79
Silurian 74–80
silver–disk pyrheliometer 88–42
simulation, mathematical 98–323
simultaneity 99–127
simultaneous events 99–128
~ observations 71–35, 86–125
single fall 87–36
~ spectrum 96–16
~ spot 89–116
~ star 96–2
single–line binary 96–15
~ spectroscopic binary 96–15
singular galaxy 98–258
~ point 94–116
~ state 99–42
singularity 94–116
naked ~ 94–117
sinuous depression 81–161
~ groove 81–167
~ rille 81–167
~ rima 81–167
sinus 81–294
~ Aestuum 81–296
~ Iridium 81–295
Sirius 92–128
~ companion 91–138
site, landing 82–14
proposed landing ~ 82–15
situation, weather 77–14
six–colo(u)r photometry 96–148
six–dimensional phase space 98–144
size distribution 82–124
~ frequency 81–249
grain ~ 82–113
~ of particles 76–114
skip distance 78–38
sky, blueness of the 77–60
bright night ~ 86–28
clear ~ 77–153
cloudy ~ 77–148
diffuse ~ radiation 77–59
~ light 77–53
light of the night ~ 79–8
night ~ 79–6
northern ~ 79–3b
overcast ~ 77–152
plane of the ~ 96–50
radiation of the ~ 77–54
southern ~ 79–3c
starry ~ 91–1a
state of the ~ 77–132
~ survey 92–11
twilight ~ 79–43
~ wave 78–13
skylight 77–53
slide 82–196
sticky ~ 87–14
~ track 82–197
sliding 82–196
slight earthquake 73–35
slipping, downward 82–198
slit, curved 88–65
oscillating ~ 88–70
slope 81–230
continental ~ 72–35
detrital ~ 81–233
moderate ~ 81–231
steep ~ 81–232
slow meteor 86–88
~ neutron absorption 94–88
~ neutron–capture process 94–88
~ nova 95–120
~ process 94–88
slump 82–199
slumping 82–198
~ of walls 82–198
Small Bear 92–97
~ Dog 92–28
~ Fox 92–101
~ Horse 92–48
~ Lion 92–60
~ Magellanic Cloud 98–214
small–scale field 89–143
~ irregularity 84–13
~ motion 88–188
SMC 98–214
smearing (out) of Fraunhofer lines 88–243
smog 76–83
smoke 97–57
~ train 87–21
smoked glass 88–33
SN event 95–131
snow 77–175
coarse–grained ~ 72–118
~ cover 77–176
~ limit 72–111
snowfield, permanent 72–119
snowline 72–111
SNR 95–133
soil characteristic 82–108
cohesive ~ 82–63
~ creep 82–152
lunar ~ 82–62
~ penetrometer 82–83
~ processor 82–79
~ sampling 82–78
~ strength 82–135
solar activity 89–1
~ antapex 98–65
~ apex 91–23, 98–64
apex of the ~ motion 98–64
ascending phase of ~ cycle 89–83
~ atmosphere 88–163
~ aureole 79–62
automatic ~ telescope 88–50
balloon–borne ~ telescope 88–52
burst of ~ plasma 84–38
burst of ~ wind 84–42
center of ~ disc 89–35
centre of ~ disc 89–35
~ constant 88–148
~ corona 79–62, 88–225
~ cosmic–ray event 89–194
~ cross 79–81
~ cycle 89–71
~ data 89–9
~ depression 79–45
descending phase of ~ cycle 89–84
~ disc 89–24
~ disk 89–24
domeless ~ refractor 88–53
~ eclipse 88–15
~ feature 89–13

~ flare 89–175
forecast of ~ activity 89–64
~ index 89–8
index of ~ activity 89–8
~ instruments 88–35
~ interplanetary loop 84–20
~ interplanetary tongue 84–20
~ limb 89–28
lower ~ atmosphere 88–166
lower layers of ~ atmosphere 88–166
~ magnetic fields 89–129
magnetized ~ plasma 84–32
~ magnetograph 88–82
~ mass 91–122
~ maximum 89–2
maximum of ~ activity 89–2
millionth of ~ disc 89–69
~ minimum 89–4
minimum of ~ activity 89–4
~ motion 98–62
~ nebula 83a–23
~ neighbo(u)rhood 98–58
~ observatory 88–7
~ oscillations 88–223a
outer ~ atmosphere 88–165
partial ~ eclipse 88–16
period of ~ rotation 88–129
permanent ~ wind 84–40
~ pertubations 83–24
~ perturbations on satellites 83–24
~ phenomena 89–13
~ planet 83–11
planet of ~ system 83–11
~ plasma 84–29
prediction of ~ activity 89–64
~ proton event 89–192
quiescent ~ wind 84–41
quiet ~ wind 84–41
~ radiation 88–134
~ rotation 88–127
~ seismology 88–223b
~ spectrograph 88–77
standard ~ motion 98–63
~ surface 88–164
~ survey 89–16
~ system 83–12
~ telescope 88–49
terrestrial effects of ~ activity 84–55
total ~ eclipse 88–17
~ tower 88–54
~ units 91–123
~ wind 88–156
~ X–ray event 89–184
solar–eclipse expedition 88–30
solar–terrestrial physics 84–56
~ relations 84–55
~ relationships 88–55
solar–type star 91–104
solar–wind burst 84–42
~ cavity 84–28
~ density 88–157
~ spectrometer 82–46
~ sputtering 82–245
~ stream 84–35a
~ velocity 88–158
solid body 83–144
~ carbon dioxide 83–144
~ crust 95–151
~ nucleus 85–39
~ particle 97–69
refractory ~ 85–49
~ rotation 98–111
solid–body rotation 98–111
solidification 83a–59
~ age 87–137
solution, stable 98–170
uniqueness of the ~ 93–82
sonic barrier 76–44
sound, anomalous 76–37
anomalous ~ propagation 76–37
anomalous propagation of ~ 76–37
~ barrier 76–44
~ energy 88–219
~ energy flux 88–221
~ field 76–40
generation of ~ 88–215
~ intensity 76–38
~ level 76–38
~ noise 88–220
normal ~ 76–35
~ probe 76–45
propagation of ~ 76–32
~ ray 76–34
~ wave 76–33
sound–emitting fireball 86–119
sounding 76–47
acoustic ~ 76–45
echo ~ 72–29
echo depth ~ 72–29
ionospheric ~ 78–19
oceanographic ~ 87–16
rocket ~ 76–48
seismic ~ 82–216
topside ~ 78–20
twilight ~ 79–41
source, compact 98–273
decametric radio ~ 83–186
double energy ~ 93–146
electron ~ 79–119
energy ~ 93–110
~ function 93–34
heat ~ 83–199
neutron ~ 94–85
~ of energy 93–110
~ of heat 83–199
quasi–stellar ~ 98–358
quasi–stellar radio ~ 98–360
South Atlantic Anomaly 75–68
~ Tropical Disturbance 83–194
Southern Cross 92–43
~ Crown 92–39
~ Fish 92–80
~ limit of eclipse 88–19
~ limiting latitude 81–106
~ polarity 89–150
~ Triangle 92–94
SPA 78–52
space 99–100
absolute ~ 99–116
allowed ~ 79–134
~ coordinate 99–124
curvature of ~ 99–115
curved ~ 99–114
~ density 86–263, 96–133
~ distribution 96–119
elliptic ~ 99–110
~ erosion 87–142
Euclidean ~ 99–107
extragalactic ~ 97–5
extraterrestrial ~ 87–73a
finite ~ 99–103
flat ~ 99–107
forbidden ~ 79–135
free ~ 76–9
~ geodesy 71–31
hyperbolic ~ 99–106
infinite ~ 99–104
interarm ~ 98–40
interfilamentary ~ 89–99
intergalactic ~ 97–7
interplanetary ~ 84–2
interstellar ~ 97–4
Lobachevski ~ 99–105
metric ~ 99–100a
Minkowski ~ 99–122
Minkowskian ~ 99–122
Newcombian ~ 99–110
non–Euclidean ~ 99–114
~ oceanography 72–69
parabolic ~ 99–107
~ reddening 97–91
Riemann('s) ~ 99–112
Riemannian ~ 99–112
six–dimensional phase ~ 98–144
spherical ~ 99–111
stellar ~ density 98–83
~ velocity 91–30
velocity ~ 98–97
~ wave 78–13
~ weathering 82–244
space–time 99–119, 99–120
Minkowski ~ 99–122
~ point 99–125
spacing, aerial 88–99
~ of aerials 88–99
~ of antennae 88–99
zero ~ 88–101
spallation 94–100
cosmic–ray ~ 82–254
~ reaction 94–100
sparks 86–84
spatial acrotriangulation 71–26
~ coordinate 99–124
~ distribution 96–119
~ resolution 88–204
~ phototriangulation 71–26

~ triangulation 71–25
~ velocity 91–30
special colo(u)r–index 83–87
~ theory of relativity 99–94
specific humidity 77–123
specimen, individual 87–42
spectral class 91–107
~ classification 91–90
Harvard system of ~ classification 91–91
~ peculiarity 91–93
~ sequence 91–94
three–dimensional ~ classification 91–106
two–dimensional ~ classification 91–105
~ type 91–108
~ variable 95–48
~ variations 95–49
spectrobologram 88–76
spectrobolometer 88–75
spectro–enregistreur des vitesses 88–57
spectrogram, dynamic 88–125
spectrograph, dynamic 88–124
horizontal ~ 88–78
meteor ~ 86–146
radio ~ 88–124
~ recording velocities 88–57
solar ~ 88–77
swept-frequency ~ 88–124
vertical ~ 88–79
spectrographic orbit 96–56
spectrography, Mössbauer 82–142
spectroheliocinematograph 88–81
spectroheliogram 88–72
spectroheliograph 88–71
spectrohelioscope 88–67
spectrometer 88–124
radio ~ 88–124
solar–wind ~ 82–46
spectrometry, non–destructive γ–ray 82–77
spectrophotometric gradient 91–112
spectroscope, prominence 88–58
solar radio ~ 88–124
spectroscopic binary 96–14
~ orbit 96–56
~ parallax 91–15
spectrum binary 96–14
composite ~ 96–19
double ~ 96–18
dynamic ~ 88–125
flash ~ 88–25
Fraunhofer ~ 88–137
integrated ~ 96–137
normal stellar ~ 91–89
~ of meteor train 86–107
principal ~ 95–115
single ~ 96–16
solar ~ 88–74a
stellar ~ 91–88
symbiotic ~ 95–117
~ variable 95–48
spectrum-luminosity relation 91–130
speed, rotation 83–32
rotational ~ 83–32
wind ~ 77–109
sphere, celestial 92–8
gas ~ 93–73
polytropic gas(eous) ~ 93–127
spherical albedo 83–91
~ atmosphere 93–24
~ galaxy 98–244
~ particles 87–29
~ space 99–111
~ symmetry 98–160
spherics 78–50
spheroid, earth 71–13
oblate ~ 98–158
spheroidal galaxy 98–244
~ system 98–159
spherules 87–29
Spica 92–129
spicule 88–195
spill, oil 72–71
spin period 83–31
~ rate 83–32
synchronous ~ 83–37
spinar 98–25
spindle–shaped galaxy 98–239
~ meteor 86–77
spiral 98–7, 98–249
Archimedean ~ 84–34
Archimedes ~ 84–34
~ arm 98–27
barred ~ 98–253
density wave theory of ~ structure 98–174
~ galaxy 98–249
~ jet 85–59
~ nebula 98–249
non–barred ~ 98–250
normal ~ 98–250
~ structure 98–8
theory of ~ structure 98–173
~ tracer 98–39
spiral–arm population 98–73
~ tracer 98–39
spirals, sequence of 98–255
spirit, wood 97–32
splash–form tektite 87–96
splitting of a comet 85–111
~ of lines by expansion 97–157
tidal ~ 85–112
Spoerer's law 89–75
sporadic coronal condensation 89–283
~ E layer 78–25
Jupiter's ~ radio emissions 83–181
~ meteor 86–23
~ rate 86–233
~ stream 86–226
spot activity 89–62
~ area 89–67
birth of ~ 89–121
development of ~ group 89–118
eastern ~ 89–114
even ~ cycle 89–77
f ~ 89–114
following ~ 89–114
formation of ~ 89–121
~ group 89–105
hot ~ 95–112
leading ~ 89–113
lifetime of a ~ group 89–123
major ~ 89–115
migration of ~ zone 89–74
odd ~ cycle 89–78
p ~ 89–113
~ penumbra 89–97
preceding ~ 89–113
~ prominence 89–252
Red ~ 83–193
return of a ~ group 88–133
single ~ 89–116
star ~ 97–62
trailer ~ 89–114
~ umbra 89–95
western ~ 89–113
White ~ 83–214
~ zone 89–54
spotted area 89–90
spray 89–258
~ prominence 89–258
spreading, ocean–floor 72–39
Spring Lake 81–302
sputtering 85–140
solar–wind ~ 82–245
Square 92–70
~ Level 92–70
ssc 75–74
SSWF 78–40
stability 93–204
coefficient of ~ 98–199
~ condition 93–205
dynamical ~ 93–207, 98–198
~ of an orbit 98–193
~ of stars 93–206
stable 93–208
~ solution 98–170
~ star 95–8
thermally ~ 93–209
stage, contracting 94–20
disk ~ 94–22
evolutionary ~ 94–28
halo ~ 94–19
intermediate ~ 94–21
nebular ~ 95–125
post–main–sequence ~ 94–68
T–Tauri ~ 94–47
standard atmosphere 76–11
~ conditions 77–3
fundamental ~ of rest 98–137
~ model 93–137
~ solar motion 98–63
~ type storm 75–72
star 91–1, 92–2

anomalous ~ 91–96
~ atlas 92–12
background ~ 96–84
Beta Persei ~ 95–77
binary ~ 96–3, 96–7
bright ~ 91–83
brightest ~ 98–304
brightest ~ of a galaxy 98–304
~ burst 94–11a
~ capture 96–73
carbon ~ 91–188
carbon–deficient ~ 91–194
carbon–poor ~ 91–194
~ catalogue 92–13
central ~ 97–148
centre of ~ 93–84
~ chain 96–77
~ chart 92–9
chemical composition of ~ 91–168
circumpolar ~ 92–2a
~ cloud 98–6
~ cluster 96–79
cluster ~ 96–82
collapsed ~ 94–111
collapsing ~ 91–164, 94–112
common ~ 91–95
companion ~ 96–39
comparison ~ 95–16
contracting ~ 91–161
cool ~ 91–156
core of ~ 93–164
~ count 98–79
dead ~ 91–167
degenerate ~ 93–180
Demon ~ 92–105
~ density 98–82
disk ~ 91–85
Dog ~ 92–128
double ~ 96–3
dwarf ~ 91–135
emission–line ~ 91–159
escaping ~ 96–160, 98–135
evening ~ 83–128
evolved ~ 94–74
exciting ~ 97–125
expanding ~ 91–162
exploding ~ 91–165
faint ~ 91–82
field ~ 96–84
first generation ~ 94–24
fixed ~ 92–4
~ flare 95–70
flare ~ 95–68
foreground ~ 96–85
~ formation 94–11
giant ~ 91–143
globular ~ cluster 96–114
halo ~ 91–84
heavy–weight ~ 91–126
helium–deficient ~ 91–192
helium–poor ~ 91–192
helium–rich ~ 91–193
high–velocity ~ 98–191
horizontal–branch ~ 91–141
hot ~ 91–155
hydrogen–deficient ~ 91–191
hydrogen–poor ~ 91–191
hyperon ~ 91–154
hyperonic ~ 91–154
imploding ~ 91–164
infrared ~ 91–157
intergalactic ~ 98–350
intermediate–mass ~ 91–125
intrinsically variable ~ 95–87
light ~ 91–124
low–mass ~ 91–124
low–velocity ~ 98–192
magnetic ~ 91–51
main–sequence ~ 91–142
~ map 92–9
massive ~ 91–126
metall-deficient ~ 91–197
metallic–line ~ 91–195
metal–line ~ 91–195
metal–poor ~ 91–197
metal–rich ~ 91–196
middle–weight ~ 91–125
morning ~ 83–127
moving ~ cluster 96–97
multiple ~ 96–5
nearby ~ 98–60
neutron ~ 91–152
nitrogen ~ 91–189
non–cluster ~ 96–84
non–elastic ~ collision 98–26
non–member ~ 96–84
non–rotating ~ 91–38
non–spherical ~ 91–86
non–stable ~ 95–9
non–stationary ~ 95–2
normal ~ 91–95
North ~ 92–121
nova–like ~ 95–118
~ number 98–80
~ of the system 98–133
old ~ 94–64
open ~ cluster 96–99
optical double ~ 96–6
outer layers of ~ 93–3
overluminous ~ 91–134
peculiar ~ 91–92
photometric binary ~ 95–73
physical double ~ 96–7
pole–on ~ 91–40
~ population 98–67
pulsating ~ 95–95
radio ~ 91–158
rapidly rotating ~ 91–39
rate of ~ deaths 94 115
rate of ~ formation 94–12
red ~ 91–97
reddened ~ 97–94
relativistic ~ cluster 96–154
revolving ~ chart 92–10
revolving ~ map 92–10
~ ring 96–78
rotating ~ 91–37
RR Lyae ~ 95–93
runaway ~ 98–135
shell ~ 93–5
shooting ~ 86–2
silicon ~ 91–190
single ~ 96–2
solar–type ~ 91–104
~ spots 97–62
stable ~ 95–8
stationary ~ 95–3
~ stream 98–98
subluminous ~ 91–133
superdense ~ 91–153
supergiant ~ 91–145
superluminous ~ 91–134
supermassive ~ 91–127
~ suspected of variability 95–5
symbiotic ~ 95–50, 96–20
~ trail 86–152
triple ~ 96–4
Tycho's ~ 95–135
~ type 91–101
U Geminorum ~ 95–112
underluminous ~ 91–133
unreddened ~ 97–95
unstable ~ 95–9
variable ~ 95–1
violent ~ formation 94–11
visual double ~ 96–8
W ~ 91–160
wide double ~ 96–9
~ with a dark companion 96–13
~ with a peculiar spectrum 91–92
~ with an extended atmosphere 93–6
~ with an extended envelope 93–7
Wolf–Rayet ~ 91–160
WR ~ 91–160
young ~ 94–63, 96–93
zero–age ~ 94–59
zero–temperature ~ 91–166
starlight 79–10
direct ~ 79–11
scattered ~ 79–12
starlike nucleus 85–38, 98–19
star–poor cluster 96–102
starquake 95–153
star–rich cluster 96–100
starry sky 91–1a
stars, age of 94–61
birth rate of ~ 94–12
early ~ 91–102
early–type ~ 91–102
evolution of ~ 94–1
formation of ~ 96–95
generation of ~ 94–23
group of ~ 92–3a
internal structure of ~ 93–71
late–type ~ 91–103
ring of ~ 96–78

EN

rotation of ~ 91–35
stability of ~ 93–206
start point 86–36
state, non–steady 98–151
~ of the sky 77–132
sea ~ 72–86
singular ~ 99–42
statistically steady ~ 98–148
steady ~ 98–140
system in a non–steady ~ 98–152
system in a steady ~ 98–141
static model 99–50
~ universe 99–50
station, automatic lunar 82–19
meteor ~ 86–158
~ of a planet 83–44
satellite ~ 71–33
seismic ~ 73–6
seismograph(ic) ~ 73–6
seismologic ~ 73–6
stationary meteor 86–34
~ model 99–52
~ point 83–45
~ position 83–45
~ potential 98–147
~ radiant 86–241
~ shell 93–11
~ star 95–3
~ universe 99–52
statistical equilibrium 93–42
~ equilibrium equations 93–44
~ parallax 91–8
statistically steady state 98–148
statistics, stellar 98–77
stay time 82–34
steady state 98–140
statistically ~ state 98–148
steady–state model 99–52
~ universe 99–52
steep slope 81–232
stellar 92–3
~ association 96–88
~ astronomy 91–2
~ atmosphere 93–2
~ brightness 91–55
~ collision 83a–16
~ content 98–172
~ content of a galaxy 98–231
~ core 93–164
~ core implosion 94–110
~ corona 93–4
~ corpse 91–167
~ cosmogony 94–7
~ dynamics 98–120
~ encounter 98–121
~ era 99–178
~ evolution 94–1
~ flare 95–70
~ hydrodynamics 98–142
~ interior 93–72
~ magnetism 95–54
~ magnitude 91–56
~ mass 91–121
~ material 93–124
~ matter 93–124
~ model 93–123
~ nebula 97–63
~ parallax 91–4
parent ~ nebula 97–63
~ population 98–67
primitive ~ nebula 97–63
rate of ~ extinction 94–115
relativistic ~ model 93–198
~ rotation 91–35
~ space density 98–83
~ spectrum 91–88
~ statistics 98–77
~ structure 93–1
~ system 96–1
~ system formed by superposition 98–165
~ type 91–101
~ variability 95–4
~ wind 93–14
step, geothermic 74–59
~ method 95–10
~ weakener 88–80
~ wedge 88–80
Stephan's Quintet 98–309
stereocomparator 91–32
stereographic projection 71–47
stereoscopic method 81–115
sticky slide 87–14
stone field 82–139
meteoritic ~ 87–60
stones, rolling 81–237
stony meteorite 87–60
stony–iron meteorite 87–80
stopping point 86–123
storage of particles 75–123
particle ~ 75–123
stored energy 89–196
storm, dust 83–255
geomagnetic ~ 75–71
gradually commencing ~ 75–78
interplanetary ~ 84–44
ionospheric ~ 78–49
Jovian noise ~ 83–185
standard–type ~ 75–72
~ sudden commencement 75–74
~ surge 72–76
~ with sudden commencement 75–73
straight fan 85–61
~ tail 85–72
~ Wall 81–185
strain 73–67
~ tensor 73–70
stratification 82–133, 97–154
thermal ~ 77–40
stratified rock 74–48
stratigraphic age 82–267
strato–cone 73–121
Stratocumulus 77–142
stratopause 76–71
stratosphere 76–70
strato–volcano 73–121
Stratus 77–139
stray Earth currents 75–6
streaks, lunar 81–168
stream, active 86–223
~ activity 86–229
age of ~ 86–245
asteroid ~ 84–100
asteroidal jet ~ 84–100
~ branch 86–240
cometary ~ 86–222
corpuscular ~ 88–160
daylight ~ 86–219
day–time ~ 86–219
density of meteor ~ 86–264
descending ~ 93–150
ecliptical ~ 86–221
gaseous ~ 95–67
index of ~ activity 86–230
interstellar ~ 86–224
inward ~ 93–150
jet ~ 77–88, 84–100
knot within meteor ~ 86–217
~ line 82–194
meteor ~ 86–216
meteor ~ structure 86–243
night–time ~ 86–218
~ of solar plasma 84–33
outward ~ 93–151
periodic ~ 86–228
permanent ~ 86–227
planetary ~ 86–220
radiant of meteor ~ 86–237
relative ~ activity 86–231
solar corpuscular ~ 84–35
solar–wind ~ 84–35a
sporadic ~ 86–226
star ~ 98–98
temporary ~ 86–225
streamer, coronal 89–287
prominence ~ 89–245
strength, bearing 82–64
cohesive ~ 85–117
effective field ~ 95–59
magnetic field ~ 84–11
soil ~ 82–135
stress 73–62
crustal ~ 82–210
principal ~ 73–64
shear ~ 73–63
tangential ~ 73–63
~ tensor 73–69
thermal ~ 86–205
striation 89–173
chromospheric ~ 88–209
~ pattern 89–174
string, cosmic 99–171c
strip, light 86–27
Strömgren parallax 91–13
structural element 98–287
~ features 98–224

~ forms 81–152
~ index of cluster 96–80
~ map 73–97
~ parameter 96–120
~ relations 82–266
~ type 98–225
structure, cell 99–171a
cellular ~ 99–171a
corotating ~ 84–17
corotating ~ of magnetic field 84–17
~ element 98–287
filamentary ~ 97–130
fine ~ 88–203
galactic ~ 98–4
hierarchic ~ 99–47
hierarchic ~ of universe 99–47
internal ~ of stars 93–71
laminar ~ 83–148
large–scale ~ 88–207
meteor stream ~ 86–243
~ of comets 85–34
~ of field 89–140
~ of meteorites 87–100
~ of Milky Way 98–4
~ of the Galaxy 98–4
post–mare ~ 81–154
pre–mare ~ 81–153
ring ~ 81–155
ring–like ~ 81–155
sector ~ 84–15
small–scale ~ 88–208
spiral ~ 98–8
stellar ~ 93–1, 93–71
structural element 98–287
subaquatic volcano 73–117
subauroral belt 79–128
subcentre 86–239
subcooled 76–106
subcooling 76–105
subdwarf 91–140
sub–Earth point 81–128
subflare 89–195
subgiant 91–151
sublimation 85–138
heat of ~ 85–139
subluminous star 91–133
sublunar point 81–129
submarine canyon 72–46
~ earthquake 73–2c
~ mountain 72–43
~ sedimentation 72–47
~ tablemount 72–44
~ volcano 73–117
submerged contour 72–32
subnuclear density 93–190
subphotospheric layer 88–168
subpolar sheets 79–51
subpulse 95–145
subsatellite point 71–30
subsidence caldera 81–182
subsoil 82–105
permanently frozen ~ 83–152
subsolar point 81–127
substance difficult to fuse 83a–55
easily fusible ~ 83a–56
non–volatile ~ 85–142
not–easily–fusible ~ 83a–55
porous ~ 82–111
volatile ~ 82–262, 83a–52
substellar point 81–130
substorm 79–144
auroral ~ 79–141
geomagnetic ~ 75–95
magnetic ~ 75–95
polar magnetic ~ 75–96
subsun 79–72
subsurface current 72–61
~ investigation 82–82
~ ocean current 72–61
subsystem 98–74
subtraction, photographic 88–94
sudden commencement 75–74
~ commencement absorption 78–47
~ cosmic noise absorption 78–42
~ disappearance 86–76, 89–279
~ disappearance of filament 89–279
~ enhancement of atmospherics 78–51
~ extinction 86–76
~ ionospheric disturbance 78–41
~ jump (in pulsar frequency) 95–154
~ phase anomaly 78–52
sulphate = sulfate 74–29
sulphide = sulfide 74–28
carbonyl ~ 97–36
hydrogen ~ 97–44
sulphite 74–28a
sum of precipitation 77–168
summer constellation 92–1a
~ Lake 81–300
summit crater 73–129
~ pit 81–244
~ pit of central peak 81–244
Sun 88–1
active ~ 89–12
apex of ~ 91–23
atmosphere of ~ 88–163
blue ~ 79–92
daily map of ~ 89–18
disturbed ~ 89–12
eclipse of ~ 88–15
general magnetic field of ~ 89–128
~ guider 88–51
mock ~ 79–71
optical ~ 88–2
~ pillar 79–78
planet of ~ 83–11
quiet ~ 89–11
Quiet ~ Year 89–10
radio ~ 88–3
rotation of ~ 88–127
~ seeker 88–51
total eclipse of ~ 88–17
undisturbed ~ 89–11
visible ~ 88–2
X–ray ~ 88–4
sun–dog 79–71
Sun–Earth relations 84–55
sungrazer 85–13
Sun–grazing comet 85–13
sunlight 77–49
sunlit aurora 79–130
~ hemisphere 83–67
~ side 75–20
sunrise 88–8
~ colo(u)rs 79–56
eclipse ends at ~ 88–21
~ terminator 83–70
Sun's declination 81–63
sunset 88–11
~ colo(u)rs 79–55
eclipse begins at ~ 88–22
~ glow 79–55
~ terminator 83–71
sunshine 77–52, 88–9
duration of ~ 88–10
~ recorder 77–61
sunspot 89–88
active ~ prominence 89–254
~ activity 89–62
apparent ~ area 89–91
area of ~ 89–67
birth of ~ 89–121
~ cycle 89–72
decay of ~ group 89–126
~ drawing 88–34
~ evolution 89–120
foreshortening of ~ 89–92
~ frequency 89–66
invisible ~ 89–159
~ model 89–102
~ number 89–65
~ penumbra 89–97
~ prominence 89–252
proper motion of ~ 89–117
~ umbra 89–95
sunspot–evolution table 89–120
sunward side 81–34
~ tail 85–78
superadiabatic gradient 93–99
superbolide 86–9
supercluster of galaxies 98–346
Local ~ of galaxies 98–347
superconducting nucleus 95–150
supercooled 76–106
supercooling 76–105
superdense star 91–153, 93–197
superfluid 95–149
supergalaxy 98–346
supergiant 91–145
~ galaxy 98–223
less luminous ~ 91–148
most luminous ~ 91–147
supergranulation 88–183
supergranule 88–182
supergravitation 99–170r

superheavy particle 99–170n
superior conjunction 83–56
~ planet 83–15
superluminous star 91–134
supermassive black hole 98–22a
~ body 98–22
~ object 98–22
~ star 91–127
supernova 95–126
~ explosion 95–131
extragalactic ~ 95–128
galactic ~ 95–127
~ of type I 95–130
~ outburst 95–131
past ~ 95–134
post- ~ 95–134
relics of ~ 95–133
remains of ~ 95–133
~ remnant 95–133
remnants of ~ explosion 95–133
~ search 95–136
Tychonic ~ 95–135
supernovae, chain of 95–132
superposed fluctuations 95–44
superposition of spectra 96–24
supersaturation 76–102
superstrong magnetic field 95–148
superunification 99–170q
superunified force 99–170r
~ interaction 99–170r
support of chromosphere 88–223
surface activity 82–231
apparent ~ density 96–132
back ~ 87–115
~ boundary coudition 93–80
~ cavity 87–118
~ condition 93–80
contact ~ 96–27
cratered ~ 81–245
critical ~ 96–27
crusted ~ 87–109
~ current 72–62
~ density 98–85
diffuse luminous ~ 79–148
Earth's ~ 71–5
equipotential ~ 71–61
~ erosion 82–241
~ feature 83–105
~ field 95–64
first type ~ 87–107
~ formation 83–105
~ fracture 81–173
front ~ 87–114
geopotential ~ 71–61
highland ~ 82–56
inter-ray ~ 81–172
~ inversion 77–38
layered ~ 83–101
lunar ~ 82–55
mare ~ 82–57
~ marking 81–148, 83–105
meteorite ~ 87–106
Moon's ~ 82–55
neutral ~ 75–16
~ nuclear reaction 91–181
~ of planets 83–97
~ of the Moon 82–55
planetary ~ 83–97
polished ~ 87–101
primary ~ 87–107
pulsating ~ 79–149
ray ~ 81–171
rear ~ 87–115
~ reconnaissance 82–32
rough ~ 83–102
~ roughness 82–106, 83–99
~ sampler 82–80
second type ~ 87–108
secondary ~ 87–108
solar ~ 88–164
solid ~ 83–98
~ temperature 93–88
~ wave 73–47
surge 89–255
bright ~ 89–257
dark ~ 89–256
flare ~ 89–190
sea ~ 72–76
storm ~ 72–76
surveillance, ice 72–85
survey, asteroid 84–94
geologic ~ of the Moon 82–50
ice ~ 72–85
magnetic ~ 75–64
photographic ~ 82–30
satellite ~ 83–26
selenologic ~ 82–50
selenophysical ~ 82–39
sky ~ 92–11
solar ~ 89–16
surveying, photographic 82–30
susceptibility, magnetic 75–27
suspected fall of meteorite 87–49
~ variable 95–5
suspended cloud 89–275
suspension current 72–49
Swan 92–44
Swan bands 85–133
~ nebula 97–143
swarm, November 86–255
sweep in position 88–105
~ in wavelength 88–111
sweep–frequency interferometer 88–110
~ spectrograph 88–124
sweeping out of gases 83a–41
swept–lobe interferometer 88–109
switch, phase–changing 88–114
Sword of Orion 92–130
Swordfish 92–46
symbiotic spectrum 95–117
~ star 95–50
symmetrical system 98–157
axially ~ 98–157
symmetry breaking 99–170s
breaking of ~ 99–170s
~ condition 98–162
cylindrical ~ 98–155
helical ~ 98–161
spherical ~ 98–160
sympathetic flare 89–179
synchrone 85–92
complete ~ 85–94
terminal ~ 85–93
synchronous rotation 81–14
~ spin 83–37
synclinal fold 73–89
syncline 73–89
syndyname 85–91
syndyne 85–91
synodic(al) month 81–46
~ revolution 81–46
synoptic(al) chart 77 11
~ map 77–11
~ map of chromosphere 89–21
~ map of corona 89–22
~ map of photosphere 89–20
~ meteorology 77–8
synthesis, aperture 88–118
cosmological ~ 99–175
nuclear ~ 94–52
~ of elements 94–52
system, axially symmetrical 98–157
collisional ~ 98–166
collisionless ~ 98–167
colo(u)r ~ 91–72, 96–140
comoving coordinate ~ 99–44
compact ~ 96–157
continental fracture ~ 73–81
Copernican ~ 99–24
cosmological ~ 99–20
current ~ 75–97
dimensions of ~ 96–64
Earth–Moon ~ 81–12
equatorial longitude ~ 83–178
extragalactic ~ 98–210
flat ~ 98–251
~ for intermediate latitudes 83–179
fracture ~ 73–79
galactic ~ 98–3
geocentric ~ 99–21
geodetic reference ~ 71–19
gravitationally bound ~ 96–156
gravitationally stable ~ 96–156
gravitationally unbound ~ 96–158
gravitationally unstable ~ 96–158
Harvard ~ of spectral classification 91–91
heliocentric ~ 99–23
~ in non–steady state 98–152
~ in steady state 98–141
inertial ~ 99–98
longitude ~ 83–177
member of ~ 98–133
mid–ocean fracture ~ 73–80
multiple–arm ~ 98–44
multiple–armed ~ 98–44

non-inertial ~ 99–147
~ of Copernicus 99–24
~ of globular clusters 98–75
~ of rays 81–170
~ of reference 98–136
~ of rings 98–291
~ of stars 96–1
~ of vortices 83a–13
planetary ~ 83–2
Ptolemaic ~ 99–22
ray ~ 81–170
satellite ~ 83–19
self–gravitating ~ 98–169
solar ~ 83–12
solar–~ meteor 86–44
spheroidal ~ 98–159
star of the ~ 98–133
stellar ~ 96–1
stellar ~ formed by superposition 98–165
synthesis ~ 88–119
two–arm ~ 98–43
UBV ~ 96–144
ultraviolet–blue–visual colo(u)r ~ 96–144
ultraviolet–blue–yellow colo(u)r ~ 96–144
unstable ~ 98–171
~ with negative total energy 96–156
~ with positive total energy 96–158
world ~ 99–20
world geodetic ~ 71–39
syzygial tide 82–273
syzygies 81–29

T

Table 92–66
~ Mountain 92–66
table, sunspot evolution 89–120
tablemount 72–44
submarine ~ 72–44
tadpole galaxy 98–281
taenite 87–93
tail, anomalous 85–75
cometary ~ 85–70
~ condensation 85–95
curved ~ 85–73
~ directed away from Sun 85–81
~ directed toward Sun 85–78
dust ~ 85–79
~ formation 85–86
ion ~ 85–82
magnetospheric ~ 75–10
main ~ 85–77
multiple ~ 85–76
~ of comet 85–70
~ of Earth 84–65
~ of galaxy 98–280
~ of Serpent 92–89
straight ~ 85–72
sunward ~ 85–78
type of cometary ~ 85–71
talus 81–234
tangent plane 96–50
tangential arc of halo 79–75
~ ray 81–169
~ stress 73–63
T–association 96–90
Tau 92–91
Taurids 86–254
Taurus 92–91
tectonic earthquake 73–2d
~ forces 73–106
~ map 73–96, 82–202
~ origin 82–200
~ process 82–201
tectonics 73–95
tektite 87–95
drop-form ~ 87–96
splash-form ~ 87–96
Tel 92–92
telegram, astronomical 85–20
Telescope 92–92
automatic solar ~ 88–50
balloon–borne solar ~ 88–52
chromospheric ~ 88–73
horizontal solar ~ 88–55
monochromatic ~ 88–38
solar ~ 88–49
tower ~ 88–54
telescopic comet 85–14a
~ meteor 86–18
Telescopium 92–92
teleseism 73–27
telluric currents 75–5
~ line 88–140a
temperate zone 77–192, 83–189
temperature 77–29
adiabatic ~ gradient 93–98
ambient ~ 77–32
brightness ~ 91–114
central ~ 93–89
colo(u)r ~ 91–109
condensation ~ 97–66
~ conditions 77–35
convective ~ gradient 93–157
Curie ~ 75–41
decrease in ~ 77–41
degeneracy ~ 93–188
dissociation ~ 91–120a
~ drop 77–41
effective ~ 91–113
electron ~ 91–116
excitation ~ 91–118
fall in ~ 77–41
fusing ~ 86–204
~ gradient 93–96
~ in the shade 77–34
~ inversion 77–36
ionization ~ 91–117
kinetic ~ 91–115
melting ~ 86–204
normal room ~ 77–31
~ profile (of atmosphere) 83–259
radiative ~ gradient 93–97
room ~ 77–31
rotation ~ 91–119
shade ~ 77–34
surface ~ 93–88
vibration ~ 91–120
temporary stream 86–225
tension 73–66
~ crack 82–206
~ fracture 82–207
water–vapour ~ 77–125
tensor, curvature 99–159
deformation ~ 73–70
distortion ~ 73–70
energy ~ 99–158
energy–momentum ~ 99–158
~ equation 99–157
fundamental ~ 99–157a
matter ~ 99–158
metric ~ 99–157a
Riemannian curvature ~ 99–159
Riemann–Christoffel ~ 99–159
strain ~ 73–70
stress ~ 73–69
term, cosmological 99–156a
differential–motion ~ 98–115
galactic rotation ~ 98–116
K– ~ 98–119
terminal moraine 72–130
~ synchrone 85–93
terminator 83–69
dawn ~ 83–70
dusk ~ 81–125
Earth's ~ 81–91
evening ~ 81–125
lunar ~ 81–123
morning ~ 83–70
sunrise ~ 83–70
sunset ~ 81–125
terra 81–190, 81–271
terrace, crater 81–229
terraced depression 81–162
~ wall 81–257
terracing, internal 81–256
terrain, chaotic 83–164
terrella 79–133
terrestrial 71–3
~ effects of solar activity 84–55
~ electricity 75–1
~ globe 71–10, 79–133
~ magnetic axis 75–47
~ magnetic field 75–44
~ magnetism 75–24
~ meteoritic crater 87–54
~ origin 87–74
~ planets 83–119
~ sphere 71–10
terrigenous deposits 72–48
~ sediments 72–48
Tertiary 74–87
tesseral harmonics 95–102

texture 82–102
flow ~ 82–191
fluidal ~ 82–191
theorem, Clairaut's 71–77
Russell–Vogt ~ 93–83
uniqueness ~ 93–83
virial ~ 98–181
Vogt–Russell ~ 93–83
theory, accretion 83a–20
capture ~ 85–99, 96–72
contraction ~ 73–77
ejection ~ 85–98
fission ~ 96–75
gauge ~ 99–170j
geomagnetic dynamo ~ 75–99
mechanical ~ of cometary forms 85–84
meteori(ti)c ~ 81–262
meteoroidal ~ 81–262
~ of cycles 97–156
~ of heat death 99–15
~ of meteors 86–159
~ of multiple formation 96–76
~ of spiral arm formation 98–173
~ of spiral structure 98–173
physical ~ of meteors 86–159
restricted ~ of relativity 99–94
separate nuclei ~ 96–76
special ~ of relativity 99–94
tidal ~ 82–305
Weizsäcker('s) turbulence ~ 83a–11
thermal demagnetization 75–40
~ dissipation 94–76
~ equilibrium 93–78
~ evolution 83a–61
~ history 74–66, 82–260
lunar ~ map 81–135
~ mapping 72–83
~ metamorphism 74–37
~ pressure 94–41
~ shock 86–206
~ stratification 77–40
~ stress 86–205
~ tides 82–299
thermally stable 93–209
thermodynamic(al) equilibrium 93–35
departure from ~ ~ 93–36
deviations from local ~ ~ 93–40
deviations from ~ ~ 93–36
local ~ ~ 93–39
thermoluminescence 82–249
thermonuclear evolution 94–50
~ reaction rate 94–53
thermoremanent magnetization 75–30
thermoresidual magnetization 75–30
thermosphere 76–59
thickness of rings 83–229
thin layer 83–149
thioformaldehyde 97–43
third integral of motion 98–186
~ quarter 81–30
thorium–lead dating 82–270
thread, prominence 89–245
Three Wise Men 92–129a
three–body collision 96–74
~ encounter 96–74
three–colo(u)r photometry 96–141
three–dimensional orbit 98–185
~ spectral classification 91–106
three–hour–range index 75–89
three–kiloparsec arm 98–37
threshold curve of the secondary component 96–40
~ of the secondary 96–40
thrust 73–65
Thuban 92–131
thunder 77–165
thunderstorm 77–159
tidal action 83a–15
~ amplitude 82–292
~ bulge 81–119, 83a–67
criterion of ~ instability 98–312
~ current 82–279
~ day 82–290
~ deformation 82–277, 83a–66
~ disintegration 85–118
~ disruption 85–118
~ dissipation 82–286
~ distortion 96–30
~ disturbance 98–314
~ effect 82–276
~ evolution 83a–65
~ field 98–196
~ force 82–289
forced ~ oscillations 82–306
~ friction 82–312
galactic ~ field 96–106
~ heat 82–287
~ instability 98–313
~ interaction 82–274, 98–311
~ oscillations 76–28
~ perturbation 98–314
~ potential 82–293
~ range 82–292
~ splitting 85–112
~ stream 82–279
~ theory 82–305
~ torque 82–288
~ wave 82–291
~ wave damage 72–77
tide, big 82–273
dead ~ 82–272
falling ~ 82–282
~ gauge 82–311
high ~ 72–90, 82–281
low ~ 72–91
neap ~ 82–272
spring ~ 82–273
syzygial ~ 82–273
total ~ 82–273
tide–generating force 82–289
tide–raising body 82–284
~ force 82–289
tides 72–89
air ~ 82–294
atmospheric ~ 82–294
crustal ~ 82–297
earth ~ 82–296
Earth–induced ~ 82–275
Earth–produced ~ 82–275
geomagnetic ~ 82–295
gravitational ~ 82–298
lunar ~ 82–308
ocean ~ 82–310
sea ~ 82–309
terrestrial ~ 82–296
thermal ~ 82–299
till 72–128
tilt, axial 83–35
time, absolute 99–117
apparent lunar ~ 82–301
arrival ~ 73–50
characteristic ~ 99–79
constant with ~ 98–150
cooling ~ 83–201, 93–201
~ coordinate 99–123
~ dilatation 99–137
dispersion ~ 96–104
dissociation ~ 96–104
dissolution ~ 96–104
echo ~ 86–291
evaporation ~ 96–109
expansion ~ 99–78
Friedmann's ~ 99–80
Friedmann's expansion ~ 99–80
geomagnetic ~ 79–132
Hubble ~ 99–79
Hubble expansion ~ 99–79
invariable in ~ 98–150
lunar ~ 82–300
mean lunar ~ 82–302
Newtonian ~ 99–118
~ of appearance 86–49
~ of meteoritic fall 87–51
~ of periastron passage 96–60
~ of relaxation 98–128
~ of transit Sun-Earth 84–47
period ~ 95–39
Planck ~ 99–170i
proper ~ 99–139
relaxation ~ 98–128
~ resolution 88–205
stay ~ 82–34
transit ~ 73–51, 84–47
travel ~ 73–51
travelling ~ 73–51
timelike coordinate 99–123
time–scale 94–65
long ~ 94–67
short ~ 94–66
Titan 83–233
Titius–Bode formula 83–3
~ law 83–3
tongue, glacier 72–125
ice ~ 72–125
interplanetary ~ 84–21

lava ~ 82–187
solar interplanetary ~ 84–20
tool, sample–collection 82–80
topographic feature 81–149
lunar ~ map 81–138
~ unit 81–149
topside sounding 78–20
tornado prominence 89–262
torque 83a–5
tidal ~ 82–288
torrid zone 77–191
torus, hydrogen 83–230
total abundance 91–177
~ albedo 83–90b
~ area 89–94
~ brightness 85–107
~ eclipse begins 81–77
~ eclipse ends 81–80
~ eclipse of Moon 81–66
~ eclipse of Sun 88–17
~ energy flux 93–116
~ extinction 97–82
~ magnitude 96–136
~ mass of a cluster 96–139
~ precipitation 77–168
~ pressure 93–93
~ radiation 88–146
~ solar eclipse 88–17
~ tide 82–273
totality, beginning of 81–77
duration of ~ 88–26
end of ~ 81–80
length of ~ 88–26
totalizator 77–171
totalizer rain–gauge 77–171
Toucan 92–95
touchdown 82–16
towards the anticenter 98–14
tower, solar 88–54
~ telescope 88–54
TrA 92–94
trace concentration 83–266
~ element 82–145
~ gas 76–88
tracer, spiral 98–39
spiral–arm ~ 98–39
track 82–251
cooling ~ 93–202
evolutionary ~ 94–29
~ density 82–253, 87–141
particle ~ 82–251
radiation ~ 87–120
slide ~ 82–197
trade winds 77–100
trades 77–100
trail echo 86–298
meteor ~ 86–153
meteor ~ on plate 86–29
star ~ 86–152
trailer 89–114
~ spot 89–114
trailing 98–28
~ arm 98–29
~ spot 89–114
~ velocity 86–154
train, day–time 86–106
drift of a ~ 86–114
duration of ~ visibility 86–111
dust ~ 86–115, 87–22
gaseous ~ 86–116
long–duration ~ 86–113
long–enduring meteor ~ 86–113
meteor ~ 86–103
meteor with ~ 86–17
noctilucent ~ 86–105
~ of a meteor 86–103
persistence of a ~ 86–111
persistent ~ 86–104
self–luminous ~ 86–108
short–duration ~ 86–112
short–enduring meteor ~ 86–112
smoke ~ 87–21
spectrum of a meteor ~ 86–107
unstable ~ 86–109
trajectory, ascending 82–9
circumlunar ~ 82–11
~ in atmosphere 86–162
interplanetary ~ 82–12
meteor ~ 86–30
~ of a meteor 86–30
particle ~ 75–118
transfer, energy 93–111
~ equation 93–32
equation of radiative ~ 93–33
heat ~ 86–202
mass ~ 96–33
mass ~ from one component to the other 96–33
~ of radiation 93–29
radiative ~ 93–29
transform, Laplace 93–54
transformation, Galilean 99–131
Galileo('s) ~ 99–131
homologous ~ 93–132
homology ~ 93–132
invariance under a ~ 99–133
Laplace ~ 93–54
Lorentz ~ 99–132
~ of energy 86–208
transient, coronal 89–185a
lunar ~ 82–153
lunar ~ phenomenon 82–153
transit 95–81
meridian ~ of the Moon 82–303
~ of a planet 83–51
~ of a planet across solar disk 83–51
~ of the Moon 82–303
satellite ~ 83–54
satellite ~ across planet's disk 83–54
shadow ~ 83–53
~ time 84–47
transition layer 72–13, 88–223c
crust–mantle ~ 72–14
phase ~ 83a–58, 99–170p
~ region 88–224
~ zone 72–13
transition–type galaxy 98–248
translunar 81–6
transmission of heat 86–202
~ window 83–264
transmitter 78–15, 86–269
~ aerial 78–16
~ antenna 78–16
~ frequency 86–273
~ power 86–270
transmitting aerial 78–16
~ antenna 78–16
transneptunian planet 83–237
transparency window 83–264
transplutonian planet 83–124
transport, energy 93–111
equation of radiative ~ 93–33
~ of magnetic field 84–48
transversal component 89–136
~ field 89–137
~ wave 73–41
trap, light 88–64
Trapezium in Orion 96–92
trapped electrons 79–111
artificial ~ radiation 75–113
geomagnetically ~ particles 75–111
geomagnetically ~ radiation 75–112
~ particle lifetime 75–122
~ particles 75–111
~ radiation 75–112
~ solar–wind gases 82–148
trapped–radiation region 75–101
trapping 75–109
~ region 75–110
travel time 73–51
travelling time 73–51
traverse 82–35
~ route 82–37
scientific ~ 82–36
~ survey 82–35
treatment, atmospheric 87–116
tree trunks 89–272
trees 89–273
tremors, earth 73–2a
trench 72–40
marine ~ 72–40
ocean ~ 72–40
Tri 92–93
Triangle 92–93
S ~ 92–94
Southern ~ 92–94
triangulation 71–23
cosmic ~ 71–28
first order ~ 71–24
primary ~ 71–24
satellite ~ 71–27
spatial ~ 71–25
Triangulum 92–93
~ Australe 92–94
trias 87–91
Triassic 74–84

Trifid Nebula 97–141
trigger 89–198
~ action 89–199
triggered volcanism 82–155
triggering 89–199
trigonometric parallax 91–5
relative ~ parallax 91–10
trilateration 71–29
triple collision 79–30, 96–74
~ conjunction 83–58
~ star 96–4
troilite 82–94
Trojan group 84–91
~ group of asteroids 84–91
Trojans 84–91
tropical cyclone 77–77
~ month 81–44
~ revolution 81–44
tropopause 76 77
troposphere 76–78
trough 73–89
potential ~ 98–176
true diameter 83–118
~ flattening 98–296
~ height 78–29
~ modulus 96–118
~ normal 71–71
~ orbit 96–53
~ path 86–31
~ radiant 86–59
~ shadow height 79–49
truncated cylinder 88–63
trunks, tree 89–272
tsunami 72–95
~ wave 72–95
T–Tauri stage 94–47
tube, cathode ray 86–293
lava ~ 82–186
magnetic ~ of force 89–145
Tuc 92–95
Tucana 92–95
turbid medium 83–251
turbidity 83–252
~ current 72–49
turbidness 83–252
turbosphere 76–72
turbulence, cosmic–scale 83a–12
large–scale ~ 83a–12
turbulent convection 93–152
turning point 98–178
tweek 78–63
twilight 79–34, 79–40
~ airglow 79–4
astronomical ~ 79–37
civil ~ 79–38
evening ~ 79–36
~ glow 79–4
morning ~ 79–35
nautical ~ 79–39
~ observation 79–42
~ phenomena 79–32
~ sky 79–43
~ sounding 79–41
twin galaxy 98–305
Twins 92–51
Heavenly ~ 92–51
two–arm galaxy 98–43
two–armed galaxy 98–43
two–blade shutter 86–149
two–colo(u)r diagram 96–147
two–component model 88–196
two–dimensonal spectral classification 91–105
two–element interferometer 88–98
two–level atom 93–45
two–photon emission 97–155
two–ribbon flare 89–187a
two–stream hypothesis 98–99
two–temperature model 88–197
Tychonic supernova 95–135
Tycho's star 95 135
type, cloud 77–135
cometary–tail ~ 85–71
~ of cometary tail 85–71
~ I population 98–71
spectral ~ 91–108
star ~ 91–101
stellar ~ 91–101
structural ~ 98–225
~ II population 98–68
~ variable 95–47
typical nova 95–111

U

U–B colo(u)r index 96–146
UBV system 96–144
ultrabasic rock 74–6
ultrashort–period variable 95–90
ultratelescopic meteor 86–20
ultraviolet excess 91–78
~ pumping 97–18a
ultraviolet–blue–visual colo(u)r system 96–144
ultraviolet–blue–yellow colo(u)r system 96–144
ultraviolet–minus–blue colo(u)r index 96–146
UMa 92–96
umbra 89–95
Moon enters ~ 81–76
Moon leaves ~ 81–81
~ of Earth 81–58
spot ~ 89–95
sunspot ~ 89–95
umbral dot 89–96
~ granule 88–180
UMi 92–97
Umkehr effect 76–126
unblanketed model 93–19
unbound system 96–158
gravitationaly ~ system 96–158
uncatalogued object 84–96
uncertain member of cluster 96–87
undae longae 73–48
~ primae 73–46
~ secundae 73–43
under–abundance 91–187
under–abundant element 91–186
underdense 86–284
underestimate 95–14
underluminous star 91–133
undersun 79–72
undisturbed photosphere 88–169
~ photospheric region 89–56
~ Sun 89–11
undriven camera 86–143
undulation, geoid 71–74
Unicorn 92–68
unidentified bands 85–134
unidirectional intensity 75–103
unification, grand 99–170u
~ of interactions 99 170w
unified force 99–170x
grand ~ force 99–170v
unilateral galaxy 98–42
unilluminated hemisphere 75–21
unipolar group 89–110
~ magnetic region 89–155a
unique creation 99–90
uniqueness of the solution 93–82
~ theorem 93–83
unit decrease of pressure 76–19
phase–changing ~ 88–115
prominence ~ 89–235
solar ~ 91–123
universe 99–1
age of the ~ 99–12
anisotropic ~ 99–65
asymptotic ~ 99–59
asymptotic ~ of first kind 99–60
Charlier's ~ 99–67
classical ~ 99–93
closed ~ 99–73
collapse of the ~ 99–181
contracting ~ 99–56
contraction of the ~ 99–180
De Sitter('s) ~ 99–69
early ~ 99–170a
Einstein('s) ~ 99–68
Einstein–De Sitter ~ 99–70
empty ~ 99–49
eternity of the ~ 99–13
evolution of the ~ 99–171e
evolving ~ 99–54
expanding ~ 99–55
expansion of the ~ 99–74
finiteness of the ~ 99–4
Friedmann's ~ 99–71
heat death of the ~ 99–14
hierarchic ~ 99–48
hierarchic structure of the ~ 99–47
homogeneous ~ 99–64
horizon of the ~ 99–10
hot ~ 99–166
infinity of the ~ 99–6

inflationary ~ 99–170b
initial ~ 99–170a
isotropic ~ 99–63
model of the ~ 99–46
monotonic ~ 99–61
monotonic ~ of first kind 99–62
Newtonian ~ 99–93
non–homogeneous ~ 99–66
non–static ~ 99–51
non–stationary ~ 99–53
~ of Einstein and De Sitter 99–70
open ~ 99–72
origin of the ~ 99–11
oscillating ~ 99–57
radius of the ~ 99–5
relativistic ~ 99–92
static ~ 99–50
stationary ~ 99–52
steady–state ~ 99–52
theory of the heat death of the ~ 99–15
unnamed crater 81–202
unreddened star 97–95
unresolved binary 96–10
unstable 93–210
gravitationally ~ system 96–158
~ star 95–9
~ system 98–171
~ train 86–109
unwinding arm 98–31
up–current 77–91
updraft 77–91
updraught 77–91
uplift or depression 73–73
upper atmosphere 76–52
~ boundary 76–58
~ chromosphere 88–192
~ mantle 72–16
~ tangential arc 79–77
upwelling 72–88
uranium method 74–96
uranium–lead age 74–101
Uranus 83–234
Urca process 94–51
Ursa Major 92–96
~ Minor 92–97
Ursids 86–258
December ~ 86–258
UV coma 85–67
UV excess 91–78

V

vacuum 76–9
~ energy 99–170k
valley 81–163
~ glacier 72–114
value, peak 76–55
Van Allen belt 75–101
inner ~ ~ 75–106
outer ~ ~ 75–105
vane, wind 77–104
vapo(u)r, water 77–119
vapo(u)rization, energy of 86–199
~ equation 86–200
intensive ~ 86–185
~ of a meteor body 86–198
variability, magnetic field 84–14
optical ~ 95–141
star suspected of ~ 95–5
stellar ~ 95–4
variable 95–1
~ base–line 88–103
~ base–line interferometer 88–107
cataclysmic ~ 95–109
Cepheid ~ 95–92
chart of the ~ 95–15
classification of ~ stars 95–46
cluster ~ 95–93
cluster–type ~ 95–93
cyclical ~ 95–108
delta Cephei type ~ 95–92
eclipsing ~ 95–74
ellipsoidal ~ 95–79
extrinsic ~ 95–86
false ~ 95–51
geometrical ~ 95–72
improper ~ 95–51
intrinsic ~ 95–87
intrinsically ~ star 95–87
irregular ~ 95–107
irregular magnetic ~ with constant polarity 95–57
irregular magnetic ~ with reversing polarity 95–56
long–period ~ 95–91
magnetic ~ 95–52
~ mass 86–167, 96–70
nebular ~ 95–85
non–dimensional ~ 93–133
nova–like ~ 95–118
occultation ~ 95–83
periodic ~ 95–88
periodic magnetic ~ 95–55
R Coronae Borealis ~ 95–83
rotational ~ 95–65
short–period ~ 95–89
semi–regular ~ 95–105
semi–regular red ~ 95–106
~ spacing interferometer 88–107
spectral ~ 95–48
spectrum ~ 95–48
~ star 95–1
suspected ~ 95–5
. .–type ~ 95–47
ultra–short ~ 95–90
veil ~ 95–83
veiled ~ 95–83
variation, brightness 95–7
center–limb ~ 88–145
centre–limb ~ 88–145
centre–to–limb ~ 88–145
daily ~ 86–137
diurnal ~ 86–137
geomagnetic ~ 75–69
light ~ 95–6
magnetic ~ 95–53
~ of geomagnetic field 75–69
seasonal ~ 86–139
secular geomagnetic ~ 75–85
spectral ~ 95–49
yearly ~ 86–138
variational method 93–57
vector, four–dimensional 99–138
magnetic ~ 89–139
~ magnetograph 88–84
Poynting ~ 93–92
propagation ~ 78–36
world ~ 99–138
Vega 92–132
Vegard–Kaplan bands 79–19
veil 83–250
cloud ~ 77–151
~ Nebula 97–137
~ of clouds 77–151
~ variable 95–83
veiled variable 95–83
veiling 95–84
veins 87–111
Vel 92–98
Vela 92–98
velocities, Maxwellian distribution of 98–96
velocity, angular 86–89
approach ~ 91–27
~ at mid–point of trail 86–97
atmospheric ~ 86–94
average ~ 86–98
characteristic ~ 76–43
coefficients of the ~ ellipsoid 98–102
~ dispersion 98–103
dispersion of ~ 98–103
distribution of residual ~ 98–92
~ ellipsoid 98–101
equatorial ~ 91–49
escape ~ 96–155
escape ~ from a system 96–155
expansion ~ 96–94
~ field 88–187
geocentric meteor ~ 86–95
geocentric ~ of meteor 86–95
heliocentric ~ 86–96
~ in the atmosphere 86–94
initial ~ 86–92
instantaneous ~ 86–99
instantaneous angular ~ 86–90
line–of–sight ~ 91–24
~ map 89–23
mean ~ 86–98
mean angular ~ 86–91
meteor ~ 86–86
negative ~ 91–27
~ of ejection 85–90
~ of entry 86–92
~ of expansion 99–75
~ of rotation 83–32

orbital ~ 96–63
~ outside the atmosphere 86–93
peculiar 91–31
positive ~ 91–26
radial ~ 91–24
recession ~ 99–75
relativistic ~ 99–136
relaxation ~ 98–129
residual ~ 98–91
rotation ~ 91–48
rotational ~ 83–32, 91–48
solar-wind ~ 88–158
~ space 98–97
space ~ 91–30
spatial ~ 91–30
trailing ~ 86–154
velocity-depth curve 73–52
~ section 73–52
velocity-distance relation 98–300
Venetian blind 88–126
Venus 83–125
crescent ~ 83–129
crescent of ~ 83–129
dichotomy of ~ 83–132
extension of horns of ~ 83–131
Venusian 83–126
vertex 96–98, 98–105
vertical 71–71
~ coelostat 88–48
~ component 75–62
deflection of the ~ 71–72
~ difference 81–140
~ distribution 76–17
~ field intensity 75–62
~ intensity 75–62
~ motion 88–185
~ photo 82–28
~ photograph 82–28
~ profile 81–228
~ profile of density 76–53
~ profile of temperature 76–53
~ spectrograph 88–79
very compact cluster of galaxies 98–335
~ faint telescopic meteor 86–20
~ low frequency emission 78–54
vesicle 82–180
vesicular 82–181
vesiculation 82–182
Vesta 84–87
vibration temperature 91–120
vibrational instability 93–214
vicinity of the Earth's orbit 84–7
videomagnetograph 88–85
violation, CP 99–170t
~ of charge parity 99–170t
violent earthquake 73–37
violet layer 83–165
Vir 92–99
Virgin 92–99
Virgo 92–99
~ cluster 98–345
~ cluster of galaxies 98–345
virial, law of 98–181
~ theorem 98–181
virtual height 78–28
viscosity 73–71
magnetic ~ 75–34
magnetic ~ of rocks 75–34
viscous magnetization 75–32
visibility 77–24
area of ~ 86–126
duration of train ~ 86–111
~ function 89–55
good ~ 77–25
~ of an occultation 81–103
~ of eclipse 81–88
poor ~ 77–26
region of ~ 87–31
zone of ~ 81–89
visible coma 85–66
~ disc 88–171
~ flare 89–181
~ hemisphere 89–27
millionths of ~ hemisphere 89–68
~ side of the Moon 81–15
~ Sun 88–2
vision, averted 86–128
extrafoveal ~ 86–128
line of ~ 91–28
parafoveal ~ 86–128
peripheral ~ 86–128
point of clearest ~ 86–127
side ~ 86–128
visual brightness 91–65
~ double star 96–8
~ light-curve 95–24
~ magnitude 91–65
~ meteor 86–16
~ polarimeter 83–82
VLF 78–54
Vogt-Russell theorem 93–83
void 99–171d
Vol 92–100
Volans 92–100
volatile 82–262
~ substance 82–262
volcanic activity 73–110
~ ash 73–142
~ bomb 73–144
~ cone 73–118, 82–160
~ dome 81–176
~ dust 73–143, 82–126
~ earthquake 73–2e
~ ejecta 82–166
~ eruption 73–112
~ explosion 73–112
~ feature 82–159
~ formation 82–159
~ marking 82–159
~ neck 73–125
~ pellets 82–169
~ rock 74–3
~ vent 73–122
volcanism along fractures 82–157
meteorite-induced ~ 82–156
triggered ~ 82–155
volcano 73–111
active ~ 73–114
burning ~ 73–114
dead ~ 73–115
dormant ~ 73–116
extinct ~ 73–115
fuming ~ 73–114
inactive ~ 73–116
quiescent ~ 73–116
shield ~ 73–119
subaquatic ~ 73–117
submarine ~ 73–117
volume density 96–133
vortex hypothesis 83a–11
~ system 83a–13
vortices, system of 83a–13
vuggy 82–181
Vul 92–101
Vulcan 83–123
Vulpecula 92–101

W

W star 91–160
wake of a meteor 86–110
walk, random 89–157
wall, concentric 81–254
crater ~ 81–251
inner crater ~ 81–252
multiple ~ 81–255
outer crater ~ 81–253
Straight ~ 81–185
terraced ~ 81–257
walled plain 81–265
waning half-moon 81–30
~ moon 81–21
~ of polar cap 83–146
warm current 72–63
~ front 77–69
washing out of Fraunhofer lines 88–243
wastage, mass 82–195
wasting, mass 81–236
water 97–26
~ Bearer 92–17
~ Carrier 92–17
~ ice 85–47
high ~ 82–281
low ~ 82–283
~ Monster 92–55
~ pollution 72–73
~ resistant 76–93
~ sampler 72–101
~ Snake 92–55
water-flow pyrheliometer 88–40
Waterman 92–17
water-stir pyrheliometer 88–41
water-vapo(u)r absorption 77–127
~ pressure 77–125
~ tension 77–125
wave, body 73–42

compressional ~ 73–46
continuous ~ 86–277
density ~ 98–175
density ~ theory 98–174
density ~ theory of spiral structure 98–174
dilatational ~ 73–46
dilational ~ 73–46
direct ~ 78–12
distortional ~ 73–43
diurnal ~ 76–30
double ~ 98–117
earthquake ~ 73–9
elastic ~ 73–44
equivolumnar ~ 73–43
extraordinary ~ 78–30
~ frequency/virtual height curve 78–23
galactic density ~ 98–175
gravitational ~ 99–164
ground ~ 78–14
infrasonic ~ 72–93
infrasonic pressure ~ 75–88
longitudinal ~ 73–45
Love ~ 73–48
ordinary ~ 78–31
P ~ 73–46
penumbral ~ 89–100
planetary ~ 72–94
pressure ~ 73–46
primary ~ 73–46
push ~ 73–46
Q ~ 73–48
rarefaction ~ 97–58
S ~ 73–43
seaquake ~ 72–95
secondary ~ 73–43
seismic ~ 73–9
semi–diurnal ~ 76–29
shear ~ 73–43
sky ~ 78–13
sound ~ 76–33
space ~ 78–13
surface ~ 73–47
~ theory 98–174
~ theory of spiral structure 98–174
tidal ~ 82–291
tidal ~ damage 72–77
transverse ~ 73–41
tsunami ~ 72–95
wavelength, Compton 99–170g
effective ~ 96–142
sweep in ~ 88–111
zero–velocity ~ 98–365
waves, activity in radio 89–6
dissipation of ~ 88–222
wavy path 86–33
waxing half–moon 81–25
~ moon 81–20
~ of polar cap 83–145
Way, Milky 98–1
weakener, step 88–80
weather 77–1
~ change 77–22
change of ~ 77–22
~ chart 77–12
~ conditions 77–2
deterioration of ~ 77–23
general ~ situation 77–15
large–scale ~ situation 77–15
~ map 77–12
~ prognosis 77–13
~ report 77–10
~ service 77–5
~ situation 77–14
weathering 74–57
space ~ 82–244
wedge of high pressure 77–86
step ~ 88–80
weight function 93–58
mean molecular ~ 93–100
method of ~ functions 93–59
Weizsäcker('s) turbulence theory 83a–11
west limb 89–33
western elongation 83–62
~ quadrature 83–66
~ sheets 79–52
Weyl's postulate 99–34
Whale 92–34
whistler 78–53
electron ~ 78–60
long ~ 78–57
nose ~ 78–59
proton ~ 78–61
short ~ 78–58
whistling atmospherics 78–53
white corona 88–237
~ dwarf 91–137
~ flare 89–182
~ hole 98–24
~ mica 74–24
White Oval 83–195
white–light corona 88–237
~ flare 89–182
~ photograph 88–37
wide double star 96–9
Widmanstätten figures 87–105
~ pattern 87–105
width, effective line 89–208
pulse ~ 86–280
Wilson effect 89–89
wind 77–89
anabatic ~ 77–92
burst of solar ~ 84–42
cyclostrophic ~ 77–95
~ direction 77–115
~ erosion 83–163
~ force 77–107
galactic ~ 98–277
geostrophic ~ 77–96
gradient ~ 77–94
gusty ~ 77–111
interstellar ~ 97–151
permanent solar ~ 84–40
prevailing ~ 77–117
quiescent solar ~ 84–41
quiet solar ~ 84–41
~ rose 77–116
solar ~ 88–156
solar ~ burst 84–42
~ speed 77–109
stellar ~ 93–14
trade ~ 77–100
~ vane 77–104
~ velocity 77–109
winding of spiral arms 98–30
outward ~ of spiral arms 98–30
windows, galactic 98–57
~ in the zone of avoidance 98–57
transmission ~ 83–264
transparency ~ 83–264
Winged Horse 92–75
winking filament 89–216
winter constellation 92–1b
Wolf 92–63
~ number 89–65
~ sunspot number 89–65
Wolf–Rayet star 91–160
wood spirit 97–32
world 99–3
end of the ~ 99–183
~ geodetic network 71–38
~ geodetic system 71–39
~ gravity network 71–81
~ line 99–154
~ Magnetic Survey 75–65
~ Meteorological Organization 77–17
~ model 99–46
~ network 77–18
origin of the ~ 99–11
~ point 99–125
~ system 99–20
~ Weather Watch 77–16
~ vector 99–138
WR star 91–160
wrinkle ridge 81–187
WWW 77–16

X

X–ray bright point 89–183a
~ burst 79–117, 89–184
~ corona 88–240
~ event 89–184
~ flare 89–183
~ galaxy 98–203
~ heliogram 88–74
~ pulsar 95–138
solar ~ event 89–184
~ Sun 88–4

Y

year, light 91–18
~ of the Quiet Sun 89–10
Quiet Sun ~ 89–10

yellow 91–98
~ colo(u)r–index 83–88
~ line 89–292
ylem 99–170
young crater 81–206
~ star 94–63, 96–93
Young's modulus 73–59

Z

zenith attraction 86–52
~ magnitude 91–63
zenithal magnitude 91–63
~ projection 71–45
zero line 89–151
~ line of magnetic field 89–151
~ phase 83–73
~ point of period–luminosity relation 95–104
~ spacing 88–101
zero–age line 94–60
~ main sequence 94–60, 96–152
~ sequence 96–152
~ star 94–59
zero–point energy 93–187
~ kinetic energy 93–187
zero–temperature model 93–186
~ model of a star 93–186
~ star 91–166
zero–velocity wave–length 98–365
z–motion 98–188
zodiac 92–5a
zodiacal band 84–63
~ cloud 84–54
~ constellation 92–5
false ~ light 84–62
~ light 79–13
~ particles 84–66
~ signs 92–6
zonal harmonic 93–55
~ harmonic functions 93–55
zone, abyssal 72–41
activity ~ 89–54
artificial radiation ~ 75–108
central ~ 81–114, 89–47
convective ~ 93–149
earthquake ~ 73–31
equatorial ~ 83–188
helium burning ~ 93–145
hydrogen burning ~ 93–144
hydrogen convective ~ 93–159
limb ~ 81–113
marginal ~ 81–113
migration of spot ~ 89–74
natural radiation ~ 75–107
occultation ~ 81–107
~ of ablation 72–123
~ of avoidance 98–56
~ of dissipation 72–124
~ of growth of glacier 72–122
~ of maximum auroral frequency 79–125
~ of occultation 81–107
~ of silence 76–42
~ of snow–supply 72–122
orogenetic ~ 73–75a
orogenic ~ 73–75a
outer radiation ~ 75–105
polar ~ 77–193
radiative ~ 93–134
royal ~ 89–54
seismic ~ 73–31
semi–convective ~ 93–160
spot ~ 89–54
temperate ~ 77–192, 83–189
torrid ~ 77–191
transition ~ 72–13
windows in ~ of avoidance 98–57
zones 83–187
Fresnel ~ 86–294
z–oscillation 98–187
Zürich classification 89–127
~ relative sunspot number 89–65

INDEX FRANÇAIS

French Index

Französischer Index

Índice francés

Índice francês

французский указатель

ALPHABET:

a	h	o	v
b	i	p	w
c	j	q	x
d	k	r	y
e	l	s	z
f	m	t	
g	n	u	

A

abaissement de la température 77–41
aberration diurne 86–57
ablation 86–182
abondance anormale 91–173
~ chimique 91–169
~ cosmique des éléments 94–107
~ d'un élément 91–169
~ isotopique 91–179, 94–106
~ métallique 91–170
~ naturelle 91–171
~ normale 91–172
~ totale 91–177
abondances des éléments 82–141
~ relatives des éléments 91–174
absorption à début brusque 78–47
~ aurorale 78–46
~ brusque du bruit cosmique 78–42
~ dans la raie 93–108
~ de vapeur d'eau 77–127
~ différentielle 97–90
~ interstellaire 97–84
~ interstellaire générale 97–87
~ interstellaire sélective 97–90
~ interstellaire totale 97–87a
~ ionosphérique polaire 78–44
~ lente des neutrons 94–88
~ libre-libre 93–106
~ lié-libre 93–107
~ lié-lié 93–108
~ neutronique lente 94–88
~ neutronique rapide 94–89
~ par la vapeur d'eau 77–127
abyssale 72–41
plaine ~ 72–42
zone ~ 72–41
abysse 72–41
accélération de la pesanteur 71–64
~ effective 85–88
~ équatoriale 88–131
accrétion 83a–33
accroissement de la masse de la Terre 87–18
~ de luminosité 86–79
~ de masse 83a–38
accumulation 83a–34
acétaldéhyde 97–42
acétonitrile 97–38
acétylène de méthyle 97–41
Achernar 92–102
achondrite 87–77
acide α-amino-propionique 87–72
~ aminé 87–70
~ formique 97–33
~ isocyanique 97–39
~ organique 87–69
acoustique atmosphérique 76–31
action atmosphérique 87–116
~ de l'atmosphère 87–116
~ de marée 83a–15
activation d'un filament 89–215
~ d'une protubérance 89–239
~ d'une protubérance quiescente 89–239
activité aurorale 79–100
~ cométaire 85–110
~ corpusculaire 89–7
~ d'éruptions 89–176
~ d'éruptions chromospériques 89–17
~ d'un courant 86–229
~ de fumerolles 82–164
~ de surface 82–231
~ des météores sporadiques 86–136
~ des taches solaires 89–62
~ du champ magnétique 89–5
~ du noyau galactique 98–271
~ éruptive 89–176, 95–71
~ galactique 98–261
~ géomagnétique 75–86
~ interne 82–158
~ lunaire 82–149
~ magnétique 89–5
~ météorique 86–134
~ optique 89–3
~ radioélectrique 89–6
~ récurrente 98–267
~ relative d'un courant 86–231
~ séismique 73–4
~ sismique 73–4
~ solaire 89–1
~ superficielle 82–231
~ volcanique 73–110, 82–154
~ volcanique déclenchée 82–155
~ volcanique déclenchée par chute météoritique 82–156
~ volcanique le long de fractures 82–157
adhésif 82–137
adsorbant 85–130
adsorbat 85–131
adsorber 85–129
adsorption 85–128
adsorptivité 85–120
advection 77–63
aérographique, carte 83–136
aérolithe 87–60
aérologie 76–122
aéronomie 76–122a
aérosol 76–89
filtre pour ~ 76–94
particule d' ~ 76–91
aérotriangulation spatiale 71–26
affaissement 82–199
afflux de gaz de l'espace intergalactique 98–356
âge absolu 82–269, 83a–70
~ caractéristique 99–79
~ d'exposition 87–139
~ d'exposition au rayonnement cosmique 87–139
~ d'un courant 86–245
~ d'un essaim météorique 86–246
~ d'une pluie météorique 86–246
~ de l'univers 99–12
~ de la Lune 81–31, 82–264
~ de la rétention gazeuse 87–135
~ de la Terre 83a–68
~ de rétention 87–135
~ de solidification 87–137
~ de strontium 82–271
~ des étoiles 94–61
~ des météorites 87–127
~ déterminé par la méthode au plomb isotopique 74–100
~ déterminé par la méthode du rapport uranium-plomb 74–101
~ déterminé par le rapport rubidium-strontium 82–271
~ géologique 74–90, 83a–69
~ par le strontium 82–271
~ relatif 82–268
~ stratigraphique 82–267
agglomération 83a–32
aggravation du temps 77–23
agitation de la mer 72–86
~ micros(é)ismique 73–54
~ sismique 73–54
agrégat moléculaire 97–65
agrégation de la matière 99–170z
Aigle 92–18
aiguilles de glace 76–112
aimantation chimique 75–31
~ de dépôt 75–33
~ détritique 75–33
~ induite 75–29
~ inverse 75–37
~ inversée 75–37
~ primaire 75–35
~ rémanente 75–28
~ résiduelle 75–28
~ secondaire 75–36
~ thermorémanente 75–30
~ thermorésiduelle 75–30
~ visqueuse 75–32
air 76–6
~ ambiant 77–4
aire calculée 89–93
~ corrigée 89–93
~ d'audibilité 86–120
~ de dispersion 87–38
~ de l'éruption 89–207
~ de la chute 87–38
~ de radiation 86–238
~ de répartition 87–38
~ de visibilité 87–31
~ des protubérances 89–233
~ des taches 89–67
~ projetée des taches 89–91
~ réelle 89–93
~ sombre 83–155
~ tachée 89–90
~ totale des taches 89–94
aires choisies 98–78

F

~ de Kapteyn 98–78
ajustement isostatique 73–100
al 91–18
a. l. 91–18
alanine 87–72
albédo de Bond 83–91
~ de Lambert 83–89
~ des nuages 83–95
~ du rayonnement cosmique 75–116
~ du rayonnement radio 83–94
~ géométrique 83–92
~ global 83–90b
~ monochromatique 83–93
~ moyen 83–90a
~ neutronique 75–116
~ radioélectrique 83–94
~ sphérique 83–91
~ total 83–90b
Albirco 92–103
albite 74–17
alcool méthylique 97–32
Alcor 92–104
aldéhyde acétique 97–42
~ formique 97–27
Algol 92–105
algolide 95–77
Algonquien 74–77
alignement de grains 97–78
alizés 77–100
allongement des cornes de Vénus 83–131
Altaïr 92–106
altération 74–57
~ en environnement cosmique 82–244
~ météorique 74–57
~ par exposition à l'espace 82–244
altimétrie par satellite 71–50
altitude au-dessus du niveau de la mer 71–51
~ d'apparition 86–39
~ de disparition 86–40
~ de la couche émissive 79–26
~ des météores 86–193
~ du point d'apparition 86–39
~ du point d'extinction 86–40
~ du point miroir 75–125
~ ellipsoïdale 71–52
~ finale 86–40
~ pression 76–16
altocumulus 77–144
altostratus 77–140
alunissage 82–13
amas à forte concentration 96–129
~ compact de galaxies 98–334
~ d'amas de galaxies 98–346
~ d'étoiles 96–79
~ de dômes 81–178
~ de galaxies 98–329
~ de la Vierge 98–345
~ double 96–81
~ en mouvement 96–97
~ galactique 96–99
~ globulaire 96–114
~ globulaire de galaxies 98–330
~ irrégulier de galaxies 98–337
~ ouvert 96–99
~ ouvert de galaxies 98–333
~ pauvre en étoiles 96–102
~ pauvre en galaxies 98–331
~ régulier de galaxies 98–336
~ riche en étoiles 96–100
~ riche en galaxies 98–332
~ stellaire 96–79
~ stellaire pauvre en métaux 96–113
~ stellaire relativiste 96–154
~ stellaire riche en métaux 96–112
~ très compact de galaxies 98–335
ambiant 76–7
amide formique 97–35
amine 87–71
aminoacide 87–70
ammoniac 97–25
amortisseur 82–24
amphibole 74–20
amplification du champ 89–156
amplitude de variation 95–38
~ de variation d'éclat 95–38
~ des marées 82–292
analyse chimique 82–75
~ de sol 82–79
~ in situ 82–136
~ minéralogique 82–53
~ par microsonde 82–143
~ pétrographique 82–52
~ sur place 82–136
ancien cratère 81–205
~ satellite 83–240
ancienne supernova 95–134
Andromède 92–14
Andromédides 86–256
anémographe 77–114
anémomètre 77–113
~ enregistreur 77–114
angle caractéristique 98–33
~ de phase 83–72
~ de position 81–71
~ héliocentrique 89–37
anhydride carbonique solide 83–144
anneau A 83–225
~ B 83–222
~ brillant 89–101
~ C 83–221
~ d'étoiles 96–78
~ de Bishop 79–88
~ de courants 75–100
~ de crêpe 83–221
~ de Saturne 83–215
~ des astéroïdes 84–70
~ équatorial de courant 75–98
~ extérieur 83–225
~ fantôme 81–242
~ gazeux 95–66
~ intérieur 83–221
~ médian 83–222
~ nébulaire 83a–10
~ sombre 83–221
~ stellaire 96–78
~ transparent 83–221
~ vu de profil 83–218
~ vu par la tranche 83–218
année de lumière 91–18
~ du Soleil Calme 89–10
année-de-lumière 91–18
année-lumière 91–18
annihilation des champs magnétiques 89–158
annulaire 97–146
anomalie de l'abondance 91–183
~ de la pesanteur 71–75
~ de la pesanteur lunaire 81–144
~ de phase à début brusque 78–52
~ gravimétrique 71–75
~ magnétique 75–67
~ sud-atlantique 75–68
anorthite 74–18
anorthosite 82–87
antapex 86–65
Antarès 92–107
Antécambrien 74–70
antenne émettrice 78–16
~ mobile 88–108
~ réceptrice 78–18
anthélie 79–83
antiapex 86–65
~ solaire 98–65
anticentre galactique 98–14
anticlinal 73–90
pli ~ 73–90
anticrépuscule 79–57
anticyclone 77–84
anticyclonique 77–85
antiqueue 85–78
antisélène 79–85
antivertex 98–106
Antlia 92–15
apex 86–64
~ du Soleil 91–23
hauteur de l' ~ 86–66
~ solaire 98–64
aphélie 83–28a
aplatissement 83a–22
~ apparent 98–295
~ d'un amas globulaire 96–123
~ de la couronne 88–247
~ de la Terre 71–8
~ polaire 83–114
~ réel 98–296
apoapside galactique 98–190
apoastre 96–61
~ galactique 98–190
~ martien 83–141
~ saturnien 83–211
apocentre galactique 98–190
apogée 81–39
apojove 83–173
apojovium 83–173

Apollon 84–81
apolune 81–41
apomars 83–141
aposaturnium 83–211
aposélène 81–41
appareil de sismologie active 82–220
~ scientifique déposé sur la Lune 82–42
~ sismique 82–219
~ sismique actif 82–220
apparition d'un bolide 86–117
~ d'un météore 86–35
~ d'une comète 85–18
point d' ~ 86–36
première ~ 85–19
première ~ d'une comète 85–19
appartenance physique 96–83
~ physique à l'amas 96–83
approche minimale 84–69
approximation de Chandrasekhar 93–51
~ de Milne-Eddington 93–50
~ non-relativiste 99–97
appulse 81–101
apside inférieur galactique 98–189
~ supérieure galactique 98–190
Apus 92–16
Aquarides 86–249
Aquarius 92–17
Aquila 92–18
Ara 92–19
arbres 89–273
en forme d' ~ 89–273
arc auroral 79–146a
~ avec structure de rayons 79–153
~ circumhorizontal 79–82
~ circumzénithal 79–74
~ crépusculaire 79–50
~ de méridien 71–21
~ homogène calme 79–143
~ montagneux 73–84
~ primaire 73–82
~ pulsant 79–147
~ rayé 79–153
~ secondaire 73–83
~ tangent au halo 79–75
~ tangent inférieur 79–76
~ tangent infralatéral 79–87
~ tangent supérieur 79–77
arche coronale 89–298
~ éruptive 89–251
arc-en-ciel 79–65
Archéozoïque 74–76
arcs intertropicaux 79–53
arène lunaire 82–112
aréo- 83–135
aréocentrique 83–138
coordonnée ~ 83–138
orbite ~ 83–138
aréographique 83–135
aréophysique 83–134
arête 81–186
argile pélagigue 72–54a
~ rouge 72–54
Ariel 83–235
Aries 92–20
arrivée sur la Terre 89–193
ascension 82–8
~ droite 81–60
~ droite de la Lune 81–60
~ droite du Soleil 81–61
aspect 83–50
~ d'une planète 83–109
association accidentelle 97–126
~ génétique 97–127
~ O 96–89
~ stellaire 96–88
~ T 96–90
associé à une éruption 89–188
assombrissement au bord 88–142
~ de gravité 91–41
~ gravitationnel 91–41
~ par gravité 91–41
~ vers le bord 88–142
astérisme 92–3a
astéroïde = planétoïde = petite planète 84–68
~ artificiel 84–69a
~ binaire 84–79
~ catalogué 84–95
~ d'approche serrée 84–78
~ de courte approche 84–78
~ de l'anneau 84–71
~ de la famille de Mars 84–74
~ de libration 84–91a
~ de Mars 84–74
~ de type Amor 84–80
~ de type Apollo 84–80a
~ double 84–79
~ en forme de cigare 84–77
~ numéroté 84–95
~ perdu 84–97
~ traversant l'orbite terrestre 84–76
~ traversant l'orbite de Mars 84–75
astre 79–9a
astronomie depuis une base lunaire 82–4
~ extragalactique 98–208
~ météorique 86–3
~ météoritique 87–6
~ stellaire 91–2
asymétrie est-ouest 89–49
~ nord-sud 89–58
asymptotique aux orbites circulaires 98–180
ataxite 87–87
~ pauvre en nickel 87–88
~ riche en nickel 87–89
Atelier du Sculpteur 92–87
atlas céleste 92–12
~ de la Lune 81–136
~ du ciel 92–12
~ lunaire 81–136
~ photométrique 88–138
atmosphère 83–241
~ adiabatique 76–12
~ d'électrons 93–28
~ de la Terre 76–5
~ d'une planète 83–243
~ du Soleil 88–163
~ électronique 93–28
~ étendue 93–9
~ extérieure 76–68
~ finie 93–23
~ grise 93–26
~ homogène 76–10
hors de l'~ 88–149
~ inférieur 76–50
~ initiale 83a–44
~ isotherme 93–25
~ libre 76–13
~ modèle 83–247
~ moyenne 76–51
~ non grise 93–27
~ oxydante 83a–47
~ planétaire 83–243
~ primitive 83–244
~ purement diffusante 93–16
~ réductrice 83a–46
~ semi-infinie 93–22
~ solaire 88–163
~ sphérique 93–24
~ standard 76–11
~ stellaire 93–2
~ stratifiée en couches planes parallèles 93–21
~ supérieure 76–52
~ terrestre 76–5
atmosphérique 83–242
atmosphériques 78–50
atome à deux niveaux 93–45
~ à plusieurs niveaux 93–46
~ primitif 99–169
~ unique 99–169
atomisation par le vent solaire 82–245
attaque chimique d'une surface polie 87–102
atténuateur à échelons 88–80
atterrissage humain 82–33
~ lunaire 82–13
~ sur la Lune 82–13
attraction zénithale 86–52
aube 79–35
auge de potentiel 98–176
augmentation brusque d'éclat 95–69
~ brusque des atmosphériques 78–51
~ brusque du niveau des atmosphériques 78–51
~ d'éclat 86–79
~ de la température 77–42
~ relativiste de la masse 99–142
auréole 79–60, 81–168
~ lunaire 79–61
~ solaire 79–62
Auriga 92–21
aurore 79–56
~ à électrons 79–110

F

~ à protons 79–108
~ artificielle 79–99
~ australe 79–106
~ avec structure à rubans 79–140
~ boréale 79–105
~ de basse latitude 79–96
~ de calotte polaire 79–98
~ de jour 79–97
~ détectée par radar 79–95
~ éclairée par le Soleil 79–130
~ flamboyante 79–157
~ polaire 79–93
~ radio 79–94
australite 87–98
Autel 92–19
auteur de la découverte 85–21
autogravitant 94–33
autogravitation 94–34
système soumis à la ~ 98–169
auto-inversion de l'aimantation 75–38
avalanches 81–237
avance du périastre 96–71
~ du périhélie 99–160
~ relativiste du périhélie 99–160
averse de météorites 87–37
~ météoritique 87–37
axe de rotation 83–33
~ de visée 91–28
~ géomagnétique 75–47
~ magnétique 75–46
~ magnétique terrestre 75–47
Azoïque 74–75
azote, premier système positif de l' 79–20
système négatif de l' ~ 79–21
azur du ciel 77–60

B

baie magnétique 75–80
baisse de la température 77–41
Balance 92–62
balayage des gaz 83a–41
~ en longueur d'onde 88–111
~ en position 88–105
Baleine 92–34
banc 72–96
bande 83–190
~ de lumière 86–27
~ de lumière zodiacale 84–63
~ équatoriale 83–191
~ homogène 79–146
~ interstellaire diffuse 97–100
~ lumineuse 86–27
~ rayée 79–150
~ sombre 83–190
~ zodiacale 84–63
bandelette de mica 88–87
bandes de fluorescence 85–132
~ de Swan 85–133
~ de Vegard-Kaplan 79–19
~ non-identifiées 85–134
~ sombres 98–288
système de ~ 79–18
banquise 72–102
barre 72–96
~ galactique 98–254
barysphère 72–17
basalte 74–9
~ lunaire 82–84
base de l'interferomètre 88–100
~ d'une protubérance 89–234
~ fixe 88–102
~ lunaire 82–17
~ lunaire habitée 82–18
~ sur la Lune 82–17
~ variable 88–103
basse-mer 82–283
bassin 81–175
~ volcanique 81–179
batholite 73–139
bathymétrie 72–28
Baudrier d'Orion 92–129a
Bélier 92–20
Bételgeuse 92–108
Bielides 86–256
bifurcation 98—5
big bang 99–167
bilan d'énergie 88–251
~ de masse 88–250
~ de radiation 88–252
~ détaillé 93–41
~ énergétique 88–251
~ thermique 77–62
bille 82–129
binaire 96–3
~ à éclipse 95–74
~ à grande séparation 96–9
~ à raies simples 96–15
~ astrométrique 96–12
~ en contact 96–29
~ non-résolue 96–10
~ photométrique 95–73
~ presque séparée 96–34
~ séparée 96–35
~ serrée 96–23
~ spectroscopique 96–14
~ spectroscopique à raies doubles 96–17
~ specroscopique à raies simples 96–15
biosphère 72–8
biotite 74–25
bleu 91–99
~ du ciel 77–60
blindage 86–171
~ magnétique 89–225
blink microscope 91–34
bloc continental 72–10
~ erratique 72–132
~ rocheux 82–130
bolide 86–11
~ brillant 86–10
~ détonant 86–119
~ géant 86–9
gros ~ 86–10
très gros ~ 86–9
~ visible de jour 86–118
bombardement par les particules 82–240
~ par météorites 82–235
bombe d'Ellerman 89–200
~ volcanique 73–144
Boötes 92–22
bord 89–29
~ apparent solaire 89–28
au ~ 89–29
au-dessus du ~ 89–30
~ brillant d'un filament 89–226
~ d'une galaxie 98–294
~ de cratère 81–221
~ de la Lune 81–95
~ éclairé 81–96
~ est 89–34
~ extérieur 83–227
~ extrême 89–32
~ lunaire 81–95
~ obscur 81–97
~ ouest 89–33
~ solaire 89–28
bosse de la courbe 95–37
boucle interplanétaire 84–21
~ magnétique 84–19
~ solaire interplanétaire 84–20
bouclier 72–11
boue 72–58
bouée automatique 72–100
~ de recherche 72–99
boule de feu primordiale 99–168
~ de poussière 86–190
boulette 82–129
bourrelet de marée 81–119
~ équatorial 71–9
~ lunaire 81–118
Boussole 92–82
Bouvier 92–22
branche ascendante 95–34
~ décroissante 95–35
~ des géantes 91–152b
~ descendante 95–35
~ du courant 86–240
~ horizontale 91–132
branchement 98–5a
bras à la traîne 98–29
~ d'Orion 98–36
~ de Persée 98–34
~ de 3 kpc 98–37
~ de trois kiloparsec 98–37
~ déroulé 98–31
~ du Sagittaire 98–35
~ en expansion 98–38
~ enroulé 98–29
~ expansif 98–38
~ galactique 98–26a
~ spiral 98–27
brassage de matière stellaire 93–162
brèche lunaire 82–85

brillance, coefficient de 83–96
facteur de ~ 83–90
brise de mer 77–97
~ de terre 77–98
brisure de la symétrie 99–170s
brouillard 77–130
bruine 77–172
bruit acoustique 88–220
~ de fond séismique 73–55
~ séismique 73–54
~ sismique 73–54
brume 83–250
~ sèche 77–27
bulbe central 98–16
~ nucléaire 98–16
bulle 82–180
à ~ 82–181
bulletin météorologique 77–10
bulleux 82–181
Burin du Graveur 92–23

C

cadastrage sélénophysique 82–40
Caelum 92–23
caillou 82–128
calculateur 85–22
caldeira = caldère 81–179
~ d'affaissement 81–181
~ d'erosion 81–183
~ d'explosion 81–180
~ de subsidence 81–182
Callisto 83–208
calme 77–110
calmes équatoriaux 77–101
~ tropicaux 77–102
zone de ~ équatoriaux 77–101
calotte aurorale 79–127
~ glaciaire 72–115
~ polaire 83–143
Cambrien 74–78
Caméléon 92–35
Camelopardalis 92–24
caméra grand champ 79–104
~ pour tout le ciel 79–104
~ tout ciel 79–104
campagne d'observation des éruptions à protons 89–177
canal d'écoulement 82–192
canaux de Mars 83–157
~ martiens 83–157
Cancer 92–25
Canes Venatici 92–26
Canis Major 92–27
~ Minor 92–28
cañon 83–154
Canopus 92–109
canyon 83–154
~ sous-marin 72–46
capacité de portance 82–64
Capella 92–110
Capricorne 92–29
Capricornus 92–29
capture 75–109, 83a–19
~ d'étoile 96–73
~ de matière intergalactique 98–355
~ de neutrons 94–84
~ de protons 94–90
~ des comètes 85–101
hypothèse de ~ 83a–18
~ neutronique 94–84
~ protonique 94–90
région de ~ 75–110
caractère d'impact 82–228
~ fini de l'univers 99–4
~ infini de l'univers 99–6
~ structural 88–248
~ topographique 81–149
~ volcanique 82–159
caractères lunaires 81–148
~ structuraux 98–224
caractéristique du sol 82–108
~ d'impact 82–228
~ lunaire 81–148
~ primaire 81–150
~ secondaire 81–151
~ topographique 81–149
carbonate 74–27
carbone radioactif 74–95
Carbonifère 74–82
carbylamine d'hydrogène 97–40
Carène 92–30
Carina 92–30
carottage 72–51
carotte 72–52
~ échantillon 82–60
carte à haute résolution 89–133
~ aérographique 83–136
~ altimétrique 81–141
~ bathymétrique 71–54
~ céleste 92–9
~ céleste mobile 92–10
~ céleste tournante 92–10
~ colorimétrique 81–134
~ d'atterrissage 72–33
~ de l'éclipse 88–28
~ de la Lune 81–133
~ de la variable 95–15
~ de l'étoile variable 95–15
~ de pesanteur 71–79
~ de profondeurs 71–54
~ de vitesses 89–23
~ des environs de l'étoile variable 95–15
~ des étoiles tournante 92–10
~ du champ magnétique 89–131
~ du ciel 92–9
~ du temps 77–12
~ en courbes de niveau 81–141
~ gravimétrique 81–143
~ héliographique 89–17
~ hydrologique 72–25
~ hypsométrique 81–141
~ journalière du Soleil 89–18
~ lunaire 81–133
~ lunaire photographique 81–137
~ lunaire physique 81–134
~ lunaire thermique 81–135
~ lunaire topographique 81–138
~ magnétique 89–131
~ météorologique 77–12
~ nautique d'atterrissage 72–33
~ océanographique 72–23
~ photométrique 81–134
~ planiciel 92–10
~ polaire 89–19
~ polarimétrique 81–134
~ prévisionnelle 77–21
~ radio 81–135a
~ radioélectrique 81–135a
~ sélénographique 81–133
~ stellaire tournante 92–10
~ structurale 82–203
~ synoptique 77–11
~ synoptique de la chromosphère 89–21
~ synoptique de la couronne 89–22
~ synoptique de la photosphère 89–20
~ tectonique 73–96, 82–202
cartographie de la Lune 81–111
~ de littoral 72–82
~ du ciel 92–11
~ lunaire 81–111
~ sélénologique 82–51
~ thermique 72–83
cas de pollution 72–72
Cassiopée 92–31
Cassiopeia 92–31
cassure 73–78
~ avec déplacement relatif 73–91
~ par collision 87–125
Castor 92–111
catabatique 77–93
catalogue d'étoiles 92–13
~ de Messier 92–12a
~ des étoiles suspectées de variabilité 95–58
~ stellaire 92–13
catastrophe produite par une météorite 87–20
~ s(é)ismique 73–40
cavité 87–117
~ creusée par une météorite 87–58
~ du vent solaire 84–28
~ géomagnétique 75–8
~ héliosphérique 88–257
~ hydromagnétique 84–22
~ interne 87–119
~ interplanétaire 84–25
~ magnétique 84–22
~ planétaire 84–24
~ sombre (autour des protubérances) 89–297

F

F

~ superficielle 87–118
ceinture active 73–32
~ d'astéroïdes 84–70
~ de radiation naturelle 75–107
~ de rayonnement 75–101
~ de rayonnement artificiel 75–108
~ de rayonnement externe 75–105
~ de rayonnement naturel 75–107
~ de Van Allen 75–101
~ extérieure 75–105
~ intérieure 75–106
~ interne 75–106
~ minaurorale 79–129
~ orogénique 73–75
~ séismique 73–30
~ sismique 73–30
~ subaurorale 79–128
cémentation 74–52
cendre volcanique 73–142
~ volcanique fine 82–126
Cénozoïque 74–73
Centaure 92–32
Centaurus 92–32
centre actif 89–60
~ d'activité 89–60
~ d'attraction 89–241
~ de haute pression 77–84
~ de l'étoile 93–84
~ de la plaque 86–156
~ du disque 89–35
~ du disque solaire 89–35
~ du disque visible 88–172
~ du système galactique 98–9
~ galactique 98–9
~ photométrique 96–46
Céphée 92–33
céphéïde 95–92
~ à courte période 95–93
~ à longue période 95–94
~ classique 95–94
Cepheus 92–33
cercle circumzénithal 79–74
~ parhélique 79–82
Cérès 84–82
Cetus 92–34
chaîne de cratères 81–241
~ de galaxies 98–328
~ de montagnes 73–85
~ stellaire 96–77
chaleur d'ablation 86–184
~ de condensation 77–155
~ de désintégration radioactive 83a–62
~ de désorption 85–124
~ de marée 82–287
~ de sublimation 85–139
~ de vaporisation 86–199
~ latente 77–157
~ originelle 74–63
~ primitive 74–63
~ primordiale 83–200
~ radioactive 83a–62
~ radiogénique 83a–62
Chamaeleon 92–35
chambre entraînée 86–142
~ fixe 86–143
~ guidée 86–142
~ non-entraînée 86–143
~ pour la détection des météores 86–145
~ pour la photographie des météores 86–141
~ stationnaire 86–143
champ à grande échelle 89–142
~ à petite échelle 89–143
~ coercitif 75–39
~ de cratères 87–56
~ de fond 89–141
~ de force galactique 98–146
~ de gravitation 98–145
~ de gravité 71–57
~ de Higgs 99–170k
~ de lave 82–184
~ de marées galactique 96–106
~ de névé 72–119
~ de pesanteur 71–57
~ de pierres 82–139
~ de rayonnement 93–30
~ de rayonnement diffus 93–31
~ de rochers 82–139
~ de vitesses 88–187
~ des courants telluriques 75–4
~ des forces de marées 98–196
~ des mouvements différentiels 98–114
~ effectif 95–59
~ fossile 91–53
~ géomagnétique 75–44
~ gravitationnel 98–145
~ gravitationnel galactique 98–146
~ interne 95–63
~ longitudinal 89–138
~ magnétique de fond 89–141
~ magnétique de Terre 75–44
~ magnétique du Soleil 89–129
~ magnétique fossile 91–53
~ magnétique galactique 98–50
~ magnétique général 89–128
~ magnétique général du Soleil 89–128
~ magnétique interplanétaire 84–9
~ magnétique interstellaire 97–106
~ magnétique lunaire 82–45
~ magnétique photosphérique 89–130
~ magnétique primitif 95–62
~ magnétique résiduel 91–53
~ magnétique terrestre 75–44
~ magnétique très intense 95–148
~ magnétisant 75–42
~ quantique 99–170m
~ sélénomagnétique 82–45
~ sonore 76–40
~ superficiel 95–64
~ tellurique 75–4
~ transversal 89–137
~ transverse 89–137
changement de coloration de la mer 72–80
~ de forme 73–68
~ de la ligne littorale 72–78
~ de phase 99–170p
~ de polarité 89–153
~ de temps 77–22
~ du climat 77–184
~ évolutif 94–27
changements diurnes 83–167
~ saisonniers 83–166
changeur de phase 88–114
charge de la Terre 75–3
chauffage de la chromosphère 88–213
~ de la couronne 88–213a
chemin de refroidissement 93–202
~ évolutif 94–29
cheminée 73–123
~ de volcan 82–162
~ du pic central 81–244
~ volcanique 73–122
cheminement 82–35
chemosphère 76–62
chercheur solaire 88–51
Cheval Ailé 92–75
Chevalet du Peintre 92–78
chevelure 85–64
~ atomique 85–67
~ de Bérénice 92–38
~ de radicaux 85–66
~ intérieure 86–65
~ moléculaire 85–65
~ ultraviolette 85–67
~ visible 85–66
Chiens de Chasse 92–26
chimie de l'atmosphère 76–79
~ de la Lune 82–74
chimiluminescence 79–31
chimiosphère 76–62
chlorinité 72–68
choc thermique 86–206
choeur de l'aube 78–62
chondre 87–76
chondrite 87–61
~ carbonée 87–63
~ non-carbonée 87–75
chromosphère 88–191
basse ~ 88–193
hauteur de la ~ 88–202
soutien de la ~ 88–223
structure fine de la ~ 88–203
~ supérieure 88–192
support de la ~ 88–223
chromosphérique 88–194
chronologie géologique 74–67
chute d'un bolide 86–117
~ d'un météore 86–5
~ d'une météorite 87–19
~ libre 94–39

lieu d'une ~ météoritique 87–50
~météoritique 87–19
moment d'une ~ météoritique 87–50
~ naturelle 82–227
~ suspectée de météorite 87–49
témoin oculaire de la ~ 87–48
~ unique 87–36
ciel austral 79–3c
~ boreal 79–3b
~ clair 77–153
~ couvert 77–152
~ crépusculaire 79–43
~ diurne 79–3a
~ étoilé 91–1a
lumière du ~ 77–53
lumière du ~ nocturne 79–8
luminescence du ~ 79–2
~ moutonné 77–143
~ nocturne 79–6
~ nocturne lumineux 86–28
~ nuageux 77–148
~ serein 77–153
cimentation 74–52
Circinus 92–36
circonstances de la chute 87–47
circulation atmosphérique 77–64
~ générale 77–65
~ méridienne 93–163
~ méridionale 93–163
~ planétaire 77–66
circumlunaire 81–5
circumstellaire 97–3
cirque lunaire 81–265
cirro-cumulus 77–143
cirro-stratus 77–141
cirrus 77–136
Ciseau du Graveur 92–23
cislunaire 81–4
classe d'éruption 89–205
~ de luminosité 91–146
~ spectrale 91–107
classification de Harvard 91–91
~ de Zürich 89–127
~ des étoiles variables 95–46
~ des galaxies 98–226
~ des météorites 87–59
~ magnétique 89–109
~ pétrographique 82–101
~ spectrale 91–90
~ spectrale bidimensionnelle 91–105
~ spectrale tridimensionnelle 91–106
clathrate 85–51
cliché météorique 86–151
climat 77–180
~ continental 77–185
~ de montagne 77–187
~ maritime 77–186
climatique 77–182
climatologie 77–179
~ par satellite 77–181
coagulation 76–119
coalescence 76–120
Cocheer 92–21
coefficient d'assombrissement 88–144
~ de brillance 83–96
~ de diffusion atmosphérique 86–287
~ de forme 86–177
~ de l'éllipsoïde des vitesses 98–102
~ de phase 83–77
~ de Poisson 73–60
~ de rugosité 82–107
~ de stabilité 98–199
~ de traînée 86–180
~ de transmission 86–201
~ de transmission calorifique 86–201
~ harmonique 71–59
cœlostat 88–46
~ horizontal 88–47
~ vertical 88–48
coeur de graphite 97—71
cogéoïde 71–17
cohérence 99–33
cohésif 82–138
coiffe d'air 86–170
~ gazeuse 86–170
collapse = écroulement = effondrement 94–109
~ de l'univers 99–181
~ final 99–182
~ gravitationnel 94–109
collecteur de poussière météorique 87–13
collection de météorites 87–8
~ de poussières météoriques 87–12
collision à trois corps 96–74
~ catastrophique 84–106
~ centrale 83a–17
~ de galaxies 98–322
~ directe 83a–17
~ entre galaxies 98–322
~ entre étoiles 83a–16
~ érosive 84–105
~ inélastique des étoiles 98–26
~ stellaire 83a–16
~ triple 79–30
colloïdal 76–97
colloïde 76–96
Colombe 92–37
colonne d'ionisation 86–282
~ ionisée 86–282
~ lumineuse 79–78
coloration de la mer 72–80
changement de ~ de la mer 72–80
Columba 92–37
Coma Berenices 92–38
comblement d'une dépression 77–83
combustion de l'hélium 94–78
~ de l'hydrogène 94–56
~ de l'oxygène 94–82
~ du carbone 94–81
~ du silicium 94–83
~ explosive de l'hélium 94–79
cométaire 85–2
comète 85–1
~ à courte période 85–5
~ à longue période 85–6
~ active 85–149
~ artificielle 85–14
~ associée à un essaim 86–259
~ de l'éclipse 85–12
découverte d'une ~ 85–16
désactivation d'une ~ 85–148
désagrégation d'une ~ 85–115
désignation d'une ~ 85–31
éjection depuis une ~ 86–262
~ elliptique 85–4
enveloppes d'une ~ 85–69
~ éteinte 85–150
~ génératrice 86–260
halos d'une ~ 85–68
~ hyperbolique 85–8
~ inactive 85–147
~ mère 86–260
~ non-périodique 85–10
~ parabolique 85–7
~ périodique 85–3
~ prédite 85–11
queue d'une ~ 85–70
~ qui rase le Soleil 85–13
~ recouverte pendant l'éclipse 85–12
retour d'une ~ 85–9
rupture d'une ~ 85–114
~ télescopique 85–14a
tête d'une ~ 85–35
comètes, essaim de 85–100
famille de ~ 85–102
groupe de ~ 85–105
nuage de ~ 85–100
nuage de ~ d'Oort 85–97
origine des ~ 85–96
physique des ~ 85–104
recherche des ~ 85–15
structure des ~ 85–34
commencement au coucher du Soleil 88–22
~ de l'éclipse 81–74
commutation de phase 88–112
compaction 74–50
compagnon 96–38
~ d'une galaxie 98–216
~ de Sirius 91–138
~ elliptique 98–217
~ elliptique d'une galaxie 98–217
~ invisible 96–41
~ obscur 96–42
~ optique 98–218
~ optique d'une galaxie 98–218
~ planétaire 96–43
~ physique 98–219
~ physique d'une galaxie 98–219
Compas 92–36
~ du Bateau 92–82
~ du Marin 92–82
compensateur d'effet Doppler-Fizeau 88–91
compensation de l'effet Doppler-Fizeau 88–90

F

~ isostatique 73–100
complètement dégénéré 93–185
complexe de base 72–12
~ fondamental 72–12
composante 96–36
~ de poussière 98–232
~ horizontale magnétique 75–60
~ longitudinale 89–135
~ lunisolaire 82–307
~ météorique 86–234
~ principale 96–37
~ secondaire 96–38
~ σ 88–89
~ transversale 89–136
~ verticale 75–62
composition chimique 91–168
~ chimique d'une étoile 91–168
~ ionique 78–10
~ isotopique 74–93, 82–140
~ moléculaire 78–11
compressibilité 73–57
concentration 96–127
~ de masses lunaires 81–145
~ de traces 83–266
~ galactique 98–76
~ vers le centre 98–76
~ vers le plan fondamental 98–76
~ vers le plan galactique 98–55
condensation 83a–26
~ centrale 85–37, 86–244
~ coronale 89–281
~ coronale permanente 89–282
~ coronale sporadique 89–283
~ dans la queue 85–95
~ dans un courant météorique 86–217
niveau de ~ 77–156
noyau de ~ 76–103
condenser, se 77–129
condition à la surface 93–80
~ au centre 93–81
~ de dégénérescence 93–181
~ de raccordement 93–142
~ de stabilité 93–205
~ de symétrie 98–162
conditions météorologiques 77–2
~ normales 77–3
~ thermiques 77–35
Conducteur de boeuf 92–22
~ de taureau 92–22
~ de troupeau 92–22
conduit 73–124
cône composé 73–121
~ d'éruption 73–118
~ d'ombre 81–90
~ de cendres 73–120
~ de débris 73–120
~ de lave 81–176a
~ de libération 76–27
~ lumineux 95–156
~ pyroclastique 73–120
~ réfléchissant 88–62
~ volcanique 73–118
configuration de la surface 83–105
~ des planètes 83–50
~ du champ 89–140
~ planétaire 83–50
confinement magnétique des rayons cosmiques 98–51
congélation perpétuelle 83–152
conglomérat 85–41
~ de glace 85–43
conglomération de la matière 99–170z
conjonction 83–55
~ inférieure 83–57
~ supérieure 83–56
~ triple 83–58
connexion accidentelle 97–126
~ génétique 97–127
consolidation du magma 82–190
constant avec temps 98–150
constante cosmique 99–29
~ cosmologique 99–29
~ de Hubble 98–302
~ de Poisson 73–60
~ de récession 98–302
~ solaire 88–148
constantes de l'ellipsoïde des vitesses 98–102
~ de Oort 98–118
constellation 92–1
~ circumploraire 92–4a
~ d'été 92–1a
~ d'hiver 92–1b
~ zodiacale 92–5
constituant atmosphérique 83–265
~ des roches 82–99
~ minéralogique 82–100
constitution interne des étoiles 93–71
contact, dernier 81–85
premier ~ 81–84
container d'échantillons 82–81
conteneur d'échantillons 82–81
contenu stellaire 98–172
~ stellaire d'une galaxie 98–231
continent 72–9, 81–191
contours isophotes 97–124
~ isophotes d'une nébuleuse 97–124
contraction 93–117
~ catastrophique 94–109
~ de l'univers 99–180
~ de Lorentz 99–134
~ des longueurs 99–134
~ d'Helmholtz 94–37
~ du noyau 94–70
~ gravifique 94–37
~ gravitationnelle 83–203, 94–37
~ quasi stationnaire 94–40
~ quasi statique 94–40
~ relativiste 99–134
contrainte 73–62
~ de cisaillement 73–63
~ de la croûte 82–210
~ principale 73–64
~ trangentielle 73–63
~ thermique 86–205, 87–27
contre-alizés 77–103
convectif 93–148
convection 93–147
~ laminaire 93–153
~ turbulente 93–152
coordonnée d'espace 99–124
~ de genre temps 99–123
~ de position 99–124
~ de temps 99–123
~ spatiale 99–124
coordonnées comobiles 99–44
~ cylindriques 98–156
~ galactocentriques 98–12
~ géomagnétiques 75–51
~ héliographiques 89–39
~ jovicentriques 83–176
~ jovigraphiques 83–175
~ planétographiques 83–8
~ sélénographiques 81–9
~ zenocentriques 83–176
~ zenographiques 83–175
coplanarité des orbites 83a–3
Corbeau 92–41
corde cosmique 99–171c
~ magnétique 89–146
cordilière 81–194
cordon protubérantiel 89–245
cornes 83–130
~ de la Lune 81–24
~ lunaires 81–24
cornet magnétosphérique 75–13
~ polaire 75–13
Corona Australis 92–39
~ Borealis 92–40
coronal 88–228
coronographe 88–59
~ à disque occulteur externe 88–60
~ à occultation externe 88–60
coronomètre K 88–66
corotation 84–17
structure en ~ 84–17
corps adsorbant 85–130
~ d'impact 82–225
~ impactant 82–225
~ météorique 86–160a
~ parental 87–123
~ préstellaire 94–15
~ producteur de marée 82–284
~ rigide 83–111
~ solide 83–110
~ supermassif 98–22
~ tournant 91–42
corpusculaire 88–155
corpuscule 88–153
~ solide 85–48
correction bolométrique 91–80
~ de champ 86–157
~ de courbure 86–62

F

~ de courbure de la surface terrestre 86–62
~ de trajectorie 82–10
~ relativiste 99–140
Corvus 92–41
cosmochimie 74–31
cosmochronologie 83a–72
cosmogonie 94–2
~ planétaire 83a–1
~ stellaire 94–7
cosmogonique 94–3
cosmogoniste 94–4
cosmologie 99–16
~ cinématique 99–26
~ newtonienne 99–25
~ non-relativiste 99–27
~ observationnelle 99–17
~ relativiste 99–28
cosmologique 99–18
cosmologiste 99–19
cosmologue 99–19
cosmos 99–2
côte 72–30
côté à l'ombre 75–21
~ éclairé 75–20
~ ensoleillé 81–34
~ illuminé 83–67
~ jour 75–20
~ nocturne 83–68
~ non-éclairé 83–68
~ nuit 75–21
couche absorbante 88–175
~ absorbante galactique 97–89
~ atmosphérique 76–49
~ convective 93–149
~ d'Heaviside 78–26
~ d'hélium 93–145
~ d'hydrogène 93–144
~ d'ozone 76–123
~ de brume 83–249
~ de Kennelly-Heaviside 78–26
~ de lave 82–175
~ de neige 77–176
~ de nuages 77–150
~ de nuages de plui 77–145
~ de photoélectrons 82–246
~ de plasma 75–11
~ de poussière 82–122
~ diffusante 83–258
~ E 78–26
~ E sporadique 78–25
~ émissive 79–17
~ extrêmement mince 83–149
~ intermédiaire 72–13
~ ionisée 78–3
~ ionosphérique 78–5
~ luminescente 79–17
~ lumineuse 79–17
~ mince 83–149
~ neutre 75–16
~ nuageuse 83–248
~ radiative 93–134
~ renversante 88–174
~ subphotosphérique 88–168
~ violette 83–165
coucher du Soleil 88–11
couches basses de l'atmosphère solaire 88–166
~ élevées de l'atmosphère solaire 88–165
~ extérieures d'une étoile 93–3
~ extérieures non-dégénérées 93–199
~ externes de l'atmosphère 88–165
~ profondes de l'atmosphère 88–166
coulage de la lave 82–188
coulée de cendre 82–174
~ de lave 73–140, 82–173
couleur intrinsèque 91–74
~ du couchant 79–55
Coupe 92–42
coupelle du pied d'atterrissage 82–23
couplage de rayonnement 99–173
couple dû à la marée 82–288
couple freinant de marée 82–288
courant actif 86–223
~ aérien 77–90
~ ascendant 77–91, 93–151
~ asymétrique 98–104
~ chaud 72–63
~ cométaire 86–222
~ corpusculaire 88–160
~ corpusculaire solaire 84–35
~ d'astéroïdes 84–100
~ d'étoiles 98–98
~ de convection 73–76
~ de jour 86–219
~ de jusant 72–65
~ de marée 82–279
~ de plasma solaire 84–33
~ de subsurface 72–61
~ de surface 72–62
~ de suspension 72–49
~ de turbidité 72–49
~ descendant 93–150
~ diurne 86–219
~ du vent solaire 84–35a
~ écliptique 86–221
~ froid 72–64
~ gazeux 95–67
~ interstellaire 86–224
~ jet 77–88
~ marin 72–60
~ marin chaud 72–63
~ marin froid 72–64
~ météorique 86–216
~ nocturne 86–218
~ océanique 72–60
~ périodique 86–228
~ permanent 86–227
~ planétaire 86–220
~ sporadique 86–226
~ stellaire 96–97, 98–98
~ surgissant 93–151
~ temporaire 86–225
~ tombant 93–150
~ vers l'extérieur 93–151
~ vers l'intérieur 93–150
~ visible de jour 86–219
courants 83–187
anneau de ~ 75–100
~ d'anneau 75–100
~ telluriques 75–5
~ telluriques vagabonds 75–6
courbe bathymétrique 71–55
~ d'abondances 94–108
~ d'éclat 95–17
~ d'égale hauteur 81–139
~ d'évolution 89–119
~ d'évolution d'un groupe 89–119
~ d'éxtinction 97–81
~ de couleur 95–21
~ de croissance 93–68
~ de croissance empirique 93–69
~ de croissance théorique 93–70
~ de lumière 95–17
~ de lumière d'un météore 86–70
~ de lumière moyenne 95–19
~ de lumière photoélectrique 95–22
~ de lumière photographique 95–23
~ de lumière visuelle 95–24
~ de niveau 71–53
~ de niveau magnétique 89–134
~ de phase 83–74
~ de polarisation 83–81
~ de rotation 98–109
~ de rougissement 97–93
~ de seuil de la secondaire 96–40
~ de vitesses radiales 95–20
~ hypsographique 81–142
~ isophote 81–92
~ moyenne d'éclat 95–19
~ période-luminosité 95–103
courbure de l'espace 99–115
~ de la surface terrestre 86–61
~ de la Terre 71–6
~ de la trajectorie 86–51
~ de lignes de force 75–121
couronne 79–60, 79–156
~ aurorale 79–137a
~ Australe 92–39
~ blanche 88–237
~ Boréale 92–40
~ d'émission 88–241a
~ d'hydrogène 83–246
~ de Fraunhofer 88–241
~ de maximum 88–244
~ de minimum 88–246
~ de rayons X 88–240
~ du Nord 92–40
~ du Sud 92–39
~ E 88–241a
~ électronique 88–242
~ en expansion 88–232
~ extérieure 88–231
~ F 88–241
~ galactique 98–49

F

~ intermédiaire 88–245
~ interne 88–229
~ K 88–242
~ lunaire 79–61
~ monochromatique 88–238
~ moyenne 88–230
~ planétaire 83–245
~ polaire 89–232
~ polaire des filaments 89–232
~ poussiéreuse 88–241
~ radio 88–239
~ solaire 79–62, 88–225
~ stellaire 93–4
~ X 88–240
couvert 77–152
~ de nuages 83–253
~ de poussière 82–120
couverture 93–20
~ de glace 72–115
Crater 92–42
cratère 73–126
~ à écoulement 81–239
~ à halo sombre 81–215
~ à raies 81–214
~ à rayons 81–214
~ à rebord élevé 81–223
~ à rebord relevé 81–223
~ à rempart d'éboulis 81–225
~ adventif 73–128
~ ancien 81–205
~ antérieur à la mare 81–207
~ antérieur aux mers 81–207
~ avec pic central 81–258a
~ central 81–217
~ circulaire 81–213
~ comblé 81–203
~ concentrique 81–216
~ d'explosion 81–238
~ d'impact 81–240
~ du type maar 81–264
~ en bol 81–219
~ en coupe 81–219
~ enseveli 81–204
~ extérieur 81–218
~ fantôme 81–209
~ irrégulier 81–220
~ jeune 81–206
~ latéral 73–128
~ lunaire 81–197
~ maar 81–264
~ météorique 87–53
~ météoritique 87–53
~ météoritique fossile 87–55
~ météoritique terrestre 87–54
~ météoroïdal 81–263
~ non-baptisé 81–202
~ non-désigné 81–202
~ parasitaire 73–128
petit ~ 81–250
~ polygonal 81–212
~ possible 81–211
~ postérieur à la mare 81–208
~ postérieur aux mers 81–208
~ postmare 81–208
~ prémare 81–207
~ primaire 81–198
~ principal 73–127, 87–57
~ probable 81–210
~ rayonnant 81–214
~ rempli 81–203
~ sans nom 81–202
~ sans rebord 81–226
~ satellite 81–200
~ secondaire 81–199
~ sommital 73–129
cratérisation 82–232
craterlet 81–250
création continue 99–91
~ de matière 99–89
~ unique 99–90
crêpes primordiales 99–171b
crépusculaire 79–40
crépuscule 79–34
~ astronomique 79–37
~ civil 79–38
~ du matin 79–35
~ du soir 79–36
~ nautique 79–39
~ vespéral 79–36
Crétacé 74–86
crête 81–186, 81–195
~ barométrique 77–87
~ de haute pression 77–87
~ de mare 81–188
~ de mer 81–188
~ de plissement 81–187
creusement d'une dépression 77–82
creux de potential 98–176
petit ~ 81–243
crevasse 81–165
crique de tension 82–206
cristal de glace 76–111
cristallisation 76–107
~ fractionnée 74–40
cristobalite 82–95
critère d'instabilité 98–312
~ d'instabilité due aux marées 98–312
~ de classification 91–110
~ de dégénérence 93–181
crochet magnétique 75–79
croissance 76–54
~ d'une planète 83a–37
~ de la calotte polaire 83–145
~ des grains 97–61
~ relativiste de la masse 99–142
croissant de Vénus 83–129
~ lunaire 81–23
croix de halo 79–79
~ de Mills 88–121
~ du Nord
~ du Sud 92–43
~ lumineuse 79–79
~ lunaire 79–80
~ solaire 79–81
croûte continentale 72–5
~ de fusion 87–110
~ lunaire 82–54
~ solide 95–151
~ terrestre 72–4
Crux 92–43
culmination de la Lune 82–303
cumulonimbus 77–137
cumulus 77–138
~ fractus 77–146
cyanoéthylène 97–30
cyanogène 97–24
cyanure d'hydrogène 97–29
~ de méthyle 97–38
cycle d'activité solaire 89–71
~ de onze ans 89–73
~ des taches solaires 89–72
~ géochimique 74–32
~ impair 89–78
~ pair 89–77
~ undécennal 89–73
cyclone 77–76
~ tropical 77–77
Cygne 92–44
Cygnus 92–44
cylindre de lumière 95–155
~ tronqué 88–63
cylindrosymétrie 98–155
cythérien 83–126

D

danger de navigation 72–97
datation 87–126
~ au plomb 74–100
~ des événements lunaires 82–265
~ isotopique 83a–73
~ par le rapport thorium-plomb 82–270
~ par les isotopes 83a–73
~ par radioactivité 87–131
~ plomb-plomb 74–100
~ radioactive 87–131
~ uranium-plomb 74–101
Dauphin 92–45
Daurade 92–46
débit d'énergie 93–109
débris de rochers 81–235
~ rocheux 81–235
début brusque 75–74
~ brusque d'orage 75–74
~ de la totalité 81–77
~ du cycle 89–82
décalage anormal vers le rouge 98–367
~ cosmologique 98–263a
~ gravifique 99–162
~ gravifique vers le rouge 99–162
~ multiple vers le rouge 98–364
~ périodique des raies 96–22
~ vers le rouge 98–297
décélération apparente 86–100

~ de l'expansion 99–82
~ réelle 86–101
décharge atmosphérique 77–161
décibel 76–39
déclenchement 89–199
déclencher 89–198
déclin de la courbe de lumière 95–35
déclinaison de la Lune 81–62
~ du Soleil 81–63
~ magnétique 75–56
décline 76–57
décomposer, se 74–65
décomposition photochimique 76–84
découplage du rayonnement 99–177
~ entre matière et rayonnement 99–177
décours de la Lune 81–21
découverte d'une comète 85–16
~ d'une météorite 87–44
décroissance 76–57
~ d'un groupe de taches 89–126
~ de la calotte polaire 83–146
~ séculaire de l'éclat 85–127
dédoublement des raies par expansion 97–157
dédoubleur de faisceau 88–88
défaut de la phase 81–28
déférent 83–48
défluidisation 82–263
déformation 73–67
~ due à la marée 82–277
~ due aux marées 83a–66
dégat par ondes de marées 72–77
dégazage 82–261, 87–136
~ d'un noyau cométaire 84–109
dégénéré 93–178
complètement ~ 93–185
fortement ~ 93–184
partiellement ~ 93–183
dégénérescence 93–172
~ complète 93–175
condition de ~ 93–181
~ de neutrons 93–191
degré de ~ 93–174
~ neutronique 93–191
~ non-relativiste 93–177
~ partielle 93–173
~ relativiste 93–176
degré de concentratin 96–128
~ de dégénérescence 93–174
~ géothermique 74–59
Deimos 83–169
déjections 82–165
délai de groupe 78–56
Delphinus 92–45
démarrage de la combustion de l'hélium 94–71
~ de la combustion de l'hydrogène 94–55
~ de la combustion du carbone 94–80
demi-diamètre apparent de la Lune 81–70
~ apparent du Soleil 81–69
~ extérieur 83–228
~ intérieur 83–226
demi-lune 81–26
Deneb 92–112
Denebola 92–113
dénombrement d'étoiles 98–79
~ de galaxies 98–343
densité centrale 93–87
~ cosmologique 98–370
~ critique 98–197
~ d'un courant météorique 86–264
~ de cratères 81–246
~ de galaxies 98–344
~ de l'ombre 81–59
~ de la distribution projetée 98–85
~ de radiation 99–85
~ de surface apparente 96–132
~ de traces 82–253
~ des radiants 86–63
~ du flux neutronique 94–86
~ du fond 96–131
~ du fond stellaire 96–131
~ du vent solaire 88–157
~ électronique 78–6, 88–233
~ élecronique linéaire 86–283
~ galactique 98–344
~ linéaire inférieure à 10^{12} cm^{-1} 86–284
~ linéaire des électrons 86–283
~ moyenne 93–86
~ moyenne de matière 99–84
~ projetée 98–85
~ réelle 96–133
~ spatiale 96–133
~ spatiale des étoiles 98–83
~ stellaire 98–82
~ subnucléaire 93–190
~ supérieure à 10^{12} cm^{-1} 86–281
~ volumique 96–133
dénudation 73–108
déphaseur 88–115
déplacement aberrant vers le rouge 98–367
~ cosmologique 98–363a
~ de la traînée 86–114
~ des zones de formation des taches 89–74
~ des zones royales 89–74
~ Doppler 91–25
~ Doppler-Fizeau 91–25
~ du radiant 86–242
~ gravitationnel vers le rouge 99–162
~ non-Doppler-Fizeau 98–363
~ périodique des raies 96–22
~ relativiste des raies spectrales 99–162
~ relativiste des raies spectrales vers le rouge 99–162
~ vers le rouge 98–297
~ vers le rouge dans le champ de gravitation 99–162
~ vers le rouge non-Doppler-Fizeau 98–363
dépolarisation 97–118
déposition 73–109
dépôt 74–47
~ annuel 87–17
dépôts abyssaux 72–50
~ de profondeurs océaniques 72–50
~ des mers profondes 87–15
~ glaciaires 72–127
~ organogéniques 72–59
~ pélagiques 72–57
~ terrigènes 72–48
dépourvu de poussière 82–119
dépression 77–76
~ à terrasses 81–162
~ continentale 81–158
~ en arc 81–160
~ linéaire 81–159
~ lunaire 81–157
petite ~ 81–243
~ sinueuse 81–161
~ solaire 79–45
dérive des continents 73–104
~ en longitude 75–120
dernier contact 81–85
~ point d'impact 87–39
~ quartier 81–30
déroulement des bras spiraux 98–30
désactivation 85–148
~ d'une comète 85–148
désagrégation 74–57
~ d'un amas 96–105
~ d'un noyau cométaire 86–261
~ du groupe de taches 89–126
~ due aux marées 85–118
~ en milieu spatial 82–244
~ radioactive 74–64
~ rotationnelle 85–119
désaimantation thermique 75–40
descendant 77–93
désert 83–161
désignation définitive 85–33
~ provisoire 85–32
désintégration d'un amas 96–103
~ d'un corps météorique 86–186
~ d'un météore 86–82
~ d'une comète 85–115
~ des astéroïdes 84–108
désintégrer, se 74–65
désionisation 86–213
désorption 85–123
dessèchement 74–51
dessication 74–51
dessin d'une planète 83–16
~ des taches 88–34
destruction 87–24
~ de la surface 87–35
~ par une météorite 87–20
détail de la surface 83–105
~ permanent 83–106

F

F

détermination de l'âge = datation 74–91, 87–126
~ de l'âge par radioactivité 87–131
~ de la masse 96–69
~ de la pesanteur 71–67
développement de la calotte polaire 83–145
déversement de pétrole 72–71
déviation d'une étoile 98–123
~ de la lumière 99–161
~ de la verticale 71–72
~ des rayons lumineux par le champ de gravitation 99–161
~ du mouvement d'une étoile 98–123
~ relativiste de la lumière 99–161
Dévonien 74–81
diagenèse 74–38
diagramme à deux couleurs 96–147
~ couleur-magnitude 96–150
~ d'évolution 96–151
~ de Hertzsprung-Russell 91–130
~ de Hubble 98–301
~ de Maunder 89–76
~ déplacement vers le rouge – magnitude apparente 98–301
~ en papillon 89–76
~ évolutif 96–151
~ luminosité-type spectral 91–130
~ masse-luminosité 91–129
~ papillon 89–76
diamants météoritiques 87–65
diamètre angulaire 97–145
~ apparent 83–117, 96–115
~ photométrique 84–93
~ réel 83–118
diastrophisme 73–72
diatrème 73–122
dichotomie de Vénus 83–132
différence de chemin optique 88–117
~ d'éclat 96–65
~ de magnitudes 91–59
~ de marche 88–117
~ verticale 81–140
différenciation chimique 83a–50
~ géochimique 74–55
~ gravitationnelle 83a–54
~ par cristallisation 74–41
~ par gravité 74–56
diffusion atmosphérique de la lumière 77–57
~ de Mie 83–76
~ des ions 86–286
~ du rayonnement cosmique 84–27
~ ionique 86–286
dilatation du temps 99–137
dilution des raies de Fraunhofer 88–243
dimension réduite par (effet de) perspective 89–92
dimensions, à quatre 99–121
~ des grains 82–113
~ du système 96–64
diminution apparente centre-bord 89–160
~ de la masse 83a–42
~ séculaire de l'éclat 85a–127
direct 83–42
direction antisolaire 75–22
~ de l'axe 83–34
~ de l'axe de rotation 84–102
~ de propagation 78–35
~ du champ magnétique 84–10
~ du vent 77–115
dirigé à l'opposé du Soleil 85–81
discontinuité de Moho 72–14
~ de Mohorovičić 72–14
~ interplanétaire 84–58
dislocation 82–212
disparition 81–98
~ brusque 89–279
~ d'un météore 86–37
~ de l'anneau 83–217
~ progressive 86–75
dispersion de matière 81–236
~ des vitesses 98–103
dispositif modifiant la longueur d'onde 88–69
~ de sélection de λ 88–69
disque d'accrétion 96–21
~ d'occultation 88–61
~ en lumière visible 88–171
~ galactique 98–52
~ lunaire 81–54
~ occulteur 88–61
~ solaire 89–24
~ visible 89–26
dissipation 94–73
~ d'ondes 88–222
~ de l'énergie 82–285
~ de marée 82–286
~ par effet de marée 82–286
~ thermique 94–76
dissociation d'un amas galactique 96–103
dissolution d'une protubérance 89–280
dissymétrie est-ouest 89–49
~ nord-sud 89–50
distance 86–288, 91–3
~ angulaire 96–52
~ au centre de l'étoile 93–85
~ au centre de la Galaxie 98–11
~ au centre du disque 89–36
~ au méridien central 89–46
~ de l'épicentre 73–19
~ de l'hypocentre 73–18
~ de la Lune à la Terre 81–13
~ des antennes 88–99
~ du foyer 73–18
~ entre les composantes 96–66
~ galactocentrique 98–11
~ géocentrique 84–3
~ héliocentrique 84–4
~ limite 86–289
~ maximale 86–289
~ photométrique 96–116
~ Terre-Lune 81–13
distorsion 73–68
~ par effet de marée 96–30
~ rotationnelle 91–44
~ rotationnelle d'une courbe de vitesse 91–44
distribué au hasard 98–93
~ d'une manière aléatoire 98–93
distribution dans l'espace 96–119
~ de la brillance 88–141
~ de la densité 96–124
~ de matière 99–8
~ des dimensions 82–124
~ des masses 84–52
~ des vitesses résiduelles 98–92
~ ellipsoïdale des vitesses 98–100
~ en latitude 89–51
~ en latitude héliographique 89–51
~ en longitude 89–48
~ en longitude héliographique 89–48
~ galactique 98–54
~ maxwellienne des vitesses 98–96
~ selon éclat 86–166
~ spatiale 96–119, 98–86
~ verticale 76–17
diviseur de faisceau 88–88
division 83–223
~ de Cassini 83–224
dôme de lave 81–176a
~ lunaire 81–177
~ volcanique 81–176
dômes, amas de 81–178
donnée géodésique 71–20
données solaires 89–9
Dorade 92–46
Dorado 92–46
dorsale barométrique 77–86
~ de haute pression 77–86
~ médio-océanique 72–45
double = binaire 96–3
~ non-résolue 96–10
~ optique 96–6
~ physique 96–7
~ semi-détachée 96–34
~ semi-séparée 96–34
~ séparée 96–35
~ visuelle 96–8
Draco 92–47
Draconides 86–252
dragage 72–53
draperies 79–154
durée de vie, à courte 89–125
à longue ~ de vie 89–124
~ d'ensoleillement 88–10
~ d'évaporation 96–109
~ d'impulsion 86–280
~ d'une éruption 89–209
~ de dissociation 96–104
~ de l'écho météorique 86–299
~ de l'éclipse 95–78
~ de la période 95–39

~ de la totalité 88–26
~ de la visibilité d'une traînée 86–111
~ de parcours 73–51
~ de propagation 73–51
~ de rotation 83–31
~ de vie d'un groupe 89–123
~ de vie d'une particule capturée 75–122
~ de vie d'une particule piégée 75–122
~ du cycle 89–87
~ du parcours d'un météore 86–50
~ du séjour 82–34
dyke 73–138
dynamique des gaz interstellaires 97–150
~ stellaire 98–120
dynamo-métamorphisme 74–36
dynamo-théorie géomagnétique 75–99

E

eau 97–26
éboulement 82–196
~ de remparts 82–198
trace d' ~ 82–197
éboulis 82–199
écarts à l'équilibre thermodynamique local 93–40
~ à l'ET 93–36
~ à l'ETL 93–40
~ par rapport à l'équilibre thermodynamique 93–36
échange d'énergie 98–125
échantillon 87–42
~ carotté 82–60
~ de carotte 82–60
~ de l'air 76–86
~ de poussière lunaire 82–121
~ lunaire 82–69
~ lunaire rapporté 82–70
échantillonage 76–85
~ de l'air 76–85
échantillonneur d'eau 72–101
échantillons, prise d' 82–72
retour d' ~ 82–71
échauffement atmosphérique 75–87
~ du corps météorique 86–174
échelle Celsius 77–33
~ centigrade 77–33
~ courte d'évolution 94–66
~ d'altitude 76–18
~ de Beaufort 77–108
~ de distance 98–303
~ de hauteur 76–18, 88–236
~ de la carte 71–49
~ de temps 94–65
~ géochronologique 74–68
~ longue d'évolution 94–67
échelon géothermique 74–59
~ hypsométrique 76–19
écho auroral 79–121
~ de traînée 86–298
~ météorique 86–296
~ météorique de courte durée 86–300
~ météorique de longue durée 86–301
~ météorique de tête 86–297
~ radar 86–295
~ radio 86–295
~ radioélectrique 86–295
~ sismique 82–223
~ sondages 72–29
éclair 77–162
~ en boule 77–163
~ en chapelet 77–164
éclat 91–54
~ absolu 85–109, 91–64
~ absolu d'un météore 86–74
~ absolu d'une comète 85–109
~ apparent 91–61
~ apparent d'un météore 86–73
~ d'un météore 86–8
~ de l'éruption 89–212
~ électrographique 91–67
~ électronographique 91–67
~ global 85–107
~ infrarouge 91—71
~ intégré 96–135
~ intrinsèque 85–107
~ limite 91–62
maximum d' ~ 86–78
~ photographique 91–65
~ photovisuel 91–69
~ radiométrique 91–70
~ réduit 85–108
~ stellaire 91–55
~ total 85–107
~ visuel 91–65
éclatement 86–82
~ d'un météore 86–82
~ du corps parental 87–124
éclipse 88–14
~ annulaire 88–18
~ annulaire de Soleil 88–15a
~ artificielle 88–227a
~ artificielle de Soleil 88–17a
~ atmosphérique 95–76
carte de l' ~ 88–28
carte de visibilité de l' ~ 88–28
~ centrale 88–16a
~ corpusculaire 88–24
~ de Lune 81–56
~ de Lune par la pénombre 81–65
~ du Soleil 88–15
fin de l' ~ 81–83
grandeur de l' ~ 81–79
~ invisible à ... 81–87
~ ionosphérique 88–23
limite australe de l' ~ 88–19
limite boréale de l' ~ 88–20
milieu de l' ~ 81–78
~ par la pénombre 81–65
~ partielle 81–79a
~ partielle de Lune 81–67
~ partielle de Soleil 88–16
~ solaire 88–15
~ totale de Lune 81–66
~ totale de Soleil 88–17
~ visible à ... 81–86
écologie 76–2
écorce de la Terre 72–4
écoulement continu 98–276
~ de cendre 82–174
~ de la glace 72–120
~ de la matière 98–275
~ de la matière du noyau galactique 98–275
~ de lave 73–130, 82–172
~ plastique 72–121
écran de projection 88–13
écrasement gravitationnel 84–72
écroulement catastrophique 94–109
~ gravitationnel 94–109
Écu de Sobiesky 92–88
effacement de raies de Fraunhofer 88–243
effet bêtatron 91–180
~ Coriolis 77–81
~ cumulatif 98–124
~ d'écran 86–171
~ d'inversion 76–126
~ d'opposition 83–75
~ de couverture 93–67
~ de marée 82–276
~ de réflexion 96–26
~ de réflexion dans les binaires 95–82
~ de sélection 98–81
~ de serre 83–263
~ Evershed 89–103
~ Götz 76–126
~ isotopique 91–178
~ Poyinting-Robertson 84–67
~ Rossiter 91–44
~ rotationnel 91–44
~ rotationnel sur la courbe de vitesse 91–44
~ terrestre de l'activité solaire 84–55
~ Wilson 89–89
efficacité lumineuse 86–210
effondrement 82–199
~ catastrophique 94–109
~ de l'univers 99–181
étoile en ~ 91–164
~ final 99–182
Grand ~ 99–182
effusion de lave 73–130
éjecta 82–165
~ d'impact 82–167
~ volcaniques 82–166
éjection 89–259, 98–179
~ associée à une éruption 89–214
~ de masse 93–15
~ de matière 94–77, 95–124

F

~ de matière du noyau galactique 98–274
~ de particules 85–89
~ de particules du Soleil 88–161
~ de plasma solaire 84–38
~ depuis une comète 86–262
~ des nuages à émission radio 98–285
~ des nuages radio 98–285
~ d'un jet 98–278
~ récurrente 98–282
~ solaire de particules 88–161
éjections 82–165
~ de choc 82–167
~ d'impact 82–167
élargissemnt de l'ombre de la Terre 81 93
élasticité 73–58
électricité atmosphérique 75–2
~ tellurique 75–1
electrojet auroral 79–113
électrons auroraux 79–109
~ piégés 79–111
~ précipités 79–118
~ relativistes 97–153
élément de structure 98–287
~ météorologique 77–28
~ sous-abondant 91–186
~ structural 98–287
~ surabondant 91–184
~ trace 82–145
éléments atmophiles 83a–29
~ chalcophiles 83a–31
~ dominants 82–144
~ du groupe du fer 94–105
~ géomagnétiques 75–54
~ lithophiles 83a–30
~ magnétiques 75–54
~ photométriques 95–25
~ réfractaires 97–64
élévation au-dessus du niveau de la mer 71–51
ellipsoïde de référence 71–18
~ des vitesses 98–101
~ international 71–14
~ local 71–15
~ terrestre 71–11
ellipticité d'un amas globulaire 96–122
élongation 83–60
~ du radiant apparent 86–67
~ du radiant vrai 86–68
~ maximale 83–63
~ occidentale 83–62
~ orientale 83–61
embryon d'une planète 83a–36
émersion 81–99
émetteur 78–15
émission atmosphérique 79–2
~ brusque de particules 88–162
~ d'extrêmement basse fréquence 78–55
~ corpusculaire 88–152
~ de lave 73–130
~ de masse 85–145
~ de particules 88–151
~ de très basse fréquence 78–54
~ du ciel nocturne 79–5
~ fissurale 73–141
~ Lα du ciel nocturne 79–9
~ post-crépusculaire 79–52
~ précrépusculaire 79–51
~ radio sporadique de Jupiter 83–181
~ simultanée de deux photons 97–155
empreintes de pas 82–38
enceinte du cratère 81–251
endogène 73–145
énergie accumulée 89–196
~ acoustique 88–219
~ atomique 93–121
~ au point zéro 93–187
~ au repos 99–144
~ au zéro absolu 93–187
~ cinétique au point zéro 93–187
~ de rotation 93–120
~ de vaporisation 86–199
~ disponible 89–197
~ du vide 99–170k
~ échangée 98–126
~ emmagasinée 89–196
~ gravitationnelle 93–119
~ non-radiative 88–218
~ nucléaire 93–122
~ sonore 88–219
enregistrement ionosphérique 78–23
enroulement à la traîne 98–28
ensoleillement 88–9
entrée dans l'atmosphère 86–161
~ de la Lune dans l'ombre 81–76
~ de la Lune dans la pénombre 81–75
enveloppe caractéristique 98–177
~ circumstellaire 93–10
~ commune d'une double 96–32
~ convective 93–136
~ d'une comète 85–69
~ de glace 97–72
~ de poussière 97–116
~ de poussière circumstellaire 97–117
~ en expansion 93–12
~ étendue 93–8
~ glacée 97–72
~ radiative 93–135
~ stationnaire 93–11
~ stellaire 93–10
environnement 76–3
Éozoïque 74–77
épaisseur des anneaux 83–229
épanchement de lave 82–171
Épée d'Orion 92–130
épeirogenèse 73–73
éphéméride des petites planètes 84–92
Épi 92–129
épicentre 73–15
épicycle 83–49
épirogenèse 73–73
époque 74–89
~ de découplage 99–177a
~ de maximum 89–81
~ de minimum 89–80
~ du passage au périastre 96–60
~ géologique 74–89
~ postglaciaire 72–109
épuisement de l'hydrogène 94–69
équateur d'illumination 81–122
~ lunaire 81–121
~ magnétique 75–50
équation d'Abel 88–235
~ d'altitude 86–194
~ d'Emden 93–130
~ de champ d'Einstein 99–156
~ de continuité 93–49
~ de couleur 91–73
~ de l'équilibré mécanique 93–76
~ de lumière 95–45
~ de luminosité 86–196
~ de masse 86–169
~ de Pogson 91–58
~ de résistance 86–179
~ de traînée 86–179
~ de transfert 93–32
~ de transfert radiatif 93–33
~ de vaporisation 86–200
~ du viriel 98–181
~ hydrostatique 93–76
~ intégrale de Schwarzschild et Milne 93–53
~ intégro-différentielle 93–52
~ pour la perte de masse 86–169
~ tensorielle 99–157
équations de l'équilibre statistique 93–44
~ du champ 99–155
Equerre 92–70
équilibre adiabatique 76–21
~ adsorptif 85–121
~ convectif 76–22, 93–79
~ de diffusion 76–23
~ général 93–43
~ hors ETL 93–43
~ hydrostatique 93–75
~ isostatique 73–99
~ isotherme 76–20
~ mécanique 93–75
~ nucléaire 94–103
~ radiatif 93–48
~ statistique 93–42, 98–149
~ thermique 93–78
~ thermodynamique 93–35
~ thermodynamique local 93–39
équivalence de la masse et de l'énergie 99–145
Equuleus 92–48
ère 74–69
~ cénozoïque 74–73
~ cosmologique 99–166a

~ d'anniquilation 99–173a
~ d'inflation 99–170d
~ de découplage 99–177a
~ de la force superunifiée 99–170a
~ de matière 99–178
~ de Planck 99–170e
~ de rayonnement 99–172
~ du plasma photonique 99–172
~ géologique 74–69
~ hadronique 99–174
~ inflationnaire 99–170d
~ inflatoire 99–170d
~ leptonique 99–176
~ mésozoïque 74–72
~ néozoïque 74–73
~ paléozoïque 74–71
~ précambrienne 74–70
~ primaire 74–71
~ stellaire 99–178
Eridan 92–94
Eridanus 92–49
érosion 73–107
~ cosmique 87–142
~ dans l'espace cosmique 87–142
~ de surface 82–241
~ éolienne 83–163
~ gazeuse 82–243
~ météoritique 82–242
~ par météorites 82–242
~ par météoroïdes 82–242
erratique 72–132
~ glaciaire 72–132
éruption 89–175, 95–69
~ à deux fils 89–187a
~ à protons 89–185
~ au delà du limbe 89–203
~ ayant une composante radio 89–180
~ chromosphérique 89–175
~ chromosphérique au bord 89–202
~ d'importance 2 89–206
~ dans le domain visible 89–181
~ derrière le bord 89–203
~ en double filet 89–187a
~ en lumière blanche 89–182
~ en rayons X 89–183
~ fissurale 73–113
~ homologue 89–178
~ lumineuse d'un météore 86–80
~ lunaire 82–247
~ par contagion 89–179
~ radio 89–180
~ radioélectrique 89–180
~ sans émission de protons 89–186
~ solaire 89–175
~ stellaire 95–70
~ sur le disque 89–204
~ visible en lumière blanche 89–182
~ volcanique 73–112, 82–170
escarpement 81–184
espace 99–100
~ absolu 99–116
~ courbe 99–114
~ d'événements 99–122
~ de Lobatchevski 99–105
~ de Newcomb 99–110
~ de Riemann 99–112
~ des phases à six dimensions 98–144
~ des vitesses 98–97
~ distancié 99–100a
~ elliptique 99–110
~ entre deux bras spiraux 98–40
~ euclidien 99–107
~ extragalactique 97–5
~ extraterrestre 87–73a
~ fini 99–103
~ hyperbolique 99–106
~ infini 99–104
~ interdit 79–135
~ interfilamentaire 89–99
~ intergalactique 97–7
~ interplanétaire 84–2
~ interstellaire 97–4
~ libre 76–9
~ métrique 99–100a
~ minkowskien 99–122
~ non-euclidien 99–114
~ parabolique 99–107
~ permis 79–134
~ plat 99–107
~ riemannien 99–112
~ sphérique 99–111
~ temps 99–119
~ temps de Minkowski 99–122
~ vide 76–9
espacement des antennes 88–99
~ fixe 88–102
~ nul 88–101
~ variable 88–103
espace-temps 99–122
espèce de nuage 77–135
essaim d'α Piscis Austrini 86–250
~ de comètes 85–100
~ diurne 86–219
~ météorique 86–216
~ nocturne 86–218
estimation d'éclat 95–11
estimer 95–12
ET 93–35
état de coulage du magma 82–189
~ de fluidité 82–189
~ de la mer 72–86
~ du ciel 77–132
~ du temps 77–14
~ non-stationnaire 98–151
~ stationnaire 98–140
~ statistiquement stationnaire 98–148
étendue de pierres 82–139
éternité de l'univers 99–13
éthyne méthylique 97–41
étincelles d'un météore 86–84
ETL 93–39
étoile 91–1
~ à atmosphère étendue 93–6
~ à azote 91–189
~ à carbone 91–188
~ à enveloppe 93–5
~ à enveloppe étendue 93–7
~ à grande vitesse 98–191
~ à hypérons 91–154
~ à neutrons 93–195
~ à particularités spectrales 91–92
~ à raies d'émission 91–159
~ à raies métalliques 91–195
~ à silicium 91–190
~ à spectre particulier 91–92
~ à spectre variable 95–48
~ à température nule 91–166
~ à vitesse de libération 96–160
~ anomale 91–96
~ au zéro absolu 91–166
~ avec un compagnon obscur 96–13
~ azotée 91–189
~ banale 91–95
~ binaire 96–3
~ brillante 91–83
~ carbonée 91–188
~ centrale 97–148
~ chaude 91–155
~ circumpolaire 92–2a
~ collapsante 94–112
~ colapsée 94–111
~ courante 91–95
~ compagnon 96–39
~ d'âge zéro 94–59
~ d'évasion 96–160
~ d'un amas 96–82
~ de champ 96–84
~ de comparaison 95–16
~ de faible luminosité 91–133
~ de faible vitesse 98–192
~ de la branche horizontale 91–141
~ de la première génération 94–24
~ de la série principale 91–142
~ de masse faible 91–124
~ de masse intermédiaire 91–125
~ de très grande masse 91–127
~ de Tycho 95–135
~ de type nova 95–118
~ de type solaire 91–104
~ de Wolf-Rayet 91–160
~ dégénérée 93–180
~ double 96–3
~ double non-résolue 96–10
~ double optique 96–6
~ double physique 96–7
~ double serrée 96–23
~ double spectroscopique 96–14
~ double spectroscopique à deux spectres 96–17
~ double spectroscopique à un seul spectre 96–15
~ double visuelle 96–8

F

~ du berger 83–128
~ du Chien 92–128
~ du disque 91–85
~ du disque galactique 91–85
~ du fond stellaire 96–84
~ du halo 91–84
~ du matin 83–127
~ du soir 83–128
~ du système 98–133
~ écroulée 94–111
~ effondrée 94–111
~ en contraction 91–161
~ en cours d'écroulement 94–112
~ en cours d'effondrement 94–112
~ en effondrement 91–164
~ en évasion 96–160
~ en expansion 91–162
~ en explosion 91–165
~ en implosion 91–164
~ en libération 98–135
~ en rotation 91–37
~ en rotation rapide 91–39
~ en voie d'échappement 98–135
~ étrangère au système 98–134
~ évoluée 94–74
~ excitatrice 97–125
~ explosive 91–165
~ faible 91–82
~ filante 86–2
~ filante traînée 86–17
~ fixe 92–4
~ froide 91–156
~ géante 91–143
~ hyperdense 93–197
~ hypéronique 91–154
~ implosive 91–164
~ infrarouge 91–157
~ instable 95–9
~ intergalactique 91–163
~ interlope 98–134
~ jeune 94–63
~ la plus brillante 98–304
~ la plus brillante de la galaxie 98–304
~ légère 91–124
~ magnétique 91–51
~ massive 91–126
~ membre de l'amas 96–82
~ membre du système 98–133
~ morte 91–167
~ multiple 96–5
~ n'appartenant pas à l'amas 96–84
~ naine 91–135
~ neutronique 91–152
~ non-membre de l'amas 96–84
~ non-membre du système 98–134
~ non-rougie 97–95
~ non-sphérique 91–86
~ non-stationnaire 95–2
~ non-tournante 91–38
~ normale 91–95
~ observée dans la direction de son pôle 91–40
~ particulière 91–92
~ pauvre en carbone 91–194
~ pauvre en hélium 91–192
~ pauvre en hydrogène 91–191
~ pauvre en métaux 91–197
~ polaire 92–121
~ principale 96–37
~ proche 98–60
~ pulsante 95–95
~ qui s'est enfuie 98–135
~ radio 91–158
~ riche en hélium 91–193
~ riche en métaux 91–196
~ rouge 91–97
~ rougie 97–94
~ sans rotation 91–38
~ simple 96–2
~ située devant 96–85
~ située en avant 96–85
~ soupçonnée variable 95–5
~ sous-lumineuse 91–133
~ stable 95–8
~ stationnaire 95–3
~ superdense 91–153
~ superlumineuse 91–134
~ supermassive 91–127
~ suspectée de variabilité 95–5
~ surlumineuse 91–134
~ symbiotique 95–50
~ tournante 91–37
~ triple 96–4
~ variable 95–1
~ vieille 94–64
~ W 91–160
~ WR 91–160
étoiles, amas d' 96–79
classification des ~ variables 95–46
~ des premiers types 91–102
~ des types avancés 91–103
nombre des ~ 98–80
~ tardives 91–103
euclidien 99–101
Europe 83–206
évanouissement des ondes courtes 78–40
évaporation d'un amas 96–108
~ d'un corps météorique 86–198
~ de trous noirs 94–118
évasion d'étoiles d'un amas 98–194
~ d'une étoile du système 96–159
événement 99–125
~ à particules de haute énergie 89–191
~ à protons 84–59
~ adjacent 99–126
~ de pollution 72–72
~ infiniment voisin 99–126
~ sismique 82–215
~ solaire à protons 89–192
~ simultané 99–128
~ solaire à rayons cosmiques 89–194
évent 73–122, 82–162
éventail 85–60
~ coronal 89–295
~ droit 85–61
en ~ 85–74
~ incurvé 85–62
~ inversé 85–63
jet d' ~ 89–296
jet en forme d' ~ 89–296
évoluer 94–75
évolution avant la séquence principale 94–26
~ chimique 82–257
~ d'un groupe des taches 89–118
~ de l'univers 99–171e
~ des étoiles 94–1
~ dynamique 96–161
~ géochimique 74–33
~ nucléaire 94–50
~ préstellaire 94–14
~ préstellaire des galaxies 94–14
~ stellaire 94–1
~ sous l'effet des marées 83a–65
~ thermique 83a–61
~ thermique passée 82–260
~ thermonucléaire 94–50
excès de couleur 91–77
~ infrarouge 91–79
~ IR 91–79
~ ultraviolet 91–78
~ UV 91–78
excursion 82–35
~ scientifique 82–36
exogène 73–146
exosphère 76–68
expansion accélérée 99–81
~ de l'univers 99–74
~ des fonds océaniques 72–39
~ hydromagnétique 88–159
~ ultrarapide 99–170c
~ ultrarapide de l'univers 99–170c
expéditeur 85–23
expédition d'éclipse 88–30
expérience de diffusion alpha 82–76
~ de Michelson 99–130
~ de Michelson et Morley 99–130
~ sismique 82–219
~ sismique active 82–220
~ sur la Lune 82–42
explosion 87–30
~ d'un météore 86–81
~ d'un noyau galactique 98–283
~ d'une comète 85–125
~ d'une nova 95–121
~ d'une supernova 95–131
~ initiale 99–167
~ nucléaire à haute altitude 75–114
~ primitive de l'univers 99–167
~ primordiale 99–167
~ volcanique 73–112
expulsion 73–132
extinction atmosphérique 77–127a
~ d'un météore 86–37

~ interstellaire 97–80
~ par la vapeur d'eau 77–126
~ progressive 86–75
~ sélective 97–83
~ soudaine 86–76
~ totale 97–82
extragalactique 98–207
extrémités de croissant 83–130
extremum de la courbe 95–37
extrusion 73–132
~ de lave 82–171

F

face antérieure de la Lune 81–15
~ arrière de la Lune 81–16
~ cachée de la Lune 81–16
~ éclairée 81–34
~ invisible de la Lune 81–16
~ opposée de la Lune 81–16
~ visible de la Lune 81–15
facteur d'expansion 99–81a
~ de contraction de Lorentz 99–135
~ de dilution 93–37
~ de forme 86–177
~ de Gaunt 93–104
~ de guillotine 93–103
~ de luminosité 86–197
~ d'échelle 99–81a
~ d'échelle cosmique 99–81a
~ géométrique 86–177
~ guillotine 93–103
facule 89–161
facules chromosphériques 89–164
~ photosphériques 89–161
~ polaires 89–162
faible lueur 79–151
faillage 73–92
~ local 82–204
faille 73–91, 82–212
~ compressive 73–94
~ de compression 73–94
~ de distension 73–93
~ directe 73–93
~ distensive 73–93
~ inverse 73–94
~ normale 73–93
~ régulière 73–93
~ renversée 73–94
faire la carte des champs magnétiques 89–132
faisceau électronique 79–116
famille de comètes 85–102
~ de Jupiter 85–103
~ des astéroïdes 84–73
fausse lumière zodiacale 84–62
feldspath 74–15
~ potassique 74–16
Femme Enchaînée 92–14
fenêtre de transparence 83–264
fenêtres de la zone d'absences 98–57
~ galactiques 98–57
fente courbe 88–65
~ oscillante 88–70
fer magnétique 82–92
~ météorique 87–83
~ météoritique 87–83
ferromagnétisme 75–26
ferro-nickel 87–90
feuillatage 74–53
feuille 83–149
feuillet 83–149
fibres de la chromosphère 88–209
fibrilles 88–209
figure d'équilibre 83–112
~ d'équilibre hydrostatique 83–112
~ d'interférence 88–120
~ d'une planète 83–109
~ de la Lune 81–116
~ de la Terre 71–7
figures de Widmanstätten 87–105
fil d'une éruption 89–187
filament 85–55, 89–224
~ d'hydrogène 89–267
~ d'une nébuleuse 97–131
~ de pénombre 89–98
~ métallique 89–268
~ oscillant 89–216
~ pénombral 89–98
filet brillant d'une éruption 89–187
~ protubérantiel 89–245
film de protubérances 89–242
filon 73–138
~ lithique 73–138
~ rocheux 73–138
filon-couche 73–135
filtre pour aérosol 76–94
~ pour poussière 76–95
fin au lever du Soleil 88–21
~ de l'éclipse partielle 81–83a
~ de la totalité 81–80
~ du cycle 89–85
~ du monde 99–183
finitude de l'univers 99–4
fission d'une comète 85–111
~ due aux marées 85–112
fissure 73–78
~ d'extension 82–207
~ de tension 82–207
flash de l'hélium 94–79
Flèche 92–84
flocule 89–167
flocules brillantes d'hydrogène 89–169
~ d'hydrogène 89–168
~ de calcium ionisé 89–170
flocculi 89–167
~ brillants d'hydrogène 89–169
~ d'hydrogène 89–168
~ de calcium ionisé 89–170
flocculus 89–167
flot 82–280
~ et jusant 72–89
flottabilité magnétique 89–147
fluctuations primordiales 99–170y
~ secondaires 95–44
~ superposées 95–44
fluide chimiquement actif 74–43
fluidité de la lave 82–188
flux 82–280
~ convectif 93–115
~ d'énergie 93–112
~ d'énergie acoustique 88–221
~ d'énergie convectif 93–115
~ d'énergie total 93–116
~ de chaleur 74–60
~ de marée 82–279
~ de masse 93–13
~ de neutrons 94–86
~ de rayonnement 93–113
~ de rayonnement émergent 93–114
~ intégré 93–47
FmU 78–33
FMU 78–37
foliation 74–53
Fomalhaut 92–116
fonction de densité 98–88
~ de diffusion 77–58
~ de distribution 98–143
~ de fréquences de masses 84–52
~ de luminosité 98–89
~ de phase 83–78
~ de saturation 93–60
~ de visibilité 89–55
~ des densités 98–88
~ des masses 96–68
~ intégro-exponentielle 93–62
~ poids 93–58
~ R(t) 99–81a
~ source 93–34
fonctions harmoniques tessérales 95–102
~ harmoniques zonales 93–55
~ sphériques primitives 95–101
fond de la mer 72–38
~ des océans 72–37
~ du cratère 81–259
~ marin 72–38
~ météorique 86–232
~ océanique 72–37
~ plat 81–260
~ séismique 73–55
~ sismique 73–55
force coercitive 75–39
~ cohésive 85–117
~ de champ effective 95–59
~ de Coriolis 77–79
~ de marée 82–289
~ de répulsion 85–87
~ déviante 77–80
~ du vent 77–107
~ électrofaible 99–170x
grande ~ unifiée 99–170v
~ interne 73–61
~ marémotrice 82–289
~ portante 82–64
~ répulsive 85–87
~ superunifiée 99–170r

~ unifiée 99–170x
forces de marée 72–92
~ tectoniques 73–106
formaldéhyde 97–27
formamide 97–35
formation 94–10
~ active 79–138
~ atmosphérique 83–107
~ circulaire 81–155
~ d'impact 82–228
~ d'un écran magnétique 89–225
~ d'une tache 89–121
~ de cratères 82–232
~ de cratères par impact 82–234
~ de failles 73–92
~ de la queue 85–86
~ de la surface 83–105
~ de raies 93–63
~ des éléments 94–52
~ des étoiles 94–11, 96–95
~ des failles locales 82–204
~ des failles régionales 82–205
~ des grains 97–60
~ des montagnes 73–74
~ du monde 99–11
~ lunaire 81–148
~ par impact 82–229
~ permanente 83–106
~ primaire 81–150
~ secondaire 81–151
~ topographique 81–149
~ violente des étoiles 94–11a
~ volcanique 82–159
forme circulaire 81–155
~ cométaire 85–83
~ d'équilibre 83–112
~ d'une planète 83–109
~ de la particule météorique 86–175
~ de nuage 77–133
~ du corps météorique 86–175
~ en anneau 98–41
~ orientée 87–113
formes aurorales 79–137
~ avec structure de rayons 79–152
~ sans structure de rayons 79–142
~ structurales 81–152
formule altimétrique 76–15
~ barométrique 76–15
~ internationale de la pesanteur 71–76
~ internationale gravimétrique 71–76
Fornax 92–50
fossé 82–132
~ d'effondrement 82–132
fosse marine 72–40
~ océanique 72–40
Fourneau 92–50
~ Chimique 92–50
foyer du tremblement de terre 73–17
fractionnement 87–138
~ des éléments chimiques 83a–51
fracture 73–78
~ au rebord d'un cratère 82–211
~ au rempart d'un cratère 82–211
~ concentrique 82–209
~ de surface 81–173
~ linéaire 82–208
fractures, système de 73–79
système de ~ continentales 73–81
système de ~ médio-océaniques 73–80
fragile 86–191
fragilité 86–192
fragment météoritique 87–43
~ d'une météorite 87–43
~ nucléaire 94–102
~ rocheux 82–68
~ vitreux 82–97
fragmentation 83a–81, 87–25
~ d'un corps météorique 86–188
~ d'un nuage cosmique 94–38
~ d'un météore 86–83
~ des astéroïdes 83a–82
~ du corps parental 87–124
~ du noyau cométaire 85–116
~ due aux marées 85–118
~ météorique 86–188
~ par collision 87–125
fragments 83a–83
frange polaire 83–151
freinage apparent 86–100
~ de la rotation 91–45
~ magnétique 91–47
~ réel 86–101
fréquence aurorale 79–122
~ critique 78–24
~ d'apparition des taches 89–66
~ d'impulsions 86–279
~ de formation des taches 89–66
~ de l'émetteur 86–273
~ de modulation 86–275
~ de récurrence 86–279
~ des cratères 81–247
~ des éruptions 89–217
~ des protubérances 89–237
~ en fonction de la taille 81–249
~ horaire 86–135
~ horaire des météores 86–135
~ maximum utilisable 78–37
~ minimum utilisable 78–33
~ porteuse 86–274
friction des marées 82–312
front 77–68
~ chaud 77–69
~ d'ionisation 97–48
~ de la coulée 82–193
~ froid 77–70
~ occlus 77–71
~ polaire 77–73
frontière de la magnétosphère 75–9
frontogénèse 77–74
frontolyse 77–75
frottement 86–173
fuite de galaxies 98–298
fumée 97–57
fumerolle 82–163
fusion 86–181

G

gain d'énergie 88–253
~ de masse 83a–38
gal 71–65
galactique 98–2
galactocentrique 98–10
distance ~ 98–11
galaxie 98–201
~ à bras multiples 98–44
~ à deux bras 98–43
~ à noyau multiple 98–319
~ à plusieurs bras 98–44
~ à rayons X 98–203
~ à un bras 98–42
~ à un seul bras 98–42
~ active 98–260
~ anomale 98–256
~ annulaire 98–259
~ barrée 98–253
~ bleue compacte 98–265
~ compacte 98–264
~ compagnon 98–216
~ condensée 98–264
~ de champ 98–340
~ de Groupe Local 98–327
~ de Haro 98–266
~ de l'amas 98–338
~ de Seyfert 98–262
~ de transition 98–248
~ désintégrée 98–317
~ distordue 98–316
~ double 98–305
~ du fond 98–340
~ du fond du ciel 98–340
~ du type N 98–263
~ du type têtard 98–281
~ ellipsoïdale 98–241
~ elliptique 98–241
~ elliptique EO 98–244
~ elliptique géante 98–243
~ elliptique naine 98–242
~ en aiguille 98–239
~ en cigare 98–239
~ en explosion 98–268
~ en fuseau 98–239
~ éruptive 98–269
~ fortement applatie 98–252
~ fragmentée 98—317
~ géante 98–222
~ infrarouge 98–205
~ irrégulière 98–237
~ jumelle 98–305
~ la plus brillante de l'amas 98–339
~ lenticulaire 98–245
~ locale 98–327
~ membre de l'amas 98–338
~ mère 98–284

~ multiple 98–308
~ naine 98–220
~ non-membre de l'amas 98–340
~ normale 98–221
~ optique 98–204
~ par la tranche 98–240
~ parente 98–284
~ particulière 98–256
~ perturbée 98–315
~ post-éruptive 98–270
~ quasistellaire 98–361
~ radio 98–206
~ régulière 98–236
~ riche en gaz 98–238
~ roentgen 98–203
~ satellite 98–216
~ singulière 98–258
~ spirale 98–249
~ spiralée 98–249
~ supergéante 98–223
~ voisine 98–211
~ vue de face 98–247
~ vue de profil 98–246
Galaxie 98–1
notre ~ 98–1
galaxies, amas de 98–293
classification des ~ 98–226
~ en action réciproque 98–310
~ en collision 98–321
~ en interaction 98–310
groupe de ~ 98–325
Groupe Local de ~ 98–326
~ interconnectées 98–306
séquence des ~ régulières 98–240
galet 82–128
~ indicateur 72–132
Ganymède 83–207
gaz baryonique 93–194
~ capturé de vent solaire 82–148
~ circumgalactique 98–351
~ congelés 85–48
~ d'hypérons 93–193
~ dans un amas de galaxies 98–352
~ de baryons 93–194
~ de neutrons 93–192
~ dégénéré 93–179
~ en trace 76–88
~ entre deux bras spiraux 98–40
~ hypéronique 93–193
~ incandescents 83a–43
~ intergalactique 98–351a
~ intergalactique d'un amas 98–352
~ intergalactique entre amas 98–353
~ interstellaire 97–9
~ neutronique 93–192
~ non-dégénéré 93–182
~ rare primordial 87–133
~ rare radiogénique 87–132
~ relativiste 97–152
géante 91–143
~ bleue 91–142a
~ brillante 91–149
~ des premiers types 91–143a
~ des types avancés 91–144a
~ normale 91–150
~ rouge 91–152a
gegenschein 84–64
gelée blanche 77–177, 83–150
Gémeaux 92–51
Gemini 92–51
Géminides 86–257
génération d'énergie 93–105
~ d'étoiles 94–23
~ stellaire 94–23
générescence d'étoiles 94–11a
géocentrique 84–3
géochimie 74–30
géochronologie 74–67
géocouronne 76–73
~ d'hydrogène neutre 76–7
géodesie lunaire 81–109
~ par satellite 71–32
~ spatiale 71–31
géodésique 99–153
géoélectricité 75–1
géoïde 71–16
~ compensé 71–17
~ régularisé 71–17
géologie 72–2
~ des planètes 83–5
~ lunaire 82–47
géomagnétisme 75–24
géométrie d'Euclide 99–108
~ euclidienne 99–108
~ non-euclidienne 99–109
~ riemannienne 99–113
géophysique 72–1
~ cosmique 84–57
géopotentiel 71–60
géosphère 72–7
géosynclinal 73–105
géotectonique 73–95
géothermie 74–58
géothermique, degré 74–59
Giacobinides 86–252
Girafe 92–24
girouette 77–104
glace 85–42
~ amorphe 85–44
~ cubique 85–46
~ d'eau 85–47
hummock de ~ 72–102
noyaux de ~ 76–104
~ sale 97–73
glaciaire 72–106
glaciation 72–104
glacier 72–112
~ alpin 72–114
~ continental 72–116
~ d'écoulement 72–125
~ de montagne 72–113
~ de vallée 72–114
glaciologie 72–103
glissement 82–196
~ de remparts 82–198
~ du sol 82–152
globe céleste 92–7
~ en fusion 83a–64
~ fondu 83a–64
~ lunaire 81–112
~ terrestre 71–10
globule 97–133
~ de poussière 98–290
gloire 79–63
golfe 81–294
~ des Iris 81–295
~ Torride 81–296
Gothlandien 74–80
goutte 76–113
~ noire 83–52
gouttelette 76–100
graben 82–132
gradient 93–94
~ absolu 91–111
~ adiabatique 93–98
~ adiabatique de température 93–98
~ adiabatique sec de température 76–14
~ convectif de température 93–157
~ de densité 98–87
~ de pression 77–106, 93–95
~ de température 93–96
~ de température radiatif 93–97
~ géothermique 74–59
~ spectrophotométrique 91–112
~ superadiabatique 93–99
~ thermique 93–96
~ vertical de température 76–14
grain 72–20
à ~ fin 82–114
à gros ~ 82–117
à ~ moyen 82–116
~ de fer 97–74
~ de glace 85–52
~ de graphite 97–75
~ de poussière 82–110
~ de silicate 97–76
~ enveloppé de glace 97–77
~ fin 82–110
~ pierreux 85–54
~ polymérique 97–77a
grains de Baily 88–32
~ de poussière interstellaire 97–59
~ interstellaires 97–59
Grand Chariot 92–96
~ Chien 92–27
~ effondrement 99–182
~ halo 79–70
~ météore 86–11
grande interaction unifiée 99–170v
~ Louche 92–96
~ marée 82–273
~ Nuage de Magellan 98–213
~ Ourse 92–96
~ Perturbation Australe 83–194

F

~ structure aplatie 99–171b
~ Tache Rouge 83–193
~ unification 99–170u
grandeur de l'éclipse 81–79
~ des particules 76–114
granit(e) 74–8
granulation faculaire 89–163
~ photosphérique 88–176
granule 88–178
~ d'ombre 88–180
~ dans l'ombre 88–180
~ faculaire 88–179
~ photosphérique 88–181
gravier 82–128
gravimètre 71–70
gravimétrie 71–63
gravitation propre 94–34
gravité 71–56
grêle 77–178
grille quart d'onde 88–126
grosses planètes 83–13
grotte de lave 82–177
groupe amine 87–71
~ astéroïdal 84–90
~ astéroïdal de commensurabilité 84–101
~ bipolaire 89–111
~ complexe 89–112
~ d'étoiles 92–3a
~ de comètes 85–105
~ de petites planètes 84–90
~ de taches 89–105
~ de taches solaires 89–105
~ du cycle précédent 89–107
~ du fer 94–105
~ du nouveau cycle 89–106
éléments du ~ du fer 94–105
~ récurrent 89–108
~ troyen 84–91
~ troyen d'astéroïdes 84–91
~ unipolaire 89–110
Grue 92–52
Grus 92–52
guide solaire 88–51
guyot 72–44

H

haie, en forme de 89–274
halo 79–67
~ chaud 98–46
~ de 46° 79–70
~ froid 98–47
~ galactique 98–45
grand ~ 79–70
~ lunaire 79–68
~ magnétique 98–48
~ phénomène de ~ 79–66
~ principal 79–69
halos d'une comète 85–68
Hamal 92–118
harmoniques tesseraux 95–102
~ zonaux 95–101
hausse de la température 77–42
haute latitude 89–53
hautement ionisé 89–289
hautes eaux de syzygie 82–273
hauteur d'ombre 79–46
~ d'ombre géométrique 79–47
~ d'une couche 78–27
~ de blindage 79–48
~ de l'apex 86–66
~ de l'atmosphère homogène 76–17a
~ de la chromosphère 88–202
~ de référence 76–18
~ des précipitations 77–158
~ équivalente 78–28
~ réelle 78–29
~ virtuelle 78–28
~ vraie 78–29
~ vraie d'ombre 79–49
haut-fond 72–96
haut-plateau 81–189
surface de ~ 82–56
héliogéophysique 84–56
héliogramme en rayons X 88–74
héliographe 77–61, 88–36
~ monochromatique 88–38
héliomètre 88–12
héliophysique 88–6
hélios(é)ismologie 88–223b
héliosphère 88–255
héliosphérique 88–256
héliostat 88–45
hémisphère diurne 83–67
~ éclairé 75–20, 83–67
~ ensoleillé 81–126
~ illuminé 81–34
millioniémes de l'~ visible 89–68
~ nocturne 75–21, 83–68
~ obscur 75–21, 83–68
~ proche 81–15
~ visible 89–27
Hercule 92–53
Hercules 92–53
Hermès 84–83
hétérosphère 76–61
hexaédrite 87–84
hiérarchie de structure 99–47
~ de structure de l'univers 99–47
hiss 78–64
histoire thermique 74–66, 82–260
holocène 72–109
holosidères 87–82
homologue 93–131
homopause 76–65
homosphère 76–60
horizon cosmique 99–10
~ cosmologique 99–179
~ de l'univers 99–10
~ des événements 99–43
~ des masses 99–41
Horloge 92–54
Horologium 92–54
hors de l'atmosphère 88–149
humidité absolue 77–122
~ atmosphérique 77–118
~ de l'air 77–118
~ relative 77–128
~ spécifique 77–123
Hyades 96–111a
Hydra 92–55
Hydre Australe 92–56
~ Femelle 92–55
~ Mâle 92–56
hydrodynamique stellaire 98–142
hydrogène interstellaire 97–46
~ métallique 83–202
~ moléculaire 97–31
~ neutre 97–47
hydrophile 76–92
hydrophobe 76–93
hydrosol 76–90
hydrosphère 72–21
hydroxyle 79–25
Hydrus 92–56
hygromètre 77–120
hypergéantes 91–147
hypocentre 73–17
hypothèse 94–6
~ catastrophique 83a–14
~ cosmogonique 94–8
~ d'éruption 89–63
~ de capture 83a–18
~ de Lambert 99–45
~ de marée 83a–14
~ des deux courants stellaires 98–99
~ des impacts 81–261
~ météorique 81–262
~ météoritique 81–262
~ météoroidale 81–262
~ nébulaire 83a–7

I

Icare 84–84
iceberg 72–117
identités de Lagrange 98–182
~ de Lagrange généralisées 98–183
île, nouvelle 72–81
illumination, équateur d' 81–122
ilménite 82–91
image photographique de la Lune 82–26
images monochromatiques 97–158
immersion 81–98
impact 82–16
~ artificiel 82–226
~ cométaire 82–230
~ d'une comète 82–230
~ d'une météorite 87–33
~ météoritique 87–33
implosion du noyau stellaire 94–110
importance d'une éruption 89–205
impulsion d'écho 86–290
impulsions 86–278

inclinaison d'une orbite 96–51
~ de l'axe 83–35
~ de l'axe de rotation sur... 91–50
~ du plan orbital 96–51
~ magnétique 75–58
~ sur l'équateur lunaire 81–37
incline 76–54
inclusion organique 87–66
index polytropique 93–128
indicateur d'amplitudes des marées 82–311
~ de distance 86–266
~ de la structure spirale 98–39
indicatrice de diffusion 77–58
indice d'activité solaire 89–8
~ de chaleur 91–81
~ de concentration 96–101
~ de couleur 91–76
~ de couleur B-V 96–145
~ de couleur composite 96–138
~ de couleur global 96–138
~ de couleur intrinsèque 91–75
~ de couleur pour le jaune 83–88
~ de couleur spécial 83–87
~ de couleur U-B 96–146
~ de l'activité d'un courant 86–230
~ de structure d'un amas 96–80
~ K 75–89
~ K d'activité géomagnétique 75–89
~ planétaire Kp 75–90
~ polytropique 93–128
~ solaire 89–8
~ structural de l'amas 98–342
~ trihoraire 75–89
~ U-B 96–146
Indien 92–57
Indus 92–57
inégalité de Schwarzschild 93–156
infinitude de l'univers 99–6
inflation 99–170c
~ cosmologique 99–170c
inhomogénéité de champ magnétique 78–9
~ de densité d'électrons 78–8
~ ionosphérique 78–7
~ macroscopique 88–206
~ oblique 88–190
inondation 72–75
insolation 77–51
instabilité 93–211
~ à l'égard des marées 98–313
~ convective 93–155
~ de rotation 83a–8
~ due aux marées 96–107
~ gravitationnelle 94–35
~ locale 93–154
~ pulsationnelle 93–213
~ rotationnelle 83a–8
~ séculaire 93–212
~ vibrationnelle 93–214
instable 93–210
instant d'alunissage 82–16
~ d'apparition 86–49
~ d'arrivée 73–50
~ d'atterrissage 82–16
~ du chute météoritique 87–51
~ du passage au périastre 96–60
instrument automatisé 82–41
~ solaire 88–35
intégrale de phase 83–79
intensité au centre de la raie Hα 89–211
~ de champ effective 95–59
~ de la pesanteur 71–62
~ directionnelle 75–103
~ du champ 84–11
~ du champ effective 95–59
~ du champ géomagnétique 75–55
~ du champ magnétique 84–11
~ horizontale 75–60
~ horizontale du champ magnétique 75–60
~ omnidirectionnelle 75–104
~ séismique 73–33
~ sismique 73–33
~ verticale 75–62
interaction atmosphérique 79–120
~ collective 98–168
~ de marées 98–311
~ due aux marées 82–274
~ électrofaible 99–170x
grande ~ unifiée 99–170v
~ superunifiée 99–170r
interface 93–141
interféromètre à antennes multiples 88–123
~ à balayge de fréquence 88–110
~ à balayage de lobe 88–109
~ à balayage de phase 88–109
~ à base variable 88–107
~ à commutation de phase 88–113
~ à deux antennes 88–98
~ à espacement variable 88–107
~ à faisceau en éventail 88–122
~ à grande base 88–104
~ à liaison radioélectrique 88–116
~ à lobe balayé 88–109
~ à lobe en éventail 88–122
~ à lobe étroit 88–121
~ à pinceau fin 88–121
~ à réseau 88–123
~ à une antenne 88–106
~ de Christiansen 88–123
~ en pinceau 88–121
~ marin 88–106
~ radio 88–97
~ radioastronomique 88–97
intergalactique 97–6
interglaciaire 72–108
intérieur de la Terre 72–3
~ des étoiles 93–72
~ stellaire 93–72
~ terrestre 72–3
interplanétaire 84–8
interruption par l'obturateur 86–150
interstellaire 97–2
intrusion 73–133
~ concordante 73–134
~ de roches ignées 82–98
~ discordante 73–137
invariable dans le temps 98–150
invariance dans une transformation 99–133
invasion des particules à haute énergie 78–43
inversion atomosphérique 77–37
~ au sol 77–38
~ de la polarité 89–153
~ de population 97–15
~ de température 77–36
~ thermique 77–36
investigation souterraine 82–82
Io 83–205
ion d'hydrogène 93–65
~ de Langevin 76–115
~ lent 76–115
~ lourd 76–115
~ négatif d'hydrogène 93–66
ionisation par les météores 86–281
rendement d' ~ 86–209
ionogramme 78–23
ionosonde 78–22
ionosphère 76–69, 78–1
~ extérieure 76–76
ionosphérique 78–2
irradiation par rayons cosmiques 82–250
irrégularité à petite échelle 84–13
~ du champ magnétique 84–12
isobathe 71–55
isocline 75–59
isodyname 75–61
isogone 75–57
isohypse 81–139
iso-intensité 75–102
isonitrile d'hydrogène 97–40
isophote 81–92
isoplèthe 96–125
isorotation 88–130
isostasie 73–98
isotope cosmogénique 82–147
~ de courte période 83a–74
~ de longue période 83a–75
~ du plomb 83a–76
~ primordial 82–146
~ produit par les rayons cosmiques 82–147
itinéraire 82–37

J

jaune 91–98
jet 85–55
~ coronal 89–287
~ coronal monochromatique 89–288
~ d'électrons 79–114
~ de matière 89–255

F

~ de matière absorbante 89–256
~ de matière accompagnant une éruption 89–190
~ de matière visible par son émission 89–257
~ du noyau galactique 98–278
~ en forme de bulbe 89–295
~ en spirale 85–59
~ incurvé 85–58
~ optique 98–279
~ rectiligne 85–57
jonction géodésique des continents 71–37
jour agité 75–93
~ calme 75–92
~ de marée 82–290
~ lunaire 81–32
~ lunaire moyen 82–304
~ magnétiquement actif 75–93
~ magnétiquement agité 75–93
~ magnétiquement calme 75–92
~ perturbé 75–94
jovicentrique, coordonnée 83–175
Jovien 83–171
jovigraphique, coordonnée 83–175
Junon 84–85
Jupiter 83–170
Jurassique 74–85
jusant 82–282

K

K coronomètre 88–66
kamacite 87–92

L

La Chèvre 92–110
La Dentelle du Cygne 97–137
lac 81–297, 83–158
~ de l'Autumne 81–301
~ de l'Été 81–300
~ de la Mort 81–298
~ de lave 82–183
~ des Rêves 81–299
~ du Printemps 81–302
laccolite 73–136
Lacerta 92–58
lacune 84–88
~ d'Hertzsprung 91–144
~ de commensurabilité 84–88
~ de Kirkwood 84–89
lame concordante 73–135
~ de compensation 88–92
~ de mica 88–87
~ intrusive 73–135
~ $\lambda/4$ 88–126
~ quart d'onde 88–126
lamelles 87–104
langue de lave 82–187
~ glaciaire 72–125
~ interplanétaire 84–21
~ solaire interplanétaire 84–20
lapilli 82–169
largeur d'impulsion 86–280
~ de la zone de silence 78–38
~ effective de la raie 89–208
L'Arrière du Bateau 92–81
laser, réflecteur 82–43
latitude aréographique 83–140
basse ~ 89–52
~ des calmes tropicaux 77–102
~ géomagnétique 75–52
haute ~ 89–53
~ héliographique 89–42
~ limite de la visibilité d'une occultation 81–104
~ limite nord 81–105
~ limite sud 81–106
~ sélénographique 81–11
lave figée 82–176
plaine de ~ 82–185
~ refroidie 82–176
tube de ~ 82–186
Le Chasseur 92–73
Le Fleuve Eridan 92–49
Le Petit du Cheval 92–48
Le Petit du Lion 92–60
Le Roi 92–33
Le Vainqueur 92–76
lentille gravitationnelle 98–256a
Leo 92–59
Leo Minor 92–60
Léonides 86–255
Lepus 92–61
levé magnétique 75–64
~ magnétique mondial 75–65
~ sélénologique 82–50
~ sélénophysique 82–39
lever du Soleil 88–8
lèvre d'évent 81–221
Lézard 92–58
liant 82–1188
~ à grain fin 82–115
libération, cône de 76–27
~ des gaz 76–24
Libra 92–62
libration 81–48
~ diurne 81–52
~ en latitude 81–50
~ en longitude 81–49
~ géométrique 81–51
nuage de ~ 84–43
nuage de ~ de la Lune 84–43
~ optique 81–51
~ parallactique 81–52
~ physique 81–53
libre de poussière 82–119
Licorne 92–68
lieu de découverte 87–52
Lièvre 92–61
ligne d'âge zéro 94–60
~ d'écoulement 82–194
~ d'égal taux de comptage 75–102
~ d'inversion des polarités 89–151
~ d'univers 99–154
~ de champ non-fermée 75–15
~ de fil à plomb 71–71
~ de force en co-rotation 75–14
~ de même profondeur 71–55
~ de séparation des polarités 89–151
~ de vision 91–28
~ géodésique 99–153
~ isocline 75–59
~ isodynamique 75–61
~ isogauss 89–134
~ isogone 75–57
~ littorale 72–31
~ neutre 89–151
lignes de Neumann 87–103
limbe solaire 89–28
limite australe 81–106
~ australe de l'éclipse 88 19
~ boréale 81–105
~ boréale de l'éclipse 88–20
~ de Chandrasekhar 93–189
~ de secteur 84–18
~ des neiges 72–111
~ supérieure 76–58
~ supérieure de l'atmosphère 76–25
Lion 92–59
lit fluvial 83–153
lithification 74–49
~ par impact 82–238
lithophiles, éléments 83a–30
lithosidérite 87–80
lithosphère 72–6
littoral 72–30
lobe à emission radio 98–286
~ de Roche 96–28
loi d'assombrissement 88–143
~ de Bode 83–3
~ de déplacement des zones royales 89–75
~ de distances planétaires 83–3
~ de Hale 89–154
~ de Hubble 98–299
~ de Kramers 93–102
~ de l'alternance de polarité 89–154
~ de polarité 89–154
~ de rotation 98–110
~ de Spoerer 89–75
~ de Titius-Bode 83–3
~ du rougissement 97–92
~ du rougissement interstellaire 97–92
~ virielle 98–181
longitude active 89–41
~ aréographique 83–139
~ au périastre 96–59
~ géomagnétique 75–53
~ héliographique 89–40
~ sélénographique 81–10
longueur angulaire 98–32
~ angulaire de la trajectoire 86–48
~ d'impulsion 86–280
~ d'onde au repos 98–365

~ d'onde Compton 99–170g
~ d'onde effective 96–142
~ de la trajectoire d'un météore 86–47
~ de Planck 99–170h
~ fondamentale de Planck 99–170h
Loup 92–63
lueur antisolaire 84–64
~ d'hydrogène 79–145
~ diffuse 79–148
faible ~ 79–151
~ pourprée 79–54
lumière cendrée 81–35
~ du ciel 77–53
~ du ciel nocturne 79–8
~ extraterrestre 79–7
fausse ~ zodiacale 84–62
~ galactique 79–14
~ galactique diffus(é)e 97–56
~ propre 96–45
~ solaire 77–49
~ stellaire 79–10
~ stellaire diffus(é)e 79–12
~ stellaire directe 79–11
~ zodiacale 79–13
luminescence atmosphérique 79–2
~ atmosphérique crépusculaire 79–4
~ atmosphérique diurne 79–3
~ atmosphérique nocturne 79–5
~ du ciel 79–3
~ du ciel crépusculaire 79–4
~ du ciel nocturne 79–5
~ Lα du ciel nocturne 79–9
~ lunaire 82–248
lumineux par lui-même 96–45
luminosité 91–128
~ d'un météore 86–195
~ maximale 86–78
~ neutrinique 94–96
lunaire 81–3
lunaison 81–18
Lune 81–2
~ croissante 81–20
de la ~ 81–3
~ décroissante 81–21
décours de la ~ 81–21
demi-~ 81–26
~ éclipsée 81–57
~ en dernier quartier 81–30
~ en premier quartier 81–25
~ jeune 81–20
nouvelle ~ 81–22
pleine ~ 81–27
~ vieille 81–21
lunette guide 88–51
Lupus 92–63
Lynx 92–64
Lyra 92–65
Lyre 92–65
Lyrides 86–248

M

maar 82–161
Machine Pneumatique 92–15
macrométéorite 87–4
macrostructure 99–9
magma 73–131
magnétisme des roches 75–43
~ induit 75–29
~ résiduel 75–28
~ stellaire 95–54
~ terrestre 75–24
magnétite 82–92
magnétogramme 75–84
magnétographe 75–83
~ du champ longitudinal 88–83
~ du vecteur champ 88–84
~ longitudinal 88–83
~ pour vecteur champ 88–84
~ solaire 88–82
~ vectoriel 88–84
magnétohéliographe 88–93
magnétoïde 98–25
magnétomètre portatif 82–44
magnétopause 75–9
magnétosphère 75–7
~ de pulsar 95–147
magnitude 91–56
~ absolue 91–64
~ absolue d'un météore 86–72
~ absolue d'une planète 83–86
~ apparente 91–61
~ apparente d'un météore 86–71
~ bolométrique 91–68
~ d'un tremblement de terre 73–34
~ électrographique 91–67
~ électronographique 91–67
~ globale 96–136
~ infrarouge 91–71
~ intégrée 96–136
~ limite 91–62
~ moyenne 83–85
~ photographique 91–66
~ photovisuelle 91–69
~ radioélectrique = radio-électrique 91–70
~ radiométrique 91–70
~ réduite au zénith 91–63
~ s(é)ismique 73–34
~ stellaire 91–56
~ totale 96–136
~ visuelle 91–65
~ zénithale 91–63
manifestation corpusculaire de l'activité 89–7
~ de l'activité solaire 89–13
~ optique de l'activité 89–3
~ radio de l'activité 89–6
~ visible d'une éruption 89–181
~ X d'une éruption 89–183
manques de Kirkwood 84–89
manteau 72–15
~ supérieur 72–16
~ terrestre 72–15
marais 81–303
~ de la Putréfaction 81–305
~ du Sommeil 81–306
marche aléatoire 89–157
~ au hasard 89–157
mare 91–266
marée basse 72–91
~ de morte eau 82–272
~ de quadrature 82–272
~ de syzygie 82–273
~ de vive-eau 82–273
~ descendante 82–282
effet de ~ 82–276
grande ~ 82–273
~ haute 82–281
~ montante 82–280
petite ~ 82–272
~ produite par la Terre 82–275
marées 72–89
~ atmosphériques 82–294
~ de la croûte 82–297
~ géomagnétiques 82–295
~ gravitationnelles 82–298
~ lunaires 82–308
~ maritimes 82–309
~ océaniques 82–310
~ terrestres 82–296
~ thermiques 82–299
marégraphe 82–311
Margarita 92–117
marge brillante d'un filament 89–226
~ continentale 73–103
Markab 92–119
marnage 82–292
Mars 83–133
martien 83–135
mascon 81–145
maser cosmique 97–13
~ interstellaire 97–14
~ naturel 97–13
masse au repos 99–143
~ cachée 99–88
~ critique 93–189
~ critique des naines blanches 93–189
~ d'air 77–67
~ d'intertie 99–148
~ d'un météore 86–163
~ d'une particule météorique 86–163
~ de Planck 99–170f
~ fondamentale de Planck 99–170f
~ gazeuse 94–32
~ globale de l'amas 96–139
~ gravitationnelle 99–149
~ inertielle 99–148
~ initiale 86–164
~ limite 93–189
~ manquante 99–87
~ originelle 86–164
perte de ~ 86–168
~ pesante 99–149

F

~ principale 87–41
~ relativiste 99–141
~ solaire 91–122
~ stellaire 91–121
~ total de l'amas 96–139
~ variable 86–167, 96–70
matériau cosmique 99–7
~ de monticule 82–103
~ détritique 81–236
~ extraterrestre 72–56
~ lunaire 82–58
~ lunaire de surface 82–59
matière absorbante 97–85
~ carbonée 87–64
~ cosmique 99–7
~ gaz-poussière 83a–25
~ intergalactique 98–349
~ interplanétaire 84–1
~ interstellaire 97–1
~ lumineuse 97–109, 99–86
~ météoritique 87–9
~ météoritique organique 87–68
~ obscure 97–108
~ poreuse 82–111
~ préstellaire 94–16
~ primordiale 99–170
~ réfractaire 83a–55
~ stellaire 93–124
matières éjectées 82–165
~ volcaniques éjectées 82–166
matrice 82–118
~ à grain fin 82–115
mauvaise visibilité 77–26
maximum aigu 86–235
~ d'éclat 86–78
~ de l'activité solaire 89–2
~ de la région 76–55
~ de la région E 76–56
~ de luminosité 86–78
~ plat 86–236
~ primaire 95–31
~ principal 95–31
~ secondaire 95–32
~ solaire 89–2
mécanisme d'ablation 86–183
~ d'excitation 79–27
~ de chauffage 88–214
~ de destruction 97–79
~ de freinage 91–46
~ de la perte de masse 86–183
~ de pompage 97–16
mélange de gaz et de rayonnement 93–125
~ de Russell 93–101
~ gaz-poussière 83a–25
membre de l'amas 96–82
~ du système 98–133
~ incertain de l'amas 96–87
~ probable de l'amas 96–86
Mensa 92–66
mer 81–266, 83–156
~ Australe 81–286
~ circulaire 81–267
~ Connue 81–289
~ de Humboldt 81–288
~ de la Connaissance 81–289
~ de la Fécondité 81–274
~ de la Sérénité 81–276
~ de la Tranquillité 81–275
~ de Moscou 81–283
~ de Smyth 81–287
~ des Crises 81–280
~ des Humeurs 81–282
~ des Inventions 81–292
~ des Nuages 81–281
~ des Pluies 81–273
~ des Vagues 81–290
~ du Froid 81–278
~ du Nectar 81–279
~ du Serpent 81–291
~ intérieure 72–26
~ intracontinentale 72–26
~ limitrophe 72–27
~ marginale 72–27, 81–285
~ méditerranée 72–26
~ Orientale 81–284
Mercure 83–120
méridien central 89–43
~ de l'aube 75–18
~ de midi 75–17
~ du coucher 75–19
~ magnétique 75–49
mésopause 76–67
mésosidérite 87–79
mésosphère 76–66
Mésozoïque 74–72
ère ~ 74–72
mesure absolue de la pesanteur 71–68
~ d'arc 71–22
~ d'émission 88–234
~ de l'âge 87–126
~ de la pesanteur 71–67
~ du degré 71–22
~ micrométrique 96–49
~ polarimétrique 83–80
~ relative de la pesanteur 71–69
Métagalaxie 98–357
métalicité 91–170
métamorphisme 74–34
~ de choc 82–237
~ de contact 74–35
~ thermique 74–37
métasomatose 74–44
métasomatisme 74–44
météore 86–1
~ à courbe de lumière irrégulière 86–85
~ appartenant à un essaim 86–24
~ artificiel 86–21
~ avec traînée 86–17
~ brillant 86–12
~ cométaire 86–45
~ de poussière 86–190
~ d'essaim 86–24
~ d'origine cométaire 86–45
~ diurne 86–22
~ double 86–131
~ du champ 86–25
~ du système solaire 86–44
~ écliptique 86–46
~ en forme de fuseau 86–77
~ faible 86–15
~ fusiforme 86–77
grand ~ 86–11
~ hyperbolique 86–43
~ interplanétaire 86–44
~ interstellaire 86–43
~ lent 86–88
~ multiple 86–133
~ n'appartenant pas à un essaim 86–25
~ ordinaire 86–14
~ photographié 86–13
~ photographié de deux stations 86–155
~ photographique 86–13
~ présentant des variations de luminosité 86–85
~ pulsant 86–85
queue d'un ~ 86–110
radio- ~ 86–267
~ rapide 86–87
sillage d'un ~ 86–110
~ sporadique 86–23
~ stationnaire 86–34
~ télescopique 86–18
~ télescopique faible 86–19
~ télescopique très faible 86–20
~ ultratélescopique 86–20
~ visuel 86–16
météores sporadiques, acitivité des 86–136
météorique 86–4
météorite 87–1
~ carbonée 87–62
~ ferreuse 87–82
fragment d'une ~ 87–43
~ métallique 87–82
~ orientée 87–112
~ parente 87–123
~ pierreuse 87–60
surface d'une ~ 87–106
météorites, classification des 87–59
ramassage des ~ 87–45
recherche des ~ 87–45
structure des ~ 87–100
météoritique 87–2
météoroïde 86–160
~ poreux 85–50
météorologie 77–6
~ dynamique 77–7
~ synoptique 77–8
méthanal 97–27
méthane-imine 97–45
méthanol 97–32
méthode à l'hélium 87–128

~ à l'uranium 74–96
~ d'Argelander 95–10
~ des degrés 95–10
~ des fonctions poids 93–59
~ du double comptage 86–132
~ itérative 93–56
~ radioactive 74–92
~ radiométrique 74–92
~ stéréoscopique 81–115
~ variationnelle 93–57
méthylacétylène 97–41
méthyle, acétylène de 97–41
méthylidyne 97–23
méthylique, éthyne 97–41
métrique 99–151
~ de Robertson-Walker 99–152a
~ de Schwarzschild 99–152
mica 74–23
~ blanc 74–24
~ noir 74–25
microbrèche lunaire 82–86
microconcentration 83–266
microcratère 81–201
microméteore 86–20
micrométéorites 87–3
micrométéorologie 77–9
microphone à fil chaud 76–46
microphotographie 87–140
micropulsations géomagnétiques 75–82
~ magnétiques 75–82
microréversibilité 93–41
Microscope 92–67
microscope à papilletement 91–34
Microscopium 92–67
microséismes 73–53
microsismes 73–53
microstructure 82–73
microtectite 87–97
migmatite 74–39
milieu 76–3
~ de l'éclipse 81–78
~ de la totalité 81–78
~ interstellaire 97–8
~ trouble 83–251
millibar 77–47
milligal 71–66
millimètre de colonne de mercure 77–48
~ de mercure 77–48
millionième du disque solaire 89–69
minéral 74–12
minimum de l'activité solaire 89–4
~ primaire 95–26
~ principal 95–26
~ secondaire 95–27
~ solaire 89–4
minuit magnétique 79–131
mirage 79–64
miroir oscillant 86–130a
~ tournant 86–130
mission spatiale vers un astéroïde 84–99
~ vers un astéroïde 84–99
~ vers un astéroïde pour ramener des échantillons 84–98
Mizar 92–120
modèle à deux composantes 88–196
~ à deux températures 88–197
~ à l'effet de serre 83–262
~ à source ponctuelle 93–138
~ à sources en couches 93–143
~ anisotrope 99–65
~ asymptotique 99–59
~ asymptotique de première espèce 99–60
~ avec l'effet de serre 93–18
~ clos 99–73
~ composite 93–140
~ cosmologique 99–46
~ cosmologique chaud 99–166
~ d'atmosphère 93–17
~ d'eau d'une comète 85–113
~ d'Eddington 93–137
~ d'Einstein 99–68
~ d'Einstein-De Sitter 99–70
~ d'étoile 93–123
~ d'étoile au zéro absolu 93–186
~ d'hydrogène 83–197
~ d'hydrogène d'une planète 83–197
~ d'une planète 83–196
~ d'univers 99–46
~ d'univers chaud 99–166
~ de Charlier 99–67
~ de De Sitter 99–69
~ de fontaine 85–85
~ de Galaxie 98–163
~ de la Terre 79–133
~ de tache 89–102
~ en contraction 99–56
~ en expansion 99–55
~ en forme de jet de fontaine 85–85
~ éolien 83–260
~ éolosphérique 83–260
~ fermé 99–73
~ hiérarchique 99–48
~ homogène 99–64
~ ionosphérique 83–261
~ isotrope 99–63
~ mathématique 98–324
~ mixte 93–140
~ monotone 99–61
~ monotone de première espèce 99–62
~ ne tenant pas compte de l'effet de serre 93–19
~ non-homogène 99–66
~ non-stationnaire 99–53
~ non-statique 99–51
~ oscillant 99–57
~ oscillant de première espèce 99–58
~ ouvert 99–72
~ planétaire 83–196
~ polytropique 93–129
~ relativiste d'étoile 93–198
~ sans l'effet de serre 93–19
~ standard 93–137
~ stationnaire 99–52
~ statique 99–50
~ stellaire 93–123
~ stellaire non-homogène 93–139
~ tenant compte de l'effet de serre 93–18
~ vide 99–49
modèles de Friedmann 99–71
modification du trait de côte 72–78
modulation électrooptique 88–86
module apparent 96–117
~ d'élasticité 73–59
~ de commande 82–6
~ de distance 91–16
~ de service 82–7
~ de Young 73–59
~ lunaire 82–5
~ vrai 96–118
mois 81–42
~ anomalistique 81–43
~ draconitique 81–47
~ sidéral 81–45
~ synodique 81–46
~ tropique 81–44
moldavite 87–99
molécule d'air 86–172
~ -fille 85–135a
~ interstellaire 97–10
~ -mère 85–135
~ parente 85–135
~ polyatomique 97–21
moment angulaire 83a–4
~ cinétique 83a–4
~ de la quantité de mouvement 83a–4
~ de rotation 83a–5
~ de rotation axiale 83a–5
~ du couple 83a–5
~ orbital 83a–6
monde 99–3
Monoceros 92–68
monopôle magnétique 99–170o
monosulfure de carbone 97–34
monoxyde de silicium 97–37
Monstre aquatique 92–55
~ Marin 92–34
mont sous-marin 72–43
~ tabulaire sous-marin 72–44
montagne lunaire 81–193
~ plissée 73–86
~ sous-marine 72–43
montée de la courbe de lumière 95–34
~ vers le maximum 95–28
moraine 72–128
~ de fond 72–129
~ frontale 72–130
~ latérale 72–131
morphologie lunaire 81–147
morsure d'une surface polie 87–102
mort thermique de l'univers 99–14
Mouche 92–69

F

mousson 77–99
moustaches 89–200
mouvement à grande échelle 88–189
~ à petite échelle 88–188
~ aléatoire 98–94
~ apparent 83–41
~ au hasard 89–157
~ de la croûte terrestre 73–72
~ de la ligne des apsides 96–71
~ de planètes 83–28
~ des protubérances vers les pôles 89–231
~ direct 83–43
~ diurne 85–29
~ diurne du radiant 86–69
~ épirogénique 73–73
~ horaire 81–72
~ horizontal 88–184
~ non-gravitationnel 85–56
~ orbital 96–62
~ ordonné 98–95
~ parallactique 91–33
~ particulier 91–31
~ perpendiculaire au plan galactique 98–188
~ planétaire 83–28
~ propre 91–19
~ propre annuel 91–20
~ propre des taches 89–117
~ propre en ascension droite 91–21
~ propre en déclinaison 91–22
~ relatif 99–99
~ relativiste du périhélie 99–160
~ rétrograde 83–47
~ solaire 98–62
~ solaire fondamental 98–66
~ solaire standard 98–63
~ suivant la composante z 98–188
~ vertical 88–185
mouvements de masse 82–195
~ internes 97–149
multipôle 75–66
Mur Droit 81–185
mur du son 76–44
Musca 92–69
muscovite 74–24

N

naine 91–135
~ blanche 91–137
~ brune 91–138a
~ noire 91–139
~ rouge 91–136
naissance d'une tache 89–121
nappe de lave 82–175
nature de l'objet 85–24
nébuleuse à émission 97–123
~ à filaments 97–129
~ à réflexion 97–128
~ America 97–140
~ Amérique du Nord 97–140
~ annulaire 97–147
~ brillante 97–121
~ cométaire 97–142
~ cométoïde 97–142
~ d'Andromède 98–215
~ d'Orion 97–136
~ de gaz et poussière 97–119
~ de la Dentelle 97–137
~ de poussière 97–115
~ diffuse 97–122
~ du Crabe 97–134
~ du Fer à Cheval 97–143
~ du Hibou 97–139
~ en fuseau 98–239
~ extragalactique 98–202
~ filamenteuse 97–129
~ galactique 97–110
~ gazeuse 97–112
~ génératrice 97–63
Grande ~ d'Orion 97–136
~ irrégulière 98–237
~ lumineuse 97–121
~ mère 97–63
~ obscure 97–132
~ Oméga 97–143
~ planétaire 97–144
~ présolaire 83a–21
~ primitive 83a–21
~ protosolaire 83a–21
~ réfléchissante 97–128
~ régulière 98–236
~ solaire 83a–23
~ spirale 98–249
~ stellaire primitive 97–63
~ Trifide 97–141
nébulosité 77–147
neck volcanique 73–125
neige 77–175
~ à gros grains 72–118
~ carbonique 83–144
~ éternelle 72–119
Néozoïque 74–73
Neptune 83–236
neutralisation mutuelle 79–29
neutron d'albedo 75–115
neutrosphère 76–64
névé 72–118
nid de nébuleuses 98–329
nifé 72–18
nimbo-stratus 77–145
niveau critique 76–26
~ de condensation 77–156
~ de la mer 72–87
~ du son 76–38
~ hypsométrique 81–142
nodosité dans une protubérance 89–244
noeud d'un courant 86–217
~ d'une protubérance 89–244
nombre de contrôle 85–30
~ de rencontres 98–122
~ de Wolf 89–65
~ des étoiles 98–80
~ des particules 76–121
~ horaire 86–135
~ horaire des météores 86–135
~ relatif 89–65
~ relatif des taches solaires 89–65
nomenclature des configurations martiennes 83–108
non-euclidien 99–102
non-lumineux 96–44
non-membre du système 98–134
non-radiative, énergie 88–218
non-relativiste 99–96
Norma 92–70
nova 95–110
~ à répétition 95–116
~ extragalactique 95–114
~ galactique 95–113
~ lente 95–120
~ naine 95–112
~ rapide 95–119
~ récurrente 95–116
~ typique 95–111
novoïde 95–118
noyau central galactique 98–18
~ cométaire 85–36
~ convectif 93–165
~ d'hélium 93–167
~ d'un amas 96–130
~ d'un météore 86–102
~ de condensation 76–103
~ de glace 85–40
~ de l'association 96–91
~ de la Galaxie 98–18
~ de la nébuleuse planétaire 97–148
~ de nifé 72–18
~ dégénéré 93–170
~ externe 72–19
~ galactique 98–18
~ galactique actif 98–272
~ interne 72–20
~ isotherme 93–168
~ non-dégénéré 93–171
~ non-homogène 93–169
~ particulier 98–20
~ quasi stellaire 98–19
~ radiatif 93–166
~ solide 85–39
~ stellaire 85–38, 93–164
~ superconducteur 95–150
~ supraconducteur 95–150
~ terrestre 72–17
noyaux de congélation 76–109
~ de cristallisation 76–108
~ de glace 76–104
~ de sublimation 76–110
~ du pic du fer 94–105
nuage 77–131
~ à émission radio 98–286
~ à maser 97–12
~ à radio 98–286
~ absorbant 97–104

~ absorbant observé lors d'une éruption 89–189
~ circumstellaire 94–48
~ coronal 89–261
~ cosmique 94–31
~ d'électrons 79–115
~ d'étoiles 98–6
~ d'hydrogène neutre 97–103
~ de comètes 85–100
~ de comètes d'Oort 85–97
~ de gaz intergalactique 98–354
~ de particules 84–31
~ de plasma 84–30
~ de poussière 83–256, 97–114
~ de poussière circumstellaire 97–117
~ d'étoiles 98–6
~ galactique primitif 94–17
~ galactique primordial 94–17
~ gazeux 97–113
Grand ~ de Magellan 98–213
~ interstellaire 97–101
~ iridescent 79–90
~ moléculaire 97–11
~ moléculaire géante 97–13a
~ nacré 79–91
~ nébulaire 94–31
~ noctilucent 79–89
~ nocturne lumineux 86–26
~ obscur 97–107
~ orographique 77–134
Petit ~ de Magellan 98–214
~ protoplanétaire 83a–24
~ radio 98–286
~ suspendu 89–275
~ zodiacal 84–54
nuages 83–254
couvert de ~ 83–253
~ de Magellan 98–212
sans ~ 77–149
voile de ~ 77–151
nuageux 77–148
nucléogenèse = nucléosynthèse 94–52
~ cosmique 99–175
~ cosmologique 99–175
~ explosive 94–57
~ non-explosive 94–58
nucléosynthèse 94–52
~ cosmique 99–175
~ cosmologique 99–175
~ explosive 94–57
~ non-explosive 94–58
~ primordiale 99–175
nuit brillante 79–59
~ étoilée 79–3d
~ lunaire 81–33
~ photométrique 79–3f
~ sans lune 79–3e
~ sans nuages 79–3g
~ seraine 79–3g
numéro de Messier 97–111

O

objet compact 98–21
~ cosmologique 98–369
~ d'apparence stellaire 98–359
~ diffus 85–25
~ effondré 98–23
~ extragalactique 98–209
~ local 98–368
~ non-catalogué 84–96
~ non-numéroté 84–96
~ non-radioémetteur 98–362
~ quasi stellaire 98–359
~ radio-calme 98–362
~ stellaire 98–359
~ supermassif 98–22
obscurcissement 95–84
~ interstellaire 97–86
observateur de météores 86–7
observation automatique des éruptions 89–221
~ cinématographique des éruptions 89–220
~ crépusculaire 79–42
~ d'une éclipse 88–226
~ de jour 83–121
~ de la surface 82–32
~ de météores 86–6
~ de satelittes 83–26
~ diurne 83–121
~ en dehors d'une éclipse 88–227
~ par échos de radar 86–265
~ par radar 86–265
~ parallèle 86–125
~ pendant une éclipse 88–226
~ photographique 86–140
~ régulière des éruptions 89–219
~ régulière du Soleil 89–16
~ quasi simultanée 71–36
~ simultanée 86–125
~ systématique des éruptions 89–219
~ systématique du Soleil 89–16
observatoire auroral 79–101
~ d'altitude 88–56
~ de haute montagne 88–56
~ ionosphérique 78–21
~ lunaire 82–20
~ sismologique 73–5
~ solaire 88–7
~ sur la Lune 82–20
obturateur à deux lames 86–149
~ à lames mobiles 86–148
occlusion 77–72
occultation 95–80
~ d'une étoile 81–94
~ lunaire 81–94
~ par la couche d'ozone 79–28
~ rasante 81–100
occulteur externe 88–60
océan 72–22
~ des Tempêtes 81–277
océanographie 72–24
~ par satellite 72–70
~ spatiale 72–69
octaédrite 87–86
Octans 92–71
Octant 92–71
Oiseau de Paradis 92–16
olivine 74–22
ombre 89–95
~ d'une tache 89–95
~ de l'ozone 76–127
~ de la Terre 79–44
ombres volantes 88–31
OMM 77–17
onde continue 86–277
~ d'espace 78–13
~ de choc sans collisions 84–49
~ de cisaillement 73–43
~ de compression 73–46
~ de densité 98–175
~ de densité galactique 98–175
~ de dilatation 73–46
~ de distension 73–46
~ de distorsion 73–43
~ de gravitation 99–164
~ de gravité 88–201
~ de longue période 73–48
~ de Love 73–48
~ de marée 82–291
~ de raréfaction 97–58
~ de rotation 73–43
~ de sol 78–14
~ de surface 73–47
~ de tempête 72–76
~ de volume 73–42
~ directe 78–12
~ diurne 76–30
~ double 98–117
~ élastique 73–44
~ extraordinaire 78–30
~ gravitationnelle 99–164
~ infrasonique 75–88
~ infrasonore 72–93
~ longitudinale 73–45
~ ordinaire 78–31
~ P 73–46
~ pénombrale 89–100
~ planétaire 72–94
~ porteuse 86–276
~ primaire 73–46
~ Q 73–48
~ S 73–43
~ secondaire 73–43
~ séismique 73–9
~ semi-diurne 76–29
~ sismique 73–9
~ sismique marine 72–95
~ sonore 76–33
~ transversale 73–41
ondulation du géoïde 71–74
opacité neutrinique 94–95
Ophiuchus 92–72
opposition 83–59

~ en ascension droite 81–68
optique atmosphérique 79–1
orage 77–159
~ à début brusque 75–73
~ à début progressif 75–78
~ avec tonnerre 77–159
~ géomagnétique 75–71
~ ionosphérique 78–49
~ magnétique 75–71
~ magnétique à début brusque 75–73
~ magnétique typique 75–72
orbite à trois dimensions 98–185
~ absolue 96–55
~ apparente 96–48
~ astrométrique 96–57
au delà de l' ~ terrestre 84–6
au voisinage de l' ~ terrestre 84–7
~ d'une planète 83–29
~ de la Lune 81–36
~ de la Terre 84–5
~ galactique 98–184
~ galactique d'une étoile 98–184
~ géocentrique 86–41
~ héliocentrique 86–42
~ lunaire 81–36
~ photocentrique 96–47
~ planétaire 83–29
~ primitive 98–200
~ réelle 96–53
~ relative 96–54
~ spectrographique 96–56
~ spectroscopique 96–56
~ terrestre 84–5
Ordovicien 74–79
Organisation Météorologique Mondiale 77–17
orientation de l'axe 83–34
origine 94–9
~ commune 94–25
~ cosmique 87–5
~ d'impact 82–229
~ de l'univers 99–11
~ des comètes 85–96
~ extraterrestre 87–73
~ tectonique 82–200
~ terrestre 87–74
Orion 92–73
Orionides 86–253
orogenèse 73–74
orthoclase 74–16
orthose 74–16
oscillalteur hydromagnétique 95–61
~ magnétohydrodynamique 95–61
oscillation à période de 5 minutes 88–186
~ en *z* 98–187
~ forcée de marée 82–306
~ perpendiculaire au plan galactique 98–187
oscillations de marée 76–28
~ libres de la Terre 73–49
~ solaires 88–223a
oscillographe 86–292
oscilloscope 86–292
ouragan 77–78
outil de prélèvement d'échantillons 82–80
Outil du Sculpteur 92–23
ovale auroral 79–124
~ blanc 83–195
oxyde 74–26
~ de carbone 97–28
~ de silicium 74–13
oxygène atomique 79–24
~ moléculaire 79–23
ozone atmosphérique 76–124
~ quantité d' ~ atmosphérique 76–125
ozonosphère 76–123

P

paléoclimat 77–188
paléomagnétisme 75–25, 83a–79
paléotempérature 72–126
Paléozoïque 74–71
ère ~ 74–71
Pallas 84–86
pallasite 87–81
Paon 92–74
paraclase 73–91
paradoxe cosmologique 99–38
~ d'Olbers 99–40
~ de Cheseaux et d'Olbers 99–40
~ de gravitation 99–39
~ de Seeliger 99–39
~ photométrique 99–40
parallaxe absolue 91–11
~ annuelle 91–6
~ de courant stellaire 91–7
~ de groupe 91–7
~ de Strömgren 91–13
~ des céphéides 91–8a
~ des sphères de Strömgren 91–13
~ dynamique 91–12
~ hypothétique 91–12
~ par les céphéides 91–8a
~ photométrique 91–14
~ relative 91–10
~ séculaire 91–9
~ spectroscopique 91–15
~ statistique 91–8
~ stellaire 91–4
~ trigonométrique 91–5
~ trigonométrique relative 91–10
paramètre de cratère 81–227
~ de décélération 99–83
~ de structure 96–120
~ photométrique 85–106
paranthélie 79–84
parantisélène 79–86
parasélène 79–73
parasites atmosphériques 78–50
parhélie 79–71
~ inférieur 79–72
parsec 91–17
~ cube 98–84
particularité 98–257
~ spectrale 91–93
particule 88–154
~ accélérée 91–182
~ aurorale 79–107
~ cœur-manteau 97–70
~ colloïdale 76–98
~ d'aérosol 76–91
~ d'air 86–172
~ géante 76–118
~ liquide 76–100
~ lunaire 82–123
~ météorique 86–160a
~ solide 76–99
~ superlourde 99–170n
particules à grain fin 82–109
~ capturées 75–111
~ capturées par le champ magnétique 75–111
~ d'Aitken 76–116
~ fines 82–109
~ interstellaires 97–68
~ lunaires fines 82–112
~ piégées 75–111
~ précipitables 76–117
~ sphériques 87–29
~ zodiacales 84–66
partie antérieure 87–114
~ arrière 87–115
~ avant 87–114
~ postérieure 87–115
passage 95–81
~ à distance minimale 84–69
~ au bord 89–31
~ au méridien central 89–45
~ au plus près 84–69
~ d'un bolide 86–117
~ d'un météore 86–5
~ d'une discontinuité interplanétaire 84–58
~ d'une planète devant le disque solaire 83–51
~ de deux galaxies 98–320
~ de l'ombre 83–53
~ du satellite 83–54
~ proche de deux galaxies 98–320
passer au méridien central 89–44
patrouille du Soleil 89–16
-pause 76–58
Pavo 92–74
paysage lunaire 81–55
Pégase 92–75
Pegasus 92–75
Pendule Astronomique 92–54
~ horizontal 71–73
pénétromètre 82–83
pénombre 89–97
~ d'une tache 89–97
~ de la Terre 81–64

pente 81–230
~ abrupte 81–232
~ continentale 72–35
~ détritique 81–233
~ modérée 81–231
pergélisol 83–152
périapside galactique 98–189
périastre 96–58
~ de Jupiter 83–172
~ galactique 98–189
~ jovien 83–172
~ martien 83–142
~ saturnien 83–210
péricentre galactique 98–189
périgée 81–38
périhélie 83–29a
périjove 83–172
périjovium 83–172
périlune 81–40
périmars 83–142
périsaturnium 83–210
périsélène 81–40
période 95–39
~ de battement 95–42
~ de onze ans 89–70
~ de pulsation 95–100
~ de recouvrement de deux cycles 89–86
~ de relaxation 98–128
~ de rotation 83–31
~ de rotation solaire 88–129
~ diurne 83–31
~ géologique 74–74
~ glaciaire 72–105, 83a–78
~ interglaciaire 72–107
~ magnétiquement agitée 75–91
~ matérielle 99–178
~ principale 95–40
~ secondaire 95–41
~ undécennale 89–70
Perle 92–117
Permien 74–83
Persée 92–76
Perséides 86–251
Perseus 92–76
persistence d'une traînée 86–111
perte d'énergie 88–254
~ de masse 86–168, 94–72
~ de matière 81–236
perturbation associée à une éruption 84–46
~ de marées 98–314
~ due à une éruption 84–46
~ géomagnétique 75–70
~ interplanétaire 84–45
~ ionosphérique 78–39
~ ionosphérique à début brusque 78–41
~ par effet de marée 98–314
perturbations magnétiques 75–70
~ mutuelles 83–25
~ mutuelles entre les satellites 83–25
~ radiales 93–215
~ solaires 83–24
~ solaires sur les satellites 83–24
pesanteur 71–56
accélération de la ~ 71–64
carte de la ~ 71–79
intensité de la ~ 71–62
measure de la ~ 71–67
measure absolue de la ~ 71–68
potentiel de ~ 71–58
Petit Chariot 92–97
~ Cheval 92–48
~ Chien 92–28
~ cratère 81–250
~ creux 81–243
~ halo 79–69
~ Lion 92–60
~ Nuage Magellan 98–214
~ Renard 92–101
petite dépression 81–243
~ Louche 92–97
~ Ourse 92–97
~ planète 84–68
petites planètes troyennes 84–91
pétrification 74–49
~ par impact 82–238
peuplement d'une galaxie 98–230
phase ascendante du cycle solaire 89–83
~ de contraction 94–20
~ de déclin du cycle solaire 89–84
~ de disque 94–22
~ de halo 94–19
~ de Hayashi 94–44
~ de la Lune 81–19
~ de retour au calme 75–77
~ d'évolution après la séquence principale 94–68
~ d'évolution après la série principale 94–68
~ disque 94–22
~ du cycle 89–79
~ évolutive 94–28
~ explosive 89–210
~ Hayashi 94–44
~ hydrodynamique 94–43
~ initiale 75–75
~ initiale d'une orage magnétique 75–75
~ intermédiaire 94–21
~ nébulaire 95–125
~ nulle 83–73
~ partielle 81–78a
~ partielle de l'éclipse 81–78a
~ principale 73–12
phases de l'éclipse 81–73
Phénix 92–77
phénomène actif 89–14
~ de courte durée 89–15
~ lunaire bref 82–153
~ lunaire transitoire 82–153
~ optique 79–33
~ solaire 89–13
phénomènes crépusculaires 79–32
~ héliogéophysiques 84–55
Phobos 83–168
Phoenix 92–77
photo-désintégration 94–101
photographie d'un météore 86–151a
~ des météores 86–140
~ en lumière blanche 88–37
~ lunaire 82–26
~ parallactique 79–102
~ séquentielle 79–103
~ verticale 82–28
photométrie à plusieurs couleurs 96–143
~ détaillée 83–104
~ en six couleurs 96–148
~ en trois couleurs 96–141
~ globale 83–103
photoneutrino 94–93
photosphère 88–167
~ calme 88–169
~ non-perturbée 88–169
photosphérique 88–170
physique de l'atmosphère 76–4
~ de la Lune 82–151
~ des comètes 85–104
~ des relations Soleil-Terre 84–56
~ planétaire 84–4
~ solaire 88–6
pic central 81–258
~ du fer 94–105
noyaux du ~ du fer 94–105
Pictor 92–78
pied d'atterrissage 82–22
piège à lumière 88–64
piégeage 75–109
zone de ~ 75–101
pierre 82–130
anguleuse 82–131
pierres roulantes 81–237
piézoglyptes 87–122
pigeonite 82–90
Pisces 92–79
Piscides 86–250
Piscis Austrinus 92–80
piton central 81–258
plage faculaire 89–164
~ faculaire obscure 89–224
~ faculaire sombre 89–224
plagioclase 74–19
plaine abyssale 72–42
~ basse 81–272
~ de lave 82–185
plan de l'anneau 83–216
~ de l'équateur 83–231
~ de Laplace 83–232
~ du ciel 96–50
~ équatorial 83–231
~ galactique 98–53
~ invariable 83–232
~ principal de Galaxie 98–53
~ tangent à la sphère céleste 96–50
planétaire 83–6

F

planète 83–1
~ d'hydrogène 83–197
~ du Soleil 83–11
~ du système solaire 83–11
~ extérieure 83–15
~ géante 83–174
~ hypothétique 83–122
~ hypothétique d'Olbers 83a–63
~ inférieure 83–14
~ intérieure 83–14
~ intramercurielle 83–123
~ jovienne 83–174
petite ~ 84–68
petite ~ de l'anneau 84–71
~ solaire 83–11
~ supérieure 83–15
~ transneptunienne 83–237
~ transplutonienne 83–124
planètes, grosses 83–13
~ majeures 83–13
~ telluriques 83–119
~ terrestres 83–119
~ troyennes 84–91
planétesimales 83a–35
planétocentrique 83–9
planétographique 83–7
planétoïde = astéroïde = petite planète 84–68
~ binaire 84–79
~ catalogué 84–95
~ de l'anneau 84–71
~ de la famille de Mars 84–74
~ numéroté 84–95
~ perdu 84–97
~ traversant l'orbite de Mars 84–75
planétologie 83–5
planétophysique 83–4
plaque continentale 73–103
~ gluante 87–14
~ mobile 88–27
plasma, couche de 75–11
~ enrichi d'hélium 84–60
~ interplanétaire 84–39
~ primordial 99–170
~ sans collision 84–50
~ solaire 84–29
~ solaire magnétisé 84–32
plasmasphère = plasmosphère 75–12, 76–76
plateau 81–192
~ continental 72–34
plate-forme continentale 72–34
Pléiades 96–110
pleine lune 21–27
plessite 87–94
pli 73–87
~ anticlinal 73–90
~ synclinal 73–89
plis 79–139
plissement 73–88
plomb commun 74–99
~ primitif 74–98
~ radiogénique 74–97
pluie 77–169
~ coronale 89–271
~ d'étoiles filantes 86–215
~ de météores nocturnes 86–218
~ météorique 86–215
~ orographique 77–166
plumes polaires 89–294
Pluton 83–239
pluviomètre 77–170
poids moléculaire moyen 93–100
point brillant 89–201
~ brillant de l'ombre 89–96
~ brillant en rayons X 89–183a
~ chaud 95–112a
~ conique 96–31
~ d'apparition 86–36
~ d'arrêt 86–123
~ d'extinction 86–38
~ d'explosion 86–122
~ d'univers 99–125
~ de chute 86–124
~ de convergence 96–98
~ de Curie 75–41
~ de disparition 86–38
~ de fusion 86–204
~ de l'espace-temps 99–125
~ de la meilleure vision 86–127
~ de rebroussement 98–178
~ de réflexion 75–124
~ de rosée 77–174
~ de rupture 86–121
~ de vision fovéale 86–127
dernier ~ d'impact 87–39
~ initial 86–36
~ interne de Lagrange 96–31
~ miroir 75–124
~ neutre 89–152
~ ombral 89–96
~ premier ~ d'impact 87–40
premier ~ de Lagrange 96–31
~ singulier 94–116
~ stationnaire 83–45
~ sublunaire 81–129
~ subsatellite 71–30
~ subsolaire 81–127
~ substellaire 81–130
~ subterrestre 81–128
points conjugués magnétiques 75–23
Poisson Austral 92–80
~ Doré 92–46
~ Volant 92–100
Poissons 92–79
Polaire 92–121
polarimètre coronal 88–66
~ photoélectrique 83–83
~ photographique 83–84
~ visuel 83–82
Polaris 92–121
polarisation interstellaire 97–105
polarité magnétique 89–148
~ N 89–149
~ négative 89–150
~ nord 89–149
~ opposée 89–155
~ positive 89–149
~ S 89–150
~ sud 89–150
pôle d'axe 83–36
~ d'inclinaison 75–45
~ de l'aimantation 75–48
~ de l'axe géomagnétique 75–48
~ de rotation 83–36
~ géomagnétique 75–48
~ magnétique terrestre 75–45
polluant 72–74, 76–81
~ atmosphérique 76–82
pollution atmosphérique 76–80
~ chimique 72–98
~ de l'air 76–80
~ de l'eau 72–73
Pollux 92–122
polyèdre 71–12
polynômes de Legendre 93–55
polytrope 93–126
pompage chimique 97–19
~ infrarouge 97–18
~ optique 97–17
~ par choc 97–20
~ par collision 97–20
~ par rayonnement 97–17
~ par rayonnement infrarouge 97–18
~ par rayonnemet ultraviolet 97–18a
~ photonique 97–17
~ ultraviolet 97–18a
pont lumineux 89–104, 98–307
~ lumineux intergalactique 98–348
population d'une galaxie 98–230
~ de l'amas 98–341
~ de type II 98–68
~ de type I 98–71
~ des bras spiraux 98–73
~ II 98–68
~ disque 98–72
~ du disque 98–72
~ du halo 98–69
~ du type halo 98–69
~ galactique 98–341
~ intermédiaire 98–70
~ stellaire 98–67
~ I 98–71
~ I extrême 98–73
pore 89–122
porosité 83–100
portée 86–288
~ du radar 86–288
~ minimum 78–38
Porteur d'eau 92–17
porteuse 86–276
position approchée 85–26
~ calculée 85–27
~ du radiant 86–60
~ héliographique 89–38
~ précise 85–28

F

~ prédite 85–27
~ sélénocentrique 81–132
postcrépusculaire, émission 79–52
postglaciaire 72–110
époque ~ 72–109
postluminescence 79–22
postmaritime, cratère 81–208
postmaximum 95–33
postnova 95–123
postulat de cohérence 99–34
~ de Weyl 99–34
~ mathématique de la relativité de l'inertie 99–36
potentiel à symétrie axiale 98–153
~ à symétrie cylindrique 98–154
~ d'absorption 85–122
~ de gravitation 96–160a
~ de gravité 71–58
~ de marée 82–293
~ de pesanteur 71–58
~ gravitationnel 96–160a
~ stationnaire 98–147
~ terrestre 71–60
Poulain 92–48
Poupe 92–81
poussée 73–65
~ de l'eau 72–88
poussière 82–109
~ circulant autour de la Terre 84–53
~ circumstellaire 97–55
~ circumterrestre 84–53
~ cosmique 97–52
~ d'impact 82–125
~ de silicate 85–53
~ galactique 97–53
~ interplanétaire 84–51
~ interstellaire 97–54
~ météorique 87–11
~ météoritique 87–10
~ volcanique 73–143, 82–126
Poussinière 96–110
pouvoir adsorbant 85–120
Praesepe 96–111
Précambrien 74–70
~ inférieur 74–75
préchauffage 86–203
précipitation aurorale d'électrons 79–112
~ d'électrons 78–48
~ orographique 77–167
précipitations 77–154
précrépusculaire, émission 79–51
précurseur 95–143
~ de tremblement de terre 73–11
prédiction de l'activité solaire 89–64
~ des ocultations 81–102
prélèvement d'échantillons du sol 82–78
~ océanographique 87–16
prémaritime, cratère 81–207
prémaximum 95–30
premier contact 81–84
~ point de Lagrange 96–31
~ quartier 81–25
prénova 95–122
pression atmosphérique 77–43
~ au niveau de la mer 77–44
~ de dégénéréscence 93–187a
~ de particules 93–90
~ de photons 93–91
~ de radiation 93–91
~ gazeuse 93–90
~ hydrostatique 93–77
~ latérale 73–65
~ non-thermique 94–42
~ normale 77–46
~ ordinaire 77–46
sous ~ normale 77–46
~ thermique 94–41
~ totale 93–93
préstellaire 94–13
corps ~ 94–15
présupernova 95–129
prévision à courte échéance 77–20
~ à longue échéance 77–19
~ de l'activité solaire 89–64
~ du temps 77–13
~ ionosphérique 78–32
~ météorologique 77–13
Primaire 74–71
principe archimédien 73–101
~ cosmologique 99–30
~ cosmologique parfait 99–32
~ cosmologique restreint 99–31
~ d'Archimède 73–101
~ d'équivalence 99–150
~ d'équivalence d'Einstein 99–150
~ d'équivalence de masse et énergie 99–145
~ d'invariance 93–61
~ de Mach 99–37
prise d'échantillons 76–85
~ d'échantillons du sol 82–78
~ de vue oblique 82–29
~ du vue panoramique 82–27
prisme d'Anderson 88–68
~ diviseur de faisceau 88–88
~ rotatif 88–68
problème cosmogonique 94–5
processus d'éjection 98–179
~ d'équilibre 94–99
~ de capture neutronique lente 94–88
~ de capture neutronique rapide 94–89
~ de rayonnement 86–211
~ de sédimentation 74–46
~ *e* 94–99
~ interne 82–258
~ lent 94–88
~ lunaire 82–150
~ métamorphique 74–34
~ *p* 94–91
~ *r* 94–89
~ radiatif 86–211
~ rapide 94–89
~ *s* 94–88
~ tectonique 82–201
~ Urca 94–51
Procyon 92–123
production d'énergie 88–216
~ d'ondes sonores 88–215
profil de température 83–259
~ vertical 81–228
~ vertical de densité 76–53
profondeur de compensation isostatique 73–102
~ de l'éclipse 95–75
~ du foyer 73–20
~ du minimum de l'éclipse 95–75
~ effective de formation des raies 93–64
~ océanique 72–41
projection azimutale 71–44
~ cartographique 71–40
~ centrale 71–48
~ conforme 71–41
~ conique 71–42
~ cylindrique 71–43
~ équiangle 71–41
~ gnomonique 71–46
~ stéréographique 41–47
~ zénithale 71–45
promontoire 81–293
propagation anormale du son 76–37
~ de l'énergie 88–217
~ de la lumière 99–129
~ des ondes radioélectriques 78–34
~ du son 76–32
propriété cinématique 98–90
~ en ondes radioélectriques 98–235
~ globale 96–134
~ intégrée 96–134
~ intrinsèque 91–100
~ optique 98–234
~ structurale 98–224
protection magnétique 89–225
Protérozoïque 74–77
proto-atmosphère 83a–44
proto-étoile 94–45
protogalaxie 94–18
protonosphère 76–75
protoplanète 83a–40
Protosoleil 83a–39
protostellaire 94–46
protubérance 89–223
~ à apparence d'éruption 89–222
~ à éventail 89–258
~ à raies métalliques 89–265
~ active 89–238
~ active commune 89–246
~ active coronale 89–247
~ active de tache 89–254
~ ascendante 89–278
~ coronale 89–264
~ d'hydrogène 89–266
~ de basse latitude 89–227
~ de haute latitude 89–229

~ de tache 89–252
~ en arche 89–270
~ en boucle 89–270
~ en coiffe 89–253
~ en entonnoir 89–269
~ en forme de bouffée 89–277
~ en forme de troncs d'arbres 89–272
~ équatoriale 89–228
~ éruptive 89–248
~ éruptive commune 89–250
~ polaire 89–230
~ quasi-éruptive 89–249
~ quiescente 89–263
~ secondaire 89–260
~ tourbillonnaire 89–262
unité de ~ 89–235
protubérances, aires journalières des 89–236
~ interactives 89–243
région à ~ actives 89–240
spectroscope pour les ~ 88–58
pseudovariable 95–51
psychromètre 77–121
puissance de crête 86–271
~ de l'émetteur 86–270
~ moyenne 86–272
pulsar 95–137
~ à rayons gamma 95–139
~ à rayons X 95–138
~ binaire 95–140a
~ de la Nébuleuse du Crabe 95–142
~ du Crabe 95–142
~ optique 95–140
pulsation isolée 95–146
~ principale 95–144
pulsations 79–158
~ adiabatiques 95–98
~ géomagnétiques 75–81
~ magnétiques 75–81
~ non-adiabatiques 95–99
~ non-radiales 95–97
~ radiales 95–96
pulvérisation 82–239, 87–28
~ d'un corps météorique 86–189
~ par le vent solaire 82–245
Puppis 92–81
pyranomètre 88–43
pyrhéliomètre 88–39
~ à circulation d'eau 88–41
~ à courant d'eau 88–40
~ à disque d'argent 88–42
pyroxène 74–21
Pyxis 92–82

Q

QSO calme 98–362
~ non-radioémetteur 98–362
~ radio-calme 98–362
Quadrantides 86–247
quadrature 83–64
~ occidentale 83–66
~ orientale 83–65
quadridimensionnel 99–121
vecteur ~ 99–138
quadrivecteur 99–138
quantité de mouvement orbital 83a–6
~ de vapeur d'eau 77–124
~ des précipitations 77–158
quartier 81–26
quasag 98–361
quasar 98–360
~ radio-calme 98–362
~ radioélectrique 98–360
~ sans rayonnement hertzien 98–362
quasi stellaire, source radio 98–360
quasi-étoile 98–360
Quaternaire 74–88
4-espace de Minkowski 99–122
queue anomale 85–75
~ cométaire 85–70
~ d'ions 85–82
~ d'un météore 86–110
~ d'une comète 85–70
~ de la magnétosphère 75–10
~ de la Terre 84–65
~ de poussière 85–79
~ dirigée vers le Soleil 85–78
~ du Serpent 92–89
~ galactique 98–280
~ incurvée 85–73
~ magnétosphérique 75–10
~ multiple 85–76
~ principale 85–77
~ rectiligne 85–75
Quintette de Stephan 98–309

R

raccourcissement perspectif 89–92
radiant 86–53
~ apparent 86–56
~ corrigé 86–58
~ d'un courant météorique 86–237
~ d'un essaim météorique 86–237
~ de groupe 86–54
~ géocentrique 86–58
~ individuel 86–55
~ météorique 86–53
~ secondaire 86–239
~ stationnaire 86–241
~ vrai 86–59
radiatif, noyau 93–166
radiation artificielle 75–113
~ artificielle capturée 75–113
~ capturée 75–112
~ capturée par le champ magnétique terrestre 75–112
~ de la Terre 77–56
~ du ciel 77–54
~ propre d'une planète 83–198
~ provoquée 75–113
~ solaire directe 77–50
radical 85–136
~ libre 85–137
radioactivité = radio-activité 74–61
radio-calme, objet 98–362
quasar ~ 98–362
radiocarbone 74–95
radioélément primitif 82–256
radiohéliogramme 88–96
radiohéliographe 88–95
radioisotope cosmogénique 82–255
radiométéore 86–267
radioscintillation 84–36
radiosource décamétrique 83–186
~ quasi stellaire 98–360
radiosources, scintillation des 84–36
radiospectrographe 88–124
rafale 77–112
raie atmosphérique 88–140a
~ circumstellaire 97–97
~ coronale 89–290
~ D 88–140
~ d'hyperfréquence interstellaire 97–98
~ d'origine circumstellaire 97–97
~ de Fraunhofer 88–139
~ de rougissement 96–149
~ déplacée vers le rouge 98–366
~ interstellaire 97–96
~ jaune 89–292
~ micro-ondes interstellaire 97–98
~ optique interstellaire 97–99
~ radio(électrique) interstellaire 97–98
~ rouge 79–15, 89–293
~ tellurique 88–140a
~ verte 79–16, 89–291
raies 79–155
~ devenant floues 88–243
~ lunaires 81–168
~ tangentielles 81–169
rainure 81–165
~ linéaire 81–166
~ sinueuse 81–167
ralentissement apparent 86–100
~ de l'expansion 99–82
~ de la rotation de la Terre 82–313
~ réel 96–101
ramassage des météorites 87–45
rapport d'abondance 91–175
~ d'équilibre 94–104
~ d'intensité 91–60
~ de Poisson 73–60
~ de masses 96–67
~ du gaz à la poussière 97–120
~ hydrogène-métaux 91–176
~ isotopique 74–94
~ masse-luminosité 98–233
~ météorologique 77–10
~ vitesse-profondeur 73–52
Ras Alhague 92–125
rayon 85–55
~ apparent de la Lune 81–70

~ apparent du Soleil 81–69
~ de l'univers 99–5, 99–81a
~ de Schwarzschild 94–113
~ équatorial 83–115
~ extérieur 83–228
~ gravitationnel 94–113
~ intérieur 83–226
le long du ~ visuel 91–29
~ moyen 83–116a
~ polaire 83–116
~ sismique 73–14
~ sonore 76–34
~ tangentiel 81–169
~ vert 79–58
~ visuel 91–28
rayonnement corpusculaire 88–150
~ d'un météore 86–207
~ de neutrinos 94–94
~ de recombinaison 86–212
~ diffus du ciel 77–59
~ diffusé par le ciel 88–136
~ dilué 93–38
~ global 77–55
~ gravitationnel 99–165
~ incident solaire 77–52
~ intégré 88–147
~ libre-lié 86–212
~ neutrinique 94–94
~ piégé 75–112
~ solaire 88–134
~ solaire direct 88–135
~ terrestre 77–56
~ total 88–146
rayons 79–155
~ lunaires 81–168
raz de marée 72–95
réaction à neutrinos 94–92
~ de capture neutronique 94–87
~ neutrinique 94–92
~ nucléaire en surface 91–181
~ photochimique 76–63
réapparition 81–99
~ de l'anneau 83–219
rebord, cratère sans 81–226
~ élevé 81–222
récepteur 78–17
récession de galaxies 98–298
recherche des comètes 85–15
~ des météorites 87–45
~ des planètes 83–238
~ des satellites 83–27
~ des supernovae 95–136
~ en profondeur 82–82
~ lunaire 82–3
récipient des échantillons 82–81
recombinaison dissociative 86–214
reconnaissance à la surface 82–32
~ orbitale 82–31
recouvrement de deux cycles 89–86
~ des spectres 96–24
recristallisation 74–42
rectification 95–18
~ des courbes de lumière 95–18
récurrence 88–132
redécouverte 85–17
réduction à l'air libre 71–78
~ au niveau de la mer 77–45
~ centre-bord du champ magnétique 89–160
~ de Faye 71–78
~ des occultations 81–108
réferentiel non-inertiel 99–147
réflecteur laser 82–43
réflexion, effet de 95–82
effet de ~ dans les binaires 95–82
reflux 82–282
réfracteur solaire sans coupole 88–53
refroidissement 83a–57
~ neutrinique 94–97
~ par emission de neutrinos 94–97
~ par neutrinos 94–97
~ par rayonnement 94–98
~ radiatif 94–98
région à protubérances actives 89–240
~ active 89–57
~ active de la couronne 89–286
~ active sur Jupiter 83–180
~ antérieure aux mers 81–269
~ aséismique 73–29
~ asismique 73–29
~ aurorale 79–126
~ basse 81–272
~ brillante 83–160
~ centrale de la Galaxie 98–13
~ centrale d'une galaxie 98–293
~ chaude 88–200
~ circumnucléaire 98–292
~ claire 83–160
~ continentale 81–271
~ coronale 89–285
~ coronale émettant la raie rouge 89–285
~ d'ablation 72–123
~ d'alimentation 72–122
~ d'alimentation d'un glacier 72–122
~ d'activité 89–57
~ d'audibilité 87–32
~ d'hydrogène ionisé 97–50
~ d'hydrogène neutre 97–49
~ de capture 75–110
~ de dispersion 87–38
~ de dispersion de la pluie météoritique 87–38
~ de fonte des neiges 72–123
~ de l'anticentre galactique 98–15
~ de l'épicentre 73–16
~ de la modulation des rayons cosmiques 84–26
~ de transition 88–224
~ de transition chromosphère-couronne 88–224
~ de visibilité 86–126
~ des mers 81–268
~ du centre galactique 98–13
~ entre deux bras spiraux 98–40
~ épicentrale 73–16
~ équatoriale 89–59
~ faculaire observée en rayonnement visible 89–165
~ faculaire observée en rayonnement X 89–166
~ froide 88–199
~ H II 97–50
~ H II compacte 97–51
~ H I 97–49
~ interdite 79–135
~ intergranulaire 88–177
~ interspiculaire 88–198
~ ionosphérique 78–4
~ M 89–61
~ magnétique 89–144
~ magnétique bipolaire 89–160a
~ magnétique unipolaire 89–155a
~ maritime 81–268
~ montagneuse 81–196
~ nucléaire 98–17
~ permise 79–134
~ photosphérique calme 89–56
~ photosphérique non-perturbée 89–56
~ polaire 81–17, 83–192
~ postérieure aux mers 81–270
~ postmare 81–270
~ prémare 81–269
~ séismique 73–28
~ sélectionnée 98–78
~ sismique 73–28
~ sombre 83–155
Règle 92–70
regmaglyptes 87–122
régolite 82–104
régression de la calotte polaire 83–146
Régulus 92–126
rejets 82–165
relation décalage vers le rouge – distance 99–76
~ diamètre-profondeur 81–248
~ diamètre-profondeur des cratères 81–248
~ distance-décalage spectral 99–76
~ luminosité-type spectral 91–130
~ magnitude-intensité 91–58
~ masse-luminosité 91–129
~ masse-rayon 93–200
~ période-luminosité 95–103
~ vitesse-distance 98–300
~ vitesse-profondeur 73–52
relations entre les phénomènes solaires et terrestres 84–55
~ Soleil-Terre 84–55
~ structurales 82–266
relativiste 99–95
relativité de l'inertie 99–35
~ générale 99–146
~ généralisée 99–146

F

~ restreinte 99–94
relaxation 98–127
~ par collisions 98–131
~ sans collisions 98–132
relevé photographique 82–30
~ sélénographique 82–50
~ sélénophysique 82–39
remontée des filaments vers les pôles 89–231
rempart à éboulis 81–224
~ concentrique 81–254
~ en terrasse 81–257
~ extérieur du cratère 81–253
~ intérieur du cratère 81–252
~ multiple 81–255
rencontre d'étoiles 98–121
~ de trois corps 96–74
rendement d'ionisation 86–209
~ énergétique 93–109
~ lumineux 86–210
renflement dû aux marées 82–278
~ galactique 98–16
~ lunaire 81–118
renforcement brusque des atmosphériques 78–51
~ d'éclat d'une plage 89–213
~ de brillance d'une plage faculaire 89–213
répartition de la brillance 88–141
~ en taille 81–249
repérage des poissons 72–84
repère 98–136
réplique 73–13
réseau chromosphérique 89–171
~ de pesanteur 71–80
~ géodésique de satellites 71–34
~ géodésique mondial 71–38
~ gravimétrique 71–80
~ gravimétrique fondamental 71–82
~ gravimétrique mondial 71–81
~ mondial 77–18
résistance aérodynamique 86–178
~ de l'air 86–179
~ du sol 82–135
résolution spatiale 88–204
~ temporelle 88–205
restes d'explosion de supernova 95–133
~ d'une supernova 95–133
retard de groupe 78–56
rétention de gaz 87–134
~ gazeuse 87–134
réticule 86–129
Réticule 92–83
~ Astronomique 92–83
Reticulum 92–83
retour d'un groupe 88–133
rétrécissement gravitationnel 83–203
rétrograde 83–46
révolution anomalistique 81–43
~ draconitique 81–47
~ sidérale 81–45
~ synodique 81–46
~ tropique 81–44
Rigel 92–127
rigidité 73–56
~ magnétique 75–117
riomètre 78–45
roc 74–1
~ lunaire 82–65
roche 74–1
~ acide 74–4
~ anguleuse 82–131
~ basique 74–5
~ cristalline 82–66
~ effusive 74–3
~ équigrenue 82–67
~ extrusive 74–3
~ feuilletée 74–54
~ ignée 74–2
~ intrusive 74–7
~ lunaire 82–65
~ magmatique 74–2
~ métamorphique 74–11
~ plutonienne 74–7
~ schisteuse 74–54
~ sédimentaire 74–10
~ stratifiée 74–48
~ ultra-basique 74–6
~ volcanique 74–3
rocher 82–130
rose des vents 77–116
rosée 77–173
rotateur 91–42
~ oblique 91–52, 95–60
~ rapide 91–43
rotation asynchrone 83–38
~ de Kepler 98–112
~ de la ligne des apsides 96–71
~ des étoiles 91–35
~ des planètes 83–30
~ différentielle 88–128
~ différentielle de la Galaxie 98–113
~ directe 83–39
~ du corps rigide 98–111
~ du corps solide 98–111
~ du Soleil 88–127
~ du solide 98–111
~ galactique 98–107
~ galactique différentielle 98–113
~ képlérienne 98–112
~ liée 81–14
~ non-synchrone 83–38
~ non-uniforme 91–36
~ planétaire 83–30
~ rétrograde 83–40
~ rigide 98–111
~ solaire 88–127
~ stellaire 91–35
~ synchrone 83–37
rotor 91–42
~ rapide 91–43
rougissement interstellaire 97–91
rugosité de la surface 82–106
rupture 82–212
~ d'un météore 86–82
~ du noyau galactique 98–318
~ par collision 84–107

S

Sac à Charbon 97–138
Sagitta 92–84
Sagittaire 92–85
Sagittarius 92–85
salinité 72–66
sans nuages 77–149
satellite 83–17
cratère ~ 81–200
~ de glace 83–23
~ Galiléen 83–204
~ irrégulier 83–21
~ le plus proche de sa planète 83–20
~ lithique 83–22
~ naturel 83–18
~ rétrograde 83–21
~ rocheux 83–22
saturation 76–101
Saturne 83–209
saturnocentrique 83–213
saturnographique 83–212
schistosité 74–53
science de l'environnement 76–2
sciences atmosphériques 76–1
~ de la Terre 71–1
~ lunaires 82–2
scintillation interplanétaire 84–37
~ interstellaire 97–109b
Scorpio 92–86
Scorpion 92–86
Scorpius 92–86
Sculpteur 92–87
Sculptor 92–87
Scutum 92–88
secondaire 96–38
Secondaire 74–72
ère ~ 74–72
secousses consécutives 73–13
~ de la croûte 95–152
~ prémonitoires 73–10
~ séismiques 73–2a
~ sismiques 73–2a
~ stellaires 95–153
secteur 85–60
~ magnétique interplanétaire 84–16
~ tournant 86–147
section efficace différentielle 94–54
~ ~ géométrique 84–103
~ ~ gravitationnelle 84–103
~ ~ partielle 94–54
~ frontale 86–176
~ géométrique 84–103
~ gravitationnelle 84–103
~ lunaire 82–5
sédiment 74–47
sédimentation 73–109

~ fine au large 72–79
~ sous-marine 72–47
sédiments biogéniques 72–59
~ de mer profonde 72–50
~ de profondeurs océaniques 72–50
~ pélagiques 72–57
~ terrigènes 72–48
séisme = séïsme = sisme 73–2a
~ d'origine météoritique 87–34
~ de dislocation 73–2d
~ de Lune 82–214
~ intermédiaire 73–24
~ local 73–25
~ lointain 73–27
~ météoritique 87–34
~ normal 73–22
~ par la chute d'une météorite 87–34
~ proche 73–26
~ profond 73–23
~ sous-marin 73–2c
~ superficiel 73–21
~ tectonique 73–2d
~ terrestre 73–2b
~ volcanique 73–2e
séismicité = sismicité 73–3
séismogramme 73–8
~ d'un impact 82–224
séismographe 73–7
séismologie = sismologie 73–1
~ active 82–217
~ lunaire 82–213
~ passive 82–218
séismologue = sismologue 73–2
séismomètre actif 82–222
~ passif 82–221
sélection, effet de 98–81
sélénocentrique 81–131
sélénochimie 82–74
sélénodésie 81–109
sélénodétique 81–110
sélénographie 81–1
sélénographique 81–8
sélénoïde 81–117
sélénologie 82–1
sélénologique 82–48
sélénologue 82–49
sélénophysique 82–151
sens de la rotation 98–108
séparateur de faisceau 88–88
séparation angulaire 96–52
~ des composantes 96–66
~ des éléments chimiques 83a–48
~ du Soleil 83a–9
séquence d'âge zéro 96–152
~ d'évolution 94–30
~ des galaxies de Hubble 98–227
~ des spirales 98–255
~ polaire 91–87
~ polaire internationale 91–87
~ polaire nord 91–87
~ principale 91–131
~ principale initiale 94–60
~ spectrale 91–94
serein 77–149
série principale 91–131
~ principale d'âge zéro 96–152
Serpens 92–89
Serpent 92–89
~ d'Eau 92–55
Serpentaire 92–72
serre, effet de 83–263
service de prévision du temps 77–5
~ du Soleil 89–16
~ météorologique 77–5
seuil du compagnon 96–40
Sextans 92–90
Sextant 92–90
sidérite 87–82
sidérolithe 87–78
sidérostat 88–44
sifflements 78–53
siffleur 78–53
~ court 78–58
~ électronique 78–60
~ en forme de nez 78–59
~ long 78–57
~ protonique 78–61
signal continu 86–277
signaux discontinus 86–278
signes du zodiaque 92–6
silence, zone de 76–42
silicate 74–14
silice 74–13
silicium, oxyde de 74–13
sill 73–135
sillage d'un météore 86–110
sillon 81–165
Silurien 74–80
simulation mathématique 98–323
simultanéité 99–127
singularité 99–42
~ nue 94–117
singulier, point 94–116
Sirius 92–128
sisme = séisme = séïsme 73–2
~ local 73–25
~ normal 73–22
~ profond 73–23
~ superficiel 73–21
sismogramme 73–8
~ d'un impact 82–224
sismographe 73–7
sismologie 73–1
~ lunaire 82–213
~ solaire 88–223b
sismologue 73–2
sismomètre actif 82–222
~ passiv 82–221
sismométrie active 82–217
site d'alunissage 82–14
~ d'alunissage choisi 82–15
~ d'alunissage proposé 82–15
~ d'alunissage sélectionné 82–15
~ d'atterrissage 82–14
~ d'atterrissage lunaire 82–14
~ d'atterrissage sur la Lune 82–14
situation météorologique 77–14
~ météorologique générale 77–15
smog 76–83
socle 72–12
sol cohérent 82–63
~ cohésif 82–63
~ lunaire 82–62
~ perpétuellement gelé 83–152
solaire 88–5
Soleil 88–1
~ actif 89–12
au voisinage du ~ 98–59
~ bleu 79–92
~ calme 89–11
~ en rayons X 88–4
~ non-perturbé 89–11
~ optique 88–2
~ perturbé 89–12
~ primitif 83a–39
~ radio 88–3
~ visible 88–2
~ X 88–4
solide réfractaire 85–49
solidification 83a–59
solution stable 98–170
somme des précipitations 77–168
son anormal 76–36
~ normal 76–35
sondage 76–47
~ acoustique 72–29
~ crépusculaire 79–41
~ en contre-haut 78–20
~ ionosphérique 78–19
~ océanographique 87–16
~ par fusées 76–48
~ par le haut 78–20
~ sismique 82–216
sonde acoustique 76–45
sondeur ionosphérique 78–22
sortie de l'ombre 81–81
~ de la pénombre 81–82
souffler 77–105
soumis à sa propre gravitation 94–33
soupe de quarks 99–169a
source compacte 98–273
~ d'électrons 79–119
~ d'énergie 93–110
~ d'énergie double 93–146
~ de chaleur 74–62
~ de neutrons 94–85
~ décamétrique 83–186
~ neutronique 94–85
~ quasi stellaire 98–358
sous-abondance 91–187
sous-dense 86–284
sous-éruption 89–195
sous-estimer 95–14
sous-géante 91–151
sous-lumineuse, étoile 91–133
sous-naine 91–140

F

F

sous-orage 79–144
~ auroral 79–141
~ géomagnétique 75–95
~ magnétique 75–95
~ magnétique polaire 75–96
sous-pulsation 95–145
sous-refroidissement 76–105
sous-sol 82–105
sous-système 98–74
soustraction photographique 88–94
soutien de la chromosphère 98–223
spallation 94–100
~ par rayons cosmiques 82–254
~ provoquée par rayons cosmiques 82–254
réaction de ~ 94–100
spatio-temporel 99–120
spécialiste de cosmogonie 94–4
~ des météorites 87–7
spécimen 87–42
spectre composite 96–19
~ d'une traînée météorique 86–107
~ de Fraunhofer 88–137
~ dédoublé 96–18
~ dynamique 88–125
~ éclair 88–25
~ global 96–137
~ principal 95–115
~ solaire 88–74a
~ stellaire 91–88
~ stellaire normal 91–89
~ symbiotique 95–117
~ unique 96–16
spectrobologramme 88–76
spectrobolomètre 88–75
spectro-enregistreur des vitesses 88–57
spectrogramme dynamique 88–125
spectrographe à balayage de fréquence 88–124
~ dymanique 88–124
~ horizontal 88–78
~ pour les météores 86–146
~ solaire 88–77
~ vertical 88–79
spectrographie (de) Mössbauer 82–142
spectro-héliocinématographe 88–81
spectrohéliogramme 88–72
spectrohéliographe 88–71
spectrohélioscope 88–67
spectromètre 88–124
~ de vent solaire 82–46
spectroscope pour les protubérances 88–58
~ protubéranciel 88–58
spectroscopie gamma non-destructive 82–77
sphère céleste 92–8
~ de gaz polytropique 93–127
~ gazeuse 93–73
~ terrestre 71–10
spéroïde aplati 98–158
~ terrestre 71–13
sphérules 87–29
~ vitreuses 82–127
spicule 88–195
spirale 98–7
~ archimédienne 84–34
~ barrée 98–253
~ d'Archimède 84–34
galaxie ~ 98–249
nébuleuse ~ 98–249
~ non-barrée 98–250
~ normale 98–250
spiralée, galaxie 98–249
spire 98–7
stabilité 93–204
~ d'une orbite 98–193
~ des étoiles 93–206
~ dynamique 93–207, 98–198
stable 93–208
thermiquement ~ 93–209
stade après la séquence principale 94–68
~ avant la séquence principale 94–49
~ d'évolution 94–28
~ de contraction 94–20
~ de disque 94–22
~ de halo 94–19
~ disque 94–22
~ hydrodynamique 94–43
~ intermédiaire 94–21
~ T-Tauri 94–47
station d'étude des météores 86–158
~ d'observations de satellites 71–33
~ d'une planète 83–44
~ lunaire automatique 82–19
~ météorique 86–158
~ séismique 73–6
~ sismique 73–6
statistique stellaire 98–77
stellaire 92–3
stéréo-comparateur 91–32
stockage de particules 75–123
stratification 82–133, 97–154
~ horizontale 82–134
~ thermique 77–40
strato-cumulus 77–142
stratopause 76–71
stratosphère 76–70
stratus 77–139
striation 89–173
stries 89–173
structure 82–102
~ à grande échelle 88–207
~ à petite échelle 88–208
~ absorbante (en Hα) 89–224
~ allongée 99–171c
~ annulaire 98–41
~ caractéristique 88–248
~ cellulaire 99–171a
~ circulaire 81–155
~ cométaire 85–34
~ d'aluminium en nid d'abeille 82–25
~ d'écoulement 82–191
~ de la Galaxie 98–4
~ de la Voie Lactée 98–4
~ des comètes 85–34
~ des météorites 87–100
~ du courant météorique 86–243
~ en arche 88–249
~ en secteur 84–15
~ en spirale 98–8
~ filamentaire 99–171c
~ filamenteuse 97–130
~ fine de la chromosphère 88–203
~ fluidale 82–191
~ hiérarchique de l'univers 99–47
~ interne 71–83
~ interne des étoiles 93–71
~ laminaire 83–148
~ postmare 81–154
~ prémare 81–153
~ solaire 89–13
~ spirale 98–8
~ spiralée 98–8
~ stellaire 93–71
~ striée 89–174
structures chromosphériques 89–172
~ élémentaires chromosphériques 88–210
~ fines 89–172
sublimation 85–138
sublunaire, point 81–129
subsatellite, point 71–30
subsolaire, point 81–127
substance adsorbée 85–131
~ dificilement fusible 83a–55
~ facilement fusible 83a–56
~ non-volatile 83a–53
~ poreuse 82–111
~ réfractaire 83a–55
~ volatile 82–262
substellaire, point 81–130
subterrestre, point 81–128
sulfate 74–29
sulfite 74–28a
sulfure 74–28
~ d'hydrogène 97–44
~ de carbonyle 97–36
superamas de galaxies 98–346
~ de la Vierge 98–347
~ Local 98–347
superdense 86–285
superfluide 95–149
supergalaxie 98–346
~ Local 98–347
supergéante 91–145
~ faible 91–148
~ la plus brillante 91–147
supergranulation 88–183
supergranule 88–182
supergravitation 99–170r
superlourde, particule 99–170n
supermassif, objet 98–22
supernova 95–126
ancienne ~ 95–134
~ de Tycho 95–135

~ de type I 95–130
~ extragalactique 95–128
~ galactique 95–127
supernovae en chaîne 95–132
superposition des spectres 96–24
superrefroidi 76–106
superrefroidissement 76–105
supersaturation 76–102
superunification 99–170q
support de la chromosphère 88–223
surabondance 91–185
surabondant, élément 91–184
surélévation du terrain 81–156
~ lunaire 81–156
surestimer 95–13
surface antérieure 87–114
~ arrière 87–115
~ avant 87–114
~ cratérée 81–245
~ critique 96–27
~ critique de Roche 96–27
~ d'une météorite 87–106
~ d'une planète 83–97
~ de contact 96–27
~ de distribution 87–38
~ de haut-plateau 82–56
~ de la Lune 82–55
~ de la Terre 71–5
~ de mare 82–57
~ de plaine 82–57
~ de raies 81–171
~ de rayons 81–171
~ du premier type 87–107
~ du second type 87–108
~ équipotentielle 71–61
~ fragmentée 82–104
~ géopotentielle 71–61
~ interraies 81–172
~ limite de salinité 72–67
~ lumineuse diffuse 79–148
~ lunaire 82–55
~ neutre 75–16
~ planétaire 83–97
~ polie 87–101
~ postérieure 87–115
~ primaire 87–107
~ pulsante 79–149
~ recouverte d'une croûte 87–109
~ rugueuse 83–102
~ secondaire 87–108
~ solaire 88–164
~ solide 83–98
~ stratifiée 83–101
~ supérieure de... 76–58
~ terrestre 71–5
surfondu 76–106
surfusion 76–105
surge 89–255
~ brillante 89–257
~ sombre 89–256
surlumineuse, étoile 91–134
surrection 89–255
~ d'une eruption 89–190
sursaut de l'éclat 85–126
~ de l'émission X 89–184
~ de rayonnement X 79–117
~ décamétrique 83–183
~ du vent solaire 84–42
~ lumineux 95–69
~ radio-électrique 83–182
~ radio-électrique de Jupiter 83–182
~ soudain 95–154
sursauts de basse fréquence 83–184
surveillance automatique des éruptions 89–221
~ des astéroïdes 84–94
~ des éruptions 89–219
~ des icebergs 72–85
~ des satellites 83–26
~ du ciel 92–11
~ météorique 86–144
susceptibilité magnétique 75–27
symbiotique, étoile 95–50
symétrie cylindrique 98–155
~ hélicoïdale 98–161
~ sphérique 98–160
synchrone 85–92
~ complète 85–94
~ terminale 85–93
synclinal 73–89
pli ~ 73–89
syndyname 85–91
synthèse d'ouverture 88–118
~ des éléments 94–52
système à bande passante 91–72
~ à bande étroite 91–72
~ à bande large 91–72
~ à collisions 98–166
~ à énergie totale négative 96–156
~ à énergie totale positive 96–158
~ à symétrie axiale 98–157
~ à synthèse d'ouverture 88–119
~ aplati 98–251
~ autogravitant 98–169
~ collisionnel 98–166
~ compact 96–157
~ copernicien 99–24
~ cosmologique 99–20
~ d'anneaux 98–291
~ d'arches 89–299
~ d'étoiles 96–1
~ de bandes 79–18
~ de cassures 73–79
~ de classification spectrale de Harvard 91–91
~ de Copernic 99–24
~ de couleurs 96–140
~ de courants 75–97
~ de fissures 73–79
~ de fractures 73–79
~ de fractures continentales 73–81
~ de fractures médio-océaniques 73–80
~ de franges d'interférence 88–120
~ de latitudes moyennes 83–179
~ de longitudes 83–177
~ de longitudes équatorial 83–178
~ de magnitudes 91–57
~ de Ptolémée 99–22
~ de raies 81–170
~ de rayons 81–170
~ de référence 98–136
~ de référence en rotation 98–138
~ de référence fondamental 97–137
~ de référence général 97–137
~ de référence local 98–139
~ de ruptures 73–79
~ de satellites 83–19
~ de tourbillons 83a–13
~ des amas globulaires 98–75
~ des anneaux de Saturne 83–220
~ des latitudes tempérées 83–179
~ II de longitudes 83–179
~ du monde 99–20
~ en état stationnaire 98–141
~ extragalactique 98–210
~ galactique 98–3
~ géocentrique 99–21
~ géodésique de référence 71–19
~ géodésique mondial 71–39
~ gravitationnellement libre 96–158
~ gravitationnellement lié 96–156
~ Harvard des spectres 91–91
~ héliocentrique 99–23
~ inertiel 99–98
~ instable 98–171
~ lié par gravitation 96–156
~ local 98–61
~ non-inertiel 99–147
~ non-lié 96–158
~ non-stationnaire 98–152
~ planétaire 83–2
~ planétocentrique 83–10
~ planétocentrique de coordonnées 83–10
~ planétocentrique de référence 83–10
~ plat 98–251
~ ptolémaïque 99–22
~ rayonnant 81–170
~ sans collisions 98–167
~ solaire 83–12
~ soumis à la autogravitation 98–169
~ soumis à sa propre gravitation 98–169
~ sphéroïdal 98–159
~ stellaire 96–1
~ stellaire formé par superposition 98–165
~ Terre-Lune 81–12
~ très aplati 98–164
~ triple 96–4
~ UBV 96–144
~ I de longitudes 83–178

systèmes de coordonnées liés à la matière 99–44
syzygies 81–29
hautes eaux de ~ 82–273
marée de ~ 82–273

T

Table 92–66
tableau d'évolution des taches 89–120
Tache Blanche 83–214
~ Claire 83–214
tache arrière 89–114
~ de la surface 83–105
~ de queue 89–114
~ de tête 89–113
~ est 89–114
Grande ~ Rouge 83–193
~ invisible 89–159
~ isolée 89–116
~ leader 89–113
~ ouest 89–113
~ permanente 83–106
~ postérieure 89–114
~ principale 89–115
~ Rouge 83–193
~ solaire 89–88
~ sombre 83–155
~ unique 89–116
taches, activité des 89–62
dessin des ~ 88–34
~ lunaires 81–148
~ stellaires 97–62
tachetures 88–210
~ brillantes 88–211
~ chromosphériques 89–172
~ sombres 88–212
taenite 87–93
taille des grains 82–113
talus 81–234
~ continental 72–36
tas 89–276
en forme de ~ 89–276
Taureau 92–91
Taurides 86–254
Taurus 92–91
taux d'émission de masse 85–146
~ d'extincion stellaire 94–115
~ de débit lumineux 86–210
~ de déclin 95–36
~ de formation des cratères 82–233
~ de formation des étoiles 94–12
~ de montée vers le maximum 95–29
~ de mortalité des étoiles 94–115
~ de production 85–143
~ de production de gaz 85–144
~ de récurrence 86–279
~ de refroidissement 93–203
~ de relaxation 98–129
~ de répétition des impulsions 86–279
~ des réactions thermonucléaires 94–53
~ d'expansion 99–75
~ d'extinction stellaire 94–115
lignes d'égal ~ de comptage 75–102
~ sporadique 86–233
tectite 87–95
~ en forme de goutte 87–96
tectonique 73–95
télégramme astronomique 85–20
Télescope 92–92
~ chromosphérique 88–73
~ solaire 88–49
~ solaire à axe optique horizontal 88–55
~ solaire automatique 88–50
~ solaire embarqué sur ballon 88–52
~ solaire horizontal 88–55
~ solaire vertical 88–54
Telescopium 92–92
téléséisme 73–27
témoin oculaire de la chute 87–48
température 77–29
~ à l'ombre 77–34
~ ambiante 77–32
~ centrale 93–89
~ cinétique 91–115
~ d'excitation 91–118
~ d'ionisation 91–117
~ de brillance monochromatique 91–114
~ de chambre 77–31
~ de condensation 97–66
~ de couleur 91–109
~ de Curie 75–41
~ de dégénérescence 93–188
~ de dissociation 91–120a
~ de fusion 86–204
~ de la pièce 77–31
~ de rotation 91–119
~ de vibration 91–120
~ effective 91–113
~ électronique 91–116
~ et pression normales 77–30
~ limite de dégénérescence 93–188
~ locale 77–31
~ normale 77–31
~ sous abri 77–34
~ superficielle 93–88
tempête de bruit 83–185
~ de poussière 83–255
~ interplanétaire 84–44
temps 77–1
~ absolu 99–117
~ caractéristique 99–79
~ d'arrivée 73–50
~ d'écho 86–291
~ d'expansion 99–78
~ d'expansion de Friedmann 99–80
~ d'expansion de Hubble 99–79
~ de désintégration 96–104
~ de dissipation 96–104
~ de Friedmann 99–80
~ de Hubble 99–79
~ de Newton 99–118
~ de parcours Soleil-Terre 84–47
~ de Planck 99–170i
~ de refroidissement 83–201
~ de relaxation 98–128
~ de trajet 73–51
~ géomagnétique 79–132
~ lunaire 82–300
~ lunaire moyen 82–302
~ lunaire vraie 82–301
~ moyen de relaxation 98–130
~ propre 99–139
teneur en chlore 72–68
~ en nickel 72–55
~ en poussières 85–80
~ en sel 72–66
tenseur d'énergie 99–158
~ d'énergie-impulsion 99–158
~ de courbure 99–159
~ de courbure de Riemann 99–159
~ de matière 99–158
~ de Riemann-Christoffel 99–159
~ des contraintes 73–69
~ des déformations 73–70
~ fondamental 99–157a
~ métrique 99–157a
tension 73–66
~ de la vapeur d'eau 77–125
terme cosmologique 99–156a
~ de mouvement différentiel 98–115
~ de rotation galactique 98–116
~ K 98–119
terminateur 83–69
~ de la Terre 81–91
~ du matin 83–70
~ du soir 81–125
~ lunaire 81–123
~ matinal 81–124
~ vesperal 83–71
terra 81–190
terrain bas 81–272
~ chaotique 83–164
~ désordonné 83–164
terrasse de cratère 81–229
~ intérieure 81–256
Terre 71–2
système ~ -Lune 81–12
terrella 79–133
terrestre 71–3
Tertiaire 74–87
tête d'une comète 85–35
~ de Cheval 97–135
~ de Méduse 92–105
~ du Serpent 92–89
texture 82–102
~ d'écoulement 82–191
~ fluidale 82–191
thallassoide 81–175
théorème d'unicité 93–83
~ de Clairaut 71–77

~ de Vogt-Russell 93–83
~ du viriel 98–181
théorie d'ondes de densité 98–174
~ de jauge 99–170j
~ de l'accrétion 83a–20
~ de l'éjection 85–98
~ de la capture 85–99
~ de la condensation en plusieurs centres 96–76
~ de la contraction 73–77
~ de la fission 96–75
~ de la formation des bras spiraux 98–173
~ de la formation multiple 96–76
~ de la mort thermique de l'univers 99–15
~ de la relativité générale 99–146
~ de la relativité généralisée 99–146
~ de la relativité restreinte 99–94
~ de la structure spirale 98–173
~ de Weizsäcker 83a–11
~ des cycles 97–156
~ des marées 82–305
~ des noyaux séparés 96–76
~ des tourbillons 83a–11
~ dynamo du champ magnétique 75–99
~ mécanique des formes cométaires 85–84
~ physique des météores 86–159
thermoluminescence 82–249
thermosphère 76–59
thio-formaldéhyde 97–43
Thuban 92–131
Titan 83–233
tjäle 83–152
tonnerre 77–165
topographie lunaire 81–146
tore d'hydrogène 83–230
totalisateur 77–171
pluviomètre ~ 77–171
totalité commence 81–77
zone de ~ 88–29
Toucan 92–95
tour solaire 88–54
trace d'éboulement 82–197
~ d'une particule 82–251
~ fossile 87–121
~ météorique 86–153
~ météorique sur photographie 86–29
~ nucléaire 87–120
~ nucléaire fossile 82–252
~ photographique du météore 86–29
~ stellaire 86–152
tracé de température 72–83
traces organiques dans les météorites 87–67
traceur de la structure spirale 98–39
traînée 86–178
~ à limière propre 86–108
~ de courte durée 86–112
~ de fumée 87–21
~ de poussière 86–115
déplacement de la ~ 86–114
durée de la visibilité d'une ~ 86–111
écho de ~ 86–298
équation de ~ 86–179
~ gazeuse 86–116
~ instable 86–109
~ lumineuse 86–108
~ lumineuse par elle-même 86–108
~ météorique 86–103
~ météorique de longue durée 86–113
~ météorique persistante 86–104
~ visible de jour 86–106
~ visible de nuit 86–105
traînées 81–168
~ sombres 98–288
trait volcanique 82–159
trajectoire apparente 86–32
~ apparente sur le ciel 86–32
~ ascendante 82–9
~ atmosphérique 86–162
~ avec distorsion 86–33
~ circumlunaire 82–11
~ d'un météore 86–30
~ d'une particule 75–118
~ dans l'atmosphère 86–162
~ interplanétaire 82–12
~ ondulante 86–33
~ réelle 86–31
trajet aléatoire 89–157
~ de refroidissement 93–202
~ évolutif 94–29
transfert d'énergie 93–111
~ de la chaleur 86–202
~ de masse d'une composante à l'autre 96–33
~ du rayonnement 93–29
transformation de Galilée 99–131
~ de l'énergie 86–208
~ de Laplace 93–54
~ de Lorentz 99–132
~ galiléenne 99–131
~ homologue 93–132
transit 95–81
transition croûte-manteau 72–14
~ de phase 83a–58, 99–170p
zone de ~ 72–13
transitoire coronal 89–185a
translunaire 81–6
transmission de chaleur 86–202
transneptunienne, planète 83–237
transport d'énergie 93–111
~ de chaleur 86–202
~ de champ magnétique 84–48
Trapèze d'Orion 96–92
traversée de l'atmosphère 87–26
tremblement de Lune 82–214
~ de terre 73–2a
~ de terre catastrophique 73–40
~ de terre continental 73–2b
~ de terre désastreux 73–38
~ de terre destructif 73–39
~ de terre faible 73–35
~ de terre intermédiaire 73–24
~ de terre local 73–25
~ de terre moyen 73–36
~ de terre profond 73–23
~ de terre sous-marin 73–2c
~ de terre superficiel 73–21
~ de terre tectonique 73–2d
~ de terre violent 73–37
~ de terre volcanique 73–2e
Triangle 92–93
~ Austral 92–94
triangulation 71–23
~ cosmique 71–28
~ du premier ordre 71–24
~ fondamentale 71–24
~ par satellite 71–27
~ spatiale 71–25
Triangulum 92–93
~ Australe 92–94
trias 74–84, 87–91
trilatération 71–29
triple collision 96–74
troïlite 82–94
Trois Rois 92–129a
troisième intégrale du mouvement 98–186
tropopause 76–77
troposphère 76–78
trou blanc 98–24
~ coronal 88–245a
~ creusé par la météorite 87–58
~ de forage 82–61
~ météoritique 87–58
~ noir 94–114
~ noir galactique 98–23a
~ noir supermassif 98–22a
trouble 83–252
tsunami 72–95
tube cathodique 86–293
~ de force en forme de corde 89–146
~ de force magnétique 89–145
~ de lave 82–186
Tucana 92–95
turbosphère 76–72
trubulence à grande échelle 83a–12
~ à l'échelle cosmique 83a–12
twik 78–63
type d'aurore 79–136
~ d'étoiles 91–101
~ de Hubble 98–229
~ de la queue cométaire 85–71
~ de structure 98–225
~ galactique de Hubble 98–229
~ morphologique 98–228
~ spectral 91–108
~ stellaire 91–101
~ structural 98–225

F

U

unicité de la solution 93–82
unification des interactions 99–170w
grande ~ 99–170u
Union Géodésique et Géophysique Internationale 71–4
unité de protubérances 89–235
unités solaires 91–123
univers 99–1
~ anisotrope 99–65
~ asymptotique 99–59
~ asymptotique de première espèce 99–60
~ chaud 99–166
~ classique 99–93
~ clos 99–73
~ d'Einstein 99–68
~ d'Einstein-De Sitter 99–70
~ d'inflation 99–170b
~ de Charlier 99–67
~ de De Sitter 99–69
~ de Friedmann 99–71
~ de Minkowski 99–122
~ en contraction 99–56
~ en évolution 99–54
~ en expansion 99–55
~ fermé 99–73
~ hiérarchique 99–48
~ hiérarchisé 99–48
~ homogène 99–64
~ inflationnaire 99–170b
~ inflatoire 99–170b
~ initial 99–170a
~ isotrope 99–63
modèle d' ~ 99–46
~ monotone 99–61
~ monotone de première espèce 99–62
~ newtonien 99–93
~ non-homogène 99–66
~ non-stationnaire 99–53
~ non-statique 99–51
~ oscillant 99–57
~ oscillant de première espèce 99–58
~ ouvert 99–72
~ primoridal 99–170a
~ relativiste 99–92
~ stationnaire 99–52
~ statique 99–50
~ vide 99–49
upwelling 72–88
Uranus 83–234
Ursa Major 92–96
~ Minor 92–97
Ursides de décembre 86–258

V

vacuole 82–180
va-et-vient latitudinal 75–119
valeurs „plancher" du champ magnétique 89–141
vallée 81–163
~ Alpine 81–164
~ des Alpes 81–164
vapeur d'eau 77–119
vaporisation, chaleur de 86–199
~ d'un corps météorique 86–198
~ intense 86–185
variabilité du champ 84–14
~ du champ magnétique 84–14
~ optique 95–141
~ stellaire 95–4
variable = étoile variable 95–1
~ à courte période 95–89
~ à éclipses 95–74
~ à éruption 95–68
~ à flare 95–68
~ à longue période 95–91
~ à obscurcissement 95–83
~ à occultation 95–83
~ à période ultracourte 95–90
~ à polarité constante 95–57
~ cataclysmique 95–109
~ cyclique 95–108
~ d'amas 95–93
~ du type ... 95–47
~ du type Algol 95–77
~ du type δ Céphée 95–92
~ du type R Coronae Borealis 95–83
~ du type RR Lyrae 95–93
~ ellipsoïdale 95–79
~ géométrique 95–72
~ extrinsèque 95–86
~ intrinsèque 95–87
~ irrégulière 95–107
~ irrégulière à polarité alternative 95–56
~ irrégulière à polarité réversible 95–56
~ magnétique 95–52
~ magnétique à polarité alternative 95–56
~ magnétique à polarité constante 95–57
~ magnétique irrégulière 95–56
~ magnétique irrégulière à polarité réversible 95–56
~ magnétique régulière 95–55
~ nébulaire 95–85
~ périodique 95–88
~ rotationnelle 95–65
~ semi-régulière 95–105
~ semi-régulière rouge 95–106
~ spectrale 95–48
variables non-dimensionnelles 93–133
~ sans dimensions 93–133
variation annuelle 86–138
~ centre-bord 88–145
~ d'éclat 95–7
~ de la période 95–43
~ de lumière 95–6
~ diurne 86–137
~ du champ géomagnétique 75–69
~ du climat 77–183
~ du spectre 95–49
~ géomagnétique 75–69
~ lumineuse 95–6
~ magnétique 95–53
~ saisonnière 86–139
~ séculaire du champ magnétique terrestre 75–85
~ séculaire géomagnétique 75–85
~ spectrale 95–49
vase 72–58
vecteur à quatre composantes 99–138
~ champ magnétique 89–139
~ de Poynting 93–92
~ de propagation 78–36
Véga 92–132
véhicule d'alunissage 82–21
~ d'atterrissage lunaire 82–21
~ d'atterrissage sur la Lune 82–21
~ de commande 82–6
~ de service 82–7
Veille Météorologique Mondiale 77–16
veines 87–111
Vela 92–98
vénitienne 88–126
vent 77–89
~ à rafales 77–111
~ anabatique 77–92
~ cyclostrophique 77–95
~ dominant 77–117
~ du gradient 77–94
~ galactique 98–277
~ géostrophique 77–96
~ interstellaire 97–151
~ solaire 88–156
~ solaire calme 84–41
~ solaire continu 84–41
~ solaire permanent 84–40
~ stellaire 93–14
vitesse du ~ 77–109
vents alizés 77–100
Vénus 83–125
vénusien 83–126
verre fumé 88–33
~ lunaire 82–96
Verseau 92–17
vertex 98–105
verticale 71–71
déviation de la ~ 71–72
vésiculaire 82–181
vésiculation 82–182
vésicule 82–180
~ de lave 82–178
vésiculeux 82–181
Vesta 84–87
vibrations propres de la Terre 73–49
vide 76–9, 99–171d
~ d'air 76–8
vidéomagnétographe 88–85

Vierge 92–99
violation de CP 99–170t
~ de la parité de charge 99–170t
~ de la symétrie 99–170s
Virgo 92–99
viscosité 73–71
~ magnétique 75–34
~ magnétique de roches 75–34
visibilité 77–24
bonne ~ 77–25
~ directe 91–29
~ d'une occultaion 81–103
~ de l'éclipse 81–88
mauvaise ~ 77–26
vision de coté 86–128
~ extrafovéale 86–128
ligne de ~ 91–28
~ périphérique 86–128
~ scotopique 86–128
visuel, rayon 91–28
vitesse à l'entrée dans l'atmosphère 86–92
~ à mi-trajectoire 86–97
~ angulaire 86–89
~ angulaire instantanée 86–90
~ angulaire moyenne 86–91
~ caractéristique 76–43
~ d'approche 91–27
~ d'éjection 85–90
~ d'entrée 86–92
~ d'expansion 96–94
~ d'un météore 86–86
~ dans l'atmosphère 86–94
~ de déclin 95–36
~ de fuite 92–26
~ de fusion 83–147
~ de libération d'un système 96–155
~ de réaction 94–53
~ de récession 99–75
~ de refroidissement 93–203
~ de relaxation 98–129
~ de rotation 83–32
~ de trace 86–154
~ du vent 77–109
~ du vent solaire 88–158
~ en dehors de l'atmosphère 86–93
~ équatoriale 91–49
~ géocentrique 86–95
~ héliocentrique 86–96
~ instantanée 86–99
~ le long du rayon visuel 91–24
~ moyenne 86–98
~ négative 91–27
~ orbitale 96–63
~ particulière 91–31
~ positive 91–26
~ radiale 91–24
~ relativiste 99–136
~ résiduelle 98–91
~ rotatoire 91–48
~ spatiale 91–30
vive-eau, marée de 82–273
Voie Lactée 98–1
voile 83–250
~ de nuages 77–151
Voile 92–98
Voiles du Navire 92–98
voisinage du Soleil 98–58
Volans 92–100
volatil 82–262
volcan 73–111
~ actif 73–114
~ bouclier 73–119
~ effusif 73–119
~ en activité 73–114
~ en repos 73–116
~ en sommeil 73–116
~ éteint 73–115
~ inactif 73–116
~ mort 73–115
~ sous-marin 73–117
~ subaquatique 73–117
Vulcain 83–123
Vulpecula 92–101

Y

ylem 99–170

Z

zenocentric 83–176
zenographic 83–175
zéro de la relation période-luminosité 95–104
zodiaque 92–5a
zone abyssale 72–41
~ active 89–54
~ antérieure aux mers 81–269
~ aurorale 79–123
~ centrale 81–114
~ continentale 81–271
~ convective 93–149
~ convective d'hydrogène 93–159
~ d'ablation 72–123
~ d'absence 97–88
~ d'apparition des taches 89–54
~ d'audibilité 76–41, 86–120
~ d'occultation 81–107
~ de calmes équatoriaux 77–101
~ de chute 87–38
~ de combustion de l'hélium 93–145
~ de combustion de l'hydrogène 93–144
~ de compression 86–170
~ de fonte des neiges 72–123
~ de formation des taches 89–54
~ de fusion glaciaire 72–124
~ de haute pression 77–84
~ de piégeage 75–101
~ de rayonnement 75–101
~ de rayonnement artificiel 75–108
~ de rayonnement externe 75–105
~ de rayonnement naturel 75–107
~ de silence 76–42
~ de transition 72–13
~ de visibilité de l'éclipse 81–89
~ du maximum de fréquence aurorale 79–125
~ équatoriale 83–188
~ extérieure 75–105
~ interne 75–106
~ marginale 81–113
~ maritime 81–268
~ orogénique 73–75a
~ polaire 77–193
~ postérieure aux mers 81–270
~ postmare 81–270
~ prémare 81–269
~ radiative 93–134
~ royale 89 54
~ séismique 73–31
~ semi-convective 93–160
~ sismique 73–31
~ tachée 89–54
~ tempérée 77–192
~ torride 77–191
zones 83–187
~ de Fresnel 86–294
déplacement des ~ royales 89–74

F

DEUTSCHER INDEX

German Index

Index allemand

Índice alemán

Índice alemão

Немецкий указатель

ALPHABET:

a	h	o	v
b	i	p	w
c	j	q	x
d	k	r	y
e	l	s	z
f	m	t	
g	n	u	

D

A

Abbildung, gnomonische 71–46
kartographische ~ 71–40
Abbremsung der Rotation 91–45
magnetische ~ 91–47
scheinbare ~ 86–100
wirkliche ~ 86–101
Abbrennen des Wasserstoffs 94–69
Abdeckscheibe 88–61
Abelsche Gleichung 88–235
Abenddämmerung 79–36
Abendmeridian 75–19
Abendrot 79–55
Abendröte 79–55
Abendstern 83–128
Abendterminator 83–71
Aberration, tägliche 86–57
Abfall 76–57
~ de Lichtkurve 95–35
Abfallgeschwindigkeit 95–36
abgewandt von der Sonne 85–81
Abgleitung 82–196
Abhang 81–230
steiler ~ 81–232
Abiotikum 74–75
Abkühlung 83a–57
Abkühlungsgeschwindigkeit 93–203
Abkühlungskurve 93–202
Abkühlungsweg 93–202
Abkühlungszeit 93–201
Ablagerung 73–109, 74–45
biogene ~ 72–59
pelagische ~ 72–57
terrigene ~ 72–48
vom Festland stammende ~ 72–48
Ablagerungsgestein 74–10
Ablagerungsmagnetisierung 75–33
Ablation 86–182
Ablationsgebiet 72–123
Ablationsprozess 86–183
Ablationswärme 86–184
Ablenkung eines Sternes 98–123
Ablenkungskraft 77–80
Abnahme 76–57
Abnehmen der Polkappe 83–146
Abplattung 83–113
~ der Korona 88–247
~ eines Kugelsternhaufens 96–123
polare ~ 83–114
scheinbare ~ 98–295
wahre ~ 98–296
Absatz 74–47
Absatzgestein 74–10
Abschiebung 82–196
Abschirmung 86–171
~ durch die Ozonschicht 79–28
magnetische ~ 89–225
Abschirmungshöhe 79–48
Abschmelzgebiet 72–123
Abschmelzgeschwindigkeit 83–147
Abschmelzung 86–182
Abschmelzungswärme 86–184
absorbierender Stoff 85–130
Absorption, allgemeine interstellare 97–87
~ durch Wasserdampf 77–127
frei-freie ~ 93–106
gebunden-freie ~ 93–107
gebunden-gebundene ~ 93–108
interstellare ~ 97–84
~ mit plötzlichem Anfang 78–47
plötzliche ~ im kosmischen Rauschen 78–42
selektive interstellare ~ 97–90
Absorptionsanstieg, plötzlicher 78–47
Absorptionsschicht, galaktische 97–89
Absorptionswolke 97–104
Abstand 91–3
~ der Komponenten 96–66
maximaler ~ 86–289
~ vom Sternmittelpunkt 93–85
~ vom Zentralmeridian 89–46
~ von der Sonnenmitte 89–36
Abstandsgesetz der Planeten 83–3
absteigend 77–93
absteigende Phase des Sonnenzyklus 89–84
Abtragung 73–108
~ durch Meteoriten 82–242
Abtrennung 83a–9
~ von der Sonne 83a–9
Abweichungen vom lokalen thermodynamischen Gleichgewicht 93–40
~ vom LTE 93–40
~ vom TE 93–36
~ vom thermodynamischen Gleichgewicht 93–36
Abyssal 72–41
Abyssalablagerungen 72–50
Abyssalsedimente 72–50
Abyssalzone 72–41
Abzweiung 98–5
Achernar 92–102
Achondrit 87–77
Achse, erdmagnetische 75–47
geomagnetische ~ 75–47
magnetische ~ 75–46
Neigung der ~ 83–35
Achsenneigung 83–35
Achsenpol 83–36
geomagnetischer ~ 75–48
Achsenrichtung 83–34
Achterschiff 92–81
Adern 87–111
adhäsiv 82–137
Adler 92–18
Adsorbat 85–131
Adsorbens 85–130
adsorbieren 85–129
adsorbierter Stoff 85–131
Adsorption 85–128
Adsorptionsgleichgewicht 85–121
Adsorptionskraft 85–120
Adsorptionspotential 85–122
Adsorptionsvermögen 85–120
Advektion 77–63
Adventivkrater 73–128
Aerolith 87–60
Aerologie 76–122
Aeronomie 76–122a
Aerosol 76–89
Aerosolfilter 76–94
Aerosolteilchen 76–91
Aerotriangulation, räumliche 71–26
Agglomeration 83a–32
Aitkenkerne 76–116
Aitkensche Teilchen 76–116
Akkretionsscheibe 96–21
Akkretionstheorie 83a–20
Akkumulation 83a–34
Akkumulationsgebiet 72–122
Aktivierung einer Protuberanz 89–139
Aktivität der Feldmeteore 86–136
~ des galaktischen Kernes 98–271
~ des Magnetfeldes 89–5
erdmagnetische ~ 75–86
galaktische ~ 98–261
geomagnetische ~ 75–86
~ im optischen Spektralbereich 89–3
~ im Radiogebiet 89–6
~ im Radiowellengebiet 89–6
korpuskulare ~ 89–7
magnetische ~ 89–5
meteorische ~ 86–134
optische ~ 89–3
rekurrierende ~ 98–267
seismische ~ 73–4
wiederholte ~ 98–267
Aktivitätsgebiet 89–57
Aktivitätszentrum 89–60
Aktivitätszone 89–54
Akustik, atmosphärische 76–31
Alanin 87–72
Albedo, Bondsche 83–91
~ der kosmischen Strahlung 75–116
geometrische ~ 83–92
Lambertsche ~ 83–89
mittlere ~ 83–90a
monochromatische ~ 83–93
sphärische ~ 83–91
Albedo-Neutron 75–115
Albireo 92–103
Albit 74–17
Algol 92–105
Algolstern 95–77
Algolveränderlicher 95–77
Algonkium 74–77
Alkor 92–104
All = Weltall = Universum 99–1
All-Sky-Kamera 79–104
Alpenquertal 81–164
Alpental 81–164
α-Aminopropionsäure 87–72
Alpha-Streuinstrument 82–76

alpiner Gletscher 72–114
Altair 92–106
Altar 92–19
Alter, absolutes 82–269
~ der Meteoriten 87–127
~ der Sterne 94–61
~ des Mondes 81–31
~ des Weltalls 99–12
~ eines Meteorschauers 86–246
~ eines Stromes 86–245
geologisches ~ 74–90
relatives ~ 82–268
stratigraphisches ~ 82–267
Alter-Null-Hauptreihe 96–152
Altersbestimmung 74–91
absolute ~ 83a–73
~ mit Hilfe von Isotopen 87–130
~ mittels Isotopen 87–130
radioaktive ~ 83a–73, 87–131
Altokumulus 77–144
Altostratus 77–140
Aluminium-Wabenblech 82–25
Ameisensäure 97–33
Ameisensäurealdehyd 97–27
Ameisensäureamid 97–35
Aminogruppe 87–71
Aminosäure 87–70
Ammoniak 97–25
Amor-ähnlicher Asteroid 84–80a
Amphibol 74–20
Amplitude 95–38
Analyse, chemische 82–75
mineralogische ~ 82–53
petrographische ~ 82–52
Anderson-Prisma 88–68
Änderung, abrupte der Pulsperiode 95–154
spektrale ~ 95–49
stündliche ~ 81–72
Änderungen, jahreszeitliche ~ 83–166
tägliche ~ 83–167
Andromeda 92–14
Andromedanebel 98–215
Andromediden 86–256
Anemograph 77–114
Anemometer 77–113
Anfang bei Sonnenuntergang 88–22
~ der Finsternis 81–74
~ der Totalität 81–77
Anfangsgeschwindigkeit 86–92
Anfangshöhe 86–39
Anfangsmasse 86–164
Anfangsphase 75–75
~ eines magnetischen Sturmes 75–75
Anfangspunkt 86–36
Anhäufung 83a–34
~ von Domen 81–178
~ von Kuppeln 81–178
Ankunft auf der Erde 89–193
Ankunftzeit 73–50
Anlauf 82–280
Annäherung, grösste 84–69
Annäherungsgeschwindigkeit 91–27
Annihilationsära 99–173a
Anomalie, gravimetrische 75–67
magnetische ~ 75–67
Südatlantische ~ 75–68
anomalistischer Monat 81–43
Anorthit 74–18
Anorthosit 82–87
Anpassungsbedingung 93–142
anregender Stern 97–125
Anregungsmechanismus 79–27
Anregungstemperatur 91–118
Ansatzzeit 73–50
Anstieg 76–54
~ der Lichtkurve 95–34
~ zum Maximum 95–28
Anstiegsdauer, relative 95–29
Antapex 86–65
~ der Sonne 98–65
Antares 92–107
Antennenabstand 88–99
Antiklinale 73–90
Antiklinalfalte 73–90
Antikline 73–90
Antipassate 77–103
antisolare Richtung 75–22
Antivertex 98–106
Antizentrum, galaktisches 98–14
antizyklonal 77–85
Antizyklone 77–84
Antlia 92–15
Anzahl der Sternbegegnungen 98–122
stündliche ~ von Meteoren 86–135
Anziehungszentrum 89–241
Äolosphärenmodell 83–260
Apareion 83–141
Apareon 83–141
Apastron 96–61
Apertursynthese 88–118
Apex 86–64
~ der Sonne 98–64
~ der Sonnenbewegung 91–23
Aphel 83–28a
Apoapside, galaktische 98–190
Apoapsis, galaktische 98–190
Apoastron 96–61
Apofokus, galaktischer 98–190
Apogalaktikum 98–190
Apogäum 81–39
Apogeum 81–39
Apojovium 83–173
Apollo 84–81
Apollo-Asteroid 84–80
Aposaturnium 83–211
Aposelen 81–41
Apsidenbewegung 96–71
Apsidenlinie, Drehung der 96–71
Apus 92–16
Aquariden 86–249
Aquarius 92–17
Äquator, magnetischer 75–50
Äquatorband 83–191
Äquatorebene 83–231
Äquatorhalbmesser 83–115
äquatoriales Gebiet 89–59
Äquatorradius 83–115
Äquatorregion 89–59
Äquatorstreifen 83–191
Äquatorwulst 71–9
Äquatorzone 83–188
Aquila 92–18
Äquipotentialfläche des Schwerefeldes 71–61
kritische ~ 96–27
äquivalente Höhe 78–28
Äquivalenthöhe 88–236
Äquivalenzprinzip 99–150
Einsteinsches ~ 99–150
~ von Masse und Energie 99–145
Ara 92–19
Ära 74–69
inflationäre ~ 99–170d
kosmologische ~ 99–166a
Plancksche ~ 99–170e
Archaikum 74–76
Archäozoikum 74–76
Archimedisches Prinzip 73–101
areiographisch 83–137
areiozentrisch 83–138
Areo- 83–135
areographisch 83–137
Areophysik 83–134
areozentrisch 83–138
Argelandermethode 95–10
Ariel 83–235
Aries 92–20
Arm, expandierender 98–38
Drei-Kiloparsec- ~ 98–37
galaktischer ~ 98–26a
laufender ~ 98–29
nachgeschleppter ~ 98–29
nachgezogener ~ 98–29
nachlaufender ~ 98–29
Orion- ~ 98–38
sich öffnender ~ 98–31
vorlaufender ~ 98–31
Art des Objektes 85–24
~ des Polarlichtes 79–136
Asche, vulkanische 73–142
Aschenfliessen 82–174
Aschenfluss 82–174
Aschenkegel 73–120
Aschenstrom 82–174
aschgraues Licht 81–35
aseismisches Gebiet 73–29
Assoziation, O- 96–89
T- ~ 96–90
Ast, absteigender 95–35
ansteigender ~ 95–34
Asterismus 92–3a
Asteroid 84–68
Amor-ähnlicher ~ 84–80a
~ der Marsfamilie 84–74
Erdbahn kreuzender ~ 84–76

katalogisierter ~ 84–95
Marsbahn kreuzender ~ 84–75
numerierter ~ 84–95
verlorener ~ 84–97
zigarrenförmiger ~ 84–77
Asteroidenfamilie 84–73
Asteroidengruppe 84–90
Trojaner-~ 84–91
Asteroidengürtel 84–70
Asteroidenmission 84–99
Asteroidenring 84–70
Asteroidenstrom 84–100
Astronomie, extragalaktische 98–208
~ von der Mondoberfläche 82–4
astronomische Dämmerung 79–37
Asymmetrie, Nord-Süd 89–50
Ost-West-~ 89–49
asymptotisch zu Kreisbahnen 98–180
Atair 92–106
Ataxit 87–87
nickelarmer ~ 87–88
nickelreicher ~ 87–89
Atlas, photometrischer 88–138
atmophile Elemente 83a–29
Atmosphäre 83–241
adiabatisch geschichtete ~ 76–12
adiabatische ~ 76–12
ausgedehnte ~ 93–9
äussere ~ 76–68
~ der Erde 76–5
~ der Sonne 88–163
endliche ~ 93–23
freie ~ 76–13
graue ~ 93–26
halbunendliche ~ 93–22
homogene ~ 76–10
~ im Anfangszustand 83a–45
isotherme ~ 77–39
kugelsymmetrische ~ 93–24
mittlere ~ 76–51
nicht-graue ~ 93–27
obere ~ 76–52
oxydierende ~ 83a–47
Physik der ~ 76–4
planparallele ~ 93–21
reduzierende ~ 83a–46
rein streuende ~ 93–16
streuende ~ 93–16
untere ~ 76–50
ursprüngliche ~ 83–244
Atmosphärendurchgang 87–26
Atmosphärengezeiten 82–294
Atmosphärengrenze 76–25
obere ~ 76–25
Atmosphärenmodell 93–17
Atmosphärenobergrenze 76–25
atmosphärisch 83–242
Atom mit vielen Energieniveaus 93–46
~ mit zwei Energieniveaus 93–45
Atomenergie 93–121
Ätzen einer polierten Fläche 87–102
Aufbau der Kometen 85–34
~ des Milchstrassensystems 98–4
hierarchischer ~ 99–47
hierarchischer ~ der Welt 99–47
innerer ~ 71–83
innerer ~ der Sterne 93–71
Aufbrechen eines Meteorkörpers 86–188
~ eines Meteors 86–83
Aufenthalt 82–34
Aufenthaltsdauer 82–34
Aufenthaltszeit 82–34
Auffüllung einer Zyklone 77–83
Aufheizung, atmosphärische 75–87
~ der Chromosphäre 88–213
~ der Korona 88–213a
Aufheizungsmechanismus 88–214
Aufhellung eines Fackelgebietes 89–213
Aufleuchten einer Nova 95–121
~ einer Plage 89–213
~ einer Supernova 95–131
~ eines Meteors 86–35
kurzzeitiges ~ 95–69
vorübergehendes ~ 82–247
Aufleuchthöhe 86–39
Aufleuchtpunkt 86–36
Auflösung einer Fleckengruppe 89–126
~ einer Protuberanz 89–280
~ eines Kometen 85–115
~ eines Kometenkerns 86–261
~ eines Meteorkörpers 86–188
~ eines offenen Sternhaufens 96–103
plötzliche ~ 89–279
räumliche ~ 88–204
~ von Haufen 98–195
zeitliche ~ 88–205
Auflösungszeit 96–104
Aufnahme des Geländes 82–30
magnetische ~ 75–64
photographische ~ im weissen Licht 88–37
selenologische ~ 82–50
Aufprall eines Meteoriten 82–236
Aufrechterhaltung der Chromosphäre 88–223
Aufsammeln meteorischen Staubes 87–12
Aufsammlungsscheibe 96–21
Aufschlag eines Meteoriten 87–33
künstlicher ~ 82–226
aufschlagender Körper 82–225
Aufschlagspunkt 86–124
Aufschlagspur 82–228
Aufschlagsseismogramm 82–224
Aufschlagsstaub 82–125
Aufschmelzung 86–181
Aufspaltung der Linien durch Expansion 97–157
aufsteigende Phase des Sonnenzyklus 89–83
Aufstieg 82–8
~ zum Maximum 95–28
Aufstiegsbahn 82–9
Aufsturz, künstlicher 82–226
aufstürzender Körper 82–225
Aufsturzhypothese 81–261
Aufsturzseismogramm 82–224
Aufsuchen von Satelliten 83–27
Aufsuchung von Kometen 85–15
Aufteilung, chemische 83a–50
Auftreffen auf die Erde 89–193
Auftrieb, magnetischer 89–147
Aufwärtsströmung 77–91
Aufwind 77–92
Aufwölbung, hakenartige 75–79
Augenruhepunkt 86–127
Augenzeuge des Falls 87–48
Aureole 79–60
Auriga 92–21
Aurora 79–93
nördliche ~ 79–105
Radar- ~ 79–95
Radio- ~ 79–94
südliche ~ 79–106
Ausbreitung der Radiowellen 78–34
~ des Lichtes 99–129
Ausbreitungsrichtung 78–35
Ausbreitungsvektor 78–36
Ausbrennen des Wasserstoffs 94–69
Ausbruch des solaren Plasmas 84–38
~ des Sonnenwindes 84–42
~ eines Kometen 85–125
wiederholter ~ 98–282
Ausbrüche bei niedrigerer Frequenz 83–184
Ausbruchsveränderlicher 95–109
Ausbuchtung, hakenartige 75–79
Ausdehnung, beschleunigte 99–81
Ausdehnungsgeschwindigkeit 99–75
Ausdehnungsrate 99–75
Ausdehnungsspalte 82–207
Ausdehnungsverzögerung 99–82
Ausdehnungszeit 99–78
Ausdrücken der Lava 82–171
Auseinanderbrechen infolge der Rotation 85–119
ausfällende Teilchen 76–117
Ausfegen von Gasen 83a–41
Ausfluss, kontinuierlicher 98–276
ununterbrochener ~ 98–276
~ von Materie aus dem galaktischen Kern 98–275
Ausgangsdaten, geodätische 71–20
Ausgasen eines Kometenkernes 84–109
ausgelöste vulkanische Tätigkeit 82–155
Ausgleich, isostatischer 73–100
Aushöhlung 87–117
innere ~ 87–119
Auslösen 89–198
Auslösung 89–199
Ausreisser 98–135
Ausrichtung von Teilchen 97–78
Ausschleuderung von Partikeln 85–89
Ausschnittsphotometrie 83–104
Aussenhalbmesser 83–228
Aussenkrater 81–218
Aussenrand 83–227

Aussenring 83–225
Aussenschichten eines Sternes 93–3
nichtentartete ~ 93–199
ausseratmosphärische Geschwindigkeit 86–93
aussergalaktisch 98–207
ausserirdischer Ursprung 87–73
ausserordentliche Welle 78–30
Ausstrahlung der Erde 77–56
Ausströmung 85–55
geradlinige ~ 85–57
gerichtete ~ 85–57
~ von Lava 73–130
Australit 87–98
Austritt des Mondes aus dem Halbschatten 81–82
~ des Mondes aus dem Kernschatten 81–81
~ des Sternes 81–99
Austrocknen 74–51
Austrocknung 74–51
Auswahleffekt 98–81
Auswaschung 73–107
Auswurf 89–259
~ aus einem Kometen 86–262
~ von Masse 95–124
~ von Materie 98–274
~ von Partikeln 85–89
~ von Radio-Wolken 98–285
Auswurfgeschwindigkeit 85–90
Auswürflinge 82–165
vulkanische ~ 82–166
Auswurfstheorie 85–98
axialsymmetrisches Potential 98–153
~ System 98–157
Azetaldehyd 97–42
Azetonitril 97–38
Azimutalprojektion 71–44
Azoikum 74–75

B

Bahn, absolute 96–55
astrometrische ~ 96–57
atmosphärische ~ 86–162
dreidimensionale ~ 98–185
~ eines Planeten 83–29
galaktische ~ eines Sternes 98–184
geozentrische ~ 86–41
heliozentrische ~ 86–42
interplanetare ~ 82–12
photozentrische ~ 96–47
räumliche ~ 86–31
relative ~ 96–54
scheinbare ~ 96–48
scheinbare ~ auf der Himmelssphäre 86–32
spektrographische ~ 96–56
spektroskopische ~ 96–56
ursprüngliche ~ 98–200
wahre ~ 96–53
Bahnbewegung 96–62
Bahndrehimpuls 83a–6
Bahngeschwindigkeit 96–63
Bahnkorrektur 82–10
Bahnkrümmung 86–51
Bahnlänge des Meteors 86–47
Bai, erdmagnetische 75–80
Bai-Störung 75–80
Balken 87–104
galaktischer ~ 98–254
balkenlose Spirale 98–250
Balkenspirale 98–253
Ballen, kleiner 82–129
Ballon-Sonnenteleskop 88–52
Band 83–190
Bande, diffuse interstellare 97–100
homogene ~ 79–146
~ mit Strahlenstruktur 79–150
Banden, unidentifizierte 85–134
Bandensystem 79–18
Bänderung 74–53
Bär, Grosser 92–96
Kleiner ~ 92–97
Bärenhüter 92–22
Baryonengas 93–194
Barysphäre 72–17
Basalt 74–9
lunarer ~ 82–84
Basislänge, feste 88–102
veränderliche ~ 88–103
Basit 74–5
Batholith 73–139
bathymetrische Karte 71–54
Bäume 89–273
Baumstämme 89–272
~ Protuberanz 89–272
Bausch 89–277
Beanspruchung 73–62
Beaufortskala der Windstärke 77–108
Beben, flaches 73–21
kontinentales ~ 73–2b
~ meteoritischen Urspungs 87–34
meteoritisches ~ 87–34
Bebenereignis 82–215
Bebenherd 73–17
Becher 92–42
Becken 81–175
bedeckt 77–152
mit Wolken ~ 83–253
Bedeckung 95–80
Reduktion der ~ 81–108
Sichtbarkeit der ~ 81–103
streifende ~ 81–100
Bedeckungsgrad 77–147
Bedeckungsveränderlicher 95–74
~ vom Algol-Typ 95–77
Bedeckungsvorausberechnungen 81–102
Bedienungsmodul 82–7
Bedingung, Fit- 93–142
Bedingungen der Finsternis 81–73
Befehlseinheit 82–6
Begegnungshypothese 83a–14
Beginn der Finsternis 81–74
~ der totalen Finsternis 81–77
~ der totalen Verfinsterung 81–77
~ der Totalität 81–77
~ des Zyklus 89–82
Begleiter 96–38
dunkler ~ 96–42
~ einer Galaxie 98–216
elliptischer ~ 98–217
elliptischer ~ einer Galaxie 98–217
optischer ~ 98–218
physischer ~ 98–219
planetarischer ~ 96–43
unsichtbarer ~ 96–41
Beimengung 76–81
Bekanntes Meer 81–289
beleuchtete Hemisphäre 81–126
Beleuchtungsäquator 81–122
Beleuchtungsstärke, Konturen gleicher 97–124
bemannte Mondbasis 82–18
benachbartes Ereignis 99–126
Benennung von Marsformationen 83–108
quasisimultane ~ 71–36
Beobachtung ausserhalb von Finsternissen 88–227
Beobachtungen, fast gleichzeitige 71–36
gleichzeitige ~ 86–125
korrespondierende ~ 86–125
photographische ~ von Meteoren 86–140
Beobachtungskosmologie 99–17
Beobachtungsnetz 82–40
selenophysikalisches ~ 82–40
weltweites ~ 77–18
Beobachtungsrichtung 91–29
in ~ 91–29
Berechner 85–22
Berg, submariner 72–43
Bergkette 81–194
Bergrücken 81–186
~ im Meer 81–188
Bergung von Proben 82–71
Berichter 85–23
Berichtigung 95–18
~ der Lichtkurven 95–18
Berührung 81–101
erste ~ 81–84
letzte ~ 81–85
Berührungsbogen 79–75
oberer ~ 79–77
seitlicher ~ 79–87
unterer ~ 79–76
Beschleunigung, äquatoriale 88–131
effektive ~ 85–88
Beschuss mit Partikeln 82–240
Besetzungsinversion 97–15
Besonderheit 98–257
spektrale ~ 91–93
Bestandteil, atmosphärischer 83–265
~ der Atmosphäre 76–87
stellarer ~ einer Galaxie 98–231

Bestrahlung durch kosmische Strahlung 82–250
Bestrahlungsalter 87–139
Betatron-Effekt 91–180
Betelgeuze 92–108
Beule 81–177
Bewegung, diastrophische 73–72
direkte ~ 83–43
epirogene ~ 73–73
epirogenetische ~ 73–73
geordnete ~ 98–95
innere ~ 97–149
nicht durch Gravitation verursachte ~ 85–56
parallaktische ~ 91–33
rechtläufige ~ 83–43
rückläufige ~ 83–47
scheinbare ~ 83–41
~ senkrecht zur galaktischen Ebene 98–188
stochastische ~ 98–94
stündliche ~ 81–72
tägliche ~ 85–29
z-~ 98–188
Bewegungsfeld, differentielles 98–114
Bewegungshaufen 96–97
Bewegungsintegral, drittes 98–186
Bewegungssternhaufen 96–97
bewölkt 77–148
Bewölkung 77–147
Bezeichnung eines Kometen 85–31
endgültige ~ 85–33
vorläufige ~ 85–32
Beziehung Durchmesser-Tiefe 81–248
genetische ~ 97–127
Masse-Leuchtkraft- ~ 91–129
Masse-Radius- ~ 93–200
Perioden-Helligkeits- ~ 95–103
solar-terrestrische ~ 84–55
Spektrum-Leuchtkraft- ~ 91–130
zufällige ~ 97–126
~ zwischen Grössenklassen und Intensitäten 91–58
Bezug, geodätischer 71–20
Bezugsellipsoid 71–18
Bezugssystem 98–136
fundamentales ~ 98–137
geodätisches ~ 71–19
lokales ~ 98–139
rotierendes ~ 98–138
Bieliden 86–256
Big Bang 99–167
Bildaufnahme, thermische 72–83
Bilder, monochromatische 97–158
Bildhauer 92–87
Bildhauerwerkstatt 92–87
Bildteiler 88–87
Bildteilerprisma 88–88
Bildung eines Flecks 89–121
Bindemittel 82–118
Binnenland-Erdbeben 73–2b
Binnenmeer 72–26
Biosphäre 72–8
Biotit 74–25
bipolares magnetisches Gebiet 89–160a
Bishopscher Ring 79–88
Blanketing 93–20
Blanketingeffekt 93–67
Blase 82–180
Blasenstruktur 99–171a
blasig 82–181
Blasigkeit 82–182
Blätterung 74–53
Blattgestein 74–54
blau 91–99
Blei, gemeines 74–99
gewöhnliches ~ 74–99
radiogenes ~ 74–97
ursprüngliches ~ 74–98
Blei-Blei-Alter 74–100
Bleiisotop 83a–76
Bleiverhältnis-Alter 74–100
Blende 88–61
Blinkkomparator 91–34
Blinkmikroskop 91–34
Blitz 77–162
Block, erratischer 72–132
loser ~ 82–130
Blöckenrand 81–224
Boden, bindiger 82–63
kohäsiver ~ 82–63
Bodeneigenschaft 82–108
Bodenfestigkeit 82–135
Bodenform, kuppelartige 81–177
Bodenfrost, ewiger 83–152
Bodeninversion 77–38
Bodenkriechen 82–152
Bodenprobennahme 82–78
Bodenschicht, lockere 82–104
Bodenunruhe, mikroseismische 73–54
Bodenuntersuchungsgerät 82–79
Bodenwelle 78–14
Bodesches Gesetz 83–3
Bogen, eruptiver 89–251
homogener ruhiger ~ 79–143
intertropischer ~ 79–53
koronaler ~ 89–298
~ mit Strahlenstruktur 79–153
primärer ~ 73–82
pulsierender ~ 79–147
sekundärer ~ 73–83
bogenförmige Vertiefung 81–160
Bogenprotuberanz 89–270
Bogenstruktur 88–249
Bogensystem 89–299
Bohrkern 72–52
Bohrloch 82–61
böiger Wind 77–111
Boje, automatische 72–100
Bolid 86–11
grosser ~ 86–10
heller ~ 86–10
Bolographenkurve 88–76
bolometrische Helligkeit 91–68
~ Korrektion 91–80
Bombe, vulkanische 73–144
Bondsche Albedo 83–91
Boötes 92–22
Böschung 81–184
~ am Fuss des Kontinentalhanges 72–36
Breite, areographische 83–140
erdmagnetische ~ 75–52
geomagnetische ~ 75–52
heliographische ~ 89–42
selenographische ~ 81–11
Breiten, in hohen 89–53
in niedrigen ~ 89–52
Breitengrenze, nördliche 81–105
südliche ~ 81–106
Breitengrenzen der Sichtbarkeit einer Bedeckung 81–104
Breitenverteilung 89–51
Bremsmechanismus 91–46
Brennpunktsnähe, galaktische 98–189
bröckelig 86–191
Bruch 73–78, 82–212
~ am Kraterrande 82–211
ebener ~ 82–208
gerader ~ 82–208
konzentrischer ~ 82–209
Bruchbildung 73–92
brüchig 86–191
Brüchigkeit 86–192
Bruchstück eines Meteoriten 87–43
Bruchstücke 83a–83
Bruchsystem 73–79
kontinentales ~ 73–81
mittelozeanisches ~ 73–80
Bucht 81–294
~ der Gluten 81–296
Buckel 95–37
~ der Kurve 95–37
Büschel, polare 89–294

C

Caelum 92–23
Caldera 81–179
Callisto 83–208
Camelopardalis 92–24
Cancer 92–25
Canes Venatici 92–26
Canis Maior 92–27
~ Minor 92–28
Capricornus 92–29
Carina 92–30
Cassinische Teilung 83–224
Cassiopeia 92–31
Celsius-Skala 77–33
Centaurus 92–32
Cepheid 95–92
klassischer ~ 95–94
Cepheiden-Parallaxe 91–8a
Cepheus 92–33
Ceres 84–82
Cetus 92–34

chalkophile Elemente 83a–31
Chamäleon 92–35
Chandrasekharsche Grenzmasse 93–189
~ Näherung 93–51
Chemie der Atmosphäre 76–79
chemisch aktive Flüssigkeit 74–43
Chemolumineszenz 79–31
Chemosphäre 76–62
Chlorgehalt 72–68
Chlormenge 72–68
Chondre 87–76
Chondrit 87–61
kohliger ~ 87–63
nichtkohliger ~ 87–75
Chromosphäre 88–191
Höhe der ~ 88–202
höhere ~ 88–192
untere ~ 88–193
Chromosphären-Teleskop 88–73
chromosphärisch 88–194
Circinus 92–36
Cirrokumulus 77–143
Cirrostratus 77–141
Cirrus 77–136
cislunar 81–4
Clairautsches Theorem 71–77
Coelostat 88–46
Columba 92–37
Coma Berenices 92–38
Compton-Wellenlänge 99–170g
Copernicanisches System 99–24
Coriolis-Effekt 77–81
Corioliskraft 77–79
Corona Australis 92–39
~ Borealis 92–40
Corvus 92–41
Crabnebel 97–134
Crater 92–42
Cristobalit 82–95
Crux 92–43
Cumulonimbus 77–137
Cumulus 77–138
Curie-Punkt 75–41
Curie-Temperatur 75–41
cusp der Magnetosphäre 75–13
~ des Poles 75–13
Cygnus 92–44

D

Damm 89–276
Dämmerung 79–34
astronomische ~ 79–37
bürgerliche ~ 79–38
nautische ~ 79–39
Dämmerungs- 79–40
Dämmerungsbeobachtung 79–42
Dämmerungsbogen 79–50
Dämmerungschor 78–62
Dämmerungserscheinungen 79–32
Dämmerungshimmel 79–43
Dämmerungsleuchten 79–4
Dämmerungssondierung 79–41
dampfender Vulkan 73–114
Dämpfer 82–24
Datierung 87–126
~ der Mondereignisse 82–265
Dauer der Bedeckung 95–78
~ der totalen Sonnenfinsternis 88–26
~ der Totalität 88–26
~ des Zyklus 89–87
Dauerfrostboden 83–152
Dauerstrich 86–277
D. B. 89–279
Deaktivierung 85–148
~ eines Kometen 85–148
Deferent 83–48
Deferenzkreis 83–48
Defluidisation 82–263
Deformation 73 67
Deformationstensor 73–70
Dehnung 73–66
Deimos 83–169
Deionisierung 86–213
Dekameterwellenausbruch 83–183
dekametrische Radioquelle 83–186
Deklination der Sonne 81–63
~ des Mondes 81–62
magnetische ~ 75–56
Delphin 92–45
Delphinus 92–45
Delta-Cephei-Stern 95–92
klassischer ~ 95–94
Delta-Cephei-Veränderlicher 95–92
Deneb 92–112
Denebola 92–113
Denudation 73–108
Depolarisation 97–118
Depression, kontinentale 81–158
lineare ~ 81–159
lunare ~ 81–157
mäanderförmige ~ 81–160
sinusförmige ~ 81–161
Desorption 85–123
Desorptionswärme 85–124
Detailphotometrie 83–104
detonierendes Meteor 86–119
Devon 74–81
Dezelerationsparameter 99–83
Dezember-Ursiden 86–258
Dezibel 76–39
Diagenese 74–38
Diagramm, Hertzsprung-Russell- 91–130
HR-~ 91–130
Hubble ~ 98–301
Masse-Leuchtkraft- ~ 91–129
Zweifarben- ~ 96–147
Diamanten, meteoritische 87–65
Diastrophismus 73–72
Diatreme 73–122
Dichotomie, Venus- 83–132
Dichte der Kernspuren 82–253
~ des Sonnenwindes 88–157
flächenhafte ~ 98–85
kosmologische ~ 98–370
kritische ~ 98–197
mittlere ~ 93–86
projizierte ~ 98–85
räumliche ~ 96–133
subnukleare ~ 93–190
wahre ~ 96–133
zentrale ~ 93–87
Dichtefunktion 98–88
Dichtegradient 98–87
Dichteverteilung 96–124
Dichtewelle 98–175
galaktische ~ 98–175
Dichtewellentheorie 98–174
Dicke der Ringe 83–229
Dienst, meteorologischer 77–5
Differentialgleichung, Emdensche 93–130
Differentiation, chemische 83a–50
geochemische ~ 74–55
differentielle Rotation 88–128
~ Rotation der Galaxie 98–113
Differenzierung im Schwerefeld 83a–54
diffus leuchtende Fläche 79–148
diffuses galaktisches Licht 97–56
~ Objekt 85–25
Diffusion der Ionen 86–286
~ der kosmischen Strahlung 84–27
Diffusionsgleichgewicht 76–23
Diffusionskoeffizient 86–287
Dimensionen des Systems 96–64
direkt 83–42
Diskontinuität, Moho- 72–14
Mohorovičić ~ 72–14
diskordante Intrusion 73–137
Dislokationsbeben 73–2d
Dissipation 94–73
~ der Energie 82–285
thermische ~ 94–76
Dissoziationstemperatur 91–120a
dissoziative Rekombination 86–214
Distanz 91–3
Dizyan 97–24
Dom 81–177
Donner 77–165
Doppelasteroid 84–79
Doppelaufnahme 86–155
photographische ~ 86–155
Doppelgalaxie 98–305
Doppelhaufen 96–81
Doppelmeteor 86–131
Doppelplanetoid 84–79
Doppelpulsar 95–140
Doppelspektrum 96–18
Doppelstern 96–3
astrometrischer ~ 96–12
enger ~ 96–23
getrennter ~ 96–35
halbgetrennter ~ 96–34
~ mit Linien beider Komponenten 96–17
~ mit Linien einer Komponente 96–15

nicht aufgelöster ~ 96–10
optischer ~ 96–6
photometrischer ~ 96–11
physischer ~ 96–7
spektroskopischer ~ 96–14
spektroskopischer ~ mit Linien beider Komponenten 96–17
spektroskopischer ~ mit Linien einer Komponente 96–15
Spektrum-~ 96–14
visueller ~ 96–8
weiter ~ 96–9
Doppelsternsystem 96–7
Doppelunterbrecher 86–149
Doppelwelle 98–117
Doppelzählung, Methode der 86–132
Doppler-Kompensation 88–90
Doppler-Kompensator 88–91
Dopplerverschiebung 91–25
Dorado 92–46
Drache 92–47
Draco 92–47
Draconiden 86–252
drakonitischer Monat 81–47
~ Umlauf 81–47
Drall 83a–4
Dränungskrater 81–239
Draperien 79–154
drehbare Sternkarte 92–10
Drehgeschwindigkeit 91–48
Drehimpuls 83a–4
Drehmoment 83a–5
Drehung der Apsidenlinie 96–71
~ der Planeten 83–30
Drehungsgeschwindigkeit 83–32
dreidimensionale Spektralklassifikation 91–106
Dreieck 92–93
Südliches ~ 92–94
Dreierstoss 79 30
Dreifachstern 96–4
Dreifachsternsystem 96–4
Dreifarbenphotometrie 96–141
Drei-Kiloparsec-Arm 98–37
Dreikörperbegegnung 96–74
Dreikörperstoss 96–74
Dreikörperzusammentreffen 96–74
dreistündliche Kennziffer 75–89
Drift in geomagnetischer Länge 75–120
longitudinale ~ 75–120
Druck, atmosphärischer 77–43
hydrostatischer ~ 93–77
~ in Meereshöhe 77–44
nichtthermischer ~ 94–42
thermischer ~ 94–41
unter normalem ~ 77–46
Druckfläche, Höhe einer 76–16
Druckgefälle 77–106
Druckgradient 93–95
Druckhöhe 76–16
Dubhe 92–114
Dumping-Elektronen 79–118
Düne 83–162
Dunkelglimmer 74–25
Dunkelnebel 97–132
Dunkelwolke 97–107
dunkler Ring 83–221
Dunst 83–250
trockner ~ 77–27
Dunstschicht 83–249
Durchgang durch den Zentralmeridian 89–45
~ einer interplanetaren Diskontinuität 84–58
Durchgangszeit durch das Periastron 96–60
Durchlässigkeitsfenster 83–264
Durchlaufen einer interplanetaren Diskontinuität 84–58
Durchmesser, linearer 83–118
photometrischer ~ 84–93
scheinbarer ~ 83–117, 96–115
wahrer ~ 83–118
Durchmischung der Sternmaterie 93–162
Durchmusterung des Himmels 92–11
selenophysikalische ~ 82–39
Dynamik von Sternhaufen 96–153
dynamische Metamorphose 74–36
~ Meteorologie 77–7
Dynamometamorphose 74–36
Dynamotheorie, geomagnetische 75–99

E

Ebbe 82–282
~ und Flut 72–89
Ebbestrom 82–282
Ebbeströmung 72–65
Ebene, galaktische 98–53
invariable ~ 83–232
Laplacesche ~ 83–232
Echo, seismisches 82–223
Echodauer 86–299
Echoimpuls 86–290
Echolotung 72–29
eckiger Stein 82–131
Eddingtonsches Modell 93–137
Edelgas, radiogenes 87–132
Effekt, Blanketing- 93–67
Evershed- ~ 89–103
Isotopie- ~ 91–178
kumulativer ~ 98–124
Mögel-Dellinger- ~ 78–40
Poynting-Robertson- ~ 84–67
Wilson- ~ 89–89
Effekte der Sonnenaktivität 84–55
effektive Linienbreite 89–208
Effusion 73–130
Effusivgestein 74–3
ehemaliger Satellit 83–240
Eichfelder, Kapteynsche 98–78
Eichtheorie 99–170j
Eidechse 92–58
Eigenbewegung 91–19
~ der Flecken 89–117
~ in Deklination 91–22
~ in Rektaszension 91–21
jährliche ~ 91–20
Eigenfarbe 91–74
Eigenfarbenindex 91–75
Eigenfarbindex 91–75
Eigengravitation 94–34
Eigenhelligkeit 85–107
Eigenlicht 96–45
leuchtend mit ~ 96–45
Eigenschaften, globale 96–134
integrale ~ 96–134
kinematische ~ 98–90
optische ~ 98–234
stellare ~ 91–100
Eigenstrahlung 83–198
~ eines Planeten 83–198
Eigenzeit 99–139
Einarmgalaxie 98–42
einarmige Galaxie 98–42
Einbruch hochenergetischer Partikeln 78–43
Einbruchkaldera 81–181
Eindeutigkeit der Lösung 93–82
Eindeutigkeitssatz 93–83
~ des Sternaufbaus 93–83
Eindringen in die Atmosphäre 86–161
Eindringgeschwindigkeit 86–92
Einfall von Gas 98–356
Einfang 75–109
~ der Kometen 85–101
langsamer ~ von Neutronen 94–88
~ von intergalaktischer Materie 98–355
~ von Neutronen 94–89
Einfangen 75–109
Einfangregion 75–110
Einfangshypothese 83a 18
Einfangstheorie 96–72
eingefangene Elektronen 79–111
~ Gase des Sonnenwindes 82–148
eingeschränktes kosmologisches Prinzip 99–31
Einhorn 92–68
Einschlagformation 82–228
Einschlaggrube 87–58
Einschlagsauswürflinge 82–167
Einschlagshypothese 81–261
Einschlagskrater 81–240
Einschlagskraterbildung 82–234
Einschliessung der kosmischen Strahlung 98–51
Einschlüsse, organische 87–66
Einstein-De Sitter Weltall 99–70
Einsteins Weltmodell 99–68
Einsteinsche Feldgleichungen 99–156
Einstein-Universum 99–68
Einstrahlung 77–51
Einsturzstaub 82–125
Einteilung, petrographische 82–101

Eintritt des Mondes in den Halbschatten 81–75
~ des Mondes in den Kernschatten 81–76
~ des Sternes 81–98
~ in die Atmosphäre 86–161
Eintrittsgeschwindigkeit 86–92
Einwirkung, atmosphärische 87–116
Einzelexemplar 87–42
Einzelfall 87–42
Einzelfleck 89–116
Einzelheit, optische 98–234
Einzelpuls 95–146
Einzelspektrum 96–16
Einzelstern 96–2
Eis 85–42
amorphes ~ 85–44
kubisches ~ 85–46
schmutziges ~ 97–73
Eisberg 72–117
Eisbergüberwachung 72–85
Eisbewegung im Gletscher 72–120
Eisdecke 72–115
Eisengruppe, Elemente der 94–105
Eisenmeteorit 87–82
Eisenpartikel 97–74
Eishügel 72–102
Eiskappe 72–115
Eiskern 85–40
Eiskonglomerat 85–43
Eiskorn 85–52
Eiskristall 76–111
Eismantel 97–72
Eisnadeln 76–112
Eisnuklei 76–109
Eissatellit 83–23
Eisschild 72–115
Eisüberwachung 72–85
Eisüberwachungsdienst 72–85
Eiszeit 83a–78
eiszeitlich 72–106
Eiszunge 72–125
Ejektion 98–179
Ejektionsvorgang 98–179
ekliptikales Meteor 86–46
Ekliptikalstrom 86–221
E-Korona 88–241a
elastische Welle 73–44
Elastizität 73–58
Elastitätsmodul 73–59
Elektrizität, atmosphärische 75–2
Elektrojet 79–114
~ in Polarlichtbreiten 79–113
Elektronen, relativistische 97–153
Elektronen-Abregnung 78–48
~ im Polarlicht 79–112
Elektronenatmosphäre 93–28
Elektronenbündel 79–116
Elektronendichte 78–6
lineare ~ 86–283
Elektronendichteinhomogenität ~ 78–8
Elektronen-Einfall ~ 78–48
Elektronenkorona 88–242
Elektronen-Pfeifstörung 78–60
Elektronenpolaricht 79–110
Elektronenquelle 79–119
Elektronenstrahl 79–116
Elektronenstrom 79–114
Elektronentemperatur 91–116
Elektronenwhistler 78–60
Elektronenwolke 79–115
elektronographische Sterngrösse 91–67
elektrooptische Modulation 88–86
elektroschwage Wechselwirkung 99–170x
elektrotellurisches Feld 75–4
Element, überhäufiges 91–184
unterhäufiges ~ 91–186
~ von zu geringer Häufigkeit 91–186
~ von zu grosser Häufigkeit 91–184
Elementarlänge 99–170h
Elementarmasse 99–170f
Elementarsturm 75–95
magnetischer ~ 75–95
polarer magnetischer ~ 75–96
Elementaufbau 94–52
Elementbildung 94–52
Elemente, atmophile 83a–29
bewegliche ~ 88–108
chalkophile ~ 83a–31
~ der Eisengruppe ~ 94–105
erdmagnetische ~ 75–54
feuerfeste ~ 97–64
geomagnetische ~ 75–54
häufige ~ 82–144
Häufigkeit chemischer ~ 82–141
lithophile ~ 83a–30
meteorologische ~ 77–28
photometrische ~ 95–25
supergranulare ~ 88–182
Elementenenstehung 94–52
Elementenhäufigkeit 82–141
Elemententstehung 94–52
Elementensythese 94–52
Elfjahreszyklus 89–73
elfjähriger Zyklus 89–73
Ellerman-Bombe 89–200
Ellipsoid, internationales 71–14
lokales ~ 71–15
ellipsoidale Galaxie 98–241
ellipsoidische Höhe 71–52
elliptische Galaxie 99–241
~ Riesengalaxie 98–243
~ Zwerggalaxie 98–242
elliptischer Komet 85–4
Elliptizität eines Kugelhaufens 96–122
Elongation 83–60
~ des scheinbaren Radianten 86–67
grösste ~ 83–63
östliche ~ 83–61
westliche ~ 83–62
Embryo eines Planeten 83a–36
Emdensche Differentialgleichung 93–130
Emission, ELF 78–55
~ in extrem niedriger Langewellenfrequenz 78–55
~ in niedriger Langewellenfrequenz 78–54
VLF ~ 78–54
Emissionskorona 88–241a
Emissionsmass 88–234
Emissionsnebel 97–123
Emissionsschicht 79–17
Emissionsstern 91–159
Empfänger 78–17
Empfangsantenne 78–18
Ende bei Sonnenaufgang 88–21
~ der Finsternis 81–83
~ der partiellen Finsternis 81–83a
~ der totalen Verfinsterung 81–80
~ der Totalität 81–80
~ der Welt 99–183
~ des Zyklus 89–85
Endhöhe 86–40
Endlichkeit des Weltalls 99–4
Endmoräne 72–130
endogen 73–145
Endpunkt 86–38
Endzusammensturz 99–182
Energie, akustische 88–219
ausgetauschte ~ 98–126
gespeicherte ~ 89–196
verfügbare 89–197
Energieabgabe 88–254
Energieausbreitung 88–217
Energieausstoss 93–109
Energieaustausch 98–125
Energiebilanz 88–251
Energieerzeugung 88–216
Energiefluss 93–112
akustischer ~ 93–161
konvektiver ~ 93–115
Energiefreisetzung 93–105
Energiegewinn 88–253
Energie-Impuls-Tensor 99–158
Energieproduktion 93–105
Energiequelle 93–110
doppelte ~ 93–146
Energiestrom 93–112
Energietensor 99–158
Energietransport 93–111
Energieumsatz 86–208
Energieverlust 88–254
entartet 93–178
teilweise ~ 93–183
vollständig ~ 93–185
Entartung 93–172
nichtrelativistische ~ 93–177
relativistische ~ 93–176
teilweise ~ 93–173
vollständige ~ 93–175
Entartungsgrad 93–174
Entartungskriterium 93–181
Entartungstemperatur 93–188
Entdecker 85–21
Entdeckung eines Kometen 85–16

D

Entfernung 86–288
~ des Mondes von der Erde 81–13
Erde-Mond ~ 81–13
galaktozentrische ~ 98–11
geozentrische ~ 84–3
photometrische ~ 96–116
~ vom Sternmittelpunkt 93–85
~ zum Zentrum der Milchstrasse 98–11
Entfernungsanzeiger 86–266
Entfernungsmodul 91–16
scheinbarer ~ 96–117
wahrer ~ 96–118
Entfernungsskala 98–303
Entgasung 82–261
Entionisierung 86–213
Entkopplung der Strahlung 99–177
Entkopplungsära 99–177a
Entladung, atmosphärische 77–161
Entmagnetisierung, thermische 75–40
Entnahme einer Luftprobe 76–85
Entstehung 94–10
~ der Elemente 94–52
~ des Weltalls 99–11
~ durch Einschlag 82–229
~ durch Einsturz 82–229
~ eines Flecks 89–121
tektonische ~ 82–200
~ von Gasblasen 82–182
~ von Staubkörnern 97–60
Entstehungsrate 85–143
Entweichen von Gasen 76–24
~ von Sternen aus einem Haufen 98–194
Entweichgeschwindigkeit 96–155
Entweichungskegel 76–27
Enveloppe, charakteristische 98–177
entwickeln sich 94–75
Entwicklung, chemische 83a–49
~ der Sterne 94–1
~ der Welt 99–171e
~ des Weltalls 99–171e
dynamische ~ 96–161
~ einer Fleckengruppe 89–118
geochemische ~ 74–33
nukleare ~ 94–50
prästellare ~ 94–14
thermische ~ 83a–61
thermonukleare ~ 94–50
Entwicklungsänderung 94–27
Entwicklungsdiagramm 96–151
Entwicklungskosmos 99–54
Entwicklungskurve 94–29
~ einer Fleckengruppe 89–119
Entwicklungsphase 94–28
~ nach der Hauptreihe 94–68
Entwicklungsreihe 94–30
Entwicklungsstadium 94–28
Entwicklungsuniversum 99–54
Entwicklungsweg 94–29
Eozoikum 74–77
Ephemeriden von kleinen Planeten 84–92
epirogene Bewegungen 73–73
Epirogenese 73–73
epirogenetische Bewegungen 73–73
Epizentralabstand 73–19
Epizentraldistanz 73–19
Epizentralentfernung 73–19
Epizentralgebiet 73–16
Epizentrum 73–15
Epizykel 83–49
Epoche 74–89
~ des Maximums 89–81
~ des Minimums 89–80
geologische ~ 74–89
e-Prozess 94–99
Equuleus 92–48
Erd- 71–3
erdabgewandte Mondseite 81–16
Erdabplattung 71–8
erdähnliche Planeten 83–119
Erdalter 83a–68
Erdaltertum 74–71
erdartige Planeten 83–119
Erdatmosphäre 76–5
ausserhalb der ~ 88–149
Erdbahn 84–5
ausserhalb der ~ 84–6
in der Nähe der ~ 84–7
jenseits der ~ 84–6
~ kreuzender Asteroid 84–76
Erdball, flüssiger 83a–64
Erdbeben 73–2a
Binnenland- ~ 73–2b
intermediäres ~ 73–24
katastrophales ~ 73–40
kontinentales ~ 73–2b
leichtes ~ 73–35
mittelstarkes ~ 73–36
mitteltiefes ~ 73–24
normales ~ 73–22
starkes ~ 73–37
submarines ~ 73–2c
tektonisches ~ 73–2d
tiefes ~ 73–23
verwüstendes ~ 73–38
vulkanisches ~ 73–2e
zerstörendes ~ 73–39
Erdbebenflutwelle 72–95
Erdbebenforscher 73–2
erdbebenfreies Gebiet 73–29
Erdbebengebiet 73–28
Erdbebengürtel 73–30
Erdbebenintensität 73–33
Erdbebenkatastrophe 73–40
Erdbebenkunde 73–1
Erdbebenmagnitude 73–34
Erdbebenschreiber 73–7
Erdbebenstation 73–6
Erdbebentätigkeit 73–4
Erdbebenwelle 73–9
Erdbebenzone 73–31
Erde 71–2
Erdelektrizität 75–1
Erdellipsoid 71–11
Erde-Mond-System 81–12
Erdferne 81–39
Erdfigur 71–7
Erdgezeiten 82–296
erdinduzierte Gezeiten 82–275
Erdinneres 72–3
Erdkern 72–17
innerer ~ 72–20
Erdkrümmung 71–6
Erdkruste 72–4
Erdkrustenbewegungen 73–72
Erdkrustengezeiten 82–297
Erdkugel 71–10
Erdladung 75–3
Erdmagnetfeld 75–44
erdmagnetische Gezeiten 82–295
Erdmagnetismus 75–24
Erdmantel 72–15
oberer ~ 72–16
Erdmittelalter 74–72
Erdnähe 81–38
Erdneuzeit 74–73
Erdoberfläche 71–5
Erdpol, magnetischer 75–45
Erdrotation, Verzögerung der 82–313
Erdschatten 81–58
Erdschweif 84–65
Erdschwingungen 73–49
freie ~ 73–49
Erdsphäroid 71–13
Erdstrahlung 77–56
Erdströme 75–5
vagabundierende ~ 75–6
Erdterminator 81–91
Erdurzeit 74–70
erdzugewandte Mondseite 81–16
Ereignis 99–125
benachbartes ~ 99–126
~ hochenergetischer Partikeln 89–191
Protonen- ~ 99–125
raum-zeitliches ~ 99–125
~ solarer kosmischer Strahlen 89–194
Ereignishorizont 99–43
Ereignisse, gleichzeitige 99–128
Ergiebigkeit 93–34
Erguss 73–132
Ergussgestein 74–3
Erhebung auf dem Mond 81–156
Erhitzung des Meteorkörpers 86–174
Erhöhung auf dem Mond 81–156
plötzliche ~ des atmosphärischen Störpegels 78–51
Erholungsphase 75–77
Erholungszeit 98–128
Eridanus 92–49
~-Fluss 92–49
Erkundung, orbitale 82–31
~ von der Umlaufbahn aus 82–31
Erlöschen eines Meteors 86–37

D

langsames ~ 86–75
plötzliches ~ 86–76
erloschener Komet 85–150
~ Vulkan 73–115
Erosion 73–107
~ der Oberfläche 82–241
~ im Weltraum 87–142
kosmische ~ 87–142
Erosionskaldera 81–183
erosiver Zusammenstoss 84–105
erratischer Block 72–132
Erscheinen eines Kometen 85–18
erstes ~ 85–19
erstes ~ eines Kometen 85–19
Erscheinung, aktive 89–14
kurzlebige ~ 89–15
vorübergehende ~ auf dem Mond 82–153
Erscheinungen, solar-terrestrische 84–55
Erschütterung der Schale 95–152
~ des Sterns 95–153
Erstarrung 83a–59
~ des Magmas 82–190
Erstarrungsalter 87–137
Erstarrungsgestein 74–2
Eruption 89–175, 95–69
~ am Sonnenrand 89–202
~ auf der Scheibe 89–204
chromosphärische ~ 89–175
~ der Klasse 2 89–206
Fläche einer ~ 89–207
Helligkeit einer ~ 89–212
Importanz der ~ 89–205
Integrallicht-~ 89–182
~ jenseits des Randes 89–203
Nicht-Protonen- ~ 89–186
Protonen- ~ 89–185
Sonnen ~ 89–175
Eruptionsauswurf 89–190
Eruptionsband 89–187
Eruptionsdauer 89–209
Eruptionsfläche 89–207
Eruptionshäufigkeit 89–217
Eruptionshypothese 89–63
Eruptionskegel 73–118
Eruptionstätigkeit 95–71
Eruptionveränderlicher 95–68
eruptive Phase 89–210
~ Protuberanz 89–248
eruptiver Bogen 89–251
Eruptivgang 73–138
Eruptivgestein 74–2
Erwärmung, atmosphärische 75–87
~ des meteorischen Körpers 86–174
erweitertes Weltpostulat 99–32
erzeugende Galaxie 98–284
erzeugender Komet 86–259
Erzeugung, kontinuierliche 99–91
~ von Materie 99–89
Erzeugungsrate 85–143
erzwungene Gezeitenschwingungen 82–306
E-Schicht 78–26
~ -Maximum 76–56
sporadische ~ 78–25
Etamin 92–115
euklidisch 99–101
Eulennebel 97–139
Europa 83–206
Event hochenergetischer Teilchen 89–191
vorübergehendes koronales ~ 89–185a
Evershed-Effekt 89–103
Ewigkeit der Welt 99–13
~ des Weltalls 99–13
Exemplar, individuelles 87–42
Exkursion 82–35
wissenschaftliche ~ 82–36
Exkursionroute 82–37
exogen 73–146
Exosphäre 76–68
expandierende Hülle 93–12
~ Korona 88–232
expandierender Kosmos 99–55
expandierendes Weltall 99–55
Expansion, beschleunigte 99–81
~ des Weltalls 99–74
hydromagnetische ~ 88–159
ultraschnelle ~ 99–170c
Expansionsfaktor 99–81a
Expansionsgeschwindigkeit 96–94
Expansionsrate 99–75
Expansionszeit des Weltalls 99–78
Friedmannsche ~ 99–80
Hubblesche ~ 99–79
Explosion 87–30
~ eines galaktischen Kernes 98–283
nukleare ~ in grosser Höhe 75–114
Explosionskaldera 81–180
Explosionskrater 81–238
Explosionsphase 89–210
explosive Nukleosynthese 94–57
explosiver Kernaufbau 94–57
Extinktion, atmosphärische 77–127a
interstellare ~ 97–80
selektive ~ 97–83
Extinktionskurve 97–81
extrafoveales Sehen 86–128
extragalaktisch 98–207
Extrusion 73–132
Extrusivgestein 74–3
Exzess, IR 91–79

F

Fächer 85–60
gekrümmter ~ 85–62
gerader ~ 85–61
koronaler ~ 89–295
umgekehrter ~ 85–63
fächerartig 85–74
Fächerbündel-Interferometer 88–122
fächerförmig 85–74
Fächerkeulen-Interferometer 88–122
Fächerprotuberanz 89–258
Fächerstrahl 89–296
Fachmann für Kosmogonie 94–4
Fackel, optische 89–165
Röntgenstrahlen- ~ 89–166
Sonnen- ~ 89–161
Fackelgebiet, photosphärisches 89–161
Fackelgranulation 89–163
Fackelgranule 88–179
Fackelgranulum 88–179
Fackeln, chromosphärische 89–164
polare ~ 89–162
Fäden 88–209
fadenähnliche Strukturen 88–209
Fadenstruktur 88–209
Faktor, Lorentzscher 99–135
Fall, freier 94–39
natürlicher ~ 82–227
Fallgebiet eines Meteoritenschauers 87–38
Fallort des Meteoriten 87–50
falsches Zodiakallicht 84–62
Falte 73–87
Falten 79–139
Faltengebirge 73–86
Faltung 73–88
Faltungsgebirge 73–86
Faltungsrücken 81–187
Faltungszone 73–75a
Farbenänderung der See 72–80
Farbenexzess 91–77
Farbenhelligkeitsdiagramm 96–150
Farbenindex 91–76
~ B-V 96–145
integraler ~ 96–138
spezieller ~ 83–87
~ U-B 96–146
Farbenindexkurve 95–21
Farbexzess 91–77
Farbgleichung 91–73
Farbindex = Farbenindex 91–76
Farbsystem 91–72
Farbtemperatur 91–109
Faser eines Nebels 97–131
Penumbra- ~ 89–98
Fasermuster 89–174
Fasern, chromosphärische 88–209
Fasernebel 97–129
Faserstruktur 97–130
Faserung 89–173
Faye-Reduktion 71–78
Federhaufenwolke 77–143
Federschichtwolke 77–141
Federwolke 77–136
feinkörnig 82–114
Feinstruktur 88–208

chromosphärische ~ 88–203
~-Magnetfeld 89–143
Feld der differentiellen Bewegung 98–114
~ der Erdströme 75–4
effektives ~ 95–59
elektrotellurisches ~ 75–4
erdmagnetisches ~ 75–44
galaktisches ~ 98–146
geomagnetisches ~ 75–44
grossmasstäbliches ~ 89–142
grossskaliges ~ 89–142
Higgssches ~ 99–170l
inneres ~ 95–63
interplanetares magnetisches ~ 84–9
kleinskaliges ~ 89–143
kleinmasstäbliches ~ 89–143
magnetisches ~ des Mondes 82–45
magnetisierendes ~ 75–42
selenomagnetisches ~ 82–45
Feldannihilation 89–158
Felder, Kapteynsche 98–78
Feldgalaxie 98–340
Feldgleichungen 99–155
Einsteinsche ~ 99–156
Feldintensität, effektive 95–59
Feldkorrektion 86–157
Feldkorrektur 86–157
Feldlinie, mitrotierende 75–14
offene ~ 75–15
Feldmeteor 86–25
Feldspat 74–15
Feldstärke 84–11
magnetische ~ 84–11
Feldstern 96–84
Feldstruktur 89–140
Feldvernichtung 89–158
Feldverstärkung 89–156
Fels 74–1
Felsblock 82–130
Felsen 74–1
Felsenmeer 82–139
Felssatellit 83–22
Felstrümmer 81–235
Fenster, galaktische 98–57
~ in der nebelfreien Zone 98–57
Fernbeben 73–27
Fernrohr 92–92
Ferromagnetismus 75–26
fester Körper 83–110
Festkörperrotation 98–111
Festland 83–159
Festlandtafel 73–103
Feststoff, feuerfester 85–49
Feuchtigkeit = Feuchte
absolute ~ 74–122
relative ~ 77–128
spezifische ~ 77–123
Feuchtigkeitsmesser 77–120
Feuerball 86–11
grosser ~ 86–10
primordialer ~ 99–168
feuerfester Stoff 83a–55
Feuerkugel 86–11
~ mit Donner 86–119
Feuerkugelerscheinung 86–117
feuerradähnliche Rotation 98–28
Fibrillen 88–209
Figur der Erde 71–7
~ eines Planeten 83–109
Filament 89–224
kosmisches ~ 99–171c
Metall- ~ 89–268
Penumbra- ~ 89–98
Wasserstoff-~ 89–267
wedelndes ~ 89–216
Filamentaktivierung 89–215
filamentartiger Nebel 97–129
Filamentkrone, polare 89–232
Filamentstruktur 97–130
Findling 72–132
Finsternis 88–14
Ende der ~ 81–83
Ende der partiellen ~ 81–83a
Grösse der ~ 81–79
ionosphärische ~ 88–23
~ ist sichtbar in ... 81–86
~ ist unsichtbar in ... 81–87
korpuskulare ~ 88–24
künstliche ~ 88–227a
Mitte der ~ 81–78
partielle ~ 81–79a
Phasen der ~ 81–73
ringförmige ~ 88–18
Sichtbarkeit der ~ 81–88
Sichtbarkeitsgebiet der ~ 81–89
totale ~ 88–17
Verlauf der ~ 81–73
zentrale ~ 88–16a
Zone der totalen ~ 88–29
Finsternisbeobachtung 88–226
Finsternisdauer 88–26
Finsterniskarte 88–28
Finsterniskomet 85–12
Firn 72–118
Firnfeld 72–119
Firnschnee 72–118
Firnschneefeld 72–119
Fische 92–79
Fischortung 72–84
Fit-Bedingung 93–142
Fixstern 92–4
F-Korona 88–241
Flachbeben 73–21
Fläche, die von Flecken bedeckte 89–90
diffus leuchtende ~ 79–148
dunkle ~ 83–155
~ einer Eruption 89–207
~ korrigiert wegen perspektivischer Verkürzung 89–93
korrigierte~ 89–93
polierte ~ 87–101
Flächendichte 96–132
scheinbare ~ 96–132
flackerndes Meteor 86–85
Flackerstern 95–68
flammendes Nordlicht 79–157
Flare 89–175
~-Aktivität 95–70
~ auf der Scheibe 89–204
~ der Importanz 2 89–206
gleichzeitiger ~ 89–179
homologer ~ 89–178
~ im weissen Licht 89–182
~ jenseits des Randes 89–203
~ mit zwei Faden 89–187a
optischer ~ 89–181
Protonen-~ 89–185
sichtbarer ~ 89–181
sympathetischer ~ 89–179
Flare-ähnliche Protuberanz 89–222
Flare-Aktivität 89–176
Flare-Auswurf 89–214
Flare-Band 89–187
Flare-Fläche 89–207
Flare-Helligkeit 89–212
Flare-Stern 95–68
Flare-Surge 89–190
Flare-Tätigkeit 89–176
Flare-Überwachung 89–219
automatische ~ 89–221
kinematographische ~ 89–220
Flare-verbunden 89–188
Flare-verbundene dunkle Wolke 89–189
Flash-Phase 89–210
Flash-Spektrum 88–25
Fleck, F- 89–114
folgender ~ 89–114
führender ~ 89–113
Grosser Roter ~ 83–193
heisser ~ 95–112a
nachfolgender ~ 89–114
östlicher ~ 89–114
P- ~ 89–113
Roter ~ 83–193
unsichtbarer ~ 89–159
vorangehender ~ 89–113
vorderer ~ 89–113
Weisser ~ 83–214
westlicher ~ 89–113
Fleckenaktivität 89–62
Fleckenbildung 89–121
Fleckenentwicklungstabelle 89–120
Fleckenfläche 89–67
korrigierte ~ 89–93
scheinbare ~ 89–91
Fleckengruppe 89–105
Fleckenhäufigkeit 89–66
Fleckenmodell 89–102
Fleckenprotuberanz 89–252
aktive ~ 89–254
Fleckenrelativzahl 89–65
Fleckentätigkeit 89–62
Fleckenzeichnung 88–34
Fleckenzone 89–54
Fleckenzonenwanderung 89–74
Gesetz der ~ 89–75

D

Fleckenzyklus 89–72
Fliege 92–69
fliegende Schatten 88–31
Fliegender Fisch 92–100
Fliessen, plastisches 72–121
Fliessenrinne 82–192
Fliesslinie 82–194
Fliesspunkt 86–204
Fliessrinne 82–192
Fliesstruktur 82–191
flimmerndes Meteor 86–85
Flocculi 89–167
Flocculus 89–167
dunkler ~ 89–224
Flocke 89–167
dunkle ~ 89–224
Flocken 89 167
Florring 83–221
Fluchtbewegung der Galaxien 98–298
Fluchtgeschwindigkeit 99–75
Flüchtige 82–262
flüchtiger Stoff 82–262
Fluchtstern 96–160
Flugdauer 86–50
Fluidaltextur 82–191
Fluidität der Lava 82–188
Fluktuationen, primitive 99–170y
Fluoreszenzbanden 85–132
Fluss akustischer Energie 88–221
~ Eridanus 92–49
Flussbett 83–153
Flüssigkeit, chemisch aktive 74–43
Flussrinne 82–192
Flusstal, submarines 72–46
Flusszustand des Magmas 82–189
Flut 82–280
taube ~ 82–272
Flutberg 82–278
Flutkraft 82–289
Flutstrom 82–280
Flutströmung 82–279
Flutwelle 82–291
Folge der regelmässigen Galaxien 98–240
~ der Spiralgalaxien 98–255
folgender Fleck 89–114
Folie 83–149
Fomalhaut 92–116
Fontänenmodell 85–85
Förderschlot 73–124
Form, orientierte 87–113
Formaldehyd 97–27
Formamid 97–35
Formation, beständige 83–106
sekundäre ~ 81–151
topographische ~ 81–149
ursprüngliche ~ 81–150
vulkanische ~ 82–159
Formel, internationale gravimetrische 71–76
Formen, aktive 79–138
~ mit Strahlen 79–152
~ mit Strahlenstruktur 79–152
~ ohne Strahlenstruktur 79–142
Formfaktor 86–177
Formparameter 86–177
Formveränderung 73–68
Fornax 92–50
Forschungsboje 72–99
Fortpflanzung des Lichtes 99–129
fossile Kernspur 82–252
~ Spur 87–121
fossiler Meteoritenkrater 87–55
Fragmentation einer kosmischen Wolke 94–38
fraktionierte Kristallisation 74–40
Fraktionierung 87–138
~ der chemischen Elemente 83a–51
Fraktokumulus 77–146
Fraunhoferkorona 88–241
Fraunhoferlinie 88–139
Fraunhoferspektrum 88–137
Frei-Gebunden-Strahlung 86–212
Freiluftreduktion 71–78
freisichtbares Meteor 86–16
freisichtiges Meteor 86–16
Fremdstern 98–134
Frequenz der Meteore 86–135
kritische ~ 78–24
Frequenzdrift-Spektrograph 88–124
Fresnelsche Zonen 86–294
Friedmannsche Expansionszeit 99–80
~ Welten 99–71
~ Weltmodelle 99–71
Front 77–68
Frontalecho 86–297
Frontogenese 77–74
Frontolyse 77–75
Frontteil 87–114
Fuchs 92–101
Fuchschen 92–101
Füchslein 92–101
führender Fleck 89–113
Fuhrmann 92–21
Fülleisen 87–94
Fumarole 82–163
Fumarolentätigkeit 82–164
Fundament 72–12
Fundamentaltensor 99–157a
Fundort 87–52
Fundstelle 87–52
Fünf-Minuten-Oszillation 88–186
Funken 86–84
Funkensprühen 86–84
Funkprognose 78–32
Funkwellenabsorption in der Polarkappe 78–44
Funkwellenausbreitung 78–34
Furche 81–165
Fusstapfen 82–38

G

Gabelung 98–5
Gal 71–65
galaktisch 98–2
galaktische Apoapside 98–190
~ Apoapsis 98–190
~ Brennpunktsferne 98–190
~ Brennpunktsnähe 98–189
~ Periapside 98–189
~ Periapsis 98–189
~ Scheibe 98–52
~ Sternpopulation 98–230
~ Urwolke 94–17
~ Verteilung 98–54
galaktischer Apofokus 98–190
~ Kern 98–18
~ Perifokus 98–189
~ Wind 98–277
galaktisches Antizentrum 98–14
~ Feld 98–146
~ Gezeitenfeld 96–106
~ Gravitationsfeld 98–146
~ Licht 79–14
~ Zentralgebiet 98–293
~ Zentrum 98–9
galaktozentrisch 98–10
galaktozentrische Koordinaten 98–12
Galaxie 98–201
aktive ~ 98–260
aufgelöste ~ 98–317
blaue kompakte ~ 98–265
~ der Lokalen Gruppe 98–327
cigarrenförmige ~ 98–239
einarmige ~ 98–42
ellipsoidale ~ 98–241
elliptische ~ 98–241
eruptive ~ 98–269
erzeugende ~ 98–284
explodierende ~ 98–268
gasreiche ~ 98–238
gestörte ~ 98–315
Haro- ~ 92–266
hellste ~ 98–339
~ im sichtbaren Licht 98–204
infrarote ~ 98–205
irreguläre ~ 98–237
Kaulquappen- ~ 98–281
kompakte ~ 98–264
kugelförmige ~ 98–244
linsenförmige ~ 98–245
lokale ~ 98–327
Markarian- ~ 98–265a
mehrarmige ~ 98–44
mehrfache ~ 98–308
~ mit mehrfachem Kern 98–319
N- ~ 98–263
~ nach der Eruption 98–270
nadelförmige ~ 98–239
nicht zum Haufen gehörende ~ 98–340
normale ~ 98–221
optische ~ 98–204
pekuliare ~ 98–256
Pol on Ansicht der ~ 98–247
Population einer ~ 98–230

quasistellare ~ 98–361
Rand einer ~ 98–294
regelmässige ~ 98–236
Schweif einer ~ 98–280
Seyfert ~ 98–262
singuläre ~ 98–258
sphäroidale ~ 98–244
spindelförmige ~ 98–239
stark abgeplattete ~ 98–252
unregelmässige ~ 98–237
verzerrte ~ 98–316
von der Kante gesehene ~ 98–246
~ von der Seite gesehen 98–246
~ vom Übergangstyp 98–248
zerfallene ~ 98–317
zweiarmige ~ 98–43
Galaxien, kollidierende 98–321
Lokale Gruppe von ~ 98–326
verbundene ~ 98–306
wechselwirkende ~ 98–310
zusammenstossende ~ 98–321
galaxienarmer Haufen 98–331
Galaxiendichte 98–344
Galazienfolge, Hubblesche 98–226
Galaxiengruppe 98–325
Lokale ~ 98–326
Galaxienhaufen 98–329
~ im Virgo 98–345
kompakter ~ 98–334
kugelförmiger ~ 98–330
offener ~ 98–333
regelmässiger ~ 98–336
reicher ~ 98–332
sehr kompakter ~ 98–335
unregelmässiger ~ 98–337
Galaxienhaufenmitglied 98–338
hellstes ~ 98–339
Galaxienhaufenpopulation 98–341
Galaxienkette 98–328
Galaxienklassifikation 98–226
Galaxienpaar 98–305
galaxienreicher Haufen 98–332
Galaxienzählung 98–343
Galaxis 98–1
Zentralgebiet der ~ 98–13
Galileischer Satellit 83–204
Galilei-Transformation 99–131
Gamma-Pulsar 95–139
Gamma-Spektroskopie, zerstörungsfreie 82–77
Gangdifferenz 88–117
Gangunterschied 88–117
Ganymed 83–207
ganztägige Welle 76–30
Gas, entartetes 93–179
intergalaktisches ~ 98–351a
intergalaktisches ~ im Haufen 98–352
intergalaktisches ~ zwischen Haufen 98–353
interstellares ~ 97–9
nichtentartetes ~ 93–182
relativistisches ~ 97–152
zirkumgalaktisches ~ 98–351
Zwischenhaufen- ~ 98–353
Gasaustreibung 82–261
Gasblase im Gestein 82–180
Gasdruck 93–90
Gasdynamik, interstellare 97–150
Gase, gefrorene 85–45
glühende ~ 83a–43
Gaserosion 82–243
Gaserzeugungsrate 85–144
Gasgehaltalter 87–135
Gaskugel 93–73
polytrope ~ 93–127
Gasmasse 94–32
Gasnebel 97–112
Gasretention 87–134
Gasretentionsalter 87–135
Gasring 95–66
Gasschweif 86–116
gas-staubförmiger Nebel 97–119
Gas-Staub-Mischung 83a–25
Gas-Staub-Verhältnis 97–120
Gasstrom 95–67
Gaswolke 97–113
intergalaktische ~ 98–354
Gaszurückhaltung 87–134
Gauntfaktor 93–104
Gebiet, aktives 83–180, 89–57
aseismisches ~ 73–29
bipolares magnetisches ~ 89–160a
~ des galaktischen Antizentrums 98–15
~ des galaktischen Zentrums 98–13
dunkles ~ 83–155
erdbebenfreies ~ 73–29
H I ~ 97–49
H II ~ 97–50
heisses ~ 88–200
Hochland ~ 81–271
intergranulares ~ 88–177
interspikulares ~ 88–198
~ ionisierten Wasserstoffs 97–50
kaltes ~ 88–199
kompaktes H II ~ 97–51
M ~ 89–61
magnetisches ~ 89–144
~ neutralen Wasserstoffs 97–49
polares ~ 83–192
post-mare ~ 81–270
prae-mare ~ 81–269
seismisches ~ 73–28
Terra ~ 81–271
Tiefland- ~ 81–272
ungeordnetes ~ 83–164
ungestörtes ~ der Photosphäre 89–56
unipolares magnetisches ~ 89–155a
Gebilde, atmosphärisches 83–107
beständiges ~ 83–106
Gebirge 81–193
Gebirgsbildung 73–74
Gebirgsbogen 73–84
Gebirgsgletscher 72–113
Gebirgskamm 81–195
Gebirgskette 81–194
Gebirgsregion 81–196
Gebirgszug 73–85
gebundene Rotation 81–14
Gefrierkerne 76–109
Gefüge 82–102
Gegend, helle 83–160
Gegendämmerung 79–57
Gegenmond 79–85
Gegenpassate 77–103
Gegenschein 84–64
Gegenschweif 85–78
gegenseitige Störungen zwischen den Satelliten 83–25
Gegensonne 79–83
Geisterkrater 81–209
Geländegestaltung 81–149
gelb 91–98
Gelbindex 83–88
Gelbfarbenindex 83–88
Gemini 92–51
Geminiden 86–257
Gemma 92–117
Generation der Sterne 94–23
Geochemie 74–30
geochemische Differentiation 74–55
Geochronologie 74–67
geochronologische Skala 74–68
Geodäsie, kosmische 71–31
Geodätische 99–153
~ Linie 99–153
~ Verbindung von Kontinenten 71–37
Geoelektrizität 75–1
Geoid 71–16
kompensiertes ~ 71–17
regularisiertes ~ 71–17
Geoidundulation 71–74
Geokorona 76–73
neutrale Wasserstoff- ~ 76–74
Geologie 72–2
planetarische ~ 83–5
geomagnetischer Hohlraum 75–8
~ Sturm 75–71
Geomagnetismus 75–24
Geometrie, Euklidische 99–108
nichteuklidische ~ 99–109
Riemannsche ~ 99–113
Geometriefaktor 86–177
Geophysik 72–1
kosmische ~ 84–57
Geopotential 71–60
Geopotentialfläche 71–61
Geosphäre 72–7
geostrophischer Wind 77–96
Geosynklinale 73–105
Geotektonik 73–95
Geothermie 74–58
Geothermik 74–58
Geowissenschaften 71–1
geozentrische Entfernung 84–3

D

Gerade Wand 81–185
Gerät, automatisiertes 82–41
~ zur Entnahme von Proben 82–80
Geröll 82–128
Geröllabhang 81–233
Geröllmasse 81–236
Gesamtdruck 93–93
Gesamtenergiefluss 93–116
Gesamtextinktion 97–82
Gesamtfarbenindex 96–138
Gesamtfläche 89–94
Gesamtgrösse 96–136
Gesamthelligkeit 85–107, 96–135
Gesamtmasse des Haufens 96–139
Gesamtspektrum 96–137
Gesamtstrahlung 88–146
Gesamtstrahlungsstrom 93–47
Geschichte, thermische 74–66
Geschwindigkeit, äquatoriale 91–49
atmosphärische ~ 86–94
augenblickliche ~ 86–99
ausseratmosphärische ~ 86–93
charakteristische ~ 76–43
~ des Sonnenwindes 88–158
heliozentrische ~ 86–96
~ im Mittelpunkt der Bahn 86–97
~ im Sehstrahl 91–24
~ in Beobachtungsrichtung 91–24
mittlere ~ 86–98
momentane ~ 86–99
negative ~ 91–27
positive ~ 91–26
relativistische ~ 99–136
Rest-~ 98–91
Geschwindigkeit-Entfernungs-Beziehung 98–300
Geschwindigkeitsabnahme 86–100
scheinbare ~ 86–100
wahre ~ 86–101
Geschwindigkeitsdispersion 98–103
Geschwindigkeitsellipsoid 98–101
Geschwindigkeitsfeld 88–187
Geschwindigkeitskarte 89–23
Geschwindigkeitsmesser, registrierender 88–57
Geschwindigkeitsraum 98–97
Geschwindigkeitsstreuung 98–103
Geschwindigkeitsverteilung 98–104
asymmetrische ~ 98–104
ellipsoidische ~ 98–100
Maxwellsche ~ 98–96
Geschwindigkeit-Tiefe-Kurve 73–52
Gesetz, Bodesches 83–3
~ der Fleckenzonenwanderung 89–75
~ der Planetenentfernungen 89a–2
Halesches ~ 89–154
Hubblesches ~ 98–299
Kramerssches ~ 93–102
Spörersches ~ 89–75
Gestalt des Meteorkörpers 86–175
Gestaltfaktor 86–177
Gestein 74–1
basisches ~ 74–5
geschiefertes ~ 74–54
gleichmässig körniges ~ 82–67
kantiges ~ 82–131
kristallines ~ 82–66
metamorphes ~ 74–11
saueres ~ 74–4
ultrabasisches ~ 74–6
vulkanisches ~ 74–3
Gesteinbruchstück 82–68
Gesteinsbestandteil 82–99
gesteinsbildendes Mineral 82–99
Gesteinsblock 82–130
Gesteinsgang 76–138
Gesteinskorn 85–54
Gesteinsmagnetismus 75–43
gestreutes Sternlicht 79–12
Gewässerkarte 72–25
Gewichtsfunktion 93–58
Gewitter 77–159
Gewitterwolke 77–137
Gezeiten 72–89
atmosphärische ~ 82–294
~ des Erdkörpers 82–296
erdmagnetische ~ 82–295
ozeanische ~ 82–310
thermische ~ 82–299
~ verursachender Körper 82–284
Gezeitenaufspaltung 85–112
Gezeitenberg 82–278
Gezeitendeformation 82–277
Gezeitendissipation 82–286
Gezeiteneffekt 82–276
Gezeitenentwicklung 83a–65
gezeitenerzeugende Kraft 82–289
Gezeitenfeld 98–196
galaktisches ~ 96–106
Gezeitenhochwasser 82–281
Gezeitenhub 82–292
Gezeitenhubmesser 82–311
Gezeitenhypothese 83a–14
Gezeiteninstabilität 96–107
Kriterium der ~ 98–312
Gezeitenkraft 82–289
Gezeitenniedrigwasser 82–283
Gezeitenphänomen 82–276
Gezeitenpotential 82–293
Gezeitenreibung 82–312
Gezeitenschwingungen 76–28
erzwungene ~ 82–306
Gezeitenstörung 98–314
Gezeitenstrom 82–279
Gezeitenströmung 82–279
Gezeitentheorie 82–305
Gezeitenwärme 82–287
Gezeitenwechselwirkung 81–120
Gezeitenwelle 82–291
Gezeitenwellenschaden 72–77
Gezeitenwirkung 83a–15
Gezeitenwoge 82–291
Gezeitenwulst 81–119
Gezeitenzerplatzen 85–118
Gezeitenzerstückelung 85–118
Giacobiniden 86–252
Gipfelkrater 73–129
Gipfelmulde 81–244
~ des Zentralkegels 81–244
Girafe 92–24
Gitterinterferometer 88–123
Glas, geschwärztes 88–33
glasartiges Fragment 82–97
Glashauseffekt 83–263
~ eines Planeten 83–109
Glashausmodell 83–262
glasiges Bruchstück 82–97
glazial 72–106
Glazialgeologie 72–103
Glazialperiode 72–105
Glaziologie 72–103
Gleichgewicht, adiabatisches 76–21
~ des Elementarprozesses 93–41
detailiertes ~ 93–41
hydrostatisches ~ 93–75
isostatisches ~ 73–99
isothermes ~ 76–20
konvektives ~ 76–22
lokales thermodynamisches ~ 93–39
LTE ~ 93–39
nicht-LTE ~ 93–43
nukleares ~ 94–103
statistisches ~ 93–42
thermisches ~ 93–78
thermodynamisches ~ 93–35
verallgemeinertes ~ 93–43
Gleichgewichtsbedingungen 93–74
Gleichgewichtsfigur 83–112
Gleichgewichtsform 83–112
Gleichgewichtsgestalt 83–112
Gleichgewichtsprozess 94–99
Gleichgewichtsverhältnis 94–104
Gleichung, Abelsche 88–235
hydrostatische ~ 93–76
Pogsonsche ~ 91–58
Gleichungen des statistischen Gleichgewichtes 93–44
gleichzeitiger Flare 89–179
Gleichzeitigkeit 99–127
Gleitbahn 82–197
Gleiten 82–198
Gletscher 72–112
alpiner ~ 72–114
Gletscherabfluss 72–120
Gletscherablagerungen 72–127
Gletscherfliessen 72–120
Gletscherkunde 72–103
Gletschernährungsgebiet 72–122
Gletscherwall 72–128
Gletscherzunge 72–125
Glied der differentiellen Bewegung 98–115
~ der galaktischen Rotation 98–116
kosmologisches ~ 99–156a
Glimmer 74–23
dunkler ~ 74–25

weisser ~ 74–24
Glimmerstreifen 88–87
Globalstrahlung 77–55
Globule 97–133
Glorie 79–63
Gluckhenne 96–110
Golf 81–294
Gotlandium 74–80
Graben 81–165
ozeanischer ~ 72–40
Grabstichel 92–23
Gradient 93–94
absoluter ~ 91–111
spektrophotometrischer ~ 91–112
thermischer ~ 74–59
überadiabatischer ~ 93–99
Gradientwind 77–94
Gradmessung 71–22
Granit 74–8
Granulation 88–176
photosphärische ~ 88–176
Granule 88–178
photosphärische ~ 88–181
Granulum 88–178
photosphärisches ~ 88–181
Graphitkern 97–71
Graphitpartikel 97–75
Gravimeter 71–70
Gravimetrie 71–63
gravimetrische Karte 81–143
gravimetrischer Netz 71–80
Gravitationsenergie 93–119
Gravitationsfeld 98–145
galaktisches ~ 98–146
Gravitationsgezeiten 82–298
Gravitationsinstabilität 94–35
Gravitationskollaps 94–109
Gravitationskontraktion 93–118
Gravitationslinse 98–256a
Gravitationsparadoxon 99–39
Gravitationspotential 96–160a
Gravitationsquerschnitt 84–104
Gravitationsradius 94–113
Gravitationsrotverschiebung 99–162
Gravitationsstrahlug 99–165
Gravitationsverdunk(e)lung 91–41
Gravitationsverschiebung 99–162
~ nach Rot 99–162
Gravitationswelle 99–164
Gravitationszusammenschrumpfung 83–203
Grenzbreite, nördliche 81–105
südliche ~ 81–106
Grenzbreiten der Sichtbarkeit einer Bedeckung 81–104
Grenze der Finsternis 88–20
nördliche ~ der Finsternis 88–20
obere ~ 76–58
obere ~ einer Schicht 76–58
Rochesche ~ 96–28
südliche ~ der Finsternis 88–19
Grenzentfernung 86–289
Grenzfläche Kruste-Mantel 72–14
Grenzgrösse 91–62
Grenzhelligkeit 91–62
Grenzkurve der Sekundärkomponente 96–40
~ des Begleiters 96–40
Grenzmasse, Chandrasekharsche 93–189
grobkörnig 82–117
gross-ausgedehnte Turbulenz 83a–12
Grösse der Finsternis 81–79
Grosse Magellansche Wolke 98–213
~ Südliche Störung 83–194
Grössenhäufigkeit 81–249
Grössenklasse 91–56
absolute ~ 86–72
absolute ~ eines Meteors 86–72
absolute ~ eines Planeten 83–86
mittlere ~ 83–85
scheinbare ~ eines Meteors 86–57
Grössenklassendifferenz 91–59
Grössenklassensystem 91–57
Grössenverteilung 82–124
Grosser Bär 92–96
~ Hund 92–27
~ Orionnebel 97–136
~ Ring 79–70
~ Roter Fleck 83–193
~ Zusammensturz 99–182
Grossion 76–115
grossmasstäbliches Feld 89–142
grossskaliges Feld 89–142
Grosswetterlage 77–15
Grundgebirge 72–12
Grundmasse 82–118
feinkörnige ~ 82–115
Grundmoräne 72–129
Grundtensor 99–157a
Gruppe, bipolare 89–111
~ des alten Zyklus 89–107
~ des neuen Zyklus 89–106
~ des vorigen Zyklus 89–107
~ kleiner Planeten 84–90
komplexe ~ 89–112
rekurrente ~ 89–108
unipolare ~ 89–110
wiederkehrende ~ 89–108
Gruppenparallaxe 91–7
Gruppenradiant 86–54
Gruppenverzögerung 78–56
Grus 92–52
lunarer ~ 82–112
Gruskorn 82–110
Guillotinefaktor 93–103
Gürtel, aktiver 73–32
~ des Orions 92–129a
~ geringer Polarhäufigkeit 79–129
~ nördlich der Südlichtzone 79–128
seismischer ~ 73–30
~ südlich der Nordlichtzone 79–128
Guyot 72–44

H

H I–Gebiet 97–49
H I-Region 97–49
H I-Wolke 97–103
H II–Gebiet 97–50
H II-Region 97–50
Haar der Berenike 92–38
Haarstern 85–1
Hadronenära 99–174
Haftplatte 87–14
Hagel 77–178
Haken im Magnetogramm 75–79
hakenartige Aufwölbung 75–79
~ Ausbuchtung 75–79
Halbkugel, sichtbare 89–27
Halbmesser, äusserer 83–228
innerer ~ 83–226
mittlerer ~ 83–116a
scheinbarer ~ 81–69
Halbmond 81–26
Halbregelmässiger 95–105
~ Veränderlicher 95–105
Halbschatten 81–64
Halbschattenfinsternis 81–65
halbtägige Welle 76–29
halbunendliche Atmosphäre 93–22
Halesches Gesetz 89–154
Halo 79–67
galaktischer ~ 98–45
heisser ~ 98–46
kühler ~ 98–47
magnetischer ~ 98–48
Haloerscheinung 79–66
Halokreuz 79–79
Halophase 94–19
Halopopulation 98–69
Halostadium 94–19
Halostern 91–84
Halskuffe 73–125
Hamal 92–118
harmonische Koeffizienten 71–59
Harogalaxie 98–266
Harvard-Klassifikation 91–91
Harvard-System der Spektralklassen 91–91
Hase 92–61
Hauch 89–277
Haufen, galaxienarmer 98–331
galaxienreicher ~ 98–332
sternarmer ~ 96–102
Haufengalaxie 98–338
hellste ~ 98–339
Haufenmitglied 96–82
Haufenstern 96–82
Haufenveränderlicher 95–93
Haufenwolke 77–138
hohe ~ 77–144
Häufigkeit, anomale 91–173
~ chemischer Elemente 82–141
~ der Flares 89–217
~ der Isotope 94–106

D

~ eines Elementes 91–169
kosmische ~ der Elemente 94–107
natürliche ~ 91–171
normale ~ 91–172
relative ~ von Elementen 91–174
stündliche ~ 86–135
totale ~ 91–177
~ von Meteoren 86–135
Häufigkeitsanomalien 91–183
Häufigkeitskurve 94–108
Häufigkeitsverhältnis 91–175
~ von Wasserstoff zu den Metallen 91–176
Hauptbeanspruchung 72–64
Hauptbogen 73–82
Hauptebene, galaktische 98–53
Hauptfleck 89–115
Hauptkomponente 96–37
Hauptkrater 73–127
Hauptmasse 87–41
Hauptmaximum 95–31
Hauptmeteorit 87–41
Hauptminimum 95–26
Hauptperiode 95–40
Hauptphase 73–12
Hauptplaneten 83–13
Hauptpuls 95–144
Hauptpulsation 95–144
Hauptreihe 91–131
Alter-Null-~ 96–152
~ für das Alter Null 96–152
Nullalter-~ 94–60
Stern der ~ 91–142
Hauptreihenstern 91–142
Hauptschweif 85–77
Hauptsequenz 91–131
Hauptspektrum 95–115
Hauptstern 96–37
Hauptstress 73–64
Haupttriangulation 71–24
Hayashi-Phase 94–44
Heaviside-Kennelly-Schicht 78–26
Heaviside-Schicht 78–26
heisse Stelle 95–112a
Helio- 88–5
Heliogeophysik 84–56
heliogeophysikalische Effekte 84–55
Heliogramm, Röntgenstrahlen- 88–74
Heliograph 77–61
monochromatischer ~ 88–38
heliographische Koordinaten 89–39
Heliometer 88–12
Helioseismologie 88–223b
Heliosphäre 88–255
heliosphärisch 88–256
heliosphärischer Hohlraum 88–257
Heliostat 88–45
heliozentrischer Winkel 89–37
heliozentrisches Weltsystem 99–23
heliumangereichertes Plasma 84–60
Heliumbrenen 94–78
explosives ~ 94–79
Heliumflash 94–79
Heliumkern 93–167
Heliummethode 87–128
~ der Altersbestimmung 87–128
Heliumschale 93–145
Heliumzündung 94–71
helle mottles 88–211
~ Nacht 79–59
~ Seite 83–67
Hellglimmer 74–24
Helligkeit 91–54
absolute ~ 91–65
absolute ~ eines Kometen 85–109
bolometrische ~ 91–68
~ der Eruption 89–212
~ eines Meteors 86–8
infrarote ~ 91–71
integrale ~ 96–135
maximale ~ 86–78
Maximum der ~ 86–78
mittlere ~ 83–85
photographische ~ 91–66
photovisuelle ~ 91–69
reduzierte ~ 85–108
scheinbare ~ 91–61
visuelle ~ 91–65
Helligkeitsabnahme, säkulare 85–127
Helligkeitsamplitude 95–38
Helligkeitsänderung 95–7
Helligkeitsanstieg 86–79
Helligkeitsausbruch einer Nova 95–121
~ eines Kometen 85–126
Helligkeitsdifferenz 96–65
Helligkeitsfaktor 83–90
Helligkeitskoeffizient 83–96
Helligkeitskurve 86–70
~ eines Meteors 86–70
Helligkeitsschätzung 95–11
Helligkeitsunterschied 96–65
Helligkeitsverteilung 86–166
Hemisphäre, abgewandte 81–16
beleuchtete ~ 83–67
dunkle ~ 83–68
erdabgewandte ~ 81–16
erdzugewandte ~ 81–15
sichtbare ~ 89–27
sonnenbeschienene ~ 75–20
unbeleuchtete ~ 75–21
zugewandte ~ 81–15
Hemmungspunkt 86–123
Hercules 92–53
Herd des Erdbebens 73–17
Herddistanz 73–18
Herdentfernung 73–18
Herdtiefe 73–20
Herkules 92–53
Herkunft 94–9
~ der Kometen 85–96
Hermes 84–83
Hertzsprung-Lücke 91–144
Hertzsprung-Russell-Diagramm 91–130
Heterosphäre 76–61
Hexaedrit 87–84
hierarchische Struktur 99–47
~ Welt 99–48
Himmel, bedeckter 77–152
bewölkter ~ 77–148
heiterer ~ 77–153
klarer ~ 77–153
Himmelsatlas 92–12
Himmelsblau 77–60
Himmelsglobus 92–7
Himmelsjäger 92–73
Himmelskarte 92–9
Himmelskörper 79–9a
Himmelskugel 92–8
Himmelsleuchten 79–2
Himmelslicht 77–53
Himmelssphäre 92–8
Himmelsstrahlung 77–54
diffuse ~ 77–59
Himmelszustand 77–132
Hinterdeck 92–81
Hintergrund, meteorischer 86–232
seismischer ~ 73–55
Hintergrunddichte 96–131
Hintergrundfeld 89–141
Hintergrundgalaxie 98–340
Hintergrund-Magnetfeld 89–141
Hintergrundstern 96–84
Hinterteil des Schiffes 92–81
hiss 78–64
Hitzdrahtmikrophon 76–46
Hoch 77–84
Hochdruckgebiet 77–84
Hochdruckkeil 77–86
Hochdruckrücken 77–87
Hochebene 81–192
hochentartet 93–184
Hochgebirgsgletscher 72–114
hochionisiert 89–289
Hochland 81–189
Hochlandoberfläche 82–56
höchste brauchbare Übertragungsfrequenz 78–37
höckeriges Material 82–103
Höhe, äquivalente 78–28
~ der Chromosphäre 88–202
~ der Emissionsschicht 79–26
~ der homogenen Atmosphäre 76–17a
~ der Meteore 86–196
~ des Anfangspunktes 86–39
~ des Apex 86–66
~ des Aufleuchtens 86–39
~ des Endpunktes 86–40
~ des Erlöschens 86–40
~ des Spiegelpunktes 75–125
~ einer Druckfläche 76–16
ellipsoidische ~ 71–52
scheinbare ~ 78–28
~ über dem Meeresniveau 71–51
wahre ~ 78–29
Höhenformel, barometrische 76–15

Höhengleiche 81–139
Höhengleichung 86–194
Höhenkarte 81–141
Höhenklima 77–187
Höhenlinie 71–53
∼ unter Wasser 72–32
Höhenlinienkarte 81–141
Höhenniveau 81–142
Höhenobservatorium 88–56
Höhenprofil 76–53
∼ der Dichte 76–53
∼ der Temperatur 76–53
Höhenschichtenkarte 81–141
Höhenschichtlinie 71–53
Höhenstation 88–56
Höhenstufe, barometrische 76–19
Höhenunterschied 81–140
Hohlraum, geomagnetischer 75–8
heliosphärischer ∼ 88–257
hydromagnetischer ∼ 84–22
interplanetarer ∼ 84–25
magnetischer ∼ 84–22
planetarer ∼ 84–24
Sonnenwind- ∼ 84–28
homogenes Weltall 99–64
homolog 93–131
Homopause 76–65
Homosphäre 76–60
Hopentfernung 78–38
Hörbarkeitsgebiet 87–32
Hörbarkeitszone 76–41
Horizont des Weltalls 99–10
kosmologischer ∼ 99–179
Horizontalast 91–132
∼-Stern 91–141
Horizontalbewegung 88–184
Horizontalcoelostat 88–47
Horizontalintensität 75–60
∼ des Magnetfeldes 75–60
Horizontalkomponente 75–60
Horizontalkreis 79–82
Horizontalpendel 71–73
Horizontalring 79–82
Horizontalspektrograph 88–78
Horizontalzölostat 88–47
Horn der Magnetosphäre 75–13
Hörner 83–130
Hörnerspitzen 83–130
∼ des Mondes 81–24
Übergreifen der Venus ∼ 83–131
Horologium 92–54
Hörweite 87–32
HR-Diagramm 91–130
Hubble-Gesetz 98–299
Hubble-Konstante 98–302
Hubble-Sequenz 98–227
Hubble-Zeit 99–79
Hufeisennebel 97–143
Hülle, ausdehnende 93–12
ausgedehnte ∼ 93–8
∼ des Kometenkerns 85–64
expandierende ∼ 93–12
gemeinsame ∼ 96–32
gemeinsame ∼ eines Doppelsterns 96–32
∼ im Strahlungsgleichgewicht 93–135
konvektive ∼ 93–136
stationäre ∼ 93–11
zirkumstellare ∼ 93–10
Hüllenstern 93–5
Humboldt-Meer 81–288
Hundstern 92–128
Hyaden 96–111a
Hydra 92–55
Hydrodynamik, stellare 98–142
hydrodynamische Phase 94–43
hydromagnetischer Oszillator 95–61
hydrophil 76–92
hydrophob 76–93
Hydrosol 76–90
Hydrosphäre 72–21
Hydroxyl 79–25
Hydroxylgruppe 97–22
Hydrus 92–56
Hygrometer 77–120
hyperbolischer Komet 85–8
∼ Raum 99–106
Hyperonengas 93–193
Hyperonenstern 91–154
Hypothese 94–6
kosmogonische ∼ 94–8
Lambertsche ∼ 99–45
Hypozentraldistanz 73–18
Hypozentrum 73–17
hypsographische Kurve 81–142

I

Identitäten, Lagrangesche 98–182
verallgemeinerte Lagrangesche ∼ 98–183
Ikarus 84–84
Ilmenit 82–91
Importanz einer Eruption 89–205
∼ eines Flares 89–205
Impulsbreite 86–280
Impulsdauer 86–280
Impulsfrequenz 86–279
Impulslänge 86–280
Impulsmoment 83a–4
Impulssignale 86–278
inaktiver Komet 85–147
Inaktivierung eines Kometen 85–148
Inder 92–57
Index der Polytrope 93–128
∼ der Sonnenaktivität 89–8
∼ der Stromaktivität 86–230
K-∼ 75–89
K-∼der geomagnetischen Aktivität 75–89
planetarischer K_p-∼ 75–90
indirektes Sehen 86–128
individueller Radiant 86–55
Indus 92–57
induzierte Magnetisierung 75–29
Inertialsystem 99–98
Inflation 99–170c
kosmologische ∼ 99–170c
inflationäre Ära 99–170d
Inflationsära 99–170d
infrarote Galaxie 98–205
Infrarotgalaxie 98–205
Infrarotstern 91–157
Infrarot-Überschuss 91–79
Infraschallwellen 75–88
inhomogenes Weltall 99–66
Inhomogenität, geneigte 88–190
ionosphärische ∼ 78–7
makroskopische ∼ 88–206
Inklination, erdmagnetische 75–58
magnetische ∼ 75–58
Pol magnetischer ∼ 75–45
Inlandgletscher 72–116
innenbürtig 82–259
Innenhalbmesser 83–226
Innenkern 72–20
Innenring 83–221
innere Korona 88–229
innerer Planet 83–14
Inneres der Erde 72–3
∼ der Sterne 93–72
Insel, neue 72–81
Insolation 77–51
instabil 93–210
Instabilität 93–211
konvektive ∼ 93–155
lokale ∼ 93–154
säkulare ∼ 93–212
vibrationelle ∼ 93–214
Integralaufnahme 88–37
integrale Magnitude 96–136
Integralexponentialfunktion 93–62
Integralgleichung, Schwarzschild-Milnesche 93–53
Integralgrösse 96–136
Integralhelligkeit 96–135
Integrallichteruption 89–182
Intergralmasse des Haufens 96–139
Integralphotometrie 83–103
integrierte Strahlung 88–147
Integro-Differentialgleichung 93–52
Intensität des erdmagnetischen Feldes 75–55
∼ eines Erdbebens 73–33
Intensitätsklasse 89–205
Intensitätsverhältnis 91–60
interaktive Protuberanzen 89–243
Interferenzfigur 88–120
Interferenzmuster 88–120
Interferometer, Fächerkeulen- 88–122
Gitter-∼ 88–123
Kliff-∼ 88–106
Langbasis- 88–104
∼ mit Phasenumschaltung 88–113
∼ mit Radioverbindung 88–116

D

~ mit schneller Frequenzänderung 88–110
~ mit schwenkbarer Richtkeule 88–109
~ mit veränderlicher Basis 88–107
~ mit veränderlicher Basislänge 88–107
scharfbündelndes ~ 88–121
starkbündelndes ~ 88–121
Swept-Lobe- ~ 88–109
Vielelement- ~ 88–123
Interferometerbasis 88–100
interfilamentarer Raum 89–99
Interfilamentraum 89–99
intergalaktisch 97–6
intergalaktische Gaswolke 98–354
~ Lichtbrücke 98–348
~ Materiebrücke 98–348
intergalaktisches Gas zwischen Haufen 98–353
interglazial 72–108
Interglazialzeit 72–107
intergranulares Gebiet 88–177
interkontinentales Meer 72–26
Internationale Union für Geodäsie und Geophysik 71–4
interplanetar 84–8
interspikular 88–198
interstellar 97–2
intertropischer Bogen 79–53
intramerkurieller Planet 83–123
Intrusion 73–133
~ der eruptiven Gesteine 82–98
diskordante ~ 73–137
konkordante ~ 73–134
magmatische ~ 82–98
transgressive ~ 73–137
Intrusivgang 73–138
Intrusivgestein 74–7
Intrusivlager 73–135
invariable Ebene 83–232
Invarianz gegenüber einer Transformation 99–133
Invarianzprinzip 93–61
Inversion, atmosphärische 77–37
Ionendiffusion 86–286
Ionenschweif 85–82
Ionenzusammensetzung 78–10
Ionisation, meteorische 86–281
Ionisationsfront 97–48
Ionisationsschlauch 86–282
Ionisationstemperatur 91–117
ionisierte Schicht 78–3
Ionisierungsausbeute 86–209
Ionogramm 78–23
Ionosonde 78–22
Ionosphäre 78–1
äussere ~ 76–76
Ionosphärenabtaster 78–22
Ionosphärenmodell 83–261
Ionosphären-Observatorium 78–21
Ionosphärenregion 78–4
Ionosphärenschicht 78–5
Ionosphären-Sondierung 78–19
Ionosphärenstörung 78–39
plötzliche ~ 78–41
Ionosphärenstrom im Polarlichtbreiten 79–113
Ionosphärensturm 78–49
Ionosphären-Vorhersage 78–32
ionosphärisch 78–2
ionosphärische Finsternis 88–23
irdischer Meteoritenkrater 87–54
~ Ursprung 87–74
irisierende Wolke 79–90
Irregularität des Magnetfeldes 84–12
kleinräumige ~ 84–13
Irrfahrt 89–157
zufällige ~ 89–157
Isobathe 71–55
Isodyname 75–61
Isogauss-Kontur 89–134
Isogauss-Linie 89–134
Isogone 75–57
Isohypse 81–139
Isokline 75–59
Isolinie magnetischer Deklination 75–57
~ magnetischer Inklination 75–59
Isophote 81–92
Isophotenlinie 81–92
Isoplethe 96–125
Isorotation 88–130
Isostasie 73–98
isostatischer Ausgleich 73–100
Isotop, durch kosmische Strahlung erzeugtes 82–147
kosmogenes ~ 82–147
kurzlebiges ~ 83a–74
langlebiges ~ 83a–75
primäres ~ 82–146
Isotopenhäufigkeit 94–106
Isotopenverhältnis 74–94
Isotopenzusammensetzung 74–93
Isotopie-Effekt 91–178
isotropes Weltall 99–63
Isozahl-Konturen 75–102
Isozyansäure 97–39
Iterationsmethode 93–56

J

Jagdhunde 92–26
Jahr der ruhigen Sonne 89–10
jahreszeitliche Änderungen 83–166
~ Variation 86–139
jährliche Variation 86–138
jährlicher Niederschlag 87–17
Jakobsstab 92–129a
Jet 98–278
optischer ~ 98–279
Jo 83–205
jovigraphische Koordinaten 83–175
jovizentrische Koordinaten 83–1765
Jungfrau 92–99
Juno 84–85
Jupiter 83–170
jupiterähnlicher Planet 83–174
Jupiterferne 83–173
Jupiternähe 83–172
Jura 74–85

K

Käfigeinschlussverbindung 85–51
Kaldera 81–179
Kalifeldspat 74–16
Kalmengürtel 77–101
Kaltfront 77–70
Kalziumflocculi 89–170
Kalziumflocken 89–170
Kamazit 87–92
Kambrium 74–78
Kamera, feste 86–143
nachgeführte ~ 86–142
nichtnachgeführte ~ 86–143
unbewegliche ~ 86–143
Kanopus 92–109
Känozoikum 74–73
känozoische Ära 74–73
kantiges Gestein 82–131
Kap 81–293
Kapella 92–110
Kapteynsche Eichfelder 98–78
~ Felder 98–78
Karbon 74–82
Karbonat 74–27
Karbonylsulfid 97–36
Karte der Chromosphäre 89–21
gravimetrische ~ 81–143
heliographische ~ 89–17
~ hoher Auflösung 89–133
hydrologische ~ 72–25
kolorimetrische ~ 81–134
magnetische ~ 75–63, 89–131
ozeanographische ~ 72–23
photometrische ~ 81–134
polarimetrische ~ 81–134
synoptische ~ 77–11
synoptische ~ der Chromosphäre 89–21
synoptische ~ der Korona 89–22
synoptische ~ der Photosphäre 89–20
tektonische ~ 73–96
Kartenmasstab 71–49
Kartenprojektion 71–40
Kartierung, thermische 72–83
Kartographie, selenologische 82–51
kartographieren das Magnetfeld 89–132
Kartographierung des Mondes 81–111
kartographische Abbildung 71–40
Kassiopeia 92–31
Kastor 92–111
katabatisch 77–93

kataklysmicher Veränderlicher 95–109
Katalog, Messier- 92–12a
~ vermuteter veränderlicher Sterne 95–58
Katastrophe, seismische 73–40
Kathodenstrahlröhre 86–293
Kaulquappen-Galaxie 98–281
Kegel, pyroklastischer 73–120
reflektierender ~ 88–62
vulkanischer ~ 73–118
Kegelprojektion 71–42
Kennzeichen, vulkanisches 82–159
Kennziffer, dreistündliche 75–89
planetarische ~ 75–90
Kentaur 92–32
Kepheus 92–33
Keplersche Rotation 98–112
Kern, aktiver galaktischer 98–272
äusserer ~ 72–19
~ der Assoziation 96–91
~ der Erde 72–17
~ der Galaxis 98–18
~ des Kometen 85–36
~ des planetarischen Nebels 97–148
~ des Sternes 93–164
entarteter ~ 93–170
fester ~ 85–39
galaktischer ~ 98–18
inhomogener ~ 93–169
isothermer ~ 93–168
konvektiver ~ 93–165
nichtentarteter ~ 93–171
pekuliarer ~ 98–20
radiativer ~ 93–166
sternförmiger ~ 85–38
supraleitender ~ 95–150
Kernaktivität einer Galaxie 98–271
Kernaufbau 94–52
explosiver ~ 94–57
nichtexplosiver ~ 94–58
Kernbereich 98–17
Kernen 72–51
Kernenergie 93–122
Kernfragment 94–102
Kerngebiet 98–17
Kerngewinnung 72–51
Kernimplosion 94–110
Kernkontraktion 94–70
Kernprobe 82–60
Kernreaktion an der Sternoberfläche 91–181
Kernschatten 81–76
Eintritt des Mondes in den ~ 81–76
Kernsplitterung durch kosmische Strahlung 82–254
Kernspur, fossile 82–252
Kernverdickung 98–16
Kernzersplitterung 82–254
~ durch kosmische Strahlung 82–254
Kette von Supernovae 95–132
Kiel des Schiffes 92–30
kinematische Kosmologie 99–26
Kirkwood-Lücken 84–89
K-Korona 88–242
Klamm 83–154
Klassifikation der Sternspektren 91–90
~ der Veränderlichen 95–46
Harvard- ~ 91–91
magnetische ~ 89–109
Züricher ~ 89–127
Klassifikationskriterium 91–110
Klassifizierung der Meteoriten 87–59
Klassifizierungskriterien 91–110
klassisches Weltall 99–93
Kleine Magellansche Wolke 98–214
Kleiner Bär 92–97
~ Hund 92–28
~ Löwe 92–60
~ Planet 84–68
~ Ring 79–69
Kleines Pferd 92–48
kleinmasstäbliches (Magnet)feld 89–143
kleinräumige Irregularität 84–13
~ Ungeordnetheit 84–13
Kliff-Interferometer 88–106
Klima 77–180
kontinentales ~ 77–185
maritimes ~ 77–185
Klimaänderung 77–184
Klimaschwankung 77–183
klimatisch 77–182
Klimatologie 77–179
Kluft 73–78
Klumpung der Materie 99–170z
Knoten 85–95
Koagulation 76–119
Koaleszenz 76–120
Koeffizienten des Geschwindigkeitsellipsoids 98–102
harmonische ~ 71–59
Koerzitivfeld 75–39
Koerzitivkraft 75–39
Kogeoid 71–17
Kohärenz 99–33
~-Postulat 99–34
Kohäsionsfestigkeit 85–117
kohäsiv 82–138
Kohlendioxid, festes 83–144
Kohlenmonoxid 97–28
Kohlensack 97–138
Kohlensäureschnee 83–144
Kohlenstoff, radioaktiver 74–95
kohlenstoffarmer Stern 91–194
Kohlenstoffbrennen 94–81
Kohlenstoffmaterie 87–64
Kohlenstoffmeteorit 87–62
Kohlenstoffmonosulfid 97–34
Kohlenstoffstern 91–188
Kohlenstoffzündung 94–80
kohliger Chondrit 87–63
kollabiertes Objekt 98–23
Kollaps 94–109
kollektive Wechselwirkung 98–168
Kollektor des meteorischen Staubes 87–13
kollidierende Galaxien 98–321
Kollisionszersplitterung 87–125
Kolloid 76–96
kolloidal 76–97
Kolloidteilchen 76–98
Koma 85–64
atomare ~ 85–67
innere ~ 85–65
molekulare ~ 85–65
sichtbare ~ 85–66
Komet 85–1
aktiver ~ 85–149
die Sonnenatmosphäre berührender ~ 85–13
elliptischer ~ 85–4
erloschener ~ 85–150
erzeugender ~ 86–259
hyperbolischer ~ 85–8
inaktiver ~ 85–147
künstlicher ~ 85–14
kurzperiodischer ~ 85–5
langperiodischer ~ 85–6
~ mit elliptischer Bahn 85–4
~ mit hyperbolischer Bahn 85–8
~ mit parabolischer Bahn 85–7
nichtaktiver ~ 85–147
nichtperiodischer ~ 85–10
parabolischer ~ 85–7
periodischer ~ 85–3
teleskopischer ~ 85–14a
vorausberechneter ~ 85–11
kometarisch 85–2
kometarischer Meteor 86–45
kometenähnlicher Nebel 97–142
Kometenaktivität 85–110
Kometenaufbau 85–34
Kometenaufspaltung 85–111
Kometeneinfang 85–101
Kometenentdeckung 85–16
Kometenenveloppen 85–68
Kometenfamilie 85–102
~ des Jupiters 85–103
Kometenform 85–83
mechanische Theorie der ~ 85–84
Kometengruppe 85–105
Kometenhalos 85–68
Kometenkern 85–36
Kometenkopf 85–35
Kometenmeteor 86–45
Kometenschweif 85–70
Kometenschweiftypus 85–71
Kometenstoss 82–230
Kometensuche 85–15
Kometentätigkeit 85–110
Kometenwolke 85–100
Oortsche ~ 85–97
Kommando-Modul 82–6
Kommensurabilitätsasteroidengruppe 84–101
Kommensurabilitätsgruppe 84–101

D

Kommensurabilitätslücke 84–88
Kompaktation 74–50
Kompaktgalaxie 98–264
blaue ~ 98–265
Kompensation für Doppler-Verschiebungen 88–90
isostatische ~ 73–100
Kompensationsplatte 88–92
Kompensationstiefe 73–102
kompensiertes Geoid 71–17
Komplanarität der Bahnen 83a–3
Komponente 96–36
atmosphärische ~ 83–265
meteorische ~des Nachthimmelleuchtens 86–234
mineralogische ~ 82–100
σ-~ 88–89
vertikale ~ 75–62
Kompressibilität 73–57
Kompressionskappe 86–170
Kompressionswelle 73–46
Kondensation 83a–26
koronale ~ 89–281
permanente koronale ~ 89–282
sporadische koronale ~ 89–283
Kondensationskern 76–103
Kondensationsniveau 77–156
Kondensationstemperatur 97–66
Kondensationswärme 77–155
kondensieren 77–129
Konfiguration der Planeten 83–50
konforme Projektion 71–41
Konglomerat 85–41
Konjunktion 83–55
dreifache ~ 83–58
obere ~ 83–56
untere ~ 83–57
konkordante Intrusion 73–134
konstant, zeitlich 98–150
Konstante, Hubblesche 99–77
kosmische ~ 99–29
kosmologische ~ 99–29
Poisson-~ 73–60
Konstellation 83–50
Planeten- ~ 83–50
Kontakt, erster 81–84
letzter ~ 81–85
Kontaktdoppelstern 96–29
Kontaktfläche 96–27
Kontaktmetamorphose 74–35
Kontinent 72–9, 83–159
Kontinentalblock 72–10
Kontinentalböschung 72–35
Kontinentaldrift 73–104
Kontinentalfuss 72–36
Kontinentalgletscher 72–116
Kontinentalhang 72–35
Kontinentalkruste 72–5
Kontinentalschelf 72–34
kontinuierliche Erzeugung 99–91
Kontinuitätsgleichung 93–49
kontrahierender Kosmos 99–56
Kontraktion 93–117
~ des Weltalls 99–180
Lorentzsche ~ 99–134
quasistationäre ~ 94–40
quasistatische ~ 94–40
Kontraktionsfaktor 99–135
Kontraktionsstadium 94–20
Kontraktionstheorie 73–77
Kontrollzahl 85–30
Kontur gleicher Beleuchtungsstärke 97–124
Isogauss- ~ 89–134
Konvektion 93–147
laminare ~ 93–153
turbulente ~ 93–152
Konvektionsgleichgewicht 93–79
Konvektionsstrom 73–76
Konvektionszone 93–149
konvektiv 93–148
Konvergenzpunkt 96–98
Konzentration 96–127
galaktische ~ 98–76
galaktische ~ zum Zentrum 98–76
galaktische ~ zur Hauptebene 98–76
~ zur galaktischen Ebene 98–55
Konzentrationsgrad 96–128
Konzentrationsindex 96–101
konzentrischer Wall 81–254
Koordinaten, galaktozentrische 98–12
geomagnetische ~ 75–51
heliographische ~ 89–39
jovigraphische ~ 83–175
jovizentrische ~ 83–176
planetographische ~ 83–8
selenographische ~ 81–9
zeitartige ~ 99–123
zylindrische ~ 98–156
Koordinatensysteme, mitbewegte 99–44
mitschwimmende ~ 99–44
Kopernikanisches System 99–24
Kopf der Schlange 92–89
~ eines Kometen 85–35
~ und Schwanz 92–89
~ und Schwanz der Schlange 92–89
Kopfteil des Streufeldes 87–39
Koplanarität der Bahnen 83a–3
Korngrösse 82–113
von grober ~ 82–117
von mittlerer ~ 82–116
Korona 79–60, 79–156
äussere ~ 88–231
E- ~ 88—241a
expandierende ~ 88–232
F-~ 88–241
galaktische ~ 98–49
innere ~ 88–229
K-~ 88–242
mittlere ~ 88–230
monochromatische ~ 88–238
weisse ~ 88–237
X-Strahlen-~ 88–240
Koronabogen 89–298
Koronagebiet 89–286
aktives ~ 89–286
grünes ~ 89–284
rotes ~ 89–285
Koronakondensation 89–281
permanente ~ 89–282
sporadische ~ 89–283
koronal 88–228
koronale aktive Protuberanz 89–247
koronaler Transient 89–185a
Koronalinie 89–290
gelbe ~ 89–292
grüne ~ 89–291
rote ~ 89–293
Koronaregen 89–271
Koronaregion, aktive 89–286
grüne ~ 89–284
rote ~ 89–285
Koronastrahl 89–287
monochromatischer ~ 89–288
Koronograph 88–59
~ mit äusserer Abblendung 88–60
~ mit äusserer Abdeckscheibe 88–60
~ mit äusserer Abdeckung 88–60
~ mit äusserer Blende 88–60
Koronometer 88–66
K-~ 88–66
Körper, aufschlagender 82–225
aufstürzender ~ 82–225
fester ~ 83–110
Gezeiten verursachender ~ 82–284
meteorischer ~ 86–160
prästellarer ~ 94–15
rotierender ~ 91–42
rotierender magnetoplasmatischer ~ 98–25
schnell rotierender ~ 91–43
starrer ~ 83–111
supermassiver ~ 98–22
Korpuskel 88–153
korpuskular 88–155
Korpuskularemission 88–152
Korpuskularstrahlung 88–150
Korpuskularstrom 88–160
Korrektur für die Krümmung 86–62
kosmischen Ursprungs 87–5
Kosmochemie 74–31
Kosmochronologie 83a–72
kosmogenes Radionuklid 82–255
Kosmogonie 94–2
~ der Planeten 83a–1
Fachmann für ~ 94–4
planetarische ~ 83a–1
kosmogonisch 94–3
Kosmogonist 94–4
Kosmologe 99–19
Kosmologie 99–16
beobachtende ~ 99–17
kinematische ~ 99–26
Newtonsche ~ 99–25
nichtrelativistische ~ 99–27
relativistische ~ 99–28

kosmologisch 99–18
kosmologische Paradoxa 99–38
kosmologisches Glied 99–156a
~ Modell 99–46
~ Objekt 98–369
~ Prinzip 99–30
Kosmos 99–2
anisotroper ~ 99–65
asymptotischer ~ 99–59
expandierender ~ 99–55
geschlossener ~ 99–73
homogener ~ 99–64
inhomogener ~ 99–66
isotroper ~ 99–63
kontrahierender ~ 99–56
leerer ~ 99–49
monotoner ~ 99–61
offener ~ 99–72
oszillierender ~ 99–57
oszillierender ~ erster Art 99–58
stationärer ~ 99–52
statischer ~ 99–50
Krabbennebel 97–134
Kraft, fluterzeugende ~ 72–92
innere ~ 73–61
repulsive ~ 85–87
tektonische ~ 73–106
Kraftröhre, magnetische 89–145
Kramerssches Gesetz 93–102
Kranich 92–52
Krater 73–126
alter ~ 81–205
äusserer ~ 81–218
junger ~ 81–206
kleiner ~ 81–250
konzentrischer ~ 81–216
kreisförmiger ~ 81–213
Maar-ähnlicher ~ 81–264
~ mit dunklem Halo 81–215
~ mit erhöhtem Rand 81–223
~ mit Rand aus Blöcken 81–225
~ mit Strahlen 81–214
~ mit Zentralberg 81–258a
möglicher ~ 81–211
post-mare ~ 81–208
prae-mare ~ 81–207
primärer ~ 81–198
randloser ~ 81–226
schüsselförmiger ~ 81–219
sekundärer ~ 81–199
tellerförmiger ~ 81–219
überfluteter ~ 81–203
unbekannter ~ 81–202
unregelmässiger ~ 81–220
vieleckiger ~ 81–212
wahrscheinlicher ~ 81–210
zugeschütteter ~ 81–204
Kraterabmessungen 81–227
Krateraussenwand 81–253
Kraterbildung 82–232
~ beim Einsturz 82–234
Kraterbildungsrate 82–233
Kraterboden 81–259
flacher ~ 81–260
Kraterdichte 81–246
Kraterfeld 87–56
Kratergrube 81–243
Kraterhäufigkeit 81–247
Kraterinnenwand 81–252
Kraterkette 81–241
Kraterparameter 81–227
Kraterrand 81–221
Kraterschlot 82–162
Kraterterrasse 81–229
Kraterwall 81–251
Kraterwand 81–251
Krebs 92–25
Krebsnebel 97–134
Krebsnebelpulsar 95–142
Kreide 74–86
kreisförmiges Meer 81–267
Kreuz des Südens 92–43
Millsches ~ 88–121
Krippe 96–111
Kristallisation 76–107
Kristallisationsdifferentiation 74–41
Kristallisationskerne 76–108
kritische Masse 93–189
~ Masse weisser Zwerge 93–189
kritisches Niveau 76–26
Krümmung der Erde 71–6
~ der Erdoberfläche 86–61
~ der Feldlinien 75–121
~ der Kraftlinien 75–121
~ des Raumes 99–115
Krümmungstensor 99–159
Kruste, feste 95–151
~-Mantel Grenzfläche 72–14
Krustenbeben 95–152
Krustenbewegungen 73–72
Krustenspannung 82–210
Kryptozoikum 74–70
K-Term 98–119
Kubikparsec = Kubikparsek 98–84
pro ~ 98–84
Kuckuckgestirn 96–110
Kugelblitz 77–163
Kügelchen 82–129
glasige ~ 82–127
Kugelfunktionen, tesserale 95–102
zonale ~ 93–55
Kugelhaufen 96–114
System der ~ 98–75
Kugelsternhaufen 96–114
Kugelsymmetrie 98–160
Kühlungszeit 83–201
kumulativer Effekt 98–124
Kumulus 77–138
künstlicher Asteroid 84–69a
Kuppel 81–177
kuppelartige Bodenform 81–177
kuppelloser Sonnenrefraktor 88–53
Kurskorrektur 82–10
Kurve, hypsographische 81–142
~ interstellarer Verfärbung 97–93
kurzdauernder Meteorschweif 86–112
kurzfristige Prognose 77–20
Kurzfristprognose 77–20
kurzlebig 89–125
Kurzperiodischer 95–89
kurzperiodischer Komet 85–5
~ Veränderlicher 95–89
Kurzwellentotalschwund 78–40
kurzzeitiges Meteorecho 86–300
Küste 72–30
Küstenaufnahme 72–82
Küstenkartierung 72–82
Küstenlinie 72–31
Küstenlinienveränderung 72–78
Küstenveränderung 72–78

L

Lacerta 92–58
Lage, heliographische 89–38
selenozentrische ~ 81–132
Lagergang 73–135
Lagerung 82–133
Lagrange-Punkt, erster 96–31
innerer ~ 96–31
Lagrangesche Identitäten 98–182
Lakkolith 73–136
laminare Konvektion 93–153
~ Struktur 83–148
Landbeben 73–2b
Landbrise 77–98
Landebein 82–22
Landefuss 82–23
Landefussdämpfer 82–23
Landeplatz 82–14
vorgesehener ~ 82–15
Landepunkt 82–14
Landevermessung, magnetische 75–64
Landklima 77–185
Landung 82–16
bemannte ~ 82–33
Landungsmoment 82–16
Landwind 77–98
Langbasis-Interferometer 88–104
langdauernder Meteorschweif 86–113
Länge, aktive 89–41
areographische ~ 83–139
~ der Meteorbahn 86–47
~ des Periastrons 96–59
erdmagnetische ~ 75–53
geomagnetische ~ 75–53
heliographische ~ 89–40
selenographische ~ 81–10
Längensystem 83–177
äquatoriales ~ 83–178
~ für mittlere Breiten 83–179
Längenverteilung 89–48
Langevin-Ion 76–115
langfristige Prognose 77–19
~ Vorhersage 77–19
Langfristprognose 77–19

D

D

langlebig 89–124
Langperiodischer 95–91
langperiodischer Komet 85–6
~ Veränderlicher 95–91
Langsamläufer 98–192
Längswelle 73–46
Lapilli 82–169
Laplacesche Ebene 83–232
Laplace-Transformation 93–54
Lappen, Rochescher 96–28
Laser-Reflektor 82–43
Laufzeit 73–51, 84–47
Lava, erkaltete 82–176
erstarrte ~ 82–176
Lavaausfluss 73–130
Lavablasen 82–178
Lavadecke 82–175
Lavadom 81–176a
Lavaebene 82–185
Lavaerguss 82–172
Lavaextrusion 82–171
Lavafeld 82–184
Lavaflut 82–173
Lavahöhle 82–177
Lavakuppe 81–176a
Lavaröhre 82–186
Lavasee 82–183
Lavastrom 73–140
Lavazunge 82–187
Lebensdauer 89–123
~ einer Fleckengruppe 89–123
~ eines Flares 89–209
~ eingefangener Partikel 75–122
Leere 99–171d
Legendre-Polynome 93–55
Legendresche Polynome 93–55
Leier 92–65
Leighton-Magnetoheliograph 88–93
Leitgeschiebe 72–132
Leo 92–59
~ Minor 92–60
Leoniden 86–255
Leptonenära 99–176
leuchtanregender Stern 97–125
Leuchten 79–151
Lα-~ des Nachthimmels 79–9
leuchtend mit Eigenlicht 96–45
leuchtende Nachtwolke 79–89
Leuchterscheinung, vorübergehende 82–247
Leuchtkraft 91–128
~ eines Meteors 86–195
Leuchtkraftfunktion 98–89
Leuchtkraftgleichung 86–196
Leuchtkraftklasse 91–146
Leuchtschweif 86–108
Leuchtstreifen 86–27
Leyer 92–65
Libra 92–62
Libration 81–48
~ in Breite 81–50
~ in Länge 81–49
optische ~ 81–51
parallaktische ~ 81–52
physische ~ 81–53
tägliche ~ 81–52
Librationsasteroid 84–91a
Librationswolke 84–43
lunare ~ 84–43
Licht, aschgraues 81–35
~ des Nachthimmels 79–8
extraterrestrisches ~ 79–7
galaktisches ~ 79–14
Lichtablenkung im Schwerefeld 99–161
Lichtausbeute 86–197
Lichtausbruch 86–80
~ eines Meteors 86–80
Lichtbrücke 84–63, 98–307
intergalaktische ~ 98–348
Lichtfalle 88–64
Lichtfortpflanzung 99–129
Lichtgleichung 95–45
Lichtjahr 91–18
Lichtkegel 95–156
Lichtkreuz 79–79
Lichtkurve 95–17
~ eines Meteors 86–70
mittlere ~ 95–19
photoelektrische ~ 95–22
photographische ~ 95–23
visuelle ~ 95–24
Lichtleistung 86–210
Lichtmaximum 86–78
Lichtsäule 79–78
Lichtschwächung, interstellare 97–80
Lichtwechsel 95–6
Lichtzentrum 96–46
Lichtzylinder 95–155
Lineareruption 73–113
Linie, D- 88–140
geodätische ~ 99–153
grüne ~ 79–16
interstellare ~ 97–96
Isogauss-~ 89–134
magnetische Null- ~ 89–151
neutrale ~ 89–151
optische interstellare ~ 97–99
rote ~ 79–15
rotverschobene ~ 98–366
zirkumstellare ~ 97–97
Linien, Neumannsche 87–103
Linienabsorption 93–108
Linienbreite, effektive 89–208
Linienentstehung 93–63
effektive Tiefe der ~ 93–64
Linienschieber 88–69
Linienverschiebung, periodische 96–22
linsenförmige Galaxie 98–245
Lithifikation 74–49
lithophile Elemente 83a–30
Lithosiderit 87–80
Lithosphäre 72–6
Lj 91–18
Lobatschewskischer Raum 99–105
Loch, galaktisches schwarzes 98–23a
koronales ~ 88–245a
schwarzes ~ 94–114
supermassives schwarzes ~ 98–22a
weisses ~ 98–24
Lokalbeben 73–25
Lokale Galaxiengruppe 98–326
~ Gruppe von Galaxien 98–326
~ Sprungbildung 82–204
~ Supergalaxie 98–347
Lokaler Superhaufen 98–347
lokales Objekt 98–368
Longitudinalfeld 89–138
Longitudinalkomponente 89–135
Londitudinalmagnetograph 88–83
Longitudinalwelle 73–45
Loopprotuberanz 89–270
Lorentz-Kontraktion 99–134
Lorentzsche Kontraktion 99–134
Lorentzscher Faktor 99–135
Lorentz-Transformation 99–132
Lorentz-Verkürzung 99–134
Lösung, chemisch aktive 74–43
stabile ~ 98–170
Lotablenkung 71–72
Lotabweichung ~ 71–72
Lotlinie 71–71
Love-Welle 73–48
Löwe 92–59
Kleiner ~ 92–60
LTE 93–39
Luchs 92–64
Lücke 84–88
Kirkwood-~ 84–89
LUF ~ 78–33
Luft 76–6
Luftdruck 77–43
Luftelektrizität 75–2
Luftfeuchte 77–118
Luftfeuchtigkeit 77–118
luftleer 76–8
Luftleuchten 79–2
Luftmasse 77–67
Luftmolekül 86–172
Luftpartikel 86–172
Luftprobe 76–86
Luftpumpe 92–15
Luftsäule 86–282
ionisierte ~ 86–282
Luftspiegelung 79–64
Luftströmung 77–90
Luftverschmutzer 76–82
Luftverschmutzung 76–80
Luftverunreinigung 76–80
Luftwiderstand 86–178
lunar 81–3
lunare Massenkonzentration 81–145
~ Schwereanomalie 81–144
lunarer Grus 82–112

lunares Material 82–58
Lunation 81–18
Lupus 92–63
Lynx 92–64
Lyra 92–65
Lyriden 86–248

M

mäanderförmige Depression 81–160
Maar 82–161
Maar-ähnlicher Krater 81–264
Maarkrater 81–264
Machsches Prinzip 99–37
Magellansche Wolke 98–212
Grosse ~ Wolke 98–213
Kleine ~ Wolke 98–214
Magma 73–131
Magmagestein 74–2
magmatische Tätigkeit 73–110
magmatisches Gestein 74–2
Magmatit 74–2
Magneteisenerz 82–92
Magnetfeld, allgemeines 89–128
~ der Erde 75–44
~ der Sonne 89–128
fossiles ~ 91–53
galaktisches ~ 98–50
Hintergrund ~ 89–141
interstellares ~ 97–106
~ kartographieren 89–132
kleinmasstäbliches ~ 89–143
photosphärisches ~ 89–130
primordiales ~ 95–62
superstarkes ~ 95–148
überstarkes ~ 95–148
übrigbleibendes ~ 95–62
ursprüngliches ~ 95–62
Magnetfelder auf der Sonne 89–129
solare ~ 89–129
Magnetfeldinhomogenität 78–9
Magnetfeldkarte 89–131
Magnetfeld-Veränderlicher 95–52
magnetisch gestörte Periode 75–91
~ gestörter Zeitabschnitt 75–91
~ konjugierte Punkte 75–23
magnetische Karte 89–131
~ Mitternacht 79–131
magnetischer Äquator 75–50
magnetisierendes Feld 75–42
magnetisiertes Sonnenplasma 84–32
Magnetisierung, bleibende 75–28
chemische ~ 75–31
induzierte ~ 75–29
inverse ~ 75–37
primäre ~ 75–35
remanente ~ 75–28
sekundäre ~ 75–36
thermoremanente ~ 75–30
umgekehrte ~ 75–37
viskose ~ 75–32
Magnetisierungsfeld 75–42
Magnetismus, Gesteins- 83a–80
induzierter ~ 75–29
remanenter ~ 75–28
Magnetit 82–92
Magnetogramm 75–84
Magnetograph 75–83
~ für das Longitudinalfeld 87–83
Magnetoheliograph 88–93
Leighton-~ 88–93
Magnetoid 98–25
Magnetometer, tragbares 82–44
Magnetopause 75–9
Magnetosphäre 75–7
~ des Pulsars 95–147
Magnetosphärengrenze 75–9
magnetosphärischer Schweif 75–10
Magnetsektor, interplanetarer 84–16
Magnitude eines Erdbebens 73–34
integrale ~ 96–136
Makrobewegung 88–189
Makrometeorit 87–4
Makrostruktur 99–9
Maler 92–78
Malerstaffelei 92–78
Mallungen 77–101
Männliche Wasserschlange 92–56
Mantel 72–15
Mare 81–266
kreisförmiges ~ 81–267
Mare-Gebiet 81–268
Mare-Kamm 81–188
Mare-Oberfläche 82–57
Mareograph 82–311
Mare-Region 81–268
Maria 81–266
maritimes Klima 77–186
Markab 92–119
Markarian-Galaxie 98–265a
Mars 83–133
Marsasteroid 84–74
Marsfamilie, Asteroid der 84–74
Marsferne 83–141
Marsformationen 83–108
Benennung von ~ 83–108
Nomenklatur von ~ 83–108
Marskanäle 83–157
Marskarte 83–136
Marsmittelpunkt- 83–138
Marsnähe 83–142
Marsphysik 83–134
Mascon 81–145
Maser, interstellarer 97–14
kosmischer ~ 97–13
natürlicher ~ 97–13
Masermolekularwolke 97–12
Maserverstärker, natürlicher 97–13
Maserwolke 97–12
Maskon 81–145
Masse des Meteorkörpers 86–163
~ des Meteors 86–163
~ des meteorischen Teilchens 86–163
fehlende ~ 99–87
relativistische ~ 99–141
schwere ~ 99–149
träge ~ 99–148
veränderliche ~ 86–167
verborgene ~ 99–88
Masse-Energie-Äquivalenzprinzip 99–145
Masse-Leuchtkraft-Beziehung 91–129
Masse-Leuchtkraft-Diagramm 91–29
Masse-Leuchtkraft-Verhältnis 98–233
Massen, ausgeworfene 82–165
Massenabnahme 86–168
Massenabnahmegleichung 86–169
Massenausfluss 93–13
Massenausstoss 93–15
Massenaustausch 96–33
Massenbestimmung 96–69
Massenbewegung 82–195
Massenbilanz 88–250
Massendichte, mittlere 99–84
Massenemission 85–145
Massenemissionsrate 85–146
Massen-Frequenz-Funktion 84–52
Massenfunktion 96–68
Massengleichung 86–169
Massenhorizont 99–41
Massenschwund 86–168
Massentransport 96–33
~ von einer Komponente zu der anderen 96–33
Massenübertragung 96–33
~ von einen Stern zum anderen 96–33
Massenverhältnis 96–67
Massenverlust 94–72
Massenverminderung 83a–42
Massenverteilung 84–52
Massenzunahme 99–142
relativistische ~ 99–142
Massenzuwachs der Erde 87–18
relativistischer ~ 99–142
Masse-Radius Beziehung 93–200
Masstab(s)höhe 76–18
Masstensor 99–157a
Material, ausserirdisches 72–56
extraterrestrisches ~ 72–56
höckeriges ~ 82–103
kosmisches ~ 99–7
lunares ~ 82–58
rundhöckerähnliches ~ 82–103
Materie, absorbierende 97–85
Auswurf von ~ 98–274
dunkle ~ 97–108
Erzeugung von ~ 99–89
gas- und staubförmige ~ 83a–25
intergalaktische ~ 98–349
interplanetare ~ 84–1
interstellare ~ 97–1
kosmische ~ 99–7
leuchtende ~ 97–109
meteoritische ~ 87–9
meteoritische organische ~ 87–68
prästellare ~ 94–16

D

stellare ~ 93–124
Materieära 99–178
Materieakkretion 83a–33
Materieaufsammlung 83a–33
Materieausstoss 94–77
~ aus dem Kometen 86–262
Materieauswurf 89–214
~ aus dem galaktischen Kern 98–274
Materiebrücke, intergalaktische 98–348
Materiedichte, mittlere 99–84
Materieerzeugung 99–89
Materiestrahl 98–278
~ aus dem galaktischen Kern 98–278
Materietensor 99–158
Materieverteilung 99–8
mathematisches Postulat der Relativität der Trägheit 99–36
maulwurfshaufenähnliche Gebilde 82–103
maximale brauchbare Übertragungsfrequenz 78–37
Maximum, breites 86–236
~ der Helligkeit 86–78
~ der Schicht 76–55
~ der Sonnentätigkeit 89–2
flaches ~ 86–236
primäres ~ 95–31
scharfes ~ 86–235
sekundäres ~ 95–32
spitzes ~ 86–235
Maximumskorona 88–244
Maxwellsche Geschwindigkeitsverteilung 98–96
Medium, interstellares 97–8
trübes ~ 83–251
Meer 81–266, 83–156
Bekanntes ~ 81–289
~ der Feuchtigkeit 81–282
~ der Fruchtbarkeit 81–274
~ der Gefahren 81–280
~ der Heiterkeit 81–276
~ der Kälte 81–278
~ der Ruhe 81–275
~ der Schlangen 81–291
~ der Sehnsucht 81–292
~ der Wellen 81–290
~ der Wolken 81–281
~ des heiteren Himmels 81–276
~ des Wissens 81–289
Humboldt- ~ 81–288
kontinentales ~ 72–26
kreisförmiges ~ 81–267
Moskau- ~ 81–283
Moskauer ~ 81–283
Nektar- ~ 81–279
Regen- ~ 81–273
Smyth- ~ 81–287
Wolken- ~ 81–281
Meerbusen 81–294
Meeresablagerung 72–47
Meeresboden 72–38
Meeresforschung 72–24
Meeresgezeiten 82–309
Meeresgrund 72–38
Meereshöhe 71–51
Druck in ~ 77–44
Meereskarte 72–23
Meeresküste 72–30
Meeresniveau 72–87
Meeresoberfläche 82–57
Meeresspiegel 72–87
Meeresstrom 72–60
kalter ~ 72–64
warmer ~ 72–63
Meeresströmung 72–60
Meerestiefenmessung 72–28
Meereswelle, seismische 72–95
Meereszustand 72–86
Mehrarmgalaxie 98–44
mehrarmige Galaxie 98–44
Mehrfachgalaxie 98–308
Mehrfachstern 96–5
Mehrfarbenphotometrie 96–143
Mensa 92–66
Meridian, magnetischer 75–49
Meridianbogen 71–21
Merkmale der Oberfläche 81–148
Merkur 83–120
Mesopause 76–67
Mesosiderit 87–79
Mesosphäre 76–66
Mesozoikum 74–72
mesozoische Ära 74–72
Messier-Katalog 92–12a
Messier-Nummer 97–111
Messung, mikrometrische 96–49
polarimetrische ~ 83–80
Metall-Filament 89–268
Metagalaxis 98–357
Metallhäufigkeit 91–170
Metallinienstern 91–195
metamorpher Prozess 74–34
metamorphes Gestein 74–11
Metamorphit 74–11
Metamorphose 74–34
dynamische ~ 74–36
thermische ~ 74–37
Metasomatose 74–44
Meteor 86–1
detonierendes ~ 86–119
ekliptikales ~ 86–46
flackerndes ~ 86–85
flimmerndes ~ 86–85
freisichtbares ~ 86–16
freisichtiges ~ 86–16
gewöhnliches ~ 86–14
helles ~ 86–12
hyperbolisches ~ 86–43
interplanetares ~ 86–44
interstellares ~ 86–43
kometarisches ~ 86–45
künstliches ~ 86–21
langsames ~ 86–88
mehrfaches ~ 86–133
~ mit Schweif 86–17
photographiertes ~ 86–13
photographisches ~ 86–13
schnelles ~ 86–87
schwaches ~ 86–15
schwaches teleskopisches ~ 86–19
Sonnensystem- ~ 86–44
spindelförmiges ~ 86–77
sporadisches ~ 86–23
stationäres ~ 86–34
teleskopisches ~ 86–18
ultrateleskopisches ~ 86–20
visuelles ~ 86–16
Meteorastronomie 86–3
Meteorbahn 86–30
Länge der ~ 86–47
schlangenförmige ~ 86–33
verwundene ~ 86–33
wellenförmige ~ 86–33
Winkellänge der ~ 86–48
Meteorbeobachter 86–7
Meteorbeobachtung 86–6
Meteorecho 86–296
kurzzeitiges ~ 86–300
langdauerndes ~ 86–301
Meteoreisen 87–83
Meteorerscheinung 86–5
Meteorexplosion 86–81
Punkt der ~ 86–122
Meteorgeschwindigkeit 86–86
geozentrische ~ 86–95
~ in der Atmosphäre 86–94
Meteorhelligkeit 86–8
absolute ~ 86–74
scheinbare ~ 86–73
Meteorhöhe 86–193
meteorisch 86–4
Meteorit 87–1
achondritischer ~ 87–77
orientierter ~ 87–112
Meteoriten- 87–2
Meteoritenastronomie 87–6
Meteoritenaufprall 82–235
Meteoritenaufschlag 82–236
Meteoritenbruchstück 87–43
Meteoriteneinschlag 82–236
durch ~ ausgelöste vulkanische Tätigkeit 82–156
Meteoriteneisen 87–83
Meteoritenerosion 82–242
Meteoritenfall 87–19
vermuteter ~ 87–49
Meteoritenforscher 87–7
Meteoritenfund 87–44
Meteoritengrube 87–58
Meteoritenhypothese 81–262
Meteoritenimpakt 87–33
Meteoritenkrater 81–263
fossiler ~ 87–55
irdischer ~ 87–54
Meteoritenkunde 87–6

Meteoritensammlung 87–8
Meteoritenschauer 87–37
Meteoritenstaub 87–10
meteoritisch 87–2
meteoritische Materie 87–9
organische ~ Materie 87–68
Meteorkamera 86–141
Meteorkern 86–102
Meteorkörper 86–160
Meteorkrater 87–53
Meteoroid 86–160
poröses ~ 85–50
Meteoroidenhypothese 81–262
Meteoroidenkrater 81–263
Meteoroidensturz 82–236
Meteorologie 77–6
dynamische ~ 77–7
synoptische ~ 77–8
Meteorologische Weltorganisation 77–17
Meteorphotographie 86–140, 86–151a
Meteorplatte 86–151
Meteorradiant 86–53
Meteorschauer 86–215
Meteorschauerradiant 86–237
Meteorschweif 86–103
beständiger ~ 86–104
instabiler ~ 86–109
kurzdauernder ~ 86–112
kurzlebiger ~ 86–112
langdauernder ~ 86–113
langlebiger ~ 86–113
Meteorschweifverschiebung 86–114
Meteorspektrograph 86–146
Meteorspur 86–153
~ auf der Platte 86–29
photographische ~ 86–153
Meteorstation 86–158
Meteorstaub 87–11
Meteorstein 87–60
Meteorstrahlung 86–207
Meteorstrom 86–216
Meteorstromdichte 86–264
Meteortrajektorie 86–30
~ in der Atmosphäre 86–30
Meteorüberwachung 86–144
Meteorzerfall 86–187
Meteorzerplatzen 86–121
Punkt des ~ 86–121
Methanal 97–27
Methanimin 97–45
Methanol 97–32
Methode der Doppelzählung 86–132
~ der Gewichtsfunktionen 93–59
radioaktive ~ 74–92
Radioecho-~ 86–265
radiometrische ~ 74–92
stereoskopische ~ 81–115
Uran-~ 74–96
Methylalkohol 97–32
Methyläthin 97–41
Methylazetylen 97–41
Methylidin 97–23
Methylzyanid 97–38
Metrik 99–151
Roberson-Walker-~ 99–152a
Schwarzschildsche ~ 99–152
mGal 71–66
M-Gebiet 89–61
Mie-Streuung 83–76
Migmatit 74–39
Mikroaufnahme 87–140
Mikrobreccie, lunare 82–86
Mikrokrater 81–201
Mikrometeor 86–20
Mikrometeorite 87–3
Mikrometeorologie 77–9
Mikroproben-Analyse 82–143
Mikropulsationen 75–82
~ des Erdfeldes 75–82
erdmagnetische ~ 75–82
geomagnetische ~ 75–82
Mikroreversibilität 93–41
Mikroscopium 92–67
Mikroseismen 73–53
mikroseismische Bodenunruhe 73–54
Mikroskop 92–67
Mikrostruktur 82–73
Mikrotektit 87–97
Mikrowellenlinie, interstellare 97–98
Milchstrasse 98–1
Milchstrassenebene 98–53
Milchstrassensystem 98–1
Millibar 77–47
Milligal 71–66
Millimeterquecksilbersäule 77–48
Millionstel der sichtbaren Halbkugel 89–68
~ der Sonnenscheibe 89–69
Mills-Kreuz-Antenne 88–121
Millssches Kreuz 88–121
Milne-Edingtonsche Näherung 93–50
Mineral 74–12
Minimum der Sonnenaktivität 89–4
~ der Sonnentätigkeit 89–4
primäres ~ 95–26
sekundäres ~ 95–27
Minimumskorona 88–246
Minkowskis vierdimensionale Welt 99–122
Minkowskische Raum-Zeit 99–122
~ Welt 99–122
Minkowskischer Raum 99–122
Mischgestein 74–39
Mischung, Gas-Staub- 83a–25
Russell-~ 93–101
~ von Gas und Strahlung 93–125
Mission zu einem Asteroiden 84–99
Missweisung 75–56
mitbewegte Koordinatensysteme 99–44
Mitglied des Sternhaufens 96–87
unsicheres ~ 96–87
wahrscheinliches ~ 96–86
Mitgliedschaft, physische 96–83
mitrotierende Feldlinie 75–14
~ Struktur des Magnetfeldes 84–17
mitschwimmende Koordinatensysteme 99–44
Mittagsmeridian 75–17
Mitte der Sonnenscheibe 89–35
mittelkörnig 82–116
Mittelmeer 72–26
mittelozeanischer Rücken 72–45
Mittelpunktdichte 93–87
Mittelpunkttemperatur 93–89
Mitte-Rand-Variation 88–145
~ der magnetischen Feldstärke 89–160
Mitternacht, magnetische 79–131
Mizar 92–120
Modell der Erde 79–133
~ der Welt 99–46
~ des Weltalls 99–46
Eddingtonsches ~ 93–137
~ einer anisotropen Welt 99–65
~ einer asymptotischen Welt 99–59
~ einer asymptotischen Welt erster Art 99–60
~ einer expandierenden Welt 99–55
~ einer geschlossenen Welt 99–73
~ einer hierarchischen Welt 99–48
~ einer homogenen Welt 99–64
~ einer inhomogenen Welt 99–66
~ einer isotropen Welt 99–63
~ einer kontrahierenden Welt 99–56
~ einer leeren Welt 99–49
~ einer monotonen Welt 99–61
~ einer nichtstationären Welt 99–53
~ einer nichtstatischen Welt 99–51
~ einer offenen Welt 99–72
~ einer oszillierenden Welt 99–57
~ einer oszillierenden Welt erster Art 99–58
~ einer stationären Welt 99–52
~ einer statischen Welt 99–50
~ eines Sternes mit Temperatur Null 93–186
galaktisches ~ 98–163
geschlossenes kosmologisches ~ 99–73
heisses kosmologisches ~ 99–166
kosmologisches ~ 99–46
mathematisches ~ 98–324
~ mit Berücksichtigung der Linienabsorption 93–18
~ mit Linienabsorption 93–18
~ mit Temperatur Null 93–186
~ ohne Berücksichtigung der Linienabsorption 93–19
offenes kosmologisches ~ 99–72
~ ohne Linienabsorption 93–19
polytropes ~ 93–129
zusammengesetztes ~ 93–140
Zwei-Komponenten-~ 88–196
Zwei-Temperatur-~ 88–197
Zwei-Zemperatur ~ der Chromosphäre 88–197

D

Modellatmosphäre 83–247
Modellatom mit vielen Energieniveaus 93–46
~ mit zwei Energieniveaus 93–45
Modellplanet 83–196
Modul, Youngscher 73–59
Service ~ 82–7
Modulation, elektrooptische 88–86
Modulationsfrequenz 86–275
Modulationsgebiet 84–26
~ der kosmischen Strahlung 84–26
Mögel-Dellinger-Effekt 78–40
Moho-Diskontinuität 72–14
Mohorovičić-Diskontinuität 72–14
Moldavit 87–99
Molekül, interstellares 97–10
vielatomiges ~ 97–21
Molekül-Aggregat 97–65
Molekulargewicht, mittleres 93–100
Molekularwolke 97–11
Molekülzusammensetzung 78–11
Moment des Gezeitenkräftepaares 82–288
~ des Kräftenpaares 83a–5
Monat 81–42
anomalistischer ~ 81–43
drakonitischer ~ 81–47
siderischer ~ 81–45
synodischer ~ 81–46
tropischer ~ 81–44
Mond 81–2, 81–3
abnehmender ~ 81–21
verfinsterter ~ 81–57
zunehmender ~ 81–20
Mondaktivität 82–149
Mondalter 81–31
Mondäquator 81–121
Mondatlas 81–136
Mondaufnahme 82–26
Mondaureole 79–61
Mondbahn 81–36
innerhalb der ~ 81–4
jenseits der ~ 81–6
Mondbasalt 82–84
Mondbasis 82–17
bemannte ~ 82–18
Mondbeben 82–214
Mondboden 82–62
Mondbreccie 82–85
Mondbrekzie 82–85
Mondchemie 82–74
Monddurchgang 82–303
Monderhöhung 81–156
Mondferne 81–41
Mondfinsternis 81–56
partielle ~ 81–67
totale ~ 81–66
Mondflare 82–247
Mondformationen 81–148
Mondforschung 82–3
Mondfoto 82–26
Mondgebirge 81–193
Mondgeodäsie 81–109
Mondgeologie 82–47
Mondgestalt 81–116
Mondgestein 82–65
Mondgezeiten 82–308
Mondglas 82–96
Mondglobus 81–112
Mondgrund 82–62
Mondhof 79–61
Mondkarte 81–133
photographische ~ 81–137
physikalische ~ 81–134
thermische ~ 81–135
topographische ~ 81–138
Mondkartographie 81–111
Mondkranz 79–61
Mondkrater 81–197
Mondkreuz 79–80
Mondkruste 82–54
Mondkugel 81–112
Mondkulmination 82–303
Mondkunde 82–2
Mondlandeaggregat 82–5
Mondlandefähre 82–21
Mondlandegerät 82–5
Mondlandschaft 81–55
Mondlandung 82–13
Mondlandungsplatz 82–14
mondlose Nacht 79–3e
Mondlumineszenz 82–248
Mondmappierung 81–111
Mondmaterial 82–58
Mondmeer 81–266
Mondmeridiandurchgang 82–303
Mondmodul 82–5
Mondmorphologie 81–147
Mondnacht 81–33
Mondnähe 81–40
Mondoberfläche 82–55
Mondoberflächenexperiment 82–42
Mondphase 81–19
Mondphotographie 82–26
Mondphysik 82–151
Mondprobe 82–69
zurückgebrachte ~ 82–70
Mondprozess 82–150
Mondrand 81–95
dunkler ~ 81–97
heller ~ 81–96
Mondring 79–68
Mondrückseite 81–16
Mondscheibe 81–54
Mondseismologie 82–213
Mondseite, beleuchtete 81–34
der Erde zugewandte ~ 81–15
erdabgewandte ~ 81–16
erdzugewandte ~ 81–15
zugewandte ~ 81–15
Mondsichel 81–23
Mond-Sonnentidenkomponente 82–307
Mondstaub 82–112
Mondstrahlen 81–168
Mondtag 81–32
mittlerer ~ 82–304
Mondteilchen 82–123
Mondterminator 81–123
Mondtiden 82–308
Mondtopographie 81–146
Mondumlaufbahn 82–11
Mondvertiefung 81–157
Mondviertel 81–26
erstes ~ 81–25
letztes ~ 81–30
Mondwulst 81–118
Mondzeit 82–300
mittelere ~ 82–302
wahre ~ 82–301
Monoceros 92–68
Monopol, magnetischer 99–170o
monotones Weltall erster Art 99–62
Monsun 77–99
Moor 81–303
Morgendämmerung 79–35
Morgenmeridian 75–18
Morgenrot 79–56
Morgenstern 83–127
Morgenterminator 81–124
morphologischer Typ 98–228
Moskauer Meer 81–283
Mössbauer-Spektrographie 82–142
Mottles 88–210
chromosphärische ~ 89–172
dunkle ~ 88–212
helle ~ 88–111
Moustaches 89–200
Mulde 73–89
Multipol 75–66
Musca 92–69
Muskovit 74–24
Muttergalaxie 98–284
Mutterkörper, meteoritischer 87–123
Muttermeteorit 87–123
Muttermolekül 85–135

N

Nachahmung, mathematische 98–323
Nachbargalaxie 98–211
Nachbeben 73–13
Nachdämmerungsemission 79–52
Nacheiszeit 72–109
nacheiszeitlich 72–110
nachfolgender Fleck 89–114
nachgeführte Kamera 86–142
nachgeschleppter Arm 98–29
nachgezogener Arm 98–29
Nachhauptreihenstadium 94–68
nachlaufender Arm 98–29
Nachleuchten 79–22
Nacht auf dem Monde 81–33
helle ~ 79–59
klare ~ 79–3g
mondlose ~ 79–3e

photometrisch klare ~ 79–3f
photometrische ~ 79–3f
wolkenfreie ~ 79–3g
Nachthemisphäre 83–68
Nachthimmel 79–6
heller ~ 86–28
Nachthimmelsleuchten 79–5
Nachthimmelslicht 79–8
Nachtschweif 86–105
Nachtseite 83–68
Nachtstrom 86–218
Nachtwolke, leuchtende 79–89
Nadiraufnahme 82–28
Nahbeben 73–26
Näherung, Chandrasekharsche 93–51
Milne-Eddingtonsche ~ 93–50
nichtrelativistische ~ 99–97
Nährgebiet 72–122
~ eines Gletschers 72–122
Nasen-Pfeifstörung 78–59
Nassbaggern 72–53
Nassbaggerung 72–53
natürlicher Satellit 83–18
Navigationsgefahr 72–97
Nebel 77–130
aussergalaktischer ~ 98–202
diffuser ~ 97–122
extragalaktischer ~ 98–202
filamentartiger ~ 97–129
galaktischer ~ 97–110
gas-staubförmiger ~ 97–119
heller ~ 97–121
Hufeisen- ~ 97–143
Kometen-ähnlicher ~ 97–142
leuchtender ~ 97–121
Netzwerk-~ 97–137
Nordamerika- ~ 97–140
Orion-~ 97–136
planetarischer ~ 97–144
regelmässiger ~ 98–236
unregelmässiger ~ 98–237
Nebelflucht 98–298
nebelfreie Zone 97–88
Nebelhaufen 98–329
Nebelring 83a–10
Nebelstadium 95–125
Nebelveränderlicher 95–85
Nebelwolke 94–31
Nebenbogen 73–83
Nebengegenmond 79–86
Nebengegensonne 79–84
Nebenkomponente 96–38
Nebenkrater 73–128
Nebenmaximum 95–32
Nebenminimum 95–27
Nebenmond 79–73
Nebenpuls 95–145
Nebenpulsation 95–145
Nebenradiant 86–239
Nebensonne 79–71
Nebularhypothese 83a–7
Neigung 76–54
~ der Achse 83–35
~ der Bahnebene 96–51
~ der Rotationsachse gegen... 91–50
~ gegen den Mondäquator 81–37
Neigungspol, magnetischer 75–45
Nektar-Meer 81–279
Neozoikum 74–73
Neptun 83–236
Netz 86–129, 92–83
gravimetrischer ~ 71–80
Netzwerk, chromosphärisches 89–171
Netzwerknebel 97–137
Neuer Stern 95–110
Neukristallisation 74–42
Neumannsche Linien 87–103
Neumond 81–22
neutrale Linie 89–151
neutraler Punkt 89–152
Neutralfläche 75–16
Neutralisation, gegenseitige 79–29
Neutralschicht 75–16
Neutrinokühlung 94–97
Neutrinoleuchtkraft 94–96
Neutrinoopazität 94–95
Neutrinoreaktion 94–92
Neutrinostrahlung 94–94
Neutron, Albedo- 75–115
Neutronenabsorption, langsame 94–88
Neutronenalbedo 75–116
Neutroneneinfang 94–84
Neutroneneinfangreaktion 94–87
Neutronenfluss 94–86
Neutronenflussdichte 94–86
Neutronengas 93–192
Neutronengasentartung 93–191
Neutronenquelle 94–85
Neutronenstern 91–152
Neutrosphäre 76–64
Newtonsche Kosmologie 99–25
~ Zeit 99–118
Newtonsches Universum 99–93
nichtaktiver Komet 85–147
nichtentartete Aussenschichten 93–199
nichtentarteter Kern 93–171
nichtentartetes Gas 93–182
nichteuklidisch 99–102
nichtflüchtiger Stoff 83a–53
nichtgebundene Rotation 83–38
nichtgraue Atmosphäre 93–27
Nichtinertialsystem 99–147
nichtkohliger Chondrit 87–75
nichtleuchtend 96–44
nichtnachgeführte Kamera 86–143
nichtnumeriertes Objekt 84–96
Nicht-Protonen-Eruption 89–186
Nicht-Protonen-Flare 89–186
nichtrelativistisch 99–96
nichtrelativistische Entartung 93–177
~ Kosmologie 99–27
nichtstationäres Universum 99–53
~ Weltmodell 99–53
nichtstatische Welt 99–51
Nichtstrahlungsenergie 88–218
Nichtstrommeteor 86–25
nichtthermischer Druck 94–42
nichtzonale Kugelfunktionen 95–102
Nickeleisen 87–90
Nickelgehalt 72–55
Niederschlag 77–154
jährlicher ~ 87–17
orographischer ~ 77–167
Niederschlagsmenge 77–158
jährliche ~ 87–17
Niederschlagssumme 77–168
Niederschlagstotalisator 77–171
Niederung 81–272
niedrigste brauchbare Übertragungsfrequenz 78–33
Niedrigwasser 82–283
Nieseln 77–172
Nife 72–18
Nifekern 72–18
Nimbostratus 77–145
Nippflut 82–272
Niveau, kritisches 76–26
Niveaufläche des Schwerefeldes 71–61
Nomenklatur von Marsformationen 83–108
Nordamerikanebel 97–140
Nordhimmel 79–3b
nördliche Aurora 79–105
Nördliche Krone 92–40
~ Wasserschlange 92–55
Nordlicht 79–105
flammendes ~ 79–157
Nordlichtdunst 79–148
Nordlichtkrone 79–156
Nordpolarsequenz 91–87
Nord-Süd-Asymmetrie 89–50
Norma 92–70
Normalatmosphäre 76–11
Normalbedingungen 77–3
normale Raumtemperatur 77–31
~ Riesen 91–150
~ Temperatur und Druck 77–30
normaler Schall 76–35
Normalnull 72–87
Nose-Whistler 78–59
Nova 95–110
extragalaktische ~ 95–114
galaktische ~ 95–113
langsame ~ 95–120
rekurrierende ~ 95–116
schnelle ~ 95–119
Tycho's ~ 95–135
typische ~ 95–111
wiederholte ~ 95–116
wiederkehrende ~ 95–116
Novaähnlicher 95–118
~ Stern 95–118
~ Veränderlicher 95–118
Novaausbruch 95–121
Nukleosynthese 94–52
explosive ~ 94–57

D

kosmologische ~ 99–175
nichtexplosive ~ 94–58
Urknall-~ 99–175
Nullgeschwindigkeitswellenlänge 98–365
Null-Linie, magnetische 89–151
Nullphase 83–73
Nullpunkt der Perioden-Leucht-kraft-Beziehung 95–104
Nullpunktsenergie 93–187
kinetische ~ 93–187
Nummer im Messier-Katalog 97–111
Messier-~ 97–111

O

O-Assoziation 96–89
oberer Planet 83–15
Oberfläche der Erde 71–5
~ eines Meteoriten 87–106
~ eines Planeten 83–97
feste ~ 83–98
geschichtete ~ 83–101
Mare-~ 82–57
mit Kratern bedeckte ~ 81–245
mit Schmelzkruste bedeckte ~ 87–109
mit Strahlen bedeckte ~ 81–171
primäre ~ 87–107
rauhe ~ 83–102
sekundäre ~ 87–108
ursprüngliche ~ 87–107
~ zwischen den Strahlen 81–172
Oberflächenaktivität 82–231
Oberflächenaushöhlung 87–118
Oberflächenbeben 73–21
Oberflächenbedingung 93–80
Oberflächenbruch 81–173
Oberflächenerkundung 82–32
Oberflächenfeld 95–64
Oberflächenformation 83–105
~ des Mondes 81–148
Oberflächengebilde 83–105
Oberflächenmaterial 82–59
lunares ~ 82–59
Oberflächenrauheit 83–99
Oberflächenrauhigkeit 83–99
Oberflächenströmung 72–62
Oberflächentemperatur 93–88
Oberflächenwelle 73–47
Oberflächenzerstörung 87–35
Objekt, diffuses 85–25
extragalaktisches ~ 98–209
kollabiertes ~ 98–23
kompaktes ~ 98–21
kosmologisches ~ 98–369
lokales ~ 98–368
nichtkatalogisiertes ~ 84–96
quasistellares ~ 98–359
supermassives ~ 98–22
Observatorium auf dem Mond 82–20
Ionosphären-~ 78–21
Octans 92–71
Ofen, Chemischer 92–50
offener Sternhaufen 96–99
offenes kosmologisches Modell 99–72
Öffnung der Spiralwindungen in Rotationsrichtung 98–30
~ der Spiralwindungen nach aussen 98–30
Okklusion 77–72
Okklusionsfront 77–71
Ökologie 76–2
Oktaedrit 87–86
Oktant 92–71
Ölauslaufverlust 72–71
Olbersches Paradoxon 99–40
Olivin 74–22
Omeganebel 97–143
Oortsche Kometenwolke 85–97
~ Konstanten 98–118
Ophiuchus 92–72
Opposition 83–59
~ in Rektaszension 81–68
Oppositionseffekt 83–75
Optik, atmosphärische 79–1
optische Erscheinung 79–33
~ Sonne 88–2
optischer Doppelstern 96–6
ordentliche Welle 78–31
Ordovizium 74–79
organische Einschüsse 87–66
~ Säure 87–69
~ Spuren in Meteoriten 87–67
organogene Sedimente 72–59
Orientierung der Drehachse 84–102
Orion 92–73
Orion-Arm 98–36
Oriongürtel 92–129a
Orioniden 86–253
Orionnebel 97–136
Grosser ~ 97–136
Orkan 77–78
tropischer ~ 77–78
orogene Zone 73–75a
Orogenese 73–74
orogenetischer Gürtel 73–75
Orogengürtel 73–75
orographische Wolke 77–134
orographischer Niederschlag 77–167
~ Regen 77–166
Ort, genäherter 85–26
genauer ~ 85–28
vorhergesagter ~ 85–27
Orthoklas 74–16
Ortsbeben 73–25
Ortskoordinate 99–124
Ostmeer 81–284
Ostrand 89–34
Ost-West-Asymmetrie 89–49
Oszillation 98–187
Fünf-Minuten-~ 88–186
~ in z-Richtung 98–187
~ senkrecht zur galaktischen Ebene 98–187
Oszillationen der Sonne 88–223a
freie ~ der Erde 73–49
Oszillator, hydromagnetischer 95–61
magnetohydrodynamischer ~ 95–61
oszillierender Spalt 88–70
oszillierendes Weltall 99–58
Oszillograph 86–292
Oszilloskop 86–292
Oval, Weisses 83–195
Oxid = Oxyd 74–26
oxydierende Atmosphäre 83a–47
Ozean 72–22
~ der Stürme 81–277
Ozeanboden 72–37
Ozeangrund 72–37
ozeanische Tiefenströmung 72–61
ozeanischer Strom 72–60
Ozeanographie 72–24
ozeanographische Karte 72–23
Ozon, atmosphärisches 76–124
Ozongehalt 76–125
Ozonosphäre 76–123
Ozonschatten 76–127
Ozonschicht 76–123

P

Paar, enges 96–23
Packeis 72–102
Paläoklima 77–188
Paläomagnetismus 75–25
Paläotemperatur 72–126
Paläozoikum 74–71
paläozoische Ära 74–71
Pallas 84–86
Pallasit 87–81
Panoramaaufnahme 82–27
Paradiesvogel 92–16
Paradoxa, kosmologische 99–38
Paradoxon, Olberssches ~ 99–40
photometrisches ~ 99–40
Seeligersches ~ 99–39
parallaktische Bewegung 91–33
~ Photographie 79–102
Parallaxe, absolute 91–11
Cepheiden-~ 91–8a
~ der Strömgrensphären 91–13
~ aus dem Strömgrenradius 91–13
dynamische ~ 91–12
hypothetische ~ 91–12
jährliche ~ 91–6
photometrische ~ 91–14
relative ~ 91–10
relative trigonometrische ~ 91–10
säkulare ~ 91–9
spektroskopische ~ 91–15
statistische ~ 91–8
trigonometrische ~ 91–5
Parallelbeobachtung 86–125
Parameter, photometrischer 85–106
Parasitärkrater 73–128
Parsec = Parsek 91–17
Partialquerschnitt 94–54

partielle Finsternis 81–79a
~ Mondfinsternis 81–67
~ Sonnenfinsternis 88–16
Partikel 88–154
feste ~ 76–99
meteorische ~ 86–160
~ mit Mantel 97–70
überschwere ~ 99–170n
Partikelausbruch 88–162
Partikelauswurf 88–161
~ aus der Sonne 88–161
Partikelbahn 75–118
Partikeleinschlag-Spur 87–120
Partikelejektion 88–161
Partikelgrösse 76–114
Partikeln, eingefangene 75–111
feinkörnige ~ 82–109
feste ~ 85–48
geomagnetisch eingefangene ~ 75–111
interstellare ~ 97–68
sphärische ~ 87–29
Partikelspeicherung 75–123
Partikelspur 82–251
Partikelstrahlung 88–150
Partikelwolke 84–31
Partikelzahl 76–121
Passate 77–100
Passatwinde 77–100
Pavo 92–74
Pegasus 92–75
Pekuliarbewegung 91–31
pekuliare Galaxie 98–256
pelagischer Ton 72–54a
Pendeluhr 92–54
Penetrometer 82–83
Penumbra 89–97
~ eines Fleckes 89–97
~-Faser 89–98
~-Filament 89–98
~-Welle 89–100
Penumbrawellen 89–100
Periapside, galaktische 98–189
Periapsis, galaktische 98–189
Periareon 83–142
Periastron 96–58
Periastrondrehung 96–71
Periastronlänge 96–59
Periastronzeit 96–60
Peridot 82–89
Perifokus, galaktischer 98–189
Perigalaktikum 98–189
Perigäum 81–38
Perigeum 81–38
Perihel 83–29a
Perihelbewegung, relativistische 99–160
Periheldrehung, relativistische 99–160
Perijovium 83–172
Periode 95–39
elfjährige ~ 89–70
geologische ~ 74–74
magnetisch gestörte ~ 75–91
sekundäre ~ 95–41
Periodenänderung 95–43
Periodendauer 95–39
Perioden-Helligkeits-Beziehung 95–103
Periodenlänge 95–39
Perioden-Leuchtkraft-Beziehung 95–103
Nullpunkt der ~ 95–104
Periodensprung 95–154
Perisaturnium 83–210
Periselen 81–40
Perl(en)schnurblitz 77–164
Perl(en)schnur-Phänomen 88–32
Perlmutterwolke 79–91
Perm 74–83
Permafrost 83–152
permanente koronale Kondensation 89–282
permanenter Sonnenwind 84–40
Perseiden 86–251
Perseus 92–76
Perseus-Arm 98–34
perspektivische Verkürzung 89–92
~ Verkürzung der Flecken 89–92
Pfau 92–74
Pfeifstörungen, atmosphärische 78–53
kurze ~ 78–58
lange ~ 78–57
"Nasen"-~ 78–59
Pfeil 92–84
Pferd, Kleines 92–48
Pferdchen 92–48
Pferdekopfnebel 97–135
Phänomen, aktives 89–14
Phantomkrater 81–209
Phase des Zyklus 89–79
eruptive ~ 89–210
Flash-~ 89–210
grösste ~ 81–78
Hayashi-~ 94–44
hydrodynamische ~ 94–43
~ nach der Hauptreihe 94–68
partielle ~ 81–78a
Phasenanomalie, plötzliche 78–52
sonnenbedingte ~ 78–52
Phasendefekt 81–28
Phasendrehinterferometer 88–109
Phasenfunktion 83–78
Phasenintegral 83–79
Phasenkoeffizient 83–77
Phasenkurve 83–74
Phasenraum, sechsdimensionaler 98–144
Phasenschaltung 88–112
Phasenschieber 88–115
Phasenübergang 83a–58
Phasenumschalter 88–114
Phasenumschaltung, Interferometer mit 88–113
Phasenumwandlung 99–170p
Phasenwinkel 83–72
Phobos 83–168
Phoenix 92–77
photochemische Reaktion 76–63
Photoelektronenschicht 82–246
Photographie, fortlaufende 79–103
parallaktische ~ 79–102
photographische Sternspur 86–152
Photolyse 76–84
Photometrie, Sechsfarben- 96–148
Photonendruck 93–91
Photoneutrino 94–93
Photospaltung 94–101
Photosphäre 88–167
ruhiges Gebiet der ~ 89–56
ungestörte ~ 88–169
Photosphärenkarte 89–20
synoptische ~ 89–20
photosphärisch 88–170
photosphärisches Magnetfeld 89–130
Phototriangulation, räumliche 71–26
Photozerlegung 94–101
~ des Kernes 94–101
Physik der Atmosphäre 76–4
~ der Kometen 85–104
~ der solar-terrestrischen Beziehungen 84–56
~ der Sonne 88–6
~ des Mondes 82–151
physikalische Theorie der Meteore 86–159
Pictor 92–78
Piezoglypten 87–122
Pigeonit 82–90
Pisces 92–79
Pisciden 86–250
Piscis Austrinus 92–80
Plage 89–164
optische ~ 89–165
Plagioklas 74–19
Plancksche Elementarlänge 99–170h
~ Elementarmasse 99–170f
~ Länge 99–170h
~ Masse 99–170f
~ Zeit 99–170i
Planet 83–1
äusserer ~ 83–15
~ der Sonne 83–11
~ des Sonnensystems 83–11
hypothetischer ~ 83–122
innerer ~ 83–14
intramerkurieller ~ 83–123
jupiterähnlicher ~ 83–174
kleiner ~ 84–68
oberer ~ 83–15
Olbersscher ~ 83a–63
sonnenferner ~ 83–15
sonnennaher ~ 83–14
transneptunischer ~ 83–237
transplutonischer ~ 83–124
unterer ~ 83–14
planetar 83–6
planetarisch 83–6
planetarische Kennziffer 75–90
planetarischer Index 75–90
~ Nebel 97–144

D

~ Strom 86–220
Planeten 83–6
erdähnliche ~ 83–119
erdartige ~ 83–119
grosse ~ 83–13
terrestrische ~ 83–119
Planatenatmosphäre 83–243
Planetenbahn 83–29
Planetenbewegung 83–28
Planetengeologie 83–5
Planetenkonstellation 83–50
Planetenkorona 83–245
Planetenkosmogonie 83a–1
Planetenmodell 83–196
Planetenoberfläche 83–97
Planetenphysik 83–4
Planetenrotation 83–30
Planetensuche 83–238
Planetensystem 83–2
Planetenzeichnung 83–16
Planetesimale 83a–35
planetographisch 83–7
planetographische Koordinaten 83–8
Planetoid 84–68
Amor-ähnlicher ~ 84–80a
~ mit dichter Annäherung 84–78
verlorener ~ 84–97
Planetoidengruppe 84–90
Planetologie 83–5
planetozentrisch 83–9
planetozentrisches System 83–10
Plasma, heliumangereichertes 84–60
interplanetares ~ 84–39
stossfreies ~ 84–50
Plasmaschicht 75–11
Plasmasphäre 75–12
Plasmastrom 84–33
solarer ~ 84–33
Plasmawolke 84–30
Platte, kontinentale 73–103
$\lambda/4$ ~ 88–126
Plattenmitte 86–156
Plattenzentrum 86–156
Plattform 73–103
kontinentale ~ 73–103
Plejaden 96–110
Plessit 87–94
Plinsen 99–171b
plötzliche Phasenanomalie 78–52
plötzliches Verschwinden 89–279
Pluto 83–239
Plutonit 74–7
Pogsonsche Gleichung 91–58
Poisson-Konstante 73–60
Pokal 92–42
Pol der geomagnetischen Achse 75–48
magnetischer ~ 75–45
magnetischer ~ der Erde 75–45
~ magnetischer Inklination 75–45
polares Gebiet 83–192
Polarfront 77–73
Polargebiet 83–192
Polarimeter, lichtelektrisches 83–83
photographisches ~ 83–84
visuelles ~ 83–82
Polarisation, interstellare 97–105
Polarisationskurve 83–81
Polarisationsmessungen 83–80
Polarität, entgegengesetzte 89–155
magnetische ~ 89–148
N-~ 89–149
negative ~ 89–150
nördliche ~ 89–149
positive ~ 89–149
S-~ 89–150
südliche ~ 89–150
Polaritätsgesetz 89–154
Polarkarte 89–19
Polarlicht 79–93
~ auf der Tagseite 79–97
~ in der Polkappe 79–98
~ in niederen Breiten 79–96
künstliches ~ 79–99
~ mit bandförmiger Struktur 79–140
sonnenbelichtetes ~ 79–130
Polarlichtabsorption 78–46
Polarlichtaktivität 79–100
Polarlichtbogen 79–146a
Polarlichtdunst 79–148
Polarlichtecho 79–121
Polarlichtelektron 79–109
Polarlichtformen 79–137
Polarlichthäufigkeit 79–122
Gürtel geringer ~ 79–129
Zone maximaler ~ 79–125
Polarlichtkappe 79–127
Polarlichtkorona 79–137a
Polarlichtobservatorium 79–101
Polarlichtoval 79–124
Polarlichtregion 79–126
Polarlicht-Substorm 79–141
Polarlichtteilchen 79–107
Polarlichtzone 79–123
Polarstern 92–121
Polarstrahlen 89–294
Pol-Aufsicht, Stern in ~ 91–40
Poldrift der Protuberanzen 89–231
Polhalbmesser 83–116
Poligonaufnahme 82–35
Poligonieren 82–35
Polkalotte 83–143
Polkappe 83–143
Zunehmen der ~ 83–145
Polkappenabsorption 78–44
Pollutant 76–81
Pollux 92–122
Polradius 83–116
Polregion 89–58
Polsequenz 91–87
internationale ~ 91–87
polwärtige Wanderung der Protuberanzen 89–231
Polyeder 71–12
Polymerenteilchen 97–77a
Polynome, Legendresche 95–101
Polytrope 93–126
~ Gaskugel 93–127
Polytropenindex 93–128
Population der galaktischen Scheibe 98–72
~ einer Galaxie 98–230
~ I 98–71
extreme ~I 98–73
intermediäre ~ 98–70
Spiralarm-~ 98–73
~ II 98–68
Pore 89–122
porös 82–181
Porosität 83–100
Position des Radianten 86–60
genäherte ~ 85–26
genaue ~ 85–28
vorausberechnete ~ 85–27
vorherberechnete ~ 85–27
vorhergesagte ~ 85—27
Positionswinkel 81–71
postglazial 72–110
post-mare Gebiet 81–270
~ Krater 81–208
~ Region 81–270
~ Struktur 81–154
Postmaximum 95–33
Postnova 95–123
Postsupernova 95–134
Postulat, Koherenz- 99–34
mathematisches ~ der Relativität der Trägheit 99–36
Weylsches ~ 99–34
Potential, axialsymmetrisches 98–153
~ mit Zylindersymmetrie 98–154
stationäres ~ 98–147
zylindersymmetrisches 98–154
Potentialmulde 98–176
Poynting-Robertson-Effekt 84–67
Poyntingvektor 93–92
prae-mare Gebiet 81–269
~ Krater 81–207
~ Region 81–269
~ Struktur 81–153
Praemaximum = Prämaximum 95–30
Praenova = Pränova 95–122
Praesepe 96–111
Praesupernova 95–129
Präkambrium 74–70
prästellar 94–13
prästellare Materie 94–16
prästellarer Körper 94–15
Präsupernova = Praesupernova 95–129
primäres Maximum 95–31
Primärkrater 81–198
Primärmagnetisierung 75–35
Primärwelle 73–46
Primärzeit 74–71

primordiales Magnetfeld 95–62
Prinzip, Archimedisches 73–101
eingeschränktes kosmologisches ~ 99–31
Machsches ~ 99–37
perfektes kosmologisches ~ 99–32
~ von Archimedes 73–101
Prisma, Anderson- 88–68
rotierendes ~ 88–68
~ zur Bildteilung 88–88
Probe des Mondmaterials 82–69
~ des Mondstaubes 82–121
Probeentnahme = Probenentnahme 76–85
Probenbehällter 82–81
Probenentnahme vom Grund 82–78
Probennahme 82–72
~ von Boden 82–78
Problem, kosmologisches 94–5
Profil, vertikales 81–228
Prognose, kurzfristige 77–20
langfristige ~ 77–19
Projektion, azimutale 71–44
gnomonische ~ 71–46
konforme ~ 71–41
stereographische ~ 71–47
winkeltreue ~ 71–41
zylindrische ~ 71–43
Projektionsschirm 88–13
Prokyon 92–123
Proterozoikum 74–77
protogalaktische Wolke 94–17
Protogalaxie 94–18
Protoneneinfang 94–90
Protonen-Ereignis 89–192
Protonen-Eruption 89–185
Protonen-Event 89–192
Protonen-Flare 89–185
Protonen-Pfeifstörung 78–61
Protonenpolarlicht 79–108
Protonensphäre 75–12
Protonenwhistler 78–61
Protonereignis 84–59
Proton-Flare-Projekt 89–177
Protonosphäre 75–12
Protoplanet 83a–40
protoplanetare Wolke 83a–24
protostellar 94–46
Protostern 96–96
Protuberanz 89–223
aktive ~ 89–238
Aktivierung einer ~ 89–239
äquatoriale ~ 89–228
Auflösung einer ~ 89–280
aufsteigende ~ 89–278
~ der Äquatorzone 89–228
~ der Polarzone 89–230
Baumstämme-~ 89–272
eruptive ~ 89–248
Flare-ähnliche ~ 89–222
gewöhnliche aktive ~ 89–246
gewöhnliche eruptive ~ 89–250
~ in Form einer kleinen Ausbuchtung 89–253
~ in hoher Breite 89–229
~ in niedriger Breite 89–227
koronale ~ 89–264
koronale aktive ~ 89–247
~ mit Metallinien 89–265
polare ~ 89–230
quasi-eruptive ~ 89–249
ruhende ~ 89–263
sekundäre ~ 89–260
Tornado-~ 89–262
Trichter-~ 89–269
Protuberanzen, interaktive 89–243
Protuberanzenbasis 89–234
Protuberanzeneinheit 89–235
Protuberanzenfaden 89–245
Protuberanzenfilm 89–242
Protuberanzenfläche 89–233
tägliche ~ 89–236
Protuberanzen-Gebiet, aktives 89–240
Protuberanzenhäufigkeit 89–237
Protuberanzenknoten 89–244
Protuberanzenspektroskop 88–58
Proxima Centauri 92–124
Prozess der Massenabnahme 86–183
e-~ 94–99
innerer ~ 82–258
p-~ 94–91
r-~ 94–89
s-~ 94–88
tektonischer ~ 82–201
Urca-~ 94–51
Pseudo-Veränderlicher 95–51
Psychrometer 77–121
Ptolemäisches Weltsystem 99–22
Pulsar 95–137
Gamma-~ 95–139
~ mit Gammastrahlung 95–139
~ mit Röntgenstrahlung 95–138
optischer ~ 95–140
Pulsarmagnetosphäre 95–147
Pulsarpaar 95–140a
Pulsation, vereinzelte 95–146
Pulsationen 79–158
adiabatische ~ 95–98
erdmagnetische ~ 75–81
geomagnetische ~ 75–81
magnetische ~ 75–81
nichtadiabatische ~ 95–99
nichtradiale ~ 95–97
radiale ~ 95–96
Pulsationsinstabilität 93–213
Pulsationsperiode 95–100
Pulsdauer 86–280
pulsierende Fläche 79–149
pulsierender Bogen 79–147
Pulverisieren 82–239
Pulverisierung 82–239
Pumpen, chemisches 97–19
~ durch infrarote Strahlung 97–18
~ durch Strahlung 97–17
~ durch ultraviolette Strahlung 97–18a
optisches ~ 97–17
Pumpmechanismus 97–16
Punkt der Meteorexplosion 86–122
~ des Erlöschens 86–38
~ des Meteorzerplatzens 86–121
~ des Zerplatzens 86–121
heller ~ 89–201
magnetisch konjugierter ~ 75–23
neutraler ~ 89–152
singulärer ~ 94–116
stationärer ~ 83–45
sublunarer ~ 81–129
subsolarer ~ 81–127
substellarer ~ 81–130
subterrestrischer ~ 81–128
Subsatelliten-~ 71–30
Punktquellenmodell 93–138
Puppis 92–81
Purpurlicht 79–54
P-Welle 73–46
Pyranometer 88–43
Pyrheliometer 88–39
Wasserrühr-~ 88–41
Wasserstrom-~ 88–40
pyroklastischer Kegel 73–120
Pyroxen 74–21
Pyxis 92–82

Q

Quadrantiden 86–247
Quadratur 83–64
östliche ~ 83–65
westliche ~ 83–66
Quantenfeld 99–170m
Quark-Suppe 99–169a
Quartär 74–88
Quasag 98–361
Quasar 98–360
radioruhiger ~ 98–362
quasi-eruptive Protuberanz 89–249
quasisimultane Beobachtungen 71–36
quasistationäre Kontraktion 94–40
quasistatische Kontraktion 94–40
quasistellare Galaxie 98–361
~ Quelle 98–358
~ Radioquelle 98–360
quasistellares Objekt 98–359
radioruhiges ~ Objekt 98–362
Quelle, dekametrische 83–186
quasistellare ~ 98–358
Querschnitt, geometrischer 84–103
stirnseitiger ~ 86–176
Querwelle 73–43
Quintett, Stephans 98–309

R

Rabe 92–41
Radar-Aurora 79–95

Radarbeobachtungen 86–265
Radarecho 86–295
Radar-Polarlicht 79–95
Radarreichweite 86–288
Radialgeschwindigkeit 91–24
Radialgeschwindigkeitskurve 95–20
Radiant 86–53
~ eines Meteorstromes 86–237
~ geozentrischer ~ 86–58
individueller ~ 86–55
korrigierter ~ 86–58
scheinbarer ~ 86–56
stationärer ~ 86–241
wahrer ~ 86–59
Radiantendichte 86–63
Radiantenort 86–60
Radiantenverschiebung 86–242
tägliche ~ 86–69
Radiationsfeld 86–238
Radiationsfläche 86–238
radiative Schicht 93–134
Radikal 85–136
freies ~ 85–137
Radikalkoma 85–66
radioaktiver Kohlenstoff 74–95
~ Zerfall 74–64
Radioaktivität 74–61
Radioalbedo 83–94
Radio-Aurora 79–94
Radioausbruch des Jupiters 83–182
Radioecho 86–295
Radioecho-Beobachtung 86–265
Radioecho-Methode 86–265
Radioemission des Jupiters 83–181
sporadische ~ des Jupiters 83–181
Radioflare 89–180
Radioflügel 98–286
Radiofrequenzalbedo 83–94
Radiofrequenz-Spektrometer 88–124
Radiogalaxie 98–206
radiogene Wärme 83a–62
radiogenes Blei 74–97
~ Edelgas 87–132
Radioheliogramm 88–96
Radioheliograph 88–95
Radiointerferometer 88–97
Radiokarbon 74–95
Radiokarte 81–135a
Radiokohlenstoff 74–95
Radiokorona 88–239
Radiometeor 86–267
radiometrische Helligkeit 91–70
Radionuklid, kosmogenes 82–255
Radio-Polarlicht 79–94
Radioquelle 98–360
dekametrische ~ 83–186
quasistellare ~ 98–360
radioruhiger Quasar 98–362
radioruhiges quasistellares Objekt 98–362
Radiosonne 88–3
Radiostern 91–158
Radio-Sterngrösse 86–268
Radiostrahlungsalbedo 83–94
Radioszintillation 84–36
interplanetare ~ 84–37
interstellare ~ 97–109b
Radiowolke 98–286
Radius, Schwarzschildscher 99–163
Raketensondierung 76–48
Rand 89–29
am ~ 89–29
äusserster ~ 89–32
~ der Sonne 89–27
~ des Mondes 81–95
dunkler ~ 81–97
~ einer Galaxie 98–294
erhöhter ~ 81–222
heller ~ 81–96
über dem ~ 89–30
Randeruption 89–202
Randflare 89–202
randloser Krater 81–226
Randmeer 72–27, 81–285
Randmoräne 72–131
Randübertritt 89–31
Randverdunkelung 88–142
Randverdunkelungsgesetz 88–143
Randverdunkelungskoeffizient 88–144
Randzone 81–113
Rapilli 82–169
Ras Alhague 92–125
Rate sporadischer Meteore 86–233
Rauch 97–57
~ mit Nebel 76–83
Rauchglas 88–33
Rauchschweif 87–21
Rauhigkeit der Oberfläche 82–106
Rauhigkeitsbeiwert 82–107
Raum 99–100
absoluter ~ 99–116
aussergalaktischer ~ 97–5
ausserirdischer ~ 87–73a
dunkler ~ 89–297
dunkler ~ um Protuberanzen 89–297
elliptischer ~ 99–110
endlicher ~ 99–103
Euklidischer ~ 99–107
extragalaktischer ~ 97–5
flacher ~ 99–107
freier ~ 76–9
gekrümmter ~ 99–114
hyperbolischer ~ 99–106
interfilamentarer ~ 89–99
intergalaktischer ~ 97–7
interplanetarer ~ 84–2
interstellarer ~ 97–4
Lobatschewskischer ~ 99–105
metrischer ~ 99–100a
Newcombscher ~ 99–110
nichteuklidischer ~ 99–114
parabolischer ~ 99–107
Riemannscher ~ 99–112
sphärischer ~ 99–111
unendlicher ~ 99–104
Raumdichte 86–263
Räume, erlaubte 79–134
verbotene ~ 79–135
Raumgeschwindigkeit 91–30
Raumkoordinate 99–124
Raumkrümmung 99–115
räumliche Triangulation 71–25
Raumozeanographie 72–69
Raumsonde zu einem Asteroiden 84–99
Raumtemperatur 77–31
normale ~ 77–31
Raumverteilung 96–119
Raumwelle 73–42
Raum-Zeit 99–119
raumzeitlich 99–120
Raumzeitpunkt 99–125
Raum-Zeit-Welt 99–122
Rauschen, akustisches 88–220
seismisches ~ 73–54
Rauschsturm 83–185
~ des Jupiters 83–185
Reaktion, photochemische 76–63
Reaktionsgeschwindigkeit 94–53
Reaktionsrate, thermonukleare 94–53
rechtläufig 83–42
rechtläufige Bewegung 83–43
Reduktion auf den Meeresspiegel 77–45
~ der Bedeckungen 81–108
~ der Sternbedeckungen 81–108
Faye-~ 71–78
reduzierende Atmosphäre 83a–46
Referenzellipsoid 71–18
Referenzsystem 98–136
geodätisches ~ 71–19
Reflexionseffekt 96–26
~ bei Doppelsternen 95–82
Reflexionsnebel 97–128
Reflexionspunkt 75–124
Regen 77–169
koronaler ~ 89–271
orographischer ~ 77–166
Regenbogen 79–65
Regenbogenbucht 81–295
Regen-Meer 81–273
Regenmesser 77–170
Regenschichtwolke 77–145
Region 89–144
asseismische ~ 73–29
erdbebenfreie ~ 73–29
H I ~ 97–49
H II ~ 97–50
magnetische ~ 89–144
Mare-~ 81–268
post-mare ~ 81–270
prae-mare ~ 81–269
seismische ~ 73–28
regionale Sprungbildung 82–205
Regolith 82–104
regularisiertes Geoid 71–17
Regulus 92–126
Reibung 86–173

Reichweite 86–288
Reif 77–177, 83–150
rein streuende Atmosphäre 93–16
Reiterlein 92–104
Rekombination, dissoziative 86–214
Rekombinationsstrahlung 86–212
Rekristallisation 74–42
Rektaszension 81–61
  ~ der Sonne 81–61
  ~ des Mondes 81–60
Rektifikation 95–18
rekurrierende Aktivität 98–267
Relativbewegung 99–99
relativistisch 99–95
relativistische Korrektur 99–140
  ~ Kosmologie 99–28
  ~ Periheldrehung 99–160
relativistischer Sternhaufen 96–154
relativistisches Gas 97–152
  ~ Weltall 99–92
Relativität der Trägheit 99–35
Relativitätstheorie, allgemeine 99–146
  spezielle ~ 99–94
Relativzahl 89–65
Relaxation 98–127
  stossfreie ~ 98–132
Relaxationsgeschwindigkeit 98–129
Relaxationszeit 98–128
  mittlere ~ 98–130
Repulsionskraft 85–87
repulsive Kraft 85–87
Repulsivkraft 85–87
Rest-Geschwindigkeit 98–91
Restmagnetisierung 75–28
Restmagnetismus 75–28
Reticulum 92–83
retrograd 83–46
retrograde Rotation 83–40
Rhegmaglypten 87–122
Richtung, antisolare 75–22
  ~ der Achse 83–34
  ~ des Magnetfeldes 84–10
  ~ entgegen der Sonne 75–22
Riefung 89–173
Riemannscher Raum 99–112
Riese = Riesenstern 91–143
  blauer ~ 91–142a
  früher ~ 91–143a
  heller ~ 91–149
  normaler ~ 91–150
  roter ~ 91–152
  ~ späten Typs 91–144a
Riesenast 91–152b
Riesengalaxie 98–222
  elliptische ~ 98–243
Riesenkern 76–118
Riesenmeteor 86–9
Riesenmolekularwolke 97–13a
Riesenplanet 83–174
Riesenstern = Riese 91–143
Riesenteilchen 76–118
Rigel 92–127
Righeit 73–56
  magnetische ~ 75–117
Rille 81–165
  lineare ~ 81–166
  sinusförmige ~ 81–167
Ring, A- 83–225
  äusserer ~ 83–225
  B-~ 83–222
  Bishopscher ~ 79–88
  C-~ 83–221
  dunkler ~ 83–221
  grosser ~ 79–70
  heller ~ 89–101
  kleiner ~ 79–69
  ~ kleiner Planeten 84–70
  mittlerer ~ 83–222
  46°- ~ 79–70
  ~ von der Kante gesehen 83–218
  22°-~ 79–69
Ring ... 97–146
Ringasteroid 84–71
Ringebene 83–216
Ringform 98–41
ringförmig 97–146
ringförmige Sonnenfinsternis 88–15a
Ringgalaxie 98–259
Ringgebirge 81–265
Ringnebel 97–147
Ringplanetoid 84–71
Ringstrom, äquatorialer 75–98
Ringströme 75–100
Ringstruktur 81–242
Ringsystem 83–220
  ~ des Saturns 83–220
Riometer 78–45
Riss 81–165
Robertson-Walker-Metrik 99–152a
Rochesche Grenze 96–28
Rochescher Lappen 96–28
rollende Steine 81–237
Röntgenburst 79–117
Röntgenfackel 89–166
Röntgenflare 89–183
Röntgengalaxie 98–203
Röntgenheliogramm 88–74
Röntgenkorona 88–240
Röntgenplage 89–166
Röntgenpulsar 95–138
Röntgensonne 88–4
Röntgenstrahlenfackel 89–166
Röntgenstrahlenflare 89–183
Röntgenstrahlen-Heliogramm 88–74
Röntgenstrahlungsausbruch 89–184
Röntgenstrahlungs-Burst 79–117
Rossbreiten 77–102
Rossiter-Effekt 91–44
Rotation der Sonne 88–127
  ~ der Sterne 91–35
  differentielle ~ 88–128
  differentielle ~ der Galaxis 98–113
  direkte ~ 83–39
  ~ eines starren Körpers 98–111
  feuerradähnliche ~ 98–28
  galaktische ~ 98–107
  gebundene ~ 81–14, 83–37
  ~ gemäss den Keplerschen Gesetzen 98–112
  Keplersche ~ 98–112
  ~ mit konvexer Seite voraus 98–28
  nachschleppende ~ 98–28
  nachziehende ~ 98–28
  nichtgebundene ~ 83–38
  rechtläufige ~ 83–39
  retrograde ~ 83–40
  rückläufige ~ 83–40
  starre ~ 98–111
  ungebundene ~ 83–38
  ungleichförmige ~ 91–36
Rotationsachse 83–33
Rotationsenergie 93–120
Rotationsgesetz 98–110
Rotationsgeschwindigkeit 91–48
Rotationsinstabilität 83a–8
Rotationskurve 98–109
Rotationsmoment 83a–5
Rotationsperiode 83–31
  ~ der Sonne 88–129
Rotationspol 83–36
Rotationssinn 98–108
Rotationssystem 83–177
  ~ I 83–178
  ~ II 83–179
Rotationstemperatur 91–119
Rotationsveränderlicher 95–65
Rotationsverzerrung 91–44
Rotationszeit 83–31
Rotationszerfall 85–119
Rotator 91–42
  schiefer ~ 91–52
Roter Fleck 83–193
roter Riese 91–152a
  ~ Ton 72–54
  ~ Zwerg 91–136
rotierender Körper 91–42
  ~ Sektor 86–147
Rötung, interstellare 97–91
Rötungslinie 96–149
Rotverschiebung 98–297
  anomale ~ 98–367
  ~ die nicht auf dem Hubble-Effekt beruht 98–363
  ~ der Linien im Schwerefeld 99–162
  kosmologische ~ 98–363a
  vielfache ~ 98–364
Rotverschiebung-Entfernungs-Beziehung 99–76
Rotverschiebung-Helligkeits-Diagramm 98–301
rotverschobene Linie 98–366
r-Prozess 94–89
RR Lyrae-Stern 95–93
RR Lyrae-Veränderlicher 95–93
Rubidium-Strontium-Alter 82–271
Rücken, mittelozeanischer 72–45

Rückfluss 82–282
rückläufig 83–46
rückläufige Bewegung 83–47
rückläufiger Satellit 83–21
Rücklaufimpuls 86–290
Rückseite 87–115
~ des Mondes 81–16
Rückstrahlimpuls 86–290
Rücktrieb 86–178
rückwärtiger Teil des Streufeldes 87–40
rückwärtiges Streufeld 87–40
rückwärts gerichteter Strahl 85–57
Ruhemasse 99–143
ruhender Vulkan 73–116
Ruhenergie 99–144
Ruh(e)wellenlänge 98–365
ruhige Photospäre 88–169
~ Sonne 89–11
ruhiger Sonnenwind 84–41
ruhiges Gebiet der Photosphäre 89–56
Ruhmasse 99–143
Rundbildaufnahme 82–27
Russellmischung 93–101
Russglas 88–33
Rutsch 82–199
Rutschen 82–198
Rutschung 82–198

D

S

Sagitta 92–84
Sagittarius 92–85
Sagittarius-Arm 98–35
säkulare Instabilität 93–212
~ Parallaxe 91–9
Säkularparallaxe 91–9
Säkularvariation des erdmagnetischen Feldes 75–85
Salinität 72–66
Salinitätsgrenzfläche 72–67
Salzgehalt 72–66
Salzhaltigkeit 72–66
Sammeln meteoritischen Staubes 87–12
Sandbank 72–96
Satellit 83–17
~ auf retrograder Bahn 83–21
aus Eis bestehender ~ 83–23
der innerste ~ 83–20
ehemaliger ~ 83–240
Galileischer ~ 83–204
natürlicher ~ 83–18
rückläufiger ~ 83–21
unregelmässiger ~ 83–21
Satellitenaltimetrie 71–50
Satellitengeodäsie 71–32
Satellitenhöhenmessung 71–50
Satellitenklimatologie 77–181
Satellitenkrater 81–200
Satellitennetz, geodätisches 71–34
Satellitenozeanographie 72–70
Satellitenstation 71–33
Satellitensystem 83–19
Satellitentriangulation 71–27
Satellitenüberwachung 83–26
Sattel 73–90
Sättigung 76–101
Sättigungsfunktion 93–60
Saturn 83–209
Saturnferne 83–211
Saturnnähe 83–210
saturnographisch 83–212
saturnozentrisch 83–213
Saturnring 83–215
Sauertoff, atomarer 79–24
molekularer ~ 79–23
Sauerstoffbrennen 94–82
Saum der Polkappe 83–151
heller ~ eines Filaments 89–226
Säure, organische 87–69
Schale, feste 95–151
Schalenquellenmodell 93–143
Schall, anomaler 76–36
normaler ~ 76–35
Schallausbreitung 76–32
anomale ~ 76–37
Schallenergie 88–219
Schallenergiefluss 88–221
Schallerzeugung 88–215
Schallfeld 76–40, 86–120
Schallmauer 76–44
Schallpegel 76–38
Schallsonde 76–45
Schallstrahl 76–34
Schallwelle 76–33
Schatten, fliegende 88–31
Schattendichte 81–59
Schattengrenze 83–69
Schattenhöhe 79–46
abgeschirmte ~ 79–49
geometrische ~ 79–47
wahre ~ 79–49
Schattenkegel 81–90
schätzen 95–12
Schauer, meteoritischer 87–37
Scheibe, auf der 88–173
galaktische ~ 98–52
sichtbare ~ 88–171
Zentrum der sichtbaren ~ 88–172
Scheibenmitte 89–35
Scheibenphase 94–22
Scheibenpopulation 98–72
Scheibenstadium 94–22
Scheibenstern 91–85
Schein 79–151
scheinbare Meteorhelligkeit 86–73
scheinbarer Radiant 86–56
Schelf, kontinentaler 72–34
Scherbeanspruchung 73–63
Scherstress 73–63
Scherungswelle 73–43
Scherwelle 73–43
Schicht, absorbierende 88–175
atmosphärische ~ 76–49
dünne ~ 83–149
E-~ 78–26
Heaviside- ~ 78–26
Heaviside-Kennelly-~ 78–26
ionisierte ~ 78–3
konvektive ~ 93–149
radiative ~ 93–134
streuende ~ 83–258
subphotosphärische 88–168
umkehrende ~ 88–174
violette ~ 83–165
Schichtgestein 74–48
Schichthöhe 78–27
Schichtlinie gleicher Zählrate 75–102
Schichtung 82–133
horizontale ~ 82–134
Schichtvulkan 73–121
Schichtwolke 77–139
hohe ~ 77–140
schiefer Rotator 91–52
Schieferung 74–53
Schiffahrtsgefahr 72–97
Schiffskompass 92–82
Schild 72–11, 92–88
Schildvulkan 73–119
Schlackenkegel 73–120
schlafender Vulkan 73–116
Schlamm 72–58
Schlange 92–89
Kopf und Schwanz der ~ 92–89
schlangenförmige Meteorbahn 86–33
Schlangenträger 92–72
schlechte Sicht 77–26
Schleiernebel 97–137
Schleife, interplanetare 84–21
magnetische ~ 84–19
solare interplanetare ~ 84–20
Schlick 72–58
Schlot 73–123
Schlotgang 73–125
Schlucht 83–154
Schmarotzerkrater 73–128
Schmelzen 86–181
Schmelzkruste 87–110
Schmelzpunkt 86–204
Schmelztemperatur 86–204
Schmetterlingsdiagramm 89–76
Schnee 77–175
vereister ~ 72–118
Schneedecke 77–176
Schneegrenze 72–111
Schnelläufer 98–191
schneller Einfang von Neutronen 94–89
Schockwelle, stossfreie 84–49
Schöpfung, einmalige 99–90
kontinuierliche ~ 99–91
Schrägaufname 82–29
Schraubensymmetrie 98–161
Schrumpfungstheorie 73–77
Schub 73–65
Schubwelle 73–43
schüsselförmiger Krater 81–219
Schusskanal 73–122

Schuttabhang 81–233
Schuttdecke 81–234
Schuttkegel 73–120
Schütze 92–85
Schwan 92–44
Schwankungen, erdmagnetische 75–69
geomagnetische ~ 75–69
sekundäre ~ 95–44
tägliche ~ 83–167
überlagerte ~ 95–44
Schwanz der Schlange 92–89
schwarzer Tropfen 83–52
~ Zwerg 91–139
Schwarzschild-Milnesche Integralgleichung 93–53
Schwarzschildsche Metrik 99–152
~ Ungleichung 93–156
schwebende Wolke 89–275
Schwebungsperiode 95–42
Schwefelwasserstoff 97–44
Schweif, am Tage sichtbarer 86–106
anomaler ~ 85–75
~ der Magnetosphäre 75–10
~ einer Galaxie 98–280
~ eines Meteors 86–103
gekrümmter ~ 85–73
gerader ~ 85–72
magnetosphärischer ~ 75–10
mehrfacher ~ 85–76
nachts sichtbarer ~ 86–105
selbstleuchtender ~ 86–108
von der Sonne abgewandter ~ 85–81
zur Sonne gerichteter ~ 85–78
Schweifansatz 86–110
Schweifbildung 85–86
Schweifecho 86–298
Schweifkondensation 85–95
Schweifspektrum 86–107
Schweifspur 86 110
Schweifstern 85–1
Schwelle 72–96
mittelozeanische ~ 72–45
Schwellenkurve der Sekundärkomponente 96–40
~ des Begleiters 96–40
Schwenkung der Stellung 88–105
schwer schmelzbarer Stoff 83a–55
Schwere 71–56
~ Masse 99–149
Schwereanomalie 71–75
lunare ~ 81–144
Schwerebeschleunigung 71–64
Schweredifferentiation 74–56
Schwerefeld 71–57
Schwereformel, internationale 71–76
Schweregrundnetz 71–82
Schwereintensität 71–62
Schwerekarte 71–79
Schwerekontraktion 83–203
Schweremessung 71–67
absolute ~ 71–68
relative ~ 71–69
Schwerenetz 71–80
Welt-~ 71–81
Schwerepotential 71–58
~ der Erde 71–60
Schwereverdunk(e)lung 96–25
Schwerewelle 88–201
Schwerezerquetschung 84–72
Schwerezusammenziehung 83–203
Schwerkraft 71–62
Schwerkraftbeschleunigung 71–64
Schwertfisch 92–46
Schwertgehänge des Orions 92–130
Schwingung zwischen den Spiegelpunkten 75–119
Schwingugstemperatur 91–120
Schwund, totaler 78–40
Schwundgebiet 72–124
Schwung 83a–4
SCNA 78–42
Scorpion 92–86
Scorpius 92–86
Sculptor 92–87
Scutum 92–88
sechsdimensionaler Phasenraum 98–144
Sechsfarben-Photometrie 96–148
Sediment 74–47
Sedimentation 73–109
~ im Schwerefeld 83a–54
marine ~ 72–47
submarine ~ 72–47
Sedimentationsmagnetisierung 75–33
Sedimentationsprozess 74–46
Sedimentationsvorgang 74–46
Sedimentbildung 74–45
Sedimente, organogene 72–59
pelagische ~ 72–57
terrigene ~ 72–48
vom Festland stammende ~ 72–48
Sedimentgestein 74–10
Sedimentverfestigung 74–50
See 81–297, 83–158
~ der Träume 81–299
~ des Frühlings 81–302
~ des Herbstes 81–301
~ des Sommers 81–300
~ des Todes 81–298
Seebeben 73–2c
Seebebenwelle 72–95
Seeboden 72–38
Seebrise 77–97
Seegang 72–86
Seekarte 72–33
Seeklima 77–186
Seeligersches Paradoxon 99–39
Seewind 77–97
Segel 92–98
~ des Schiffes 92–98
Segelkarte 72–33
Sehen, extrafoveales 86–128
indirektes ~ 86–128
seitliches ~ 86–128
Sehstrahl 91–28
auf dem ~ 91–29
seismische Katastrofe 73–40
~ Sondierung 82–216
~ Welle 73–9
~ Zone 73–31
seismischer Strahl 73–14
~ Versuch 82–219
Seismizität 73–3
Seismogramm 73–8
Seismograph 73–7
Seismologe 73–2
Seismologie 73–1
aktive ~ 82–217
passive ~ 82–218
seismologische Station 73–6
seismologisches Observatorium 73–5
Seismometer 82–222
aktives ~ 82–222
passives ~ 82–221
Seite, dunkle 75–21
sonnenbeschienene ~ 75–20
Seitenkrater 73–128
Seitenmoräne 72–131
Sektor 85–60
interplanetarer ~ 84–16
Magnet-~ 84–16
~ mit beweglichem Unterbrecher 86–148
unterbrechender ~ 86–148
Sektorgrenze 84–18
Sektorstruktur 84–15
Sektorunterbrechung 86–150
sekundäre Formation 81–151
~ Welle 73–43
sekundärer Krater 81–199
Sekundärformation 81–151
Sekundärkomponente 96–38
dunkle ~ 96–42
Grenzkurve der ~ 96–40
Sekundärkrater 81–199
Sekundärmagnetisierung 75–36
Sekundärzeit 74–72
selbstgravitierend 94–33
selbstleuchtend 96–45
Selbstumkehr der Magnetisierung 75–38
Selected Areas 98–78
Selenochemie 82–74
Selenodäsie 81–109
selenodätisch 81–110
Selenographie 81–1
selenographisch 81–8
Selenoid 81–117
Selenologe 82–49
Selenologie 82–1
selenologisch 82–48
selenologische Kartographie 82–51
selenophysikalische Aufnahme 82–39
selenophysikalisches Beobachtungsnetz 82–40
selenozentrisch 81–131
selenozentrische Lage 81–132
semikonvektive Zone 93–160

D

Sendeatenne 78–16
Sendefrequenz 86–273
Sendeleistung 86–270
mittlere ~ 86–272
Sender 78–15
Senkrechtaufnahme 82–28
Senkungskaldera 81–182
Sequenz der Spiralgalaxien 98–255
Sequenzphotographie 79–103
Serpens 92–89
Sextans 92–90
Sextant 92–90
Seyfert-Galaxie 98–262
Sicht 77–24
gute ~ 77–25
schlechte ~ 77–26
sichtbare Scheibe 89–26
~ Sonnenscheibe 89–26
Sichtbarkeit der Finsternis 81–88
~ der Sternbedeckung 81–103
Sichtbakeitsdauer 86–111
~ eines Meteorschweifs 86–111
Sichtbarkeitsfunktion 89–55
Sichtbarkeitsgebiet 86–126
~ der Finsternis 81–89
Sichtweite 87–31
SID 78–41
siderischer Monat 81–45
~ Umlauf 81–45
Siderit 87–82
Siderolith 87–78
Siderostat 88–44
Siebengestirn 96–110
Silberscheibenpyrheliometer 88–42
Silikat 74–14
Silikatpartikel 97–76
Silikatstaub 85–53
Siliziumbrennen 94–83
Siliziumdioxid 74–13
Siliziummonoxid 97–37
Silizium-Stern 91–190
Silur 74–80
Simulation, mathematische 98–323
Simultanbeobachtungen 71–35
singuläre Galaxie 98–258
singulärer Punkt 94–116
Singularität 99–42
nackte ~ 94–117
sinusförmige Depression 81–161
~ Rille 81–167
~ Vertiefung 81–161
Sirius 92–128
~ B 91–138
Sirius-Begleiter 91–138
Skala, geochronologische 74–68
Skalenfaktor 99–81a
Skalenhöhe 76–18, 88–236
Skorpion 92–86
Smog 76–83
Smyth-Meer 81–287
Sobieskischer Schild 92–88
Sockel 72–12
solar 88–5
solarer Plasmastrom 84–33
Solarkonstante 88–148
solar-terrestrische Beziehungen 84–55
~ Erscheinungen 84–55
Sommersternbild 92–1a
Sondierung 76–47
Ionosphären-~ 78–19
seismische ~ 82–216
~ von oben 78–20
Sonne 88–1
aktive ~ 89–12
blaue ~ 79–92
gestörte ~ 89–12
optische ~ 88–2
ruhige ~ 89–11
sichtbare ~ 88–2
ungestörte ~ 89–11
zur ~ gerichteter Schweif 85–78
Sonnen- 88–5
Sonnenaktivität 89–1
Index der ~ 89–8
Maximum der ~ 89–2
Minimum der ~ 89–4
terrestrische Effekte der ~ 84–55
Voraussage der ~ 89–64
Zyklus der ~ 89–71
Sonnenapex 98–64
Sonnenatmosphäre 88–163
äussere ~ 88–165
höhere Schichten der ~ 88–165
tiefere ~ 88–166
tiefere Schichten der ~ 88–166
Sonnenaufgang 88–8
Sonnenaureole 79–62
sonnenbedingte Phasenanomalie 78–52
sonnenbelichtetes Polarlicht 79–130
sonnenbeschienene Hemisphäre 83—67
Sonnenbewegung 98–62
fundamentale ~ 98–66
Standard-~ 98–63
~ um galaktisches Zentrum 98–66
Sonnendaten 89–9
Sonnendepression 79–45
Sonenneinheiten 91–123
Sonneneinstrahlung 77–52
Sonnenerscheinung 89–13
Sonneneruption 89–175
Sonneneruptionseffekt im atmosphärischen Störpegel 78–51
~ im kosmischen Rauschen 78–42
Sonnenfackel 89–161
sonnenferner Planet 83–15
Sonnenfinsternis 88–15
künstliche ~ 88–17a
partielle ~ 88–16
ringförmige ~ 88–15a
totale ~ 88–17
zentrale ~ 88–160
Sonnenfinsternisexpedition 88–30
Sonnenfleck 89–88
Sonnenfleckenaktivität 89–62
Sonnenfleckengruppe 89–105
Sonnenfleckenrelativzahl 89–65
Sonnenfleckenzyklus 89–72
Sonnenindex 89–8
Sonneninstrumente 88–35
Sonnenkorona 79–62, 88–225
Sonnenkorpuskularstrom 84–35
Sonnenkranz 79–62
Sonnenkreuz 79–81
Sonnenlicht 77–49
Sonnenmagnetograph 88–82
Sonnenmasse 91–122
Sonnenmaximum 89–2
Sonnenminimum 89–4
sonnennaher Planet 83–14
Sonnennebel 83a–23
Sonnenoberfläche 88–164
Sonnenobservatorium 88–7
Sonnenphänomen 89–13
Sonnenphysik 88–6
Sonnenplanet 83–11
Sonnenplasma 84–29
magnetisiertes ~ 84–32
Sonnenprojektionsschirm 88–13
Sonnenrand 89–28
Sonnenrefraktor, kuppelloser 88–53
Sonnenrotation 88–127
Sonnenscheibe 89–24
Mitte der ~ 89–35
sichtbare ~ 89–26
Sonnenschein 88–9
Sonnenscheindauer 88–10
Sonnenscheinschreiber 77–61
Sonnenschwingungen 88–223a
Sonnenseismologie 88–223b
Sonnenspektrograph 88–77
Sonnenspektrum 88–74a
Sonnenstrahlung 88–134
direkte ~ 77–50
Sonnensucher 88–51
Sonnensystem 83–12
Sonnensystem-Meteor 86–44
Sonnentätigkeit = Sonnenaktivität 89–1
Maximum der ~ 89–2
Minimum der ~ 89–4
Voraussage der ~ 89–64
Sonnenteleskop 88–49
automatisches ~ 88–50
Ballon-~ 88–52
horizontales ~ 88–55
Sonnenüberwachung 89–16
Sonnenumgebung 98–58
in der ~ 98–59
Sonnenuntergang 88–11
Sonnenurnebel 83a–21
Sonnenwind 88–156
permanenter ~ 84–40
ruhiger ~ 84–41
~-Hohlraum 84–28
Sonnenwindspektrometer 82–46
Sonnenwindstrom 84–35a
Sonnenwindzerstäubung 82–245

Sonnenzyklus 89–71
Spallation 94–100
Spallationsreaktion 94–100
Spalt, gekrümmter 88–65
oszillierender ~ 88–70
Spalte 73–78
~ am Kraterrande 82–211
Spaltenausbruch 73–113
Spaltenerguss 73–141
Spalteneruption 73–113
Spaltensystem 73–79
Spaltungstheorie 96–75
Spaltungsvorgang 94–100
Spannung, thermische 87–27
Spannungsriss 82–206
Spannungstensor 73–69
Speicherung von Partikeln 75–123
Spektralklasse 91–107
Spektralklassifikation 91–90
zweidimensionale ~ 91–105
Spektrallinie, atmosphärische 88–140a
tellurische ~ 88–140a
Spektralsequenz 91–94
Spektraltyp 91–108
Spektraltypus 91–108
Spektrobologramm 88–76
Spektrobolometer 88–75
Spektrogramm, dynamisches 88–125
Spektrograph, dynamischer 88–124
Frequenzdrift-~ 88–124
Spektrographie, Mössbauer- 82–142
Spektroheliogramm 88–72
Spektroheliograph 88–71
Spektroheliokinematograph 88–81
Spektrohelioskop 88–67
Spektrum, dynamisches 88–125
Flash ~ 88–25
gemischtes ~ 96–19
integrales ~ 96–137
symbiotisches ~ 95–117
überlagertes ~ 96–19
zusammengesetztes ~ 96–19
Spektrum-Doppelstern 96–14
Spektrum-Leuchtkraft-Beziehung 91–130
Spektrumveränderlicher 95–48
Spherics 78–50
sphärische Albedo 83–90
sphärischer Raum 99–111
Sphäroid, abgeplattetes 98–158
sphäroidale Galaxie 98–244
sphäroidisches System 98–159
Spiegel, beweglicher 86–130
rotierender ~ 86–130
vibrierender ~ 86–130a
Spiegelpunkt 75–124
Spika 92–129
Spikule 88–195
Spikulum 88–195
Spinar 98–25
spindelförmige Galaxie 98–239
spindelförmiges Meteor 86–77
Spiralarm 98–27
Spiralarm-Indikator 98–39
Spiralarm-Population 98–73
Spirale 98–7, 98–249
Archimedische ~ 84–34
balkenlose ~ 98–250
normale ~ 98–250
spiralförmiger Strahl 85–59
Spiralgalaxie 98–249
~ mit Balken 98–253
Spiralnebel 98–249
Spiralstruktur 98–8
Theorie der ~ 98–173
Spitze der Schicht 76–55
Spitzenleistung 86–271
Spoerersches Gesetz 89–75
sporadische koronale Kondensation 89–283
~ Radioemission des Jupiters 83–181
sporadisches Meteor 86–23
Spray 89–258
Springflut 82–273
Springtide 82–273
Spritzfleckprotuberanz 89–255
Spritzprotuberanz 89–255
dunkle ~ 89–256
helle ~ 89–257
Sprödigkeit 86–192
Sprung 73–91
gewöhnlicher ~ 73–93
Sprungbildung 73–92
lokale ~ 82–204
regionale ~ 82–205
Sprungentfernung 78–38
Spur, fossile 87–121
Spurbildungsgeschwindigkeit 86–154
Spurendichte 87–141
Spurenelement 82–145
Spurengas 76–88
Spurenkonzentration 83–266
Spurgeschwindigkeit 86–154
stabil 93–208
thermisch ~ 93–209
Stabilität 93–204
~ der Sterne 93–206
dynamische ~ 93–207, 98–198
~ einer Bahn 98–193
Stabilitätsbedingung 93–205
Stabilitätskoeffizient 98–199
Stadium, hydrodynamisches 94–43
T-Tauri ~ 94–47
Standard-Atmosphäre 76–11
Standardmodell 93–137
Standard-Sonnenbewegung 98–63
stark abgeplattetes System 98–164
Stärke des Erdbebens 73–34
~ des Schattens 81–59
starre Rotation 98–111
starrer Körper 83–111
Starrheit 73–56
Station, automatische 82–19
automatische ~ auf dem Mond 82–19
seismologische ~ 73–6
stationärer Punkt 83–45
stationäres Potential 98–147
Staub, die Erde umkreisender 84–53
galaktischer ~ 97–53
interplanetarer ~ 84–51
interstellarer ~ 97–54
kosmischer ~ 97–52
meteorischer ~ 87–11
meteoritischer ~ 87–10
vulkanischer ~ 73–143
zirkumstellarer ~ 97–55
zirkumterrestrischer ~ 84–53
Staubball 86–190
staubbedeckt 82–120
Staubfilter 76–95
staubfrei 82–119
Staubgehalt 85–80
Staubglobule 98–290
Staubhülle 97–116
zirkumstellare ~ 97–117
Staubkollektor 87–13
Staubkomponente 98–232
Staubkörner, interstellare 97–59
Staubmeteor 86–190
Staubnebel 97–115
Gas- und ~ 97–119
Staubpartikeln, interstellare 97–68
Staubschicht 82–122
Staubschweif 85–79
Staubstreifen 98–289
Staubsturm 83–255
Staubverdichtung 98–290
Staubwolke 83–256
zirkumstellare ~ 97–117
Stauregen 77–166
Steifigkeit, magnetische 75–117
Steilabfall 81–184
Steilheit des Abstieges 95–36
~ des Anstieges 95–29
Steilwand 81–184
Stein, eckiger 82–131
Steinbock 92–29
Steine, rollende 81–237
Steinmeteorit 87–60
Steinschutt 81–236
Stellarastronomie 91–2
Stellardynamik 98–120
stellare Materie 93–124
stellarer Urnebel 97–63
Stellarstatistik 98–77
Stephans Quintett 98–309
stereographische Projektion 71–47
Stereokomparator 91–32
Stern 91–1, 92–3, 95–47
abgestorbener ~ 91–167
alter ~ 94–64
anomaler ~ 91–96
anregender ~ 97–125
begleitender ~ 96–39
δ Cephei-~ 95–92
Delta-Cephei-~ 95–92

D

~ der ersten Generation 94–24
~ der Hauptreihe 91–142
~ der Hauptsequenz 91–142
der hellste ~ 98–304
der hellste ~ einer Galaxie 98–304
~ der Scheibenpopulation 91–85
der Veränderlichkeit verdächtiger ~ 95–5
~ des Systems 98–133
entarteter ~ 93–180
entweichender ~ 96–160
entwickelter ~ 94–74
expandierender ~ 91–162
explodierender ~ 91–165
~ frühen Typs 91–102
~ grosser Masse 91–126
heisser ~ 91–155
heliumarmer ~ 91–192
heliumreicher ~ 91–193
heller ~ 91–83
Horizontalast- ~ 91–141
~ in Pol-Aufsicht 91–40
in sich zusammenstürzender ~ 94–112
instabiler ~ 95–9
intergalaktischer ~ 91–163
junger ~ 94–63
~ kleiner Masse 91–124
kohlenstoffarmer ~ 91–194
kollabierender ~ 91–164
kollabierter ~ 94–111
kontrahierender ~ 91–161
kühler ~ 91–156
leuchtanregender ~ 97–125
magnetischer ~ 91–51
massearmer ~ 91–124
massereicher ~ 91–126
metallarmer ~ 91–197
metallreicher ~ 91–196
~ mit Besonderheiten 91–92
~ mit Besonderheiten im Spektrum 91–92
~ mit einem unsichtbaren Begleiter 96–13
~ mit einer ausgedehnten Atmosphäre 93–6
~ mit einer ausgedehnten Hülle 93–7
~ mit Emissionslinien 91–159
~ mit geringer Geschwindigkeit 98–192
~ mit Hülle 93–5
~ mit pekuliarem Spektrum 91–92
~ mit Rotationsachse in Richtung des Sehstrahls 91–40
~ mit Temperatur Null 91–166
~ mit veränderlichem Spektrum 95–48
~ mittlerer Masse 91–125
naher ~ 98–60
Neuer ~ 95–110
nicht zum Haufen gehörender ~ 96–84
nichtrotierender ~ 91–38
nichtsphärischer ~ 91–86
nichtstationärer ~ 95–2
nichtverfärbter ~ 97–95
normaler ~ 91–95
novaähnlicher ~ 95–118
pulsierender ~ 95–95
R Coronae Borealis ~ 95–83
rot verfärbter ~ 97–94
roter ~ 91–97
rotierender ~ 91–37
RR Lyrae-~ 95–93
schnell rotierender ~ 91–39
schwacher ~ 91–82
Silizum- ~ 91–190
~ späten Typs 91–103
stationärer ~ 95–3
stabiler ~ 95–8
supermassiver ~ 91–127
symbiotischer ~ 96–20
systemfremder ~ 98–134
Tychonischer ~ 95–135
überdichter ~ 91–153
überleuchtkräftiger ~ 91–134
überluminoser ~ 91–134
unterleuchtkräftiger ~ 91–133
unterluminoser ~ 91–133
varänderlicher ~ 95–1
~ vom Alter Null 94–59
~ vom Spektraltyp der Sonne 91–104
~ von grosser Masse 81–126
~ von kleiner Masse 91–124
W-~ 91–160
wasserstoffarmer ~ 91–191
Wolf-Rayet-~ 91–160
~ zu geringer Leuchtkraft 91–133
~ zu hoher Leuchtkraft 91–134
zusammenfallender ~ 91–164
Sternära 99–178
sternarmer Haufen 96–102
Sternassoziation 96–88
Sternatlas 92–12
Sternatmosphäre 93–2
Sternaufbau 93–1
Sternbeben 95–153
Sternbedeckung 81–94
~ durch den Mond 81–94
Reduktion der ~ 81–108
Sichtbarkeit der ~ 81–103
Voraussage der ~ 81–102
Sternbedeckungszone 81–107
Sternbegegnungen 98–121
Anzahl der ~ 98–122
Sternbegleiter 96–39
Sternbild 92–1
Tierkreis-~ 92–5
Sterndichte 98–82
räumliche ~ 98–83
Sterne, Entwicklung der 94–1
Generation der ~ 94–23
Sterneinfang 96–73
Sternentstehung 94–11
heftige ~ 94–11a
Sternentstehungsrate 94–12
Sternentwicklung 94–1
Sterneruption 95–70
Sternflecken 97–62
sternförmiger Kern 85–38
Sterngehalt 98–231
~ einer Galaxie 98–231
Sterngeneration 94–23
Sterngrösse = Sternhelligkeit = Grössenklasse = Magnitudo 91–56
absolute ~ 91–64
absolute ~ eines Planeten 83–86
bolometrische ~ 91–68
elektronographische ~ 91–67
~ im Zenit 91–63
~ in der mittleren Opposition 83–85
infrarote ~ 91–71
mittlere ~ 83–85
photographische ~ 91–66
photovisuelle ~ 91–69
Radio- ~ 86–268
radiometrische ~ 91–70
scheinbare ~ 91–61
visuelle ~ 91–65
Sternhaufen 96–79
galaktischer ~ 96–99
kugelförmiger ~ 96–114
metallarmer ~ 96–113
metallreicher ~ 96–112
~ mit hoher Konzentration 96–129
offener ~ 96–99
relativistischer ~ 96–154
Sternhaufenauflösung 96–103
Sternhaufenkern 96–130
Sternhaufenmitglied 96–82
Sternhelligkeit 91–55
absolute ~ 91–64
scheinbare ~ 91–61
Sternhimmel 91–1a
Sterninhalt 98–172
Sterninneres 93–72
Sternkarte 92–9
drehbare ~ 92–10
Sternkatalog 92–13
Sternkette 96–77
Sternkorona 93–4
Sternkosmogonie 94–7
Sternlicht 79–10
direktes ~ 79–11
gestreutes ~ 79–12
Sternmagnetismus 95–54
Sternmasse 91–121
Sternmaterie 93–124
Sternmittelpunkt 93–84
Sternmodell 93–123
inhomogenes ~ 93–139
relativistisches ~ 93–198
Sternnacht 79–3d

D

Sternnebel 97–63
Sternpaar 96–3
getrenntes ~ 96–35
halbgetrenntes ~ 96–34
physisches ~ 96–7
weites ~ 96–9
Sternparallaxe 91–4
Sternpopulation 98–67
galaktische ~ 98–230
Typ I-~ 98–71
Typ II-~ 98–68
sternreicher Haufen 96–100
Sternring 96–78
Sternrotation 91–35
Sternschnuppe 86–2
Sternschnuppenfall 86–215
Sternschnuppenschwarm 86–215
Sternspektrum 91–88
normales ~ 91–89
Sternspur 86–152
photographische ~ 86–152
Sternsterberate 94–115
Sternstrom 98–98
Sternstromparallaxe 91–7
Sternstruktur 93–1
Sternsystem 96–1
durch Superposition gebildetes ~ 98–165
lokales ~ 98–61
Sterntyp 91–101
früher ~ 91–102
später ~ 91–103
Sternurnebel 97–63
Sternveränderlichkeit 95–4
Sternwind 93–14
Sternwolke 98–6
Sternzahl 98–80
Sternzählung 98–79
Sternzentrum 93–84
Sternzusammenstoss 83a–16
Steuermodul 82–6
Stickstoffbanden, erste positive 79–20
negative ~ 79–21
Stickstoffstern 91–189
Stier 92–91
Stillstand eines Planeten 83–44
Stirnmoräne 72–130
stirnseitiger Querschnitt 86–176
Stoff, absorbierender 85–130
absorbierter ~ 85–131
feuerfester ~ 83a–55
flüchtiger ~ 82–262, 85–141
leicht schmelzbarer ~ 83a–56
nichtflüchtiger ~ 83a–53
poröser ~ 82–111
schwer schmelzbarer ~ 83a–55
schwer schmelzender ~ 83a–55
Störpegel, atmosphärischer 78–50
Störung, Bai- 75–80
einer Eruption zugeordnete ~ 84–46
Grosse Südliche ~ 83–194
interplanetare ~ 84–45
ionosphärische ~ 78–39
plötzliche ionosphärische ~ 78–41
Südliche Tropische ~ 83–194
Störungen, atmosphärische 78–50
~ durch die Sonne 83–24
erdmagnetische ~ 75–70
gegenseitige ~ 83–25
gegenseitige ~ zwischen den Satelliten 83–25
magnetische ~ 75–70
radiale ~ 93–215
Störungsgebiet, Südliches 83–194
Stoss zwischen Sternen 98–26
unelastischer ~ zwischen Sternen 98–26
Stossauswürflinge 82–167
stossfreie Schockwelle 84–49
stossfreies Plasma 84–50
~ System 98–167
Stosshypothese 83a–14
Stossmetamorphose 82–237
Stosspumpen 97–20
Stossrelaxation 98–131
Stosssystem 98–166
Stossversteinerung 82–238
Strahl 85–55
gekrümmter ~ 85–58
grüner ~ 79–58
rückwärts gerichteter ~ 85–57
seismischer ~ 73–14
spiralförmiger ~ 85–59
tangentialer ~ 81–169
Strahlen 79–155
~ auf dem Mond 81–168
polare ~ 89–294
Strahlenkegel 95–156
Strahlenkranz 79–60
Strahlenoberfläche 81–171
Strahlenstruktur, Formen ohne 79–142
Strahlensystem 81–170
Strahlenteilungsprisma 88–88
Strahlstrom 77–88
Strahlströmung 77–88
Strahlteiler 88–88
Strahlung, andere Energieform als 88–218
eingefangene ~ 75–112
geomagnetisch eingefangene ~ 75–112
integrierte ~ 88–147
künstliche eingefangene ~ 75–113
verdünnte ~ 93–38
Strahlungsära 99–172
Strahlungsdichte 99–85
Strahlungsdruck 93–91
Strahlungsempfänger mit extrem schmaler Richtkeule 88–121
Strahlungsfeld 93–30
diffuses ~ 93–31
Strahlugsfluss 93–113
austretender ~ 93–114
Strahlungsgleichgewicht 88–252
Strahlungsgürtel 75–101
äusserer ~ 75–105
innerer ~ 75–106
künstlicher ~ 75–108
natürlicher ~ 75–107
Van Allenscher ~ 75–101
Strahlungskopplung 99–173
Strahlungskühlung 94–98
Strahlungsprozess 86–211
Strahlungspunkt 86–53
Strahlungsquelle, kompakte 98–273
Strahlungstemperatur 91–114
~ für eine gegebene Wellenlänge 91–114
Strahlungstransport 93–29
Strahlungstransportgleichung 93–33
Strahlungsvorgang 86–211
Strahlungszone 75–108
künstliche ~ 75–108
Strain 73–67
Strang, magnetischer 89–146
stratigraphisches Alter 82–267
Stratokumulus 77–142
Stratopause 76–71
Stratosphäre 76–70
Stratovulkan 73–121
Stratus 77–139
Streifen 83–190
dunkle ~ 98–288
Stress 73–62
Stresstensor 73–69
Streudiagramm 77–58
streuende Atmosphäre 93–16
rein ~ Atmosphäre 93–16
~ Schicht 83–258
Streufeld eines Meteoritenschauers 87–38
rückwärtiges ~ 87–40
Streufunktion 77–58
Streuindikatrix 77–58
Streuinstrument, Alpha-~ 82–76
Streustrahlung des Himmels 88–136
Streuung des Lichtes 77–57
atmosphärische ~ des Lichtes 77–57
Mie-~ 83–76
Strom, absinkender 93–150
aufsteigender ~ 93–151
~ des Sonnenplasmas 84–33
interstellarer ~ 86–224
kometarischer ~ 86–222
ozeanischer ~ 72–60
periodischer ~ 86–228
permanenter ~ 86–227
planetarischer ~ 86–220
sporadischer ~ 86–226
tätiger ~ 86–223
temporärer ~ 86–225
Stromaktivität 86–229
Index der ~ 86–230
relative ~ 86–231
Ströme, tellurische 75–5
Stromlinie 82–194
Strommeteor 86–24

D

Stromradiant 86–237
Stromsystem 75–97
Strömungen 83–187
Strömungsfront 82–193
Strömungslinie 82–194
Stromzweig 86–240
Strontiumalter 82–271
Struktur 82–102
~ der Meteoriten 87–100
~ der Galaxis 98–4
~ des Meteorstromes 86–243
fadenähnliche ~ 88–209
faserige ~ 97–130
grossräumige ~ 99–171
laminare ~ 83–148
makroskopische ~ 88–207
mikroskopische ~ 88–208
mitrotierende ~ 84–17
mitrotierende ~ des Magnetfeldes 84–17
post-mare ~ 81–154
prae-mare ~ 81–153
Strukturbeziehungen 82–266
Struktureigenschaften 98–224
Struktureinzelheiten 98–224
~ im Radiowellenbereich 98–235
Strukturelement 98–287
Strukturen 81–152
Strukturformen 81–152
Strukturindex eines Galaxienhaufens 98–342
~ eines Sternhaufens 96–80
Strukturkarte 73–97, 82–203
Strukturmerkmal 88–248
Strukturparameter 96–120
Strukturtyp 98–225
Stufe, geothermische 74–59
Stufenkeil 88–80
Stufenschätzungsmethode 95–10
Sturm, allmählich beginnender 75–78
~ des Standard-Typs 75–72
erdmagnetischer ~ 75–71
geomagnetischer ~ 75–71
interplanetarer ~ 84–44
~ mit plötzlichem Beginn 75–73
Sturmanfang, plötzlicher 75–74
Sturmflut 72–76
subaquatischer Vulkan 73–117
Subflare 89–195
Sublimation 85–138
Sublimationskerne 76–110
Sublimationswärme 85–139
sublunarer Punkt 81–129
submarine Sedimentation 72–47
submariner Berg 72–43
~ Vulkan 73–117
submarines Erdbeben 73–2c
subnukleare Dichte 93–190
subphotosphärische Schicht 88–168
Subsatelliten-Punkt 71–30
subsolarer Punkt 81–127
Substanz, poröse 82–111
substellarer Punkt 81–130
Substorm 79–144
subterrestrischer Punkt 81–128
Subtraktion, photographische 88–94
Suche nach Meteoriten 87–45
Supernova- ~ 95–136
Südatlantische Anomalie 75–68
Südhimmel 79–3c
südliche Aurora 79–106
~ Krone 92–39
~ Polarität 89–150
~ tropische Störung 83–194
~ Wasserschlange 92–56
Südlicher Fisch 92–80
Südliches Dreieck 92–94
~ Störungsgebiet 83–194
Südlicht 79–106
Südmeer 81–286
Sulfat 74–29
Sulfid 74–28
Sulfit 74–28a
Sumpf 81–303
~ der Epidemien 81–304
~ des Schlafes 81–306
~ des Zerfalls 81–305
Supergalaxie 98–346
Lokale ~ 98–347
Vigro-~ 98–347
Supergalaxienhaufen 98–346
supergranulares Element 88–182
Supergranulation 88–183
Supergranule 88–182
Supergranulum 88–182
Supergravitation 99–170r
Superhaufen 98–346
Lokaler ~ 98–347
supermassiver Körper 98–22
~ Stern 91–127
supermassives Objekt 98–22
~ schwarzes Loch 98–22a
Supernova 95–126
ehemalige 95–134
extragalaktische ~ 95–128
galaktische ~ 95–127
~-Suche 95–136
Tycho's ~ 95–135
Überreste der ~ 95–133
~ vom Typ I 95–130
Supernovaausbruch 95–131
Supernovarest 95–133
supervereinheitliche Wechselwirkung 99–170r
Supervereinheitlichung 99–170q
Supraflüssigkeit 95–149
Supraleitungskern 95–150
Surge 89–255
dunkle ~ 89–256
helle ~ 89–257
Suspensionsstrom 72–49
Suszeptibilität 75–27
magnetische ~ 75–27
Swanbanden 85–133
S-Welle 73–43
Swept-Lobe-Interferometer 88–109
symbiotischer Stern 96–20
symbiotisches Spektrum 95–117
Symmetrie, sphärische 98–160
zylindrische ~ 98–155
Symmetriebedingung 98–162
Symmetriebrechung 99–170s
Synchrone 85–92
letzte ~ 85–93
vollständige ~ 85–94
Syndyname 85–91
Synklinale 73–89
Synklinalfalte 73–89
Synkline 73–89
synodischer Monat 81–46
~ Umlauf 81–46
synoptische Karte 77–11
~ Photosphärenkarte 89–20
Synthesesystem 88–119
System, abgeplattetes 98–251
axialsymmetrisches ~ 98–157
Copernicanisches ~ 99–24
~ der Kugelhaufen 98–75
durch Stösse beherrschtes ~ 98–166
extragalaktisches ~ 98–210
flaches ~ 98–251
galaktisches ~ 98–3
gebundenes ~ 96–156
geozentrisches ~ 99–21
~ in einem nichtstationären Zustand 98–152
~ im stationären Zustand 98–141
instabiles ~ 98–171
kompaktes ~ 96–157
Kopernikanisches ~ 99–24
lokales ~ 98–61
~ mit Eigengravitation 98–169
~ mit negativer Gesamtenergie 96–156
~ mit positiver Gesamtenergie 96–158
nichtgebundenes ~ 96–158
planetozentrisches ~ 83–10
planetozentrisches ~ von Koordinaten 83–10
Ptolemäisches ~ 99–22
selbstgravitierendes ~ 98–169
sphäroidisches ~ 98–159
stark abgeplattetes ~ 98–164
stossfreies ~ 98–167
U, B, V- ~ 96–144
~ von Ringen 98–291
~ von Sternen 96–1
systemfremder Stern 98–134
Systemkonstanten 95–25
Systemmitglied 98–133
Systemstern 98–133
Syzygien 81–29
Syzygienhochwasser 82–273
Szintillation, interstellare 97–109a

D

T

Taenit 87–93
Tafelberg 92–66
Tag, aktiver 75–93
~ auf dem Monde 81–32
gestörter ~ 75–94
magnetisch aktiver ~ 75–93
magnetisch ruhiger ~ 75–92
ruhiger ~ 75–92
Tagesbeobachtung 83–121
Tagesbolid 86–118
Tageshimmel 79–3a
Tageskarte der Sonne 89–18
Tagesmeteor 86–22
Tages-Meteorstrom 86–219
Tagesseite 83–67
Tagesstrom 86–219
Taghimmel 79–3a
Taghimmelsleuchten 79–3
tägliche Verschiebung des Radianten 86–69
Tagschweif 86–106
Tagseite 83–67
~ des Mondes 81–34
Tagstrom 86–219
Tal 81–163
Talgletscher 72–114
Tangentialebene an der Himmelssphäre 96–50
T-Assoziation 96–90
Tastfrequenz 86–279
Tätigkeit, ausgelöste vulkanische 82–155
~ der Oberfläche 82–231
Flare-~ 89–176
innere ~ 82–158
magnetische ~ 73–110
vulkanische ~ 82–154
vulkanische ~ längs der Brüche 82–157
Tau 77–173
Taube 92–37
taube Flut 82–272
Taupunkt 77–174
Tauriden 86–254
Taurus 92–91
TE 93–35
Teilchen 88–154
Aitkensche ~ 76–116
ausfällende ~ 76–117
beschleunigtes ~ 91–182
eisüberzogenes ~ 97–77
festes ~ 85–48, 97–69
flüssiges ~ 76–100
meteorisches ~ 86–160
~ mit Mantel 97–70
überschweres ~ 99–170n
Teilchenausbruch 88–162
Teilchenbahn 75–118
Teilchendruck 93–90
Teilchenemission 88–151
Teilchenspur 87–120
Teilchenstrahlung 88–150
Teilradiant 86–239
Teilsturm, magnetischer 75–95
polarer magnetischer ~ 75–96
Teilung 83–223
Tektit 87–95
Tektonik 73–95
tektonische Karte 73–96
~ Kräfte 73–106
tektonischer Prozess 82–201
tektonisches Erdbeben 73–2d
Telegramm, astronomisches 85–20
Telemeteor 86–18
schwaches ~ 86–19
sehr schwaches ~ 86–20
Telescopium 92–92
teleskopischer Komet 85–14a
tellerförmiger Krater 81–219
tellurische Spektrallinie 88–140a
~ Ströme 75–5
Temperatur 77–29
Curie-~ 75–41
effektive ~ 91–113
~ im Schatten 77–34
kinetische ~ 91–115
zentrale ~ 93–89
Temperaturabfall 77–41
Temperaturabnahme 77–41
Temperaturgefälle 93–96
Temperaturgradient 93–96
adiabatischer ~ 93–98
konvektiver ~ 93–157
radiativer ~ 93–97
trockenadiabatischer ~ 76–14
Temperaturinversion 77–36
Temperaturprofil 83–259
~ der Atmosphäre 83–259
Temperaturschichtung 77–40
Temperaturverhältnisse 77–35
Temperaturzunahme 77–42
Tensionsspalte 82–207
Tensor, metrischer 99–157a
Riemann-Christoffelscher ~ 99–159
Tensorgleichung 99–157
Terminator 83–69
~ des Erdschattens 81–91
Terra 81–190
Terra-Gebiet 81–271
Terrain, chaotisches 83–164
Terrainbeschaffenheit 81–149
Terrasse, innere 81–256
Terrassendepression 81–162
terrassenförmige Vertiefung 81–162
~ Wand 81–257
Terrassenwall 81–257
Terrella 79–133
terrestrisch 71–3
terrestrische Effekte der Sonnenaktivität 84–55
~ Planeten 83–119
Tertiär 74–87
tesserale Kugelfunktionen 95–102
Textur 82–102
Theorem, Vogt-Russell- 93–83
Theorie der Bildung getrennter Kerne 96–76
~ der Spiralarmentstehung 98–173
~ der Spiralstruktur 98–173
~ des Wärmetodes des Weltalls 99–15
~ "gauge" 99–170j
~ getrennter Kerne 96–76
mechanische ~ der Kometenformen 85–84
physikalische ~ der Meteore 86–159
thermisch stabil 93–209
thermische Geschichte 74–66
~ Metamorphose 74–37
Thermolumineszenz 82–249
Thermometamorphose 74–37
thermonukleare Reaktionsrate 94–53
Thermosphäre 76–59
Thioformaldehyd 97–43
Thorium-Blei-Datierung 82–270
Thuban 92–131
Tiden 72–89
Tidenhochwasser 82–281
Tidenhub 82–292
Tidenstieg 82–292
Tidenstrom 82–279
Tidentag 82–290
Tief 77–76
Tiefbeben 73–23
Tiefdruckgebiet 77–76
Tiefe der isostatischen Ausgleichsfläche 73–102
~ des Minimums 95–75
~ des Minimums der Finsternis 95–75
effektive ~ der Linienentstehung 93–64
isostatische ~ 73–102
Tiefebene 81–272
Tiefenkarte 71–54
Tiefenkurve 71–55
Tiefenlinie 71–55
Tiefenmessung 72–28
Tiefenströmung 72–61
ozeanische ~ 72–61
Tiefenstufe, geothermische 74–59
Tieflandgebiet 81–272
Tiefscholle 82–132
Tiefsee 72–41
Tiefseeablagerung 72–50
Tiefseeberg 72–43
abgestumpfter ~ 72–44
Tiefsee-Ebene 72–42
Tiefseegraben 72–40
Tiefseehügel 72–43
Tiefseekuppe 72–43
Tiefseelotung 72–29
Tiefseerinne 72–40
Tiefseesedimente 87–15
Tierkreis 92–5a

D

Tierkreislicht 84–61
Tierkreis-Sternbild 92–5
Tierkreiszeichen 92–6
Tisch 92–66
Titan 83–233
Titius-Bodesche Regel 83–3
Tochtermolekül 85–135a
Ton, pelagischer 73–54a
roter ~ 72–54
topographische Mondkarte 81–138
Tornado-Protuberanz 89–262
Totalalbedo 83–90b
totale Mondfinsternis 81–66
~ Sonnenfinsternis 88–17
Totalisator 77–171
Niederschlags- ~ 77–171
Totalität, Anfang der 81–77
Beginn der ~ 81–77
Ende der ~ 81–80
Totalitätszone 88–29
toter Vulkan 73–115
Trabant 83–17
tragbares Magnetometer 82–44
träge Masse 99–148
Tragen der Chromosphäre 88–223
Träger 86–276
Trägerfrequenz 86–274
Trägerwelle 86–276
Tragfähigkeit 82–64
Trägheit, Relativität der 99–35
Tragkraft 82–64
Transformation 99–131
Galilei-~ 99–131
Galileische ~ 99–131
homologe ~ 93–132
Invarianz gegenüber einer ~ 99–133
Lorentz-~ 99–132
transgressive Intrusion 73–137
Transient, koronaler 89–185a
translunar 81–6
transneptunischer Planet 83–237
Transparenzfenster 83–264
Transpluto 83–124
transplutonischer Planet 83–124
Transport des Magnetfeldes 84–48
Transportgleichung 93–32
Transversalfeld 89–137
Transversalkomponente 89–136
Transversalwelle 73–41
Trapez im Orion 96–92
Treibhausmodell 83–262
Trennung der chemischen Elemente 83a–48
Trennungslinie der Polaritäten 89–151
Triade 87–91
Triangulation 71–23
~ erster Ordnung 71–24
kosmische ~ 71–28
räumliche ~ 71–25
Triangulum 92–93
~ Australe 92–94
Trias 74–84
Trichter 89–269
Trichter-Protuberanz 89–269
Trifidnebel 97–141
Trilateration 71–29
trockenadiabatischer Temperaturgradient 76–14
Trockeneis 83–144
Troilit 82–94
Trojaner 84–91
~ Asteroidengruppe 84–91
Tröpfchen 76–100
Tropfen 76–113
schwarzer ~ 83–52
Tropfentektit 87–96
tropischer Umlauf 81–44
Tropopause 76–77
Troposphäre 76–78
Trübestrom 72–49
Trübheit 83–252
Trübung 83–252
Trümmerfeld 82–139
Tsunami 72–95
Tucana 92–95
Tukan 92–95
Turbosphäre 76–72
Turbulenz mit kosmischen Ausmass 83a–12
Turbulenztheorie, Weizsäckersche 83a–11
Turmteleskop 88–54
tweek 78–63
Tychonischer Stern 95–135
Tycho's Supernova 95–135
Typ, Hubblescher 98–229
morphologischer ~ 98–228
Typ-I-Sternpopulation 98–71
Typ-II-Sternpopulation 98–68
Typus = Typ
Hubblescher ~ einer Galaxie 98–229

U

überadiabatischer Gradient 93–99
Überbolid 86–9
Überdeckung von Spektren 96–24
~ von zwei Zyklen 89–86
überdicht 86–285
überdichter Stern 91–153
überfluteter Krater 81–203
Überflutung 72–75
Übergangsschicht 88–224
Übergangstyp der Korona 88–245
Übergangszone 72–13
Übergreifen der Hörnerspitzen 83–131
überhäufiges Element 91–184
Überhäufigkeit 91–185
überlagertes Spektrum 96–19
Überlagerung von Spektren 96–24
Überlappung von zwei Zyklen 89–86
überleuchtkräftiger Stern 91–134
überluminoser Stern 91–134
Überreste der Supernova 95–133
Überriese 91–145
Überriesen, hellste 91–147
schwächere ~ 91–148
Überriesengalaxie 98–223
Übersättigung 76–102
überschätzen 95–13
Überschuss, infraroter 91–79
Infrarot-~ 91–79
ultravioletter ~ 91–78
Ultraviolett- ~ 91–78
Überschwemmung 72–75
überschwere Partikel 99–170n
überschweres Teilchen 99–170n
Übersicht, photographische 82–30
Übertragungsfrequenz, höchste brauchbare 78–37
maximale nutzbare ~ 78–37
niedrigste brauchbare ~ 78–33
Überwachen von Satelliten 83–26
Überwachung der Eruptionen 89–219
~ der Flares 89–219
Flare-~ 89–219
~ von Asteroiden 84–94
Überwachungskamera 86–145
übrigbleibendes Magnetfeld 95–62
U, B, V-System 96–144
ultrabasisches Gestein 74–6
Ultrabasit 74–6
ultraschnelle Expansion 99–170c
~ Expansion des Weltalls 99–170c
Ultraviolettkoma 85–67
Ultraviolett-Überschuss 91–78
Umbra 89–95
Umbragranule 88–180
Umbragranulum 88–180
Umbrapunkt 89–96
heller ~ 89–96
Umdrehungsachse 83–33
Umdrehungsgeschwindigkeit 83–32
umgebend 76–7
Umgebung der Sonne 98–58
Umgebungs- 76–7
Umgebungskarte des Veränderlichen 95–15
Umgebungsluft 77–4
Umgebungstemperatur 77–32
Umgrenzungslinien gleicher Zählrate 75–102
Umkehreffekt 76–126
umkehrende Schicht 88–174
Umkehrung der magnetischen Polarität 89–153
Umkristallisierung 74–42
Umlauf, anomalistischer 81–43
drakonitischer ~ 81–47
siderischer ~ 81–45
tropischer ~ 81–44
Umstände des Falls 87–47
Umwandlung der Energie 86–208
Umwelt 76–3
Umweltwissenschaft 76–2
unbewegliche Kamera 86–143
„undae longae“ 73–48

„undae primae“ 73–46
„undae secundae“ 73–43
Unebenheit der Oberfläche 82–106
unelastischer Stoss zwischen Sternen 98–26
unendlicher Raum 99–104
Unendlichkeit des Weltalls 99–6
Ungeordnetheit des Magnetfeldes 84–12
kleinräumige ~ 84–13
ungerader Zyklus 89–78
ungeradzahliger Zyklus 89–78
ungestörte Photosphäre 88–169
ungestörtes Gebiet der Photosphäre 89–56
Ungleichung, Schwarzschild- 93–156
unidentifizierte Banden 85–134
unipolares magnetisches Gebiet 89–155a
Universum = All = Weltall 99–1
De Sitter-~ 99–69
Einstein-De Sitter- ~ 99–70
frühes ~ 99–168
geschlossenes ~ 99–73
heisses ~ 99–166
inflationäres ~ 99–170b
leeres ~ 99–49
nichtstationäres ~ 99–53
nichtstatisches ~ 99–51
offenes ~ 99–72
oszillierendes 99–57
sich entwickelndes ~ 99–54
stationäres ~ 99–52
statisches ~ 99–50
unregelmässige Galaxie 98–237
unregelmässiger magnetischer Veränderlicher mit konstanter Polarität 95–57
~ magnetischer Veränderlicher mit wechselnder Polarität 95–56
~ Nebel 98–237
~ Veränderlicher mit konstanter Polarität 95–57
~ Veränderlicher mit wechselnder Polarität 95–56
Unregelmässigkeit des Magnetfeldes 84–12
untätiger Vulkan 73–116
unterbrechender Sektor 86–148
unterdicht 86–284
Untergrund 82–105
unterhäufiges Element 91–186
Unterhäufigkeit 91–187
unterkühlt 76–106
Unterkühlung 76–105
unterleuchtkräftiger Stern 91–133
unterluminoser Stern 91–133
Unterriesen 91–151
unterschätzen 95–14
unterseeischer Vulkan 73–117
Untersonne 79–72
Untersuchung an Ort und Stelle 82–136
in-situ ~ 82–136
~ unterhalb der Oberfläche 82–82
Untersystem 98–74
Unterwasserbaggerung 72–53
Unterzwerg 91–140
Untiefe 72–96
unveränderlich in der Zeit 98–150
zeitlich ~ 98–150
Uran-Blei-Alter 74–101
Uranus 83–234
Uratmosphäre 83a–44
Uratom 99–169
Urca-Prozess 94–51
Uredelgas 87–133
Urexplosion 99–167
Urisotop 82–146
Urknall 99–167
Urknall-Nukleosynthese 99–175
Urkosmos 99–170a
Urmaterie 99–170
Urnebel 83a–21
stellarer ~ 97–63
Urnuklid 82–256
radioaktives ~ 82–256
Urplanet 83a–40
Urplasma 99–170
Ursa Major 92–96
~ Minor 92–97
Ursiden, Dezember 86–258
Ursonne 83a–39
Ursprung 94–9
ausserirdischer ~ 87–73
~ der Kometen 85–96
~ der Welt 99–11
gemeinsamer ~ 94–25
irdischer ~ 87–74
tektonischer ~ 82–200
ursprüngliche Wärme 74–63
ursprüngliches Magnetfeld 95–62
Ursprungs, extraterrestrischen 87–5
kosmischen ~ 87–5
Ursprungskomet 86–260
Urstern 94–45
Urstoff 99–170
Urwärme 83–200
Urweltall 99–170a
Urwolke, galaktische 94–17
Urzeit der Erde 74–70
UV-Exzess 91–78

V

vagabundierende Erdströme 75–6
Vakuum 76–9
Vakuumenergie 99–170k
Van-Allen-Gürtel 75–101
Van-Allenscher Strahlungsgürtel 75–101
Variabilität des Magnetfeldes 84–14
Variation, jahreszeitliche 86–139
jährliche ~ 86–138
tägliche ~ 86–137
~ des Erdmagnetfeldes 75–69
erdmagnetische ~ 75–69
geomagnetische ~ 75–69
magnetische ~ 95–53
Variationsmethode 93–57
Vegard-Kaplan-Banden 79–19
Vektor, magnetischer 89–139
Poynting-~ 93–92
vierdimensionaler ~ 99–138
Vektormagnetograph 88–84
Vela 92–98
Venus 83–125
Venus-Dichotomie 83–132
Venussichel 83–129
Veränderliche, dimensionslose 93–133
Veränderlicher = veränderlicher Stern 95–1
Delta-Cephei-~ 95–92
~ der...Klasse 95–47
eigentlicher ~ 95–87
ellipsoidischer ~ 95–79
geometrischer ~ 95–72
halbregelmässiger ~ 95–105
kataklysmisch(er)- 95–109
kurzperiodischer ~ 95–89
langperiodischer ~ 95–91
Magnetfeld-~ 95–52
~ mit ultrakurzer Periode 95–90
novaähnlicher ~ 95–118
periodischer ~ 95–88
periodischer magnetischer ~ 95–55
physischer ~ 95–87
Pseudo-~ 95–51
regelmässiger magnetischer – 95–55
roter halbregelmässig(er) ~ 95–106
RR Lyrae- ~ 95–93
scheinbar(er) ~ 95–86
~ Stern 95–1
uneigentlicher ~ 95–86
unregelmässig(er) ~ 95–107
unregelmässiger magnetischer ~ 95–57
verdächtiger ~ 95–5
vermuteter ~ 95–5
~ vom ... Typus 95–47
zyklisch(er) ~ 95–108
Varänderlichkeit, optische 95–141
Veränderungen, spektrale 95–49
verborgene Masse 99–88
verbotene Räume 79–135
Verbreiterung des Ozeanbodens 72–39
Verdampfen eines Meteorkörpers 86–198
Verdampfung eines Meteorkörpers 86–198
~ eines Sternhaufens 96–108
intensive ~ 86–185
lebhafte ~ 86–185
~ schwarzer Löcher 94–118
Verdampfungsdauer 96–109
Verdampfungsenergie 86–199
Verdampfungsgleichung 86–200
Verdampfungswärme 86–199
Verdampfungszeit 96–109
Verdichtung innerhalb eines Meteorstromes 86–217

D

zentrale ~ 85–37, 96–126
Verdichtungskappe 86–170
Verdichtungswelle 73–46
Verdichtungszone 86–170
Verdrängung 74–44
Verdunklung, interstellare 97–86
verdünnte Strahlung 93–38
Verdünnungsfaktor 93–37
Verdünnungswelle 97–58
Vereinheitlichung der Wechselwirkungen 99–170w
grosse ~ 99–170u
Vereisung 72–104
Vereisungsperiode 72–105
Verfärbung, interstellare 97–91
Kurve interstellarer ~ 97–93
Verfärbungsgesetz 97–92
Verfärbungskurve 97–93
Verfärbungslinie 96–149
verfinsterter Mond 81–57
Verfinsterung der Atmosphäre 95–76
Verformung 73–68
Verformungstensor 73–70
Vergleichsstern 95–16
Vergletscherung 72–104
Vergrösserung der Masse 83a–38
~ des Erdschattens 81–93
Verhältnis der Beleuchtungsstärken 91–60
~ Durchmesser zu Tiefe 81–248
Masse-Leuchtkraft-~ 98–233
Poissonsches ~ 73–60
Verhüllung 95–84
Verhüllungsveränderlicher 95–83
Verkittung 74–52
~ von Gesteinen 74–52
Verkürzung, Lorentz- 99–134
perspektivische ~ 89–92
perspektivische ~ der Flecken 89–92
relativistische ~ 99–134
Verlauf der Finsternis 81–73
Verletzung der CP-Invarianz 99–170t
~ der Ladungsparität 99–170t
verlorener Asteroid 84–97
Vermessung, magnetische 75–64
magnetische ~ der Erde 75–65
photographische ~ 82–30
selenologische ~ 82–50
selenophysikalische ~ 82–39
Verschiebung des Radianten 86–242
longitudinale ~ 75–120
Verschlammung in äusserem Schelf 72–79
Verschlickung in äusserem Schelf 72–79
Verschmierung der Fraunhoferlinien 88–243
Verschmutzung, chemische 72–98
Verschmutzugsereignis 72–72
Verschwinden 81–98
~ des Ringes 83–217
plötzliches ~ 89–279
Versteinerung 74–49
Versuch, aktiver seismischer 82–220
Michelson-Morley-~ 99–130
Michelsonscher ~ 99–130
seismischer ~ 82–219
Verteilung der Geschwindigkeitskomponenten 98–92
ellipsoidische ~ der Sterngeschwindigkeiten 98–100
galaktische ~ 98–54
~ im Raum 96–119
~ nach Breite 89–51
~ nach heliographischer Breite 89–51
~ nach heliographischer Länge 89–48
räumliche ~ 96–119
~ von Restgeschwindigkeiten 98–92
Verteilungsfunktion 98–143
Vertex 96–98, 98–105
Vertiefung, bogenförmige 81–160
~ einer Zyklone 77–82
kontinentale ~ 81–158
lineare ~ 81–159
sinusförmige ~ 81–161
terrassenförmige ~ 81–162
Vertikalbewegung 88–185
Vertikalcoelostat 88–48
Vertikale 71–71
~ Komponente 75–62
Vertikalintensität 75–62
Vertikalspektrograph 88–79
Vertikalverteilung 76–17
Vertikalzölostat 88–48
Verunreinigung, chemische 72–98
Verunreinigungsfall 72–72
Verunreinigungsstoff 72–74
Verwerfen 73–92
Verwerfung 73–91
antithetische ~ 73–94
gegensinnige ~ 73–94
gewöhnliche ~ 73–93
inverse ~ 73–94
normale ~ 73–93
widersinnige ~ 73–94
Verwerfungsbildung 73–92
lokale ~ 82–204
Verwischung der Fraunhoferlinien 88–243
Verwitterung 74–57
~ im kosmischen Raum 82–244
verwundene Meteorbahn 86–33
Verwurf 73–91
verzerrte Galaxie 98–316
Verzerrung 73–68
Verzerrungstensor 73–70
Verzögerung der Erdrotation 82–313
~ der Expansion 99–82
scheinbare ~ 86–100
wahre ~ 86–101
Verzögerungsparameter 99–83
Verzweigung 98–5a
Vesta 84–87
vibrationelle Instabilität 93–214
vibrierender Spiegel 86–130
Videomagnetograph 88–85
vielatomiges Molekül 97–21
vieleckiger Krater 81–212
Vielelement-Interferometer 88–123
Vielfachinterferometer 88–123
vierdimensional 99–121
vierdimensionaler Vektor 99–138
Vierervektor 99–138
Viertel, erstes 81–25
letztes ~ 81–30
Viertelwellenlängenplatte 88–126
Viertelwellenplatte 88–126
violette Schicht 83–165
Virgo 92–99
Virgo-Haufen 98–345
Virgo-Supergalaxie 98–347
Virialgleichung 98–181
Virialsatz 98–181
Visionslinie 91–28
Visionsradius 91–28
Viskosität 73–71
magnetische ~ 75–34
Vogt-Russell-Theorem 93–83
Volans 92–100
Vollmond 81–27
vorausberechnete Position 85–27
vorausberechneter Komet 85–11
Voraussage de Sonnenaktivität 89–64
~ der Sonnentätigkeit 89–64
~ der Sternbedeckungen 81–102
Vorbeben 73–10
Vorbeigang zweier Galaxien 98–320
Vordämmerungsemission 79–51
Vordergrundstern 96–85
Vorderseite 87–114
~ des Mondes 81–15
Vorgang, innerer 82–258
Vorhauptreihenentwicklung 94–26
Vorhauptreihenstadium 94–49
vorherberechnete Position 85–27
vorhergesagter Ort 85–27
Vorhersage, kurzfristige 77–20
langfristige ~ 77–19
Vorhersagekarte 77–21
Vorläufer 95–143
~ des Erdbebens 73–11
Vortrieb 73–65
Vorübergang 95–81
~ des Schattens 83–53
~ des Schattens eines Trabanten 83–53
~ eines Planeten vor der Sonne 83–51
~ eines Planeten vor der Sonnenscheibe 83–51
~ eines Satelliten 83–54
~ eines Trabanten 83–54
~ eines Trabanten vor der Planetenscheibe 83–54
~ zweier Galaxien 98–320
vorübergehende Leuchterscheinung 82–247

vorübergehendes Aufleuchten 82–247
~ koronales Event 89–185a
Vorwärmung 86–203
Vulkan 73–111
dampfender ~ 73–114
erloschener ~ 73–115
geschichteter ~ 73–121
ruhender ~ 73–116
schlafender ~ 73–116
subaquatischer ~ 73–117
submariner ~ 73–117
tätiger ~ 73–114
toter ~ 73–115
untätiger ~ 73–116
Vulkanausbruch 73–112
Vulkanbecken 81–179
Vulkandom 81–176
Vulkaneruption 73–112
Vulkanesse 82–162
vulkanische Asche 82–168
vulkanischer Kegel 73–118
Vulkankegel 82–160
Vulkankuppe 81–176
Vulkanschlot 82–162
Vulpecula 92–101

W

Waage 92–62
Wabenblech, Aluminium- 82–25
Wachsen eines Planeten 83a–37
Wachstum der Staubkörner 97–61
Wachstumskurve 93–68
empirische ~ 93–69
theoretische ~ 93–70
wahrscheinlicher Krater 81–210
Walfisch 92–34
Wall, konzentrischer 81–254
mehrfacher ~ 81–255
Wallebene 81–265
Wand, Gerade 81–185
Lange ~ 81–185
Wandelstern 83–1
Wanderung der Aktivitätszonen 89–74
~ der Kontinente 73–104
polwärtige ~ der Protuberanzen 89–231
Wärme. latende 77–157
radiogene ~ 83a–62
ursprüngliche ~ 74–63
Wärmebilanz 77–62
~ der Atmosphäre 77–62
Wärmefluss 74–60
Wärmehaushalt 77–62
~ der Atmosphäre 77–62
Wärmeindex 91–81
Wärmequelle 74–62, 83–199
Wärmeschock 86–206
Wärmespannung 86–205
Wärmestoss 86–206
Wärmestrom 72–63
Wärmetod der Welt 99–14
~ des Weltalls 99–14
Wärmetransport 86–202
Wärmeübergang 86–202
Wärmeübertragung 86–202
Wärmeübertragungskoeffizient 86–201
Warmfront 77–69
Warmzeit 72–107
Wasser 97–26
wasserabstossend 76–93
wasserabweisend 76–93
wasseraffin 76–92
Wasserauftrieb 72–88
Wasserdampf 77–119
Wasseradampfabsorption 77–127
Wasserdampfdruck 77–125
Wasserdampfextinktion 77–126
Wasserdampfgehalt 77–124
Wasserdampfspannung 77–125
Wassereis 85–47
wasserfeindlich 76–93
wasserfreundlich 76–92
Wassermann 92–17
Wassermodell 85–113
~ eines Kometen 85–113
Wasserprobeentnahmegerät 72–101
Wasserprobenehmer 72–101
Wasserrühr-Pyrheliometer 88–41
Wasserschlange, Männliche 92–56
Nördliche ~ 92–55
Weibliche ~ 92–55
Wasserstoff, interstellarer 97–46
metallischer ~ 83–202
molekularer ~ 97–31
neutraler ~ 97–47
wasserstoffarmer Stern 91–191
Wasserstoffbrennen 94–56
Wasserstoff-Filament 89–267
Wasserstoff-Flocculi 89–168
Wasserstoff-Flocken 89–168
helle ~ 89–169
Wasserstoffion 93–65
negatives ~ 93–66
Wasserstoffisozyanid 97–40
Wasserstoffkarbylamin 97–40
Wasserstoffkonvektionszone 93–159
Wasserstoffkorona 83–245
Wasserstoffleuchten 79–145
Wasserstoff-Metall-Verhältnis 91–176
Wasserstoffmodell 83–197
~ eines Planeten 83–197
Wasserstoffoxid 97–26
Wasserstoffplanet 83–197
Wasserstoffprotuberanz 89–266
Wasserstoffschale 93–144
Wasserstofftorus 83–230
Wasserstoff-Zündung 94–55
Wasserstoffzyanid 97–29
Wasserstrom-Pyrheliometer 88–40
Wassertiefe, Linie gleicher 71–55
Wasserverunreinigung 72–73
Wechsel der magnetischen Polarität 89–153
wechselwirkende Galaxien 98–310
Wechselwirkung, atmosphärische 79–120
elektroschwache ~ 99–170x
grosse vereinheitliche ~ 99–170v
kollektive ~ 98–168
supervereinheitliche ~ 99–170r
vereinheitliche ~ 99–170x
Wechselwirkungen, Vereinheitlichung der 99–170w
wedelndes Filament 89–216
Wega 92–132
Wegunterschied 88–117
wehen 77–105
Weisser Fleck 83–214
~ Zwerg 91–137
Weisses Oval 83–195
Weizsäckersche Turbulenztheorie 83a–11
Welle, ausserordentliche 78–30
direkte ~ 78–12
elastische ~ 73–44
ganztägige ~ 76–30
halbtägige ~ 76–29
kontinuierliche ~ 86–277
Love- ~ 73–48
ordentliche ~ 78–31
P- ~ 73–46
planetarische ~ 72–94
primäre ~ 73–46
Q- ~ 73–48
S- ~ 73–43
seismische ~ 73–9
sekundäre ~ 73–43
Wellendissipation 88–222
wellenförmige Meteorbahn 86–33
Wellenlänge, Compton- 99–170g
effektive ~ 96–142
Wellenlängenänderung 88–111
schnelle ~ 88–111
Welt 99–3
Ende der ~ 99–183
hierarchische ~ 99–48
Minkowskis vierdimensionale ~ 99–122
Minkowskische ~ 99–122
Modell einer anisotropen ~ 99–65
Modell einer asymptotischen ~ 99–59
Modell einer expandierenden ~ 99–55
Modell einer geschlossenen ~ 99–73
Modell einer hierarchischen ~ 99–48
Modell einer homogenen ~ 99–64
Modell einer isotropen ~ 99–63
Modell einer leeren ~ 99–49
Modell einer offenen ~ 99–72
Modell einer oszillierenden ~ 99–57
nichtstatische ~ 99–51
Raum-Zeit- ~ 99–122
stationäre ~ 99–52
Weltall = All = Universum 99–1
anisotropes ~ 99–65
asymptotisches ~ 99–59

D

asymptotisches ~ ester Art 99–60
Charliersches ~ 99–67
De Sitter-~ 99–69
Einstein-De Sitter ~ 99–70
Einstein's ~ 99–68
expandierendes 99–55
geschlossenes ~ 99–73
heisses ~ 99–166
hierarchisches ~ 99–48
homogenes ~ 99–64
inflationäres ~ 99–170b
inhomogenes ~ 99–66
isotropes ~ 99–63
kontrahierendes ~ 99–56
monotones ~ 99–61
monotones ~ erster Art 99–62
offenes ~ 99–72
oszillierendes ~ 99–57
oszillierendes ~ erster Art 99–58
relativistisches ~ 99–92
sich entwickelndes ~ 99–54
stationäres ~ 99–52
Weltalter 99–12
Weltbild 99–20
Welten, Friedmannsche 99–71
Weltenstehung 99–11
Weltlinie 99–154
Weltmodell 99–46
anisotropes ~ 99–65
asymptotisches ~ 99–59
asymptotisches ~ erster Art 99–60
Charliersches ~ 99–67
De Sitter-~ 99–69
Einsteins ~ 99–68
expandierendes ~ 99–55
Friedmannsches ~ 99–71
heisses ~ 99–166
inhomogenes ~ 99–66
isotropes ~ 99–63
kontrahierendes ~ 99–56
nichtstationäres ~ 99–53
oszillierendes ~ 99–57
stationäres ~ 99–52
~ von Einstein und De Sitter 99–70
Weltnetz, geodätisches 71–38
Weltorganisation für Meteorologie 77–17
Weltpostulat 99–30
erweitertes ~ 99–32
Weltpunkt 99–125
unendlich benachbarter ~ 99–126
Weltradius 99–5
Welt-Schwerenetz 71–81
Weltsystem 99–20
Copernicanisches ~ 99–24
geodätisches ~ 71–39
heliozentrisches ~ 99–23
Kopernikanisches ~ 99–24
Ptolemäisches ~ 99–22
Weltuntergang 99–183
Weltwetterwacht 77–16
Wendepunkt 98–178
westliche Quadratur 83–66
Westrand 89–33
Wetter 77–1
Wetteränderung 77–22
Wetterbedingungen 77–2
Wetterbericht 77–10
Wetterdienst 77–5
Wetterkarte 77–12
Wetterlage 77–14
Wetterprognose 77–13
Wetterverhältnisse 77–2
Wetteverschlechterung 77–23
Wettervorhersage 77–13
Weylsches Postulat 99–34
Whistler 78–53
kurzer ~ 78–58
langer ~ 78–57
Widder 92–20
Widerstandsbeiwert 86–180
Widerstandsgesetz 86–179
Widmannstättensche Figuren 87–105
Wiederentdeckung 85–17
Wiedererscheinen des Ringes 83–219
wiederholte Nova 95–116
Wiederholungsfrequenz 86–279
Wiederkehr 88–132
~ der Gruppe 88–133
~ eines Kometen 85–9
wiederkehrende Nova 95–116
Wilson-Effekt 89–89
Wind 77–89
aufsteigender ~ 77–92
böiger ~ 77–111
galaktischer ~ 98–277
geostrophischer ~ 77–96
interstellarer ~ 97–151
solarer ~ 88–156
stellarer ~ 93–14
vorherschender ~ 77–117
zyklostrophischer ~ 77–95
Winderosion 83–163
Windfahne 77–104
Windgeschwindigkeit 77–109
Windmesser 77–113
Windrichtung 77–115
Windrose 77–116
Windschreiber 77–114
Windstärke 77–107
Windstille 77–110
Windstoss 77–112
Windungslänge 98–32
Winkel, charakteristischer 98–33
heliozentrischer ~ 89–37
Winkelabstand 96–52
Winkeldurchmesser 83–117, 96–115
Winkelentfernung 96–52
Winkelgeschwindigkeit 86–89
augenblickliche ~ 86–90
mittlere ~ 86–91
momentane ~ 86–90
Winkellänge der Meteorbahn 86–48
Winkelmass 92–70
Wintersternbild 92–1b
Wirbelhypothese 83a–11
Wirbelsturm, tropischer 77–78
Wirbelsystem 83a–13
Wirkungsquerschnitt, differentieller 94–54
partieller ~ 94–54
Wissenschaften, atmosphärische 76–1
Wolf 92–63
Wolf-Rayet-Stern 91–160
Wolke 77–131
dunkle ~ 89–189
Flare-~ 89–189
Flare-verbundene dunkle ~ 89–189
H I ~ 97–103
interstellare ~ 97–101
irisierende ~ 79–90
koronale ~ 89–261
kosmische ~ 94–31
~ neutralen Wasserstoffes 97–103
protogalaktische~ 94–17
protoplanetare ~ 83a–24
schwebende ~ 89–275
zirkumstellare ~ 97–102
Wolken 83–254
Wolkenalbedo 83–95
Wolkenart 77–135
Wolkenform 77–133
wolkenfrei 77–149
wolkenlos 77–149
Wolken-Meer 81–281
Wolkenschicht 77–150, 83–248
Wolkenschleier 77–151
wolkig 77–148
WR-Stern 91–160
Wulst, zentraler 98–16
Wüste 83–161

X

X-Strahlen-Burst 89–184
X-Strahlen-Flare 89–183
X-Strahlen-Heliogramm 88–74
X-Strahlen-Korona 88–240
X-Strahlen-Punkt, heller 89–183a

Y

Youngscher Modul 73–59

Z

Zähigkeit 73–71
Zahl, Poissonsche 73–60
Zählrate für den gesamten Raumwinkel 75–104
~ für eine bestimmte Richtung 75–103
Zaunreihe 89–274
z-Bewegung 98–188
Zehrgebiet 72–124
~ eines Gletschers 72–124
Zeichnung eines Planeten 83–16

Zeit, absolute 99–117
charakteristische ~ 99–79
~ des Durchganges durch das Periastron 96–60
geomagnetische ~ 79–132
Hubblesche ~ 99–79
Newtonsche ~ 99–118
unveränderlich in der ~ 98–150
Zeitabschnitt, magnetisch gestörter 75–91
Zeitalter 74–69
zeitartige Koordinate 99–123
Zeitauflösung 88–205
Zeitdehnung 99–137
Zeitdilatation 99–137
Zeitkoordinate 99–123
zeitlich konstant 98–150
~ unveränderlich 98–150
zeitliche Auflösung 88–205
Zeitpunkt des Erscheinens 86–49
~ des Falls 87–51
Zeitrechnung, geologische 74–67
Zeitskala 94–65
kurze ~ 94–66
lange ~ 94–67
Zellenstruktur 99–171a
Zementation 74–52
Zementierung 74–52
Zenitattraktion 86–52
Zenithelligkeit 91–63
Zenitprojektion 71–45
Zentaur 92–32
Zentigradskala 77–33
Zentralberg 81–258
Zentraldichte 93–87
zentraler Wulst 98–16
~ Zusammenstoss 83a–17
Zentralgebiet der Galaxis 98–13
~ einer Galaxie 98–293
galaktisches ~ 98–293
Zentralintensität der Hα-Linie 89–211
Zentralkegel 81–258
Zentralkörper 98–16
Zentralkrater 81–217
Zentralmeridian 89–43
Abstand vom ~ 89–46
durch den ~ gehen 89–44
Durchgang durch den ~ 89–45
Zentralmeridiandurchgang 89–45
Zentralprojektion 71–48
Zentralstern 97–148
Zentraltemperatur 93–89
Zentralzone 81–114, 89–47
Zentrum der sichtbaren Scheibe 88–172
~ der Sonnentätigkeit 89–60
~ des Milchstrassensystems 98–9
galaktisches ~ 98–9
Zentrumsbedingungen 93–81
Zerbrechen eines Meteorkörpers 86–188
Zerbröckeln eines Meteors 86–83
Zerbröcklung eines Meteors 86–83
Zerdrücken durch Gravitation 84–72
Zerfall einer Fleckengruppe 89–126
~ eines Kometen 85–115
~ eines Kometenkerns 86–261
~ durch Stoss 84–107
radioaktiver ~ 74–64
zerfallen 74–65
Zerkleinerung eines Kometenkernes 85–116
~ eines Meteorkörpers 86–188
Zerplatzen der Astroiden 84–108
~ eines Kometen 85–114
~ eines Meteors 86–82
Zerreissen des galaktischen Kernes 98–318
Zerrung 73–66
zersetzen 74–65
Zersplitterung 83a–81
~ der Asteroiden 83a–82
~ des Mutterkörpers 87–124
~ durch kosmische Strahlung 82–254
Kern-~ 94–100
Zerspringen eines Meteors 86–82
Zerstäubung 82–239, 85–140
~ durch den Sonnenwind 82–245
~ eines Meteorkörpers 86–189
Zerstörung 87–24
~ durch einen Meteoriten 87–20
~ eines Kometen 85–114
~ eines Sternhaufens 96–105
zerstörungsfreie Gamma-Spektroskopie 82–77
Zerstörungsmechanismus 97–79
Zerstrahlungsära 99–173a
Zerstückelung des Kometenkernes 85–116
~ eines Meteorkörpers 86–188
Zertrümmerung 87–25
Zielpunkt der Sonnenbewegung 91–23
zigarrenförmiger Asteroid 84–77
Zimmertemperatur 77–31
Zirkel 92–36
Zirkulation, allgemeine 77–65
atmosphärische ~ 77–64
meridionale ~ 93–163
planetare ~ 77–66
zirkumgalaktisches Gas 98–351
zirkumlunar 81–5
zirkumnuklearer Bereich 98–292
Zirkumpolarstern 92–2a
Zirkumpolarsternbild 92–4a
zirkumstellar 97–3
zirkumstellare Hülle 93–10
~ Linie 97–97
~ Wolke 94–48
zirkumstellarer Staub 97–55
zirkumterrestrischer Staub 84–53
Zirkumzenitalbogen 79–74
Zirkumzenitalkreis 79–74
Zodiakalband 84–63
Zodiakallicht 79–13, 84–61
falsches ~ 84–62
Zodiakallichtteilchen 84–66
Zodiakalwolke 84–54
Zölostat 88–46
Zone der Aktivität 89–54
~ der totalen Finsternis 88–29
~ des Heliumbrennens 93–145
~ des Schweigens 76–42
~ des Wasserstoffbrennens 93–144
Fresnelsche ~ 86–294
gemässigte ~ 77–192
halbkonvektive ~ 93–160
heisse ~ 77–191
~ maximaler Polarlichthäufigkeit 79–125
nebelfreie ~ 97–88
orogene ~ 73–75a
polare ~ 77–193
seismische ~ 73–31
semikonvektive ~ 93–160
Zonen 83–187
Fresnelsche ~ 86–294
Zonenwanderung 89–74
zufällig verteilt 98–93
zufällige Beziehung 97–126
~ Irrfahrt 89–157
zufälliger Zusammenhang 97–126
Zufallsbewegung 98–94
zugeschütteter Krater 81–204
Zunehmen der Polkappe 83–145
zunehmender Mond 81–20
Züricher Klassifikation 89–127
Zusammenbruch 94–109
Zusammenfliessen 76–120
zusammengesetztes Modell 93–140
~ Spektrum 96–19
Zusammenhang, genetischer 97–127
zufälliger ~ 97–126
Zusammensetzung, chemische 91–168
isotopische ~ 82–140
Zusammenstoss der Sterne 83a–16
direkter ~ 83a–17
erosiver ~ 84–105
katastrophischer ~ 84–106
~ von Galaxien 98–322
zentraler ~ 83a–17
zusammenstossende Galaxien 98–321
Zusammensturz des Weltalls 99–181
Zustand, im nichtstationären 98–151
stationärer ~ 98–140
statistisch stationärer ~ 98–148
Zweiarmgalaxie 98–43
zweiarmige Galaxie 98–43
Zweibänderflare 89–187a
zweidimensionale Spektralklassifikation 91–105
Zweielement-Interferometer 88–98
Zweifarben-Diagramm 96–147
Zweig eines Stroms 86–240
Zwei-Komponenten-Modell 88–196
Zwei-Photonen-Ausstrahlung 97–155
Zwei-Photonen-Emission 97–155
Zweistromtheorie 98–99
Zwei-Temperaturen-Modell 88–197

D

D

Zwerg 91–135
brauner ~ 91–138a
roter ~ 91–136
schwarzer ~ 91–139
weisser ~ 91–137
Zwerggalaxie 98–220
elliptische ~ 98–242
Zwergnova 95–112
Zwergstern 91–135
Zwielicht, abendliches 79–36
Zwillinge 92–51
zwischen Armen 98–40
Zwischen-Arm-Gebiet 98–40
Zwischen-Arm-Raum 98–40
Zwischeneiszeit 72–107
zwischeneiszeitlich 72–108
Zwischenhaufen-Gas 98–353
Zwischenphase 94–21
Zwischenpopulation 98–70
Zwischenraum 88–99
fester ~ 88–102
ohne ~ 88–101
veränderlicher ~ 88–103
Zwischenschicht 72–13
Zwischenstadium 94–21
Zyan 97–24
Zyanazetylen 97–30
Zyklentheorie 97–156
zyklisch Veränderlicher 95–108
zyklischer Veränderlicher 95–108
Zyklone 77–76
tropische ~ 77–77
zyklostrophischer Wind 77–95
Zyklus der Sonnenaktivität 89–71
Beginn des ~ 89–82
elfjähriger ~ 89–73
Ende des ~ 89–85
geochemischer ~ 74–32
gerader ~ 89–77
geradzahliger ~ 89–77
Phase des ~ 89–79
ungerader ~ 89–78
ungeradzahliger ~ 89–78
Zylinder, abgestumpfter 88–63
Zylinderkoordinaten 98–156
Zylinderprojektion 71–43
Zylindersymmetrie 98–155
zylindrische Koordinaten 98–156
~ Projektion 71–43
~ Symmetrie 98–155

ÍNDICE ESPAÑOL

Spanish Index

Index espagnol

Spanischer Index

Índice espanhol

Испанский указатель

ALFABETO:

a	h	ñ	u
b	i	o	v
c	j	p	w
ch	k	q	x
d	l	r	y
e	ll	rr	z
f	m	s	
g	n	t	

A

abanico 85–60
~ coronal 89–295
~ curvado 85–62
~ curvo 85–62
en forma de ~ 85–74
~ invertido 85–63
~ recto 85–61
aberración diurna 86–57
abertura volcánica 73–122
abismo 72–41
ablación 86–182
absorción auroral 78–46
~ brusca del ruido cósmico 78–42
~ de casquete polar 78–44
~ de comienzo brusco 78–47
~ de línea 93–108
~ diferencial 97–90
~ interestelar 97–84
~ interestelar general 97–87
~ interestelar selectiva 97–90
~ interestelar total 97–87a
~ ionosférica polar 78–44
~ lenta de neutrones 94–88
~ libre-libre 93–106
~ ligada-libre 93–107
~ ligada-ligada 93–108
~ por vapor de agua 77–127
~ rápida de neutrones 94–89
abultamiento central 98–16
~ debido a las mareas 81–119
~ ecuatorial 71–9
~ lunar 81–118
abundancia anómala 91–173
~ cósmica de los elementos 94–107
~ de los elementos 82–141
~ de metales 91–170
~ de un elemento 91–169
~ isotópica 91–179
~ metálica 91–170
~ natural 91–171
~ normal 91–172
~ química 91–169
~ total 91–177
abundancias relativas de los elementos 91–174
acanaladura 81–165
acción de disparo 89–199
~ de la atmósfera 87–116
~ de marea 83a–15
aceleración ecuatorial 88–131
~ efectiva 85–88
~ de la gravedad 71–64
~ negativa aparente 86—100
~ negative real 86–101
acetaldehído 97–42
acetonitrilo 97–38
ácido α-aminopropiónico 87–72
~ amínico 87–70
~ cianhídrico 97–29
~ fórmico 97–33
~ isociánico 97–39
~ orgánico 87–69
~ sulfhídrico 97–44
acondrito 87–77
acoplamiento de radiación 99–173
~ por radiación 99–173
acrecimiento 83a–33
acreción 83a–33
activación de un filamento 89–215
~ de una protuberancia 89–239
actividad auroral 79–100
~ cometaria 85–110
~ corpuscular 89–7
~ de erupciones 89–176
~ de fulguraciones 89–176
~ de las manchas 89–62
~ de las manchas solares 89–62
~ de los meteoros 86–136
~ de los meteoros esporádicos 86–136
~ de una corriente 86–229
~ del núcleo galáctico 98–271
~ en radiación radio 89–6
~ eruptiva 95–71
~ fumarólica 82–164
~ galáctica 98–261
~ geomagnética 75–86
~ interna 82–158
~ lunar 82–149
~ magnética 89–5
~ meteórica 86–134
~ óptica 89–3
~ radioeléctrica 89–6
~ recurrente 98–267
~ relativa de la corriente 86–231
~ sísmica 73–4
~ solar 89–1
~ superficial 82–231
~ volcánica 73–110, 82–154
Acuáridas 86–249
Acuario 92–17
acumulación 83a–34
acústica atmosférica 76–31
achatamiento 83–113
~ aparente 98–295
~ de la corona 88–247
~ de un cúmulo globular 96–123
~ polar 83–114
~ real 98–296
~ terrestre 71–8
Achernar 92–102
adhesivo 82–137
adsorbato 85–131
adsorbente 85–130
adsorber 85–129
adsorción 85–128
advección 77–63
aerolito 87–60
aerología 76–122
aeronomía 76–122a
aerosol 76–89
aerotriangulación espacial 71–26
afelio 83–28a
afloramiento 72–88
agitación del mar 72–86
~ microsísmica 73–54
aglomeración 83a–32
~ de la materia 99–170z
aglutinación 74–52
agotamiento del hidrógeno 94–69
agregado molecular 97–65
agua 97–26
aguas llenas 82–281
~ muertas 82–272
~ vivas 82–273
Águila 92–18
agujas de hielo 76–112
agujero blanco 98–24
~ coronal 88–245a
~ crateroide 81–250
~ meteorítico 87–58
~ negro 94–114
~ negro galáctico 98–23a
~ negro supermasivo 98–22a
~ taladrado 82–61
aire 76–6
~ ambiente 77–4
ajuste isostático 73–100
AL 91–18
alanina 87–72
alargamiento de la sombra de la Tierra 81–93
alba 79–35
albedo de Bond 83–91
~ de Lambert 83–89
~ de las nubes 83–95
~ de neutrones 75–116
~ de radiofrecuencia 83–94
~ de rayos cósmicos 75–116
~ esférico 83–91
~ geométrico 83–92
~ medio 83–90a
~ monocromático 83–93
~ total 83–90b
Albireo 92–103
albita 74–17
alcance del radar 86–288
alcohol metílico 97–32
Alcor 92–104
aldehído acético 97–42
~ fórmico 97–27
Algol 92–105
Algónquico 74–77
alineamiento de granos 97–78
alisios 77–100
almacenamiento de partículas 75–123
alta 77–84
~ atmósfera 76–52
~ cromosfera 88–192
~ mar 82–281
Altair 92–106
altamente degenerado 93–184
~ ionizado 89–289
Altar 92–19

ES

ES

alternación de polaridad magnética 89–153
altimetría por satélites 71–50
altiplanicie 81–189
altiplano 81–189
altitud de extinción 86–40
~ de los meteoros 86–193
~ del punto de aparición 86–39
~ del punto de reflectión 75–125
~ sobre el elipsoide de referencia 71–52
~ sobre el nivel del mar 71–51
altocúmulo 77–144
altoestrato 77–140
altostrato 77–140
altura a escala 76–18
~ de aparición 86–39
~ de blindaje 79–48
~ de desaparición 86–40
~ de escala 88–236
~ de extinción 86–40
~ de la atmósfera homogénea 76–17a
~ de la capa luminosa 79–26
~ de la cromosfera 88–202
~ de la sombra 79–46
~ de los meteoros 86–193
~ de presión 76–16
~ de una capa 78–27
~ del ápex 86–66
~ del punto de aparición 86–39
~ del punto de desaparición 86–40
~ del punto de extinción 86–40
~ del punto de reflexión 75–125
~ del punto espejo 75–125
~ equivalente 78–28
~ geométrica de la sombra 79–47
~ real 78–29
~ real de la sombra 79–49
~ virtual 78–28
alunizador 82–21
alunizaje 82–13
álveo 83–153
amanecer 79–35
amarillo 91–98
ambiente 76–7
amino 87–71
aminoácido 87–70
amoníaco 97–25
amortiguador 82–24
~ del pie de alunizaje 82–23
amplificación del campo 89–156
~ del campo magnético 89–156
amplitud 95–38
~ de las mareas 82–292
~ de variación de brillo 95–38
anagaláctico 98–207
análisis de microprueba 82–143
~ in situ 82–136
~ mineralógico 82–53
~ petrográfico 82–52
~ químico 82–75
anchura = ancho
~ de la zona de silencio 78–38
~ de pulso 86–280
~ efectiva de la línea 89–208
Andrómeda 92–14
Andromédidas 86–256
anemógrafo 77–114
anemómetro 77–113
~ registrator 77–114
anfíbol 74–20
ángulo característico 98–33
~ de fase 83–72
~ de posición 81–71
~ heliocéntrico 89–37
anhídrido carbónico sólido 83–144
anillo A 83–225
~ asteroidal 84–70
~ B 83–222
~ brillante 89–101
~ C 83–221
~ central 83–222
~ de asteroides 74–40
~ de Bishop 79–88
~ de canto 83–218
~ de estrellas 96–78
~ de Saturno 83–215
~ estelar 96–78
~ exterior 83–225
~ externo 83–225
~ falso 81–242
~ fantasma 81–242
~ gaseoso 95–66
~ interior 83–221
~ medio 83–222
~ nebular 83a–10
~ nebuloso 83–221
~ obscuro 83–221
~ visto de canto 83–218
aniquilación de campos magnéticos 89–158
anochecer 79–36
anomalía brusca de fase 78–52
~ de la gravedad 71–75
~ gravimétrica 71–75
~ gravitacional lunar 81–144
~ magnética 75–67
~ sudatlántica 75–68
anomálias de abundancia 91–183
anortita 74–18
anortosita 82–87
Antares 92–107
antena emisora 78–16
~ en cruz de Mills 88–121
~ receptora 78–18
~ transmisora 78–16
antiápex 86–65
~ solar 98–65
anticentro galáctico 98–14
anticiclón 77–84
anticiclónico 77–85
anticlinal 73–90
anticola 85–78
anticrepúsculo 79–57
antigua supernova 95–134
antiguo satélite 83–240
antihelio 79–83
antiselena 79–85
antiselene 79–85
antivértice 98–106
Antlia 92–15
anular 97–146
Año del Sol calmo 89–10
~ del Sol inactivo 89–10
~ del Sol quieto 89–10
~ luz 91–18
apápside galáctico 98–190
apareo 83–141
aparición de un cometa 85–18
~ de un metoro 86–35
apastro 96–61
ápex 86–64
~ del Sol 91–23
~ solar 91–23
aplanamiento 98–295
~ aparente 98–295
~ del la Tierra 71—8
~ real 98–296
aplastamiento gravitacional 84–72
apoápside galáctico 98–190
ap(o)astro 96–61
apocentro galáctico 98–190
~ joviano 83–173
~ saturnal 83–211
apocintio 81–41
apofoco galáctico 98–190
apogaláctico 98–190
apogeo 81–39
apojove 83–173
apojoviano 83–173
Apolo 84–81
apolunio 81–41
aposaturno 83–211
aprisionamiento 75–109
aproximación cercana 84–69
~ de Chandrasekhar 93–51
~ de Milne-Eddington 93–50
~ no relativista 99–97
apulso 81–101
Apus 92–16
Aquarius 92–17
Aquila 92–18
Ara 92–19
árboles 89–273
en forma de ~ 89–273
arborescente 89–273
arcilla pelágica 72–54a
~ roja 72–54
arco auroral 79–146a
~ circuncenital 79–74
~ con estructura de rayos 79–153
~ con rayos 79–153
~ coronal 89–298
~ crepuscular 79–50
~ de meridiano 71–21
~ de montañas 73–84

~ eruptivo 89–251
~ estriado 79–153
~ homogéneo quieto 79–143
~ iris 79–65
~ montañoso 73–84
~ primario 73–82
~ pulsante 79–147
~ secundario 73–83
~ tangencial de halo 79–75
~ tangencial inferior 79–76
~ tangencial infralateral 79–87
~ tangencial superior 79–77
arcos 89–270
~ intertropicales 79–53
área aparente de las manchas 89–91
~ corregida 89–93
~ de alta presión 77–84
~ de audibilidad 86–120
~ de caída 87–38
~ de impacto 87–38
~ de impacto de la lluvia meteorítica 87–38
~ de las manchas 89–67
~ de las protuberancias 89–233
~ de radiación 86–238
~ de una fulguración 89–207
~ de visibilidad 86–126
~ diaria de protuberancias 89–236
~ intergranular 88–177
~ manchada 89–90
~ reducida 89–93
~ total 89–94
áreas de Kapteyn 98–78
~ seleccionadas 98–78
areo- 83–135
areocéntrico 83–138
areofísica 83–134
areográfico 83–137
Ariel 83–235
Aries 92–20
armónicos zonales 93–55
Arqueozóico 74–76
arrastre 86–178
arrebol vespertino 79–55
arrolamiento 98–28
ascendente 76–54
ascensión recta de la Luna 81–60
~ recta del Sol 81–61
ascenso 82–8
asentamiento de tierras 82–152
asimetría este-oeste 89–49
~ norte-sur 89–50
asintótico a las órbitas circulares 98–180
asociación accidental 97–126
~ estelar 96–88
~ genética 97–127
~ O 96–89
~ T 96–90
asociado a una fulguración 89–188
aspecto 83–50
~ topográfico 81–149
aspereza de superficie 83–99
~ superficial 82–106
asterismo 92–3a
asteroide 84–68
~ Amor 84–80
~ artificial 84–69a
~ binario 84–79
~ catalogado 84–95
~ de aproximación cercana 84–78
~ de aproximación cercana a la Tierra 84–78
~ de libración 84–91a
~ del anillo 84–71
~ del cinturón 84–71
~ en forma de cigarro 84–77
~ marciano 84–74
~ numerado 84–95
~ perdido 84–97
~ que cruza la órbita de la Tierra 84–76
~ que cruza la órbita de Marte 84–75
~ que cruza la órbita terrestre 84–76
~ tipo Amor 84–80
~ tipo Apollo 84–80a
asteroides troyanos 84–91
astillado por radiación cósmica 82–254
astillamiento 84–100
astro 79–9a
astrolito 87–60
astronomía de base lunar 82–4
~ estelar 91–2
~ extragaláctica 98–208
~ hecha desde la Luna 83–4
~ meteórica 86–3
~ meteorítica 87–6
ataque químico 87–102
ataxito 87–87
~ pobre en níquel 87–88
~ rico en níquel 87–89
atenuador escalonado 88–80
aterramiento en mar arbierto 72–79
aterrizaje tripulado 82–33
atlas celeste 92–12
~ de la Luna 81–136
~ del cielo 92–12
~ fotométrico 88–138
~ lunar 81–136
atmósfera 83–241
~ adiabática 76–12
baja ~ 76–50
~ de electrones 93–28
~ de la Tierra 76–5
~ de un planeta 83–243
~ del Sol 88–163
~ difusora 93–16
~ electrónica 93–28
~ esférica 93–24
~ estelar 93–2
~ estratificada 93–21
~ estratificada en capas plano-paralelas 93–21
~ extendida 93–9
~ externa 76–68
~ finita 93–23
~ gris 93–26
~ homogénea 76–10
~ inicial 83a–44
~ isotérmica 77–39
~ libre 76–13
~ media 76–51
~ no gris 93–27
~ oxidante 83a–47
~ patrón 76–11
~ planetaria 83–243
~ plano-paralela 93–21
~ primitiva 83–244
~ puramente difusora 93–16
~ reductora 83a–46
~ semiinfinita 93–22
~ solar 88–163
~ terrestre 76–5
~ tipo 76–11
atmosférico 83–242
atmosféricos 78–50
atomización por viento solar 82–245
átomo de dos niveles 93–45
~ de niveles múltiples 93–46
~ primitivo 99–169
atracción zenital 86–52
aumento al máximo 95–28
~ brusco de atmosféricos 78–51
~ de brillo 86–79, 95–28
~ de masa 83a–38
~ de masa de la Tierra 87–18
~ de temperatura 77–42
~ relativista de massa 99–142
aureola 79–60
~ solar 79–62
Auriga 92–21
aurora 79–56
~ artificial 79–99
~ austral 79–106
~ boreal 79–105
~ de baja latitud 79–96
~ de casquete polar 79–98
~ de estructura de cinta 79–140
~ diurna 79–97
~ electrónica 79–110
~ flameante 79–157
~ iluminada por el Sol 79–130
~ permanente 79–5
~ polar 79–93
~ protónica 79–108
~ radar 79–95
~ radio 79–94
Austrálidas 86–250
australita 87–98
autogravitación 94–34
autogravitante 94–33
autogravitatorio 94–33
autoinversión de la imanación 75–38
~ de la magnetización 75–38
autoluminoso 96–45
avance del periastro 96–71
~ relativista del perihelio 99–160

Ave del Paraíso 92–16
~ Fénix 92–77
Azóico 74–75
azul 91–99
~ del cielo 77–60

B

bahía 81–294
~ del Arco Iris 81–295
~ magnética 75–80
~ Obscura 97–135
baja atmósfera 76–49
~ cromosfera 88–193
bajamar 82–283
bajío 72–96
balance calorífico 77–62
~ calorífico de la atmósfera 77–62
~ de energía 88–251
~ de masa 88–250
~ de radiación 88–252
~ detallado 93–41
Balanza 92–62
Ballena 92–34
banco 72–96
banda 83–190
~ con rayos 79–150
~ de audibilidad 96–120
~ de fulguración 89–187
~ de la luz zodiacal 84–63
~ de polvo 98–289
~ de visibilidad 86–126
~ difusa interestelar 97–100
~ ecuatorial 83–191
~ estriada 79–150
~ homogénea 79–146
~ luminosa 86–27
~ obscura 83–190
~ zodiacal 84–63
bandas de fluorescencia 85–132
~ de sombra 88–31
~ de Swan 85–133
~ de Vegard-Kaplan 79–19
~ negativas del nitrógeno 79–21
~ no identificadas 85–134
~ obscuras 98–288
~ positivas del nitrógeno 79–20
barisfera 72–17
barra 72–96
~ galáctica 98–254
barrera del sonido 76–44
~ sónica 76–44
barrido en longitud de onda 88–111
~ en posición 88–105
basalto 74–9
~ lunar 82–84
basamento 72–12
base de una protuberancia 89–234
~ del interferómetro 88–100
~ en la Luna 82–17
~ lunar 82–17
~ lunar tripulada 82–18
batería de cámaras 86–144
batiente continental 72–36
batimetría 72–28
batolito 73–139
Betelgeuse 92–108
Biélidas 86–256
bifurcación 98–5
big bang 99–167
bigotes 89–200
binaria 96–3, 96–7
~ astrométrica 96–12
~ cercana 96–23
~ con espectro simple 96–15
~ de contacto 96–29
~ de eclipse 95–74
~ de eclipse tipo Algol 95–77
~ eclipsante 95–77
~ eclipsante tipo Algol 95–77
~ espectral 96–14
~ espectroscópica 96–14
~ espectroscópica con espectro simple 96–15
~ espectroscópica con líneas dobles 96–17
~ espectroscópica con lineas simples 96–15
~ fotométrica 96–11
~ irresoluble 96–10
~ no resuelta 96–10
~ próxima 96–23
~ semiseparada 96–34
~ separada 96–35
biosfera 72–8
biotita 74–25
„blanketing" 93–20
blindaje 86–171
~ magnético 89–225
~ por la capa de ozono 79–28
bloque 82–130
~ anguloso 82–131
~ continental 72–10
~ de rocas 82–130
~ errático 72–132
boca volcánica 73–122
bocanada 89–277
bola de fuego 99–168
~ de fuego primordial 99–168
~ de polvo 86–190
boletín meteorológico 77–10
bólido 86–11
~ brillante 86–10
~ detonante 86–119
~ gigante 86–9
~ grande 86–10
~ muy grande 86–9
~ visible de día 86–118
Bolsa de Carbón 97–138
bomba de Ellerman 89–200
~ volcánica 73–144
bombardeo de partículas 82–240
~ meteorítico 82–235
~ meteoroidal 82–235
~ por partículas 82–240
bombeo colisional 97–20
~ infrarrojo 97–18
~ óptico 97–17
~ por choque 97–20
~ químico 97–19
~ radiativo 97–17
~ ultravioleta 97–18a
Boötes 92–22
borde brillante de un filamento 89–226
~ de bloques 81–224
~ de la Luna 81–95
~ de los casquetes polares 83–151
~ de una galaxia 98–294
~ del cráter 81–221
~ exterior 83–227
~ iluminado 81–96
~ levantado 81–122
~ lunar 81–95
~ obscuro 81–97
bóveda obscura 89–297
~ obscura alrededor de las protuberancias 89–297
boya automática 72–100
~ de investigación 72–99
Boyero 92–22
brazo de Orión 98–36
~ de Perseo 98–34
~ de Sagitario 98–35
~ de tres kiloparsec 98–37
~ de 3kpc 98–37
~ desenrollado 98–31
~ en expansión 98–38
~ enrollado 98–29
~ espiral 98–27
~ galáctico 98–26a
brazos, entre los 98–40
brecha lunar 82–85
brillo 91–54
~ absoluto 85–109
~ absoluto de un cometa 85–109
~ aparente 91–61
~ aparente de un meteoro 86–73
~ bolométrico 91–68
~ de una erupción 89–212
~ de una fulguración 89–212
~ de un meteoro 86–8
~ electrográfico 91–67
~ estelar 91–55
~ fotográfico 91–66
~ fotovisual 91–69
~ infrarrojo 91–71
~ integrado 96–135
~ integral 96–135
~ intrínseco 85–107
~ límite 91–62
~ radiométrico 91–40
~ reducido 85–108
~ total 85–107, 96–135
~ visual 91–65
brisa de mar 77–97
~ de tierra 77–98

ES

Brújula 92–82
bruma 83–250
~ seca 77–27
brusca emisión de rayos X 79–117
bucle coronal 89–298
~ interplanetario 84–21
~ magnético 84–19
~ solar-interplanetario 84–20
bucles 89–270
buena visibilidad 77–25
bulbo central 98–16
~ coronal 89–295
burbujas de lava 82–178
Buril 92–23
buscador del Sol 88–51
búsqueda de cometas 85–15
~ de meteoritos 87–45
~ de planetas 83–238
~ de satélites 83–27
~ de supernovas 95–136

C

Caballete del Pintor 92–78
Caballito 92–48
Caballo Menor 92–48
cabellera 85–64
Cabellera de Berenice 92–38
Cabeza de Caballo 97–135
~ de la Serpiente 92–89
~ y Cola 92–89
cabeza de un cometa 83–35
cabo 81–293
cadena de cráteres 81–241
~ de estrellas 96–77
~ de galaxias 98–328
~ de montañas 81–194
~ de supernovas 95–132
~ estelar 96–77
~ montañosa 81–194
Caelum 92–23
caída 76–57
~ aislada 87–36
~ de gas 98–356
~ de gas del espacio intergaláctico 98–356
~ de temperatura 77–41
~ de un bólido 86–117
~ de un meteorito 87–19
~ de un meteorito aislado 87–36
~ de un meteoro 86–5
~ libre 94–39
~ meteorítica 87–19
~ natural 82–227
~ supuesta de un meteorito 87–49
calculista 85–22
caldera 81–179
~ colapsada 81–181
~ de erosión 81–183
~ de explosión 81–180
~ de hundimiento 81–181
~ de subsidencia 81–182
calentamiento atmosférico 75–87
~ de la corona 88–213a
~ de la cromosfera 88–213
~ de la partícula meteórica 86–174
~ del cuerpo meteórico 86–174
calima 77–27
Calisto 83–208
calma 77–110
calmas ecuatoriales 77–101
~ tropicales 77–102
calor de ablación 86–184
~ de condensación 77–155
~ de desorción 85–124
~ de marea 82–287
~ de sublimación 85–139
~ de vaporización 86–199
~ fósil 74–63
~ latente 77–157
~ originario 74–63
~ primordial 83–200
~ radiactivo 83a–62
~ radiogénico 83a–62
calota de hielo 72–115
camacita 87–92
Camaleón 92–35
cámara de patrullaje 86–145
~ de vigilancia 86–145
~ fija 86–143
~ fotográfica de cielo global 79–104
~ guiada 86–142
~ meteórica 86–141
~ móvil 86–142
~ no guiada 86–143
~ para la detección de meteoros 86–145
~ para la fotografía de los meteoros 86–141
cambiador de fase 88–115
cambio de forma 73–68
~ de la línea de la costa 72–78
~ de período 95–43
~ del clima 77–184
~ del tiempo 77–22
~ evolutivo 94–27
cambios diurnos 83–167
~ estacionales 83–166
Cambriano 74–78
Cámbrico 74–78
Camelopardalis 92–24
Camino de Santiago 98–1
campo coercitivo 75–39
~ cuántico 99–170m
~ de corrientes telúricas 75–4
~ de cráteres 87–56
~ de fondo 89–141
~ de gran escala 89–142
~ de gravedad 71–57
~ de hielo flotante 72–102
~ de Higgs 99–1701
~ de lava 82–184
~ de macroestructura 89–142
~ de marea 98–196
~ de marea galáctico 96–106
~ de microestructura 89–143
~ de movimientos diferenciales 98–114
~ de nevero 72–119
~ de pequeña escala 89–143
~ de radiación 93–30
~ de radiación difusa 93–31
~ de rocas 82–139
~ de velocidades 88–187
~ de ventisquero 72–119
~ effectivo 95–59
~ electrotelúrico 75–4
~ galáctico 98–146
~ geomagnético 75–44
~ gravitacional 98–145
~ gravitacional galáctico 98–146
~ imanador 75–42
~ interno 95–63
~ longitudinal 89–138
~ magnético de fondo 89–141
~ magnético de gran escala 89–142
~ magnético de la Tierra 75–44
~ magnético de macroestructura 89–142
~ magnético de microestructura 89–143
~ magnético de pequeña escala 89–143
~ magnético fósil 91–53
~ magnético fotosférico 89–130
~ magnético galáctico 98–50
~ magnético general 89–128
~ magnético general del Sol 89–128
~ magnético interplanetario 84–9
~ magnético interestelar 97–106
~ magnético lunar 82–45
~ magnético muy intenso 95–148
~ magnético primitivo 95–62
~ magnético remanente 95–62
~ magnético residual 91–53
~ magnético solar 89–129
~ magnético terrestre 75–44
~ magnético ultraintenso 95–148
~ magnetizante 75–42
~ selenomagnético 82–45
~ sonoro 76–40
~ superficial 95–64
~ transversal 89–137
Can Mayor 92–27
~ Menor 92–28
canal de flujo 82–192
canales de Marte 83–157
Cancer 92–25
Canes Venatici 92–26
Cangrejo 92–25
Canis Major 92–27
~ Minor 92–28
Canopus 92–109
cantidad de nubes 77–147
~ de ozono en la atmósfera 76–125
~ de precipitaciones 77–158
~ de vapor de agua 77–124

~ del cloro 72–68
~ del sal 72–66
cantos rodados 81–237
cañon 83–154
~ submarino 72–46
capa absorbente 88–175
~ atmosférica 76–49
~ concordante 73–135
~ convectiva 93–149
~ de bruma 83–249
~ de combustión de helio 93–145
~ de combustión de hidrógeno 93–144
~ de detrito fragmentario 82–104
~ de difusión 83–258
~ de helio 93–145
~ de hidrógeno 93–144
~ de inversión 88–174
~ de lava 82–175
~ de nieve 77–176
~ de nubes 77–150, 83–248
~ de nubes lluviosas 77–145
~ de ozono 76–123
~ de polvo 82–122
~ de transición 72–13
~ difusora 83–258
~ E 78–26
~ E esporádica 78–25
~ emisiva 79–17
~ fotoelectrónica 82–246
~ galáctica absorbente 97–89
~ Heaviside-Kennelly 78–26
~ intrusiva 73–135
~ inversora 88–174
~ ionizada 78–3
~ ionosférica 78–5
~ luminosa 79–17
~ radiativa 93–134
~ semiconvectica 93–160
~ subfotosférica 88–168
~ violeta 83–165
capacidad portadora 82–64
~ sustentadora 82–64
capas bajas de la atmósfera solar 88–166
~ envolventes de un cometa 85–69
~ exteriores de una estrella 93–3
~ exteriores no degeneradas 93–199
~ superiores de la atmósfera solar 88–165
Capella 92–110
Capricornio 92–29
Capricornus 92–29
captura 75–109, 83a–19
~ de cometas 85–101
~ de estrellas 96–73
~ de la materia intergaláctica 98–355
~ de neutrones 94–84
~ de protones 94–90
~ neutrónica 94–84
~ protónica 94–90
cara iluminada 81–34
~ oculta de la Luna 81–16
~ opuesta de la Luna 81–16
~ visible de la Luna 81–15
carácter volcánico 82–159
característica de la superficie 83–105
~ del suelo 82–108
~ en ondas de radio 98–235
~ estructural 88–248
~ óptica 98–234
~ permanente 83–106
~ primaria 81–150
~ radio 98–235
~ secundaria 81–151
características activas 79–138
~ estructurales 98–224
~ lunares 81–148
carbonato 74–27
Carbonífero 74–82
carbono radiactivo 74–95
carga de la Tierra 75–3
Carina 92–30
Carnero 92–20
Carro 92–96
carta batimétrica 71–54
~ celeste 92–9
~ celeste giratoria 92–10
~ de arrumbamiento 72–33
~ de la variable 95–15
~ de predicción 77–21
~ de recalar 72–33
~ del cielo 92–9
~ del eclipse 88–28
~ del tiempo 77–12
~ física lunar 81–134
~ gravimétrica 71–79
~ hipsométrica 81–141
~ lunar 81–133
~ magnética 75–63
~ sinóptica 77–11
~ tectónica 73–96
~ térmica lunar 81–135
cartografía de la Luna 81–111
~ del litoral 72–82
~ lunar 81–111
~ selenológica 82–51
~ térmica 72–83
cartografiado selenológico 82–50
casco coronal 89–295
Casiopea 92–31
caso de contaminación 72–72
casquete auroral 79–127
~ de compresión 86–170
~ glaciar 72–115
~ polar 83–143
Cassiopeia 92–31
Cástor 92–111
catabático 77–93
catálogo de estrellas 92–13
~ de estrellas sospechosas de variabilidad 95–58
~ de Messier 92–12a
~ estelar 92–13
~ de supuestas variables 95–58
catástrofe causada por un meteorito 87–20
~ sísmica 73–40
caverna de lava 82–177
cavidad 87–117
~ del viento solar 84–28
~ geomagnética 75–8
~ heliosférica 88–257
~ hidromagnética 84–22
~ interna 87–119
~ interplanetaria 84–25
~ magnética 84–22
~ planetaria 84–24
~ superficial 87–118
cefeida 95–92
~ clásica 95–94
~ de corto período 95–93
~ de largo período 95–94
Cefeo 92–33
celóstato 88–46
~ horizontal 88–47
~ vertical 88–48
célula granular 88–178
~ supergranular 88–182
cementación 74–52
ceniza volcánica 73–142, 82–168
Cenozoico 74–73
Centauro 92–32
Centaurus 92–32
centellas de un meteoro 86–84
centelleo interestelar 97–109a
~ radio 84–36
~ radio interestelar 97–109b
~ radio interplanetario 84–37
centro activo 89–60
~ de actividad 89–60
~ de atracción 89–241
~ de la estrella 93–84
~ de la Galaxia 98–9
~ de la placa 86–156
~ del disco lunar aparente 81–128
~ del disco solar 89–35
~ del disco visible 88–172
~ fotométrico 96–46
~ galáctico 98–9
Cepheus 92–33
cerco 89–274
Ceres 84–82
Cetus 92–34
cianoacetileno 97–30
cianógeno 97–24
cianuro de hidrógeno 97–29
~ de metilo 97–38
ciclo de actividad solar 89–71
~ de las manchas 89–72
~ de las manchas solares 89–72
~ de once años 89–73
~ geoquímico 74–32
~ solar 89–71
~ solar impar 89–78
~ solar par 89–77
~ undecenal 89–73

ciclón 77–76
~ tropical 77–77
cielo austral 79–3c
~ boreal 79–3b
~ claro 77–153
~ crepuscular 79–43
~ cubierto 77–152
~ despejado 77–153
~ diurno 79–3a
~ estrellado 91–1a
~ nocturno 79–6
~ nocturno luminoso 86–28
~ nuboso 77–148
ciencia ambiental 76–2
ciencias atmosféricas 76–1
~ lunares 82–2
cilindro de luz 95–155
~ truncado 88–63
Cincel 92–23
cinta de fulguración 89–187
cinturón activo 73–32
~ de asteroides 84–70
~ de Orión 92–129a
~ de radiación 75–101
~ de radiación artificial 75–108
~ de radiación exterior 75–105
~ de radiación interior 75–106
~ de radiación natural 75–107
~ de Van Allen 75–101
~ minauroral 79–129
~ sísmico 73–30
~ subauroral 79–128
Circinus 92–36
circo lunar 81–265
circulación atmosférica 77–64
~ general 77–65
~ meridional 93–163
~ planetaria 77–66
círculo azimutal 79–82
~ circuncenital 79–74
~ deferente 83–48
~ horizontal 79–82
circumestelar 97–3
circunlunar 81–5
circunstancias de la caída 87–47
~ del eclipse 81–73
cirro 77–136
cirrocúmulo 77–143
cirroestrato 77–141
cirrostrato 77–141
cislunar 81–4
Cisne 92–44
claror atmosférico 79–2
~ atmosférico crepuscular 79–4
~ atmosférico diurno 79–3
~ atmosférico nocturno 79–5
clase de luminosidad 91–146
~ de una fulguración 89–205
~ espectral 91–107
clasificación de estrellas variables 95–46
~ de galaxias 98–226
~ de Harvard 91–91
~ de los meteoritos 87–59
~ de Zürich 89–127
~ espectral 91–90
~ espectral bidimensional 91–105
~ espectral tridimensional 91–106
~ magnética 89–109
~ petrográfica 82–101
clatrato 85–51
clima 77–180
~ continental 77–185
~ de montaña 77–187
~ marítimo 77–186
climático 77–182
climatología 77–179
~ desde satélite 77–181
cloricidad 72–68
clorinidad 72–68
coagulación 76–119
coalescencia 76–120
Cochero 92–21
coeficiente de arrastre 86–180
~ de aspereza 82–107
~ de brillo 83–96
~ de difusión atmosférica 86–287
~ de estabilidad 98–199
~ de fase 83–77
~ de forma 86–177
~ de obscurecimiento 88–144
~ de Poisson 73–60
~ de resistencia 86–180
~ de rugosidad 82–107
~ de transmisión del calor 86–201
coeficientes armónicos 71–59
~ de Legendre 95–101
~ del elipsoide de velocidades 98–102
coherencia 99–33
cohesivo 82–138
cola anómala 85–75
~ cometaria 85–70
~ curvada 85–73
~ de humo 87–21
~ de iones 85–82
~ de la Serpiente 92–90
~ de la Tierre 84–65
~ de polvo 85–79
~ de un cometa 85–70
~ de un meteoro 86–110
~ dirigida hacia el Sol 85–78
~ en dirección opuesta al Sol 85–81
~ galáctica 98–280
~ iónica 85–82
~ magnética de la Tierra 75–10
~ magnetosférica 75–10
~ múltiple 85–76
~ principal 85–77
~ recta 85–72
colada de lava 73–130, 82–173
colapso 94–109
~ del universo 99–181
~ final 99–182
Gran ~ 99–182
~ gravitacional 94–109
colección de meteoritos 87–8
~ de muestras 82–72
~ de muestras de suelos 82–78
colisión catastrófica 84–106
~ central 83a–17
~ estelar 83a–16
~ de estrellas 83a–16
~ de galaxias 98–322
~ de tres cuerpos 96–74
~ erosiva 84–105
~ frontal 83a–17
~ inelástica de estrellas 98–26
~ triple 79–30
coloidal 76–97
coloide 76–96
color intrínseco 91–74
colores de atardecer 79–55
~ de la aurora 79–56
Columba 92–37
columna de protuberancia 89–245
~ ionizada 86–282
~ luminosa 79–78
coma 85–64
~ atómica 85–67
~ Berenices 92–38
~ de radicales 85–66
~ interior 85–65
~ molecular 85–65
~ ultravioleta 85–67
~ visible 85–66
combustión del carbono 94–81
~ del helio 94–78
~ del hidrógeno 94–56
~ del oxígeno 94–82
~ del silicio 94–83
~ explosiva de helio 94–79
cometa 85–1
~ activo 85–149
~ artificial 85–14
~ asociado a un enjambre 86–259
~ de corto período 85–5
~ de eclipse 85–12
~ de largo período 85–6
~ de órbita elíptica 85–4
~ de órbita hiperbólica 85–8
~ de órbita parabólica 85–7
~ elíptico 85–4
~ extinto 85–150
~ generador 86–260
~ hiperbólico 85–8
~ inactivo 85–147
~ no periódico 85–10
~ padre 86–260
~ parabólico 85–7
~ periódico 85–3
~ predicho 85–11
~ rasante al Sol 85–13
~ telescópico 85–14a
~ vivo 85–149
cometario 85–2
comienzo brusco 75–74
~ del ciclo 89–82

~ del eclipse 81–74
~ del eclipse total 81–77
~ rápido 75–74
compactación 74–50
compañera = compañero 96–38
~ de Sirio 91–138
~ de una galaxia 98–216
~ elíptica 98–217
~ elíptica de una galaxia 98–217
~ física 98–219
~ física de una galaxia 98–219
~ invisible 96–41
~ o(b)scura 96–42
~ óptica 98–218
~ óptica de una galaxia 98–218
~ planetaria 96–43
compañero = compañera 96–38
~ de Sirio 91–138
Compás 92–36
compensación isostática 73–100
~ para los desplazamientos Doppler 88–90
compensador Doppler 88–91
complejo basal 72–12
completamente degenerado 93–185
componente 96–36
~ atmosférico 76–87
~ de polvo 98–232
~ horizontal 75–60
~ invisible 96–41
~ longitudinal 89–135
~ lunisolar 83–307
~ meteórico 86–234
~ mineralógico 82–100
~ obscuro 96–42
~ primario 96–37
~ principal 96–37
~ σ 88–89
~ transversal 89–136
~ vertical 75–62
composición iónica 78–10
~ isotópica 74–93, 82–140
~ molecular 78–11
~ química 91–168
~ química de una estrella 91–168
compresibilidad 73–57
comunicador 85–23
comunicante 85–23
concentración 96–127
~ de masa lunar 81–145
~ de trazas 83–266
~ de vestigios 83–266
~ galáctica 98–76
~ hacia el centro 98–76
~ hacia el plano galáctico 98–55
~ hacia el plano principal 98–76
condensación 83a–26
~ central 85–37, 96–126
~ coronal 89–281
~ coronal esporádica 89–283
~ coronal permanente 89–282
~ de una protuberancia 89–244
~ en la cola 85–95
~ en una corriente meteórica 86–217
condensar(se) 77–129
condición de ajuste 93–142
~ de contorno en el centro 93–81
~ de empalme 93–142
~ de estabilidad 93–205
~ de simetría 98–162
~ en el centro 93–81
~ en la superficie 93–80
condiciones de equilibrio 93–74
~ del tiempo 77–2
~ meteorológicas 77–2
~ normales 77–3
~ térmicas 77–35
condrito = condrita 87–61
~ carbonáceo 87–63
~ no carbonáceo 87–75
cóndrulo 87–76
conducto 73–124
configuración del campo 89–140
~ planetaria 83–50
confinamiento magnético de los rayos cósmicos 98–51
conglomerado 85–41
~ de hielo 85–43
conjunción 83–55
~ inferior 83–57
~ superior 83–56
~ triple 83–58
conjunto de domos 81–179
conmutación de fase 88–112
conmutador de fase 88–114
cono compuesto 73–121
~ de capas 73–121
~ de cenizas 73–120
~ de erupción 73–118
~ de escape 76–27
~ de escorias 73–120
~ de eyección 73–120
~ de luz 95–156
~ de sombra 81–90
~ mixto 73–121
~ piroclástico 73–120
~ reflectante 88–62
~ volcánico 82–160
consolidación 74–50
~ del magma 82–190
constante cósmica 99–29
~ cosmológica 99–29
~ de expansión 99–77
~ de Hubble 98–302
~ de Oort 98–118
~ de Poisson 73–60
~ de recesión 98–302
~ solar 88–148
constantes del elipsoide de velocidades 98–102
constelación 83–50, 92–1
~ circumpolar 92–4a
~ de invierno 91–1b
~ de verano 92–1a
~ zodiacal 92–5
constitución interna de las estrellas 93–71
constituyente atmosférico 83–265
~ mineralógico 82–100
contacto, primer 81–84
último ~ 81–85
contaminación 76–80
~ atmosférica 76–80
~ del agua 72–73
~ química 72–98
contaminante 72–74, 76–81
~ atmosférico 76–82
contenedor de muestras 82–81
contenido de níquel 72–55
~ de polvo 85–80
~ estelar 98–172
~ estelar de una galaxia 98–231
conteos estelares 98–79
continente 72–9, 81–191
contorno de igual contaje 75–102
~ de isoconteos 75–102
~ de isogauss 89–134
~ de nivel 81–139
~ isofotal 97–124
~ isofotal de una nebulosa 97–124
contraalisios 77–103
contrabrillo zodiacal 84–64
contracción 93–117
~ catastrófica 94–109
~ c(u)asiestacionaria 94–40
~ c(u)asiestática 94–40
~ de Helmholtz 94–37
~ de Lorentz 99–134
~ del núcleo 94–70
~ del universo 99–180
~ gravitacional 83–203, 93–118
~ Helmholtz 94–37
~ longitudinal 99–134
~ relativista 99–134
convección 93–147
~ laminar 93–153
~ turbulenta 93–152
convectivo 93–148
coordenada de espacio 99–124
~ de tiempo 99–123
~ espacial 99–124
~ temporal 99–123
coordenadas cilíndricas 98–156
~ galactocéntricas 98–12
~ geomagnéticas 75–51
~ heliográficas 89–39
~ jovicéntricas 83–176
~ jovigráficas 83–175
~ planetográficas 83–8
~ selenográficas 81–9
Copa 92–42
coplanareidad de órbitas 83a–3
corchete 75–79
cordillera 73–85, 81–186
~ de plegamiento 81–187

ES

~ marina 81–188
corona 79–60, 79–156
~ auroral 79–137a
~ Austral 92–39
~ Australis 92–39
~ blanca 88–237
~ Boreal 92–40
~ Borealis 92–40
~ de electrones 88–242
~ de emisión 88–241a
~ de Fraunhofer 88–241
~ de hidrógeno 83–246
~ de luz blanca 88–237
~ de máximo 88–244
~ de mínimo 88–246
~ de tipo intermedio 88–245
~ E 88–241a
~ en expansión 88–232
~ en rayos X 88–240
~ estelar 93–4
~ exterior 88–231
~ F 88–241
~ galáctica 98–49
~ interior 88–229
~ intermedia 88–245
~ interna 88–229
~ K 88–242
~ lunar 79–61
~ máxima 88–244
~ media 88–230
~ mínima 88–246
~ monocromática 88–238
~ planetaria 83–245
~ polar de los filamentos 89–232
~ radioeléctrica 88–239
~ solar 79–62, 88–225
coronal 88–228
coronógrafo 88–59
~ con disco externo 88–60
~ ocultado externamente 88–60
coronómetro K 88–66
coros matutinos 78–62
corpuscular 88–155
corpúsculo 88–153
corpúsculos sólidos 85–48
corrección bolométrica 91–80
~ de campo 86–157
~ de la trayectoria 82–10
~ por la curvatura 86–62
~ por la curvatura de la superficie terrestre 86–62
~ relativista 99–140
corriente activa 86–223
~ anular ecuatorial 75–98
~ ascendente 77–91, 93–151
~ cálida 72–63
~ cometaria 86–222
~ coronal 89–287
~ corpuscular 88–160
~ corpuscular solar 84–35
~ de aire 77–90
~ de asteroides 84–100
~ de cenizas 82–174
~ de convección 73–76
~ de plasma solar 84–33
~ de reflujo 72–65
~ de suspensión 72–49
~ de turbidez 72–49
~ de viento solar 84–35a
~ descendente 93–150
~ diurna 86–219
~ eclíptica 86–221
~ en chorro 77–88
~ esporádica 86–226
~ estelar 98–98
~ fría 72–64
~ gaseosa 95–67
~ interestelar 86–224
~ marina 72–60
~ marina cálida 72–63
~ marina fría 72–64
~ meteórica 86–216
~ nocturna 86–218
~ oceánica 72–60
~ periódica 86–228
~ permanente 86–227
~ planetaria 86–220
~ sub-superficial 72–61
~ superficial 72–62
~ temporal 86–225
corrientes 83–187
~ anulares 75–100
~ de anillo 75–100
~ telúricas 75–5
~ telúricas parásitas 75–6
~ telúricas vagabundas 75–6
corrimiento de la zona de actividad 89–74
~ de paredes 82–198
~ Doppler 91–25
~ gravitatorio 99–162
~ gravitatorio hacia el rojo 99–162
~ hacia el rojo 98–297
~ relativista 99–162
~ relativista de las rayas espectrales 99–162
~ relativista del perihelio 99–160
corteza continental 72–5
~ de la Luna 82–54
~ de la Tierra 72–4
~ lunar 82–54
~ sólida 95–151
~ terrestre 72–4
Corvus 92–41
cosmocronologia 83a–72
cosmogonía 94–2
~ estelar 94–7
~ planetaria 83a–1
cosmogónico 94–3
cosmogonista 94–4
cosmología 99–16
~ cinemática 99–26
~ newtoniana 99–25
~ no relativista 99–27
~ observacional 99–17
~ relativista 99–28
cosmológico 99–18
cosmólogo 99–19
cosmoquímica 74–31
cosmos 99–2
costa 72–30
costra de la Tierra 72–4
~ producida por la fusión 87–110
~ terrestre 72–4
Cráter 92–42
cráter 73–126
~ adventicio 73–128
~ anterior al mar 81–207
~ antiguo 81–205
~ bordeado de bloques 81–225
~ central 81–217
~ circular 81–213
~ con el borde de bloques 81–225
~ con el borde levantado 81–223
~ con pico central 81–258a
~ con rayos 81–214
~ concéntrico 81–216
~ de cumbre 73–129
~ de drenaje 81–239
~ de explosión 81–238
~ de halo oscuro 81–215
~ de impacto 81–240
~ en forma de plato 81–219
~ en forma de taza 81–219
~ enterrado 81–204
~ externo 81–218
falso ~ 81–209
~ fantasma 81–209
~ innominado 81–202
~ inundado 81–203
~ irregular 81–220
~ joven 81–206
~ juvenil 81–206
~ lateral 73–128
~ lunar 81–197
~ maar 81–264
~ meteórico 81–263
~ meteorítico 81–263, 87–53
~ meteorítico fósil 87–55
~ meteorítico terrestre 87–54
~ meteoroidal 81–263
~ parasitario 73–128
~ pequeño 81–250
~ poligonal 81–212
~ posible 81–211
~ posterior al mar 81–208
~ postmarítimo 81–208
~ premarítimo 81–207
~ primario 81–198
~ principal 73–127, 87–57
~ probable 81–210
~ radiado 81–214
~ reciente 81–206
~ satélite 81–200
~ secundario 81–199

ES

~ sepultado 81–204
~ sin borde 81–226
~ sin nombre 81–202
~ soterrado 81–204
~ viejo 81–205
craterito 94–10
craterización 82–232
~ por impacto 82–234
creación 94–10
~ continua 99–91
~ de estrellas 96–95
~ de materia 99–89
~ única 99–90
creciente 81–23
~ de Venus 83–129
~ lunar 81–23
crecimiento de granos 97–61
~ den un planeta 83a–37
~ del casquete polar 83–145
~ planetario 83a–37
crepuscular 79–40
crepúsculo 79–34
~ astronómico 79–37
~ civil 79–38
~ matutino 79–35
~ náutico 79–39
~ vespertino 79–36
cresta 81–195
~ de alta presión 77–87
~ de mar(e) 81–188
~ de montaña 81–195
~ de plegamiento 81–187
~ medio-oceánica 72–45
~ montañosa 81–186
Cretáceo 74–86
Criptozoico 74–70
cristal ahumado 88–33
~ de hielo 76–111
cristalización 76–107
~ fraccionada 74–40
cristobalita 82–95
criterio de degeneración 93–181
~ de inestabilidad por marea 98–312
criterios de clasificación 91–110
cromosfera 88–191
alta ~ 88–192
baja ~ 88–193
cromosférico 88–194
cronología geológica 74–67
Crux 92–43
Cruz 92–43
~ de halo 79–79
~ de Mills 88–121
~ del Norte 92–44
~ del Sur 92–43
~ luminosa 79–79
~ lunar 79–80
~ solar 79–81
cruzamiento de dos galaxias 98–320
cruzar el meridiano central 89–44
Cuadrántidas 86–247
cuadratura 83–64
~ occidental 83–66
~ oriental 83–65
cuadridimensional 99–121
cuadrivector 99–138
cuadro de evolución de los grupos 89–120
cuarto 81–26
~ creciente 81–25
~ menguante 81–30
primer ~ 81–25
tercer ~ 81–30
último ~ 81–30
cuásag 98–361
cuásar 98–360
~ inactivo en radio 98–362
~ quiescente en radio 98–362
~ silencioso 98–362
Cuaternario 74–88
cuatrovector 99–138
cubierto 77–152
cielo ~ 77–152
~ de nubes 83–253
~ de polvo 82–120
cuello magnetosférico 75–13
~ volcánico 73–125
cuenca 81–175
cuenta de galaxias 98–343
cuerda cósmica 99–171c
~ magnética 86–146
cuerno magnetosférico 75–13
cuernos 83–130
cuerpo causante de marea 82–284
~ celeste 79–9a
~ giratorio 91–42
~ impactante 82–225
~ madre 87–123
~ magnetoplasmático 98–25
~ magnetoplasmático rotatorio 98–25
~ meteórico 86–160a
~ preestelar 94–15
~ rígido 83–111
~ rotante 91–42
~ sólido 83–110
~ supermasivo 98–22
Cuervo 92–41
cueva de lava 82–177
culminación de la Luna 82–303
cúmulo 77–138
~ abierto de galaxias 98–333
~ compacto de galaxias 98–334
~ de alta concentración 96–129
~ de cúmulos 98–346
~ de cúmulos de galaxias 98–346
~ de domos 81–178
~ de estrellas relativista 96–154
~ de galaxias 98–329
~ de galaxias de la Virgen 98–345
~ de la Virgen 98–345
~ de nebulosas 98–329
~ doble 96–81
~ en movimiento 96–97
~ estelar 96–79
~ estelar abierto 96–99
~ estelar pobre 96–102
~ estelar rico 96–100
~ galáctico 96–99
~ globular 96–114
~ globular de galaxias 98–330
~ irregular de galaxias 98–337
~ muy compacto de galaxias 98–335
~ pobre de galaxias 98–331
~ pobre en metales 96–113
~ regular de galaxias 98–336
~ rico de galaxias 98–332
~ rico en metales 96–112
cúmulonimbo 77–137
cuña anticiclónica 77–86
~ de alta presión 77–86
~ escalonada 88–80
cupla de marea 82–288
curva batimétrica 71–55
~ de abundancia 94–108
~ de brillo 95–17
~ de color 95–21
~ de crecimiento 93–68
~ de crecimiento empírica 93–69
~ de crecimiento teórica 93–70
~ de desarollo de un grupo 89–119
~ de enrojecimiento 97–93
~ de evolución 89–119
~ de extinción 97–81
~ de fase 83–74
~ de luz 95–17
~ de luz de un meteoro 86–70
~ de nivel 71–53
~ de nivel sumergida 72–32
~ de polarización 83–81
~ de rotación 98–109
~ de velocidad-profundidad 73–52
~ de velocidades radiales 95–20
~ fotoeléctrica de luz 95–22
~ fotográfica de luz 95–23
~ isofota 81–92
~ media de luz 95–19
~ período-luminosidad 95–103
~ umbral de la secundaria 96–40
~ visual de luz 95–24
curvatura de la superficie terrestre 86–61
~ de la Tierra 71–6
~ de la trayectoria 86–51
~ de líneas de campo 75–121
~ de líneas de fuerza 75–121
~ del espacio 99–115
~ terrestre 71–6
cúspides 83–130
~ de la Luna 81–24
Cygnus 92–44

CH

Chamaeleon 92–35
chimenea 73–123
~ volcánica 82–162
chispas 86–84

~ de un meteoro 86–84
choque térmico 86–206
~ triple 79–30
chorro 85–55
~ brillante 89–257
~ curvo 85–58
~ de electrones 79–114
~ desde un núcleo galáctico 98–278
~ en espiral 85–59
~ galáctico 98–278
~ óptico 98–279
~ o(b)scuro 89–256
~ rectilíneo 85–57

D

daño por marea 72–77
~ por ondas de marea 72–77
datación 83a–71
~ de sucesos lunares 82–265
~ isotópica 83a–73
~ por isótopos 83a–73
~ por torio-plomo 82–270
~ radiactiva 83a–73
datos solares 89–9
datum geodésico 71–20
deceleración aparente 86–100
~ real 86–101
decibel 76–39
decibelio 76–39
declinación del Sol 81–63
~ magnética 75–56
declive 81–184
decrecimiento del casquete polar 83–146
defecto de fase 81–28
deferente 83–48
círculo ~ 83–48
deficiencia en abundancia 91–187
deformación 73–67
~ de marea 82–277
~ debida a las mareas 83a–66
~ por mareas 83a–66
degeneración 93–172
~ de neutrones 93–191
~ del gas de neutrones 93–191
~ no relativista 93–177
~ parcial 93–173
~ relativista 93–176
~ total 93–175
degenerado 93–178
altamente ~ 93–184
completamente ~ 93–185
parcialmente ~ 93–183
totalmente ~ 93–185
Deimos 83–169
Delfín 92–45
Delphinus 92–45
Deneb 92–112
Denébola 92–113
densidad central 93–87
~ cosmológica 98–370
~ crítica 98–197
~ de cráteres 81–246
~ de galaxias 98–344
~ de la sombra 81–59
~ de radiación 99–85
~ de radiantes 86–63
~ de trazas 82–253
~ de una corriente meteórica 86–264
~ del flujo neutrónico 94–86
~ del fondo 96–131
~ del viento solar 88–157
~ electrónica 78–6
~ electrónica lineal 86–283
~ en el centro 93–87
~ espacial 86–263, 96–133
~ estelar 98–82
~ estelar espacial 98–83
~ galáctica 98–344
~ lineal inferior a 10^{12} cm^{-1} 86–284
~ media 93–86
~ media de la materia 99–84
~ media de masa 99–84
~ proyectada 98–85
~ real 96–133
~ subnuclear 93–190
~ superficial 98–85
~ superficial aparente 96–132
~ superior a 10^{12} el . cm^{-1} 86–285
denudación 73–108
depolarización 97–118
depósito 74–47
~ anual 87–17
~ de partículas 75–123
~ de sedimentos 73–109
depósitos abisales 72–50
~ biogénicos 72–59
~ glaciales 72–127
~ en el fondo del mar 87–15
~ pelágicos 72–57
~ terrígenos 72–48
depresión 77–76
~ arqueada 81–160
~ aterrazada 81–162
~ continental 81–158
~ del pico central 81–244
~ en arco 81–160
~ en terrazas 81–162
~ lineal 81–159
~ lunar 81–157
~ sinuosa 81–161
~ solar 79–45
deriva continental 73–104
~ de los continentes 73–104
~ en longitud geomagnética 75–120
~ longitudinal 75–120
derrame continuo 98–276
~ de lava 73–130
~ de materia desde el núcleo galáctico 98–275
~ de petróleo 72–71
derrubio 81–234
desaceleración de la expansión 99–82
desacoplamiento de radiación 99–177
~ entre materia y radiación 99–177
desactivación 85–148
~ de un cometa 85–148
desagregación 74–57
~ de un cúmulo 96–105
desaparición 81–98
~ brusca 89–279
~ del un meteoro 86–37
~ del anillo 83–217
~ gradual 86–75
~ instantánea 86–76
desarrollo de un grupo de manchas 89–118
descarga atmosférica 77–161
descendente 77–93
descoloración del mar 72–80
descomponerse 74–65
descubridor 85–21
descubrimiento de un cometa 85–16
~ de un meteorito 87–44
desdibujamiento de las líneas de Fraunhofer 88–243
desdoblamiento de la líneas por expansión 97–157
desecación 74–51
desembarco tripulado 82–33
desenrollamiento 98–30
~ de los brazos espirales 98–30
desfasador 88–115
desfiladero submarino 72–46
desfluidización 82–263
desgasificación 82–261, 87–136
~ del núcleo cometario 84–109
desgaste de masa 81–236
desierto 83–161
designación de un cometa 85–31
~ definitiva 85–33
~ preliminar 85–32
~ previa 85–32
desigualdad de Schwarzschild 93–156
desimantación térmica 75–40
desintegración de cúmulos 98–195
[illegible] cometa 85–115
~ de un cuerpo meteórico 86–186
~ de un cúmulo galáctico 96–103
~ de un meteoro 86–187
~ del núcleo cometario 86–261
~ por efecto de marea 85–118
~ radiactiva 74–64
~ rotacional 85–119
desintegrar(se) 74–65
desionización 86–213
deslizamiento 82–196
~ de paredes 82–198
~ en frecuencia 95–154
desmagnetización térmica 75–40
desorción 85–123
despejado 77–149
desplazamiento anómalo hacia el rojo 98–367
~ asimétrico 98–104

~ cosmológico 98–363a
~ de la estela 86–114
~ de la zona de las manchas 89–74
~ de las protuberancias hacia el polo 89–231
~ del radiante 86–242
~ Doppler 91–25
~ gravitacional hacia el rojo 99–162
~ hacia el rojo 98–297
~ hacia el rojo no debido a velocidad 98–363
~ múltiple hacia el rojo 98–364
~ periódico de líneas 96–22
desplome 82–199
destello lunar 82–247
~ primigenio 99–168
destrucción 87–24
~ de la superficie 87–35
~ por marea 72—77
~ por ondas de marea 72–77
~ por un meteorito 87–20
desvanecimiento repentino 78–40
~ repentino de ondas cortas 78–40
desviación de la vertical 71–72
~ de una estrella 98–123
~ relativista de la luz 99–161
~ relativista de los rayos luminosos 99–161
desviaciones del equilibrio termodinámico 93–36
~ del equilibrio termodinámico local 93–40
~ del ET 93–36
~ del ETL 93–40
detalle de la superficie 83–105
~ estructural 88–248
~ lunar 81–148
~ permanente 83–106
~ topográfico 81–149
determinación de la edad 74–91
~ de la edad por el método del plomo 74–100
~ de la edad por la relación uranio-plomo 74–101
~ de la edad por radiactividad 87–131
~ de la gravedad 71–67
~ de la masa 96–69
~ isotópica de la edad 87–130
detrito 81–234
Devónico 74–81
día activo 75–93
~ de calma 75–92
~ de calma magnética 75–92
~ de mareas 82–290
~ lunar 81–32
~ lunar medio 82–304
~ magnéticamente activo 75–93
~ magnéticamente tranquilo 75–92
~ perturbado 75–94
diagénesis 74–38
diagrama color-color 96–147
~ color-magnitud 96–150
~ de evolución 96–151
~ de Hertzsprung-Russell 91–130
~ de Hubble 98–301
~ de mariposa 89–76
~ de Maunder 89–76
~ en dos colores 96–147
~ HR 91–130
~ magnitud aparente-desplazamiento hacia el rojo 98–301
~ masa-luminosidad 91–129
diamantes meteoríticos 87–65
diámetro angular 96–115
~ aparente 96–115
~ fotométrico 84–93
~ lineal 83–118
~ verdadero 83–118
diastrofismo 73–72
dibujo de las manchas solares 88–34
~ de un planeta 83–16
dicotomía de Venus 83–132
diferencia de camino óptico 88–117
~ de brillo 96–65
~ de magnitudes 91–59
~ de recorrido óptico 88–117
~ vertical 81–140
diferenciación geoquímica 74–55
~ gravitacional 74–56
~ por cristalización 74–41
~ química 83a–50
difusión atmosférica de la luz 77–57
~ de la luz 77–57
~ de los iones 86–286
~ de Mie 83–76
~ de rayos cósmicos 84–27
~ iónica 86–286
dilatación del tiempo 99–137
dimensiones del sistema 96–64
de cuatro ~ 99–121
dinámica de cúmulos estelares 96–153
~ del gas interestelar 97–150
~ estelar 98–120
dinamometamorfismo 74–36
dinamoteoría geomagnética 75–99
dióxido de silicio 74–13
dique 73–138
dirección antisolar 75–22
~ de la visual 91–28
~ de propagación 78–35
~ del campo 84–10
~ del campo magnético 84–10
~ del eje 83–34
~ del viento 77–115
en la ~ de la visual 91–29
~ opuesta al Sol 75–22
directo 83–42
disco de acreción 96–21
~ de acrecimiento 96–21
~ de la Luna 81–54
~ de ocultamiento 88–61
~ del Sol 89–24
en el ~ 88–173
~ galáctico 98–52
~ lunar 81–54
~ ocultador 88–61
~ ocultante 88–61
sobre el ~ 88–173
~ solar 89–24
~ visible 88–171
discontinuidad de Mohorovičić 72–14
~ Moho 72–14
disgregación de un cúmulo 96–105
~ de un cúmulo galáctico 96–103
disipación 94–73
~ de la energía 82–285
~ de marea 82–286
~ de ondas 88–222
~ térmica 94–76
dislocación 73–91
~ cortical 82–212
disminución de masa 83a–42
~ de temperatura 77–41
~ secular del brillo 85–127
disolución de un cúmulo galáctico 96–103
~ de un grupo de manchas 89–126
~ de una protuberancia 89–280
disparar 89–198
dispersión de las velocidades 98–103
dispositivo de desplazamiento de la raya 88–69
disrupción de un meteoro 86–82
disrupción de un cometa 85–114
~ de un meteoro 86–82
~ del cuerpo original 87–124
~ por colisión 87–125
~ por marea 85–118
distancia 86–288, 91–3
~ al centro del disco 89–36
~ al centro galáctico 98–11
~ al meridiano central 89–46
~ angular 96–52
~ de la Luna a la Tierra 81–13
~ del epicentro 73–19
~ del hipocentro 73–18
~ desde el centro de la estrella 93–85
~ entre los componentes 96–66
~ epicentral 73–19
~ focal 73–18
~ fotométrica 96–116
~ galactocéntrica 98–11
~ geocéntrica 84–3
~ heliocéntrica 84–4
hipocéntral 73–18
~ límite 86–289
~ máxima 86–289
~ Tierra-Luna 81–13
distorsión 73–68
~ de marea 96–30
~ rotacional 91–44
distribución de brillo 88–141
~ de densidad 96–124
~ de la materia 99–8

ES

~ de la velocidad con la profundidad 73–52
~ de las masas 84–52
~ de tamaños 82–124
~ de velocidades residuales 98–92
~ elipsoidal del velocidades 98–100
~ en el espacio 96–119
~ en latitud 89–51
~ en latitud heliográfica 89–51
~ en longitud 89–48
~ en longitud heliográfica 89–48
~ espacial 96–119, 98–86
~ galáctica 98–54
~ latitudinal 89–51
~ longitudinal 89–48
~ maxwelliana de velocidades 98–96
~ según el brillo 86–166
~ según la masa 86–165
~ vertical 76–17
distribuido al azar 98–93
~ aleatoriamente 98–93
disturbio asociado a una fulguración 84–46
~ interplanetario 84–45
disyunción 73–78
división 83–223
~ de Cassini 83–224
divisor de haz 88–88
doble 96–3
~ de contacto 96–29
~ de gran separación 96–9
~ física 96–7
~ óptica 96–6
~ visual 96–8
domo de lava 81–176a
~ lunar 81–177
~ volcánico 81–176
Dorada 92–46
Dorado 92–46
dorsal centro-oceánica 72–45
~ de alta presión 77—87
Draco 92–47
Dracónidas 86–252
dragado 72–53
dragaje 72–53
Dragón 92–47
dromocrónica 73–52
Dubhe 92–114
duna 83–162
duración 89–125
de corta ~ 89–125
~ de la insolación 88–10
~ de la totalidad 88–26
~ de la visibilidad de una estela 86–111
de larga ~ 89–124
~ de pulsos 86–280
~ de rotación 83–31
~ de un grupo 89–123
~ de una fulguración 89–209
~ de una partícula 75–122
~ del ciclo 89–87
~ del eclipse 95–78
~ del eco 86–299
~ del radioeco 86–299
~ del recorrido 73–51
~ del recorrido meteórico 86–50

E

eclipse 88–14
~ anular 88–18
~ anular del Sol 88–15a
~ artificial 88–227a
~ artificial del Sol 88–17a
~ atmosférico 95–76
~ central 88–16a
~ corpuscular 88–24
~ de Luna 81–56
~ de Luna por la penumbra 81–65
~ del Sol 88–15
~ es invisible en ... 81–87
~ es visible en ... 81–86
~ ionosférico 88–23
~ lunar 81–56
~ parcial 81–79a
~ parcial de Luna 81–67
~ parcial del Sol 88–16
~ penumbral de Luna 81–65
~ solar 88–15
~ total comienza 81–77
~ total de Luna 81–66
~ total del Sol 88–17
eco auroral 79–121
~ de estela 86–298
~ de radar 86–295
~ meteórico 86–296
~ meteórico de corta duración 86–300
~ meteórico de larga duración 86–301
~ meteórico frontal 86–297
~ sísmico 82–223
ecología 76–2
ecosondeos 72–29
ecuación de Abel 88–235
~ de altura 86–194
~ de color 91–73
~ de continuidad 93–49
~ de Emden 93–130
~ de la luminosidad 86–196
~ de la pérdida de masa 86–169
~ de la resistencia 86–179
~ de luz 95–45
~ de masa 86–169
~ de Pogson 91–58
~ de vaporización 86–200
~ de estroncio 82–271
~ del transporte 93–32
~ del transporte radiativo 93–33
~ del virial 98–181
~ hidrostática 93–76
~ integral de Schwarzschild y Milne 93–53
~ integro-diferencial 93–52
~ tensorial 99–157
ecuaciones del campo 99–155
~ del campo de Einstein 99–156
~ del equilibrio estadístico 93–44
ecuador de iluminación 81–122
~ lunar 81–121
~ magnético 75–50
edad absoluta 82–269, 83a–70
~ característica 99–79
~ de estroncio 82–271
~ de exposición a los rayos cósmicos 87–139
~ de irradiación 87–139
~ de la Luna 81–31, 82–264
~ de la Tierra 83a–68
~ de las estrellas 94–61
~ de los meteoritos 87–127
~ de radiación 87–139
~ de retención de gas 87–135
~ de solidificación 87–137
~ de un enjambre 86–246
~ de una corriente 86–245
~ de una lluvia meteórica 86–246
~ del universo 99–12
~ estratigráfica 82–267
~ geológica 74–90, 83a–69
~ por relación plomo-plomo 74–100
~ por relación uranio-plomo 74–101
~ relativa 82–268
~ rubidio-estroncio 82–271
efecto acumulativo 98–124
~ betatrón 91–180
~ blanketing 93–67
~ Coriolis 77–81
~ de invernadero 83–263
~ de inversión 76–126
~ de marea 82–276
~ de oposición 83–75
~ de pantalla 86–171
~ de reflexión 96–26
~ de reflexión en binarias 95–82
~ de selección 98–81
~ de velo 95–84
~ Evershed 89–103
~ isotópico 91–178
~ Poynting-Robertson 84–67
~ Rossiter 91–44
~ Umkehr 76–126
~ Wilson 89–89
efectos terrestres de la actividad solar 84–55
efemérides de los pequeños planetas 84–92
eficacia luminosa 86–210
eficiencia de ionización 86–209
~ luminosa 86–210
efusión de lava 73–130, 82–172
eje de rotación 83–33
~ geomagnético 75–47
~ magnético 75–46
~ magnético de la tierra 75–47

ES

ejemplar individual 87–42
elasticidad 73–58
electricidad atmosférica 75–2
~ de la Tierra 75–1
~ terrestre 75–1
electrochorro auroral 79–113
electrón auroral 79–109
electrones atrapados 79–111
~ desechados 79–118
~ relativistas 97–153
elemento deficiente 91–186
~ estructural 98–287
~ granular 88–178
~ meteorológico 77–28
~ movible 88–108
~ subabundante 91–186
~ superabundante 91–184
~ supergranular 88–182
elementos atmófilos 83a–29
~ calcófilos 83a–31
~ de trazas 82–145
~ del grupo del hierro 94–105
~ del pico del hierro 94–105
~ fotométricos 95–25
~ geomagnéticos 75–5
~ litófilos 83a–30
~ magnéticos 75–54
~ mayores 82–144
~ refractarios 97–64
elevación lunar 81–156
elipsoide de referencia 71–18
~ de velocidades 98–101
~ internacional 71–14
~ local 71–15
~ terrestre 71–11
elipticidad de un cúmulo globular 96–122
elongación 83–60
~ del radiante aparente 86–67
~ del radiante verdadero 86–68
~ este 83–61
máxima ~ 83–63
~ occidental 83–62
~ oeste 83–62
~ oriental 83–61
embrión de un planeta 83a–36
embudo 89–269
emersión 81–99
emisión brusca de partículas 88–162
~ corpuscular 88–152
~ de dos fotones 97–155
~ de extremadamente baja frecuencia 78–55
~ de lava 73–130
~ de masa 85–145
~ de muy baja frecuencia 78–54
~ de partículas 88–151
~ Lα del cielo nocturno 79–9
~ postcrepuscular 79–52
~ precrepuscular 79–51
emisor 78–15
empeoramiento del tiempo 77–23
empuje 73–65
enana 91–135
~ blanca 91–137
~ marrón 91–138a
~ negra 91–139
~ parda 91–138a
~ roja 91–136
encima del limbo 89–30
encuentro de tres cuerpos 96–74
~ estelar 98–121
~ triple 96–74
endógeno 73–145
energía acumulada 89–196
~ acústica 88–219
~ almacenada 89–196
~ atómica 93–121
~ cinética del punto cero 93–187
~ de vaporización 86–199
~ del vacío 99–170k
~ disponible 89–197
~ en reposo 99–144
~ gravitacional 93–119
~ intercambiada 98–126
~ no radiativa 88–218
~ nuclear 93–122
~ rotacional 93–120
enfriamiento 83a–57
~ neutrínico 94–97
~ radiativo 94–98
Enjambre de Orión 92–130
enjambre diurno 86–219
~ meteórico 86–216
~ nocturno 86–218
enlace geodésico intercontinental 71–37
enrojecimiento estelar 97–91
enrollamiento 98–28
~ de los brazos espirales 98–28
entrada de la Luna en la sombra 81–76
~ en la atmósfera 86–161
enturbiamiento 83–252
envoltura circumestelar 93–10
~ circumestelar de polvo 97–117
~ circumestelar pulverulenta 97–117
~ convectiva 93–136
~ de polvo 97–116
~ en expansión 93–12
~ estacionaria 93–11
~ extendida 93–8
~ pulverulenta 97–116
~ radiativa 93–135
envolturas de un cometa 85–69
envolvente característica 98–177
~ común 96–32
~ común de una binaria 96–32
Eozoico 74–77
epicentro 73–15
epiciclo 83–49
epirogénesis 73–73
epirogenia 73–73
época 74–89
~ de máximo 89–81
~ de mínimo 89–80
~ del paso por el periastro 96–60
~ del periastro 96–60
~ geológica 74–89
~ postglacial 72–109
equilibrio adiabático 76–21
~ convectivo 76–22, 93–79
~ de adsorción 85–121
~ de masa 88–250
~ de radiación 88–252
~ detallado 93–41
~ diferente del ETL 93–43
~ difusivo 76–23
~ energético 88–251
~ estadístico 93–42, 98–149
~ generalizado 93–43
~ hidrostático 93–75
~ isostático 73–99
~ isotérmico 76–20
~ nuclear 94–103
~ radiativo 93–48
~ térmico 93–78
~ térmico de la atmósfera 77–62
~ termodinámico 93–35
~ termodinámico local 93–39
equivalencia de masa y energía 99–145
Equuleus 92–48
era 74–69
~ cosmológica 99–166a
~ de aniquilación 99–173a
~ de inflación 99–170d
~ de inflación cosmológica 99–170d
~ de la fuerza superunificada 99–170e
~ de Planck 99–170e
~ de desacoplamiento 99–177a
~ del plasma radiativo 99–172
~ estelar 99–178
~ geológica 74–69
~ hadrónica 99–174
~ kenozoica 74–73
~ leptónica 99–176
~ mesozoica 74–72
~ paleozoica 74–71
~ precámbrica 74–70
~ radiativa 99–172
Erídano 92–49
Eridanus 92–49
erosión 73–107
~ cósmica 87–142
~ en el espacio cósmico 87–142
~ eólica 83–163
~ gaseosa 82–243
~ meteorítica 82–242
~ por meteoroides 82–242
~ superficial 82–241
erupción 95–69
~ a protones 89–185
~ aprotónica 89–186
~ con dos cintas 89–187a
~ de fisura 73–113
~ de clase 2 89–206
~ de importancia 2 89–206

~ detrás del limbo 89–203
~ en el disco 89–204
~ en el limbo 89–102
~ en luz blanca 89–182
~ en rayos X 89–183
~ estelar 95–70
~ fisural 73–113
~ homóloga 89–178
~ longitudinal 73–113
~ luminosa de un meteoro 86–80
~ óptica 89–181
~ protónica 89–185
~ radio(-eléctrica) 89–180
~ simpatética 89–179
~ sin protones 89–186
~ solar 89–175
~ visible 89–181
~ volcánica 73–112, 82–170
escala Beaufort 77–108
~ centesimal 77–33
~ centígrada 77–33
~ de alturas 76–18
~ de Celsio 77–33
~ de distancias 98–303
~ de la carta 71–49
~ de tiempo 94–65
~ de tiempo corta 94–66
~ de tiempo larga 94–67
~ de vientos Beaufort 77–108
~ del mapa 71–49
~ geocronológica 74–68
escalón barométrico 76–19
escape de estrellas de un cúmulo 98–194
~ de gases 76–24
~ de una estrella del sistema 96–159
escarcha 77–177, 83–150
escarpa 81–184
escisión de un cometa 85–111
~ por marea 85–112
Escorpión 92–86
escorzo de las manchas 89–92
Escuadra 92–70
escudo 72–11, 92–88
~ de Sobieski 92–88
Escultor 92–87
esfera celeste 92–8
~ de gas politrópica 93–127
~ gaseosa 93–73
~ terrestre 71–10
esféricos 78–50
esferoide achatado 98–158
~ terrestre 71–13
esférulas 87–29
~ vítreas 82–127
esfuerzo 73–62
~ cortante 73–63
~ de cizalla 73–63
~ interno 73–62
~ principal 73–64
~ térmico 86–205
esmog 76–83
espaciamiento cero 88–101
~ de antenas 88–99
~ fijo 88–102
~ variable 88–103
espacio 99–100
~ absoluto 99–116
~ curvo 99–114
~ de las fases hexadimensional 98–144
~ de las fases 6-dimensional 98–144
~ de Lobachevski 99–105
~ de Minkowski 99–122
~ de Newcomb 99–110
~ de Riemann 99–112
~ de sucesos 99–122
~ de velocidades 98–97
~ elíptico 99–110
~ esférico 99–111
~ euclídeo 99–107
~ extragaláctico 97–5
~ extraterrestre 87–73a
~ finito 99–103
~ hiperbólico 99–106
~ infinito 99–104
~ interestelar 97–4
~ interfilamentario 89–99
~ intergaláctico 97–7
~ interplanetario 84–2
~ libre 76–9
~ métrico 99–100a
~ no euclídeo 99–114
~ parabólico 99–107
~ permitido 79–134
~ plano 99–107
~ prohibido 79–135
~ riemanniano 99–112
espaciotemporal 99–120
espacio-tiempo 99–119
~ de Minkowski 99–122
especialista en meteoritos 87–7
especie de nube 77–135
espectro compuesto 96–19
~ de Fraunhofer 88–137
~ de una estela meteórica 86–107
~ dinámico 88–125
~ doble 96–18
~ estelar 91–88
~ estelar normal 91–89
~ integrado 96–137
~ integral 96–137
~ principal 95–115
~ relámpago 88–25
~ simbiótico 95–117
~ simple 96–16
espectrobologrma 88–76
espectrobolómetro 88–75
espectrografía Mössbauer 82–142
espectrógrafo horizontal 88–78
~ para meteoros 86–146
~ solar 88–77
~ vertical 88–79
espectrograma dinámico 88–125
espectroheliocinematógrafo 88–81
espectroheliógrafo 88–71
espectroheliograma 88–72
espectrohelioscopio 88–67
espectrometría gamma no destructiva 82–77
espectrómetro 88–124
~ del viento solar 82–46
espectroscopio de protuberancias 88–58
espejismo 79–64
espejo giratorio 86–130
~ oscilante 86–130a
~ rotatorio 86–130
espesor de los anillos 83–229
espícula 88–195
Espiga 92–129
espiral 98–7, 98–249
~ Arquimediana 84–34
~ barreada 98–253
~ de Arquímedes 84–34
~ no barreada 98–250
~ normal 98–250
esquistosidad 74–53
estabilidad 93–204
~ de las estrellas 93–206
~ de una órbita 98–193
~ dinámica 93–207, 98–198
estable 93–208
estación de satélites 71–33
~ de un planeta 83–44
~ lunar automática 82–19
~ meteórica 86–158
~ para la observación de meteoros 86–158
~ sísmica 73–6
~ sismográfica 73–6
~ sismológica 73–6
estadio contractivo 94–20
~ de disco 94–22
~ evolutivo 94–28
~ de preenana 94–49
~ disco 94–22
~ Hayashi 94–44
~ hidrodinámico 94–43
~ intermedio 94–21
~ postenana 94–68
~ postsecuencia principal 94–68
~ preenana 94–49
~ presecuencia principal 94–49
~ T Tauri 94–47
estadística estelar 98–77
estado de fluidez del magma 82–189
~ de la mar 72–86
~ del cielo 77–132
~ del tiempo 77–14
~ del tiempo a gran escala 77–15
~ estacionario 98–140
~ estadísticamente estacionario 98–148
~ meteorológico general 77–15
~ no estacionario 98–151
estallido de rayos X 79–117
estela con luz propia 86–108

~ de corta duración 86–112
~ de larga duración 86–113
~ de polvo 86–115
~ de un meteoro 86–103
~ gaseosa 86–116
~ inestable 86–109
~ luminosa 86–108
~ meteórica 86–103
~ meteórica persistente 86–104
~ muy corta 86–110
~ visible de día 86–106
~ visible de noche 86–105
estelar 92–3
estereocomparador 91–32
estimación de brillo 95–11
estimar 95–12
estratificación 82–133, 97–154
~ horizontal 82–134
~ térmica 77–40
estrato 77–139
~ galáctico absorbente 97–89
~ violáceo 83–165
estratocúmulo 77–142
estratopausa 76–71
estratosfera 76–70
estrella 91–1
~ aislada 96–2
~ anómala 91–96
~ binaria 96–7
~ binaria cerrada 96–23
~ binaria fotométrica 95–73
~ brillante 91–83
~ caliente 91–155
~ central 97–148
~ circumpolar 92–2a
~ colapsada 94–111
~ colapsante 91–164
~ compañera 96–39
~ común 91–95
~ con atmósfera extendida 93–6
~ con envoltura extendida 93–7
~ con espectro peculiar 91–92
~ con líneas de emisión 91–159
~ con líneas metálicas 91–195
~ con pecularidades espectrales 91–92
~ con un compañero invisible 96–13
~ de alta velocidad 98–191
~ de baja velocidad 98–192
~ de campo 96–84
~ de carbono 91–188
~ de comparación 95–16
~ de edad cero 94–59
~ de envoltura 93–5
~ de fondo 96–84
~ de hiperones 91–154
~ de la mañana 83–127
~ de la rama horizontal 91–141
~ de la secuencia principal 91–142
~ de la tarde 83–128
~ de masa intermedia 91–125
~ de masa pequeña 91–124
~ de neutrones 91–152, 93–195
~ de nitrógeno 91–189
~ de primer plano 96–85
~ de primera generación 94–24
~ de silicio 91–190
~ de temperatura cero 91–166
~ de tipo solar 91–104
~ de Tycho 95–135
~ débil 91–82
~ deficiente en carbono 91–194
~ deficiente en helio 91–192
~ deficiente en hidrógeno 91–191
~ deficiente en metales 91–197
~ degenerada 93–180
~ del alba 83–127
~ del cúmulo 96–82
~ del disco 91–85
~ del halo 91–84
~ del sistema 98–133
~ doble 96–3
~ doble de gran separación 96–9
~ doble física 96–7
~ doble óptica 96–6
~ doble visual 96–8
~ en colapso 91–164, 94–112
~ en contracción 91–161
~ en expansión 91–162
~ en explosión 91–165
~ en fuga 98–135
~ en implosión 91–164
~ en rotación 91–37
~ en rotación rápida 91–39
~ enana 91–135
~ enrojecida 97–94
~ escapante 98–135
~ estable 95–8
~ estacionaria 95–3
~ evolucionada 94–74
~ excitadora 97–125
~ excitante 97–125
~ explosiva 91–165
~ fija 92–4
~ fría 91–156
~ fugada 96–160
~ fugaz 86–2
~ fulgurante 95–68
~ gigante 91–143
~ hiperónica 93–196
~ implosiva 91–164
~ inestable 95–9
~ infrarroja 91–157
~ intergaláctica 91–163
~ intrínsecamente variable 95–87
~ intrusa 98–134
~ joven 94–93
~ liviana 91–124
~ magnética 91–51
~ más brillante 98–304
~ más brillante de una galaxia 98–304
~ masiva 91–126
~ muerta 91–167
~ múltiple 96–5
~ neutrónica 91–152
~ no enrojecida 97–95
~ no esférica 91–86
~ no estable 95–9
~ no estacionaria 95–2
~ no perteneciente al cúmulo 96–84
~ no perteneciente al sistema 98–134
~ no rotante 71–38
~ normal 91–95
~ peculiar 91–92
~ perteneciente al sistema 98–133
~ pobre en carbono 91–194
~ pobre en helio 91–192
~ pobre en hidrógeno 91–191
~ pobre en metales 91–197
~ polar 92–121
~ primaria 96–37
~ principal 96–37
~ próxima 98–60
~ pulsante 95–95
~ que escapa 98–135
~ ráfaga 95–68
~ rica en helio 91–193
~ rica en metales 91–196
~ roja 91–97
~ rotante 91–37
~ RR Lyrae 95–93
~ simbiótica 95–50
~ simple 96–2
~ sin rotación 91–38
~ sospechosa de variabilidad 95–5
~ subluminosa 91–133
~ superdensa 91–153
~ superluminosa 91–134
~ supermasiva 91–127
~ triple 96–4
~ variable 95–1
~ vecina 98–60
~ vespertina 83–128
~ vieja 94–64
~ vista por el polo 91–40
~ Wolf-Rayet 91–160
~ WR 91–160
estrellas de los primeros tipos 91–102
~ de los últimos tipos 91–103
~ de tipos tardíos 91–103
~ de tipos tempranos 91–102
estriación 89–173
estriado cromosférico 88–209
estrías 79–155, 89–173
~ cromosféricas 88–209
estructura a gran escala 99–171
~ anterior a los mares 81–153
~ anular 81–155
~ celular 99–171a
~ cometaria 85–34
~ corrotacional 84–17
~ de estriación 89–174
~ de estrias 89–174
~ de la corriente meteórica 86–243
~ de la Galáxia 98–4

~ de la Vía Láctea 98–4
~ de los cometas 85–34
~ de los meteoritos 87–100
~ del campo 89–140
~ en arco 88–249
~ en corrotación 84–17
~ en gran escala 88–207, 99–171
~ en pequeña escala 88–208
~ en sector 84–15
~ espiral 98–8
~ estelar 93–1
~ filamentosa 97–130
~ fina de la cromosfera 88–203
~ galáctica 98–4
~ interna 71–83
~ jerárquica 99–47
~ jerárquica del universo 99–47
~ laminar 83–148
~ postmarítima 81–154
~ premarítima 81–153
estructuras lunares 81–148
Etamín 92–115
etapa nebular 95–125
eternidad del universo 99–13
euclideano 99–101
euclídeo 99–101
no ~ 99–102
Europa 83–206
evaporación de agujeros negros 94–118
~ de un cúmulo 96–108
~ de una partícula meteórica 86–198
evolución anterior a la secuencia principal 94–26
~ de las estrellas 94–1
~ de un grupo de manchas 89–118
~ debida a las mareas 83a–65
~ del universo 99–171e
~ dinámica 96–161
~ estelar 94–1
~ geoquímica 74–33
~ nuclear 94–50
~ por mareas 83a–65
~ postenana 94–68
~ posterior a secuencia principal 94–68
~ preenana 94–26
~ preestelar 94–14
~ preestelar de las galaxias 94–14
~ presecuencia principal 94–26
~ química 82–257
~ térmica 83a–61
~ termonuclear 94–50
evolucionar 94–75
exceso de color 91–77
~ infrarrojo 91–79
~ IR 91–79
~ ultravioleta 91–78
~ UV 91–78
exhaustión del hidrógeno 94–69
exógeno 73–146
exosfera 76–68
expansión acelerada 99–81
~ del universo 99–74
~ hidromagnética 88–159
~ ultrarrápida 99–170c
~ ultrarrápida del universo 99–170c
expedición para observar un eclipse del Sol 88–30
experimento de Michelson 99–130
~ en la superficie lunar 82–42
~ sísmico 82–219
~ sísmico activo 82–220
exploración científica 82–36
~ de reconocimiento 82–35
explosión 87–30
~ de brillo 85–126
~ de plasma solar 84–38
~ de un cometa 85–125
~ de un meteoro 86–81
~ de un núcleo galáctico 98–283
~ de una nova 95–121
~ de una supernova 95–131
~ de viento solar 84–42
Gran ~ 99–167
~ inicial 99–167
~ nuclear a gran altura 75–114
~ primigenia 99–167
~ primordial 99–167
~ volcánica 73–112
expulsión 73–132
~ de gases 83a–41
~ de la materia 98–274
~ de la materia desde el núcleo galáctico 98–274
~ de lava 82–271
~ de masa 95–124
ex-supernova 95–134
extensión del fondo oceánico 72–39
extinción atmosférica 77–127a
~ de un meteoro 86–37
~ del vapor de agua 77–126
~ gradual 86–75
~ instantánea 86–76
~ interestelar 97–80
~ selectiva 97–83
~ total 97–82
extracción de testigos 72–51
extragaláctico 98–207
extrusión 73–132
~ de lava 82–171
eyección 89–259, 98–179
~ de la materia 98–274
~ de la materia desde el núcleo galáctico 98–274
~ de las radionubes 98–285
~ de masa 93–15
~ de materia brillante 89–257
~ de materia o(b)scura 89–256
~ de partículas 85–89
~ de partículas por el Sol 88–161
~ del plasma solar 84–38
~ por un cometa 86–262
~ recurrente 98–282
eyectos 82–165
~ de impacto 82–167
~ de volcánicos 82–166

F

factor de brillo 83–90
~ de contracción de Lorentz 99–135
~ de dilución 93–37
~ de escala 99–81a
~ de expansión cósmica 99–81a
~ de forma 86–177
~ de Gaunt 93–104
~ de guillotina 93–103
~ de luminosidad 86–197
~ geométrico 86–177
fácula 89–161
~ en rayos X 89–166
~ óptica 86–165
fáculas cromosféricas 89–164
~ fotosféricas 89–161
~ polares 89–162
faja de polvo 98–289
~ luminosa 86–27
falda 72–36
falla 73–91, 82–212
~ de compresión 73–94
~ de distensión 73–93
~ directa 73–93
~ inversa 73–94
~ invertida 73–94
~ normal 73–93
fallamiento 73–92
falsa variable 95–51
falso sol 79–71
familia de cometas 85–102
~ de Júpiter 85–103
~ de los asteroides 84–73
fango 72–58
fase ascendente del ciclo solar 89–3
~ cero 83–73
~ creciente 89–83
~ creciente de la Luna 81–23
~ creciente del ciclo solar 89–83
~ de contracción 94–20
~ de disco 94–22
~ de evolución 94–28
~ de flash 89–210
~ de halo 94–19
~ de Hayashi 94–44
~ de la Luna 81–19
~ de restablecimiento 75–77
~ declinante 89–84
~ declinante del ciclo solar 89–84
~ del ciclo 89–79
~ descendente del ciclo solar 89–84
~ disco 94–22
~ eruptiva 89–210
~ explosiva 89–210
~ halo 94–19
~ Hayashi 94–44
~ hidrodinámica 94–43

ES

~ inicial 75–75
~ inicial de una tormenta magnética 75–75
~ intermedia 94–21
~ máxima del eclipse 81–78
~ nebular 95–125
~ parcial 81–78a
~ parcial del eclipse 81–78a
~ principal 73–12, 75–76
~ T-Tauri 94–47
~ volátil 82–262
fases del eclipse 81–73
faultening local 82–204
faz iluminada de la Luna 81–34
~ invisible de la Luna 81–16
~ visible de la Luna 81–15
fechado de sucesos lunares 82–265
~ torio-plomo 82–270
feldespato 74–15
~ potásico 74–16
Fénix 92–77
fenómeno activo 89–14
~ de corta duración 89–15
~ de halo 79–66
~ de vida corta 89–15
~ luminoso transitorio 82–247
~ lunar transitorio 82–153
~ óptico 79–33
~ solar 89–13
~ transitorio coronal 89–185a
fenómenos crepusculares 79–32
ferromagnetismo 75–26
ferroníquel 87–90
fibra de protuberancia 89–245
fibrilla de penumbra 89–98
fibrillas 88–209
~ cromosféricas 88–209
figura de equilibrio 83–112
~ de interferencia 88–120
~ de la Luna 81–116
~ de la Tierra 71–7
~ de un planeta 83–109
figuras de Widmanstaetten 87–105
filamento 85–55, 89–224
~ de hidrógeno 89–267
~ de penumbra 89–98
~ de una nebulosa 97–131
~ metálico 89–268
~ oscilante 89–216
~ tremolante 89–216
filete de protuberancia 89–245
filón de inyección 73–138
~ de roca 73–138
filtro de aerosol 76–94
~ de polvo 76–95
fin de la totalidad 81–80
~ del ciclo 89–85
~ del eclipse 81–83
~ del eclipse a la salida del Sol 88–21
~ del eclipse parcial 81–83a
~ del eclipse total 81–80
~ del mundo 99–183
finitud del universo 99–4
finos lunares 82–112
fisica atmosférica 76–4
~ de la Luna 82–151
~ de las relaciones Sol-Tierra 84–56
~ de los cometas 85–104
~ de los planetas 83–4
~ planetaria 83–4
~ solar 88–6
fisura 73–78
Flecha 92–84
flóculo 89–167
~ o(b)scuro 89–224
~ o(b)scuro de hidrógeno 89–224
flóculos 89–167
~ brillantes de hidrógeno 89–169
~ de calcio 89–170
~ de hidrógeno 89–168
flotación magnética 89–147
fluctuaciones primordiales 99–170y
~ secundarias 95–44
~ sobrepuestas 95–44
fluidez de la lava 82–188
fluido químicamente activo 74–43
flujo 82–280
~ contínuo hacia afuera 98–276
~ convectivo 93–115
~ de calor 74–60
~ de cenizas 82–174
~ de energía 93–112
~ de energía acústica 88–221
~ de energía convectivo 93–115
~ de fisura 73–141
~ de lava 73–140
~ de marea 82–279
~ de masa 93–13
~ de masa hacia el exterior 93–13
~ de materia desde el núcleos galáctico 98–275
~ de materia hacia afuera 98–275
~ de neutrones 94–86
~ de radiación 93–113
~ fisural 73–141
~ hacia afuera 98–275
~ integrado 93–47
~ plástico 72–121
~ radiativo emergente 93–114
~ total de energía 93–116
~ y reflujo 72–89
Fobos 83–168
foco del terremoto 73–17
~ sísmico 73–17
foliación 74–53
Fomalhaut 92–116
fondo de un cráter 81–259
~ del mar 72–38
~ del oceáno 72–37
~ llano 81–260
~ meteórico 86–232
~ oceánico 72–37
forma anular 81–155
~ de equilibrio 83–112
~ de equilibrio hidrostático 83–112
~ de la Luna 81–116
~ de la partícula meteórica 86–175
~ de la Tierra 71–7
~ de nube 77–133
~ de un cometa 85–83
~ de un planeta 83–109
~ del cuerpo meteórico 86–175
en ~ de montículo 89–276
en ~ de nube suspendida 89–275
en ~ de soplo 89–277
~ orientada 87–113
formación 94–10
~ atmosférica 83–107
~ de cráteres 82–232
~ de cráteres por impacto 82–234
~ de elementos 94–52
~ de estrellas 94–11, 96–95
~ de fallas 73–92
~ de granos 97–60
~ de la cola 85–86
~ de la superficie 83–105
~ de las líneas 93–63
~ de las montañas 73–74
~ de una falla local 82–204
~ de una falla regional 82–205
~ de una mancha 89–121
~ permanente 83–106
~ por impacto 82–229
~ secundaria 81–151
~ topográfica 81–149
~ violenta de estrellas 94–11a
~ volcánica 82–159
formaciones lunares 81–148
formaldehído 97–27
formamida 97–35
formas aurorales 79–137
~ con estructura de rayos 79–152
~ con rayos 79–152
~ estructurales 81–152
~ sin estructura de rayos 79–142
~ sin rayos 79–142
fórmula barométrica de la altura 76–15
~ de nivelación barométrica 76–15
~ de Titius-Bode 83–3
~ internacional de la gravedad 71–76
Fornax 92–50
fosa 83–130
~ marina 72–40
~ oceánica 72–40
foto lunar 82–26
fotocentro 96–46
fotodesintegración 94–101
~ del núcleo 94–101
fotografía de la Luna 82–26
~ de meteoros 86–140
~ de un meteoro 86–151a
~ en luz blanca 88–37
~ en secuencia 79–103
~ lunar 82–26
~ oblícua 82–29
~ panorámica 82–27

ES

~ paraláctica 79–102
~ vertical 82–28
fotolisis 76–84
fotometría detallada 83–104
~ en seis colores 96–148
~ en tres colores 96–141
~ integrada 83–103
~ multicolor 96–143
fotoneutrino 94–93
fotosfera 88–167
~ inactiva 88–169
~ no perturbada 88–169
~ tranquila 88–169
fotosférico 88–170
fototriangulación espacial 71–26
fraccionamiento 87–138
~ de los elementos químicos 83a–51
~ de un cometa 85–114
fractocúmulo 77–146
fractura 73–78
~ concéntrica 82–209
~ de borde de cráter 82–211
~ de tensión 82–207
~ lineal 82–208
~ superficial 81–173
fracturación 73–92
frágil 86–191
fragilidad 86–192
fragmentación 83a–81, 87–25
~ de asteroides 83a–82
~ de un cuerpo meteórico 86–188
~ de un meteoro 86–83
~ de una nube cósmica 94–38
fragmento de un meteorito 87–43
~ nuclear 94–102
~ rocoso 82–68
~ vítreo 82–97
fragmentos 83a–83
~ de roca 81–235
franja 83–190
~ activa 73–32
~ de polvo 98–289
~ ecuatorial 83–191
~ orogénica 73–75
~ oscura 83–190
~ polar 83–151
~ sísmica 73–30
franjas 79–154
frecuencia crítica 78–24
~ de auroras 79–122
~ de cráteres 81–247
~ de las fulguraciones 89–217
~ de las manchas 89–66
~ de las protuberancias 89–237
~ de los meteoros esporádicos 86–233
~ de los pulsos 86–279
~ de recurrencia 86–279
~ de tamaño 81–249
~ del transmisor 86–273
~ horaria de los meteoros 86–135
~ moduladora 86–275
~ portadora 86–274
frenado de la rotación 91–45
~ magnético 91–47
frente 77–68
~ cálido 77–69
~ de flujo 82–193
~ de ionización 97–48
~ frío 77–70
~ ocluido 77–71
~ polar 77–73
fricción 86–173
~ de la marea 82–312
frontera magnetosférica 75–9
frontogénesis 77–74
frontólisis 77–75
fuente compacta 98–273
~ compacta de radiación 98–273
~ cuasi-estelar 98–358
~ de calor 74–62
~ de electrones 79–119
~ de energía 93–110
~ de neutrones 94–85
~ decamétrica 83–186
~ doble de energía 93–146
~ neutrónica 94–85
~ radioeléctrica cuasi-estelar 98–360
fuera de la atmósfera terrestre 88–149
fuertemente ionizado 89–289
fuerza coercitiva 75–39
~ cohesiva 85–117
~ de Coriolis 77–79
~ de la gravedad 71–62
~ de marea 82–289
~ de repulsión 85–87
~ del viento 77–107
~ desviadora 77–80
~ electrodébil 99–170x
~ interna 73–61
~ repulsiva 85–87
~ superunificada 99–170r
~ unificada 99–170x
fuerzas de marea 72–92
~ tectónicas 73–106
fuga de las galaxias 98–298
~ de una estrella del sistema 96–159
fulguración 95–69
~ a protones 89–185
~ aprotónica 89–186
~ con dos cintas 89–187a
~ cromosférica 89–175
~ de clase 2 89–206
~ de importancia 2 89–206
~ detrás del limbo 89–203
~ en el disco 89–204
~ en el limbo 89–202
~ en luz blance 89–182
~ en rayos X 89–183
~ estelar 95–70
~ homóloga 89–178
~ óptica 89–181
~ protónica 89–185
~ radioeléctrica 89–180
~ simpatética 89–179
~ sin protones 89–186
~ solar 89–175
~ visible 89–181
fumarola 82–163
función de densidad 98–88
~ de difusión 77–58
~ de distribución 98–143
~ de fase 83–78
~ de frecuencia de las masas 84–52
~ de las masas 96–68
~ de luminosidad 98–89
~ de masa 96–68
~ de peso 93–58
~ de saturación 93–60
~ de visibilidad 89–55
~ fuente 93–34
~ integro-exponencial 93–62
~ R(t) 99–81a
funciones esféricas teserales 95–102
fusión 86–181

G

gal 71–65
galáctico 98–2
galactoncéntrico 98–10
Galaxia 98–1
galaxia 98–201
~ activa 98–260
~ altamente achatada 98–252
~ anular 98–259
~ azul compacta 98–265
~ barreada 98–253
~ compacta 98–264
~ con barra 98–253
~ con brazo único 98–42
~ con núcleo multiple 98–319
~ cuasi-estelar 98–361
~ de barra 98–253
~ de brazos múltiples 98–44
~ de campo 98–340
~ de dos brazos 98–43
~ de fondo 98–340
~ de rayos X 98–203
~ de tipo N 98–263
~ de transición 98–248
~ de un solo brazo 98–42
~ de varios brazos 98–44
~ deformada 98–316
~ del cúmulo 98–338
~ del Grupo Local 98–327
~ desintegrada 98–317
~ después de explosión 98–270
~ disgregada 98–317
~ distorsionada 98–316
~ doble 98–305
~ elipsoidal 98–241
~ elíptica 98–241
~ elíptica enana 98–242
~ elíptica gigante 98–243
~ en colisión 98–321

ES

~ en explosión 98–268
~ en forma de huso 98–239
~ en radiación X 98–203
~ enana 98–220
~ eruptiva 98–269
~ esférica 98–244
~ esferoidal 98–244
~ espiral 98–249
~ fusiforme 98–239
~ gigante 98–222
~ Haro 98–266
~ infrarroja 98–205
~ irregular 98–237
~ lenticular 98–245
~ local 98–327
~ madre 98–284
~ Markarian 98–265a
~ más brillante del cúmulo 98–339
~ múltiple 98–308
~ no barreada 98–250
~ no perteneciente al cúmulo 98–340
~ normal 98–221
~ observable en infrarojo 98–205
~ observable en luz 98–204
~ óptica 98–204
~ peculiar 98–256
~ perturbada 98–315
~ posteruptiva 98–270
~ regular 98–236
~ rica en gas 98–238
~ Seyfert 98–262
~ singular 98–258
~ supergigante 98–223
~ tipo EO 98–244
~ tipo renacuajo 98–281
~ vecina 98–211
~ vista de canto 98–246
~ vista de frente 98–247
~ vista de perfil 98–246
galaxias colisionantes 98–321
~ en colisión 98–321
~ interactuantes 98–310
~ interconectadas 98–306
ganancia de energía 88–253
gancho magnético 75–79
Ganímedes 83–207
gas bariónico 93–194
~ circungaláctico 98–351
~ de bariones 93–194
~ de hiperones 93–193
~ de neutrones 93–192
~ degenerado 93–179
~ entrecúmulos 98–353
~ hiperónico 93–193
~ interestelar 97–9
~ intergaláctico 98–351a
~ intergaláctico en cúmulo 98–352
~ intergaláctico entre cúmulos 98–353
~ intracumular 98–352
~ neutrónico 93–192
~ no degenerado 93–182
~ noble primordial 87–133
~ noble radiogénico 87–132
~ relativista 97–152
~ residual 76–88
gases atrapados del viento solar 82–148
~ congelados 85–45
~ incandescentes 83a–43
gatillar 89–198
Gemelos 92–51
Gemini 92–51
Gemínidas 86–257
Gemma 92–117
generación de energía 88–216
~ de estrellas 94–23
~ de sonido 88–215
~ estelar 94–23
geociencias 71–1
geocorona 76–73
~ de hidrógeno neutro 76–74
geocronología 74–67
geodesia espacial 71–31
~ lunar 81–109
~ por satélite 71–32
geodésica 99–153
geofísica 72–1
~ cósmica 84–57
geoide 71–16
~ compensado 71–17
geología 72–2
~ lunar 82–47
~ planetaria 83–5
geomagnetismo 75–24
geometría de Riemann 99–113
~ euclídea 99–108
~ no euclídea 99–109
geopotencial 71–60
geoquímica 74–30
geosfera 72–7
geosinclinal 73–105
geotectónica 73–95
geotermia 74–58
Giacobínidas 86–252
gigante 91–143
~ azul 91–142a
~ brillante 91–149
~ de tipos tardíos 91–144a
~ de tipos tempranos 91–143a
~ luminosa 91–149
~ normal 91–150
~ roja 91–152a
galaciación 72–104
glacial 72–106
glaciar 72–112
~ alpino 72–114
~ continental 72–116
~ de montaña 72–113
glaciología 72–103
globo celeste 92–7
~ derretido 83a–64
~ fundido 83a–64
~ lunar 81–112
~ terrestre 71–10
glóbulo 97–133
~ de polvo 98–290
gloria 79–63
golfo 81–294
~ tórrido 81–296
gota 76–113
~ negra 83–52
gotita 76–100
Gotlandiano 74–80
graben 82–132
gradiente 93–94
~ absoluto 91–111
~ adiabático de temperatura 93–98
~ adiabático seco 76–14
~ bárico 77–106
~ barométrico 77–106
~ de densidad 98–87
~ de presión 77–106
~ de temperatura 93–96
~ de temperatura convectivo 93–157
~ espectrofotométrico 91–112
~ radiativo de temperatura 93–97
~ superadiabático 93–99
~ térmico 93–96
grado de concentración 96–128
~ de degeneración 93–174
~ geotérmico 74–59
Gran Colapso 99–182
~ Explosión 99–167
~ interacción unificada 99–170v
~ Mancha Roja 83–193
~ Nube Magallánica 98–213
~ Perturbación Austral 83–194
~ unificación 97–170u
grandes planetas 83–13
granito 74–8
granizo 77–178
grano de grafito 97–75
~ de hielo 85–52
~ de hierro 97–74
~ de roca 85–54
~ de silicato 97–76
~ de tamaño medio 82–116
~ fino 82–114
~ grueso 82–117
~ polimérico 97–77a
~ recubierto de hielo 97–77
granos de „finos" 82–110
~ de materia interestelar 97–59
~ interestelares 97–59
granulación facular 89–163
~ fotosférica 88–176
granulado medio 82–116
gránulo 88–178
~ de la umbra 88–180
~ facular 88–179
~ fotosférico 88–181
~ umbral 88–180
gránulos 82–129
~ faculares 89–163
gravedad 71–56
~ propia 94–34

gravimetría 71–63
gravímetro 71–70
gravitación propia 94–34
grieta 81–165
~ de tensión 82–206
Grulla 92–52
grupo amino 87–71
~ asteroidal 84–90
~ asteroidal de conmensurabilidad 84–101
~ bipolar 89–111
~ complejo 89–112
~ de cometas 85–105
~ de conmensurabilidad 84–101
~ de galaxias 98–325
~ de los asteroides 84–90
~ de los planetas menores 84–90
~ de manchas 89–105
~ de manchas solares 89–105
~ del ciclo anterior 89–107
~ del ciclo precedente 89–107
~ del nuevo ciclo 89–106
~ del viejo ciclo 89–107
~ Local de galaxias 98–326
~ recurrente 89–108
~ troyano 84–91
~ unipolar 89–110
Grus 92–52
guijarro 82–128
guyot 72–44

H

hacer el mapa del campo magnético 89–132
hallazgo de un meteorito 87–44
halo 79–67
~ caliente 98–46
~ frío 98–47
~ galáctico 98–45
~ grande 79–70
~ lunar 79–68
~ magnético 98–48
~ primario 79–69
~ principal 79–69
~ secundario 79–70
halos de un cometa 85–68
Hamal 92–118
haz de electrones 79–116
~ electrónico 79–116
helero 72–112
heliofísica 88–6
heliogeofísica 84–56
heliógrafo 77–61, 88–36
~ monocromático 88–38
heliograma en rayos X 88–74
heliómetro 88–12
heliosfera 88–255
heliosférico 88–256
heliosismología 88–223b
helióstato 88–45
hemisferio alumbrado por el Sol 75–20
~ de dia 83–67
~ diurno 83–67
~ iluminado 81–126
~ invisible 81–16
~ no iluminado 83–68
~ nocturno 75–21
~ o(b)scuro 83–68
~ visible 89–27
hendidura 81–165
~ lineal 81–166
Hércules 92–53
Hermes 84–83
herramienta recogedora de muestras 82–80
heterogeneidad de la densidad electrónica 78–8
~ del campo magnético 78–9
~ ionosférica 78–7
heterosfera 76–61
hexaedrito 87–84
Híadas 96–111a
Híades 96–111a
Hidra Austral 92–56
~ Hembra 92–55
~ Macho 92–56
hidrodinámica estelar 98–142
hidrofílico 76–92
hidrófilo 76–92
hidrofóbico 76–93
hidrófobo 76–93
hidrógeno interestelar 97–46
~ metálico 83–202
~ molecular 97–31
~ neutro 97–47
hidrosfera 72–21
hidrosol 76–90
hidroxilo 79–25
hielo 85–42
~ amorfo 85–44
campo de ~ flotante 72–102
~ cúbico 85–46
~ de agua 85–47
~ flotante 72–102
~ seco 83–144
~ sucio 97–73
hierro meteórico 87–83
~ meteorítico 87–83
higrómetro 77–120
hipocentro 73–17
hipótesis 94–6
~ catastrófica 83a–14
~ cosmogónica 94–8
~ de impacto 81–261
~ de la captura 83a–18
~ de la erupción 89–63
~ de Lambert 99–45
~ de las dos corrientes 98–99
~ de los torbellinos 83a–11
~ de los vórtices 83a–11
~ de Weizsäcker 83a–11
~ del choque 83a–14
~ del encuentro 83a–14
~ meteorítica 81–262
~ meteoroidal 81–262
~ nebular 83a–7
historia térmica 74–66, 82–260
hoja de plasma 75–11
~ neutra 75–16
hojuelas primordiales 99–171b
holoceno 72–109
homólogo 93–131
homopausa 76–65
homosfera 76–60
horizonte cósmico 99–10
~ cosmológico 99–179
~ de las masas 99–41
~ de los sucesos 99–43
~ del universo 99–10
Horno 92–50
Horologium 92–54
hoya de la cumbre 81–244
hoyo meteorítico 87–58
hoyuelo 81–243
huellas del pie 82–38
humedad absoluta 77–122
~ atmosférica 77–118
~ del aire 77–118
~ específica 77–123
~ relativa 77–128
humo 97–57
hundimiento 82–199
huracán 77–78
~ tropical 77–78
Hydra 92–55
Hydrus 92–56

I

Ícaro 84–84
iceberg 72–117
identidades lagrangianas 98–182
~ generalizadas 98–183
ignición del carbono 94–80
~ del helio 94–71
~ del hidrógeno 94–55
ilmenita 82–91
imagen de interferencia 88–120
imágenes monocromáticas 97–158
imanación inducida 75–29
~ inversa 75–37
~ primaria 75–35
~ química 75–31
~ residual 75–28
~ secundaria 75–36
~ sedimentaria 75–33
~ termorremanente 75–30
~ termorresidual 75–30
~ viscosa 75–32
immersión 81–98
impacto 82–16
~ artificial 82–226
~ cometario 82–230
~ de un meteorito 87–33
~ meteorítico 82–236

ES

~ meteoroidal 82–236
impetu angular 83a–4
implosión del núcleo estelar 94–110
importancia de una fulguración 89–205
inclinación de la órbita 96–51
~ del eje 83–35
~ del eje de rotación 91–50
~ del plano orbital 96–51
~ magnética 75–58
~ sobre el ecuador lunar 81–37
inclusión orgánica 87–66
incremento de luminosidad 86–79
~ de masa 83a–38
indicador de distancia 86–266
~ de estructura espiral 98–39
~ de mareas 82–311
indicatriz de difusión 77–58
indice de actividad de una corriente 86–230
~ de actividad solar 89–8
~ de azul menos amarillo 96–145
~ de calor 91–81
~ de color 91–76
~ de color amarillo 83–88
~ de color B–V 96–145
~ de color compuesto 96–138
~ de color especial 83–87
~ de color integral 96–138
~ de color U–B 96–146
~ de concentración 96–101
~ de ultravioleta menos azul 96–146
~ estructural de un cúmulo 96–80
~ intrínseco de color 91–75
~ K de actividad magnética 75–89
~ magnético planetario 75–90
~ planetario K_p 75–90
~ politrópico 93–128
~ solar 89–8
~ tri-horario 75–89
Indio 92–57
Indus 92–57
inestabilidad 93–211
~ convectiva 93–155
~ de marea 96–107
~ de pulsación 93–213
~ de rotación 83a–8
~ de vibración 93–214
~ gravitacional 94–35
~ local 93–154
~ por marea 98–313
~ pulsante 93–213
~ rotacional 93–216
~ secular 93–212
~ vibracional 93–214
inestable 93–210
infinidad del universo 99–6
infinitud del universo 99–6
inflación 99–170c
~ cosmológica 99–170c
inhomogeneidad inclinada 88–190
~ macroscópica 88–206
insolación 77–51
instante de alunizaje 82–16
~ de aparición 86–49
~ de caída 87–51
~ del paso por el periastro 96–60
instrumento automatizado 82–41
~ para dispersión alfa 82–76
instrumentos solares 88–35
integral de fase 83–79
tercera ~ del movimiento 98–186
intemperización 74–57
intensidad de la gravedad 71–62
~ del campo magnético 84–11
~ del seísmo 73–33
~ del terremoto 73–33
~ efectiva del campo 95–59
~ en el centro de la línea 89–211
~ horizontal 75–60
~ horizontal del campo magnético 75–60
~ omnidireccional 75–104
~ sísmica 73–33
~ unidireccional 75–103
~ vertical 75–62
intensificación del brillo de una fácula 89–213
interacción atmosférica 79–120
~ colectiva 98–168
~ de marea 81–120, 82–274
~ electrodébil 99–170x
gran ~ unificada 99–170v
~ superunificada 99–170r
intercambio de energía 98–125
interestelar 97–2
interface 93–141
interfase de salinidad 72–67
interferencias estáticas 78–50
interferómetro con barrido de fase 88–109
~ con barrido de lóbulo 88–109
~ de acantilado 88–106
~ de antenas múltiples 88–123
~ de barrido en frecuencia 88–110
~ de base varible 88–107
~ de Christiansen 88–123
~ de conmutación de fase 88–113
~ de dos elementos 88–98
~ de enlace radioeléctrico 88–116
~ de haz en abanico 88–122
~ de haz en pincel 88–121
~ de larga base 88–104
~ de línea-base larga 88–104
~ de lóbulo en abanico 88–122
~ de lóbulo móvil 88–109
~ de red 88–123
~ marítimo 88–106
~ superdireccional 88–121
intergaláctico 97–6
interglaciación 72–107
interglacial 72–108
interior de estrellas 93–72
~ de la Tierra 72–3
~ estelar 93–72
interplanetario 84–8
interrupción por el obturador 86–150
intervalo de salto 78–38
~ geotérmico 74–59
intrusión 73–133
~ concordante 73–134
~ discordante 73–137
~ ígnea 82–98
~ magmática 82–98
inundación 72–75
invariable con el tiempo 98–150
~ en el tiempo 98–150
invariancia respecto de una transformación 99–133
inversión atmosférica 77–37
~ de poblaciones 97–15
~ de polaridad magnética 89–153
~ de temperatura 77–36
~ junto al suelo 77–38
~ térmica 77–36
investigación del subsuelo 82–82
~ lunar 82–3
~ subsuperficial 82–82
Io 83–205
ion de hidrógeno 93–65
~ grande 76–115
~ Langevin 76–115
~ lento 76–115
~ negativo del hidrógeno 93–66
ionización por los meteoros 86–281
ionizado, altamente 89–289
ionograma 78–23
ionosfera 76–69
~ exterior 76–76
ionosférico 78–2
ionosonda 78–22
irradiación de rayos cósmicos 82–250
~ por radiación cósmica 82–250
irregularidad a pequeña escala 84–13
~ del campo magnetico 84–12
isla nueva 72–81
isobata 71–55
isocianuro de hidrógeno 97–40
isoclina 75–59
isodinámica 75–61
isofota 81–92
isogona 75–57
isohipsa 81–139
isonitrilo de hidrógeno 97–40
isopleta 96–125
isorrotación 88–130
isostasia 73–98
isótopo cosmogénico 82–147
~ de largo período 83a–75
~ de vida corta 83a–74
~ del plomo 83a–76
~ primordial 82–146
~ producido por rayos cósmicos 82–147

ES

J

Jirafa 92–24
joroba 95–37
~ de la curva 95–37
joviano 93–171
Juno 84–85
Júpiter 83–170
Jurásico 74–85

K

kamacita 87–92
Kenozoico 74–73

L

La Cabra 92–110
La Espiga 92–129
Lacerta 92–58
lacolito 73–136
lado alejado de la Luna 81–16
~ de día 75–20
~ de noche 83–68
~ del Sol 75–20
~ diurno 75–20
~ iluminado 81–126
~ obscuro 83–68
~ visible de la Luna 81–15
Lagarto 92–58
lago 81–297, 83–158
~ de la Muerte 81–298
~ de lava 82–183
~ de los Sueños 81–299
~ de Otoño 81–301
~ de Primavera 81–302
~ de Verano 81–300
laguna 84–88
~ de conmensurabilidad 84–88
~ de Hertzsprung 91–144
lagunas de Kirkwood 84–89
lámina 83–149
~ de mica 88–87
~ de plasma 75–11
~ de un cuarto de onda 88–126
~ $\lambda/4$ 88–126
laminillas 87–104
Las Siete Cabrillas 96–110
lapilli 82–169
latitud alta 89–53
~ areográfica 83–140
de ~ alta 89–53
de ~ baja 89–52
~ geomagnética 75–52
~ heliográfica 89–42
~ límite austral 81–106
~ límite boreal 81–105
~ límite norte 81–105
~ selenográfica 81–11
~ sur 81–106
latitudes de las calmas tropicales 77–102
~ límites de visibilidad de una ocultación 81–104
lava enfriada 82–176
~ solidificada 82–176
lazos 89–270
Lebreles 92–26
lecho fluvial 83–153
~ marítimo 72–38
~ oceánico 72–37
lengua de hielo 72–125
~ de lava 82–187
~ de un glaciar 72–125
~ interplanetaria 84–21
~ solar-interplanetaria 84–20
lente gravitacional 98–256a
Leo 92–59
~ Minor 92–60
León 92–59
~ Menor 92–60
Leónidas 86–255
Lepus 92–61
levantamiento de satélites 83–26
~ fotográfico 82–30
~ magnético 75–64
~ magnético mundial 75–65
~ selenofísico 82–39
~ selenológico 82–50
ley de alteración de la polaridad 89–154
~ de Bode 83–3
~ de distancias planetarias 83a–2
~ de enrojecimiento 97–92
~ de Hale 89–154
~ de Hubble 98–299
~ de Kramers 93–102
~ de la resistencia 86–179
~ de las latitudes 89–75
~ de o(b)scurecimiento 88–143
~ de polaridad 89–154
~ de rotación 98–110
~ de Spoerer 89–75
~ de Titius (y) Bode 83a–2
liberación de energía 93–109
Libra 92–62
libración 81–48
~ diurna 81–52
~ en latitud 81–50
~ en longitud 81–49
~ física 81–53
~ geométrica 81–51
~ óptica 81–51
~ paraláctica 81–52
libre de polvo 82–119
Liebre 92–61
limbo, encima del 89–30
~ este 89–34
~ extremo 89–32
~ oeste 89–33
sobre el ~ 89–29
~ solar 89–28
límite austral del eclipse 88–19
~ boreal del eclipse 88–20
~ de Chandrasekhar 93–189
~ de las nieves 72–111
~ de masa 93–189
~ del sector 84–18
~ norte del eclipse 88–20
~ superior de la atmósfera 76–25
~ superior de una capa 76–58
~ sur del eclipse 88–19
limo 72–58
Lince 92–64
línea amarilla 89–292
~ atmosférica 88–140a
~ circumestelar 97–97
~ coronal 89–290
~ D 88–140
~ de campo abierta 75–15
~ de enrojecimiento 96–149
~ de flujo 82–194
~ de Fraunhofer 88–139
~ de fuerza corrotacional 75–14
~ de isogauss 89–134
~ de la costa 72–31
~ de la visual 91–28
~ de microondas interestelar 97–98
~ de nivel 81–139
~ de separación de polaridades opuestas 89–151
~ de sombra 83–69
~ de universo 99–154
~ desplazada hacia el rojo 98–366
~ geodésica 99–155
~ interestelar 97–96
~ interestelar en radio 97–98
~ interestelar óptica 97–99
~ isoclina 79–59
~ isoclinal 75–59
~ isodinámica 75–61
~ isogona 75–57
~ neutra 89–151
~ roja 79–15
~ telúrica 88–140a
~ verde 79–16, 89–291
línea-base fija 88–102
~ variable 88–103
líneas de igual intensidad 75–102
Lira 92–65
Líridas 86–248
litificación 74–49
~ por impacto 82–238
litoral 72–30
litosfera 72–6
litosiderito 87–80
Lobo 92–63
lóbulo de Roche 96–28
localización de la zona piscícola 72–84
lodo 72–58
longitud activa 89–41
~ angular 98–32
~ angular de la trayectoria 86–48
~ areográfica 83–139
~ de la trayectoria 86–47
~ de onda de Compton 99–170g
~ de onda de velocidad cero 98–365

ES

~ de onda efectiva 96–142
~ de onda en reposo 98–365
~ del periastro 96–59
~ elemental de Planck 99–170h
~ de Planck 99–170h
~ geomagnética 75–53
~ heliográfica 89–40
~ selenográfica 81–10
lucero de la mañana 83–127
~ de la tarde 83–128
~ del alba 83–127
~ miguero 83–127
lugar de caída 87–50
~ de caída de un meteorito 87–50
~ de hallazgo 87–52
luminescencia atmosférica 79–2
~ crepuscular 79–4
~ diurna 79–3
~ Lα del cielo nocturno 79–9
~ lunar 82–248
~ nocturna 79–5
luminosidad 91–128
~ crepuscular 79–4
~ de un meteoro 86–195
~ débil 79–151
~ del aire 79–2
~ del cielo nocturno 79–8
~ del hidrógeno 79–145
~ diurna del aire 79–3
~ neutrínica 94–96
~ nocturna del aire 79–5
Luna 81–2
~ creciente 81–20
~ decreciente 81–21
~ eclipsada 81–57
~ llena 81–27
media ~ 81–26
~ menguante 81–21
~ nueva 81–22
lunación 81–18
lunamoto 82–214
lunar 81–3
lunicéntrico 81–131
Lupus 92–63
luz antisolar 84–64
~ cenicienta 81–35
con ~ propia 96–45
~ del cielo 77–53
~ del cielo nocturno 79–8
~ estelar 79–10
~ estelar difusa 79–12
~ estelar directa 79–11
~ extraterrestre 79–7
~ galáctica 79–14
~ galáctica difusa 97–56
~ purpúrea 79–54
sin ~ propia 96–44
~ solar 77–49
~ zodiacal 79–13
~ zodiacal falsa 84–62
Lynx 92–64
Lyra 92–65

Ll

llegada a la Tierra 89–193
llenado de una depresión 77–83
llovizna 77–172
lluvia 77–169
~ coronal 89–271
~ de estrellas 86–215
~ de estrellas fugaces 86–215
~ de meteoritos 87–37
~ meteórica 86–215
~ meteorítica 87–37
~ orográfica 77–166

M

maar 82–161
macroestructura 99–9
macrometeorito 87–4
magma 73–131
magnetismo de las rocas 75–43
~ estelar 95–54
~ inducido 75–29
~ remanente 75–28
~ residual 75–28
~ terrestre 75–24
magnetita 82–92
magnetización detrítica 75–33
~ inducida 75–29
~ remanente 75–28
~ residual 75–28
~ sedimentaria 75–33
~ termoremanente 75–30
~ termoresidual 75–30
~ viscosa 75–32
magnetógrafo 75–83
~ del campo longitudinal 88–83
~ longitudinal 88–83
~ solar 88–82
~ vectorial 88–84
magnetograma 75–84
magnetoheliógrafo 88–93
~ de Leighton 88–93
magnetoide 98–25
magnetómetro portátil 82–44
magnetopausa 75–9
magnetosfera 75–7
~ de un pulsar 95–147
magnitud absoluta 91–64
~ absoluta de un meteoro 86–72
~ absoluta de un planeta 83–86
~ aparente 91–61
~ aparente de un meteoro 86–71
~ bolométrica 91–68
~ cenital 91–63
~ de un terremoto 73–34
~ del eclipse 81–79
~ electrográfica 91–67
~ en oposición media 83–85
~ estelar 91–56
~ fotográfica 91–66
~ fotovisual 91–69
~ infrarroja 91–71
~ integrada 96–136
~ integral 96–136
~ límite 91–62
~ media 83–85
~ radiométrica 91–70
~ reducida al cenit 91–63
~ sísmica 73–34
~ total 96–136
~ visual 91–65
~ zenital 91–63
mala visibilidad 77–26
mancha aislada 89–116
~ Blanca 83–214
~ caliente 95–112a
~ del este 89–114
~ del oeste 89–113
~ delantera 89–113
Gran ~ Roja 83–193
~ f 89–114
~ invisible 89–159
~ p 89–113
~ precedente 89–113
~ principal 89–115
~ Roja 83–193
~ siguiente 89–114
~ solar 89–88
~ trasera 89–114
~ zaguera 89–114
manchas estelares 97–62
manto de hielo 97–72
~ de la Tierra 72–15
~ superior 72–16
~ terrestre 72–15
mapa colorimétrico 81–134
~ de alta resolución 89–133
~ de campos magnéticos 89–131
~ de gravedad 71–79
~ de la Luna 81–133
~ de las manchas solares 88–34
~ de nivel constante 81–141
~ de velocidades 89–23
~ del Marte 83–136
~ del tiempo 77–12
~ diario del Sol 89–18
~ estructural 73–97
~ físico lunar 81–134
~ fotográfico de la Luna 81–137
~ fotográfico lunar 81–137
~ fotométrico 81–134
~ gravimétrico 81–143
~ heliográfico 89–17
~ hidrográfico 72–25
~ hipsométrico 81–141
~ lunar 81–133
~ magnético 89–131
~ oceanográfico 72–23
~ polar 89–19
~ polarimétrico 81–134
~ radio 81–135a
~ sinóptico 77–11
~ sinóptico de la corona 89–22

ES

~ sinóptico de la cromosfera 89–21
~ sinóptico de la fotosfera 89–20
~ tectónico 82–202
~ térmico lunar 81–135
~ topográfico de la Luna 81–138
~ topográfico lunar 81–138
mapeo de la Luna 81–111
~ selenológico 82–51
mapificación coestra 72–82
~ de temperatura 72–83
Máquina Neumática 92–15
mar 81–266, 83–156
alta ~ 72–90
~ Austral 81–286
~ circular 81–267
~ Conocido 81–289
~ continental 72–26
~ de Humboldt 81–288
~ de la Fecundidad 81–274
~ de la Ingenuidad 81–292
~ de la Serenidad 81–276
~ de la Tranquilidad 81–275
~ de las Crisis 81–280
~ de las Lluvias 81–273
~ de las Nubes 81–281
~ de las Ondas 81–290
~ de las Serpientes 81–291
~ de los Fríos 81–278
~ de los Humores 81–282
~ de los Néctares 81–279
~ de Moscú 81–283
~ de Smyth 81–287
~ del Conocimiento 81–289
~ del Frío 81–278
~ del Néctar 81–279
~ interior 72–26
~ marginal 72–27, 81–285
~ mediterráneo 72–26
~ Oriental 81–284
marciano 83–135
marea alta 82–281
~ baja 72–91
~ entrante 82–280
~ equinoccial 82–273
~ menguante 82–282
~ muerta 82–272
~ vacilante 82–282
mareas 72–89
~ altas 82–273
~ atmosféricas 82–294
~ bajas 82–272
~ de cuadratura 82–272
~ de la corteza 82–297
~ de mar 82–309
~ de sicigias 82–273
~ geomagnéticas 82–295
~ gravitacionales 82–298
~ inducidas por la Tierra 82–275
~ lunares 82–308
~ oceánicas 82–310
~ térmicas 82–299
~ terrestres 82–296
~ vivas 82–273
marégrafo 82–311
marejada 72–76
maremoto 73–2c
mareógrafo 82–311
mareómetro 82–311
Markab 92–119
Marte 83–133
masa ausente 99–87
~ crítica 93–189
~ crítica de las enanas blancas 93–189
~ de aire 77–67
~ de la partícula meteórica 86–163
~ de Planck 99–170f
~ del cuerpo meteórico 86–163
~ del Sol 91–122
~ elemental de Planck 99–170f
~ en reposo 99–143
~ escondida 99–88
~ estelar 91–121
~ faltante 99–87
~ gaseosa 94–32
~ gravitatoria 99–149
~ inercial 99–148
~ inicial 86–164
~ integral del cúmulo 96–139
~ meteórica 86–163
~ original 86–164
~ principal 87–41
~ relativista 99–141
~ solar 91–122
~ total del cúmulo 96–139
~ variable 86–167, 96–70
mascon 81–145
máser cósmico 97–13
~ interestelar 97–14
~ natural 97–13
materia absorbente 97–85
~ adsorbente 85–130
~ adsorbida 85–131
~ ausente 99–87
~ carbonácea 87–64
~ carbónica 87–64
~ cósmica 99–7
~ de gas y polvo 83a–25
~ estelar 93–124
~ faltante 99–87
~ interestelar 97–1
~ intergaláctica 98–349
~ interplanetaria 84–1
~ luminosa 97–109
~ meteorítica 87–9
~ orgánica meteorítica 87–68
~ oscura 97–108
~ pre-estelar 94–16
~ primigenia 99–170
material cósmico 99–7
~ de la superficie lunar 82–59
~ de montículo 82–103
dificil de fundir 83a–54
~ extraterrestre 72–56
~ eyectado 82–165
~ lunar 82–58
~ refractario 85–49
materiales detríticos 81–236
~ expulsados 82–165
~ proyectados 82–165
~ volcánicos proyectados 82–166
matriz 82–118
~ de grano fino 82–115
máxima elongación 83–63
~ frecuencia utilizable 78–37
máximo 76–55
~ agudo 86–235
~ brillo 86–78
~ de la actividad solar 89–2
~ de la capa E 76–56
~ de una capa 76–55
~ llano 86–236
~ primario 95–31
~ principal 95–31
~ secundario 95–32
~ solar 89–2
mecanismo de bombeo 97–16
~ de calentamiento 88–214
~ de destrucción 97–79
~ de excitación 79–27
~ de frenado 91–46
media luna 81–26
medianoche magnética 79–131
medición absoluta de la gravedad 71–68
~ de la gravedad 71–67
~ del grado 71–22
~ micrométrica 96–49
~ polarimétrica 83–80
~ relativa de la gravedad 71–69
medida de la emisión 88–234
~ de la gravedad 71–67
medidor relativo de opacidad ionosférica 78–45
medio ambiente 76–3
~ del eclipse 81–78
~ interestelar 97–8
~ turbio 83–251
membrecía del cúmulo 96–83
Mensa 92–66
Mercurio 83–120
meridiano central 89–43
~ de mediodía 75–17
~ del amanecer 75–18
~ del anochecer 75–19
~ magnético 75–49
~ matutino 75–18
~ vespertino 75–19
mes 81–42
~ anomalístico 81–43
~ draconítico 81–47
~ lunar 81–18
~ sidéreo 81–45
~ sinódico 81–46
~ trópico 81–44
Mesa 92–66

ES

mesa submarina 72–44
meseta 81–192
~ continental 73–103
mesopausa 76–67
mesosfera 76–66
mesosiderito 87–79
Mesozoico 74–72
metagalaxia 98–357
metamorfismo 74–34
~ de contacto 74–35
~ de choque 82–287
~ dinámico 74–36
~ térmico 74–37
metanal 97–27
metanamida 97–35
metanimina 97–45
metanol 97–32
metasomatismo 74–44
metasomatosis 74–44
meteórico 86–4
meteóritico 87–2
meteorito 87–1
~ acondrítico 87–77
~ carbonáceo 87–62
~ férrico 87–82
mayor ~ 87–41
~ orientado 87–112
~ original 87–123
~ pétreo 87–60
meteorización 74–57
~ espacial 82–244
meteoro 86–1
~ artificial 86–21
~ brillante 86–12
~ cometario 86–43
~ con rastro 86–17
~ con variaciones irregulares de brillo 86–85
~ de enjambre 86–24
~ de origen cometario 86–45
~ de polvo 86–190
~ débil 86–15
~ del sistema solar 86–44
~ diurno 86–22
~ doble 86–131
~ eclíptico 86–46
~ esporádico 86–23
~ estacionario 86–34
~ fotografiado 86–13
~ fotografiado desde dos estaciones 86–155
~ fotográfico 86–13
~ fusiforme 86–77
~ hiperbólico 86–43
~ interestelar 86–43
~ interplanetario 86–44
~ lento 86–88
~ múltiple 86–133
~ no perteneciente a un enjambre 86–25
~ ordinario 86–14
~ pulsante 86–5
~ rápido 86–87
~ solitario 86–25
~ telescópico 86–18
~ telescópico débil 86–19
~ telescópico muy débil 86–20
~ ultratelescópico 86–20
~ visible 86–16
~ visual 86–16
meteoroide 86–160
~ poroso 85–50
meteorología 77–6
~ dinámica 77–7
~ sinóptica 77–8
metilacetileno 97–41
metilideno 97–23
método de Argelander 95–10
~ de iteración 93–56
~ de las funciones de peso 93–59
~ de pasos 95–10
~ del doble conteo 86–132
~ del helio 87–128
~ del uranio 74–96
~ estereoscópico 81–115
~ radiactivo 74–92
~ radiométrico 74–92
~ variacional 93–57
métrica 99–151
~ de Robertson-Walker 99–152a
~ de Schwarzschild 99–152
mezcla de gas y polvo 83a–25
~ de gas y radiación 93–125
~ de la materia estelar 93–162
~ de Russell 93–101
mFU 78–33
MFU 78–37
mica 74–23
~ blanca 74–24
~ negra 74–25
microbrecha lunar 82–86
microcráter 81–201
microestructura 82–73
microfotografia 87–140
micrófono de hilo caliente 76–46
micrometeoritos 87–3
micrometeoro 86–20
micrometeorología 77–9
micropulsaciones geomagnéticas 75–82
~ magnéticas 75–82
microrreversibilidad 93–41
Microscopio 92–67
microscopio „blink“ 91–34
~ de centelleo 91–34
~ de salto 91–34
Microscopium 92–67
microseísmost 73–53
microsismos 73–53
microtectita 87–97
miembro del cúmulo 96–82
~ del sistema 98–133
~ incierto del cúmulo 96–87
~ más brillante del cúmulo 98–339
~ probable del cúmulo 96–86
migmatita 74–39
migración de protuberancias hacia el polo 89–231
milibar 77–47
miligal 71–66
milímetro de la columna de mercurio 77–48
millonésima de disco solar 89–69
millonésimas de hemisferio visible 89–68
mineral 74–12
~ que constituye la roca 82–99
~ que forma la roca 82–99
mínima frecuencia utilizable 78–33
mínimo de la actividad solar 89–4
~ primario 95–26
~ principal 95–26
~ secundario 95–27
~ solar 89–4
misión a un asteroide 84–99
~ a un asteroide con retorno de muestras 84–98
mitad del disco de Venus 83–132
Mizar 92–120
modelo abierto 99–72
~ anisótropo 99–65
~ asintótico 99–59
~ asintótico de primera especie 99–60
~ cerrado 99–73
~ compuesto 93–140
~ con blanketing 93–18
~ cosmológico 99–46
~ cosmológico caliente 99–166
~ de agua 85–113
~ de agua de cometas 85–113
~ de atmósfera 83–247, 93–17
~ de Charlier 99–67
~ de De Sitter 99–69
~ de dos componentes 88–196
~ de dos temperaturas 88–197
~ de Eddington 93–137
~ de Einstein 99–68
~ de Einstein y De Sitter 99–70
~ de estrella 93–123
~ de estrella de temperatura nula 93–186
~ de fuente en capa 93–143
~ de fuente puntual 93–138
~ de hidrógeno 83–197
~ de hidrógeno de un planeta 83–197
~ de invernadero 83–262
~ de la Tierra 79–133
~ de manchas 89–102
~ de surtidor 85–85
~ de temperatura nula 93–186
~ de un planeta 83–196
~ del universo 99–46
~ del universo en contracción 99–56
~ del universo en expansión 99–55
~ en contracción 99–56
~ en expansión 99–55
~ eolosférico 83–260

~ estacionario 99–52
~ estándar 93–137
~ estático 99–50
~ estelar 93–123
~ estelar inhomogéneo 93–139
~ estelar no homogéneo 93–139
~ estelar relativista 93–198
~ galáctico 98–163
~ homogéneo 99–64
~ inhomogéneo 99–66
~ ionosférico 83–261
~ isótropo 99–63
~ jerárquico 99–48
~ matemático 98–324
~ monótono 99–61
~ monótono de primera especie 99–62
~ no estacionario 99–53
~ no estático 99–51
~ no homogéneo 99–66
~ oscilatorio 99–57
~ oscilatorio de primera especie 99–58
~ politrópico 93–129
~ sin blanketing 93–19
~ vacío 99–49
modelos de Friedmann 99–71
modificación del litoral 72–78
modulación electro-óptica 88–86
módulo aparente 96–117
~ aparente de distancia 96–117
~ de alunizaje 82–5
~ de comando 82–6
~ de distancia 91–16
~ de elasticidad 73–59
~ de mando 82–6
~ de servicio 82–7
~ de Young 73–59
~ lunar 82–5
~ verdadero 96–118
moldavita 87–99
mole errática 72–132
molécula de aire 86–172
~ hija 85–135a
~ interestelar ~ 97–10
~ madre 85–135
~ poliatómica 97–21
momento angular 83a–4
~ angular orbital 83a–6
~ cinético 83a–4
~ de caída de un meteorito 87–51
~ de la cantidad de movimiento 83a–4
~ de par 83a–5
~ del paso por el periastro 96–60
~ rotacional 83a–5
monitoraje de fulguraciones 89–219
~ de satélites 83–26
monitoreo de fulguraciones 89–219
Monoceros 92–68
monopolo magnético 99–170o
monosulfuro de carbono 97–34
monóxido de carbono 97–28
~ de silicio 97–37
Montaña de la Mesa 92–66
~ de plegamiento 77–86
~ plegada 73–86
~ submarina 72–43
montañas lunares 81–193
monte submarino 72–43
~ tabular 72–44
montículo 89–276
en forma de ~ 89–276
monzón 77–99
morfología lunar 81–147
morrena = morena 72–128
~ de fondo 72–129
~ frontal 72–130
~ lateral 72–131
~ terminal 72–130
Mosca 92–69
moscovita 74–24
motas 88–210
~ brillantes 88–211
~ cromosféricas 89–172
~ o(b)scuras 88–212
moteado 88–210
~ cromosférico 89–172
movimiento a gran escala 88–189
~ a pequeña escala 88–188
~ al azar 89–157, 98–94
~ aleatorio 98–94
~ aparente 83–41
~ apsidal 96–71
~ casual 98–94
~ de masa 82–195
~ de planetas 83–28
~ del glaciar 72–120
~ diario 85–29
~ directo 83–43
~ diurno 86–69
~ diurno del radiante 86–69
~ en gran escala 88–189
~ en pequeña escala 88–188
~ en *z* 98–188
~ epirogénico 73–73
~ horario 81–72
~ horizontal 88–184
~ no gravitacional 85–56
~ no gravitatorio 85–56
~ orbital 96–62
~ ordenado 98–95
~ paraláctico 91–33
~ peculiar 91–31
~ perpendicular al plano galáctico 98–188
~ planetario 83–28
~ propio 91–19
~ propio anual 91–20
~ propio de las manchas 89–117
~ propio en ascensión recta 91–21
~ propio en declinación 91–22
~ relativo 99–99
~ retrógrado 83–47
~ solar 98–62
~ solar básico 98–66
~ solar estándar 98–63
~ vertical 88–185
movimientos de la corteza terrestre 73–72
~ internos 97–149
muerte térmica del universo 99–14
muestra de aire 76–86
~ de material lunar 82–69
~ de polvo lunar 82–121
~ del núcleo 82–60
~ individual 87–42
~ lunar 82–69
~ lunar devuelta 82–70
~ obtenida con taladro 72–52
muestreador de superficie 82–80
muestreo 76–85
~ de aire 76–85
multipolo 75–66
mundo 99–3
Muro Recto 81–185
Musca 92–69

N

nacimiento de una mancha 89–121
naturaleza del objeto 85–24
nebulosa anular 97–147
~ brillante 97–121
~ cometaria 97–142
~ cometoide 97–142
~ con filamentos 97–129
~ de Andrómeda 98–215
~ de emisión 97–123
~ de gas y polvo 97–119
~ de Orión 97–136
~ de polvo 97–115
~ de reflexión 97–128
~ del Búho 97–139
~ del Cangrejo 97–134
~ del Encaje 97–137
~ del Velo 97–137
~ difusa 97–122
~ espiral 98–249
~ extragaláctica 98–202
~ filamentosa 97–129
~ galáctica 97–110
~ gaseosa 97–112
~ gaseoso-polvorienta 97–119
~ generadora 97–63
~ Herradura 97–143
~ irregular 98–237
~ luminosa 97–121
~ Norteamérica 97–140
~ Omega 97–143
~ oscura 97–132
~ planetaria 97–144
~ preestelar 97–63
~ presolar 83a–21

ES

O

~ fotográfica de meteoros 86–140
~ fuera del eclipse 88–227
~ regular del Sol 89–16
~ simultánea 86–125
observaciones c(u)asisimultáneas 71–36
~ con radar 86–265
~ simultáneas 71–35
observador de meteoros 86–7
observatorio auroral 79–101
~ de alta montaña 88–56
~ de gran altitud 88–56
~ ionosférico 78–21
~ lunar 82–20
~ sismológico 73–5
~ solar 88–7
obturador de dos láminas 86–149
~ de lámina móvil 86–148
~ giratorio 86–147
océano 72–22
~ de las Tempestades 81–277
~ de las Tormentas 81–277
oceanografia 72–24
~ espacial 72–69
~ por satélite 72–70
oclusión 77–72
octaedrito 87–86
Octans 92–71
Octante 92–71
ocultación 95–80
~ de una estrella por la Luna 81–94
~ lunar 81–94
~ por la capa de ozono 79–28
~ rasante 81–100
Ofiuco 92–72
ola ciclónica 72–76
~ de marea 82–278
oleada 89–255
~ de fulguración 89–190
olivina 74–22
OMM 77–17
onda continua 86–277
~ de cizalla 73–43
~ de compresión 73–46
~ de choque sin colisiones 84–49
~ de densidad 98–175
~ de densidad galáctica 98–175
~ de distorsión 73–43
~ de gravedad 88–201
~ de gravitación 99–164
~ de Love 73–48
~ de marea 82–291
~ de presión 73–46
~ de rarefacción 97–58
~ de terremoto 73–9
~ de volumen 73–42
~ del espacio 78–13
~ directa 78–12
~ diurna 76–30
~ doble 98–117
~ elástica 73–44
~ equivolumétrica 73–43
~ extraordinaria 78–30
~ gravitacional 99–164
~ infrasónica 72–93
~ larga 73–48
~ longitudinal 73–45
~ Love 73–48
~ ordinaria 78–31
~ P 73–46
~ planetaria 72–94
~ portadora 86–276
~ primaria 73–46
~ Q 73–48
~ reflejada 78–13
~ S 73–43
~ secundaria 73–43
~ semidiurna 76–29
~ sísmica 73–9
~ sísmica de maremoto 72–95
~ sonora 76–33
~ superficial 73–47
~ terrestre 78–14
~ transversal 73–41
ondas infrasónicas de presión 75–88
~ penumbrales 89–100
ondulación del geoide 71–74
opacidad neutrínica 94–95
Ophiuchus 92–72
oposición 83–59
~ en ascensión recta 81–68
óptica atmosférica 79–1
órbita absoluta 96–55
~ aparente 96–48
~ areocéntrica 83–138
~ astrométrica 96–57
~ de la Luna 81–36
~ de la Tierra 84–5
~ de un planeta 83–29
~ espectrográfica 96–56
~ espectroscópica 96–56
~ fotocéntrica 96–47
~ galáctica de una estrella 98–184
~ geocéntrica 86–41
~ heliocéntrica 86–42
~ lunar 81–36
~ original 98–200
~ planetaria 83–29
~ primitiva 98–200
~ relativa 96–54
~ terrestre 84–5
~ tridimensional 98–185
~ verdadera 96–53
Ordoviciano 74–79
Organización Meteorológica Mundial 77–17
orientación del eje de rotación 84–102
origen 94–9
~ común 94–25
~ de los cometas 85–96
de ~ cósmico 87–5
de ~ extraterrestre 87–5
~ del mundo 99–11
~ del universo 99–11
~ extraterrestre 87–73
~ por impacto 82–229
~ tectónico 82–200
~ terrestre 87–74
Orion 92–73
Orión 92–73
Oriónidas 86–253
orogénesis 73–74
orogenia 73–74
ortoclasa 74–16
Osa Mayor 92–96
~ Menor 92–97
oscilación perpendicular al plano galáctico 98–187
~ en z 98–187
oscilaciones de 5 minutos 88–186
~ de marea 76–28
~ de marea inducidas 82–306
~ forzadas de las mareas 82–306
~ libres de la Tierra 73–49
~ solares 88–223a
oscilador hidromagnético 95–61
~ magnetohidrodinámico 95–61
oscilógrafo 86–292
osciloscopio 86–292
oscurecimiento = obscurecimiento 88–142
~ del limbo 88–142
óvalo auroral 79–124
Óvalo blanco 83–195
oxhidrilo 79–25
óxido 74–26
~ de hidrógeno 97–26
oxidrilo 97–22
oxígeno atómico 79–24
~ molecular 79–23
ozono atmosférico 76–124
ozonosfera 76–123

P

paisaje lunar 81–55
Palas 84–86
paleoclima 77–188
paleomagnetismo 75–25, 83a–79
paleotemperatura 72–126
paleotraza nuclear 82–252
Paleozoico 74–71
pallasito 87–81
Paloma 92–37
panal de aluminio 82–25
pantalla 88–13
pantano 81–303
~ de la Decadencia 81–305
~ de la Putrefacción 81–305
~ de las Epidemias 81–304
~ del Sueño 81–306
par de marea 82–288
paraclasa 73–91
paradoja de Chesaux y Olbers 99–40
~ de Olbers 99–40
~ de Seeliger 99–39
~ fotométrica 99–40

ES

~ gravitatoria 99–39
paradojas cosmológicas 99–38
paralaje absoluta 91–11
~ anual 91–6
~ cefeídica 91–8a
~ de cúmulo 91–7
~ de esferas de Strömgren 91–13
~ de grupo 91–7
~ de Strömgren 91–13
~ dinámica 91–12
~ espectroscópica 91–15
~ estadística 91–8
~ estelar 91–4
~ fotométrica 91–14
~ relativa 91–10
~ secular 91–9
~ trigonométrica 91–5
~ trigonométrica relativa 91–10
parámetro de desaceleración 99–83
~ estructural 96–120
parámetros de un cráter 81–227
~ fotométricos 85–106
paranthelio 79–84
parantiselena 79–86
parantiselene 79–86
paraselena 79–73
paraselene 79–73
pared aterrazada 81–257
~ con terrazas 81–257
~ concéntrica 81–254
~ del cráter 81–251
~ externa del cráter 81–253
~ interna del cráter 81–252
~ múltiple 81–255
~ Recta 81–185
parhelio 79–71
~ inferior 79–72
parsec 91–17
~ cúbico 98–84
parte posterior de la Luna 81–16
partícula 88–154
~ acelerada 91–182
~ auroral 79–107
~ coloidal 76–98
~ de aerosol 76–91
~ de aire 86–172
~ del núcleo-manto 97–70
~ gigante 76–118
~ líquida 76–100
~ lunar 82–123
~ meteórica 86–160a
~ sólida 76–99, 97–69
partículas Aitken 76–116
~ atrapadas 75–111
~ capturadas 75–111
~ de "finos" 82–109
~ esféricas 87–29
~ finas 82–109
~ geomagnéticamente atrapadas 75–111
~ interestelares 97–68
~ precipitantes 76–117
~ sólidas 85–48
~ superpesadas 99–107n
~ zodiacales 84–66
pasaje 95–81
~ próximo de dos galaxias 98–320
paseo al azar 89–157
paso de la Luna por el meridiano 82–303
~ de un bólido 86–117
~ de un meteoro 86–5
~ de un planeta por el disco solar 83–51
~ de un satélite 83–54
~ de un satélite a través del disco de un planeta 83–54
~ de una discontinuidad interplanetaria 84–58
~ por el meridiano central 89–45
~ por la atmósfera 87–26
~ sobre el limbo 89–31
pata del alunizador 82–22
patrullaje automático de fulguraciones 89–221
~ cinematográfico de fulguraciones 89–220
~ de asteroides 84–94
~ de fulguraciones 89–219
~ de satélites 83–26
~ meteórico 86–144
~ solar 89–16
pausa 76–58
Pavo 92–74
~ Real 92–74
Peces 92–79
peculiaridad 98–257
~ espectral 91–93
Pegaso 92–75
Pegasus 92–75
película cinematográfica de protuberancias 89–242
pendiente 81–230
~ del derrubio 81–233
~ detrítico 81–233
~ empinada 81–232
~ fuerte 81–232
~ moderada 81–231
péndulo horizontal 71–73
penetración de las partículas energéticas 78–43
penetrómetro 82–83
~ de suelos 82–83
penumbra 89–97
~ de la mancha 89–97
~ de la Tierra 81–64
peña 74–1
pequeña depresión 81–243
~ Nube Magallánica 98–214
pequeño planeta 84–68
pérdida de energía 88–254
~ de masa 86–168
perfil de altura 76–53
~ de densidad 76–53
~ de humedad 76–53
~ de temperatura 83–259
~ del viento 76–53
~ vertical 81–228
periápside galáctico 98–189
periáreo 83–142
periastro 96–58
pericentro galáctico 98–189
~ joviano 83–172
~ saturnal 83–210
pericintio 81–40
perifoco galáctico 98–189
perigaláctico 98–189
perigeo 81–38
perihelio 83–29a
perijove 83–172
perijoviano 83–172
perilunio 81–40
período 95–39
~ de la rotación del Sol 88–129
~ de las pulsaciones 95–100
~ de once años 89–70
~ de pulsación 95–42
~ de rotación 83–31
~ de superposición de dos ciclos 89–86
~ geológico 74–74, 83a–77
~ glacial 72–105, 83a–78
~ glaciario 72–105
~ interglacial 72–107
~ magnéticamente perturbado 75–91
~ primario 95–40
~ secundario 95–41
~ sidéreo 81–45
~ sinódico 81–46
~ undecenal 89–70
perisaturno 83–210
perlas de Baily 88–32
Pérmico 74–83
Perros de Caza 92–26
Perseidas 86–251
Perseo 92–76
Perseus 92–76
persiana veneciana 88–126
persistencia de una estela 86–111
pertenencia física al cúmulo 96–83
Perturbación Austral 83–194
perturbación asociada a una fulguración 84–46
~ de marea 98–314
~ interplanetaria 84–45
~ ionosférica 78–39
~ ionosférica repentina 78–41
~ Sur Tropical 83–194
perturbaciones geomagnéticas 75–70
~ magnéticas 75–70
~ mutuas 83–25
~ mutuas entre satélites 83–25
~ radiales 93–215
~ solares 83–24
~ solares en los satélites 83–24
Pesebre 96–111

ES

peso molecular medio 93–100
petrificación 74–49
~ por impacto 82–238
Pez Austral 92–80
~ Volador 92–100
Phoenix 92–77
pico central 81–258
Pictor 92–78
pie del vehículo de alunizaje 82–22
piedras rodantes 81–237
piezoglyptos 87–122
pigeonita 82–90
Pintor 92–78
piranómetro 88–43
pirheliómetro 88–39
~ de circulación de agua 88–41
~ de corriente de agua 88–40
~ de disco de plata 88–42
piroxeno 74–21
pisadas 82–38
Pisces 92–79
Piscis Austrinus 92–80
pista de deslizamiento 82–197
placa adhesiva 87–14
~ compensadora 88–92
~ de un cuarto de onda 88–126
~ meteórica 86–151
~ móvil 88–27
plagioclasa 74–19
plancha delgada 83–149
planeta 83–1
~ de hidrógeno 83–197
~ del sistema solar 83–11
~ del Sol 83–11
~ exterior 83–15
~ gigante 83–174
~ hipotético 83–122
~ hipotético de Olbers 83a–63
~ inferior 83–14
~ interior 83–14
~ intramercurial 83–123
~ joviano 83–174
~ menor 84–68
pequeño ~ 84–68
~ solar 83–11
~ superior 83–15
~ transneptuniano 83–237
~ transplutoniano 83–124
~ X 83–124
planetario 83–6
planetas grandes 83–13
~ mayores 83–13
~ principales 83–13
~ telúricos 83–119
~ terrestres 83–119
planetesimales 83a–35
planetilla 84–68
planetocéntrico 83–9
planetográfico 83–7
planetoide 84–68
~ de la familia de Marte 84–74
~ de Marte 84–74
~ del anillo 84–71
~ doble 84–79
~ hecho por el hombre 84–69a
~ marciano 84–74
~ numerado 84–95
~ tipo Apolo 84–80a
planetología 83–5
planicie abisal 72–42
~ de lava 82–185
planicies 81–272
planisferio 92–10
plano de anillo 83–216
~ del cielo 96–50
~ del ecuador 82–231
~ ecuatorial 83–231
~ galáctico 98–53
~ invariable 83–232
~ Laplaciano 83–232
~ principal de la Galaxia 98–53
~ tangencial a la esfera celeste 96–50
plasma enriquecido en hélio 84–60
~ interplanetario 84–39
~ primordial 99–170
~ sin colisiones 84–50
~ solar 84–29
~ solar magnetizado 84–32
plasmasfera = plasmaesfera 76–76
plataforma continental 72–34
playa en rayos X 89–166
~ óptica 89–165
playas 89–164
pleamar 82–281
plegamiento 73–88
plena mar 82–281
plenilunio 81–27
plessita 87–94
Pléyades = Pléyadas 96–110
pliegue 73–87
~ anticlinal 73–90
~ sinclinal 73–89
pliegues 79–139
plomada 71–71
plomo común 74–99
~ ordinario 74–99
~ original 74–98
~ primitivo 74–98
~ radiogénico 74–97
plumas polares 89–294
Plutón 83–239
pluviómetro 77–170
~ totalizador 77–171
población de los brazos espirales 98–73
~ de una galaxia 98–230
~ del cúmulo 98–341
~ del disco 98–72
~ del halo 98–69
~ del tipo II 98–68
~ del tipo I 98–71
~ II 98–68
~ estelar 98–67
~ galáctica 98–230
~ intermedia 98–70
~ I 98–71
~ I extrema 98–73
poder adsorbente 85–120
~ de adsorción 85–120
Polar 92–121
polaridad magnética 89–148
~ magnética negativa 89–150
~ magnética positiva 89–149
~ norte 89–149
~ opuesta 89–155
~ sur 89–150
polarímetro coronal 88–66
~ fotoeléctrico 83–83
~ fotográfico 83–84
~ visual 83–82
polarización interestelar 97–105
poliedro 71–12
polinomios de Legendre 93–55
politropo 93–126
polo de la inclinación magnética 75–45
~ de rotación 83–36
~ del eje 83–36
~ del eje geomagnético 75–48
~ geomagnético 75–48
~ magnético terrestre 75–45
polución del aire 76–80
Pólux 92–122
polvo circumestelar 97–55
~ circunterrestre 84–53
~ cósmico 97–52
~ de impactos 82–125
~ de silicato 85–53
~ galáctico 97–53
~ interestelar 97–54
~ interplanetario 84–51
~ meteórico 87–11
~ meteorítico 87–10
~ volcánico 73–143, 82–126
Popa 92–81
poro 89–122
porosidad 83–100
portadora 86–276
posición aproximada 85–26
~ calculada 85–27
~ del radiante 86–60
~ estacionaria 83–45
~ heliográfica 89–38
~ precisa 85–28
~ predicha 85–27
~ selenocéntrica 81–132
postglacial 72–110
postluminiscencia 79–22
postmáximo 95–33
postnova 95–123
postulado de coherencia 99–34
~ de Weyl 99–34
~ matemático de la relatividad de inercia 99–145
potencia cohesiva 85–117
~ de pico 86–271
~ del transmisor 86–270
~ media 86–272

potencial axisimétrico 98–153
~ de adsorción 85–122
~ de gravedad 71–58
~ de marea 82–293
~ de simetría cilíndrica 98–154
~ estacionario 98–147
~ gravitacional 96–160a
~ gravitatorio 96–160a
~ terrestre 71–60
precalentamiento 86–203
Precámbrico 74–70
~ inferior 74–75
precipitación de electrones 78–48
~ de electrones aurorales 79–112
~ orográfica 77–167
~ total 77–168
precipitaciones 77–154
~ atmosféricas 77–154
precursor 95–143
~ de un terremoto 73–11
predicción a corto plazo 77–20
~ a largo plazo 77–19
~ de la actividad solar 89–64
~ de ocultaciones 81–102
~ del tiempo 77–13
~ ionosférica 78–32
predicciones de las posiciones de pequeños planetas 84–92
preestelar 94–13
premáximo 95–30
premonitores 73–10
prenova 95–122
Presepio 96–111
presión al nivel del mar 77–44
~ atmosférica 77–43
~ de degeneración 93–187a
~ de fotones 93–91
~ de la radiación 93–91
~ de partículas 93–90
~ del gas 93–90
~ del vapor de agua 77–125
~ hidrostática 93–77
~ lateral 73–65
~ no térmica 94–42
~ normal 77–46
~ térmica 94–41
~ total 93–93
presupernova 95–129
previsión a corto plazo 77–20
~ a largo plazo 77–19
~ de la actividad solar 89–64
~ del tiempo 77–13
primaria 73–46
Primario 74–71
primer contacto 81–84
~ punto Lagrangiano 96–31
primera aparición 85–19
~ aparición de un cometa 85–19
primeras bandas positivas del nitrógeno 79–20
principio Arquimediano 73–101
~ cosmológico 99–30
~ cosmológico perfecto 99–32
~ cosmológico restringido 99–31
~ de Arquímedes 73–101
~ de equivalencia 99–150
~ de invariancia 93–61
~ de la equivalencia de masa y energía 99–145
~ de la totalidad 81–77
~ de Mach 99–37
~ del eclipse 81–74
~ del mundo 99–11
prisma de Anderson 88–68
~ rotante 88–68
~ separador de haces 88–88
problema cosmogónico 94–5
procesador de suelos 82–79
proceso de ablación 86–183
~ de equilibrio 94–99
~ de eyección 98–179
~ de radiación 86–211
~ *e* 94–99
~ endógeno 82–258
~ interno 82–258
~ lento de captura de neutrones 94–88
~ lunar 82–150
~ metamórfico 74–34
~ *p* 94–91
~ *r* 94–89
~ radiativo 86–211
~ rápido de captura de neutrones 94–89
~ *s* 94–88
~ sedimentario 74–46
~ tectónico 82–201
~ Urca 94–51
Proción 92–123
producción de energía 93–105
profundidad de compensación 73–102
~ de compensación isostática 73–102
~ del eclipse 95–75
~ del foco 73–20
~ efectiva de formación de las líneas 93–64
~ focal 73–20
profundización de una depresión 77–82
prognóstico a corto plazo 77–20
~ a largo plazo 77–19
~ del tiempo 77–13
prolongación de los cuernos de Venus 83–131
promontorio 81–293
propagación anómala del sonido 76–37
~ de la energía 88–217
~ de la luz 99–129
~ de las ondas 78–34
~ de las ondas de radio 78–34
~ del sonido 76–32
propiedad en ondas de radio 98–235
~ intrínseca 91–100
~ óptica 98–234
propiedades cinemáticas 98–90
~ estructurales 98–224
~ globales 96–134
~ integrales 96–134
propino 97–41
proporción isotópica 74–94
protección magnética 89–225
Proterozóico 74–77
protoatmósfera 83a–44
protoestelar 94–46
protoestrella 94–45
protogalaxia 94–18
protonoesfera 75–12
protonósfera 75–12
protoplaneta 83a–40
Protosol 83a–39
protuberancia 89–223
~ activa 89–238
~ activa coronal 89–247
~ activa de mancha 89–254
~ activa ordinaria 89–246
~ ascendente 89–278
~ c(u)asi-eruptiva 89–249
~ con líneas metálicas 89–265
~ coronal 89–264
~ de arco 89–270
~ de casquete 89–253
~ de embudo 89–269
~ de hidrógeno 89–266
~ de las mareas 81–119
~ de latitud alta 89–229
~ de latitud baja 89–227
~ de mancha 89–252
~ de tipo fulguración 89–222
~ de tornado 89–262
~ de troncos de árboles 89–272
~ ecuatorial 89–228
~ en abanico 89–258
~ en chorro 89–255
~ en forma de arco 89–270
~ en forma de casquete 89–253
~ en forma de embudo 89–269
~ en forma de tornado 89–262
~ en forma de troncos de árboles 89–272
~ eruptiva 89–248
~ eruptiva ordinaria 89–250
~ lunar 81–118
~ polar 89–230
~ quiescente 89–263
~ secundaria 89–260
~ semejante a una fulguración 89–222
protuberancias interactivas 89–243
Próxima Centauri 92–124
proximidades del Sol 98–58
proyección acimutal 71–44
~ cartográfica 71–40
~ cenital 71–45
~ central 71–48
~ cilíndrica 71–43
~ conforme 71–41
~ cónica 71–42

~ estereográfica 71–47
~ gnomónica 71–46
~ zenital 71–45
proyecto de fulguraciones protónicas 89–177
(p)seudocráter 81–250
(p)seudovariable 95–51
psicrómetro 77–121
puente brillante 89–104, 98–307
~ intergaláctico 98–348
puesta del Sol 88–11
pulsación aislada 95–146
~ principal 95–144
pulsaciones 79–158
~ adiabáticas 95–98
~ geomagnéticas 75–81
~ magnéticas 75–81
~ no adiabáticas 95–99
~ no radiales 95–97
~ radiales 95–96
pulsar 95–137
~ binario 95–140a
~ de rayos gamma 95–139
~ de rayos X 95–138
~ del Cangrejo 95–142
~ óptico 95–140
pulso aislado 95–146
~ de retorno 86–290
~ del eco 86–290
~ principal 95–144
pulverización 82–239, 87–28
~ de un cuerpo meteórico 86–189
~ por viento solar 82–245
punto brillante 89–201
~ brillante en rayos X 89–183a
~ brillante umbral 89–96
~ cero de la relación período-luminosidad 95–104
~ cónico 96–31
~ de aparición 86–36
~ de caída 86–124
~ de convergencia 96–98
~ de Curie 75–41
~ de desaparición 86–38
~ de disrupción 86–121
~ de explosión 86–122
~ de extinción 86–38
~ de fusión 86–204
~ de impacto 86–124
~ de parada 86–123
~ de reflexión 75–124
~ de regresión 98–178
~ de retorno 98–178
~ de rocío 77–174
~ de umbra 89–96
~ de visión foveal 86–127
~ del espacio-tiempo 99–125
~ del universo 99–125
~ espejo 75–124
~ estacionario 83–45
~ extremo de la región de dispersión 87–39
~ interior de Lagrange 96–31
~ más próximo de la región de dispersión 87–40
~ neutro 89–152
primer ~ Lagrangiano 96–31
~ singular 94–116
~ subestelar 81–130
~ sublunar 81–129
~ subsatélite 71–30
~ subsolar 81–127
~ subterrestre 81–128
~ umbral 89–96
puntos magnéticamente conjugados 75–23
~ magnetoconjugados 75–23
Puñal de Orión 92–130
Puppis 92–81
Pyxis 92–82

Q

QSO inactivo en radio 98–362
QSO quiescente en radio 98–362
quebramiento de simetría 99–170s
quebradizo 86–191
quemazón de helio 94–78
~ de hidrógeno 94–56
~ del carbono 94–81
~ del oxígeno 94–82
~ del silicio 94–83
~ explosiva del helio 94–79
Quilla 92–30
química de la atmósfera 76–79
~ lunar 82–74
quimioluminiscencia 79–31
quimiosfera 76–62
Quinteto de Stephan 98–309

R

RA 89–57
racha 77–112
radiación artificial atrapada 75–113
~ capturada 75–112
~ corpuscular 88–150
~ de la Tierra 77–56
~ de recombinación 86–212
~ de un meteoro 86–207
~ del cielo 77–54
~ difusa del cielo 77–59
~ diluída 93–38
~ geomagnéticamente atrapada 75–112
~ global 77–55
~ gravitacional 99–165
~ integral 88–147
~ libre-ligado 86–212
~ neutrínica 94–94
~ propia de un planeta 83–198
~ solar 88–134
~ solar directa 77–50
~ solar incidente 77–52
~ terrestre 77–56
~ total 88–146
radiactividad 74–61
radiante 86–53
~ aparente 86–56
~ corregido 86–58
~ de grupo 86–54
~ de una corriente meteórica 86–237
~ de una lluvia meteórica 86–237
~ estacionario 86–241
~ geocéntrico 86–58
~ individual 86–55
~ meteórico 86–53
~ secundario 86–239
~ verdadero 86–59
radical 85–136
~ libre 85–137
radio aparente de la Luna 81–70
~ de Schwarzschild 94–113
~ aparente de Sol 81–69
~ del universo 99–5
~ ecuatorial 83–115
~ exterior 83–228
~ externo 83–228
~ gravitacional 94–113
~ interior 83–226
~ interno 83–226
~ medio 83–116a
~ polar 83–116
radioalbedo 83–94
radiocarbono 74–95
radiocentelleo 84–36
~ interestelar 97–109b
~ interplanetario 84–37
radioeco 86–295
radioemisiones esporádicas de Júpiter 83–181
radioespectrógrafo 88–124
~ de barrido 88–124
~ dinámico 88–124
radioespectrómetro 88–124
radioestrella 91–158
radiofuente cuasi-estelar 98–360
~ decamétrica 83–186
radiogalaxia 98–206
radioheliógrafo 88–95
radioheliograma 88–96
radiointerferómetro 88–97
radiolóbulo 98–286
radiomagnitud 86–268
radiometeoro 86–267
radiómetro de lóbulo estrecho 88–121
radionube 98–286
radionucl(e)ido cosmogénico 82–255
~ primitivo 82–256
~ primordial 82–256
~ producido por bombardeo 82–255
radiosol 88–3
ráfaga 77–112, 95–69
rama ascendente 95–34
~ creciente 95–34
~ de las gigantes 91–152b

ES

~ de una corriente 86–240
~ decreciente 95–35
~ descendente 95–35
~ horizontal 91–132
ramificación 98–5a
rango de las mareas 82–292
ranura 81–165
~ curvada 88–65
~ oscilante 88–70
~ sinuosa 81–167
rapidez de crecimiento hacia el máximo 95–29
Ras Alhague 92–125
rasgo de impacto 82–228
~ primaro 81–150
~ secundario 81–151
~ volcánico 82–159
rasgos lunares 81–148
rastro de un meteoro 86–30
rayas de Neumann 87–103
rayo 77–162, 85–55
~ coronal 89–287
~ coronal monocromático 89–288
~ curvado 85–58
~ de un abanico coronal 89–296
~ de un bulbo coronal 89–296
~ en bola 77–163
~ en forma de rosario 77–164
~ rectilíneo 85–57
~ rosario 77–164
~ sísmico 73–14
~ sonoro 76–34
~ tangencial 81–169
~ verde 79–58
~ visual 91–28
rayos 79–155
~ lunares 81–168
~ polares 89–294
razón de abundancia 91–175
~ de iluminancias 91–60
~ de intensidades 91–60
~ del gas al polvo 97–120
~ hidrógeno-metales 91–176
~ masa-luminosidad 98–233
reacción de astillado 94–100
~ de astillamiento 94–100
~ de captura de neutrones 94–87
~ de espalación 94–100
~ de neutrinos 94–92
~ fotoquímica 76–63
~ neutrínica 94–92
~ nuclear superficial 91–181
reaparición 81–99
~ del anillo 83–219
rebote latitudinal 75–119
receptor 78–17
recesión de las galaxias 98–298
recipiente de muestras 82–81
recogedor de muestras 72–101
~ de muestras de agua 72–101
recolección de polvo meteórico 87–12
recolector de polvo meteórico 87–13
recombinación disociativa 86–214
reconocimiento de hielos 72–85
~ de la superficie 82–32
~ desde la órbita 82–31
~ en la superficie 82–32
~ fotográfico 82–30
~ magnético 75–64
~ magnético mundial 75–65
~ orbital 82–31
~ selenofísico 82–39
~ selenológico 82–50
recristalización 74–42
rectificación 95–18
~ de la curva de luz 95–18
recuento de galaxias 98–343
recuentos estelares 98–79
recurrencia 88–132
red de gravedad 71–80
~ de vigilancia selenofísica 82–40
~ geodésica de satélites 71–34
~ geodésica mundial 71–38
~ gravimétrica 71–80
~ gravimétrica fundamental 71–82
~ gravimétrica mundial 71–81
~ monitora selenofísica 82–40
~ mundial 77–18
redescubrimiento 85–17
reducción al aire libre 71–78
~ al nivel de mar 77–45
~ de ocultaciones 81–108
~ Faye 71–78
referencia geodésica 71–20
referencial 98–136
reflujo 82–282
refractor solar sin cúpula 88–53
región abisal 72–41
~ activa 89–57
~ activa coronal 89–286
~ activa de Júpiter 83–180
~ anterior a los mares 81–269
~ asísmica 73–29
~ auroral 79–126
~ baja 81–272
~ brillante 83–160
~ caliente 88–200
~ central de la Galaxia 98–13
~ central de una galaxia 98–293
~ circunnuclear 98–292
~ clara 83–160
~ continental 81–271
~ de actividad 89–57
~ de acumulación de un glaciar 72–122
~ de aprisionamiento 75–110
~ de audibilidad 87–32
~ de auroras 79–126
~ de caída 87–38
~ de captura 75–110
~ de dispersión de los meteoritos 87–38
~ de fusión de un glaciar 72–124
~ de hidrógeno ionizado 97–50
~ de hidrógeno neutro 97–49
~ de mares 81–268
~ de modulación 84–25
~ de modulación de la radiación cósmica 84–25
~ de protuberancias activas 89–240
~ de terremoto 73–28
~ de transición 88–224
~ de transición entre cromosfera y corona 88–224
~ de visibilidad 86–126
~ del anticentro galáctico 98–15
~ del centro galáctico 98–13
~ del epicentro 73–16
~ del núcleo 98–17
~ ecuatorial 89–59
~ epicentral 73–16
~ exterior de radiación 75–105
~ fotosférica no pertubada 89–56
~ fría 88–199
~ H II 97–50
~ H II compacta 97–51
~ H I 97–49
~ interespicular 88–198
~ intergranular 88–177
~ interior de radiación 75–106
~ ionosférica 78–4
~ M 89–61
~ magnética 89–144
~ magnética bipolar 89–160a
~ magnética unipolar 89–155a
~ marítima 81–268
~ montañosa 81–196
~ nuclear 98–17
~ o(b)scura 83–155
~ polar 81–17, 83–192
~ posible 79–134
~ posterior a los mares 81–270
~ postmarítima 81–270
~ premarítima 81–269
~ prohibida 79–135
~ roja de la corona 89–285
~ sin terremotos 73–29
~ sísmica 73–28
~ solar activa 89–57
~ verde de la corona 89–284
registrador de luz solar 77–61
~ espectroscópico de velocidades 88–57
~ ionosférico 78–22
regmaglyptos 87–122
regolito 82–104
Régulo 92–126
relación accidental 97–126
~ de abundancia 91–175
~ de equilibrio 94–104
~ de intensidades 91–60
~ de las masas 96–67
~ de Poisson 73–60
~ del gas al polvo 97–120
~ diámetro-profundidad 81–248

ES

~ entre corrimiento hacia el rojo y la distancia 99–76
~ entre los fenómenos solares y terrestres 84–55
~ espectro-luminosidad 91–130
~ genética 97–127
~ hidrógeno-metales 91–176
~ isotópica 74–94
~ magnitud-intensidad 91–58
~ masa-luminosidad 91–129
~ masa-radio 93–200
~ período-luminosidad 95–103
~ velocidad-distancia 98–300
relaciones estructurales 82–266
~ heliogeofísicas 84–55
~ Sol-Tierra 84–55
relajación = relajamiento 98–127
~ por colisiones 98–131
~ sin colisiones 98–132
relatividad de la inercia 99–35
~ especial 99–94
~ general 99–146
~ restringida 99–94
relativista 99–95
relativístico 99–95
relevamiento celeste 92–11
~ del ciclo 92–11
selenológico 82–50
Reloj 92–54
~ de Péndulo 92–54
remoción de masa 81–236
rendija curvada 88–65
~ oscilante 88–70
rendimento energético 93–109
réplicas 73–13
resistencia aerodinámica 86–178
~ del aire 86–178
~ del suelo 82–135
resolución espacial 88–204
~ en el tiempo 88–205
~ temporal 88–205
resplandor 79–151
~ residual 79–22
restos de una supernova 95–133
~ orgánicos en meteoritos 87–67
retardación de la expansión 99–82
retardo de grupo 78–56
~ de la rotación terrestre 82–313
retención de gas 87–134
reticulado cromosférico 89–171
retículo 86–129, 92–83
Reticulum 92–83
retorno de muestras 82–71
~ de un cometa 85–9
~ de un grupo de manchas 88–133
retrógrado 83–46
retroreflector 82—43
~ laser 82–43
revolución anomalística 81–43
~ draconítica 81–47
~ sideral 81–45
~ sinódica 81–46
~ trópica 81–44
riesgos de la navegación 72–97
Rigel 92–127
rigidez 73–56
~ magnética 75–117
Río Erídano 92–49
riómetro 78–45
ritmo de aumento hacia el máximo 95–29
roca 74–1
~ ácida 74–4
~ básica 74–5
~ cristalina 82–66
~ efusiva 74–3
~ equigranular 82–67
~ esquistosa 74–54
~ estratificada 74–48
~ extrusiva 74–3
~ foliada 74–54
~ ígnea 74–2
~ intrusiva 74–7
~ lamelar 74–54
~ lunar 82–65
~ magmática 74–2
~ metamórfica 74–11
~ plutónica 74–7
~ sedimentaria 74–10
~ ultrabásica 74–6
~ volcánica 74–3
rocío 77–173
rosa de los vientos 77–116
rotación asincrónica 83–38
~ de cuerpo rígido 98–111
~ de Kepler 98–112
~ de la línea de los ápsides 96–71
~ de las estrellas 91–35
~ de planetas 83–30
~ del Sol 88–127
~ diferencial 88–128
~ diferencial de la Galaxia 98–113
~ directa 83–39
~ estelar 91–35
~ forzada 81–14
~ galáctica 98–107
~ galáctica diferencial 98–113
~ kepleriana 98–112
~ ligada 81–14
~ no sincrónica 83–38
~ no uniforme 91–36
~ planetaria 83–30
~ retrógrada 83–40
~ rígida 98–111
~ sincrónica 83–37
~ solar 88–127
rotor 91–42
~ oblícuo 91–52, 95–60
~ rápido 91 43
rotura 73–78
rugosidad superficial 82–106
ruido acústico 88–220
~ sísmico 73–54
~ sísmico de fondo 73–55
ruptura de un cometa 85–111
~ de un núcleo galáctico 98–318
~ por colisión 84–107
ruta de exploración 82–37

S

sacar muestras 76–85
Saco de Carbón 97–138
sacudida premonitora 73–10
Sagitario 92–85
Sagitta 92–84
Sagittarius 92–85
salida de la Luna de la penumbra 81–82
~ del Sol 88–8
~ de la Luna de la sombra 81–80
salinidad 72–66
salto brusco 95–154
satélite 83–17
antiguo ~ 83–240
~ de hielo 83–23
~ Galileano 83–204
~ irregular 83–21
~ más interior 83–20
~ natural 83–18
~ retrógrado 83–21
~ rocoso 83–22
saturación 76–101
Saturno 83–209
saturnocéntrico 83–213
saturnográfico 83–212
Scorpio 92–86
Scorpius 92–86
Sculptor 92–87
Scutum 92–88
sección eficaz diferencial 94–54
~ frontal 86–176
~ geométrica 84–103
~ gravitacional 84–104
sector 85–60
~ giratorio 86–147
~ magnético interplanetario 84–16
secuencia boreal 91–87
~ de edad cero 94–60
~ de espirales 98–255
~ de galaxias de Hubble 98–227
~ de galaxias regulares 98–240
~ espectral 91–94
~ evolutiva 94–30
~ polar 91–87
~ polar internacional 91–87
~ principal 91–131
~ principal de edad cero 96–152
secundaria 73–43
Secundario 74–72
sedimentación 74–45
~ en alta mar 72–79
~ en mar abierto 72–79
~ submarina 72–47
sedimento 74–47
sedimentos abisales 72–50
~ en el fondo de mar 87–15

ES

~ en region abis(m)al 72–50
~ marinos 87–15
~ marinos profundos 72–50
~ organógenos 72–59
~ pelágicos 72–57
~ terrígenos 72–48
seguimiento automático de fulguraciones 89–221
~ cinematográfico de fulguraciones 89–220
~ de fulguraciones 89–219
seísmo = sismo 73–2a
~ cercano 73–26
~ lejano 73–27
~ local 73–25
~ meteorítico 87–34
~ normal 73–22
~ superficial 73–21
seismograma 73–8
seismógrafo 73–7
seismología 73–1
selenocéntrico 81–131
selenodesia 81–109
selenodésico 81–110
selenofísica 82–151
selenografía 81–1
selenográfico 81–8
selenoide 81–117
selenología 82–1
selenológico 82–48
selenólogo 82–49
selenoquímica 82–74
semidiámetro aparente de la Luna 81–70
~ aparente del Sol 81–69
~ ecuatorial 83–115
~ exterior 83–228
~ externo 83–228
~ interior 83–226
~ interno 83–226
sentido de la rotación 98–108
señales intermitentes 86–278
~ pulsadas 86–278
separación de los elementos químicos 83a–48
~ desde el Sol 83a–9
~ entre antenas 88–99
separador de haces 88–88
sereno 77–149
Serpens 92–89
Serpentario 92–72
Serpiente 92–89
Cabeza de la ~ 92–90
Cabeza y Cola de la ~ 92–90
Cola de la ~ 92–90
~ de Agua 92–55
servicio meteorológico 77–5
seto 89–274
en forma de ~ 89–274
seudocrater 81–250
seudovariable 95–51
Sextans 92–90
Sextante 92–90
sicigias = cizigias 81–29
siderito 87–82
siderolito 87–78
sideróstato 88–44
sierra 73–85
signos del zodíaco 92–6
silbido corto 78–58
~ de nariz 78–59
~ electrónico 78–60
~ en forma de nariz 78–59
~ largo 78–57
~ prolongado 78–57
~ protónico 78–61
silbidos 78–53
~ atmosféricos 78–53
silicato 74–14
sílice 74–13
sill 73–135
Silúrico 74–80
simetría cilíndrica 98–155
~ esférica 98–160
~ heliocoidal 98–161
simulación matemática 98–323
simultaneidad 99–127
sinclinal 73–89
síncrona 85–92
~ completa 85–94
~ terminal 85–93
sindinama 85–91
singularidad 99–42
~ desnuda 94–117
síntesis de abertura 88–118
~ de elementos 94–52
~ nuclear 94–52
Sirio 92–128
Sirius 92–128
sismicidad 73–3
sismógrafo 73–7
sismograma 73–8
~ de impacto 82–224
sismología 73–1
~ activa 82–217
~ de la Luna 82–213
~ lunar 82–213
~ pasiva 82–218
~ solar 88–223b
sismólogo 73–2
sismómetro activo 82–222
~ pasivo 82–221
sistema aplanado 98–251
~ aplastado 98–251
~ autogravitante 98–169
~ autogravitatorio 98–169
~ axisimétrico 98–157
~ colisional 98–166
~ compacto 96–157
~ con colisiones 98–166
~ con energía total negativa 96–156
~ con energía total positiva 96–158
~ copernicano 99–24
~ cosmológico 99–20
~ de anillos 98–291
~ de arcos coronales 89–299
~ de bandas 79–18
~ de bucles 89–299
~ de clasificación espectral 91–91
~ de clasificación espectral de Harvard 91–91
~ de colores 91–72, 96–140
~ de Copérnico 99–24
~ de corrientes 75–97
~ de cúmulos globulares 98–75
~ de estrellas 96–1
~ de fracturas 73–79, 81–174
~ de fracturas continentales 73–81
~ de fracturas oceánicas 73–80
~ de franjas de interferencia 88–120
~ de longitudes 83–177
~ de los anillos de Saturno 83–220
~ de magnitudes 91–57
~ de rayos 81–170
~ de referencia 98–136
~ de referencia en rotación 98–138
~ de referencia rotante 98–138
~ de roturas 73–79
~ de satélites 83–19
~ de síntesis 88–119
~ de torbellinos 83a–13
~ de vórtices 83a–13
~ del mundo 99–20
~ II de longitudes 83–179
~ ecuatorial de longitudes 83–178
~ en estado estacionario 98–141
~ en estado no estacionario 98–152
~ esferoidal 98–159
~ estelar 96–1
~ estelar formado por superposición 98–165
~ extragaláctico 98–210
~ fundamental de referencia 97–137
~ galáctico 98–3
~ geocéntrico 99–21
~ geodésico de referencia 71–19
~ goedésico mundial 71–39
~ gravitatoriamente libre 96–158
~ gravitatoriamente ligado 96–156
~ heliocéntrico 99–23
~ inercial 99–98
~ inestable 98–171
~ ligado por gravitación 96–156
~ local 98–61
~ local de referencia 98–139
~ muy achatado 98–164
~ no inercial 99–147
~ para latitudes intermedias 83–179
~ planetario 83–2
~ planetocéntrico 83–10
~ planetocéntrico de coordenadas 83–10
~ planetocéntrico de referencia 83–10
~ ptolemaico 99–22
~ sin colisiones 98–167
~ solar 83–12
~ Tierra-Luna 81–12
~ triple 96–4

~ U, B, V 96–144
~ I de longitudes 83–178
sistemas de coordenadas comóviles 99–44
~ de referencia invariablemente unidos a... 99–44
sitio de alunizaje 82–14
~ del hallazgo de un meteorito 87–52
~ propuesto de alunizaje 82–15
situación general meteorológica 77–15
~ meteorológica 77–14
smog 76–83
sobreenfriado 76–106
sobreenfriamiento 76–105
sobreestimar 95–13
sobrefusión 76–105
sobreposición de los espectros 96–24
sobresalto en ondas decamétricas 83–183
~ en radioondas de Júpiter 83–182
~ en rayos X 89–184
sobresaltos de baja frecuencia 83–184
sobresaturación 76–102
Sol 88–1
~ activo 89–12
Año del ~ calmo 89–10
Año del ~ quieto 89–10
~ azul 79–92
~ calmo 89–11
~ en rayos X 88–4
falso ~ 79–71
~ inactivo 89–11
~ no perturbado 89–11
~ óptico 88–2
~ perturbado 89–12
~ quieto 89–11
~ radio 88–3
~ tranquilo 89–11
~ visible 88–2
~ X 88–4
solar 88–5
solidificación 83a–59
solución estable 98–170
sombra de la Tierra 79–44
~ de ozono 76–127
sombras volantes 88–31
sonda acústica 76–45
~ de suelos 82–83
sondaje descendente 78–20
sondeo 76–47
~ con taladro 72–51
~ crepuscular 79–41
~ descendente 78–20
~ ionosférico 78–19
~ por cohete 76–48
~ sísmico 82–216
sondeos acústicos 72–29
~ marinos 87–16
~ por eco 72–29
sonido anómalo 76–36
~ normal 76–35
sopa de quarks 99–169a
soplar 77–105
soplido 78–64
soplo 89–277
en forma de ~ 89–277
soporte de la cromosfera 88–223
Spica 92–129
subabundancia 91–187
subenana 91–140
subenfriado 76–106
subenfriamiento 76–105
suberupción 89–195
subestimar 95–14
subfulguración 89–195
subfundido 76–106
subfusión 76–105
subgigantes 91–151
subida 76–54
~ al máximo 95–28
~ de temperatura 77–42
sublimación 85–138
subpulsación 95–145
subpulso 95–145
subsistema 98–74
substancia difícil de fundir 83a–55
~ fácilmente fundible 83a–56
~ no volátil 83a–53
~ porosa 82–111
~ refractaria 83a–55
~ volátil 83a–52
subsuelo 82–105
subtormenta 79–144
~ auroral 79–141
subtracción fotográfica 88–99
sucesión de las galaxias de Hubble 98–227
suceso 99–125
~ adyacente 99–126
~ de contaminación 72–72
~ de partículas energéticas 89–191
~ en rayos X 89–184
~ infinitamente próximo 99–126
~ protónico 84–59
~ sísmico 82–215
~ solar con eyección de protones 89–192
~ solar de rayos cósmicos 89–194
sucesos simultáneos 99–128
suelo cohesivo 82–63
~ consistente 82–63
~ lunar 82–62
~ marítimo 72–38
~ oceánico 72–37
sulfato 74–29
sulfito 74–28a
sulfuro 74–28
~ de carbonilo 97–36
~ de hidrógeno 97–44
suma de las precipitaciones 77–168
superabundancia 91–185
supercúmulo de galaxias 98–346
~ de galaxias Local 98–347
superficie alta 82–56
~ anterior 87–114
~ con costra 87–109
~ con cráteres 81–245
~ craterizada 81–245
~ crítica 96–27
~ de altiplanicie 82–56
~ de contacto 96–27
~ de la Luna 82–55
~ de la Tierra 71–5
~ de mar 82–57
~ de rayos 81–171
~ de un meteorito 87–106
~ de un planeta 83–97
~ en capas 83–101
~ entre rayos 81–172
~ equipotencial 71–61
~ estratificada 83–101
~ frontal 87–114
~ geopotencial 71–61
~ luminosa difusa 79–148
~ lunar 82–55
~ neutra 75–16
~ original 87–107
~ planetaria 83–97
~ posterior 87–115
~ primaria 87–107
~ pulida 87–101
~ pulsante 79–149
~ rayada 81–171
~ rugosa 83–102
~ secundaria 87–108
~ solar 88–164
~ sólida 83–98
~ terrestre 71–5
superfluido 95–149
supergalaxia 98–346
~ de la Virgen 98–347
Supergalaxia Local 98–347
supergigante 91–145
supergigantes brillantes 91–147
~ débiles 91–148
~ luminosas 91–147
~ más brillantes 91–147
~ más luminosas 91–147
~ subluminosas 91–148
supergranulación 88–183
supergranulo 88–182
supergravitación 99–170r
supernova 95–126
antigua ~ 95–134
~ de tipo I 95–130
~ de Tycho 95–135
~ extragaláctica 95–128
~ galáctica 95–127
superposición de dos ciclos 89–86
período de ~ 89–86
~ de los espectros 96–24
superunificación 99–170q
supuesta variable 95–5
surco 81–165
surtidor 89–255
~ brillante 89–257
~ de fulguración 89–190

ES

~ o(b)scuro 89–256
susceptibilidad magnética 75–27
sustancia no volátil 85–142
~ volátil 85–141

T

tabla de evolución de los grupos 89–120
teanita 87–93
Tahalí de Orión 92–129a
talud 81–234
Taller del Escultor 92–87
tamaño de grano 82–113
~ de partículas 76–114
tasa de crecimiento hacia el máximo 95–29
~ de decrecimiento 95–36
~ de desaparición de estrellas 94–115
~ de emisión de masa 85–146
~ de expansión 99–75
~ de formación 85–143
~ de formación de estrellas 94–12
~ de producción 85–143
~ de producción de gases 85–144
~ de reacción termonuclear 94–53
~ de recesión 99–75
~ de recurrencia 86–279
~ de relajación 98–129
~ de repetición de pulsos 86–279
Táuridas 86–254
Taurus 92–91
tectita 87–95
~ en forma de gota 87–96
tectónica 73–95
techo de la atmósfera 76–25
telegrama astronómico 85–20
Telescopio 92–92
~ cromosférico 88–73
~ solar 88–49
~ solar automático 88–50
~ solar horizontal 88–55
~ solar llevado por un globo 88–52
Telescopium 92–92
temblor de la corteza 95–152
~ de la estrella 95–153
~ de tierra 73–2a
temblores secundarios 73–13
temperatura 77–29
~ a la sombra 77–34
~ ambiente 77–32
~ central 93–89
~ cinética 91–115
~ de brillo 91–114
~ de brillo a la longitud de onda λ 91–114
~ de color 91–109
~ de condensación 97–66
~ de Curie 75–41
~ de degeneración 93–188
~ de disociación 91–120a
~ de excitación 91–118
~ de fusión 86–204
~ de ionización 91–117
~ de rotación 91–119
~ de vibración 91–120
~ efectiva 91–113
~ electrónica 91–116
~ en el centro 93–89
~ normal 77–31
~ superficial 93–88
~ y presión normales 77–30
tempestad geomagnética 75–71
~ ionosférica 78–49
tensión 73–66
~ cortical 82–210
~ térmica 87–27
tensor de curvatura 99–159
~ de deformaciones 73–70
~ de energía 99–158
~ de esfuerzos 73–69
~ de materia 99–158
~ de Riemann 99–159
~ energía-impetu 99–158
~ fundamental 99–157a
~ métrico 99–157a
teorema de Clairaut 71–77
~ de Russell-Vogt 93–83
~ de unicidad 93–83
~ de Vogt-Russell 93–83
~ del virial 98–181
teoría calibracional 99–170j
~ de captura 85–99, 96–72
~ de eyección 85–98
~ de formación múltiple 96–76
~ de la acreción 83a–20
~ de la contracción 73–77
~ de la estructura espiral 98–173
~ de la fisión 96–75
~ de la muerte térmica del universo 99–15
~ de la relatividad especial 99–94
~ de la relatividad restringida 99–94
~ de las mareas 82–305
~ de los brazos espirales 98–173
~ de los cíclos 97–156
~ de medida 99–170j
~ de núcleos separados 96–76
~ de ondas de densidad 98–174
~ física de los meteoros 86–159
~ general de la relatividad 99–146
~ geomagnética de la dinamo 75–99
Terciario 79–133
térmicamente estable 93–209
terminador 81–91, 83–69
~ de la Tierra 81–91
~ lunar 81–123
~ matutino 81–124
~ terrestre 81–91
~ vespertino 81–125
término cosmológico 99–156a
~ de la rotación galáctica 98–116
~ del movimiento diferencial 98–115
~ K 98–119
termoluminiscencia 82–249
termometamorfismo 74–37
termosfera 76–59
terra 81–190
terraplén 89–276
terraza del cráter 81–229
terrazas internas 81–256
terrella 79–133
terremoto 73–2a
~ catastrófico 73–40
~ cercano 73–26
~ continental 73–2b
~ de dislocación 73–2d
~ de foco profundo 73–23
~ de profundidad intermedia 73–24
~ débil 73–35
~ desastroso 73–38
~ destructor 73–39
~ intenso 73–37
~ intermedio 73–24
~ lejano 73–27
~ local 73–25
~ medio 73–36
~ muy intenso 73–38
~ normal 73–22
~ profundo 73–23
~ submarino 73–2c
~ superficial 73–21
~ tectónico 73–2d
~ terrestre 73–2b
~ volcánico 73–2e
terreno caótico 83–164
~ desordenando 83–164
terrestre 71–3
testigo 72–52
testimonio visual de la caída 87–48
tetradimensional 99–121
tetravector 99–138
textura 82–102
~ de flujo 82–191
~ fluidal 82–191
Thuban 93–131
tiempo 77–1
~ absoluto 99–117
~ característico 99–79
~ de aparición 86–49
~ de desintegración 96–104
~ de disgregación 96–104
~ de disolución 96–104
~ de eco 86–291
~ de enfriamiento 83–201
~ de evaporación 96–109
~ de expansión 99–78
~ de Friedmann 99–80
~ de Hubble 99–79
~ de llegada 73–50
~ de permanencia 82–34
~ de Planck 99–170i
~ de propagación 73–51
~ de recorrido 73–51
~ de relajación 98–128
~ de vida de una partícula atrapada 75–122

ES

~ del paso por periastro 96—60
~ del periastro 96–60
~ del período 95–39
~ del tránsito 84–47
~ del tránsito Sol-Tierra 84–47
~ geomagnético 79–132
~ lunar 82–300
~ lunar aparente 82–301
~ lunar medio 82–302
~ medio de relajación 98–130
~ newtoniano 99–118
~ propio 99–139
Tierra 71–2
~ alta 81–189
~ permanente congelada 83–152
tierras altas 82–56
~ bajas 81–272
tioformaldehído 97–43
tipo de aurora 79–136
~ de estrellas 91–101
~ de Hubble 98–229
~ de la cola cometaria 85–71
~ de variable 95–47
~ espectral 91–108
~ estelar 91–101
~ estructural 98–225
~ galáctico de Hubble 98–229
~ morfológico 98–228
tira de mica 88–87
Titán 83–233
titileo interplanetario en radio 84–37
~ radio 84–36
toma de contacto 82–16
tomamuestras de agua 72–101
topografía lunar 81–146
tormenta 77–159
~ de comienzo brusco 75–73
~ de comienzo gradual 75–78
~ de partículas 88–162
~ de polvo 83–255
~ de ruido de Júpiter 83–185
~ de trueno 77–159
~ geomagnética 75–71
~ geomagnética secundaria 75–95
~ interplanetaria 84–44
~ magnética 75–71
~ magnética de comienzo brusco 75–73
~ magnética polar secundaria 75–96
~ magnética secundaria 75–95
~ magnética típica 75–72
Toro 92–91
toro de hidrógeno 83–230
torre solar 88–54
torrente de lava 73–140
totalmente degenerado 93–185
trampa para la luz 88–64
transferencia de calor 86–202
transformación de energía 86–208
~ de Galileo 99–131
~ de Laplace 93–54
~ de Lorentz 99–132
~ galileana 99–131
~ homológica 93–132
~ laplaciana 93–54
transición corteza-manto 72–14
~ de fase 83a–58, 99–170p
tránsito 95–81
~ de la Luna 82–303
~ de la sombra 83–53
~ de un planeta sobre el disco solar 83–51
~ del satélite 83–54
~ lunar 82–303
transitorio coronal 89–185a
translunar 81–6
transmisión de calor 86–202
transmisor 78–15
transporte de energía 93–111
~ de masa 96–33
~ de masa de una componente a otra 96–33
~ de radiación 93–29
~ del campo magnético 84–48
~ radiativo 93–29
Trapecio de Orión 96–92
tratamiento con ácido 87–102
travesía científica 82–36
~ de levantamiento 82–35
trayectoria aparente 86–32
~ aparente en la esfera celeste 86–32
~ ascendente 82–9
~ atmosférica 86–162
~ circunlunar 82–11
~ de enfriamiento 93–202
~ de un meteoro 86–30
~ de una partícula 75–118
~ distorsionada 86–33
~ en la atmósfera 86–162
~ evolutiva 94–29
~ interplanetaria 82–12
~ ondulada 86–33
~ verdadera 86–31
traza de enfriamiento 93–202
~ de una partícula 82–251
~ estelar 86–152
~ evolutiva 94–29
~ fósil 87–121
~ fotográfica de un meteoro 86–29
~ meteórica 86–153
~ meteórica sobre la placa fotográfica 86–29
~ nuclear 87–120
~ nuclear fósil 82–252
trazador de estructura espiral 98–39
Tres Marías 92–129a
tríada 87–91
triangulación 71–23
~ cósmica 71–28
~ de primer orden 71–24
~ espacial 71–25
~ por satélites 71–27
~ primaria 71–24
Triángulo 92–93
~ Austral 92–94
Triangulum 92–93
~ Australe 92–94
Triásico 74–84
trilateración 71–29
troilita 82–94
troncos 89–278
tropopausa 76–77
troposfera 76–78
Troyanos 84–91
trueno 77–165
tsunami 72–95
tubo de fuerza magnética 89–145
~ de lava 82–186
~ de rayos catódicos 86–293
Tucán 92–95
Tucana 92–95
turbiedad 83–252
turbosfera 76–72
turbulencia a escala cósmica 83a–12
~ a gran escala 83a–12
tvik 78–63

U

último contacto 81–85
umbra 89–95
~ de la mancha 89–95
umbral de la compañera 96–40
~ de la secundaria 96–40
unicidad de la solución 93–82
Unicornio 92–68
unidad de protuberancia 89–235
~ topográfica 81–149
unidades solares 91–123
unificación de las interacciones 99–170w
gran ~ 99–170u
Unión Internacional de Geodesia y Geofísica 71–4
universo 99–1
~ abierto 99–72
~ anisótropo 99–65
~ asintótico 99–59
~ asintótico de primera especie 99–60
~ caliente 99–166
~ cerrado 99–73
~ clásico 99–93
~ de Charlier 99–67
~ de De Sitter 99–69
~ de Einstein 99–68
~ de Einstein y De Sitter 99–70
~ en contracción 99–56
~ en evolución 99–54
~ en expansión 99–55
~ estacionario 99–52
~ estático 99–50
~ homogéneo 99–64
~ inflacionario 99–170b
~ inicial 99–170a
~ isótropo 99–63
~ jerárquico 99–48

ES

~ monótono 99–61
~ monótono de primera especie 99–62
~ newtoniano 99–93
~ no estacionario 99–53
~ no estático 99–51
~ no homogéneo 99–66
~ oscilatorio 99–57
~ oscilatorio de primera especie 99–58
~ primordial 99–170a
~ relativista 99–92
~ vacío 99–49
universos de Friedmann 99–71
Urano 83–234
Ursa Major 92–96
~ Minor 92–97
Úrsidas de diciembre 86–258

ES

V

vacío 76–8, 76–9, 99–171d
valle 81–163
~ Alpino 81–164
vapor de agua 77–119
vaporización de un cuerpo meteórico 86–198
~ intensa 86–185
variabilidad del campo magnético 84–14
~ estelar 95–4
~ óptica 95–141
variable 95–1
~ Algol 95–77
~ cataclísmica 95–109
~ cefeida 95–92
~ cíclica 95–108
~ de corto período 95–89
~ de cúmulo 95–93
~ de eclipse 95–74
~ de espectro 95–48
~ de fulguraciones 95–68
~ de largo período 95–91
~ de período ultracorto 95–90
~ de tipo cúmulo 95–93
~ de velo 95–83
~ eclipsante 95–74
~ elipsoidal 95–79
~ espectral 95–48
~ espectroscópica 95–48
~ extrínseca 95–86
falsa ~ 95–51
~ geométrica 95–72
~ intrínseca 95–87
~ irregular 95–107
~ irregular con polaridad alterna(nte) 95–56
~ irregular con polaridad constante 95–57
~ magnética 95–52
~ magnética periódica 95–55
~ magnética regular 95–55
~ nebular 95–85
~ parecida a nova 95–118
~ periódica 95–88
~ por ocultación 95–83
~ roja semirregular 95–106
~ rotacional 95–65
~ semirregular 95–105
~ sospechosa 95–5
supuesta ~ 95–5
~ tipo... 95–47
~ tipo δ Cephei 95–93
~ tipo R Coronae Borealis 95–83
~ velada 95–83
variables adimensionales 93–133
variación anual 86–138
~ centro-borde 88–145
~ centro-limbo 88–145
~ de brillo 95–7
~ de luz 95–6
~ del campo magnético del centro al limbo 89–160
~ del clima 77–183
~ diurna 86–137
~ estacional 86–139
~ geomagnética secular 75–85
~ secular del campo geomagnético 75–85
variaciones espectrales 95–49
~ geomagnéticas 75–69
~ magnéticas 95–53
vecindad de la órbita terrestre 84–7
en la ~ solar 98–59
~ solar 98–59
vector de Poynting 93–92
~ de propagación 78–36
~ magnético 89–139
Vega 92–132
vehículo de alunizaje 82–21
Vela 92–98
veleta 77–104
velo 83–250
~ auroral 79–148
~ de nubes 77–151
velocidad angular 86–89
~ angular instantánea 86–90
~ angular media 86–91
~ atmosférica 86–94
~ característica 76–43
~ de craterización 82–233
~ de aproximación 91–27
~ de decrecimiento 95–36
~ de desaparición de estrellas 94–115
~ de enfriamiento 93–203
~ de entrada en la atmósfera 86–92
~ de escape desde un sistema 96–155
~ de expansión 96–94
~ de eyección 85–90
~ de formación de cráteres 82–233
~ de formación de estrellas 94–12
~ de fusión 83–147
~ de giro 83–32
~ de reacción termonuclear 94–53
~ de recesión 91–26
~ de relajación 98–129
~ de rotación 91–48
~ de un meteoro 86–86
~ del viento 77–109
~ del viento solar 88–158
~ ecuatorial 91–49
~ en la atmósfera 86–94
~ en la traza 86–154
~ en la visual 91–24
~ espacial 91–30
~ fuera de la atmósfera 86–93
~ geocéntrica de un meteoro 89–95
~ heliocéntrica 86–96
~ inicial 86–92
~ instantánea 86–99
~ media 86–98
~ negativa 91–27
~ orbital 96–63
~ positiva 91–26
~ radial 91–24
~ relativista 99–136
~ residual 98–91
~ rotacional 91–48
ventana de transmisión 83–264
~ de transparencia 83–264
ventanas de la zona de ausencia 98–57
~ galácticas 98–57
ventisquero 72–118
campo de ~ 72–119
Venus 83–125
venusino 83–126
vertical 71–71
~ astronómica 71–71
vértice 96–98, 98–105
vertido de petróleo 72–71
vesícula 82–180
vesiculación 82–182
vesicular 82–181
vesiculoso 82–181
Vesta 84–87
vetas 87–111
Vía Láctea 98–1
vibraciones libres de la Tierra 73–49
vida corta 89–125
de ~ larga 89–124
~ de una partícula atrapada 75–122
~ larga 89–124
videomagnetógrafo 88–85
vidrio ahumado 88–33
~ lunar 82–96
viento 77–89
~ a rachas 77–111
~ anabático 77–92
~ ascendente 77–92
~ ciclostrófico 77–95
~ del gradiente 77–94
~ dominante 77–117
~ estelar 93–14
~ galáctico 98–277
~ geostrófico 77–96
~ interestelar 97–151

~ predominante 77–117
~ solar 88–156
~ solar en calma 84–41
~ solar permanente 84–40
~ solar quiescente 84–41
vientos alisios 77–100
~ contraalisios 77–103
vigilancia automática de fulguraciones 89–221
~ cinematográfica de fulguraciones 89–220
~ de fulguraciones 89–219
~ de satélites 83–26
~ meteórica 86–144
~ Meteorológica Mundial 77–16
violación de la paridad de carga 99–170t
~ de simetría 99–170s
virazón 77–97
Virgen 92–99
Virgo 92–99
viscosidad 73–71
~ magnética 75–34
visibilidad 77–24
buena ~ 77–25
~ de una ocultación 81–103
~ del eclipse 81–88
mala ~ 77–26
visión extrafoveal 86–128
~ parafoveal 86–128
~ periférica 86–128
visual 91–28
en la ~ 91–29
en la dirección de la ~ 91–29
Volans 92–100
volátil 82–262
volcán 73–111
~ activo 73–114
~ apagado 73–115
~ de capas 73–121
~ de lava 73–119
~ durmiente 73–116
~ en actividad 73–114
~ extinguido 73–115
~ extinto 73–115
~ Hawaiano 73–119
~ inactivo 73–116
~ submarino 73–117
~ suboceánico 73–117
~ vivo 73–114
volcanismo a lo largo de fracturas 82–157
~ inducido 82–155
~ inducido por meteoritos 82–156
Vulcano 83–123
Vulpécula 92–101
Vulpeja 92–101

Z

zanja de potencial 98–176
zócalo continental 72–34
zodíaco 92–5a
zona auroral 79–123
~ central 81–114
~ común 93–141
~ convectiva 93–149
~ convectiva de hidrógeno 93–159
~ de ablación 72–123
~ de actividad 89–54
~ de alimentación 72–122
~ de audibilidad 76–41, 86–120
~ de ausencia 97–88, 98–56
~ de calmas ecuatoriales 77–101
~ de combustión de helio 93–145
~ de combustión de hidrógeno 93–144
~ de compresión 86–170
~ de crecimiento 72–122
~ de disipación 72–124
~ de helio 93–145
~ de hidrógeno 93–144
~ de manchas 89–54
~ de máxima frecuencia auroral 79–125
~ de ocultación 81–107
~ de radiación 75–101
~ de radiación atrapada 75–101
~ de radiación natural 75–107
~ de silencio 76–42
~ de totalidad 88–29
~ de transición 72–13
~ de visibilidad 81–89
~ de visibilidad del eclipse 81–89
~ del borde 81–113
~ del limbo 81–113
~ ecuatorial 83–188
~ exterior de radiación 75–105
~ interior de radiación 75–106
~ marginal 81–113
~ orogénica 73–75a
~ polar 77–193
~ radiativa 93–134
~ real 89–54
~ semiconvectiva 93–160
~ sísmica 73–31
~ templada 77–192
~ tórrida 77–191
zonas 83–187
~ de Fresnel 86–294
Zorra 92–101

ES

ÍNDICE PORTUGUÊS

Portuguese Index

Index portugais

Portugiesischer Index

Índice portugués

Португальский указатель

ALFABETO:

a	h	o	v
b	i	p	w
c	j	q	x
d	k	r	y
e	l	s	z
f	m	t	
g	n	u	

A

abaixamento da temperatura 77–41
abalos da crosta 95–152
~ da estrela 95–153
aberração diurna 86–57
abissal 72–41
ablação 86–182
absorção auroral 78–46
~ brusca de ruído cósmico 78–42
~ da calota polar 78–44
~ de começo brusco 78–47
~ de início súbito 78–47
~ de linha 93–108
~ de raia 93–108
~ devido às auroras 78–46
~ diferencial 97–90
~ do vapor de água 77–127
~ inter(e)stelar 97–84
~ inter(e)stelar geral 97–87
~ inter(e)stelar seletiva 97–90
~ inter(e)stelar total 97–87a
~ ionosférica polar 78–44
~ lenta de nêutrons 94–88
~ ligada-ligada 93–108
~ ligada-livre 93–107
~ livre-livre 93–106
~ por vapor de água 77–127
~ rápida de nêutrons 94–89
~ súbita do ruído cósmico 78–42
abundância anômala 91–173
~ cósmica dos elementos 94–107
~ de isótopos 94–106
~ de um elemento 91–169
~ isotópica 91–179
~ metálica 91–170
~ natural 91–171
~ normal 91–172
~ química 91–169
~ total 91–177
abundâncias elementares 82–141
~ relativas dos elementos 91–174
ação da atmosfera 87–116
~ de degasar 82–261
aceleração da gravidade 71–64
~ efetiva 85–88
~ equatorial 88–131
acetaldeído 97–42
acetileno de metilo 97–41
acetonitrilo 97–38
achado de um meteorito 87–44
achatamento 83–113
~ aparente 98–295
~ da coroa 88–247
~ da Terra 71–8
~ de um cúmulo globular 96–123
~ polar 83–114
~ por perspectiva 89–92
~ real 98–296
ácido aminado 87–70
~ α-aminopropiônico 87–72
~ cianídrico 97–29
~ fórmico 97–33
~ isociánico 97–39
~ orgânico 87–69
~ sulfídrico 97–44
acima do limbo 89–30
acondrito 87–77
acontecimento 99–125
~ adjacente 99–126
acontecimentos simultâneos 99–128
acoplamento da radiação 99–173
~ por radiação 99–173
acrescimento 83a–33
acréscimo = acresção 83a–33
teoria de ~ 83a–20
acumulação 83a–34
~ de partículas 75–123
acústica atmosférica 76–31
adesivo 82–137
adsorção 85–128
adsorvato 85–131
adsorvente 85–130
adsorver 85–129
advecção 77–63
aerólito 87–60
aerologia 76–122
aeronomia 76–122a
aerossol 76–89
aerotriangulação espacial 71–26
afastamentos do ET 93–36
~ do ETL 93–40
afélio 83–28a
afloramento 72–88
afluxo de gás 98–356
~ del gás do espaço intergaláctico 98–356
afundamento 82–199
agitação do mar 72–86
~ microssísmica 73–54
aglomeração 83a–32
~ da matéria 99–170z
algomerado = cúmulo = enxame 96–79
~ aberto de galáxias 98–333
~ compacto de galáxias 98–334
~ de galáxias 98–329
~ de grande concentração 96–129
~ de Virgem 98–345
~ duplo 96–81
~ estelar 96–79
~ globular 96–114
~ globular de galáxias 98–330
~ irregular 98–337
~ móvel 96–97
~ muito compacto de galáxias 98–335
~ pobre de galáxias 98–331
~ pobre de estrelas 96–102
~ pobre em metais 96–113
~ regular de galáxias 98–336
~ rico de galáxias 98–332
~ rico em metais 96–112
aglutinação 74–52
agravamento do tempo 77–23
agregado molecular 97–65
água 97–26
~ rasa 72–96
Águia 92–18
agulhas de gêlo 76–112
ajustamento isostático 73–100
alameda 89–274
com forma de ~ 89–274
alanina 87–72
alargamento da sombra da Terra 81–93
albedo das nuvens 83–95
~ de Bond 83–91
~ de Lambert 83–89
~ de radiação cósmica 75–116
~ de radiofreqüências 83–94
~ esférico 83–91
~ geométrico 83–92
~ médio 83–90a
~ monocrómatico 83–93
~ neutrônico 75–116
~ total 83–90b
Albíreo 92–103
albita 74–17
alcance 86–288
~ do radar 86–288
~ mínimo 78–38
álcool metílico 97–32
Alcor 92–104
aldeído acético 97–42
~ fórmico 97–27
além da órbita terrestre 84–6
Alfeca 92–117
Algol 92–105
Algonquiano 74–77
alinhamento de grãos 97–78
alísios 77–100
alta 77–84
Altair 92–106
altamente ionizado 89–289
Altar 92–19
alteração pelo tempo 74–57
altimetria por satélites 71–50
altiplano 81–189
altitude de pressão 76–16
~ de uma camada 78–27
~ do ponto espelhento 75–125
~ dos meteoros 86–193
~ real 78–29
~ sobre o elipsóide de referência 71–52
~ sobre o nível do mar 71–51
~ virtual 78–28
alto-cúmulo 77–144
alto-estrato 77–140
altura da atmosfera homogênea 79–26
~ da camada emissora 79–26
~ da cromosfera 88–202
~ da sombra 79–46
~ de aparição 86–39
~ de abrigo 79–48
~ de blindagem 79–48
~ de desaparição 86–40
~ de ecala 88–236

P

~ de extinção 86–40
~ de referência 76–18
~ de uma camada 78–27
~ do ápex 86–66
~ do ponto de aparição 86–39
~ do ponto de desaparição 86–40
~ do ponto de extinção 86–40
~ dos meteoros 86–193
~ equivalente 78–28
~ geométrica da sombra 79–47
~ real 78–29
~ verdadeira 78–29
~ verdadeira da sombra 79–49
~ virtual 78–28
alumínio alveolado 82–25
~ celular 82–25
~ em forma de favos de colmeira 82–25
alunagem 82–13
alunissagem 82–13
álveo 83–153
amanhecer 79–35
amarelo 91–98
ambiente 76–3, 76–7
amino 87–71
aminoácido 87–70
amoníaco 97–25
amortecedor 82–24
~ de pouso 82–23
amostra de material lunar 82–69
~ de poeira lunar 82–121
~ do ar 76–86
~ do núcleo 82–60
~ lunar 82–69
~ lunar de retorno 82–70
~ lunar devolvida 82–70
titar ~ 76–85
amostrador de água 72–101
amostragem 76–85
~ do ar 76–85
~ do solo 82–78
amplificação do campo magnético 89–156
amplitude 95–38
~ da maré 82–292
~ da variação de brilho 95–38
anã 91–135
~ branca 91–137
estrela ~ 91–135
~ marrom 91–138a
~ negra 91–139
nova ~ 95–112
~ parda 91–138a
~ preta 91–139
~ vermelha 91–136
anagaláctico 98–207
análise de microsonda 82–143
~ local 82–136
~ mineralógica 82–53
~ petrográfica 82–52
~ química 82–75
Andromeda 92–14
Andrômeda 92–14
Andromédidas 86–256
Andromedídios 86–256
anéis de Saturno 83–215
anel A 83–225
~ asteroidal 84–70
~ B 83–222
~ brilhante 89–101
~ C 83–221
~ central 83–222
~ de Bishop 79–88
~ de crepe 83–221
~ de estrelas 96–78
~ de Saturno 83–215
~ dos asteróides 84–70
~ elementar 81–209
~ estelar 96–78
~ exterior 83–225
~ fantasma 81–242
~ gasoso 95–66
~ interior 83–221
~ lunar 81–265
~ médio 83–222
~ nebular 83a–10
~ transparente 83–221
~ visto de perfil 83–218
anelar 97–146
anemógrafo 77–114
anemômetro 77–113
~ registrador 77–114
anfibólio 74–20
ângulo característico 98–33
~ de fase 83–72
~ de posição 81–71
~ heliocêntrico 89–37
anidrido carbônico sólido 83–144
aniquilação de campos magnéticos 89–158
Ano do Sol calmo 89–10
~ do Sol quieto 89–10
ano-luz 91–18
anomalia brusca de fase 78–52
~ da gravidade 71–75
~ do Atlântico Sul 75–68
~ gravimétrica 71–75
~ gravitacional lunar 81–144
~ magnética 75–67
~ súbita de fase 78–52
anomalias de abundância 91–183
anortita 74–18
anortosita 82–87
anos-luz 91–18
Antares 92–107
antélio 79–83
antena cruzada Mills 88–121
~ de emissão 78–16
~ de recepção 78–18
~ emissora 78–16
~ receptora 78–18
~ transmissora 78–16
antiápex 86–65
~ solar 98–65
anticauda 85–78
anticentro galáctico 98–14
anticiclone 77–84
anticiclônico 77–85
anticlinal 73–90
anticrepúsculo 79–57
antigo satélite 83–240
anti-helio 79–83
anti-selene 79–85
antisselênio 79–85
antivértice 98–106
Antlia 92–15
anular 97–146
aparecimento de um bólido 86–117
~ de um cometa 85–18
~ de um meteoro 86–5
apáreon 83–141
aparição de um cometa 85–18
~ de um meteoro 86–35
ápex 86–64
~ do Sol 91–23
~ solar 91–23
apoapside galáctica 98–190
apoastro 96–61
apocentro galáctico 98–190
apocíntio 81–41
apofoco galáctico 98–190
apogaláctico 98–190
apogeu 81–39
apojove 83–173
Apolo 84–81
apolúnio 81–41
apossatúrnio 83–211
apossaturno 83–211
aposselênio 81–41
aprisionamento 75–109
aprofundamento de uma depressão 77–82
aproximação de Chandrasekhar 93–51
~ de Milne-Eddington 93–50
~ máxima 84–69
~ não-relativista 99–97
apulso 81–101
Apus 92–16
Aquáridas 86–249
Aquarídeos 86–249
Aquário 92–17
Aquarius 92–17
aquecimento atmosférico 75–87
~ da coroa 88–213a
~ da cromosfera 88–213
~ da partícula meteórica 86–174
~ do corpo meteórico 86–174
Aquernar 92–102
Aquila 92–18
ar 76–6
~ ambiente 77–4
Ara 92–19
arco auroral 79–146a
~ circum-horizontal 79–82
~ circunzenital 79–74
~ com estrutura raiada 79–153
~ coronal 89–298

P

~ crepuscular 79–50
~ de meridiano 71–21
~ eruptivo 89–251
~ homogêneo calmo 79–143
~ interplanetário 84–21
~ magnético 84–19
~ montanhoso 73–84
~ primário 73–82
~ pulsante 79–147
~ secundário 73–83
~ solar interplanetário 84–20
~ tangencial do halo 79–75
~ tangencial inferior 79–76
~ tangencial infralateral 79–87
~ tangencial superior 79–77
arco-íris 79–65
arcos intertropicais 79–53
área aparente das manchas 89–91
~ com manchas 89–90
~ corrigida de perspectiva 89–93
~ da zona de silêncio 78–38
~ das manchas 89–67
~ das protuberâncias 89–233
~ de alta pressão 77–84
~ de audibilidade 86–120
~ de queda 87–38
~ de radiação 86–238
~ de uma erupção 89–207
~ de visibilidade 86–126
~ do impacto 87–38
~ do impacto da chuva mereorítica 87–38
~ intergranular 88–177
~ manchada 89–90
~ reduzida 89–93
~ total (das manchas) 89–94
áreas de Kapteyn 98–78
~ de plagas 89–164
~ de praias 89–164
~ diárias das protuberâncias 89–236
~ escolhidas 98–78
~ selecionadas 98–78
areo- 83–135
areocêntrico 83–138
areofísica 83–134
areográfico 83–137
argila pelágica 72–54a
~ vermelha 72–54
Ariel 83–235
Aries 92–20
armadilha de luz 88–64
armazenamento de partículas 75–123
Arqueiro 92–85
Arqueozóico 74–76
arrasto (atmosférico) 86–178
arrefecimento 83a–57
~ neutrínico 94–97
~ por neutrinos 94–97
~ por radiação 94–98
~ radiativo 94–98
árvores 89–273
com forma de ~ 89–273
ascensão para o máximo 95–28
~ reta da Lua 81–60
aspe(c)to dos planetas 83–50
~ lunar 81–148
~ topográfico 81–149
~ vulcanico 82–159
aspectos lunares 81–148
aspereza de superfície 83–99
~ superficial 83–99
ASRC 78–42
assimetria este-oeste 89–49
~ norte-sul 89–50
assintótico às órbitas circulares 98–180
assoalho oceânico 72–37
assobiadores 78–53
assobio curto 78–58
~ longo 78–57
assobios 78–53
associação acidental 97–126
~ estelar 96–88
~ genética 97–127
~ O 96–89
~ T 96–90
associado à erupção 89–188
~ à fulguração 89–188
asterismo 92–3a
asteróide 84–68
~ binário 84–79
~ catalogado 84–95
~ cruzante à órbita da Terra 84–76
~ cruzante à órbita de Marte 84–75
~ da família de Marte 84–74
~ de grande aproximação 84–78
~ de grande aproximação à Terra 84–78
~ de Marte 84–74
~ do anel 84–71
~ do cinturão 84–71
~ em forma de charuto 84–77
~ em libração 84–91a
~ marciano 84–74
~ numerado 84–95
~ perdido 84–97
~ rasante 84–78
~ rasante à Terra 84–78
~ tipo Amor 84–80
~ tipo Apolo 84–80a
asteróides troianos 84–91
astro 79–9a
astronomia de base lunar 82–4
~ com base lunar 82–4
~ estelar 91–2
~ extragaláctica 98–208
~ meteórica 86–3
~ meteorítica 87–6
ataxita = ataxito 87–87
~ pobre de níquel 87–88
~ pobre em níquel 87–88
~ rica de níquel 87–89
Atelier do Escultor 92–87
atenuador escalonado 88–80
ativação de um filamento 89–215
~ de uma protuberância 89–239
~ de uma protuberância quiescente 89–239
atividade auroral 79–100
~ cometária 85–110
~ corpuscular 89–7
~ das manchas 89–62
~ das manchas solares 89–62
~ de erupções 95–71
~ de fulgurações 89–176
~ de uma corrente 86–229
~ do núcleo galáctico 98–271
~ dos meteoros esporádicos 86–136
~ em erupcões 89–176
~ em fulgurações 89–176
~ em ondas radioelé(c)tricas 89–6
~ eruptiva 89–176
~ fulgurante 89–176, 95–71
~ fumarólica 82–164
~ galáctica 98–261
~ geomagnética 75–86
~ interna 82–158
~ lunar 82–149
~ magnética 89–5
~ meteórica 86–134
~ ó(p)tica 89–3
~ rádio 89–6
~ recorrente 98–267
~ relativa da corrente 86–231
~ sísmica 73–4
~ solar 89–1
~ superficial 82–231
~ vulcânica 73–110
atlas celeste 92–12
~ da Lua 81–136
~ do céu 92–12
~ fotométrico 88–138
~ lunar 81–136
atmosfera 83–241
~ adiabática 76–12
alta ~ 76–52
baixa ~ 76–50
~ cinza 93–26
~ da Terra 76–5
~ de camadas plano-paralelas 93–21
~ de difusão 93–16
~ de difusão pura 93–16
~ de elé(c)trons 93–28
~ de um planeta 83–243
~ difusora 93–16
~ do Sol 88–163
~ ele(c)trônica 93–28
~ esférica 93–24
~ estelar 93–2
~ extensa 93–9
~ exterior 76–68
~ finita 93–23
~ homogênea 76–10
~ inicial 83a–44
~ isotérmica 77–39
~ livre 76–13
~ média 76–51

P

~ não-cinza 93–27
~ oxidante 83a–47
~ padrão 76–11
~ planetária 83–243
~ plano-paralela 93–21
~ primitiva 83–244
~ redutora 83a–46
~ semi-infinita 93–22
~ solar 88–163
~ solar exterior 88–165
~ solar inferior 88–166
~ standard 76–11
~ superior 76–52
~ terrestre 76–5
atmosférico 83–242
atmosféricos 78–50
átomo de dois níveis 93–45
~ de níveis múltiplos 93–46
~ primitivo 99–169
atração zenital 86–52
atraso da rotação da Terra 82–313
atrito 86–173
~ das marés 82–312
aumento 76–54
~ brusco de brilho 95–69
~ brusco dos atmosféricos 78–51
~ da massa 83a–38
~ da massa da Terra 87–18
~ da temperatura 77–42
~ de brilho 86–79
~ de brilho de uma plaga 89–213
~ de luminosidade 86–79
~ relativista da massa 99–142
~ relativístico da massa 99–142
~ súbito dos atmosféricos 78–51
auréola 79–60
~ lunar 79–61
~ solar 79–62
Auriga 92–21
aurora 79–35, 79–56
~ artificial 79–99
~ austral 79–106
~ boreal 79–105
~ com estrutura de fita 79–140
~ de baixa latitude 79–96
~ de calota polar 79–98
~ de elé(c)trons 79–110
~ de prótons 79–108
~ diurna 79–97
~ flamejante 79–157
~ iluminada pelo Sol 79–130
~ permanente 79–5
~ polar 79–93
~ radar 79–95
~ rádio 79–94
Austrálidas 86–250
australita = australito 87–98
auto-gravitação 94–34
auto-gravitante 94–33
auto-inversão de magnetização 75–38
autoluminoso 96–45
autor da descoberta 85–21
auto-reversão da magnetização 75–38
avanço do periastro 96–71
~ relativista do periélio 99–160
Ave-do-paraíso 92–16
avermelhamento inter(e)stelar 97–91
Azóico 74–75
azul 91–99
~ do céu 77–60

B

bacia 81–175
baía 81–294
~ do Arco-Iris 81–295
~ dos Ardores 81–296
~ dos Calores 81–296
~ magnética 75–80
~ Tórrida 81–296
baixa atmosfera 76–50
~ cromosfera 88–193
baixa-mar 72–91, 82–283
baixio 72–96
Balança 92–62
balanceamento detalhado 93–41
~ pormenorizado 93–41
balanço de calor 77–62
~ de calor da atmosfera 77–62
~ detalhado 93–41
~ pormenorizado 93–41
~ térmico 77–62
Baleia 92–34
banda 83–190
~ da luz zodiacal 84–63
~ de poeira 98–289
~ difusa inter(e)stelar 97–100
~ equatorial 83–191
~ luminosa 86–27
~ zodiacal 84–63
bandas de fluorescência 85–132
~ de Swan 85–133
~ de Vegard-Kaplan 79–19
~ escuras 98–288
~ não-identificadas 85–134
~ negativas do nitrogênio 79–21
banqueta submarina 72–34
barisfera 72–17
barra galáctica 98–254
barranco 83–154
barreira do som 76–44
~ sonora 76–44
basalto 74–9
~ lunar 82–84
basamento 72–12
base de uma protuberância 89–234
~ do interferômetro 88–100
~ lunar 82–17
~ lunar habitada 82–18
~ lunar tripulada 82–18
batimetria 72–28
batólito 73–139
Betelgeuse 92–108
Bielídios 86–256
bifurcação 98–5
big-bang 99–167
bigodes 89–200
bigue-bangue 99–167
binária 96–3, 96–7
~ afastada 96–9
~ astrométrica 96–12
~ cerrada 96–23
~ com raias duplas 96–17
~ com raias simples 96–15
~ de contacto 96–29
~ desligada 96–35
~ destacada 96–35
~ eclipsante 95–74
~ eclipsante tipo Algol 95–77
~ em conta(c)to 96–29
~ espe(c)troscópica 96–14
~ espe(c)troscópica com raias duplas 96–17
~ espe(c)troscópica com raias simples 96–15
~ estreita 96–23
~ fotométrica 95–73
~ não-resolúvel 96–10
~ não-resolvida 96–10
~ semidesligada 96–34
~ semi-separada 96–34
~ separada 96–35
~ visual 96–8
binário de maré 82–288
biosfera 72–8
biotita 74–25
blindagem 86–171
~ magnética 89–225
bloco anguloso 82–131
~ continental 72–10
~ de rocha 82–130
~ errático 72–132
boa visibilidade 77–25
bóia automática 72–100
~ de pesquisa 72–99
Boiadeiro 92–22
Boieiro 92–22
bojo devido à maré 83a–67
~ equatorial 71–9
~ galáctico 98–16
~ lunar 81–118
bola de fogo 77–163
~ de fogo primordial 99–168
~ de poeira 86–190
boletim do tempo 77–10
~ meteorológico 77–10
bolhas de lava 82–178
bólido 86–11
~ brilhante 86–10
~ detonante 86–119
~ gigante 86–9
~ grande 86–10
~ muito grande 86–9
~ produzindo efeitos sonoros 86–119
~ sonoro 86–119
~ visível de dia 86–118

P

Bomba de Ar 92–15
~ de Ellerman 89–200
~ Pneumática 92–15
~ vulcânica 73–144
bombardeamento de partículas 82–240
~ meteorítico 82–235
~ meteoroidal 82–235
~ por partículas 82–240
bombardeio meteorítico 82–235
~ meteoroidal 82–235
bombeamento colisional 97–20
~ infravermelho 97–18
~ ó(p)tico 97–17
~ químico 97–19
~ radiativo 97–17
~ ultravioleta 97–18a
Boötes 92–22
borda com blocos de pedra 81–224
~ da cratera 81–221
~ de uma galáxia 98–294
~ levantada 81–222
bordo brilhante de um filamento 89–226
~ claro 81–96
~ com blocos 81–224
~ da cratera 81–221
~ da Lua 81–95
~ do leste 89–34
~ do oeste 89–33
~ escuro 81–97
~ exterior 83–227
~ extremo 89–32
~ leste 89–34
~ levantado 81–222
~ lunar 81–95
~ oeste 89–33
braço de Orion 98–36
~ de Perseu 98–34
~ de Sagitário 98–35
~ de três kiloparsecs 98–37
~ de 3 kpc 98–37
~ desenrolado 98–31
~ desenroscado 98–31
~ em expansão 98–38
~ enrolado 98–29
~ enroscado 98–29
~ espiral 98–27
~ galáctico 98–26a
brecha lunar 82–85
brilhância absoluta 85–109
~ de uma fulguração 89–212
~ global 85–107
~ intrínseca 85–107
~ reduzida 85–108
~ total 85–107
brilho 91–54
~ absoluto 85–109
~ absoluto de um meteoro 86–74
~ anti-solar 84–64
~ aparente 91–61
~ aparente de um meteoro 86–73
~ atmosférico 79–2
~ atmosférico crepuscular 79–4
~ atmosférico diurno 79–3
~ bolométrico 91–68
~ de um meteoro 86–8
~ de uma erupção 89–212
~ ele(c)trográfico 91–61
~ estelar 91–55
~ fotográfico 91–66
~ fotovisual 91–69
~ global 85–107
~ infravermelho 91–71
~ integral 96–135
~ limite 91–62
~ máximo 86–78
~ radiométrico 91–70
~ reduzido 85–108
~ terreal 81–35
~ terreal na Lua 81–35
~ total 85–107
~ visual 91–65
brisa marítima 77–97
~ terrestre 77–98
brugalhau 81–237
bruma 83–250
~ com polu(i)ção 76–83
~ seca 77–27
buraco branco 98–24
~ coronal 88–245a
~ do impacto 87–58
~ negro 94–114
~ negro galáctico 98–23a
~ negro supermassivo 98–22a
~ no pico central 81–244
burgalhau 81–237
Buril 92–23
busca de cometas 85–15
~ de meteoritos 87–45
~ de satélites 83–27
~ de supernovas 95–136
buscador solar 88–51
Bússola 92–82

C

Cabeça da Serpente 92–90
~ de Cavalo 97–135
~ de um cometa 85–35
~ e Cauda 92–90
~ e Cauda da Serpente 92–90
cabeleira 85–64
~ de Berenice 92–38
cabo 81–293
Cabra 92–29, 92–110
Caçador 92–73
Grande ~ 92–73
cadeia de crateras 81–241
~ de estrelas 96–77
~ de galáxias 98–328
~ de montanhas 81–194
~ de supernovas 95–132
~ estelar 96–77
Caelum 92–23
Cães Caçadores 92–26
~ de Caça 92–26
calculista 85–22
calderia 81–179
~ de colapso 81–181
~ de erosão 81–183
~ de explosão 81–180
~ de subsidência 81–182
calhau 82–128
Calisto 83–208
calmaria 77–110
calmarias equatoriais 77–101
~ tropicais 77–102
calor da maré 82–287
~ de ablação 86–184
~ de condensação 71–155
~ de dessorção 85–124
~ de sublimação 85–139
~ de vaporização 86–199
~ de devido à radioatividade 83a–62
~ fóssil 74–63
~ latente 77–157
~ original 74–63
~ primordial 83–200
~ radiogênico 83a–62
calota auroral 79–127
~ de compressão 86–170
~ glacial 72–115
~ polar 83–143
camacita 87–92
camada absorvente 88–175
~ atmosférica 76–49
~ concordante 73–135
~ convectiva 93–149
~ de bruma 83–249
~ de combustão de hélio 93–145
~ de combustão de hidrogênio 93–144
~ de detritos fragmentários 82–104
~ de espalhamento 83–258
~ de inversão 88–174
~ de lava 82–175
~ de nuvens 77–150, 83–248
~ de ozone 76–123
~ de ozônio 76–123
~ de plasma 75–11
~ de poeira 82–122
~ de transição 72–13, 88–224
~ delgada 83–149
~ difusora 83–258
~ E 78–26
~ E esporádica 78–25
~ emissora 79–17
~ fotoele(c)trônica 82–246
~ galáctica absorvente 97–89
~ intermediária 72–13
~ Heaviside-Kennelly 78–26
~ intrusiva 73–135
~ ionizada 78–3
~ ionosférica 78–5
~ luminescente do ar 79–17
~ luminosa 79–17
~ neutra 75–16

P

~ radiativa 93–134
~ semiconvectiva 93–160
~ subfotosférica 88–168
~ violeta 83–165
camadas exteriores de uma estrela 93–3
~ exteriores não-degeneradas 93–199
~ mais baixas da atmosfera solar 88–166
~ mais elevadas da atmosfera solar 88–165
~ ocidentais 79–52
~ subpolares 79–51
Camaleão 92–35
câmara de patrulhamento 86–145
~ de vigilância 86–145
~ fotográfica fixa 86–143
~ fotográfica guiada 86–142
~ fotográfica para todo o céu 79–104
~ meteórica 86–141
~ móvel 86–142
~ não-guiada 86–143
~ não-móvel 86–143
Cambriano 74–78
Camelopardalis 92–24
câmera = câmara 86–145
campo coerci(ti)vo 75–39
~ das correntes teluricas 75–4
~ de conjunto 89–142
~ de crateras 87–56
~ de forças de maré 98–196
~ de fundo 89–141
~ de grande escala 89–142
~ de gravidade 71–57
~ de Higgs 99–1701
~ de lava 82–184
~ de maré galáctico 96–106
~ de nevado 72–119
~ de pedras 82–139
~ de pequena escala 89–143
~ de radiação 93–30
~ de radiação difusa 93–31
~ de rochas 82–139
~ de som 76–40
~ de velocidades 88–187
~ dos movimentos diferenciais 98–114
~ efetivo 95–59
~ ele(c)trotelúrico 75–4
~ galáctico 98–146
~ geomagnético 75–44
~ gravífico 98–145
~ gravitacional 98–145
~ gravitacional galáctico 98–146
~ iman(iz)ante 75–42
~ interno 95–63
~ longitudinal 89–138
~ magnético da Terra 75–44
~ magnético de fundo 89–141
~ magnético de pequena escala 89–143
~ magnético fóssil 91–53
~ magnético fotosférico 89–130
~ magnético galáctico 98–50
~ magnético geral 89–128
~ magnético geral do Sol 89–128
~ magnético interplanetário 84–9
~ magnético inter(e)stelar 97–106
~ magnético local 89–143
~ magnético lunar 82–45
~ magnético muito intenso 95–148
~ magnético primitivo 95–62
~ magnético remanescente 95–62
~ magnético superintenso 95–148
~ magnético terrestre 75–44
~ magnetizante 75–42
~ quântico 99–170m
~ selenomagnético 82–45
~ sonoro 76–40
~ superficial 95–64
~ transversal 89–137
campos magnéticos solares 89–129
canais de Marte 83–157
canal de fluxo 82–192
Cancer 92–25
Canes Venatici 92–26
canhão 83–154
~ submarino 72–46
cânion submarino 72–46
Canis Major 92–27
~ Minor 92–28
Canopéia 92–109
Canopo 92–109
Canopus 92–109
Cão Maior 92–27
~ Menor 92–28
capa de gelo 72–115
capacidade de suporte 82–64
Capella 92–110
Capricórnio 92–29
Capricornus 92–29
captura 75–109
~ da matéria intergaláctica 98–355
~ de cometas 85–101
~ de estrela 96–73
~ de nêutrons 94–84
~ de prótons 94–90
~ neutrônica 94–84
característica ativa 89–14
~ da superfície 83–105
~ de impacto 82–228
~ do solo 82–108
~ lunar 81–148
~ ó(p)tica 98–234
~ permanente 83–106
~ rádio 98–235
~ solar 89–13
características estruturais 98–224
Caranguejo 92–25
carbonato 74–27
Carbonífero 74–82
carbono radiativo 74–95
Carena 92–30
carga da Terra 75–3
Carina 92–30
Carneiro 92–20
carta batimétrica 71–54
~ celeste 92–9
~ celeste giratória 92–10
~ colorimétrica 81–134
~ da variável 95–15
~ das velocidades 89–23
~ de alta resolução 89–133
~ de campos magnéticos 89–131
~ de grande resolução 89–133
~ de marear 72–33
~ de previsão 77–21
~ diária do Sol 89–18
~ do céu 92–9
~ do eclipse 88–28
~ do tempo 77–12
~ física lunar 81–134
~ fotométrica 81–134
~ gravimétrica 71–79
~ heliográfica 89–17
~ hidrográfica 72–25
~ hipsométrica 81–141
~ lunar 81–133
~ magnética 75–63, 89–131
~ náutica 72–33
~ oceánográfica 72–23
~ polar 89–19
~ polarimétrica 81–134
~ sinó(p)tica 77–11
~ sinó(p)tica da coroa 89–22
~ sinó(p)tica da cromosfera 89–21
~ sinó(p)tica da fotosfera 89–20
~ tectônica 73–96
~ térmica lunar 81–135
cartografia do litoral 72–82
~ lunar 81–111
Cassiopéia 92–31
Castor 92–111
catabático 77–94
catálogo das estrelas suspeitas de variabilidade 95–58
~ das variáveis suspeitas 95–58
~ de estrelas 92–13
~ de Messier 92–12a
~ estelar 92–13
catástrofe pela queda de um meteorito 87–20
cata-vento 77–104
cauda anômala 85–75
~ cometária 85–70
~ curvada 85–73
~ da magnetosfera 75–10
~ da Serpente 92–89
~ da Terra 84–65
~ de fumaça 87–21
~ de fumo 87–21
~ de um meteoro 86–110
~ de pó 85–78
~ de poeira 87–22
~ de um cometa 85–70
~ dirigida para o Sol 85–78

~ em direção oposta ao Sol 85–81
~ galáctica 98–280
~ iônica 85–82
~ magnética da Terra 75–10
~ múltipla 85–76
~ principal 85–77
~ reta 85–72
~ retilínea 85–72
cauterização 87–102
Cavalete do Pintor 92–78
Cavalinho 92–48
cavamento de uma depressão 77–82
caverna de lava 82–177
cavidade 87–117
~ do vento solar 84–28
~ geomágnetica 75–8, 84–23
~ heliosférica 88–257
~ hidromagnética 84–22
~ interior 89–119
~ interplanetária 84–25
~ magnética 84–22
~ planetária 84–24
~ superficial 87–118
cefeida 95–92
~ clássica 95–94
~ de curto período 95–93
~ de longo período 95–94
Cefeu 92–33
celóstato 88–46
~ horizontal 88–47
~ vertical 88–48
cementação 74–52
Cenozóico 74–73
Centauro 92–32
Centaurus 92–32
centro ativo 89–60
~ da chapa fotográfica 86–156
~ da estrela 93–84
~ da Via Láctea 98–9
~ de atividade 89–60
~ de atração 89–241
~ do disco 89–35
~ do disco aparente da Lua 81–128
~ do disco solar 89–35
~ do disco visível 88–172
~ do sistema galáctico 98–9
~ fotométrico 96–46
~ galáctico 98–9
Cepheus 92–33
Ceres 84–82
Cetus 92–34
céu austral 79–3c
~ boreal 79–3b
~ claro 77–153
~ crepuscular 79–43
~ diurno 79–3a
~ encoberto 77–152
~ estrelado 91–1a
~ limpo 77–153
~ noturno 79–6
~ noturno luminoso 86–28
~ nublado 77–148
chã 81–272
terra ~ 81–272
Chamaeleon 92–35
chaminé 73–123
~ vulcânica 73–122
chapa adesiva 87–14
~ delgada 83–149
~ fotográfica móvel 88–27
~ meteórica 86–151
chegada à Terra 89–193
chispas de um meteoro 86–84
choque térmico 86–206
~ triplo 79–30
choques premonitores 73–10
chumbo comum 74–99
~ original 74–98
~ primitivo 74–98
~ radiogênico 74–97
chuva 77–169
~ coronal 89–271
~ de estrelas cadentes 86–215
~ de meteoritos 87–37
~ de meteoros 86–215
~ meteorítica 87–37
~ orográfica 77–166
chuvisco 77–172
cianeto de metila 97–38
cianoacetileno 97–30
cianogênio 97–24
ciclo da atividade solar 89–71
~ das manchas 89–72
~ das manchas solares 89–72
~ de onze anos 89–73
~ geoquímico 74–32
~ solar 89–71
~ solar impar 89–78
~ solar par 89–77
~ undecenal 89–73
ciclone 77–76
~ tropical 77–77
ciência ambiental 76–2
~ dos meteoritos 87–6
ciências atmosféricas 76–1
~ da Terra 71–1
~ lunares 82–2
cilindro de luz 95–155
~ truncado 88–63
cintilação inter(e)stelar 97–109a
~ inter(e)stelar das radiofontes 97–109b
~ interplanetária das radiofontes 84–37
cinturão ativo 73–32
~ de asteróides 84–70
~ de Órion 92–129a
~ de radiação 75–101
~ de radiação artificial 75–108
~ de radiação interno 75–106
~ de radiação natural 75–107
~ de Van Allen 75–101
~ dos pequenos planetas 84–70
~ externo 75–105
~ interno 75–106
~ minauroral 79–129
~ orogênico 73–75
~ sísmico 73–30
~ subauroral 79–128
cinza vulcânica 73–142, 82–168
Cinzel 92–23
Circinus 92–36
circo lunar 81–265
circulação atmosférica 77–64
~ geral 77–65
~ meridional 93–163
~ planetária 77–66
círculo circunzenital 79–74
~ de iluminação 83–69
~ par(h)élico 79–82
circum-estelar 97–3
circum-lunar 81–5
circunlunar 81–5
circunstâncias da queda 87–47
~ do eclipse 81–73
circunstelar 97–3
cirro 77–136
cirro-cúmulo 77–143
cirro-estrato 77–141
cisão causada por maré 85–112
~ de maré 85–112
~ de um cometa 85–111
~ de um meteoro 86–82
cislunar 81–4
Cisne 92–44
clarão atmosférico 79–2
~ fraco 79–151
classe de luminosidade 91–146
~ de uma erupção 89–205
~ espe(c)tral 91–107
classificação das estrelas variáveis 95–46
~ de galáxias 98–226
~ de Harvard 91–91
~ de Zurich 89–127
~ dos meteoritos 87–59
~ espe(c)tral 91–90
~ espe(c)tral bidimensional 91–105
~ espe(c)tral de Harvard 91–91
~ espe(c)tral tridimensional 91–106
~ magnética 89–109
~ petrográfica 82–101
clatrato 85–51
clima 77–180
~ continental 77–185
~ de montanha 77–187
~ marítimo 77–186
climático 77–182
climatologia 77–179
~ por satélites 77–181
clorinidade 72–68
coagulação 76–119
coalescência 76–120
coberto de núvens 83–253
~ de poeira 82–120
cobertura 93–20
~ de neve 77–176

Cocheiro 92–21
coeficiente de arrasto 86–180
~ de brilhância 83–96
~ de brilho 88–96
~ de difusão atmosférica 86–287
~ de diluição 93–37
~ de estabilidade 98–199
~ de fase 83–77
~ de forma 86–177
~ de obscurecimento 88–144
~ de resistencia 86–180
~ de rugosidade 82–107
~ de transmissão de calor 86–201
coeficientes de Legendre 93–55
~ do elipsóide de velocidades 98–102
~ harmônicos 71–59
coerência 99–33
coesivo 82–138
coifa de compressão 86–170
colapso 94–109
~ do universo 99–181
~ final 99–182
grande ~ 99–182
~ gravitacional 94–109
colchete 75–79
coleção de amostras 82–72
~ de amostras de solo 82–78
~ de meteoritos 87–8
coleta de amostras de solo 82–78
~ de poeira meteórica 87–12
coletor de poeira meteórica 87–13
colisão catastrófica 84–106
~ central 83a–17
~ de estrelas 83a–16
~ de galáxias 98–322
~ de três corpos 96–74
~ erosiva 84–105
~ estelar 83a–16
~ frontal 83a–17
~ inelástica de estrelas 98–26
~ tripla 96–74
coloidal 76–97
colóide 76–96
Columba 92–37
coluna ionizada 86–282
~ luminosa 79–78
~ solar 79–78
coma 85–64
~ atômica 85–67
~ Berenices 92–38
~ de radicais 85–66
~ interior 85–65
~ molecular 85–65
~ ultravioleta 85–67
~ visível 85–66
combustão do hidrogênio 94–56
~ do oxigênio 94–82
~ do silício 94–83
começo brusco 75–74
~ da totalidade 81–77
~ do eclipse 81–74
~ do eclipse total 81–77
cometa 85–1
~ artificial 85–14
~ associado a um enxame 86–259
~ ativo 85–149
~ com órbita elí(p)tica 85–4
~ com órbita hiperbólica 85–8
~ com órbita parabólica 85–7
~ de curto período 85–5
~ de eclipse 85–12
~ de longo período 85–6
~ elíptico 85–4
~ elítico 85–4
~ extinto 85–150
~ gerador 86–260
~ hiperbólico 85–8
~ inativo 85–147
~ não-periódico 85–10
~ pai do enxame 86–260
~ parabólico 85–7
~ periódico 85–3
~ predito 85–11
~ previsto 85–11
~ rasante ao Sol 85–13
~ telescópico 85–14a
~ vivo 85–149
cometário 85–2
compactação 74–50
companheira 96–38
~ de Sírio 91–138
~ de uma galáxia 98–216
~ elíptica 98–217
~ escura 96–42
~ física 98–219
~ invisível 96–41
~ obscura 96–42
~ ó(p)tica 98–218
~ planetária 96–43
companheiro 96–38
~ de uma galáxia 98–216
~ elíptico 98–217
~ físico 98–219
~ físico de uma galáxia 98–219
~ ó(p)tico 98–218
~ ó(p)tico de uma galáxia 98–218
Compasso 92–36
compensação do efeito Doppler 88–90
~ isostática 73–100
compensador Doppler 88–91
completamente degenerado 93–185
complexo basal 72–12
componente 96–36
~ atmosférico 76–87, 83–265
~ de poeira 98–232
~ horizontal do campo magnético 75–60
~ longitudinal 89–135
~ lunissolar 82–307
~ meteórico da radiação do céu noturno 86–234
~ primário 96–37
~ principal 96–37
~ secundário 96–38
~ σ 88–89
~ transversal 89–136
~ vertical 75–62
composição iônica 78–10
~ isotópica 74–93
~ molecular 78–11
~ química 91–168
~ química de uma estrela 91–168
compressibilidade 73–57
comprimento angular 98–32
~ angular da trajetória 88–48
~ da onda Compton 99–170g
~ da trajetória de um meteoro 86–47
~ de onda de velocidade nula 98–365
~ de onda efetivo 96–142
~ de onda em repouso 98–365
~ de Planck 99–170h
~ elementar de Planck 99–170h
comunicador 85–23
comutação de fase 88–112
comutador de fase 88–114
concentração 96–127
~ de massa lunar 81–145
~ de traços 83–266
~ de vestígios 83–266
~ em direção ao centro 98–76
em direção ao plano fundamental 98–76
~ galáctica 98–76
~ para o plano galáctico 98–55
condensação 83a–26
~ central 85–37, 96–126
~ coronal 89–281
~ coronal esporádica 89–283
~ coronal permanente 89–282
~ de uma corrente meteórica 86–217
~ em uma corrente meteórica 86–217
~ na cauda 85–95
condensar(-se) 77–129
condição à superfície 93–80
~ de ajuste 93–142
~ de estabilidade 93–205
~ de simetria 98–162
~ na superfície 93–80
~ no centro 93–81
condições de equilíbrio 93–74
~ de temperatura 77–35
~ do tempo 77–2
~ meteorológicas 77—2
~ meteorológicas gerais 77–15
~ normais 77–3
condrito 87–61
~ carbonáceo 87–63
~ não-carbonáceo 87–75
côndrulo 87–76
conduto 73–124
~ vulcânico 82–162
cone de cinza 73–120
~ de ejeção 73–120
~ de erupção 73–118
~ de escapamento 76–27

~ de escape 76–27
~ de estratos 73–121
~ de explosão 73–120
~ de luz 95–156
~ de sombra 81–90
~ estraficado 73–121
~ luminoso 95–156
~ piroclástico 73–120
~ refletor 88–62
~ vulcânico 73–118
conexão acidental 97–126
~ genética 97–127
configuração do campo 89–140
~ estriada 89–174
~ planetária 83–50
confinamento magnético dos raios cósmicos 98–51
conglomerado 85–41
~ de gelo 85–43
conjugado da maré 82–288
conjunção 83–55
~ inferior 83–57
~ superior 83–56
~ tripla 83–58
consolidação do magma 82–190
constante cósmica 99–29
~ cosmológica 99–29
~ da recessão 98–302
~ de expansão 99–77
~ de Hubble 98–302
~ de Poisson 73–60
~ no tempo 98–150
~ solar 88–148
constantes de Oort 98–118
~ do elipsóide de velocidades 98–102
constelação 92–1
~ circumpolar 92–4a
~ de inverno 92–1b
~ de verão 92–1a
~ zodiacal 92–5
constituicão interna das estrelas 93–71
constituinte atmosférico 76–87
~ mineral de uma rocha 82–99
~ mineralógico 82–100
contagem de estrelas 98–79
~ de galáxias 98–343
contenedor de amostras 82–81
conteúdo de ozônio atmosférico 76–125
~ de pó 85–80
~ do cloro 72–68
~ do sal 72–66
~ em níquel 72–55
~ estelar 98–172
~ estelar de uma galáxia 98–231
continente 72–9, 81–191
contorno da costa 72–31
~ de nível 81–139
~ de isofotas 81–92
~ isofótico 81–92
~ isogáussico 89–134
contornos de igual contagem 75–102
~ de igual intensidade 75–102
~ isófotos 97–124
~ isófotos de uma nebulosa 97–124
contra-alísios 77–103
contrabrilho zodiacal 84–64
contração 93–117, 94–36
~ catastrófica 94–109
~ de comprimentos 99–134
~ de Lorentz 99–134
~ devida à gravitação 94–37
~ do núcleo 94–70
~ do universo 99–180
~ gravitacional 93–118
~ quase-estacionária 94–40
~ quase-estática 94–40
~ relativista 99–134
~ relativística 99–134
convecção 93–147
~ laminar 93–153
~ turbulenta 93–152
convectivo 93–148
conversor de fase 88–114
convexidade equatorial 71–9
coordenada do espaço 99–124
~ do tempo 99–123
~ espacial 99–124
~ tempo 99–123
~ temporal 99–123
coordenadas areocêntricas 83–138
~ cilíndricas 98–156
~ galactocêntricas 98–12
~ geomagnéticas 75–51
~ heliográficas 89–39
~ jovicêntricas 83–176
~ jovigráficas 83–175
~ planetográficas 83–8
~ selenográficas 81–9
coplanaridade das órbitas 83a–3
cor intrínseca 91–74
corda cósmica 99–171c
~ magnética 89–146
cordilheira 73–85
~ dobrada 81–187
cores do (Sol) nascente 79–56
~ do (Sol) poente 79–55
coro de alvorada 78–62
~ de aurora 78–62
coroa 79–156, 79–60
~ Austral 92–39
~ Boreal 92–40
~ branca 88–237
~ de elé(c)trons 88–242
~ de emissão 88–241a
~ de Fraunhofer 88–241
~ de hidrogênio 83–246
~ de máximo 88–244
~ de mínimo 88–246
~ de raios X 88–240
~ E 88–241a
~ em expansão 88–232
~ em luz branca 88–237
~ estelar 93–4
~ externa 88–231
~ F 88–241
~ galáctica 98–49
~ intermediária 88–245
~ interna 88–229
~ K 88–242
~ lunar 79–61
~ máxima 88–244
~ média 88–230
~ mínima 88–246
~ monocromática 88–238
~ planetária 83–245
~ polar 89–232
~ polar de filamentos 89–232
radio ~ 88–239
~ Roentgen 88–240
~ solar 79–62, 88–225
Corona Australis 92–39
~ Borealis 92–40
coronal 88–228
coronógrafo 88–59
~ de ocultação exterior 88–60
~ ocultado exteriormente 88–60
coronómetro 88–66
K ~ 88–66
corpo adsorvente 85–130
~ celeste 79–9a
~ de impacto 82–225
~ em rotação 91–42
~ girante de magnetoplasma 98–25
~ meteórico 86–160a
~ origem do meteorito 87–123
~ parente do meteorito 87–123
~ pré-estelar 94–15
~ produtor de maré 82–284
~ rígido 83–111
~ rotante 91–42
~ rotatório de magnetoplasma 98–25
~ sólido 83–110
~ supermassivo 98–22
corpuscular 88–155
corpúsculo 88–153
corpúsculos sólidos 85–48
correção bolométrica 91–80
~ de campo 86–157
~ de curvatura 86–62
~ de trajetória 82–10
~ de trajetória do vôo 82–10
~ para a curvatura 86–62
~ para a curvatura da superfície da Terra 86–62
~ relativista 99–140
~ relativística 99–140
corrediça 82–196
corrente anular equatorial 75–98
~ ascendente 77–91, 93–151
~ ativa 86–223
~ atmosférica 77–90
~ cometária 86–222
~ coronal 89–287
~ corpuscular 88–160
~ corpuscular solar 84–35
~ de asteróides 84–100

P

P

~ de convecção 73–76
~ de estrelas 98–98
~ de ja(c)to 77–88
~ de plasma solar 84–33
~ de refluxo 72–65
~ de suspensão 72–49
~ de turbidez 72–49
~ de vento solar 84–35a
~ descendete 93–150
~ diurna 86–219
~ eclíptica 86–221
~ esporádica 86–226
~ estelar 98–98
~ fria 72–64
~ gasosa 95–67
~ inter(e)stelar 86–224
~ marinha 72–60
~ marinha fria 72–64
~ marinha quente 72–63
~ meteórica 86–216
~ noturna 86–218
~ oceânica 72–60
~ periódica 86–228
~ permanente 86–227
~ planetária 86–220
~ quente 72–63
~ subsuperficial 72–61
~ superficial 72–62
~ temporária 86–225
correntes anulares 75–100
~ telúricas 75–5
~ telúricas vagabundas 75–6
corte pelo obturador 86–150
cortinas 79–154
Corvo 92–41
Corvus 92–41
cosmo 99–2
cosmocronologia 83a–72
cosmogonia 94–2
~ estelar 94–7
~ planetária 83a–1
cosmogônico 94–3
cosmogonista 94–4
cosmologia 99–16
~ cinemática 99–26
~ não-relativista 99–27
~ newtoniana 99–25
~ observacional 99–17
~ relativista 99–28
cosmológico 99–18
cosmologista 99–19
cosmólogo 99–19
cosmoquímica 74–31
cosmos 99–2
costa 72–30
cova 81–243
~ meteorítica 87–58
pequena ~ 81–244
pequena ~ de pico central 81–244
Crater 92–42
cratera 73–126
~ adventícia 73–128
~ alagada 81–203
~ anterior à mar 81–207
~ antiga 81–205
~ central 81–217
~ cheia 81–203
~ circular 81–213
~ com bordos formados por blocos 81–225
~ com bordos levantados 81–223
~ com halo escuro 81–215
~ com pico central 81–258a
~ com raias 81–214
~ concêntrica 81–216
~ de bordos com blocos 81–225
~ de drenagem 81–239
~ de explosão 81–238
~ de impacto 81–240
~ de raios 81–214
~ de taça 81–219
~ de vértice 73–129
~ em forma de prato 81–219
~ enterrada 81–204
~ externa 81–218
~ falsa 81–209
~ fantasma 81–209
~ inominada 81–202
~ inundada 81–203
~ irregular 81–220
~ jovem 81–206
~ lateral 73–128
~ lunar 81–197
~ maar 81–264
~ meteórica 81–263
~ meteorítica 81–263, 87–53
~ meteorítica fóssil 87–55
~ meteorítica terrestre 87–54
~ meteoroidal 81–263
~ nova 81–206
~ parasita 73–128
pequena ~ 81–250
~ poligonal 81–212
~ pós-mares 81–208
~ pós-marítima 81–208
~ possível 81–211
~ posterior à mar 81–208
~ precedente à mar 81–207
~ pré-mares 81–207
~ pré-marítima 81–207
~ primária 81–198
~ principal 73–127, 87–57
~ provável 81–210
~ raiada 81–214
~ satélite 81–200
~ secundária 73–128, 81–199
~ sem bordos 81–226
~ sem nome 81–202
~ tipo maar 81–264
craterização 82–232
craterleta 81–250
crepuscular 79–40
crepúsculo 79–34
~ astronômico 79–37
~ civil 79–38
~ matutino 79–35
~ náutico 79–39
~ vespertino 79–36
crescente de Vênus 83–129
~ lunar 81–23
crescimento da calota polar 83–145
~ de grãos 97–61
~ de um planeta 83a–37
~ para o máximo 95–28
~ planetário 83a–37
Cretáceo 74–86
criação contínua 99–91
~ da matéria 99–89
~ das estrelas 96–95
~ única 99–90
Criptozóico 74–70
crista 81–195
~ anticiclônica 77–87
~ de dobra 81–187
~ de dobramento 81–187
~ de mar 81–188
~ de médio-oceânica 77–45
~ de montanhas 81–195
~ montanhosa 81–195
cristal de gelo 76–111
cristalização 76–107
~ fracionada 74–40
cristobalita 82–95
critério de degeneração 93–181
~ de degenerescência 93–181
~ de instabilidade de maré 98–312
critérios de classificação 91–110
crochê magnético 75–79
cromosfera 88–191
alta ~ 88–192
cromosférico 88–194
cronologia geológica 74–67
crosta continental 72–5
~ da Lua 82–54
~ da Terra 72–4
~ de fusão 87–110
~ lunar 82–54
~ sólida 95–151
~ terrestre 72–4
Crux 92–43
cruz de halo 79–79
~ do Norte 92–44
~ do Sul 92–43
~ luminosa 79–79
~ lunar 79–80
~ solar 79–81
Cruzeiro 92–43
~ do Sul 92–43
culminação da Lua 82–303
cúmulo 77–138
~ aberto 96–99
~ aberto de galáxias 98–333
~ compacto de galáxias 98–334
~ de galáxias 98–329
~ de galáxias de Virgem 98–345
~ de grande concentração 96–129

~ de nebulosas 98–329
~ de Virgem 98–345
~ dos cúmulos de galáxias 98–346
~ duplo 98–81
~ em movimento 96–97
~ estelar 96–79
~ estelar rico 96–100
~ galáctico 96–99
~ globular 96–114
~ globular de galáxias 98–330
~ irregular de galáxias 98–337
~ móvel 96–97
~ muito compacto de galáxias 98–335
~ pobre de estrelas 96–102
~ pobre de galáxias 98–331
~ pobre em galáxias 98–331
~ pobre em metais 96–113
~ regular de galáxias 98–336
~ rico 96–100
~ rico de estrelas 96–100
~ rico de galáxias 98–332
~ rico em estrelas 96–100
~ rico em galáxias 98–332
cúmulo-nimbo 77–137
cunha de alta pressão 77–86
cúpula de lava 81–176a
~ vulcânica 81–176
curva batimétrica 71–55
~ de abundância 94–108
~ de avermelhamento 97–93
~ de brilho 95–17
~ de cor 95–21
~ de crescimento 93–68
~ de evolução 89–119
~ de evolução de um grupo 89–119
~ de extinção 97–81
~ de fase 83–74
~ de igual profundidade 71–55
~ de luz 95–17
~ de luz de um meteoro 86–70
~ de luz fotoelé(c)trica 95–22
~ de luz fotográfica 95–23
~ de luz média 95–19
~ de luz visual 95–24
~ de nível 71–53
~ de nível submergida 72–32
~ del nível submersa 72–32
~ de polarização 83–81
~ de rotação 98–109
~ de um meteoro 86–70
~ de velocidades radiais 95–20
~ empírica de crescimento 93–69
~ geodésica 99–153
~ hipsográfica 81–142
~ isoclinal 75–59
~ limiar do componente secundário 96–40
~ média de brilho 95–19
~ período-luminosidade 95–103
~ teórica de crescimento 93–70
~ visual de luz 95–24
curvatura da Superfície da Terra 86–61
curvatura de superfície da Terra 86–61
~ da Terra 71–6
~ da trajetória 86–51
~ das linhas de força 75–121
~ do espaço 99–115
~ terrestre 71–6
cúspide magnetosférica 75–13
cúspides lunares 81–24
Cygnus 92–44

D

dados solares 89–9
dano por ondas de maré 72–77
dar início 89–198
datação 83a–71
~ absoluta 83a–73
~ de eventos lunares 82–265
~ isotópica 87–130
~ por isótopos 83a–73
~ radi(o)ativa 83a–73
~ tório-chumbo 82–270
datando por meio da relação chumbo--tório 82–270
datum geodésico 71–20
decaimento de um grupo de manchas 89–126
decair 74–65
decibel 76–39
declinação do Sol 81–63
~ magnética 75–56
declínio 76–57
~ de temperatura 77–41
declive 81–230
~ abrupto 81–232
~ detrítico 81–233
~ moderato 81–231
decrescimento da calota polar 83–146
decréscimo de massa 83a–42
defasador 88–115
defeito de fase 81–28
deferente 83–48
deficiência em abundância 91–187
deflexão da luz pelo campo de gravidade 99–161
~ de uma estrela 98–123
~ relativista da luz 99–161
~ do movimento de uma estrela 98–123
defluidação 82–263
deformação 73–67
~ de maré 82–277
~ devida à maré 83a–66
~ por maré 83a–66
~ de um núcleo cometário 84–109
degasação do núcleo cometário 84–109
degasificação 87–136
degeneração 93–172
degenerado 93–178
altamente ~ 93–184
completamente ~ 93–185
fortemente ~ 93–184
parcialmente ~ 93–183
totalmente ~ 93–185
degenerescência 93–172
~ completa 93–175
~ de nêutrons 93–191
~ do gás de nêutrons 93–191
~ não-relativista 93–177
~ parcial 93–173
~ relativista 93–176
~ total 93–175
Deimos 83–169
deionização 86–213
Delfim 92–45
Delphinus 92–45
Deneb 92–112
Denébola 92–113
densidade central 93–87
~ cosmológica 98–370
~ crítica 98–197
~ da radiação 99–85
~ da sombra 81–59
~ de crateras 81–246
~ de fundo 96–131
~ de fundo estelar 96–131
~ de galáxias 98–344
~ de traços 82–253
~ de uma corrente meteórica 86–264
~ do fluxo neutrônico 94–86
~ do vento solar 88–157
~ dos radiantes 86–63
~ dos traços 87–141
~ ele(c)trônica 78–6
~ ele(c)trônica linear 86–283
~ espacial 86–263, 96–133
~ espacial das estrelas 98–83
~ estelar 98–82
~ galáctica 98–344
~ inferior a 10^{12} el . cm^{-1} 86–284
~ média 93–86
~ média da matéria 99–84
~ projetada 98–85
~ real 96–133
~ subnuclear 93–190
~ superficial 98–85
~ superficial aparente 96–132
~ superior a 10^{12} el . cm^{-1} 86–285
denudação = desnudamento 73–108
deposição 73–109
depósito anual 87–17
depósitos abissais 72–50
~ biogênicos 72–59
~ glaciais 72–127
~ marinhos 87–15
~ na zona abissal 72–50
~ pelágicos 72–57
~ terrígenos 72–48
depressão 77–76
~ arqueada 81–160
~ continental 81–158
~ de terrações 81–162
~ linear 81–159

P

~ lunar 81–157
pequena ~ 81–243
~ sinuosa 81–161
~ solar 79–45
~ terraceada 81–162
deriva continental 73–104
~ dos continentes 73–104
~ em longitude 75–120
~ longitudinal 75–120
derrame de óleo 72–71
desaceleração aparente 86–100
~ da expansão 99–82
~ real 86–101
desacoplamento de radiação 99–177
~ entre matéria e radiação 99–177
desagregação de um cometa 85–114
~ de um cúmulo 96–105
~ do núcleo de um cometa 86–261
~ espacial 82–244
~ rotacional 85–119
desaparecimento = desaparição 81–98
~ do anel 83–217
~ súbito 86–76
~ súbito das ondas curtas 78–40
~ súbito de um filamento 89–279
desaparição = desaparecimento 81–98
~ de um meteoro 86–37
~ gradual 86–75
desativação 85–148
~ de um cometa 85–148
descarga atmosférica 77–161
~ vulcânica 82–162
descendente 77–93
descoberta de um cometa 85–16
~ de um meteorito 87–44
descobridor 85–21
descoloração do mar 72–80
desdobramento das raias por expansão 97–157
desenho das manchas 88–34
~ de um planeta 83–16
desenrolamento 98–30
~ dos braços espirais 98–30
desenroscamento 98–30
~ dos braços espirais 98–30
desenvolvimento de un grupo de manchas 89–118
deserto 83–161
desfasador 88–115
desfluidação 82–263
desfluidificação 82–263
desfluidização 82–263
desgaste de massa 81–236
designação de um cometa 85–31
~ definitiva 85–33
~ preliminar 85–32
~ provisória 85–32
desigualdade de Schwarzschild 93–156
desintegração de cúmulos 98–195
~ de um cometa 85–115
~ de um corpo meteórico 86–186
~ de um cúmulo galáctico 96–103
~ de um meteoro 86–187
~ do núcleo cometário 85–116
~ dos asteróides 84–108
~ radioativa 74–64
desintegrar 74–65
desionização 86–213
deslizamento 82–196
~ de paredes 82–198
~ do solo 82–152
deslocador de raia 88–69
deslocamento 73–91
~ anômalo para o vermelho 98–367
~ assimétrico 98–104
~ cosmológico 98–363a
~ da crosta 82–212
~ da zona de manchas 89–74
~ das raias para o vermelho 99–162
~ discrepante para o vermelho 98–367
~ do radiante 86–242
~ do rast(r)o 86–114
~ Doppler 91–25
~ gravitacional para o vermelho 99–162
~ múltiplo para o vermelho 98–364
~ para o vermelho 98–287
~ para o vermelho não-proveniente de velocidade 98–363
~ periódico das raias 96–22
~ relativista das raias para o vermelho 99–162
desmagnetização térmica 75–40
desmemabramento de um cúmulo galáctico 96–103
desmoronamento 82–198
desnudamento = denudação 73–108
de(s)polarização 97–118
desprendimento de gás 82–261
dessecação 74–51
dessorção 85–123
destruição 87–24
~ da superfície 87–35
~ por ondas de marés 72–77
~ por um meteorito 87–20
desviador de raia 88–69
desvio anômalo para o vermelho 98–367
~ assimétrico 98–104
~ cosmológico 98–363a
~ da luz pelo campo de gravidade 99–161
~ da vertical 71–72
~ discrepante para o vermelho 98–367
~ do movimento de uma estrela 98–123
~ Doppler 91–25
~ gravitacional para o vermelho 99–162
~ múltiplo para o vermelho 98–364
~ para o vermelho 98–297
~ relativista da luz 99–161
desvios do equilíbrio termodinânico 93–36
~ do equilíbrio termodinâmico local 93–40
~ do ET 93–36
~ do ETL 93–40
detalhe da superfície 83–105
~ lunar 81–148
~ permanente 83–106
determinação da gravidade 71–67
~ da idade 74–91, 87–126, 94–62
~ da idade isotópica 87–130
~ da idade por isótopos 87–130
~ da idade por radioatividade 87–131
~ da intensidade de gravidade 71–67
~ de massa 96–69
detrito 81–234
detritos fragmentários 82–104
Devoniano 74–81
dia ativo 75–93
~ calmo 75–92
~ de maré 82–290
~ lunar 81–32, 81–290
~ lunar médio 82–304
~ magnéticamente ativo 75–93
~ magneticamente calmo 75–92
~ magneticamente quieto 75–92
~ perturbado 75–94
~ quieto 74–92
diagênese 74–38
diagrama a duas cores 96–147
~ bicolor 96–147
~ cor-magnitude 96–150
~ de borboleta 89–76
~ de evolução 96–151
~ de Hertzsprung-Russell 91–130
~ de Hubble 98–301
~ de interferência 88–120
~ de Maunder 89–76
~ deslocamento para o vermelho – – magnitude 98–301
~ desvio para o vermelho-magnitude 98–301
~ em borboleta 89–76
~ evolutivo 96–151
~ H-R 91–130
~ magnitude-desvio para o vermelho 98–301
~ massa-luminosidade 91–129
diamantes meteoríticos 87–65
diâmetro angular 83–117, 97–145
~ aparente 83–117, 96–115
~ fotométrico 84–93
~ linear 83–118
~ real 83–118
~ verdadeiro 83–118
diastrofismo 73–72
diatrema 73–122
dicotomia de Vênus 83–132
diferença de brilho 96–65
~ de caminho ó(p)tico 88–117
~ de magnitudes 91–59

P

~ de trajeto 88–117
~ vertical 81–140
diferenciação geoquímica 74–55
~ gravitacional 74–56
~ por cristalização 74–41
~ química 83a–50
difusão atmosférica da luz 77–57
~ da luz 77–57
~ de Mie 83–76
~ dos íons 86–286
~ dos raios cósmicos 84–27
~ iônica 86–286
dilatação do tempo 99–137
dimensão das partículas 76–114
dimensões do sistema 96–64
diminuicão secular do brilho 85–126
dinâmica do gas inter(e)stelar 97–150
~ dos cúmulos estelares 96–153
~ estelar 98–120
dióxido de carbono sólido 83–144
~ de silício 74–13
dique 73–138
direcão anti-solar 75–22
~ de propagação 78–35
~ do campo magnético 84–10
~ do eixo 83–34
~ do vento 77–115
direto 83–42
dirupção de um cometa 85–114
~ por efeito de maré 85–118
disco da Lua 81–54
~ de acresção 96–21
~ de acrescimento 96–21
~ de ocultação 88–61
~ do Sol 89–24
~ galáctico 98–52
~ lunar 81–54
no ~ 88–173
~ ocultante 88–61
~ solar 89–24
~ visível 88–171
discontinuidade de Mohorovičić 72–14
~ Moho 72–14
disjunção 73–78
disparar 89–198
dispersão de velocidades 98–103
dispositivo de alunissagem 82–21
~ de pouso lunar 82–21
dissipação 94–73
~ da energía 82–285
~ da maré 82–286
~ de ondas 88–222
~ de um cúmulo galáctico 96–103
~ térmica 94–76
dissociação de um cúmulo galáctico 96–103
dissolução de uma protuberância 89–280
distância 86–288
~ angular 96–52
~ ao centro da Galáxia 98–11
~ ao centro do disco 89–36
~ ao meridiano central 89–46
~ da Lua à Terra 81–13
~ de antenas 88–99
~ do centro da estrela 93–85
~ do epicentro 73–19
~ do foco 73–18
~ entre os componentes 96–66
~ epicentral 73–19
~ focal 73–18
~ fotométrica 96–116
~ galactocêntrica 98–11
~ geocêntrica 84–3
~ heliocêntrica 84–4
~ hipocentral 73–18
~ máxima 86–289
~ Terra-Lua 81–13
distorcão 73–68
~ de maré 96–30
~ rotacional 91–44
~ rotacional da curva de velocidade 91–44
distribuição da densidade 96–124
~ da matéria 99–8
~ das massas 86–165
~ das velocidades residuais 98–92
~ de brilhância 88–141
~ de brilho 86–166
~ de tamanhos 82–124
~ elipsoidal das velocidades 98–100
~ em latitude 89–51
~ em latitude heliográfica 89–51
~ em longitude 89–48
~ em longitude heliográfica 89–48
~ espacial 96–119
~ galáctica 98–54
~ latitudinal 89–51
~ longitudinal 89–48
~ maxwelliana das velocidades 98–96
~ no espaço 98–86
~ segundo a massa 86–165
~ vertical 76–17
distribuido aleatoriamente 98–93
~ ao acaso 98–93
~ casualmente 98–93
distúrbio ionosférico 78–39
~ ionosférico súbito 78–41
divisão 83–223
~ de Cassini 83–224
dobra 73–87
~ anticlinal 73–90
~ sinclinal 73–89
dobramento 73–88
dobras 79–139
domo de lava 81–176a
~ lunar 81–177
~ vulcânico 81–176
domos, aglomerado de 81–179
~ escuros 89–297
~ escuros em torno das protuberâncias 89–297
Dorado 92–46
dorsal 81–186
~ anticiclônica 77–87
~ marinha 81–188
~ marítima 81–188
~ (médio-)oceânica 72–45
Dourado 92–46
Draco 92–47
Dracónidas 86–252
Dragão 92–47
dragagem 72–53
Dubhe 92–114
duna 83–162
dupla 96–3
~ astrométrica 96–12
~ cerrada 95–23
~ com espe(c)tro simples 96–15
~ com raias simples 96–15
~ de conta(c)to 96–29
~ de grande separação 96–9
~ destacada 96–35
~ em conta(c)to 96–29
~ espe(c)troscópica 96–14
~ espe(c)troscópica com linhas duplas 96–17
~ espe(c)troscópica com raias simples 96–15
~ estreita 95–23
~ fotométrica 96–11
~ não-resolúvel 96–10
~ não-resolvida 96–10
~ ó(p)tica 96–6
~ semidesligada 96–34
~ semidestacada 96–34
~ separada 96–35
~ visual 96–8
duração da evaporação 96–109
~ da insolação 88–10
~ da rotação 83–31
~ da totalidade 88–26
~ da visibilidade do rast(r)o 86–111
de longa ~ 89–124
de pequena ~ 89–125
~ de um grupo de manchas 89–123
~ de uma erupção 89–209
~ de vida de um grupo de manchas 89–123
~ do ciclo 89–87
~ do desmembramento 96–104
~ do eclipse 95–78
~ do eclipse total 88–26
~ do eco 86–299
~ do percurso 73–51
~ do percurso de um meteoro 86–50
~ do período 95–39
~ de radioeco 86–299
~ dos pulsos 86–280
duto 73–124

E

eclipse 88–14
~ anular 88–18
~ anular do Sol 88–15a
~ artificial 88–227a

P

~ artificial do Sol 88–17a
~ atmosférico 95–76
~ central 88–16a
~ começa ao pôr do Sol 88–22
~ corpuscular 88–24
~ da Lua 81–56
~ do Sol 88–15
~ é invisível em ... 81–87
~ é visível em ... 81–86
~ ionosférico 88–23
~ lunar 81–56
~ parcial 81–79a
~ parcial da Lua 81–67
~ parcial do Sol 88–16
~ penumbral da Lua 81–65
~ solar 88–15
~ termina ao nascer do Sol 88–21
~ total começa 81–77
~ total da Lua 81–66
~ total do Sol 88–17
eco auroral 79–121
~ de rast(r)o 86–298
~ meteórico 86–296
~ meteórico de curta duração 86–300
~ meteórico de longa duração 86–301
~ meteórico frontal 86–297
~ rádio 86–295
~ radioelé(c)trico 86–295
~ sísmico 82–223
ecologia 76–2
eco-sondagens 72–29
efeito bétatron 91–180
~ Coriolis 77–81
~ cumulativo 98–124
~ de cobertura 93–67
~ de estufa 83–263
~ Götz 76–126
~ de inversão 76–126
~ de maré 82–276
~ de oposição 83–75
~ de reflaxão 96–26
~ de reflaxão nas binárias 95–82
~ de seleção 98–81
~ de Umkehr 76–126
~ estufa 83–263
~ Evershed 89–103
~ isotópico 91–178
~ Poynting-Robertson 84–67
~ Rossiter 91–44
~ Wilson 89–89
efeitos terrestres da atividade solar 84–55
efemérides dos pequenos planetas 84–92
eficiência luminosa 86–210
efluxo contínuo 98–276
~ da matéria 98–275
~ da matéria do núcleo galáctico 98–275
~ de massa 93–13
efusão de lava 82–172
eixo de rotação 83–33
~ geomagnético 75–47
~ magnético 75–46
~ magnético da Terra 75–47
ejeção 89–259
~ da matéria 89–214
~ da matéria absorvente 89–256
~ da matéria brilhante 89–257
~ da matéria do núcleo galáctico 98–274
~ das radionuvens 98–285
~ de massa 95–124
~ de partículas 85–89
~ de partículas do Sol 88–161
~ de plasma solar 84–38
~ de um cometa 86–262
~ recorrente 98–282
ejetamentos 82–165
ejetólitos 82–169
ejetos 82–165
~ de impacto 82–167
~ vulcânicos 82–166
elasticidade 73–58
ele(c)tricidade atmosférica 75–2
~ terrestre 75–1
ele(c)troja(c)to auroral 79–113
elé(c)tron auroral 79–109
elé(c)trons aprisionados 79–111
~ precipitados 79–118
~ relativistas 97–153
~ relativísticos 97–153
~ retidos 79–111
elemento deficiente 91–186
~ em deficiência 91–186
~ em traços 82–145
~ estrutural 98–287
~ granular 88–178
~ meteorológico 77–28
~ móvel 88–108
~ subabundante 91–186
~ superabundante 91–184
~ supergranular 88–182
elementos atmófilos 83a–29
~ calcófilos 83a–31
~ do grupo do ferro 94–105
~ fotométricos 95–25
~ geomagnéticos 75–54
~ litófilos 83a–30
~ magnéticos 75–54
~ principais 82–144
~ refratários 97–64
elevação 76–54
~ da temperatura 77–42
~ de gelo 72–102
~ lunar 81–156
elipsóide das velocidades 98–101
~ de referência 71–18
~ internacional 71–14
~ local 71–15
~ terrestre 71–11
elipticidade de um cúmulo globular 96–122
elmo coronal 89–295
elongação 83–60
~ do radiante aparente 86–67
~ do radiante verdadeiro 86–68
máxima ~ 83–63
~ ocidental 83–62
~ oriental 83–61
embrião de um planeta 83a–36
emersão 81–99
emissão atmosférica 79–2
~ atmosférica crepuscular 79–4
~ atmosférica noturna 79–5
~ brusca de partículas 88–162
~ brusca de raios X 79–117
~ corpuscular 88–152
~ de dois fótons 97–155
~ de freqüência extremamente baixa 78–55
~ de freqüência muito baixa 78–54
~ de lava 73–130
~ de massa 85–145
~ de partículas 88–151
~ de Lα do céu noturno 79–9
~ pós-crepuscular 79–52
~ pré-crepuscular 79–51
~ simultânea 97–155
~ simultânea do dois fótons 97–155
emissões de rádio esporádicas de Júpiter 83–181
emissor 78–15
empuxo 73–65
enchente 82–280
~ da maré 82–280
enchimento de uma depressão 77–83
encoberto 77–152
encontro de estrelas 98–121
~ de três corpos 96–74
~ estelar 98–121
encurtamento das manchas 89–92
~ magnético 89–160
~ magnético por perspectiva 89–160
endógeno 73–145
energia acumulada 89–196
~ acústica 88–219
~ armazenada 89–196
~ atômica 93–121
~ cinética do ponto zero 93–187
~ de repouso 99–144
~ de rotação 93–120
~ de vaporização 86–199
~ disponível 89–197
~ do ponto zero 93–187
~ do vácuo 99–170k
~ em repouso 99–144
~ gravitacional 93–119
~ não-radiativa 88–218
~ nuclear 93–122
~ rotacional 93–120
~ trocada 98–126
enfraquecimento brusco 78–40
~ brusco das ondas curtas 78–40
engatilhamento 89–199
enlace geodésico intercontinental 71–37

P

enrolamento dos braços espirais 98–28
enroscamento 98–28
~ dos braços espirais 98–28
entrada da Lua na penumbra 81–75
~ da Lua na sombra 81–76
~ na atmosfera 86–161
entre os braços 98–40
envelope comum (de uma binária) 96–32
~ convectivo 93–136
~ extenso 93–8
~ radiativo 93–135
envoltória circunstelar 93–10
~ convectiva 93–136
~ em expansão 93–12
estrela com ~ 93–5
~ estacionária 93–11
~ radiativa 93–135
enxame = aglomerado = cúmulo 96–79
~ de estrelas relativista 96–154
~ de grande concentração 96–129
~ diurno 86–219
~ estelar 96–79
~ globular 96–114
~ meteórico 86–216
~ móvel de estrelas 96–97
~ noturno 86–218
~ pobre em metais 96–113
~ rico em metais 96–112
Eozóico 74–77
epicentro 73–15
epiciclo 83–49
epirogênese 73–73
epirogenia 73–73
época 74–89
~ de máximo 89–81
~ de mínimo 89–80
~ do periastro 96–60
~ geológica 74–89
~ glacial 83a–78
~ holocena 72–109
~ pós-glacial 72–109
equação da luz 95–45
~ da perda de massa 86–169
~ de Abel 88–235
~ de altura 86–194
~ de continuidade 93–49
~ de cor 91–73
~ de Emden 91–130
~ de luminosidade 86–196
~ de massa 86–169
~ de Pogson 91–58
~ de resistência 86–179
~ de transferência 93–32
~ de transferência radiativa 93–33
~ de transporte 93–32
~ de transporte radiativo 93–33
~ de vaporização 86–200
~ do arrasto 86–179
~ do equilíbrio hidrostático 93–76
~ do virial 98–181
~ hidrostática 93–76
~ integral de Schwarzschild e Milne 93–53
~ integro-diferencial 93–52
~ tensorial 99–157
equações de campo 99–155
~ de campo de Einstein 99–156
~ de equilíbrio estatístico 93–44
equador de iluminação 81–122
~ lunar 81–121
~ magnético 75–50
eqüiescalares de igual contagem 75–102
equilíbrio adiabático 76–21
~ convectico 76–22
~ de adsorção 85–121
~ de difusão 76–23
~ de energia 88–251
~ de massa 88–250
~ de radiação 88–252
~ diferente do ETL 93–43
~ energético 88–251
~ estatístico 93–42
~ estatístico nuclear 94–103
~ fora do ETL 93–43
~ generalizado 93–43
~ hidrostático 93–75
~ isostático 73–99
~ isotérmico 76–20
~ nuclear 94–103
~ pormenorizado 93–41
~ radiativo 93–48
~ térmico 93–78
~ termodinâmico 93–35
~ termodinâmico local 93–39
equivalência de massa e energia 99–145
Equuleus 92–48
era 74–69
~ cenozóica 74–73
~ cosmológica 99–166a
~ da inflação 99–170d
~ da inflação cosmológica 99–170d
~ da radiação 99–172
~ de aniquilação 99–173a
~ de desmaterialização 99–173a
~ de Planck 99–170e
~ da supergravitação 99–170e
~ do desacoplamento 99–177a
~ do plasma fotónico 99–172
~ estelar 99–178
~ hadrônica 99–174
~ leptônica 99–176
~ mesozóica 74–72
~ neozóica 74–73
~ paleozóica 74–71
~ pré-cambriana 74–70
Erídano 92–49
Eridanus 92–49
erosão 73–107
~ cósmica 87–142
~ da superfície 82–241
~ eóli(c)a 83–163
~ gasosa 82–243
~ meteorítica 82–242
~ meteoroidal 82–242
~ no espaço cósmico 87–142
erupção = fulguração 95–69
~ branca 89–182
~ com duas fitas 89–187a
~ cromosférica 89–175
~ de classe 2 89–206
~ de importância 2 89–206
~ de plasma solar 84–38
~ de raios X 89–183
~ de prótons 89–185
~ de vento solar 84–42
~ em luz branca 89–182
~ em ondas decamétricas 83–183
~ em radiofreqüências de Júpiter 83–182
~ em raios X 89–183
~ estelar 95–70
~ fissural 73–113
~ homóloga 89–178
~ longitudinal 73–113
~ luminosa de um meteoro 86–80
~ lunar 82–247
~ não-protônica 89–186
~ no bordo 89–202
~ no disco 89–204
~ no limbo 89–202
~ ó(p)tica 89–181
~ por detrás do bordo 89–203
~ por detrás do limbo 89–203
~ protônica 89–185
~ rádio 89–180
~ simpatética 89–179
~ sobre o disco 89–204
~ solar 89–175
~ visível 89–181
~ vulcânica 73–112, 82–170
erupcões de baixa freqüência 83–184
escala barométrica 76–19
~ Beaufort 77–108
~ centésimal 77–33
~ centígrada 77–33
~ curta de tempo 94–66
~ da carta 71–49
~ de altura 76–18
~ de distâncias 98–303
~ de tempo 94–65
~ do mapa 71–49
~ dos ventos (de) Beaufort 77–108
~ geocronológica 74–68
~ longa de tempo 94–67
escalão hipsométrico 76–19
escapamento de gases 76–24
escape de gás 82–261
~ de uma estrela do sistema 96–159
escarpa 81–184
escoamento contínuo 98–276
~ da matéria 98–275
~ da matéria do núcleo galáctico 98–275
Escorpião 92–86
escorregadela de paredes 82–198

P

escorregamento 82–196
~ de paredes 82–198
escudo 72–11
~ de Sobieski 92–88
Escultor 92–87
escurecimento de limbo 88–142
~ inter(e)stelar 97–86
~ por gravidade 96–25
esfera celeste 92–8
~ gasosa 93–73
~ gasosa politrópica 93–127
~ terrestre 71–10
esferóide achatado 98–158
~ terrestre 71–13
esférulas 87–29
~ vitrificadas 82–127
esfoliação 74–53
esforço 73–62
~ de cisalhamento 73–63
~ de corte 73–63
~ principal 73–64
~ térmico 86–205
esgotamento do hidrogênio 94–69
esguicho 89–258
esmagamento gravitacional 84–72
espaçamento entre antenas 88–99
~ nulo 88–101
~ variável 88–103
espaço 99–100
~ absoluto 99–116
~ curvo 99–114
~ das velocidades 98–97
~ de fases a seis dimensões 98–144
~ de Lobatchevski 99–105
~ de Minkowski 99–122
~ de Newcomb 99–110
~ de Riemann 99–112
~ dos eventos 99–122
~ dos acontecimentos 99–122
~ elíptico 99–110
~ entre os braços 98–40
~ esférico 99–111
~ euclidiano 99–107
~ extragaláctico 97–5
~ extrarrestre 87–73a
~ finito 99–103
~ hiperbólico 99–106
~ infinito 99–104
~ interdito 79–135
~ inter(e)stelar 97–4
~ interfilamentar 89–99
~ intergaláctico 97–7
~ interplanetário 84–2
~ livre 76–9
~ métrico 99–100a
~ não-euclidiano 99–114
~ parabólico 99–107
~ permitido 79–134
~ plano 99–107
~ proibido 79–135
~ riemanniano 99–112
espaço-tempo 99–119
~ de Minkowski 99–122
espaço-temporal 99–120
Espada do Órion 92–130
espalação 94–100
~ por raios cósmicos 82–254
espalhamento (atmosférico) da luz 77–57
~ de Mie 83–76
especialista em meteoritos 87–7
espécie de nuvem 77–135
espécime(n) individual 87–42
espe(c)tro composto 96–19
~ de Fraunhofer 88–137
~ dinâmico 88–125
~ do rast(r)o de um meteoro 86–107
~ duplo 96–18
~ estelar 91–88
~ estelar normal 91–89
~ global 96–137
~ integral 96–137
~ principal 95–115
~ simbiótico 95–117
~ simples 96–16
espe(c)trobolograma 88–76
espe(c)trobolômetro 88–75
espe(c)trografia Mössbauer 82–142
espe(c)trógrafo dinâmico 88–124
~ horizontal 88–78
~ para meteoros 86–146
~ solar 88–77
~ vertical 88–79
espe(c)trograma dinâmico 88–125
espe(c)tro(h)eliocinematógrafo 88–81
espe(c)tro(h)eliógrafo 88–71
espe(c)tro(h)eliograma 88–72
espe(c)tro(h)elioscópio 88–67
espe(c)trometria não-destrutiva de raios gama 82–77
espe(c)trômetro 88–124
~ de vento solar 82–46
espe(c)troregistrador de velocidades 88–57
espe(c)tro relâmpago 88–25
espe(c)troscópio de protuberâncias 88–58
espelho giratório 86–130
~ oscilante 86–130
espessura dos anéis 83–229
espícula 88–195
Espiga 92–129
espiral 98–7, 98–249
~ arquimediana 84–34
~ barrada 98–253
~ de Arquimedes 84–34
~ não-barrada 98–250
~ normal 98–250
Esquadro 92–70
estabilidade 93–204
~ das estrelas 93–206
~ de uma órbita 98–193
~ dinâmica 93–207, 98–198
estação de satélites 71–33
~ de um planeta 83–44
~ lunar automática 82–19
~ meteórica 86–158
~ para a observação de meteoros 86–158
~ sísmica 73–6
~ sismográfica 73–6
~ sismológica 73–6
estádio = estágio 94–19, 94–20
~ de contração 94–20
~ de disco 94–22
~ evolutivo 94–28
~ intermediário 94–21
~ pré-seqüência principal 94–49
~ pré-série principal 94–49
estado de fluência 82–189
~ de fluidez 82–189
~ do céu 77–132
~ do mar 72–86
~ do tempo 77–14
~ do tempo de grande escala 77–15
em ~ não-estacionário 98–151
~ estacionário 98–140
~ estatísticamente estacionário 98–148
estágio = estádio 94–19, 94–20
~ anterior à seqüência principal 94–49
~ anterior à série principal 94–49
~ de contração 94–20
~ de disco 94–22
~ de halo 94–19
~ de Hayashi 94–44
~ evolutivo 94–28
~ Hayashi 94–44
~ hidrodinâmico 94–43
~ intermediário 94–21
~ pós-seqüência principal 94–68
~ posterior à seqüência principal 94–68
~ T-Tauri 94–47
estatística estelar 98–77
estável 93–208
termicamente ~ 93–209
esteira de um meteoro 86–110
estelar 92–3
estelho de hélio 94–79
estereo-comparador 91–32
estimar 95–12
estimativa de brilho 95–11
estratificação 82–133, 97–154
~ horizontal 82–134
~ térmica 77–40
estrato 77–139
estrato-cúmulo 77–142
estrato-nimbo 77–145
estratopausa 76–71
estratosfera 76–70
estratovulcão 73–121
estrela 91–1
~ a temperatura zero 91–166
~ a zero absoluto 91–166
~ anesférica 91–86

P

~ anômala 91–96
~ avermelhada 97–94
~ binária 96–7
~ brilhante 91–83
~ cadente 86–2
~ carbonada 91–188
~ central 97–148
~ circumpolar 92–2a
~ colapsada 94–111
~ colapsante 91–164
~ com atmosfera extensa 93–6
~ com envelope 93–5
~ com envelope extenso 93–7
~ com envoltória 93–5
~ com espe(c)tro peculiar 91–92
~ com invólucro 93–5
~ com invólucro extenso 93–7
~ com linhas de emissão 91–159
~ com linhas metálicas 91–195
~ com o pólo à vista 91–40
~ com peculiaridades espe(c)trais 91–92
~ com raias de emissão 91–159
~ com raias metálicas 91–195
~ com uma companheira invisível 96–13
~ companheira 96–39
~ comum 91–95
~ d'alva 83–127
~ da manhã 83–127
~ da primeira geração 94–24
~ da seqüência principal 91–142
~ da tarde 83–128
~ de alta velocidade 98–191
~ de baixa velocidade 98–192
~ de carbono 91–88
~ de comparação 95–16
~ de escape 96–160
~ de híperons 91–154
~ de idade nula 94–59
~ de idade zero 94–59
~ de massa intermédia 91–125
~ de massa intermediária 91–125
~ de massa moderada 91–125
~ de nêutrons 93–195
~ de nitrogênio 91–189
~ de pequena massa 91–124
~ de primeiro plano 96–85
~ de rotação rápida 91–39
~ de silício 91–190
~ de tipo solar 91–104
~ de Tycho 95–135
~ deficiente em carbono 91–194
~ deficiente em hélio 91–192
~ deficiente em hidrogênio 91–191
~ deficiente em metais 91–197
~ degenerada 93–180
~ do aglomerado 96–82
~ do campo 96–84
~ do cúmulo 96–82
~ do disco 91–85
~ do fundo 96–84
~ do halo 91–84
~ do pastor 83–128
~ do ramo horizontal 91–141
~ do sistema 98–133
~ dupla 96–3
~ dupla de grande separação 96–9
~ dupla espe(c)troscópica 96–14
~ dupla física 96–7
~ dupla ó(p)tica 96–6
~ dupla visual 96–8
~ em afastamento 96–160
~ em colapso 91–164, 94–112
~ em contração 91–161
~ em escape 98–135
~ em expansão 91–162
~ em explosão 91–165
~ em fuga 98–135
~ em implosão 91–164
~ em rotação 91–37
~ em rotação rápida 91–39
~ escapante 96–160
~ estacionária 95–3
~ estável 95–8
~ estranha 98–134
~ evoluída 94–74
~ excitadora 97–125
~ excitante 97–125
~ explosiva 92–165
~ filante 86–2
~ fixa 92–4
~ fraca 91–82
~ fria 91–156
~ fugaz 86–2
~ fugitiva 96–160
~ fulgurante 95–68
~ gigante 91–143
~ hiperônica 91–154
~ implosiva 91–164
~ infravermelha 91–157
~ instável 95–9
~ intergaláctica 91–163
~ intrinsicamente variável 95–87
~ intrusa 98–134
~ jovem 96–93
~ leve 92–124
~ magnética 91–51
~ mais brilhante 98–304
~ mais brilhante de uma galáxia 98–304
~ massiva 91–126
~ matutina 83–127
~ membro do cúmulo 96–82
~ morta 91–167
~ múltipla 96–5
~ não-avermelhada 97–95
~ não-esférica 91–86
~ não-estacionária 95–2
~ não-membro do cúmulo 96–84
~ não-pertencente ao cúmulo 96–84
~ não-pertencente ao sistema 98–134
~ não-rotante 92–38
~ neutrônica 91–152
~ nitrogenada 91–189
~ normal 91–95
~ parecida à nova 95–118
~ peculiar 91–92
~ pertencente ao cúmulo 96–82
~ pertencente ao sistema 98–133
~ pobre em carbono 91–194
~ pobre em hélio 91–192
~ pobre em hidrogênio 91–191
~ pobre em metais 91–197
~ poco massiva 91–124
~ polar 92–121
~ preposta 96–85
~ primária 96–37
~ principal 96–37
~ próxima 98–60
~ pulsante 95–95
~ que escapa 98–135
~ quente 91–155
~ rica em hélio 91–193
~ rica em metais 91–196
~ rotante 91–37
~ sem rotação 91–38
~ simbiótica 95–50
~ simples 96–2
~ subluminosa 91–133
~ superdensa 91–153
~ superluminosa 91–134
~ supermassiva 91–127
~ suspeita de variabilidade 95–5
~ tripla 96–4
~ tríplice 95–4
~ única 96–2
~ variável 95–1
~ velha 94–64
~ vermelha 91–97
~ Vespertina 83–128
~ vista do pólo 91–40
~ vizinha 98–60
~ Wolf-Rayet 91–160
~ WR 91–160
estrelas anteriores 91–102
~ de tipo avançado 91–103
~ de tipo tardio 91–103
~ dos primeiros tipos 91–102
~ dos tipos finais 91–103
~ dos tipos iniciais 91–102
~ dos últimos tipos 91–103
~ posteriores 91–103
estriação 89–173
~ cromosférica 88–209
estriamento 89–173
estrias cromosféricas 88–209
estrutura anelar 81–155, 98–41
~ anular 81–155, 98–41
~ característica 88–248
~ celular 99–171a
~ cometária 85–34
~ corrotacional 84–17
~ da corrente meteórica 86–243
~ da Galáxia 98–4
~ da Via Láctea 98–4

P

~ de grande escala 88–207
~ de pequena escala 88–208
~ de setor 84–15
~ do campo 89–140
~ dos cometas 85–34
~ dos meteoritos 87–100
~ em arco 88–249
~ em corrotação 84–17
~ em forma de favos de colmeia 82–25
~ em grande escala 99–171
~ espiral 98–8
~ estelar 93–71
~ estriada 88–209
~ fibrilar 88–209
~ filamentar 97–130
~ filamentosa 97–130
~ fina da cromosfera 88–203
~ galáctica 98–4
~ hierárquica 99–47
~ hierárquica do universo 99–47
~ interna 71–83
~ interna das estrelas 93–71
~ laminar 83–148
~ pré-mares 81–153
~ pré-marítima 81–153
~ pós-mares 81–154
~ pós-marítima 81–154
~ raiada 79–150, 79–153
~ setorial 84–15
equilíbrio termodinâmico local 93–39
esva(ne)cimento súbito (das ondas curtas) 78–40
ET 93–35
Etamin 92–115
eternidade do universo 99–13
ETL 93–39
euclidiano 99–101
Europa 83–206
evaporação de buracos negros 94–118
~ de um cúmulo 96–108
~ de uma partícula meteórica 86–198
evento 99–125
~ a partículas energéticas 89–191
~ adjacente 99–126
~ de poluição 72–72
~ de protons 84–59
~ de raios cósmicos solares 89–194
~ em raios X 89–184
~ poluente 72–72
~ protônico 84–59
~ rádio de Júpiter 83–182
~ simultâneo 99–128
~ sísmico 82–215
~ solar de prótons 89–192
~ solar em raios X 89–184
~ transitório coronal 89–185a
evolução anterior à seqüência principal 94–26
~ das estrelas 94–1
~ de um grupo de manchas 89–118
~ devida a marés 83a–65
~ dinâmica 96–161
~ do universo 99–171e
~ estelar 94–1
~ geoquímica 74–33
~ nuclear 94–50
~ por efeito de marés 83a–65
~ pré-estelar 94–14
~ pré-estelar das galáxias 94–14
~ pré-seqüência principal 94–26
~ pré-série principal 94–26
~ química 82–257
~ térmica 83a–61
~ termonuclear 94–50
evoluir 94–75
evolver 94–75
exaustão do hidrogênio 94–69
excesso de cor 91–77
~ de infravermelho 91–79
~ de IV 91–79
~ de ultravioleta 91–78
~ infravermelho 91–79
~ IV 91–79
~ ultravioleta 91–78
~ UV 91–78
excursão 82–35
~ científica 82–36
exógeno 73–146
exosfera 76–68
expansão acelerada 99–81
~ do assoalho oceânico 72–39
~ do fundo oceânico 72–39
~ do universo 99–74
~ hidromagnética 88–159
~ ultra-rápida 99–170c
~ ultra-rápida do universo 99–170c
expedição de eclipse solar 88–30
experiência de Michelson 99–130
~ na superfície lunar 82–42
~ sísmica 82–219
~ sísmica ativa 82–220
experimento sísmico 82–219
~ sísmico ativo 82–220
explosão 87–30
~ de brilho 85–126
~ de partículas 88–162
~ de raios X 79–117
~ de um cometa 85–125
~ de um meteoro 86–81
~ de um núcleo galáctico 98–283
~ de uma nova 95–121
~ de uma supernova 95–131
~ de vento solar 84–42
~ inicial 99–167
~ nuclear de alta altitude 75–114
~ primordial 99–167
~ vulcânica 73–112
expulsão 73–132
~ da matéria 98–274
~ da matéria do núcleo galáctico 98–274
~ de gases 83a–41
~ de lava 82–171
ex-supernova 95–134
extensão da maré 82–292
extenuação das linhas de Fraunhofer 88–243
exterior à atmosfera terrestre 88–149
extinção atmosférica 77–127a
~ de um meteoro 86–37
~ inter(e)stelar 97–80
~ por vapor de água 77–126
~ progressiva 86–75
~ seletiva 97–83
~ total 97–82
extragaláctico 98–207
extrusão 73–132
~ de lava 82–171

F

face distante da Lua 81–16
~ frontal da Lua 81–15
~ iluminada 81–34
~ não-visível 81–16
~ oculta da Lua 81–16
~ traseira 81–16
~ traseira da Lua 81–16
~ visível 81–15
fácula 89–161
fáculas fotosféricas 89–161
~ cromosféricas 89–164
~ em raios X 89–166
~ ó(p)ticas 89–165
~ polares 89–162
faíscas 86–84
~ de um meteoro 86–84
faixa 83–190
~ ativa 73–32
~ com estrutura raiada 79–150
~ de poeira 98–289
~ de totalidade 88–29
~ equatorial 83–191
~ escura 83–190
~ homogênea 79–146
~ luminosa 86–27
~ sísmica 73–30
~ zodiacal 84–63
faixas escuras 98–288
~ sombreadas 88–31
falha 73–91
~ de compressão 73–94
~ de distensão 73–93
~ de Hertzsprung 91–144
~ direta 73–93
~ inversa 73–94
~ invertida 73–94
~ normal 73–93
falhamento local 82–204
~ regional 82–205
falsa luz zodiacal 84–62
família de cometas 85–102
~ de Júpiter 85–103
~ dos asteróides 84–73
fase ascendente do ciclo solar 89–83

P

~ crescente 89–83
~ crescente da Lua 81–23
~ da Lua 81–19
~ de contração 94–20
~ de disco 94–22
~ de halo 94–19
~ de recuperação 75–77
~ descendente do ciclo solar 89–84
~ do ciclo 89–79
~ eruptiva 89–210
~ evolutiva 94–28
~ explosiva 89–210
~ "flash" 89–210
~ Hayashi 94–44
~ hidrodinâmica 94–43
~ inicial 75–75
~ inicial de uma tempestade magnética 75–75
~ intermediária 94–21
~ máxima do eclipse 81–78
~ nebular 95–125
~ nula 83–73
~ parcial 81–78a
~ parcial do eclipse 81–78a
~ pós-seqüência principal 94–68
~ principal 73–12, 75–76
~ T Tauri 44–47
~ zero 83–73
fases do eclipse 81–73
fator da expansão cósmica 99–81a
~ de brilhância 83–90
~ de brilho 83–90
~ de contração de Lorentz 99–135
~ de diluição 93–37
~ de forma 86–177
~ de Gaunt 93–104
~ de guilhotina 93–103
~ de luminosidade 86–197
~ geométrico 86–177
~ R(*t*) 99–81a
feixe ele(c)trônico 79–116
feldspato 74–15
~ potássico 74–16
fenda 81–165
~ curva 88–65
~ curvada 88–65
~ de tensão 82–206
~ oscilante 88–70
Fênix 92–77
fenômeno 89–14
~ ativo 89–14
~ de curta duração 89–15
~ de halo 79–66
~ lunar transiente 82–153
~ ó(p)tico 79–33
~ solar 89–13
~ transitório coronal 89–185a
fenômenos crepusculares 79–32
ferramenta de coleta de amostras 82–80
ferro meteórico 87–83
~ meteorítico 87–83
ferromagnetismo 75–26
ferro-níquel 87–90
fibra de uma protuberância 89–245
fibrila penumbral 89–98
figura da Lua 81–116
~ da Terra 71–7
~ de equilíbrio 83–112
~ de equilíbrio hidrostático 83–112
~ de impacto 82–228
~ de uma planeta 83–109
figuras de Widmanstaetten 87–105
filamento 85–55, 89–224
~ da penumbra 89–98
~ de hidrogênio 89–267
~ de uma erupção 89–187
~ de uma nebulosa 97–131
~ de uma protuberância 89–245
~ metálico 89–268
~ oscilante 89–216
~ tremulante 89–216
filme cinematográfico de protuberâncias 89–242
~ de protuberâncias 89–242
filtro de aerossol 76–94
~ de poeiras 76–95
fim da totalidade 81–80
~ da trajetória aparente 86–38
~ do ciclo 89–85
~ do eclipse 81–83
~ do eclipse parcial 81–83a
~ do eclipse total 81–80
~ do mundo 99–183
finitude do universo 99–4
física atmosférica 76–4
~ da Lua 82–151
~ das relações Sol-Terra 84–56
~ dos cometas 85–104
~ dos planetas 83–4
~ lunar 82–151
~ planetária 83–4
~ solar 88–6
fissura 73–78
fita eruptiva 89–187
Flecha 92–84
flóculo 89–167
flóculos 89–167
~ brilhantes de hidrogênio 89–169
~ de cálcio 89–170
~ de hidrogênio 89–168
~ escuros 89–224
~ escuros de hidrogênio 89–224
flotação magnética 89–147
fluência de lava 82–188
fluidez da lava 82–188
fluido quimicamente ativo 74–43
flutuação climática 77–183
flutuações primordiais 99–170y
~ secundárias 95–44
~ sobrepostas 95–44
~ superpostas 95–44
fluxo 82–279
~ convectivo 93–15
~ da maré 82–279
~ de calor 74–60
~ de cinza 82–174
~ de energia 93–112
~ de energia acústica 88–221
~ de energia convectivo 93–115
~ de energia sonora 88–221
~ de geleira 72–120
~ de lava 73–140
~ de massa 93–13
~ de nêutrons 94–86
~ de radiação 93–113
~ e refluxo 72–89
~ fissural 73–141
~ glaciário 72–120
~ integrado 93–47
~ plástico 72–121
~ radiativo emergente 93–114
~ total de energia 93–116
FmU 78–33
FMU 78–37
Fobos 83–168
foco do sismo 73–17
~ do terremoto 73–17
folha de mica 88–87
Fomalhaut 92–116
fonte compacta 98–273
~ de calor 74–62
~ de elé(c)trons 79–119
~ de energia 93–110
~ de nêutrons 94–85
~ de rádio quase-estelar 98–360
~ decamétria 83–186
~ dupla de energia 93–146
~ neutrônica 94–85
~ quase-estelar 98–358
fora da atmosfera terrestre 88–149
força coercitiva 75–39
~ coerciva 75–39
~ da gravidade 71–62
~ de coesão 85–117
~ de Coriolis 77–79
~ de maré 82–289
~ de repulsão 85–87
~ desviante 77–80
~ do vento 77–107
~ electrofraca 99–170x
~ geradora da maré 82–289
~ interna 73–61
~ portante 82–64
~ repulsiva 85–87
~ superunificada 99–170r
~ unificada 99–170x
forças de maré 72–92
~ tectônicas 73–106
forma anelar 81–155
~ anular 81–155
~ da Lua 81–116
~ da partícula meteórica 86–175
~ da Terra 71–7
~ de nuvem 77–133
~ de um cometa 85–83
~ do corpo meteórico 86–175

em ~ de leque 85–74
~ orientada 87–113
~ superficial 83–105
formação 94–10
~ ativa 79–138
~ atmosférica 83–107
~ da cauda 85–86
~ da superfície 83–105
~ das estrelas 94–11
~ das falhas 73–92
~ de crateras 82–232
~ de crateras por impacto 82–234
~ de estrelas 96–95
~ de falha local 82–204
~ de falha regional 82–205
~ de grãos 97–60
~ de impacto 82–228
~ de linhas 93–63
~ de montanhas 73–74
~ de raias 93–63
~ de uma mancha 89–121
~ do mundo 99–11
~ dos elementos 94–52
~ estelar 94–11
~ estrutural 88–248
~ lunar 81–148
~ permanente 83–106
~ primária 81–150
~ secundária 81–151
~ topográfica 81–149
~ violenta de estrelas 94–11a
formações ativas 79–138
~ lunares 81–148
formaldeído 97—27
formamida 97–35
formas com estrutura raiada 79–152
~ de aurora 79–137
~ estruturais 81–152
~ sem estrutura raiada 79–142
fórmula altimétrica 76–15
~ barométrica 76–15
~ de Bode 83–3
~ de nivelamento barométrico 76–15
~ de Titius-Bode 83–3
~ internacional da gravidade 71–76
~ internacional gravimétrica 71–76
Fornalha 92–50
Fornax 92–50
Forno 92–50
fortemente degenerado 93–184
fossa 82–132
~ marítima 72–40
~ oceânica 72–40
foto da Lua 82–26
fotocentro 96–46
fotodesintegração 94–101
~ do núcleo 94–101
fotografia da Lua 82–26
~ de meteoros 86–140
~ de seqüência 79–103
~ de um meteoro 86–151a
~ em luz branca 88–37
~ lunar 82–26
~ oblíqua 82–29
~ panorâmica 82–27
~ paralá(c)tica 79–102
~ vertical 82–28
fotólise 76–84
fotometria em seis cores 96–148
~ detalhada 83–104
~ em três cores 96–141
~ integrada 83–103
~ multicolor 96–143
~ multicor 96–143
~ pormenorizada 83–104
fotoneutrino 94–93
fotosfera 88–167
~ calma 88–169
~ não-perturbada 88–169
fotosférico 88–170
fototriangulação espacial 71–26
fracionamento 87–138
~ dos elementos químicos 83a–51
fra(c)to-cúmulo 77–146
frágil 86–191
fragilidadc 86–192
fragmentação 83a–81
~ de um corpo meteórico 86–188
~ de um meteoro 86–83
~ de uma nuvem cósmica 94–38
~ dos asteróides 84–108
fragmento de rocha 82–68
~ de um meteorito 87–43
~ nuclear 94–102
~ principal 87–41
~ vitrificado 82–97
fragmentos 83a–83
~ de rocha 81–235
franja polar 83–151
fratura 81–165
~ concêntrica 82–209
~ da margem da cratera 82–211
~ de tensão 82–207
~ linear 82–208
~ plana 82–208
~ superficial 81–173
frenagem da rotação 91–45
~ magnética 91–47
fre(n)amento magnético 91–47
frente 77–68
~ de ionização 97–48
~ do fluxo 82–193
~ fria 77–70
~ oclusa 77–71
~ polar 77–73
~ quente 77–69
freqüência crítica 78–24
~ da ocorrência de auroras 79–122
~ da onda portadora 86–274
~ de auroras 79–122
~ de crateras 81–247
~ de erupções 89–217
~ de fulgurações 89–217
~ de manchas 89–66
~ de protuberâncias 89–237
~ de recorrência 86–279
~ de tamanho 81–249
~ do transmissor 86–273
~ dos meteoros esporádicos 86–233
~ dos pulsos 86–279
~ horária de meteoros 86–135
~ máxima utilizável 78–37
~ mínima utilizável 78–33
~ moduladora 86–275
fricção das marés 82–312
fronteira da magnetosfera 75–9
frontogênese 77–74
frontólise 77–75
fuga das galáxias 98–298
~ de estrelas de um cúmulo 98–194
~ de uma estrela do sistema 96–159
fugitiva 96–160
fulguração 89–175, 95–69
~ com duas fitas 89–187a
~ de clase 2 89–206
~ de importância 2 89–206
~ de prótons 89–185
~ de raios X 89–183
~ em luz branca 89–182
~ em raios X 89–183
~ estelar 95–70
~ homóloga 89–178
~ não-protônica 89–186
~ no bordo 89–202
~ no disco 89–204
~ no limbo 89–202
~ ó(p)tica 89–181
~ por detrás do bordo 89–203
~ por detrás do limbo 89–203
~ protônica 89–185
~ rádio 89–180
~ sem prótons 89–186
~ simpatética 89–179
~ sobre o disco 89–204
~ visível 89–181
fumaça 97–57
fumarola 82–163
fumo 97–57
função da freqüência de massa 84–52
~ das massas 96–68
~ de densidade 98–88
~ de difusão 77–58
~ de distribuičao 98–143
~ de espalhamento 77–58
~ de fase 83–78
~ de luminosidade 98–89
~ de massa 96–68
~ de peso 93–58
~ de ponderação 93–58
~ de saturação 93–60
~ de visibilidade 89–55
~ fonte 93–34
~ integro-exponencial 93–62
~ origem 93–34
~ peso 93–58
~ saturação 93–60

P

funções esféricas zonais 95–101
~ harmônicas não-zonais 95–102
~ harmônicas tesserais 95–102
fundo de uma cratera 81–259
~ do oceano 72–37
~ marítimo 72–38
~ meteórico 86–232
~ oceânico 72–37
~ plano 81–260
~ sísmico 73–55
funil 89–269
protuberância de ~ 89–269
furacão 72–51, 77–78
~ tropical 77–78
furo 82–61
fusão 86–181

G

gal 71–65
galáctico 98–2
galactocêntrico 98–10
Galáxia 98–1
galáxia 98–1, 98–201
~ anã 98–220
~ anular 98–259
~ ativa 98–260
~ azul de Haro 98–266
~ barrada 98–253
~ com núcleo múltiplo 98–319
~ com um braço 98–42
~ compacta 98–264
~ compacta azul 98–265
~ de dois braços 98–43
~ de Haro 98–266
~ de múltiplos braços 98–44
~ de raios X 98–203
~ de Seyfort 98–262
~ de transição 98–248
~ de um braço 98–42
~ de um só braço 98–42
~ de vários braços 98–44
~ deformada 98–316
~ depois de uma erupção 98–270
~ desintegrada 98–317
~ distorcida 98–316
~ do campo 98–340
~ do cúmulo 98–338
~ do Grupo Local 98–327
~ do tipo N 98–263
~ dupla 98–305
~ elipsoidal 98–241
~ elíptica 98–241
~ elíptica anã 98–242
~ elíptica gigante 98–243
~ em explosão 98–268
~ em forma de fuso 98–239
~ em raios X 98–203
~ eruptiva 98–269
~ esférica 98–244
~ esferoidal 98–244
~ espiral 98–249
~ espiral normal 98–250
~ explosiva 98–268
~ fusiforme 98–239
~ gigante 98–222
~ infravermelha 98–205
~ irregular 98–237
~ lenticular 98–245
~ local 98–327
~ mãe 98–284
~ mais brilhante de cúmulo 98–339
~ muito achatada 98–252
~ múltipla 98–308
~ não-membro do cúmulo 98–340
~ não-pertencente ao cúmulo 98–340
~ normal 98–221
~ observável em luz 98–204
~ observável no infravermelho 98–205
~ ó(p)tica 98–204
~ peculiar 98–256
~ perturbada 98–315
~ pós-eruptiva 98–270
~ quase-estelar 98–361
~ regular 98–236
~ rica em gás 98–238
~ singular 98–258
~ supergigante 98–223
~ tipo girino 98–281
~ vísivel 98–204
~ vista de frente 98–247
~ vista de perfil 98–246
~ vizinha 98–211
galáxias em colisão 98–321
~ interativas 98–310
~ interconectadas 98–306
ganho de energia 88–253
Ganimedes 83–207
garganta vulcânica 73–125
gás bariônico 93–194
~ circungaláctico 98–351
~ de bárions 93–194
~ de híperons 93–193
~ de nêutrons 93–192
~ degenerado 93–179
~ em um cúmulo 98–352
~ entre os braços 98–40
~ hiperônico 93–193
~ intercúmulo 98–353
~ inter(e)stelar 97–9
~ intergaláctico em cúmulo 98–352
~ intergaláctico entre cúmulos 98–353
~ intracúmulo 98–352
~ não-degenerado 93–182
~ neutrônico 93–192
~ nobre primordial 87–133
~ nobre radiogênico 87–132
~ relativista 97–152
~ relativístico 97–152
gases captados do vento solar 82–148
~ congelados 85–45
~ incandescentes 83a–43
geada 77–177
~ branca 77–177
geleira 72–112
~ alpina 72–114
~ continental 72–116
~ de montanha 72–113
gelo 85–42
~ amorfo 85–44
~ cúbico 85–46
~ de água 85–47
~ flutuante 72–102
~ seco 83–144
~ sujo 97–73
Gêmeos 92–51
Gemini 92–51
Gemínidas 86–257
Geminídeos 86–257
Gemma 92–117
gênese 94–52
~ dos elementos 94–52
geociências 79–1
geocorona 76–73
~ de hidrogênio neutro 76–74
geocronologia 74–67
geodesia espacial 71–31
~ lunar 81–109
~ por satélite 71–32
geodésica 99–153
geofisica 72–1
~ cósmica 84–57
geóide 71–16
~ compensado 71–17
~ regularizado 71–17
geologia 72–2
~ de gelo 72–103
~ dos planetas 83–5
~ lunar 82–47
~ planetária 83–5
geomagnetismo 75–24
geometria euclidiana 99–108
~ não-euclidiana 99–109
~ riemanniana 99–113
geopotencial 71–60
geoquímica 74–30
geosfera 72–7
geossinclinal 73–105
geotectônica 73–95
geotermia 74–58
geração de energia 93–105
~ de estrelas 94–23
~ estelar 94–23
Giacobínidas 86–252
Giacobinídeos 86–252
gigante 91–143
~ azul 91–142a
estrela ~ 91–143
~ vermelha 91–152a
gigantes brilhantes 91–149
~ dos tipos finais 91–144a
~ dos tipos iniciais 91–143a
~ luminosas 91–149
~ normais 91–150
Girafa 92–24

P

glaciação 72–104
glacial 72–106
glaciar 72–112
~ de montanha 72–113
~ de vale 72–114
glaciário 72–106
glaciologia 72–103
globo celeste 92–7
~ derretido 83a–64
~ fundido 83a–64
~ lunar 81–112
~ terrestre 71–10
glóbulo 97–133
~ de poeira 98–290
glória 79–63
Golfinho 92–45
golfo 81–294
Golfo do Acro-Iris 81–295
~Tórrido 81–296
gota 76–113
~ negra 83–52
gotícula 76–100
Gotlândico 74–80
graben 82–132
gradiente 93–94
~ absoluto 91–111
~ adiabático de temperatura 93–98
~ adiabático seco 76–14
~ bárico 77–106
~ barométrico 77–106
~ de densidade 98–87
~ de pressão 93–95
~ de temperatura 93–96
~ de temperatura convectivo 93–157
~ espe(c)trofotométrico 91–112
~ geotérmico 74–59
~ radiativo de temperatura 93–97
~ superadiabático 93–99
~ térmico 93–96
~ vertical de temperatura 76–14
Grande Caçador 92–73
~ força unificada 99–170v
~ halo 79–70
~ interação unificada 99–170v
~ Mancha Vermelha 83–193
~ Nuvem de Magalhães 98–213
~ Perturbação Austral 83–194
~ planeta 83–13
~ unificação 99–170u
grandeza do eclipse 81–79
granito 74–8
granizo 77–178
granulacão facular 89–163
~ fina 82–114
~ fotosférica 88–176
~ média 82–116
grânulo 88–178
~ da umbra 88–180
~ facular 88–179
~ fotosférico 88–181
grão de ferro 97–74
~ de fines 82–110
~ de gelo 85–52
~ de grafite 97–75
~ de polímero 97–77a
~ de silicato 97–76
~ fino 82–110
~ recoberto de gelo 97–77
~ rochoso 85–54
grãos de Baily 88–32
~ de poeira inter(e)stelar 97–59
~ grossos 82–117
~ inter(e)stelares 97–59
grau de concentração 96–128
~ de degenerescência 93–174
~ geotérmico 74–59
gravidade 71–56
gravimetria 71–63
gravímetro 71–70
gravitação própria 94–34
greta 81–165
grimpa 77–104
Grou 92–52
Grua 92–52
grupamento amino 87–71
grupo asteroidal 84–90
~ asteroidal de comensurabilidade 84–101
~ bipolar 89–111
~ complexo 89–112
~ de asteróides 84–90
~ de comensurabilidade 84–101
~ de cometas 85–105
~ de galáxias 98–325
~ de manchas 89–105
~ de manchas solares 89–105
~ de pequenos planetas 84–90
~ do ciclo anterior 89–107
~ do ciclo precedente 89–107
~ do novo ciclo 89–106
~ Local de Galáxias 98–326
~ recorrente 89–108
~ troiano 84–91
~ unipolar 89–110
Grus 92–52
guyot 72–44

H

halo 79–67
~ de 46° 79–70
~ de 22° 79–69
~ frio 98–47
~ galáctico 98–45
grande ~ 79–70
~ lunar 79–68
~ magnético 98–48
pequeno ~ 79–69
~ quente 98–46
halos de um cometa 85–68
Hamal 92–118
harmônicas tesserais 95–102
~ zonais 93–55
heliofísica 88–6
heli(o) 88–5
helio-geofísica 84–56
heliógrafo 77–61, 88–36
~ monocromático 88–38
heliograma em raios X 88–74
heliômetro 88–12
heliosfera 88–255
heliosférico 88–256
heliosismologia 88–223b
helióstato 88–45
hemisfério diurno 83–67
~ do lado da Terra 81–15
~ escuro 83–68
~ iluminado 81–126
~ iluminado pelo Sol 75–20
~ não-iluminado 75–21
~ noturno 83–68
~ visível 89–27
~ voltado para a Terra 81–15
Hercules 92–53
Hércules 92–53
Hermes 84–83
heterogeneidade da densidade ele(c)trónica 78–8
~ do campo magnético 78–9
~ inclinada 88–190
~ ionosférica 78–7
heterosfera 76–61
hexaédrita 87–84
Híades 96–111a
Hidra Austral 92–56
~ Fêmea 92–55
~ Macho 92–56
hidrodinâmica estelar 98–142
hidrófilo 76–92
hidrofóbico 76–93
hidrófobo 76–93
hidrogênio inter(e)stelar 97–46
~ metálico 83–202
~ molecular 97–31
~ neutro 97–47
hidrosfera 72–21
hidrossol 76–90
hidroxila 97–22
hierarquia de estrutura 99–47
~ de estrutura do Universo 99–47
higrômetro 77–120
hipergigantes 91–147
hipocentro 73–17
hipótese 94–6
~ catastrófica 83a–14
~ cosmogônica 94–8
~ das duas correntes 98–99
~ de captura 83a–18
~ de colisão 83a–14
~ de erupção 89–63
~ de impacto 81–261
~ de Lambert 99–45
~ de maré 83a–14
~ de Weizsäcker 83a–11
~ dos turbilhões 83a–11
~ dos vórtices 83a–11

~ meteorítica 81–262
~ meteoroídica 81–262
~ nebular 83a–7
hiss 78–64
história térmica 74–66, 83a–60
holoceno 72–109
homólogo 93–131
homopausa 76–65
homosfera 76–60
hora de aparição 86–49
horizontle cósmico 99–10
~ cosmológico 99–179
~ de acontecimentos 99–43
~ de eventos 99–43
~ de massas 99–41
~ do universo 99–10
Horologium 92–54
Hydra 92–55
Hydrus 92–56

I

Ícaro 84–84
iceberg 72–117
idade absoluta 82–269, 83a–70
~ característica 99–79
~ chumbo-chumbo 74–100
~ da Lua 81–31
~ da Terra 83a–68
~ das estrelas 94–61
~ de estrôncio 82–271
~ de exposição aos raios cósmicos 87–139
~ de radiação 87–139
~ de razão isotópica do chumbo 74–100
~ de retencão de gás 87–135
~ de rubídio-estrôncio 82–271
~ de solidificação 87–137
~ de um enxame de meteoros 86–246
~ de uma corrente 86–245
~ do universo 99–12
~ dos meteoritos 87–127
~ estratigráfica 82–267
~ geológica 74–90
~ relativa 82–268
~ urânio-chumbo 74–101
identidades de Lagrange 98–182
~ lagrangianas 98–182
~ lagrangianas generalizadas 98–183
ignicão do carbônio 94–80
~ do carbono 94–80
~ do hélio 94–71
~ do hidrogênio 94–55
ilha nova 72–81
ilmenita 82–91
imagens monocromáticas 97–158
imantação inducida 75–29
~ residual 75–29
imersão 81–98
impacto 82–16
~ artificial 82–226
~ cometário 82–230
~ de um meteorito 87–33
~ meteorítico 82–236
~ meteoroidal 82–236
~ natural 82–227
implosão do núcleo estelar 94–110
importância de uma erupção 89–205
inclinação axial 83–35
~ da órbita 96–51
~ do eixo 83–35
~ do eixo de rotação 91–50
~ do plano orbital 96–51
~ do ramo ascendente 95–29
~ magnética 75–58
~ sobre o equador lunar 81–37
inclusão orgânica 87–66
incremento de massa 83a–38
indicador da estrutura espiral 98–40
~ da opacidade ionosférica relativa 78–45
~ de alcance 86–266
indicatriz de difusão 77–58
~ de espalhamento 77–58
índice B–V 96–145
~ azul menos amarelo 96–145
~ de atividade da corrente 86–230
~ de atividade solar 89–8
~ de calor 91–81
~ de concentração 96–101
~ de cor 91–76
~ de cor B–V 96–145
~ de cor amarelo 83–88
~ de cor azul menos amarelo 96–145
~ de cor azul menos visual 96–145
~ de cor composto 96–138
~ de cor especial 83–87
~ de cor global 96–138
~ de cor integral 96–138
~ de cor intrínseco 91–75
~ de cor para o amarelo 83–88
~ de cor U–B 96–146
~ de cor ultravioleta menos azul 96–146
~ estrutural de um cúmulo 96–80
~ K da atividade geomagnética 75–89
~ planetário K_p 75–90
~ politrópico 93–128
~ solar 89–8
~ triorário 75–89
~ U–B 96–146
~ ultravioleta menos azul 96–146
Índio 92–57
Indus 92–57
infinidade do universo 99–6
inflação 99–170c
~ cosmológica 99–170c
início brusco 75–74
~ de trajetória aparente 86–36
~ do ciclo 89–82
~ súbito da tempestade 75–74
inomogeneidades macroscópicas 88–206
insolação 88–9
instabilidade 93–211
~ de maré 96–107
~ de pulsação 93–213
~ de rotação 83a–8, 93–216
~ de vibração 93–214
~ gravitacional 94–35
~ local 93–154
~ pulsacional 93–213
~ rotacional 83a–8, 93–216
~ secular 93–212
~ vibracional 93–214
instante da passagem pelo periastro 96–60
~ da queda de um meteorito 87–51
~ de alunissagem 82–16
~ de aparição 86–49
~ de chegada 73–50
instável 93–210
instrumento automatizado 82–41
~ de espalhamento alfa 82–76
instrumentos solares 88–35
integral de fase 83–79
terceira ~ do movimento 98–186
intemperismo 74–57
intensidade central da raia Hα 89–211
~ da gravidade 71–62
~ de terremoto 73–33
~ do campo geomagnético 75–55
~ do campo magnético 84–11
~ do som 76–38
~ horizontal 75–60
~ horizontal do campo magnético 75–60
~ o(m)nidirecional 75–104
~ sísmica 73–33
~ unidirecional 75–103
~ vertical 75–62
interação atmosférica 79–120
~ coletiva 98–168
~ de maré 81–120, 98–311
~ ele(c)trofraca 99–170x
~ superunificada 99–170r
interestelar = interstelar 97–2
interface 93–141
~ de salinidade 72–67
interferências estáticas 78–50
interferômetro a comutação de fase 88–113
~ a feixe de lapiz 88–121
~ a feixe estreito 88–121
~ a multi-elementos 88–123
~ a varredura de fase 88–109
~ a varredura em fase 88–109
~ a varredura em freqüência 88–110
~ a varrimento em fase 88–109
~ com antenas múltiplas 88–123
~ de base variável 88–107
~ de Christiansen 88–123
~ de dois elementos 88–98
~ de lóbulo em leque 88–122
~ de lóbulo estreito 88–121

P

~ de longa linha de base 88–104
~ em rede 88–123
~ marítimo 88–106
~ multionda 88–123
~ por ligação radioelé(c)trica 88–116
~ por varredura de lóbulo 88–109
~ produzindo feixe em leque 88–122
interfilamentar 89–99
espaço ~ 89–99
intergaláctico 97–6
interglacial 72–108
interglaciário 77–190
interior da Terra 72–3
~ das estrelas 93–72
~ estelar 93–72
interplanetário 84–8
interrupção pelo obturador 86–150
inter(e)stelar 97–2
interzona 93–141
intrusão 73–133
~ concordante 73–134
~ discordante 73–137
~ ígnea 82–98
~ magmática 82–98
inundação 72–75
invariância em uma transformação 99–133
invariante no tempo 98–150
invariável com o tempo 98–150
invasão de partículas de alta energia 78–43
inversão atmosférica 77–37
~ da polaridade magnética 89–153
~ de população 97–15
~ de temperatura 77–36
~ no centro da raia Hα 89–211
~ no solo 77–38
~ superficial 77–38
investigação subsuperficial 82–82
invólucro característico 98–177
~ circum-estelar 93–10
~ circum-estelar de pó 97–117
~ circunstelar 93–10
~ comum 96–32
~ comum de uma binária 96–32
~ convectivo 93–136
~ de pó 97–116
~ de poeira 97–116
~ em expansão 93–12
~ estacionário 93–11
~ extenso 93–8
~ radiativo 93–135
invólucros de um cometa 85–69
Io 83–205
íon de hidrogênio 93–65
~ grande 76–115
~ Langevin 76–115
~ negativo de hidrogênio 93–66
ionização pelos meteoros 86–281
ionizado 89–289
altamente ~ 89–289
ionograma 78–23
ionosfera 76–69, 78–1
~ exterior 76–76
ionosférico 78–2
ionossonda 78–22
irradiação de raios cósmicos 82–250
~ por raios cósmicos 82–250
irregularidade de pequena escala 84–13
~ do campo magnético 84–12
isóbata 71–55
isocianeto de hidrogênio 97–40
isóclina 75–59
isodinâmica 75–61
isofota 81–92
isofotas de uma nebulosa 97–124
isógona 75–57
isoípsa 81–139
isonitrilo de hidrogênio 97–40
isopleta 96–125
isorrotação 88–130
isostasia 73–98
isótopo cosmogênico 82–147
~ de vida curta 83a–74
~ de vida longa 83a–75
~ do chumbo 83a–76
~ primordial 82–146
~ produzido por raios cósmicos 82–147
itinerário 82–37
~ da travesia 82–37

J

ja(c)to 85–55
~ auroral 79–113
~ curvado 85–58
~ do núcleo galáctico 98–278
~ ele(c)trônico 79–114
~ em espiral 85–59
~ ó(p)tico 98–279
~ retilíneo 85–57
janela de transmissão 83–264
~ veneziana 88–126
janelas galácticas 98–57
~ na zona de ausência 98–57
jaspes 88–210
~ brilhantes 88–211
~ escuros 88–212
~ obscuros 88–212
joviano 83–171
Juno 84–85
Júpiter 83–170
Jurássico 74–85

K

K-coronômetro 88–66
kamacita 87–92

L

Lacerta 92–58
lacólito = lacolito 73–136, 82–179
laço interplanetário 84–21
~ magnético 84–19
~ solar interplanetário 84–20
lacuna 84–88
~ de comensurabilidade 84–88
~ de Hertzsprung 91–144
lacunas de Kirkwood 84–89
lado da noite 75–21
~ distante 81–16
~ do dia 75–20
~ do Sol 75–20
~ escuro 75–21
~ frontal da Lua 81–15
~ iluminado 81–34
~ iluminado pelo Sol 75–20
~ não-visível 81–16
~ oposto da Lua 81–16
~ traseiro 81–16
~ traseiro da Lua 81–16
~ visível 81–15
~ voltado para o Sol 81–34
Lagartixa 92–58
Lagarto 92–58
lago 81–297
~ da Morte 81–298
~ da Primavera 81–302
~ de lava 82–183
~ do Outono 81–301
~ do Verão 81–300
~ dos Sonhos 81–299
lama 72–58
lamelas 87–104
lâmina 83–149
~ de compensação 88–92
~ de mica 88–87
~ de um quarto de onda 88–126
~ neutra 75–16
~ quatro de onda 88–126
lamínula de compensação 88–92
~ de mica 88–87
lâmpada de raios catódicos 86–293
lampejo de hélio 94–79
lapíli 82–169
largura da zona de silêncio 78–38
~ do pulso 86–280
~ efetiva de uma raia 89–208
latitude areográfica 83–140
de alta ~ 89–53
de baixa ~ 89–52
~ geomagnética 75–52
~ heliográfica 89–42
~ limite norte 81–105
~ limite sul 81–106
~ selenográfica 81–11
latitudes limites da visibilidade 81–104
lava solidificada 82–176
Leão 92–59
~ Menor 92–60
Lebre 92–61
lei das distancias planetárias 83a–2
~ das latitudes 89–75
~ de alternação de polaridade 89–154

P

~ de avermelhamento 97–92
~ de Hale 89–154
~ de Hubble 98–299
~ de Kramers 93–102
~ de obscurecimento 88–143
~ de polaridade 89–154
~ de resistência 86–179
~ de rotação 98–110
~ de Spoerer 89–75
~ de Titius e Bode 83a–2
leito do oceano 72–37
~ marítimo 72–38
lente gravitacional 98–256a
Leo 92–59
~ Minor 92–60
Leônidas 86–255
Leonídeos 86–255
Lepus 92–61
leque 85–60
~ coronal 89–295
~ curvado 85–62
~ invertido 85–63
~ reto 85–61
em forma de ~ 85–74
levantamento celeste 92–11
~ de asteróides 84–94
~ de geleiras 72–85
~ de satélites 83–26
~ de travessia 82–35
~ do céu 92–11
~ fotográfico 82–30
~ geológico da Lua 82–50
~ magnético 75–64
~ magnético mundial 75–65
~ poligonal 82–35
~ selenofísico 82–39
~ selenológico 82–50
liber(t)ação de energia 93–109
Libra 92–62
libração 81–48
~ diurna 81–52
~ em latitude 81–50
~ em longitude 81–49
~ física 81–53
~ geométrica 81–51
~ ó(p)tica 81–51
~ paralá(c)tica 81–52
limbo, acima do 89–30
~ brilhante 81–96
~ da Lua 81–95
~ do leste 89–34
~ do oeste 89–33
~ do Sol 89–28
~ escuro 81–97
~ extremo 89–32
~ iluminado 81–96
~ leste 89–34
no ~ 89–29
~ oeste 89–33
~ solar 89–28
limiar do secundário 96–40
curva ~ do componente secundário 96–40
limite de Chandrasekhar 93–189
~ de neve 72–111
~ do setor 84–18
~ norte do eclipse 88–20
~ sul do eclipse 88–19
~ superior 76–58
~ superior da atmosfera 76–25
limpo 77–149
Lince 92–64
língua da geleira 72–125
~ de gelo 72–125
~ de lava 82–187
~ glaciária 72–125
~ interplanetária 84–21
~ solar interplanetária 84–20
linha amarela 89–292
~ atmosférica 88–140a
~ circum-estelar 97–97
~ circunstelar 97–97
~ coronal 89–290
~ D 88–140
~ de arrefecimento 93–202
~ de avermelhamento 96–149
~ de base fixa 88–102
~ de base variável 88–103
~ de campo aberta 75–15
~ de corrente 82–194
~ de força corrotativa 75–14
~ de Fraunhofer 88–139
~ de idade zero 94–60
~ de neve 72–111
~ de nível 81–139
~ de prumo 71–71
~ de separação de polaridades opostas 89–151
~ de universo 99–154
~ de visada 91–28
~ de visão 91–28
~ deslocada para o vermelho 98–366
~ do fluxo 82–194
~ evolutiva 94–29
~ geodésica 99–153
~ inter(e)stelar de microondas 97–98
~ inter(e)stelar de rádio 97–98
~ inter(e)stelar ó(p)tica 97–99
~ isogônica 75–57
~ litorânea 72–31
na ~ de visada 91–29
~ neutra 89–151
~ telúrica 88–140a
~ verde 79–16, 89–291
~ vermelha 79–15, 89–293
linhas de Neumann 87–103
Lira 92–65
Líridas 86–248
Lirídeos 86–248
litificação 74–49
~ por impacto 82–238
litoral 72–30
litosfera 72–6
litossiderito 87–80
livre de poeira 82–119
Lobo 92–63
lóbulo de Roche 96–28
locação de peixes 72–84
~ de zona piscosa 72–84
local da alun(iss)agem 82–14
~ da descoberta 87–52
~ da queda de um meteorito 87–50
~ de encontro 87–52
~ de pouso 82–14
~ de pouso na Lua 82–14
~ proposto da alunagem 82–15
localizador do Sol 88–51
lodo 72–58
longitude areográfica 83–139
~ ativa 89–41
~ do periastro 96–59
~ geomagnética 75–53
~ heliográfica 89–40
~ selenográfica 81–10
Lua 81–2
~ cheia 81–27
~ crescente 81–20
~ eclipsada 81–57
meia ~ 81–26
~ minguante 81–21
~ nova 81–22
luminescência anticrepuscular 79–57
~ crepuscular do ar 79–4
~ diurna do ar 79–3
~ do ar 79–2
~ Lα do céu noturno 79–9
~ lunar 82–248
~ noturna do ar 79–5
luminosidade 91–128
~ de neutrino 94–96
~ de um meteoro 86–195
~ do céu noturno 79–8
~ do hidrogênio 79–145
~ neutrínica 94–96
luminoso por si próprio 96–45
lunação 81–18
lunamoto 82–214
lunar 81–3
lunicêntrico 81–131
Lupus 92–63
luz cinzenta 81–35
com ~ própria 96–45
~ do céu 77–53
~ do céu noturno 79–8
~ estelar 79–10
~ estelar difusa 79–12
~ estelar direta 79–11
~ estelar espalhada 79–12
~ extraterrestre 79–7
~ galáctica 79–14
~ galáctica difusa 97–56
~ purpúrea 79–54
sem ~ própria 96–44
~ solar 77–49, 88–9
~ zodiacal 79–13, 84–61

P

~ zodiacal falsa 84–62
Lynx 92–64
Lyra 92–65

M

má visibilidade 77–26
maar 82–161
macroestrutura 99–9
macrometeorito 87–4
magma 73–131
magnetismo das rochas 75–43
~ estelar 95–54
~ inducido 75–29
~ residual 75–28
~ terrestre 75–24
magnetita 82–92
magnetização deposicional 75–33
~ detrítica 75–33
~ induzida 75–29
~ inversa 75–37
~ primária 75–35
~ química 75–31
~ remanente 75–28
~ remanescente 75–28
~ residual 75–28
~ reversa 75–37
~ secundária 75–36
~ sedimentar 75–33
~ termorremanente 75–30
~ termorremanescente 75–30
~ viscosa 75–32
magnetógrafo 75–83
~ do campo longitudinal 88–83
~ longitudinal 88–83
~ solar 88–82
~ ve(c)torial 88–84
magnetograma 75–84
magneto(h)eliógrafo 88–93
~ de Leighton 88–93
magnetóide 98–25
magnetômetro portátil 82–44
magnetopausa 75–9
magnetosfera 75–7
~ de um pulsar 95–147
magnitude absoluta 91–64
~ absoluta de um meteoro 86–72
~ absoluta de um planeta 83–86
~ aparente 91–61
~ aparente de um meteoro 86–71
~ bolométrica 91–68
~ de um terremoto 73–34
~ ele(c)tronográfica 91–67
~ estelar 91–56
~ fotográfica 91–66
~ fotovisual 91–69
~ global 96–136
~ infravermelha 91–71
~ integral 96–136
~ limite 91–62
~ média 83–85
~ média na oposição 83–85
~ radiométrica 91–70
~ reduzida ao zênite 91–63
~ sísmica 73–34
~ total 96–136
~ visual 91–65
~ zenital 91–63
maior meteorito 87–41
Mancha Branca 83–214
~ Clara 83–214
~ Vermelha 83–193
mancha da frente 89–113
~ de cauda 89–114
~ de este 89–114
~ de oeste 89–113
~ *f* 89–114
~ invisível 89–159
~ isolada 89–116
~ *p* 89–113
~ precedente 89–113
~ principal 89–115
~ quente 95–112a
~ seguinte 89–114
~ solar 89–88
~ Vermelha 83–193
manchas estelares 97–62
manto da Terra 72–15
~ de gelo 97–72
~ superior 72–16
~ terrestre 72–15
mapa batimétrico 71–54
~ da gravidade 71–79
~ da Lua 81–133
~ da variável 95–15
~ de curvas de nível 81–141
~ de linhas de nível 81–141
~ de Marte 83–136
~ do tempo 77–12
~ estrutural 73–97
~ físico lunar 81–134
~ fotográfico da Lua 81–137
~ fotográfico lunar 81–137
~ gravimétrico 81–143
~ hipsométrico 81–141
~ lunar 81–133
~ rádio 81–135a
~ sinó(p)tico 77–11
~ te(c)tônico 73–96
~ térmico lunar 81–135
~ topográfico da Lua 81–138
~ topográfico lunar 81–138
mapeamento da costa 72–82
~ da Lua 81–111
~ lunar 81–111
~ selenológico 82–51
~ térmico 72–83
Máquina de Ar 92–15
~ Pneumática 92–15
máquina fotográfica para todo o céu 79–104
mar 81–266, 83–156
~ Austral 81–286
baixa ~ 82–283
~ circular 81–267
~ Conhecido 81–289
~ continental 72–26
~ da Engenhosidade 81–292
~ da Fecundidade 81–274
~ da Fertilidade 81–274
~ da Margem 81–285
~ da Serenidade 81–276
~ da Tranqüilidade 81–275
~ da Umidade 81–282
~ das Chuvas 81–273
~ das Crises 81–280
~ das Nuvens 81–281
~ das Ondas 81–290
~ das Serpentes 81–291
~ de Humboldt 81–288
~ de Moscou 81–283
~ de Smyth 81–287
~ do Engenho 81–292
~ do Frio 81–278
~ do Néctar 81–279
~ dos Humores 81–282
~ dos Néctares 81–279
~ interior 72–26
~ marginal 72–27, 81 285
~ mediterrâneo 72–26
~ Moscovita 81–283
~ Oriental 81–284
~ Úmido 81–282
marca permanente 83–106
~ primária 81–150
~ secundária 81–151
~ superficial 83–105
~ vulcânica 82–159
marcas superficiais 81–148
marciano 83–135
maré alta 72–90
~ alta de sizígias 82–273
~ baixa 82–283
~ cheia 82–281
~ de águas mortas 82–272
~ de águas vivas 82–273
~ de quadratura 82–272
~ de quarto 82–272
~ de sizígias 82–273
~ descendente 82–282
~ montante 82–280
~ morta 82–272
~ vazia 72–91
maré-cheia 72–90
marégrafo 82–311
marêmetro 82–311
maremoto 73–2c
mareógrafo 82–311
mareômetro 82–311
marés 72–89
~ atmosféricas 82–294
~ da crosta 82–297
~ geomagnéticas 82–295
~ gravitacionais 82–298
~ induzidas pela Terra 82–275
~ lunares 82–308

P

~ marítimas 82–309
~ oceânicas 82–310
~ sólidas 82–296
~ térmicas 82–299
~ terrestres 82–296
maria 81–266
Markab 92–119
Marte 83–133
mascon 81–145
maser cósmico 97–13
~ inter(e)stelar 97–14
~ natural 97–13
massa ausente 99–87
~ crítica 93–189
~ crítica das anãs brancas 93–189
~ da partícula meteórica 86–163
~ de ar 77–67
~ de Chandrasekhar 93–189
~ de Planck 99–170f
~ de repouso 99–143
~ do meteoro 86–163
~ do Sol 91–122
~ elementar de Planck 99–170f
~ em repouso 99–143
~ escondida 99–88
~ estelar 91–121
~ faltante 99–87
~ gasosa 94–32
~ global do cúmulo 96–139
~ gravitacional 99–149
~ inercial 99–148
~ inicial 86–164
~ integral do cúmulo 96–139
~ limite 93–189
~ meteórica 86–163
~ original 86–164
~ principal 87–41
~ relativista 99–141
~ relativística 99–141
~ solar 91–122
~ total do cúmulo 96–139
~ variável 86–167, 96–70
matacão 82–130
matéria absorvente 97–85
~ adsorvida 85–131
~ ausente 99–87
~ cósmica 99–7
~ estelar 93–124
~ inter(e)stelar 97–1
~ intergaláctica 98–349
~ interplanetária 84–1
~ luminosa 97–109
~ meteorítica 87–9
~ obscura 97–108
~ orgânica meteorítica 87–68
~ pré-estelar 94–16
~ primordial 99–170
materiais detríticos 81–236
~ vulcânicos ejetados 82–166
material carbonâceo 87–64
~ carbônico 87–64
~ cósmico 99–7
~ da superfície lunar 82–59
~ de cômoros 82–103
~ estelar 93–124
~ extraterrestre 72–56
~ lunar 82–58
~ pré-estelar 94–16
~ refratário 83a–55
matriz 82–118
~ de granulação fina 82–115
máximo 76–55
~ achatado 86–236
~ agudo 86–235
~ da atividade solar 89–2
~ da camada E 76–56
~ de uma camada 76–55
pós- ~ 95–33
pré- ~ 95–30
~ primário 95–31
~ principal 95–31
~ secundário 95–32
~ solar 89–2
mecanismo de aquecimento 88–214
~ de bombeamento 97–16
~ de destruição 97–79
~ de excitação 79–27
~ de frenagem 91–46
~ de travagem 91–46
medição absoluta da gravidade 71–68
~ da gravidade 71–67
~ do grau 71–22
~ relativa da gravidade 71–69
medida absoluta da gravitade 71–68
~ da gravidade 71–67
~ de emissão 88–234
~ micrométrica 96–49
~ relativa da gravidade 71–69
medidas polarimétricas 83–80
medidor da opacidade ionosférica 78–45
meia-lua 81–26
meia-noite magnética 79–131
meio ambiente 76–3
~ do eclipse 81–78
~ inter(e)stelar 97–8
~ turvo 83–251
membro do aglomerado 96–82
~ do cúmulo 96–82
~ do sistema 98–133
~ incerto do aglomerado 96–87
~ incerto do cúmulo 96–87
~ provável do aglomerado 96–86
~ provável do cúmulo 96–86
Mensa 92–66
Mercúrio 83–120
meridiano central 89–43
~ do meio-dia 75–17
~ do nascente 75–18
~ do poente 75–19
~ magnético 75–49
mês 81–42
~ anomalístico 81–43
~ dracônico 81–47
~ lunar 81–18
~ nódico 81–47
~ sideral 81–45
~ sinódico 81–46
~ trópico 81–44
Mesa 92–66
mesa submarina 72–44
mesopausa 76–67
mesosfera 76–66
mesossiderito 87–79
Mesozóico 74–72
metade do disco de Vênus 83–132
metagaláxia 98–357
metalicidade 91–170
metamorfismo 74–34
~ de choque 82–237
~ de conta(c)to 74–35
~ dinâmico 74–36
~ térmico 74–37
metanal 97–27
metanamida 97–35
metanimina 97–45
metanol 97–32
metassomatismo 74–44
metassomatose 74–44
meteórico 86–4
meteorítico 87–2
meteorito 87–1
~ acondrítico 87–77
~ carbonáceo 87–62
~ férrico 87–82
~ ferrolítico 87–80
~ lítico 87–60
maior ~ 87–41
~ orientado 87–112
~ original 87–123
meteorização 74–57
~ espacial 82–244
meteoro 86–1
~ artificial 86–21
~ brilhante 86–12
~ com esteira 86–17
~ com rast(r)o 86–17
~ com rast(r)o fusiforme 86–77
~ com variações luminosas 86–85
~ cometário 86–45
~ de chuva 86–24
~ de enxame 86–24
~ de luz pulsante 86–85
~ de origem cometária 86–45
~ de pó 86–190
~ diurno 86–22
~ do sistema solar 86–44
~ duplo 86–131
~ eclíptico 86–46
~ esporádico 86–23
~ estacionário 86–34
~ fotografado 86–13
~ fotografado de dois locais 86–155
~ fotográfico 86–13
~ fraco 86–15
~ fusiforme 86–77
~ inter(e)stelar 86–43

P

~ interplanetário 86–44
~ hiperbólico 86–43
~ lento 86–88
~ múltiplo 86–133
~ não-pertencente a um enxame 86–25
~ ordinário 86–14
~ pertencente a um enxame 86–24
~ rápido 86–87
~ telescópico 86–18
~ telescópico fraco 86–19
~ telescópico muito fraco 86–20
~ ultratelescópico 86–20
~ visível 86–16
~ visível a olho nu 86–16
~ visual 86–14
meteoróide 86–160
~ poroso 85–50
meteorologia 77–6
~ dinâmica 77–7
~ sinó(p)tica 77–8
metilacetileno 97–41
metilidino 97–23
método das funções de peso 93–59
~ de Argelander 95–10
~ de contagem dupla 86–132
~ do hélio (de determinação da idade) 87–128
~ do urânio 74–96
~ dos passos 95–10
~ estereoscópico 81–115
~ iterativo 93–56
~ radi(o)ativo 74–92
~ radiométrico 74–92
~ variacional 93–57
métrica 99–151
~ de Robertson-Walker 99–152a
~ de Schwarzschild 99–152
mica 74–23
~ branca 74–24
~ preta 74–25
microbrecha lunar 82–86
microcratera 81–201
microestrutura 82–73
microfone de fio quente 76–46
microfografia 87–140
micrometeoritos 87–3
micrometeoro 86–20
micrometeorologia 77–9
micropulsações geomagnéticas 75–82
~ magnéticas 75–82
microrreversibilidade 93–41
Microscópio 92–67
~ de cintilação 91–34
Microscopium 92–67
microssismos 73–53
microtectita = microtectito 87–97
migmatito 74–39
migração da zona ativa 89–74
~ de protuberâncias para os pólos 89–231
milibar 77–47
miligal 71–66
milímetro de mercúrio 77–48
~ da coluna de mercúrio 77–48
milionésimo do disco solar 89–69
milionésimos do hemisfério visível 89–68
mineral 74–12
~ constituinte de rocha 82–99
mínimo da atividade solar 89–4
~ primário 95–26
~ principal 95–26
~ secundário 95–27
~ solar 89–4
miragem 79–64
missão a um asteróide 84–99
missão a um asteróide com retorno de amostrais 84–98
mistura da matéria estelar 93–162
~ de gás e poeira 83a–25
~ de gás e radiação 93–125
~ de Russell 93–101
Mizar 92–120
modelo aberto 99–72
~ anisotrópico 99–65
~ anisótropo 99–65
~ assintótico 99–59
~ assintótico de primeira espécie 99–60
~ composto 93–140
~ cosmológico 99–46
~ cosmológico quente 99–166
~ da estrela 93–123
~ da mancha 89–102
~ da Terra 79–133
~ de água de cometas 85–113
~ de atmosfera 83–247, 93–17
~ de Charlier 99–67
~ de cobertura 93–18
~ de De Sitter 99–69
~ de dois componentes 88–196
~ de duas temperaturas 88–197
~ de Eddington 93–137
~ de Einstein 99–68
~ de Einstein e De Sitter 99–70
~ de estufa 83–262
~ de fonte em camada 93–143
~ de fonte pontual 93–138
~ de hidrogênio 83–197
~ de hidrogênio de um planeta 83–197
~ de temperatura zero 93–186
~ de temperatura zero de uma estrela 93–186
~ de um planeta 83–196
~ de universo 99–46
~ descoberto 93–19
~ em chafariz 85–85
~ em contração 99–56
~ em expansão 99–55
~ eolosférico 83–260
~ estacionário 99–52
~ estático 99–50
~ estelar 93–123
~ estelar inomogêneo 93–139
~ estelar não-homogêneo 93–139
~ estelar relativista 93–198
~ estelar relativístico 93–198
~ fechado 99–73
~ galáctico 98–163
~ hierárquico 99–48
~ homogêneo 99–64
~ inomogêneo 99–66
~ ionosférico 82–261
~ isotrópico 99–63
~ isótropo 99–63
~ matemático 98–324
~ monotônico 99–61
~ monotônico de primeira espécie 99–62
~ não-estacionário 99–53
~ não-estático 99–51
~ não-homogêneo 99–66
~ não-isotrópico 99–65
~ não-isótropo 99–65
~ padrão 93–137
~ politrópico 93–129
~ pulsante 99–57
~ pulsante de primeira espécie 99–58
~ sem cobertura 93–18
~ vazio 99–49
modelos de Friedmann 99–71
modificação da linha litorânea 72–78
~ do contorno da costa 72–78
modulação ele(c)troó(p)tica 88–86
módulo aparente 96–117
~ aparente de distancia 96–117
~ de alun(iss)agem 82–5
~ de comando 82–6
~ de distância 91–16
~ de elasticidade 73–59
~ de serviço 82–7
~ de Young 73–59
~ lunar 82–5
~ verdadeiro 96–118
moldavita = moldavito 87–99
molécula de ar 86–172
~ -filha 85–135a
~ inter(e)stelar 97–10
~ -mãe 85–135
~ poliatômica 97–21
momento angular 83a–4
~ cinético 83a–4
~ de rotação 83a–5
~ orbital 83a–6
monção 77–99
monitoração de satélites 83–26
monitoramento de erupções 89–219
Monoceros 92–68
monopolo magnético 99–170o
monossulfeto de carbono 97–34
monóxido de carbono 97–28
~ de silício 97–37
montanha submarina 72–43
montanhas de dobramento 73–86
~ dobradas 73–86

P

~ lunares 81–193
monte de gelo flutuante 72–117
~ submarino 72–43
morena 72–128
~ de fundo 72–129
~ frontal 72–130
~ lateral 72–131
~ terminal 72–130
morfologia lunar 81–147
morte térmica do universo
99–14
Mosca 92–69
moscovita 74–24
movimento aleatório 89–157
~ ao acaso 98–94
~ aparente 83–41
~ apsidal 96–71
~ casual 98–94
~ de grande escala 88–189
~ de massa 82–195
~ de pequena escala 88–188
~ de planetas 83–28
~ diário 85–29
~ diário do radiante 86–69
~ direto 83–43
~ em *z* 98–188
~ entre pontos espelhentos 75–119
~ epirogenético 73–73
~ glaciário 72–120
~ horário 81–72
~ horizontal 88–184
~ não-gravitacional 85–56
~ orbital 96–62
~ ordenado 98–95
~ paralá(c)tico 91–33
~ peculiar 91–31
~ planetário 83–28
~ próprio 91–19
~ próprio anual 91–20
~ próprio das manchas 89–117
~ próprio em ascensão reta 91–21
~ próprio em declinação 91–22
~ relativista do periélio 99–160
~ relativo 99–99
~ retrógrado 83–47
~ solar 98–62
~ solar básico 98–66
~ solar padrão 98–63
~ vertical 88–185
~ *z* 98–188
movimentos crustais 73–72
~ da crosta terrestre 73–72
~ internos 97–149
mudança climática 77–184
~ da Lua 81–19
~ de fase 83a–58
~ de forma 73–68
~ do clima 77–184
~ do tempo 77–22
~ evolutiva 94–27
multipolo 75–66
mundo 99–3
Musca 92–69
musicais atmosféricos 78–63

N

não-euclidiano 99–102
não-luminoso 96–44
não-relativista 99–96
não-relativístico 99–96
nascer do Sol 88–8
nascimento de uma mancha 89–121
natureza do objeto 85–24
Nebulosa América do Norte 97–140
~ anular 97–147
~ brilhante 97–121
~ cometária 97–142
~ cometóide 97–142
~ da Coruja 97–139
~ de Andrômeda 98–215
~ de emissão 97–123
~ de filamentos 97–129
~ de gás e pó 97–119
~ de gás e poeira 97–119
~ de Órion 97–136
~ de pó 97–115
~ de poeira 97–115
~ de reflexão 97–128
~ de Renda 97–137
~ difusa 97–122
~ do Caranguejo 97–134
~ do Laço 97–137
~ do Mocho 97–139
~ do Véu 97–137
~ em fuso 98–239
~ escura 97–132
~ espiral 98–249
~ extragaláctica 98–202
~ Ferradura 97–143
~ filamentar 97–129
~ filamentosa 97–129
~ fusiforme 98–239
~ galáctica 97–110
~ gasosa 97–112
~ irregular 98–237
~ luminosa 97–121
~ mãe 97–63
~ obscura 97–132
~ Ômega 97–143
~ planetária 97–144
~ pré-estelar 97–63
~ pré-solar 83a–21
~ primitiva 83a–21, 97–63
~ proto-solar 83a–21
~ refletora 97–128
~ regular 98–236
~ solar 83a–23
~ Trífida 97–141
nebulosidade 77–147
Neozóico 74–73
Netuno 83–236
neutralização mútua 79–29
nêutron de albedo 75–115
neutrosfera 76–64
nevado 72–118
neve 77–175
~ eterna 72–119
~ granulada 72–118
~ granular 72–118
~ granular endurecida 72–118
névoa 83–250
~ seca 77–27
nevoeiro 77–130
nife 72–18
nimbo de erupção 89–189
~ escuro 89–189
nimbo-estrato 77–145
nível crítico 76–26
~ de condensação 77–156
~ do mar 72–87
~ do som 76–38
~ hipsométrico 81–142
nó de uma protuberância 89–244
noite clara 79–59
~ de céu estrelado 79–3d
~ fotométrica 79–3f
~ fotometricamente clara 79–3f
~ lunar 81–33
~ sem luar 79–3e
~ sem nuvens 79–3g
nome de um cometa 85–31
nomenclatura das formações marcianas
83–108
~ dos detalhes de Marte 83–108
Norma 92–70
nova 95–110
~ extragaláctica 95–114
~ galáctica 95–113
~ lenta 95–120
pós- ~ 95–123
pré- ~ 95–122
~ rápida 95–119
~ recorrente 95–116
~ típica 95–111
novae 95–110
novas 95–110
novilúnio 81–22
novóide 95–118
nublado 83–253
núcleo central 72–18
~ cometário 85–36
~ convectivo 93–165
~ da associação 96–91
~ da Galáxia 98–18
~ da nebulosa planetária 97–148
~ da Terra 72–17
~ de condensação 76–103
~ de cúmulo 96–130
~ de furação 72–52
~ de gelo 85–40
~ de grafite 97–71
~ de hélio 93–167
~ de um cómeta 85–36
~ de um meteoro 86–102

P

~ de uma estrela 93–164
~ degenerado 93–170
~ estelar 85–38, 93–164
~ esteliforme 98–19
~ externo 72–19
~ galáctico 98–18
~ galáctico ativo 98–272
~ gigante 76–118
~ inomogêneo 93–169
~ interno 72–20
~ isotérmico 93–168
~ não-degenerado 93–171
~ não-homogéneo 93–169
~ peculiar 98–20
~ quase-estelar 98–19
~ radiativo 93–166
~ sólido 85–39
~ supercondutor 95–150
~ terrestre 72–17
nucleogênese 94–52
~ cosmológica 99–175
~ explosiva 94–57
~ não-explosiva 94–58
~ primordial 99 175
núcleos (de) Aitken 76–116
~ de congelamento 76–109
~ de cristalização 76–108
~ de gelo 76–104
~ de sublimação 76–110
nucleossíntese 94–52
~ cosmológica 99–175
~ explosiva 94–57
~ não-explosiva 94–58
~ primordial 99–175
número de controle 85–30
~ de encontros 98–122
~ de estrelas 98–80
~ de Messier 97–111
~ de meteoros por hora 86–135
~ de partículas 76–121
~ de verificação 85–30
~ de Wolf 89–65
~ relativo de manchas 89–65
nuvem 77–131
~ absorvente 97–104
~ circum-estelar 94–48
~ circunstelar 94–48
~ coronal 89–261
~ cósmica 94–31
~ cumuliforme 77–138
~ de cometas 85–100
~ de cometas de Oort 85–97
~ de elé(c)trons 79–115
~ de estrelas 98–6
~ de gás 97–113
~ de gás intergaláctico 98–354
~ de hidrogênio neutro 97–103
~ de libração 84–43
~ de libração lunar 84–43
~ de Oort 85–97
~ de partículas 84–31
~ de plasma 84–30
~ de pó 83–256
~ de poeira 83–256, 97–114
~ de poeira circum-estelar 97–117
~ escura 97–107
~ estelar 98–6
~ galáctica primordial 94–17
Grande ~ de Magalhães 98–213
~ H I 97–103
~ inter(e)stelar 97–101
~ iridescente 79–90
~ luminosa noturna 79–89
~ maser 97–12
~ molecular 97–11
~ molecular gigante 97–13a
~ nacarada 79–91
~ noctilúcia 86–26
~ noctiluzente 79–89
~ obscura 97–107
~ orográfica 77–134
Pequena ~ de Magalhães 98–214
~ protogaláctica 94–17
~ protoplanetária 83a–24
~ suspensa 89–275
~ zodiacal 84–54
nuvens 83–254
camada de ~ 83–248
coberto de ~ 83–253
~ de estrelas 98–6
~ de Magalhães 98–212
sem ~ 77–149

O

objeto colapsado 98–23
~ compacto 98–21
~ cosmológico 98–369
~ difuso 85–25
~ extragaláctico 98–209
~ local 98–368
~ não-catalogado 84–96
~ não-numerado 84–96
~ quase-estelar 98–359
~ supermassivo 98–22
~ tipo Apolo 84–80a
obscurecimento = escurecimento 88–142, 95–84
~ de bordo 88–142
~ de limbo 88–142
~ gravitacional 91–41
~ inter(e)stelar 97–86
~ por gravidade 96–25
observação de eclipse 88–226
~ de meteoros 86–6
~ diurna 83–121
~ do crepúsculo 79–42
~ fora do eclipse 88–227
~ regular do Sol 89–16
~ simultânea 86–125
observações por radar 86–265
~ quase-simultâneas 71–36
~ simultâneas 71–35
observador de meteors 86–7
observatório auroral 79–101
~ de aurora 79–101
~ de grande altitude 88–56
~ ionosférico 78–21
~ lunar 82–20
~ sismológico 73–5
~ solar 88–7
obturador de duas lâminas 89–149
~ de lâmina móvel 86–148
~ rotativo 86–147
ocaso do Sol 88–11
oceano 72–22
~ das Tempestades 81–277
oceanografia 72–24
~ espacial 72–69
~ por satélite 72–70
oclusão 77–72
octaedrita 87–86
Octans 92–71
Octante 92–71
ocultacão 95–80
~ de uma estrela pela Lua 81–94
~ lunar 81–94
~ pela camada de ozônio 79–28
~ rasante 81–100
Ofiúco 92–72
Oitante 92–71
olivina 74–22
OMM 77–17
onda contínua 86–277
~ de choque acolisional 84–49
~ de choque sem colisões 84–49
~ de cisalhamento 73–43
~ de compressão 73–46
~ de densidade 98–175
~ de densidade galáctica 98–175
~ de distorção 73–43
~ de gravidade 88–201
~ de gravitação 99–164
~ de Love 73–48
~ de maré 82–291
~ de pressão 73–46
~ de rarefação 97–58
~ de tempestade 72–76
~ de terremoto 73–9
~ de volume 73–42
~ direta 78–12
~ distorcida 73–43
~ diurna 76–30
~ dupla 98–117
~ elástica 73–44
~ eqüivolumétrica 73–43
~ espacial 78–13
~ extraordinária 78–30
~ gravitacional 99–164
~ infra-sônica 72–93
~ lateral 73–43
~ longitudinal 73–45
~ no solo 78–14
~ ordinária 78–31
~ P 73–46
~ planetária 72–94

~ portadora 86–276
~ primária 73–46
~ Q 73–48
~ rotacional 73–43
~ S 73–43
~ secundária 73–43
~ semidiurna 76–29
~ sísmica 73–9
~ sísmica do maremoto 72–95
~ sonora 76–33
~ superficial 73–47
~ terrestre 78–14
~ transversal 73–41
ondas infra-sônicas de pressão 75–88
~ na penumbra 89–100
~ penumbrais 89–100
ondulação do geóide 71–74
opacidade neutrínica 94–95
Ophiuchus 92–72
oposição 83–59
~ em ascensaõ reta 81–68
ó(p)tica atmosférica 79–1
órbita absoluta 96–55
~ aparente 96–48
~ astrométrica 96–57
~ da Lua 81–36
~ da Terra 84–5
~ de um planeta 83–29
~ espe(c)troscópica 96–56
~ fotocêntrica 96–47
~ galáctica de uma estrela 98–184
~ geocêntrica 86–41
~ heliocêntrica 86–42
~ lunar 81–36
~ planetária 83–29
~ primitiva 98–200
~ relativa 96–54
~ terrestre 84–5
~ tridimensional 98–185
~ verdadeira 96–53
Ordoviciano 74–79
Organização Meteorológica Mundial 77–17
orientação do eixo de rotação 84–102
origem 94–9
~ comum 94–25
~ cósmica 87–5
~ de impacto 82–229
~ do mundo 99–11
~ do universo 99–11
~ dos cometas 85–96
~ extraterrestre 87–73
~ tectônica 82–200
~ terrestre 87–74
Órion 92–73
Oriônidas 86–253
Orionídeos 86–253
orla da calota polar 83–151
orogênese 73–74
orogenia 73–74
ortoclásio 74–16
orvalho 77–173
oscilação latitudinal 75–119
~ perpendicular ao plano galáctico 98–187
~ *z* 98–187
oscilações de cinco minutos 88–186
~ de maré 76–28
~ forçadas da maré 82–306
~ livres da Terra 73–49
~ solares 88–223a
oscilador hidromagnético 95–61
~ magneto-idrodinâmico 95–61
oscilógrafo 86–292
osciloscópio 86–292
oval auroral 79–124
~ branca 83–195
óxido 74–26
~ de hidrogênio 97–26
oxidrila 79–25
oxigênio atômico 79–24
~ molecular 79–23
ozone atmosférico 76–124
ozônio atmosférico 76–124
ozonosfera 76–123

P

padrão de espalhamento 77–58
paisagem lunar 81–55
Palas 84–86
paleoclima 77–188
paleomagnetismo 75–25
paleotemperatura 72–126
pallasito 87–81
panorama lunar 81–55
panquecas primordiais 99–171b
pântano 81–303
~ da Decadência 81–305
~ da Putrefação 81–305
~ das Epidemias 81–304
~ do Sono 81–306
papa-ceia 83–127
paráclase 73–91
paradoxo de Chesaux e de Olbers 99–40
~ de gravitação 99–39
~ de Olbers 99–40
~ de Seeliger 99–39
~ fotométrico 99–40
~ gravitacional 99–39
paradoxos cosmológicos 99–38
paralaxe absoluta 91–11
~ ânua 91–6
~ anual 91–6
~ cefêidica 91–8a
~ das esferas de Strömgren 91–13
~ de cefeidas 91–8a
~ de corrente estelar 91–7
~ de cúmulo móvel 91–7
~ de grupo 91–7
~ de Strömgren 91–13
~ dinâmica 91–12
~ espe(c)troscópica 91–15
~ estatística 91–8
~ estelar 91–4
~ fotométrica 91–14
~ relativa 91–10
~ secular 91–9
~ trigonométrica 91–5
~ trigonométrica relativa 91–10
parâmetro de desaceleração 99–83
~ estrutural 96–120
parâmetros de uma cratera 81–227
~ fotométricos 85–106
parantélio 79–84
paranti-helio 79–84
paranti-selene 79–86
parantisselênio 79–86
parasitas atmosféricos 78–50
parasselene 79–73
parasselênio 79–73
parcialmente degenerado 93–183
parede concêntrica 81–254
~ de cratera 81–251
~ Direita 81–185
~ em forma de terraço 81–257
~ externa de cratera 81–253
~ interna de cratera 81–252
~ múltipla 81–255
~ Reta 81–185
parélio 79–71
~ inferior 79–72
parsec 91–17
parsec cúbico 98–84
por ~ 98–84
partícula 88–154
~ acelerada 91–182
~ auroral 79–107
~ coloidal 76–98
~ de aerossol 76–91
~ de ar 86–172
~ do núcleo-manto 97–70
~ gigante 76–118
~ líquida 76–100
~ lunar 82–123
~ meteórica 86–160a
~ sólida 76–99, 97–69
~ superpesada 99–170n
partículas Aitken 76–116
~ aprisionadas 75–111
~ de "fines" 82–109
~ esféricas 87–29
~ finas 82–109
~ inter(e)stelares 97–68
~ precipitáveis 76–117
~ retidas 75–111
~ retidas geomagneticamente 75–111
~ sólidas 85–48
~ zodiacais 84–66
passagem 95–81
~ da sombra 83–53
~ de discontinuidade interplanetária 84–58
~ de duas galáxias 98–320
~ de um bólido 86–117
~ de um meteoro 86–5

P

P

~ de um planeta diante do disco solar 83–51
~ do satélite 83–54
~ no limbo 89–31
~ pela atmosfera 87–26
~ pelo meridiano central 89–45
passar pelo meridiano central 89–44
passo geotérmico 74–59
patrulha automática das erupções 89–220
~ cinematográfica das erupções 89–220
~ de erupções 89–219
~ de fulgurações 89–219
~ de satélites 83–26
~ dos asteróides 84–94
patrulhamento de fulgurações 89–219
~ dos asteróides 84–94
~ dos meteoros 86–144
~ dos satélites 83–26
pausa 76–58
Pavão 92–74
Pavo 92–74
peculiaridade 98–257
~ espe(c)tral 91–93
pedra meteórica 87–60
~ meteorítica 87–60
pedras roladas 81–237
pedregulho 82–130
pedrinhas vulcânicas 82–169
pegadas 82–38
Pégaso 92–75
Pegasus 92–75
Peixe Austral 92–80
Peixes 92–79
Peixe-voador 92–100
pelota 82–129
pêndulo horizontal 71–73
penetrômetro 82–83
~ de solo 82–83
~ de superfície 82–83
penumbra 89–97
~ da mancha 89–97
~ da Terra 81–64
Pequena Nuvem de Magalhães 98–214
pequeno planeta 84–68
percurso aleatório 89–157
perda de energia 88–254
~ de massa 86–168
perfil de temperatura 83–259
~ de temperatura da atmosfera 83–259
~ vertical 81–228
~ vertical da densidade 76–53
~ vertical da temperatura 76–53
~ vertical da umidade 76–53
perfuração 72–51
periapside galáctica 98–189
periáreon 83–142
periastro 96–58
~ joviano 83–172
pericentro galáctico 98–189
pericíntio 81–40
periélio 83–29a
perifoco galáctico 98–189
perigaláctico 98–189
perigeu 81–38
perijove 83–172
perilúnio 81–40
período 95–39
~ arqueano 74–75
~ da rotação solar 88–129
~ de batimento 95–42
~ de onze anos 89–70
~ de pulsação 95–100
~ de rotação 83–31
~ de superposição de dois ciclos 89–86
~ geológico 74–74, 83a–77
~ glacial 72–105
~ glaciário 77–189
~ interglacial 72–107
~ interglaciário 72–107
~ perturbado magneticamente 75–91
~ principal 95–40
~ secundário 95–41
~ undecenal 89–70
perissatúrnio 83–210
perissaturno 83–210
perisselênio 81–40
Permiano 74–83
perna para pouso 82–22
Pérola 92–117
Perseidas 86–251
Perseideos 86–251
Perseu 92–76
Perseus 92–76
pertenência física ao cúmulo 96–82
perturbação associada a uma fulguração 84–46
~ de maré 98–314
~ interplanetária 84–45
~ ionosférica 78–39
~ ionosférica brusca 78–41
~ ionosférica de início repentino 78–41
perturbações geomagnéticas 75–70
~ mútuas 83–25
~ mútuas entre satélites 83–25
~ radiais 93–215
~ solares 83–24
~ solares em satélites 83–24
peso molecular médio 93–100
pesquisa de satélites 83–26
~ lunar 82–3
petrificação 74–49
~ por impacto 82–238
Phoenix 92–77
pico central 81–258
~ submarino 72–43
Pictor 92–78
piezoglitas 87–122
pigeonita 82–90
pintas 88–210
~ brilhantes 88–211
~ cromosféricas 89–172
~ escuras 88–212
Pintor 92–78
Cavalete do ~ 92–78
piranômetro 88–43
pireliômetro 88–39
~ a circulação de água 88–41
~ a corrente de água 88–40
~ a disco de prata 88–42
piroxênio 74–21
Pisces 92–79
~ Australídeos 86–250
Piscis Austrinus 92–80
~ Volans 92–100
piso de uma cratera 81–259
pista de deslizamento 82–197
placa adesiva 87–14
~ continental 73–103
~ meteórica 86–151
~ quarto de onda 88–126
plaga em raios X 89–166
~ ó(p)tica 89–165
plagas 89–164
plagioclásio 74 19
planalto 81–189
~ submarino 72–44
planeta 83–1
~ do sistema solar 83–11
~ do Sol 83–11
~ exterior 83–15
~ externo 83–15
~ gigante 83–174
~ hipotético 83–122
~ hipotético de Olbers 83a–63
~ inferior 83–14
~ interior 83–14
~ intramercurial 83–123
~ joviano 83–174
~ modelo 83–196
pequeno ~ 84–68
~ solar 83–11
~ superior 83–15
~ transnetuniano 83–237
~ transplutoniano 83–124
~ X 83–124
planetário 83–6
planetas telúricos 83–119
~ terrestres 83–119
planetesimais 83a–35
planetocêntrico 83–9
planetofísica 83–4
planetográfico 83–7
planetóide 84–68
~ catalogado 84–95
~ do anel 84–71
~ do cinturão 84–71
~ duplo 84–79
~ feito pelo homem 84–69a
~ numerado 84–95
~ tipo Amor 84–80
~ tipo Apolo 84–80a

planetologia 83–5
planície abissal 72–42
~ de lava 82–185
planícies 81–272
planisfério celeste 92–10
plano de Laplace 83–232
~ do anel 83–216
~ do céu 96–50
~ do equador 83–231
~ equatorial 83–231
~ galáctico 98–53
~ invariável 83–232
~ tangente à esfera celeste 96–50
plasma acolisional 84–50
~ enriquecido de hélio 84–60
~ interplanetário 84–39
~ primordial 99–170
~ sem colisões 84–50
~ solar 84–29
~ solar magnetizado 84–32
plasmosfera 75–12, 76–76
plataforma continental 72–34
platô = plató 81–192
Plêiades 96–110
plenilúnio 81–27
plessita 87–94
plumas 89–294
Plutão 83–239
pluviômetro 77–170
~ totalizador 77–171
pó circum-estelar 97–55
~ circunestelar 97–55
~ circunterrestre 84–53
~ cósmico 97–52
~ de impacto 82–125
~ de silicato 85–53
~ fino lunar 82–112
~ galáctico 97–53
~ inter(e)stelar 97–54
~ interplanetário 84–51
livre de ~ 82–119
~ meteórico 87–11
~ meteorítico 87–10
~ orbitante ao redor da Terra 84–53
~ vulcânico 73–143
poder adsorvente 85–120
~ de adsorção 85–120
~ de coesão 85–117
poeira circum-estelar 97–55
~ circunstelar 97–55
~ circunterrestre 84–53
~ cósmica 97–52
~ de impacto 82–125
~ de silicato 85–53
~ fina lunar 82–112
~ galáctica 97–53
~ inter(e)stelar 97–54
~ interplanetária 84–51
livre de ~ 82–119
~ meteórica 87–11
~ meteorítica 87–10
~ orbitante 84–53
~ orbitante ao redor da Terra 84–53
~ vulcânica 73–143
Polar 92–121
Estrela ~ 92–121
polaridade magnética 89–148
~ magnética negativa 89–150
~ magnética positiva 89–149
~ norte 89–149
~ oposta 89–155
~ sul 89–150
polarímetro fotoelé(c)trico 83–83
~ fotográfico 83–84
~ visual 83–82
polarização inter(e)stelar 97–105
poliedro 71–12
polinômios de Legendre 93–55
politrópica 93–126
pólo de rotação 83–36
~ do eixo 83–36
~ do eixo geomagnético 75–48
~ geomagnético 75–48
~ magnético da Terra 75–45
~ magnético de inclinação 75–45
poluente 72–74
~ atmosférico 76–82
poluição = polução 76–80
~ atmosférica 76–80
~ da água 72–73
~ do ar 76–80
~ química 72–98
Pólux 92–122
Pomba 92–37
ponta magnetosférica 75–13
pontas 83–130
ponte brilhante 89–104, 98–307
~ intergaláctica 98–348
~ luminosa 89–104, 98–307
ponto brilhante 89–201
~ brilhante em raios X 89–183a
~ cônico 96–31
~ de apariação 86–36
~ de cisão 86–121
~ de convergência 96–98
~ de Curie 75–41
~ de explosão 86–122
~ de extinção 86–38
~ de fusão 86–204
~ de impacto 86–124
~ de máxima visão 86–127
~ de orvalho 77–174
~ de parada 86–123
~ de queda 86–124
~ de reflexão 75–124
~ de regressão 98–178
~ de retorno 98–178
~ de retrocesso 98–178
~ de ruptura 86–121
~ de sombra 89–96
~ de umbra 89–96
~ de universo 99–125
~ de visão foveal 86–127
~ do espaço-tempo 99–125
~ espelhento 75–124
~ estacionário 83–45
~ interior de Lagrange 96–31
~ neutro 89–152
primeiro ~ de Lagrange 96–31
~ singular 94–116
~ subestelar 81–130
~ sublunar 81–129
~ subsatélite 71–30
~ subsolar 81–127
~ subterrestre 81–128
~ umbral 89–96
pontos magnéticamente conjugados 75–23
~ magnetoconjugados 75–23
Popa 92–81
população de tipo II 96–68
~ de tipo I 98–71
~ de uma galáxia 98–230
~ do cúmulo 98–341
~ do disco 98–72
~ do halo 98–69
~ dos braços espirais 98–73
~ estelar 98–67
~ galáctica 98–230
~ intermediária 98–70
~ I extrema 98–73
pôr do Sol 88–11
pormenor da superfície 83–105
~ estrutural 88–248
~ permanente 83–106
poro 89–122
porosidade 83–100
poroso 82–181
portadora 86–276
pós-glacial 72–110
posição aproximada 85–26
~ calculada 85–27
~ do radiante 86–60
~ estacionária 83–45
~ heliográfica 89–38
~ precisa 85–28
~ predita 85–27
~ prevista 85–27
~ relativa dos planetas 83–50
~ selenocêntrica 81–132
pós-luminescência 79–22
pós-máximo 95–33
pós-nova 95–123
posterior a mar, cratera 81–208
postulado de coerência 99–34
~ de Weyl 99–34
potência do transmissor 86–270
~ máxima 86–271
~ média 86–272
potencial axialmente simétrico 98–153
~ da gravidade 71–58
~ da maré 82–293
~ de adsorção 85–122
~ de gravitação 96–160a
~ de simetria cilíndrica 98–154
~ estacionário 98–147

~ gravitacional 96–160a
~ terrestre 71–60
Potro 92–48
pouso controlado por ser humano 82–33
praia em raios X 89–166
~ ó(p)tica 89–165
praias = plagas 89–164
preamar 72–90, 82–281
preaquecimento 86–203
Pré-cambriano 74–70
~ inferior 74–75
precedente a mar, cratera 81–207
precipitação 77–158
~ atmosférica 77–154
~ auroral ele(c)trônica 79–112
~ de elé(c)trons 78–48
~ ele(c)trônica 78–48
~ orográfica 77–167
~ total 77–168
precursor 95–143
~ do terremoto 73–11
predição da atividade solar 89–64
~ das ocultações 81–102
pré-estelar 94–13
pré-máximo 95–30
pré-nova 95–122
Presépio 96–111
pressão ao nível do mar 77–44
~ atérmica 94–42
~ atmosférica 77–43
~ da radiação 93–91
~ de partículas 93–90
~ do gás 93–90
~ dos fótons 93–91
~ gasosa 93–90
~ hidrostática 93–77
~ lateral 73–65
~ não-térmica 94–42
~ normal 77–46
~ térmica 94–41
~ total 93–93
pré-supernova 95–129
previsão a curto prazo 77–20
~ a longo prazo 77–19
~ da atividade solar 89–64
~ das ocultações 81–102
~ do tempo 77–13
~ ionosférica 78–32
Primário 74–71
primário 96–37
primeira aparição de um cometa 85–19
primeiras bandas positivas do nitrogênio 79–20
primeiro conta(c)to 81–84
primeiro quarto 81–25
princípio arquimediano 73–101
~ cosmológico 99–30
~ cosmológico perfeito 99–32
~ cosmológico restrito 99–31
~ de Arquimedes 73–101
~ de eqüivalência 99–150
~ de eqüivalência de Einstein 99–150
~ de invariânica 93–61
~ de Mach 99–37
~ do eclipse 81–74
prisma de Anderson 88–68
~ rotativo 88–68
~ separador do feixe luminoso 88–88
problema cosmogônico 94–5
processador de solo 82–79
processo de ablação 86–183
~ de ejeção 98–179
~ de equilíbrio 94–99
~ de radiação 86–211
~ *e* 94–99
~ endógeno 82–258
~ interno 82–258
~ lento de captura de nêutrons 94–88
~ lunar 82–150
~ metamórfico 74–34
~ *p* 94–91
~ *r* 94–89
~ rápido de captura de nêutrons 94–89
~ *s* 94–88
~ sedimentar 74–46
~ tectônico 82–201
~ Urca 94–51
Prociâo 92–123
Prócion 92–123
procura de cometas 85–15
~ de planetas 83–238
~ de satélites 83–27
produção de energia 88–216
~ de som 88–215
profundidade da compensação 73–102
~ da compensação isostática 73–102
~ do eclipse 95–75
~ do foco 73–20
~ efetiva da formação das linhas 93–64
~ focal 73–20
prognóstico 73–11
~ a curto prazo 77–20
~ a longo prazo 77–19
~ do tempo 77–13
projeção azimutal 71–44
~ cartográfica 71–40
~ central 71–48
~ cilíndrica 71–43
~ conforme 71–41
~ cônica 71–42
~ esterográfica 71–47
~ gnomônica 71–46
~ zenital 71–45
projeto de erupções protônicas 89–177
~ de fulgurações de prótons 89–177
prolongamento do crescente de Vênus 83–131
promontório 81–293
propagação anômala do som 76–37
~ da luz 99–129
~ das ondas radioelé(c)tricas 78–34
~ de energia 88–217
~ do som 76–32
propriedade intrínseca 91–100
propriedades cinemáticas 98–90
~ globais 96–134
~ integrais 96–134
Proterozóico 74–77
proto-atmosfera 83a–44
proto-estelar 94–46
proto-estrela 94–45
protogaláxia 94–18
protonosfera 76–75
protoplaneta 83a–40
Protosol 83a–39
protuberância 89–223, 94–37
~ ascendente 89–278
~ ativa 89–238
~ ativa comum 89–246
~ ativa coronal 89–247
~ ativa de mancha 89–254
~ central 98–16
~ com forma de tronco 89–272
~ coronal 89–264
~ da curva 95–37
~ da maré 81–119
~ da maré alta 82–278
~ de alta latitude 89–229
~ de anel 89–270
~ de arco 89–270
~ de baixa latitude 89–227
~ de calota 89–253
~ de funil 89–269
~ de hidrogênio 89–266
~ de linhas metálicas 89–265
~ de mancha 89–252
~ de tronco de árvore 89–272
~ em forma de anel 89–270
~ em forma de arco 89–270
~ em forma de funil 89–269
~ em leque 89–258
~ equatorial 89–228
~ eruptiva 89–248
~ eruptiva comum 89–250
~ lunar 81–118
~ polar 89–230
~ quase-eruptiva 89–249
~ quiscente 89–263
~ secundária 89–260
~ semelhante a erupção 89–222
~ turbilionária 89–262
protuberâncias interativas 89–243
Próxima Centauri 92–124
proximidade da órbita da Terra 84–7
proximidades do Sol 98–58
próximo à órbita terrestre 84–7
pseudovariável 95–51
psicrómetro 77–121
pulsações 79–158
~ adiabáticas 95–98
~ geomagnéticas 75–81
~ magnéticas 75–81
~ naõ-adiabáticas 95–99

P

~ não-radiais 95–97
~ radiais 95–96
pulsar 95–137
~ binário 95–97
~ de raios gama 95–139
~ de raios X 95–138
~ do Caranguejo 95–142
~ ó(p)tico 95–140
pulso de formação estelar 94–11a
~ do eco 86–290
~ isolado 95–146
~ principal 95–144
pulverização 82–239, 87–28
~ de um corpo meteórico 86–189
~ pelo vento solar 82–245
Puppis 92–81
Pyxis 92–82

Q

QSO calmo 98–362
~ radiocalmo 98–362
Quadrântidas 86–247
Quadrantídeos 86–247
quadratura 83–64
~ ocidental 83–66
~ oriental 83–65
quadri(di)mensional 99–121
quadrive(c)tor 99–138
quadro de evolução das manchas 89–120
~ de projeção 88–13
quantidade de movimento angular 83a–4
~ de movimento de rotação 83a–5
~ de movimento orbital 83a–6
~ de nuvens 77–147
~ de precipitação 77–158
~ de vapor de água 77–124
quarto crescente 81–25
~ minguante 81–30
primeiro ~ 81–25
terceiro ~ 81–30
último ~ 81–30
quasag 98–361
quasar 98–360
~ calmo 98–362
~ radiocalmo 98–362
Quaternário 74–88
quebra de simetria 99–170s
quebradiço 86–191
queda 76–57
~ de gás (do espaço intergaláctico) 98–356
~ de temperatura 77–41
~ de um bólido 86–117
~ de um meteorito 87–19
~ de um meteoro 86–5
~ livre 94–39
~ meteorítica 87–19
~ natural 82–227
~ pressuposta de um meteorito 87–49
~ única 87–36
queima do carbono 94–81
~ do hélio 94–78
~ do hidrogênio 94–56
~ do oxigênio 94–82
~ do silício 94–83
~ explosiva do hélio 94–79
Querena 92–30
Quilha 92–30
química da atmosfera 76–79
~ lunar 82–74
quimi(o)luminescência 79–31
quimiosfera 76–62
Quinteto de Stephan 98–309

R

radiação aprisionada 75–112
~ artificial aprisionada 75–113
~ artificial retida 75–113
~ corpuscular 88–150
~ da Terra 77–56
~ de partículas 88–150
~ de recombinação 86–212
~ de um meteoro 86–207
~ difusa do céu 77–59, 88–136
~ diluída 93–38
~ do céu 77–54
~ global 77–55
~ gravitacional 99–165
~ incidente solar 77–52
~ integrada 88–147
~ livre-ligado 86–212
~ neutrínica 94–94
~ própria de um planeta 83–198
~ retida 75–112
~ retida geomagneticamente 75–112
~ solar 88–134
~ solar direta 77–50, 88–135
~ terrestre 77–56
~ total 88–146
radiante 86–53
~ aparente 86–56
~ corrigido 86–58
~ de grupo 86–54
~ de um enxame meteórico 86–237
~ de uma corrente meteórica 86–237
~ estacionário 86–241
~ geocêntrico 86–58
~ individual 86–55
~ meteórico 86–53
~ secundário 86–239
~ verdadeiro 86–59
radical 85–136
~ livre 85–137
radiestrela 91–158
radioalbedo 83–94
radioatividade 74–61
radioaurora 79–94
radiocintilação 84–36
~ interplanetária 84–37
radiocoroa 88–239
radioeco 86–295
radioeliógrafo 88–95
radioeliograma 88–96
radioespe(c)trógrafo 88–124
radioespe(c)trômetro 88–124
radioespe(c)troscópio solar 88–124
radioestrela 91–158
radiofonte 98–360
~ decamétrica 83–186
~ quase-estelar 98–360
radiogaláxia 98–206
radiointerferômetro 88–97
radiolóbulo 98–286
radiomagnitude 86–268
radiomapa 81–135a
radiometeoro 86–267
radionuclídeo cosmogênico 82–255
~ primordial 82–256
~ produzido por bombardeio 82–255
radionuvem 98–286
rádio-Sol 88–3
radiossol 88–3
raia amarela = linha amarela 89–292
~ atmosférica 88–140a
~ circum-estelar 97–97
~ circunstelar 97–97
~ coronal 89–290
~ D 88–140
~ inter(e)stelar 97–96
~ inter(e)stelar en microondas 97–98
~ inter(e)stelar de rádio 97–98
~ inter(e)stelar ó(p)tica 97–99
~ tangencial 81–169
~ telúrica 88–140a
~ verde 89–291
~ vermelha 89–293
raias lunares 81–168
raio 77–162
~ aparente 81–70
~ aparente da Lua 81–70
~ aparente do Sol 81–69
~ coronal 89 287
~ coronal monocromático 89–288
~ de leque coronal 89–296
~ de Schwarzschild 99–163
~ de visada 91–28
~ do universo 99–5
~ equatorial 83–115
~ exterior 83–228
~ externo 83–228
~ gravitacional 94–113, 99–163
~ interior 83–226
~ interno 83–226
~ médio 83–116a
~ polar 83–116
~ sísmico 73–14
~ sonoro 76–34
~ tangencial 81–169
~ verde 79–58
raios 79–155
~ lunares 81–168
~ polares 89–294
rajada 77–112
ramificação 98–5a
ramo ascendente 95–34

P

~ das gigantes 91–152b
~ de uma corrente 86–240
~ descendente 95–35
~ horizontal 91–132
ranhura 81–165
~ linear 81–166
~ sinuosa 81–167
Raposa 92–101
Raposinha 92–101
Ras Alhague 92–125
rastejamento (encosta abaixo) 82–195
rast(r)o de curta duração 86–112
~ de fumaça 87–21
~ de longa duração 86–113
~ de poeira 86–115
~ de um meteoro 86–29
~ de uma estrela 86–152
~ gasoso 86–116
~ instável 86–109
~ luminoso 86–108
~ meteórico 86–103
~ meteórico persistente 86–104
~ noctilucente 86–105
~ nóctilúcio 86–105
~ visível de dia 86–106
rastro = rasto 86–103
razão das massas 96–67
~ de abundância 91–175
~ de equilíbrio 94–104
~ de expansão 99–75
~ de intensidades 91–60
~ de isótopos 74–94
~ de Poisson 73–60
~ entre gás e pó 97–120
~ gás-poeira 97–120
~ hidrogênio-metais 91–176
~ isotópica 74–94
~ massa-luminosidade 98–233
reação de captura de nêutrons 94–87
~ de espalação 94–100
~ de neutrino 94–92
~ fotoquímica 76–63
~ nuclear superficial 91–181
reaparecimento 81–99
~ do anel 83–219
reaparicão 81–99
~ do anel 83–219
receptor 78–17
recessão das galáxias 98–298
recolha de poeira meteórica 87–12
recolhedor de amostras de água 72–101
~ de poeira meteórica 87–13
recolhimento de poeira meteórica 87–12
recombinação dissociativa 86–214
reconhecimento da superfície 82–32
~ na superficie 82–32
~ orbital 82–31
~ superficial 82–32
recorrência 88–132
recristalização 74–42
rede cromosférica 89–171
~ de gravidade 71–80
~ de referênca selenofísica 82–40
~ de vigilancia lunar 84–40
~ geodésica de satélites 71–34
~ geodésica mundial 71–38
~ gravimétrica 71–80
~ gravimétrica básica 71–82
~ gravimétrica fundamental 71–82
~ gravimétrica mundial 71–81
~ mundial 77–18
redescoberta 85–17
redução ao ar livre 71–78
~ ao nível do mar 77–45
~ aparente de velocidade 86–100
~ das ocultações 81–108
~ de Faye 71–78
~ real de velocidade 86–101
redutor em escada 88–80
referência géodesica 71–20
referencial 98–136
~ em rotação 98–138
~ padrão 98–137
~ rotante 98–138
refluxo da maré 82–282
reforço brusco dos atmosféricos 78–51
~ súbito dos atmosféricos 78–51
refrator solar sem cúpula 88–53
região anterior aos mares 81–269
~ assísmica 73–29
~ ativa 89–57
~ ativa de protuberâncias 89–240
~ ativa em Júpiter 83–180
~ auroral 79–126
~ baixa 81–272
~ brilhante 83–160
~ calma da fotosfera 89–56
~ central da Galáxia 98–13
~ central de uma galáxia 98–293
~ circum-nuclear 98–292
~ circunuclear 98–292
~ clara 83–160
~ continental 81–271
~ coronal ativa 89–286
~ de ablação 72–123
~ de acumulação de uma geleira 72–122
~ de alta pressão 77–84
~ de aprisionamento 75–110
~ de atividade 89–57
~ de audibilidade 87–32
~ de auroras 79–126
~ de calmarias tropicais 77–102
~ de escape 76–68
~ de hidrogênio ionizado 97–50
~ de hidrogênio neutro 97–49
~ de mares 81–268
~ de modulação 84–26
~ de modulação da radiação cósmica 84–26
~ de visibilidade 87–31
~ de retenção 75–110
~ de terremoto 73–28
~ de transição 88–224
~ de transiçâo cromosfera-coroa 88–224
~ do anticentro galáctico 98–15
~ do centro galáctico 98–13
~ do impacto 87–38
~ do impacto dos meteoritos 87–38
~ do núcleo 98–17
~ epicentral 73–16
~ epicêntrica 73–16
~ equatorial 89–59
~ escura 83–155
~ fria 88–199
~ H II 97–50
~ H II compacta 97–51
~ H I 97–49
~ interdita 79–135
~ intergranular 88–177
~ interspicular 88–198
~ ionosférica 78–4
~ M 89–61
~ magnética 89–144
~ magnética bipolar 89–160a
~ magnética unipolar 89–155a
~ marítima 81–268
~ montanhosa 81–196
~ nuclear 98–17
~ permitida 79–134
~ polar 81–17, 89–58
~ pós-mares 81–270
~ pós-marítima 81–270
~ posterior aos mares 81–269
~ pré-mares 81–269
~ pré-marítima 81–269
~ proibida 79–135
~ quente 88–200
~ sem terremotos 73–29
~ solar ativa 89–57
~ sísmica 73–28
~ verde da coroa 89–284
~ vermelha da coroa 89–285
registrador de brilho solar 77–61
~ ionosférico 78–22
registrar o campo magnético 89–132
registro ionosférico 78–23
regmaglitas 87–122
regolito 82–104
Régua 92–70
~ paraláctica 92–70
Régulo 92–126
relação acidental 97–126
~ das massas 96–67
~ de equilíbrio 94–104
~ de Hubble 99–76
~ de intensidades 91–60
~ desvio para o vermelho-distância 99–76
~ diâmetro-profundidade 81–248
~ genética 97–127
~ Hubble 99–76
~ luminosidade-tipo espe(c)tral 91–130

~ magnitude-desvio para o vermelho 98–301
~ magnitude-intensidade 91–58
~ massa-luminosidade 91–129
~ massa-raio 93–200
~ período-luminosidade 95–103
~ velocidade-distância 98–300
~ velocidade-profundidade 73–52
relações estruturais 82–266
~ helio-geofísicas 84–55
~ solares-terrestres 84–55
~ Sol-Terra 84–55
relâmpago 77–162
~ em bola 77–163
~ em cadeia 77–164
~ em contas 77–164
~ em rosário 77–164
relatividade da inércia 99–35
~ especial 99–94
~ geral 99–146
relativista 99–95
relativístico 99–95
relaxação 98–127
~ acolisional 98–132
~ colisional 98–131
~ por colisões 98–131
~ sem colisões 98–132
Relógio 92–54
remanescentes de uma supernova 95–133
rendimento de ionização 86–209
~ energético 93–109
~ luminoso 86–210
réplicas 73–13
resfriamento 83a–57
~ neutrínico 94–97
~ por neutrinos 94–97
~ por radiação 94–98
~ radiativo 94–98
resistência aerodinâmica 86–178
~ do ar 86–178
~ do solo 82–135
resolução espacial 88–204
~ temporal 88–205
restos da explosão de uma supernova 95–133
~ de uma supernova 95–133
retardação da expansão 99–82
retardo da expansão 99–82
~ de grupo 78–56
retenção 75–109
~ de gás 87–134
Retículo 92–83
retículo 86–129
Reticulum 92–83
retificação 95–18
~ da curva de luz 95–18
retorno de amostras 82–71
~ de um cometa 85–9
~ de um grupo de manchas 88–133
retrógrado 83–46
retrorrefletor 82–43
~ laser 82–43
revolução anomalística 81–43
~ dracônica 81–47
~ sideral 81–45
~ sinódica 81–46
~ trópica 81–44
Rigel 92–127
rigidez 73–56
~ magnética 75–117
rima 81–165
riômetro 78–45
riscos de navegação 72–97
rocha 74–1
~ ácida 74–4
~ angulosa 82–131
~ básica 74–5
~ cristalina 82–66
~ efusiva 74–3
~ eqüigranular 82–67
~ estratificada 74–48
~ extrusiva 74–3
~ folheada 74–54
~ ígnea 74–2
~ instrusiva 74–7
~ lunar 82–65
~ magmática 74–2
~ metamórfica 74–11
~ plutônica 74–7
~ sedimentar 74–10
~ ultrabásica 74–6
~ vulcânica 74–3
~ xistácea 74–54
~ xistóide 74–54
rosa dos ventos 77–116
rota cruzada 82–37
~ de excursão 82–37
rotação assíncrona 83–38
~ da linha dos apsides 96–71
~ das estrelas 91–35
~ de corpo rígido 98–111
~ de Kepler 98–112
~ diferencial 88–128
~ diferencial da Galáxia 98–113
~ direta 83–39
~ dol Sol 88–127
~ dos planetas 83–30
~ estelar 91–35
~ forçada 81–14
~ galáctica 98–107
~ galáctica diferencial 98–113
~ kepleriana 98–112
~ ligada 81–14
~ não-sincrónica 83–38
~ não-uniforme 91–36
~ planetária 83–30
~ retrógrada 83–40
~ rígida 98–111
~ síncrona 83–37
~ sincrônica 83–37
~ solar 88–127
~ vinculada 81–14
rotor 91–42
~ inclinado 95–60
~ oblíquo 91–52
~ rápido 91–43
rugosidade superficial 82–106
ruído acústico 88–220
~ de fundo 73–55
~ de fundo sísmico 73–55
~ sísmico 73–54
ruptura 73–78
~ colisional 84–107
~ de um cometa 85–114
~ de um meteoro 86–82
~ de um núcleo 98–318
~ de um núcleo galáctico 98–318
~ do corpo original 87–124
~ do corpo parente 87–124
~ por colisão 87–125
~ por efeito de maré 85–118
~ por maré 85–118

S

Saco de Carvão 97–138
Sagitário 92–85
Sagitta 92–82
Sagittarius 92–85
saída da Lua da penumbra 81–82
~ da Lua da sombra 81–81
Saipé 92–46
saliência da maré alta 82–278
salinidade 72–66
salto brusco 95–154
~ repentino 95–154
satélite 83–17
antigo ~ 83–240
~ de gelo 83–23
~ Galileano 83–204
~ irregular 83–21
~ mais interior 83–20
~ natural 83–18
~ retrógrado 83–21
~ rochoso 83–22
saturação 76–101
saturnicêntrico 83–213
saturnigráfico 83–212
Saturno 83–209
Scorpio 92–86
Scorpius 92–86
Sculptor 92–87
Scutum 92–88
se(c)ção de choque diferencial 94–54
~ de choque geométrica 84–103
~ de choque gravitacional 84–104
~ de choque parcial 94–54
~ eficaz diferencial 94–54
~ eficaz geométrica 84–103
~ eficaz gravitacional 84–104
~ frontal 86–176
~ geométrica 84–103
~ gravitacional 84–104
se(c)tor 85–60
~ curvado 85–62
~ invertido 85–63

~ magnético interplanetário 84–16
~ reto 85–61
secundário 96–38
Secundário 74–72
sedimentação 73–109
~ em mar alto 72–79
~ oceânica 72–79
~ submarina 72–47
sedimento 74–47
sedimentos abissais 72–50
~ marinhos 87–15
~ marinhos profundos 72–50
~ na zona abissal 72–50
~ organogênicos 72–59
~ pelágicos 72–57
~ terrígenos 72–48
seixo 82–128
selenocêntrico 81–131
selenodésia 81–109
selenodésico 81–110
selenofísica 82–151
selenografia 81–1
selenográfico 81–8
selenóide 81–117
selenologia 82–1
selenológico 82–48
selenólogo 82–49
selenoquímica 82–74
sem luz própria 96–44
semidiâmetro aparente da Lua 81–70
~ aparente do Sol 81–69
~ equatorial 83–115
~ exterior 83–228
~ interior 83–226
sentido da rotação 98–108
separação angular 96–52
~ do Sol 83a–9
~ dos compontentes 96–66
~ dos elementos químicos 83a–48
separador de feixe luminoso 88–88
seqüência boreal 91–87
~ das espirais 98–255
~ das galáxias 98–240
~ das galáxias regulares 98–240
~ de fotografias 79–103
~ de galáxias de Hubble 98–227
~ de idade zero 96–152
~ espe(c)tral 91–94
~ evolutiva 94–30
~ polar internacional 91–87
~ polar norte 91–87
~ principal 91–131
~ principal de idade zero 94–60
Serpens 92–89
Serpentário 92–72
Serpente 92–89
serra 73–85
serviço meteorológico 77–5
Seta 92–84
Sete-cabrinhas 96–110
Sete-estrelo 96–110
Sextans 92–90
Sextante 92–90
siderito 87–82
siderólito 87–78
sideróstato 88–44
signos do zodíaco 92–6
sílica 74–13
silicato 74–14
sill 73–135
Siluriano 74–80
silvo curto 78–58
~ de nariz 78–59
~ ele(c)trônico 78–60
~ em forma de nariz 78–59
~ longo 78–57
~ protônico 78–61
silvos atmosféricos 78–53
simetria cilíndrica 98–155
~ esférica 98–160
~ helicoidal 98–161
simulação matemática 98–323
simultaneidade 99–127
sinais pulsados 86–278
~ pulsantes 86–278
sinclinal 73–89
síncrona 85–92
~ completa 85–94
~ terminal 85–93
sindínama 85–91
singularidade 94–116, 99–42
~ desnuda 94–117
síntese de abertura 88–118
~ dos elementos 94–52
Sírio 92–128
Sirius 92–128
sismicidade 73–3
sismo 73–2a
~ continental 73–2b
~ distante 73–27
~ intermediário 73–24
~ local 73–25
~ meteorítico 87–34
~ normal 73–22
~ profundo 73–23
~ próximo 73–26
~ submarino 73–2c
~ superficial 73–21
~ tectônico 73–2d
~ vulcânico 73–2e
sismógrafo 73–7
sismograma 73–8
~ de impacto 82–224
sismologia 73–1
~ ativa 82–217
~ lunar 82–213
~ passiva 82–218
~ solar 88–223b
sismólogo 73–2
sismômetro 73–7
~ ativo 82–222
~ passivo 82–221
sismos da crosta 95–152
~ estelares 95–153
sistema achatado 98–251
~ acolisional 98–167
~ autogravitante 98–169
~ chato 98–251
~ colisional 98–166
~ com colisões 98–166
~ com energia total negativa 96–156
~ com energia total positiva 96–158
~ compacto 96–157
~ copernicano 99–24
~ cosmológico 99–20
~ de anéis 98–291
~ de arcos 89–299
~ de bandas 79–18
~ de classificação espe(c)tral 91–91
~ de classificação espe(c)tral de Harvard 91–91
~ de Copernico 99—24
~ de cores 91–72, 96–140
~ de correntes 75–97
~ de estrelas 96–1
~ de fissuras 73–79
~ de franjas de interferência 88–120
~ de fraturas 73–79, 81–174
~ de fraturas continentais 73–81
~ de fraturas oceânicas 73–80
~ de gravitação própria 98–169
~ de longitudes 83–178
~ de longitudes I 83–177
~ de longitudes II 83–179
~ de magnitudes 91–57
~ de raios 81–170
~ de referência 98–136
~ de referência em rotação 98–138
~ de referênica geral 98–137
~ de referência rotante 98–138
~ de rupturas 73–79
~ de satélites 83–19
~ de simetria axial 98–157
~ de sintetização 88–119
~ de turbilhões 83a–13
~ de vórtices 83a–13
~ do mundo 99–20
~ dos anéis de Saturno 83–220
~ dos cúmulos globulares 98–75
~ em estado estacionário 98–141
~ equatorial de longitudes 83–178
~ esferoidal 98–159
~ estacionário 98–41
~ estável 96–156
~ estável gravitacionalmente 96–156
~ estelar 96–1
~ estelar formado por sobreposição 98–165
~ estelar formado por superposição 98–165
~ extragaláctico 98–210
~ fundamental de referência 98–137
~ galáctico 98–3
~ geocêntrico 99–21
~ geodésico de referência 71–19
~ geodésico mundial 71–39

P

~ gravitacionalmente livre 96–158
~ héliocêntrico 99–23
~ inercial 99–98
~ instável 98–171
~ instável gravitacionalmente 96–158
~ ligado por gravitação 96–156
~ local 98–61
~ local de referência 98–139
~ muito achatado 98–164
~ não-estacionário 98–152
~ não-inercial 99–147
~ para latitudes intermédias 83–179
~ planetário 83–2
~ planetocêntrico 83–10
~ planetocêntrico 83–10
~ planetocêntrico de coordenadas 83–10
~ planetocêntrico de referência 83–10
~ ptolemaico 99–22
~ sem colisões 98–167
~ sintetizador 88–119
~ solar 83–12
~ Terra-Lua 81–12
~ triplo 96–4
~ U, B, V 96–144
sistemas de coordenadas associados à matéria 99–44
sítio de alunagem 82–14
~ de pouso 82–14
~ de pouso na Lua 82–14
~ proposto de pouso 82–15
situação meteorológica 77–14
~ meteorológica geral 77–15
sizígia 81–29
sizígias = sizígios 81–29
smog 76–83
sobredenso 86–285
sobreestimar 95–13
sobreposição dos espe(c)tros 96–24
soergimento 72–88
Sol 88–1
~ ativo 89–12
~ azul 79–92
~ calmo 89–11
~ em raios X 88–4
~ falso 79–71
~ ó(p)tico 88–2
~ perturbado 89–12
~ primitivo 83a–39
~ quieto 89–11
~ rádio 88–3
~ visível 88–2
~ X 88–4
solar 88–5
solidificação 83a–59
solo coerente 82–63
~ coesivo 82–63
~ do mar 72–38
~ do oceano 72–37
~ lunar 82–62
~ marítimo 72–38
~ oceânico 72–37
~ permanentemente gelado 83–152
solução estável 98–170
som anômalo 76–36
~ normal 76–35
soma das precipitacões 77–168
sombra da Terra 79–44
~ da mancha 89–98
~ de ozônio 76–127
sombras volantes 88–31
sonda acústica 76–45
~ de solos 82–83
sondagem 76–47
~ descendente 78–20
~ do crepúsculo 79–41
~ ionosférica 78–19
~ por foguete 76–48
~ sísmica 82–216
sondagens acústicas 72–29
~ oceanográficas 87–16
sopa de quarks 99–169a
soprar 77–105
sopro 89–277
com forma de ~ 89–277
Spica 92–129
subabundância 91–187
subanã 91–140
subdenso 86–284
sub-erupção 89–195
subestimar 95–14
subgigantes 91–151
subida 76–54, 82–8
~ da temperatura 77–42
~ para o máximo 95–28
sublimação 85–138
subpulso 95–145
sub-resfriado 76–106
sub-resfriamento 76–105
subsistema 98–74
subsolo 82–105
su(b)stância dificilmente fusível 83a–55
~ facilmente fusível 83a–56
~ não-volátil 83a–53
~ porosa 82–111
~ refratária 83a–55
~ sólida refratária 85–49
~ volátil 82–262
subtempestade 79–144
~ auroral 79–141
~ magnética 75–95
~ magnética polar 75–96
subtração fotográfica 88–94
sulco 81–165
sulfato 74–29
sulfeto 74–28
~ de carbonila 97–36
~ de hidrogênio 97–44
sulfito 74–28a
sulfureto 74–28
superabundância 91–185
superaglomerado de galáxias 98–346
~ local 98–347
supercúmulo de galáxias 98–346
~ local de galáxias 98–347
superdenso 86–285
superestimar 95–13
superfície anterior 87–114
~ áspera 83–102
~ coberta com crateras 81–245
~ com crateras 81–245
~ crítica 96–27
~ da Lua 82–55
~ da Terra 71–5
~ de contacto 96–27
~ de planalto 82–56
~ de raios 81–171
~ de um meteorito 87–106
~ de um planeta 83–97
~ do mar 82–57
~ em camadas 83–101
~ encrostada 87–109
~ entre raias 81–172
~ eqüipotencial 71–61
~ eqüipotencial da gravidade 71–61
~ estratificada 83–101
~ frontal 87–114
~ geopotencial 71–61
~ luminosa difusa 79–148
~ lunar 82–55
~ neutra 75–16
~ original 87–107
~ planetária 83–97
~ polida 87–101
~ posterior 87–115
~ primária 87–107
~ pulsante 79–149
~ raiada 81–171
~ secundária 87–108
~ solar 88–164
~ sólida 83–98
~ terrestre 71–5
superfluido 95–149
superfundido 76–106
superfusão 76–105
supergaláxia 98–346
~ da Virgem 98–347
~ local 98–347
supergigante 91–145
~ brilhante 91–147
~ fraca 91–148
~ luminosa 91–147
~ mais brilhante 91–147
~ mais luminosa 91–147
~ menos brilhante 91–148
~ menos luminosa 91–148
~ subluminosa 91–148
supergranulação 88–183
supergrânulo 88–182
supergravitação 99–170r
supernova 95–126
~ de Tycho 95–135
~ extragaláctica 95–128
~ galáctica 95–127
~ passada 95–134
~ tipo I 95–130

P

superposição de dois ciclos 89–86
~ dos espe(c)tros 96–24
super-resfriado 76–106
super-resfriamento 76–105
supersaturação 76–102
superunificação 99–170q
suporte da cromosfera 88–223
surgimento 89–255
~ brilhante 89–257
~ de uma erupção 89–190
~ escuro 89–256
susce(p)tibilidade magnética 75–27

P

T

Taça 92–42
t(a)enita 87–93
talude 81–234
~ continental 72–35
~ continental de inclinação suave 72–36
tálus 81–234
tamanho das partículas 76–114
~ dos grãos 82–113
Táuridas 86–254
Taurídeos 86–254
Taurus 92–91
taxa da craterização 82–233
~ da formação de crateras 82–233
~ da formação estelar 94–12
~ de arrefecimento 93–203
~ de decréscimo 95–36
~ de desaparecimento da estrelas 94–115
~ de emissão de massa 85–146
~ de expansão 99–75
~ de extinção estelar 94–115
~ de formação 85–143
~ de formação das estrelas 94–12
~ de fusão 83–147
~ de morte das estrelas 94–115
~ de nascimento das estrelas 94–12
~ de produção 85–143
~ de produção de gases 85–144
~ de reação termonuclear 94–53
~ de recessão 99–75
~ de recorrência 86–279
~ de relaxação 98–129
~ de repetição de pulsos 86–279
~ de resfriamento 93–203
~ de subida para o máximo 95–29
~ de variação vertical de temperatura 76–14
tectita 87–95
~ em forma de gota 87–96
tela 88–13
telegrama astronômico 85–20
Telescópio 92–92
telescopio cromosférico 88–73
~ solar 88–49
~ solar automático 88–50
~ solar colocado em um balão 88–52
~ solar horizontal 88–55
Telescopium 92–92
temperatura 77–29
~ à sombra 77–34
~ ambiente 77–32
~ ambiente normal 77–31
~ central 93–89
~ cinética 91–115
~ Curie 75–41
~ da sala 77–31
~ de brilho monocromático 91–114
~ de condensação 97–66
~ de cor 91–109
~ de Curie 75–41
~ de degenerescência 93–188
~ de dissociação 91–120a
~ de excitação 91–118
~ de fusão 86–204
~ de ionização 91–117
~ de rotação 91–119
~ de sala 77–31
~ de vibração 91–120
~ do ar 77–34
~ do ar à sombra 77–34
~ e pressão normais 77–30
~ efetiva 91–113
~ ele(c)trônica 91–116
~ no abrigo 77–34
~ no centro 93–89
~ superficial 93–88
tempestade com trovoadas 77–159
~ de início gradual 75–78
~ de início súbito 75–73
~ de poeira 83–255
~ de ruídos em Júpiter 83–185
~ geomagnética 75–71
~ interplanetária 84–44
~ ionosférica 78–49
~ típica 75–72
tempo 77–1
~ absoluto 99–117
~ característico 99–79
~ da passagem pelo periastro 96–60
~ de arrefecimento 83–201
~ de chegada 73–50
~ de desintegração 96–104
~ de dissociação 96–104
~ de evaporação 96–109
~ de expansão 99–78
~ de Friedmann 99–80
~ de Hubble 99–79
~ de percurso 73–51
~ de período 95–39
~ de permanência 82–34
~ de Planck 99–170i
~ de recepção do eco 86–291
~ de relaxação 98–128
~ de resfriamento 83–201
~ de trânsito 84–47
~ de trânsito Sol-Terra 84–47
~ de vida da partícula aprisionada 75–122
~ do eco 86–291
~ do Sol descoberto 77–51
~ geomagnético 79–132
~ lunar 82–300
~ lunar aparente 82–301
~ lunar médio 82–302
~ médio de relaxação 98–130
~ newtoniano 99–118
~ próprio 99–139
tensão 73–66
~ crustal 82–210
~ da crosta 82–210
~ do vapor de água 77–125
~ térmica 87–27
tensor das deformações 73–70
~ de curvatura 99–159
~ de energia 99–158
~ de matéria 99–158
~ de Riemann 99–159
~ dos esforços 73–69
~ energia-momento 99–158
~ fundamental 99–157a
~ métrico 99–157a
teor de ozônio na atmosfera 76–125
teorema da unicidade 93–83
~ de Clairaut 71–77
~ de Russell-Vogt 93–83
~ do virial 98–181
teoria da captura 96–72
~ da divisão 96–75
~ da estrutura espiral 98–173
~ da fissão 96–75
~ da formação múltipla 96–76
~ da morte térmica do universo 99–15
~ da relatividade generalizada 99–146
~ da relatividade restrita 99–94
~ das marés 82–305
~ das ondas de densidade 98–174
~ de acréscimo 83a–20
~ de calibração 99–170j
~ de captura 85–99
~ de contração 73–77
~ de ejeção 85–98
~ de formação dos braços espirais 98–173
~ de medida 99–170j
~ dínamo do campo geomagnético 75–99
~ dínamo geomagnética 75–99
~ dos braços espirais 98–173
~ dos ciclos 97–156
~ dos núcleos separados 96–76
~ física dos meteoros 86–159
~ gauge 99–170j
~ mecânica da forma dos cometas 85–84
Terciário 74–87
termicamente estável 93–209
terminadoiro = terminadouro 83–69
~ da manhã 81–124

~ da tarde 83–71
~ da Terra 81–91
~ do anoitecer 81–125
~ do nascente 81–124
~ do poente 81–125
~ lunar 81–123
~ matutino 83–70
~ terrestre 81–91
~ vespertino 83–71
termo cosmológico 99–156a
~ da rotação galáctica 98–116
~ do movimento diferencial 98–115
~ K 98–119
termoluminescência 82–249
termosfera 76–59
Terra 71–2
terra 81–190
~ chã 81–272
terraço de cratera 81–229
~ interno 81–256
terras altas 81–189
~ baixas 81–272
terrela 79–133
terremoto 73–2a
~ catastrófico 73–40
~ continental 73–2b
~ de deslocamento 73–2d
~ de profundidade intermediária 73–24
~ desastroso 73–38
~ destruidor 73–39
~ distante 73–27
~ fraco 73–35
~ intenso 73–37
~ local 73–25
~ médio 73–36
~ muito intenso 73–38
~ normal 73–22
~ profundo 73–23
~ próximo 73–26
~ submarino 73–2c
~ superficial 73–21
~ tectônico 73–2d
~ vulcânico 73–2e
terreno caótico 83–164
~ desordenado 83–164
terrestre 71–3
testemunha ocular de queda 87–48
testemunho 72–51
tetradimensional 99–121
tetrave(c)tor 99–138
textura 82–102
~ de fluxo 82–191
~ fluidal 82–191
tioformaldeído 97–43
tipo da cauda cometária 85–71
~ de aurora 79–136
~ de estrelas 91–101
~ de Hubble 98–229
~ espe(c)tral 91–108
~ estelar 91–101
~ estrutural 98–225
~ galáctico de Hubble 98–229
~ morfológico 98–228
tirar amostra 76–85
~ amostra do ar 76–85
Titã 83–233
topografia lunar 81–146
toro de hidrogênio 83–230
torque 83a–5
~ da maré 82–288
torre solar 88–54
torrente corpuscular 88–160
totalmente degenerado 93–185
Touro 92–91
traçado da temperatura 72–83
traçador da estrutura espiral 98–39
traço de arrefecimento 93–202
~ de escorregamento 82–197
~ de gás 76–88
~ de idade nula 94–60
~ de idade zero 94–60
~ de partícula 82–251
~ de radiação 87–120
~ evolutivo 94–29
~ fóssil 87–121
~ nuclear 87–120
~ nuclear fóssil 82–252
traços de elemento 82–145
elemento em ~ 82–145
~ orgânicos em meteoritos 87–67
trajetória aparente 86–32
~ aparente no céu 86–32
~ ascendente 82–9
~ circunlunar 82–11
~ de arrefecimento 93–201
~ de resfriamento 93–201
~ de um meteoro 86–30
~ de um meteoro na atmosfera 86–30
~ de uma partícula 75–118
~ distorcida 86–33
~ evolutiva 94–29
~ interplanetária 82–12
~ na atmosfera 86–162
~ ondulada 86–33
~ real 86–31
transferência da radiaçâo 93–29
~ de calor 86–202
~ de energia 93–111
~ de massa 96–33
~ de massa de um componente para outro 96–33
~ radiativa 93–29
transformação de energia 86–208
~ de Galileu 99–131
~ de Laplace 93–54
~ de Lorentz 99–132
~ galileana 99–131
~ homóloga 93–132
transição crosta-manto 72–14
~ de fase 83a–58, 99–170p
transiente coronal 89–185a
trânsito 95–81
~ da Lua 82–303
~ de um planeta diante do Sol 83–51
~ de um satélite 83–54
~ lunar 82–303
transitório coronal 89–185a
translunar 81–6
transmissão de calor 86–202
transmissor 78–15, 86–269
transporte da radiação 93–29
~ de energia 93–111
~ de massa 96–33
~ do campo magnético 84–48
~ radiativo 93–29
Trapézio de Órion 96–92
travessia 82–35
~ científica 82–36
levantamento de ~ 82–35
trecho ascendente 95–34
~ descendente 95–35
tremor da costra 95–152
~ da estrella 95–153
~ de terra 73–2a
tremores secundários 73–13
Três Reis Magos 92–130
~ Irmãs 92–129a
~ Marias 92–129a
tríade 87–91
Triádico 74–84
triangulação 71–23
~ cósmica 71–28
~ de 1ª ordem 71–24
~ espacial 71–25
~ fundamental 71–24
~ por satélites 71–27
Triângulo 92–93
~ Austral 92–94
Triangulum 92–93
~ Austral 92–94
Triás(s)ico 74–84
trilateração 71–29
troca de energia 98–125
troilita 82–94
tropopausa 76–77
troposfera 76–78
trovão 77–165
trovoada 77–159
tsunami 72–95
Tubã 92–131
tubo de força magnética 89–145
~ de lava 82–187
~ de raios catódicos 86–293
Tucana 92–95
Tucano 92–95
turbosfera 76–72
turbulência cósmica 83a–12
~ em escala cósmica 83a–12
~ em grande escala 83a–12
turvação 83–252
turvamento 83–252

P

U

último conta(c)to 81–85
umbra 89–95
~ da mancha 89–95
umidade absoluta 77–122
~ atmosférica 77–118
~ do ar 77–118
~ específica 77–123
~ relativa 77–128
União Internacional de Geodesia e Geofísica 71–4
Unicórnio 92–68
unidade de mudança de fase 88–115
~ de protuberância 89–235
~ topográfica 81–149
unidades solares 81–123
unificação das interações 99–170w
universo 99–1
~ aberto 99–72
~ anisotrópico 99–65
~ assimptótico 99–59
~ assintótico 99–59
~ clássico 99–93
~ de Charlier 99–67
~ de De Sitter 99–69
~ de Einstein 99–68
~ de Friedmann 99–71
~ em contração 99–56
~ em evolução 99–54
~ em expansão 99–55
~ estacionário 99–52
~ estático 99–50
~ fechado 99–73
~ hierárquico 99–48
~ homogêneo 99–64
~ inflacionário 99–170b
~ inomogêneo 99–66
~ isotrópico 99–63
~ isótropo 99–63
~ monotônico 99–61
~ monotônico de primeira espécie 99–62
~ não-estacionário 99–53
~ não-estático 99–51
~ não-homogêneo 99–66
~ não-isotrópico 99–65
~ não-isótropo 99–65
~ newtoniano 99–93
~ oscilante 99–57
~ primordial 99–170a
~ pulsante 99–57
~ pulsante de primeira espécie 99–58
~ quente 99–166
~ relativista 99–92
~ relativístico 99–92
~ vazio 99–49
universos de Friedmann 99–71
Urano 83–234
Ursa Maior 92–96
~ Major 92–96
~ Menor 92–97
~ Minor 92–97
Úrsidas de dezembro 86–258
Ursídeos 86–258

V

vácuo 76–9
vaga de tempestade 72–76
vala de potencial 98–176
vale 81–163
~ Alpino 81–164
~ dos Alpes 81–164
valor máximo 76–55
~ máximo da camada E 76–56
~ máximo de uma camada 76–55
vapor da água 77–119
vaporização 86–198
~ de um corpo meteórico 86–198
~ intensa 86–185
variabilidade do campo 84–14
~ do campo magnético 84–14
~ estelar 95–4
~ ó(p)tica 95–141
variação anual 86–138
~ centro-ao-limbo 88–145
~ centro-limbo 88–145
~ de brilho 95–7
~ de luz 95–6
~ diária 86–137
~ do clima 77–183
~ do período 95–43
~ estacional 86–139
~ luminosa 95–6
~ sazonal 86–139
~ secular do campo geomagnético 75–85
~ secular geomagnética 75–85
variações diurnas 83–167
~ do campo geomagnético 75–69
~ do espe(c)tro 95–49
~ espe(c)trais 95–49
~ estacionais 83–166
~ geomagnéticas 75–69
~ magnéticas 95–53
~ sazonais 83–166
variáveis adimensionais 93–133
variável 95–1
~ cataclísmica 95–109
~ cíclica 95–108
~ de cúmulo 95–93
~ de curto período 95–89
~ de eclipse 95–74
~ de erupções 95–68
~ de longo período 95–91
~ de período ultracurto 95–90
~ de velamento 95–83
~ eclipsante 95–74
~ elipsoidal 95–79
~ eruptiva 95–68
~ espe(c)tral 95–48
~ espe(c)troscópica 95–48
~ extrínseca 95–86
~ fulgurante 95–68
~ geométrica 95–72
~ imprópria 95–51
~ intrínseca 95–87
~ irregular 95–107
~ magnética 95–52
~ magnética irregular de polaridade alternada 95–56
~ magnética irregular de polaridade constante 95–57
~ magnética regular 95–55
~ nebular 95–85
~ periódica 95–88
~ por obscurecimento 95–83
~ por ocultação 95–83
~ rotacional 95–65
~ semi-regular 95–105
~ semi-regular vermelha 95–106
~ suspeita 95–5
~ tipo ... 95–47
~ tipo Algol 95–77
~ tipo cefeida 95–92
~ tipo R Coronae Borealis 95–83
~ tipo RR Lyrae 95–93
varredura em comprimento de onda 88–111
~ em posição 88–105
vazamento de óleo 72–71
vazio 76–8, 99–171d
ve(c)tor de Poynting 93–92
~ de propagação 78–36
~ magnético 89–139
Vega 92–132
veio mineral discordante 73–138
veios 87–111
Vela 92–98
velamento 95–84
Velas 92–98
velocidade angular 86–89
~ angular instantânea 86–90
~ angular média 86–91
~ atmosférica 86–94
~ característica 76–43
~ da entrada na atmosfera 86–92
~ de afastamento 91–26
~ de aproximação 91–27
~ de ejeção 85–90
~ de escape de um sistema 96–155
~ de expansão 96–94
~ de fusão 83–147
~ de recessão 91–26, 99–75
~ de relaxação 98–129
~ de rotação 83–32
~ do vento 77–109
~ do vento solar 88–158
~ de um meteoro 86–86
~ equatorial 91–49
~ espacial 91–30
~ fora da atmosfera 86–93
~ geocêntrica 86–95
~ geocêntrica de um meteoro 86–95
~ heliocêntrica 86–96

P

~ instantânea 86–99
~ média 86–98
~ na atmosfera 86–94
~ na linha de visão 91–24
~ na linha de visada 91–24
~ negativa 91–27
~ no ponto médio da trajetória 86–97
~ orbital 96–63
~ positiva 91–26
~ radial 91–24
~ relativista 99–136
~ relativística 99–136
~ residual 98–91
~ rotacional 91–48
~ sobre o rast(r)o 86–154
~ sobre o rast(r)o na chapa fotográfica 86–154
vento 77–89
~ anabático 77–92
~ ascendente 77–92
~ ciclostrófico 77–95
~ de rajadas 77–111
~ do gradiente 77–94
~ estelar 93–14
~ galáctico 98–277
~ geostrófico 77–96
~ gradiente 77–94
~ inter(e)stelar 97–151
~ predominante 77–117
~ solar 88–156
~ solar calmo 84–41
~ solar permanente 84–40
~ solar quiescente 84–41
ventos alísios 77–100
~ anti-alísios 77–103
~ contra-alísios 77–103
Vênus 83–125
venusiano 83–126
vertical 71–71
vértice 96–98, 98–105
vesícula 82–180
vesiculação 82–182
vesicular 82–181
vesiculoso 82–181
Vésper 83–128
Véspero 83–128
Vesta 84–87
Véu da Noiva 97–137
véu 83–250
~ de nuvens 77–151
Via Láctea 98–1
Via-látea 98–1
vibrações livres da Terra 73–49
vida da partícula aprisionada 75–122
~ da partícula retida 75–122
~ de um grupo de manchas 89–123
videomagnetógrafo 88–85
vidro enfumaçado 88–33
~ fosco 88–33
~ lunar 82–96
vigilância de iceberges 72–85
~ de meteoros 86–144
~ Meteorológica Mundial 77–16
violação da paridade de carga 99–170t
~ de simetria 99–170s
Virgem 92–99
Virgo 92–99
visão extrafoveal 86–128
~ indireta 86–128
~ periférica 86–128
viscosidade 73–71
~ magnética 75–34
visibilidade 77–24
boa ~ 77–25
~ de uma ocultação 81–103
~ do eclipse 81–88
má ~ 77–26
vizinhança do Sol 98–59
na ~ do Sol 98–59
~ solar 98–58
VMM 77–16
Volans 92–100
volátil 82–262
su(b)stância ~ 82–262
vulcanismo ao longo de fraturas 82–157
~ induzido 82–155
~ induzido por meteorito 82–156
~ producido por queda de um meteorito 82–156
Vulcano 83–123
Vulcão 83–123
vulcão 73–111
~ ativo 73–114
~ dormente 73–116
~ escudo 73–119
~ extinto 73–115
~ inativo 73–116
~ submarino 73–117
~ suboceânico 73–117
~ tipo Hawai 73–119

X

xistácea, rocha 74–54
xistocidade 74–53
xistóide, rocha 74–54

Z

zero da relação período-luminosidade 95–104
zodíaco 92–5a
zona abissal 72–41
~ ativa 89–54
~ auroral 79–123
~ central 81–114, 89–47
~ convectiva 93–149
~ convectiva do hidrogênio 93–159
~ de ablação 72–123
~ de alimentação glaciária 72–122
~ de audibilidade 76–41
~ de ausência 98–56
~ de calmarias equatoriais 77–101
~ de combustão do hélio 93–145
~ de combustão do hidrogênio 93–144
~ de compressão 86–170
~ de convecção 93–149
~ de convecção do hidrogênio 93–159
~ de crescimento 72–122
~ de dissipacão 71–124
~ de exclusão 98–56
~ de freqüência máxima de auroras 79–125
~ de fusão glaciária 72–124
~ de hélio 93–145
~ de hidrogênio 93–144
~ de manchas 89–54
~ de ocultação 81–107
~ de queima do hélio 93–145
~ de queima do hidrogênio 93–144
~ de radiação 75–101
~ de radiação aprisionada 75–101
~ de radiação artificial 75–108
~ de radiação externa 75–105
~ de radiação interna 75–106
~ de radiação natural 75–107
~ de silêncio 76–42
~ de visibilidade 81–89
~ de visibilidade do eclipse 81–89
~ do limbo 81–113
~ equatorial 83–188
~ marginal 81–113
~ orogênica 73–75a
~ orogenética 73–75a
~ polar 77–193
~ radiativa 93–134
~ régia 89–54
~ semiconvectiva 93–160
~ sísmica 73–31
~ temperada 77–192, 83–189
~ tórrida 77–191
zonas 83–187
~ de Fresnel 86–294

РУССКИЙ УКАЗАТЕЛЬ

Russian Index

Index russe

Russischer Index

Índice ruso

Índice russo

АЛФАВИТ:

а	б	в	г	д	е	ё	ж	з
и	й	к	л	м	н	о	п	р
с	т	у	ф	х	ц	ч	ш	щ
ъ	ы	ь	э	ю	я			

A

абелево уравнение 88–235
аберрация, суточная 86–57
абиссаль 72–41
абиссальная зона 72–41
~ равнина 72–41
абляция 86–182
абсорбция = абсорпция = поглощение
~ в линии 93–108
~ водяным паром 77–127
свободно-свободная ~ 93–106
связанно-свободная ~ 93–107
авроральная электроструя 79–113
авроральный электрон 79–109
австралит 87–98
автоматизированный инструмент 82–41
агломерация 83a–32
агрегат, молекулярный 97–65
адвекция 77–63
адронная эра 99–174
адсорбат 85–131
адсорбент 85–130
адсорбированное вещество 85–131
адсорбироватъ 85–129
адсорбируемость 85–120
адсорбирующее вещество 85–130
адсорбционная способность 85–120
адсорбция 85–128
азой 74–75
азойская группа 74–75
~ эра 74–75
айсберг 72–117
Аквариды 86–249
аккреция 83a–33
аккумуляция 83a–34
подводная ~ 72–47
активизация волокна 89–215
~ протуберанца 89–239
активная область 89–57
активность в оптическом диапазоне 89–3
~ в радиоволнах 89–6
~ в радиодиапазоне 89–6
внутренняя ~ 82–158
вспышечная ~ 89–176, 95–71
вулканическая ~ 73–110, 82–154
галактическая ~ 98–261
~ галактического ядра 98–271
геомагнитная ~ 75–86
кометная ~ 85–110
корпускулярная ~ 89–7
лунная ~ 82–149
магнитная ~ 89–5
~ магнитных полей 89–5
метеорная ~ 86–134
~ метеорного фона 86–136
поверхностная ~ 82–231
повторяющаяся ~ 98–267
~ полярных сияний 79–100
~ потока 86–229
рекуррентная ~ 98–267
сейсмическая ~ 73–4
~ Солнца 89–1
солнечная ~ 89–1
~ солнечных пятен 89–62
фоновая ~ 86–136
фумарольная ~ 82–164
акустика, атмосферная 76–31
аланин 87–72
Алголь 95–77, 92–105
алмазы, метеоритные 87–65
альбедный нейтрон 75–115
альбедо Бонда 83–91
~ в области радиочастот 83–94
видимое ~ 83–90
геометрическое ~ 83–92
истинное ~ 83–89
~ космических лучей 75–116
ламбертово ~ 83–89
монохроматическое ~ 83–93
нейтронное ~ 75–116
~ нейтронов 75–116
~ облаков 83–95
плоское ~ 83–89
~ по Бонду 83–91
~ по Ламберту 83–89
полное ~ 83–90b
среднее ~ 83–90a
сферическое ~ 83–91
Альбирео 92–103
альбит 74–17
Алькор 92–104
альпийский ледник 72–114
Альтаир 92–106
альтиметрия, спутниковая 71–50
альтокумулус 77–144
альтостратус 77–140
α-аминопропионовая кислота 87–72
алюминий, сотовый 82–25
ячеистый ~ 82–25
амид муравьиной кислоты 97–35
аминогруппа 87–71
аминокислота 87–70
аммиак 97–25
амортизатор 82–24
амплитуда 95–38
~ изменения блеска 95–38
~ прилива 82–292
амфибол 74–20
анагалактический 98–207
анализ микроскопических проб 82–143
минералогический ~ 82–53
~ на месте 82–136
петрографический ~ 82–52
химический ~ 82–75
Андромеда 92–14
Андромедиды 86–256
анемограф 77–114
анемометр 77–113
самопишущий ~ 77–114
аннигиляционная эра 99–173a
аннигиляция магнитных полей 89–158
аномалии в содержании 91–183
~ обилия 91–183
аномалистический месяц 81–43
~ период обращения 81–43
аномалия, гравиметрическая 71–75
лунная гравитационная ~ 81–144
магнитная ~ 75–67
~ обилия 91–183
~ силы тяжести 71–75
югоатлантическая ~ 75–68
южноатлантическая ~ 75–68
анортит 74–18
анортозит 82–87
анталголь 95–93
Антарес 92–107
антелий 79–83
антенна, передающая 78–16
приёмная ~ 78–18
антиапекс 86–65
солнечный ~ 98–65
антивертекс 98–106
антигелий 79–83
антиклиналь 73–90
антиклинальная складка 73–90
антипассаты 77–103
антиселена 79–85
антисолнечное направление 75–22
антицентр 98–14
галактический ~ 98–14
антициклон 77–84
антициклонический 77–85
апекс 86–64
~ движения Солнца 91–23
солнечный ~ 91–23, 98–64
~ Солнца 91–23
апоапсида, галактическая 98–190
апоарий 83–141
апоастр 96–61
апоастрий 96–61
апогалактий 98–190
апогей 81–39
апогелий 83–28a
апойовий 83–173
апомарсий 83–141
Аполлон 84–81
апол уний 81–41
апосатурний 83–211
апоселений 81–41
апоцентр, галактический 98–190
аппарат для посадки на Луну 82–21
лунный посадочный ~ 82–21
аппульс 81–101
апсидальное движение 96–71
арео- 83–135
ареографический 83–137
ареоцентрический 83–138
Ариэль 83–235
арка, эруптивная 89–251
арковидная депрессия 81–160
арочная система 89–299
археозой 74–76

археозойская группа 74–76
~ эра 74–76
архимедов принцип 73–101
архимедова спираль 84–34
асейсмичный район 73–29
асимметрия, восточно-западная 89–49
северно-южная ~ 89–50
асимптотический к круговым орбитам 98–180
асинхронное вращение 83–38
аспекты планет 83–50
ассоциация, звёздная 96–88
О-~ 96–89
случайная ~ 97–126
Т-~ 96–90
астеризм 92–3а
астероид 84–68
~ в форме сигары 84–77
двойной ~ 84–79
искусственный ~ 84–69а
каталогизированный ~ 84–95
~ кольца 84–71
либрирующий ~ 84–91а
~ Марса 84–74
нумерованный ~ 84–95
~ пересекающий орбиту Земли 84–76
~ пересекающий орбиту Марса 84–75
потерянный ~ 84–97
~ семейства Марса 84–74
сигарообразный ~ 84–77
~ тесного сближения 84–78
~ типа Амур 84–80
~ типа Аполлон 84–80а
астрономия базирующаяся на наблюдениях с Луны 82–4
внегалактическая ~ 98–208
звёздная ~ 91–2
метеорная ~ 86–3
атаксит 87–87
бедный никелем ~ 87–88
богатый никелем ~ 87–89
атлас, звёздный 92–12
лунный ~ 81–136
~ Луны 81–136
~ неба 92–12
фотометрический ~ 88–138
атмосфера 83–241
адиабатическая ~ 76–12
верхняя ~ 76–52
восстановительная ~ 83а–46
звёздная ~ 93–2
~ Земли 76–5
земная ~ 76–5
изотермическая ~ 93–25
конечная ~ 93–23
~ конечной толщины 93–23
несерая ~ 93–27
нижняя ~ 76–50
однородная ~ 76–10
окислительная~ 83а–47
первичная ~ 83–244
первоначальная ~ 83а–44
планетная ~ 83–243
~ планеты 83–243
плоско-параллельная ~ 93–21
полубесконечная ~ 93–22
протяжённая ~ 93–9
ранняя ~ 83а–45
рассеивающая ~ 93–16
свободная ~ 76–13
серая ~ 93–26
солнечная ~ 88–163
~ Солнца 88–163
~ составленная из электронов 93–28
~ состоящая из электронов 93–28
средняя ~ 76–51
стандартная ~ 76–11
сферическая ~ 93–24
чисто рассеивающая ~ 93–16
электронная ~ 93–28
атмосферики 78–50
длинные свистящие ~ 78–57
короткие свистящие ~ 78–58
свистящие ~ 78–53
свистящие ~ в форме носа 78–59
атмосферный 83–242
атмофилы 83а–29
атмофильные элементы 83а–29
атом, двухуровенный 93–45
первоначальный ~ 99–169
~ с несколькими уровнями 93–46
атомизация солнечным ветром 82–245
афелий 83–28а
афтершок 73–13
Ахернар 92–102
ахондрит 87–77
ацетальдегид 97–42
ацетонитрил 97–38
аэрозоль 76–89
аэролит 87–60
аэрология 76–122
аэрономия 76–122а
аэротриангуляция, пространственная 71–26

Б

бабочки Маундера 89–76
база интерферометра 88–100
лунная ~ 82–17
~ на Луне 82–17
неизменяющаяся ~ 88–102
обитаемая лунная ~ 82–18
переменная ~ 88–103
базальт 74–9
лунный ~ 82–84
базис интерферометра 88–100
~ протуберанца 89–234
базит 74–5
баланс, детальный 93–41
~ массы 88–250
радиационный ~ 88–252
тепловой ~ 77–62
тепловой ~ атмосферы 77–62
энергетический ~ 88–251
~ энергии 88–251
балансирование, детальное 93–41
балдж 98–16
балки 87–104
балл вспышки 89–205
~ землетрясения 73–34
балльность землетрясения 73–33
банка 72–96
барионный газ 93–194
барический градиент 77–106
барометрическая высота 76–16
~ формула 76–15
барьер, звуковой 76–44
бассейн 81–175
батиметрическая карта 71–54
~ линия 71–55
батиметрия 72–28
батолит 73–139
башня, солнечная 88–54
бегущие тени 88–31
бедное звёздное скопление 96–102
безветрие 77–110
безвоздушное пространство 76–9
безвоздушный 76–8
безлунная ночь 79–3е
безоблачно 77–149
безразмерные переменные 93–133
безымянный кратер 81–202
Белое пятно 83–214
белый овал 83–195
берег 72–30
береговая линия моря 72–31
бесконечность вселенной 99–6
беспорядочная поверхность 83–164
беспорядочное движение 89–157
бесстолкновительная релаксация 98–132
~ система 98–167
~ударная волна 84–49
бетатронный эффект 91–180
Бетельгейзе 92–108
биг-бэнг 99–167
Биелиды 86–256
биосфера 72–8
биотит 74–25
биполярная группа 89–111
~ магнитная область 89–160а
бифуркация 98–5
благородный газ 87–133
первичный ~ газ 87–133
блеск 91–54
абсолютный ~ 85–109
абсолютный ~ кометы 85–109
абсолютный ~ метеора 86–74
болометрический ~ 91–68
видимый ~ звезды 91–61
визуальный ~ 91–65
~ звезды 91–55

R

интегральный ~ 85–107
инфракрасный ~ 91–71
максимальный ~ 86–78
~ метеора 86–8
общий ~ 96–135
предельный ~ 91–62
приведённый ~ 85–108
собственный ~ 85–107
сумарный ~ 96–135
фотовизуальный ~ 91–69
фотографический ~ 91–66
Близнецы 92–51
близполюсное созвездие 92–4a
блинк-компаратор 91–34
блинк-микроскоп 91–34
блин, первичный 99–171b
блок 82–130
~ с острыми углами 82–131
блуждающие токи 75–6
богатое звёздное скопление 96–100
болид 86–11
большой ~ 86–10
детонирующий ~ 86–119
дневной ~ 86–118
звуковой ~ 86–119
очень большой ~ 86–9
яркий ~ 86–10
болото 81–303
~ Гниения 81–305
~ Гнилое 81–305
~ Сна 81–306
~ Сонное 81–306
~ Эпидемий 81–304
Большая Медведица 92–96
~ туманность Ориона 97–136
Большое красное пятно 83–193
~ Магелланово Облако 98–213
~ объединение 99–170u
~ южное возмущение 83–194
Большой Ковш 92–96
~ коллапс 99–182
~ Пёс 92–27
бомба, вулканическая 73–144
~ Эллермана 89–200
бомбардировка метеорными телами 82–235
~ метеороидами 82–235
~ частицами 82–240
борозда 81–165
извилистая ~ 81–167
прямолинейная ~ 81–166
брекчия, лунная 82–85
бриз, береговой 77–98
дневной ~ 77–97
морской ~ 77–97
ночной ~ 77–98
бугристый материал 82–103
буй, автоматический 72–100
исследовательский ~ 72–99
булыжник 82–128
бурное образование звёзд 94–11a
буровой образец 82–60
буря, геомагнитная 75–71
ионосферная ~ 78–49
магнитная ~ 75–71
межпланетная ~ 84–44
~ обычного типа 75–72
пылевая ~ 83–255
~ с внезапным началом 75–73
~ с постепенным началом 75–78
шумовая ~ на Юпитере 83–185
бухта, магнитная 75–80
бухтообразное возвущение 75–80
бюллетень погоды 77–10

В

вакуумная энергия 99–170k
вал, внешний 81–253
внутренний ~ 81–252
выступающий ~ 81–222
концентрический ~ 81–254
~ кратера 81–251
~ кратный ~ 81–255
~ состоящий из глыб 81–224
террасовый ~ 81–257
валун 82–130
ледниковый ~ 72–132
эрратический ~ 72–132
валун-индикатор 72–132
вариации спектра 95–49
вариация, вековая 75–85
геомагнитная ~ 75–69
годичная ~ 86–138
сезонная ~ 86–139
суточная ~ 86–137
~ элементов земного магнетизма 75–69
вдоль луча зрения 91–29
Вега 92–132
веерообразный 85–74
~ выброс 85–62
вековые вариации геомагнитного поля 75–85
вектор магнитного поля 89–139
мировой ~ 99–138
~ Пойнтинга 93–92
~ распространения 78–36
~ Умова-Пойнтинга 93–92
четырёхмерный ~ 99–138
великое объединение 99–170u
~ объединённое взаимодействие 99–170v
величина, абсолютная 91–64
аболютная звёздная ~ 91–64
аболютная ~ метеора 86–72
аболютная ~ планеты 83–86
болометрическая ~ 91–68
болометрическая звёздная ~ 91–68
видимая ~ 91–61
видмая звёздная ~ 91–61
видимая звёздная ~ метеора 86–71
визуальная звёздная ~ 91–65
звёздная ~ 91–56
звёздная ~ в зените 91–63
звёздная ~ в средней оппозиции 83–85
звёздная ~ по радионаблюдениям 86–268
~ землетрясения 73–34
зенитная звёздная ~ 91–63
интегральная звёздная ~ 96–136
инфракрасная звёздная ~ 91–71
~ наибольшей фазы 81–79
общая звёздная ~ 96–136
предельная звёздная ~ 91–62
радиозвёздная ~ 86–268
радиометрическая звёздная ~ 91–70
средняя звёздная ~ 83–85
суммарная звёздная ~ 96–136
фотовизуальная звёздная ~ 91–69
фотографическая звёздная ~ 91–66
электрографическая звёздная ~ 91–67
элетронографическая звёздная ~ 91–67
Венера 83–125
венерианский 83–126
венец 79–60
~ вокруг Луны 79–61
~ вокруг Солнца 79–62
веретенообразная галактика 98–239
веретенообразный метеор 86–77
вертекс 98–105
верхние слои атмосферы 76–52
~ слои солнечной атмосферы 88–165
верхняя граница атмосферы 76–25
вершина, атмосферы 76–25
подводная ~ 72–43
вес, средний молекулярный 93–100
Веста 84–87
Весы 92–62
ветвь, восходящая 95–34
галактическая ~ 98–26a
~ гигантов 91–152b
горизонтальная ~ 91–132
закрученная галактическая ~ 98–29
закручивающаяся галактическая ~ 98–29
нисходящая ~ 95–35
нисходящая ~ цикла 89–84
~ Ориона 98–36
~ Персея 98–34
~ потока 86–240
раскручиваемая ~ 98–31
раскручивающаяся ~ 98–31
~ роста активности (солнечного цикла) 89–83
~ спада активности (солнечного цикла) 89–84
спиральная ~ 98–27
~ Стрельца 98–35
ветер 77–89
анабатический ~ 77–92
восходящий ~ 77–92

R

галактический ~ 98–277
геострофический ~ 77–96
градиентный ~ 77–94
звёздный ~ 93–14
межзвёздный ~ 97–151
порывистый ~ 77–111
постоянный солнечный ~ 84–40
преобладающий ~ 77–117
солнечный ~ 88–156
спокойный солнечный ~ 84–41
циклострофический ~ 77–95
вечность вселенной 99–13
вещество, адсорбирующее 85–130
внеземное ~ 72–56
вяжущее ~ 82–118
газово-пылевое ~ 83а–25
дозвездное ~ 94–16
звёздное ~ 93–124
~ звезды 93–124
космическое ~ 99–7
легкоплавкое ~ 83а–56
летучее ~ 82–262
лунное ~ 82–58
межзвёздное ~ 97–1
межпланетное ~ 84–1
мелкозернистое вяжущее ~ 82–115
метеоритное ~ 87–9
метеоритное органическое ~ 87–68
нелетучее ~ 83а–53
первичное ~ 99–170
поглощающее ~ 97–85
пористое ~ 82–111
светящееся ~ 99–86
тёмное ~ 97–108
тугоплавкое ~ 83а–55
углеродистое ~ 87–64
углистое ~87–64
цементирующее ~ 82–118
взаимная нейтрализация 79–29
взаимодействие, атмосферное 79–120
коллективное ~ 98–168
объединённое ~ 99–170х
приливное ~ 81–120, 98–311
~ с атмосферой 79–120
сверхобъединённое ~ 99–170г
электрослабое ~ 99–170х
взаимодействующие галактики 98–310
~ протуберанцы 89–243
взаимосвязанные галактики 98–306
взброс 73–94
взвешенное облако 89–275
вздутие, приливное 81–119
экваториальное ~ 71–9
взрыв 87–30, 95–131
Большой ~ 99–167
высотный ~ 75–114
~ галактического ядра 98–283
~ метеора 86–81
~ новой 95–121
~ новой звезды 95–121
первоначальный ~ 99–167
~ сверхновой 95–131
~ сверхновой звезды 95–131
взрывная гипотеза 89–63
~ фаза 89–210
взятие донной пробы 87–16
вибрирующее зеркало 86–130а
вид облака 77–135
видеомагнитограф 88–85
видимость 77–24
~ затмения 81–88
плохая ~ 77–26
~покрытия 81–103
хорошая ~ 77–25
видманштеттеновы фигуры 87–105
винтовая симметрия 98–161
вириальная теорема 98–181
висящее облако 89–275
вихревая теория 83а–11
включение, органическое 87–66
влажность 77–118
абсолютная ~ 77–122
атмосферная ~ 77–118
~ воздуха 77–118
относительная ~ 77–128
удельная ~ 77–123
ВМО 77–17
вне земной атмосферы 88–149
внегалактический 98–207
внедрение 73–133
внезапная фазовая аномалия 78–52
внезапное исчезновение 89–279
~ исчезновение волокна 89–279
~ начало магнитной бури 75–74
~ повышение атмосфериков 78–51
~ поглощение коротких волн 78–40
~ поглощение космического шума 78–42
~ усиление атмосфериков 78–51
внезапный выброс частиц 88–162
внезатменный коронограф 88–59
внеземное вещество 72–56
~ происхождение 87–73
~ пространство 87–73а
внеземной материал 72–56
внешний вал 81–253
~ склон вала 81–253
внутренний вал 81–252
~ склон вала 81–252
внутритропические дуги 79–53
вода, малая 82–283
меженная ~ 82–283
низкая ~ 82–283
полная ~ 82–281
Водолей 92–17
водород, межзвёздный 97–46
металлический ~ 83–202
молекулярный ~ 97–31
неионизованный ~ 97–47
нейтральный ~ 97–47
сернистый ~ 97–44
возврат 88–132
~ группы пятен 88–133
возвратный выброс 89–255
~ протуберанец 89–255
возвращение 88–132
~ группы пятен 88–133
~ кометы 85–9
возвышение, лунное 81–156
~ лунной местности 81–156
возгонка 85–138
возгорание вспышки 89–210
воздействие атмосферы 87–116
приливное ~ 83а–15
воздух 76–6
воздушное сопротивление 86–178
возмущение, внезапное ионосферное 78–41
геомагнитное ~ 75–70
ионосферное ~ 78–39
магнитное ~ 75–70
межпланетное ~ 84–45
~ обусловленное вспышкой 84–46
приливное ~ 98–314
Южное тропическое ~ 83–194
возмущения, взаимные 83–25
взаимные ~ в движении спутников 83–25
~ вызванные Солнцем 83–24
радиальные ~ 93–215
возмущённое Солнце 89–12
возникновение 94–9
~ пятна 89–121
Возничий 92–21
возраст, абсолютный 82–269
~вселенной 99–12
геологический ~ 74–90
~ затвердевания 87–137
~ звёзд 94–61
~ Земли 83а–68
~ Луны 81–31, 82–264
~ метеоритов 87–127
~ определённый по количеству остаточных газов 87–135
относительный ~ 82–268
~ по радиогенному стронцию 82–271
~ по соотношению изотопов свинца 74–100
~ потока 86–245
~ роя 86–246
рубидий-стронциевый ~ 82–271
свинцово-свинцовый ~ 74–100
стратиграфический ~ 82–267
ураново-свинцовый ~ 74–101
характерный ~ 99–79
экспозиционный ~ 87–139
волдыри, лавовые 82–178
Волк 92–63
волна, бесстолкновительная 84–49
бесстолкновительная ударная ~ 84–49
вторичная ~ 73–43
галактическая ~ плотности 98–175
галактическая ~ сжатия 98–175
гравитационная ~ 99–164

двойная ~ 98–117
звуковая ~ 76–33
~ землетрясения 73–9
земная ~ 78–14
инфразвуковая ~ 72–93
~ Лява 73–48
незатухающая ~ 86–277
необыкновенная ~ 78–30
несущая ~ 86–276
объёмная ~ 73–42
обыкновенная ~ 78–31
P ~ 73–46
первичная ~ 73–46
планетарная ~ 72–94
~ плотности 98–175
поверхностная ~ 73–47
полусуточная ~ 76–29
поперечная ~ 73–41
поперечная ~ землетрясения 73–43
приземная ~ 78–14
приливная ~ 82–291
продольная ~ 73–45
продольная ~ землетрясения 73–46
пространственная ~ 78–13
прямая ~ 78–12
Q ~ 73–48
~ разрежения 97–58
S ~ 73–43
~ сдвига 73–43
сейсмическая ~ моретрясения 72–95
~ сжатия 73–46
сейсмическая продольная ~ 73–46
суточная ~ 76–30
~ тсунами 72–95
~ тяготения 88–201
упругая ~ 73–44
~ цунами 72–95
волнение 72–86
волнистость геоида 71–74
волновая теория спиральной структуры 98–174
волны, инфразвуковые 75–88
~ полутени 89–100
волокна, хромосферные 88–209
волокнистая структура 97–130
Волокнистая туманность 97–137
волокно 89–224
водородное ~ 89–267
металлическое ~ 89–268
мигающее ~ 89–216
осциллирующее ~ 89–216
полутеневое ~ 89–98
~ полутени 89–98
~ протуберанца 89–245
~туманности 97–131
волоконная структура 89–174
волоконца, хромосферные 89–173
волоконцевая структура 89–174
Волопас 92–22
Волосы Вероники 92–38
Ворон 92–41
воронка 87–58, 89–269
воронкообразный протуберанец 89–269
восприимчивость, магнитная 75–27
восход Солнца 88–8
водходящая ветвь солнечного цикла 89–83
восхождение, прямое 81–60
впадина 81–175
вершинная ~ 81–244
вершинная ~ центральной горки 81–244
внутриматериковая ~ 81–158
вулканическая ~ 81–179
глубоководная ~ 72–40
лунная ~ 81–157
вращательная дисторсия 91–44
вращающееся зеркало 86–130
~ тело 91–42
вращение, вынужденное 81–14
галактическое ~ 98–107
дифференциальное ~ 88–128
дифференциальное ~ Галактики 98–113
~ звёзд 91–35
кеплеровское ~ 98–112
~ линии апсид 96–71
неоднородное ~ 91–36
несвободное ~ 81–14
~ планет 83–30
планетное ~ 83–30
обратное ~ 83–40
прямое ~ 83–39
синхронное ~ 83–37
солнечное ~ 88–127
~ Солнца 88–127
твердотельное ~ 98–111
временное лунное явление 82–153
время, абсолютное 99–117
~ вступления волны 73–50
геомагнитное ~ 79–132
~ до маскимума 95–30
~ жизни захваченной частицы 75–122
истинное лунное ~ 82–301
лунное ~ 82–300
ньютоновское ~ 99–118
~ охлаждения 83–201
~ падения метеорита 87–51
~ Планка 99–170i
~ после максимума 95–33
~ появления 86–49
~ пребывания 82–34
~ прихода 73–50
~ пробега 84–47
~ пробега от Солнца к Земли 84–47
~ пробега сейсмической волны 73–51
~ распада 96–104
~ расширения 99–78
~ релаксации 98–128
собственное ~ 99–139
среднее лунное ~ 82–302
среднее ~ релаксации 98–130
фридмановское ~ 99–80
фридмановское ~ расширения 99–80
хабблoвское ~ 99–79
характеристическое ~ 99–79
вселенная 99–1
анизотропная ~ 99–65
асимптотическая ~ 99–59
асимптотическая ~ первого рода 99–60
горячая ~ 99–166
~ Де Ситтера 99–69
закрытая ~ 99–73
замкнутая ~ 99–73
иерархическая ~ 99–48
изотропная ~ 99–63
классическая ~ 99–93
монотонная ~ 99–61
монотонная ~ первого рода 99–62
неоднородная ~ 99–66
нестатическая ~ 99–51
нестационарная ~ 99–53
ньютоновская ~ 99–93
однородная ~ 99–64
осциллирующая ~ 99–57
осциллирующая ~ первого рода 99–58
открытая ~ 99–72
первоначальная ~ 99–170a
пульсирующая ~ 99–57
пустая ~ 99–49
раздувающаяся ~ 99–170b
ранняя ~ 99–170a
расширяющаяся ~ 99–55
релятивистская ~ 99–92
сжимающаяся ~ 99–56
статическая ~ 99–50
стационарная ~ 99–52
~ Фридмана 99–71
~ Шарлье 99–67
эволюционирующая ~ 99–54
~ Эйнштейна 99–68
~ Эйнштейна-Де Ситтера 99–70
всемирная геодезическая система 71–39
~ метеорологическая организация 77–17
~ служба погоды 77–16
всенаправленная интенсивность 75–104
ВСП 77–16
всплеск, декаметровый 83–183
низкочастотный ~ 83–184
~ рентгеновского излучения 79–117
~ солннечного ветра 84–42
вспышечная активность 89–176, 95–71
~ лента 89–187
вспышечноподобный протуберанец 89–222

R

вспышечный лимб 89–189
вспышка 89–175, 95–69
~ балла 2 89–206
~ блеска 85–126
~ в белом свете 89–182
~ в видимом диапазоне 89–181
~ в радиоизлучении 89–180
~ в рентгеновских лучах 89–183
гелиевая ~ 94–79
гомологическая ~ 89–178
двухленточная ~ 89–187а
двухниточная ~ 89–187а
залимбовая ~ 89–203
звёздная ~ 95–70
~ звезды 95–70
~ кометы 85–125
лимбовая ~ 89–202
лунная ~ 82–247
~ метеора 86–80
~ на диске 89–204
~ на краю диска 89–202
непротонная ~ 89–186
~ новой 95–121
~ новой звезды 95–121
оптическая ~ 89–181
протонная ~ 89–185
рентгеновская ~ 89–183
~ сверхновой 95–131
~ сверхновой звезды 95–131
симпатическая ~ 89–179
солнечная ~ 89–175
~ солнечного ветра 84–42
~ солнечных космических лучей 89–194
хромосферная ~ 89–175
вспышкоподобный протуберанец 89–222
вступление Луны в полутень 81–75
~ Луны в тень 81–76
вторжение частиц высоких энергий 78–43
~ электронов 78–48
~ электронов полярного сияния 79–112
вуалирование 95–84
вуалируемая переменная 95–83
~ переменная звезда 95–83
вуаль 83–250
вулкан 73–111, 83–123
бездействующий ~ 73–116
действующий ~ 73–114
мёртвый ~ 73–115
наслоенный ~ 73–121
недействующий ~ 73–116
подводный ~ 73–117
потухший ~ 73–115
смешанный ~ 73–121
спящий ~ 73–116
субмаринный ~ 73–117
уснувший ~ 73–116
щитовидный ~ 73–119
щитовой ~ 73–119
вулканизм вдоль трещин 82–157
~ вызванный ... 82–155
~ вызванный падением метеорита 82–156
вулканит 74–3
вулканическая порода 74–3
вулканическое извержение 82–170
вход в атмосферу 86–161
~ в плотные слои атмосферы 86–161
выбрасывание 98–179
~ вещества 98–274
выброс 85–55, 89–259
быстрый вспышечный ~ 89–258
веерообразный ~ 85–60, 89–258
~ вещества 89–214, 94–77
внезапный ~ частиц 88–162
возвратный ~ 89–255
вспышечный ~ 89–214
вспышечный возвратный ~ 89–190
~ из кометы 86–262
искривлённый ~ 85–58
кривящийся ~ 85–58
~ массы 93–15
~ материи 98–274
~ материи из галактического ядра 98–274
обращённый веерообразный ~ 85–63
оптический ~ 98–279
послевспышечный ~ 89–190
~ протонов 84–59
прямолинейный ~ 85–57
прямой веерообразный ~ 85–61
~ радиоизлучающих облаков 98–285
рекуррентный ~ 98–282
~ солнечной плазмы 84–38
спиральный ~ 85–59
~ струи 98–278
~ струи из галактического ядра 98–278
струйчатый ~ 89–258
тёмный возвратный ~ 89–256
~ частиц 85–89
яркий возвратный ~ 89–257
выбросы 82–165
вулканические ~ 82–166
выветривание 74–57
пространственное ~ 82–244
выделение газов 87–136
~ газов из кометного ядра 84–109
~ энергии 93–105
выигрыш энергии 88–253
вылет звёзд из скопления 98–194
выметание газов 83а–41
вынужденное вращение 81–14
выпуклость, приливная 82–278
экваториальная ~ 71–9
вырождение 93–172
~ нейтройнного газа 93–191
нейтронное ~ 93–191
нерелятивистское ~ 93–177
полное ~ 93–175
релятивистское ~ 93–176
частичное ~ 93–173
вырожденный 93–178
~ в высокой степени 93–184
полностью ~ 93–185
частично ~ 93–183
высверленное отверстие 82–61
высокогорная обсерватория 88–56
~ станция 88–56
высокогорье 81–189, 82–56
высокоионизованный 89–289
высококучевые облака 77–144
высокослоистое облако 77–140
высокоширотный 89–53
высота апекса 86–66
барометрическая ~ 76–16
геометрическая ~ тени 79–47
действительная ~ 78–29
действующая ~ 78–28
~ зеркальной точки 75–125
~ излучающего слоя 79–26
истинная ~ 78–29
истинная ~ тени 79–49
~ исчезновения 86–40
масштабная ~ 76–18
~ метеоров 86–193
~ над уровнем моря 71–51
~ над эллипсоидом 71–52
~ однородной атмосферы 76–17а
~ по давлению 76–16
~ погасания 86–40
~ появления метеора 86–39
предельная ~ атмосферы 76–25
приведённая ~ 76–18
реальная ~ 78–29
~ слоя 78–27
~ слоя свечения 79–26
~ тени 79–46
~ точки исчезновения 86–40
~ точки появления 86–39
~ хромосферы 88–202
эквивалентная ~ 78–28
~ экранирования 79–48
экранированная ~ тени 79–49
эллипсоидическая ~ 71–52
высотно-частотная характеристика 78–23
выступ, лунный 81–118
межпланетный ~ 84–21
приливный ~ 81–119
экваториальный ~ 71–9
выступающий вал 81–222
высушивание 74–51
высыпание электронов 79–112
высыпанные электроны 79–118
высыпающиеся частицы 76–117
выталкивание лавы 82–171
выход Луны из полутени 81–82
~ Луны из тени 81–81
~ на орбиту 82–8

~ энергии 93–109
вычислитель 85–22
вычитание, фотографическое 88–94
вяжущее вещество 82–118
вязкий 82–138
вязкость 73–71
магнитная ~ 75–34
магнитная ~ горных пород 75–34
вязкотекучесть 73–71

Г

газ, барионный 93–194
вырожденный ~ 93–179
гиперонный ~ 93–193
замерзший ~ 85–45
замороженный ~ 85–45
захваченный ~ солнечного ветра 82–148
межгалактический ~ 98–351а
межгалактический ~ внутри скопления 98–352
~ между скоплениями галактик 98–353
межзвёздный ~ 97–9
нейтронный ~ 93–192
окологалактический ~ 98–351
радиогенный ~ 87–132
радиогенный благородный ~ 87–132
раскалённый ~ 83а–43
релятивистский ~ 97–152
угарный ~ 97–28
газово-пылевая туманность 97–119
газодинамика, межзвёздная 97–150
гайот 72–44
гал 71–65
Галактика 98–1
галактика 98–201
активная ~ 98–260
~ Аро 98–266
веретенообразная ~ 98–239
взрывающаяся ~ 98–268
~ видимая плашмя 98–247
~ видимая с ребра 98–246
возмущённая ~ 98–315
~ гигант 98–222
гигантская ~ 98–222
гигантская эллиптическая ~ 98–243
головастиковая ~ 98–281
головастикообразная ~ 98–281
голубая компактная ~ 98–265
двойная ~ 98–305
деформированная ~ 98–316
двухрукавная ~ 98–43
иглообразная ~ 98–239
инфракрасная ~ 98–205
искажённая ~ 98–316
~ карлик 98–220
карликовая ~ 98–220
карликовая эллиптическая ~ 98–242
квазизвёздная ~ 98–361
кольцевая ~ 98–259
кольцеобразная ~ 98–259
компактная ~ 98–264
кратная ~ 98–308
линзовая ~ 98–245
линзовидная ~ 98–245
линзообразная ~ 98–245
местная ~ 98–327
~ Местной Группы 98–327
~ Маркаряна 98–265а
многорукавная ~ 98–44
N ~ 98–263
~ наблюдаемая в плане 98–247
~ наблюдаемая с ребра 98–246
наша ~ 98–1
~ не принадлежащая скоплению 98–340
неправильная ~ 98–237
нормальная ~ 98–221
обычная ~ 98–221
однорукавная ~ 98–42
оптическая ~ 98–204
особенная ~ 98–258
пекулярная ~ 98–256
пересечённая спиральная ~ 98–253
~ переходного типа 98–248
~ поля 98–340
послеэруптивная ~ 98–270
постэруптивная ~ 98–270
правильная ~ 98–236
разрушенная ~ 98–317
распавшаяся ~ 98–317
рентгеновская ~ 98–203
родительская ~ 98–284
~ с большим содержанием газа 98–238
~ с кратным ядром 98–319
~ сверхгигант 98–223
сверхгигантская ~ 98–223
~ Сейферта 98–262
сейфертовская ~ 98–262
сигарообразная ~ 98–239
сильно сплюснутая ~ 98–252
сингулярная ~ 98–258
~ скопления 98–338
соседняя ~ 98–211
спиральная ~ 98–249
спиральная ~ с перемычкой 98–253
~ спутник 98–216
сферическая ~ 98–244
сфероидальная ~ 98–244
~ типа голова-хвост 98–281
~ типа N 98–263
~ фона 98–340
чечевицеобразная ~ 98–245
эллипсоидальная ~ 98–241
эллиптическая ~ 98–241
эллиптическая ~ спутник 98–217
эруптивная ~ 98–269
ярчайшая ~ 98–339
галактики, взаимодействующие 98–310взаимосвязанные ~ 98–306
сталкивающиеся ~ 98–321
галактическая ветвь 98–26а
~ волна плотности 98–175
~ чёрная дыра 98–23а
галактический 98–2
~ ветер 98–277
~ рукав 98–26а
галактическое вращение 98–107
~ гало 98–45
галактоцентрический 98–10
Галилеев спутник 83–204
галилеевский спутник 83–204
галилеевское преобразование 99–131
гало 79–67
большое ~ 79–70
~ вокруг Луны 79–68
галактическое ~ 98–45
горячее ~ 98–46
лунное ~ 79–68
магнитное ~ 98–48
малое ~ 79–69
холодное ~ 98–47
галосы 79–67
~ кометы 85–68
галька 82–128
гамма-пульсар 95–139
гамма-спектроскопия 82–77
недеструктивная ~ 82–77
неразрушающая ~ 82–77
Ганимед 83–207
гарвардская спектральная классификация 91–91
гармоники, зональные 95–101
незональные ~ 95–102
тессеральные ~ 95–102
гауптовский поправочный множитель 93–104
гексаэдрит 87–84
гелиевая вспышка 94–79
гелиевый метод 87–128
~ метод определения возраста 87–128
гелио- 88–5
гелиогеофизика 84–56
гелиограмма в рентгеновских лучах 88–74
рентгеновская ~ 88–74
гелиограф 77–61, 88–36
монохроматический ~ 88–38
гелиодолгота 89–40
гелиометр 88–12
гелиосейсмология 88–223b
гелиостат 88–45
гелиосфера 88–255
гелиосферная полость 88–257
гелиосферный 88–256
гелиофизика 88–6
гелиоцентрическая система 99–23

R

гелиоцентрическое расстояние 84–4
гелиоширота 89–42
Геминиды 86–257
Гемма 92–117
генерация звука 88–215
~ энергии 93–105
геоблок 72–11
геодезическая система относимости 71–19
геодезия, космическая 71–31
спутниковая ~ 71–32
геоид 71–16
компенсированный ~ 71–17
регуляризированный ~ 71–17
геоисторическая шкала 74–68
геокорона 76–73
~ из нейтрального водорода 76–74
геологический возраст 74–90
~ период 74–74
геологическое летоисчисление 74–67
геология 72–2
~ планет 83–5
планетная ~ 83–5
геомагнетизм 75–24
геомагнитное время 79–132
геомагнитные вариации 75–69
~ возмущения 75–70
геометрическая высота тени 79–47
геометрия Евклида 99–108
евклидова ~ 99–108
неевклидова ~ 99–109
риманова ~ 99–113
~ Эвклида 99–108
эвклидова ~ 99–108
геопотенциал 71–60
геопотенциальная поверхность 71–61
геосинклиналь 73–105
геострофический ветер 77–96
геосфера 72–7
геотектоника 73–95
геотермика 74–58
геотермия 74–58
геофизика 72–1
космическая ~ 84–57
геохимический цикл 74–32
геохимия 74–30
геохронологическая шкала 74–68
геохронология 74–67
геоцентрическая система 99–21
геоцентрическое расстояние 84–3
геоэлектричество 75–1
Геркулес 92–53
Гермес 84–83
гетеросфера 76–61
Гиады 96–111а
гибель группы пятен 89–126
гигант 91–143
голубой ~ 91–142а
красный ~ 91–152а
нормальный ~ 91–150
~ поздних спектральных классов 91–144а
~ ранних спектральных классов 91–143а
яркий ~ 91–149
гигантская галактика 98–222
~ эллиптическая галактика 98–243
гигантское молекулярное облако 97–13а
гигрометр 77–120
гидируемая камера 86–142
Гидра 92–55
гидродинамика, звёздная 98–142
гидрозоли 76–90
гидроксил 79–25
гидриксильная группа 97–22
гидрологическая карта 72–25
гидростатическое давление 93–77
гидросфера 72–21
гидрофильный 76–92
гидрофобный 76–93
гильотинный множитель 93–103
гипербазит 74–6
гипотеза 94–6
взрывная ~ 89–63
~ двух потоков 98–99
~ захвата 85–99
~ извержения 85–98
катастрофическая ~ 83а–14
~ контракции 73–77
контракционная ~ 73–77
космологическая ~ 94–8
~ Ламберта 99–45
метеоритная ~ 81–262
метеороидная ~ 81–262
небулярная ~ 83а–7
~ падения 81–261
приливная ~ 83а–14
~ сжатия 73–77
~ столкновсний 81–261
~ столкновения 83а–14
~ увлечения 85–99
гипоцентр 73–17
гипоцентральное расстояние 73–18
гипсометрический уровень 81–142
гл 71–65
главная последовательность 91–131
глетчер 72–112
глины, красные 72–54
красные глубоководные ~ 72–54
пелагические ~ 72–54а
глобула 97–133
пылевая ~ 98–290
глобус, лунный 81–112
небесный ~ 92–7
глория 79–63
глубина залегания очага 73–20
~затмения 95–75
~ изостатической компенсации 73–102
~ очага 73–20
~ фокуса 73–20
эффективная ~ испускания излучения 93–64
эффективная ~ образования линий 93–64
глубинная порода 74–7
глубинного происхождения 82–259
глубоководная впадина 72–40
глубоководный промер 72–28
глыба 82–130
континентальная ~ 72–11
ледниковая ~ 72–132
гляциальный 72–106
гляциология 72–103
Год Спокойного Солнца 89–10
световой ~ 91–18
голая особенность 94–117
Голова Змеи 92–89
голова, кометная 85–35
~ кометы 85–35
головастиковая галактика 98–281
головастикообразная галактика 98–281
головная часть области рассеяния 87–39
голубая компактная галактика 98–265
голубой 91–99
~ цвет неба 77–60
Голубь 92–37
голыш 82–128
гомологическая вспышка 89–178
гомологический 93–131
гомопауза 76–65
гомосфера 76–60
Гончие Псы 92–26
~ Собаки 92–26
гора, ледяная 72–117
плосковершинная подводная ~ 72–44
подводная ~ 72–43
горб кривой 95–37
~ кривой блеска 95–37
лунный ~ 81–118
приливный ~ 81–119, 83а–67
горение водорода 94–56
взрывное ~ гелия 94–79
~ гелия 94–78
~ кислорода 94–82
~ кремия 94–83
~ углерода 94–81
горизонт вселенной 99–10
космологический ~ 99–179
~ масс 99–41
~ событий 99–43
горизонталь 71–53, 81–139
гипсометрическая ~ 81–142
подводная ~ 72–32
горка, центральная 81–258
горловина вулкана 73–122
горообразование 73–74
горы, лунные 81–193
складкообразные ~ 73–86
складчатые ~ 73–86
горячая вселенная 99–166
горячее гало 98–46

R

~ место 95–112a
~ пятно 95–112a
грабен 82–132
гравиметр 71–70
гравиметрическая аномалия 71–75
гравиметрия 71–63
гравитационная волна 99–164
~ линза 98–256a
гравитационно несвязанная система 96–158
~ связанная система 96–156
гравитационное потемнение 91–41
гравитационный потенциал 96–160a
~ радиус 99–163
гравитирующий 94–33
град 77–178
градиент 93–94
абсолютный ~ 91–111
адиабатический температурный ~ 93–98
барический ~ 77–106
геотермический ~ 74–59
~ давления 77–106, 93–95
конвективный температурный ~ 93–157
~ плотности 98–87
сверхадиабатический ~ 93–99
спектрофотометрический ~ 91–112
сухоадиабатический ~ 76–14
сухоадиабатический ветрикальный ~ 76–14
температурный ~ 93–96
~ температуры 93–96
~ температуры при лучистом равновесии 93–97
градиентный ветер 77–94
гранит 74–8
граница, верхняя 76–58
верхняя ~ атмосферы 76–25
верхняя ~ слоя 76–58
~ земной тени 81–91
~ кора-мантия 72–14
~ магнитосферы 75–9
~ Мохо 72–14
~ Мохоровичича 72–14
~ раздела солённости 72–67
северная ~ затмения 88–20
~сектора 84–18
секторная ~ 84–18
снеговая ~ 72–111
~ тени 83–69
южная ~ затмения 88–19
граничный слой солён(н)ости 72–67
гранула 88–178
~ мелкодисперсного материала 82–110
стеклянная ~ 82–127
~ тени 88–180
факельная ~ 88–179
фотосферная ~ 88–181
грануляция, факельная 89–163
фотосферная ~ 88–176
график изменения скорости с глубиной 73–52
гребень, барический 77–87
~ высокого давления 77–87
горный ~ 81–186
~ моря 81–188
~ повышенного давления 77–87
складчатый ~ 81–187
гроза 77–159
гром 77–165
грот, лавовый 82–177
грубозернистый 82–117
грунт 81–190
лунный ~ 82–62
морской ~ 72–38
связный ~ 82–63
группа, азойская 74–75
археозойская ~ 74–76
~ астероидов 84–90
биполярная ~ 89–111
~ галактик 98–325
гидроксильная ~ 97–22
докембрийская ~ 74–70
~ звёзд 92–3a
~ комет 85–105
комплексная ~ 89–112
~ малых планет 84–90
Местная ~ галактик 98–326
~ нового цикла 89–106
палеозойская ~ 74–71
~ планетоидов 84–90
~ предыдущего цикла 89–107
протерозойская ~ 74–77
~ пятен 89–105
рекуррентная ~ 89–108
сложная ~ 89–112
~ соизмеримых астероидов 84–101
~ солнечных пятен 89–105
~ старого цикла 89–107
~ Троянцев 84–91
униполярная ~ 89–110

Д

давление, атмосферное 77–43
боковое ~ 73–65
~ водяного пара 77–125
~ воздуха 77–43
~ вырождения 93–187a
~ газа 93–90
газовое ~ 93–90
гидростатическое ~ 93–77
~ излучения 93–91
лучистое ~ 93–91
~ на уровне моря 77–44
нетепловое ~ 94–42
полное ~ 93–93
~ приведённое к уровню моря 77–44
радиационное ~ 93–91
тепловое ~ 94–41
~ фотонов 93–91
~ частиц 93–90
дайка 73–138
дальность 86–288
~ действия 86–288
~ действия радиолокатора 86–288
максимальная ~ 86–289
предельная ~ 86–289
данные, исходные геодезические 71–20
солнечные ~ 89–9
датирование 87–126
~ лунных событий 82–265
~ по радиоизотопам 87–131
радиоактивное ~ 87–131
даты, исходные геодезические 71–20
движение, беспорядочное 98–94
вертикальное ~ 88–185
вертикальное ~ континентов 73–73
видимое ~ 83–41
горизонтальное ~ 88–184
горизонтальное ~ континентов 73–104
~ земной коры 73–72
~ континентов 73–104
крупномасштабное ~ 88–188
~ ледника 72–120
~ линии апсид 96–71
~ масс 82–195
~ материков 73–104
мелкомасштабное ~ 88–188
негравитационное ~ 85–56
обратное ~ 83–47
орбитальное ~ 96–62
основное солнечное ~ 98–66
относительное ~ 99–99
параллактическое ~ 91–33
пекулярное ~ 91–31
~ перпендикулярное плоскости Галактики 98–188
~ планет 83–28
планетное ~ 83–28
попятное ~ 83–47
прямое ~ 83–43
релятивистское ~ перигелия 99–160
собственное ~ 91–19
собственное годичное ~ 91–20
собственное ~ по прямому восхождению 91–21
собственное ~ по склонению 91–22
собственное ~ пятен 89–117
солнечное ~ 98–62
~ Солнца 98–62
~ Солнца вокруг центра Галактики 98–66
стандартное солнечное ~ 98–63
суточное ~ 85–29
тектоническое ~ 73–72
упорядоченное ~ 98–95
хаотическое ~ 98–94
часовое ~ 81–72
эпейрогеническое ~ 73–73
двойная 96–7
астрометрическая ~ 96–12

R

~ волна 98–117
контактная ~ 96–29
неразрешённая ~ 96–10
неразрешимая ~ 96–10
оптическая ~ 96–6
полуразделённая ~ 96–34
разделённая ~ 96–35
~ с двойными линиями в спектре 96–17
~ с одиночными линиями в спектре 96–15
спектральная ~ 96–14
спектрально- ~ 96–14
спекроскопическая ~ 96–14
физическая ~ 96–7
фотометрическая ~ 95–73
широкая ~ 96–9
двуокись кремния 74–13
твёрдая ~ углерода 83–144
двухленточная вспышка 89–187а
двухлопастный обтюратор 86–149
двухниточная вспышка 89–187а
двухрукавная галактика 98–43
двухуровенный атом 93–45
двухфотонное излучение 97–155
Дева 92–99
девон 74–81
девонский период 74–81
дегазация 82-261
~ кометного ядра 84–109
дезактивизация кометы 85–148
дезинтеграция, вращательная 85–119
~ кометного ядра 85–116
~ кометы 85–115
деионизация 86–213
Деймос 83–169
действие, спусковое 89–199
триггерное ~ 89–199
действующий вулкан 73–114
Декабрьские Урсиды 86–258
декаметровый всплеск 83–183
~ радиовсплеск 83–183
деление 83–223
~ галактического ядра 98–318
~ Кассини 83–224
делитель пучка 88–88
Дельфин 92–45
Денеб 92–112
Денебола 92–113
денудация 73–108
день, активный 75–93
возмущённый ~ 75–94
лунный ~ 81–32
~ на Луне 81–32
спокойный ~ 75–92
деполяризация 97–118
депрессия 77–76
арковидная ~ 81–160
внутриматериковая 81–158
извилистая ~ 81–161
линейная ~ 81–159
лунная ~ 81–157
террасовая ~ 81–162
деревья 89–273
десорбция 85–123
деталь в радиоизлучении 98–235
~ на поверхности 83–105
оптическая ~ 98–234
постоянная ~ 83–106
структурная ~ 98–224
~ структуры 98–287
детальное балансирование 93–41
~ равновесие 93–41
~ уравновешивание 93–41
детальный баланс 93–41
детонирующий болид 86–119
дефект фазы 81–28
деферент 83–48
дефлюидизация 82–263
деформация, относительная 73–67
приливная ~ 82–277, 83а–66
деформирование 73–68
децибел 76–39
деятельность, вулканическая 73–110
магматическая ~ 73–110
пятнообразовательная ~ 89–62
региональная сбросовая ~ 82–205
Джакобиниды 86–252
диагенез 74–38
диаграмма бабочек 89–76
~ Герцшпрунга-Рессела 91–130
двухцветная ~ 96–147
~ красное смещение-звёздная величина 98–301
~ масса-светимость 91–129
~ показатель цвета-звёздная величина 96–150
~ развития 96–151
~ спектр-абсолютная величина 91–130
~ спектр-светимость 91–130
~ Хаббла 98–301
~ цвет-величина 96–150
~ широта-время 89–76
эволюционная ~ 96–151
диаметр, видимый 83–117
истинный ~ 83–118
линейный ~ 83–118
угловой ~ 96–115, 97–145
фотометрический ~ 84–93
диастрофизм 73–72
диатрема 82–162
дилютированное излучение 93–38
динамика, звёздная 98–120
~ звёздных скоплений 96–153
динамометаморфизм 74–36
диск, аккреционный 96–21
видимый ~ 88–171
галактический ~ 98–52
затмевающий ~ 88–61
лунный ~ 81–54
~ Луны 81–54
оптический ~ 88–2
солнечный ~ 89–24
~ Солнца 89–24
дискордантная интрузия 73–137
дислокация 82–212
дисперсия скоростей 98–103
диссипация 94–73
~ волн 88–222
приливная ~ 82–286
термическая ~ 94–76
~ энергии 82–285
диссоциационная температура 91–120а
дисторсия, вращательная 91–44
приливная ~ 96–30
дифференциональное вращение 88–128
дифференциация, геохимическая 74–55
гравитационная ~ 83а–54
кристаллизационная ~ 74–41
химическая ~ 83а–50
диффузия, ионная 86–286
~ ионов 86–286
~ космических лучей 84–27
диффузная светящаяся поверхность 79–148
дихотомия Венеры 83–132
дициан 97–24
длина в угловой мере 98–32
~ волны для нулевой скорости 98–365
~ импульса 86–280
~ периода 95–39
~ Планка 99–170h
~ пути метеора 86–47
угловая ~ пути 86–48
элементарная ~ Планка 99–170h
эффективная ~ волны 96–142
длительность импульса 86–280
~ периода 95–39
~ полёта 86–50
дневная полусфера 83–67
~ сторона 83–67
дневное небо 79–3а
дно кратера 81–259
морское ~ 72–38
~ моря 72–38
~ океана 72–37
океаническое ~ 72–37
океанское ~ 72–37
плоское ~ 81–260
догалактическое облако 94–17
дождемер 77–170
суммарный ~ 77–171
дождь 77–169
звёздный ~ 86–215
корональный ~ 89–271
метеоритный ~ 87–37
~ метеоритов 87–37
метеорный ~ 86–215
орографический ~ 77–166
дозвёздная эволюция 94–14
~ эволюция галактик 94–14
дозвёздное вещество 94–16
~ тело 94–15

R

дозвёздный 94–13
докембрий 74–70
докембрийская эра 74–70
долгоживущий 89–124
долгосрочный прогноз 77–19
долгота, активная 89–41
ареографическая ~ 83–139
гелиографическая ~ 89–40
геомагнитная ~ 75–53
~ периастра 96–59
селенографическая ~ 81–10
долготный дрейф 75–120
долина 81–163
~ Альп 81–164
долинный ледник 72–114
доморская структура 81–153
донная морена 72–129
досолнечная туманность 83а–21
доставка образца 82–71
дочерняя молекула 85–135а
драгирование 72–53
драгировка 72–53
Дракон 92–47
Дракониды 86–252
драконический месяц 81–47
~ период обращения 81–47
драпри 79–154
дрейф, асимметрический 98–104
асимметричный ~ 98–104
долготный ~ 75–120
~ континентов 73–104
~ материков 73–104
~ метеорного следа 86–114
~ по долготе 75–120
полярный ~ протуберанцев 89–231
широтный ~ 89–75
дробинка 82–129
дробление 83а–81, 87–25
~ астероидов 83а–82
~ космического облака 94–38
~ метеора 86–83
~ метеорного тела 86–188
Дубге 92–114
дуга, верхняя касательная 79–77
вторичная ~ 73–83
касательная ~ гало 79–75
корональная ~ 89–298
~ меридиана 71–21
нижняя касательная ~ 79–76
однородная ~ 79–143
окологоризонтальная ~ 79–82
околозенитная ~ 79–74
первичная ~ 73–82
побочная нижняя касательная ~ 79–87
пульсирующая ~ 79–147
~ с лучевой структурой 79–153
сумеречная ~ 79–50
дуги, внутритропические 79–53
дугообразное понижение 81–160
дугоограздный горный хребет 73–84
дуть 77–105
дым 97–57
дымка 77–27, 83–250
дымчатое стекло 88–33
дыра, белая 98–24
галактическая чёрная ~ 98–23а
корональная ~ 88–245а
сверхмассивная чёрная ~ 98–22а
чёрная ~ 94–114
дюна 83–162

Е

Евклида 99–101
геометрия ~ 99–108
евклидов 99–101
евклидова геометрия 99–108
Европа 83–206
единицы, солнечные 91–123
~ Солнца 91–123
Единорог 92–68
единственность решения 93–82
единый акт творения 99–90
Е-корона 88–241
естественная распространённость 91–171
естественный спутник 83–18

Ж

жаркая зона 77–191
жгут магнитного поля 89–146
железо, метеорное 87–83
никелистое ~ 87–90
железняк, магнитный 82–92
титанистый ~ 82–91
жёлоб, океанский глубоководный 72–40
жёлтый 91–98
Жеребёнок 92–48
жерло вулкана 73–122
жерловина 73–125
Жертвеник 92–19
жёсткое тело 83–111
жёсткость 73–56
магнитная ~ 75–117
Живописец 92–78
жидкость, сверхтекучая 95–149
химически активная ~ 74–43
жилки 87–111
Жираф 92–24
Журавль 92–52

З

заатмосферное свечение 79–7
забор проб лунного грунта 82–78
зависимость звёздной величины от освещённости 91–58
~ красное смещение-расстояние 99–76
~ масса-радиус 93–200
~ масса-светимость 91–129
~ период-светимость 95–103
~ скорость-расстояние 98–300
~ спектр-светимость 91–130
загораживаемая переменная 95–83
~ переменная звёзда 95–83
загораживание 86–171
загорание водорода 94–55
~ гелия 94–71
~ углерода 94–80
загрязнение атмосферы 76–80
~ воды 72–73
~ воздуха 76–80
химическое ~ 72–98
загрязнитель 76–81
~ атмосферы 76–82
загрязняющее вещество 72–74
задерживающая сила 75–39
задержка, групповая 78–56
заиление в открытом море 72–79
заиливание во внешней зоне шельфа 72–79
закись 74–26
закон Архимеда 73–101
архимедов ~ 73–101
~ Боде 83–3
~ вращения 98–110
~ зон 89–75
~ Крамерса 93–102
~ красного смещения 98–299
~ межзвёздного покраснения 97–92
~ планетных расстояний 83а–2
~ потемнения 88–143
~ сопротивления 86–179
~ Тициуса-Боде 83а–2
~ Хаббла 98–299
~ Хэйлла 89–154
~ чередования полярности 89–154
~ Шпёрера 89–75
закрученный галактический рукав 98–29
закручивание ветвей 98–28
закручивающийся галактический рукав 98–29
закрытая вселенная 99–73
закрытие звезды 81–98
залежь, интрузивная 73–135
залив 81–294
~ Волнений 81–296
~ Зноя 81–296
~ Радуги 81–295
залимбовая вспышка 89–203
залунный 81–6
замагниченная солнечная плазма 84–32
замедление, видимое 86–100
~ вращения Земли 82–313
~ времени 99–137
истинное ~ 86–101
~ расширения 99–82
~ течения времени 99–137
замерзшие газы 85–45
замирание коротких волн 78–40

R

замкнутая вселенная 99–73
замороженные газы 85–45
запаздывание эха 86–291
запись магнитографа 75–84
заполнение депрессии 77–83
~ циклона 77–83
запустить 89–198
запятненная площадь 89–67
зарисовка Солнца 88–34
зародыш планеты 83а–36
заря, вечерняя 79–55
утренняя ~ 79–35, 79–56
заряд Земли 75–3
электрический ~ Земли 75–3
засолённость 72–66
затвердевание 83а–59
затвердение 83а–59
~ магмы 82–190
затемнение озонным слоем 79–28
затишье 77–110
затмение 88–14
атмосферное ~ 95–76
~ видимо в ... 81–86
ионосферное ~ 88–23
искусственное ~ 88–227а
искусственное ~ Солнца 88–17а
кольцевое ~ 88–18
кольцеобразное ~ 88–18
кольцеобразное ~ Солнца 88–15а
корпускулярное ~ 88–24
лунное ~ 81–56
~ Луны 81–56
~ невидимо в ... 81–87
полное лунное ~ 81–66
полное ~ Луны 81–66
полное ~ начинается 81–77
полное солнечное ~ 88–17
полутеневое лунное ~ 81–65
солнечное ~ 88–15
~ Солнца 88–15
центральное ~ 88–16а
частное ~ 81–79а
частное ~ Луны 81–67
частное лунное ~ 81–67
частное солнечное ~ 88–16
затменная переменная 95–74
затмившаяся Лина 81–57
затменно-переменная 95–74
затопление 72–75
захват 83а–19
быстрый ~ нейтронов 94–89
~ газа 98–356
~ газа из межгалактического пространства 98–356
~ звезды 96–73
~ комет 85–101
медленный ~ нейтронов 94–88
~ межгалактической материи 98–355
нейтронный ~ 94–84
~ нейтронов 94–84
протонный ~ 94–90
~ протонов 94–90
~ частиц 75–109
захваченная радиация 75–112
~ радиация искусственного происхождения 75–113
захваченные газы солнечного ветра 82–148
заход Солнца 88–11
Заяц 92–61
звезда 92–1
азотная ~ 91–189
аномальная ~ 91–96
астрометрическая двойная ~ 96–12
~ бедная металлами 91–197
близкая ~ 98–60
близполюсная ~ 92–2а
~ большой массы 91–126
бывшая новая ~ 95–123
быстрая новая ~ 95–119
быстровращающаяся ~ 91–39
быстродвижущаяся ~ 98–191
вечерняя ~ 83–128
взрывающаяся ~ 91–165
взрывная переменная ~ 95–109
~ видимая с полюса 91–40
визуально-двойная ~ 96–8
внегалактическая новая ~ 95–114
внегалактическая сверхновая ~ 95–128
возбуждающая ~ 97–125
~ Вольфа-Райе 91–160
вращающаяся ~ 91–37
вспыхивающая ~ 95–68
~ вылетевшая из системы 98–135
вырожденная ~ 93–180
галактическая новая ~ 95–113
галактическая сверхновая ~ 95–127
~ гало 91–84
~ гигант 91–143
гиперонная ~ 91–154
главная ~ 96–37
~ главной последовательности 91–142
~ горизонтальной ветви 91–141
горячая ~ 91–155
двойная ~ 96–3
~ диска 91–85
загораживаемая переменная ~ 95–83
~ заподозренная в переменности 95–5
зарождающаяся ~ 94–45
затмеваемая переменная ~ 95–83
затменная переменная ~ 95–74
затменно-переменная ~ 95–74
~ избыточной светимости 91–134
инфракрасная ~ 91–157
~ карлик 91–135
карликовая ~ 91–135
карликовая новая ~ 95–112
катаклизмическая переменная ~ 95–109
коллапсирующая ~ 91–164
короткопериодическая переменная ~ 95–89
красная ~ 91–97
красная полуправильная переменная ~ 95–106
кратная ~ 96–5
кремниевая ~ 91–190
магнитная ~ 91–51
магнитная переменная ~ 95–52
магнитно-переменная ~ 95–52
~ малой массы 91–124
массивная ~ 91–126
медленная ~ 98–192
медленная новая ~ 95–120
межгалактическая ~ 91–163
мёртвая ~ 91–167
металлическая ~ 91–195
молодая ~ 94–63
~ населения гало 91–84
~ населения диска 91–85
~ не принадлежащая системе 98–134
~ не принадлежащая скоплению 96–84
~ не является членом скопления 96–84
небулярная переменная ~ 95–85
невращающаяся ~ 91–38
недосветимая ~ 91–133
нейтронная ~ 91–152
неподвижная ~ 92–4
непокрасневшая ~ 97–95
неправильная переменная ~ 95–107
неразрешимая двойная ~ 96–10
несобственная переменная ~ 95–86
нестационарная ~ 95–2
несферическая ~ 91–86
нефизическая переменная ~ 95–86
неустойчивая ~ 95–9
новая ~ 95–110
новая ~ до вспышки 95–122
~ нормальная ~ 91–95
~ нулевого возраста 94–59
~ нулевой температуры 91–166
обычная ~ 91–95
одиночная ~ 96–2
околополюсная ~ 92–2а
околополярная ~ 92–2а
оптическая двойная ~ 96–6
освобождающаяся ~ 96–160
падающая ~ 86–2
пекулярная ~ 91–92
~ первого поколения 94–24
~ переднего фона 96–85
передняя ~ 96–85
переменная ~ 95–1
переменная ~ класса ... 95–47
переменная ~ типа ... 95–47
переменная ~ типа скоплений 95–93
пересветимая ~ 91–134

периодическая магнитно-переменная ~ 95–55
повторная новая ~ 95–116
~ повышенной светимости 91–134
подозреваемая в переменности 95–5
~ поздних спектральных классов 91–103
~ поздних спектральных типов 91–103
покрасневшая ~ 97–94
полуправильная переменная ~ 95–105
~ поля 96–84
~ пониженной светимости 91–133
~ принадлежащая системе 98–134
~ принадлежащая скоплению 96–84
~ промежуточной массы 91–125
проэволюционировавшая ~ 94–74
псевдопеременная ~ 95–51
пульсирующая ~ 95–95
~ ранних спектральных классов 91–102
~ ранних спектральных типов 91–102
ранняя ~ 91–102
расширяющаяся ~ 91–162
~ с большим содержанием гелия 91–193
~ с большим содержанием металлов 91–196
~ с дефицитом металлов 91–197
~ с избыточной светимостью 91–134
~ с малой скоростью 98–192
~ с низким содержанием водорода 91–191
~ с низким содержанием гелия 91–192
~ с низким содержанием углерода 91–194
~ с оболочкой 93–5
~ с пекулярным спектром 91–92
~ с протяжённой атмосферой 93–6
~ с протяжённой оболочкой 93–7
~ с тёмным спутником 96–13
~ с эмиссионными линиями 91–159
~ сверхкороткопериодическая переменная ~ 95–90
сверхмассивная ~ 91–127
сверхновая ~ 95–126
сверхновая ~ после вспышки 95–134
сверхплотная ~ 91–153
сжатая ~ 91–164
сжимающаяся ~ 91–161
симбиотическая ~ 95–50
~ системы 98–133
сколлапсировавшая ~ 94–111
~ скопления 96–82
слабая ~ 91–82
собственная переменная ~ 95–87
~ солнечного типа 91–104
спавшая ~ 91–164
спадающая ~ 91–164
спектрально-двойная ~ 96–14
спектрально-переменная ~ 95–48
спектроскопическая двойная ~ 96–14
~ спутник 96–39
~ сравнения 95–16
~ средней массы 91–125
старая ~ 94–64
стационарная ~ 95–3
тесная двойная ~ 96–23
~ типа Вольфа-Райе 91–160
~ типа дельта Цефея 95–92
~ типа RR Лиры 95–93
~ типа Солнца 91–104
~ типа U Близнецов 95–112
типичная новая ~ 95–111
~ Тихо-Браге 95–135
тройная ~ 96–4
убегающая ~ 98–135
угасшая ~ 91–167
углеродная ~ 91–188
умершая ~ 91–167
ускользающая ~ 96–160
устойчивая ~ 95–8
утренняя ~ 83–127
уходящая ~ 96–160
физическая двойная ~ 96–7
физическая переменная ~ 95–87
~ фона 96–84
фотометрическая двойная ~ 96–11
холодная ~ 91–156
центральная ~ 96–37, 97–148
циклическая переменная ~ 95–108
широкая двойная ~ 96–9
эволюционировавшая ~ 94–74
экранируемая переменная ~ 95–83
эруптивная ~ 95–68
яркая ~ 91–83
ярчайшая ~ галактики 98–304
звёздная вспышка 95–70
~ гидродинамика 98–142
~ динамика 98–120
~ ночь 79–3d
~ эволюция 94–1
звёздное вещество 93–124
~ население 98–67
~ небо 91–1a
~ поколение 94–23
~ содержание 98–172
звёздные недра 93–72
звёздный 92–3
~ атлас 92–12
~ ветер 93–14
~ дождь 86–215
звездообразное ядро 85–38
звёздообразование 94–11, 96–95
бурное ~ 94–11a
звездоподобное ядро 98–19
звездотрясение 95–153
звуковая волна 76–33
~ энергия 88–219
звуковой барьер 76–44
землеведение 71–1
землетрясение 73–2a
близкое ~ 73–26
~ вулканического происхождения 73–2e
вулканическое ~ 73–2e
глубокое ~ 73–23
глубокофокусное ~ 73–23
далекое ~ 73–27
дислокационное ~ 73–2d
катастрофическое ~ 73–40
локальное ~ 73–25
материковое ~ 73–2b
мелкое ~ 73–21
местное ~ 73–25
~ на суше 73–2b
неглубокое ~ 73–21
нормальное ~ 73–22
обвальное ~ 73–2d
опустошительное ~ 73–38
подводное ~ 73–2c
провальное ~ 73–2d
промежуточное ~ 73–24
~ промежуточной глубины 73–24
разрушительное ~ 73–38
~ с материковым очагом 73–2b
сильное ~ 73–37
слабое ~ 73–35
среднее ~ 73–24
тектоническое ~ 73–2d
удалённое ~ 73–27
умеренное ~ 73–36
уничтожающее ~ 73–39
Земля 71–2
земная волна 78–14
~ кривизна 71–6
земной 71–3
земные проявления солнечной деятельности 84–55
зенитное притяжение 86–52
зеркало, вибрирующее 86–130a
вращающееся ~ 86–130
зерно мелочи 82–110
змеевидная траектория 86–33
Змееносец 92–72
Змея 92–89
зодиак 92–5a
зодиакальное созвездие 92–5
зодиакальные знаки 92–6
~ обозначения 92–6
зодиакальный свет 84–61
Золотая Рыба 92–46
зона, абиссальная 72–41
~ абляции 72–123
внутренняя ~ захваченной радиации 75–106
~ горения водорода 93–144
~ горения гелия 93–145
жаркая ~ 77–191

~ затишья 77–101
~ захваченной радиации 75–101
~ землетрясения 73–31
~ избегания 97–88
искусственная ~ радиации 75–108
конвективная ~ 93–149
конвективная водородная ~ 93–159
королевская ~ 89–54
краевая ~ 81–113
краевая ~ Луны 81–113
~ лучистого равновесия 93–134
~ максимальной повторяемости 79–125
~ максимума полярных сияний 79–125
~ минимума полярных синяний 79–129
~ молчания 76–42
орогеническая ~ 73–75a
~ перехода 72–13
переходная ~ 72–13
~ питания леднкиков 72–122
подвижная ~ 73–75a
~ покрытия 81–107
полуконвективная ~ 93–160
полярная ~ 77–193
~ полярных сияний 79–123
~ пятен 89–54
~ пятнообразования 89–54
сейсмическая ~ 73–31
~ слышимости 76–41
~ таяния 72–124
тропическая ~ 77–191
умеренная ~ 77–192, 83–189
центральная ~ 81–114, 89–47
экваториальная ~ 83–188
экваториальная ~ затишья 77–101
~ экваториальных штилей 77–101
зональные гармоники 95–101
зонд, акустический 76–45
зондирование 76–47
~ ионосферы 78–19
ракетное ~ 76–48
~ сверху 78–20
сейсмическое ~ 82–216
сумеречное ~ 79–41
зоны 83–187
~ Френеля 86–294
зрение, периферическое 86–128

И

иглообразная галактика 98–239
иглы, ледяные 76–112
иерархическая вселенная 99–48
~ структура 99–47
~ структура вселенной 99–47
избирательная экстинкция 97–83
избыток инфракрасного излучения 91–79
инфракрасный ~ 91–79
~ обилия 91–185
~ содержания 91–185
ультрафиолетовый ~ 91–78
УФ ~ 91–78
~ цвета 91–77
извержение 89–259
~ вулкана 73–112
вулканическое ~ 73–112
~ из региональной трещины 73–113
линейное ~ 73–113
трещинное ~ 73–113
~ частиц 85–89
экструзивное ~ 73–132
изверженный материал 82–165
известковый полевой шпат 82–87
извилистая борозда 81–167
~ депрессия 81–161
излиб 73–87
изгородь 89–274
живаяа ~ 89–274
излившаяся порода 74–3
излияние 85–55
~ из трещины 73–141
искривлённое ~ 85–58
~ лавы 73–130
прямолинейное ~ 85–57
спиральное ~ 85–59
трещинное ~ 73–141
излом 73–78
концентрический ~ 82–209
~ растяжения 82–207
ровный ~ 82–208
излучение, гравитационное 99–165
двухфотонное ~ 97–155
дилютированное ~ 93–38
диффузное ~ неба 88–136
~ Земли 77–56
~ и газ 93–125
импульсное ~ 86–278
интегральное ~ 88–147
корпускулярное ~ 88–150, 88–152
~ метеора 86–207
~ неба 77–54
нейтринное ~ 94–94
непрерывное ~ 86–277
полное ~ 88–146
прямое солнечное ~ 88–135
рассеянное ~ 88–136
расслабленное ~ 93–38
рекомбинационное ~ 86–212
свободно-связанное ~ 86–212
собственное ~ планеты 83–198
солнечное ~ 88–134
~ Солнца 88–134
изменение береговой линии 72–78
~ блеска 95–7
~ климата 77–184
~ магнитной полярности 89–153
~ от центра к краю 88–145
~ периода 95–43
~ погоды 77–22
скачкообразное ~ периода 95–154
~ формы 73–68
~ цвета моря 72–80
эволюционное ~ 94–27
изменения, вековые магнитные 75–85
магнитные ~ 95–53
световые ~ 95–6
сезонные ~ 83–166
спектральные ~ 95–49
суточные ~ 83–167
~ яркости 95–6
измерение силы тяжести, абсолютное 71–68
абсолютное ~ ускорения силы тяжести 71–68
~ абсолютной величины силы тяжести 71–68
~ возраста 87–126
~ глубин 72–28
~ глубин эхолотом 72–29
градусное ~ 71–22
миркометрическое ~ 96–49
относительное ~ силы тяжести 71–69
относительное ~ ускорения силы тяжести 71–69
~ относительной величины силы тяжести 71–69
поляриметрическое ~ 83–80
~ силы тяжести 71–67
спутниковое ~ высоты 71–50
измерительный прибор относительной непрозрачности ионосферы 78–45
изначальное магнитное поле 95–62
изначальный радионуклид 82–256
изобата 71–55
изображение, интерференционное 88–120
изображения, монохроматические 97–158
изогаусса 89–134
изогипса 71–53
изогона 75–57
изогональная 75–57
изодинама 75–61
изоклина 75–59
изолиния магнитного поля 89–134
~ плотности 96–125
~ скорости счёта 75–102
изонитрил водорода 97–40
изоплета 96–125
изоротация 88–130
изостазия 73–98
изостатическая компенсация 73–100
изостатическое выравнивание 73–100
~ равновесие 73–99
изотермическая атмосфера 93–25
изотоп, догложивущий 83a–75
изначальный ~ 82–146
короткоживущий ~ 83a–74
космогенный ~ 82–147
~ образованный космическими лучами 82–147
первичный ~ 82–146

первичный радиоактивный ~ 82–256
примордиальный ~ 82–146
~ свинца 83а–76
изотопический состав 87–129
~ эффект 91–178
изотопическое определение возраста 87–130
изотропная вселенная 99–63
изотопный состав 82–140
изофота 81–92
изофоты туманности 97–124
изучение Луны 82–3
Икар 84–84
ил 72–58
глубоководный ~ 87–15
илем 99–170
ильменит 82–91
иммерсия 81–98
имплозия звёздного ядра 94–110
импульс, главный 95–144
изолированный ~ 95–146
отражённый ~ 86–290
импульсы 86–278
инвариантность относительно преобразования 99–133
инверсия, атмосферная 77–37
~ населённостей 97–15
~ населённостей уровней 97–15
приземная ~ 77–38
температурная ~ 77–36
Индеец 92–57
индекс активности потока 86–230
К ~ геомагнитной активности 75–89
~ концентрации 96–101
планетарный ~ 75–90
планетарный ~ геомагнитной активности 75–90
~ политропы 93–128
~ солнечной активности 89–8
структурный ~ скопления 96–80
трёхчасовой ~ активности 75–89
индикатор дальности 86–266
~ спиральной структуры 98–39
индикатриса рассеяния 77–58
индуцированная намагниченность 75–29
иней 77–177, 83–150
инсоляция 77–51, 88–9
интеграл движения, третий 98–186
фазовый ~ 83–79
интегральная звёздная величина 96–136
~ яркость 96–135
интенсивность в линии Hα 89–211
всенаправленная ~ 75–104
~ землетрясения 73–33
однонаправленная ~ 75–103
~ силы тяжести 71–62
центральная ~ линии Hα 89–211
интервал перекрытия двух циклов 89–86
интергляциальный 72–108
интерференционная картина 88–120
интерференционное изображение 88–120
интерферометр, двухантенный 88–98
длиннобазисный ~ 88–104
крестообразный ~ 88–121
многоантенный ~ 88–123
многозеркальный ~ 88–123
многолучевой ~ 88–123
могоэлементный ~ 88–123
морской ~ 88–106
наскальный ~ 88–106
~ Райла 88–113
~ с большой базой 88–104
~ с веерообразной диаграммой направленности 88–122
~ с карандашной диаграммой направленности 88–121
~ с карандашным нучком 88–121
~ с качающейся диаграммой 88–109
~ с переключением фазы 88–113
~ с переменной базой 88–107
~ с радиосвязью 88–116
~ с фазовой перестройкой 88–109
~ с частотной развёрткой 88–110
~ со сканированием по частоте 88–110
интрамеркуриальная планета 83–123
интрузивная залежь 73–135
интрузия 73–133
дискордантная ~ 73–137
конкордантная ~ 73–134
магматическая ~ 82–98
несогласная ~ 73–137
пирогенная ~ 82–98
пластовая ~ 73–135
согласная ~ 73–134
инфляционная стадия 99–170d
инфляция 99–170c
инфразвуковые волны 75–88
инфракрасная звезда 91–157
Ио 83–205
иови- 83–171
иовиграфические координаты 83–175
иовицентрические координаты 83–176
ион, водородный 93–65
отрицательный ~ водорода 93–66
~ Ланжевена 76–115
тяжёлый ~ 76–115
ионизационная температура 91–117
ионизация, метеорная 86–281
ионизованный слой 78–3
ионная диффузия 86–286
ионный состав 78–10
ионограмма 78–23
ионосфера 78–1
внешняя ~ 76–76
ионосферное возмущение 78–39
ионосферный 78–2
~ прогноз 78–32
~ самописец 78–22
иризирующее облако 79–90
искажённая галактика 98–316
искатель, солнечный 88–51
~ Солнца 88–51
искривленный веерообразный выброс 85–62
искры метеора 86–84
искусственная луна 88–61
искусственное затмение Солнца 88–17а
искусственный радиационный пояс 75–108
испарение, интенсивное 86–185
~ метеорного тела 86–198
~ метеорной частицы 86–198
~ скопления 96–108
~ чёрных дер 94–118
исправление за кривизну земной поверхности 86–62
испускание массы 85–145
~ частиц 88–161
~ частиц из Солнца 88–161
испыление 82–239
исследования Луны 82–3
исследователь метеоритов 87–7
исследовательский буй 72–99
истечение, веерообразное 85–60
~ вещества 93–13
~ вещества из галактического ядра 98–275
~ из трещины 73–141
искривлённое веерообразное ~ 85–62
~ лавы 82–172
~ массы 93–13
непрерывное ~ 98–276
обращенное веерообразное ~ 85–63
перевернутое веерообразное ~ 85–63
прямое веерообразное ~ 85–61
трещинное ~ 73–141
истинный диаметр 83–118
история, тепловая 74–66, 83а–60
термическая ~ 74–66, 83а–60
источник декаметрового излучения 83–186
квазизвёздный ~ 98–358
квазизвёздный ~ излучения 98–358
компактный ~ 98–273
~ нейтронов 94–85
~ тепла 74–62
~ тепловой энергии 74–62
~ электронов 79–119
~ энергии 93–110
истощение массы 81–236
~ породы 81–236
истощившаяся комета 85–150
исходные геодезические данные 71–20

R

~ геодезические даты 71–20
исчезновение звезды 81–98
~ колец 83–217
~ метеора 86–37
исчерпание водорода 94–69

К

каверна 87–117
лавовая ~ 82–177
кавернозный 82–181
кайма полярной шапки 83–151
яркая ~ волокна 89–226
кайнозой 74–73
кайнозойская группа 74–73
~ эра 74–73
калибровочная теория 99–170j
калийный полевой шпат 74–16
Каллисто 83–208
кальдера 81–179
~взрыва 81–180
~обрушения 81–181
~опускания 81–182
~оседания 81–181
эрозионная ~ 81–183
камасит 87–92
каменноугольная система 74–82
каменноугольный период 74–82
камень, лунный 82–65
камера всего неба 79–104
гидируемая ~ 86–142
метеорная ~ 86–141
негидируемая ~ 86–143
неподвижная ~ 86–143
патрульная ~ 86–145
подвижная ~ 86–142
фиксированная ~ 86–143
камни, катящиеся 81–237
канал вулкана 73–124
~ потока 82–192
каналы Марса 83–157
марсианские ~ 83–157
кандидат в переменные 95–5
Канопус 92–109
каньон 83–154
подводный ~ 72–46
Капелла 92–110
капелька 76–100
капля 76–113
чёрная ~ 83–52
карбиламин водорода 97–40
карбон 74–82
карбонат 74–27
карбонил-сульфид 97–36
карлик 91–135
белый ~ 91–137
звезда ~ 91–135
коричневый ~ 91–138a
красный ~ 91–136
чёрный ~ 91–139
карликовая галактика 98–220
~ звезда 91–135
~ новая 95–112
~ новая звезда 95–112
карта, батиметрическая 71–54
~ высот 81–141
гелиографическая ~ 89–17
гидрологическая ~ 72–25
гипсометрическая ~ 81–141
~ глубин 71–54
~ горизонталей 81–141
гравиметрическая ~ 71–79, 81–143
ежедневная ~ Солнца 89–18
звёздная ~ 92–9
~ изолиний силы тяжести 81–143
изомагнитная ~ 75–63
~ изофот радиоизлучения 81–135a
колориметрическая ~ 81–134
лунная ~ 81–133
~ Луны 81–133
магнитная ~ 75–63
~ магнитного поля 89–131
~ Марса 83–136
~ неба 92–9
океанографическая ~ 72–23
~ окрестностей переменной звезды 95–15
~ переменной звезды 95–15
~ погоды 77–12
подвижная звёздная ~ 92–10
подвижная ~ неба 92–10
поляризационная ~ 81–134
полярная ~ 89–19
прогностическая ~ 77–21
путевая ~ 72–33
путевая береговая ~ 72–33
~ рельефа 81–141
~ с высоким разрешением 89–133
~ силы тяжести 71–79
синоптическая ~ 77–11
синоптическая ~ короны 89–22
синоптическая ~ хромосферы 89–21
~ синоптическая ~ фотосферы 89–20
~ скоростей 89–23
~ солнечного затмения 88–28
структурная ~ 73–97
тектоническая ~ 73–96
тепловая ~ Луны 81–135
термическая ~ Луны 81–135
топографическая ~ Луны 81–138
физическая ~ 81–134
фотографическая ~ Луны 81–137
фотометрическая ~ 81–134
~ элементов земного магнетизма 75–63
картина, интерференционная 88–120
картирование, береговое 72–82
~ Луны 81–111
селенологическое ~ 82–51
термическое ~ 72–83
картографирование Луны 81–111
~ температуры 72–83
картография, лунная 81–111
касательная дуга (к) гало 79–75
касп, полярный 75–13
Кассиопея 92–31
Кастор 92–111
катабатический 77–93
катаклизмическая переменная 95–109
каталог звёзд 92–13
~ звёзд подозренных в переменности 95–58
звёздный ~ 92–13
~ Мессье 92–12a
катастрофа, сейсмическая 73–40
катастрофическое землетрясение 73–40
катящиеся камни 81–237
Квадрантиды 86–247
квадратура 83–64
восточная ~ 83–65
западная ~ 83–66
квазаг 98–361
квазар 98–360
радиоспокойный ~ 98–262
квазизвёздная галактика 98–361
квазизвёздный источник 98–358
~ источник излучения 98–358
~ объект 98–359
~ объект без радиоизлучения 98–362
~ радиоисточник 98–360
квазисинхронные наблюдения 71–36
квазистатическое сжатие 94–40
квазистационарное сжатие 94–40
квантовое поле 99–170m
кварковый суп 99–169a
Квинтет Стефана 98–309
кембрий 74–78
кембрийская система 74–78
кембрийский период 74–78
кеплеровское вращение 98–112
керн 72–52
Киль 92–30
кильватер 86–110
кинофильм протуберанцев 89–242
кислород, атомный 79–24
молекулярный ~ 79–23
кислота, α-аминопропионовая 87–72
изоциановая ~ 97–39
муравьиная ~ 97–33
органическая ~ 87–69
Кит 92–34
класс светимости 91–146
спектральный ~ 91–107
классификация галактик 98–226
гарвардская спектральная ~ 91–91
двумерная спектральная ~ 91–105
магнитная ~ 89–109
~ метеоритов 87–59
~ переменных звёзд 95–46
петрографическая ~ 82–101
спектральная ~ 91–90
трёхмерная спектральная ~ 91–106
цюрихская ~ 89–127

R

классическая вселенная 99–93
клатрат 85–51
клевок 87–33
клейкая пластинка 87–14
клефт, магнитосферный 75–13
климат 77–180
горный ~ 77–187
континентальный ~ 77–185
материковый ~ 77–185
морской ~ 77–186
приморской ~ 77–186
климатический 77–182
климатология 77–179
спутниковая ~ 77–181
клин высокого давления 77–86
клуб дыма 89–277
коагуляция 76–119
коалесценция 76–120
когезионная прочность 85–117
когеоид 71–17
когерентность 99–33
Козерог 92–29
когдыри, лавовые 82–178
колебания, адиабатические 95–98
вторичные ~ 95–44
вынужденные приливные ~ 82–306
геомагнитные ~ 75–81
z ~ 98–187
~ климата 77–183
~ между зеркальными точками 75–119
неадиабатические ~ 95–99
нерадиальные ~ 95–97
~ перпендикулярные плоскости Галактики 98–187
приливные ~ 76–28
радиальные ~ 95–96
свободные ~ Земли 73–49
солнечные ~ 88–223a
колебательная температура 91–120
колеблющаяся щель 88–70
количество водяного пара 77–124
~ метеоров в один час 86–135
~ облаков 77–147
~ осадков 77–158
~ хлора 72–68
коллапс 94–109
~ вселенной 99–181
гравитационный ~ 94–109
~ звёздного ядра 94–110
коллапсирующая звезда 91–164
коллективное взаимодействие 98–168
коллекция метеоритов 87–8
коллоид 76–96
коллоидальный = коллоидный 76–97
колонка 72–52
колонна, ионизированная 86–282
колориметрическая карта 81–134
колор-индекс 91–76
кольцевое затмение 88–18
кольцевой 97–146
кольцеобразная галактика 98–259
кольцеобразное затмение 88–18
кольцеобразный 97–146
кольцо A 83–225
~ астероидов 84–70
~ B 83–222
~ Бишопа 79–88
большое ~ 79–70
~ видимое с ребра 83–218
внешнее ~ 83–225
~ газовое ~ 95–66
~ звёзд 96–78
звёздное ~ 96–78
креповое ~ 83–221
малое ~ 79–69
~ призрак 81–242
~ C 83–221
~ Сатурна 83–215
светлое ~ 89–101
среднее ~ 83–222
туманное ~ 83a–10
яркое ~ 89–101
кома 85–64
атомарная ~ 85–67
видимая ~ 85–66
внутренняя ~ 85–65
молекулярная ~ 85–65
свободнорадикальная ~ 85–66
~ свободных радикалов 85–66
ультрафиолетовая ~ 85–67
фотохимическая ~ 85–65
комета 85–1
активная ~ 85–149
гиперболическая ~ 85–8
действующая ~ 85–149
долгопериодическая ~ 85–6
задевающая Солнце ~ 85–13
затменная ~ 85–12
искусственная ~ 85–14
истощившаяся ~ 85–150
короткопериодическая ~ 85–5
материнская ~ 86–260
неактивная ~ 85–147
непериодическая ~ 85–10
~ открытая при затмении 85–12
параболическая ~ 85–7
периодическая ~ 85–3
предвычисленная ~ 85–11
предсказанная ~ 85–11
родительская ~ 86–259
~ родоначальница 86–260
~ с гиперболической орбитой 85–8
~ с параболической орбитой 85–7
~ с эллиптической орбитой 85–4
~ связанная с роём 86–259
телескопическая ~ 85–14a
эллиптическая ~ 85–4
кометарная туманность 97–142
кометная активность 85–110
~ форма 85–83
кометный 85–2
~ метеор 86–45
кометообразная туманность 97–142
комнатная температура 77–31
нормальная комнатная ~ 77–31
комок, пылевой 86–190
компактный источник 98–273
Компас 92–82
компенсатор, доп(п)леровский 88–91
~ доп(п)леровских сдвигов 88–91
компенсация доп(п)леровских сдвигов 88–90
изостатическая ~ 73–100
компенсирующая пластинка 88–92
компланарность орбит 83a–3
комплексная группа 89–112
компонент = компонента 96–36
вертикальный ~ напряженности 75–62
главный ~ 96–37
горизонтальный ~ напряженности 75–60
минералогический ~ 82–100
поперечный ~ 89–136
продольный ~ 89–135
пылевой ~ 98–232
σ-~ 88–89
комптоновская длина волны 99–170g
конвективная водородная зона 93–159
~ зона 93–149
конвективный 93–148
~ температурный градиент 93–157
конвекция 93–147
ламинарная ~ 93–153
турбулентная ~ 93–152
конгломерат 85–41
ледяной ~ 85–43
конденсация 83a–26
корональная ~ 89–281
неизменная корональная ~ 89–282
перманентная корональная ~ 89–282
спорадическая корональная ~ 89–283
конденсировать 77–129
конденсироваться 77–129
конец видимого пути 86–38
~ затмения 81–83
~ затмения при восходе 88–21
~ мира 99–183
~ полного затмения 81–80
~ полутеневого затмения 81–82
~ цикла 89–85
~ частного затмения 81–83a
конечность вселенной 99–4
Конская голова 97–135
конские широты 77–102
консолидация магмы 82–190
контакт, первый 81–84
последний ~ 81–85
четвёртый ~ 81–85
контактный метаморфизм 74–35
контейнер для образцов 82–81
континент 72–9, 83–159
континентальная отмель 72–34

R

~ платформа 72–34
континентальный склон 72–35
конус вулкана 73–118
вулканический ~ 73–118
лавовый ~ 81–176а
насыпной 73–120
отражающий ~ 88–62
пепловый ~ 73–120
пирокластический ~ 73–120
~ потер 76–27
световой ~ 95–156
сложный вулканический ~ 73–121
~ тени 81–90
туфовый ~ 73–120
~ убегания 76–27
~ ускользания 76–27
эксплозивный ~ 73–120
конфигурация планет 83–50
~ поля 89–140
концентрация 96–127
галактическая ~ 98–76
галактическая ~ к основной плоскости 98–76
галактическая ~ к центру 98–76
~ к плоскости галактики 98–55
~ масс в коре Луны 81–145
незначительная ~ 83–266
электронная ~ 78–6
координата времени 99–123
временная ~ 99–123
координаты, галактоцентрические 98–12
гелиографические ~ 89–39
геомагнитные ~ 75–51
иовиграфические ~ 83–175
иовицентрические ~ 83–176
планетографические ~ 83–8
планетоцентрические ~ 83–9
пространственные ~ 99–124
селенографические ~ 81–9
цилиндрические ~ 98–156
юпитерографические ~ 83–175
юпитероцентрические ~ 83–176
кора Земли 72–4
земная ~ 72–4
континентальная ~ 72–5
лунная ~ 82–54
~ Луны 82–54
материковая ~ 72–5
~ плавления 87–110
твёрдая ~ 95–151
коридоры видимости 98–57
галактические ~ видимости 98–57
кориолисова сила 77–79
Корма 92–81
королевская зона 89–54
корона 79–156
авроральная ~ 79–137а
белая ~ 88–237
~ в белом свете 88–237
~ в период максимума 88–244
~ в период минимума 88–246
внешняя ~ 88–231
внутренняя ~ 88–229
водородная ~ 83–246
галактическая ~ 98–49
E ~ 88–241a
звёздная ~ 93–4
излучающая ~ 88–241a
K ~ 88–242
максимальная ~ 88–244
минимальная ~ 88–246
монохроматическая ~ 88–238
планетная ~ 83–245
промежуточная ~ 88–245
расширающаяся ~ 88–232
рентгеновская ~ 88–240
солнечная ~ 88–225
средная ~ 88–230
F ~ 88–241
фраунгоферова ~ 88–241
электронная ~ 88–242
эмиссионная ~ 88–241a
эмиссионная линейчатая ~ 88–241a
корональная дуга 89–298
~ дыра 89–245a
~ конденсация 89–281
корональное транзиентное явление 89–185a
корональный 88–228
~ дождь 89–271
~ протуберанец 89–264
коронограф 88–59
внезатменный ~ 88–59
~ с внешним затмением 88–60
коронометр 88–66
K ~ 88–66
коротирующая структура 84–17
короткоживущий 89–125
~ изотоп 83a–74
короткопериодическая комета 85–5
коротрясение 95–152
корпускула 88–153
корпускулярная активность 89–7
корпускулярное затмение 88–24
~ излучение 88–150
корпускулярный 88–155
коррекция траектории 82–10
корыто, потенциальное 98–176
космическая геодезия 71–31
~ геофизика 84–57
~ материя 99–7
~ струна 99–171c
космического происхождения 87–5
космическое вещество 99–7
~ облако 94–31
космогенный изотоп 82–147
космогонист 94–4
космогоническая гипотеза 94–8
космогонический 94–3
космогония 94–2
звёздная ~ 94–7
планетная ~ 83a–1
~ планетной системы 83a–1
космолог 99–19
космологическая постоянная 99–29
~ система 99–20
космологический 99–18
~ горизонт 99–179
~ член 99–156a
космологическое красное смещение 98–363a
космология 99–16
кинематическая ~ 99–26
наблюдательная ~ 99–17
нерелятивистская ~ 99–27
ньютоновская ~ 99–25
релятивистская ~ 99–28
космос 99–2
космохимия 74–31
космохронология 83a–72
котловина 81–175
коэрцитивная сила 75–39
коэффициент атмосферной диффузии 86–287
~ дилюции 93–37
~ потемнения 88–144
~ Пуассона 73–60
~ светимости 86–197
~ сокращения 99–135
~ сокращения Лоренца 99–135
~ сопротивления 86–180
~ теплопередачи 86–201
~ увлечения 86–180
~ устойчивости 98–199
фазовый ~ 83–77
~ фазы 83–77
~ формы 86–177
~ шереховатости 82–107
~ эллипсоида скоростей 98–102
~ эффективности излучения 86–197
~ яркости 83–96
Краб 97–134
Крабовидная туманность 97–134
краевая зона 81–113
~ зона Луны 81–113
край, внешний 83–227
восточный ~ 89–34
~ галактики 98–294
~ диска Солнца 89–28
западный ~ 89–33
~ кратера 81–221
~ Луны 81–95
поднятый ~ 81–222
светлый ~ 81–96
~ солнечного диска 89–28
~ состоящий из глыб 81–224
тёмный ~ 81–97
экстремальный ~ 89–32
Красное пятно 83–193
кратер без вала 81–226
безымянный ~ 81–202
боковой ~ 73–128
вероятный ~ 81–210
вершинный ~ 73–129

R

~ взрывного происхождения 81–238
взрывной ~ 81–238
внешний ~ 81–218
внутренний ~ 81–217
возможный ~ 81–211
~ возникший до образования морей 81–207
~ возникший после образования морей 81–208
вторичный ~ 81–199
вулканический ~ 73–126
главный ~ 73–127, 87–57
древний ~ 81–205
древний метеоритный ~ 87–55
дренированный ~ 81–239
заполненный ~ 81–203
затопленный ~ 81–203
земной метеоритный ~ 87–54
концентрический ~ 81–216
круглый ~ 81–213
лунный ~ 81–197
лучевой ~ 81–214
лучистый ~ 81–214
маар ~ 81–264
мелкий ~ 81–250
метеоритный ~ 87–53
метеорный ~ 87–53
метеороидный ~ 81–263
микроскопический ~ 81–201
молодой ~ 81–206
небольшой ~ 81–250
неправильный ~ 81–220
~ образованный при истечении лавы 81–239
паразитический ~ 73–128
первичный ~ 81–198
побочный ~ 81–199
погруженный ~ 81–204
поздний ~ 81–199
полигональный ~ 81–212
реликтовый ~ 81–209
~ с валом из глыб 81–225
~ с выступающим валом 81–223
~ с плоским дном 81–219
~ с тёмным кольцом 81–215
~ с тройным валом 81–255
~ с центральной горкой 81–258а
~ спутник 81–200
терминальный ~ 73–129
ударный ~ 81–240
~ фантом 81–209
центральный ~ 81–217
чашечный ~ 81–219
кратер-лунка 81–250
кратерная цепочка 81–241
~ цепь 81–241
кратер-призрак 81–209
кратер-фантом 81–209
краткосрочный прогноз 77–20
кратная галактика 98–308
~ звезда 96–5
кратный вал 81–255
кратчайшая 99–153
~ линия 99–153
кремнезём 74–13
кремниевая звезда 91–190
крепповое кольцо 83–221
Крест 92–43
крест, лунный 79–80
~ Миллса 88–121
световой ~ 79–79
солнечный ~ 79–81
крестообразный интерферометр 88–121
кривая блеска 95–17
~ блеска метеора 86–70
визуальная кривая ~ 95–24
~ вращения 98–109
~ изменения блеска 95–17
~ изменения показателя цвета 95–21
~ изменения цвета 95–21
~ лучевых скоростей 95–20
~ ослабления света 97–81
~ охлаждения 93–202
~ период-светимость 95–103
~ покраснения 97–93
предельная ~ спутника 96–40
~ равной освещённости 81–92
~ равной яркости 81–92
~ развития 89–119
~ развития группы пятен 89–119
~ распространённости 94–108
~ распространённости элементов 94–108
~ роста 93–68
средняя ~ блеска 95–19
теоретическая ~ роста 93–70
фазовая ~ 83–74
фотографическая ~ блеска 95–23
фотоэлектрическая ~ блеска 95–22
~ хода поляризации 83–81
~ экстинкции 97–81
эмпирическая ~ роста 93–69
кривизна земной поверхности 71–6
~ пространства 99–115
~ пути 86–51
~ силовых линий 75–121
кривые равного счёта 75–102
криптозой 74–70
кристалл, ледяной 76–111
кристаллизационная дифференциация 74–41
кристаллизация 76–107
фракционная ~ 74–40
кристобалит 82–95
критерий вырождения 93–181
~ классификации 91–110
классификационный ~ 91–110
~ приливной неустойчивости 98–312
крона волокон 89–232
полярная ~ волокон 89–232
кроше 75–79
круг, большой 79–70
горизонтальный ~ 79–82
малый ~ 79–69
паргелический ~ 79–82
круглое море 81–267
круговое море 81–267
крупнозернистый 82–117
крупномасштабная структура 88–207
крупномасштабное движение 88–189
крюк, магнитный 75–79
крючок, магнитный 75–79
кряж 73–85
горный ~ 73–85
кульминация Луны 82–303
кумулонимбус 77–138
купол 81–177
вулканический ~ 81–176
лавовый ~ 81–176а
ледниковый ~ 72–115
лунный ~ 81–177
купола, тёмные 89–297
тёмные ~ вокруг протуберанцев 89–297
курс 82–35
кучево-дождевое облако 77–137

Л

лава, застывшая 82–176
затвердевшая ~ 82–176
отвердевшая ~ 82–176
лавовая пещера 82–177
~ трубка 82–186
лавовые волдыри 82–178
~ пузыри 82–178
лавовый грот 82–177
~ конус 81–176а
~ купол 81–176
~ язык 82–187
лайншифтер 88–69
лакколит 73–136
ламинарная конвекция 93–153
ландшафт, лунный 81–55
лапилли 82–169
лапласово преобразование 93–54
Лебедь 92–44
Лев 92–59
Малый ~ 92–60
легкоплавкое вещество 83а–56
лёд 85–42
аморфный ~ 85–44
водяной ~ 85–47
грязный ~ 97–73
загрязнёный ~ 97–73
зернистый ~ 72–118
кубический ~ 85–46
паковый ~ 72–102
плотный ~ 72–102
ледник 72–112
альпийский ~ 72–114
~ альпийского типа 72–114

выводной ~ 72–125
горный ~ 72–113
долинный ~ 72–114
континентальный ~ 72–116
материковый ~ 72–116
покровный ~ 72–116
разгрузовочный ~ 72–125
стоковый ~ 72–125
ледниковая глыба 72–132
~ шапка 72–115
ледниковый 72–106
~ валун 72–132
~ купол 72–115
~ отторженец 72–132
~ период 77–189
~ покров 72–115
~ сток 72–120
~ язык 72–125
ледяная гора 72–117
ледяной = ледяный
~ кристалл 76–111
~ спутник 83–23
ледяные иглы 76–112
~ ядра 76–104
Лейтоновский магнитогелиограф 88–93
лента, вспышечная 89–187
~ вспышки 89–187
~ протуберанца 89–245
яркая ~ вспышки 89–187
Леониды 86–255
лепесток Роша 96–28
лептонная эра 99–176
летоисчисление, геологическое 74–67
Летучая Рыба 92–100
летучее 82–262
~ вещество 82–262
либрационное облако 84–43
~ облако Луны 84–43
либрация 81–48
геометрическая ~ 81–51
оптическая ~ 81–51
параллактическая ~ 81–52
~ по долготе 81–49
~ по широте 81–50
суточная ~ 81–52
физическая ~ 81–53
либрирующий астероид 84–91а
лидер 89–113
лимб, восточный 89–34
западный ~ 89–33
~ Луны 81–95
светлый ~ 81–96
солнечный ~ 89–28
тёмный ~ 81–97
тёмный ~ вспышки 89–189
экстремальный ~ 89–32
линейная депрессия 81–159
линейный диаметр 83–118
линза, гравитационная 98–256а
линзовая галактика 98–245
линзовидная галактика 98–245
линзообразная галактика 98–245
линии Неймана 87–103
неймановы ~ 87–103
линия, атмосферная 88–140а
атмосферная спектральная ~ 88–140а
батиметрическая ~ 71–55
вертикальная ~ 71–71
геодезическая ~ 99–153
D ~ 88–140
жёлтая ~ 89–292
зелёная ~ 89–291
зелёная спектральная ~ 79–16
корональная ~ 89–290
красная ~ 89–293
красная спектральная ~ 79–15
кратчайшая ~ 99–153
межзвёздная ~ 97–96
микрковолновая межзвёздная ~ 97–98
мировая ~ 99–154
нейтральная ~ 89–151
нулевая ~ магнитного поля 89–151
~ нулевого возраста 94–60
околозвёздная ~ 97–97
оптическая межзвёздная ~ 97–99
~ отвеса 71–71
отвесная ~ 71–71
~ покраснения 96–149
~ потока 82–194
~ равного магнитного склонения 75–57
~ равной напряжённости 75–61
~ равных высот 71–53
~ равных глубин 71–55
~ равных значений наклонения 75–59
~ раздела полярностей 89–151
разомкнутая силовая ~ 75–15
силовая ~ участвующая во вращении 75–14
~ смещённая к красному концу спектра 98–366
снеговая ~ 72–111
теллурическая ~ 88–140а
~ течения 82–194
фраунгоферова ~ 88–139
липкий 82–137
Лира 92–65
Лириды 86–248
Лисичка 92–101
лист 83–149
листоватость 74–53
литификация 74–49
ударная ~ 82–238
литосидерит 87–80
литосфера 72–6
литофилы 83а–30
литофильные элементы 83а–30
лоб течения 82–193
ловушка, световая 88–64
ложе океана 72–37
ложное Солнце 79–71
ложный зодиакальный свет 84–62
ломкий 86–191
ломкость 86–192
ЛТР 93–39
не ~ 93–43
отсутствие ~ 93–43
Луна 81–2
~ в земной тени 81–57
~ в первой четверти 81–25
затмившаяся ~ 81–57
искусственная ~ 88–61
ложная ~ 79–73
~ на ущербе 81–21
нарастающая ~ 81–20
новая ~ 81–22
полная ~ 81–27
прибывающая ~ 81–20
растущая ~ 81–20
убывающая ~ 81–21
ущерблённая ~ 81–21
ущербная ~ 81–21
лунация 81–18
лунка 81–243
метеоритная ~ 87–58
лунная база 82–17
~ брекчия 82–85
~ впадина 81–157
~ всыпшка 82–247
~ геодезия 81–109
~ геология 82–47
~ депрессия 81–157
~ карта 81–133
~ картография 81–111
~ люминесценция 82–248
~ орбита 81–36
~ почва 82–62
~ станция 82–17
~ фаза 81–19
лунное вещество 82–58
~ время 82–300
~ гало 79–68
~ затмение 81–56
~ картирование 81–111
~ понижение 81–137
лунно-солнечная составляющая 82–307
лунно-солнечный компонент 82–307
лунные горы 81–193
~ лучи 81–168
~ образования 81–148
~ сутки 81–32
лунный 81–3
~ атлас 81–136
~ базальт 82–84
~ выступ 81–118
~ глобус 81–112
~ горб 81–118
~ грунт 82–62
~ день 81–32
~ диск 81–54
~ кратер 81–197

~ крест 79–80
~ купол 81–177
~ ландшафт 81–55
~ маскон 81–145
~ материал 82–58
~ образец 82–69
~ пейзаж 81–55
~ почвенный слой 82–62
~ серп 81–23
~ терминатор 81–123
~ цирк 81–265
~ шар 81–112
~ экватор 81–121
лунотрясение 82–214
луч, аномальный звуковой 76–36
звуковой ~ 76–34
зелёный ~ 79–58
~ зрения 91–28
касательный ~ 81–169
корональный ~ 89–287
монохроматический корональный ~ 89–288
нормальный звуковой ~ 76–35
~ опахала 89–296
опахальный ~ 89–296
сейсмический ~ 73–14
тангенциальный ~ 81–169
лучевая поверхность 81–171
~ система 81–170
~ скорость 91–24
лучевой кратер 81–214
лучи 79–155
лунные ~ 81–168
полярные ~ 89–294
светлые ~ 81–168
лучистая оболочка 93–135
лучистый кратер 81–214
люк 84–88
люки Кирквуда 84–89
люминесценция, лунная 82–248

М

маар 82–161
~ кратер 81–264
Магеллановы Облака 98–212
магма 73–131
магматическая деятельность 73–110
~ интрузия 82–98
магнетизм горных пород 83а–80
звёздный ~ 95–54
земной ~ 75–24
остаточный ~ 75–28
~ пород 75–43
магнетит 82–92
магнитная активность 89–5
~ восприимчивость 75–27
~ жёсткоть 75–117
~ звезда 91–51
~ классификация 89–109
~ плавучесть 89–147
~ подъёмная сила 89–147
магнитно-возмущённый период 75–91
магнитное гало 98–48
магнитносопряжённые точки 75–23
магнитные возмущения 75–70
~ изменения 95–53
магнитный железняк 82–92
~ крюк 75–79
~ крючок 75–79
~ монополь 99–170о
магнитогелиограф 88–93
лейтоновский ~ 88–93
магнитограмма 75–84
магнитограф 75–83
векторный ~ 88–84
~ (для) продольного поля 88–83
солнечный ~ 88–82
магнитоид 98–25
магнитометр, переносный 82–44
портативный ~ 82–44
магнитопауза 75–9
магнитоплазменное тело 98–25
вращающееся ~ тело 98–25
магнитосфера 75–7
~ пульсара 95–147
магнитосферный клефт 75–13
магнитуда землетрясения 73–34
мазер, естественный 97–13
космический ~ 97–13
межзвёздный ~ 97–14
макрометеорит 87–4
макросиноптическое положение 77–15
макроструктура 99–9
максвелловское распределение скоростей 98–96
максимальная корона 88–244
~ применимая частота 78–37
максимум 76–55
вторичный ~ 95–32
главный ~ 95–31
острый ~ 86–235
плоский ~ 86–236
пологий ~ 86–236
~ слоя 76–55
~ слоя E 76–56
~ солнечной активности 89–2
солнечный ~ 89–2
Малая Медведица 92–97
малое кольцо 79–69
Малое Магелланово Облако 98–214
маломасштабная неоднородность 84–13
Малый Ковш 92–97
~ Конь 92–48
~ Лев 92–60
~ Пёс 92–28
мантия 72–15
верхняя ~ 72–16
~ Земли 72–15
мареограф 82–311
Маркаб 92–119
Марс 83–133
марсианские каналы 83–157
марсианский 83–135
марсов 83–135
маршрутный обзор 82–36
~ траверс 82–36
маскон 81–145
лунный ~ 81–145
масса, воздушная 77–67
~ газа 94–32
газовая ~ 94–32
главная ~ 87–41
гравитационная ~ 99–149
звёздная ~ 91–121
~ звезды 91–121
инертная ~ 99–148
инерционная ~ 99–148
критическая ~ 93–189
критическая ~ белых карликов 93–189
метеорная ~ 86–163
~ метеорной частицы 86–163
наибольшая ~ 87–41
начальная ~ 86–164
недостающая ~ 99–87
общая ~ скопления 96–139
основная ~ 82–118
переменная ~ 86–167
~ Планка 99–170f
планковская ~ 99–170f
~ покоя 99–143
полная ~ скопления 96–139
релятивистская ~ 99–141
скрытая ~ 99–88
~ Солнца 91–122
солнечная ~ 91–122
суммарная ~ скопления 96–139
тяжёлая ~ 99–149
массив, континентальный 72–10
массивная звезда 91–126
масштаб времени 94–65
~ карты 71–49
масштабная высота 76–18
масштабный фактор 99–81а
материал изверженный соударением 82–167
~ лунной поверхности 82–59
лунный ~ 82–58
мелкодисперсный лунный ~ 82–112
ударом изверженный ~ 82–167
материк 72–9, 81–191, 83–159
материковая кора 72–5
~ область 81–271
~ поверхность 82–56
материковое землетрясение 73–2b
материковый климат 77–185
~ ледник 72–116
~ район 81–271
материнская порода 82–118
материнское тело 87–123
материя, дозвёздная 94–16
космическая ~ 99–7
межгалактическая ~ 98–349
межзвёздная ~ 97–1

межпланетная ~ 84–1
маятник, горизонтальный 71–73
мгал 71–66
мгла, голубая 83–165
межантенное расстояние 88–99
межгалактическая материя 98–349
межгалактический 97–6
межгранульный промежуток 88–177
Международный геофизический и геодезический союз 71–4
~ Полярный Ряд 91–87
~ эллипсоид 71–14
межзвёздная газодинамика 97–150
~ линия 97–96
~ материя 97–1
межзвёздное вещество 97–1
~ мерцание радиоисточников 97–109b
межзвёздный 97–2
~ ветер 97–151
~ водород 97–46
~ газ 97–9
межледниковый 72–108
межледниковье 72–107
межлучевая поверхность 81–172
межпланетная материя 84–1
~ петля 84–21
межпланетное вещество 84–1
~ возмущение 84–45
~ пространство 84–2
межпланетный 84–8
~ выступ 84–21
межспикульный промежуток 88–198
мезозой 74–72
мезозойская группа 74–72
~ эра 74–72
мезопауза 76–67
мезосидерит 87–79
мезосфера 76–66
мел 74–86
мелкий кратер 81–250
мелководье 72–96
мелкодисперсный материал 82–109
мелкое землетрясение 73–21
мелкозернистый 82–114
мелкомасштабное движение 88–188
меловой период 74–86
мелочь, лунная 82–112
мель 72–96
мера эмиссии 88–234
мерзлота, вечная 83–152
меридиан, вечерний 75–19
магнитный ~ 75–49
полуденный ~ 75–17
утренний ~ 75–18
центральный ~ 89–43
меридиональная циркуляция 93–163
Меркурий 83–120
мероопределение 99–151
мёртвая звезда 91–167
мерцание, межзвёздное 97–109a
межзвёздное ~ радиоисточников 97–109b
межпланетное ~ радиоисточников 84–37
~ радиоисточников 84–36
мёссбауэровская спектрография 82–142
местная галактика 98–327
место находки 87–52
~ находки метеорита 87–52
~ обнаружения 87–52
~ обнаружения метеорита 87–52
~ падения метеорита 87–50
~ посадки 82–14
предполагаемое ~ посадки 82–15
~ прилунения 82–14
~ удара 86–124
месяц 81–42
аномалистический ~ 81–43
драконический ~ 81–47
звёздный ~ 81–45
лунный ~ 81–46
сидерический ~ 81–45
синодический ~ 81–46
тропический ~ 81–44
метагалактика 98–357
металл, сернистый 74–28
металлический водород 83–202
металличность 91–170
метаморфизм 74–34
динамический ~ 74–36
контактный ~ 74–35
термальный ~ 74–37
ударный ~ 82–237
метаморфический процесс 74–34
метанимин 97–45
метанол 97–32
метасоматизм 74–44
метасоматоз 74–44
метенил 97–23
метеор 86–1
базисный ~ 86–155
быстрый ~ 86–87
веретенообразный ~ 86–77
визуальный ~ 86–16
галактический ~ 86–43
гиперболический ~ 86–43
двойной ~ 86–131
дневной ~ 86–22
искусственный ~ 86–21
~ кометного происхождения 86–45
кометный ~ 86–45
кратный ~ 86–133
медленный ~ 86–88
межзвёздный ~ 86–43
межланетный ~ 86–44
мерцающий ~ 86–85
~ наблюдаемый радиолокатором 86–267
~ не принадлежащий (к) потоку 86–25
обычный ~ 86–14
~ потока 86–24
поточный ~ 86–24
~ принадлежащий к потоку 86–24
пылевой ~ 86–190
слабый ~ 86–15
слабый телескопический ~ 86–19
~ снятый из двух пунктов 86–155
~ со следом 86–17
~ солнечной системы 86–44
спорадический ~ 86–23
стационарный ~ 86–34
сфотографированный ~ 86–13
телескопический ~ 86–18
фотографический ~ 86–13
эклиптикальный ~ 86–46
яркий ~ 86–12
метеорит 87–1
железный ~ 87–82
каменножелезный ~ 87–80
каменный ~ 87–60
ориентированный ~ 87–112
углистый ~ 87–62
метеоритика 87–6
метеоритная лунка 87–58
метеоритное вещество 87–9
~ железо 87–83
~ органическое вещество 87–68
метеоритные алмазы 87–65
метеоритный 87–2
~ дождь 87–37
~ кратер 81–263
метеоритолог 87–7
метеорная активность 86–134
~ ионизация 86–281
~ камера 86–141
~ частица 86–160a
метеорный 86–4
~ дождь 86–215
~ поток 86–216
~ рой 86–217
метеороид 86–160
пористый ~ 85–50
метеороидный кратер 81–263
метеорология 77–6
динамическая ~ 77–7
синоптическая ~ 77–8
метеорщик 86–7
метеоусловия 77–2
метиламин 97–43
метилацетилен 97–41
метиловый спирт 97–32
метилцианид 97–38
метилэтин 97–41
метин 97–23
метод, вариационный 93–57
~ весовых функций 93–59
~ двойного счёта 86–132
~ итераций 93–56
радиологический ~ 74–92
радиологический ~ определения возраста 74–92
стереоскопический ~ 81–115

R

торий-свинцовый ~ определения возраста 82–270
урановый ~ 74–96
метрика 99–151
~ Робертсона-Уокера 99–152a
шварцшильдовская ~ 99–152
метрический тензор 99–157a
метрическое пространство 99–100a
механизм возбуждения 79–27
~ деструкции 97–79
~ нагрева 88–214
~ накачки 97–16
~ разрушения 97–79
~ торможения 91–46
механическая теория кометных форм 85–84
Меч Ориона 92–130
мигающее волокно 89–216
мигматит 74–39
миделево сечение 86–176
мидель 86–176
микроанализ 82–143
микробарограф, электромагнитный 76–46
микробрекчия, лунная 82–86
микробухта 75–79
микроволновая межзвёздная линия 97–98
микрогаз 76–88
микроколебания 73–53
микрократер 81–201
микрометеор 86–20
микрометеориты 87–3
микрометеорология 77–9
микрометрическое измерение 96–49
микропульсации геомагнитного поля 75–82
геомагнитные ~ 75–82
микросейсмы 73–53
Микроскоп 92–67
микроскопический кратер 81–201
микроснимок 87–140
микроструктура 82–73
микротектит 87–97
микрофотография 87–140
микрофотоснимок 87–140
миллибар 77–47
миллигал 71–66
миллиметр ртутного столба 77–48
миллионная доля солнечного диска 89–69
миллионные доли видимой полусферы 89–68
минерал 74–12
породообразующий ~ 82–99
минерализация 72–66
минералогическая составная часть 82–100
минимальная корона 88–246
~ применимая частота 78–33
минимум, вторичный 95–27
главный ~ 95–26
~ солнечной активности 89–4
солнечный ~ 89–4
мир 99–3
четырёхмерный ~ 99–122
мираж 79–64
мировая гравиметрическая сеть 71–81
~ линия 99–154
~ магнитная съёмка 75–65
Мицар 92–120
Млечный Путь 98–1
многоантенный интерферометр 88–123
многозеркальный интерферометр 88–123
многократное красное смещение 98–364
многолучевой интерферометр 88–123
моногорукавная галактика 98–44
многочлены Лежандра 95–101
многоэлементный интерферометр 88–123
множитель, гаунтовский поправочный 93–104
гильотинный ~ 93–103
моделирование, математическое 98–323
модель, анизотропная 99–65
асимптотическая ~ 99–59
асимптотическая ~ первого рода 99–60
атмосферная ~ 83–247
~ атмосферы 83–247, 93–17
~ без экранирования 93–19
водородная ~ 83–197
водородная ~ планеты 83–197
водяная ~ 85–113
~ вселенной 99–46
~ Галактики 98–163
горячая ~ вселенной 99–166
горячая космологическая ~ 99–166
~ двойного источника 93–146
двухкомпонентная ~ 88–196
двухтемпературная ~ 88–197
двухтемпературная ~ хромосферы 88–197
~ Де Ситтера 99–69
закрытая ~ 99–73
замкнутая ~ 99–73
звёздная ~ 93–123
~ звезды 93–123
~ звезды с нулевой температурой 93–186
~ Земли 79–133
иерархическая ~ 99–48
изотропная ~ 99–63
ионосферная ~ 83–261
космологическая ~ 99–46
математическая ~ 98–324
монотонная ~ 99–61
монотонная ~ первого рода 99–62
неоднородная ~ 99–66
неоднородная ~ звезды 93–139
нестационарная ~ 99–53
однородная ~ 99–64
осциллирующая ~ 99–57
осциллирующая ~ первого рода 99–58
открытая ~ 99–72
парниковая ~ 83–262
~ планеты 83–196
политропная ~ 93–129
пульсирующая ~ 99–57
пустая ~ 99–49
~ пятна 89–102
раздувающаяся космологическая ~ 99–170b
расширяющаяся ~ 99–55
релятивистская ~ 93–198
~ с двумя источниками 93–146
~ с источниками в оболочке 93–143
~ с точечным источником 93–138
~ с экранированием 93–18
сжимающаяся ~ 99–56
составная ~ 93–140
стандартная ~ 93–137
статическая ~ 99–50
стационарная ~ 99–52
тепличная ~ 83–262
фонтанная ~ 85–85
~ Фридмана 99–71
~ Шарлье 99–67
~ Эддингтона 93–137
~ Эйнштейна 99–68
~ Эйнштейна–Де Ситтера 99–70
эолосферная ~ 83–260
модуль, видимый 96–117
видимый ~ расстояния 96–117
истинный ~ расстояния 96–118
командный ~ 82–6
~ продольной упругости 73–59
~ расстояния 91–16
~ упругости 73–59
~ Юнга 73–59
модуляция, электроническая 88–86
молдавит 87–99
молекула воздуха 86–172
дочерняя ~ 85–135a
материнская ~ 85–135
межзвёздная ~ 97–10
многоатомная ~ 97–21
родительская ~ 85–135
молекулярный агрегат 97–65
молния 77–162
чёточная ~ 77–164
шаровая ~ 77–163
момент вращения 83a–5
~ количества движения 83a–4
~ наибольшей фазы 81–78
орбитальный ~ 83a–6
орбитальный ~ количества движения 83a–6
~ посадки 82–16
~ появления 86–49

R

приливный ~ вращения 82–288
~ прохождения через периастр 96–60
угловой ~ 83a–4
моноокись кремния 97–37
монополь, магнитный 99–170o
моносульфид углерода 97–34
монотонная вселенная 99–61
монохроматический корональный луч 89–288
море 81–266, 83–156
~ Влажности 81–282
внутреннее ~ 72–26
внутриконтинентальное ~ 72–26
внутриматериковое ~ 72–26
~ Волн 81–290
~ Восточное 81–284
~ Гумбольдта 81–288
~ Дождей 81–273
~ Змеи 81–291
~ Изобилия 81–274
каменное ~ 82–139
краевое ~ 72–27, 81–285
~ Кризисов 81–280
круглое ~ 81–267
круговое ~ 81–267
~ Мечты 81–292
~ Москвы 81–283
~ Нектара 81–279
~ Облаков 81–281
окраинное ~ 72–27
~ Плодородия 81–274
~ Познаний 81–289
~ Познанное 81–289
~ Смита 81–287
~ Спокойствия 81–275
средиземное ~ 72–26
~ Холода 81–278
эпиконтинентальное ~ 72–27
~ Южное 81–286
~ Ясности 81–276
морена 72–128
береговая ~ 72–131
боковая ~ 72–131
донная ~ 72–129
конечная ~ 72–130
краевая ~ 72–130
основная ~ 72–129
поддонная ~ 72–129
моретрясение 73–2c
морось 77–172
морской грунт 72–38
~ интерферометр 88–106
~ район 81–268
морфологический тип 98–228
морфология, лунная 81–147
~ лунных образований 81–147
мост, межгалактический 98–348
межгалактический яркий ~ 98–348
светлый ~ 89–104
яркий ~ 98–307
мощность вспышки 89–205
~ передатчика 86–270
пиковая ~ 86–271
средняя ~ 86–272
МПЧ 78–37
мульда 73–89
сбросовая ~ 82–132
мультиполь 75–66
муравьиная кислота 97–33
муравьиный альдегид 97–27
мусковит 74–24
муссон 77–99
мутность 83–252
Муха 92–69
мыс 81–293

Н

наблюдатель метеоров 86–7
наблюдательная космология 99–17
наблюдение, внезатменное 88–227
дневное ~ 83–121
затменное ~ 88–226
квазисинхронное ~ 71–36
корреспондирующее ~ 86–125
~ метеоров 86–6
параллельное ~ 86–125
почти одновременное ~ 71–36
радиолокационное ~ 86–265
~ радиоэхо 86–265
синхронное ~ 71–35
~ спутников 83–26
~ сумерск 79–42
наводнение 72–75
нагон воды 72–88
нагорье 81–189
нагрев атмосферы 75–87
~ короны 88–213a
предварительный ~ 86–203
~ хромосферы 88–213
надвиг 73–94
надэра, протерозойская 74–77
наземный 71–3
наивысший уровень прилива 72–89
наименьшее расстояние связи 78–38
накачка, инфракрасная 97–18
~ инфракрасным излучением 97–18
оптическая ~ 97–17
радиативная ~ 97–17
~ столкновениями 97–20
столкновительная ~ 97–20
ультрафиолетовая ~ 97–18a
фотонная ~ 97–17
химическая ~ 97–19
наклон 81–230
значительный ~ 81–232
~ к лунному экватору 81–37
крутой ~ 81–232
~ оси 83–35
~ оси вращения к ... 91–50
умеренный ~ 81–231
наклонение, магнитное 75–58
~ оси вращения к ... 91–50
~ плоскости орбиты 96–51
наклонная неоднородность 88–190
накопление 83a–34
~ частиц 75–123
наложение спектров 96–24
намагничение, вторичное 75–36
вязкое ~ 75–32
индуцированное ~ 75–29
обратное ~ 75–37
ориентационное ~ 75–33
остаточное ~ 75–28
первичное ~ 75–35
термоостаточное ~ 75–30
химическое ~ 75–31
намагниченная солнечная плазма 84–32
намагниченность = намагничение 75–36
намывание 73–109
нанос 73–109
напластование 82–133
направление ветра 77–115
~ вращения 98–108
~ магнитного поля 84–10
~ оси 83–34
~ оси вращения 84–102
~ от Солнца 75–22
~ поля 84–10
~ распространения 78–35
направленный от Солнца 85–81
напряжение 73–62
~ в коре 82–210
~ в породах коры 82–210
главное ~ 73–64
касательное ~ 73–63
~ сдвига 73–63
сдвигающее ~ 73–63
~ силы тяжести 71–62
срезывающее ~ 73–63
тангенциальное ~ 73–63
тепловое ~ 87–27
термическое ~ 87–27
напряжённость геомагнитного поля 75–55
~ магнитного поля 84–11
~ магнитного поля Земли 75–55
эффективная ~ магнитного поля 95–59
нарастание полярной шапки 83–145
нарастающая Луна 81–20
наращивание 83a–33
нарушение зарядовой чётности 99–170t
разрывное ~ 73–92
~ симметрии 99–170s
население второго типа 98–68
~ галактики 98–230
~ галактического диска 98–72
галактическое ~ 98–230
~ гало 98–69
звёздное ~ 98–67
~ первого типа 98–71
~ спиральных рукавов 98–73

R

~ типа II 98–68
~ типа I 98–67
экстремное ~ первого типа 98–341
населённость скопления 98–341
наскальный интерферометр 88–106
наслоение 82–133
Насос 92–15
насыпной конус 73–120
насыщение 76–101
натекание газа 98–356
~ газа из галактического пространства 98–356
натяжение 73–66
Наугольник 92–70
наука о землетрясениях 73–1
~ о ледниках 72–103
~ об окружающей среде 76–2
науки, атмосферные 76–1
~ о Земли 71–1
~ о Луне 82–2
находка метеорита 87–44
начало вселенной 99–11
~ затмения 81–74
~ полного затмения 81–77
~ расширения 99–167
~ цикла 89–82
неадиабатические колебания 95–99
неактивная комета 85–147
небо, дневное 79–3a
звёздное ~ 91–1a
ночное ~ 79–6
облачное ~ 77–148
насмурное ~ 77–152
северное ~ 79–3b
сумеречное ~ 79–43
южное ~ 79–3c
яркое ночное ~ 86–28
ясное ~ 77–153
небулярная гипотеза 83a–7
невозмущённое Солнце 89–11
невращающаяся звезда 91–38
невырожденный газ 93–182
негидируемая камера 86–143
неглубокое землетрясение 73–21
негравитационное движение 85–56
недействующий вулкан 73–116
недеструктивная γ-спектроскопия 82–77
недооценить 95–14
недоп(п)леровское красное смещение 98–363
недосветимая звезда 91–133
недостаток обилия 91–187
недостаточность содержания 91–187
недостающаяся масса 99–87
недостоверный член скопления 96–87
недоуплотнённый 86–284
недоценить 95–14
недра, звёздные 93–72
~ Земли 72–3
земные ~ 72–3
неевклидов 99–102
незатухающая волна 86–277
незональные гармоники 95–102
неизменная корональная конденсация 89–282
неизменный во времени 98–150
неизменяемая плоскость 83–232
неизменяемый во времени 98–150
неизменяющийся с временем 98–150
неинерциальная система 99–147
неионизованный водород 97–47
неймановы линии 87–103
нейтрализация, взаимная 79–29
нейтральный водород 97–47
нейтринное излучение 94–94
нейтрон, альбедный 75–115
нейтронная звезда 91–152
нейтронное альбедо 75–116
~ вырождение 93–191
нейтронный газ 93–192
~ захват 94–84
~ источник 94–85
~ поток 94–86
нейтросфера 76–64
некаталогизированный объект 84–96
некк 73–125
вулканический ~ 73–125
нелетучее вещество 83a–53
нелучевая энергия 88–218
ненасыщенный 86–284
ненумерованный объект 84–96
необыкновенная волна 78–30
неоднородная вселенная 99–66
неоднородная модель звезды 93–139
неоднородность, ионосферная 78–7
~ магнитного поля 78–9
макроскопическая ~ 88–206
маломасштабная ~ 84–13
мелкомасштабная ~ 84–13
наклонная ~ 88–190
~ электронной плотности 78–8
неозой 74–73
неотож(д)ествлённые полосы 85–134
непериодическая комета 85–10
неподвижная звезда 92–4
непокрасневшая звезда 97–95
неправильная галактика 98–237
неправильная переменная 95–107
неправильный кратер 81–220
непрерывное излучение 86–277
непрозрачность, нейтринная 94–95
непротонная вспышка 89–186
Нептун 83–236
неравенство Шварцшильда 93–156
нерадиальные колебания 95–97
неразрешённая двойная ~ 96–10
неразрешимая двойная 96–10
неразрушающая гамма-спектроскопия 82–77
нерастворимый 76–93
нерелятивистская космология 99–27
нерелятивистский 99–96
нерелятивистское вырождение 93–177
~ приближение 99–97
неровность поверхности 82–106
несамосветящийся 96–44
несветящийся 96–44
несвободное вращение 81–14
несейсмичный район 73–29
нессрая атмосфера 93–27
несобственная переменная 95–86
несогласная интрузия 73–137
нестабильность, конвективная 93–155
локальная ~ 93–154
нестатическая вселенная 99–51
нестационарная вселенная 99–53
~ звезда 95–2
несущая волна 86–276
~ способность 82–64
несферическая звезда 91–86
нетепловое давление 94–42
неустойчивая звезда 95–9
неустойчивость 93–211
вековая ~ 93–212
вибрационная ~ 93–214
вращательная ~ 93–216
гравитационная ~ 94–35
динамическая ~ 93–207
колебательная ~ 93–214
конвективная ~ 93–155
локальная ~ 93–154
приливная ~ 98–313
пульсационная ~ 93–213
ротационная ~ 83a–8
неусточивый 93–210
нефизическая переменная 95–86
нечлен скопления 96–84
~ скопления галактик 98–340
неэвклидов 99–102
низина 81–272
низкочастотные всплески 83–184
низкоширотный 89–52
низменность 81–272
низменная область 81–272
низменный район 81–272
никелистое железо 87–90
нимбостратус 77–145
нисходящая ветвь 95–35
~ фаза солнечного цикла 89–84
нисходящий 77–93
нитеподобные структуры 88–209
нифе 72–18
ядро ~ 72–18
новая = новая звезда 95–110
бывшая ~ 95–123
быстрая ~ 95–119
внегалактическая ~ 95–114
галактическая ~ 95–113
~ до вспышки 95–122
~ звезда = новая 95–110
медленная ~ 95–120
обыкновенная ~ 95–111
повторная ~ 95–116
типичная ~ 95–111

новолуние 81–22
новоподобная переменная 95–118
нога, посадочная 82–22
номенклатура марсианских образований 83–108
номер по каталогу Мессье 97–111
нормальная звезда 91–95
нормальные температура и давление 77–30
ночь, безлунная 79–3e
безоблачная ~ 79–3g
звёздная ~ 79–3d
лунная ~ 81–33
~ на Луне 81–33
светлая ~ 79–59
фотометрическая ~ 79–3f
фотометрически ясная ~ 79–3f
ясная ~ 79–3g
нуклеогенез(ис) 94–52
взрывной ~ 94–57
невзрывной ~ 94–58
первичный ~ 99–175
нуклеосинтез 94–52
взрывной ~ 94–57
космологический ~ 99–175
невзрывной ~ 94–58
первичный ~ 99–175
нуль-пункт зависимости период-светимость 95–104
нумерованный астероид 84–95
ньютоновская вселенная 99–93
~ космология 99–25
ньютоновское время 99–118
нэк = некк 73–125
вулканический ~ 73–125

O

О-ассоциация 96–89
обвал 82–199
обвальное землетрясение 73–2d
обезвоживание 74–51
обезгаживание 82–261
обеспыленный 82–119
обзор, маршрутный 82–35
~ наблюдений малых планет 84–94
научный ~ 82–36
~ неба 92–11
~ небесной сферы 92–11
фотографический ~ 82–30
обилие, аномальное 91–173
космическое ~ элементов 94–107
нормальное ~ 91–172
полное ~ 91–177
~ химических элементов 82–141
~ элемента 91–169
обитаемая лунная база 82–18
~ лунная станция 82–18
облака 83–254
Магеллановы ~ 98–212
облако 77–131
взвешенное ~ 89–275
висящее ~ 89–275
высокое слоистое ~ 77–140
высококучевое ~ 77–144
высокослоистое ~ 77–140
газовое ~ 97–113
гигантское молекулярное ~ 97–13a
догалактическое ~ 94–17
~ дыма 89–277
звёздное ~ 98–6
зодиакальное ~ 84–54
иризирующее ~ 79–90
~ комет 85–97
кометное ~ 85–100
корональное ~ 89–261
космическое ~ 94–31
кучево-дождевое ~ 77–137
кучевое ~ 77–138
либрационное ~ 84–43
либрационное ~ Луны 84–43
мазерное ~ 97–12
~ межгалактического газа 98–354
межзвёздное ~ 97–101
молекулярное ~ 97–11
~ нейтрального водорода 97–103
ночное светящееся ~ 79–89
околозвёздное ~ 94–48
околозвёздное пылевое ~ 97–117
~ Оорта 85–97
орографическое ~ 77–134
перистое ~ 77–136
перисто-кучевое ~ 77–143
перисто-слоистое ~ 77–141
перламутровое ~ 79–91
плазменное ~ 84–30
~ плазмы 84–30
поглощающее ~ 97–104
протогалактическое ~ 94–17
протопланетное ~ 83a–24
пылевое ~ 83–256, 87–23
разорванно-кучевое ~ 77–146
серебристое ~ 79–89
слоисто-дождевое ~ 77–145
слоистое ~ 77–139
слоисто-кучевое ~ 77–142
тёмное ~ 97–107
~ частиц 84–31
электронное ~ 79–115
~ Эпика-Оорта 85–97
область абляции ледника 72–123
авроральная ~ 79–126
~ аккумуляции 72–122
активная ~ 89–57
активная ~ на Юпитере 83–180
~ активности 89–57
~ активных протуберанцев 89–240
~ антицентра 98–15
асейсмичная ~ 73–29
биполярная магнитная ~ 89–160a
~ B 72–16
~ видимости 86–126
~ видимости затмения 81–89
внешняя ~ атмосферы 76–68
~ возникшая до образования морей 81–269
~ возникшая после образования морей 81–270
~ высокого давления 77–84
галактическая ~ антицентра 98–15
~ галактического центра 98–13
горячая ~ 88–200
~ задержки 86–123
запрещённая ~ 79–135
~ захвата 75–110
~ захвата частиц 75–110
зелёная корональная ~ 89–284
~ ионизованного водорода 97–50
~ ионосферы 78–4
~ H II 97–50
компактная ~ H II 97–51
континентальная ~ 81–271
корональная активная ~ 89–286
красная корональная ~ 89–285
M ~ 89–61
магнитная ~ 89–144
~ магнитного поля 89–144
материковая ~ 81–271
межгранульная ~ 88–177
~ между спиральными рукавами 98–40
межрукавная ~ 98–40
межспикульная ~ 88–198
~ модуляции 84–26
~ модуляции космических лучей 85–26
морская ~ 81–268
~ накопления ледника 72–122
невозмущённая ~ 89–56
невозмущённая ~ фотосферы 89–56
~ нейтрального водорода 97–49
~ низкого давления 77–76
околоядерная ~ 98–292
~ падения метеоритного дождя 87–38
переходная ~ 88–224
~ питания 72–122
~ повышенной концентрации масс 81–145
полярная ~ 81–17
~ полярных сияний 79–126
промежуточная ~ 72–13
разрешённая ~ 79–134
~ рассеяния 76–68
~ рассеяния метеоритного дождя 87–38
светлая ~ 83–160
сейсмическая ~ 73–28
~ слышимости звука 86–120
солнечная активная ~ 89–57
~ стаивания ледника 72–124
тёмная ~ 83–155
униполярная магнитная ~ 89–155a
холодная ~ 88–199
центральная ~ галактики 98–13

экваториальная ~ 89–59
~ эпицентра 73–16
ядерная ~ 98–17
~ ядра 98–17
облачно 77–148
облачная пелена 77–151
облачность 77–147
облачный 83–253
~ слой 77–150
обледенение, континентальное 72–104
обломки 83а–83
~ пород 81–235
обломок горной породы 82–68
~ метеорита 87–43
метеоритный ~ 87–43
окатный ~ 82–128
~ породы 82–68
стекловатый ~ 82–97
обломочная порода 82–85
облучение космическими лучами 82–250
~ Солнцем 88–9
обмен энергии 98–125
обобщённое равновесие 93–43
ободок полярной шапки 83–151
обозначение кометы 85–31
окончательное ~ 85–33
предварительное ~ 85–32
обозначения, зодиакальные 92–6
обозрение астероидов 84–94
~ неба 92–11
селенологическое ~ 82–50
фотографическое ~ 82–30
оболочка в конвективном равновесии 93–136
~ в лучистом равновесии 93–135
водная ~ Земли 72–21
водородная ~ 93–144
водородная ~ Земли 76–74
гелиевая ~ 93–145
~ головы кометы 85–64
~ звезды 93–10
конвективная ~ 93–136
ледяная ~ 97–72
лучистая ~ 93–135
общая ~ 96–32
общая ~ двойной звезды 96–32
околозвёздная пылевая ~ 97–117
плазменная ~ 75–11
протяжённая ~ 93–8
пылевая ~ 97–116
расширяющаяся ~ 93–12
стационарная ~ 93–11
оболочки кометы 85–69
оборудование для сбора метероной пыли 87–13
обработка, атмосферная 87–116
~ покрытий 81–108
образец из пробоотборника 82–60
~ керна 82–60
~ лунного вещества 82–69
~ лунного грунта 82–70
~ лунного грунта, доставленный на Землю 82–70
~ лунной пыли 82–121
лунный ~ 82–69
привезённый лунный ~ 82–70
образование 94–10
активное ~ 89–14
атмосферное ~ 83–107
бурное ~ звёзд 94–11а
~ вещества 99–89
вулканическое ~ 82–159
~ звёзд 94–11
короткоживущее ~ 89–15
~ кратеров 82–232
~ кратеров ударом 82–234
~ линий 93–63
~ материи 99–89
местное ~ сбросов 82–204
~ на поверхности 83–105
облачное ~ в хвосте 85–95
постоянное ~ 83–106
~ пузырьков во время застывания 82–182
~ пылинок 97–60
~ пятна 89–121
региональное ~ сбросов 82–205
~ сбросов 73–92
солнечное ~ 89–13
структурное ~ 88–248
ударное ~ 82–228
ударное ~ кратеров 82–234
~ фронта 77–74
~ хвоста 85–86
~ элементов 94–52
образования лунной поверхности 81–148
лунные ~ 81–148
обратное вращение 83–40
~ движение 83–47
обратный 83–46
обращающий слой 88–174
обращение, драконическое 81–47
~ магнитной полярности 89–153
сидерическое ~ 81–45
синодическое ~ 81–46
обрыв 81–184
обсерватория, высокогорная 88–56
ионосферная ~ 78–21
лунная ~ 82–20
~ на Луне 82–20
~ полярных сияний 79–101
сейсмологическая ~ 73–5
солнечная ~ 88–7
обследование льдов 72–85
обстоятельства затмения 81–73
~ падения 87–47
обтюратор 86–147
двухлопастный ~ 86–149
~ с подвижным крылом 86–148
общая масса скопления 96–139
~ теория относительности 99–146
объединение, великое 99–170u
~ взаимодействий 99–170w
сверхвеликое ~ 99–170q
объединённое взаимодействие 99–170x
объект, близкий 98–368
внегалактический ~ 98–209
диффузный ~ 85–25
квазизвёздный ~ 98–359
компактный ~ 98–21
космологический ~ 98–369
некаталогизированный ~ 84–96
ненумерованный ~ 84–96
сверхмассивный ~ 98–22
сколлапсировавший ~ 98–23
сколлапсированный ~ 98–23
спокойный квазизвёздный ~ 98–362
~ типа Амур 84–80
тихий квазизвёздный ~ 98–362
объёмная волна 73–42
~ плотность 96–133
обыкновенная волна 78–31
обычная звезда 91–95
овал, белый 83–195
~ полярных сияний 79–124
Овен 92–20
огибающая, характеристическая 98–177
огненный шар 99–168
огнеупорные элементы 97–64
одиннадцатилетний период 89–70
одиночная звезда 96–2
одновременность 99–127
однозначность решения 93–82
однонаправленная интенсивность 75–103
одноокись кремния 97–37
однородная вселенная 99–64
однорукавная галактика 98–42
односернистый углерод 97–34
озеро 81–297, 83–158
~ Весны 81–302
кратерное ~ 82–161
лавовое ~ 82–183
~ Лета 81–300
~ Осени 81–301
~ Смерти 81–298
~ Сновидений 81–299
озон, атмосферный 76–124
озонная тень 75–129
озоносфера 76–123
окаймление волокна, яркое 89–226
окаменение 74–49
~ в результате удара 82–238
окатный обломок 82–128
океан 72–22
~ Бурь 81–277
глубокий ~ 72–41
океанографическая карта 72–23
океанография 72–24
космическая ~ 72–69
спутниковая ~ 72–70
океанский глубоководный жёлоб 72–40

R

окисел 74–26
окись 74–26
~ водорода 97–26
~ углерода 97–28
окклюзия 77–72
окклюдированный фронт 77–71
окна видимости 98–57
окно прозрачности 83–264
окологалактический газ 98–351
окологоризонтальная дуга 79–82
околозвёздный 97–3
околозенитная дуга 79–74
околозенитный круг 79–74
окололунный 81–5
~ ореол 79–61
околополюсное созвездие 92–4а
околополярное созвездие 92–4а
околосолнечный ореол 79–62
околоядерная область 98–292
окраинное море 72–27
окрестности Солнца 98–58
окружающая среда 76–3
окружающий 76–7
~ воздух 77–4
оксид 74–26
Октант 92–71
октаэдрит 87–86
оледенение 72–104
оливин 74–22
ондуляция геоида 71–74
ОНЧ 78–54
опасность для судоходства 72–97
навигационная ~ 72–97
опахало 89–295
опахальный луч 89–296
оползание 82–195
~ стенок 82–198
оползень 82–196
опора, посадочная 82–23
оппозиция 83–59
определение возраста 74–91, 94–62
~ возраста по радиоактивности 87–131
~ возраста с помощью изотопов 83а–73
изотопическое ~ возраста 87–130
~ массы 96–69
~ рыболовного места 72–84
~ силы тяжести 71–67
~ хода рыбы 72–84
оптика, атмосферная 79–1
оптическая вспышка 89–181
~ двойная звезда 96–6
~ галактика 98–204
~ особенность 98–234
~ струя 98–279
опыт Майкельсона 99–130
~ Майкельсона-Морли 99–130
оранжерейный эффект 83–263
орбита, абсолютная 96–55
астрометрическая ~ 96–57
видимая ~ 96–48
галактическая ~ звезды 98–184
гелиоцентрическая ~ 86–42
геоцентрическая ~ 86–41
~ Земли 84–5
земная ~ 84–5
истинная ~ 96–53
~ Луны 81–36
лунная ~ 81–36
относительная ~ 96–54
планетная ~ 83–29
~ планеты 83–29
первичная ~ 98–200
пространственная ~ 98–185
спектрографическая ~ 96–56
спектроскопическая ~ 96–56
трёхмерная ~ 98–185
фотоцентрическая ~ 96–47
орбитальное движение 96–62
органическая кислота 87–69
органическое включение 87–66
ордовик 74–79
ордовикский период 74–79
Орел 92–18
ореол 79–60
~ Луны 79–61
окололунный ~ 79–61
околосолнечный ~ 79–62
~ Солнца 79–62
ориентация частиц 97–78
преимущественная ~ частиц 97–78
Орион 92–73
Ориониды 86–253
Орионов рукав 98–36
орогенез 73–74
орогеническая зона 73–75а
орогенический пояс 73–75
орографический дождь 77–166
ортоклаз 74–16
орудие для собирания образцов 82–80
осадки 77–154
абиссальные ~ 72–50
атмосферные ~ 77–154
биогенные ~ 72–59
глубоководные ~ 72–50
морские ~ 72–57
~ образованные с суши 72–48
океанические ~ 72–57
орографические ~ 77–167
пелагические ~ 72–57
терригенные ~ 72–48
осадконакопление 74–45
осадкообразование 74–45
морское ~ 72–47
осадок 74–47
глубоководный ~ 87–15
годовой ~ 87–17
осаждение 73–109
освобождающаяся звезда 96–160
освобождение звезд из скопления 98–194
~ звезды из системы 96–159
оседание пластов 82–199
осколки 83а–83
осколок метеорита 87–43
~ ядра 94–102
ослабитель, ступенчатый 88–80
ослабление света, межзвёздное 97–80
селективное ~ 97–83
основание протуберанца 89–234
основная порода 74–5
особая точка 94–116
особенность 94–116
вулканическая ~ 82–159
древняя характерная ~ 81–150
~ лунного рельефа 81–148
новейшая ~ 81–151
новейшая характерная ~ 81–151
~ поверхности 83–105
подлинная ~ 91–100
поздняя характерная ~ 81–151
спектральная ~ 91–93
структурная ~ 98–224
осолён(н)ость 72–66
остатки в метеоритах, органические 87–67
остаток вспышки сверхновой 95–133
~ сверхновой 95–133
~ сверхновой звезды 95–133
остаточный магнетизм 75–28
остров, новый 72–81
остывание 83а–57
осциллирующая вселенная 99–57
осциллирующее волокно 89–216
осциллограф 86–292
осциллоскоп 86–292
осциллятор, гидромагнитный 95–61
магнитогидродинамический ~ 95–61
осцилляции между зеркальными точками 75–119
пятиминутные ~ 88–186
солнечные ~ 88–223а
осыпь 81–234
ось вращения 83–33
~ геодиполя 75–47
земная магнитная ~ 75–47
магнитная ~ 75–46
магнитная ~ Земли 75–47
отбор грунтовых проб 72–51
~ керна 72–51
~ образца грунта 82–78
~ проб 76–85
~ проб воздуха 76–85
отверде(ва)ние 83а–59
~ лавы 82–190
отверстие, высверленное 82–61
отвесная линия 71–71
отдача, световая 86–210
отделение 83а–9
~ от Солнца 83а–9
отклонение, вертикальное 71–72
гравитационное ~ света 99–161
~ звезды 98–123

~ луча света в гравитационном поле 99–161
~ магнитной стрелки 75–56
~ от локального термодинамического равновесия 93–40
~ от ЛТР 93–40
~ от термодинамического равновесия 93–36
~ от ТР 93–36
~ отвеса 71–72
откос 81–184
открыватель 85–21
~ кометы 85–21
открытая вселенная 99–72
открытие звезды 81–99
~ кометы 85–16
отлив 82–282
отложение 74–47
~ осадочных материалов 73–109
отложения, абиссальные 71–50
биогенные ~ 72–59
глубинные морские ~ 72–50
глубоководные ~ 87–15
гляциальные ~ 72–127
ледниковые ~ 72–127
органические ~ 72–59
пелагические ~ 72–57
терригенные ~ 72–48
отмель 72–96
континентальная ~ 72–34
относительное движение 99–99
относительность инерции 99–35
относительный возраст 82–268
отношение водорода к металлам 91–176
~ газа и пыли 97–120
~ диаметра к глубине 81–248
изотопное ~ 74–94
~ изотопов 74–94
~ интенсивностей 91–60
~ масс 96–67
~ масс газа к пыли 97–120
~ масса-светимость 98–233
~ освещённостей 91–60
~ Пуассона 73–60
равновесное ~ 94–104
~ содержаний 91–175
отношения, структурные 82–266
ОТО 99–146
отправитель 85–23
отражатель 82–43
лазерный ~ 82–43
отражающий конус 88–62
отражённый импульс 86–290
~ сигнал 86–290
отрицательный ион водорода 93–66
отрыв излучения от вещества 99–177
отсек, командный 82–6
лунный ~ 82–5
служебный ~ 82–7
~ со вспомогательными системами 82–7
~ управления 82–6
отстой 74–47
отсутствие ЛТР 93–43
отталкивательная сила 85–87
отталкивающая сила 85–87
отторженец, ледниковый 72–132
охлаждение 83а–57
~ излучением 94–98
нейтринное ~ 94–97
радиативное ~ 94–98
оценивать 95–12
оценить 95–12
оценка блеска 95–11
очаг землетрясения 73–17
сейсмический ~ 73–17
очевидец падения 87–48

П

Павлин 92–74
падающая звезда 86–2
падение 76–57
~ болида 86–117
вековое ~ блеска 85–127
естественное ~ 82–227
~ кометы 82–230
~ метеора 86–5
~ метеорита 87–19
одиночное ~ метеорита 87–36
подозреваемое ~ метеорита 87–49
предполагаемое ~ метеорита 87–49
слободное ~ 94–39
~ температуры 77–41
паковый лёд 72–102
палеозой 74–71
палеозойская эра 74–71
палеоклимат 77–188
палеомагнетизм 75–25
палеотемпература 72–126
Паллада 84–86
палласит 87–81
пар, водяной 77–119
пара, визуальная 96–8
тесная ~ 96–23
физическая ~ 96–7
широкая ~ 96–9
парадокс, гравитационный 99–39
~ Зеелигера 99–39
космологический ~ 99–38
~ Неймана-Зеелигера 99–39
~ Ольберса 99–40
фотометрический ~ 99–40
~ Шезо-Ольберса 99–40
паразитический кратер 73–128
параклас 73–91
параллакс, абсолютный 91–11
вековой ~ 91–9
~ вычисленный по цефеидам 91–8а
годичный ~ 91–6
групповой ~ 91–7
динамический ~ 91–12
звёздный ~ 91–4
~ звезды 91–4
относительный ~ 91–10
относительный тригонометрический ~ 91–10
спектральный ~ 91–15
спектроскопический ~ 91–15
статистический ~ 91–8
~ Стрёмгрена 91–13
~ сферы Стрёмгрена 91–13
тригонометрический ~ 91–5
фотометрический ~ 91–14
~ цефеид 91–8а
цефеидный ~ 91–8а
~ Штрёмгрена 91–13
параллактическая либрация 81–52
параллактическое движение 91–33
параллельное наблюдение 86–125
параметр замедления 99–83
~ расширения Хаббла 98–302
структурный ~ 96–120
параметры кратера 81–227
фотометрические ~ 85–106
парантгелий 79–84
парантиселена 79–86
паргелий 79–71
паргелический круг 79–82
парниковый эффект 83–263
парсек 91–17
кубический ~ 98–84
Паруса 92–98
пасмурно 77–152
пассатные ветры 77–100
пассаты 77–100
патруль, автоматический 89–221
~ вспышек 89–219
метеорный ~ 86–144
фотографический ~ вспышек 89–220
Пегас 92–75
пейсаж, лунный 81–55
пекулярная галактика 98–256
~ звезда 91–92
пекулярное движение 91–31
~ ядро 98–20
пекулярность 98–257
пелагические осадки 72–57
~ отложения 72–57
пелена облаков 77–151
облачная ~ 77–151
пенетрометр 82–83
пепел, вулканический 73–142
пепельный свет Луны 81–35
пепловый конус 73–120
первичная волна 73–46
~ дуга 73–82
~ орбита 98–200
первичное вещество 99–170
первичные блины 99–171b
~ флуктуации 99–170у
первичный благородный газ 87–133
~ изотоп 82–146
~ кратер 81–198

R

~ нуклеосинтез 99–175
~ радиоактивный изотоп 82–256
первоатом 99–169
первоначальная атмосфера 83а–44
~ вселенная 99–170а
первоначальный атом 99–169
первооткрыватель 85–21
~ кометы 85–21
передатчик 78–15
передающая антенна 78–16
передлунный 81–4
передморская структура 81–153
передняя звезда 96–85
переключатель фазы 88–114
переключение фазы 88–112
перекристаллизация 74–42
перекрытие спектров 96–24
переменная 95–1
~ база 88–103
взрывная ~ 95–109
вращающаяся ~ 95–65
вуалируемая ~ 95–83
геометрическая ~ 95–72
долгопериодическая ~ 95–91
залораживаемая ~ 95–83
затмеваемая ~ 95–83
затменная ~ 95–74
затменная ~ типа Алголя 95–77
~ звезда = переменная 95–1
катаклизмическая ~ 95–109
~ класса ... 95–47
короткопериодическая ~ 95–89
красная полуправильная ~ 95–106
магнитная ~ 95–52
~ масса 86–167
небулярная ~ 95–85
неправильная ~ 95–107
неправильная магнитная ~ с переменной полярностью 95–56
неправильная магнитная ~ с постоянной полярностью 95–57
неправильная ~ с переменной полярностью 95–56
неправильная ~ с постоянной полярностью 95–57
несобственная ~ 95–86
нефизическая ~ 95–86
новоподобная ~ 95–118
периодическая ~ 95–88
полуправильная ~ 95–105
сверхкороткопериодическая ~ 95–90
собственная ~ 95–87
спектральная ~ 95–48
~ типа ... 95–47
~ типа дельта Цефея 95–92
~ типа R Coronae Borealis 95–83
~ типа скоплений 95–93
физическая ~ 95–87
экранируемая ~ 95–83
эллипсоидальная ~ 95–79
эруптивная ~ 95–68
переменность, звёздная 95–4
~ магнитного поля 84–14
оптическая ~ 95–141
~ поля 84–14
перемешивание звёздного вещества 93–162
перемещение материков 73–104
горизонтальное ~ материков 73–104
перемычка, галактическая 98–254
яркая ~ в пятне 89–104
перенасыщение 76–102
перенос излучения 93–29
~ магнитного поля 84–48
~ массы 96–33
~ массы с одной компоненты на другую 96–33
~ тепла 86–202
~ энергии 93–111
переносный магнитометр 82–44
переоткрытие 85–17
переохлаждение 76–105
переохлаждённый 76–106
переоценить 95–13
перерыв от обтюратора 86–150
пересветимая звезда 91–134
пересечённая спиральная галактика 98–253
пересыщение 76–102
пересыщенный 86–285
переуплотненный 86–285
переход, фазовый 83а–58, 99–170р
переходная область 72–13
периапсида, галактическая 98–189
периарий 83–142
периастр 96–58
периастрий 96–58
перигалактий 98–189
перигей 81–38
перигелий 83–29а
перийовий 83–172
перилуний 81–40
перимарсий 83–142
период 95–39
~ биений 95–42
вращательный ~ 83–31
~ вращения 83–31
~ вращения Солнца 88–129
вторичный ~ 95–41
геологический ~ 74–74
главный ~ 95–40
гляциальный ~ 72–105
девонский ~ 74–81
драконический ~ обращения 81–47
звёздный ~ обращения 81–45
каменноугольный ~ 74–82
кембрийский ~ 74–78
ледниковый ~ 72–105
магнитно-возмущённый ~ 75–91
межледниковый ~ 72–107
меловой ~ 74–86
~ наложения соседних циклов 89–86
одиннадцатилетний ~ 89–70
ордовикский ~ 74–79
~ осевого вращения 83–31
основной ~ 95–40
~ перекрытия двух циклов 89–86
пермский ~ 74–83
~ пульсаций 95–100
~ релаксации 98–128
силурийский ~ 74–80
синодический ~ 81–46
синодический ~ обращения 81–46
третичный ~ 74–87
четвертичный ~ 74–88
юрский ~ 74–85
периодическая комета 85–3
~ магнитно-переменная звезда 95–55
перисатурний 83–210
периселений 81–40
перисто-кучевые облака 77–143
перисто-слоистые облака 77–141
периферическое зрение 86–128
перицентр, галактический 98–189
перламутровое облако 79–91
перманентная корональная конденсация 89–282
пермь 74–83
Персеев рукав 98–34
Персеиды 86–251
Персей 92–76
петельный протуберанец 89–270
петлеобразный протуберанец 89–270
петли 89–270
Петля в Лебеде 97–137
магнитная ~ 84–19
солнечная межпланетная ~ 84–20
петрографическая классификация 82–101
Печь 92–50
пещера, лавовая 82–177
пиджонит 82–90
пижонит 82–90
пик слоя 76–55
пиковая мощность 86–271
пиранометр 88–43
пиргелиометр 88–39
водоструйный ~ 88–40
водяной ~ с перемешиванием 88–41
~ с текущей водой 88–40
~ с серебряным диском 88–42
пирогенная интрузия 82–98
пирокластический конус 73–120
пироксен 74–21
Писциды 86–250
пк 91–17
плавление 86–181
плавучесть, магнитная 89–147
плагиоклаз 74–19
плазма, бесстолкновительная 84–50

замагниченная солнечная ~ 84–32
межпланетная ~ 84–39
намагниченная ~ 84–32
~ обогащённая гелием 84–60
солнечная ~ 84–29
плазменная оболочка 75–11
плазменный слой 75–11
плазмосфера 75–12
планета 83–1
большая ~ 83–13
верхняя ~ 83–15
внешняя ~ 83–15
внутренняя ~ 83–14
водородная ~ 83–197
~ гигант 83–174
гипотетическая ~ 83a–63
гипотетическая ~ Ольберса 83a–63
заплутоновая ~ 83–124
земная ~ 83–119
~ земного типа 83–119
~ земной группы 83–119
интрамеркуриальная ~ 83–123
малая ~ 84–68
малая ~ кольца 84–71
нижняя ~ 83–14
солнечная ~ 83–11
~ солнечной системы 83–11
~ Солнца 83–11
~ спутник 96–43
~ типа Земли 83–119
~ типа Юпитера 83–174
транснептуновая ~ 83–237
трансплутоновая ~ 83–124
планетарная волна 72–94
планетарный индекс 75–90
планетезимали 83a–35
планетная конфигурация 83–50
~ корона 83–245
~ космогония 83a–1
~ орбита 83–29
планетный 83–6
планетографический 83–7
планетоид = астероид = малая планета 84–68
двойной ~ 84–79
~ пересекающий орбиту Земли 84–76
~ пересекающий орбиту Марса 84–75
~ тесно сближающийся 84–78
~ тесного сближения 84–78
планетология 83–5
планетофизика 83–4
планетоцентрический 83–9
планеты, большие 83–13
главные ~ 83–13
планковская масса 99–170f
~ эра 99–170e
пластинка в четверть волны 88–126
клейкая ~ 87–14
компенсирующая ~ 88–92
метеорная ~ 86–151
тонкая ~ 83–149
четвертьволновая ~ 88–126
плато 81–192
платформа, континентальная 72–34
материковая ~ 73–103
плашмя, галактика видимая 98–247
плессит 87–94
Плеяды 96–110
плоская возвышенность ~ 81–192
плосковершинная подводная гора 72–44
плоскогорье 82–56
плоское дно 81–260
плоскость Галактики 98–53
галактическая ~ 98–53
картинная ~ 96–50
~ кольца 83–216
~ Лапласа 83–232
неизменяемая ~ 83–232
основная ~ Галактики 95–53
~ перпендикулярная к лучу зрения 96–50
~ экватора 83–231
экваториальная ~ 83–231
плотность, видимая поверхностная 96–132
~ галактик 98–344
~ звёзд 98–82
~ излучения 99–85
истинная ~ 96–133
космологическая ~ 98–370
~ кратеров 81–246
критическая ~ 98–197
линейная электронная ~ 86–283
линейная ~ электронов 86–283
~ метеорного потока 86–264
~ нейтронного потока 94–86
объёмная ~ 96–133
поверхностная ~ 98–85
погонная электронная ~ 86–283
проекционная ~ 98–85
пространственная ~ 86–263
пространственная ~ звёзд 98–83
пространственная ~ метеорного потока 86–264
~ радиантов 86–63
~ следов 87–141
~ солнечного ветра 88–157
средняя ~ 93–86
средняя ~ вещества во вселенной 99–84
средняя ~ массы 99–84
субъядерная ~ 93–190
~ тени 81–59
~ треков 82–253
~ фона 96–131
центральная ~ 93–87
электронная ~ 78–6
площади протуберанцев 89–236
ежедневные ~ протуберанцев 89–236
избранные ~ 98–78
среднесуточные ~ протуберанцев 89–236
площадки, избранные 98–78
~ Каптейна 98–78
площадь, видимая 89–91
~ вспышки 89–207
запятнённая ~ 89–90
общая ~ 89–94
~ протуберанцев 89–233
~ пятен 89–67
~ пятен исправленная за перспективное сокращение 89–93
~ радиации 86–238
~ с пятнами 89–90
суммарная ~ пятен 89–94
Плутон 83–239
плутоническая порода 74–7
побочный кратер 73–128
поверка 95–18
~ кривой блеска 95–18
поверхностная волна 73–47
поверхность, вторичная 87–108
~ второго рода 87–108
геопотенциальная ~ 71–61
гористая ~ 82–56
~ Земли 71–5
земная ~ 71–5
контактная ~ 96–27
критическая ~ 96–27
критическая ~ Роша 96–27
критическая эквипотенциальная ~ 96–27
лобовая ~ 87–114
лунная ~ 82–55
~ Луны 82–55
лучевая ~ 81–171
~ моря 82–57
межлучевая ~ 81–172
~ метеорита 87–106
~ Мохоровичича 72–14
первичная ~ 87–107
~ первого рода 87–107
передняя ~ 87–114
планетная ~ 83–97
~ покрытая кратерами 81–245
~ покрытая лучами 81–171
полированная ~ 87–101
промежуточная ~ 93–141
пульсирующая ~ 79–149
~ с корой плавления 87–109
слоистая ~ 83–101
~ Солнца 88–164
твёрдая ~ 83–98
тыловая ~ 87–115
шероховатая ~ 83–102
эквипотенциальная ~ поля тяжести 71–61
повреждение приливной волной 72–77
повторная новая 95–116
~ новая звезда 95–116
повторяемость полярных сияний 79–122

R

повышение температуры 77–42
погасание метеора 86–37
мгновенное ~ 86–76
плавное ~ 86–75
постепенное ~ 86–75
резкое ~ 86–76
тупое ~ 86–75
поглощающее вещество 97–85
поглощение, авроральное 78–46
~ в линии 93–108
~ в полярной шапке 78–44
внезапное ~ коротких волн 78–40
внезапное ~ космического шума 78–42
~ водяным паром 77–127
дифференциальное ~ 97–90
избирательное межзвёздное ~ 97–90
межзвёздное ~ 97–84
общее межзвёздное ~ 97–87
поверхностное ~ 85–128
полное межзвёздное ~ 97–87a
~ с внезапным началом 78–47
свободно-свободное ~ 93–106
связанно-свободное ~ 93–107
погода 77–1
погружение Солнца под горизонт 79–45
погруженный кратер 81–204
подвижная зона 73–75a
подводная горизонталь 72–32
подводное землетрясение 73–2c
подводный вулкан 73–117
~ каньон 72–46
подгрунт 82–105
поддержание хромосферы 88–223
подзвёздная точка 81–130
подземная точка 81–128
подимпульс 95–145
подлунная точка 81–129
подножие, континентальное 72–36
материковое ~ 72–36
поднятый край 81–222
подповерхностное исследование 82–82
подробности лунного рельефа 81–148
подробность, постоянная 83–106
~ рельефа 83–105
подсистема 98–74
подсолнечная точка 81–127
подспутниковая точка 71–30
подсчёт галактик 98–343
~ звёзд 98–79
подушка, газовая 86–170
подфотосферный слой 88–168
подъём 76–54, 82–8
~ к максимуму 95–28
~ прилива 82–292
поиск комет 85–15
~ метеоритов 87–45
~ планет 83–238
~ сверхновых звёзд 95–136
~ спутников 83–27
показатель желтизны 83–88
интегральный ~ цвета 96–138
истинный ~ цвета 91–75
~ политропы 93–128
собственный ~ цвета 91–75
специальный ~ цвета 83–87
структурный ~ скопления 98–342
~ тепла 91–81
тепловой ~ 91–81
~ цвета 91–76
~ цвета B-V 96–145
~ цвета в системе B, V 96–145
~ цвета U-B 96–146
поколение звёзд 94–23
звёздное ~ 94–23
покрасневшая звезда 97–94
покраснение, межзвёздное 97–91
покров, лавовый 82–175
ледниковый ~ 72–115
снежный ~ 77–176
покровный ледник 72–116
покрытие 95–80
~ звезды Луной 81–94
касательное ~ 81–100
покрытый облаками 83–253
~ пылью 82–120
поле, акустическое 87–32
внутреннее ~ 95–63
галактическое ~ 98–146
галактическое гравитационное ~ 98–146
галактическое магнитное ~ 98–50
галактическое приливное ~ 96–106
геомагнитное ~ 75–44
гравитационное ~ 98–145
~ дифференциальных движений 98–114
~ диффузного излучения 93–31
звуковое ~ 76–40
земное магнитное ~ 75–44
~ земных электрических токов 75–4
~ излучения 93–30
изначальное магнитное ~ 95–62
каменное ~ 82–139
~ камней 82–139
квантовое ~ 99–170m
коэрцитивное ~ 75–39
кратерное ~ 87–56
крупномасштабное ~ 89–142
лавовое ~ 82–184
магнитное ~ Земли 75–44
межзвёздное магнитное ~ 97–106
межпланетное магнитное ~ 84–9
мелкомасштабное ~ 89–143
~ намагничивания 75–42
общее магнитное ~ 89–128
общее магнитное ~ Солнца 89–128
оптическое ~ 87–31
очень сильное магнитное ~ 95–148
первоначальное магнитное ~ 95–62
поверхностное ~ 95–64
поперечное ~ 89–137
~ приливных сил 98–196
продольное ~ 89–138
реликтовое магнитное ~ 91–53
сверхсильное магнитное ~ 95–148
селеномагнитное ~ 82–45
~ силы тяжести 71–57
~ скоростей 88–187
солнечное магнитное ~ 89–129
~ тяготения 98–145
фирновое ~ 72–119
флоккульное ~ 89–164
флоккульное ~ в рентгеновских лучах 89–166
фоновое ~ 89–141
фоновое магнитное ~ 89–141
фотосферное магнитное ~ 89–130
~ Хиггса 99–170l
электротеллурическое ~ 75–4
полевой шпат 74–15
полёт болида 86–117
~ к астероиду 84–99
~ к астероиду с возвращением образцов 84–98
~ к малой планете 84–99
~ метеора 86–5
ползучесть грунта 82–152
полигональный кратер 81–212
полиномы Лежандра 93–55
политропа 93–126
политропная модель 93–129
полиэдр 71–12
Поллукс 92–122
полная вода 82–281
~ Луна 81–27
~ масса скопления 96–139
~ фаза Луны 81–27
полное вырождение 93–175
~ давление 93–93
~ лунное затмение 81–66
~ солнечное затмение 88–17
полнолуние 81–27
полночь, магнитная 79–131
полный вектор магнитно поля 89–139
~ прилив 82–281
половина диска Луны 81–26
пологий максимум 86–236
положение, гелиографическое 89–38
общее синоптическое ~ 77–15
предвычисленное ~ 85–27
предсказанное ~ 85–27
приблизительное ~ 85–26
~ радианта 86–60
селеноцентрическое ~ 81–132
синоптическое ~ 77–14
точное ~ 85–28
полоса 83–190
диффузная межзвёздная ~ 97–100
зодиакальная ~ 84–63
однородная ~ 79–146
~ полного затмения 88–29
~ полной фазы 88–29

R

пылевая ~ 98–289
~ с лучевой структурой 79–150
светящаяся ~ 86–27
тёмная ~ 83–190
экваториальная ~ 83–191
полоска из слюды 88–87
~ поляроидной мозаики 88–87
полость 87–117
~ в минерале 82–180
внутренняя ~ 87–119
гелиосферная ~ 88–257
геомагнитная ~ 75–8
геомагнитная ~ в потоке солнечной плазмы 75–8
гидромагнитная ~ 84–22
магнитная ~ 84–22
межпланетная ~ 84–25
планетная ~ 84–24
поверхностная ~ 87–118
~ солнечного ветра 84–28
полосчатость 74–53
полосы Вегарда-Каплана 79–19
неотож(д)ествлённые ~ 85–134
~ отрицательной системы азота 79–21
~ Свана 85–133
тёмные ~ 98–288
флуоресцентные ~ 85–132
полубесконечная атмосфера 93–22
полуденный меридиан 75–17
полудиаметр Луны, видимый 81–70
видимый ~ Солнца 81–69
полуконвективная зона 93–160
полумесяц 81–26
полуразделённая двойная 96–34
полусуточная волна 76–29
полусфера, видимая 89–27
дневная ~ 83–67
неосвещённая ~ 83–68
ночная ~ 83–68
освещенная ~ 83–67
освещённая ~ Земли 75–20
полутеневое волокно 89–98
~ лунное затмение 81–65
полутень 89–97
~ Земли 81–64
земная ~ 81–64
~ пятна 89–97
полушарие, видимое 89–27
видимое ~ Луны 81–15
дневное ~ 83–67
дневное ~ Земли 75–20
неосвещённое ~ 83–68
ночное ~ 83–68
освещённое ~ 83–67
~ освещённое Солнцем 75–20
северное ~ 79–3b
южное ~ 79–3c
южное ~ неба 79–3c
полюс вращения 83–36
геомагнитный ~ 75–45
магнитный ~ Земли 75–45
~ наклонения 75–45
~ однородного намагничения 75–48
~ оси 83–36
~ оси вращения 83–36
поляризация, межзвёзднаия 97–105
поляриметр, визуальный 83–82
фотографический ~ 83–84
фотоэлектрический ~ 83–83
поляриметрическое измерение 83–80
Полярная 92–121
полярная зона 77–193
~область 83–192
~ шапка 83–143
полярное сияние 79–93
~ сияние в радиодиапазоне 79–94
полярность, магнитная 89–148
отрицательная ~ 89–150
положительная ~ 89–149
противоположная ~ 89–155
северная ~ 89–149
южная ~ 89–150
полярный венец волокон 89–232
помехи, атмосферные 78–50
помутнение 83–252
понижение, дугообразное 81–160
извилистое ~ 81–161
линейное ~ 81–159
лунное ~ 81–157
попадание метеорита 87–20
поперечная волна 73–41
поправка, болометрическая 91–80
~ за кривизну земной поверхности 86–62
~ к формуле Крамерса 93–103
~ поля 86–157
релятивистская ~ 99–140
попятное движение 83–47
попятный 83–46
пора 89–122
пористое вещество 82–111
пористость 83–100
пористый 82–181
порог разрешения спутника 96–40
порода 74–1
вулканическая ~ 74–3
глубинная ~ 74–7
горная ~ 74–1
гранитная ~ 82–67
излившаяся ~ 74–3
интрузивная ~ 74–7
кислая ~ 74–4
кристаллическая ~ 82–66
лунная ~ 82–65
магматическая ~ 74–2
метаморфическая ~ 74–11
обломочная ~ 82–85
осадочная ~ 74–10
основная ~ 74–5
плутоническая ~ 74–7
поверхностная рыхлая ~ 82–104
покрывающая ~ 82–104
равномерно зернистая ~ 82–67
рассланцованная ~ 74–54
расслоенная ~ 74–54
ультраосновная ~ 74–6
эффузивная ~ 74–3
породообразующий минерал 82–99
породосоставляющая часть 82–99
портативный магнитометр 82–44
порыв ветра 77–112
порывистый ветер 77–111
посадка на Луну 82–13
~ на поверхность Луны 82–13
пилотируемая ~ 82–33
посадочная нога 82–22
~ опора 82–23
последовательность, главная 91–131
главная ~ нулевого возраста 94–60
~ нулевого возраста 96–152
~ правильных галактик 98–240
спектральная ~ 91–94
~ спиральных галактик 98–255
хаббловская ~ галактик 98–227
эволюционная ~ 94–30
последующее пятно 89–114
послеледниковый 72–110
послеледниковье 72–109
послеморская структура 81–154
послесвечение 79–22
послесумеречная эмиссия 79–52
послеэруптивная галактика 98–270
постгляциальный 72–110
постоянная, космическая 99–29
космологическая ~ 99–29
~ Пуассона 73–60
солнечная ~ 88–148
~ Хаббла 98–302
хаббловская ~ 99–77
постоянные Оорта 98–118
~ Стокса 71–59
Стоксовы ~ 71–59
построить карту магнитного поля 89–132
постулат Вейля 99–34
~ когерентности 99–34
математический ~ относительности инерции 99–36
постэруптивная галактика 98–270
потемнение, гравитационное 91–41
~ диска к краю 88–142
~ к краю 88–142
потенциал адсорбции 85–122
адсорбционный ~ 85–122
гравитационный ~ 96–160a
~ Земли 71–60
приливный ~ 82–293
~ с осевой симметрией 98–153
~ с цилиндрической симметрией 98–154
~ силы тяжести 71–58
стационарный ~ 98–147
потенциальное корыто 98–176
потеря массы 86–168, 94–72
~ энергии 88–254

R

поток, активный метеорный 86–223
~ астероидов 84–100
воздушный ~ 77–90
восходящий ~ 77–91
восходящий конвективный ~ 93–151
временный ~ 86–225
выходящий ~ излучения 93–114
~ газа 95–67
дневной ~ 86–219
~ звёзд 98–98
звёздный ~ 98–98
~ звуковой энергии 88–221
~ излучения 93–113
интегральный ~ 93–47
кометный ~ 86–222
конвективный ~ энергии 93–115
конвекционный ~ 73–76
~ корпускул 88–160
корпускулярный ~ 88–160
лавовый ~ 73–140, 82–173
~ малых планет 84–100
межзвёздный ~ 86–224
местный звёздный ~ 96–97
метеорный ~ 86–216
мутьевой ~ 72–49
нисходящий конвективный ~ 93–150
ночной ~ 86–218
~ пепла 82–174
падающий ~ 93–150
периодический ~ 86–228
~ планетного происхождения 86–220
полный ~ 93–47
полный ~ энергии 93–116
постоянный ~ 86–227
~ солнечного ветра 84–35а
~ солнечной плазмы 84–33
солнечный корпускулярный ~ 84–35
спорадический ~ 86–226
суспензионный ~ 72–49
~ тепла 74–60
тепловой ~ 74–60
эклиптикальный ~ 86–221
~ энергии 93–112
потолок атмосферы 76–25
поточный метеор 86–24
потухший вулкан 73–115
почва, лунная 82–62
почвенный слой, лунный 82–62
появление звезды 81–99
~ кольца 83–219
~ кометы 85–18
~ метеора 86–35
первое ~ кометы 85–19
~ частиц высокой энергии 89–191
~ энергичных частиц 89–191
поярчание флоккула 89–213
пояс 83–190
активный ~ 73–32
~ астероидов 84–70
~ ван Аллена-Вернова 75–101
~ видимости покрытия 81–107
внешний радиационный ~ 75–105
внутренний радиационный ~ 75–106
второй радиационный ~ 75–105
естественный радиационный ~ 75–107
~ землетрясения 73–30
искусственный радиационный ~ 75–108
~ малых планет 84–70
~ Ориона 92–129а
орогенический ~ 73–75
первый радиационный ~ 75–106
подвижный ~ 73–75
~ радиации искусственного происхождения 75–108
радиационный ~ 75–101
радиационный ~ Земли 75–101
сейсмический ~ 73–30
субавроральный ~ 79–128
субполярный ~ 79–128
экваториальный ~ 83–191
ППШ 78–44
праатом 99–169
правило Тициуса-Боде 83–3
правильная галактика 98–236
превращение, фазовое 83а–58
~ фаз 83а–58
~ энергии 86–208
предварительное исследование поверхности 82–32
предвестник 95–143, 73–11
~ землетрясения 73–11
предвычисление покрытий 81–102
~ положений малых планет 84–92
предвычисленная комета 85–11
предвычисленное положение 85–27
предельная высота атмосферы 76–25
~ дальность 86–289
~ звёздная величина 91–62
предлунный 81–4
предморской район 81–269
предновая 95–122
предполагаемое падение метеорита 87–49
предсверхновая 95–129
предсказание, долгосрочное 77–19
краткосрочное ~ 77–20
~ погоды 77–13
~ покрытий 81–102
~ солнечной активности 89–64
предсказанная комета 85–11
предсумеречная эмиссия 79–51
предшественник 95–143
преимущественная ориентация частиц 97–78
преобладающие элементы 82–144
преобладающий ветер 77–117
преобразование, галилеевское 99–131
~ Галилея 99–131
гомологическое ~ 93–132
~ Лапласа 93–54
~ Лоренца 99–132
преходящий лунный процесс 82–153
приближение Милна-Эддингтона 93–50
нерелятивистское ~ 99–97
~ Чандрасекара 93–51
приблизительное положение 85–26
прибой, штормовой 72–76
прибор испытывающий грунт 82–79
прибывающая Луна 81–20
приведение к уровню моря 77–45
приведённая высота 76–18
приводить в действие 89–198
приёмная антена 78–18
приёминик 78–17
приземление 82–16
приземная волна 74–14
~ инверсия 77–38
призма Андерсона 88–68
вращающаяся ~ 88–68
~ разделяющая пучок 88–88
светоделительная ~ 88–88
признак, вулканический 82–159
прилив 82–280
~ вызванный Землей 82–275
~ и отлив 72–89
квадратурный ~ 82–272
штормовой ~ 72–76
приливная выпуклость 82–278
~ деформация 83а–66
~ неустройчивость 98–313
~ эволюция 83а–65
приливное взаимодействие 82–274
приливные силы 72–92
приливный горб 82–278
~ потенциал 82–293
приливомер 82–311
приливообразующее тело 82–284
приливообразующие силы 72–92
приливо-отливные силы 72–92
приливы 72–89
атмосферные ~ 82–294
~ земной коры 82–297
земные ~ 82–296
геомагнитные ~ 82–295
гравитационные ~ 82–298
лунные ~ 82–308
морские ~ 82–309
океанские ~ 82–310
сизигийные ~ 82–273
термические ~ 82–299
прилунение 82–13
примордиальный изотоп 82–146
принадлежность, физическая 96–83
принцип, архимедов 73–101
~ инвариантности 93–61
космологический ~ 99–30
~ Маха 99–37

ограниченный космологический ~ 99–31
совершенный космологический ~ 99–32
~ эквивалентности 99–150
~ эквивалентности массы и энергии 99–145
приращение массы 99–142
природа объекта 85–24
прирост массы 83a–38
~ массы Земли 87–18
пристающий 82–137
притяжение, зенитное 86–52
собственное ~ 94–34
приход на Землю 89–193
проба воздуха 76–86
лунная ~ 82–69
лунная ~ доставленная на Землю 82–70
пробел 84–88
~ Герцшпрунга 91–144
~ соизмеримости 84–88
проблема, космогоническая 94–5
пробоотбиратель 82–80
~ воды 72–101
пробоотборник 82–80
провал взрыва 82–161
провальное землетрясение 73–2d
прогноз, долгосрочный 77–19
ионосферный ~ 78–32
краткосрочный ~ 77–20
~ погоды 77–13
~ солнечной активности 89–64
прогностическая карта 77–21
продолжительность видимости следа 86–111
~ вспышки 89–209
~ жизни вспышки 89–209
~ жизни группы пятен 89–123
~ затмения 95–78
~ импульса 86–280
~ испарения 96–109
~ метеорного эха 86–299
~ полёта 86–50
~ полного затмения 88–26
~ солнечного сияния 88–10
~ цикла 89–87
продольный компонент 89–135
продукты извержения 82–165
проект „протонная вспышка" 89–177
проекция, азимутальная 71–44
гномоническая ~ 71–46
зенитная ~ 71–45
картографическая ~ 71–40
коническая ~ 71–42
конформная ~ 71–41
стереографическая ~ 71–47
центральная ~ 71–48
цилиндрическая ~ 71–43
происхождение 94–9
~ атомных ядер 94–52
внеземное ~ 87–73
~ вселенной 99–11
земное ~ 87–74
~ комет 85–96
~ мира 99–11
совместное ~ 94–25
тектоническое ~ 82–200
ударное ~ 82–229
~ ударом 82–229
происшествие загрязнения 72–72
прокладка 82–134
Проксима Центавра 92–124
пролёт болида 86–117
~ метеора 86–5
~ через атмосферу 87–26
промежутки Кирквуда 84–89
промежуток 84–88
межгранульный ~ 88–177
межспикульный ~ 88–198
~ соизмеримости 84–88
промежуточная корона 88–245
промежуточное землетрясение 73–24
промер глубины 72–28
эхолотный ~ 72–29
прослойка, пылевая 98–289
пространственная волна 78–13
пространственно-временной 99–120
пространственное выветривание 82–244
пространство 99–100
абсолютное ~ 99–116
безвоздушное ~ 76–9
бесконечное ~ 99–104
внегалактическое ~ 97–5
внеземное ~ 87–73a
гиперболическое ~ 99–106
~ Евклида 99–107
запрещённое ~ 79–135
искривлённое ~ 99–114
конечное ~ 99–103
~ Лобачевского 99–105
межгалактическое ~ 97–7
~ между волокнами 89–99
межзвёздное ~ 97–4
межпланетное ~ 84–2
метрическое ~ 99–100a
~ Минковского 99–122
неевклидово ~ 99–114
~ Ньюкома 99–110
параболическое ~ 99–107
плоское ~ 99–107
разрешённое ~ 79–134
~ Римана 99–112
риманово ~ 99–112
свободное ~ 76–9
~ скоростей 98–97
сферическое ~ 99–111
шестимерное фазовое ~ 98–144
~ Эвклида 99–107
эллиптическое ~ 99–110
пространство-время 99–119
~ Минковского 99–122
протерозой 74–77
протерозойская группа 74–77
противолуна 79–85
побочная ~ 79–86
противоположная полярность 89–155
противосияние 84–64
противосолнце 79–83
побочное ~ 79–84
противостояние 83–59
~ по прямому восхождению 81–68
противосумерки 79–57
протоатмосфера 83a–44
протогалактика 94–18
протогалактическое облако 94–17
протозвезда 94–45
протозвёздный 94–46
протонная вспышка 89–185
протоносфера 76–75
протопланета 83a–40
протопланетное облако 83a–24
протосолнечная туманность 83a–21
Протосолнце 83a–39
протуберанец 89–223
активный ~ 89–238
активный ~ типа солнечных пятен 89–254
водородный ~ 89–266
возвратный ~ 89–255
воронкообразный ~ 89–269
вспышечноподобный ~ 89–222
вспышкоподобный ~ 89–222
вторичный ~ 89–260
высокоширотный ~ 89–229
квазиэруптивный ~ 89–249
колпачковый ~ 89–253
корональный ~ 89–264
корональный активный ~ 89–247
металлический ~ 89–265
низкоширотный ~ 89–227
обычный активный ~ 89–246
обычный эруптивный ~ 89–250
петельный ~ 89–270
петлеобразный ~ 89–270
поднимающийся ~ 89–278
~ полярной зоны 89–230
полярный ~ 89–230
спокойный ~ 89–263
тёмный возвратный ~ 89–256
~ типа колпак 89–253
~ типа солнечных пятен 89–252
~ торнадо 89–262
экваториальный ~ 89–228
эруптивный ~ 89–248
протуберанц-единица 89–235
протуберанц-спектроскоп 88–58
протуберанцы, взаимодействующие 89–243
протяжённая оболочка 93–8
профиль, вертикальный 81–228
высотный ~ 76–53
температурный ~ 83–259
температурный ~ атмосферы 83–259

R

проходить через центральный меридиан 89–44
прохождение 95–81
близкое ~ двух галактик 98–320
~ Луны через меридиан 82–303
~ разрыва в межпланетной среде 84–58
~ планеты по диску Солнца 83–51
~ спутника 83–54
~ спутника по диску планеты 83–54
~ тени 83–53
~ тени спутника по диску планеты 83–53
~ через край диска 89–31
~ через центральный меридиан 89–45
процесс абляции 86–183
внутренний ~ 82–258
~ выбрасывания 98–179
e ~ 94–99
~ излучения 86–211
лунный ~ 82–150
~ на Луне 82–150
~ осадкообразования 74–46
осадочный ~ 74–46
p ~ 94–91
преходящий лунный ~ 82–153
равновесный ~ 94–99
радиационный ~ 86–211
r ~ 94–89
s ~ 94–88
тектонический ~ 82–201
~ уноса массы 86–183
эндогенный ~ 82–258
Процион 92–123
прочность грунта 82–135
когезионная ~ 85–117
~ почвы 82–135
проэволюционировавшая звезда 94–74
проявление солнечной активности 89–13
прояснено 77–149
прямая волна 78–12
~ Стена 81–185
прямое восхождение Солнца 81–61
~ вращение 83–39
~ движение 83–43
~ солнечное излучение 88–135
прямой 83–42
прямолинейная борозда 81–166
прямолинейный выброс 85–57
пс 91–17
псевдопеременная звезда 95–51
психрометр 77–121
птолемеева система 99–22
пузырок 82–180
пузырчатый 82–181
пульверизация 82–239
пульсар 95–137
~ в Крабовидной туманности 95–142
гамма ~ 95–139
двойной ~ 95–140a
оптический ~ 95–140
рентгеновский ~ 95–138
пульсации 79–158
адиабатические ~ 95–98
~ геомагнитного поля 75–81
геомагнитные ~ 75–81
неадиабатические ~ 95–99
нерадиальные ~ 95–97
радиальные ~ 95–96
пульсационная неустойчивость 93–213
пульсирующая дуга 79–147
~ звезда 95–95
~ поверхность 79–149
пурпурный свет 79–54
пустая вселенная 99–49
пустота 99–171d
пустыня 83–161
путевая карта 72–33
путь в атмосфере 86–162
видимый ~ 86–32
видимый ~ на сфере 86–32
истинный ~ 86–31
~ метеора 86–30
~ обвала 82–197
~ развития 94–29
~ скольжения 82–197
эволюционный ~ 94–29
пучок, электронный 79–116
~ электронов 79–116
пылевая буря 83–255
~ глобула 98–290
~ оболочка 97–116
~ составляющая 98–232
~ туманность 97–115
пылевое облако 83–256
~ сгущение 98–290
пылевой комок 86–190
~ компонент 98–232
~ метеор 86–190
пылесодержание 85–80
пылинка, графитовая 97–75
железная ~ 97–74
силикатная ~ 97–76
пылинки, межзвёздные 97–59
пыль, вулканическая 73–143
галактическая ~ 97–53
~ движущаяся вокруг Земли 84–53
космическая ~ 97–52
межзвёздная ~ 97–54
межпланетная ~ 84–51
метеоритная ~ 87–10
метеорная ~ 87–11
~ образованная ударом 82–125
околозвёздная ~ 97–55
силикатная ~ 85–53
~ ударного происхождения 82–125
пьезоглипты 87–122
пятиминутные колебания 88–186
пятка посадочной ноги 82–23
пятно, Большое 83–193
ведущее ~ 89–113
главное ~ 89–115
головное ~ 89–113
горячее ~ 95–112a
заднее ~ 89–114
звёздное ~ 97–62
Красное ~ 83–193
невидимое ~ 89–159
одиночное ~ 89–116
последующее ~ 89–114
следующее ~ 89–114
солнечное ~ 89–88
хвостовое ~ 89–114
пятнообразовательная деятельность 89–62

Р

равнина, абиссальная 72–42
лавовая ~ 82–185
равновесие, адиабатическое 76–21
адсорбционное ~ 85–121
гидростатическое ~ 93–75
детальное ~ 93–41
диффузионное ~ 76–23
изостатическое ~ 73–99
изотермическое ~ 76–20
конвективное ~ 76–22, 93–79
локальное термодинамическое ~ 93–39
лучистое ~ 93–48
обобщённое ~ 93–43
статистическое ~ 93–42, 98–149
тепловое ~ 93–78
термодинамическое ~ 93–35
ядерное ~ 94–103
равновесное отношение 94–104
радиальные возмущения 93–215
~ колебания 95–96
~ пульсации 95–96
радиант 86–53, 96–98
видимий ~ 86–56
геоцентрический ~ 86–58
групповой ~ 86–54
индивидуальный ~ 86–55
исправленный ~ 86–58
истинный ~ 86–59
~ метеорного потока 86–237
метеорный ~ 86–53
побочный ~ 86–239
стационарный ~ 86–241
радиативное охлаждение 94–98
радиационная эра 99–172
радиационный процесс 86–211
радиация, захваченная геомагнитным полем 75–112
корпускулярная ~ 88–150
~ небесного свода 77–54
прямая солнечная ~ 88–135
суммарная ~ 77–55
радикал 85–136
свободный ~ 85–137
радиоаврора 79–94

радиоактивное датирование 87–131
~ тепло 83a–62
радиоактивность 74–61
радиоальбедо 83–94
радиовсплеск, декаметровый 83–183
~ на Юпитере 83–182
радиовспышка 89–180
радиогалактика 98–206
радиогелиограмма 88–96
радиогелиограф 88–95
радиогенное тепло 83a–62
радиогенный благородный газ 87–132
радиозвезда 91–158
радиозвёздная величина 86–268
радиоизлучение Юпитера, спорадическое 83–181
радиоинтерферометр 88–97
многоэлементный ~ 88–123
~ с переменной базой 88–107
~ Христиансена 88–123
радиоисточник декаметрового излучения 83–186
квазизвёздный ~ 98–360
радиокарта 81–135a
радиокомпонент 98–286
радиокорона 88–239
радиолиния, межзвёздная 97–98
радиологический метод 74–92
~ метод определения возраста 74–92
радиолокационные наблюдения 86–265
радиомерцание 84–36
радиометеор 86–267
радиометрическая звёздная величина 91–70
радионуклид, космогенический 82–255
радиооблако 98–286
радиоособенность 98–235
радиосолнце 88–3
радиоспектрометр 88–124
радиоспектроскоп, солнечный 88–124
радиоуглерод 74–95
радиоэхо 86–295
радиус 81–69
видимый ~ Луны 81–70
видимый ~ Солнца 81–69
внешний ~ 83–228
внутренний ~ 83–226
~ вселенной 99–5
гравитационный ~ 94–113
~ Луны 81–70
полярный ~ 83–116
~ Солнца 81–69
средний ~ 83–116a
~ Шварцшильда 94–113
шварцшильдовский ~ 94–113
экваториальный ~ 83–115
радуга 79–65
разбегание галактик 98–298
разведка, орбитальная 82–31
развертка в положении 88–105
развёртывание ветвей 98–30
разветвление 98–5a
развитие вселенной 99–171
~ группы пятен 89–118
~ до главной последовательности 94–26
развязывание излучения 99–177
разгрузочный ледник 72–125
раздавление, гравитационное 84–72
раздвигание дна океана 72–39
раздвоение 98–5
разделение химических элементов 83a–48
раздробление малых планет 84–108
раздробленный слой 82–104
раздувание космологическое 99–170c
сверхбыстрое космологическое ~ 99–170c
раздувающаяся вселенная 99–170b
разлёт галактик 98–298
разлистование 74–53
разлитие нефти 72–71
различие по высоте 81–140
разлом 81–165
поверхностный ~ 81–173
размагничивание, тепловое 75–40
термическое ~ 75–40
размазывание фраунгоферовых линий 88–243
размер зёрен 82–113
~ частиц 76–114
размеры системы 96–64
размыв 73–107
размывание фронта 77–75
разность блеска 96–65
~ звёздных величин 91–59
~ путей 88–117
~ хода лучей 88–117
~ яркости 96–65
разогрев метеорного тела 86–174
разомкнутая силовая линия 75–15
разорванно-кучевое облако 77–146
разрез, скоростной 73–52
разрешение, временное 88–205
пространственное ~ 88–204
разрешённая область 79–134
разрешённое пространство 79–134
разрушение 87–24
~ галактического скопления 96–103
~ гравитационное ~ 84–72
~ кометного ядра 85–116
~ кометы 85–115
~ метеора 86–82
~ материнского тела 87–124
~ от метеорита 87–20
~ поверхности 87–35
~ при ударении 84–107
приливное ~ 85–118
~ скоплений 98–195
столкновительное ~ 87–125
~ ударением 84–107
~ ударом 84–107
разрушенная галактика 98–317
разрыв 73–78, 82–212
~ галактического ядра 98–318
~ Герцшпрунга 91–144
~ растяжения 82–207
~ ударением 84–107
разряд, атмосферный 77–161
район, асейсмичный ~ 73–29
гористый ~ 81–196
материковый ~ 81–271
морской ~ 81–268
несейсмичный ~ 73–29
низменный ~ 81–272
послеморской ~ 81–270
предморской ~ 81–269
сейсмический ~ 73–28
Райская Птица 92–16
Рак 92–25
Ракообразная туманность 97–134
ранняя вселенная 99–170a
~ звезда 91–102
Рас-Альхаг 92–125
раскаленные газы 83a–43
раскалывание 94–100
раскол 81–165
раскручение рукавов 98–30
раскрученный рукав 98–31
раскручивание рукавов 98–30
раскручиванный рукав 98–31
раскручивающаяся ветвь 98–31
распавшаяся галактика 98–317
распад группы пятен 89–126
~ кометного ядра 86–261
~ кометы 85–114
~ метеора 86–187
~ метеорного тела 86–186
~ метеорной частицы 86–186
~ открытого скопления 96–103
~ при столкновении 84–107
радиоактивный ~ 74–64
~ родительского тела 87–124
~ скоплений 98–195
столкновительный ~ 87–125
распадаться 74–65
расплавление 86–181
расплавленнй земной шар 83a–64
распределение, асимметричное 98–104
асимметрическое ~ скоростей 98–104
~ в пространстве 96–119
вертикальное ~ 76–17
галактическое ~ 98–54
долготное ~ 89–48
максвелловское ~ скоростей 98–96
~ материи 99–8
~ остаточных скоростей 98–92
~ плотности 96–124
~ по гелиографической долготе 89–48
~ по массе 84–52, 86–165
~ по размерам 82–124
~ по широте 89–51

R

~ по яркости 86–166
пространственное ~ 98–86
широтное ~ 89–51
эллипсоидальное ~ скоростей 98–100
~ яркости 88–141
распределённый, случайно 98–93
распространение, аномальное 76–37
~ звука 76–32
~ радиоволн 78–34
~ света 99–129
~ энергии 88–217
распространённость, естественная 91–171
~ элементов 82–141
~ изотопов 91–179, 94–106
космическая ~ элементов 94–107
относительная ~ элементов 91–174
~ элемента 91–169
распыление 82–239, 85–140
~ метеорного тела 86–189
~ метеорной частицы 86–189
~ солнечным ветром 82–245
рассасывание протуберанца 89–280
рассвет 79–35
расселина 73–78
рассеяние Ми 83–76
~ света 77–57
~ света в атмосфере 77–57
~ энергии 82–285
рассеянное излучение неба 88–136
рассеянный свет неба 77–59
расслабленнное излучение 93–38
рассланцованная порода 74–54
расслоение 74–53
расслоенная порода 74–54
расстояние 86–288, 91–3
~ антенн 88–99
галактоцентрическое ~ 98–11
гелиоцентрическое ~ 84–4
геоцентрическое ~ 84–3
~ действия радиолокатора 86–288
~ до очага землетрясения 73–18
~ до центра Галактики 98–11
межантенное ~ 88–99
~ между антеннами 88–99
~ между компонентами 96–66
наименьшее ~ связи 78–38
нулевое ~ антенн 88–101
~ от Земли до Луны 81–13
~ от центра диска 89–36
~ от центра звезды 93–85
~ от центрального меридиана 89–46
~ от эпицентра 73–19
переменное ~ антенн 88–103
постоянное ~ антенн 88–102
угловое ~ 96–52
угловое ~ между компонентами 96–52
фотометрическое ~ 96–116
эпицентральное ~ 73–19
растрескивание космическими лучами 82–254
растущая Луна 81–20
растяжение 73 67
расхождение по высоте 81–140
расширение вселенной 99–74
гидромагнитное ~ 88–159
~ земной тени 81–93
~ морского дна 72–39
ускоренное ~ 99–81
расширяющаяся вселенная 99–55
~ звезда 91–162
~ оболочка 93–12
~ корона 88–232
расширяющийся рукав 98–38
расщелина 81–165
расщепление кометы 85–111
~ линий вследствие расширения 97–157
приливное ~ 85–111
реакция захвата нейтрона 94–87
нейтринная ~ 94–92
~ нейтронного захвата 94–87
~ скалывания 94 100
ядерная ~ на поверхности 91–181
регмаглипты 87–122
реголит 82–104
лунный ~ 82–104
Регул 92–126
регуляризированный геоид 71–17
редукция за свободный воздух 71–78
~ наблюдений покрытий 81–108
~ Фая 71–78
Резец 92–23
рекогносцировка, орбитальная 82–31
~ поверхности 82–32
рекомбинационное излучение 86–212
рекомбинация, диссоциативная 86–214
рекристаллизация 74 42
ректификация 95–18
рекуррентная группа 89–108
рекуррентность группы пятен 88–133
рекуррентный выброс 98–282
релаксация 98 127
бесстолкновительная ~ 98–132
~ за счёт столкновений 98–131
столкновительная ~ 98–131
реликтовое магнитное поле 95–62
реликтовый кратер 81–209
релятивистская вселенная 99–92
~ космология 99–28
~ масса 99–141
релятивистский 99–95
~ газ 97–152
релятивистское вырождение 93–176
~ движение перигелия 99–160
~ приращение массы 99–142
~ сокращение 99–134
рентгеновская вспышка 89–183
~ галактика 98–203
~ гелиограмма 88 74
~ корона 88–240
рентгеновский пульсар 95–138
ресселова смесь 93–101
ретрорефлектор, лазерный 82–43
рефрактор без башни 88–53
решение, устойчивое 98–170
Ригель 92–127
риманова геометрия 99–113
риманово пространство 99–112
риометр 78–45
рисунок планеты 83–16
ров, сбросовый 82–132
рог магнитосферы 75–13
рога 83–130
~ Луны 81–24
~ серпа 83–130
~ серпа Луны 81–24
родительская галактика 98–284
~ комета 86–259
~ молекула 85–135
рождение пятна 89–121
роза ветров 77–116
рой, дневной 86–219
метеорный ~ 86–217
~ метеорных тел 86–217
ночной ~ 86–218
роса 77–173
рост планеты 83а–37
~ пылевых частиц 97–61
ротатор 91–42
быстрый ~ 91–43
наклонный ~ 95–60
ротационная неустойчивость 83а–8
рубидий-стронциевый возраст 82–271
рукав, галактический 98–26а
закрученный галактический ~ 98–29
закручивающийся галактический ~ 98–29
Орионов ~ 98–36
расширяющийся ~ 98–38
спиральный ~ 98–27
~ Стрельца 98–35
3 кпс ~ 98–37
русло 83–153
Рыбы 92–79
Рысь 92–64
Ряд, Международный Полярный 91–87
Северный Полярный ~ 91–87

С

самогравитация 94–34
самогравитирующая система 98–169
самогравитирующий 94–33
самообращение намагниченности 75–38
самописец, ионосферный 78–22
самопишущий анемометр 77–114
самосветящееся небесное тело 79–9а
самосветящийся 96–45
~ след 86–108

самосжатие, гравитационное 94–37
катастрофическое ~ 94–109
Сатурн 83–209
сатурнографический 83–212
сатурноцентрический 83–213
сближение, звёздное 98–121
теснейшее ~ 84–69
сбор метеоритов 87–45
~ метеорной пыли 87–12
сбрасывание, местное 82–204
региональное ~ 82–205
сброс 73–93
~ вещества 94–77
~ материи 95–124
нисходящий ~ 73–93
нормальный ~ 73–93
обратный ~ 73–94
обращённый ~ 73–94
обыкновенный ~ 73–93
простой ~ 73–93
сбросовая деятельность 73–92
местная ~ деятельность 82–204
~ мульда 82–132
сбросовый ров 82–132
сверхадиабатический градиент 93–99
сверхболид 86–9
сверхбыстрое космологическое раздувание 99–170c
сверхвеликое объединение 99–170g
Сверхгалактика 98–347
сверхгалактика 98–346
~ Девы 98–347
Местная ~ 98–347
сверхгигант 91–145
сверхгигантская галактика 98–223
сверхгиганты, менее яркие 91–148
наиболее яркие ~ 91–147
слабые ~ 91–148
ярчайшие ~ 91–147
сверхгранула 88–182
сверхмассивная звезда 91–127
~ чёрная дыра 98–22a
сверхмассивное тело 98–22
сверхмассивный объект 98–22
сверхновая = сверхновая звезда 95–126
бывшая ~ 95–134
внегалактическая ~ 95–128
галактическая ~ 95–127
~ до вспышки 95–129
~ звезда 95–126
~ после вспышки 95–134
~ типа I 95–130
~ Тихо Браге 95–135
сверхобъединение 99–170q
сверхобъединённое взаимодействие 99–170r
сверхплотная звезда 91–153
сверхсильное магнитное поле 95–148
сверхскопление галактик 98–346
Местное ~ 98–347
сверхтекучая жидкость 95–149
сверхтяжёлая частица 99–170n
свет, внеатмосферный 79–7
зодиакальный ~ 84–61
галактический ~ 79–14
диффузный галактический ~ 97–56
~ звёзд 79–10
звёздный ~ 79–10
ложный зодиакальный ~ 84–62
~ неба 77–53
~ ночного неба 79–8
пепельный ~ Луны 81–35
прямой ~ звёзд 79–11
пурпурный ~ 79–54
рассеянный ~ звёзд 79–12
рассеянный ~ неба 77–59
солнечный ~ 88–9
светило 79–9a
светимость 91–128
~ метеора 86–195
нейтринная ~ 94–96
светлый мост 89–104
световая ловушка 88–64
световой год 91–18
~ крест 79–79
~ столб 79–78
световые изменения 95–6
светящаяся материя 99–86
светящееся вещество 97–109
светящийся 96–45
свечение 79–151
авроральное ~ 79–146a
атмосферное ~ 79–2
атмосферное ~ дневного неба 79–3
атмосферное ~ ночного неба 79–5
атмосферное ~ сумеречного неба 79–4
~ атмосферы 79–2
водородное ~ 79–145
дневное ~ атмосферы 79–3
заатмосферное ~ 79–7
Lα ~ ночного неба 79–9
~ неба 77–53
ночное ~ атмосферы 79–5
собственное ~ атмосферы 79–2
сумеречное ~ атмосферы 79–4
свинец, обыкновенный 74–99
первоначальный ~ 74–98
~ радиоактивного происхождения 74–97
радиогенный ~ 74–97
свинцово-свинцовый возраст 74–100
свист, протонный 78–61
электронный ~ 78–60
свисты 78–53
свистящие 78–53
~ атмосферики 78–53
свободное падение 94–39
свободно-свободная абсорбция 93–106
свободно-связанное излучение 86–212
свободные колебания 73–49
~ колебания Земли 73–49
свободный радикал 85–137
свойства, интегральные 96–134
кинематические ~ 98–90
собственные ~ звезды 91–100
суммарные ~ 96–134
связанно-свободное поглощение 93–107
связанный со вспышкой 89–188
связи, солнечно-земные 84–55
связная почва 82–63
связный 82–138
~ грунт 82–63
связующий 82–138
связь, генетическая 97–127
~ излучения с веществом 99–173
~ между солнечными и земными явлениями 84–55
~ посредством радиации 99–173
радиационная ~ 99–173
случайная ~ 97–126
~ Солнце-Земля 84–55
сгусток протуберанца 89–244
сгущение 96–127
пылевое ~ 98–290
центральное ~ 85–37, 86–244
сдвиг 73–91
~ коры 82–212
сдвигающее напряжение 73–63
сдувание расплавленного слоя 86–182
Северная Корона 92–40
северное небо 79–3b
~ полушарие неба 79–3b
Северный Полярный Ряд 91–87
седиментация 74–45
седло 73–90
сезонные изменения 83–166
сейсма, метеоритная 87–34
сеймическая волна 72–95
~ зона 73–31
~ катастрофа 73–40
~ область 73–28
~ станция 73–6
сейсмический луч 73–14
~ очаг 73–17
~ пояс 73–30
~ район 73–28
~ фон 73–55
~ шум 73–54
сейсмическое эхо 82–223
сейсмичность 73–3
сейсмограмма 73–8
~ падения 82–224
~ падения тела на Луну 82–224
сейсмограф 73–7
сейсмозапись 73–8
сейсмолог 73–2
сейсмология 73–1
активная ~ 82–217
лунная ~ 82–213
~ Луны 82–213
пассивная ~ 82–218
сейсмометр, активный 82–222
пассивный ~ 82–221

R

сейсмоприёмник 73–7
сейфертовская галактика 98–262
Секстант 92–90
сектор, вращающийся 86–147
межпланетный ~ 84–16
межпланетный магнитный ~ 84–16
секторная граница 84–18
~ структура 84–15
селенографическая долгота 81–10
~ широта 81–11
селенографические координаты 81–9
селенографический 81–8
селенография 81–1
селенодезический 81–110
селенодезия 81–109
селеноид 81–117
селенолог 82–49
селенологическая съёмка 82–50
селенологический 82–48
селенологическое картирование 82–51
~ обозрение 82–50
селенология 82–1
селеномагнитное поле 82–45
селенофизика 82–151
селенофизическая сеть регистрации 82–40
~ съёмка 82–39
селенохимия 82–74
селеноцентрический 81–131
семейство астероидов 84–73
~ комет 85–102
~ комет Юпитера 85–103
~ малых планет 84–73
~ Юпитера 85–103
сердцевина, углеродная 97–71
серебристое облако 79–89
середина затмения 81–78
сернистокислая соль 74–28а
сернистый водород 97–44
~ металл 74–28
сернокислая соль 74–29
сероводород 97–44
серп Венеры 83–129
лунный ~ 81–23
~ Луны 81–23
серповидная фаза Луны 81–23
сетка 86–129, 92–83
хромосферная ~ 89–171
сеть, геодезическая спутниковая 71–34
гравиметрическая ~ 71–80
мировая ~ 77–18
мировая геодезическая ~ 71–38
мировая гравиметрическая ~ 71–81
опорная гравиметрическая ~ 71–82
селенофизическая ~ регистрации 82–40
~ силы тяжести 71–80
сечение столкновения 84–103
геометрическое ~ столкновения 84–103
гравитационное ~ 84–104
гравитационное ~ столкновения 84–104
дифференциальное ~ 94–54
лобовое ~ 86–176
миделево ~ 86–176
парциальное ~ 94–54
сжатие 93–117, 94–36
видимое ~ 98–295
~ вселенной 99–180
гравитационное ~ 93–118
~ Земли 71–8
истинное ~ 98–296
катастрофическое ~ 94–109
квазистатическое ~ 94–40
квазистационарное ~ 94–40
~ короны 88–247
полярное ~ 83–114
~ шарового скопления 96–123
~ ядра 94–70
сжатый сфериод 98–158
сжимаемость 73–57
сжимающаяся вселенная 99–56
~ звезда 91–161
~ модель 99–56
сигарообразная галактика 98–239
сигарообразный астероид 84–77
сигнал, отражённый 86–290
сидерит 87–82
сидерический месяц 81–45
сидерическое обращение 81–45
сидеролит 87–78
сидеростат 88–44
сизигии 81–29
сизигийный прилив 82–273
сила ветра 77–107
внутренняя ~ 73–61
задерживающая ~ 75–39
~ Кориолиса 77–79
кориолисова ~ 77–79
коэрцитивная ~ 75–39
магнитная подъёмная ~ 86–147
максимальная ~ света 86–78
отклоняющая ~ 77–80
~ отталкивания 85–87
отталкивательная ~ 85–87
отталкивающая ~ 85–87
приливная ~ 82–289
приливообразующая ~ 82–289
приливо-отливная ~ 82–289
тектоническая ~ 73–106
~ тяжести 71–56
силикат 74–14
силикатная пылинка 97–76
~ пыль 85–53
силл 73–135
сильно сплюснутая галактика 98–252
симбиотическая звезда 95–50
симметрия, винтовая 98–161
сферическая ~ 98–160
цилиндрическая ~ 98–155
симпатическая вспышка 89–179
сингулярная галактика 98–258
сингулярное состояние 99–42
сингулярность 94–116
голая ~ 94–117
синдинама 85–91
синева неба 77–60
синклиналь 73–89
синклинальная складка 73–89
синодический месяц 81–46
синоптическая карта 77–11
~ метеорология 77–8
синтез, апертурный 88–118
~ апертуры 88–118
взрывной ядерный ~ 94–57
космологический ~ ядер 99–175
невзрывной ядерный ~ 94–58
~ элементов 94–52
ядерный ~ 94–52
синхрона 85–92
концевая ~ 85–93
полная ~ 85–94
синхронное вращение 83–37
Сириус 92–128
система апертурного синтеза 88–119
арочная ~ 89–299
бесстолкновительная ~ 98–167
~ в нестационарном состоянии 98–152
~ в стационарном состоянии 98–141
~ вихрей 83а–13
внегалактическая ~ 98–210
вращающаяся ~ отсчёта 98–138
галактическая ~ 98–3
гарвардская ~ спектральный классов 91–91
гелиоцентрическая ~ 99–23
геодезическая ~ относимости 71–19
геоцентрическая ~ 99–21
гравитационно несвязанная ~ 96–158
гравитационно связанная ~ 96–156
гравитирующая ~ 98–169
девонская ~ 74–81
~ для умеренных широт 83–179
~ долгот 83–177
~ долгот II 83–179
~ долгот для умеренных широт 83–179
~ долгот I 83–178
~ дуг 89–299
~ звёзд 96–1
звёздная ~ 96–1
звёздная ~ образованная наложением 98–165
~ звёздных величин 91–57
~ Земля-Луна 81–12
~ изломов 73–79
инерциальная ~ отсчёта 99–98
каменноугольная ~ 74–82
кембрийская ~ 74–78
~ колец 98–291
~ колец Сатурна 83–220

R

компактная ~ 96–157
континентальная ~ разломов 73–81
~ координат 98–136
~ Коперника 99–24
коперникова ~ 99–24
космологическая ~ 99–20
лучевая ~ 81–170
меловая ~ 74–86
местная ~ 98–61
местная ~ отсчёта 98–139
~ мира 99–20
мировая ~ 99–20
неинерциальная ~ 99–147
неустойчивая ~ 98–171
океаническая ~ 73–80
океанская ~ разломов 73–80
ордовикская ~ 74–79
~ отсчёта 98–136
пермская ~ 74–83
планетная ~ 83–2
планетоцентрическая ~ 83–10
планетоцентрическая ~ координат 83–10
плоская ~ 98–251
~ полос 79–18
птоломеева ~ 99–22
~ Птоломея 99–22
~ разломов 81–174
~ разрывов 73–79
~ с осевой симметрией 98–157
~ с отрицательной полной энергией 96–156
~ с положительной полной энергией 96–158
самогравитирующая ~ 98–169
~ сатурновых колец 83–220
силурийская ~ 74–80
сильно сжатая ~ 98–164
солнечная ~ 83–12
сопутствующая координатная ~ 99–44
~ спутников 83–19
спутниковая ~ 83–19
срединно-океанская ~ разломов 73–80
столкновительная ~ 98–166
сфероидальная ~ 98–159
третичная ~ 74–87
~ трещин 81–174
триасовая ~ 74–84
тройная ~ 96–4
~ U, B, V 96–144
фундаментальная ~ отсчёта 98–137
~ цветов 96–140
цветовая ~ 91–72
четвертичная ~ 74–88
~ шаровых скоплений 98–75
экваториальная ~ долгот 83–178
~ электрических токов 75–97
юрская ~ 74–85
сияние, искусственное полярное 79–99
низкоширотное полярное ~ 79–96
пламенное полярное ~ 79–157
полярное ~ 79–93
полярное ~ в виде пламени 79–157
полярное ~ в полярной шапке 79–98
полярное ~ излучающее радиоволны 79–94
полярное ~ на дневной стороне 79–97
полярное ~ наблюдённое с помощью радиолокатора 79–95
полярное ~ освещённое Солнцем 79–130
полярное ~ с ленточной структурой 79–140
протонное полярное ~ 79–108
радарное полярное ~ 79–95
северное полярное ~ 79–105
солнечное ~ 77–52
электронное полярное ~ 79–110
южное полярное ~ 79–106
скала 74–1
лунная ~ 82–65
~ с острыми углами 82–131
скалывание 94–100
~ ядра космическими лучами 82–254
сканирование по длине волны 88–111
~ по небу 88–105
скат 81–234
скважина, буровая 82–61
скважность 83–100
складка 73–87
антиклинальная ~ 73–90
синклинальная ~ 73–89
складки 79–139
складкообразные горы 73–86
складкообразование 73–88
складчатость 73–88
складчатые горы 73–86
складчатый гребень 81–187
~ хребет 81–187
склон 81–230
внутренний ~ вала 81–252
континентальный ~ 72–35
крутой ~ 81–232
материковый ~ 72–35
~ покрытый обломками 81–233
пологий ~ 81–231
склонение Луны 81–62
магнитное ~ 75–56
~ Солнца 81–63
сколлапсировавшая звезда 94–111
сколлапсировавший объект 98–23
сколлапсированный объект 98–23
скольжение 82–196
~ пепла 82–174
скопление 83а–32
бедное ~ галактик 98–331
~ бедное звёздами 96–102
бедное звёздное ~ 96–102
богатое ~ галактик 98–332
~ богатое звёздами 96–100
богатое звёздное ~ 96–100
~ в Деве 98–345
~ внегалактических туманностей 98–329
~ галактик 98–329
~ галактик в Деве 98–345
галактическое звёздное ~ 96–99
движущееся звёздное ~ 96–97
двойное ~ 96–81
~ звёзд 96–79
звёздное ~ 96–79
компактное ~ галактик 98–334
~ куполов 81–178
неправильное ~ галактик 98–337
открытое звёздное ~ 96–99
очень компактное ~ галактик 98–335
правильное ~ галактик 98–336
рассеянное ~ галактик 98–333
рассеянное звёздное ~ 96–99
релятивистское звёздное ~ 96–154
~ с высоким содержанием металлов 96–112
~ с низким содержанием металлов 96–113
~ с сильным сгущением к центру 96–129
~ скоплений 98–346
шаровое ~ галактик 98–330
шаровое звёздное ~ 96–114
шарообразное звёздное ~ 96–114
скорость, атмосферная 86–94
~ в атмосфере 86–94
~ в середине пути 86–97
~ ветра 77–109
~ влёта в атмосферу 86–92
внеатмосферная ~ 86–93
вращательная ~ 91–48
~ вращения 91–48
~ входа 86–92
~ выброса 85–90
~ выделения 85–143
~ выделения газов 85–144
~ газовыделения 85–144
гелиоцентрическая ~ 86–96
геоцентрическая ~ 86–95
~ звёздообразования 94–12
~ извержения 85–90
~ испускания массы 85–146
лучевая ~ 91–24
мгновенная ~ 86–99
мгновенная угловая ~ 86–90
~ метеора 86–86
начальная ~ 86–92
~ образования 85–143
~ образования кратеров 82–233
орбитальная ~ 96–63
остаточная ~ 98–91
отрицательная ~ 91–27
~ охлаждения 93–203
~ падания блеска 95–36
~ перемещения изображения 86–154

R

~ повторения импульсов 86–279
~ подъёма к максимуму 95–29
положительная ~ 91–26
~ приближения 91–27
пространственная ~ 91–30
~ расширения 96–94
~ релаксации 98–129
релятивистская ~ 99–136
~ солнечного ветра 88–158
средняя ~ 86–98
средняя угловая ~ 86–91
~ таяния 83–147
~ термоядерной реакции 94–53
~ убегания из системы 96–155
угловая ~ 86–89
~ удаления 91–26
~ умирания звезд 94–115
~ ускользания из системы 96–155
характеристическая ~ 76–43
экваториальная ~ 91–49
~ эмиссии массы 85–146
Скорпион 92–86
скручивание галактических рукавов 98–28
скрытая масса 99–88
Скульптор 92–87
скучивание вещества 99–170z
слабая звезда 91–82
сланцеватость 74–53
след, весьма кратковременный 86–110
газовый ~ 86–116
длительный ~ 86–113
дневной ~ 86–106
долговременный ~ 86–113
древний ~ 87–121
дымовой ~ 87–21
звёздный ~ 86–152
~ звезды 86–152
~ звезды на пластинке 86–152
кратковременный ~ 86–112
~ метеора 86–103
~ метеора на пластинке 86–29
метеорный ~ 86–103
неустойчивый ~ 86–109
ночной ~ 86–105
~ охлаждения 93–202
первичный ядерный ~ 82–252
пылевой ~ 86–115, 87–22
~ радиационного разрушения 87–120
самосветящийся ~ 86–108
светящийся ~ 86–108
стойкий ~ 86–104
стойкий ~ метеора 86–104
~ частицы 82–251
следы ног 82–38
слежение за спутниками 83–26
сложная группа 89–112
слои атмосферы, верхние 76–52
верхние ~ солнечной атмосферы 88–165
внешние ~ звезды 93–3
наружные ~ звезды 93–3
невырожденные внешние ~ 93–199
нижние ~ солнечной атмосферы 88–166
слоистая поверхность 83–101
~ порода 74–48
слоисто-дождевые облака 77–145
слоисто-кучевые облака 77–142
слоистость 82–133
горизонтальная ~ 82–134
слоистые облака 77–139
слой, атмосферный 76–49
~ атмосферы 76–49
галактический поглощающий ~ 97–89
~ дымки 83–249
~ E 78–26
излучающий ~ 79–17
ионосферный ~ 78–5
~ Кеннелли-Хевисайда 78–26
нейтральный ~ 75–16
облачный ~ 77–150, 83–248
обращающий ~ 88–174
озонный ~ 76–123
переходный ~ 72–13
переходный ~ между хромосферой и короной 88–224
плазменный ~ 75–11
~ повышения вверх 76–54
~ повышения температуры вверх 76–54
поглощающий ~ 88–175
подфотосферный ~ 88–168
~ понижения вверх 76–57
~ понижения температуры вверх 76–57
пылевой ~ 82–122
~ пыли 82–122
рассеивающий ~ 83–258
~ свечения 79–17
спорадический ~ E 78–25
~ тумана 83–249
фиолетовый ~ 83–165
фотоэлектронный ~ 82–246
~ Хевисайда 78–26
служба наблюдения за льдами 72–85
~ погоды 77–5
~ прогнозов 77–5
~ Солнца 89–16
случай загрязнения 72–72
случайная связь 97–126
случайно распределённый 98–93
случайное блуждание 89–157
слюда 74–23
белая ~ 74–24
обыкновенная ~ 74–24
чёрная ~ 74–25
смежное событие 99–126
сместитель линий 88–69
смесь газа и пыли 83a–25
~ излучения и газа 93–125
смешанный вулкан 73–121
смещение, аномальное красное 98–367
гравитационное красное ~ 99–162
доп(п)леровское ~ 91–25
~ зоны пятен 89–74
~ зоны пятнообразования 89–74
космологическое красное ~ 98–363a
красное ~ 98–297
многократное ~ 98–364
недоп(п)леровское красное ~ 98–363
периодическое ~ линий 96–22
~ протуберанцев к полюсам 89–231
~ радианта 86–242
суточное ~ радианта 86–69
смог 76–83
смыв 73–107
смывание 73–107
снег 77–175
снеговая граница 72–111
~ линия 72–111
снежный покров 77–176
снимок метеора на пластинке 86–29
метеорный ~ 86–151a
панорамный ~ 82–27
широкоугольный ~ 82–27
снос 73–108
СНЧ 78–55
собирание образцов 82–72
собрание метеоритов 87–8
собственное время 99–139
~ движение 91–19
~ движение по прямому восхождению 91–21
~ движение пятен 89–117
~ излучение планеты 83–198
собственный показатель цвета 91–75
событие 99–125
~ в рентгене 89–184
~ в рентгеновских лучах 89–184
~ в солнечном рентгене 89–184
~ в солнечных космических лучах 89–194
одновременное ~ 99–128
протонное ~ 89–192
сейсмическое ~ 82–215
совершенный космологический принцип 99–32
совращающаяся структура 84–17
~ структура магнитного поля 84–17
содержание, аномальное 91–173
звёздное ~ 98–172
~ металлов 91–170
~ никеля 72–55
нормальное ~ 91–172
~ озона в атмосфере 76–125
общее ~ 91–177
относительное ~ элементов 91–174
~ хлора 72–68
~ элемента 91–169
соединение 83–55
верхнее ~ 83–56

R

геодезическое ~ материков 71–37
нижнее ~ 83–57
тройное ~ 83–58
созвездие 92–1
близполюсное ~ 92–4a
~ зимнего неба 92–1b
зодиакальное ~ 92–5
~ летнего неба 92–1a
околополюсное ~ 92–4a
околополярное ~ 92–4a
сокращение длин 99–134
лоренцево ~ 99–134
магнитное ~ 89–160
перспективное ~ пятен 89–92
релятивистское ~ 99–134
~ пятен 89–92
солёность 72–66
солнечная активная область 89–57
~ активность 89–1
~ башня 88–54
~ вспышка 89–175
~ планета 83–11
~ постоянная 88–148
~ радиация 88–134
~ сейсмология 88–223b
~ система 83–12
~ туманность 83a–23
~ экспедиция 88–30
солнечное затмение 88–15
~ излучение 88–134
~ образование 89–13
~ пятно 89–88
~ сияние 77–52
солнечно-земные связи 84–55
солнечные данные 89–9
~ инструменты 88–35
~ колебания 88–223a
~ осцилляции 88–223a
солнечный 88–5
~ ветер 88–156
~ диск 89–24
~ искатель 88–51
~ корпускулярный поток 84–35
~ крест 79–81
~ магнитограф 88–82
~ максимум 89–2
~ минумум 89–4
~ свет 77–52
~ телескоп 88–49
~ фотогид 88–51
Солнце 88–1
активное ~ 89–12
~ в радиоволнах 88–3
возмущённое ~ 89–12
голубое ~ 79–92
ложное ~ 79–71
невозмущённое ~ 89–11
нижнее ~ 79–72
нижнее побочное ~ 79–72
оптическое ~ 88–2
побочное ~ 79–71
рентгеновское ~ 88–4
спокойное ~ 89–11
соль, сернистая 74–28
сернокислая ~ 74–29
углекислая ~ 74–27
сообщение о погоде 77–10
соотношение масса-светимость 91–129
~ скорость-глубина 73–52
сопротивление, воздушное 86–178
лобовое ~ 86–178
сопутствующая координатная система 99–44
состав, звёздный 98–231
звёздный ~ галактики 98–231
изотопический ~ 87–129
изотопный ~ 87–129
молекулярный ~ 78–11
химический ~ 91–168
составление селенологической карты 82–51
составляющая 96–36
атмосферная ~ 76–87
атмосферная ~ свечения ночного неба 79–5
горизонтальная ~ магнитного поля 75–60
лунно-солнечная ~ 82–307
малая ~ атмосферы 76–88
метеорная ~ свечения ночного неба 86–234
плоская ~ 98–71
поперечная ~ 89–136
промежуточная ~ 98–70
пылевая ~ 98–232
сферическая ~ 98–69
состояние моря 72–86
~ неба 77–132
нестационарное ~ 98–151
общее ~ погоды 77–15
особое ~ 99–42
~ погоды 77–14
сингулярное ~ 99–42
сотовый алюминий 82–25
спавшая звезда 91–164
спадающая звезда 91–164
спектр вспышки 88–25
главный ~ 95–115
двойной ~ 96–18
динамический ~ 88–125
звёздный ~ 91–88
~ звезды 91–88
интегральный ~ 96–137
комбинированный ~ 95–117
~ метеорного следа 86–107
неискажённый звёздный ~ 91–89
нормальный звёздный ~ 91–89
одиночный ~ 96–16
симбиотический ~ 95–117
сложный ~ 96–19
суммарный ~ 96–137
фраунгоферов ~ 88–137
спектральная двойная 96–14
~ классификация 91–90
~ особенность 91–93
~ последовательность 91–94
спектрально-двойная 96–14
~ звезда 96–14
~ с двойными линиями в спектре 96–17
~ с одиночными линиями в спектре 96–15
спектрально-переменная 95–48
~ звезда 95–48
спектральные изменения 95–49
спектральный класс 91–107
~ параллакс 91–15
~ тип 91–108
спектроболограмма 88–76
спектроболометр 88–75
спектрогелиограмма 88–72
спектрогелиграф 88–71
спектрогелиокинематограф 88–81
спектрогелископ 88–67
спектрограмма, динамическая 88–125
спектрограф, вертикальный 88–79
горизонтальный ~ 88–78
динамический ~ 88–124
метеорный ~ 86–146
солнечный ~ 88–77
спектрография, мессбауэровская 82–142
ядерная резонансная ~ 82–142
спектрометр, динамический 88–124
~ солнечного ветра 82–46
спектрометрия, гамма 82–77
недеструктивная гамма ~ 82–77
спекторегистратор скоростей 88–57
спектроскопическая двойная 96–14
~ двойная звезда 96–14
~ орбита 96–56
спектроскопия, гамма 82–77
неразрушающая гамма ~ 82–77
спектрофотометрический градиент 91–112
специальная теория относительности 99–94
специальный показатель цвета 83–87
Спика 92–129
спикула 88–195
спираль 98–7, 98–249
~ Архимеда 84–34
архимедова ~ 84–34
~ без перемычки 98–250
нормальная ~ 98–250
обычная ~ 98–250
спиральная галактика 98–249
~ галактика с перемычкой 98–253
спиральный выброс 85–59
~ рукав 98–27
спирт, метиловый 97–32
сплюснутость Земли 71–8
~ короны 88–247
~ шарового скопления 96–123
сплющенность 83–113

R

сплющивание 83а–22
спокойное Солнце 89–11
спокойный день 75–92
~ квазизвёздный объект 98–362
~ солнечный ветер 84–41
сползание 82–195
спорадическая конденсация 89–283
спорадический метеор 86–23
~ поток 86–226
спорадическое радиоизлучение Юпитера 83–181
способ степеней 95–10
степенной ~ 95–10
способность, несущая 82–64
СПР 91–87
спуск 89–199
спусковое действие 89–199
спутник 83–17, 96–38
ближайший ~ 83–20
бывший ~ 83–240
~ галактики 98–216
галилеевский ~ 83–204
естественный ~ 83–18
~ из каменных пород 83–22
~ из скальных пород 83–22
иррегулярный ~ 83–21
каменный ~ 83–22
кратер- ~ 81–200
ледяной ~ 83–23
невидимый ~ 96–41
обратный ~ 83–21
оптический ~ галактики 98–218
планета- ~ 96–43
планетарный ~ 96–43
планетный ~ 96–43
~ с обратным движением 83–21
~ Сириуса 91–138
тёмный ~ 96–42
физический ~ галактики 98–219
эллиптический ~ галактики 98–217
спутниковая альтиметрия 71–50
~ климатология 77–181
~ система 83–19
спящий вулкан 73–116
среда, межзвёздная 97–8
мутная ~ 83–251
окружающая ~ 76–3
цементирующая ~ 82–118
средиземное море 72–26
срединно-океанический хребет 72–45
среднее время релаксации 98–130
среднезернистый 82–116
средний молекулярный вес 93–100
~ радиус 83–116а
срезывающее напряжение 73–63
срок испарения 96–109
стабильность 93–204
стадия гало 94–19
дисковая ~ 94–22
~ до главной последовательности 94–49
гидродинамическая ~ 94–43
инфляционная ~ 99–170d
~ космологического раздувания 99–170d
небулярная ~ 95–125
~ после главной последовательности 94–68
промежуточная ~ 94–21
~ раздувания 99–170d
~ сжатия 94–20
~ Т-Тельца 94–47
~ эволюции 94–28
эволюционная ~ Хаяши 94–44
сталкивающиеся галактики 98–321
стандартная модель 93–137
станция, автоматическая 82–19
автоматическая лунная ~ 82–19
высокогорная ~ 88–56
лунная ~ 82–17
метеорная ~ 86–158
обитаемая лунная ~ 82–18
сейсмическая ~ 73–6
сейсмографическая ~ 73–6
спутниковая ~ 71–33
старая звезда 94–64
статистика, звёздная 98–77
статистически стационарное состояние 98–148
статистический параллакс 91–8
статистическое равновесие 93–42
статическая вселенная 99–50
~ модель 99–50
стационарная вселенная 99–52
~ звезда 95–3
~ модель 99–52
~ оболочка 93–11
стационарное состояние 98–140
стационарный метеор 86–34
~ потенциал 98–147
стволы деревьев 89–272
стекание ледника 72–120
стекло, дымчатое 88–33
лунное ~ 82–96
стекловатый обломок 82–97
стекловидный фрагмент 82–97
стеклянные гранулы 82–127
~ сферулы 82–127
~ шарики 82–127
стена, внутренняя 81–252
~ кратера 81–251
Прямая ~ 81–185
степень вырождения 93–174
~ концентрации 96–128
~ облачности 77–147
~ сгущения 96–128
стереокомпаратор 91–32
СТО 99–94
Стожары 96–110
стойкий след 86–104
сток, ледниковый 72–120
стоковый ледник 72–125
Стоксовы постоянные 71–59
столб, световой 79–78
солнечный ~ 79–78
столкновение в лоб 83а–17
~ галактик 98–322
звёздное ~ 83а–16
катастрофическое ~ 84–106
~ кометы 82–230
неупругое ~ 98–26
неупругое ~ звёзд 98–26
тройное ~ 79–30
центральное ~ 83а–17
эрозионное ~ 84–105
столкновительная накачка 97–20
~ система 98–166
столкновительный распад 87–125
Столовая Гора 92–66
сторона, ближайшая 81–15
видимая ~ Луны 81–15
дневная ~ 81–34
задняя ~ Луны 81–16
~ Земли обращенная к Солнцу 75–20
невидимая ~ Луны 81–16
неосвещённая ~ 75–21
ночная ~ 75–21
обратная ~ Луны 81–16
обращённая к Земле ~ Луны 81–15
~ обращённая к Солнцу 81–34
освещённая ~ 81–126
~ освещённая Солнцем 81–34
передняя ~ 81–15
солнечная ~ 83–67
теневая ~ 83–68
теневая ~ Земли 75–21
стояние планеты 83–44
стратиграфический возраст 82–267
стратификация 97–154
температурная ~ 77–40
стратовулкан 73–121
стратокумулус 77–142
стратопауза 76–71
стратосфера 76–70
стратус 77–139
Стрела 92–84
Стрелец 92–85
стресс 73–62
~ в породах коры 82–210
строение 82–102
внутреннее ~ 71–83
внутреннее ~ звёзд 93–71
внутреннее террасовое ~ 81–256
~ Галактики 98–4
~ звёзд 93–1
~ комет 85–34
структура 82–102
арочная ~ 88–249
волокнистая ~ 97–130
волоконцевая ~ 89–174
доморская ~ 81–153
иерархическая ~ 99–47
кольцевая ~ 81–155
~ комет 85–34

R

коротирующая ~ 84–17
коротирующая ~ магнитного поля 84–17
крупномасштабная ~ 88–207
ламинарная ~ 83–148
мелкомасштабная ~ 88–208
~ метеоритов 87–100
~ метеорного потока 86–243
~ Млечного Пути 98–4
предморская ~ 81–153
послеморская ~ 81–154
секторная ~ 84–15
секторная ~ магнитного поля 84–15
слоистая ~ 83–148
совместно вращающаяся ~ 84–17
совращающаяся ~ 84–17
спиральная ~ 98–8
тонкая ~ хромосферы 88–203
ячеистая ~ 99–171a
структурные формы 81–152
структурный индекс скопления 96–80
~ параметр 96–120
~ тип 98–225
струна, космическая 99–171c
струя 89–245
оптическая ~ 98–279
~ протуберанца 89–245
~ электронов 79–114
ступень, барическая 76–19
геотермическая ~ 74–59
ступенчатый ослабитель 88–80
субавроральный пояс 79–128
суббуря 79–144
авроральная ~ 79–141
магнитная ~ 75–95
полярная магнитная ~ 75–96
субвспышка 89–195
субгиганты 91–151
субимпульс 95–145
субкарлик 91–140
сублимация 85–138
субполярный пояс 79–128
субсистема 98–74
субтропические широты 77–102
субъядро 72–20
сульфат 74–29
сульфид 74–28
карбонил-~ 97–36
карбонильный ~ 97–36
сульфит 74–28a
сумеречная дуга 79–50
сумеречное зондирование 79–41
~ небо 79–43
~ свечение атмосферы 79–4
сумеречные явления 79–32
сумеречный 79–40
сумерки 79–34
астрономические ~ 79–37
вечерние ~ 79–36
гражданские ~ 79–38
навигационные ~ 79–39
утренние ~ 79–35
сумма осадков 77–168
суммарная звёздная величина 96–136
~ масса скопления 96–139
суммарные свойства 96–134
сумрак 79–36
суп, кварковый 99–169a
супергранула 88–182
супергранулярный элемент 88–182
супергрануляция 88–183
суспензионный поток 72–49
сутки, лунные 81–32
приливные ~ 82–290
средние лунные ~ 82–304
суточная волна 76–30
~либрация 81–52
суточное движение 85–29
суточные изменения 83–167
суфляр 82–163
сухая мгла 77–27
сфера, небесная 92–8
политропная газообразная ~ 93–127
сферическая галактика 98–244
~симметрия 98–160
сфероид, земной 71–13
сфероидальная галактика 98–244
~система 98–159
сферулы 87–29
стеклянные ~ 82–127
съёмка, геомагнитная 75–64
~льдов 72–85
магнитная ~ 75–64
маршрутная ~ 82–35
Мировая магнитная ~ 75–65
~ морских берегов 72–82
научная ~ 82–36
селенологическая ~ 82–50
селенофизическая ~ 82–39
фотографическая ~ 82–30

Т

таблица развития солнечных пятен 89–120
таласоид 81–175
тангенциальное напряжение 73–63
Т-ассоциация 96–90
Тауриды 86–254
твёрдая кора 95–151
~ частица 83a–27
твердотельное вращение 98–111
твик 78–63
творение материи 99–89
непрерывное ~ 99–91
текстура 82–102
~течения 82–191
флуидальная ~ 82–191
тектит 87–95
каплеобразный ~ 87–96
тектоника 73–95
тектоническая карта 73–96
тектонические движения 73–72
~ силы 73–106
тектоническое землетрясение 73–2d
текучесть лавы 82–188
телеграмма, астрономическая 85–20
телеметеор 86–18
очень слабый ~ 86–20
слабый ~ 86–19
телескоп 92–92
автоматический солнечный ~ 88–50
башенный ~ 88–54
горизонтальный солнечный ~ 88–55
солнечный ~ 88–49
солнечный ~ без купола 88–53
солнечный ~ на баллоне 88–52
стратосферный солнечный ~ 88–52
хромосферный ~ 88–73
телескопическая комета 85–14a
Телец 92–91
теллурическая линия 88–140a
теллурические токи 75–5
тело, абсолютно твёрдое 83–111
вращающееся ~ 91–42
вращающееся магнитоплазменное ~ 98–25
дозвёздное ~ 94–15
жёсткое ~ 83–111
исходное ~ 87–123
материнское ~ 87–123
метеоритное родительское ~ 87–123
метеорное ~ 86–160a
небесное ~ 79–9a
приливообразующее ~ 82–284
самосветящееся небесное ~ 79–9a
сверхмассивное ~ 98–22
твёрдое ~ 83–110
ударяющее ~ 82–225
упавшее ~ 82–225
тёмный край 81–97
темп звездообразования 94–12
~умирания звёзд 94–115
температура 77–29
~в тени 77–34
~возбуждения 91–118
вращательная ~ 91–119
~вырождения 93–188
диссоциационная ~ 91–120a
ионизационная ~ 91–117
кинетическая ~ 91–115
колебательная ~ 91–120
комнатная ~ 77–31
~конденсации 97–66
окружающая ~ 77–32
~плавления 86–204
поверхностная ~ 93–88
~размагничения 75–41
~таяния 86–204
цветовая ~ 91–109
центральная ~ 93–89
электронная ~ 91–116
эффективная ~ 91–113

R

яркостная ~ 91–114
температурная инверсия 77–36
~стратификация 77–40
температурный градиент 93–96
теневая сторона 83–68
тензор деформации 73–70
~кривизны 99–159
~материи 99–158
метрический ~ 99–157а
~напряжения 73–69
~Римана 99–159
~Римана-Кристоффеля 99–159
фундаментальный ~ 99–157а
~энергии 99–158
~энергии-импульса 99–158
тени, бегущие 88–31
тень Земли 81–58
земная ~ 81–58
озонная ~ 76–127
~пятна 89–95
теорема вириала 98–181
~Клеро 71–77
~о вириале 98–181
~однозначности 93–83
~Рессела-Фогта 93–83
теоретическая кривая роста 93–70
теория аккреции 83а–20
~Веицзеккера 83а–11
вихревая ~ 83а–11
~геомагнитного динамо 75–99
~деления 96–75
~захвата 96–72
~контракции 73–77
контракционная ~ 73–77
механическая ~ кометных форм 85–84
~наращивания 83а–20
~образования отдельных ядер 96–76
общая ~ относительности 99–146
~приливов 82–305
~сжатия 73–77
~спиральной структуры 98–173
~тепловой смерти вселенной 99–15
турбулентная ~ 83а–11
физическая ~ метеоров 86–159
~циклов 97–156
частная ~ относительности 99–94
тепличная модель 83–262
тепличный эффект 83–263
тепло конденсации 77–155
первичное ~ 83–200
радиоактивное ~ 83а–62
радиогенное ~ 83а–62
скрытое ~ 77–157
тепловая история 83–60
~карта Луны 81–135
~смерть вселенной 99–14
~эволюция 83а–61
тепловое давление 94–41
~напряжение 87–27
~равновесие 93–78
тепловой баланс атмосферы 77–62
~показатель 91–81
~поток 74–60
~удар 86–206
теплопередача 86–202
теплота абляции 86–184
~вызванная приливом 82–287
~десорбции 85–124
~испарения 86–199
~конденсации 77–155
скрытая ~ 77–157
~сублимации 85–139
удельная ~ испарения 86–199
терелла 79–133
термальный метаморфизм 74–37
терминальный кратер 73–129
терминатор 83–69
вечерний ~ 81–125
земной ~ 81–91
лунный ~ 81–123
~ на рассвете 81–124
~ при восходе Солнца 81–124
~ при заходящем Солнце 81–125
сумеречный ~ 81–125
утренний ~ 81–124
термическая диссипация 94–76
~ история 83а–60
~ эволюция 83а–61
термические приливы 82–299
термическое напряжение 87–27
термодинамическое равновесие 93–35
термолюминесценция 82–249
термометаморфизм 74–37
термоостаточная намагниченность 75–30
термосфера 76–59
термоядерная эволюция 94–50
терра 81–190
терраса, кратерная 81–229
террасированная депрессия 81–162
террасовое понижение 81–162
террасовый вал 81–257
терригенные осадки 72–48
~ отложения 72–48
тесная двойная звезда 96–23
~ пара 96–23
теснейшее сближение 84–69
тессеральные гармоники 95–102
течение, восходящее 77–91
конвективное ~ 73–76
~ ледника 72–120
морское ~ 72–60
океаническое ~ 72–60
отливное ~ 72–65
~ пепла 82–174
пластическое ~ 72–121
поверхностное ~ 72–62
~ под поверхностью 72–61
приливное ~ 82–279
приливно-отливное ~ 82–279
струйное ~ 77–88
тёплое ~ 72–63
тёплое морское ~ 72–63
холодное ~ 72–64
холодное морское ~ 72–64
тиоформальдегид 97–43
тип звёзд 91–101
~ кометного хвоста 85–71
морфологический ~ 98–228
~ полярного сияния 79–136
спектральный ~ 91–108
структурный ~ 98–225
типичная новая 95–111
~ новая звезда 95–111
Титан 83–233
титанистый железняк 82–91
тихий квазизвёздный объект 98–362
тождества Лагранжа 98–182
обобщённые ~ Лагранжа 98–183
ток, экваториальный кольцевой 75–98
токи, земные 75–5
земные электрические ~ 75–5
кольцевые ~ 75–100
теллурические ~ 75–5
толчки, предварительные 73–10
толчок землетрясения 73–2а
затухающий ~ землетрясения 73–13
последующий ~ землетрясения 73–13
толщина кольца 83–229
тонкая пластинка 83–149
тонкий слой 83–149
тонкозернистые частицы лунного грунта 82–112
тонкозернистый 82–114
топографический элемент 81–149
топография, лунная 81–146
~ Луны 81–146
тор, водородный 83–230
торий-свинцовый метод определения возраста 82–270
торможение, аэродинамическое 86–178
видимое ~ 86–100
~ вращения 91–45
истинное ~ 86–101
магнитное ~ 91–47
торос 72–102
точка взрыва 86–122
внутренняя ~ Лагранжа 96–31
~ возврата 98–178
~ задержки 86–123
зеркальная ~ 75–124
~ исчезновения 86–38
коническая ~ 96–31
~ Кюри 75–41
мировая ~ 99–125
нейтральная ~ 89–152
~ отражения 75–124
первая критическая ~ 96–31
первая ~ Лагранжа 96–31
~ погасания 86–38
подзвёздная ~ 81–130
подземная ~ 81–128

R

подлунная ~ 81–129
подсолнечная ~ 81–127
подспутниковая ~ 71–30
~ появления 86–36
~ пространства-времени 99–125
~ распада 86–121
~ росы 77–174
~ стояния 83–45
~ тени пятна 89–96
~ фиксации 86–127
яркая ~ 89–201
яркая рентгеновская ~ 89–183a
яркая ~ тени 89–96
траверс 82–35
маршрутный ~ 82–36
научный ~ 82–36
травление полированной поверхности 87–102
траектория, атмосферная 86–162
~ в атмосфере Земли 86–162
~ вокруг Луны 82–11
волнистая ~ 86–33
восходящая ~ 82–9
змеевидная ~ 86–33
истинная ~ 86–31
межпланетная ~ 82–12
~ метеора 86–30
~ метеора в атмосфере 86–30
~ облёта Луны 82–11
~ подъёма 82–9
селеноцентрическая ~ 82–11
~ частицы 75–118
транзиент, корональный 89–185a
транслунный 81–6
транснептуновая планета 83–237
трансплутоновая планета 83–124
Трапеция Ориона 96–92
трасса маршрута 82–37
~ прохождения 82–37
трек 82–251
древний ~ 87–121
~ охлаждения 93–202
первичный ~ ядра 82–252
~ радиационного разрушения 87–120
~ развития 94–29
~ частицы 82–251
эволюционный ~ 94–29
трение 86–173
приливное ~ 82–312
~ приливной волны 82–312
третий интеграл движения 98–186
третичный период 74–87
Треугольник 92–93
Южный ~ 92–94
трёхмерная орбита 98–185
~ спектральная классификация 91–106
Трёхраздельная туманность 97–141
трёхчасовой индекс активности 75–89
трещина 81–165
~ вдоль края кратера 82–211
концентрическая ~ 82–209
поверхностная ~ 81–173
~ растяжения 82–206
трещинное извержение 73–113
триада 87–91
триангуляция 71–23
космическая ~ 71–28
основная ~ 71–24
~ первого класса 71–24
пространственная ~ 71–25
спутниковая ~ 71–27
Триас 74–84
триасовый период 74–84
триггер 89–199
триггерное действие 89–199
тригонометрический параллакс 91–5
трилатерация 71–29
троилит 82–94
тройная звезда 96–4
Тройная туманность 97–141
тройное сближение 96–74
~ соударение 96–74
~ столкновение 96–74
тропическая зона 77–191
тропический месяц 81–44
~ период обращения 81–44
тропопауза 76–77
тропосфера 76–78
Троянцы 84–91
трубка взрыва 73–123
катодная ~ 86–293
катодно-лучевая ~ 86–293
лавовая ~ 82–186
магнитная силовая ~ 89–145
электронно-лучевая ~ 86–293
трясение коры 95–152
Тубан 92–131
тугоплавкое вещество 83a–55
Тукан 92–95
туман 77–130
туманное кольцо 83a–10
Туманность Андромеды 98–215
Большая ~ Ориона 97–136
внегалактическая ~ 98–202
волокнистая ~ 97–129, 97–137
газовая ~ 97–112
газово-пылевая ~ 97–119
галактическая ~ 97–110
диффузная ~ 97–122
дозвёздная ~ 97–63
досолнечная ~ 83a–21
кольцевая ~ 97–147
кольцеобразная ~ 97–147
кометарная ~ 97–142
кометообразная ~ 97–142
Крабовидная ~ 97–134
материнская ~ 97–63
неправильная ~ 98–237
~ Омега 97–143
отражательная ~ 97–128
отражающая ~ 97–128
планетарная ~ 97–144
~ Подкова 97–143
правильная ~ 98–236
протосолнечная ~ 83a–21
пылевая ~ 97–115
Ракообразная ~ 97–134
родительская ~ 97–63
~ Рыбачья Сеть 97–137
~ с эмиссионным спектром 97–123
светлая ~ 97–121
~ Северная Америка 97–140
~ Сова 97–139
солнечная ~ 83a–23
спиральная ~ 98–249
тёмная ~ 97–132
Трёхраздельная ~ 97–141
Тройная ~ 97–141
эмиссионная ~ 97–123
яркая ~ 97–121
туманные образования 83–250
турбосфера 76–72
турбулентная конвекция 93–152
турбулентность большого масштаба 83a–12
~ космического масштаба 83a–12
туфовый конус 73–120
тыловая часть области рассеяния 87–40
тэнит 87–93
тяжёлая масса 99–149
тяжесть 71–56

У

убегание газов 76–24
~ газов из атмосферы 76–24
убегающая звезда 98–135
убывание полярной шапки 83–146
убывающая Луна 81–21
убыль массы 83a–42, 86–168
увеличение земной тени 81–93
~ массы 83a–38
~ массы Земли 87–18
увлечение 86–178
угарный газ 97–28
угасшая звезда 91–167
углекислая соль 74–27
углекислота, твёрдая 83–144
углерод, односернистый 97–34
радиоактивный ~ 74–95
углеродистое вещество 87–64
углеродная звезда 91–188
~ сердцевина 97–71
углистое вещество 87–64
углистый метеорит 87–62
угловой диаметр 97–145
~ момент 83a–4
углубление депрессии 77–82
маленькое ~ 81–243
~ циклона 77–82
угол, гелиоцентрический 89–37
~ наклона оси 83–35
позиционный ~ 81–71

~ положения 81–71
фазовый ~ 83–72
~ фазы 83–72
характеристический ~ 98–33
Угольный мешок 97–138
удаление, взаимное 98–298
взаимное ~ галактик 98–298
~ галактик 98–298
удар землетрясения 73–2а
затухающий ~ землетрясения 73–13
искусственный ~ 82–226
~ метеорита 87–33
~ метеорита о поверхность Земли 87–33
~ метеорного тела 82–236
~ метеороида 82–236
последующий ~ землетрясения 73–13
тепловой ~ 86–206
ударная литификация 82–238
ударное образование 82–228
ударный кратер 81–240
~ метаморфизм 82–237
ударяющее тело 82–225
удержание газа 87–134
~ космических лучей магнитным полем 98–51
магнитное ~ космического излучения 98–51
удерживание газа 87–134
удлинение рогов 83–131
~ рогов Венеры 83–131
~ рогов серпа Венеры 83–131
узел в хвосте 85–95
~ протуберанца 89–244
узелки 88–210
тёмные ~ 88–212
хромосферные ~ 89–172
яркие ~ 88–211
узлы хромосферной сетки 89–172
указатель расстояния 86–266
уклонение отвеса 71–72
уклонения геоида 71–74
уксусный альдегид 97–42
ультрабазит 74–6
ультрателеметеор 86–20
ультрафиолетовый избыток 91–78
~ показатель цвета 96–146
уменьшение массы 83а–42, 86–168
умеренная зона 77–192, 83–189
умеренный наклон 81–231
умершая звезда 91–167
умкер–эффект 76–126
ундуляция геоида 71–74
униполярная группа 89–110
~ магнитная область 89–155а
уничтожающее землетрясение 73–39
унос массы 86–182
уплотнение 83а–26
~ осадков 74–50
центральное ~ 85–37, 96–126
уплощение 83а–22
упорядоченное движение 98–95
упорядоченность частиц 97–78
упругая волна 73–44
упругость 73–58
уравнение, абелево 88–235
~ Абеля 88–235
~ высоты 86–194
~ гидростатики 93–76
~ гидростатического равновесия 93–76
гидростатическое ~ 93–76
интегральное ~ Шварцшильда--Милна 93–53
интегро-дифференциальное ~ 93–52
~ испарения 86–200
~ лучистого переноса 93–33
~ непрерывности 93–49
~ переноса 93–32
~ переноса излучения 93–33
~ Погсона 91–58
~ поля 99–155
~ потери массы 86–169
~ светимости 86–196
световое ~ 95–45
~ стационарности 93–44
тензорное ~ 99–157
~ торможения 86–179
~ убыли массы 86–169
~ цвета 91–73
цветовое ~ 91–73
~ Эмдена 93–130
эйнштейновское ~ поля 99–156
уравновешивание, детальное 93–41
ураган 77–78
тропический ~ 77–78
Уран 83–234
ураново-свинцовый возраст 74–101
урановый метод 74–96
Урка процесс 94–51
уровень звука 76–38
~ сылы звука 76–38
~ конденсации 77–156
критический ~ 76–26
~ моря 72–87
усадка, гравитационная 83–203
усиление блеска 86–79
внезапное ~ атмосфериков 78–51
~ магнитного поля 89–156
~ поля 89–156
ускользание газов 76–24
~ газов из атмосферы 76–24
~ звёзд из скопления 98–194
ускользающая звезда 96–160
ускорение силы тяжести 71–64
экваториальное ~ 88–131
эффективное ~ 85–88
ускоренное расширение 99–81
условие в центре 93–81
граничное ~ в центре 93–81
граничное ~ на поверхности 93–80
~ на поверхности 93–80
~ симметрии 98–162
~ согласования 93–142
~ сшивки 93–142
~ устойчивости 93–205
условия, метеорологические 77–2
~ равновесия 93–74
стандартные ~ 77–3
стандартные атмосферные ~ 77–3
температурные ~ 77–35
уснувший вулкан 73–116
устойчивая звезда 95–8
устойчивое решение 98–170
устойчивость 93–204
динамическая ~ 98–198
~ звёзд 93–206
~ орбиты 98–193
устойчивый 93–208
термически ~ 93–209
усы 89–200
усыхание 74–51
утёс с острыми углами 82–131
утечка нефти 72–71
утолщение, центральное 98–16
утренние сумерки 79–35
~ хоры 78–62
утренний меридиан 75–18
утренняя заря 79–35
~ звезда 83–127
уход звезды из системы 96–159
уходящая звезда 96–160
ущелье, узкое 83–154
ущерблённая Луна 81–21
ущербная Луна 81–21
ухудшение погоды 77–23
участок траектории 82–12
межпланетный ~ траектории 82–12

Ф

фаза, взрывная 89–210
~ возрастания 89–83
~ восстановления 75–77
восходящая ~ солнечного цикла 89–83
~ вплеска 89–210
~ гало 94–19
главная ~ 75–76
дисковая ~ 94–22
лунная ~ 81–19
~ Луны 81–19
начальная ~ 75–75
начальная ~ магнитной бури 75–75
нисходящая ~ солнечного цикла 89–84
нулевая ~ 83–73
основная ~ 73–12
полная ~ Луны 81–27
промежуточная ~ 94–21
~ развития 94–28
~ роста (активности) солнечного цикла 89–83
~ сжатия 94–20

~ спада (активности) солнечного цикла 89–84
текучая ~ магмы 82–189
~ цикла 89–79
частная ~ 81–78a
частная ~ затмения 81–78a
эволюционная ~ 94–28
фазовая кривая 83–74
фазовое превращение 83a–58
~ пространство 98–144
фазовращатель 88–115
фазовый коэффициент 83–77
~ переход 83a–58, 99–170p
~ угол 83–62
фазосдвигающее устройство 88–115
файербол 99–168
факел 89–161
факелы, полярные 89–162
фотосферные ~ 89–161
хромосферные ~ 89–164
факельная гранула 88–179
фактор дилюции 93–37
масштабный ~ 99–81a
~ разрежения 93–37
~ R(t) 99–81
яркостный ~ 83–90
Феникс 92–77
ферромагнетизм 75–26
фибриллы 88–209
фигура, гидростатическая 83–112
~ гидростатического равновесия 83–112
~ Земли 71–7
~ Луны 81–116
~ планеты 83–109
~ равновесия 83–112
фигуры, видманштеттеновы 87–105
физика атмосферы 76–4
~ Земли 72–1
~ комет 85–104
~ Луны 82–151
~ Марса 83–134
солнечная ~ 88–6
~ солнечно-земных связей 84–56
~ Солнца 88–6
физическая двойная звезда 96–7
~ либрация 81–53
физический спутник 98–219
~ спутник галактики 98–219
физическое членство 96–83
фильтр, аэрозольный 76–94
~ для пыли 76–95
фиолетовый слой 83–165
фирн 72–118
фирновый снег 72–118
флоккул 89–167
тёмный ~ 89–224
тёмный водородный ~ 89–224
флоккулы 89–167
~ в оптическом диапазоне 89–165
водородные ~ 89–168
кальциевые ~ 89–170
рентгеновские ~ 89–166
яркие водородные ~ 89–169
флоккульное поле 89–165
~ поле в оптическом диапазоне 89–165
флуктуации, вторичные 95–44
первичные ~ 99–170y
флюгер 77–104
Фобос 83–168
фокус землетрясения 73–17
Фомальгаут 92–116
фон, метеорный 86–232
сейсмический ~ 73–55
фоновая активность 86–136
фоновое поле 89–141
~ магнитное поле 89–141
форма, кольцевая 98–41
кометная ~ 85–83
~ кометы 85–83
~ метеорного тела 86–175
облачная ~ 77–133
ориентированная ~ 87–113
~ планеты 83–109
формальдегид 97–27
формамид 97–35
формирование 94–10
формула, барометрическая 76–15
международная гравиметрическая ~ 71–76
~ Погсона 91–58
формы без лучевой структуры 79–142
лунные ~ 81–148
~ полярных сияний 79–137
~ с лучевой структурой 79–152
структурные ~ 81–152
форшоки 73–10
фотовизуальная величина 91–69
фотогелиограф 88–36
фотогид, солнечный 88–51
фотографирование метеоров 86–140
фотографическая карта Луны 81–137
~ кривая блеска 95–23
~ сьёмка 82–30
фотографическое вычитание 88–94
~ обозрение 82–30
фотография в белом свете 88–37
~ в интегральном свете 88–37
~ в наклонном направлении 82–29
~ в отвесном направлении 82–28
горизонтальная ~ 82–28
~ лунной поверхности 82–26
~ Луны 82–26
метеорная ~ 86–140
параллактическая ~ 79–102
~ под углом наклона 82–29
последовательная ~ 79–103
фотодезинтеграция 94–101
фотокамера для небесного свода 79–104
фотолиз 76–84
фотометрическая двойная 95–73
~ ночь 79–3f
фотометрически ясная ночь 79–3f
фотометрический диаметр 84–93
фотометрическое расстояние 96–116
фотометрия, детальная 83–104
интегральная ~ 83–103
многоцветная ~ 96–143
трёхцветная ~ 96–141
шестицветная ~ 96–148
фотонейтрино 94–93
фотопластинка, движущаяся 88–27
фоторасщепление 94–101
~ ядра 94–101
фотосфера 88–167
невозмущённая ~ 88–169
спокойная ~ 88–169
фотосферное магнитное поле 89–130
фотосферные факелы 89–161
фотосферный 88–170
фотохимическая реакция 76–63
фотохимическое разложение 76–84
фотоцентр 96–46
фотоцентрическая орбита 96–47
фотоэлектрическая кривая блеска 95–22
фотоэлектрический поляриметр 83–83
фотоэлектронный слой 82–246
фрагментация 87–25
фрагменты 83a–83
фракционирование 87–138
~ химических элементов 83a–51
фракционная кристаллизация 74–40
фрактокумулус 77–146
фраунгоферова корона 88–241
~ линия 88–139
фридмановское время 99–80
~ время расширения 99–80
фронт 77–68
~ ионизации 97–48
ионизационный ~ 97–48
окклюдированный ~ 77–71
~ окклюзии 77–71
полярный ~ 77–73
~ потока 82–193
тёплый ~ 77–69
холодный ~ 77–70
фронтогенез 77–74
фронтолиз 77–75
фронтообразование 77–74
фумарола 82–163
фумарольная активность 82–164
фундамент, кристаллический 72–12
фундаментальная система отсчёта 98–137
функция, весовая 93–58
~ видимости 89–55
зональная гармоническая ~ 93–55
интегрально-показательная ~ 93–62
~ источника 93–34
~ масс 96–68
~ насыщения 93–60
~ отдачи 93–34

R

~ плотности 98–88
~ распределения 98–143
~ рассеяния 77–58
~ светимости 98–89
фазовая ~ 83–78

Х

хаббловская последовательность галактик 98–227
~ постоянная 99–77
хаббловский тип галактики 98–229
хаббловское время 99–79
халькофильные элементы 83а–31
халькофилы 83а–31
Хамаль 92–118
Хамелеон 92–35
хаотическая местность 83–164
хаотическое движение 98–94
характер местности 81–149
характеристика, высотно-частнотная 78–23
~ грунта 82–108
характеристическая огибающая 98–177
~ скорость 76–43
характеристический угол 98–33
характерное время 99–79
хвост, аномальный 85–75
~ галактики 98–280
главный ~ 85–77
~ Земли 84–65
~ Змеи 92–90
изогнутый ~ 85–73
ионный ~ 85–82
кометный ~ 85–70
~ кометы 85–70
кратный ~ 85–76
~ магнитосферы 75–10
~ метеора 86–110
многолучевой ~ 85–76
~ направленный к Солнцу 85–78
~ направленный от Солнца 85–81
прямой ~ 85–72
пылевой ~ 85–79
хвостовое пятно 89–114
хемилюминесценция 79–31
хемисфера 76–62
химическая дифференциация 83а–50
~ накачка 97–19
~ намагниченность 75–31
~ эволюция 82–257
химический состав 91–168
химическое загрязнение 72–98
химия атмосферы 76–79
лунная ~ 82–74
~ Луны 82–74
хис 78–64
холм 89–276
холодная звезда 91–156
холодное течение 72–64
холодный фронт 77–70
хондра 87–76
хондрит 87–61
неуглистый ~ 87–75
углистый ~ 87–63
хоры, утренние 78–62
хребет 81–186
горный ~ 73–85
морской ~ 81–188
складчатый ~ 81–187
срединно-океанический ~ 72–45
хромосфера 88–191
верхняя ~ 88–192
нижняя ~ 88–193
хромосферная вспышка 89–175
хромосферные волокна 88–209
~ волоконца 89–173
~ факелы 89–164
хромосферный 88–194
~ телескоп 88–73
хрупкий 86–191
хрупкость 86–192

Ц

цвет, истинный 91–74
цветовое уравнение 91–73
целостат 88–46
вертикальный ~ 88–48
горизонтальный ~ 88–47
цементация 74–52
~ осадков 74–52
цементирующая среда 82–118
Центавр 92–32
центр активности 89–60
активный ~ 89–60
~ видимого диска 88–172
~ видимого диска Луны 81–128
~ Галактики 98–9
галактический ~ 98–9
~ диска 81–128, 89–35
~ звезды 93–84
~ пластинки 86–156
~ притяжения 89–241
~ солнечного диска 89–35
~ фиксации 86–127
центральная вершина 81–258
~ горка 81–258
~ звезда 96–37
~ зона 81–114, 89–47
~ интенсивность (в) линии 89–211
~ область галактики 98–293
центральное затмение 88–16а
~ сгущение 85–37, 86–244
~ столкновение 83а–17
~ уплотнение 85–37, 96–126
центральный кратер 81–217
~ меридиан 89–43
цепочка галактик 98–328
звёздная ~ 96–77
цепь галактик 98–328
горная ~ 81–194
кратерная ~ 81–241
~ кратеров 81–241
~ сверхновых 95–132
Церера 84–82
цефеида 95–92
долгопериодическая ~ 95–94
классическая ~ 95–94
короткопериодическая ~ 95–93
цефеидный параллакс 91–8а
Цефей 92–33
цианацетилен 97–30
цианид водорода 97–29
водородный ~ 97–29
цианистый метил 97–38
цизлунный 81–4
цикл, геохимический 74–32
нечётный ~ 89–78
одиннадцатилетний ~ 89–73
~ пятнообразователной деятельности 89–72
~ солнечной активности 89–71
солнечный ~ 89–71
~ солнечных пятен 89–72
чётный ~ 89–77
циклическая переменная 95–108
~ переменная звезда 95–108
циклон 77–76
тропический ~ 77–77
циклострофический ветер 77–95
цилиндр, ионизованный 86–282
косо срезанный ~ 88–63
световой ~ 95–155
усечённый ~ 88–63
цилиндрическая симметрия 98–155
цилиндрические координаты 98–156
цирк, лунный 81–265
Циркуль 92–36
циркуляция, атмосферная 77–64
меридиональная ~ 93–163
общая ~ 77–65
планетарная ~ 77–66
циррокумулус 77–143
цирростратус 77–141
циррус 77–136
цоколь 72–12
цунами 72–95
волна ~ 72–95
цюрихская классификация 89–127
цюрихское число пятен 89–65

Ч

часовое движение 81–72
~ изменение 81–72
частица 88–154
авроральная ~ 79–107
~ аэрозольная 76–91
~ воздуха 86–172
гигантская ~ 76–118
жидкая ~ 76–100
каменистая ~ 85–54
коллоидная ~ 76–98
ледяная ~ 85–52
лунная ~ 82–123

метеорная ~ 86–160a
полимерная ~ 97–77a
~ с ледяной оболочкой 97–77
сверхтяжёлая ~ 99–170n
~ содержащая ядро и оболочку 97–70
твёрдая ~ 97–69
ускоренная ~ 91–182
~ ядро-оболочка 97–70
частицы Айткена 76–116
высыпающиеся ~ 76–117
~ захваченные геомагнитным полем 75–111
~ зодиакального облака 84–66
зодиакальные ~ 84–66
межзвёздные ~ 97–68
межзвёздные пылевые ~ 97–59
~ мелочи 82–109
мельчайшие ~ 82–109
сферические ~ 87–29
частичное вырождение 93–173
частная теория относительности 99–94
~ фаза 81–78a
~ фаза затмения 81–78a
частное затмение 81–79a
~ затмение Луны 81–67
~ лунное затмение 81–67
~ солнечное затмение 88–16
частота величин 81–249
~ импульсов 86–279
~ кратеров 81–247
критическая ~ 78–24
максимальная применимая ~ 78–37
~ метеоров 86–135
минимальная применимая ~ 78–33
~ модуляции 86–275
несущая ~ 86–274
~ образования кратеров 82–233
~ передатчика 86–273
~ повторения импульсов 86–279
~ появления всплесков 89–217
~ появления полярных сияний 79–122
~ протуберанцев 89–237
~ размеров 81–249
~ солнечных пятен 89–66
~ спорадических метеоров 86–233
часть ядра, внешняя 72–19
минералогическая составная ~ 82–100
Часы 92–54
Чаша 92–42
чашечный кратер 81–219
чёрная дыра 94–114
чёрный карлик 91–139
чертёж Солнца 88–34
черты местности 81–148
топографические ~ 81–149
характерные ~ местности 81–148
четвертичная система 74–88
четвертичный период 74–88
четверть 81–26
первая ~ 81–25
первая или последняя ~ 83–132
последняя ~ 81–30
третья ~ 81–30
чётки Бейли 88–32
чётный цикл 89–77
чёточная молния 77–164
четырёхмерный 99–121
~ мир 99–122
четырёхразмерный 99–121
число Вольфа 89–65
~ звёзд 98–80
контрольное ~ 85–30
относительное ~ 89–65
относительное ~ солнечных пятен 89–65
~ сближений 98–122
цюрихское ~ пятен 89–65
часовое ~ 86–135
часовое ~ метеоров 86–135
~ частиц 76–121
чисто рассеивающая атмосфера 93–16
член, возможный 96–86
возможный ~ скопления 96–86
~ галактического вращения 98–116
дифференциальный ~ 98–115
K- ~ 98–119
космологический ~ 99–156a
недостоверный ~ скопления 96–87
неопределённый ~ скопления 96–87
неуверенный ~ скопления 96–87
~ системы 98–133
~ скопления 96–82
~ скопления галактик 98–338
членство, физическое 96–83

Ш

шапка, ледниковая 72–115
полярная ~ 83–143
~ полярных сияний 79–127
~ сжатого воздуха 86–170
шар, газовой 93–73
земной ~ 71–10
лунный ~ 81–112
огненный ~ 99–168
политропный газообразный ~ 93–127
шарик 82–129
шарики 87–29
стеклянные ~ 82–127
шаровая молния 77–163
шварцшильдовская метрика 99–152
шварцшильдовский радиус 94–113
шельф 72–34
шероховатая поверхность 83–102
шероховатость поверхности 83–99
шестимерное фазовое пространство 98–144
шестицветная фотометрия 96–148
ширина импульса 86–280
эффективная ~ линии 89–208
широкая двойная звезда 96–9
широкоугольная фотография 82–27
широта, ареографическая 83–140
гелиографическая ~ 89–42
геомагнитная ~ 75–52
предельная ~ видимости покрытия 81–104
северная предельная ~ 81–105
селенографическая ~ 81–11
южная предельная ~ 81–106
широтный дрейф 89–75
широты, конские 77–102
субтропические ~ 77–102
шкала Бофорта 77–108
~ времени 94–65
~ высот 76–18, 88–236
геоисторическая ~ 74–68
геохронологическая ~ 74–68
длинная ~ времени 94–67
короткая ~ времени 94–66
~ расстояний 98–303
стоградусная ~ 77–33
~ Цельсия 77–33
шлейф, геомагнитный 75–10
шлем 89–295
шпат, известковый полевой 82–87
калийный полевой ~ 74–16
полевой ~ 74–15, 82–93
штиль 77–110
штормовой прибой 72–76
~ прилив 72–76
шум, акустический 88–220
сейсмический ~ 73–54

Щ

щель 83–223
~ Кассини 83–224
колеблющаяся ~ 88–70
кривая ~ 88–65
~ между кольцами 83–223
щёточки, полярные 89–294
щит 72–11, 92–88
щитовидный вулкан 73–119
щитовой вулкан 73–119

Э

эвклидов 99–101
эвклидова геометрия 99–108
эволюционировавшая звезда 94–74
эволюционировать 94–75
эволюционирующая вселенная 99–54
эволюционная диаграмма 96–151
~ последовательность 94–30
~ фаза 94–28
эволюционное изменение 94–27
эволюционный путь 94–29
~ трек 94–29
эволюция вселенной 99–171e
геохимическая ~ 74–33
динамическая ~ 96–161

R

~ до главной последовательности 94–26
дозвёздная ~ 94–14
дозвёздная ~ галактик 94–14
~ звёзд 94–1
звёздная ~ 94–1
приливная ~ 83а–65
тепловая ~ 83а–61
термическая ~ 83а–61
термоядерная ~ 94–50
химическая ~ 82–257
ядерная ~ 94–50
эжекция частиц 88–161
~ частиц из Солнца 88–161
эйнштейновские уравнения 99–156
~ уравнения поля 99–156
экватор, лунный 81–121
магнитный ~ 75–50
~ освещения 81–122
экваториальная зона 83–188
~ область 89–59
экваториальный пояс 83–191
~ протуберанец 89–228
~ радиус 83–115
экземпляр, индивидуальный 87–42
экзогенный 73–146
экзосфера 76–68
эклиптический метеор 86–46
~ поток 86–221
экология 76–2
экран для наблюдения Солнца 88–13
экранирование 86–171, 93–20
~ линиями 93–20
магнитное ~ 89–225
~ межзвёздным веществом 97–86
~ озоном 79–28
экранированная высота тени 79–49
экранировка 86–171
экранируемая переменная 95–83
~ переменная звезда 95–83
экспедиция для наблюдения солнечного затмения 88–30
солнечная ~ 88–30
эксперимент, активный 82–220
активный сейсмический ~ 82–220
~ на лунной повехности 82–42
~ рассеяния альфа-частиц 82–76
сейсмический ~ 82–219
эксплозивный конус 73–120
экспозиционный возраст 87–139
экстинкция, атмосферная 77–127а
~ водяным паром 77–126
избирательная ~ 97–83
межзвёздная ~ 97–80
полная ~ 97–82
экстремальное население первого типа 98–73
экструзивное извержение 73–132
экструзия 73–132, 82–171
электричество, атмосферное 75–2
электрографическая звёздная величина 91–67
электронная концентрация 78–6
~ корона 88–242
~ плотность 88–233
электронное облако 79–115
электронный свист 78–60
электронографическая звёздная величина 91–67
электроны, высыпанные 79–118
захваченные ~ 79–111
релятивистские ~ 97–153
электрооптическая модуляция 88–86
электрослабое взаимодействие 99–170x
электроструя, авроральная 79–113
~ полярного сияния 79–113
электротеллурическое поле 75–4
элемент, метеорологиеский 77–28
подвижный ~ 88–108
рассеянный ~ 82–145
~ с избыточным содержанием 91–184
~ с недостатком обилия 91–186
~ с повышенным содержанием 91–184
~ с пониженным содержанием 91–186
структурный ~ 98–287
супергранулярный ~ 88–182
топографический ~ 81–149
элементарная длина Планка 99–170h
~ масса Планка 99–170f
элементы, атмофильные 83а–29
геомагнитные ~ 75–54
главные ~ 82–144
~ группы железа 94–105
~ железного пика 94–105
литофильные ~ 83а–30
магнитные ~ 75–54
огнеупорные ~ 97–64
~ ориентирования 71–20
преобладающие ~ 82–144
фотометрические ~ 95–25
халькофильные ~ 83а–31
эллипс рассеяния 87–38
эллипсоид, земной 71–11
международный ~ 71–14
местный ~ 71–15
~ относимости 71–18
отсчётный ~ 71–18
референц- ~ 71–18
~ скоростей 98–101
~ сравнения 71–18
эллипсоидальная галактика 98–241
~ переменная 95–79
эллипсоидальное распределение скоростей 98–100
эллиптическая галактика 98–241
~ галактика-спутник 98–217
~ комета 85–4
эллиптическое пространство 99–110
эллиптический спутник 98–217
~ спутник галактики 98–217
эллиптичность шарового скопления 96–122
элонгация 83–60
~ видимого радианта 86–67
восточная ~ 83–61
западная ~ 83–62
~ истинного радианта 86–58
наибольшая ~ 83–63
эмиссионная звезда 91–159
~ кайма волокна 89–226
~ туманность 97–123
эмиссия исключительно низких частот 78–55
корпускулярная ~ 88–152
~ массы 85–145
~ ОНЧ 78–54
~ очень низких частот 78–54
послесумеречная ~ 79–52
предсумеречная ~ 79–51
~ сверхнизких частот 78–55
~ частиц 88–151
эмпирическая кривая роста 93–69
эндогенный 73–145
энергетический баланс 88–251
~ выход 93–109
энергия, акустическая 88–219
атомная ~ 93–121
~ вакуума 99–170k
вакуумная ~ 99–170k
вращательная ~ 93–120
~ вращения 93–120
гравитационная ~ 93–119
запасённая ~ 89–196
звуковая ~ 88–219
имеющаяся ~ 89–197
~ испарения 86–199
накопленная ~ 89–196
нелучевая ~ 88–218
нулевая ~ 93–187
обмениваемая ~ 98–126
~ покоя 99–144
~ при нулевой температуре 93–187
ядерная ~ 93–122
энерговыделение 93–105
эолосферная модель 83–260
эпейрогенез 73–73
эпейрогеническое движение 73–73
эпейрофорез 73–104
эпицентр 73–15
эпицентральное расстояние 73–19
эпицикл 83–49
эпоха 74–89
геологическая ~ 74–89
~ максимума 89–81
~ минимума 89–80
~ отрыва излучения от вещества 99–177а
~ наложения двух циклов 89–86
~ перекрытия двух циклов 89–86
послеледниковая ~ 72–109
~ развязывания 99–177а
~ разделения излучения 99–177а

эра 74–69
адронная ~ 99–174
азойская ~ 74–75
~ аннигиляции 99–173a
аннигиляционная ~ 99–173a
археозойская ~ 74–76
~ вещества 99–178
докембрийская ~ 74–70
звёздная ~ 99–178
~ излучения 99–172
кайнозойская ~ 74–73
космологическая ~ 99–166a
лептонная ~ 99–176
мезозойская ~ 74–72
~ отрыва излучения 99–177a
планковская ~ 99–170e
~ радиации 99–172
радиационная ~ 99–172
~ радиационно-доминированной плазмы 99–172
~ развязывания излучения 99–177a
~ разделения излучения и вещества 99–177a
~ раздувания 99–170d
~ фотонной плазмы 99–172
Эридан 92–49
эрозионное столкновение 84–105
эрозия 73–107, 82–243
~ в космическом пространстве 87–142
ветровая ~ 83–163
газовая ~ 82–243
метеоритная ~ 82–242
метеороидная ~ 82–242
поверхностная ~ 82–241
эрратический валун 72–132
эруптивная арка 89–251
~ галактика 98–269
~ звезда 95–68
~ переменная 95–68
эрупция 95–69
Этамин 92–115
этап гало 94–19
дисковый ~ 94–22
~ до главной последовательности 94–49
~ после главной последовательности 94–68
промежуточный ~ 94–21
~ сжатия 94–20
эфемериды малых планет 84–92
еффект, бетатронный 91–180
~ Вильсона 89–89
вращательный ~ 91–44
~ Деллинджера 78–40
избирательный ~ 98–81
изотопический ~ 91–178
~ Кориолиса 77–81
кумулятивный ~ 98–124
~ обращения 76–126
оранжерейный ~ 83–263
~ отражения 96–26
~ отражения в двойных 95–82
парниковый ~ 83–263
~ Пойнтинга-Робертсона 84–67
покровный ~ 93–67
приливный ~ 82–276
~ противостояния 83–75
тепличный ~ 83–263
умкер ~ 76–126
~ Эвершеда 89–103
~ экранирования 93–67
эффективная длина волны 96–142
~ ширина линии 89–208
эффективность излучения 86–210
~ ионизации 86–209
световая ~ 86–210
эффузивная порода 74–3
эффузия 73–130, 82–172
эхо, головное 86–297
головное ~ метеора 86–297
длительное метеорное ~ 86–301
кратковременное метеорное ~ 86–300
метеорное ~ 86–296
~ от следа 86–298
~ полярного сияния 79–121
сейсмическое ~ 82–223
эхолотирование 72–29
эхолотный промер 72–29
эхо-сигнал 86–295

Ю

Южная Гидра 92–56
~ граница затмения 88–19
~ Корона 92–39
~ предельная широта 81–106
~ Рыба 92–80
южное небо 79–3c
~ полярное сияние 79–106
~ тропическое возмущение 83–194
Южный Змей 92 56
~ Крест 92–43
~ Треугольник 92–94
Юнона 84–85
Юпитер 83–170
юпитерианский 83–171
юпитеров 83–171
юпитерографические координаты 83–175
юпитероцентрические координаты 83–176
юра 74–85
юрская система 74–85
юрский период 74–85

Я

явление гало 79–66
корональное транзиентное ~ 89–185a
лунное временное ~ 82–153
оптическое ~ 79–33
сейсмическое ~ 82–215
протонное ~ 89–192
солнечное ~ 89–13
явления, сумеречные 79–32
ядерная резонанская спектрография 82–142
ядерное равновесие 94–103
ядерный синтез 94–52
ядра Айткена 76–116
~ замерзания 76–109
льдообразующие ~ 76–104
ледяные ~ 76–104
~ сублимации 76–110
ядро, активное 98-272
активное галактическое ~ 98–272
~ ассоциации 96–91
~ в лучистом равновесии 93–166
внешнее ~ 72–19
внешнее ~ Земли 72–19
внутреннее ~ 72–20
внутреннее ~ Земли 72–20
вырожденное ~ 93–170
~ Галактики 98–18
галактическое ~ 98–18
гелиевое ~ 93–167
гигантское ~ 76–118
графитовое ~ 97–71
звёздное ~ 93–164
звездообразное ~ 85–38
звездоподобное ~ 98–19
~ звезды 93–164
~ Земли 72–17
земное ~ 72–17
изотермическое ~ 93–168
кометное ~ 85-36
~ кометы 85–36
конвективное ~ 93–165
~ конденсации 97–67
~ кристаллизации 76–108
ледяное ~ 85–40
~ метеора 86–102
невырожденное ~ 93–171
неоднородное ~ 93–169
~ нифе 72–18
пекулярное ~ 98–20
~ планетарной туманности 97–148
~ пятна 89–95
сверхпроводящее ~ 95–150
~ скопления 96–130
твёрдое ~ 85–39
язык, лавовый 82-187
ледниковый ~ 72–125
яма, метеоритная 87–58
ямочка 81–243
ЯМР-спектрография 82–142
яркая звезда 91–83
~ кайма волокна 89–226
яркие гиганты 91-149
яркий возвратный выброс 89–257
~ возвратный протуберанец 89–257
~ мост 98–307
яркое ночное небо 86–28

R

яркостный фактор 83–90
яркость 91–54
∼ вспышки 89–212
звёздная ∼ 91–55
∼ звезды 91–56
∼ метеора 86–8
ярчайшая галактика 98–339
∼ звезда галактики 98–304
ярчайший член скопления 98–339
Ясли 96–111
ясно 77–153
ясное небо 77–153
ячеистая структура 99–171a
ячеистый 82–181
∼ алюминий 82–25
Ящерица 92–58

R